CISM COURSES AND LECTURES

This series presents lecture notes, monographs, edited works and proceedings in the field of the Mechanics, Engineering, Computer Science and Applied Mathematics.
Purpose of the series is to make known in the international scientific and technical community results obtained in some of the activities organized by CISM, the International Centre for Mechanical Sciences.

INTERNATIONAL CENTRE FOR MECHANICAL SCIENCES

COURSES AND LECTURES - No. 437

AMST'02

ADVANCED MANUFACTURING SYSTEMS AND TECHNOLOGY

PROCEEDINGS OF THE
SIXTH INTERNATIONAL CONFERENCE

EDITED BY

ELSO KULIANIC
UNIVERSITY OF UDINE

Springer-Verlag Wien GmbH

This book was printed with the contribution of Consorzio
Universitario del Friuli and Regione Friuli Venezia Giulia

This volume contains 547 illustrations

Originally published by Springer-Verlag Wien New York in 2002
Softcover reprint of the hardcover 1st edition 2002

SPIN 10882242

In order to make this volume available as economically and as rapidly as possible the authors' typescripts have been reproduced in the original forms. This method unfortunately has its typographical limitations but it is hoped that they in no way distract the reader.

ISBN 978-3-7091-2557-1 ISBN 978-3-7091-2555-7 (eBook)
DOI 10.1007/978-3-7091-2555-7

MATERIALS SCIENCE AND THE SCIENCE
OF MANUFACTURING - INCREASING PRODUCTIVITY
MAKING PRODUCTS MORE RELIABLE AND LESS EXPENSIVE

ORGANIZERS

University of Udine – Faculty of Engineering – Department of
Electrical, Managerial and Mechanical Engineering – Italy
Centre International des Sciences Mecaniques, CISM – Udine – Italy
University of Rijeka – College of Engineering – Croatia

CONFERENCE VENUE

University of Udine – PALAZZO ANTONINI
Via Tarcisio Petracco 8, Udine, Italy

PREFACE

Manufacturing a product is not difficult, the difficulty consists in manufacturing a product of high quality, at a low cost and rapidly.

Progress in materials and manufacturing sciences and technology, as well as in intelligent machining systems, promotes an increase in productivity and a decrease of costs and production time without compromising the reliability of the products. In the new global and national condition the competitiveness of products depends in turn on their quality, which is becoming increasingly relevant.

The International Conference on Advanced Manufacturing Systems and Technology – AMST is held every third year. The First International Conference on Advanced Manufacturing Systems and Technology AMST'87 was held in Opatija (Croatia) in October 1987. The Second International Conference on Advanced Manufacturing Systems and Technology AMST'90 was held in Trento (Italy) in June 1990. The Third, Fourth an Fifth Conferences on Advanced Manufacturing Systems and Technology were all held in Udine (Italy) as follows: AMST'93 in April 1993, AMST'96 in September 1996 and AMST'99 in June 1999.

The Sixth International Conference on Advanced Manufacturing Systems and Technology - AMST'02, which was held in Udine in June 2002, aimed at presenting up-to-date information on the latest developments – research results and experience – in the field of machining of conventional and advanced materials, machine tools and flexible manufacturing systems, forming, nonconventional processes, robotics, measurement and control, quality, design and ecodesign, rapid prototyping, rapid tooling and manufacturing, materials and mechanics, thus providing an international forum for a beneficial exchange of ideas and furthering a favourable cooperation between research and industry.

Elso Kuljanic

HONOUR COMMITTEE

R. TONDO, President of Giunta Regione Autonoma Friuli – Venezia Giulia
M. STRASSOLDO DI GRAFFEMBERGO, President of Provincia di Udine
F. HONSELL, Rector of the University of Udine
D. RUKAVINA, Rector of the University of Rijeka
A. STELLA, Dean of the Faculty of Engineering, University of Udine
S. CECOTTI, Mayor of Udine
A. VALDUGA, President of the Associazione Industriali della Provincia di Udine
G. FRAU, President of Consorzio Universitario del Friuli
B. SCHREFLER, General Secretary of CISM
G. BENEDETTI, Danieli & C.
M. FANTONI, Fantoni Spa
A. PITTINI, Gruppo Pittini
R. SNAIDERO, Snaidero R. Spa

SCIENTIFIC COMMITTEE

E. KULJANIC, (Chairman), University of Udine, Italy
N. ALBERTI, University of Palermo, Italy
P. BARIANI, University of Padova, Italy
A. BUGINI, University of Bergamo, Italy
R. CEBALO, University of Zagreb, Croatia
G. CHRISSOLOURIS, University of Patras, Greece
B. FRANKOVIC, University of Rijeka, Croatia
M.F. DE VRIES, University of Wisconsin Madison, U.S.A.
R. IPPOLITO, Polytechnic of Torino, Italy
F. JOVANE, Polytechnic of Milano, Italy
I. KATAVIC, University of Rijeka, Croatia
H.J.J. KALS, University of Twente, The Netherlands
F. KLOCKE, T.H. Aachen, Germany
F. LE MAITRE, Ecole Nationale Superieure de Mechanique, France
E. LENZ, Technion, Israel
R. LEVI, Polytechnic of Torino, Italy
B. LINDSTROM, Royal Institute of Technology, Sweden
G. MARINSEK, Centro Ricerche Fiat, Italy
J.A. Mc GEOUGH, University of Edimburg, UK
M.E. MERCHANT, IAMS, Ohio, U.S.A.
G.F.MICHELETTI, Polytechnic of Torino, Italy
B. MILCIC, INAS, Zagreb, Croatia
T. NAKAGAWA, University Of Tokyo, Japan
S. NOTO LA DIEGA, University of Palermo, Italy
J. PEKLENIK, University of Ljubljana, Slovenia
H. SCHULZ, T.H. Darmstadt, Germany

G. SPUR, T.U. Berlin, Germany
N.P. SUH, MIT, Mass., U.S.A.
H.K. TONSHOFF, University of Hannover, Germany
B.F. von TURKOVICH, University of Vermont, U.S.A.
K. UEDA, Kobe University, Japan
A. VILLA, Polytechnic of Torino, Italy

ORGANIZING COMMITEE

E. KULJANIC (Chairman)
F. MIANI (Secretary)
G. BIANCHI, C. BANDERA, F. DE BONA, S. FILIPPI, A. GASPARETTO,
M. GIOVAGNONI, S. MIANI, G. CUKOR, M. SORTINO

SUPPORTING ORGANIZATIONS

Giunta Regione Autonoma Friuli – Venezia Giulia, Provincia di Udine, Comune di Udine, University of Udine, Consorzio Universitario del Friuli, Danieli & C., Fantoni Spa, Gruppo Pittini, Snaidero R. Spa, Fondazione Cassa di Risparmio di Udine e Pordenone, Comitato per la promozione degli studi tecnico scientifici, Centro Convegni ed Accoglienza, Pietro Rosa T.B.M., FAMUP, ONRIFO Office of Naval Research International Field Office, US Navy.

CONTENTS

Machine Tools and Flexible Manufacturing Systems

Forming

Nonconventional Processes

Quality, Design and Ecodesign

Rapid Prototyping, Tooling and Manufacturing

Materials and Mechanics

THE DYNAMIC CLUSTER STRUCTURES: A NEW MANUFACTURING PARADIGM FOR PRODUCTION OF HIGH-TECH PRODUCTS

J. Peklenik

Department of Control and Manufacturing Systems, University of Ljubljana, Slovenia

KEYWORDS: cluster structure, elementary work unit, production unit, networking, education, subject.

ABSTRACT. The contribution discusses a new approach how to organize the development and manufacturing of high-tech (HT)-products by small and medium size enterprises. The paper introduces the concept of a dynamic production cluster (DPC), explaining its structure and operation. The importance of the cooperation between research and development units (RDU) and production units is stressed. The market analysis represents the basis for forming the dynamic production cluster. The role of continuous education of the subject on various levels is explained.

1 MOTIVATION AND OBJECTIVES

Since the introduction of computers into manufacturing technology and the production of various products in 1952 [1], the trend of ever increasing application of these devices has become the leading direction in the industrial development. Modern computer equipment and software development have provided the means and methods for an unprecedented advancement of the production of goods for global markets.

It is, however, necessary to point out that these developments in manufacturing have decisively contributed to the improvements and competitiveness of the companies that are financially and from the managerial viewpoint capable of investing and implementing these advanced production means and methods.

On the other hand, national manufacturing industries reveal a very large number of small and medium size production systems. These firms have, in comparison with large companies, some apparent disadvantages. Let us review and analyze the reasons for this conclusion.

In order to become or to remain competitive on the markets, the manufacturers of products must be able to continuously execute a number of vital activities required in the production of goods.

The analysis of the markets represents the first and very important step, providing the answer to the question, whether a product, planned or improved, will have competitive chances on the markets. Only if the result of the inquiry is positive, the product development should be introduced. An important step in this procedure is making and testing the prototype. After this step has been assessed as successful, there follows the design of the product for manufacturing, planning, programming, fabrication of components, assembly and testing. The on-time delivery of the products to the customers is an imperative!

Published in: E.Kuljanic (Ed.) *Advanced Manufacturing Systems and Technology*,
CISM Courses and Lectures No. 437, Springer Wien New York, 2002.

All these steps must be implemented by applying the best and the most advanced work methods and technologies in order to offer to the markets a competitive product of high quality.

The complexity of the work methods or/and procedures, as well as the complexity of modern manufacturing technologies require highly knowledgeable and competent subjects on various levels, as well as the financial means in order to implement the production with advanced machinery processes, information technologies (IT), etc. to accomplish the objectives. The level of investments into the subjects as well as the equipment and information technologies is so high that only well designed, well equipped and managed factories of a certain critical size can be successful in this complex world of manufacturing.

The chances for small and medium size enterprises to become successful are very small due to the fact that:

- they are hardly to develop an innovative product based on new and advanced research accomplishments and be competitive;
- manufacturing technologies and work methods supported by IT do not only require the financial means, but also advanced knowledge and creativity on a wide spectrum of activities;
- marketing of these firms is usually not capable of providing reliable answers about the chances of the products on the markets;
- the risk capital for support of these firms is very difficult to find.

A large and economically strong company can, of course, meet these requirements, but a small or medium size firm is definitely not in the position to fulfill all the conditions related to the marketing development, fabrication and selling. On the other hand, a large company is usually not flexible enough to meet very randomly variable input orders, quickly enough. In addition, the organizational structure of a factory is fixed and difficult to change. Also the structure of subjects working in the factory and their knowledge does not always meet the levels required for the development of innovative products, their manufacturing and management and control of the factory on various levels.

Due to the fact that e.g. in the EU the small and medium size companies (SME) employ about 66 % of all employees and generate 56 % GNP. The most important objective is to find new ways and means of how to improve the productivity and competitiveness of SME on the global markets with innovative products of high quality.

In order to accomplish these objectives, it is essential to develop a new manufacturing paradigm, different from the over ninety years old F.W. Taylor paradigm [2] still in operation. The condition for such a development is a very close cooperation between marketing as an economic entity, research and development, and the production systems.

2 THE EXISTING STRUCTURES OF PRODUCTION SYSTEMS

In order to establish a basis for the comparison between classical manufacturing structures, the proposed dynamic production clusters might be useful to reassess the basic manufacturing structures operating today.

2.1 STRUCTURE OF A CLASSICAL PRODUCTION SYSTEM

Figure 1 illustrates a cybernetic structure of a factory system exhibiting three levels of activities, which are vital for the manufacturing of HT-products [3]. The first level is focused on the strategic formulation of the production policy and on the decision making process. The second level is responsible for the managerial activities in general and for planning development, marketing, commercial activities and manufacturing standards in particular. On the third level – the manufacturing processes necessary for the fabrication of products are implemented.

The majority of the production systems (factories) today is structured as shown in Figure 1.

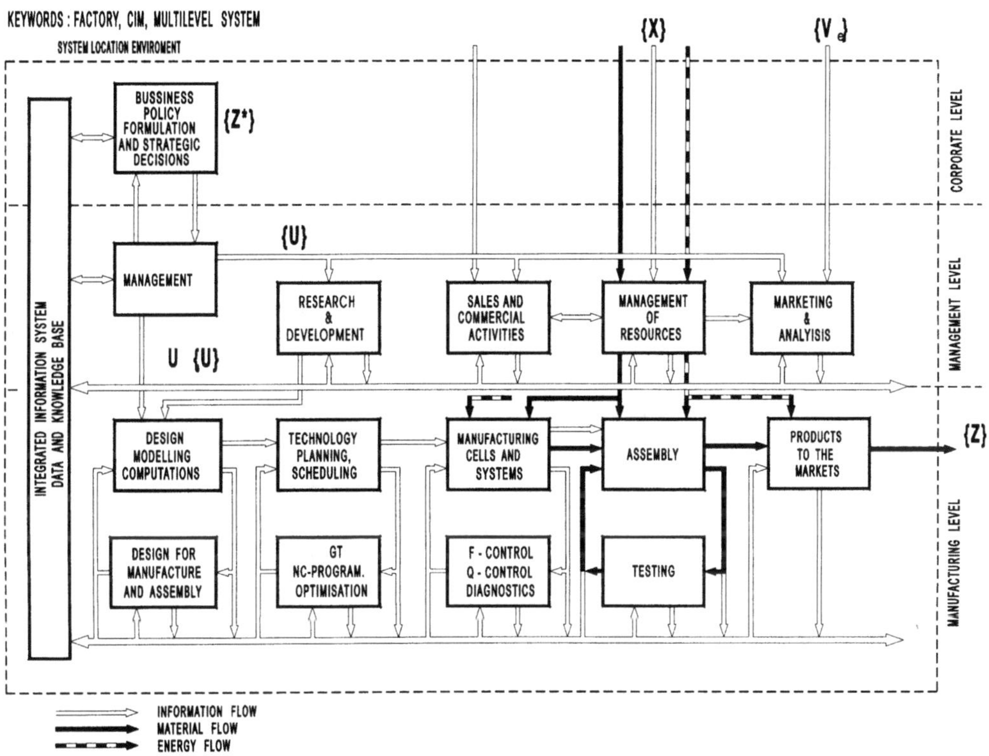

FIGURE 1. Cybernetic structure of a factory

2.2 CLUSTER STRUCTURE OF A PRODUCTION SYSTEM

The last two decades exhibit a strong influence of the information technology on manufacturing. Not only the CNC-machinery and equipment for fabrication of components have been intensively affected, also the work processes and procedures such as design, planning, scheduling, production control decision making, marketing etc. have been computerized to great extent. This has led to more effective implementation of the human work and faster adaptation to changes of manufacturing technologies.

The efficiency of the work force and management in factories have increased tremendously due to the introduction of local and global networks (e.g. LAN, INTERNET, INTRANET). These developments offer new opportunities in the search for a new paradigm in this field. We know that the globalization of markets has forced the industrial enterprises into greater production flexibility. The response and delivery time for the orders is a decisive factor in this highly competitive environment. That means that the structure of the production system must exhibit a very high adaptability in terms of a quick fabrication adaptation regarding the

- type of products A, B, ... K
- number of orders N_A, N_B ... N_K
- number of customers C (C_1, C_2 ... C_n)
- time of incoming orders T_A T_B T_K

Figure 2 illustrates the stochastic structure of the incoming orders for products A, B ... K and customers C1, C2, ...CH. Also the timing of orders varies randomly.

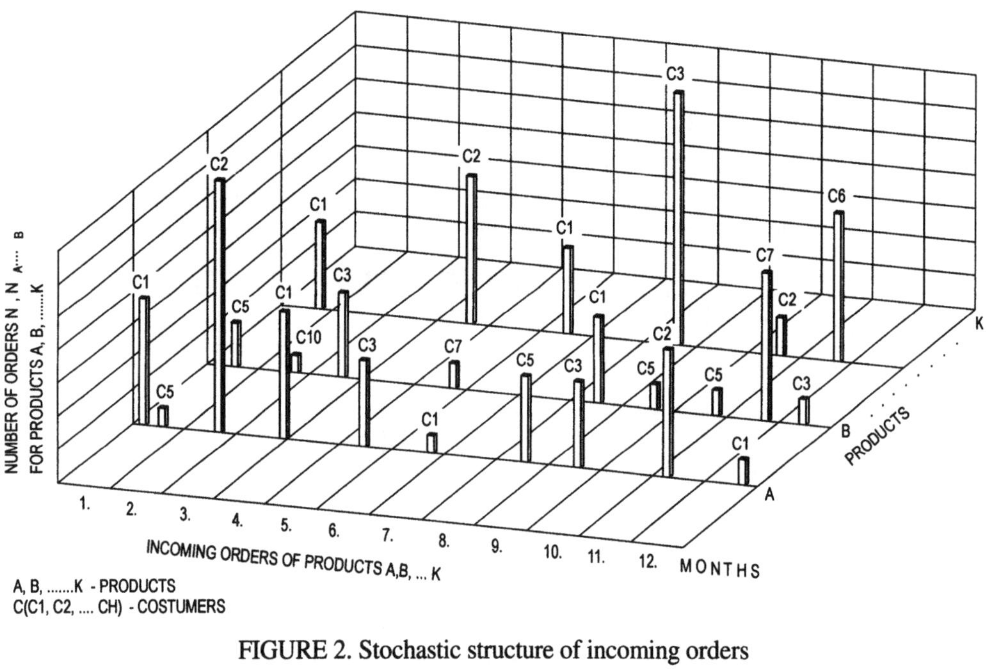

FIGURE 2. Stochastic structure of incoming orders

These stochastic characteristics of incoming orders generate problems in the planning and control of fabrication activities and require a highly adaptable system.

Several years ago the formation of a production system as a cluster of several enterprises appeared on the industrial scene. Figure 3 yields a cluster structure of a production system [4-7]. The main manufacturer (MM), who is the cluster leader, conducts the marketing research for a product to be developed and manufactured. In case that the analysis reveals a positive answer, the MM initiates the development and product prototype. Usually the product development takes place in his own factory. Only in exceptional cases an external R&D-unit is employed. There follows a selection of partners or cooperators K_1, K_2 ... K_v, and the information networking of the cluster units by LAN,

INTERNET, INTRANET, ect. Transport logistics also represents an integrative means in forming the cluster. Of course, the principle of the virtual enterprise plays an important role in the integration of the decentralized production units [4,7].

This type of structuring provides a move effective production, better control, higher flexibility and adaptability of the units integrated into the cluster. MM is responsible for all the important decisions and the coordination of the activities within the cluster.

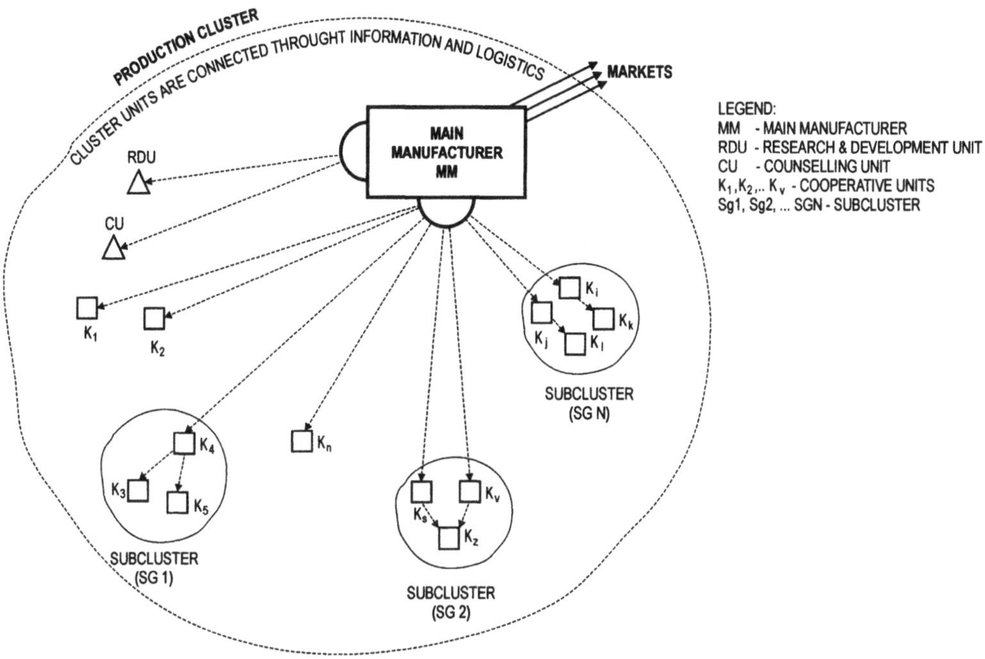

FIGURE 3. Cluster structure of a production system

3 GENERIC MODEL OF A DYNAMIC PRODUCTION CLUSTER, DPC

Our brief survey of the existing production systems indicates that the manufacturing paradigm is slowly changing, turning away from the F. W. Taylor approach [2] to the more adaptable cluster structure [5]. However, it should be stated that some deficiencies could be observed in the cluster structure of a production system as shown in Figure 3.

With the globalisation of markets, the competitiveness of the SME may considerably be affected when compared with the large international companies (LIC). The criteria, which have to be considered, are:

- marketing
- development of HT-products
- innovations related to the HT-products and manufacturing technologies
- product quality
- usage of IT technology
- human resources

TABLE 1. Comparison of the relevant criteria for a successful operation of productions systems

LIC-Large International Corporations	SME-Small & Medium Size manufacturers
MARKETING	
very professional and worldwide	limited and more or less localized
HT-PRODUCT DEVELOPMENT	
top level due to human resources and financial means	limited due to insufficient knowledge, financial and human resources
INNOVATIONS RELATED TO PRODUCTS & MANUFACTURING TECHNOLOGIES	
strongly supported, promoted and financially rewarded. Advanced research on related problems, also support from research institutions	with strong intentions to get on the market and stay there. Innovations depend on the creativity and knowledge of the owners or the employees. Financial limitations regarding investments into new technologies impedes the innovation process. Research institutions are only very seldom asked for help.
QUALITY OF PRODUCTS	
high priority pursuit without a compromise	is properly pursued only if knowledge and financial means are adequate
USAGE OF INFORMATION TECHNOLOGY	
in manufacturing activities and various other processes is intensive and wide spread. There are almost no limitations	depends upon the available IT-knowledge which is rather limited in SME. There are also financial restrictions
HUMAN RESOURCES	
selection of high quality human resources is top priority. Financial questions are not of great importance	high quality human resources are exceptional and rarely selected by the company owner. There are financial limitations.

Table 1 compares these criteria in terms of LIC and SME.

The analysis of the criteria for a successful operation of the SME manufacturers shows that the only way for these enterprises to become competitive in a certain niche against the LIC, is as follows:

1. The SME must cooperate with a R&D-unit or the institution which is responsible for research and development in the field of SME interest;
2. The human resources of SME must get the opportunities to increase the knowledge required for the activities related to marketing, HT-product development, quality and usage of IT in manufacturing. This means that additional education of the subjects is necessary;
3. The state must create a policy which will enable the SME manufacturers to invest into the infrastructure, particularly into the software and hardware of the information and control systems.

On the basis of this analysis a generic model of a DPC has been developed [3]. Figure 4 shows its conceptual model. It consists of three subsystems:

- the production units PU_1, PU_2 … PU_n.
- research & development unit (RDU) and
- a virtual unit for coordination, control and marketing (CCMU).

The selection of various units depends upon the HT-product to be manufactured. That means:

a) market anticipations for a particular HT-product;
b) the production capacities of the cooperating PU_n-units, including their technological capabilities, must be well-known;
c) the knowledge of the subjects (managers, designers, operators etc) on various levels of activities is required the knowledge on information and computing technologies is particularly important;
d) capabilities of PU to invest and
e) RDU dealing with research and development in the field or related field of the proposed HT-product.

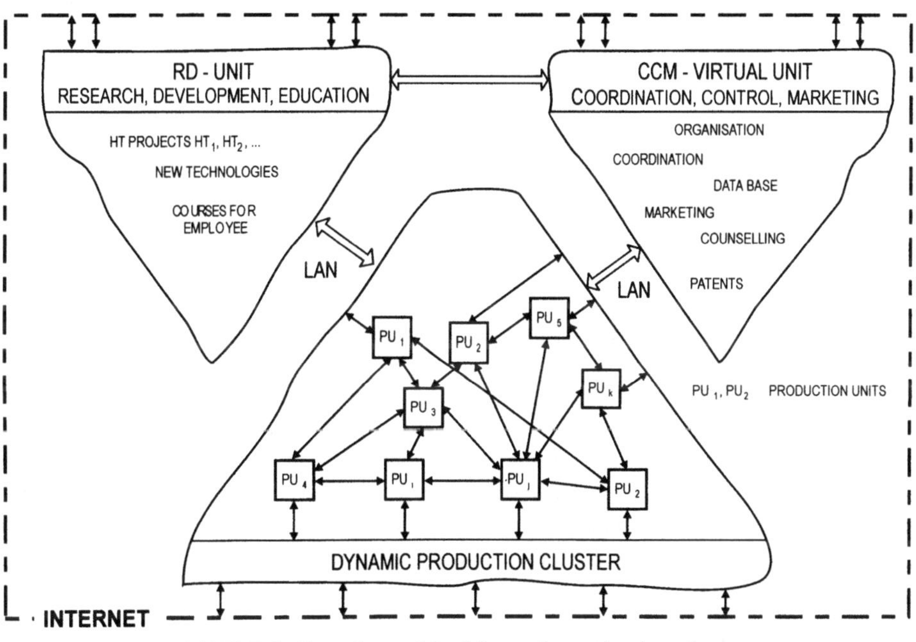

FIGURE 4. Generic model of dynamic production cluster

Based on this information, knowledge, related market and economic analysis, the formation of DPC for a certain HT-product can be implemented.

It should be stressed that, if the HT-product is changed, the selection and integration of cooperaing units PU_1, PU_2 … PU_n, as well as RDU must be modified in order to get optimal cluster structure for the manufacturing of this particular product.

3.1 STRUCTURE OF THE PRODUCTION UNIT, PU

A production enterprise consists of one or more production units PU1, PU2 ... PUk. They are equipped with proper machinery, subjects (operators, group leaders, managers etc) internal transportation equipment, LAN-local area networks and perhaps the INTERNET connection. A typical structure of PU is shown in Figure 5

In order to get a unified basis for structuring the PU and integrating it into the cluster, the concept of the elementary work system EWS has been proposed [9].

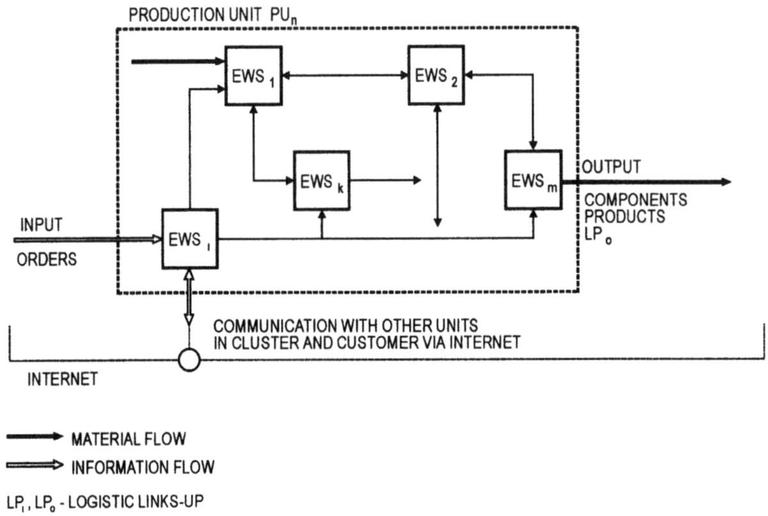

FIGURE 5. Structure of production unit, PU

Figure 6 reveals the structure of EWS. It consists of the process P (cutting, drawing, computing, welding, etc.) the process implementation device PID (lathe, drawing table, computer, welding device, etc.) and the subject (operator, program designer, manager, etc). The block presentation of the EWS enables the use of the control system theory for the description of the PID and process behaviors. The subjects can be characterized by assessing his competence [10] for the job. His task is to determine the process conditions, to supervision the process and to communicate with other subjects.

In case that the PID and/or the process P are computer controlled, a control system is introduced and connected via LAN with other workplaces for programming, control and integration, Figure 7.

3.2 RESEARCH & DEVELOPMENT UNIT, RDU

The development of a HT-product represents a very complex problem for a SME manufacturer. The knowledge of the entrepreneur about the newest accomplishments in scientific and technological research is usually lacking. This means that to develop a new HT-product he has to get in touch with an institution doing research in the field in which he is interested. However, the mechanism and the procedure required to establish such a link, should be developed and tested. The state with its development policy can, in this case, be very helpful.

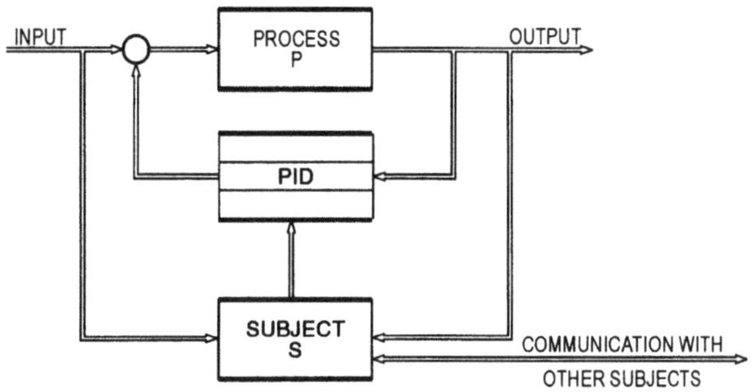

FIGURE 6. Elementary work system EWS

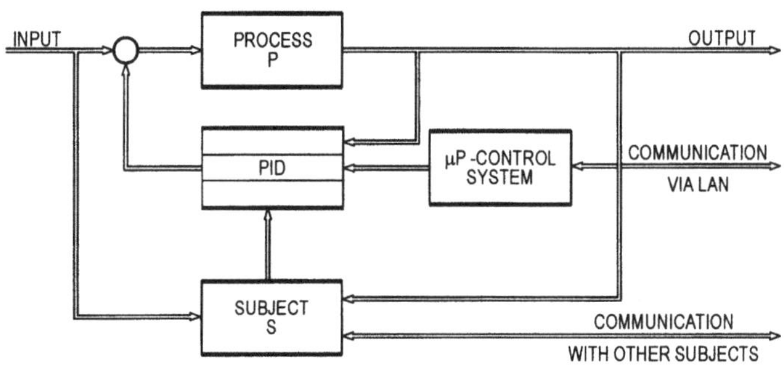

FIGURE 7. Extended EWS

To implement an active cooperation between the PU and RDU a mixed team should be formed in order to develop a new product or/and to improve an existing one.

The same sort of cooperation applies also when new manufacturing and IT technologies have to be introduced in the PU. The education of employees is vital for a successful operation of production clusters. The RDU should organize short courses for e.g.:

- general work and communication with computers and networks;
- computer aided design CAD;
- technological process planning CAPP;
- NC-programming;
- computer aided planning and production control;
- quality control CAQ;
- etc.

There is another important problem which the RDU must address by creating cooperation with the production units. The development of an innovative HT-product requires creativity and competence of the subjects working on the project.

Figure 8 reveals the work system for the development of a HT-product in cooperation between PU and RDUs. It is necessary to form a PU-team supervised by a competent researcher and instructors from the RDU. The project is specified and the PU-team searches for solutions, supported by advice from the supervisor and the instructors. The PU-team should have the necessary knowledge as well as creative ideas contributing to the solution of the problem. This type of project approach has been developed at the Department of Control and Manufacturing System, University of Ljubljana and has proved very successful [11].

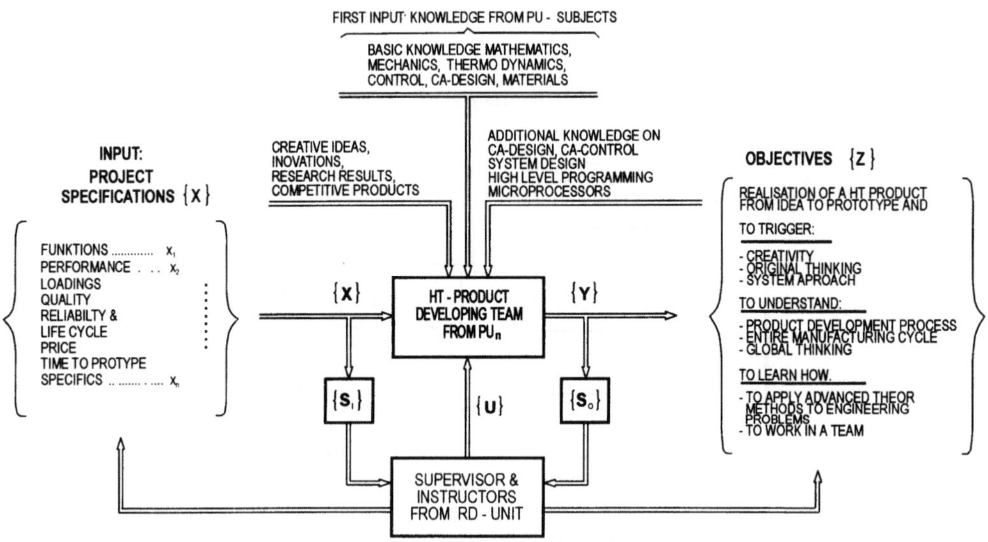

FIGURE 8. Work system consisting a PU-team and RDU instructors

3.3 VIRTUAL COORDINATION, CONTROL AND MARKETING UNIT, CCMU

The formation of a dynamic production cluster DPC is possible only if the marketing analysis indicates successful selling of a product. The operation of the DPC may only be promising and acceptable if the work, control and logistics are properly organized and managed.

The question to be answered is: "How should the structure and operation of the CCMU be formulated in order to implement its integrative and coordinative role in the cluster?"

The information which is vital for the operation of the DPC and has to be at disposal, is listed below:

- survey and analysis of competitive products;
- detailed information about the manufacturing technologies and capacities, available in PU;
- data on materials, machinery elements, electrical and electronic elements etc., required for the HT-product, in addition, the costs, suppliers, delivery times, etc.;

- information on the suppliers of manufacturing machinery and devices, tools, accessories etc;
- information on transport logistics;
- etc.

The organization of the CCMU is virtual due to the fact that the individual PU_s possess a lot of information, and so do the subjects with knowledge required for the functions of coordination, control and marketing. In addition the contributions, particularly in marketing, are coming from the RDU. The localization of the active subjects of the CMMU and relevant data bases can be widespread, but interconnected via computer networks. The communication between the decision making subjects on different locations is vital for the operation of DPC.

4 FORMATION AND OPERATION OF DPC

Based on the presentation and explanations of the DPC concept, its formation as a cybernetic structure will be briefly discussed, Figure 9.

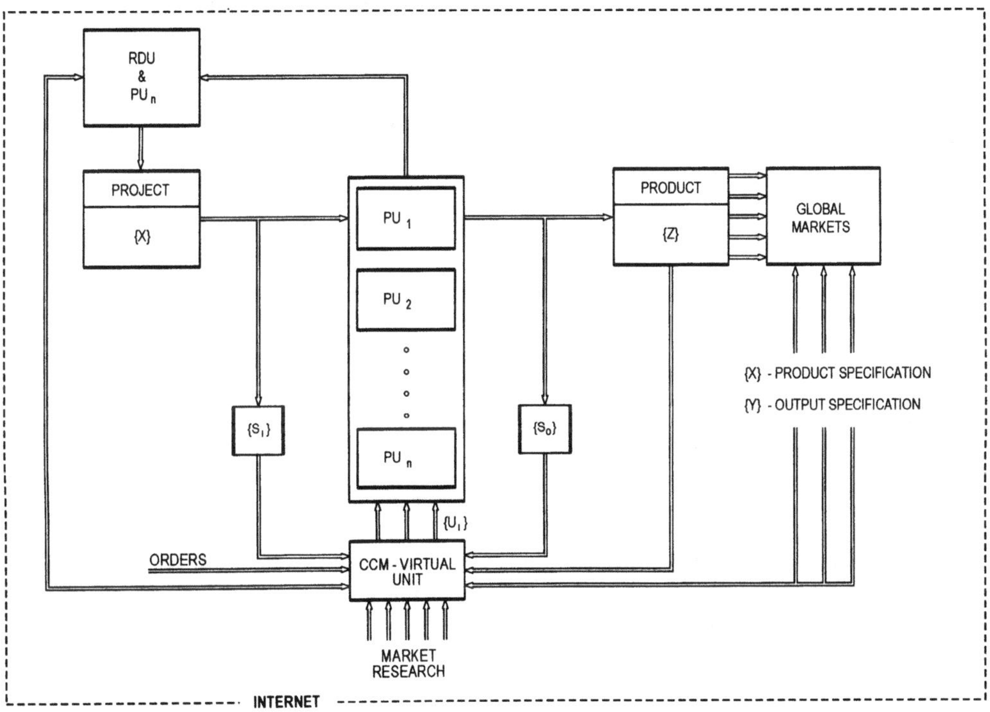

FIGURE 9. Cybernetic structure of a dynamic production cluster DPC

The virtual unit of CCM, establishes what chances a certain HT-product may have. A team of production unit PUn consults a relevant RDU about the development of a HT-product vaguely conceived by the PUn. when the decision of this team is acceptable, the virtual CCMU starts the market research. If the results are promising, the PUn and RDU form a team in order to develop this HT-product. The organizational team structure is shown in Figure 9. The specifications of the project are defined as a set of requirements, the technical documentation prepared and the prototype manufactured.

After these steps have been taken the formation of the rest of the DPC is implemented. The selection of the production units PU1, PU2 ... PUn is based on the economic interest of the PU to become partners on equal footing. Of course, the condition is that the criteria specified in previous discussion are met. That is to say that they have:

- adequate manufacturing means and technologies;
- adequate fabrication capacities in terms of delivery dates, quantity and quality;
- capable subjects; and
- an acceptable location.

When forming a DPC it is important that the cybernetic and control system methods are applied in its structuring. The inputs and outputs of the cooperating units must be well known in order to control them as well as the entire cluster.

Illustrative example of a DPC. In order to test the proposed concept of DPC, in terms of economic and organizational suitability, the following simple example has been initiated.

Manufacturers need for their products various sheet components with the following variations, Table 2.

TABLE 2. Example data

Type	Characteristics		
dimensions	small	medium	large size
materials	A	B	C
sheet thickness	< 2 mm	$2 \div 8$ mm	> 8
parts/order	< 5	$5 \div 30$	> 30
shapes	simple	medium	complex

The cutting technologies employed for the fabrication of sheet parts as shown in Table 2 are:

- mechanical cutting with plate shears, piercing machines;
- electrical arc, flame cutting;
- water jet;
- laser beam.

Five manufacturers PU1, PU2 ... PU5 have been selected for the test. They have CNC machines with the above cutting technologies available. The individual machines are not fully employed. The actual time in which the sheet components are produced varies between 20 % and 70 %, considering the eight-hour work time.

The aim is to increase the actual work time to 100 % and later to increase it to two to three times. And the prospect is very promising.

5 CONCLUSION

The structure of the dynamic production cluster (DPC) represents an innovative approach in the forming and realization of all activities required for the manufacture of a HT-product by SME enterprises. The formation of the DPC is based on market research, executed with the help of a virtual coordination control and marketing unit (CCMU).

The cooperation between the PU and RDU is vital for the innovation of the HT-products based on research in a certain field. By introducing this concept, the transfer of knowledge between PU and RDU will become very intensive and very efficient. The flexibility of the DPC is high and the results of the adaptation of the cluster manufacturing structure to the market fluctuations will be extraordinary.

In addition to these positive influences, further continuous education of the subjects of SME will contribute to a higher technical culture in the industrial environment.

REFERENCES

1. Cincinnati Milacron Report, 1952.
2. Taylor F. W. (1907): Trans. An. Soc. Mech. Eng.
3. Peklenik J. (1988): Fertigungskybernetik – eine Wissenschaffliche Disziplin für die Produktionstechnik, Sonderdruck der Technischen Universität Berlin 1988, pp. 1-25.
4. Sihn W., Gehr F. (19998) Manufacturing in Networks – Competitive Advantages for Virtual Enterprises, Proc. of the 31. CIRP-Int. Seminar on Manufacturing Systems, Berkley, California, 1998, pp. 76-80.
5. Becker M., Crostack H. A., Saar M. (1998): Process Networks Engineering, A model based approach to design and improvement of processes in decentralized manufacturing systems, Proc. of the 31. CIRP-Int. Seminar on Manufacturing Systems, Berkley; California, 1998, pp. 105-110.
6. Martnes P. (1996): Virtuelle Unternehmen, Eine Organisationsstruktur für die Zukunft Wirtschaftwwissenschaftliches Studium, Vol. 6, 1996, pp. 280-285.
7. Mehler B.H., Reinhart G. (1998): Building the Virtual Enterprise – Manufacturing in Decentralized Networks, Proc. 31th CIRP Seminar on Manufacturing Systems, Berkley, California, 1998, pp. 81.86.
8. Wright P.K. (1995): Principles of Open Architecture in Manufacturing, Journal of Manufacturing Systems, Vol. 14, No. 3, 1995, pp. 187-202.
9. Peklenik J. (1994): Theoretical Models of Manufacturing Systems – A Critical Assessment, Proc. of the 27. CIRP-Int. Seminar, Vol. 23 (1994), No.2, pp. 1-9.
10. Peklenik J. (1996): A New Approach to Education of University Graduates for Complex World of Manufacturing, Manufacturing Education for the 21st Century, Vol. III, pp. 11-16, SME, San Diego.
11. Vovk Avšič M., Sluga A., Peklenik J. (2001): Modelling of a Subject Competence, Proc. 3rd Int. Workshop on emergent Synthesis, IWES '01, Bled, Slovenia, 12-13.3., pp. 207-216

RECENT DEVELOPEMENT AND TRENDS IN TOOL CONDITION MONITORING

E. Kulianic, M. Sortino

Department of Electrical, Management and Mechanical Engineering, University of Udine, Italy

KEYWORDS: Intelligent machining systems, tool monitoring, sensors

ABSTRACT: The trend in manufacturing is towards the intelligent machining systems. The productivity and quality of the output of such a system, as well as the machining cost, depend on many factors. One of the most significant factors for tool condition monitoring in machining is the sensor system. This paper reviews recent developments and trends in tool condition monitoring, among them are: the motivation for application of these systems in turning, milling and drilling, optical sensor systems, electrical resistance and temperature sensor systems, radioactive sensors, cutting force and torque measuring systems, effective power and motor current measuring systems, vibration sensors systems and acoustic emission sensors. A multiple sensor approach and its application in industrial conditions are discussed. Further investigations needed in tool condition monitoring and process monitoring are presented.

1 INTRODUCTION

The effectiveness of new machining systems has been lower than expected. The weakest part of the machining system is the tool, i.e. the cutting edge. Sometimes we do not realize that the cutting edge is working probably in the most severe conditions within the whole technical field. The specific cutting force in turning hardened steel could be more than 5000N and the temperature is around 1000°C, when two surfaces are sliding one over the other without any lubricant. Namely, the coolant vaporizes before entering between the tool and the chip or between contact surfaces of the tool and the workpiece. The problem is how long the tool will be able to cut, i.e. what the tool wear rate or the tool life will be. The productivity, the machining cost, the quality, the integrity of the machined surface, and the profit strongly depend on tool wear or tool life.

Not only the tool breakage is very important for the effectiveness of the machining system but the detection of tool wear as well. When the tool is worn the machining process has to be stopped and the tool must be changed. The machining system is not producing when the tool or the tool tip is changed. Since the operator is not available on a new machining system to determine when the tool is worn, there is a need for reliable and simple tool wear sensor systems. The chip form can change with the increase of tool wear, which could be very important in turning, since the long chip can stop the turning process in long-cut cases. The chip control system has to be used in such conditions. In milling, the chip is usually short and does not stop the milling process. From the author's experience a long chip can be obtained in face milling stainless steel with carbide cutter, even if the width of the workpiece is only 50 mm and the cutter diameter is 100mm. The chip could be more than 500 mm long due to the welding of stainless steel chips.

Published in: E.Kuljanic (Ed.) *Advanced Manufacturing Systems and Technology*, CISM Courses and Lectures No. 437, Springer Wien New York, 2002.

Tool wear sensor system and, for turning in some cases, the chip control system are the minimum requirements for a successful functioning of a machining system. Advanced sensor design, coupled with signal processing technologies, provides improved information about the process condition thus enabling process optimization and control. It is not adequate to have information relating solely to tool wear or to tool condition. Additionally, capabilities such as in-process quality control and machine tool diagnostics are the requirements of the future, i.e. there is a shift from monitoring the tool condition to monitoring the process conditions and the resulting part quality.

Many different sensors are available for monitoring the machining systems, Figure 1 [1]. However, the most common sensors used in industrial conditions are force, power and acoustic emission sensors. The most important sensor is the one which ensures a successful and reliable machining process.

In order to obtain almost an optimum performance of the machining system, there is a need for monitoring the following parts of the system: tools or tooling - state of wear; lubrication; workpiece dimensions and geometry; surface roughness; tolerances and surface integrity; the process itself, chip formation; cutting temperature and energy consumption. In practice, it is quite difficult to obtain the optimum. Therefore, it is important to identify the optimum and to come as close to it as possible.

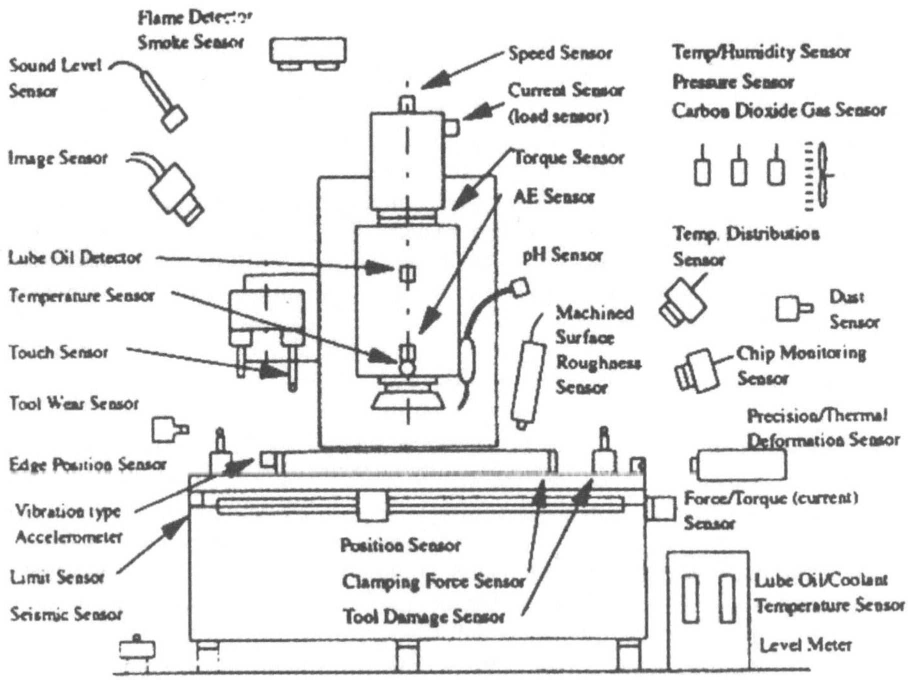

FIGURE 1. Sensors for Machining system monitoring

There is a great amount of information relating to this topic. The first comprehensive survey was published by Micheletti et.al. [2] in 1976 on tool wear sensors for cutting operations. A critical review of sensors for unmanned machining systems was published by Tlusty and Andrews [3] in 1983. Development and trends in monitoring and control of machining process were published by Tönshoff et al. [4] in 1988. Other surveys had been published on various aspects of tool and process monitoring [5,6,7,8]. A comprehensive review was published by Byrne et al [9] in 1995. The review database of tool condition monitoring was published by CIRP – Teti [10] in 1995. In 1999 the author of this paper partly covered the concept of intelligent tool [11].

The aim of this work is to investigate the recent development and trends in tool condition monitoring in machining process including turning, milling and drilling with an emphasis on progressive tool wear, tool chipping, catastrophic tool failure, tool fracture and collision.

2 TOOL WEAR SENSORS

The tool wear sensors can detect the signal out from the tool, the workpiece and the machining process. According to Micheletti et al. [2], there are continuous and intermittent systems, and direct and indirect measuring system. The continuous measuring sensor systems use measured data from the machining process, and the intermittent measuring systems take data only during intervals, i.e. when the tool is not in contact with the workpiece.

2.1 DIRECT MEASURING SYSTEM

Direct systems measure actual quantity of the tool wear, i.e. the flank wear land VB (average or maximum), crater wear, Figure 2, or the plastic deformation of the cutting edge, Figure 3 [12]. According to the author's research, the plastic deformation of the cutting edge can occur in hard turning with carbides tool only few seconds after the start of turning.

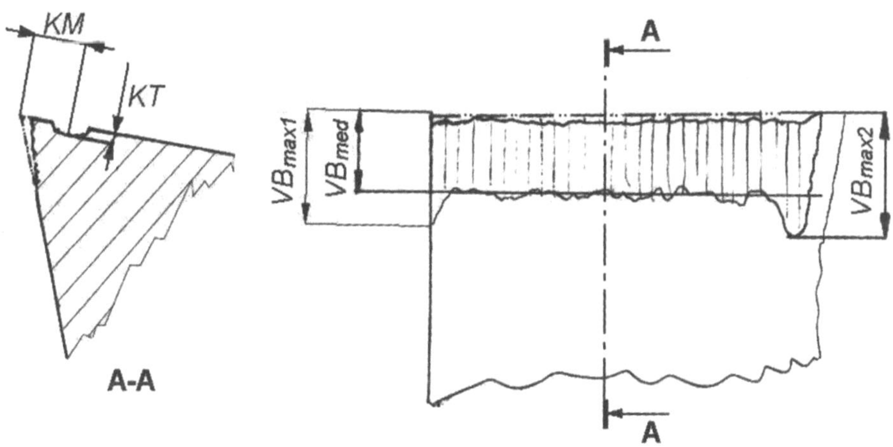

FIGURE 2. Geometry of tool wear

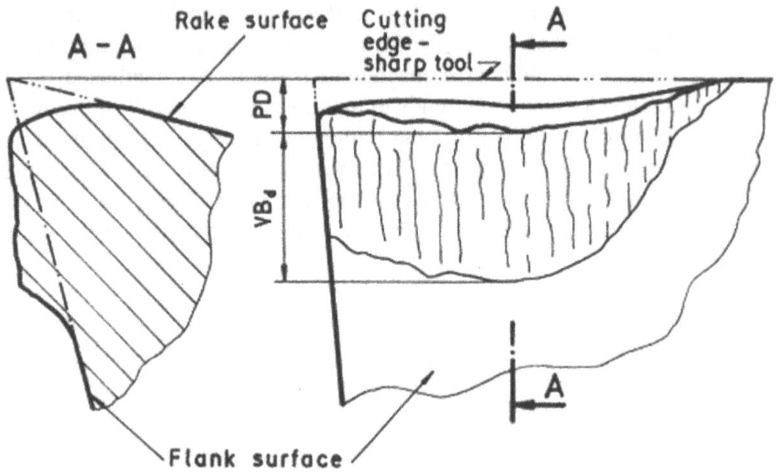

FIGURE 3. Plastic deformation of the carbides cutting edge

OPTICAL SENSOR SYSTEM

Optical sensor systems can be applied to cutting edge when the cutting edge is not in contact with the workpiece. Recent direct tool wear monitoring systems are most frequently based on machine vision by camera. Determining tool wear from processed images of wear zone of the cutting tool has been investigated for more than thirty years.

A research for optical detection of tool wear is going on at the University of Udine – DIEGM, in order to develop an automatic system for tool wear measurement. The basic system for tool wear measurement used in the research is illustrated in Figure 4.

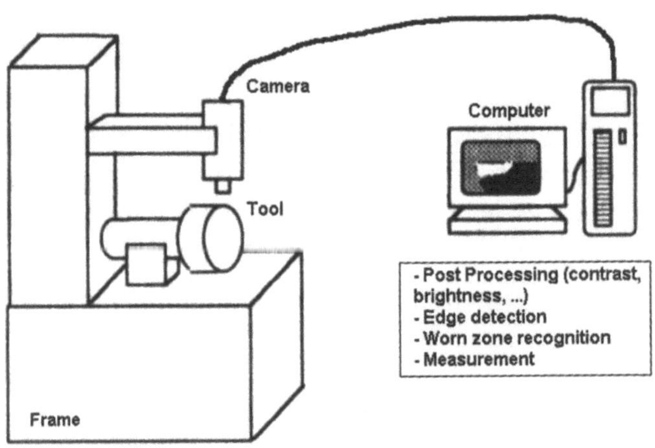

FIGURE 4. Basic System for tool wear measurement

The image of the tool wear is digitalized and processed by a computer. The algorithm used to detect the worn zone from the image is a high-pass bidimensional filter that enhances the edges. In this way a new direction algorithm will be applied for tool worn area detection. It seems that the system is a promising one. The obtained data will be used for tool condition monitoring automatically.

Kurada and Bradley [13] have published a review of machine vision sensors for tool condition monitoring. They covered the basic principles, instrumentation and various image-processing schemes used in the development of vision based on tool condition monitoring system –TCMS. Wong et al. [14] showed that there is a good correlation between tool wear and intensity of laser scatter pattern of the reflected laser ray. The system consisted of a laser focused on the machined workpiece surface in such a way that its reflected ray was captured through a digital camera. The deduced surface roughness was related to the tool wear. It was found that it was very difficult to determine the tool wear by observing the roughness of the machined surface.

According to Pfeifer and Wiegers [15] optical direct measurements are very promising for tool wear monitoring. They proposed a method for adaptive illumination of the tool wear area to obtain optimised camera images. In such a way several machined vision based on monitoring systems have been realized [16,17]. Optical sensors systems can be used for workpiece surface quality measurement, i.e. laser focus system for circle form error, eccentricity and surface roughness, and stray-light sensor system for circle form error and for surface roughness measurements. Also, a laser triangulation sensor can be used to sense grinding wheel profiles during grinding. In general, optical sensor systems are sensitive to chips and dirt and suitable solutions have to be applied.

ELECTRICAL RESISTANCE AND TEMPERATURE SENSOR SYSTEMS

The electrical resistance sensors can be used to detect flank wear land of cutting tools.
Some solutions were found in this field in the early seventies [8, 9, 10].
In the nineties an important research, in this field, was done in cooperation among the Chair on Manufacturing Technology at the Laboratory of Machine Tool and Production Engineering (WZL) at the Technical University of Aachen, Fraunhofer Institute for Coating and Surface Technology (FhG-IST) in Braschwerg Fh-IPT in Aachen, and Fh-IZM in Berlin. The result of the research was an intelligent tool [18,19]. The sensor approach is based on thin film sensor integration into the coating of the coated carbides insert for turning operation. The concept of intelligent tool is given in Figure 5.

In this way it is possible to detect the tool wear and the cutting tool temperature during cutting. The position and the dimensions of the wear zone are monitored on-line using geometrical, adapted mesh of strip conductors. The electrical resistance of this structure, during the signal processing correlated with their location, gives information of the wear geometry on the coated carbides insert. Also, it gives information of the temperature close to the cutting edge.

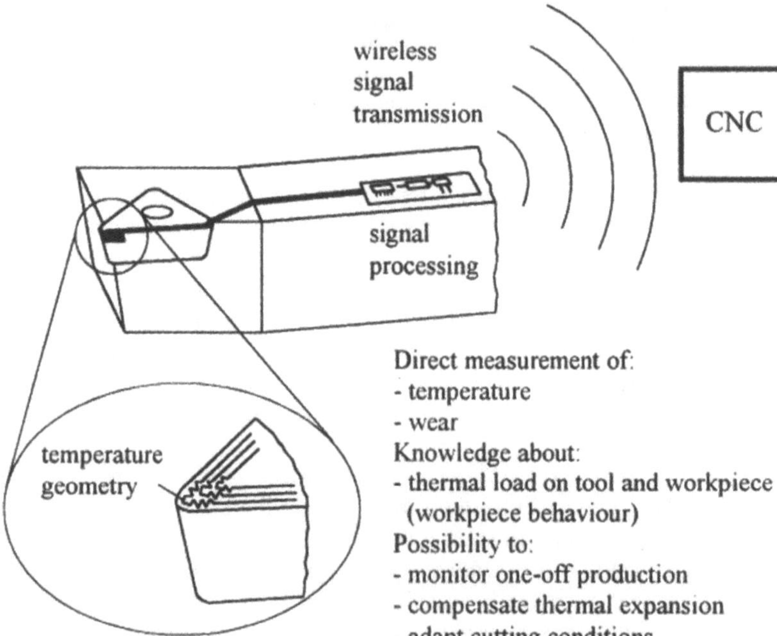

FIGURE 5. Concept of intelligent tool

The advantages of such sensors are that they can operate on-line during machining, and there are no problems with dust, chips, coolant fluid, sensor fixing, and positioning. Therefore, to make such sensors, which have to work in very difficult conditions close to the cutting edge – high temperature (more than 900°C) and high specific cutting force, the solutions have to be found in different fields such as thin-film and surface technology, micro-structuring, bonding technology and cutting process monitoring.

The first prototype of the force sensor integrated in a tool for turning using strain gages was done at the University of Pisa [20]. The sensor system is given in Figure 6.

The advantages of the sensor system are as follows: no modification of the machine tool, negligible reduction of static and dynamic stiffness of the tool, no modification of the external dimensions of the tool, quick and easy installation on the machine tool, adaptability to workshop conditions and acceptable cost.

Since the tool cost is low in comparison to other cost in machining, the intelligent tool or tool condition monitoring system, even if more expensive than a conventional tool, could be a practical solution for tool and process monitoring in the future.

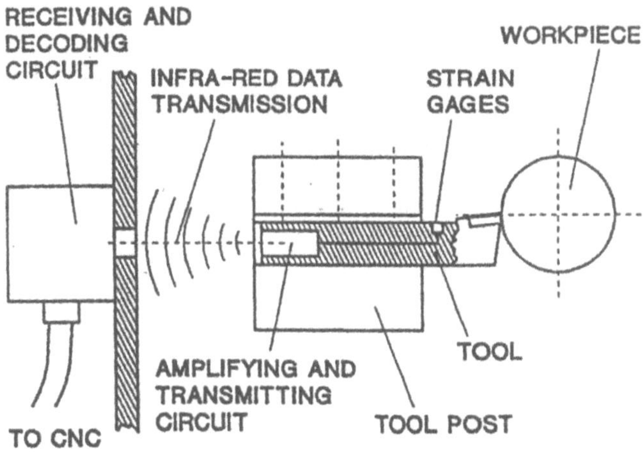

FIGURE 6. Schematic view of the sensor system with the strain gages

RADIOACTIVE SENSOR

The research on radioactive sensors started approximately fifty years ago. Merchant et al. [21] developed a method for rapid tool life testing with radioactive cutting tool. Opitz and Hake [22] made a wear analysis of carbide tools by means of radioisotopes. After these publications a lot of research was going on in this direction.

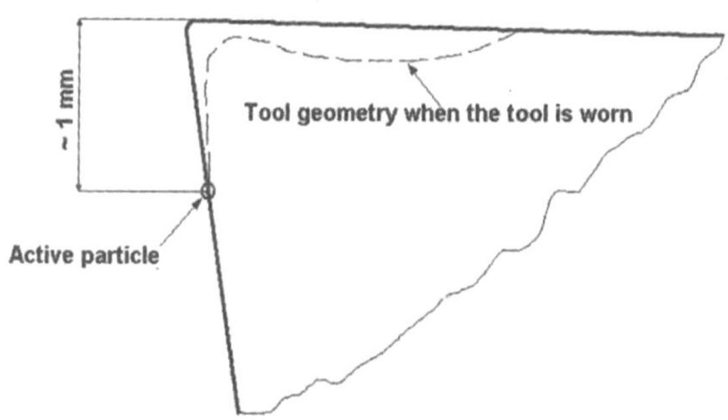

FIGURE 7. Radioactive sensor

At MIT Cook et al. [23] developed the micro isotope tool wear sensing method in 1975. The method consisted in placing a small radioactive particle, with an activity 10^{-8} Curie, beyond the cutting edge on a given distance corresponding to tool wear criterion for example, VB=1mm for rough turning, Figure 7. When the tool wear reaches VB=1 mm, the radioactive particle will be worn out too, and a drop of radioactivity will be noted. This method is very simple. However, it did not find industrial application due to radioactivity. Also, other radioactive sensors were not applied in industrial condition, for the same reason.

2.1 INDIRECT MEASURING SYSTEMS

The tool wear, tool breakage and some important process parameters can be determined by applying indirect measuring processes. The main characteristics of the indirect methods are the continuity of measuring. These processes are generally less accurate than direct methods.
In this section the measured suitable variables for industrial application such as cutting force and/or torque, electrical power and motor current, vibrations and acoustic emission, will be discussed.

CUTTING FORCES AND TORQUE

The cutting forces and/or torque measuring systems are the most important for indirect continuous measuring monitoring. Intensive research has been done to correlate cutting force with tool wear since 1966 Micheletti [24].

The conditions, in this field, have completely changed in the meantime. The most important characteristic of the new dynamometers is the wireless signal transmission. This possibility leads us to a rotating dynamometer.
The Kistler rotating dynamometer, Figure 8, is suitable for three forces and torque measurement.

FIGURE 8. Kistler rotating dynamometer

The advantage of this dynamometer is that there are no constraints of the dimensions of the workpiece in comparison to the plate dynamometer. The rotating dynamometer can be mounted on the spindle of a milling machine in production, and the workpiece can be of any dimensions. This dynamometer is suitable for measurements in milling, drilling and taping during machining in industrial conditions. The rotating dynamometer is composed of a rotor and a stator. The rotor is clamped between the spindle and the tool and the stator are fixed on the spindle housing near the rotor.

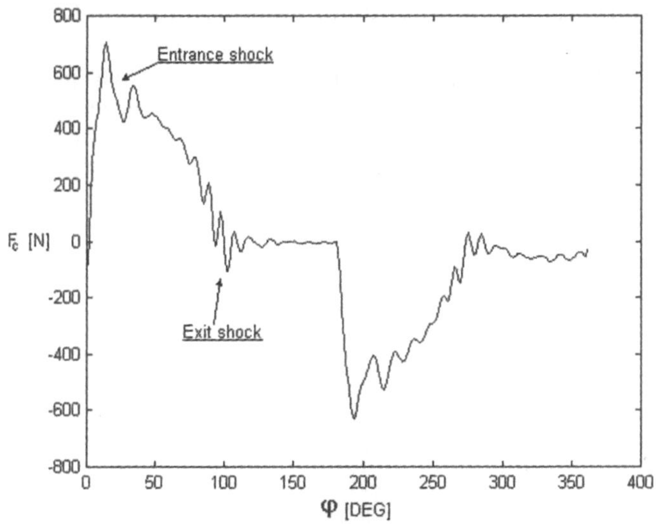

FIGURE 9. Measured cutting force in milling

The measured cutting force F_c in milling, by means of the rotating Kistler dynamometer, is given in Figure 9, and the corresponding torque is illustrated in Figure 10. The investigation of cutting forces and torque in milling with the rotating dynamometer is presented in [25]. The disadvantage of dynamometers for force based monitoring is high cost and a lack of overload protection. The overload can occur in case of tool collision.

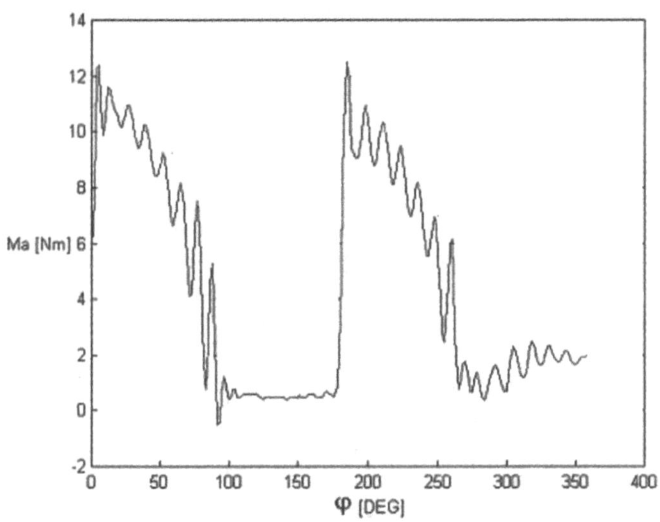

FIGURE 10. Measured torque in milling

MACHINE INTEGRATED FORCE AND TORQUE MEASURING SYSTEMS

For practical reasons, instead of using dynamometers, force measuring plates or rings are installed in the spindle of the machine tool or between the turret disc and the turret housing. In the latter the cross-sensitivity could be greater than the one in dynamometers. For example, Kistler commercial piezoelectric quartz force sensors, for use in machine structure, are available.

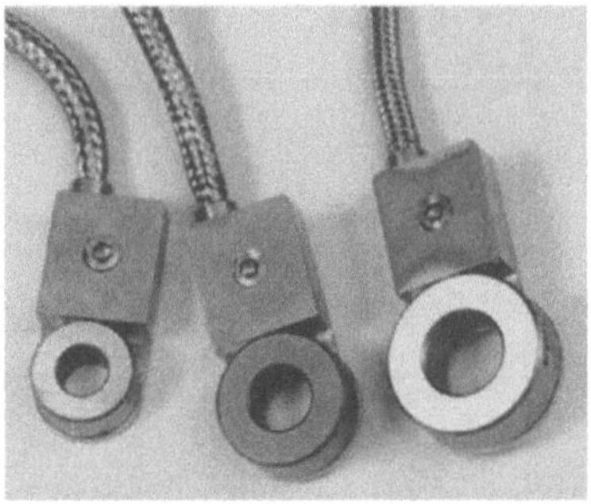

FIGURE 11. Prometec flexible measuring plates.

In Prometec flexible measuring plates – FMP [25], 1D or 3D wedge-type measuring elements are applied. Ready-made standard distance plates are available for all commercial tool turrets. These sensors are over-load protected. An important characteristic of the sensors is high stiffness.

Montronix presents force sensors with integrated charge amplifiers – ICA, Figure 12, which are designed specifically for fixed tool applications, for example, the lathe [27]. Also, they can be used in other machining operations, for instance, in broaching. These force sensors are available in retrofittable versions as well as in more traditional versions designed to be embedded within the machine structure. The characteristics of the force sensors with ICA are as follows: fast response time to detect force transient, improved reliability and easier maintenance in industrial conditions.

Pins, extension sensors, measurement of displace to detect the displacement or bending of tool, force measuring bearing, are well covered in Byrne et al. [9]. Pins are extension sensors, which generally possess a low level of sensitivity. They are suitable only for breakage identification during rough machining.

Measuring the displacement or bending of tool in the direct vicinity of the cutting point represents a good approach for tool monitoring. However, the risk of disturbances due to chips, dirt and cooling lubricant is very high. Therefore, the use of such sensors is not recommended particularly when an automatic tool changing is applied. Force measuring bearing usually reduces the stiffness of the spindle. Since the spindle stiffness is very important it might be used in special cases.

Recent research on the application of sensors integrated motor spindle for monitoring a flexible machining center was done by O'Donnel et al. [28]. A method for pre-processing of sensor signals is detailed and the performance of the sensor integrated motor spindle is examined with regards to realization of the process forces, detection of collision and spindle condition monitoring in drilling.

EFFECTIVE POWER AND MOTOR CURRENT MEASURING SYSTEMS

One of the simplest sensors is the sensor to measure the current or effective power of the main spindle or the feed drives. The disadvantages of the sensors are as follows [1]: when the current or power of the feed drives is measured, the problems occur due to the friction in guide ways or in other places contained in the measuring signal, whereby this component is usually greater than the component of the signal, which is of relevance to the process and may fluctuate strongly in accordance with both the lubrication state and the transverse rate [29]. Due to inert masses the output signal has a low-pass filter characteristic. The tool breakage is detected after the consequent damage has occurred, due to the integrating characteristic of the measured parameters.

The spindle power is proportional to the main cutting force in the direction of the primary motion, which is the least wear sensitive parameter. In spite of this, the first author [30] has found out that there is an increase in the sum of the spindle power and feed power of approximately 20 to 22%, when the face milling cutter is worn in comparison to the power when the cutter is sharp, Figure 12. Since the difference is significant, the power sensor can be used in rough face milling. Brons [30] came to similar conclusion. Such sensors can be used to detect collision, tool breakage and tool wear if adequately large enough cross-section areas of the cut are used.

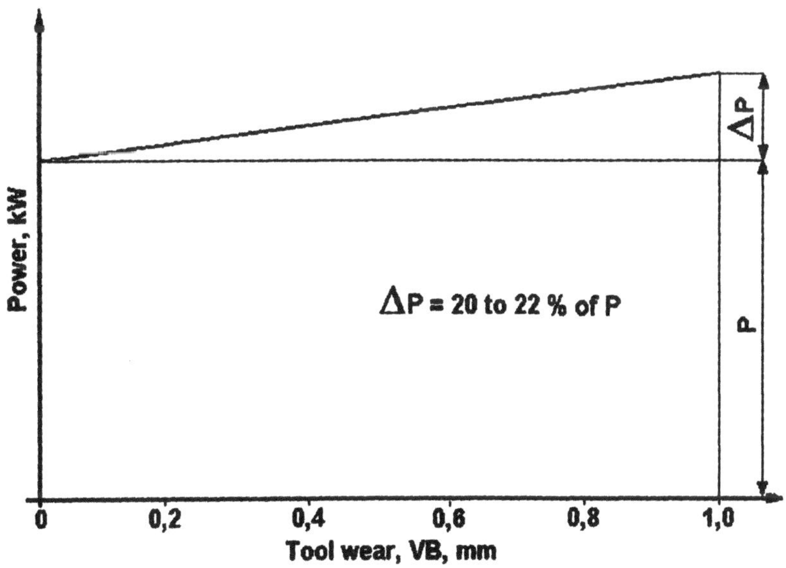

FIGURE 12. The increase of power – spindle and feed - due to tool wear

VIBRATION SENSOR SYSTEM

Oscillation of cutting forces induces vibrations of the machine structure, and these vibrations change due to tool wear. Vibrations are connected to natural frequencies of machining system, therefore, they are connected to system dynamics. Vibrations can be measured either directly by using a position transducer or indirectly by using an accelerometer. Widely used vibration sensors are accelerometers. Typical accelerometers for tool condition monitoring are given in Figure 13 and Figure 14.

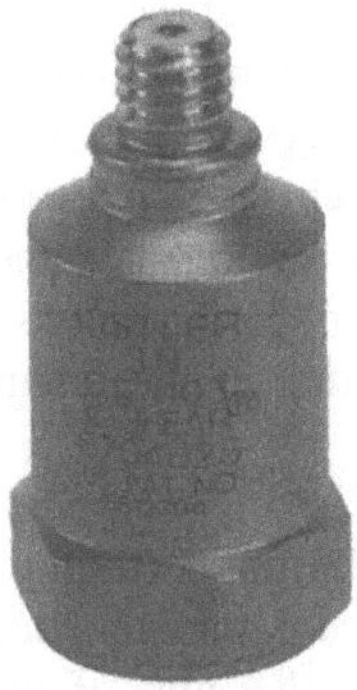

FIGURE 13. Single direction accelerometer - Kistler

Vibration models as a function of tool wear and cutting parameters have been developed in order to understand the basic principles of the phenomenon and they have been used to obtain information for tool condition monitoring. Measurement of vibration and signals analysis could be used for tool condition monitoring. Both systems for tool breakage detection and tool wear monitoring have been developed [9]. Breakage detection can be accomplished by detecting signal peaks that occur during tool breakage. Clearly, tool wear monitoring is a more complex task than tool breakage detection.

FIGURE 14. Multiple direction accelerometer - Kistler

After acquisition and conditioning, the vibration signal is post-processed in order to extract features parameters to be used for tool condition identification. Several approaches have been tempted to detect tool wear state by feature parameters using different indicators: wavelet packet transform [32], time-series modelling [33], Fourier frequency transform, signal root mean square [34], etc. The wavelet packet transform seems to be very useful.

Recently F.Klocke et al. [35], have developed a system for tool condition monitoring in mould and die milling finishing using a tri-axial accelerometer mounted very near the tool. Thus it was proved that the system reliability is high. Compared to the dynamometers and the acoustic emission sensors, the vibration sensors are less expensive and the results are more reliable.

ACOUSTIC EMISSION SENSOR SYSTEM

Acoustic emission for tool condition monitoring is a widespread research activity. Acoustic emissions are generated in the chip formation zone, thus information on tool condition and cutting dynamics are embedded into acoustic emission. However, there are other sources of acoustic emission, for example, the impact of the chip at the workpiece, friction on the rake face and the flank, crack formation and propagation and chip breakage. Several analytical models [36,37] for acoustic emission generation have been developed considering the chip deformation mechanism [38], the cutting parameters [39-41], tool and workpiece geometry, the type of workpiece material [42], and surface characteristic of machined workpiece [43][44].

In the past, several systems for tool breakage detection based on acoustic emission had been proposed [9][45][46]. The development of an acoustic emission-based monitoring system for tool breakage detection is quite simple. When the tool breaks, high-power pulse of acoustic emission is released, as shown in Figure 15.

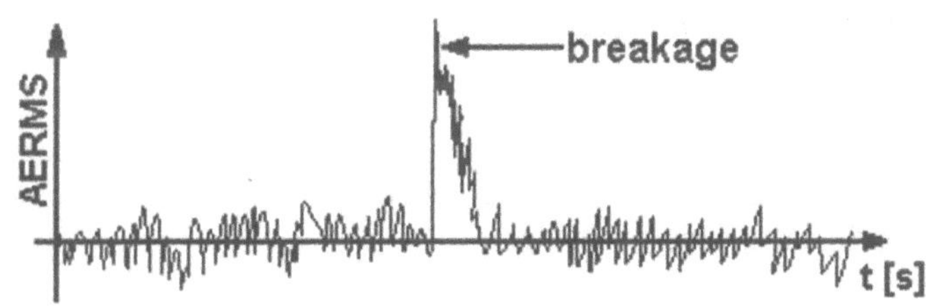

FIGURE 15. Acoustic emission tool breakage example

Nowadays the industry requires on-line tool wear estimators to avoid, if possible, tool breakage. Thus, in recent years, research has been focused on application of acoustic emission for tool wear estimation.

The acoustic emission signal is acquired by using sensors that must be placed very near the acoustic emission source to reduce sensibility of the environment noise. The acoustic emission sensor with integrated charge amplifier is given in Figure 16.

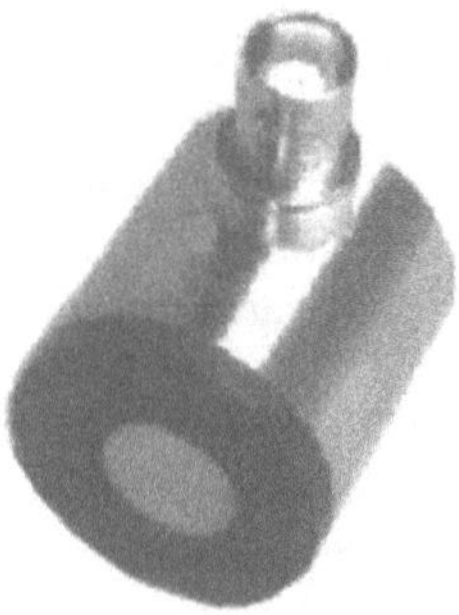

FIGURE 16. Acoustic emission sensor with integrated charge amplifier – Vallen Systeme

The acoustic emission sensor signal is processed in order to extract features parameters for tool state characterization. These feature parameters should reassume all the information on tool state contained in acoustic emission signal. A large number of parameters can be used for this purpose: root mean square (RMS) [47-49], auto-regressive time-series model parameters [50], Fourier frequency transform coefficients [51], skew and kurtosis [52], wavelet packet transform [53,54], etc. The correlation between feature parameters and tool condition is not simple. Thus, self-organizing information processing system should be used. Common used methods for tool state identification are: neural networks, polynomial networks (GMDH), fuzzy classifiers, etc.

The state of the art regarding generation, detection, storage, processing and characterization of acoustic emission sensor has been presented by R.Teti and P.Buonadonna in the CIRP collaborative work "Round Robin on AE Monitoring of Machining" [55]. Acoustic emission signals have been acquired in three different laboratories and the obtained data have been investigated using state of the art methods in six different laboratories. A standard procedure to facilitate industrial application of acoustic emission in tool condition monitoring during machining has been proposed.

Another important branch of research activities covers the application of acoustic emission sensor coupled with other sensors in order to obtain a multiple sensor monitoring system, which seems to be efficient and more reliable.

3 MULTIPLE SENSOR APPROACH

The need for a reliable tool condition monitoring system over a wide range of industrial application is taking research efforts towards a multiple sensor approach. By using multiple sensors, more indicators can be used for tool condition identification. Therefore, the monitoring system can be more reliable and more accurate. An integrated multi-sensor, i.e. acoustic emission and vibration sensor is shown in Figure 17.

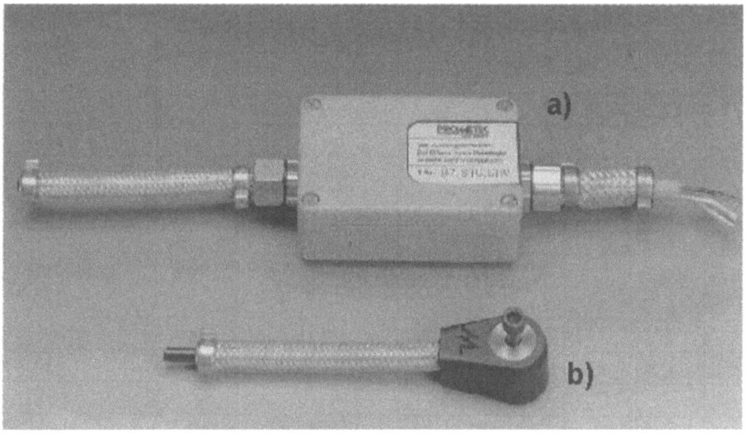

FIGURE 17. a) charge amplifier, b) acoustic emission and vibration sensor – Prometec

The main problem of single sensor system is its extreme specialization that makes it useless when the system is used in different conditions. By using more sensors at the same time, the system can be provided with self-learning capabilities to adapt itself dynamically to different conditions.

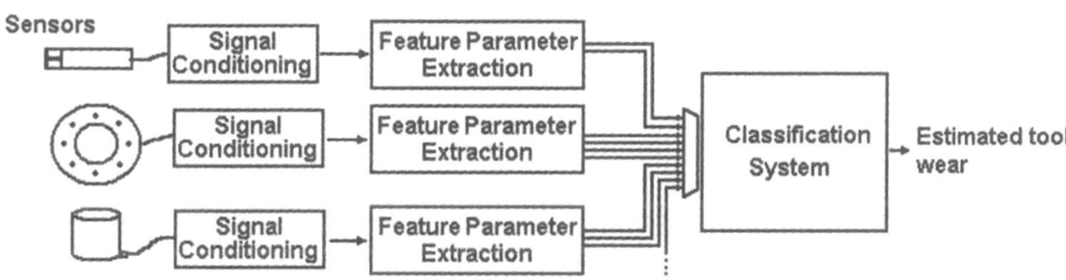

FIGURE 18. Multiple sensor monitoring system

Figure 18 shows a typical multi-sensor monitoring system. The data obtained by different sensors are conditioned and features parameters are extracted independently from every channel. The acquired parameters constitute the input of a tool wear classification system. The result could be the estimated tool wear. The classification system can be realized by using neural networks, polynomial networks, etc.

Intensive research has been done in coupling different sensors. In table 1, different sensor combinations of the sensors application are given, where: CF is Cutting Force, VI is Vibrations, AE is Acoustic Emissions, EP is Electrical Power, IP is Image Processing, NN is Neural Networks, FC is Fuzzy Classifier, GMDH is polynomial networks - Group Method Data Handling, SOM is Self Organizing Map – Unsupervised Neural Network, AR is autoregressive parameter extraction.

TABLE 1. State of the art of different approaches on multiple sensor monitoring systems

On-line Sensors				Off-line sensors	Decision Making	Paper
CF	VI	AE	EP	IP		
x	x	x				[66]
		x	x		Algorithm	[62]
x		x		Tool Geometry		[57]
x	x	x			Fuzzy Reasoning, NN	[64]
		x	x		NN	[60]
x		x			GMDH	[61]
x	x	x	x			[67]
	x			X		[58]
x	x				Data fusion tests	[63]
x		x			NN	[56]
x	x				SOM	[65]
x	x		x		AR	[59]

It is obvious that the multiple sensor monitoring systems are more reliable than single sensor systems. The question is which sensors combination will provide the required information at the minimum cost.

4 CONCLUSION

After more than thirty years of the tool condition monitoring research we still do not have reliable and simple solutions for industrial application. In accordance with the considerations presented in this paper, we may draw the following conclusions.

The application of the tool condition monitoring systems contributes to improvement in the stability of the machining process, tool wear detection, prevention and detection of tool breakage, reduction of non-productive time, optimization of machining process, in-process quality control, etc.

One possible solution for an intelligent tool system could be a new sensor system based on thin film sensor integration in the coating of a coated carbide insert.

Rotating dynamometer as a sensors is more practical for milling, drilling and taping than the plate dynamometer, since it is mounted on the spindle of the machine tool independently of the size of the workpiece.

Furthermore, vibration sensor are less expensive and yield more reliable results than dynamometers and acoustic emission sensors.

Since the cost of intelligent tool or tool condition monitoring system will be much lower than the cost of an intelligent machining system, the application of the intelligent tool or tool condition monitoring system will not significantly increase the machining cost on advanced intelligent machining systems.

It is not adequate to have information based solely on the tool wear or on the condition of the tool. There is an equal need for monitoring the machining process conditions as well as the resulting part quality.

There is also a need for suitable decision making strategy. The sensors and sensor systems must become integrated as a part of the intelligent machining system not only for the increase of the effectiveness of the machining system, but also for data acquisition, for example, machinability data acquisition, probably, by applying the first author's Integrated Machinability Testing Concept [68].

It will be of a further interest to investigate multiple sensor monitoring systems regarding reliability and flexibility, practical application and cost.

REFERENCES

1. T.Moriwaki, (1993), Proc. First Workshop on Tool Condition Monitoring (TCM), D.A.Dornfeld,G.Byrne Editors, CIRP

2. G.F.Micheletti, W.Koenig, H.R. Victor, (1976), In Process Tool Wear Sensors for Cutting Operations, Annals of CIRP, 25-2, 483-496

3. J.Tlusty, G.C.Andrews, (1983), A Critical Review of Sensors for Unmanned Machining, Annals of CIRP, 32-2, 563-572

4. H.K.Toenshoff, J.P.Wulfsberg, H.J.J.Kals, C.A. van Luttervelt, (1988), Development and trends in Monitoring and Control of Machining Processes, Annals of CIRP, 37-2, 611-622

5. D.A.Dornfeld, W.Koenig, G.Kettler, (1993), Present State of Tools and Process Monitoring in Cutting, Proc. New Developments in Cutting, VDI Berichte NR 988, Duesseldorf, 363-376

6. K.Iwata, (1988), Sensing Technologies for Improving the Machine Tool Function, Proc. 3rdInternational Machine Tool Engineer's Conference, JMTBA, Tokyo, 87-109

7. T.Hoshi, (1990), Automatic Tool Failure Monitoring in Drilling and Thread Tapping, Proc. III International Conference on Automatic Supervision, Monitoring and Adaptive Control in Manufacturing, CIRP, Rydzyna, Poland, 41-58

8. D.E.Dimla, (2000), Sensor Signals for Tool.Wear Monitoring in Metal Cutting Operations- a Review of Methods, International Journal of Machine Tools and Manufacture, Vol. 40/8, 1073-1098

9. G.Byrne, D.Dornfeld, I.Inasaki, G.Ketteler, W.Konig, R.Teti, (1995), Tool Condition Monitoring (TCM) – The Status of research and industrial application, Annals of the CIRP, Vol 44/2/1995, 541-567

10. R.Teti, (1995), A Review of Tool Condition Monitoring Literature Database, Annals of CIRP, 44-2

11. E.Kuljanic, (1999), Machining: the Present and the Future, Advanced Manufacturing Systems and Technology, CISM Courses and Lectures No. 406, Kuljanic, E. (Ed.), Springer Wien New York, 11-24
12. E.Kuljanic, (1992), Macro Plastic Deformation of Cutting Edge - A Method for Maximum Utilization of Cutting Tool , Annals of CIRP, 41-1, 151-154
13. S.Kurada, C.Bradley, (1997), A Review of Machine Visions Sensors for Tool Condition Monitoring, Computers in Industry, 34-1, 55-72
14. Y.S.Wong, A.Y.C.Nee, X.Q.Li,C.Reisdorf, (1997), Tool Condition Monitoring Using Laser Scatter Pattern, Journal of Materials Processing Technology, 63, 205-210
15. T. Pfeifer, L.Wiegers, (2000): Reliable tool wear monitoring by optimised image and illumination control in machine vision. Measurement 28, 209-218
16. D.Spath, W.Weis, V.Huntrup, (1995), Dynamische maschinenintegrirte Werkzeugvermessung, VDI-Spezial Werkzenge, Springer Verlag, Heidelberg, 1995, 62-64
17. T.Pfeifer, T.Steffen, L.Wiegers, (1996), Optische Erfassung des Schneidenvershaisses an spaneden Bearbeitungswerlezengen, IX Oberflachenkolloquium, Technische Universitaet Chemmitz, Tagungsband
18. F.Klocke, M.Rehse, (1997), Intelligent Tools Through Integrated Micro SyStems, Production Engineering, Vol. IV/2, 65-68
19. T.Loehken, (1996), Unpublished Technical Report, Fraunhofer Institute for Coating and Surfaces Technology, Braunshweig
20. M.Santochi, G.Dini, G.Tantussi, (1997), A Sensor-Integrated Tool for Cutting Force Monitoring, Annals of CIRP, 46-1, 49-56.
21. M.E.Merchant, et al., (1953), Radioactive Cutting Tools for Rapid Tool Life Testing, Trans. ASME 75, p. 549
22. H.Opitz,O.Hake, (1956), Wear Analysis of Hard Metal Cutting Turning Tools by Means of Radioisotopes, Microtecnic, 10/1
23. N.H.Cook, K.Subramanian, S.Basile, (1975), Microisotope Tool Wear Sensing Method, Report MIT
24. G.F.Micheletti, (1966), Relationship Between Cutting Forces and Tool Wear in Steel Turning, 7th MTDR Conference, Birmingham
25. E.Kuljanic, M.Sortino, F.Miani, (2002), Application of Rotating Dynamometer for Cutting Forces Measurement in Milling, Advanced Manufacturing Systems and Technology, CISM Courses and Lectures, E.Kuljanic (Ed.), Springer Wien New York.
26. N.N., (2002), Flexible Measuring Plates – FMP, Brochure of the Prometec Company, Aachen
27. Montronix – Force Sensors with Integrated Charge Amplifier, 2002, brochure of the Montronix Company, Raleigh
28. G.E.O'Donnell, K.Kelly, G.Byrne, Use of Sensors Integrated Motor Spindle for Monitoring of a Flexible Machining Centre, Intelligent Computation in Manufacturing Engineering,2
29. M.Weck, (1989), Werkzeugmaschinen, Band 2, Konstruktion and Berechnung, VDI-Verlag, Duesseldorf

30. E.Kuljanic, (1972), Study of Wear in Single-Tooth and Multi-Tooth Milling, Doctor of Philosophy Disertation, University of Cincinnati, Ohio, Congress Library, USA

31. S.Brons, (1994), Application of Motor Current Sensing for Indirect Measurement of Cutting Forces – Possibilities and Limitations, doctoral thesis N° 21, Royal Institute of Technology, KTH, Stockholm

32. G.Y.Luo, D. Osypiw, M.Irle, (2000), Real-time Condition by Significant and Natural Frequencies Analysis of Vibration Signal with Wavelet Filter and Autocorrelation Enhancement, Journal of Sound and Vibration, Vol.236/4, 413-430

33. J.C.Chen, W-L.Chen, (1999), A Tool Breakage Detection System Using an Accelerometer Sensor, Journal of Intelligent Manufacturing, Vol. 10/2, pp 187-197

34. T.I.El-Wardany, M.A.Elbestawi, (1996), Tool Condition Monitoring in Drilling Using Vibration Signature Analysis, International Journal of Machine Tools and Manufacture, Vol. 36/6, 687-711

35. F.Klocke, H.-H.Kratz, (2001), Process Monitoring in Mould and Die Milling with Accelerometer Signals, Production Engineering, Vol. VIII/2, 45-48

36. R.Y.Chiou, S.Y.Liang, (2000), Analysis of Acoustic Emission in Chatter Vibration with Tool Wear Effect in Turning, International Journal of Machine Tools and Manufacture, Vol.40/7, 927-941

37. I.Grabec, E.Govekar, E.Susic, B.Antolovic, (1998), Monitoring Manufacturing Processes by Utilizing Empirical Modelling, Ultrasonics, Vol. 36/1, 263-271

38. J.Barry, G.Byrne, D.Lennon, (2001), Observations on Chip Formation and Acoustic Emission in Machining Ti-6Al-4V alloy, International Journal of Machine Tools and Manufacture, Vol. 41/7, 1055-1070

39. D.P.Saini, Y.J.Park, (1996), A Quantitative Model of Acoustic Emissions in Orthogonal Cutting Operations, Journal of Materials Processing Technology, Vol. 58/4, 343-350

40. J.Barry, G.Byrne, (2001), Study on Acoustic Emission in Machining Hardened Steels Part 1: Acoustic Emission During Saw-Tooth Chip Formation, Proceedings of the I MECH E Part B Journal of Engineering Manufacture, Vol. 215/11, 1549-1559

41. J.Barry, G.Byrne, (2001), Study on Acoustic Emission in Machining Hardened Steels Part 2: Acoustic Emission During Continuous Chip Formation with a Non-Overlapping Cutting Arrangement, Proceedings of the I MECH E Part B Journal of Engineering Manufacture, Vol. 215/11, 1561-1570

42. R.Teti, P.Buonadonna, Process Monitoring Of Composite Materials Machining Through Neural Networks, Intelligent Computation in Manufacturing Engineering, Vol. 2, 377-381

43. P.Wilkinson, R.L.Reuben, J.D.C.Jones, J.S.Barton, D.P.Hand, T.A.Carolan, S.R.Kidd, (1999), Tool Wear Prediction From Acoustic Emission And Surface Characteristics Via An Artificial Neural Network, Mechanical Systems and Signal Processing, Vol. 13/6, 995-966

44. H.K.Tonshoff, M.Jung, S.Mannel, W.Rietz, (2000), Using Acoustic Emission Signals For Monitoring Of Production Processes, Ultrasonics, Vol. 37/10, 681-686

45. K.Jemielniak, O.Otman, (1998), Tool Failure Detection Based On Analysis Of Acoustic Emission, Journal of Materials Processing Technology, Vol. 76/1, 192-197

46. I.Tansel, M.Trujillo, A.Nedbouyan, C.Velez, W.-Y.Bao, T.T.Arkan, B.Tansel, (1998), Micro-End-Milling-III. Wear Estimation And Tool Breakage Detection Using Acoustic Emission Signals, International Journal of Machine Tools and Manufacture, Vol. 38/12, 1449-1466

47. T.A.Carolan, S.R.Kidd, D.P.Hand, S.J.Wilcox, P.Wilkinson, J.S.Barton, J.D.C.Jones, R.L.Reuben, (1997), Acoustic Emission Monitoring of Tool Wear During the Face Milling of Steels and Aluminium Alloys using a Fibre Optic Sensorpart 2: Frequency Analysis, Proceedings of the I MECH E Part B Journal of Engineering Manufacture, Vol. 211/4, 311-319

48. J.-S.Kim, M.-C.Kang, B.-J.Ryu, Ji Young-Kwon, (1999), Development Of An On-Line Tool-Life Monitoring System Using Acoustic Emission Signals In Gear Shaping, International Journal of Machine Tools and Manufacture, Vol. 39/11, 1761-1777

49. R.Y.Chiou, S.Y.Liang, (2000), Analysis Of Acoustic Emission In Chatter Vibration With Tool Wear Effect In Turning, International Journal of Machine Tools and Manufacture, Vol.40/7, 927-941

50. H.V.Ravindra, Y.G.Srinivasa, R.Krishnamurthy, (1997), Acoustic Emission For Tool Condition Monitoring In Metal Cutting, Wear, Vol. 212/1, 78-84

51. T.A.Carolan, S.R.Kidd, D.P.Hand, S.J.Wilcox, P.Wilkinson, J.S.Barton, J.D.C.Jones, R.L.Reuben, (1997), Acoustic Emission Monitoring of Tool Wear During the Face Milling of Steels and Aluminium Alloys using a Fibre Optic Sensorpart 1: Energy Analysis, Proceedings of the I MECH E Part B Journal of Engineering Manufacture, Vol. 211/4, 299-309

52. K.Jemalniak, (2001), Some Aspects Of Acoustic Emission Signal Pre-Processing, Journal of Materials Processing Technology, Vol.109/3, 242-247

53. L.Xiaoli, Y.Yingxue, Y.Zhejun, (1997), On-Line Tool Condition Monitoring System With Wavelet Fuzzy Neural Network, Journal of Intelligent Manufacturing, Vol. 8/4, 271-276

54. L.Xiaoli, Y.Zhejun, (1998), Tool Wear Monitoring With Wavelet Packet Transform-Fuzzy Clustering Method, Wear, Vol. 219/2, pp-145-154

55. R.Teti, P.Buonadonna, (2000), Round Robin On Acoustic Emission Monitoring In Machining, Annals of the CIRP, Vol. , 47-69

56. K.-H. Lou, C.-J. Lin, (1996), An Intelligent Sensor Fusion System For Tool Monitoring On A Machining Center, Proceedings of the 1999 IEEE/SICE/RSJ International Conference on Multisensor Fusion and Integration for Intelligent Systems, 208-214

57. S.J.Wilcox, R.L.Reuben, P.Souquet, (1997), The Use Of Cutting Force And Acoustic Emission Signals For The Monitoring Of Tool Insert Geometry During Rough Face Milling, International Journal of Machine Tools and Manufacture, Vol. 37/4, 481-494

58. B.Bahr, S.Montavalli, T.Arfi, (1997), Sensor Fusion For Monitoring Machine Tool Conditions, International Journal of Computer Integrated Manufacturing, Vol. 10/5, 314-323

59. Kumar S.A., Ravindra H.V., Srinivasa H.G., (1997), In-Process Tool Wear Monitoring Through Time Series Modeling And Pattern Recognition, International Journal of Production Research, Vol. 35/3, 739-751

60. Y.Quan, M.Zhou, Z.Luo, (1998), On-Line Robust Identification Of Tool-Wear Via Multi-Sensor Neural-Network Fusion, Engineering Applications of Artificial Intelligence, Vol. 11/6, 717-722

61. H.Henderson, J.Simunek, R. Angulo-Jaramillo, M.G.Schaap, J.-P. Vandervaere, M.T.van Genuchten, C.L.Jiaa, D.A.Dornfeld, (1998), A Self-Organizing Approach To The Prediction And Detection Of Tool Wear, ISA Transactions, Vol. 37/4, 239-255

62. X.Li, S.Dong, Z.Yuan, (1999), Discrete Wavelet Transform For Tool Breakage Monitoring, International Journal of Machine Tools and Manufacture, Vol. 39/12, 1935-1944

63. S.-L.Chen, Y.W. Jen, (2000), Data Fusion Neural Network For Tool Condition Monitoring In CNC Milling Machining, International Journal of Machine Tools and Manufacture, Vol.40/3, 381-400

64. A.Hamrol, (2000), Process Diagnostics As A Means Of Improving The Efficiency Of Quality Control, Production Planning and Control, Vol. 11/8, 797-805

65. C.Scheffer, P.S.Heyns, (2001), Wear Monitoring In Turning Operations Using Vibration And Strain Measurements, Mechanical systems and Signal Processing, Vol. 15/6, 1185-1202

66. M.C.Kang, J.S.Kim, J.H.Kim, (2001), A Monitoring Technique Using A Multi-Sensor In High Speed Machining, Journal of Materials Processing Technology, Vol.113/1, 331-336

67. W.Hu, A.Starr, A.Leung, (2001), A Multisensor-Based System For Manufacturing Process Monitoring, Proceedings of the I MECH E Part B Journal of Engineering Manufacture, Vol. 215/9, 1165-1175

68. E.Kuljanic, (1996), Machinability Testing in 21st Century – Integrated Machinability Testing Concept, Advanced Manufacturing System and Technology, CISM Courses and Lectures 372, E.Kuljanic (Ed.), SpringerWienNewYork, 23-36

COMPREHENSIVE SIMULATION OF MANUFACTURING PROCESSES

H. K. Tönshoff, M. Clausen, K. Tracht

Institute of Production Engineering and Machine Tools, University of Hannover, Germany

KEYWORDS: virtual manufacturing, cutting simulation, structural dynamics.

ABSTRACT. This paper presents a new approach to predict the manufacturing accuracy of given workpiece geometries by comprehensive numerical simulation of the whole metal cutting production process. The intention of the simulation is to determine dynamical influences of the machine tool structure, the metal cutting process and the CNC/PLC on the precision of the workpiece. The simulation of metal-cutting manufacturing consists of four elements, which depend on each other: a CAM module, an elasto-kinematic model of the machine tool structure, a model of the cutting process, and a simulation of the CNC/PLC. After the simulation of the manufacturing process, the actual workpiece geometry is well known. A comparison of this geometry and the nominal workpiece geometry is the basis for an optimization of the NC-Code. The result is an improved manufacturing precision. One application of this simulation system is to obtain higher manufacturing accuracies on weak machine tool structures.

1 INTRODUCTION

In up to date CAM-systems numerical control code for machine tools is generated automatically. If this NC-Code is used, the resulting workpiece differs from the CAD-model. Unaccounted inaccuracies between the real system and its model cause deviations above the design tolerance. These deviations can be due to wear, vibrations of the machine tool with respect to the tool, milling cutter deflection and thermodynamic influences. These influences are not modelled completely yet [1]. This is the starting point of the department of virtual manufacturing at the Institute of Production Engineering and Machine Tools at the University of Hannover, founded by the lower saxony ministry for science and culture. This paper shows how the process simulation, the machine tool simulation and the NC-program simulation will be linked and merged in one comprehensive system.

To enable the virtual machine tool to create and optimise the NC-program, a CAD-model of the real machine tool plus further boundary conditions (type of machine control, tool data etc.) is the input data. The comprehensive model is the basis for a simulation of the cutting process and optimisation of the NC-program.

It is almost impossible to specify all influences in one comprehensive model. There is no comprehensive perception of the interaction between single influences and the workpiece, as a huge number of interactions occurs (cutting tool deflection, cutting behaviour, damping properties of the machine tool, thermodynamic influences, etc.). Therefore well known models of single influences can be integrated into the simulation model. The interaction between the influences can then be investigated to figure out if some influences can be neglected. The quality of the model and the

Published in: E.Kuljanic (Ed.) *Advanced Manufacturing Systems and Technology*,
CISM Courses and Lectures No. 437, Springer Wien New York, 2002.

simulation will be approved by comparing results of the simulation and machined components. Optimisation criteria are dimensional stability, finish quality, machining time and costs.

Influences can be classified in four groups: machine tool (kinematics, dynamics, capacity, etc.), tool (geometry, wear, cutting material, etc.), workpiece (material properties, surface properties, etc.) and cutting process (forces, cutting parameter, arising temperatures, etc.).

These groups will be integrated into the cutting simulation system (CutS), developed at the Institute of Production Engineering and Machine Tools. The comprehensive simulation system consists of separate modules, each containing several models. The general principle of modelling and simulation and the modules will be described in the following.

2 GENERAL PRINCIPLE OF MODELLING AND SIMULATION

A model is a simplified representation of a system. Simulation is the process of imitating (appearance, effect) important aspects of the behavior of the system in real time, compressed time, or expanded time by constructing and experimenting using the system model. Studying the behavior of the system by modelling and simulation is necessity in the case of the machine tool system. A computer simulation model can be tested and proofed in two different ways. Figure 1 shows the paths for model verification.

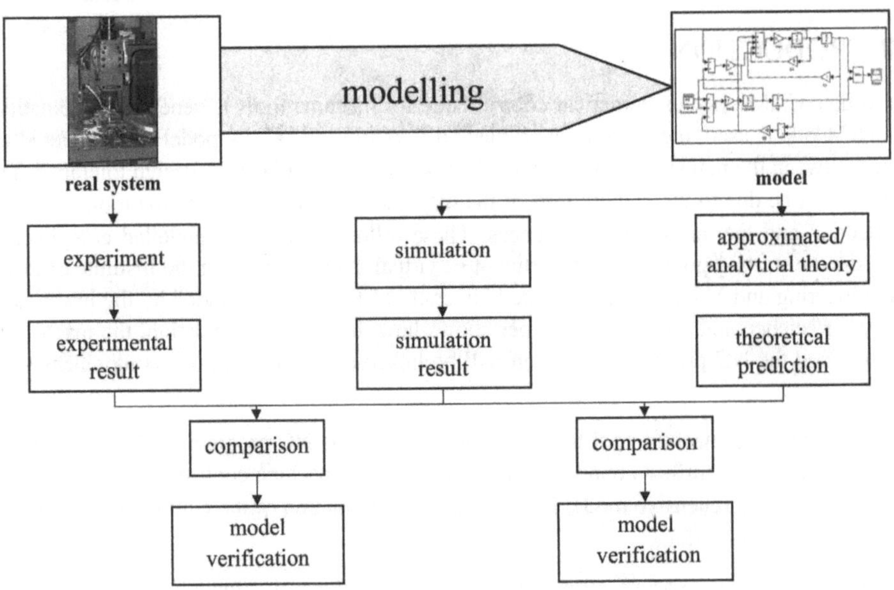

FIGURE 1. Model verification

At first the real system, in this example the machine tool, has to be represented by a model. This model can be expressed in a approximated or analytical theory and a simulation. The theory and the simulation result have to be compared with the experimental result. If the simulation result is not close enough to the experimental result, the model must be improved. That means, modelling is an iterative process. This is the procedure of specified cutting simulation system.

To develop a comprehensive simulation of manufacturing processes the described procedure must be repeated several times. So it is necessary to generate a modular simulation program, that will easily allow changes of single models. Up to date approaches try to include all physical behaviors in one closed system of equations. With these models the verification can be done like in figure 1, but if there are deviations at the results there are no chances to find out the reasons.

The new approach is to model complex systems in a modularized way. This will give the opportunity to include well known but inexact models and substitute them if there are new perceptions. In the following modelling of the appearing processes and the structure of the simulation system is explained and an example model is shown.

3 MODELLING THE CUTTING PROCESS

The cutting process is the most significant part of the manufacturing process. A lot of researchers are engaged with the modelling and simulation in the scope of cutting, chip formation, and the contact between tool and workpiece. Various existing models specify the shear zone and the shear plane respectively while cutting or strain process is in progress [2]. These models determine the present strains and temperatures as well as the graded index and contribute to the determination of workpiece stresses during the cutting process.

To calculate cutting forces in CutS, an analytical approach was chosen. The implemented module of cutting simulation is based on the theories of Merchant, Shaw, Oxley and the semi-analytical approach of Armarego [3][4][5]. The models are enlarged by theories of tribology. These theories are necessary to describe tool wear [6]. The cutting forces themselves are influenced by the friction coefficient and the resulting tool wear. To test models that are influenced by microscopic material properties FEA will be used also. Most approaches concentrate on macroscopic relations and calculate the median of appearing stresses. The new model integrates works that investigate plastic and elastic stresses and rheology models which can be imported through solids subdivided into parts.

With the input parameters, speed, width of cut, cutting velocity, tool angles for all directions etc. it is possible to calculate the cutting forces, temperatures and strains. Coupled with dynamic and static behavior of the machine tool, it is possible to predict the momentary surface formation point. The exactness of the prediction depends on the models that are chosen.

The forces, temperatures, friction coefficients, and strains are calculated three dimensional and time-dependent because of the dynamic process behavior that is assumed in the simulation model. As an example the prediction of cutting forces is shown exemplary. In the following the cutting force prediction module is presented.

PREDICTING CUTTING FORCES

In this part, as an example, the prediction of turning process cutting forces is described. Figure 2 visualize the parameters of turning.

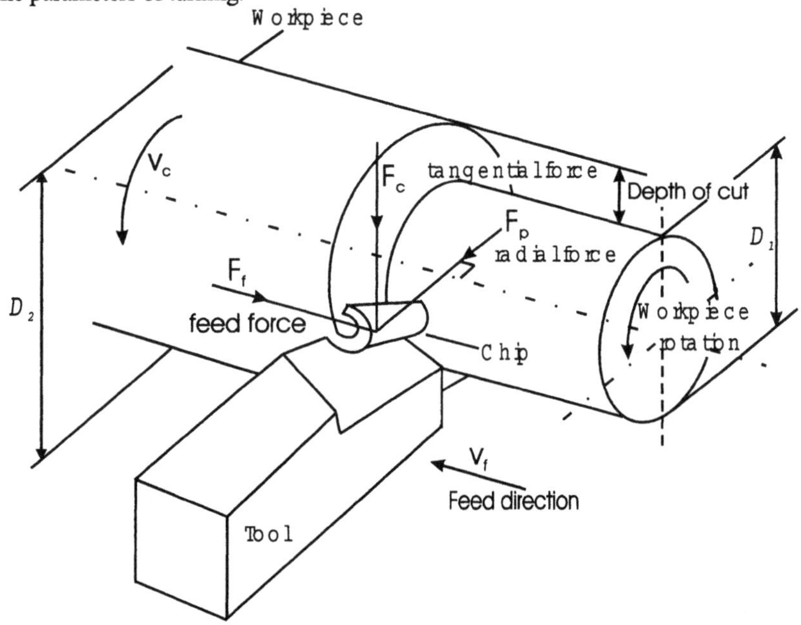

FIGURE 2. Parameters of turning

The workpiece diameter is reduced at the difference between D_2 and D_1.. This is called the depth of cut. The Rotation is v_c and the cutting feed is v_f. The geometric circumstances are explained in the figures 3 and 4. In turning processes the cutting velocity has an inclination angle κ (oblique cutting angle), and thus the directions of shear, friction, chip flow, and resulting cutting force vectors have components in all three Cartesian coordinates (x,y,z).

The x axis is perpendicular to the cutting edge but lies on the cut surface, y is aligned with the cutting edge and z is perpendicular to the xy plane. The forces exist in all three directions. The important planes are the shear plane, the rake face, the cut surface xy, the normal plane xz and the velocity plane. The angle between the shear and xy planes is called the normal shear angle ϕ_n. The shear velocity lies on the shear plane but makes an oblique shear angle ϕ_κ with the vector normal to the cutting edge on the normal plane. The sheared chip moves over the rake face plane with a chip flow angle η measured form a vector on the rake face but normal to the cutting edge. The friction force between the z axis and normal vector on the rake face is defined as the normal rake angle γ_n. The friction force on the rake face ($\vec{F}_{T\gamma}$) and normal force to the rake ($\vec{F}_v$) form the resultant cutting force $\vec{F}_z$ with a friction angle of ρ. The resultant force vector ($\vec{F}_z$) has an acute projection angle of ϕ_κ with the normal plane, which in turn has an in-plane angle of $\phi_n + \gamma_n$ with the normal force F_v. ϕ_n is the angle between the x axis and the projection of $\vec{F}_z$ on the normal plane.

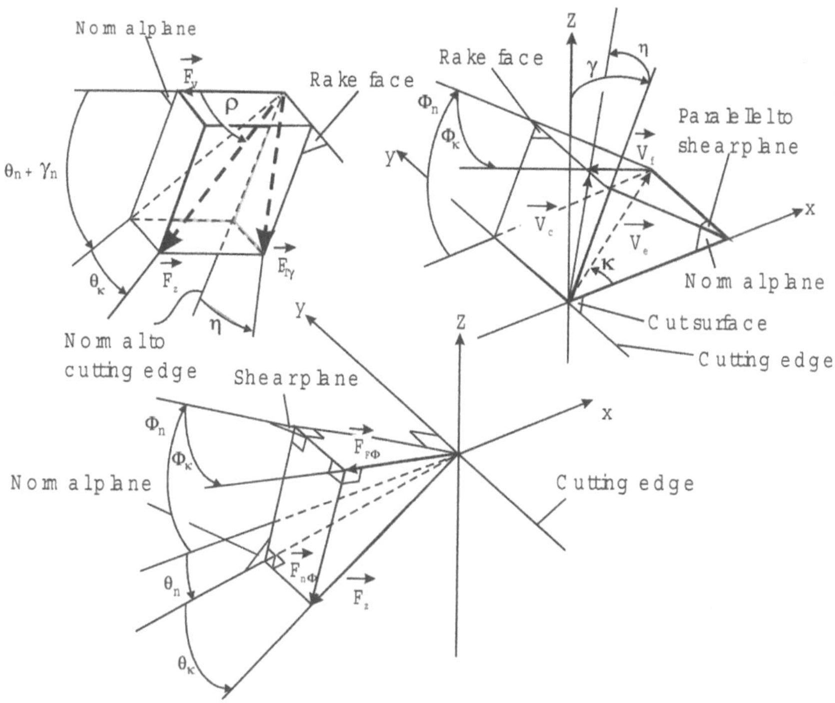

FIGURE 3. Shear, force, velocity diagrams

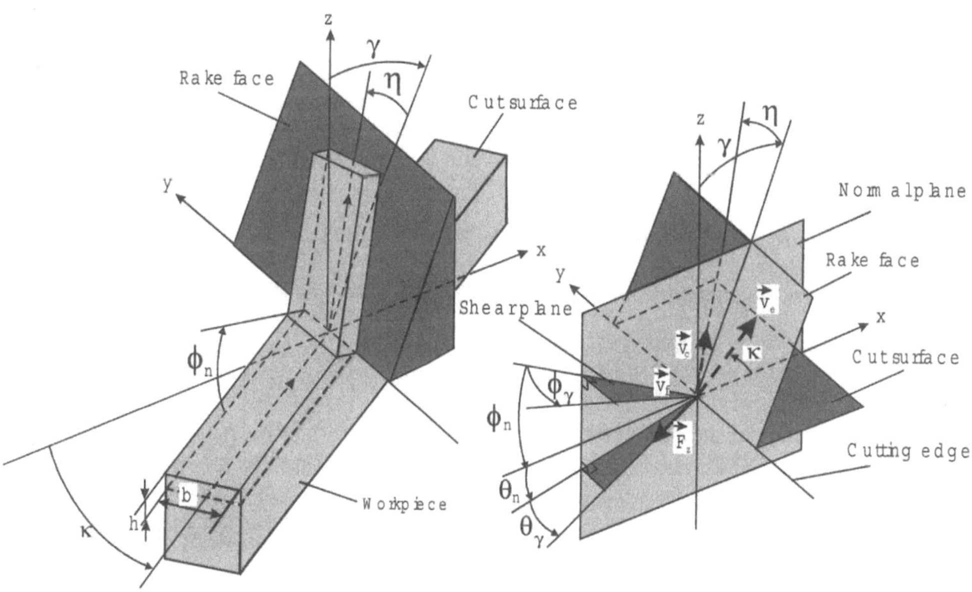

FIGURE 4. Geometrie of oblique cutting

The outcome of this is the following geometric relations:

$$F_{T\gamma} = F_z \sin\rho = F_z \frac{\sin\Theta_\kappa}{\sin\eta} \rightarrow \sin\Theta_\kappa = \sin\gamma \sin\eta \tag{1}$$

$$F_{T\gamma} = F_v \tan\rho = F_v \frac{\tan(\Theta_n + \gamma_n)}{\cos\eta} \rightarrow \tan(\Theta_n + \gamma_n) = \tan\rho \cos\eta \tag{2}$$

The velocities of chip ($\vec{v}_c$), shear ($\vec{v}_f$), and cutting ($\vec{v}_e$) form the velocity plane as shown in figure 4. Each velocity vector can be defined by its Cartesian components:

$$\overrightarrow{v_e} = \begin{bmatrix} v_e \cos\kappa \\ v_e \sin\kappa \\ 0 \end{bmatrix} \tag{3}$$

$$\overrightarrow{v_c} = \begin{bmatrix} v_c \cos\eta \sin\gamma_n \\ v_c \sin\eta \\ v_c \cos\eta \cos\gamma_n \end{bmatrix} \tag{4}$$

$$\overrightarrow{v_f} = \begin{bmatrix} -v_f \cos\Phi_i \cos\Phi_n \\ -v_f \sin\Phi_i \\ v_f \cos\Phi_i \sin\Phi_n \end{bmatrix} \tag{5}$$

By elimination V, V_c and V_s from the velocity relation

$$\overrightarrow{v_f} = \overrightarrow{v_c} - \overrightarrow{v_e} \tag{6}$$

the geometric relation between the shear and the chip flow directions can be obtained.

$$\tan\eta = \frac{\tan\kappa \cos(\Phi_n - \gamma_n) - \cos\gamma_n \tan\Phi_\kappa}{\sin\Phi_n} \tag{7}$$

Now the geometric relationships are defined. To predict the cutting force, it is necessary to make some assumptions. From the geometry, the shear force is expressed as a projection of $\vec{F}$ in the directions of shear.

$$F_{F\Phi} = F_z[\cos(\Theta_n + \Phi_n)\cos\Theta_\kappa \cos\Phi_\kappa + \sin\Theta_\kappa \sin\Phi_\kappa] \tag{8}$$

Another possibility is to take the force as a product of shear stress and shear plane area.

$$F_{F\Phi} = \tau_s A_s = \tau\left(\frac{b}{\cos\kappa}\right)\left(\frac{h}{\sin\Phi_n}\right) \tag{9}$$

where A_s, b and h are the shear area, the width of cut, and the depth of cut (uncut chip thickness), respectively. By equation the two shear force expressions the resultant force is derived:

$$F_z = \frac{\tau_s bh}{[\cos(\Theta_n + \Phi_n)\cos\Theta_\kappa \cos\Phi_\kappa + \sin\Theta_\kappa \sin\Phi_\kappa]\cos\kappa \sin\Phi_n} \tag{10}$$

The assumptions regards the shear angle and the ship flow angle. There are up to 50 different theories [2] for prediction these angles. The same problem appears with the material constants.

Until now there are no assumptions about friction or thermal effects. This is done through the Coulomb friction law which integrates the slip-stick effect. By including the energy input of the system it is possible to calculate the current temperatures. Even if the theories are chosen, the equations can only be solved by iteration. To enable calculations of high dynamic processes, like cutting, the times step size has to be variable. This becomes more important if the machine tool dynamic is integrated into the simulation.

CUTTING PROCESS MODULE

The cutting force prediction module is divided into single model parts, as shown in figure 5. A workpiece material model, the cutter geometry and the cutting parameters plus the decision which cutting process are chosen (turning or milling). Than the cutting force prediction module calculates the cutting force surface with a chosen shear angle, friction and temperature model. The cutting force prediction module interacts with the machine tool model, and the environment temperature. The environment temperature interacts with the heat dissipation caused by the cutting process. The machine tool needs the cutting tool edge location and the appearing forces to calculate the machine tool movement and the cutting tool deflection.

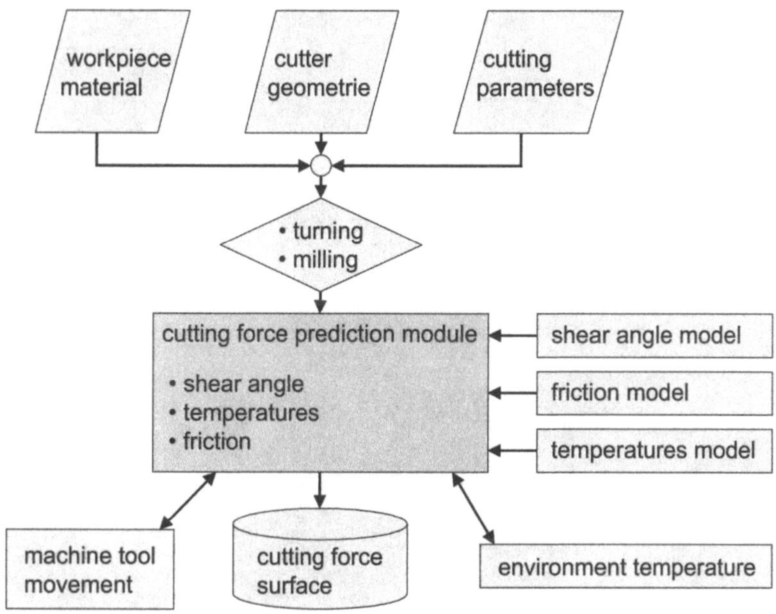

FIGURE 5. Modelling the cutting process

4 MODELLING THE MACHINE TOOL

The machine tool design is very important for the dynamical behavior of the machine tool structure. Also it is necessary to include the elasticity of the clamping and the cutting tool into the machine tool model.

The machine tool model is separated into different parts of the real machine tool. One part is called the machine tool structure model, it represents the machine structure. Other parts are models of the workpiece, the cutting tool and the numerical control system. All these parts form the machine tool module.

Figure 6 is a principle overview of the machine tool module. The dataflow will be shown in the following. The CAD model which represents geometric data of the machine tool is extended by information about material properties, damping properties of the parts and their connections. This information is stored in a database and has been developed by experiments during the design and assembly of the machine tool. A software tool is under development to extend the CAD model with definitions of sliding contact bearings, bearings, spindle and drives, that means all moveable parts of the machine tool. The datamodel is now prepared for a meshing with subpart-elements, which are based on the superelement technique [7]. The subpart-elements were developed during the proceeding of the project. With these subpart-elements it is possible to create a multi-body system to calculate the dynamic behavior of the machine tool structure.

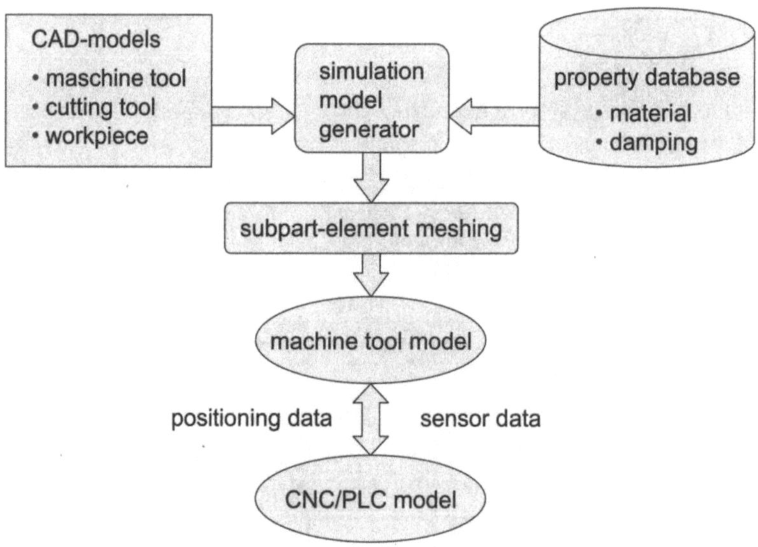

FIGURE 6. Machine tool module

A separate model of the cutting tool enables the possibility of tool changes during the manufacturing process. The structure of the tool is treated like the machine tool structure and is coupled with the tool fitting. The cutting tool can be a milling or a turning tool, depending on the selected machine tool structure. The cutter deflection is calculated time dependent during the dynamic simulation of the cutting process. This is achieved by the implemented multi-body system.

The workpiece is modeled in the geometric shape of the raw part. The clampings fix it to the clamping plate with a static force, that has to be entered before calculations. The subpart-elements mesh is more detailed than the machine tool structure to raise accuracy of the prediction. It is possible to choose a constant property allocation or a statistical allocation for material design. The first one represents the median which is determined by tension tests and the second one is a closer approximation of real circumstances determined by grinding patterns.

The workpiece simulation will include a model of the elastomechanic properties to determine the system response of force effects and a model of the material properties to determine the surface properties and reactions to thermal influences.

A control unit model is necessary to simulate the machine movements. So it is possible to send NC-code to the application, and NC-commands will be executed step by step. The NC-commands will be translated and transferred to the virtual spindle drives and this will force the movement of the dependent machine tool parts.

The machine tool module interacts with the process module and the NC-program. This way the dynamical behavior of the machine while cutting is calculated. The result of the calculation is used to recalculate the cutting forces and predict the surface properties.

The graphical representation and the aim of the simulations system are described in the next chapter.

5 THE SIMULATION SYSTEM

The machine tool module and the process module and their connection are implemented in the simulation system. Output parameters are input parameters for the depending module. Calculation results are visualized in the graphical user interface. The visualization is done during or after the simulation process.

The input parameters to simulate the machining process are shown in figure 7. The machine-tool on which the machining should be processed, must be well known. The CAD-model, the material, damping and elastic properties of the machine tool have to be defined before simulation can start. The raw part of the workpiece has to be known and the cutting tools that are in use need similar datasets as the machine tool. Now the machine tool module preprocesses the dataset as described in the chapter 'modelling the machine tool' (extend CAD-data, mesh, etc.).

After preprocessing the machine tool dataset, the simulation system expects the NC-program. It is possible to create the program online on the CNC/PLC interface like on a real machine tool control unit. Furthermore a CAM system can be used to create an NC-Program nearly automatically.

When the NC-program is imported or coded manually, the simulation system is ready to start. The machine tool module and the process module are interacting while processing the NC-blocks step by step. After processing the NC-code, the workpiece surface is well known.

The cutting sequence and the differences between target and actual contour are visualized. Additionally the machining time and workpiece tolerances are calculated and plotted as a result. If the simulation result is not close enough to the planned workpiece surface, form, dimension and/or machining time, an optimization process can be started. If the machining strategy can be improved, the reprogramming of the NC-program is necessary. An automatic change is not implemented yet.

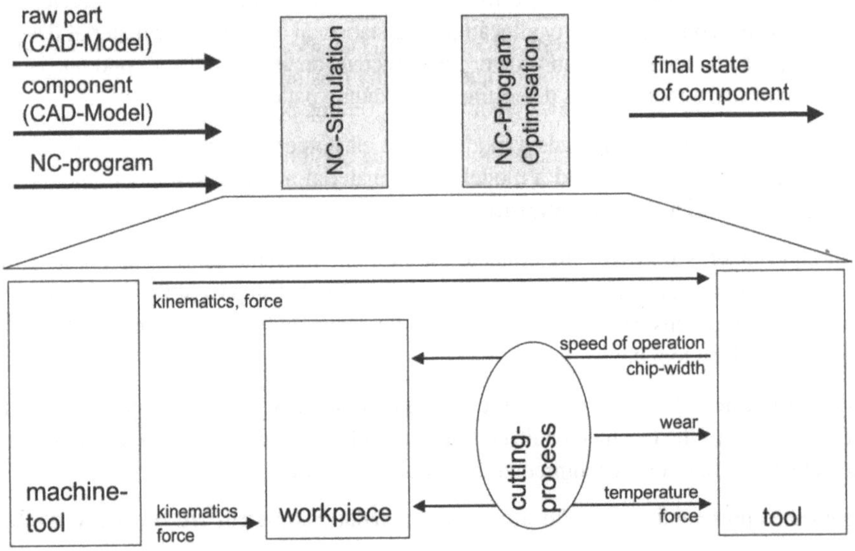

FIGURE 7. Cutting simulation system (CutS)

6 CONCLUSION AND OUTLOOK

This article presented the systematical proceeding to simulate cutting processes in one comprehensive model. The model will be part of the cutting simulation system (CutS), which is currently under development. With a suchlike system it is possible to simulate the machine tool behavior under machining conditions. The presented modelling, simulation and verification process will be continued. The quality of the model and the simulation will be approved by comparing results of the simulation and machined components. An experimental machine tool is under development for this purpose.

The simulation of machine tool behavior has two considerable benefits. The start-up times of cutting processes will be shortened by reducing costly tests on real machine tools. The tests will be performed on a virtual machine tool. The system can be used in the design phase to predict properties of the designed part. In this stage changes can be done with mean effort. This will reduce failure costs in progress planning.

REFERENCES

1. Luttervelt, C.A. v.,, et.al.: Present Situation and Future Trends in Modelling of Machining Operations, Keynote Paper, Annals of the CIRP, Vol. 47/2, 1998.

2. Luttervelt van, C.A.: Typology of models and simulations of cutting operations, 4. CIRP International Workshop Modelling of machining operations, 17-18 August 2001, pp. 41-46, 2001
3. Shaw, M.C.; Cook, N.H.; Smith, P.A.: The basic mechanics of three dimensional cutting operations, Trans. ASME, Vol 74/6, 1952.
4. Oxley, P.L.B.: Mechanics of Machining: An Analytical Approach to Assessing Machinability, Ellis Horwood, Chichester, 1989.
5. Armarego, E.J.A., Deshpande, N.P.: Computerised end-milling force predictions with cutting models allowing for eccentricity and cutter deflection, Annals of the CIRP, Vol. 40/1, pp. 25-29, 1991
6. Schmidt, J., Walter, U.: Verschleiß- und Belastungscharakterisierung im tribologischen System Fräsen, Konferenz-Einzelbericht zum 10. Internationalen Kolloquium "Tribology-Solving Friction and Wear Problems" in Esslingen, Bd. 1 (1996), S. 15-22. (in German)
7. Maißer, P., Tuan, P. A., Härtel, T. Freudenberg, H., Asch, M., Borgwardt, P., Kotzias, B., (1993), Modellierung und Simulation hybrider Mehrkörpersysteme mittels Superelementtechnik, Technischer Bericht im Auftrag der deutschen Agentur für Raumfahrtangelegenheiten Förderprojekt-Nr 50 IB 9206, Institut für Mechatronik, Chemnitz (in German)

KNOWLEDGE MANAGEMENT - AN ESSENTIAL CONTRIBUTION TO COMPANY'S SUCCESS

H. Schulz

Institute of Production Management, Technology and Machine Tools (PTW),
Darmstadt University of Technology, Germany

KEYWORDS: Knowledge management, best practise, competition leadership

ABSTRACT: Today knowledge management must be understood in this way that knowledge essentially contributes to company's success.
Successful companies use new and specific methods of knowledge management with the objective to become faster, better and cheaper than the competitors. Knowledge is the key for this and not only the product, the capital and the personnel of the company. Good knowledge management correlates with company's success.

1 INTRODUCTION

Today knowledge management must be understood in this way that knowledge essentially contributes to company's success.
The introduction and use of information and communication (IT)-technology alone is not sufficient. To only look on hard figures for company's success like financial indicators, e. g. profit, marketshare etc., leads to a wrong point of view. A holistic consideration shows that the so-called weak aspects, which cannot be fixed by real figures, e. g. customers relationship, patents, special strong facts of the company and its products essentially guarantee the future life of the company.
Therefore, knowledge is the understanding of relations and causalities, and is essential in making operations effective, building business processes or predicting the outcomes of business models.
Knowledge-management is a conscious and systematic decision making about the best use of scarce resources under uncertainty to achieve lasting improvements in an organization's performance.
So modern knowledge-management leads to a new kind of thinking in a holistic dimension.

2 THE BRIDGE TO COMPANY'S SUCCESS

PTW and McKinsey created a survey which very clearly shows the correlation between a good knowledge management and company's success. This is a very new point of consideration.
We analysed companies of different industrial sectors (fig. 1), roughly equally throughout Europe, the US and Japan.

Published in: E.Kuljanic (Ed.) *Advanced Manufacturing Systems and Technology*,
CISM Courses and Lectures No. 437, Springer Wien New York, 2002.

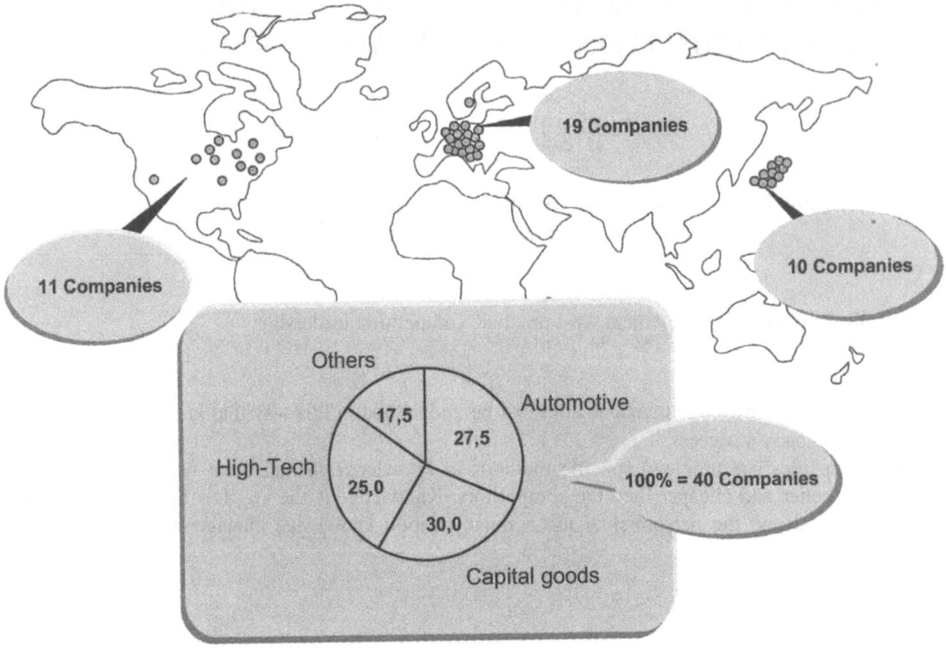

FIGURE 1: Participants of Knowledge Management Survey

As a result of more than 400 interviews with 100 specific questions a database with almost 50.000 entries was generated. At each company we conducted 8 managers of different management levels: R & D, procurement, production, marketing, sales and the chief executive for knowledge-management. A special performance indicator reflects the company's process and financial performance (fig. 2).

Using this indicator we categorized the top 15 companies as "more successful" and the bottom 15 as "less successful". Figure 3 shows the remarkable differences of the performance figures of the more and the less successful companies.

Analyzing the state-of-the-art and the actual literature we found a basic set of 139 knowledge management techniques. Regarding this in the survey we got a fundamental statement considering the relationship between a good knowledge management and the company's success.

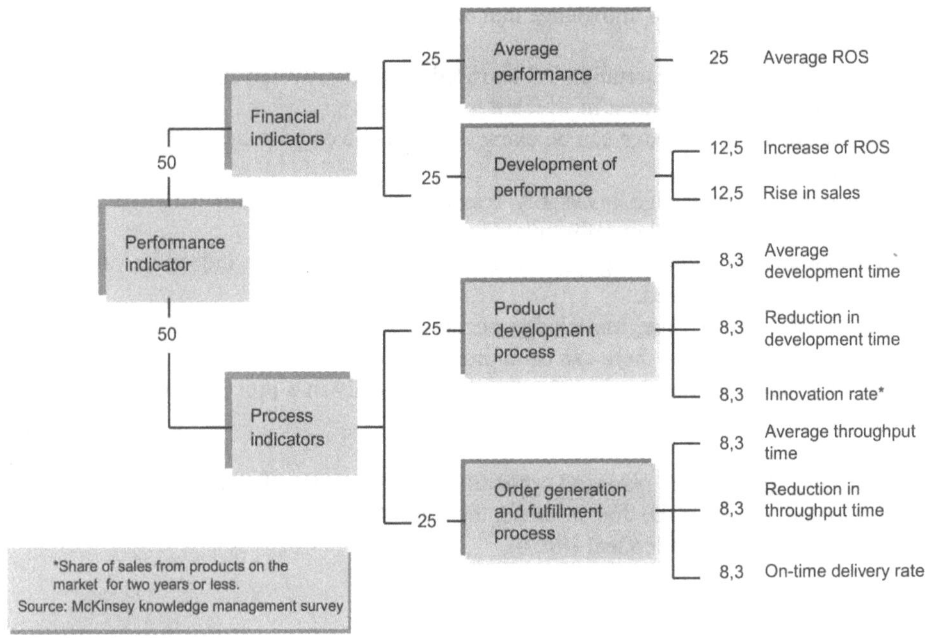

FIGURE 2: Performance indicator

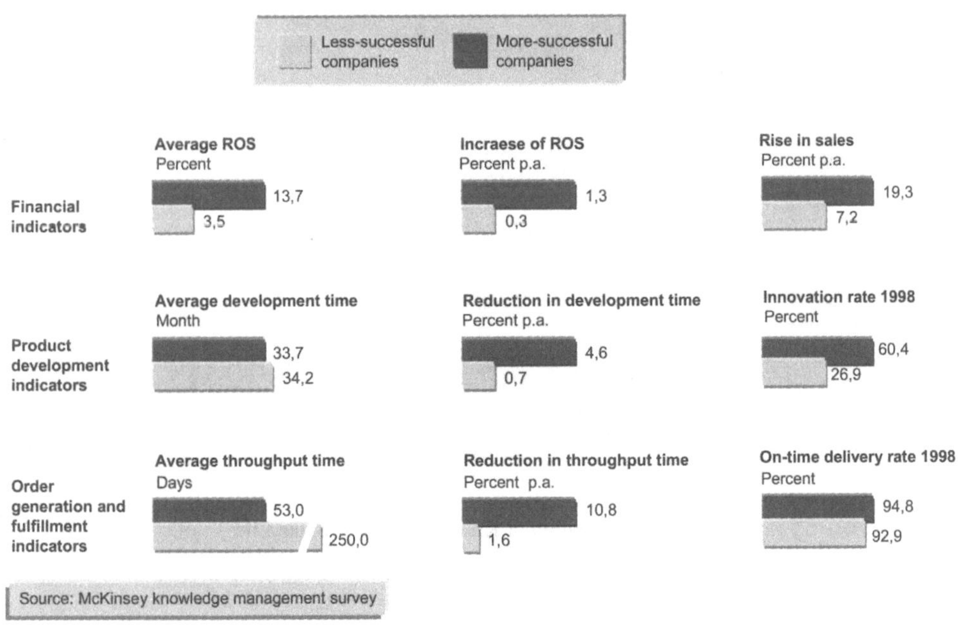

FIGURE 3: Performance figures, 1995-1998

We defined six characteristics of knowledge that distinguish from other aspects:

- *Subjective* – the interpretation of knowledge is heavily dependent on individual's background and the context in which it is used
- *Transferable* – knowledge can be extracted from one context and profitably applied in a new one
- *Embedded* – knowledge invariably resides in a static and often buried form that cannot easily be moved or reformulated
- *Self-reinforcing* – knowledge does not use value when shared, indeed its value grows when widely distributed
- *Perishable* – over time, knowledge becomes outdated, especially for an individual organization, although there can be unpredictable volatility
- *Spontaneous* – knowledge can develop unpredictably in a process that cannot always be controlled
-

Naturally, knowledge has to be managed optimally in all functions of a company. In this paper it is demonstrated on R & D, for instance, because this function is responsible for the determination of the costs of a product (fig. 4).

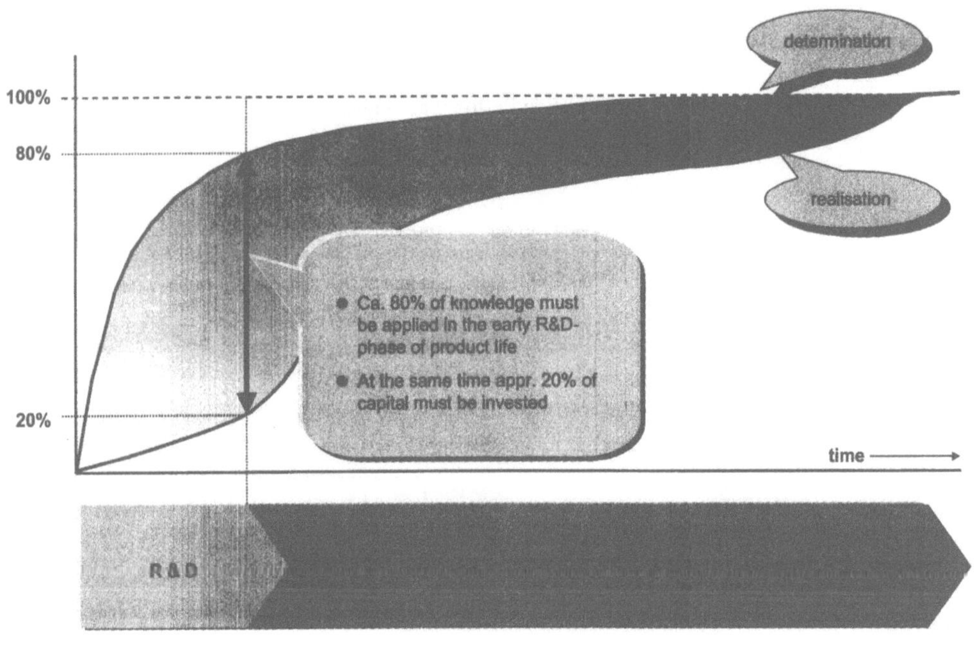

FIGURE 4: Application of Knowledge and Product Life

Product life becomes shorter and shorter and creates a complexity pit (fig. 5).

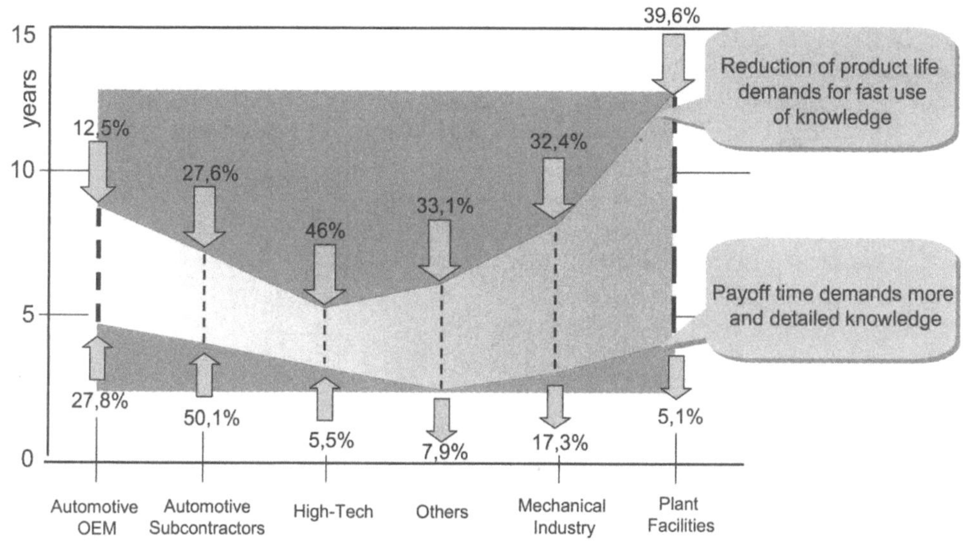

FIGURE 5: 10 Years Changes of Product Life and Payoff Time

Therefore, we have three centralized sections of problems (fig. 6).

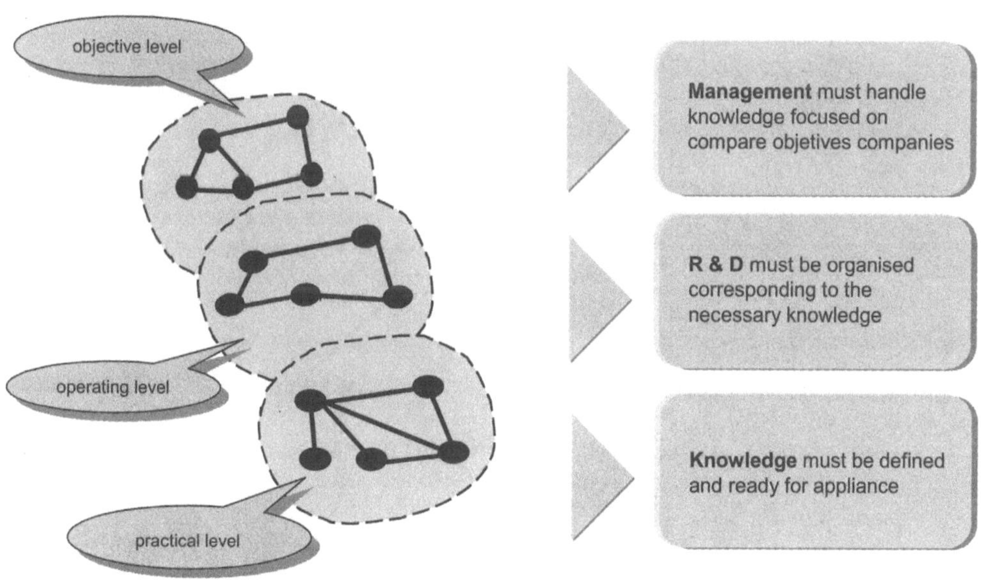

FIGURE 6: Levels of Knowledge Management in R&D

At the practical level knowledge must be created, transferred and applied (fig. 7).

	creation of knowledge	transfer of knowledge	application of knowledge
individual	Spontaneous	Embedded	Subjective
group	Self- reinforcing	Transferable	Perishable

FIGURE 7: Knowledge Processes of Practical Level

At the operating level we found three instruments, which could be individually configured (fig. 8).

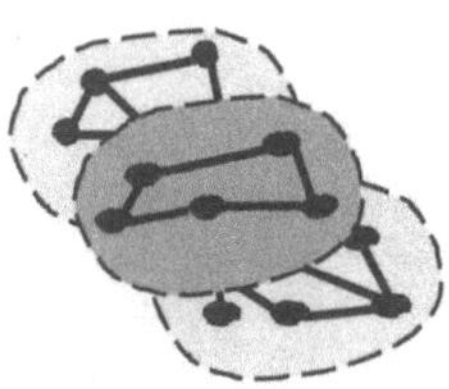

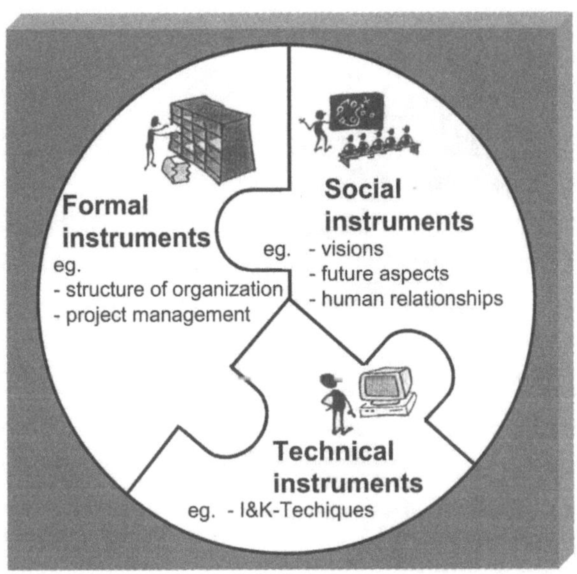

FIGURE 8: Methods of Operating Level

At the objective level a feed back control process must be installed (fig. 9).

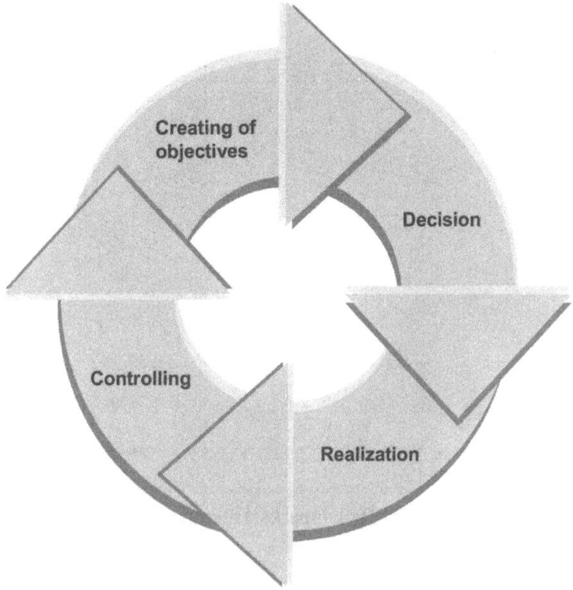

FIGURE 9: Feedback Process of Objective Level

But only the very close correlation of the general objectives with benchmark data leads to the company's specific targets (fig. 10).

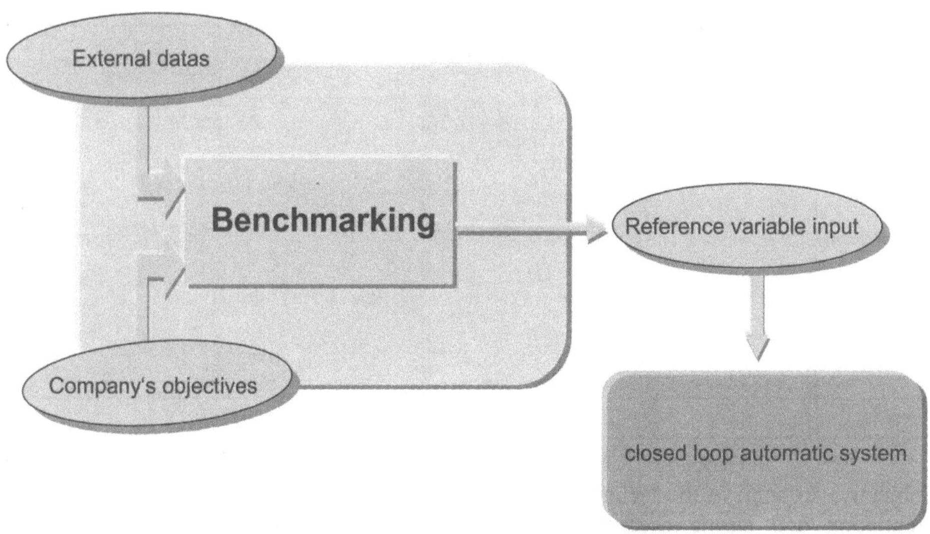

FIGURE 10: Introduction of Benchmarks

The comparism with the benchmarks allows to create sections of activity with different priorities. Industrial and global sectors must be regarded (fig. 11, 12).

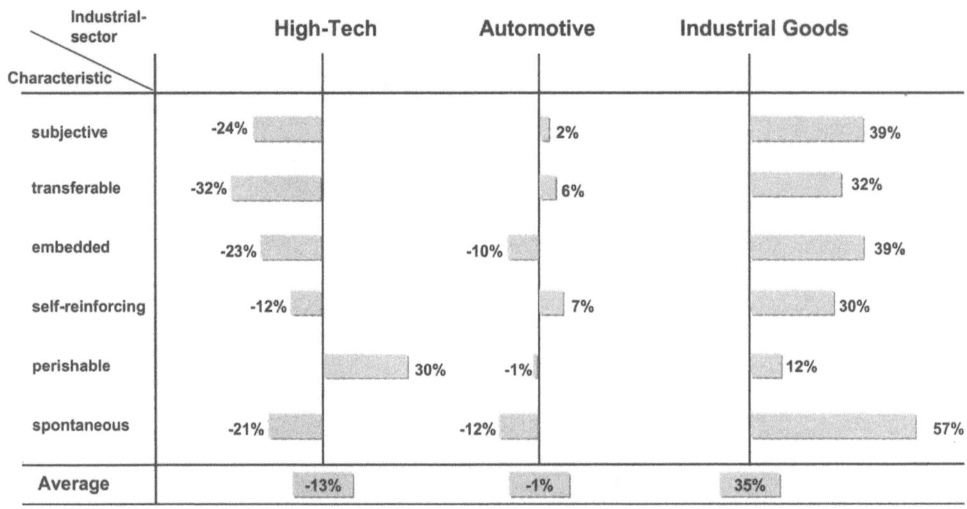

FIGURE 11: Divergencies of Benchmarks Tolerances of Different Industrial Sectors

An example, a machine tool manufacturer with a turnover of appr. 500 Mio Euro, shows the efficiency of a good knowledge management.

FIGURE 12: Global Divergencies of Benchmarks Tolerances

The objective to achieve the technology leadership forces to use knowledge as a factor for differentiation and strong competition. The average annual growth demands for new organizational structures and processes. The growing personal number leads to the necessity to compensate the knowledge gap. The globalisation demands for changes in the management and communication sector. The use of a special normalized analysis model (fig. 13), developed at PTW, allows to find the priorities for the activities (fig. 14).

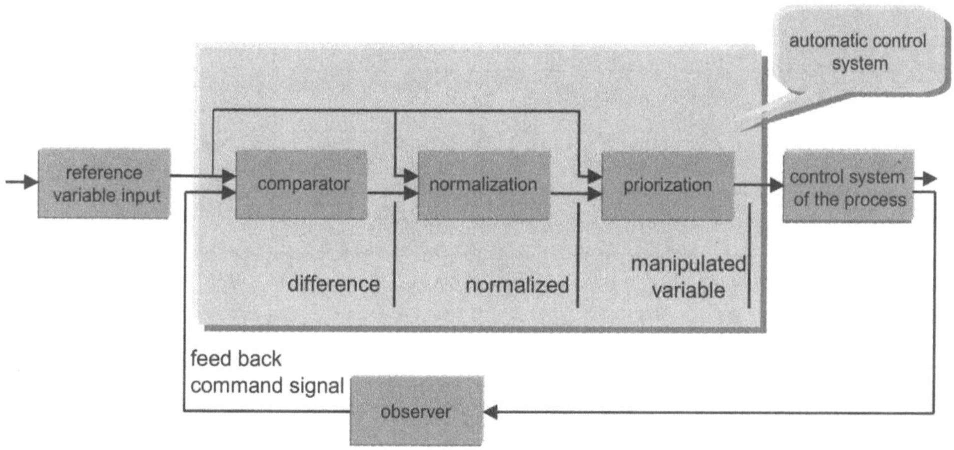

FIGURE 13: Closed Loop System

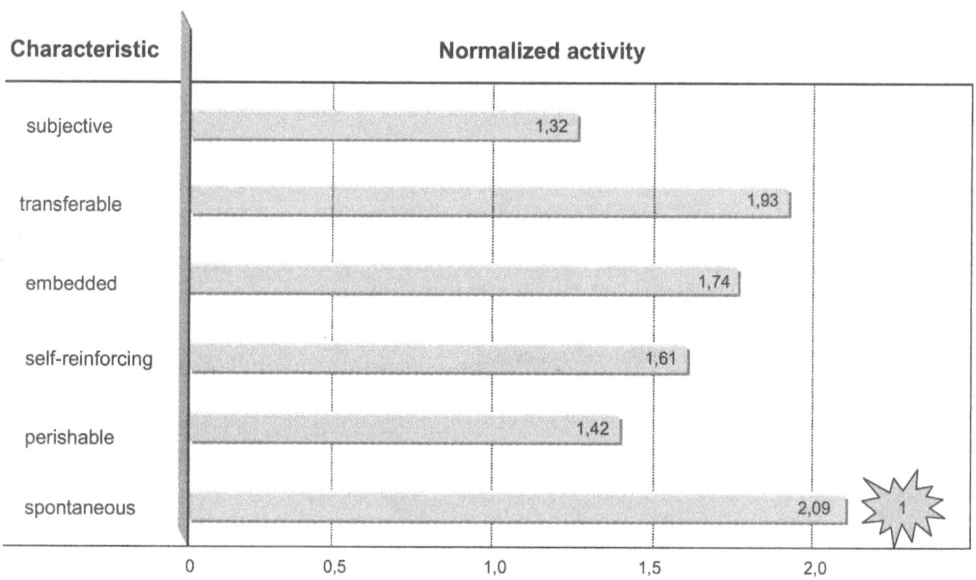

FIGURE 14: Priorization of the Specific Necessary Activities

Based on this it is possible to derivate and to introduce the necessary relevant strategies (fig. 15).

	task	Best practice solution
1	Installation of a competition-oriented knowledge in purchasing department	Detailed comparision by external benchmarking
2	Installation of a competition-oriented knowledge base in R&D	Permanent analysis of costs and process performance
3	Access to external information in product marketing	Information services (internal and external) by internet
4	Installation of a market oriented knowledge base	Cooperation with external service-organisations
	• • • • •	• • • • •

FIGURE 15: Examples of Activities Considering Spontaneous Knowledge

3 CONCLUSION

Successful companies use new and specific methods of knowledge management with the objective to become faster, better and cheaper than the competitors. Knowledge is the key for this and not only the product, the capital and the personnel of the company. Good knowledge management correlates with company's success.

4 REFERENCES

1. Kluge, J., e.a., Knowledge Unplugged, Palgrave publishers Ltd. N.Y., USA, 2001
2. Hauschild S., e.a., Creating a Knowledge culture, McKinsey quarterly, 2001-1
3. Stoffels, A., Wissensorientiertes Management der Produkt- und Prozessentwicklung Diss. TU Darmstadt, 2001
4. Metternich, J., Wissen als Grundlage von Wettbewerbsstrategien, Diss. TU Darmstadt, 2001
5. Bendler, A., Wissensmanagement im internationalen Vertrieb produzierender Unternehmen, Diss. TU Darmstadt 2001
6. Rabrenovic, O., Die Rolle der Organisation des Wissensmanagements im Unternehmen, Diss. TU Darmstadt 2000
7. Schulz, H., Will Platform Concepts reduce Costs of Machine Tools? Computer Integrated Manufacturing and High Speed Machining, HUPS, Zagreb, 2001

INDUSTRIAL EXPERIMENTS – THEORY AND PRACTICE

R. Levi[1], D. Romano[2]

[1] Department of Production Systems and Economics, Polytechnic of Torino, Italy
[2] Department of Mechanics, University of Cagliari, Italy

KEYWORDS: Design of Experiments, Principles, Industrial Problems.

ABSTRACT. Basic, time proven design principles are a key factor for successful industrial experiments today, as they were in the past. One must also take into proper considerations practical aspects and human factors, easier said than done with some characters – and some situations. Simple guidelines may help in the difficult task of selecting a compromise between abstract formulations and real world constraints, a recurrent problem with no general solution.

1 INTRODUCTORY RUMINATIONS

On a side you've got abstract principles, theoretical considerations, and on the other one dirt, mistakes, practical constraints; in a nutshell "All is fair in love and war", and that sure applies to experimental work at shop floor level. No platoon leader ignores that with the smoke of the first shot fired in anger up and away go all elaborate theories and abstract strategies of military school, all he's left to fend off with are guts, cohesion between mates and leadership, and precious little else. On the factory floor life's seldom at stake, reputation however is; important decisions must be taken on the spot, armchair strategist's comfort is denied, and consequences are to be faced squarely. There's nothing wrong with such a situation, though; once the initial shock is over, and you've come to appreciate that it's also piranhas besides catfish in the stream, you can play the game, and find some fun, too. The script may well range in between kiddy cartoons and westerns, perhaps with a fair sprinkling of comics.

Let us recall a few incidents collected over some four decades. In the early fifties here applied statistics and DOE were a non starter in mechanical engineering, control charts were just showing up in QC; probability paper had to be ordered abroad, or painstakingly hand made on the drawing board. No access to modern computing facilities, non-existent or priced sky high; clumsy, noisy electromechanical calculating machines of Fisher – Yates vintage were the rule for all sort of computations. Printing calculators were slow, their rigid, inconvenient format entailed anyway plenty of transcription and input work, a sure source of errors; everybody feared the dreaded moment a sum of squares would turn up negative, sure sign you'd better start all over again from scratch. In linear algebra inversion of a 6×6 matrix even with the abbreviated Gauss-Doolittle method entailed half a day's hard work, even when the mandatory check via identity matrix gave eventually the green light.

Sure enough you'd better think twice about numerical problems, the price for the wrong choice could be stiff. Then electronic calculators came, and digital computers; nothing like what even kids have today, in the late sixties a HP 2116B was as large as a kitchen refrigerator, had a 4k memory (that's right, a measly 4096; no mega, no giga) and cost would top a quarter million € in today's term. Input/output were via a 10 car/s teletype, programs were recorded on punched

Published in: E.Kuljanic (Ed.) *Advanced Manufacturing Systems and Technology*,
CISM Courses and Lectures No. 437, Springer Wien New York, 2002.

paper tape in ASCII code; sometimes either punch or reading head went awry like voting machines in the last presidential election in Florida, and you were in a big mess. Homemade software was the rule, diagnostics were cryptic at best if at all existent, and to flush out bugs Sherlock Holmes' flair was needed – or else.

What hardly changed with time is the interaction with the other guy, the chap we have to cooperate (or fight) with if meaningful results are to turn up. There's a whole rogue gallery of them, let's take a guided tour, taking a leaf or two off a well-worn diary; by the way, we're not talking about fiction, and don't believe this experience is an exceptional one.

Be on the lookout for the guy who's a notch too sharp, or just a plain cheat; in the course of a comprehensive round robin test program on material testing methods one cropped up who thought he could just as well dispense with a lot of test work and feed us with fake results, who'd notice anyway. Well, the moment I smelled the rat I felt personally stung, and started plotting result after result on control charts; sure enough something pointed straight in his direction, yet on that evidence only we couldn't nab the man. So we fed him some samples taken from a lot with peculiar properties he wasn't aware of, and sure enough a properly concocted set of "results" soon came back – which blissfully ignored facts real tests would have clearly shown. That's all we needed to throw to the round file whatever came from that lab, label it as an unreliable source, and exclude it altogether from the final report, duly published; the man didn't have the gall of asking how we came to that conclusion.

Lazy people are a nuisance, however I've found at my expenses many years ago that idiots or simply careless people may be even worse; there's plenty of them around nowadays, an old adage says the mother of fools is always pregnant. In the course of an extensive experimental investigation on surface properties performed for a large industrial concern we couldn't make head or tails of results on many tests, some aspects simply couldn't be explained. After double checking and recalibrating the setup, and having the inexplicable results confirmed by replication, error in identification of experimental material – supplied by the customer under his responsibility – came to be the only possible explanation, and after days and weeks of inconclusive work we faced the man with evidence at hand. At first he outright rejected even considering that possibility, then pushed to the wall he shrugged and admitted that, after all, a mistake might have occurred, but why make all that fuss … A strong temptation of doing summary justice here and there was resisted out of sheer laziness, after all getting rid of all idiots would be such a formidable task as to dwarf those of Herculean fame.

There's hardly a more damaging character than the big shot who believes he knows everything, and acts accordingly. He may let you work, but on his conditions; no DOE please, it won't work, just do what I tell you and that's all you're expected to do, I already know whatever we need to know about this stuff so your work here isn't really needed, is it? Unless you manage to perform behind his back a meaningful test program, meaningless results is what you get saddled with according to his recipe, and he'll not fail to throw in for good measure he already knew it. Wasted time and plenty of frustration is what you get working under these conditions; you end up wiser, and learn how to deal with such fellows – one of them is one too many.

Sulfuric, vicious characters may prove to be hard to deal with, and rather unpleasant too. The manager of a large pipe mill we were called into to perform badly needed tests simply wouldn't tolerate us around his machinery, and we were warned against crossing him; no way of interfering with production in his fiefdom, where machines kept breaking apart. The task was by no means easy, meaningful signals of a few mV had to be extracted from strain gage circuitry under strong electromagnetic fields originated by electrical machinery in the MW power range, signal to noise ratio in the beginning was abysmally low; heat, dust, noise and vibration galore. No problem, we were graciously permitted to work on Saturday night and Sunday as late as we wished, provided we packed up and disappeared before Monday's morning shift – after that it would be a mere 300 km car ride back home. Everybody slept in the car but the driver; we took turns at the wheel and still remember how hard we struggled against falling asleep while driving in the night, thoroughly done. By the way, results were obtained, a nasty source of fatigue failure was tracked down and remedial action pointed out, so the case was closed. Full credit for the team's survival goes to Heavens, of course.

There's plenty of fine, cooperative people around it's a pleasure to work with, I've made a number of good friends among them; however they don't make headlines, it's a case of dog bites postman.

2 BASIC PRINCIPLES

Let's restrict to where it belongs the Aristotelian/deterministic frame of mind, we know experiments only can provide proper answers to some questions; problem number one is, how to exploit efficiently time and resources available. By the way, the logic of knowledge buildup established by classic Greek philosophers hardly changed over the last two dozen of centuries, the revolving sequence "thesis → hypothesis → synthesis" still holds true today. No matter what the field, some sort of "a priori" knowledge is often available, if only at the level of gut feeling. A classic knowledge fusion device, Bayes' theorem, may then be exploited, giving at the same time a wide berth to the infamous postulate, hardly beyond criticism.

Exploitation of a priori knowledge, even if scanty, is particularly needed at planning stage, where practical decision must be taken, such as which factors and responses – and within which ranges - should be considered, which materials, instrument systems, tooling & machines are to be used, usw. A priori knowledge invariably affects our choices, which are best made taking into account the strength of available evidence, that is giving nature a fair chance of disproving our expectations whenever based upon flawed information. A cyclic nature is therefore typical of the cognitive process, since on the ground of priori knowledge, that is tentative model No. 1, we design the first experiment and analyze its results. Synthesis or fusion of results with previously held concepts leads to an a posteriori model No. 2 (hopefully improved), which may in turn provide the foundations for a second experiment, and so on. When analyzing experimental results an all too common whine sounds like "if only I'd known that", therefore it makes sense to adopt damage control devices such as keeping some spare resources just in case, or if at all possible a sequential approach.

We'll deal here only with that subset of experiments in which factors may be made to take selected levels within a given range, and responses observed accordingly. A test entails therefore controlled, substantial stimulation of the system under consideration, and recording selected output signals, as opposed to passive observation typical of such fields as astronomy and meteorology, in which planned system stimulation is generally out of question.

Two major problems appear in experimental investigation, namely one due to the finite amount of resources and the second to the well-established fact that if lady luck is as blind as a bat, jinx's got an uncomfortably sharp sight. Let's examine how we can tackle both, with a clear understanding that sometimes failures are unavoidable, it's a fact of life we've got to accept.

Given r factors (either quantitative or not) a r-dimensional space may be defined accordingly, where the subset under consideration, that is the sample space, is defined by the ranges pertaining to variables. Standardization makes sample space a segment if $r = 1$, a square if $r = 2$, a cube if $r = 3$, and a hypercube if $r \geq 4$. Within the sample space experimental design is defined by coordinates of the n treatment combinations, corresponding to test points, whose number n is defined on technical and economic grounds.

Common sense would suggest spreading test points evenly over sample space, avoiding overcrowding as well as voids. Thus the old and inefficient criterion of "one factor at a time", that is exploring sample space only along coordinate axis, is disposed of once and for all by simple space filling logic, since it leads to ignore systematically large regions without a shred of justification. Let us recall a sentence R.A. Fisher wrote with his characteristic style in the mid 'twenties, debunking forever what was then accepted as common wisdom,

"No aphorism is more frequently repeated in connection with field trials, than that we must ask Nature few questions or, ideally, one question at a time. The writer is convinced that this view is wholly mistaken. Nature, he suggests, will best respond to a logical and carefully thought out questionnaire; indeed, if we ask her a single question, she will often refuse to answer until some other topic has been discussed."

That is, attempts of explaining a complex real world in terms of overly simplified linear, additive models are necessarily bound to fail, as they cannot account for interactions and non-linearity. A no nonsense approach to experimental investigation addresses squarely the issue of higher order terms, to be dropped eventually should they prove negligible.

Everything's fine as long as $n \gg r$, and exhaustive exploration of sample space is possible; difficulties do however arise whenever the number of tests is limited, and/or stationarity is far from granted. Estimates of order of magnitude of main effects, with substantial uncertainties, may be obtained as long as $n > r$, below that level from investigation one slips into gambling. As a good start, and simple too, it's worth exploring every vertex of sample space, that is taking $n = 2^r$, corresponding to two-level factorial designs; some replications and a center point are worth adding whenever possible. Second order effects of both single factors and their interactions may be investigated taking $n = 3^r$, at the price however of a substantially larger number of tests.

No matter which design is selected, it necessarily represents a filter between a variegated real world and the finite set of conclusions which may be supported by a finite sized experiment, a fact which requires careful consideration. In the elementary case of $r = 1$, $n = 2$, sample space is a segment whose end points only are accessed; no inference may be drawn in between, nor is an estimate of error available. At two levels only, our "green tinted spectacles" of Kant's paradigm cater only for green (linear) models, a good starting point in many instances, a far cry from accurate description in other cases.

Let's stick to the elementary case of $r = 1$, with sample space boiling down to a unit segment of a straight line. The n test points allotted may be spread out evenly without replication, spaced 1/(n 1) apart, or piled up in two heaps sized n/2 at both ends of segment, or any other pattern. In the first case no degrees of freedom are left for the estimate of pure error, and $n - 2$ are allotted to lack of fit, the opposite being true in the second case; as for the third, a mixture of the former two would apply. Again, in the first case a polynomial up to the $n - 2$ degree can be fitted, while in the second uncertainty in the estimate of slope would be lowest; there's no way of having both at the same time, just as you cannot have your pie and eat it. Either extreme options are justified in particular circumstances; as a rule they are not, and a sensible balance between estimation capabilities in several directions appears more reasonable a choice.

In a given sample space, for any combination of r and n , should model's form be known, identification of an "optimal" experimental design is a simple enough mathematical exercise in terms of a selected objective function, e.g. minimization of uncertainty in parameter estimates. Such a result should not be construed into proving the superiority of the so derived design over any other one, for in fact the opposite might well be true, a truism the freshman may be uncomfortable with; hardly so whoever's got a modicum of practical experience under his belt. First of all, optimal designs are necessarily tailor fitted to a given model, and fare distinctly worse than general purpose designs if actual model's form turns out to differ from the assumed one, that is they're seldom robust to model changes. Furthermore, they are prone to dictate test point concentration on sample space borders, to the expense of inside regions that may well be close to actual operating range, where dependable information is a key requirement. Last but not least, in industrial work several responses are considered as a rule, the case of a single one being an exception; a model adequate for one may be ways off the mark for another. Optimal models look great on the blackboard; they're however seldom worth considering in applied work, where solid, reliable tools are preferred to fragile analytical niceties.

Sometimes resources hardly permit reaching at all the goal, forget about how; targets must be soberly redefined in terms of what's at hand, larger uncertainties and less strict significance levels accepted out of sheer necessity – and made to be accepted to all involved parties, easier said than done. A thin budget entail cuts in replications, reduced effect separation capability, and larger risk of error when interpreting results. In order to address the situation adequately, proper attention must be paid to an often overlooked facet of the problem, since unbeknown to us something devilish may actually be working towards our demise; according to a corollary of that fundamental axiom familiar in every lab - Murphy's law - "Mother Nature is a bitch".

That is, besides the set of factors under control, or at least known, there's a whole array of others uncontrolled and even ignored at all, which may well play havoc with our experiment by causing on responses effects we may blissfully ascribe to the wrong cause. Not only may thus uncertainty be blown out of proportion, but also inherently false results may be arrived at. Not to take proper precautions would be irresponsible just as surfing the net with no anti-virus worth speaking of; the experimental design must therefore pack proper defensive devices, along with those aimed at the main goal.

3 STATISTICAL TOOLBOX

Application of statistical devices may lead to positive results in experimental work – just as in any other endeavor – only if guided by common sense, a commodity whose demand may well exceed supply in not a few instances. Let us consider a few items of a checklist best kept in mind at preliminary stages, hopefully before getting definitely committed.

Identify main goals and scope of investigation right at the beginning – and, do so in a properly organized meeting, with all interested parties in attendance, with a properly drafted agenda and minutes initialed by whoever's taking part in the exercise, first and foremost the guy who's going to foot the bill. It may be not that easy, especially under emergency conditions when results are needed urgently, but it's definitely worth the effort; in other words, the price of failing to do so may be rather stiff. Resources, expectations, factors and responses, who's going to do what, must be spelled out beforehand, not as an afterthought. To start without a clear plan "in order to save time" is a sure recipe for misunderstandings, bickering and eventual failure.

Insist on quantitative responses if at all possible – just recall Lord Kelvin's adage. People argue easily about words, properly documented figures leave little room for discussion. If qualitative responses must be dealt with, make a point of insisting on scales with universally agreed reference terms, as e.g. in Mohs or Welzenbach scales. Needless to say, quantitative measurements are a farce if gages and instruments are not properly calibrated by competent technicians, and traceable to the relevant standards; in case of doubt do check right away.

Replication – Andrews went as far as affirming that the three most important R in experimental work are .... Replication, Replication, Replication! (Snee et al., 1985). The man was worth its salt, scatter/noise/error is a key component of system's signature, the premier way of estimating it is by replication – no mere repetition - of actual tests. Precision of estimates derived from n replicates increases by a factor proportional to $\sqrt{n}$, distribution of averages tends to normal form no matter how individual elements are distributed, and above all unbiased estimate of error may be obtained, a key factor in all tests. Repeatability is assessed, and precision of estimates may be made to match requirements, given of course the needed resources.

Randomization – horror stories abound on the devastating effects of superposition over a "regular" factor of a lurking variable, no one in his senses would add his name to an already long list of absurdities derived according to the infamous principle "post hoc ergo propter hoc". Again, R. A. Fisher showed that protection at reasonable cost is readily available, just allot randomly experimental material, time order of tests and whatever else may conceivably affect test outcome. By assigning to each and every treatment combination the same probability of being allotted to any given time span, operator and machine a major portion of bias is averaged out, the remaining part becoming a legitimate component of error. Randomization's like insurance, in that by paying a moderate price in terms of planning complexity coverage is obtained against potentially devastating occurrences. It may not be that easy, situations may dictate otherwise; in field work, you may face an angry Count Almaviva with as much leeway as the hapless Don Basilio had at the business end of the gun's barrel, and behave accordingly.

Blocks – as a rule tests performed in the same day, on the same setup, on the same lot are more uniform, and enable comparisons more sensitive than when a wide range of situations is involved. On the other hand, tests must be performed on different days, on different machines and by different operators, if only to ensure general validity of results. The overall design may however be split up in a patchwork of subsets, or blocks, inherently more uniform. By accommodating contrasts to be privileged in comparisons within blocks, precision and generality are preserved, while between blocks comparisons evaluate overall variability, a term to be reckoned with. The price's paid mainly in terms of a more complex organization of tests.

Alias & confounding – nice as it would be estimating effects unaffected by all factors but one, time and resources often do not permit it. When the number of tests allowed exceeds but marginally the number of factors considered, aliases must be accepted as a fact of life. That is, estimates of single and combined effects are more or less inextricably mixed with some unwanted terms, usually corresponding to interactions of an order high enough to make it plausible to assume their expected value is small enough not to upset the apple cart. Some all important terms may be privileged at the expenses of others, or if need be a few additional tests may be added later on, if added precision is worth the price. What's really needed is to make it clear right from the beginning to the executive in charge, whose statistical knowledge may be a wee bit rusty, that savings in tests come to a price he must be ready to pay in full, and that in DOE just as in any other endeavor there are no free lunches.

Resolution – a nickname for confounding and aliasing, typical of fractional replication of factorial designs. Resolution III designs entails confounding single effects with two-factor (double) interactions, acceptable at screening level only when identification of the vital few out of the trivial many is the main goal. At least Resolution V is required for routine work, single effects being confounded with quadruple interactions, and double interactions with triple. Any available hint about the presence of interactive effects helps in selecting resolution level; as a rule in presence of several effects of comparable order of magnitude at least some sizable interactions are to be expected, as inadequacies of simple additive models are exposed.

Sequential experimentation and model validation – it makes sense to consider experimentation aimed at model building, and at model validation, as separate entities. Ideally the former should cater for model identification, and parameter estimation, while the latter must put model in jeopardy, expose its weak spots and describe shortcomings. When validation work is aimed at model improvement, sequential experimentation steps in, where every stage is planned according to the outcome of the previous one, and provides the grounds for the next set of tests. When at all feasible such a strategy may lead to substantial savings, as after conclusion of most large scale testing information is gleaned on the advantages of following another way – if only we'd known it beforehand. Dynamic rescheduling and reprogramming testing in the light of fresh information is the buzzword, not to be confused with jaywalking – that is, hopefully so. And, in any case, never empty your magazine with the first volley, few people ever regretted saving a lump of test resources for a second round.

TABLE 1. Main requirements and methods in DOE

Requirements	Methods
Clear identification of goals, priorities, and resources	Statistical and specific techniques appropriate for proper selection of: factors and levels, responses, testing equipment and procedure, field of application
Protection against effects of factors uncontrolled or ignored	Experimental design: material & test instruments control, blocks, randomization, ANOVA, ANCOVA
Systematic error prevention	Blocks (balanced if possible), proper techniques, randomization, replication
Univocal results	Quantitative responses and standard test procedures; otherwise, use scales with recognized reference terms
Error & noise evaluation	Replication, randomization, residual analysis
Accuracy of estimates, adequacy to stated purposes	Resolution, replication, randomization, orthogonal comparisons, ANOVA, ANCOVA

4 BASIC EXPERIMENTAL DESIGNS

Let us recall the main features of typical experimental designs, starting from the simpler ones catering for first order models, where additive effects only are considered and any interaction among factors is ruled out. More involved structures afford as a rule better precision, and enhanced protection against lurking factors; these benefits are paid for in terms of more complex management of experimental work, and increased vulnerability to such facts of life as mistakes and missing data, which may occur owing to factors beyond experimenter's control.

Simple randomized designs – straightforward, adaptable structures, with no restrictions concerning allotment of experimental material to treatments, nor replications. Typical features are fair precision, and robustness to missing data, the loss being restricted to the corresponding degrees of freedom for error. Given k treatments and n replications, of the nk-1 degrees of freedom corrected for the mean k-1 are available for estimates of effects, while k(n-1) go to the estimate of error.

Randomized blocks – enhanced precision thanks to superior homogeneity which may be achieved within blocks, as typical of smaller subsets of a large entity. Thus in tire testing on the road a block may correspond to a vehicle, or an axle thereof; in machinability testing to a given machine tool, in manufacturing experiments to a shift, and so on. Randomization is restricted in that treatments must all appear within every block, no particular limits apply to replication. Given k treatments and b blocks, for a single replication of the bk-1 degrees of freedom corrected for the mean k-1 are for treatments, b-1 correspond to blocks, and to the estimate of error (b-1)(k-1).

Incomplete blocks – a necessity whenever block size cannot accommodate every treatment, and therefore contrasts may not be restricted exclusively within blocks. When total number of tests divided by number of tests per block equals treatments, every contrast appears once in a block, and the balanced design caters for uniform precision of estimates. If need be unbalanced designs may be resorted to, accepting their inherent limitations.

Latin squares – an efficient design providing for elimination of the effects of two potentially damaging factors, thanks to a two-dimensional randomized, balanced square pattern. Levels of such factors must be the same, and equal to the number of treatment combinations. Original applications were in agriculture, aimed at elimination of effects of fertility gradients in two directions; extension to other fields soon followed. For instance, on a wear testing machine which can accommodate four specimen at a time, four different materials may be tested in as many test runs, allotting to all materials different locations on the machine in every run. Given k treatments, the k^2-1 degrees of freedom corrected for the mean are split up into k – 1 for treatments, rows and columns, the remaining (k-1)(k-2) corresponding to error; thus for the example the split is 15 = 3+3+3+6.

Graeco-Latin squares – an extension of the concept of Latin squares to the case in which three distinct effects of as many disturbing factors must be eliminated, with the added limitation that the number of levels of the three factors must be the same, and equal to the number of treatment combinations. Given the basic Latin square layout, the third factor is taken care of by associating it with treatment combinations, every level being associated once only with any one row, column and treatment. As in the case of Latin squares, no interaction among disturbing factors is admitted.

Youden squares – quite similar to Latin squares, with one limitation somehow relaxed. One disturbing factor only is required to have as many levels as treatment combinations, the number of replications however must be equal to the number of levels of the remaining disturbing factor.

Restricted randomization is mandatory for rows, columns and treatment order, just as in the case of Latin squares.
A design may be preferable to another in a given situation, the reverse being true in another case; there's no such thing as a hard and fast selection rule, some kind of compromise is inevitable just as in day to day life. The main properties listed in Table 2 provide some criteria for selection.

TABLE 2. Main properties of some basic experimental designs

Characteristic	Design		
	Simple Randomized	Randomized Blocks	Latin Squares
No. of replications	No limitation	Same for every treatment	Same for every treatment
Test preparation	Simple	Not very complex	Rather complex
No. of treatments	No limitation	Limit: block size	Range 3 – 12
Degrees of freedom for error	Maximum	Reduced by No. of blocks	Reduced by No. of rows & columns
Protection against lack of uniformity	Poor	Good in one direction	Good in two directions
Control of heteroschedasticity	Simple	Feasible	Rather complex
Missing data	No problem	Limited damage	Large damage
Errors in test organization	Rather robust	Quite vulnerable	Highly vulnerable

5 FACTORIAL DESIGNS

Simple designs described above are appropriate whenever responses are expected to be linear and effects simply additive; in real world it may well be otherwise, and R. A. Fisher's observation quoted above applies. Having first defined the problem, he devised an appropriate solution under the form of factorial designs, first termed complex experiments (Fisher 1935, Yates 1937). A major stumbling block stood in the way under the form of interaction, that is variation in one factor's effect associated with a shift in the levels of one or several other factors. By the way, let us stress that the concept doesn't involve causation, which may be implied by specific mechanistic considerations only. Thus interaction entails a multiplicative action originated by the association of factors, whose action when taken one at a time may be substantially smaller (or larger), as frequently observed in field trials, pharmacology and many other fields.

Two factor (or more) interactions corresponds to non additivity, while single factor interaction amounts to non-linearity, both common occurrences whenever a sizeable range is covered. The opposite concept, that is absence of interaction, corresponds to a linear, additive model, a not unreasonable starting assumption at exploratory level, hardly a safe one for an in depth investigation. The "one factor at a time" approach is inevitably bound to fail when interactions are present, as their effect cannot be explained in terms of single factor effects only; owing also to the low efficiency and inadequate coverage of sample space such an approach has nowadays but marginal applications, being superseded by inherently superior designs.

Factorial designs – a strikingly efficient tool for systematic analysis of single and combined effects of several factors – either quantitative or qualitative - on experimental response(s). A complete factorial design with r factors at k levels entails the k^r treatment combinations corresponding to all combinations of factors and levels. For $k = 2$, a rather common choice particularly at exploratory level, the model must accommodate all significant terms among r effects of single factors, $r!/2(r-2)!$ two factor interactions, $r!/3!(r-3)!$ three factor interactions, and generally $r!/q!(r-q)!$ interactions of q-th order ($q \leq r$), up to a grand total of as many as 2^r-1 effects whose estimates are all orthogonal. In the general case of m replications of r factors each at k_i levels, the number of tests required, equal to $m\times k_1\times k_2 \times\ldots\times k_r$, gets fairly large should r and/or the k_i exceed moderate values.

Fractional factorial designs – a solution for accommodating many factors in a design of moderate size, assuming that some higher order interactions may be disregarded at all, and other interactions safely confounded with the estimates of effects under consideration. The option is particularly attractive at exploratory level, since it caters for expeditious examination of a large number of factors, and further tests may be planned on the most important ones should more detailed information be required. A subset only of the complete set of combinations of factors and levels is considered; thus with e.g. r factors each at $k = 2$ levels a one half fraction entails 2^{r-1} tests at as many treatment combinations, and in general a fraction of order $1/2^p$ entails 2^{r-p} tests. Savings in time and resources thus achieved are paid for in terms of complete loss of information on at least one effect – selected as a rule among higher order interactions, expected to be negligible – and confounding all estimated effects with one or several aliases, besides the general loss in precision entailed by the smaller number of tests. The degree of entanglement among estimates is identified by the resolution of the design, that is the sum of the orders of effects confounded. Thus a Resolution III design entail confusion between the effects of single factors and second order interaction, acceptable at exploratory level only, while with a Resolution V design single effects are confounded only with fourth and higher order interactions, and second order estimations with third and higher order ones, a frequent compromise between cost and precision. As a rule error estimates are obtained from higher order interactions or external information, as there are no d.f. for error.

Saturated designs – pared down (almost) to the bone fractional factorial replications or similar plans with a number n of treatment combinations barely exceeding the number r of factors, typically $n = r + 1$; estimates of effects of single factor are orthogonal, but all aliased with second and higher order interactions since these designs are Resolution III. Fractional factorial designs of the 2^{r-p} series are frequently adopted in this context, proper of preliminary, exploratory investigations.

Supersaturated designs – two levels only, more factors than treatment combinations ($r > n$); r and n are both even in order to cater for estimation of single effects, albeit not independently of course. Factors exceed orthogonal comparisons, and entanglement among single factor effects cannot be avoided completely; confounding with a number of second and higher order interactions also occurs. Applications cover mainly preliminary screening work, in which many factors must be considered although but a few of them are expected to have substantial effects, dwarfing those of interactions. Single replication being the rule with most fractional designs, in absence of external estimates of error normal plots of effects and scree plots may be relied upon to provide reference terms of some sort.

A broad range of options is provided by these designs, as shown by considering what may be done given eight tests corresponding to as many treatment comparisons (but for the first case), with different tradeoffs between precision and comprehensiveness. The experimenter may e.g. select one among the following lines, to estimate the single effects of:

- two factors and their single interaction independently, averaged over two replications, with four degree of freedom for the estimate of error;
- three factors and all their four interactions (three double, one triple) independently, with a single replication, no d.f. for the estimate of error;
- four factors and three double interactions (out of the six double, four triple and one quadruple) with a 1/2 replication, all remaining interactions being assumed negligible, all estimates having one alias, no d.f. for error;
- five factors and two double interactions with a 1/4 replication, all other interactions being assumed negligible, all estimates having three aliases, no d.f. for error;
- six factors and one double interaction with a 1/8 replication, all other interactions being assumed negligible, all estimates having seven aliases, no d.f. for error;
- seven factors with a 1/16 replication, all interactions being assumed negligible, all estimates having fifteen aliases, no d.f. for error.

Any one of these options may be preferable in a given context, and less than ideal in another; there's no such thing as a "best" design, just as a shoe you've got to select it according to intended use, and it doesn't hurt to check it, too.

Definition of a full factorial design is straightforward, as opposed to a fractional replication thereof, especially if many factors are considered and fractions are small; published catalogs of fractional designs or ad hoc software should be used. A realistic evaluation of costs versus benefits of alternative options is a must in order to reach as objective a decision as possible, and steer clear of gambling, unless you really wish to do so. In case of doubt keep in mind Andrews' leit motiv – Replication, Replication, Replication! – and play it safe. Not only does replication increase precision by averaging out random errors, but thanks to the central limit theorem means tend to normal form no matter how individual responses are distributed; since most statistical tests assume normality, use on unreplicated values may well lead to sheer nonsense.

A typical application of fractional factorials in industry deals with robust design, aimed at meeting specific requirements in terms of means, scatter and insensitivity to external disturbances beyond operator's control, within given technical and economic constraints. Introduced some forty years ago by Morrison, "rediscovered" and widely advertised by Taguchi over two decades later, this methodology is aimed at identifying target values and related tolerances for the parameters under consideration, in order to meet the robustness required at the lowest cost. Experimental investigation is almost always required to obtain relevant information; exploration of sample space with both numerical and physical tests leads usually to substantial savings in time and resources.

6 CONCLUDING REMARKS

Shifting reliance, from blind trust into almost religious beliefs beyond criticism to experimental evidence open to rational discussion, was the bold step taken by the founding fathers of modern science of Rinascimento, a turning point of the history of mankind. Oddly enough, that developments in error theory and statistics had obviously to lead to experimental design, an accepted fact today as it was a hotly debated upon subject three quarters of a century ago, is still argued among educated people who should know better. Applications of DOE in industry often take place after quite a bit of haggling, and authority must be brought to bear to push aside all sorts of lame excuses purporting to show why it won't work in that particular case. When results speak for themselves it's another story, and success finds itself son of many fathers, as usual. Contrary to common wisdom, higher education does not appear to affect favorably the attitude toward DOE, as it is not uncommon to find technicians far more open to fruitful cooperation than university graduates. An open mind and willingness to learn by doing are not privileges depending upon higher education; indeed some universities have curricula which, if not actually designed to stifle innovative minds, at any rate do so quite effectively.

ACKNOWLEDGMENT

This lecture is presented under the auspices of Pro-Enbis, a thematic network of the EC Fifth Framework Program dedicated to furthering the use of statistics in industry and commerce.

REFERENCES

1. Box, G.E.P., Hunter, W.G., Hunter J.S., (1978), Statistics for Experimenters: an Introduction to Design, Data Analysis and Model Building, J. Wiley, New York.
2. Cochran W.G., Cox, G.M., (1950), Experimental Designs, J. Wiley, New York.
3. Davies, O.L., (1954), The Design and Analysis of Industrial Experiments, Oliver & Boyd, Edinburgh.
4. Fisher, R.A., (1925), Statistical Methods for Research Workers, Oliver & Boyd, Edinburgh.
5. Fisher, R.A., (1926), The arrangement of field experiments, J. Min. Agr. Gr. Brit., 33, 505-513.
6. Fisher, R.A., (1935), The Design of Experiments, Oliver & Boyd, Edinburgh.
7. Morrison, S.J., (1957), The Study of Variability in Engineering Design, Appl. Statist., Royal Statistical Society, 6, No. 2, 133-138.
8. Morrison, S.J., (1998), Variance Synthesis Revisited, Quality Engineering, 11, No. 1, 149-155.
9. Myers, R.H., Khuri, A.I., Vining, G., (1992), Response Surface Alternatives to the Taguchi Robust Parameter Design Approach, The American Statistician, 46, No. 2, 131-139.
10. Romano, D., Bertagnolio, A., Levi, R., (1999), Dynamic analysis of production processes by planned experiments, Proc. 5th Int. Conf. Advanced Mfg. Systems and Technology - AMST'99, 97-106.
11. Romano, D., Alosi, F., Barbato, G., Levi, R., (1999), Simultaneous Parameter and Tolerance Design by Sequential Numerical Experiments, Proc. Int. Conf. on Quality Manufacturing, Rand Afrikaans University and University of Stellenbosch, 16-23.
12. Snee, R.D., Hare, L.B., Trout, J.R., (1985), Experiments in Industry: Design, Analysis and Interpretation of Results, American Society for Quality Control, Milwaukee.
13. Taguchi, G., (1987), System of Experimental Design: Engineering Methods to Optimize Quality and Minimize Cost,: UNIPUB/Kraus International, White Plains.
14. Yates, F., (1937), The Design and Analysis of Factorial Experiments, Technical Communication 35, Imperial Bureau of Soil Science, Harpenden.

DISASSEMBLY FOR RECYCLING, MAINTENANCE AND REMANUFACTURING: STATE OF THE ART AND PERSPECTIVES

M. Santochi, G. Dini, F. Failli

Department of Mechanical, Nuclear and Production Engineering, University of Pisa, Italy

KEYWORDS: Disassembly, Recycling, Maintenance, Remanufacturing

ABSTRACT. The continuous progress in manufacturing techniques and the increasing ecological consciousness have led to a constant refinement in natural resources exploitation. In particular, the advisability of a longer and more reliable life cycle of products and the reintroduction of mere materials in the production cycle have created new research fields in maintenance, remanufacturing and recycling, respectively. All these subjects have a common need: the complete or partial disassembly of the product in order to obtain more efficient and profitable results. The aim of this paper is therefore to analyse the state of the art of the research in disassembly techniques applied in these fields, emphasising the problems to be solved and the perspectives in the next years.

1 INTRODUCTION

In the last years, due to the spreading of sustainable development concepts, disassembly has been considered as a key issue in activities like maintenance, remanufacturing and recycling [1], which play a fundamental role, respectively, in extending product life, in decreasing the environmental impact of new products, in closing the materials life cycle. A lot of work has been done by many researchers to develop methods and tools to make disassembly a promising, profitable and widespread activity. In this context, at least three actors must be considered:

- *producers,* who are pushed to use Design for Disassembly (DFD) techniques [2] or try to embed Life Cycle Units (see section 3.3) [3] in their products or use innovative tools in maintenance (e.g. augmented reality [4]);
- *users,* who start to consider the new concept of selling use instead of selling products [5] with the consequence of a completely different way to conceive the market and their habits; this practice, allowing a total control (i.e. correct maintenance, strict monitoring, final disposal, etc.) on a huge number of products, could be crucial in remanufacturing as well as in the general optimisation of resources.

Published in: E.Kuljanic (Ed.) *Advanced Manufacturing Systems and Technology*, CISM Courses and Lectures No. 437, Springer Wien New York, 2002.

- *dismantlers,* who are interested in decreasing the dismantling costs and speeding up the disassembly process of end-of-life goods: the development of innovative dedicated tools is one of the ways to solve their problems.

2 GENERAL ASPECTS OF DISASSEMBLY PROCESSES

The following list of relevant issues should be considered in order to outline the complex scenario of disassembly (Table 1):

- In maintenance only standard (i.e. non destructive, see below) disassembly operation can be normally performed, since the main target is the restoration of a working condition in machines or facilities.
- If the aim of disassembly is a mere material recovery, the destruction of parts during dismantling is often profitable to speed up the process and to increase the profits. This can be necessary in particular for damaged products, where the status of the joints may be altered (rust, oxidation, deformations...), or in presence of drive fits, welded or glued joints. In Figure 1, an example of profit increased by using destructive operations is shown [6].
- A very important issue in disassembly for remanufacturing and maintenance is the knowledge of the level of wear and damage affecting the various components of the product. Unfortunately these data are usually unknown since they depend on the product life, its storage after discarding and its transport to the disassembly plant

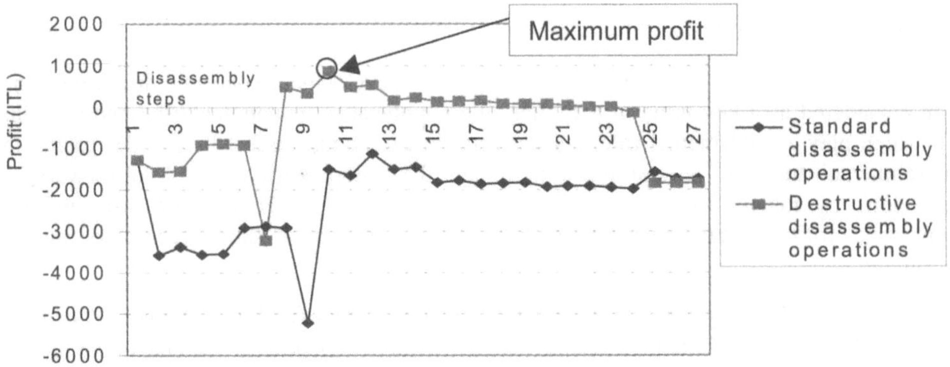

FIGURE 1. Effects on the general profit by using destructive disassembly operations [6]

- Almost all widely diffused products are not designed to make the disassembly process easier: only functionality and assembly constraints are mostly considered. Traditional joining methods are unsuitable for fast disconnection, materials are chosen only to maximise the performances and to decrease production costs [1].

- At present the market for remanufactured components is limited.
- End-of-life goods have to be transported to the disassembly plant with high costs owing to unsolved logistic problems.
- Disassembly for recycling process can be economically profitable only with a high plant input volume. So the only available categories of products are currently cars, household appliances (TV sets, washing machines, refrigerators, dishwashers...), PCs, monitors, printers.
- Legislation can largely influence the planning and the results of a disassembly process because some constraints can be imposed. Typical examples are: removal of dangerous or toxic substances (oils, refrigeration fluids, acids...); incentives for using recycled materials [1]; EU guidelines on End-of-Life Vehicles (ELV); EU guideline draft on Waste of Electrical and Electronic Equipment (WEEE) [7].

The main result of the described problems is that the creation of an effective collection, within a complete disassembly and recycling network, has a relevant delay and disassembly issues are not yet fully considered. However many signals let look at disassembly as an unavoidable technology and the future challenges are the decreasing of disassembly costs and rationalisation of disassembly activities.

TABLE 1. Relevant issues in disassembly process: level of importance in different application fields (⊗: low; ⊗⊗: medium; ⊗⊗⊗: high)

Related aspects	**Maintenance**	**Remanufacturing**	**Recycling**
Use of a destructive disassembly	⊗	⊗⊗	⊗⊗⊗
Product status (level of wear, damages, etc.)	⊗⊗⊗	⊗⊗⊗	⊗⊗
Dedicated product design (DFD)	⊗⊗⊗	⊗⊗⊗	⊗⊗⊗
Limited market dimension	⊗	⊗⊗⊗	⊗⊗⊗
Logistic problems	⊗	⊗⊗	⊗⊗⊗
Volume of products to be disassembled	⊗	⊗⊗	⊗⊗⊗
Legislative constraints	⊗	⊗⊗	⊗⊗⊗

2.1 DISASSEMBLY OPERATIONS, TOOLS AND EQUIPMENT

From a general point of view, a disassembly operation is "an action which separates (or enables to separate) two or more components of a product". This definition therefore includes both the simple detachment of two parts and the preliminary operations such as unfastening, elimination of connections, creation of holes or cavities for improving accessibility to the parts to be disassembled, etc. Disassembly operations can be divided in the following three categories:

- *Standard operations*: they are the most intuitive, representing the reverse of the assembly operations. They need standard tools and equipment, similar to those ones used in assembly. These operations are *non destructive* and therefore assure the integrity of parts after disassembling. Due to this property they are almost the only possibility for a maintenance intervention but, considered as recycling actions, their execution can be difficult to be performed or unprofitable (depending on the kind of connection).
- *Destructive operations*: they perform the physical (partial or complete) destruction of one or more parts in order to obtain the successive separation or to improve the accessibility to the other parts. These operations do not preserve the integrity of parts but they are usually faster than standard operations. Often they are performed to avoid the disassembly of other components economically not relevant. Destructive operations can be also necessary in the case of non reversible connections (e.g.: welded joints) or part alterations (e.g. oxidation, damages on screw heads, etc.).
- *General purpose operations*: they are all the operations not included in the previous categories, not strictly related to disassembly, but usually requested for completing the process (e.g.: handling, presenting a product or a single part in a desired orientation and in a given position, etc.). They usually require special tools and equipment.

Table 2 summarises the main disassembly operations, giving a short description, some examples of applications and a brief list of problems which can be encountered during disassembling.

Tools and equipment required in a disassembly process are usually the same used in general purpose activities such as pliers for separating objects, wrenches and spanners for unscrewing operations, cutters, shears and grinders for destructive cutting operations. Table 2 also shows the various kinds of tools usually adopted to perform the different disassembly operations. However, particularly in disassembly for recycling, the use of traditional tools does not allow to achieve sufficient profits [5]: therefore dedicated tools to speed up and rationalise the disassembly of widely diffused goods have to be developed, as described in section 3.4.

3 THE STATE OF THE ART IN DISASSEMBLY

In the following paragraphs some of the most recent developments in the research on disassembly are presented.

3.1 PRODUCT DESIGN

A correct approach to disassembly is based on a design oriented to solve the problems before they occur. Using DFD techniques [8, 9] or other related DFx procedures, such as Design for Environment (DFE) and Design for Recycling (DFR), the best results in terms of recycling easiness and environmental impact can be achieved. DFD techniques are aspects of concurrent engineering process, where different and often antithetic requirements have to be harmonised in a single product.

The most important DFD criteria deal with:

- *Innovative fastening techniques*: during the disassembly phase the detachment of the various parts of the product has to be easy and quick. In particular the separation of parts made by different materials has to be performed as easier as possible. In the case of material recovery, the integrity of the various parts is not important and breakable joints can be adopted. An innovative and interesting approach to the material separation problem named "active disassembly" [10, 11] has been developed. The product structure is studied to identify the zone where a separation action will be more effective to obtain pure and recyclable materials. During the production phase one or more simple and cheap devices, able to facilitate the separation, are embedded in the product. They remain "sleeping" during the whole working life of the product. Only during the disassembly phase they are activated to obtain the separation of desired parts. An example is the insertion of a Ni-Cr wire around the screen of a CRT. During the dismantling of the TV set, a current flows through the wire with the consequence of a heat generation and the CRT breakage in two parts with different glass properties, more efficiently recyclable.
- *Materials homogeneity*: the use of a minimum number of different materials in the product can increase the purity of the recovered materials. Also the use of painted parts should be minimised. Although the presence in the products of different materials is very often necessary, the designer should try to cluster the parts made of the same, or compatible, materials. In this way the dismantling of the products can include a lower number of operations because the homogeneous subassemblies do not need any further disassembly.
- *Material selection*: metals and glass are the most recyclable materials and some plastics (PET, for example) are quite easily recyclable. Thermosetting resins cannot be recycled but only shredded and reused for different purposes, for example as fillers. In any case every plastic part should be identified by a proper symbol indicating the material.
- *Product structure*: a very important aspect related to the product structure is the accessibility of parts. Particularly in maintenance, where standard disassembly operations are mostly used, it deeply influences the disassembly cost and time.
- *Toxic substances*: the elimination of toxic substances should be made as easy as possible because they must be always separated from the rest of the product, independently from the selected disassembly cycle. For instance, if a lubrication or cooling circuit has to be included in the machine, the drain plug should be easily reachable by the worker.
- *Handling and grasping*: in many cases the disassembly of a component can be made easier and more reliable by including shapes suitable for their grasping and handling. This is particularly important in automated disassembly.

In these last years different commercial DFx software has been presented. A widespread application is DFMA (Design for Manufacturing and Assembly) [2]. This tool, together with another software for DFE, can be used to implement DFD techniques. Although these applications need a high volume of input data concerning the product, their interface is extremely user-friendly and the designer can quickly obtain useful suggestions (also from the economical point of view). Another interesting tool is Design for Service (DFS), specifically oriented to maintenance.

TABLE 2. Classification of the main disassembly operations

	Operation	Definition	Example of application	Problems	Typical Tools
Standard	Separating	Detachment of parts	Separation after removing fasteners or connections	Corrosion of parts, deformations, accessibility	Pliers, chisels, chucks, grippers, special tools
Standard	Rotating	Rotating a part in order to disengage it	Bayonet coupling, generic extraction of parts from the product	Corrosion of parts, deformations, accessibility	Pliers, spanners
Standard	Pulling/ pushing	Pulling or pushing a part	Separation of bearings from shafts, hammering a part for extraction	Corrosion, accessibility, force needed to separate parts	Punches, hammers, presses, extractors
Standard	Unscrewing	Disassembling threaded connections	Unscrewing screws, nuts and bolts	Corrosion, accessibility, damage to the nuts or screw heads	Screwdrivers, ratchets, wrenches, spanners, Allen keys
Standard	Deforming	Deforming a part in order to disengage it	Deforming a snap fit, enlarging a Seeger ring	Accessibility	Pliers
Standard	Emptying	Removing fluids from tanks or cavities	Removing CFC fluids from refrigerators	Accessibility, toxic agents	Tanks, hoses, funnels, pumps
Destructive	Drilling	Making holes in order to remove fasteners	Destruction of screw heads or screw bodies	Accessibility, damage to the parts	Portable drills or drilling machine tools
Destructive	Cutting	Making a cut to separate parts or improve accessibility	Cutting the metal sheet of a cover to obtain a fast access inside a product	Damage to the parts	Cutters, shears, grinders, hacksaws, flame torches, laser or water jet equipment
Destructive	Milling	Milling to separate parts, remove fasteners or improve accessibility	Milling rivet heads, local shredding of parts	Accessibility, damage to the parts	Standard and portable milling units, equipped with mills or routers
Destructive	Shredding/ crushing	Local elimination of parts or subassemblies to dispose or recycle materials	Shredding of printed circuit boards, cryogenic crushing	Impure recycled material	Mills, crushers, special equipment, cryogenic systems
Destructive	Bending	Bending permanently a part to improve accessibility	Quick opening of covers	Graspability, stiffness of components	Pliers, special equipment
General purpose	Recognition	Recognition of interesting materials or components during disassembly	Separation of various kinds of plastics or metals	Lack of information about materials or components	Artificial vision systems, dedicated instruments and sensors
General purpose	Grasping	Grasping a part in order to handle it	Extracting a part from its cover	Deformations, accessibility, weight	Standard grippers (fingered, vacuum-type, magnetic), special equipment
General purpose	Moving/ Orienting	Moving/orienting a part or a product	Moving products between workstations or orienting them to improve their accessibility	Weight, dimension, shape	Conveyor belts, special platforms, pick-and-place devices, robots
General purpose	Cleaning/ coat removal	Cleaning/purifying a part or a product	Removing dust from CRTs, removing paint from parts	Accessibility, chemical compatibility	Manual or automated brushes
General purpose	Blocking	Give a stable position to a product under disassembly	Blocking and referring a product by a chuck	Surfaces availability, stiffness of components	Chucks

3.2 DISASSEMBLY PLANNING SOFTWARE TOOLS

The main reasons for the development of disassembly planning software tools can be synthesised in the need for finding optimal solutions among many alternatives and the possibility of integrating it in a CAE environment (interaction with other computer-aided activities such as simulation, scheduling, etc.).

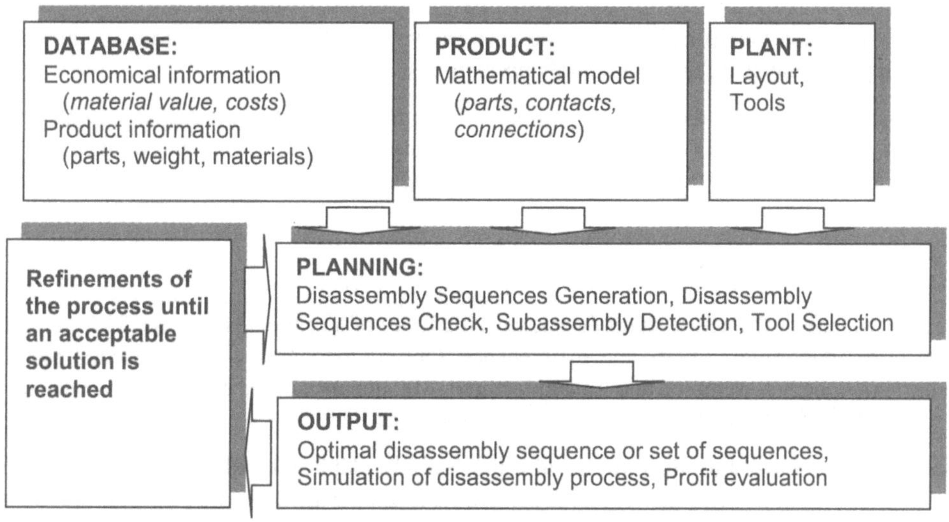

FIGURE 2. Flow chart of a general disassembly planning system

In Figure 2 the typical structure of a such kind of a system is shown. Using these tools, where a model of the product is stored together with a series of related information (such as the number of parts, their weight, kind of joints, materials, market values, etc.) a disassembly sequence can be tested to evaluate the achievable profit. The most complete systems are able to suggest one or more optimal sequences to the planner. If used to simulate the disassembly of a product during the design phase, these systems represent, together with DFD techniques, powerful concurrent engineering tools, allowing the production of easily and profitably recyclable products.

Most disassembly planning systems have been developed in recycling research area. Both traditional and innovative methods have been proposed to implement these systems: expert systems, neural networks [12], ant colony systems [13], genetic algorithms, fuzzy logic [14], Petri nets [15, 16, 17], wave propagation approach [18, 19].

An interesting approach to the typical problems of disassembly has been recently proposed in [20, 21] and named *reactive planning*. Very often the conditions/characteristics of a product are unknown, therefore predetermined plans may be unrealistic [22]. While the usual predictive planning assumes to know everything about the product to be disassembled, including the expected deterioration, reactive planning starts from a predicted plan and adapts the decisions to the actual conditions of the product. The adaptation is carried out dynamically during the recycling process when the data on the product become available.

Another example of disassembly planning system for recycling problems is DIRECT (DIsassembly for RECycling Tool) proposed in [6]. It performs:

- an integration with CAD: avoiding a tedious insertion of all geometrical information for product model creation;
- an automated detection of subassemblies having relevant recycling properties (e.g. homogeneity of materials);
- an automated identification of all the possible ways to disassemble a product: allowing a complete analysis of the disassembly possibilities.

The system includes a database containing information about the materials present in the product, the market data (energy cost, labour cost, material values, available tools, etc.). The results are automatically displayed as shown in Figure 1, where the profit evolution of a disassembly sequence is reported.

3.3 LIFE CYCLE UNITS

The knowledge of the actual status (wear, working time) of a product is very useful to optimise both remanufacturing of parts and maintenance. Only components with a sufficient residual value should be disassembled and remanufactured. Maintenance could be performed only when really useful, with a general saving of time and resources.

Unfortunately, when a product arrives at a disassembly plant, normally no information about its actual working life and general wear are available and in the same way no information are available for maintenance (apart from a rough working time estimation). To overcome this lack of information, a category of devices named Life Cycle Units (LCU) has been proposed [3]. They are based on one or more sensors able to monitor a physical phenomenon (e.g. temperature, humidity) concerning a component of the product, and to store the collected information in on-board memory or communicate them to an external elaboration unit.

Different prototypes have been developed: monitoring of the RPM fluctuation in a fan cooler to predict the bearing wear [3]; determination of residual loading capacity of accumulators [3]. In [23] a device named Electronic Data Log (EDL) has been used to monitor the life of an electric motor in terms of starts and stops of the motor, accumulated run time, motor temperature and power consumption in each individual use cycle (peak and average values).

In a near future, these units could be used also to control actuators for a simultaneous and selective disassembly. For instance, a magnetostrictive actuator embedded in a connection device [3] could release automatically the joint, driven by an external computer where a specific disassembly plan has been stored.

3.4 INNOVATIVE DISASSEMBLY TOOLS AND EQUIPMENT

The great variety of geometrical shapes to be treated, their unknown status and the need for very short working times, have led to the development of new families of tools (e.g. tools forming new acting surfaces). The next list reports, for some relevant disassembly operation, innovative examples of disassembly tools and equipment proposed in these recent years:

- *Separating operation*: a special device dedicated to this operation is the *splitting tool* described in [24]. It is a simple mechanism allowing an entry strength or impulse (given by an hammer, for example) to be divided into two forces that execute the separation of the parts.

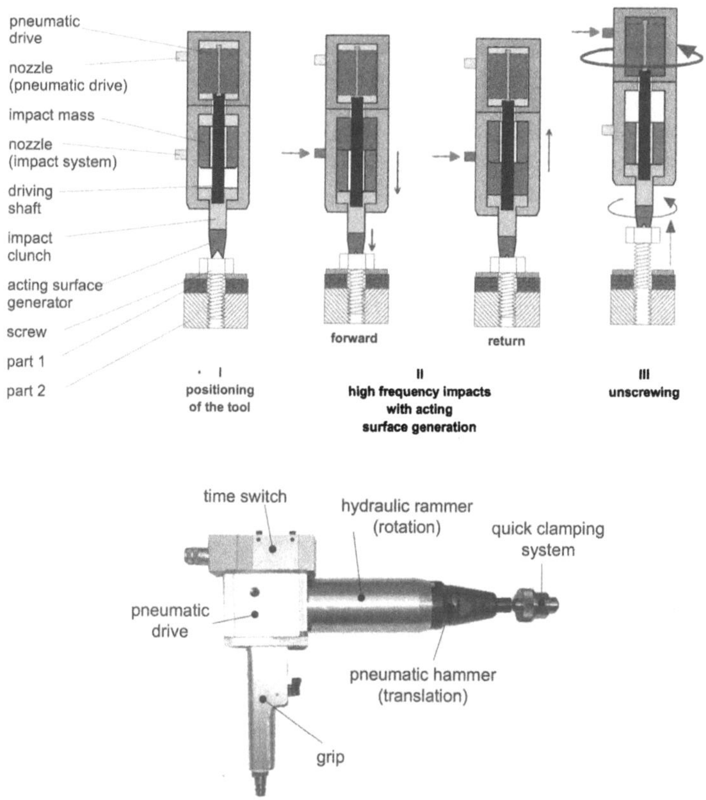

FIGURE 3. Special disassembly tool for fast unscrewing [25, 26]

- *Unscrewing operation*: a screwdriver using a device capable of creating acting surfaces on the head of the object to be unscrewed is described in [25, 26]. It is particularly useful to speed up the unscrewing operation or to allow the unscrewing of damaged joints. The basic idea is generated by the difficulty encountered in disassembly oxidised screws using standard screwdrivers. In this disassembly tool (Figure 3) a special screwdriver is used to imprint the screw head by creating a new slot and unscrewing the part. This tool can be used also as an universal screwdriver being able to create proper acting surfaces on each kind of screw heads.
  Another tool named *drilldriver* is proposed in [24]. If working points are available on the screw to be disassembled, it works as a standard screwdriver, with the possibility to grip on several types of screws (cross recessed, slotted, etc.); if the working points are not available,

a short drilling process is started (at low rotational speed) in order to create drill chips to be used to transmit torque during unscrewing. If this last process does not succeed (due to corrosion, for instance), the complete destruction of the joining element is obtained by using the drill at a high rotational speed;

- *Grasping operation*: a special device, named *drillgripper* for handling components without working points is described in [24]. The tool consists of a drill and a gripper. The drill is used to make a hole on the workpiece and to create a working point. The gripper, represented by an expandable elastomer, penetrates the hole and grasps the workpiece.

During the disassembly process, both manual and automatic, the parts to be disassembled have to be reached in a short time, so their accessibility assumes a great importance for the profitability of the process. In addition ergonomics and safety have to be considered in manual disassembly. Thus special manipulators have been developed for products having high weight and/or dimensions. In Figure 4 an example of such devices is shown [27].

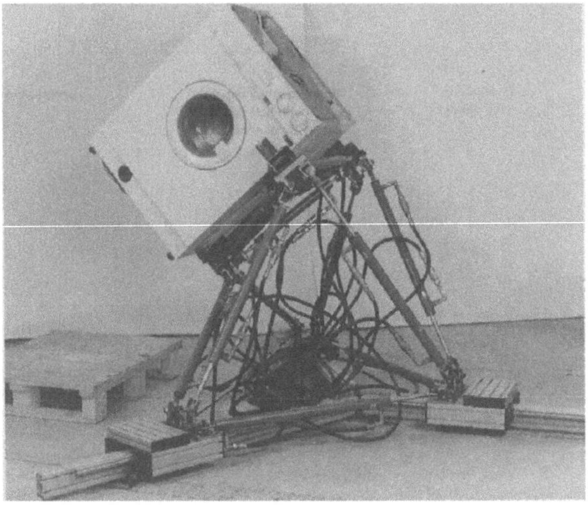

FIGURE 4. Manipulation equipment for ergonomic household appliances orienting [27]

3.5 DISASSEMBLY AUTOMATION

The difficulties (mainly the unpredictability) affecting the disassembly process make its automation quite difficult. Thus, most examples of disassembly plants are manual or only partially automated: only the easiest and repetitive operations are performed by robots, sometime guided by automatic vision systems. For the complex and variable operations, human operators are employed.

Some attempts of automation have been made to disassemble minidisks [28], industrial electric motors [29] and cars [30] in a prototypical disassembly cell based on a vision guided robot. In these cases the variability of the input can be drastically reduced (the products are very similar to each other and the damaged ones are rare) and the automation of the process can be realised.

In the field of household appliances, a complete automation is more difficult to achieve, mainly due to a higher variety of shapes and internal layouts. In [31] a refrigerator disassembly plant with a six axes robot is described. The robot is used to cut the cooling circuit by a shear and to extract the compressor by a gripper. The difficulties related to the variety of models are overcome by teaching action points in a previous station in the disassembly line. In [32] similar concepts are applied to the disassembly of air conditioners. The robot is guided by an automatic vision system and the final disassembly of the compressor is performed in a manual station. TV sets are disassembled in the plant described in [33]; also in this case manual (TV loading, components disassembly) and automated operations (rear cover cutting and removal, cleaning, CRT breaking and removal) are present.

4 A CASE STUDY: A ROBOTIZED CELL FOR AUTOMATED DISASSEMBLY OF COMPRESSORS

At the Dept. of Mechanical, Nuclear and Production Engineering of the University of Pisa several projects in disassembly have started in the last years both in manual and in automated operations. One of the most meaningful examples of automation is a feasibility study regarding the automated disassembly of compressors usually present in refrigerators. The reasons for such research are:

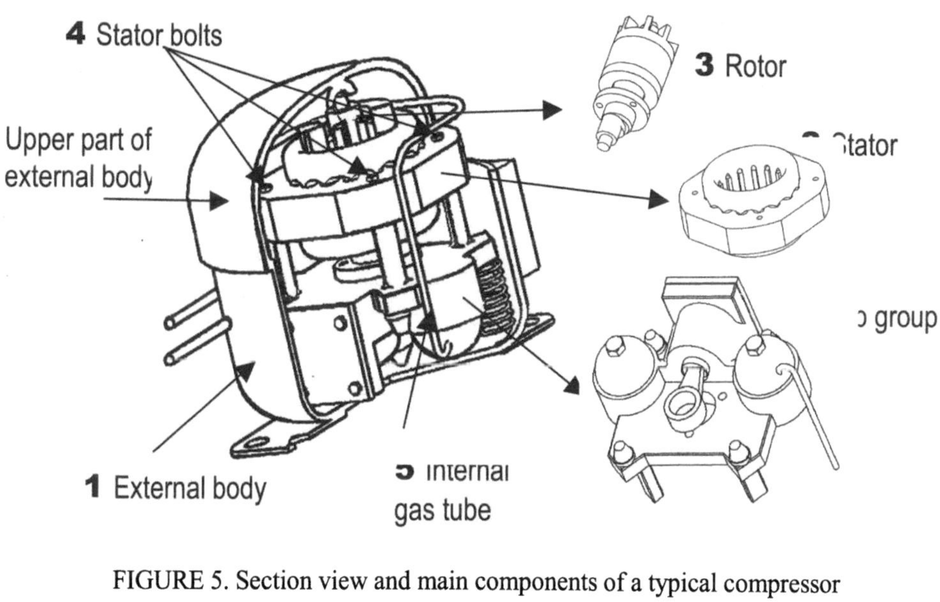

FIGURE 5. Section view and main components of a typical compressor

- at present compressors do not have a wide market as reusable components and on the other side most refrigerators are recycled through cryogenic crushing and separation of different materials; nevertheless some problems of impurity still remain on the ferrous fraction;
- from the point of view of their material content, compressors may represent a source of profit, since significant percentages of copper and aluminium are present in addition to steel and cast iron;

- even if some types of compressors differing in some features are currently present on the market, their overall standardisation may allows a high degree of automation. In this study two widespread types have been considered.

A typical example of compressor with its main components is shown in Figure 5.

The design of the robotized cell started from a set of manual disassembly tests in order to optimise the disassembly sequence and to find the optimal solutions for destructive operations, like cutting of the external body and bolts. It must be considered that actual compressors have not been designed following the DFD rules. The cell is intended to receive compressors fed through a conveyor, with electrical supply cables and gas tubes already cut to a fixed length. A simplified layout of the cell is shown in Figure 6. A robot and a simple pick-and-place manipulator are present.

The selected sequence for one of the two possible types of compressors processed in the plant is shown in Figure 7 and it is based on the following steps:

a) The compressor on the conveyor is located through a sensor, the conveyor stops and the compressor is identified through a vision system. It is possible to select the proper sequence as a function of the compressor type, to check the external conditions and to transmit to the robot the data necessary for a correct grasping;

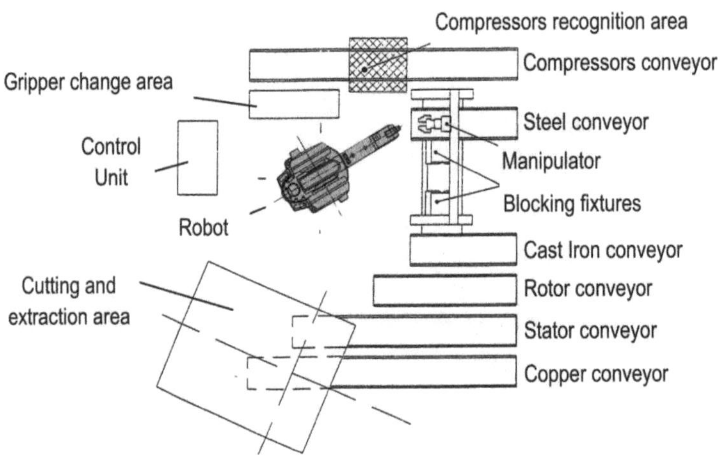

FIGURE 6. Disassembly cell layout

b) The compressor is grasped by the robot and positioned in a proper blocking fixture; then the robot, after a gripper change, uses a cutting tool to cut the upper part of the external body 1 (see Figure 5). Data from the vision system are used for the tool path. During this operation the part is held by a vacuum cup and then moved to the proper output conveyor through the arm of the manipulator.

c) Another special tool mounted on the robot wrist cuts the internal gas tube 5 whose position is constant and does not require the use of a vision system. A further special tool cuts the electrical connector on the external body.

d) The internal part of the compressor is extracted by the robot through another gripper and the external body is moved from its position to the output conveyor by the manipulator. Then the robot moves the internal part to the blocking fixture.
e) A milling operation of the four bolts 4 is then necessary to separate the stator from the other components: this is made through a proper tool mounted on the robot wrist.
f) The stator 2 is extracted and positioned on the cutting station for the separation of the copper winding.
g) The subgroup remaining is then positioned in the same blocking fixture after a 180° rotation. This is necessary since the milling operation of a connecting element can be made only from this side and through the unavoidable help of the vision system. Another milling operation of connecting elements is then required to separate the pump group 6 from the rotor.
h) Finally the robot moves the cast iron pump group 6 and the rotor 3 to the proper output conveyors.

On the stator previously positioned on the cutting station, the following operations are performed while the robot is performing the previous actions:

a) A part of the copper winding is separated through a cutting operation performed by a properly designed shears; during such operation the cut part falls into the proper conveyor.
b) In the same position the remaining part of the winding is extracted from the other side through a properly designed tool and released into the conveyor;
c) Finally the stator 2 is deposited into the proper conveyor by a simple pick-and-place device.

A similar sequence is possible in the cell for the other type of the compressor. A series of different grippers, tools, a proper design of the blocking elements and the presence of the vision system give the cell sufficient flexibility.

While the robot is a general purpose articulated 6 d.o.f. one, the manipulator was the result of a careful design. In order to avoid complex gripper magazines, a solution based on a rotating turret was preferred. Both the vacuum cup and the fingered gripper are mounted on the turret. The flexibility required by the disassembly sequences has been obtained through properly designed active surfaces of the hydraulic vice of the blocking fixture.

All the operations required to separate the copper winding from the iron stator are here performed in a time shorter than the robot and manipulator disassembly time, so that they do not affect the total treatment time for each compressor.

An economical analysis has been made to evaluate the feasibility of the cell. The analysis considers the total cost of the cell, the total disassembly time (about 5 minutes), the operative costs including two persons for supervisory tasks and loading/unloading operations. Possible profits have also considered under the hypothesis of selling copper, aluminium and ferrous content within the recycling market. The result is that a profit of about 1.2 Euro can be achieved for each compressor. Also the financial analysis has given encouraging results in terms of return of investment.

The research is going on and a prototype of a multipurpose gripper is under construction. By such a gripper the following operation will be performed: 1) moving of the whole compressor from the input conveyor to the blocking fixture, 2) cutting of the internal gas tube, 3) extraction of internal part of the compressor (electric motor and pump).

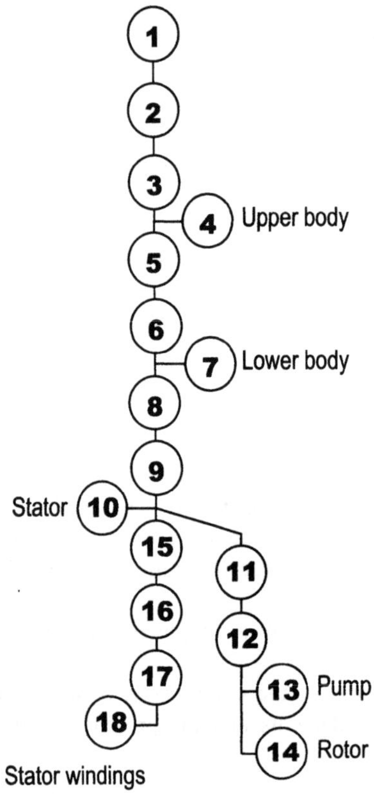

Disassembly sequence no.1	
Disassembly steps	**Operations**
1. Compressor recognition	Recognition
2. Positioning on the blocking fixt.	Moving/ blocking
3. Body cutting and opening	Cutting
4. Upper body recovery	Moving
5. Internal gas tubes cutting	Cutting
6. Motor and pump extraction	Separating / moving
7. External body recovery	Moving
8. Positioning on the blocking fixt.	Moving/ blocking
9. Milling of four bolts	Milling
10. Extraction of the stator	Separating / moving
11. Repositioning of group 3-6	Rotating / moving / blocking
12. Milling of connecting elements	Milling
13. Extraction of the pump	Separating / moving
14. Extraction of the rotor	Separating / moving
15. Positioning on the cutting stat.	Moving/ blocking
16. Cutting of copper winding	Cutting
17. Extraction of remaining copper	Separating
18. Moving stator on the conveyor	Moving

Total recovered materials (Kg)			
Copper	Cast Iron	Steel	Aluminium
0.85	1.9	5.5	0.5

FIGURE 7. Disassembly sequence of the compressor

5 CONCLUDING REMARKS AND PERSPECTIVES

Disassembly is likely to be a strategic technology in the next future in the fields of maintenance, remanufacturing and recycling. Even if a lot of work has been done in different aspects of this wide and multidisciplinary field, some open issues are still present and many problems have to be solved. The most relevant from a scientific point of view, representing possible and interesting directions and perspectives for research activities in the next years could be:

- development of new solutions to be used in DFD such as new fastening techniques, solutions for easy identification of materials, toxic agents, etc.;
- development of efficient disassembly planning systems able to find optimal solutions in terms of maximum profit and disassembly depth and able to interact with the disassembly plant;
- automation of disassembly operations through innovative, modular and flexible solutions (development of fixtures, equipment and sensors);
- development of dedicated tools for improving the feasibility of disassembly operations and reducing disassembly times.

In addition to these concepts, the results of a survey conducted consulting more than 100 German companies involved in disassembly of electronic devices can be also considered [34]. In particular, an interesting aspect of this survey is represented by an investigation about the main problems occurring in disassembly and the demands derived from them for further research and developments. It appears that the most important problems is the huge variety of electronic devices to be disassembled and the biggest demand for further research and development is in the field of product design.

Furthermore, it is advisable that a greater amount of work should be done in legislative matter which will play a fundamental and driving role in research and development activity.

6 ACKNOWLEDGEMENTS

The authors would like to express their sincere thanks to Prof. G. Tantussi and Dr. M. Braglia for their technical support and to Dr. M. Letari for his active contribution in this work.

7 REFERENCES

1. Jovane, F., Alting, L., Armillotta, A., Eversheim, W., Feldmann, K., Seliger, L., Roth, N. (1993) A Key Issue in Product Life Cycle: Disassembly, Annals of the CIRP Vol. 42/2/1993, pp. 651-658
2. http://www.dfma.com/
3. Grudzien, W., Middendorf, A., Reichl, H., Seliger, G. (2000) A micro system tool for improved maintenance, quality assurance and recycling Proceedings of the CAPE 2000, 7-9 August, 2000, Edinburgh, Scotland, p 391-399
4. Lu, S.C.Y., Shpitalni, M., Gadh, R., (1999) Virtual and Augmented Reality Technologies for Product Realization, Annals of the CIRP, Vol48/2, pp.471-495
5. Seliger, G., Basdere, B., Schumann, S., Wapler, M. (2000) Technological Conditions for Selling Use instead of Selling Products Proceedings of the 7th International CIRP Seminar on Life Cycle Engineering, Tokyo, Japan, 2000. p. 17-22
6. Dini, G., Failli, F., Santochi, M., A Disassembly Planning Software System For The Optimization Of Recycling Processes, Production Planning & Control, Taylor & Francis Ltd. London, 2001, Vol. 12, No. 1, pp. 2-12.
7. Hesselbach, J., Herrmann, C., Ohlendorf, M., Graf, R. (2001) Approach of Substance Flow Oriented Closed Loop Supply Chain Management in the Electrical and Electronic Equipment Industry Proceedings of EcoDesign 2001: 2nd International Symposium on Environmentally Conscious Design and Inverse Manufacturing, December 11-15, 2001 Tokyo, Japan

8. Boothroyd, G., Alting, L., (1992) Design for Assembly and Disassembly, Annals of the CIRP Vol. 41/2/1992 pp. 625-636
9. Alting, L., Legarth, J.B. (1995) Life Cycle Engineering and Design, Annals of the CIRP Vol. 44/2/1995, p. 569-580
10. Masui, K., Mizuhara, K., Ishii, K., Rose, C.M. (1999) Development of products embedded disassembly process based on end-of-life strategies, Proceedings First International Symposium on Environmentally Conscious Design and Inverse Manufacturing. IEEE Comput. Soc, Los Alamitos, CA, USA; 1999; xxx+1021 pp. p.570-5
11. Suga, T. (1999) Disassemblability assessment for IM, Proceedings First International Symposium on Environmentally Conscious Design and Inverse Manufacturing. IEEE Comput. Soc, Los Alamitos, CA, USA; 1999; xxx+1021 pp. p.580-
12. Hsin Hao Huang, Wang, M.H., Johnson, M.R. (2000) Disassembly sequence generation using a neural network approach, Journal of Manufacturing Systems. Vol.19, no.2; 2000; pp.73-82
13. Failli, F,. Dini, G., (2001) Optimization of Disassembly Sequences for Recycling of End-of-Life Products by Using a Colony of Ant-Like Agents, Proceedings of the 14th International Conference on Industrial and Engineering Applications of Artificial Intelligence and Expert Systems IEA/AIE 2001 Budapest, Hungary, pp. 632-639
14. Chevron, D., Binder, Z., Horacek, P., Perret, R., (1997) Disassembling process modelling and operations planning under imprecise operation time Proceedings of the 13th World Congress, International Federation of Automatic Control. Vol.L. Systems Engineering and Management. Pergamon, Oxford, UK; 1997; x+420 pp. p.367-72
15. Moore, K.E., Gungor, A., Gupta, S.M. (1998) Disassembly Petri net generation in the presence of XOR precedence relationships SMC'98 Conference Proceedings. 1998 IEEE International Conference on Systems, Man, and Cybernetics IEEE, New York, NY, USA; 1998; pp. 13-18 vol.1
16. Moore, K.E., Gungor, A., Gupta, S.M. (1998) A Petri net approach to disassembly process planning Computers-&-Industrial-Engineering. vol.35, no.1-2; Oct. 1998; pp. 165-8
17. Salomonski, N., Zussman, E. (1999) On-line predictive model for disassembly process planning adaptation Robotics-and-Computer-Integrated-Manufacturing. vol.15, no.3; June 1999; pp.211-20
18. Srinivasan, H., Gadh, R., (1998) Complexity reduction in geometric selective disassembly using the wave propagation abstraction, Proceeding 1998 IEEE International Conference on Robotics and Automation IEEE, New York, NY, USA; 1998; 4 vol. lxv+3744 pp. p.1478-83 vol.2
19. Srinivasan, H., Figueroa, R., Gadh, R. (1999) Selective disassembly for virtual prototyping as applied to de-manufacturing, Robotics-and-Computer-Integrated-Manufacturing. vol.15, no.3; June 1999; p.231-45
20. Geiger,D., Zussmann, E. (1996) Probabilistic reactive disassembly planning Annals of the CIRP Vol 45/1 pp. 49-52
21. Scholz-Reiter, B., Scharke, H. 1997 Implementation and Testing of a Reactive Disassembly Planner Life Cycle Networks, F.-L. Krause & G. Seliger1997 Chapman & Hall, Section 32
22. Tang, Y., MengChu, Z., Zussman, E., Caudill, R. (2000) Disassembly modeling, planning and application: a review Proceedings 2000 ICRA. Millennium Conference. IEEE International Conference on Robotics and Automation. Symposia Proceedings, IEEE, Piscataway, NJ, USA; 2000; Vol.3 pp.2197-202
23. Klausner, M., Grimm, W., M., Hendrickson, C., (1998) Reuse of Electric Motors in Consumer Products - Design and Analysis of an Electronic Data Log, Journal of Industrial Ecology, Volume 2, No. 2, p. 89-102
24. Feldmann, K., Trautner, S., Meedt, O., (1998), Innovative Disassembly Strategies Based on Flexible Partial Destructive Tools, proc. of Intelligent Assembly and Disassembly IAD98, 193, 1-6.

25. Seliger, G., Keil, T., Rebafka, U., Stenzel, A. (2001) Flexible Disassembly Tools Proceedings of the 2001 IEEE International Symposium on Electronics & the Environment, Denver, Colorado, pp. 30-35
26. Wagner, M., Seliger, G., (1996), Modeling of Geometry-Independent End Effectors for Flexible Disassembly Tools, Proc. of the 3rd Int. CIRP Sem. on Life Cycle Engineering, Zurich, Verlag, 219-228.
27. Uhlmann, E., Seliger, G., Haertwig, J.P., Keil, T. (2000) A Pilot System For The Disassembly Of Home Appliances Using New Tools And Concepts, Proceedings of the Third World Congress on Intelligent Manufacturing Processes & Systems, Cambridge, MA - June 28-30, 2000, p. 453-456
28. Gschwendtner et al (1995) Robotized disassembly of minidisks, 26th Int. Symposium on Industrial robots
29. Karlsson, B. (1998) Automatized disassembly of electrical industrial motors, SPIE Proceedings vol. 3517
30. Schraft et. al. (1995) Robotic dismantling. Experimental results in automobile disassembly, 26th Int. Symposium on Industrial robots
31. Kotera, Y., Hirasawa, E., Sakitani, H. (1999) Integrated Recycle Plant for Electric Home Appliances - Automated Primary Disassembling for Refrigerators, Proceedings First International Symposium on Environmentally Conscious Design and Inverse Manufacturing, IEEE comp. Soc, Los Alamitos, CA, USA, 1999, pp. 268-272
32. Uchiyama, Y., Fujisawa, R., Oda, Y., Hirasawa, E. (1999) Air conditioner and Washing Machine Primary Disassembly Process, Proceedings First International Symposium on Environmentally Conscious Design and Inverse Manufacturing, IEEE comp. Soc, Los Alamitos, CA, USA, 1999, pp. 258-262
33. Kotera, Y., Sato, S. (1997)An Integrated Recycling Process for Electric Home Appliances, Mitsubishi - Electric - Advance, Sept. 1997, Vol. 80, pp. 23-26
34. Hesselbach, J., Kuhn, M., (1998) Survey of Processes at Disassembly Companies, IFAC Intelligent Assembly and Disassembly, Bled, Slovenia, pp.99-102.

ANALYSIS OF SOME INNOVATIVE AND FLEXIBLE SHEET FORMING PROCESSES

N. Alberti[1], F. Micari[2]

[1] Department of Mechanical Technology and Production, University of Palermo, Italy
[2] Department of Mechanical Engineering, University of Calabria, Italy

KEYWORDS: Sheet metal forming, Hydro-forming, Incremental forming

ABSTRACT. In the field of sheet forming, innovative and flexible processes that do not impose the use of expensive conventional equipments and/or do not require time consuming set-up operations have become, nowadays, a rather promising research topic. Two different main lines are currently followed: the former is based on the development of new stamping processes based on the utilization of flexible media, while the latter is aimed to develop so called progressive forming processes (spinning and incremental forming). Both hydro-forming and incremental forming permit a very relevant reduction of the tooling and set-up costs and improve process flexibility. These items are discussed in detail in the paper; furthermore some experiments aimed to the characterization of material formability in the mentioned processes are presented.

1 INTRODUCTION

The growing request for customer oriented products leads to a deep change in manufacturing systems. In particular, the capability of a rapid conversion of the production towards new products and the possibility to carry out a quick variation of manufacturing capability in order to match market requirements, are definitely crucial competitive factors for modern manufacturing industries.

In this way, the development of innovative technologies within manufacturing is highly requested. In particular, as far as forming processes are concerned, innovative processes that do not impose the use of expensive conventional equipments and/or do not require time consuming set-up operations have become, nowadays, a rather promising research field. In fact, such technologies give rise to flexibility improvement in manufacturing processes, and, as a consequence, enable a competitive advantage for industries in the modern manufacturing environment.

The availability of different production alternatives is a quite relevant improvement both from an economical point of view and from a technological one. Thus, innovative sheet stamping processes can be considered as useful manufacturing opportunities that can lead to economical advantages with respect to traditional processes, mostly because of the reduction of tooling costs.

Moreover, in the modern competitive scenarios, the request of flexible production technologies for small batches production is becoming more and more important since manufacturing industries are greatly interested in small market segments where technological capacity and flexibility are crucial.

Published in: E.Kuljanic (Ed.) *Advanced Manufacturing Systems and Technology*, CISM Courses and Lectures No. 437, Springer Wien New York, 2002.

The research in the field of innovative stamping processes follows two different evolution directions. The former is based on the development of new stamping processes based on the utilization of flexible media, such as elastomers (rubber forming processes), liquids (hydro-forming processes) or gases, with the aim to substitute one of the dies of the traditional stamping processes. The latter is aimed to develop so called progressive forming in which the final shape of the component is determined by the relative movement of a geometrically simple and small punch with respect to the blank (spinning and incremental forming), rather than by the shape of the dies. Such processes should be carried out on CNC machines, where it is possible to assign and control the punch movement according to fixed paths.

Both hydro-forming and incremental forming permit a very relevant reduction of the tooling and set-up costs and improve process flexibility; on the other hand they are quite "young" and under development processes, which require a considerable research effort.

In the last few years the authors have been involved in several research projects both on hydro-forming and on incremental forming. In this paper some of the most relevant obtained results are reported and discussed in detail. In particular the former part of the paper is focused on hydro-forming: the process of hydro-mechanical deep drawing is taken into account and some experimental tests aimed to the investigation of material formability are shown, together with a detailed numerical simulation of the process. Moreover a simple and economic experimental apparatus to evaluate material formability in tube hydro-forming is presented with an extensive experimental investigation, carried out to investigate material formability over a wide range.

In the latter part of the paper, in turn, some results of a wide research activity on incremental forming will be presented. Several incremental forming tests were carried out on 1050-O and 6082-T6 Aluminum Alloy sheets utilizing a properly designed fixture mounted on a 3-axis controlled CNC milling machine. The tests were aimed to the production of complex shape components. Here again, a finite element analysis of the forming process was carried out, in order to investigate the process mechanics and to evaluate the forming limits.

2 HYDRO-FORMING

2.1 SHEET HYDRO-FORMING

The application of sheet hydro-forming in the modern automotive industry has had a very large increase in the last decade [1]. It is worth pointing out some of the most relevant advantages offered by sheet hydro-forming. This technology allows to strongly reduce both the costs due to the manufacturing and the maintenance of the dies and the ones linked to set-up times. Such reduction is not only due to the avoidance of one of the dies; furthermore the pressure distribution due to the fluid action over the sheet is more gradual and distributed, thus permitting to construct the remaining die in less expensive materials than the conventional ones.

Other advantages supplied by hydro-forming are a remarkable increase of the limiting drawing ratio in deep drawing, a more uniform thickness distribution on the stamped sheet, a safer residual stresses distribution and finally a better surface quality than using other traditional processes [2-6].

The former item was carefully investigated by the authors in a couple of recent papers [7-8]: a simple experimental equipment was developed to carry out hydro-mechanical deep drawing operations on a typical deep drawing steel and on AA6082T6 Aluminum Alloy sheets. The aim of the research was to investigate the influence on the process mechanics of some of the most important operating parameters (ram speed, blank-holder pressure), in order to obtain the highest deep drawing ratio.

A simple and schematic sketch of the experiment is reported in Figure 1. The rigid punch and the drawn cup penetrate into an enclosed liquid chamber, generating an hydrostatic pressure of the liquid which pushes the drawn component against the punch: thus fluid pressurization is obtained through the movement of the punch. The main difference between the experiment and the industrial sheet hydro-forming operation is that in the latter the hydrostatic pressure of the fluid replaces the action of the punch and it is obtained by means of external working media sources, such as pumps or pressure intensifier systems.

The pressure value is controlled through a maximum pressure valve; furthermore as the punch goes down, a thin fluid film flows under the flange. According to the above considerations, it is quite easy to conclude that process mechanics is strongly affected by the blank-holder force value: such parameter in fact determines to some extent the thickness of the fluid film and then the value of fluid pressure in the chamber due to the punch penetration.

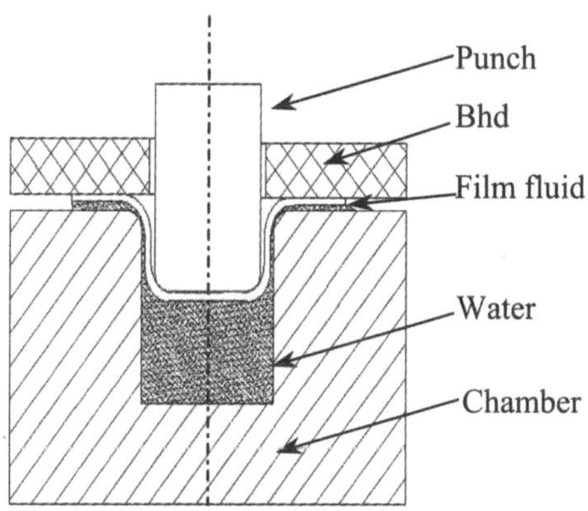

FIGURE 1. Hydro-mechanical deep drawing

The process conditions which allow the highest deep drawing ratio were searched for. From the theoretical point of view, the idea that hydro-mechanical deep drawing should permit a relevant increase of such ratio is based on the observation that in this process the sheet is strongly pushed by the fluid action against the punch and consequently large frictional forces occur at the punch-sheet interface. These forces are very helpful, since they cooperate to apply the drawing force to the sheet: in other words the drawing force is not applied just on the bottom of the cup, as occurs in conventional deep drawing, but a relevant contribution is due to the above mentioned friction forces. By this way a relevant reduction of the axial stresses acting on the cup walls is achieved, moving away the danger of necking and tearing.

Such considerations are confirmed by the comparison of typical fractures in the traditional process and in hydro-mechanical deep drawing. In the former case fracture typically occurs at the punch radius, i.e. at the corner between the bottom and the wall of the cup, due to the above mentioned tensile axial stresses; in the latter case when material formability is exceeded, fracture is generally localized close to the chamber radius, i.e. at the corner between the flange and the wall of the cup.

The theoretical considerations were confirmed by the experiments: in particular the highest "safe" deep drawing ratio utilizing steel blanks was β=2.62, i.e. quite a very high value with respect to the typical ones in conventional deep drawing. As well the highest "safe" deep drawing ratio on AA6082T6 blanks was β=1.83, again a very high and promising value.

On the other hand sheet hydro-forming is characterized by forming times much larger than traditional operations; as a consequence such technology cannot be taken into account for large scale productions, but only for small batches.

Furthermore the suitability of sheet hydro-forming depends on the complexity of the part to be stamped: if, for instance, it is characterized by small fillet radii, very high fluid pressures could be necessary, requiring large and expensive equipments and determining relevant problems from the point of view of material safety.

In this sense another experiment was developed by the authors [9]. The study was focused on the investigation of material formability, the latter being measured through the capability of the blank to deform according to the punch shape due to the action of the fluid pressure. Thus the tests were carried out at the varying both of the geometry of the front part of the punch and of the blank-holder force which really determines the pressure of the fluid inside the chamber.

As far as the former issue is concerned, three different punches were utilized as shown in Figure 2. The punches are characterized by a different geometry of the nose, as regards both the number of the cylindrical sections (one or two) and the heights of the sections themselves. All over the tests both the corner and the fillet radii of the punches were maintained constant and equal to 1mm, while the diameter of the blank was equal to 70mm. The tests were carried out on the deep drawing steel ASTM A622.

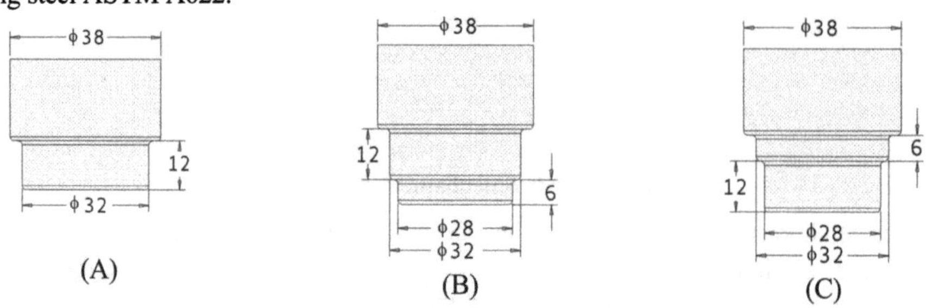

FIGURE 2. The utilized punches

Figures 3 (a and b) shows for instance the cups obtained utilizing the A-type punch and fixing the blank-holder force to a value corresponding to an initial blank-holder pressure (Bhp_i) equal to 2% or 6% of the initial yield stress of the blank material (196MPa). Actually the increase of the BHF permits to increase the fluid pressure inside the die chamber and consequently allows a better overlapping of the sheet metal over the punch shape, even if the corner radius is not completely filled. The above considerations are confirmed by the next Figure 4 which shows the fluid pressure histories during the two tests: the pressures were measured by means of an high pressure fluid sensor based on the strain gage technology.

FIGURE 3. A-type cups (a:Bhp_i=2%σ_0; b:Bhp_i= 6%σ_0)

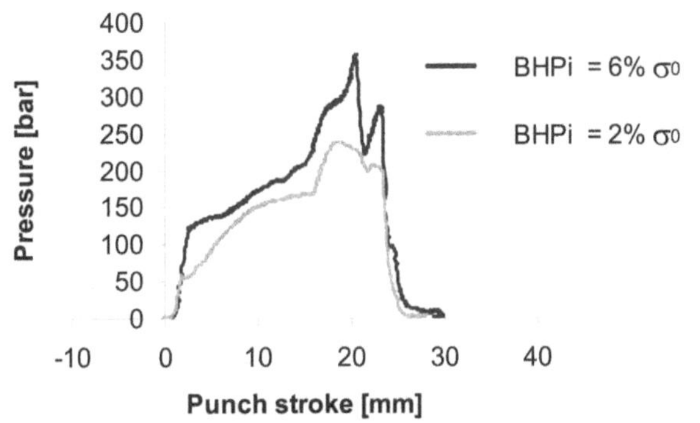

FIGURE 4. Fluid pressure vs. stroke curves

The hydro-mechanical deep drawing operation here addressed was numerically simulated by the implicit elastic-plastic MSC-AutoForge commercial code. The effect of the fluid pressure was taken into account applying the above reported fluid pressure history on the lower surface of the sheet, i.e. the one in contact with the fluid inside the die chamber. Figure 5 (a and b) reports a particular of the deformed meshes corresponding to the two investigated process conditions: actually the numerical analysis is able to predict the better overlapping of the punch shape obtained with the higher BHF and the incomplete filling of the corner radius. The effectiveness of the numerical simulation is confirmed as well by the very good prediction of the punch load during the process, as shown in the next Figure 6.

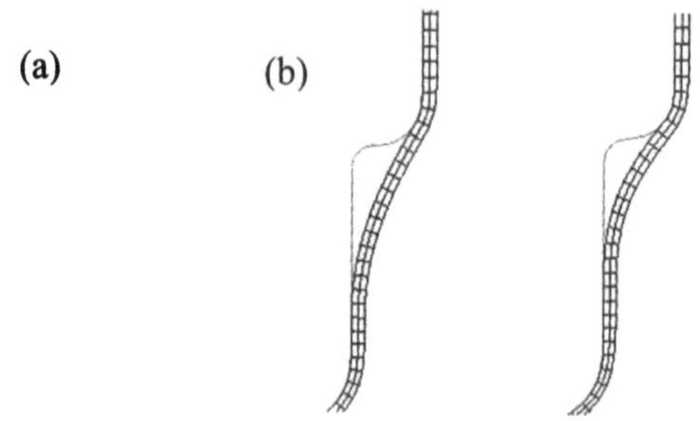

FIGURE 5. Deformed meshes (a: $Bhp_i=2\%\sigma_0$; b: $Bhp_i=6\%\sigma_0$)

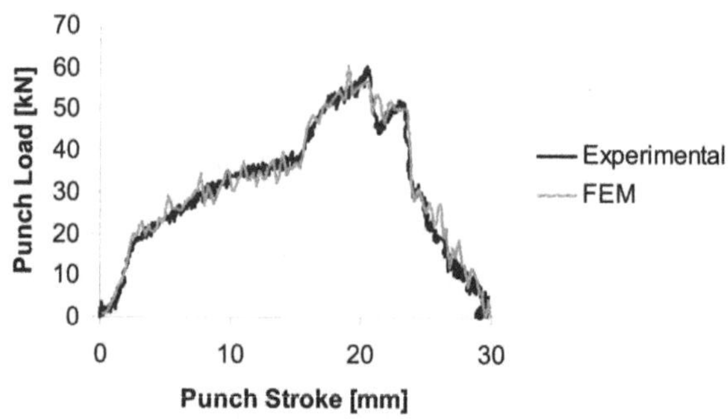

FIGURE 6. Comparison of the predicted and the experimental punch load ($Bhp_i=6\%\ \sigma_0$)

Further details on the experiments at the varying of the punch geometry and, as a consequence, of the heaviness of the forming conditions can be found out in reference [9].

2.2 TUBE HYDRO-FORMING

In the most recent years, tube hydro-forming has become an economic and industrially suitable alternative to various stamping processes. Tube hydro-forming enables in some cases the manufacture of complex parts, starting from hollow components, in just one operation, while traditionally a whole sequence of stamping and welding processes would have been necessary; thus both the tooling costs, due to fewer parts, and other secondary operations can be reduced [10-15].

Again, together with the above mentioned advantages, some relevant drawbacks must be mentioned as well: tube hydro-forming is characterized by slow cycle time as compared to conventional stamping operations and requires very expensive equipment. Moreover there is still a relevant lack of knowledge about tube hydro-forming in particular as far as material formability is concerned.

A simple and economic experimental device to evaluate material formability in tube hydro-forming was proposed by the authors [16]. Simplicity and cost effectiveness represented the most relevant constraints to the design of the experiment: thus no external pressure source were util-

ized, because the fluid pressure is obtained through the movement of the punch. The tests were carried out placing the tube into a die characterized by a proper expansion zone and filling the tube completely with the fluid (Figure 7). Then the movement of the punch was able to supply both the axial feeding of the tube material and the internal pressure, the latter being controlled by means of a maximum pressure valve. In this way, at the varying of the expansion zone geometry and the limiting pressure value, material formability was investigated over a wide range.
The tests were carried out on AISI1015 steel tubes. The diameter of the tubes was 40 mm, while two different thickness values (namely 1.0 and 1.5mm) were taken into account. Three different dies were designed and constructed, characterized by different expansion zones for the tube. The latter task was achieved utilizing three different values of the maximum diameter of the die cavity (D) and of the semi-cone die angle (α), see Figure 7. The values of the above geometrical parameters for each one of the three dies are summarized in Table1; in the latter column of this Table, the largest theoretical hoop strain occurring in the expanded tube is reported for each die geometry.

TABLE 1. Geometry of the utilized dies

Die	Maximum Inner Diameter (mm)	Semi-cone angle	Maximum hoop strain
1	48	20	0.18
2	56	20	0.34
3	64	30	0.47

Fluid pressure inside the chamber was measured and controlled. The former task was achieved by means of a strain gage pressure sensor, while the latter was carried out by means of a maximum pressure valve. The working ranges of the sensor and the valve were 0-100 MPa and 10-50 MPa respectively.

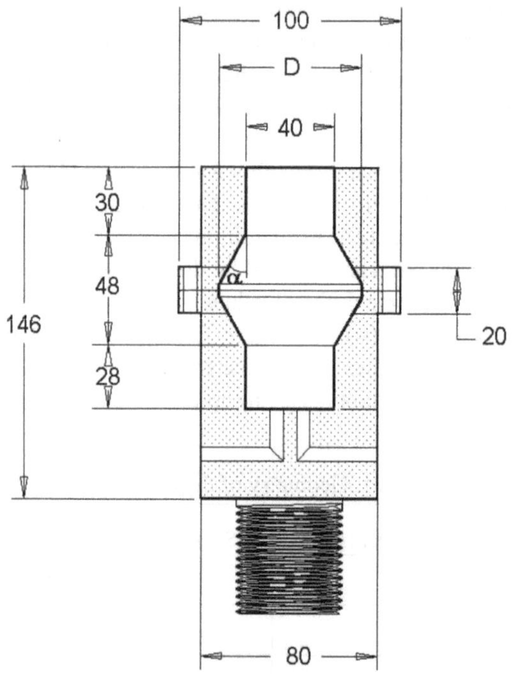

FIGURE 7. Die sketch (all the dimensions are in mm)

The above described equipment was mounted in a 1000kN hydraulic testing machine; a PC based data acquisition system was set-up to obtain a continuous monitoring of the most relevant process variables, i.e. the punch stroke, the axial load and the fluid pressure inside the chamber. As mentioned the tests were carried out varying the expansion zone geometry (i.e. utilizing the three dies described above) and the value of the fluid pressure inside the chamber. Depending on the test conditions it was possible to observe a wide variety of flow defects, as well as ductile fractures on the tubes: Figure 8 reports some typical defects, namely wrinkling, buckling and bursting.

FIGURE 8. Some defects observed during the tests

Therefore the developed experiment, even if very cheap and simple, permitted a wide investigation of material formability in tube hydro-forming. Taking into account, for instance, the set of tests on 1.5mm thick tubes, the experimental evidences are reported in Table 2: the result of each test is simply indicated as "sound" or "defective", depending whether the fluid pressure was able to ensure the expansion of the free surface of the tube up to completely fill the die cavity or, on the contrary, a shape defect occurred. No ductile fractures were, in fact, observed on 1.5mm thick tubes within the investigated pressure range.

TABLE 2. Test results (tube thickness:1.5mm)

Fluid Press. [MPa]	Die n.1	Die n.2	Die n.3
40	Sound	Sound	Defective
35	Sound	Defective	Defective
30	Defective	Defective	Defective

Tube hydro-forming was simulated as well utilizing the implicit elastic-plastic MSC-AutoForge commercial code. An axi-symmetric analysis was developed taking into account the geometry of the parts. The numerical model was able to follow the process mechanics and to predict either the insurgence of flow defects or the complete filling of the die cavity as confirmed in Figure 9.

FIGURE 9. Experimental and numerical results for different testing conditions: a) die n.1, pressure 30MPa, thickness 1.5mm (wrinkling occurrence); b) die n.2, pressure 40MPa, thickness 1.5mm (sound part); c) die n.3, pressure 40MPa, thickness 1.5mm (defective part)

The set of tests on 1.0mm thick tubes showed, in turn, the insurgence of ductile fractures, i.e. bursting, for the heaviest testing conditions (die n.3 and highest fluid pressure value). Therefore, in order to predict the insurgence of fractures, the numerical simulations were carried out applying a ductile fracture criterion as well. In particular the Cockroft and Latham fracture criterion was implemented by means of user subroutine [16-17].

3 INCREMENTAL FORMING

Incremental forming is based on the idea to progressively deform a sheet utilizing a small and geometrically simple tool which moves along an assigned trajectory controlled by computer. The tool deforms the material through a bending and/or stretching deformation mechanics. In this way no conventional punches and dies are required and the final shape of the sheet is determined by the trajectory assigned to the tool only. In other words, in incremental forming material deformation does not depend on the punch and the die shapes (as usually occurs in metal stamping), but on the relative movement of the punch with respect to the blank. The final shape is obtained through the "sum" of local deformations induced by the punch. Thus the effective definition of the tool path on a CNC machine represents one of the most relevant issues to be investigated [18-22]. According to the above considerations the advantages offered by incremental forming in terms of process flexibility are quite evident. Nevertheless incremental forming is typically a very slow process. Therefore industrial suitability of such technology can be assessed only for very small batch productions. A very promising field of application for progressive forming may be, on the contrary, rapid prototyping. In the last few years the authors carried out an extensive basic research activity on incremental forming [9]. The tests were carried out on 6082-T6 and 1050-O Aluminum Alloys 1.0mm thick sheets utilizing a properly designed fixture mounted on a 3-axis CNC milling machine (Figure 10 a,b,c).

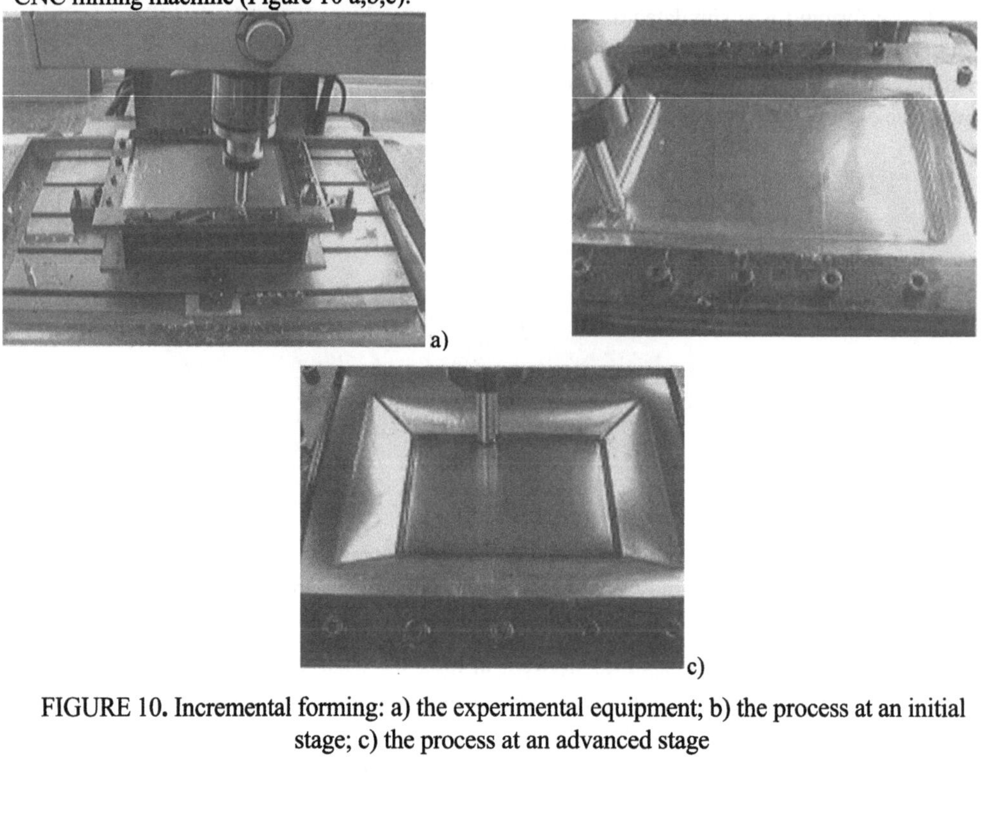

a) b) c)

FIGURE 10. Incremental forming: a) the experimental equipment; b) the process at an initial stage; c) the process at an advanced stage

At each test the blank, which had square or rectangular shape, was clamped at the boundary; in this way the punch movement determined an almost pure stretching deformation mechanics. The tests were aimed to the production of different complex shape components, such as conical and cross shells. Some of the obtained shapes are reported in Figure 11.

a)

b)

c)

FIGURE 11. Some typical shapes obtained through incremental forming

A detailed description of the experimental investigations may be found out in references [9] and [23]. Here it is worth pointing out, more in detail, the process mechanics which characterizes incremental forming. Actually, in this process, the applied deformation is strongly localized: step by step, the moving punch applies plastic deformation on the material through a local biaxial stretching mechanism, i.e. to sheet thickness cost. Therefore material formability in incremental forming is limited only by the maximum thinning that the material is capable to undergo before fracture.

According to the above considerations the forming limit diagram for incremental forming is, for the same material, quite different from the conventional one. The latter statement has been recently demonstrated by the authors, as reported in reference [24].

Finally numerical simulation of incremental forming requires a fully three-dimensional analysis since no symmetry conditions can be assumed in the progressive interaction between the punch path and the clamped sheet. Thus an explicit dynamic model was utilized by the commercial code PC Dynaform. The simultaneous contact of several shell elements with the tool surface allowed the proper modeling of the progressive deformation impressed to the sheet metal. In the next Figure 12 the typical obtained deformed meshes reproducing the investigated shell shapes are reported, confirming that the numerical simulation is a powerful tool in order to investigate the process mechanics of the incremental forming operations.

FIGURE 12. Typical deformed meshes obtained through numerical simulation

4 ACKNOWLEDGEMENTS

This work has been made using MURST (Italian Ministry for University and Scientific Research) funds.

REFERENCES

1. Schmoeckel, D., Hielscher, C., Huber, R., (1999), Metal Forming of Tubes and Sheets with Liquid and Other Flexible Media. Annals of CIRP, Vol 48/2, 497-513.
2. Vollertsen, F., Prange, T., Sander, M., (1999), Hydroforming: needs, developments, and perspectives. Proceedings of the 6th ICTP, Vol 2, 1197-1210.
3. Kleiner, M., Homberg, W., Brosius, A., (1999), Process and Control of Sheet Metal Hydroforming. Proceedings of the 6th ICTP, Vol 2, 1243-1252.
4. Hein, P., Geiger, M., (1999), Advanced process control strategies for the hydroforming of sheet metal pairs. Proceedings of the 6th ICTP, Vol 2, 1267-1272.
5. Tirosh, J., Konvalina, P., (1995), On the hydrodynamic deep-drawing process. Int. J. Mech. Sci., Vol 27, 595-607.
6. Danckert, J., Nielsen, K.B., (2000), Hydromechanical deep drawing with uniform pressure on the flange. Annals of CIRP, Vol 49/1, 217-220.
7. Filice, L., Fratini, L., Micari, F., (1999), Experimental analysis of the hydroforming process. Proceedings of the 4th AITEM Conference, 371-378.
8. Filice, L., Fratini, L., Micari, F., (2000), Hydro-mechanical deep drawing of AA6082T6 sheets. Proceedings of the 7th Saxon Conference on Forming Technology, 187-198.
9. Filice, L., Fratini, L., Micari, F., (2001), New trends in sheet metal forming processes. Proceedings of PRIME 2001 Conference, 143-148.
10. Ahmetoglu, M., Altan, T., (2000), Tube Hydroforming: State-of-the-art and Future Trends. J. of Mat. Proc. Technology, Vol 98, 25-33.
11. Sokolowski, T., Gerke, K., Ahmetoglu, M., Altan, T., (2000), Evaluation of Tube Formability and Material Characteristics: Hydraulic Bulge Testing of Tubes. J. of Mat. Proc. Technology, Vol 98, 34-40.
12. Leitermann, W., Von Zengen, K.H., Dick., P., (1999), Innovative Forming Technologies for Space Frames. Proceedings of the 6th ICTP, Vol 2, 1165-1170.
13. Schmoeckel, D., Hessler, C., Engel, B., (1992), Pressure control in hydraulic tube forming. Annals of the CIRP, Vol 41/1, 311-314.
14. Ghouati, O., Baida, M., Lenoir, H., Gelin, J.C., (1999), 3D Numerical Simulation of Tube Hydroforming. Proceedings of the 6th ICTP, Vol 2, 1211-1216.

15. Gelin J.C., Ghouati O., Labergère C., (2000), From Optimal Design to Control of Process in Sheet Forming and Tube Hydroforming. Proceedings of ECCOMAS 2000.
16. Filice L., Fratini L., Micari F., (2001), A simple experiment to characterize material formability in tube hydroforming. Annals of CIRP, Vol 50/1, 181-184.
17. Cockroft, M. G., Latham, D. J., (1968), Ductility and the Workability of Metals. J. Inst. Metals, Vol 96, 33-39.
18. Sawada, T., Matsubara, S., Sakamoto, M., Fukuhara, G., (1999), Deformation analysis for stretch forming of sheet metal with CNC machine tool. Proceedings of the 6th ICTP, Vol 2, 1501-1504.
19. Kitazawa, K., Nakajima, A., (1999), Cylindrical incremental drawing of sheet metals by CNC incremental forming process. Proceedings of the 6th ICTP, Vol 2, 1495-1500.
20. Iseki, H., (1999), A simple deformation analysis for incremental bulging of sheet metal using high speed water jet. Proceedings of the 6th ICTP, Vol 2, 1483-1488.
21. Tanaka, S., Nakamura, T., Hayakawa, K., (1999), Incremental sheet metal forming using elastic tools, 2000. Proceedings of the 6th ICTP, Vol 2, 1477-1482.
22. Iseki H., (2000), As experimental and theoretical study on a forming limit curve in incremental forming of sheet metal using spherical roller. Proceedings of Metal Forming 2000, 557-562.
23. Filice L., Fratini L., Pantano F., (2001), CNC incremental forming of Aluminum alloy sheets. Proceedings of the 5th AITEM Conference, 473-482.
24. Filice L., Fratini L., Micari F., (2002), Analysis of material formability in incremental forming. Accepted for publication on the Annals of Cirp, Vol 51/1.

IMPROVEMENT OF THE SURFACE TEXTURE QUALITY OF MILLED DIES AND MOLDS

A. Moisan[1], M. Boujelbene[1], B. Brenier[2], A. Fabre[2]

[1] LaBoMaP laboratory, ENSAM Cluny, France
[2] MécaSurf laboratory, ENSAM Aix-en-Provence, France

KEYWORDS: Multi-axis milling, 3D surface texture, tool path, continuity of tangency, dies and molds.

ABSTRACT. One of the goals of multi-axis milling for dies and molds is to achieve good accuracy together with an improvement in surface finish, in order to minimize polishing operations, still necessary to obtain the surface finish required by the automotive industry and the plastic injection sector. This present study concerns the multi-axis milling of a specific mold steel by a ball nose carbide tool coated with Ti(CN). Surface finish analysis given by 3D measurements and data processing techniques shows that micro-geometrical defaults generated by a C^1 continuous tool path are definitely less important than those obtained by classical tool paths driven via linear interpolation.

1 INTRODUCTION

Die and mold manufacturing represents a significant area of application of cast steels and alloy steels [1]. Machining and polishing of dies/molds represent approximately two thirds of their total manufacturing costs [2][3].

Since 1960, molds and dies making is of growing interest. Machining of complex surfaces with ball or bull nose tools is today done with algorithms written in the 70s. Modern CAD-CAM (computer-aided design, computer-aided manufacturing) software, provide paths with reasonable accuracy; collisions between tool and part are avoided (but, sometimes, not the spindle and part). Generally optimized paths are now quickly computed, and at the end of the processes ([4][5][6][7][8]), the machine tool receives a list of numerous statement programming linear trajectories, always shorter rather than longer (some microns) [9].

Due to the improvement of machine tools, we have started this study in order to check the quality of the machined part (accuracy, roughness) using long paths with respect of tangency continuity and appropriate cutting conditions as a result of the tool-work material combination behavior. The main goal of this task is to reduce the subsequent "polishing" length of time for molds and dies.

2 TECHNICAL AND ECONOMIC ANALYSIS OF MOLD MANUFACTURING PROCESSES

Technical and economic analysis of mold manufacturing processes can be achieved only by working on real examples. The injection market exhibits a wide variety of aspects, various sizes of pieces, various kinds of molds. Molds for plastic injection can be made from different grades of HB 300 or 400 pre-hardened steels.

Published in: E.Kuljanic (Ed.) *Advanced Manufacturing Systems and Technology*, CISM Courses and Lectures No. 437, Springer Wien New York, 2002.

Transport (automotive industry, industrial vehicles, etc.) use represents more than a third of injected plastics needs. The automotive industry presents both representative and homogeneous characteristics. Injected parts are visible pieces for which surface finishing is of prime importance, with large dimensions, made into polypropylene. Injection series quantities are around hundreds of thousands, into molds with single or more cavities, made of HB 300 pre-hardened mold steels.

The analysis of mold steels sold around the world shows that the automotive industry is representative of numerous other small markets. Indeed, mold steels represent around 30 % of the tool steels market [10][11]. HB 300 pre-hardened mold steels, like those used by the automotive industry, represent around 70 % of the mold steels market [12][13]. Assuming that mold making for the automotive industry is sufficiently representative, this mold making process and sector is considered to represent the various stages involved in mold making. (see Figure 1).

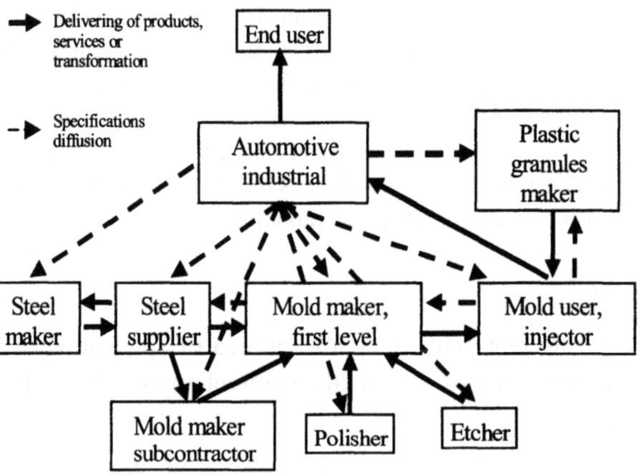

FIGURE 1. Stages involved in mold designing and making [10].

Amongst those involved in mold designing and manufacturing, the mold maker has a central position.

Operation sharing:

- Design (plastic piece definition): Automotive industrial;
- Studies, simulations (mold definition): Injector and mold maker first level;
- First rough machining (cube machining, first cavity, etc…): Steel supplier;
- Rough machining (milling, EDM, drilling, deep drilling): Mold maker subcontractor;
- Half fine and fine machining (Milling, EDM): Mold maker;
- Adjusting, polishing: Mold maker;
- Injection tests: Mold maker;
- Retouching: Mold maker;
- Surface finishing: Polisher;
- Qualifying injection: Mold maker, Injector and customer.

First rough machining is generally achieved by the steel supplier, while the mold maker subcontractor takes charge of the rough machining of the cavities, drilling, etc. Semi-fine and fine milling as well as adjustments are achieved by the first level mold maker.

Figure 2a gives a cost analysis of the mold making process [12][13] and Figure 2b shows the importance of plastic injection molds (73 %) for the mold industry [14].

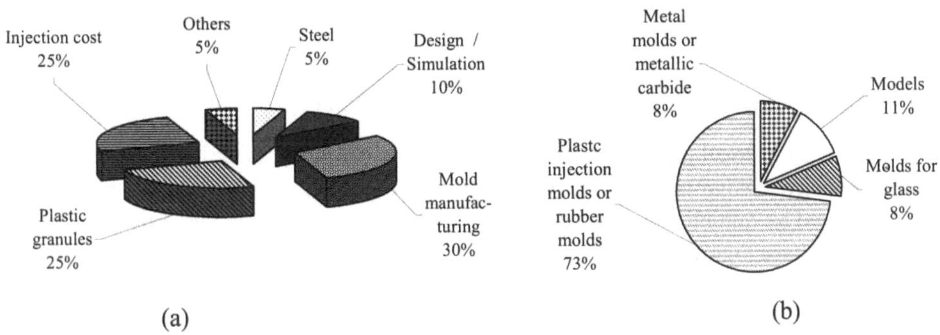

FIGURE 2. (a) Analysis of machining costs of a plastic injection mold, (b) Product type partitions.

3 MULTI-AXIS MILLING OF COMPLEX SURFACES AND TRAJECTORY APPROXIMATION

Major CAD-CAM software can be used to produce previously designed complex shapes; generally, using surfaces are known as parametric surfaces:

- Bézier surface:

If P_{ij} poles, a Bézier surface equals:

$$S(u,v)=\sum_{i=0}^{n}\sum_{j=0}^{m}P_{ij}B_i^n(u)B_j^m(v) \quad (1)$$

where $$B_i^k(u)=\begin{pmatrix} i \\ k \end{pmatrix}u^i(1-u)^{k-i} \text{ (Bernstein polynomials),} \quad (2)$$

- B-Spline surface:

$$S(u,v)=\sum_{i=0}^{p}\sum_{j=0}^{q}P_{ij}N_i^m(u)N_j^n(v) \quad (3)$$

where $N_i^m(u)$, B-Splines , m polynomials degree, $u_0\langle u_1\langle\ldots\langle u_{n+m}$.

$$\begin{cases} \sum_i N_i^m(u)=1 \\ N_i^m(u)\geq 0 \\ N_i^m(u)=0 \text{ if } u\notin [u_i,u_{i+m+1}] \end{cases} \quad (4)$$

$$N_i^m(u)=\frac{(u-u_i)}{(u_{i+m}-u_i)}N_i^{m-1}(u)+\frac{(u_{i+m+1}-u)}{(u_{i+m+1}-u_{i+1})}N_{i+1}^{m-1}(u) \quad (5)$$

$$N_i^0(u)=1 \text{ if } u\in [u_i,u_{i+1}]$$

- NURBS Surface (Non Uniform Rational B-Spline):

$$S(u,v) = \frac{\sum_{i=0}^{p}\sum_{i=0}^{q} w_{ij} P_{ij} N_i^m(u) N_j^n(v)}{\sum_{i=0}^{p}\sum_{j=0}^{q} w_{ij} N_i^m(u) N_j^n(v)} \tag{6}$$

where w_{ij} are the weights allocated respectively to the poles P_{ij} [15].

They are meshed in small plane faces and the tool path is generated using linear trajectories, with simple continuity, linking points computed as intersections of faces with principal planes. Accuracy is achieved by minimizing chord error [16] (distance between exact curve path and segment approximation) (see Figure 3).

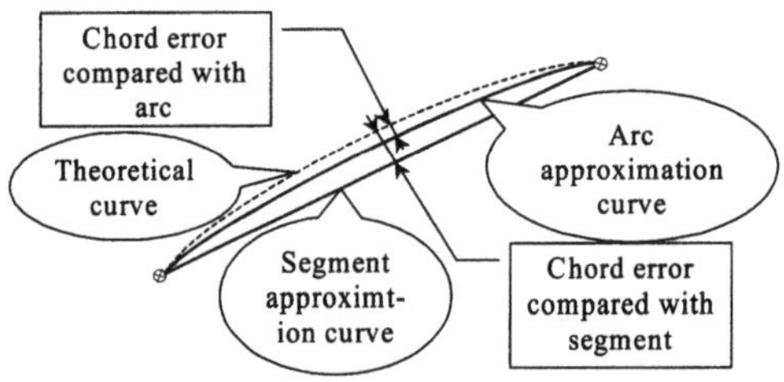

FIGURE 3. Chord error.

Usual machines tools know only linear and circular trajectories and a surprising phenomenon occurs frequently. Complex surfaces elaborated by the designer are approximated by a series of short segments often not machinable due to overhead and overshooting problems.

Segments are programmed by machine software which is then able to define complex curves machinable by NC milling machines. Original surfaces are first approximated, then constructed again to be machinable. Original information from designer is lost. Naturally cutting information (feed per tooth f_z, cutting speed V_c, radial engagement a_e, etc...) [17][18][19][20] is not taken into account during the optimal path computation. Trajectories with a mathematical approximation reduced to zero produce a part which requires a time-consuming polishing process taking some hours.

LINEAR TRAJECTORIES. Linear trajectory algorithms are well known in literature. The tool is driven from its center, extremities or by a point on its surface. In this case, the theoretical path is programmed with a chord tolerance and the N.C. computer does the correction on the path normal, if it exists. Mold or die end milling by a ball or bull nose tool is often done by millions of linear trajectories, always hard to check before real machining. The finished surface is generally "faceted" and 3D movement is possible.

LINEAR AND CIRCULAR TRAJECTORIES. Rarely are algorithms combining linear and circular trajectories published. Paths are approximated by a series of segments and circles with respect to continuity of tangency at their ends. Elementary trajectories are naturally computed to set in the tolerance areas. The NC milling program becomes a succession of ISO statement G1 (linear interpolation) and G2 / G3 (circular interpolation) moves.

4 EXPERIMENTAL STUDY

Tested molds materials are a new redesigned mold steel grade SP300 (Table 1) (Usinor Industeel, France), and an aluminum alloy, AlCu4Mg, with high machinability. In this study, the work surface (ellipsoid 40-20 mm) (see Figure 4a) is machined by a Widia ball nose carbide tool coated with hard coatings such as TiCN, with 2 teeth, and diameter (D) 16 mm. For each experiment, the numerical approximation of the surfaces is estimated with a theoretical chord error of 0.02 mm.

TABLE 1. Chemical analysis of tested mold steel SP300

% ponderal						HB
C	Cr	Mn	Mo	S	Ca, V, B	
0.25	1.3	1.3	0.4	0.02	-	300

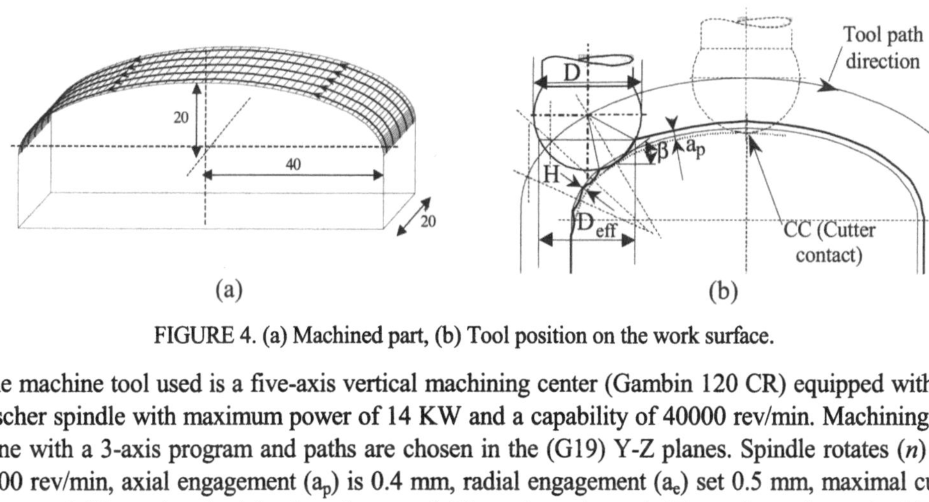

(a) (b)

FIGURE 4. (a) Machined part, (b) Tool position on the work surface.

The machine tool used is a five-axis vertical machining center (Gambin 120 CR) equipped with a Fischer spindle with maximum power of 14 KW and a capability of 40000 rev/min. Machining is done with a 3-axis program and paths are chosen in the (G19) Y-Z planes. Spindle rotates (n) at 3000 rev/min, axial engagement (a_p) is 0.4 mm, radial engagement (a_e) set 0.5 mm, maximal cutting speed ($V_{ceff\,max}$) and minimal cutting speed ($V_{ceff\,min}$), as shown in Figure 5, are determined by:

$$V_{ceff\ \max} = \frac{\pi.n.D}{1000} \tag{7}$$

$$V_{ceff\ \min} = \frac{2.\pi.n.\sqrt{D.a_p - a_p^{\ 2}}}{1000} \tag{8}$$

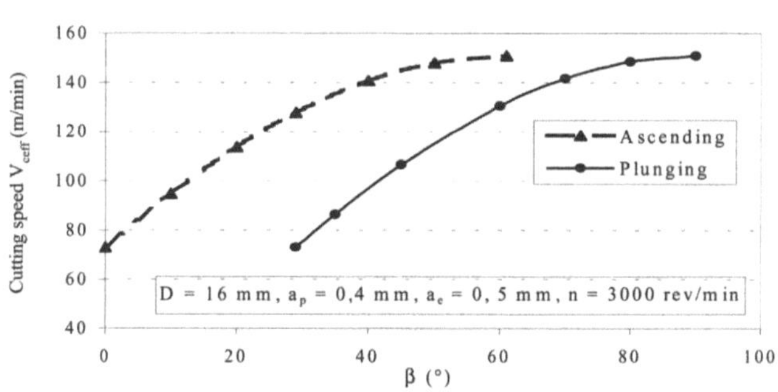

FIGURE 5. Cutting speed V_{ceff} variation (ascending and plunging).

The milling tool path is the trajectory of the cutter center, which is also known as the cutter location CL path, and is given by the NC part program. The cutter contact CC path is the tangential trajectory between the ball-end cutter and the part surface. It is shifted from the tool path by a distance equal to the cutter radius (see Figure 4b). When the cutter moves in parallel trajectories, as shown in Figure 4b, scallops are created on the finished surface. The distance between the parallel trajectories is the CC path interval, which depends on the local curvature of the surface, the size of the cutter, and the allowable scallop height remaining on the surface after the machining operation.

5 RESULTS

5.1 MICROGEOMETRICAL TEXTURE OF THE SURFACE

This study aims to characterize the machined surfaces with the use of the SOMICRONIC-SURFASCAN three-dimentional profilometer. The 3D analysis enlarges the possibilites of knowledge and representation of the surface texture [21].

Result shows (see Figure 6) the associated surface by the least squares fit criteria. The effect of linear trajectories pictured with small planes of length equal to the length of programmed linear segments (see Figure 6a) is clearly observed. On the other surface, machined with respect of continuity of tangency, facetisation is not visible (see Figure 6b).

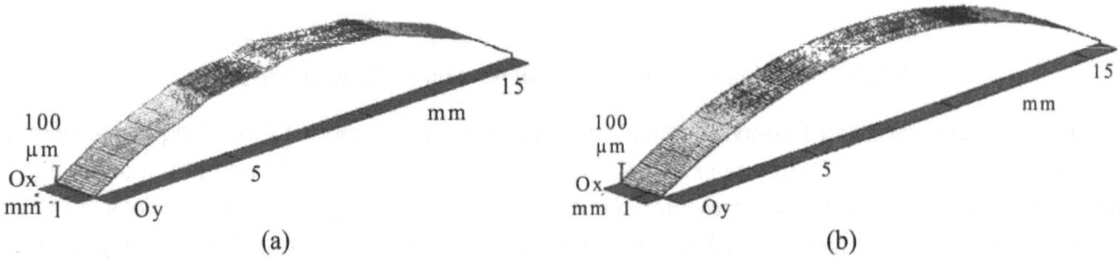

FIGURE 6. Surface texture of the workpiece, (a) Linear interpolation, (b) Continuity of tangency.

The tridimensional parameters of the texture surface: arithmetical mean surface Sa, root mean square deviation Sq and skewness of height distribution Ssq (see Figure 7a) of the workpiece compare the continuity of tangency method with programmed linear interpolation.

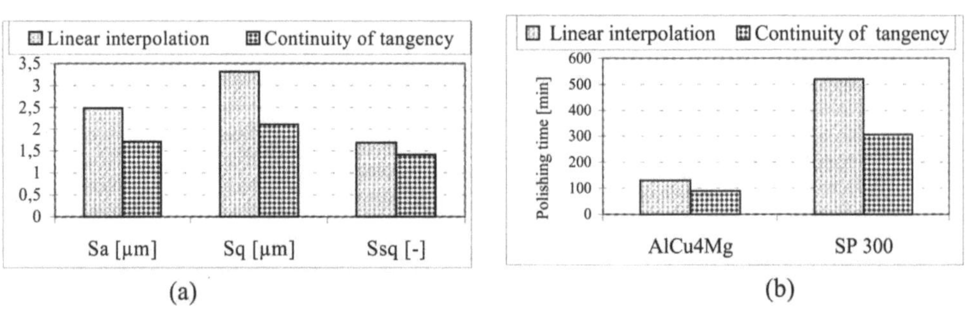

FIGURE 7. Workpiece one-way, linear interpolation and continuity of tangency, (a) 3D surface texture parameters. (b) Polishing time.

5.2 POLISHING TIME

At the end of the milling process, surface texture of the workpiece was measured in the various interpolation cases.

Then, a manual polishing was done up to a finite surface chosen as polished mirror, by means of abrasive papers (of grading from 320 to 1200) and diamond paste.

As previous results on convex ellipsoid surfaces show that a strong reduction in manual polishing time is obtained when the part is machined from a path respecting the continuity of tangency (see Figure 7b, 8). Using the time for polishing the same part machined with the classic linear interpolation program as a reference, the reduction in time is 31 % with the aluminum alloy part to 41 % with the steel SP300 part.

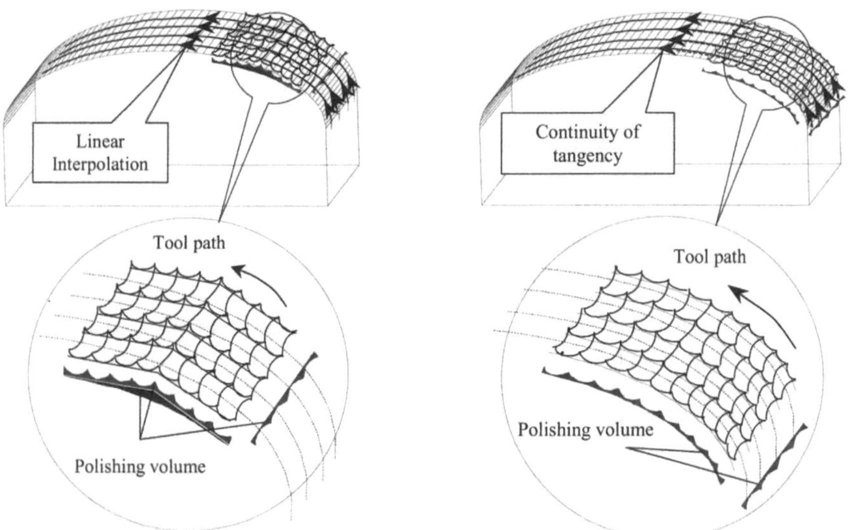

FIGURE 8. Workpiece one-way, linear interpolation and continuity of tangency.

6 CONCLUSION

It is obvious that polishing will still need to be done on molds of this kind in the future, particularly when a mirror finish is needed. The benefits in choosing paths programmed with trajectories with respect to the continuity of tangency, even with the constraint of circular interpolation in principal planes, have been clearly shown.

In addition to a lower occupation of the machine tool memory and a shorter, more readable program, the accuracy of the part and the 3D surface roughness are improved compared with those obtained with paths based on linear approximation (even in 3D).

The improved surface quality together with the higher precision obtained by using a program with continuity of tangency lead to a decrease in polishing time from 31 % in the case of AlCu4Mg, to 41 % in the case of SP 300 mold steel grade for plastic injection.

REFERENCES

1. Fallböhmer P., Rodriguez C.A., Özel T., Altan T., (2000), High-speed machining of cast iron and alloy steels for die and mold manufacturing. J. of Materials Processing Technology 98, 104-115.
2. Elbestawi M.A., Chen L., Becze C.E., El-Wardany T.I.., (1997), High-speed milling of dies and molds in their hardened state. Annals of the CIRP, Vol 46/1, 57-62.
3. Boujelbene M., Fabre A., Brenier B., Moisan A., (2001), Improvement of the surface finish of multi-axis milled surfaces. Advances in Manufacturing Technology-XV. (Ed.), Professional Engineering Publishing UK, ISBN 1 86058 325 3, 119-124.
4. Bézier P., (1984), Courbes et surfaces, Mathématiques et CAO. Vol 4, (Ed.) Hermès, Paris.
5. Seon Hwang J., Chien Chang T., (1998), Three-axis Machining of compound surfaces using flat and filleted end mills. Computer-Aided Design, Vol 30/8, 641-647.
6. Chih-Ching L., (1998), A new approach to CNC path generation. Computer-Aided Design, Vol 30/8, 649-655.
7. Farouki R.T., Manjunathaiah J., Nicholas D., Yuan G.F., Jee S., (1998), Variable-feedrate CNC interpolators for constant material removal rates along Pythagorean-hodograph curves. Computer-Aided Design, Vol 30/8, 631-640.
8. Lin Y.J., Lee T.S., (1999), An adaptative tool path generation algorithm for precision surface machining. Computer-Aided Design, 237-247.
9. Boujelbene M., Brenier B., Fabre A., Moisan A., Hadj Sassi B., (2000), Interactions FAO et fraisage multi-axes de surfaces complexes via des opérations de surfaçage. Actes 3èmes Journées de Mécanique et d'Ingénierie JMI 2000, 28-30 avril, Sfax, Tunisie.
10. Le Calvez C., Rech J., Dessoly M., Moisan A., (2001), Steels for Plastic Injection Molds: Rough milling characterisation and optimisation. Proc. of the CIRP-2nd International Conference and Exhibition on Design and Production of Dies and Molds, June 21-23, Kusadasi, Turkey.
11. Usinor Industeel, (1999), Internal reports, France.
12. Le Calvez C., Rech J., Moisan A., Dessoly M., (2001), Rough Milling Characterization and Optimization of Steels for Plastic Injection Molds. CIRP Workshop on Machining of Dies and Molds, 24 January, Paris, France.
13. Altan T., Lilly B., (2001), Manufacturing of Dies and Molds. Annals of the CIRP, Vol 50/2, 405-423.
14. Duthoit M., Tremare R., (1999), L'industrie française des moules et modèles – Savoir faire et haute technologie dans les PMI. Le 4 pages des statistiques industrielles. N° 124, 1-4.
15. Piegl L., (1991), On Nurbs: A Survey. IEEE Computer Graphics & Applications, Vol 20, 55-71.
16. Boujelbene M., Brenier B., Fabre A., Moisan A., (2001), Contribution to complex surface finishing by milling ; interactions tool-part. The 3rd International conference on metal cutting and high speed machining, June 27-29, Metz, France, Vol 1, 163-166.
17. Moisan A., (1998), L'usinage, Conception en Mécanique Industrielle. Ch. 12. 6, (Ed.), Dunod.
18. Altintas Y., (2001), Manufacturing Automation, Metal cutting Mechanics, Machine tool vibrations and CNC Design. (Ed.), Combridge University Press.
19. Schulz H., Hock St., (1995), High speed Milling of dies and Moulds – Cutting conditions and Technology. Annals of the CIRP, Vol 44/1, 35-38.
20. Boujelbene M., Brenier B., Fabre A., Moisan A., (2000), Etat de surface des pièces de formes complexes fraisées en bout sur centre d'usinage. TCMM (Ed.), Editura Technica, Bucharesti, Romania, ISBN 973-31-1492-8, N° 41, 215-220.
21. De Chiffre L., Lonardo P., Trumpold H., Lucca D.A., Goch G., Brown C.A., Raja J., Hansen H.N., (2000), Quantitative Characterization of Surface texture. Annals of the CIRP, Vol 49/2, 635-652.

MACHINING ABRASION-RESISTANT HARD-FACINGS WITH PCBN TOOLS

E. J. Brookes, R. D. James, C. Shutt, C. J. Taylor

Department of Engineering, University of Hull, United Kingdom

KEYWORDS: Milling, Hard-facing, Quick-stop

ABSTRACT: Iron-based hard-facings are a popular choice for abrasion resistant materials in industries as diverse as mining and food processing. Alloys forming chromium carbides dispersed in a relatively soft matrix are particularly effective because these carbides are usually present as relatively large micro-constituents, which present large surface areas to the abrasive material. Hard-facing materials are usually deposited onto a substrate by welding to provide wear-resistant layers several millimetres thick. Subsequent machining is usually necessary to achieve required standards of dimensional accuracy and surface finish. However, effective abrasion-resistant materials are, by their nature, very difficult to machine other than by grinding. Cutting tools based on PCBN present the possibility of turning and milling as more cost-effective alternatives. The work reported here included industrial and laboratory-based cutting trials together with a detailed examination of both the tool:work-piece and tool:chip interface acquired using a novel milling 'quick-stop' device, developed at the University of Hull.

1 INTRODUCTION

Abrasion-resistant materials are important in industries such as mining and food processing, where excessive wear of critical components will adversely affect equipment performance and operating costs [1]. Iron-based hard-facing materials are commonly used because of their excellent wear-resistance, which is attributable to hard carbides dispersed in a softer matrix [2]. Alloys forming chromium carbides are particularly effective because these carbides are usually present as relatively large micro-constituents, which offer large surface areas to the passage of abrasive material. Hard-facing materials are usually applied to a substrate by arc welding to provide a wear-resistant layer several millimetres thick. The welded layer almost invariably requires subsequent machining and, until quite recently, grinding was the only effective process. However, ultra-hard cutting tools, based on polycrystalline cubic boron nitride (PCBN), now offer the possibility of turning and milling as economic alternatives to abrasive machining.

A particular advantage of PCBN tools is that, unlike diamond, they are suitable for machining hard ferrous materials. Their superior performance over more conventional tool materials, particularly in terms of higher material removal rate, has been shown in many areas [3]. PCBN tools can successfully machine hardened alloy steels, cobalt-based alloys, nickel-based alloys and tool steels. Furthermore, cast irons, including nickel/chromium cast iron, can be machined at very high speeds due largely to the ability of PCBN to retain its of strength at high temperature. PCBN tools can also be used to machine very tough materials including Ni-hard [4], tungsten carbide and other engineering ceramics [5].

PCBN tooling products were intended for use with difficult-to-machine ferrous work-pieces but not specifically to cope with carbide-containing hard-facings. Lack of understanding of the machining process with these materials, especially the work-piece material deformation process and tool:work-piece interaction, has restricted the application of ultra-hard tools in this

Published in: E.Kuljanic (Ed.) *Advanced Manufacturing Systems and Technology*, CISM Courses and Lectures No. 437, Springer Wien New York, 2002.

field. In earlier work, the authors [6] reported an application of ultra-hard tools in turning of wear-resistant welded hard-facings. The hard-facing materials used were of high hardness (HRC 55-60) and high carbide content, which made machining very difficult with tungsten carbide tools. Furthermore, the ridged surface layer, resulting from the weld-deposition process, together with micro-structural in-homogeneity and varying hardness added to the difficulties by creating unstable loading conditions on the tool inserts [7].
The work reported here is concerned with assessing the suitability of high-CBN content cutting tool materials for both turning and milling iron-based hard-faced work-pieces. The work included cutting trials and generation of samples, using 'quick-stop' devices, for detailed examination of the tool:work-piece and tool:chip interfaces. Quick-stop devices for turning processes are well established and have been in use for many years. However, preservation of the work-piece:chip interface in milling operations is far more difficult. In the work reported here, the authors describe the development of a novel quick-stop device and its use in the investigation of tool:work-piece interactions when milling an iron-based hard-facing material with PCBN tools.

2 EXPERIMENTAL DETAILS

2.1 TOOL MATERIAL

For the turning tests, the PCBN tools used were solid 7mm diameter circular inserts. In the milling tests, solid, 'square' inserts were used with an edge length of 12.5mm and a corner radius of 3.5mm. The PCBN turning and milling inserts were manufactured from material with a high volume fraction of CBN and had a very dense structure. The measured hardness of the CBN was HK2 36.5GPa.

2.2 WORK-PIECE MATERIAL

The microstructure of hard-facing material used is complex. The material used is representative of the Fe-Cr-C alloys suitable for weld deposition on a mild steel substrate. In the welding process, solidification begins with the formation of primary $(Cr, Fe)_7C_3$ carbides, the residual liquid eventually decomposing by a ternary eutectic reaction into a mixture of austenite and more $(Cr, Fe)_7C_3$ [8]. As can be seen from the micro-section, a main feature for the material is the high volume fraction of large-sized primary carbides (8-15μm in diameter) exhibiting columnar growth with hexagonal cross section. Hardness measurements revealed that the bulk hardness of the hard-facing was approximately HRC 57 but the Knoop hardness of the large carbides is about HK50g 14GPa.

2.3 MACHINE TOOL AND WORK-PIECE DETAILS

LABORATORY TESTS

Turning tests were carried out in a laboratory with work-pieces prepared from a chromium-carbide-based hard-facing layer, nominally 6mm thick, deposited on a mild steel bar (ϕ100, L300mm). The hardness of the welded layer was HRC 55-58. The samples were pre-turned, to remove the rough welded skin, before the controlled cutting tests commenced. The tests were conducted, without a coolant, using a Churchill 'Computurn' 290 CNC lathe. The cutting conditions were S 70m/min, F 0.25mm/rev, DOC 0.65mm. Stages in the wear of the flank and rake faces of the cutting tool were monitored and cutting stopped when an insert ceased to cut effectively. Milling tests in the laboratory were carried out on a Kryle V535 CNC vertical milling machine on flat samples of hard-faced mild steel with a two-flute 32mm diameter end mill. The cutting inserts in both materials were 12.5mm square with 3.5mm corner radii and a thickness of 5mm. Cutting conditions were S 3000rev/min, F 600mm/min and the DOC 0.8mm.

FIELD TESTS

Field tests were conducted in a factory producing hard-faced components on a regular basis. Turning tests were carried out using a Warner & Swasey Universal 315 CNC lathe at S70m/min, F0.25mm/rev and DOC 2mm; this depth was chosen to ensure that cutting occurred below the rough surface layer. The work-pieces used in the milling field tests were helical components with diameters between 200-400 mm. The components had, typically, a 6mm thick weld-deposited layer on top of a mild steel substrate. As in the case of the turned parts, the components had rough, undulating surfaces as a result of the welding process. The milling was carried out on an Ingersoll Bohle BW120 horizontal 4 axis CNC machining centre. The tool holder was a 63mm diameter end-milling cutter with provision for 4 'square' inserts. The cutting conditions were S909 rev/min, F0.06-0.12 mm/tooth and DOC 1.5mm.

2.4 MILLING QUICK-STOP TESTS

An attempt to produce milling quick-stop samples by accelerating the workpiece material away from the tool/workpiece interface using an explosive device has been reported, however the disadvantages include the need to develop a very high force, a very small sample size and the very critical requirement to time the explosion to occur at a precise position [9]. A reduction in the mass of either the tool or the portion of the workpiece, which is detached from the bulk, would translate into a reduction in the forces required to achieve a separation rate comparable to the explosive methods used in turning - a rationale employed by Byrne *et al* [10] for a turning quick-stop. In the work reported here, this approach has been adapted and a low mass workpiece has been developed in which the forces generated during machining fractures the workpiece with the chip intact. Pre-machined into a flat bar, with a reduced section at one end, the workpiece is clamped to the machine bed along the 'back', whilst a spring-loaded device to the right of the spindle can pivot away from the milling cutter (Figure 1(a)). Cutting proceeds from left to right, and material is continuously removed from the bar until a narrow bridge-shaped section is formed. The cutting forces generated during machining are supported until, ultimately, the reduced bridge section fractures. At that point, the unclamped portion is rapidly accelerated away from the tool:workpiece interface, shown in Figure 1(b).

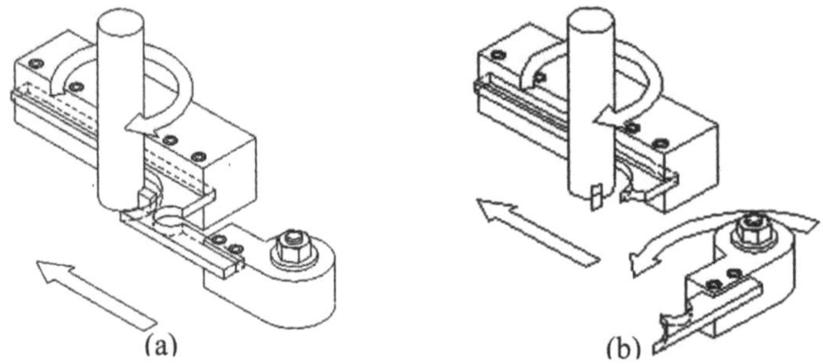

FIGURE 1. Geometry of the milling quick-stop, (a) prior to fracture, (b) after fracture

3 OBSERVATIONS

3.1 TURNING

In the turning trials, high cutting loads and temperatures were clearly being generated. The tip of the PCBN cutting tool glowed and red-hot chips were observed even when new inserts were used. With increased cutting time and consequential tool wear, the temperature increased still further. On-site measurement using an infrared optical pyrometer showed that local temperatures in the region of 750-800°C were developed during the cutting process. The cutting tool life achieved was approximately ten times longer than that of the standard tungsten carbide tools.

3.2 MILLING

The milling field tests were started at S909 rev/min, F0.06 mm/tooth and DOC 1.5mm. In the process of cutting, it was observed that the tool inserts glowed red-hot and the chips formed were highly segmented. At these conditions, the PCBN tool inserts gave better performance in terms of the cutting tool life, as judged by the time an insert lasts before losing its ability to cut effectively. The average cutting life of the PCBN tools was found about three to four times that of the standard whiskered ceramic tools [11].

3.3 TOOL WEAR MECHANISMS

Typical tool wear that developed during milling for each tool:work-piece combination is shown in Figure 2. Details of the tool wear that occurred during turning have been described in detail by the authors elsewhere [12]. In general, as edge chipping occurs a loss of performance results and, with significant loss of tool geometry, the feed force increases during cutting.

The primary wear mode of the PCBN tool insert, shown in Figure 2(a), was by edge chipping on a relatively small scale. It is suggested that this small-scale chipping gradually removed the tool material from the cutting edge probably as a result of the impact forces developed during milling. Chipping was predominant in the early stages of cutting and was concentrated at the lower edge of the chamfer. A more detailed examination of worn tools showed that, within the zone of contact on the flank face, cracks have propagated around the primary CBN particles and wear of the near edge region occurred due to the removal of individual grains or aggregates. The consequence is that individual particles have separated from the bulk of the tool material. The flaking process on the rake face, however, was the result of transgranular fracture i.e. cleavage cracks which have propagated across the CBN particles. A high magnification view of the worn, fractured surface of PCBN, Figure 2(b), features both fractured and loose particles. This suggests that the PCBN ultimately failed by mixed modes: both transgranular individual particle failure, andintergranular failure through the particle boundaries. This observation is in general agreement with earlier work outlining wear mechanisms in PCBN [13]. A smeared layer of work-piece material was present on the flank face of the PCBN, but no such layer was observed on the whiskered ceramic tools. However, this layer does not appear to directly contribute to the wear process in this tool:workpiece pairing.

(a)

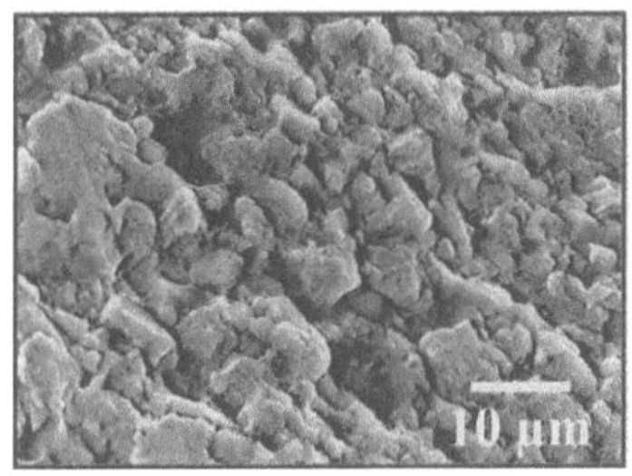

(b)

FIGURE 2. (a) PCBN tool wear, (b) transgranular and intergranular fracture of PCBN particles

3.4 DEFORMATION OF THE WORKPIECE

An earlier paper by the authors [13] showed that the strain imposed during chip formation in turning operations resulted in the production of discontinuous chips. The hard-facing chips were saw-toothed, semicircular and severely serrated on the side near the minor cutting edge. The tool face side was regular and no apparent flow of the workpiece was observed. Saw-tooth chips of varying segment size were formed, influenced by the microstructural heterogeneity of the work-piece. The result of the 'quick stop' test for PCBN tooling is reproduced in Figure 3, and shows a clean detachment of the cutting tool with the retention of the chip to the workpiece. Clearly both the morphology and the quantity of the carbides inhibit the chip formation process and the deformation within the chip segment. Large, columnar primary carbides can be seen, roughly perpendicular to the cutting edge and hexagonal grains, with defects in the form of inclusions at the centre, are observed roughly parallel to the cutting edge. Cleavage fracture of the large carbides ahead of the cutting edge, parallel to the cutting direction, occurred and in some cases the carbide and matrix separated along the boundary of the carbide and the chromium-depleted zone. After fracture, fragments of the carbide have moved into the chip in the subsequent chip formation process and no other movement, e.g. rotation, of these large fractured segments of the carbide was observed. Cracking and bending of columnar carbides was evident near the surface region and it is clear that plastic deformation of the carbides has occurred during the cutting process and where the columnar primary carbide particles were close packed, the carbides were severely bent and cracked.

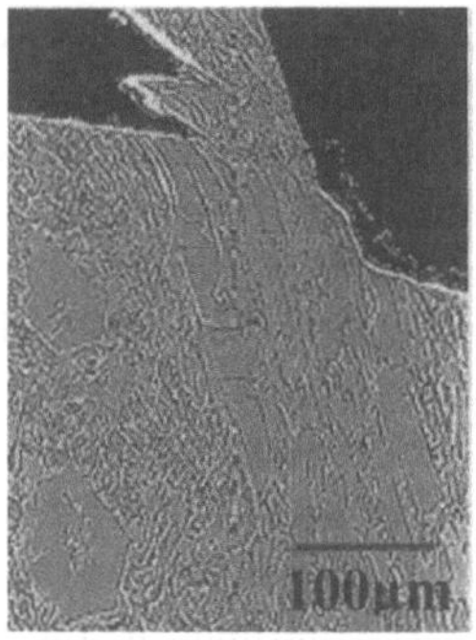

FIGURE 3. Turning quick-stop of welded hard-facing with PCBN tooling

An example of a milling quick-stop produced using novel technique is shown in Figure 4. In this case tungsten carbide has been used to cut titanium alloy Ti-6Al-4V, to demonstrate the potential of this method to produce a well-formed interface and clean detachment of the tool. Figure 4(a) shows the long, continuous and severely deformed chip, typical of materials which undergo extensive plastic deformation during machining, still attached to the workpiece. The characteristic narrow shear zone produced when cutting titanium alloys can be seen in the metallographic section shown in Figure 4(b)

A milling quick-stop produced during milling welded hardfacing with PCBN tooling is shown in Figure 5. As in the case of turning, the hard-facing chips were saw-toothed, semicircular and severely serrated and the tool face side was smooth. The chip was extremely friable and became segmented and detached while still on the machine (Figure 5(a). Clearly the compressed material ahead of the tool has failed in a brittle manner (Figure 5(b)) with cleavage fracture of the large carbides ahead of the cutting edge and significant failure of the carbide matrix interface. Fragments of the carbide have moved into the chip, as was seen in the turning quick-stop.

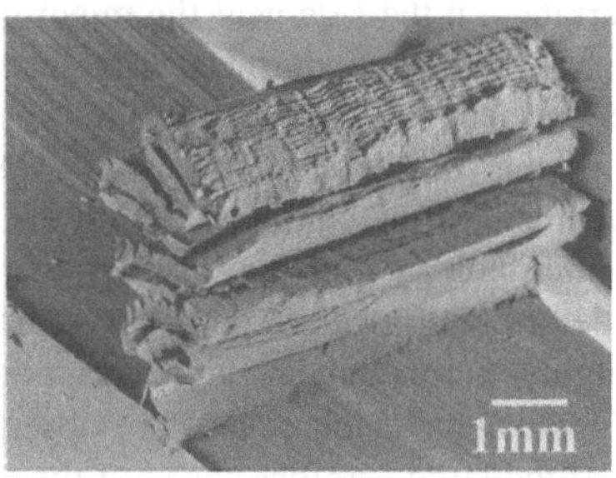

(a)

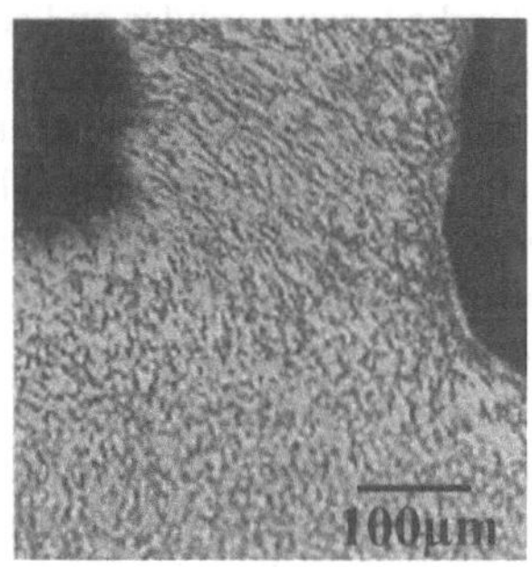

(b)

FIGURE 4. Quick-stop chip produced when milling Ti-6Al-4V with carbide;
(a) SEM micrograph, (b) etched section

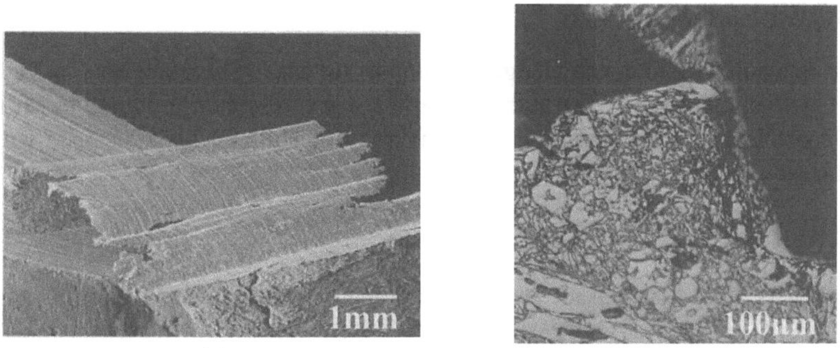

(a) (b)

FIGURE 5. Quick-stop chip produced when milling hardfacing with PCBN; (a) SEM micrograph, (b) etched section

4.0 DISCUSSION

The poor machinability of the carbide-containing hard-facing materials is related to their microstructure and, specifically, the behaviour of the carbides in the cutting process. The presence of significant numbers of large carbides in a relatively stiff matrix will result in high, but intermittent, forces on the edge of the cutting tool, which leads to damage. Thus, the main mechanical requirements of a successful tool material for this application will be high hardness, to withstand the abrasive carbide particles, combined with toughness to withstand the variations in cutting forces.

Whilst the performance of PCBN is impressive in both turning and milling applications, there remains much to be done in optimising its use for particular applications. For example, in the milling field tests, although the material removal rate could be increased to twice that of the standard whiskered ceramic material, under these conditions so much heat was transferred to the tool holder that thermal expansion resulted in the PCBN inserts becoming loose. A further important consideration for many potential users is the cost of PCBN tools, which are typically five times more expensive than whiskered ceramic equivalents.

5.0 CONCLUSIONS

- PCBN tools can be used successfully for turning and milling carbide-containing welded hard-facings,
- In both turning and milling operations, damage to the PCBN tools was mainly by progressive chipping and small-scale flaking.
- The excellent performance of PCBN is mainly due to its higher hardness, fracture toughness and thermal conductivity.
- Despite the established performance advantages, the current high cost of PCBN tooling remains a barrier to its wider usage
- A novel quick-stop device has been developed which facilitates investigation of material removal processes in milling operations

ACKNOWLEDGEMENTS

The authors would like to acknowledge the support for Mr C Shutt and Mr C J Taylor through EPSRC CASE studentships. The authors would also express their gratitude to Mr R W Swain of the Department of Engineering at the University of Hull for his assistance with the work reported here.

REFERENCES

1. Gregory, E., (1980), Materials for Weld-surfacing and Hard-facing, The Welding Institute, 11-22.
2. Menon, R., (1995), New Developments in Hard-facing Alloys, *The Welding Journal*, 75, No. 2, 43-49.
3. Roebuck B., (1995), Technologies and Markets for Cutting Tool Materials, National Physical Laboratory
4. Heath P., (1986), 14th Structure, Property and Application of Polycrystalline Cubic Boron Nitride, North American Manufacturing Research Conference Proceedings, 66-80.
5. Kitagawa T. and Maekawa K., (1990), Plasma Hot Machining for New Engineering Materials, *Wear*, 139, 251-267.
6. Brookes E., James R., Ren X., (1999), Hard turning with PCBN Tooling, AMST '99 5th International Conference on Advanced Manufacturing Systems, June 1999, Udine, Italy, 107-115.
7. Ren X., James R., Brookes E., (1999), Wear Characteristics of CBN Tools in Machining Welded Hard-facing Materials, *Journal of Superhard Materials*, 1, 36-45.
8. Atamert S., Bhadeshia H. K. D. II, (1990), Mater. Sci. Eng. 130, 101-111.
9. Wager J. G. and Brown R. H., (1980), A New Quick-stop Device for Milling and Grinding, *Annals of the CIRP*, 15-18.
10. Byrne G., Heinzel J., Barry J., Eitzenberger H., (1997), Chip Root Tests Using PCD and MCD Cutting Tools, *Industrial Diamond Review*, 57, 96-101.
11. Brookes E., James R., Ren X., (1998), International Conference on Behaviour of Materials in Machining: Opportunities and Prospects for Improved Operations, Stratford on Avon (U), 189-198.
12. Brookes E., James R., Ren X., (2000), Turning and Milling Welded Hard-facings with PCBN Tooling, Intertech 2000 Conference, Vancouver, Canada.
13. Brookes C., Hooper R., (1981), Towards Improved Performance of Tool Materials', Proceedings of the International Conference Organized Jointly by The National Physical Laboratory and The Metals Society, Teddington, 32- 36.

MACHINING PROCESS IMPROVEMENT BY PRACTICAL TESTS IN SHOP FLOOR

N. L. Coppini[1], E. A. Baptista[1]

[1]Post-graduate Program in Production Engineering, Methodist University of Piracicaba, Brasil

KEYWORDS: Machining, Optimisation, Process.

ABSTRACT. The machining process optimisation always received special attention of process engineers. Nowadays however, the competitiveness increase asks for additional technologies that result in competitive capacity improvement. Good effects can be achieved when the process optimisation is accomplished in shop floor, based in technical/economical constraints, using parts that are in manufacturing. In this way, the productive line is not broke off and the machining process can be optimised. That is the conclusion of this work, which presents a practical methodology to optimize machining process in shop floor, based in the related constraints, joined with test results of turning and drilling process optimisation.

1 INTRODUCTION

The utilization of modern tools, modern fixtures systems, modern machines, raw material reasoning and acquiring cutting parameters by the traditional way, such as manufacturer's catalogue, could not be enough to guarantee the needed increase of competitiveness of industries that use the machining process in an expressive form. Even if the data from the reliable tests in statistics terms is used [1], on an industrial application these values may need to be optimised considering the scenarios characteristics.

The traditional way normally used to make a practical tool replacement by some "better or new generation tool" one is to compare both tools through cutting performance and cost/benefit tests. These tests normally are not based on optimised cutting condition and, because of this, they could give wrong results.

Can the performance of the tool in use still be improved? And the newer tool, can be optimised or not? This problem happen because is unusual the use of the Maximum Efficiency Interval – MEI, in process development or in process optimisation. The refereed interval is formed by the minimum cost speed, V_{cmc}, and the maximum cost speed, V_{cmxp}, see Figure 1. Know if the cutting speed in use, or the optimised cutting speed, belongs to MEI, can represent a great competitive distinguish, to avoid unnecessary costs that occur when the cutting speed below the V_{cmc} or above the V_{cmxp} is used.

When MEI is applied in the development or in process optimisation, the different scenarios have to be considered such as: mass production; flexible production with Group Technology - GT and flexible production less GT. The challenge is to determine which cutting speed from MEI will be used, as a reference, in the process optimisation of specific scenery. There are same cases that the V_{cmc} is indicated, however, there are another cases that V_{cmxp} shows better results.

Published in: E.Kuljanic (Ed.) *Advanced Manufacturing Systems and Technology*,
CISM Courses and Lectures No. 437, Springer Wien New York, 2002.

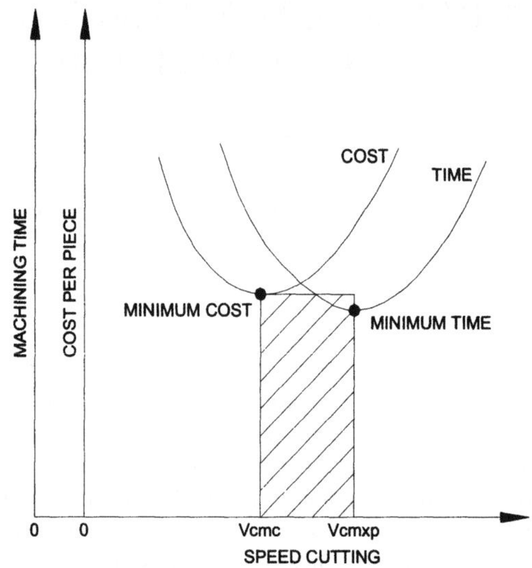

FIGURE 1. Maximum Efficient Interval, MEI.

To find the final results will be necessary to accomplish machining tests in shop floor, to determine the x coefficient and the k constant of Taylor Life's Equation – see Equation 1,. Joined with an industrial atmosphere analysis, which allow that the reference to optimisation will be specified. An Expert System, ES, to optimize the machining process based on the described until here, can be very helpful, allowing that the all steps of the optimisation could be accomplished by machine operator, in real time with the occurrence of the process.

$$T = \frac{K}{V_c^{\,x}} \tag{1}$$

Where T (min) is the tool life , k is the constant e x is the coefficient of the Taylor's equation. V_c is cutting speed (m/min).

Techniques and information that allow the machining process optimisation occurs in shop floor, without stop the production, respecting the MEI and the productive scenery are presented in this work.

The optimisation methodology is showed together with the scenario's classification and their information, followed by a short introduction about the TOES (Tool Optmisation Expert System) and their application.

2 OPTIMISATION METHODOLOGY

The machining process optimisation methodology its based on a model proposed by Pallerosi & Coppini [2], which consists in the x and k values determination in industrial atmosphere, following the steps described below:

a) adopt start operational conditions according to the traditional method in use, such as manufacture tool catalogues or previous experience;

b) execute the cutting process of the programmed workpieces batch with a constant cutting speed and carry on until the tool life end, according with a previously defined criterion. Take note of the tool life value, which can be expressed in time, cutting length or number of pieces;

c) modify the cutting speed value at the most 20% greater than the previously cutting speed used, if this is not possible in function of the process constraints, a value 20% smaller have to be adopted [3];

d) execute the cutting process with the second cutting speed until obtain a second value to the tool life, according with the same previous criterion, and keep the second value of the tool life;

e) determine through the Equations 2 and 3 the x and k values, and through the Equations 4 and 5, the minimum cost speed V_{cmc} and the maximum production speed V_{cmxp} values, which are the MEI;

$$x = \frac{\log\left(\frac{Z_{t1}}{Z_{t2}}\right)}{\log\left(\frac{V_{c2}}{V_{c1}}\right)} + 1 \qquad (2)$$

$$k = Z_{t1} \cdot t_c \cdot V_{c1}^{\;x} \qquad (3)$$

Where Z_{t1} and Z_{t2} are tool life in number of workpieces, for the first and second cutting speed (V_{c1} and V_{c2}), and t_c is the cutting time (min).

f) the calculated cutting speed values through the Equations 4 and 5 should belong to the speed interval used on the machining tests, with a tolerance of ±10%. In case this condition is not attended, a new speed interval should be adopted and the tests should be repeated until that this condition is satisfied. Sometime, mainly in function of cutting time, V_{cmxp} can be bigger then the maximum cutting speed available in the machine tool. So, this second cutting speed must be considered as the reference instead of V_{cmxp}.

For reasons that will be renowned in this work, the V_{cmc} and V_{cmxp} determination was not supported by the statistics proceedings. Off course, that the simple tool life determination, based in the tests above described, given us a reasonable reliability. Whatever, the cutting speed optimize based on the V_{cmc} and V_{cmxp} must be validated during their application, in function of the specific scenery as showed bellow.

$$V_{cmxp} = \sqrt[x]{\frac{k}{(x-1)\cdot t_{ft}}} \tag{4}$$

$$V_{cmc} = \left\{ \frac{k\cdot(Sh+Sm)}{60\cdot(x-1)\cdot\left[K_{ft} + \frac{(Sh+Sm)}{60}\cdot t_{ft}\right]} \right\}^{\frac{1}{x}} \tag{5}$$

Where the t_{ft} is the tool replacement time (min), K_{ft} is the tool cost per tool life (U$), and Sh is the man/hour cost (U$), Sm is the machine/hour cost.

3 SCENARIES AND INFORMATIONS ABOUT THE PRODUCTIVE ENVIRONMENT

3.1 SCENARIES CLASSIFICATION

The scenarios were classified by the batch size, the workpieces diversity, the material type and geometric forms. Thus, the following scenarios are identified as:

a) mass production – this scenery has a big batch size, which allow that the machining tests, to x and k determination, be accomplished a lot of times, as much as necessary. This is the ideal scenery to apply the optimisation methodology proposed in this work. In it, the MEI can be defined more accurate and the statistical validation can be done easily;

b) flexible production where workpieces have the same material and similar geometry – this scenery although be flexible, consent that all workpieces be treated like just one batch. In this case, the total workpieces quantity, permits that the tests be accomplished like the mass production;

c) flexible production where workpieces have the same material, but different geometries – this scenery will be treated in the same way of the last scenery. However, in function of the difference among geometries, the results can be influenced by the tool wear down variation. Although of these distortion possibilities on the result beings, this scenery will be treated like the last scenario. But, in this case, the tool life must be expressed in time units, due the difference among the geometries. The statistical validation can be followed by a bigger dispersion when the cutting speed optimised will be used. This dispersion occurs in function of the sequence of the different geometries in the batch, although of the pair tool/material be keep constant;

d) flexible production with material's and geometrie's workpieces different, however using the same tool to cut all the workpieces for the same operation – this scenery is more complicated when compared with others described before, because the sequence of the workpieces can vary during the tests running and it is possible to cut different workpieces by one tool life.

However, to manage the number of tools, the portion of tool life shared among different workpieces must be calculated. It can be used x and k values from the literature. In this case, the reliability is not so good. But also, they can by obtained in the machining process development, using the optimisation methodology proposed in this work, which is more expensive, however, more trustful. In function of the tool wear down variation, caused by the different material in process, the results can be show some variation. Thus, this approach given us conditions to specify the number of the tool that will be necessary to machine all batch of different workpieces, and the operator need to report the real number of the tool spend. The statistical validation, in this case, is damaged, so, the results must be used like an orientation to the operator. This means that the operator is free about the responsibility related with tool replacement moment.

3.2 INFORMATIONS ABOUT THE PRODUCTIVE ENVIRONMENT

Joined with the MEI some information about the productive environment, which has some influence over the cutting speed, must be analyzed before the optimised cutting speed choose, this information are:

a) setup time – when this time cause a delay in the productive line, the machine can become a neck, in this situation, the V_{cmxp} is indicated to be reference to optimisation;

b) tool replacement time – the V_{cmxp} is inversely proportional the tool replacement time, so, as much this value down, the V_{cmxp} value will increase, see the Equation 4;

c) tool cost – when the tool cost is expensive, the use of the V_{cmxp} is not indicated, due to the tool cost increase;

d) cost system – the Activity Based Cost – ABC cost system, allow that the V_{cmc} be calculate trustfully;

e) standard tool – when a standard tool is used, which can be found easily with the supplier, there is no problem, but, when a special tool is used, which the buy ask for previous programming, the use of the V_{cmxp} is not indicated.

f) tool stock – this information is essential when the special tool is used. Is valid to remember that the use of V_{cmxp} spend more tool;

g) kinds of workpieces and their quantities – used to classified the productive line, like mass or flexible production;

h) family code – used to arrange different workpieces into just one batch;

i) material – used to scenery classification.

This information can be used to built rules and consequently, Expert Systems, which allow that the machining process optimisation be done reliable.

4 TOOL OPTIMISATION EXPERT SYSTEM - TOES

To aid machining process optimisation the TOES was developed [4], which is an expert system to accomplish all calculus required to MEI determination, and off course, also choice the reference cutting speed, to optimisation, respecting the productive scenery.

Simulation, laboratory and industrial tests [5, 6], made the TOES performance validation. In all of then, their effectiveness was verified. The TOES also has a database with the optimisation results, which can be used to adopt cutting parameter in machining process similar [7].

5 APPLICATION

How described, some application tests were done, but in this work, only the last one will be present.

The tests happened in an automobile industry placed in Brazil. The workpiece choose is called "base tube", which is used in buses and trucks. The Figure 2, shows a draw of the referred workpiece, eminence the optimised area (dimension of Ø100 x 69, Ø110 x 53 e Ø115 x 22.5 mm).

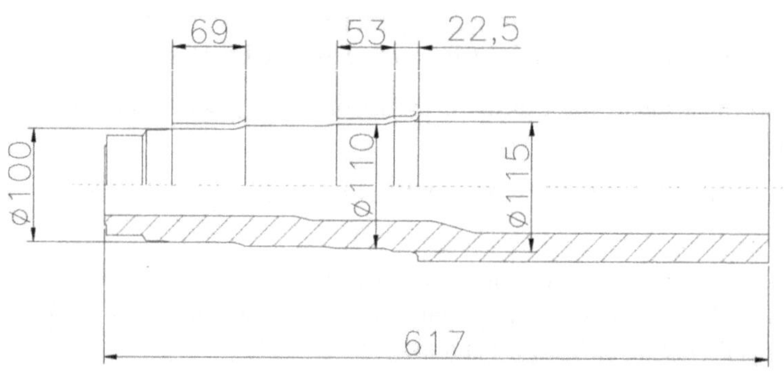

FIGURE 2. Workpiece.

The tests consists in the finishing operation, which have a fine dimensional tolerance, controlled by the operator by the micrometer and roughness both equipment are Mitutoyo. The cutting parameters in use to the operator were adopted like the first cutting conditions. The tool life criterion was the superficial roughness, which was the criterion in use by operator. Another informations are: workpiece material: ST 52 3N with 83HRB hardness; Machine: horizontal lathe INDEX: Model GU 800; Tool: insert TCMT 16 T3 08 UM, 4025 class – Sandvik.

The place, where the workpiece is manufactured, is a flexible cell and two another pieces are made in it. However, in this test only one piece was appraised. All informations about the scenery configuration were obtained with the area supervisor, and all of then were imputed into the TOES. The first cutting speed, V_{c1}, was 300 m/min and the second, V_{c2}, was 360 m/min. Two life values were obtained to each condition, see the table 1.

TABLE 1. Test's results.

	Tool life $V_{c1} = 300$ m/min (Z_t)	Tool Life $V_{c2} = 360$ m/min (Z_t)
Test 1	29	17
Test 2	27	17
Average	28	17

After impute the test results into the TOES, the following information were showed from it: 1) scenery: mass production; 2) MEI: V_{cmxp} = 1067 m/min, V_{cmc} = 387 m/min; 3) cutting speed to optimisation reference: V_{cmc}. Thus, it is recommended to use 390 m/min as the optimised cutting speed. This means that the industry was using a cutting speed less than V_{cmc} and for this reason, loosing in terms of competitiveness. The loss by each piece manufactured using 300m/min instead of 390 m/min can be determined and showed to be around US0,05/piece. Table 2 shows the advantages obtained by the methodology application.

TABLE 2. Productivity analyses.

Item	$V_{c1} = 300$ m/min	$V_{coptimised} = 360$ m/min
Cutting time / workpiece (min)	2,25	1,72
Manufacturing time / workpiece (min)	4,63	4,1
Tool life (workpieces)	28	14
Hourly production (workpieces / h)	13 (12,95)	15 (14,63)
Machining cost (US$)	1,25	1,20

The hourly production increases about 13%, and despite rise of the consumed tool, the machining cost per workpiece is about 4% smaller, although of the cutting time be about 30% smaller.

6 CONCLUSIONS

Based on the results obtained in this work it can conclude that:

a) the cutting parameters indicated by the tool manufacturer's catalogue can not suggest the best economical conditions, as it was showed in the test during on line shop floor application ;

b) the MEI, joined with the shop floor information's, allow to optimise the machining process with better reliability;

c) the shop floor information accomplished to the scenery classification and choice of the optimised cutting speed, showed to be very appropriate;

d) the proposed methodology to optimize cutting speed, can be reproduced by an Expert System. TOES make easily the use of the proposed optimisation method and allow to rich very good results.

e) Based on the authors experience, normally the methodology application could introduce productivity improvement and more profit during cutting process.

f) The methodology and the use of TOES are not very intrusive for real time applications in shop floor, as to say, during production by cutting process normally running.

ACKNOWLEDGMENTS

The authors would like to thank FAPESP and Sandvik Coromant do Brasil Ind. e Com.

REFERENCES

1. Kuljanic, E., Cukor, G., (2001), Machinability testing and optimization of machining process. In.: International Scientific Conference on Production Engineering, 7 – CIM 2001, 2001, Croatia, Croatian Association of Production Engineering, Zagreb.
2. Pallerosi, C. A., Coppini, N. L., (1991), Durability of cutting tools under true conditions. Proceedings of CANCAM 91 – Canadian Congress in Applied Mechanics, pp. 173-175. Canada, 1991.
3. Pallerosi, C. A., Coppini, N. L., (1975), Cutting tool durability in the metal machining. (in Portuguese – "Durabilidade de ferramentas de corte na usinagem dos metais"). Metalurgia, São Paulo, n.215, p. 645-649, out. 1975.
4. Baptista, E. A.,, Expert system development to optimise the machining process (in Portuguese – "Desenvolvimento de um sistema especialista para a otimização do processo de usinagem".) Santa Bárbara d'Oeste: FEMP, UNIMEP, 2000. Dissertação (Mestrado em Engenharia de Produção) – Faculdade de Engenharia Mecânica e de Produção, Universidade Metodista de Piracicaba, 2000, 194p.
5. Coppini, N. L., Coelho, R. T., Baptista, E. A., (1999), Tool Optimisation Expert System. **In:** Integrity Reliability and Failure – an International Conference, 1999, Portugal, University of Porto.
6. Baptista, E. A., Coppini, N. L., (2001), Machining process optimisation aided by an expert system (in Portuguese – "Otimização do processo de usinagem com auxílio de sistema especialista.") In.: Congresso Brasileiro de Engenharia de Fabricação 1 – COBEF, 2001, Curitiba – Brasil. Associação Brasileira de Ciências Mecânicas – ABCM. 2001.
7. Ribeiro, M. V., (1999), Cutting conditions optimisation aided by computer during the process development (in Portuguese – "Otimização das condições de corte assistida por computador durante o desenvolvimento do processo"). Campinas: FEM, UNICAMP, 1999. Tese (Doutorado em Engenharia Mecânica) – Faculdade de Engenharia Mecânica, Universidade Estadual de Campinas, 1999, 143p.

CRANKSHAFT MANUFACTURING ON MACHINING CENTRES – TURN-MILLING AND DEEP HOLE DRILLING

K. Weinert, H. Löbbe

Department of Machining Technology, University of Dortmund, Germany

KEYWORDS: Cutting, Process chain, Crankshaft

ABSTRACT. Large crankshafts are used in the field of stationary power generation and ship motor construction. Due to the strong global market pressure and the very high technical requirements on the component, economical and reliable production technologies have to be used. This treatise presents optimisation possibilities for machining large forged crankshafts on machining centres with respect to the complete process chain in the field of cutting technology. Investigations in special cutting technologies such as turn-milling of grooves to reduce tension and of the crankpins as well as deep hole drilling with single-edge gundrills for the purpose of machining oil holes are presented. A reduction of the production time, an increase of the process security at the required quality and a higher flexibility regarding the complete process chain can be achieved simultaneously.

1 INTRODUCTION

Large forged crankshafts are used in the field of ship engine manufacturing, stationary power plants and off-road vehicles. For the complete machining of forged crankshafts different machining operations such as turning, turn-milling, milling, drilling, deep hole drilling, tapping, grinding and deburring are necessary. Machining large forged crankshafts leads to specific problems due to the component and the employed processes. Problems are caused by the extremly large size of the crankshafts and the balance error during the single process steps. These important factors necessarily demand a minimal amount of chucking and a low rotational speed of the workpiece. The amount of chucking time, cost intensive manual work, e.g. transport and set-up time, and the damage risk can be reduced. Therefor the complete process chain in the field of cutting technology has to be kept in mind. Key aspects are the substitution of process steps by using combined manufacturing methods such as turn-milling, investigations in drilling, machining of form elements, minimization of burr formation and the development of cooling lubricant with a low class of water endangerment. The process chain can be reduced as shown in Figure 1.

2 TECHNOLOGIES

2.1 TURN-MILLING

To ensure the exploitation of the potential of modern cutting materials when machining large crankshafts the use of rotating tools is necessary. This leads to the use of the turn-milling technology instead of turning. Orthogonal turn-milling is mostly used for external machining of shafts.

Published in: E.Kuljanic (Ed.) *Advanced Manufacturing Systems and Technology*,
CISM Courses and Lectures No. 437, Springer Wien New York, 2002.

turn-milling parallel to the axis is used for machining parts clamped in a chuck [1]. Scientific research almost exclusively focusses on machining rotationally symmetric work pieces. This allows a high number of revolutions of the work piece. When machining large, eccentric work pieces this is not possible which leads to different boundary conditions.

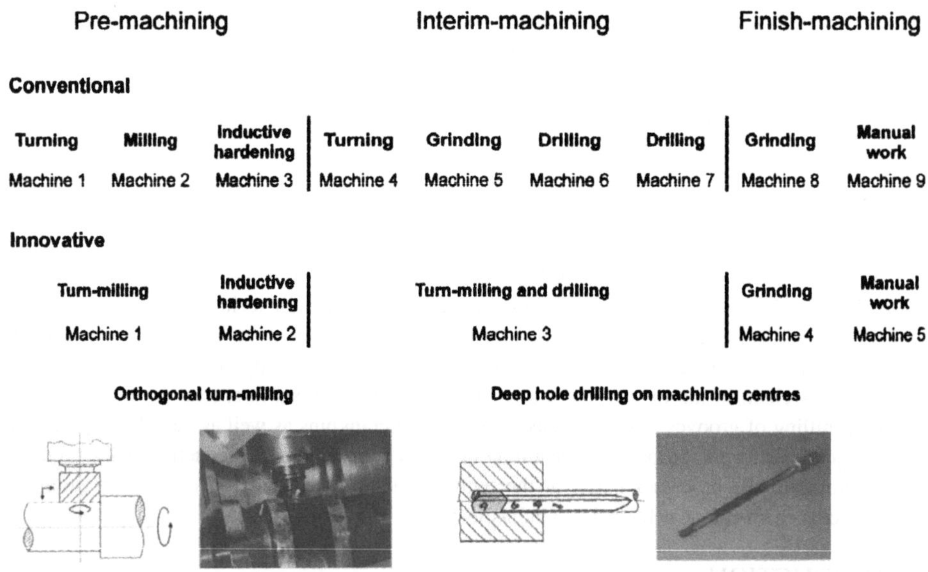

FIGURE 1: Process chain configuration

2.2 DEEP HOLE DRILLING

Generally, there are three different types of deep hole drilling technology: BTA-, ejector- and single-edge gun drilling. Besides large drilling depth, deep hole drilling can be used to machine bore holes with a high quality. For diameters up to 32 mm single-edge gun drilling is mostly used [2]. The use of deep hole drilling technology on machining centres is for example desirable to enable a complete machining of work pieces on one machine tool.

To use deep hole drilling tools on machining centres special requirements have to be fulfilled. A high coolant pressure and flow rate is necessary to transport the chips out of the deep bore hole. Furthermore, water based cutting fluids instead of special deep hole drilling oil must be used because of the operative conditions. In addition, guidance for the spot-drilling with the single-edge gun drill has to be realised by a pilot hole, since a spot-drilling bush with periphery is normally not available in machining centres. Especially the problem of guiding the single-edge gun drill when starting the hole on hardened and sloped surfaces is still not solved. On conventional bench drilling machines accordingly shaped spot-drilling bushes are normally used. These have to be designed and chosen for every special case.

3 RESULTS

3.1 TURN-MILLING OF ANNEALED CRANKSHAFTS

At the crankpin and at the internal sides of the crankwebs there is a large overmeasure on the forging blank which has to be rough machined. Depending on the forging die and the actual crankpin the overmeasure varies by a large amount. Turn-milling parallel to the axis with extremely big disk milling cutters is conventionally used. This leads to a high material removal rate because of the very big tools. Furthermore a low number of revolutions of the crankshafts is possible, which is advantageous because of the balance error. However, for this special machining task the crankshaft has to be chucked separately on an appropriate machine tool. Furthermore the landed costs are very high, so that this strategy is only profitable when machining a high amount of the same crankshaft type in big batch sizes.

To reduce the necessary amount of chucking and consequently the non-productive time this machining step can be done by orthogonal turn-milling on a turn-milling centre. Using this strategy is advantageous when machining small batch sizes or a small total amount of the same crankshaft type, e. g. job lots, single work packages, prototypes or rush-orders.

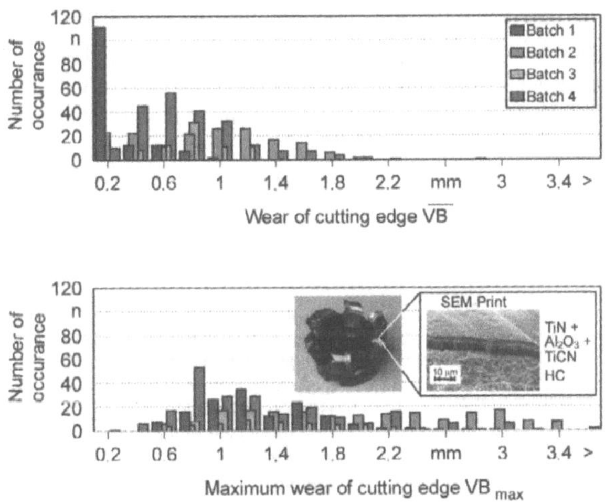

Workpiece mat.:	34CrNiMo6	
Tool:	Torus milling cutter	
Indexable tip:	HC; Geometry C; Ø 16 mm $TiCN+Al_2O_3+TiN$	
	Crankweb	***Crankweb***
Cutting speed:	v_c = 251 m/min	v_c = 151 m/min
Feed rate:	f = 4 mm	f = $3^1/_3$ mm
Depth of cut:	a_p = 3-4 mm	
Working engagement:	$a_e \approx$ 6-15 mm	

FIGURE 2: Wear of the cutting edge after machining two complete cranks

To reach an acceptable essential operation time using the orthogonal turn milling, a high material removal rate must be achieved. Therefore high feed rates and high depths of cut and working engagements have to be realised, which leads to a very high mechanical tool load. The choice of appropriate indexable inserts here is very important. A solid wedge and the coating system $TiCN+Al_2O_3+TiN$ proved to be best.

The rough milling of two complete cranks, in this case the crankpins and the inner sides of the crankwebs, can be realised with a high process security (Figure 2). The wear of cutting edge varies due to variation in heat treatment and the wear of the forging die. This leads to different material properties and working engagements between the single batches.

3.2 MACHINING OF GROOVES TO REDUCE TENSION

Due to the sharp-edged transition, peak stress-loads occur within the crankpin where it enters the crankweb. To achieve a stress reduction in this area, grooves are machined into the crankpin. For this purpose, orthogonal turn-milling using tools having an appropriate geometry is employed. Compared to conventional turning with a special machine and HSS-tools, the essential operation time can be reduced significantly (Figure 3). The rough machining is carried out with a standard tool and round indexable inserts. The two stress reduction grooves of one crankpin are simultaneously finished using up and down milling. This requires the use of ultra-accurate tools as well as an individual selection of the indexable inserts for the single insert seats. Thereby an even wear progression, a high process security and satisfactory surface qualities can be achieved. The surface quality is improved significantly and consequently the manual work is reduced by half.

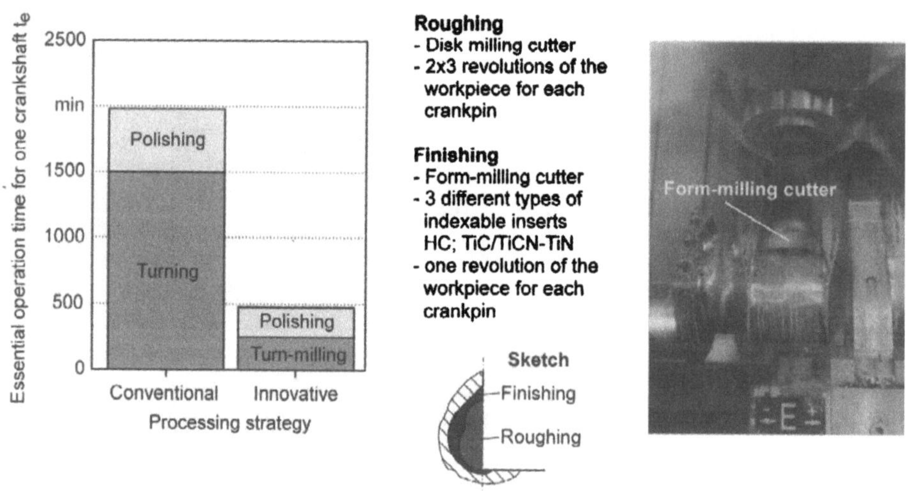

FIGURE 3: Machining of grooves to reduce tension

3.3 MACHINING OF HOLES IN ANNEALED CRANKSHAFTS

For the greasing of the crank pin and middle bearings oil supply holes have to be machined into the crankshafts (Figure 4). Dependent on the construction of the crankshaft the oil supply is for example realised through the front of the crankshaft. In this case each half of the crankshaft is supplied through the respective shaft end with the aid of a branched oil supply hole system. Another possibility is the supply of every single crank pin through the adjacent middle bearing. The problems when machining the oil hole system are nearly the same independent of the construction. Most critical are the deep holes and the spot-drilling on sloped and hardened surfaces.

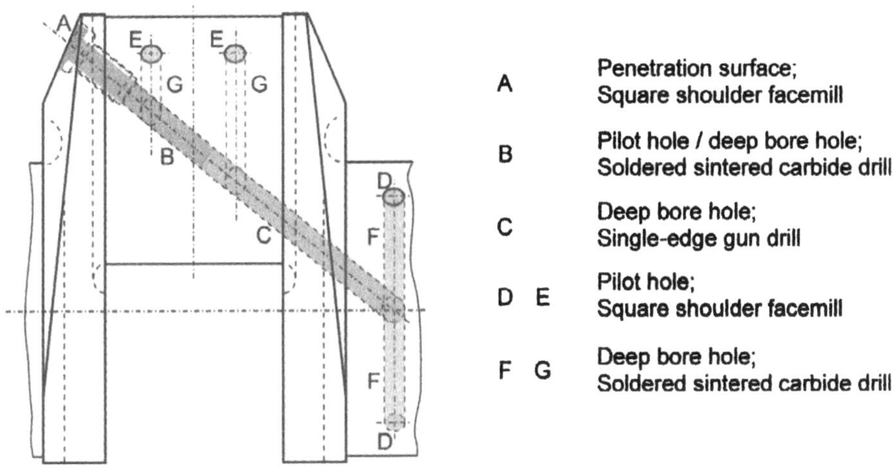

FIGURE 4: Machining of oil supply holes

For machining deep holes on machining centres, single-edge gun drilling is an appropriate technology to produce them efficiently with high quality. However, soldered cemented carbide drills can be used to machine drilling depths up to l/D = 7 at significantly higher feed speeds. Furthermore, for using single-edge gun drills on machining centres, a pilot hole for the guidance during spot drilling is also necessary. Twist drills are used for this, so that a tool change must be carried out in any case. Thus, the use of soldered cemented carbide drills was tested. These were constructed to drill up to l/D = 7.5. If the complete drilling depth is less than this value, oil supply holes can alternatively be machined completely in this way without using single-edge gun drills.

In this special usecase the diameter of the pilot hole machined with the soldered cemented carbide drill (D = 14.5 mm; l = 100 mm) was maximally 25 μm above the nominal diameter and is thus suited to be used as a pilot hole. Generally spoken, the diameter of the pilot should be as close as possible to the nominal diameter [3]. For the second part of the bore hole (l = 100-250 mm) the use of deep hole drilling technology is necessary, since the relative drilling depth is l/D = 17.5.

Using this combined machining strategy and simultaneously employing optimised process parameters, the essential operation time can be reduced by approximately 30% (Figure 5). The quality of the hole deteriorates within the first section, where the soldered cemented carbide drill is used, compared to the section machined with the single-edge gun drill. However, the hole still clearly exceeds the quality required.

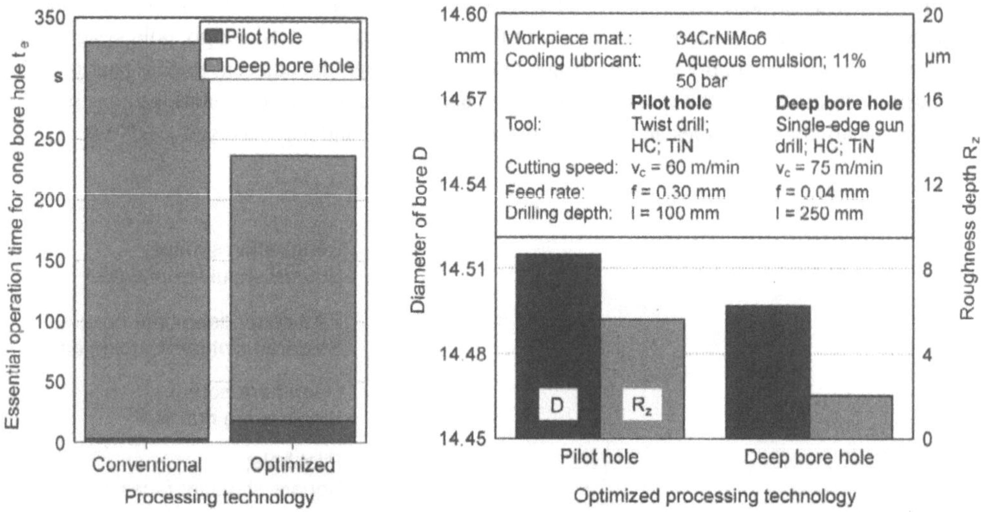

FIGURE 5: Milling of pilot holes in hardened surfaces

3.4 MACHINING OF PILOT HOLES IN HARDENED AND SLOPED SURFACES

The machining of pilot holes to guide the single-edge gun drill in hardened and sloped surfaces is problematic. A cemented carbide drill is pushed away in this case. This is why the helical milling with a square shoulder facemill is used. The coolant is externally supplied emulsion to remove the chips out of the bore hole.

During the experiments, the pilot holes could be machined in a wide range of parameters. A variation of the diameter of bore is possible by radius correction. All bore holes are conical towards the end (Figure 6). Using down milling, the process is acoustical more quiet and there is less vibration. Furthermore the range of diameter in this case is smaller and absolute run-out of the centre is significantly smaller compared to up milling. Totally, down milling is favourable to up milling in this case. Using this technology, process secure single-edge gun drilling is possible even in hardened and sloped surfaces on machining centres.

3.5 COOLING LUBRICANT

Using deep hole drill technology there is a high demand on the coolant. At the guiding pads a good lubrication has to be ensured. On machining centres this has to be realised by water based cutting fluids, since other operations are carried out here as well. Based on a standard product, a water based cutting fluid was developed, which fulfilled the technical effort. The newly developed emulsion has a low pH-value, which leads to a high skin compatibility. Another goal was to reach as low a class of water endangerment as possible while retaining universal applicability of the coolant. This goal was reached through further development.

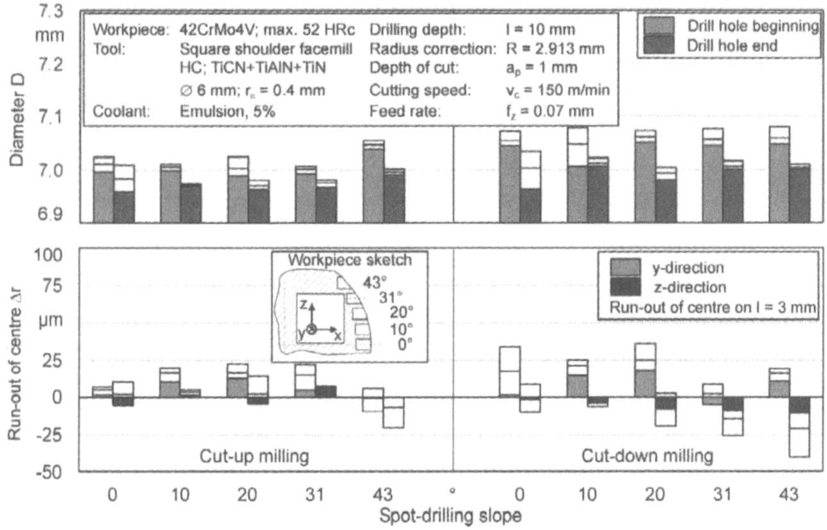

FIGURE 6: Processing technology for machining oil supply holes

4 CONCLUSION

By using the orthogonal turn-milling for machining large forged crankshafts the process chain can be shortened, costs can be reduced and the flexibility rises. The rough machining of annealed crankshafts can be performed in an economic way and with a high process security. The use of ultra-accurate tools makes the machining of form elements profitable.

Oil supply holes in non hardened crankshafts can be machined with a combined processing technology so that the demanded quality can be reached in an economic way. Using helical milling in hardened and sloped surfaces to machine pilot holes enables to transfer the deep hole drilling technology from the bench drilling machine to the machining centre. This leads to a time advantage and a higher flexibility.

REFERENCES

1. Daniel, A.: Oberflächengestalt beim achsparallelen Innen- und Außendrehfräsen. Dissertation TH Darmstadt, 1994
2. VDI-Richtlinie 3208: Richtwerte für das Tiefbohren mit Einlippenbohrern. März 1996, Beuth-Verlag, Berlin
3. Löbbe, H.; Webber, O.: Einlippentiefbohren auf BAZ – Bohrungsqualität und Prozessüberwachung. In „Spanende Fertigung“, 3. Edition, Publisher: K. Weinert, Essen, Vulkan-Verlag 2001, pp. 101-113

A CONTRIBUTION TO THE STRAIN-HARDENING PROCESS ANALYSIS OF HARDENED STEEL DURING HIGH-SPEED MACHINING

S. Dolinšek[1], S. Ekinović[2]

[1] Faculty of Mechanical Engineering, University of Ljubljana, Slovenia
[2] Faculty of Mechanical Engineering in Zenica, University of Sarajevo, BIH

KEYWORDS: High-speed machining, hardened steel, chip formation.

ABSTRACT. High-speed machining of hard materials, which is nowadays widely applied in the tool manufacturing industry, has several advantages over conventional machining. However, in practical machining one of the important tasks is still to define the values of the cutting speed limit, where we enter into the high-speed region. In the paper some of the results of the investigation of the strain hardening process of hardened steel (52 HRC) in milling with a speed range from 50 to 1500 m/min are presented. The influence of the cutting speed into the strain-hardening rate is identified by micrographs and measurement of micro hardness, and the chip shape is also discussed. On basis of the results obtained it is possible to predict the searched speed limit when the cutting process enters into the high-speed region.

1. INTRODUCTION

High-speed machining, like the cutting of hard materials, at higher speeds are those technologies, which are recently being increasingly applied in the production industries especially in mould producing. These machining processes are characterised by their high productivity, good surface finish quality and higher dimensional tolerances. Additionally these technologies make it possible to execute final machining operations without consequent grinding or similar finishing operations. Among different high-speed machining processes the most commonly applied is high-speed milling.

The newest investigations in this segment of processes are focused on the four characteristic directions: mechanisms of tool wear [1][2], quality of surface finish [3][4], mechanisms of chip formations [5][6][7], and problems of machining of materials in their hardened state (hard machining) [3][8][9][10][11]. All these researches are mainly focused on the one common goal, to establish the possibilities of optimal use of the high-speed machining in practical applications. One of the most important tasks, therefore, is to establish the regime of the cutting speeds that correspond to the level when the process enters into the high-speed range.

In the present work an analysis of the chip formation mechanism; plastic behaviour of the machined material and analysis of the chip shape produced is used as the principal approach to investigate those problems. In examples of the high-speed machining of hard materials, the main common chip types appear in the form of the saw-tooth chip. This kind of chip is the result of some interrelated mechanisms such as localised shear, adiabatic shear and also as a catastrophic shear in the form of extensive cracks [8]. Chips are formed segmentally and the morphology of their formation is a result of the changeable cutting conditions, which mainly depend on:

Published in: E.Kuljanic (Ed.) *Advanced Manufacturing Systems and Technology*,
CISM Courses and Lectures No. 437, Springer Wien New York, 2002.

- Mechanical, thermal and thermo-chemical characteristics of the work material,
- Cutting conditions,
- Changes in sliding characteristics at the primary zone,
- Changes in tribological circumstances at the tool-chip interface,
- The possible interactions between the primary and secondary shear zone and the dynamic behaviours of the machine-tool system and its linkage with the cutting process.

2. EXPERIMENTAL WORK

The experimental work has been carried out at the Faculty of Mechanical Engineering, University of Ljubljana. The machining was conducted on the milling machine, type Moriseiki-Frontier. The machining conditions were as follows: cutter diameter D = 80 mm, depth of cut d = 2mm, tooth feed f_z = 0.1 mm, cutter type SUM-UFO-4000, cutting insert SFKN 12T3 A2TN - AC230, cutting speed v_c = 50-1500 m/min, $\gamma = 27^\circ$, $\lambda = -7^\circ$, $\chi = 45^\circ$.

The chemical composition of the investigated steel grade is shown in Table 1, and Figure 1 shows the microstructures and results of Vickers micro-hardness measurements of the machined steels.

TABLE 1. Chemical composition of the investigated steel

Steel grade	Chemical composition [%]									
	C	Si	Mn	P	S	Cr	Mo	V	Cu	Al
X63CrMoV51	0.62	1.0	0.59	0.017	0.004	5.46	1.21	0.46	0.26	0.028

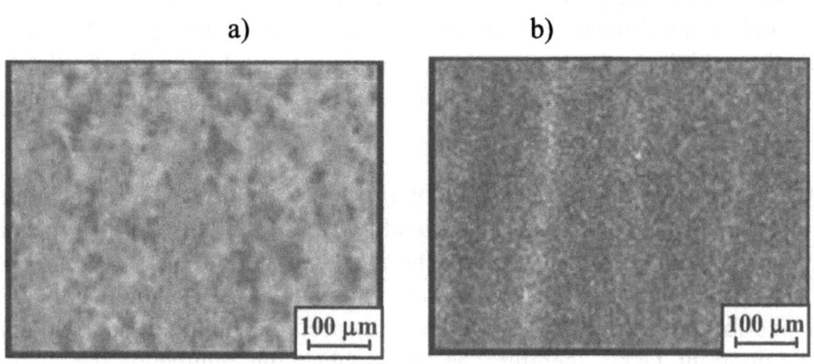

FIGURE 1. Microstructure of the machined steel grades:
a) X63CrMoV51- annealed, 282 HV,
b) X63CrMoV51- tempered, 629 HV.

The microstructure of X63CrMoV51 in annealed state consists of ferrite with spheroidal carbides and low quantity of perlite due to incomplete annealing. The microstructure of the same steel in tempered state is martensite with the characteristic eutectoid carbides and some retained austenite due to "yielding" after quenching. Steel in the annealed state was machined at the cutting speeds of between the 150 – 1500 m/min, and in tempered state at speeds 50, 150, 300 and 1500 m/min.

3. RESULTS AND DISCUSSION

All the experimental results are presented in Figure 2 in the form of a diagram of the micro-hardness HV and cutting speed magnitude; the diagram is illustrated by the microstructures of the chip shape produced. On the left hand side the figure presents the microphotographs of the initial microstructure for the steel at the annealed (292 HV) and tempered state 629 HV - 52 HRc).

In machining of the above mentioned steel at the annealed state with a speed of v_c = 150 m/min a continuous type of chip is produced with the distinctive level of strain hardening, microphotograph No.1, Figure 2. The average micro-hardness of this chip is 470 HV, which gives us the coefficient of strain hardening of n = 470/282 = 1.67. The texture of the grains and also the friction layer (secondary deformation) at the inner side of the chip can also be clearly noticed. The shear angle is changeable (unstable), and consequently waviness on the other side of chip can be observed. Changes at the shear angle can be explained by the instability of the deformation processes and tribological conditions in the chip-tool contacts at the rake face. Instabilities of the deformation process in this steel grade are the consequences of the differences in deformability of the ductile ferrite base and hard and fragile carbides. Tribological circumstances in the chip-tool contact are changeable due to the stick-slip effect and high contact strains, which lead to the instability of the deformations at the secondary shear zone.

In machining at a speed of 1500 m/min the chips are segmented as the consequence of the localized shear deformation, microphotograph No.2, Figure 2. The average micro-hardness at the boundaries of the segmented chips is 457 HV which gives us strain hardening coefficient n = 1.62. On the inner side of the segments the hardness of the chip was measured as 380 HV. Therefore the material within the chip segment is less deformed. The white layer, however - characteristic of the predominant local shear deformation and thermal softening - is not evident. In the relation between the chip shape and the mechanism of formation it might be ascertained that for this kind of annealed steel the level of speed v_c = 1500 m/min corresponds to the region of high-speed cutting.

In machining of these kinds of steel grade in tempered state at a speed of v_c = 50 m/min the mechanism of the chip shape is quite similar to the circumstances of machining the annealed grade at a speed of v_c = 150 m/min, microphotograph No.3, Figure 2. The main distinctive difference is in the appearance of the white layer on the inner side of the chip, which is the consequence of the thermal softening of the material. The average measured micro-hardness is 660 HV, which in relation to the initial values (629 HV) gives us a low level of strain hardening n = 1.05. The structure is characteristically identical, i.e. with uniform plastic deformation.

In machining at a speed of v_c = 150 m/min, the chip is segmented and has a typical saw tooth shape, microphotograph No.4, Figure 2. Clearly visible is the white layer both on the inner part of chip and also between the segments. Therefore, the appearance of the thermal softening and mechanism of deforming is present. The average measured value of the white layer is 756 HV, consequently the strain hardening coefficient is n = 1.2. However, on the inner part of the segments the average hardness is only 632 HV, which shows as a complete un-deformed structure in relation to the initial state of the workpiece material. On basis of the above findings it can be concluded that the speed of v_c = 150 m/min is already of the range in which we can talk about high-speed machining.

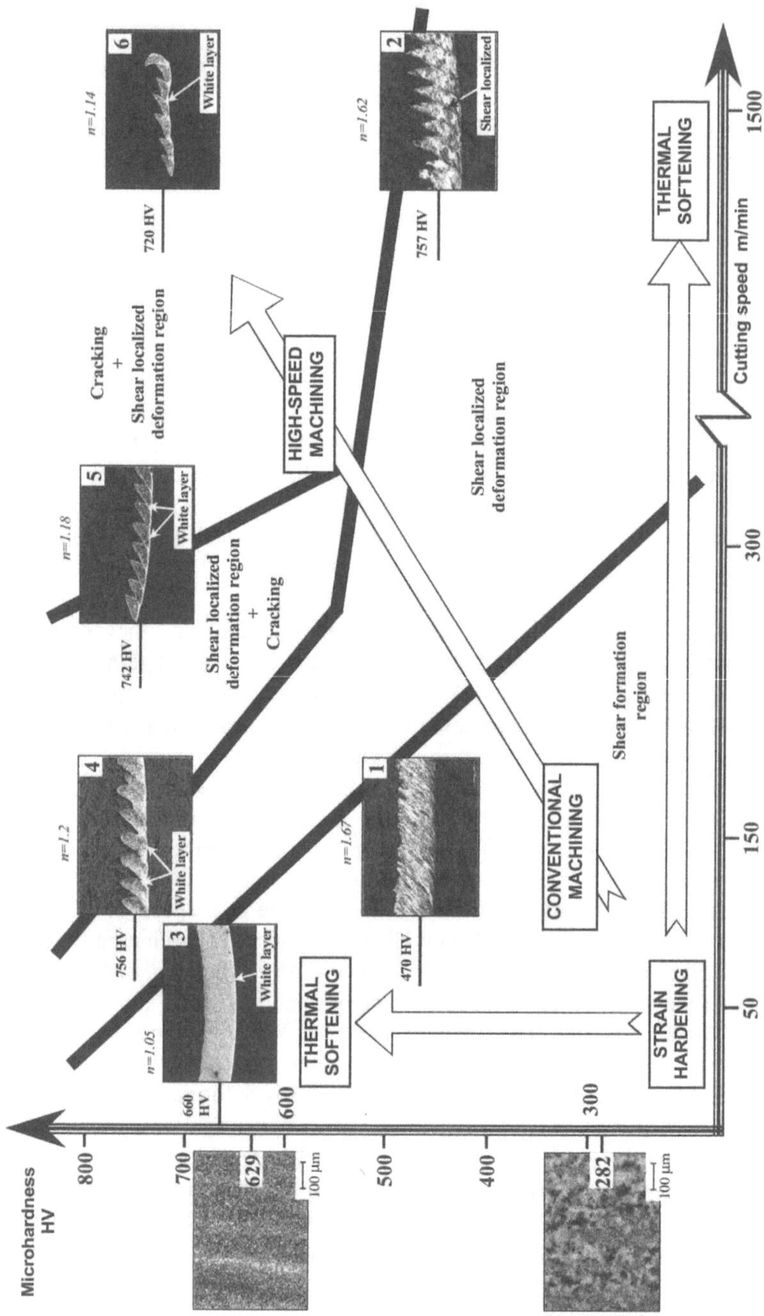

FIGURE 2. Chip morphology in conventional and high-speed regions in relation to the cutting speed and micro-hardness.

In machining at a speed of v_c = 300 m/min the chip is even more strongly segmented, with lower thickness and with the segments of lower dimensions in comparison with the lower speeds, see microphotograph No. 5, Figure 2. At the same time the thickness of the white layer is lower and its average measured micro-hardness is 742 HV, consequently n = 720/629 = 1.18. The micro-hardness between segments is 640 HV.

As a final ascertainment, with the cutting speed of v_c = 1500 m/min the chip shape is heavily segmented with even lower thickness and magnitude of the segments, microphotograph No. 6, Figure 2. Moreover the thickness of the white layer is lower compared to the machining with the lower speeds, its average micro-hardness is 720 HV and strain hardening coefficient n 0 720/629 = 1.14. Micro-hardness within the segments is 618 HV.

4. CONCLUSION

In the paper the experimental investigations of the chip shape and measurement of the chip micro-hardness serve as a basis for the analysis of the process of plastic deformation in machining of the steel X63CrMoV51 at the annealed and tempered conditions. Cutting has been performed by the milling process t a cutting speed within the range of v_c = 50 – 1500 m/min. In relation to the results obtained some main conclusions can be defined:

- In machining with lower cutting speeds and lower workpiece hardness, strain hardening of the steel is dominant.
- At the higher cutting speeds and higher hardness of the work material thermal softening of the material and initiation of cracking is dominant, and the chip is mainly shown as being of segmented type.
- On the basis of the analysis of the chips produced, some characteristic regions of strain hardening can be defined in Fig 2 as:
  a) region of the strain hardening,
  b) transitional region with shear localised deformation with lower influence of the thermal softening,
  c) and region with prevailing initiation of cracking and influence of thermal softening.
- On the basis of the presented analysis some distinctions within the cutting speed region defined as conventional or high speed machining can be ascertained.

ACKNOWLEDGMENTS

The authors are grateful to the Slovenian Science Foundation for providing financial support for this research.

REFERENCES

1. Becze, C.E., Clayton, P., Chen, L., El-Wardany, T.I., Elbestawi, M.A., (2000), High-speed five-axis milling of hardened tool steel, Internationl Journal of Machine Tools & Manufacture, Vol.40, 869-885.

2. Dolinšek, S., Kopač, J., (1999), Mechanism and types of tool wear; some particularities in using advanced cutting materials and newest machining processes, Proceedings of the 8th International Scientific Conference on Achievements in Mechanical & Materials Engineering, 185-188.
3. El-Wardany, T.I., Kishawy, H.A., Elbestawi, M.A., (2000), Surface Integrity of die Material in High Speed Hard Machining, Part 1: Micro-graphical Analysis, Transactions of the ASME, Journal of Manufacturing Science and Engineering, Vol.122, 620-631.,
4. El-Wardany, T.I., Kishawy, H.A., Elbestawi, M.A., (2000), Surface Integrity of die Material in High Speed Hard Machining, Part 2: Micro-hardness Variations and Residual Stresses, Transactions of the ASME, Journal of Manufacturing Science and Engineering, Vol.122, 632-641.
5. Ning, Y., Rahman, M., Wong, Y.S., (2001), Investigation of chip formation in high-speed end milling, Journal of Materials Processing Technology, Vol.113, 360-367.
6. Fallbohmer, P., Rodriguez, C.A., Ozel, T., Altan, T., (2000), High-speed machining of cast iron and alloy steels for die and mold manufacturing, Journal of Materials Processing Technology, Vol.98, 104-115.
7. Vyas, A., Shaw, M.c., (1999), Mechanics of Saw-Tooth Chip Formation in Metal Cutting, Transactions of the ASME, Journal of Manufacturing Science and Engineering, Vol.121, 163-172.
8. Poulachon, G., Moisan, A.L., (2000), Hard-Turning: Chip Formation Mechanisms and Metallurgical Aspects, Transactions of the ASME, Journal of Manufacturing Science and Engineering, Vol.122, 406-412.
9. Schulz, H., Kneisel, T., (1994) Turn Milling of Hardened Steel – an Alternative to Turning, Annals of the CIRP, Vol43, 93-96.
10. Elbestawi, M.A., Chen, L., Besze, C.E., El-Wardany, T.I., (1997), High-speed Milling of Dies and Moulds in Their Hardened State, Annals of the CIRP, Vol.46, 57-62.
11. Tonshoff, H.K., Arendt, C., Ben Amor, R., (2000), Cutting of Hardened Steel, Annals of the CIRP, Vol.49, 547-566.

TOOL FLANK WEAR PREDICTION USING THE FORCE-TIME MEASUREMENTS IN TURNING

G. Cukor[1], E. Kuljanić[2]

[1] Department of Production Engineering, Faculty of Engineering, University of Rijeka, Croatia
[2] Department of Mechanical Engineering, University of Udine, Italy

KEYWORDS: Tool flank wear, Adhesion wear, Cutting force, Turning

ABSTRACT: The paper proposes a mathematical model to predict the tool flank wear during the turning process. The thrust component of cutting force has been found to correlate with the gradual flank wear progress. Prediction of tool flank wear has been achieved by combining the thrust force-time measurements and developed adhesion flank wear model. The model was verified by comparing the predicted values with the measured values. A strong correlation was obtained between these two sets of values.

1 INTRODUCTION

The most undesirable yet inherent characteristic of machining processes is wear of the cutting tool. The tool wear is a complex phenomenon that manifests itself in different and varied ways. Cutting tool wear can be classified into several types as follows:

1. adhesive wear associated with shear plane deformation,
2. abrasive wear resulting from hard particles cutting action,
3. diffusion wear occurring at high temperatures,
4. fracture wear such as chipping due to fatigue, and
5. other mechanisms.

Tool wear processes generally occur in combination with the predominant wear mode (Fig. 1) that is dependent upon the cutting conditions (principally cutting speed v_c), workpiece/tool material and the tool insert geometry. The more predominantly modes of cutting tool wear in turning operations are flank and crater wear. Flank wear has been traditionally emphasized more than crater wear because of the more direct influence that flank wear has on the accuracy of the workpiece dimension, as well as surface quality.

Flank wear arises due to both adhesive and abrasive wear mechanisms from the intense rubbing action of the two surfaces in contact, i.e. the clearance face of the cutting tool and the newly machined surface of the workpiece. Its rate of increase at the beginning of the tool life is rapid, settling down to a steady state then accelerating rapidly again at the end of tool life.

Enough work has been done in the past to correlate various parameters with the progress of tool flank wear [1, 2]. One of the most promising techniques for its indirect detection appears to be the measurement and use of cutting force, since it can be successfully correlated to flank wear [3]. This can largely be attributed to the fact that cutting force signals are highly sensitive carriers of information about the status of the machining process and, hence, they are suitable alternatives to tool flank wear monitoring.

Published in: E.Kuljanic (Ed.) *Advanced Manufacturing Systems and Technology*, CISM Courses and Lectures No. 437, Springer Wien New York, 2002.

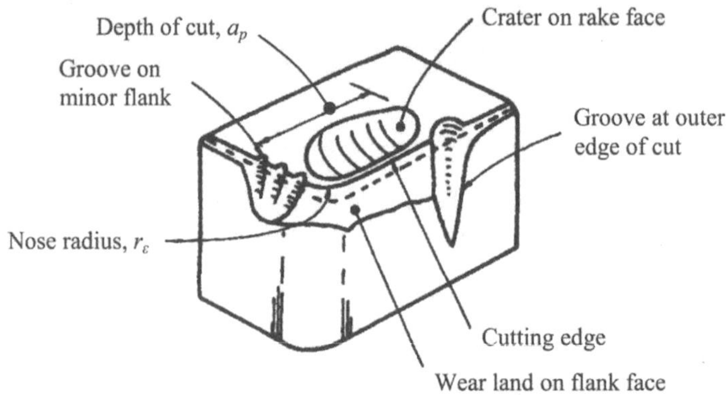

FIGURE 1. Various wear and fracture surfaces on a worn cutting tool

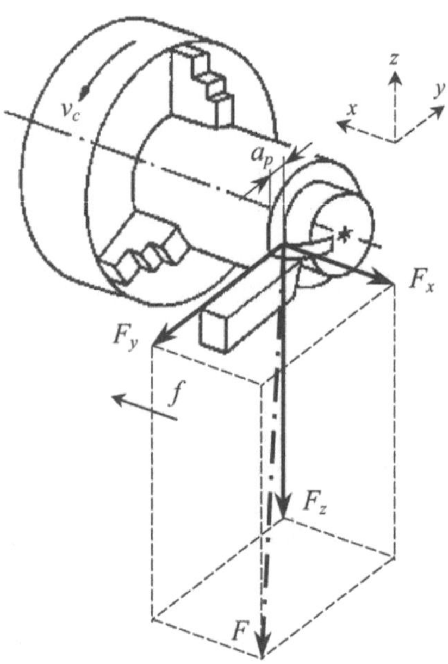

FIGURE 2. Cutting force components

In turning, it is convenient to consider the resulting cutting force F as a three-component system: the feed (axial or x-axis) component F_x, the thrust (radial or y-axis) component F_y and the main cutting force (tangential or z-axis) component F_z (Fig. 2). Although it has been widely established that variation in the cutting force can represent an accurate and reliable approach to assess tool flank wear and failure [4,5], disagreement still exists as to which cutting force component (or combination) is the most sensitive and reliable.

The aim of this study was to develop a reliable mathematical model for predicting the flank wear of turning tool. The thrust force (F_y) in time domain has been modeled from the experimental data and it was introduced in developed adhesion flank wear model. This relationship was then used for prediction of tool flank wear. A strong correlation was obtained between predicted and the measured flank wears data.

2 ADHESION FLANK WEAR MODEL

In this investigation, adhesion has been considered as the predominant flank wear mechanism. When two mating surfaces are brought together, they touch at the tips of the higher asperities. The real area of contact supports the load, and increasing the normal load increases the amount of deformation and contact area, thus increasing the number of asperities that support the load. According to the fundamental law of adhesion wear [6], the possible adhesion flank wear volume V_{ad} under normal load F after sliding distance l is given by:

$$V_{ad} = K_{ad} \frac{F\, l}{H_B}, \tag{1}$$

where H_B is the Brinnel hardness value of the wearing material and K_{ad} is called the wear coefficient for adhesive wear. Wear coefficient K_{ad} is a principal value for a friction pair to describe its wear rate. The physical meaning of K_{ad} is the wear volume fraction at the plastic contact zone, and it is strongly affected by the material properties and the geometry of the zone in compression and shearing. It varies between 10^{-7} and 10^{-2} depending on the operating conditions and material properties [6]. It should be recognized that a wear coefficient K_{ad} is not a constant value but it is a possible value in the range of adhesive wear rate.

In order to predict the tool flank wear under different conditions, it is essential to build the cutting force model accounting for the influence of cutting speed v_c, feed f (and depth of cut a_p if vary), along the duration of machining t. Cukor [7] had developed a model-building method for the non-linear modeling of machining processes. The method is well suited for this application since:

- data storage requirements for this method are small,
- the computation is efficient,
- subjective prior information can be made use of, and
- it allows adaptive estimation of model parameters.

The basic m-order polynomial model is:

$$r = e^{b_0} \prod_{i=1}^{n} \xi_i^{\sum_{j=1}^{m} b_{m(i-1)+j} (\ln \xi_i)^{j-1}}, \tag{2}$$

where r is the measured response, ξ_i are the input variables, n is the number of input variables and $b_{0,1,2,...}$ are the estimated model parameters. Accordingly, for a_p = *const.*, the proposed model of the thrust force F_y is

$$F_y = e^{b_0} v_c^{\,b_1 + b_2 \ln v_c} f^{\,b_3 + b_4 \ln f} t^{\,b_5}. \tag{3}$$

Since $dl = v_c\, dt$, Eq. (1) can be rewritten as

$$\frac{dV_{ad}}{dt} = \frac{K_{ad}}{H_B} e^{b_0} v_c^{\,1 + b_1 + b_2 \ln v_c} f^{\,b_3 + b_4 \ln f} t^{\,b_5}. \tag{4}$$

Tool flank wear geometry can be simplified as presented in Fig. 3. Considering ΔABC, where α is the clearance angle and γ is the rake angle,

$$ds = \frac{\tan \alpha}{1 - \tan \alpha \tan \gamma} dV_b.$$

For the entering angle κ, the width of cut is $b = a_p/\cos \kappa$, and the volume of the worn out tool material from the flank face at any instant during a time interval of Δt is dV_{fw}, which is given by

$$dV_{fw} = b\, ds \left(V_b + \frac{1}{2} dV_b \right) = \frac{a_p dV_b \tan \alpha}{\cos \kappa \left(1 - \tan \alpha \tan \gamma\right)} \left[V_b + \frac{1}{2} dV_b \right].$$

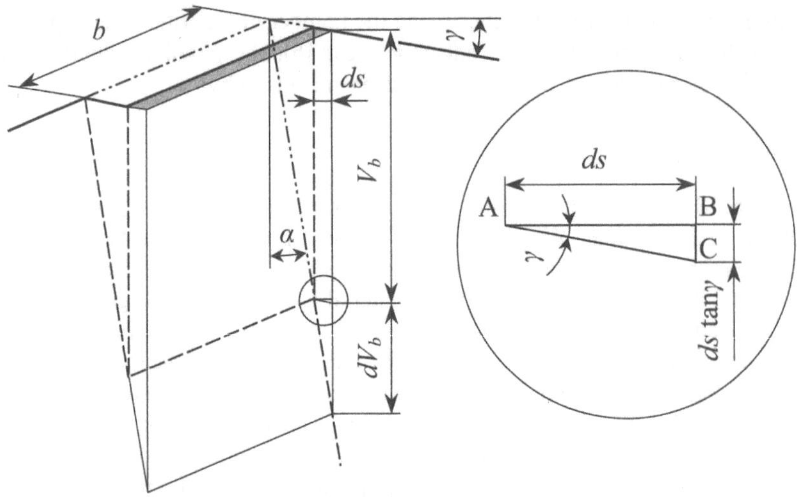

FIGURE 3. Tool flank wear geometry

Neglecting the dV_b^2 term,

$$dV_{fw} = \frac{a_p \tan\alpha}{\cos\kappa\,(1-\tan\alpha\tan\gamma)} V_b dV_b,$$

and upon differentiating, the value of the flank wear volume rate is obtained as:

$$\frac{dV_{fw}}{dt} = \frac{a_p \tan\alpha}{\cos\kappa\,(1-\tan\alpha\tan\gamma)} V_b \frac{dV_b}{dt}. \tag{5}$$

Since the adhesion flank wear volume rate dV_{ad}/dt is the same as flank wear volume rate dV_{fw}/dt, equating Eqs. (4) and (5) yields

$$V_b \frac{dV_b}{dt} = \frac{K_{ad}\cos\kappa\,(1-\tan\alpha\tan\gamma)}{H_B\, a_p \tan\alpha} e^{b_0} v_c^{\,1+b_1+b_2\ln v_c} f^{\,b_3+b_4\ln f} t^{b_5}.$$

Integrating both sides,

$$\int_0^{V_b} V_b dV_b = \frac{K_{ad}\cos\kappa\,(1-\tan\alpha\tan\gamma)}{H_B\, a_p \tan\alpha} e^{b_0} v_c^{\,1+b_1+b_2\ln v_c} f^{\,b_3+b_4\ln f} \int_0^t t^{b_5} dt,$$

therefore,

$$\frac{1}{2} V_b^2 = \frac{K_{ad}\cos\kappa\,(1-\tan\alpha\tan\gamma)}{H_B\, a_p \tan\alpha} e^{b_0} v_c^{\,1+b_1+b_2\ln v_c} f^{\,b_3+b_4\ln f} \frac{t^{1+b_5}}{1+b_5}.$$

Finally, a mathematical model for flank wear prediction is

$$V_b = \left[\frac{2K_{ad}\cos\kappa\,(1-\tan\alpha\tan\gamma)}{H_B\, a_p \tan\alpha\,(1+b_5)} e^{b_0} v_c^{\,1+b_1+b_2\ln v_c} f^{\,b_3+b_4\ln f} t^{1+b_5}\right]^{0,5}. \tag{6}$$

3 VALIDATION OF THE PROPOSED MODEL

The aim of this section is to validate the predictive mathematical model. For this purpose the experimental data appear in Dimla et al. [8] was used. Details of the experimental set up were as follows: machine tool - Lang Swing J6, workpiece material - EN24 BS 970, hardness - 255 HB, tool holder - Sandvik SSBCR 2020 K12, insert - SCMT 12 04 08 UM, tool material - Sandvik Coromant P25 4025, $\kappa = 75°$, $\alpha = 7°$, $\gamma = 0°$, $v_c = 275$ and 300 m/min, $f = 0{,}1$, 0,2 and 0,3 mm/rev, $a_p = 2$ mm.

Due to limitation of space only one but typical plot for experiments is presented. Fig. 4 shows the three components of cutting force and flank wear for different durations of machining for one particular cutting condition. There is an increase in all components of cutting force with progressive increase in flank wear. The fluctuations in force values can be attributed to random variations in material properties. Close analysis of Fig. 4 indicates a good correlation between flank wear V_b, feed force F_x and thrust force F_y.

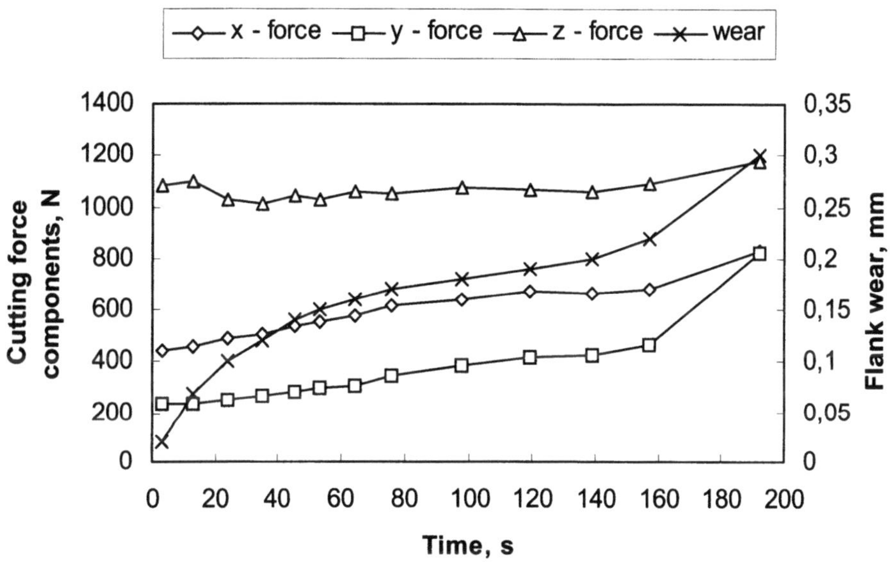

FIGURE 4. Cutting force components and flank wear with time ($v_c = 275$ m/min, $f = 0{,}3$ mm/rev)

The thrust force model derived above, Eq. (3), was solved to estimate the unknown model parameters. For this purpose the method of least squares was used, therefore

$$F_y = e^{-33{,}8897} v_c^{\,12{,}09768-0{,}80147\ln v_c} f^{\,4{,}155551+0{,}979588\ln f} t^{\,0{,}160592}. \tag{7}$$

For the given friction pair $K_{ad} = 1{,}05\text{E-}10$ [6], hence the following mathematical model of thrust force-wear characteristics, for $a_p = 2$ mm, can be obtained:

$$V_b = \left(7{,}183 \cdot 10^{-24} v_c^{\,13{,}09768-0{,}80147\ln v_c} f^{\,4{,}155551+0{,}979588\ln f} t^{\,1{,}160592}\right)^{0{,}5}. \tag{8}$$

Once the mathematical model was developed, the input parameters, i.e. cutting speed, feed and time, were fed to the Eq. (8) and the output variable, tool flank wear, was predicted for the experimental data. A few flank wear-time plots (Figs. 5-8) were drawn comparing the predicted flank wear and the measured flank wear. A good correlation was obtained between these two sets of values. Thus it can be stated that the developed model predicted the values of flank wear reliably over the selected range of input cutting parameters.

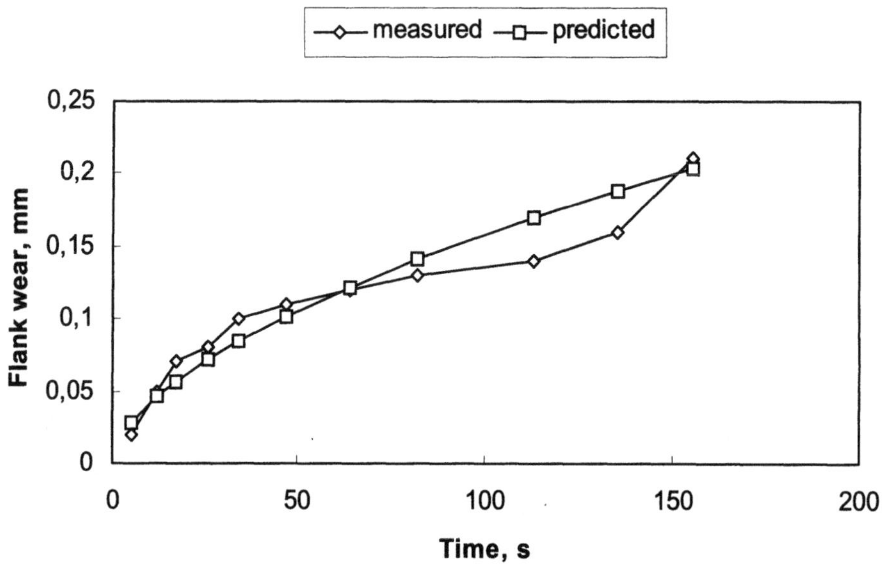

FIGURE 5. Flank wear-time plot (v_c = 300 m/min and f = 0,1 mm/rev)

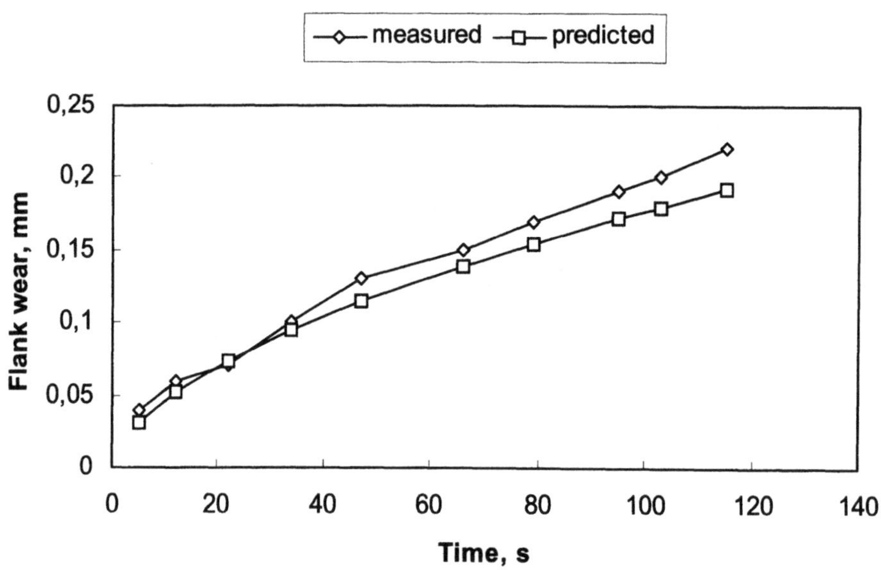

FIGURE 6. Flank wear-time plot (v_c = 300 m/min and f = 0,2 mm/rev)

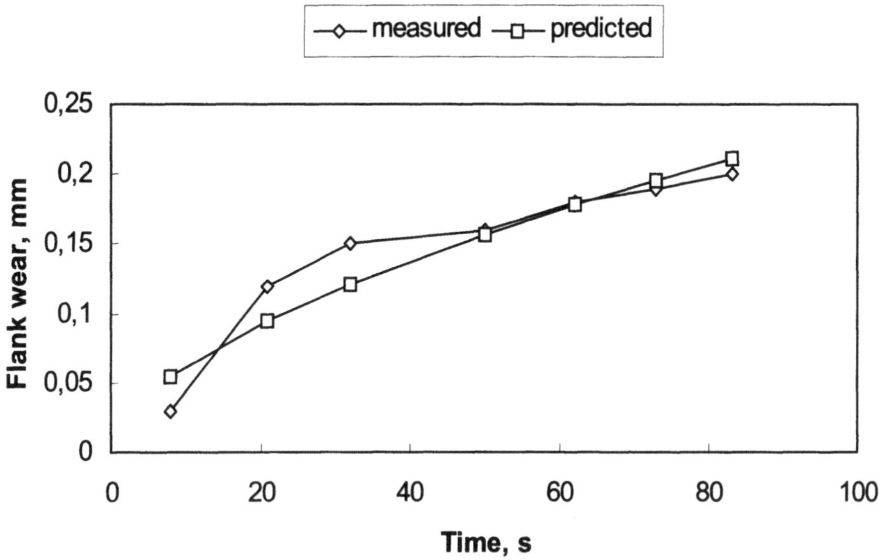

FIGURE 7. Flank wear-time plot (v_c = 300 m/min and f = 0,3 mm/rev)

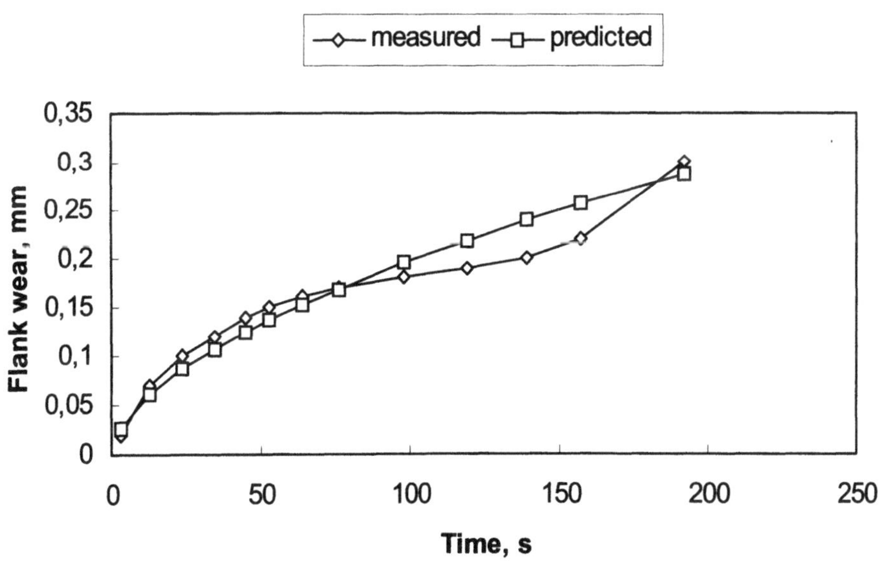

FIGURE 8. Flank wear-time plot (v_c = 275 m/min and f = 0,3 mm/rev)

4 CONCLUSIONS

The basic aim of this study on the investigation of the possibility of indirect prediction of the tool flank wear has been fulfilled. In the present work, a predictive mathematical model has been developed. Flank wear has been deduced indirectly by measuring the thrust force (F_y) in time domain and implementing the derived adhesion flank wear model. The following conclusions can be made.

- For monitoring of tool flank wear, a mathematical models of both thrust force (F_y) and adhesive flank wear have been developed successfully.
- Comparison of predicted and measured values of tool flank wears shows that strong agreement has been achieved between them. Therefore the developed model is recommended to be used for predicting the flank wear.
- As an extension of this study, an effective control strategy can be implemented for on-line monitoring and control during turning process using the mathematical model presented.

REFERENCES

1. G. Byrne, D. Dornfeld, I. Inasaki, G. Ketteler, W. König, R. Teti, Tool condition monitoring (TCM) – the status of research and industrial application, Annals of CIRP 44 (2) (1995) 541-567.
2. D.E. Dimla Sr., Sensor signals for tool wear monitoring in metal cutting operations – a review of methods, Int. J. Mach. Tools Manufact. 40 (2000) 1073-1098.
3. S.E. Oraby, D.R. Hayhurst, Development of models for tool wear force relationships in metal cutting, Int.J. Mech. Sci. 33 (2) (1991) 125-138.
4. H.V. Ravindra, Y.G. Srinivasa, R. Krishnamurthy, Modelling tool wear based on cutting forces in turning, Wear 169 (1993) 25-32.
5. S.K. Choudhury, K.K. Kishore, Tool wear measurement in turning using force ratio, Int. J. Mach. Tools Manufact. 40 (2000) 899-909.
6. B. Bhushan, Modern tribology handbook (2001) by CRC Press LLC.
7. G. Cukor, Optimization of machining process for advanced machining systems, Ph.D. thesis, Faculty of Engineering, University of Rijeka (1999).
8. D.E. Dimla Sr., P.M. Lister, On-line metal cutting tool condition monitoring. I: force and vibration analyses, Int. J. Mach. Tools Manufact. 40 (2000) 739-768.

METAL MATRIX COMPOSITES - MMC - TURNING: COMPARISON OF TOOL MATERIALS

C. Borsellino[1], S. Lo Casto[2], E. Lo Valvo[2], V. F. Ruisi[2]

[1] Department of Industrial Chemistry and Materials Engineering, University of Messina, Italy
[2] Department of Technology and Mechanical Production, University of Palermo, Italy

KEYWORDS: Turning, MMC, Tool wear.

ABSTRACT. The main properties required to the modern cutting tools to be employed in production, are both high wear resistance and chemical stability. In this last years, in the machining of MMCs several kinds of coated tools have been developed; they are characterized by high hardness to resist to the relevant strength and abrasive action of the reinforce fibers or particles, arranged inside the matrix. In turning of MMC the employ of Tungsten Carbide tools coated with Polycrystalline Diamond (PCD) have shown to be very effective, whereas they are very expensive thus increasing the production costs.
In order to reduce the tool cost, in the present work, some other tool materials, characterized by hardness values similar to PCDs, like coated Cermet tools and CBN (Cubic Boron Nitrite), are employed in the cut of a MMC. To evaluate the possibility of employing these tools in substitute of PCDs, tool life data have been compared taking into account both their cost and the surfaces quality obtained.

1 INTRODUCTION

The family of material classified as MMCs comprises a very broad range of advanced composites of great importance to both industrial and aerospace applications. However, the development and use of MMCs are still in their infancy when compared to monolithic materials or even PMC systems. Therefore, only a handful of applications have been designed and produced, but these are illustrative of the potential of MMCs. The primary advantage of advanced composites is their ability to provide higher mechanical properties and tailored physical properties at less weight than that of conventional materials. The use of metals as matrices imparts important properties to the resultant composites. High matrix strength and elastic modulus impart high composite shear and transverse strength and stiffness compared with monolithic metals, MMCs have higher strength-to-density and stiffness-to-density ratios, better fatigue resistance and superior temperature properties, lower CTEs, and better wear resistance. In the case of MMCs, these advantages are available for high-temperature applications (>316°) where PMCs are inadequate. In general, it can be seen that the strength of the composite is usually better than for unreinforced aluminum, and the modulus (a measure of the stiffness) is improved significantly. Furthermore, because metals are more electrically and thermally conductive, MMCs can be used in heat dissipation and transmission applications. In addition MMCs show higher temperature capability, fire resistance, higher transverse stiffness and strength than PMCs and also no moisture absorption, higher electric and thermal conductivity, better radiation resistance, no outgassing in a service environment, and manufacturing with conventional metalworking equipment.
The addition of a reinforcement to metal can be ceramic in nature with high strength and high stiffness retention at high temperature, or it can have physical attributes such as high heat transfer

Published in: E.Kuljanic (Ed.) *Advanced Manufacturing Systems and Technology*,
CISM Courses and Lectures No. 437, Springer Wien New York, 2002.

of low thermal expansion characteristics, or it can exhibit high wear resistance. Being very small in size, generally these reinforcements do not suffer from the flaws and defects generally attributed to bulk materials of the same chemistry. The presence of reinforcements can lead on the converse some disadvantages due to the higher material and fabrication costs, lack of material databases, and limited service experience. In addition the costs of design, analysis, and quality control are greater [1-4].

From the workability point of view, the reinforce inside the matrix causes very strong abrasive action on tools during the cutting operations. In figure 1 is shown the SEM image of the MMC material employed in the current investigation. In this image it's possible to recognize the presence of the reinforcement particles inside the metal matrix. This characteristic leads to very short tool life, bad surface quality, built up edge and high power needed for the cut [5,6].

FIGURE 1. SEM Image

In turning operation the employ of Tungsten Carbide tools coated with Polycrystalline Diamond have shown to be very effective, due to their high hardness, whereas they are very expensive thus increasing the production costs.

In order to reduce the tool cost, in the present work, some other tool materials, characterized by hardness values similar to PCDs, like coated Cermet tools and CBN (Cubic Boron Nitrite), are employed in the cut of a MMC.

To evaluate the possibility of employing these tools in substitute of PCDs, tool life data have been compared taking into account both their cost and the surfaces quality obtained.

2 TOOL WEAR EXPERIMENTS

2.1 EXPERIMENTAL SET-UP

The cutting testes were conducted on a Boeringer DM550/1000 lathe. The workpiece employed was an MMC 6061/Al_2O_3/10p (Duralcan® W6A.10A-T6) with an external diameter of 228 mm. Its characteristics are reported in the following:

Chemical composition: Si = 0,6%, Fe = 0,13%, Cu = 0,28%, Mn = 0,007%, Mg = 1,03%, Cr = 0,1%, Sn = 0,004%

Hardness: $HBN_{(2,5/62,5/30)} = 50$ [daN/mm^2]

Tensile strength: $R = 352\ N/mm^2$

A piece of this material, approximately 800 mm long, was fixed between chuck and tailstock. The commercially available materials selected for the tests, according to the insert number TCMW 16 03 08, were as follows:

- Cermet HT 3015 K10 in the following called "**A**"
- Cubic Boron Nitride - CBN (CB20) in the following called "**B**"
- Polycrystalline Diamond - PCD in the following called "**C**"

Each material was tested three times in continuous dry turning with the following cutting condition:

- geometry:
  - rake angle $\gamma = 0°$
  - clearance angle $\alpha = 7°$
  - side cutting edge angle $\Psi = 0°$
  - inclination angle $\lambda = 0°$

- parameters:
  - depth of cut: 1,5 mm
  - feeds: 0,14 mm/rev; 0,28 mm/rev
  - speeds: 5,5 m/s; 7,8 m/s

At the end of each cut, flank wear was evaluated by acquiring the image of the primary edge employing a processing image system as shown in figure 2. It is composed by a CCD camera (Sony AVC-D50CE monochrome, high resolution) connected with a PC, properly equipped to store and process the acquired images.

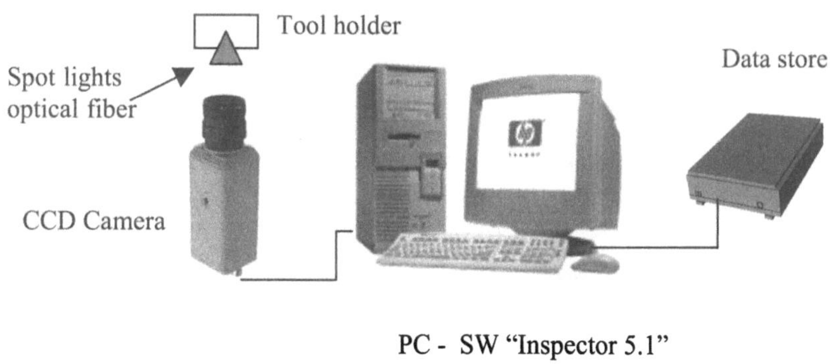

FIGURE 2. The image processing system employed for the tool wear evaluation.

On the acquired image, the value of flank wear was measured employing the commercial software "Inspector 5.1" by Matrox. This measure was performed ad fixed time interval until the value of $VB_B = 0,2$ mm was reached, for each test.

2.2 EXPERIMENTAL RESULTS

For all the couples of cutting parameters above mentioned, the flank wear trends are shown in figure 3 for the three kind of tools employed, called A, B, C.

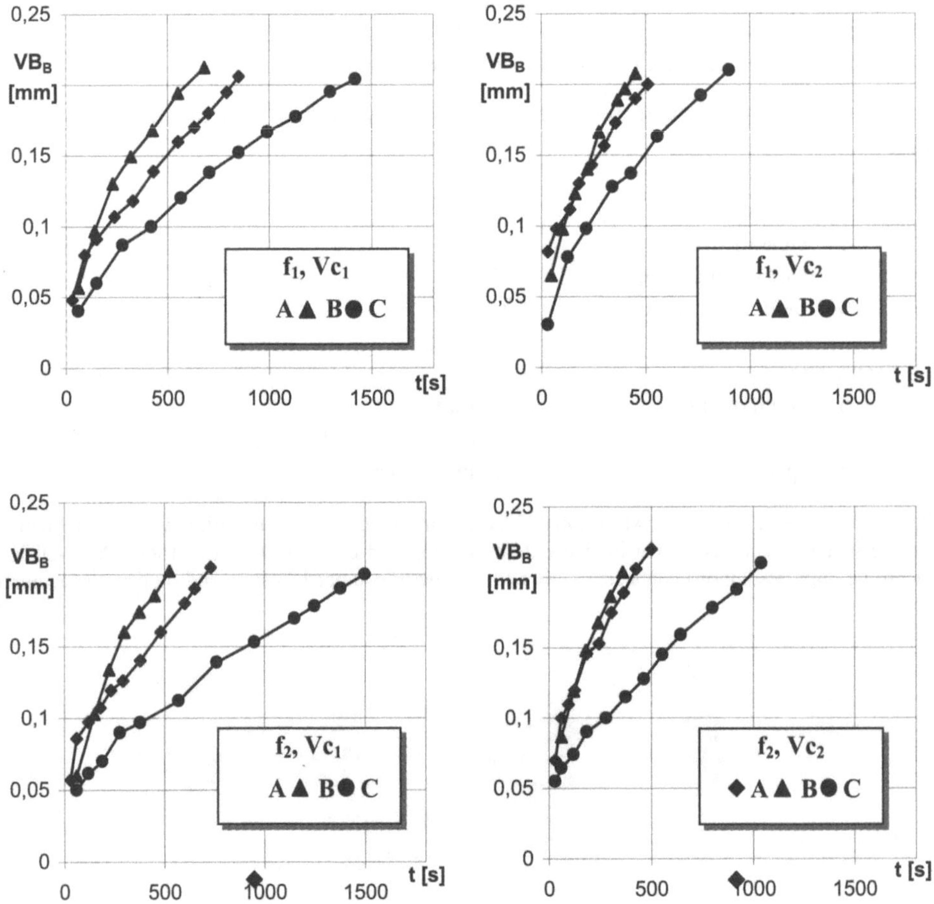

FIGURE 3. Flank wear trends.

Among the wear criteria, only the value of VB_B has been taken into account, because the crater wear was negligible.

Since the workpiece material during the cut have a tendency to weld on the tool tip, after the cut, it was necessary to perform a slight mechanical removal to clear the primary cutting edge allowing the correct evaluation of the flank wear. The flank wear along the cutting edges was regular as shown in figure 4.

FIGURE 4. The SEM image of a PCD tool – X100.

The lifetime is evaluated on the base of a flank wear limit VB_B = 0,2 mm. In table 1, are summarized their values, for all the above-mentioned tools and cutting parameters.

TABLE 1. Tool life

Tool	A		B		C	
V_C f	0,14 [mm/re	0,28 [mm/re	0,14 [mm/re	0,28 [mm/re	0,14 [mm/re	0,28 [mm/re
5,5 [m/s]	850	730	660	525	1420	1500
7,8 [m/s]	510	490	450	360	890	1000

At the end of each cut also the roughness of the workpiece was measured on six longitudinal profiles in different radial positions, by means of a profilometer Handysurf E-35A by Zeiss. In the following table 2 the values of workpiece roughness at the end of the tool life are summarized.

TABLE 2. Workpiece roughness [μm]

Tool	A		B		C	
V_C f	0,14 [mm/re	0,28 [mm/re	0,14 [mm/re	0,28 [mm/re	0,14 [mm/re	0,28 [mm/re
5,5 [m/s]	2,2	5,7	1,15	3,46	1,62	4,12
7,8 [m/s]	2	4,8	1,14	2	1,7	3,6

In the figure 5, the trend of the Ra parameter is reported for the tools employed.

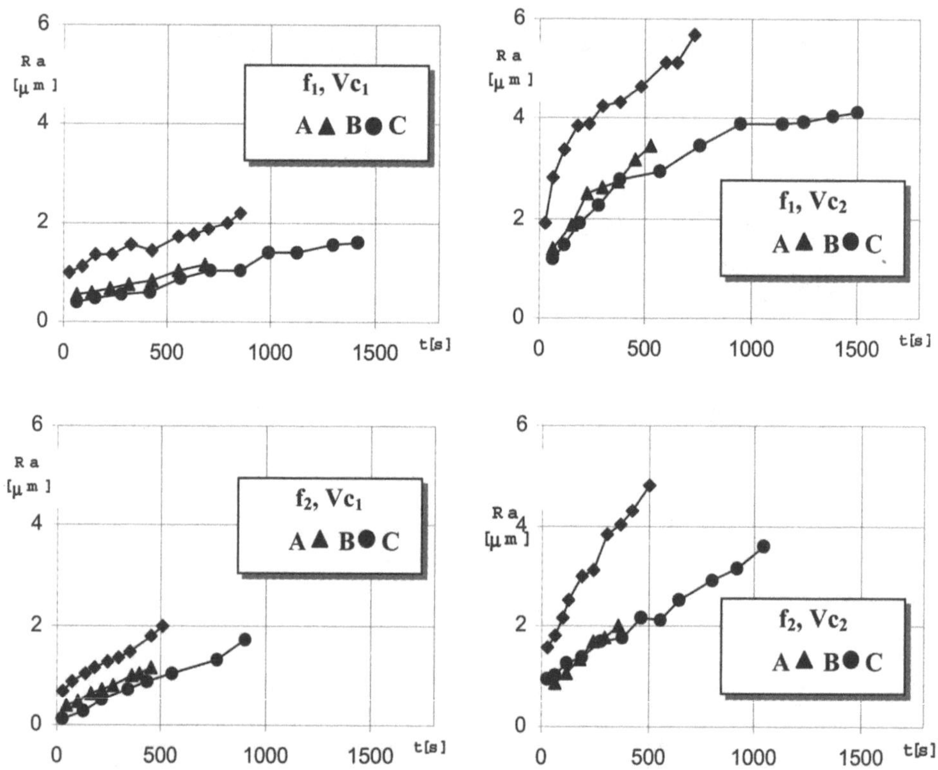

FIGURE 5. Workpiece roughness trends.

3 DISCUSSION OF THE RESULTS

During the cutting of MMC the main responsible of the wear is the direct contact between the reinforce particles inside the workpiece material and the primary cutting edge and the flank, that causes several grooves along the flank wear, as shown, in figure 4 [7].

This kind of abrasion require the correct choice of the cutting parameters in order to avoid the removal of the coating thin from the tool [8], caused by the strong shear actions that are generated between the coating and the substrate of the tool during the cut.

For the employed cutting parameters, examining the tool life data reported in table 1, and the workpiece roughness data, in table2, we can draw that the best performance are obtained by the PCD tools (C).

The flank wear trends show that, at increasing the cutting speed, the tool life for all the examined tools decreases, whereas at increasing the feed, only PCDs tool life tends to increase [9,10].

This behavior can be explained by supposing the following dynamics: at the same cutting speed, at increasing the feed, both the mechanical stress and the abrasive effect of the alumina particles grow, since a greater quantity of material is cut; on the converse, from the observation of figure 6 a, b, where the SEM image of the chips for the two feeds adopted is reported, we can draw that the opposite effect is generated from alumina particles at increasing the feed. In particular, in figure 6a, we can observe that the reinforce particles, during the cut at lower feed, tend to thicken along well-defined directions generating a very strong abrasion on the primary edge and flank. For higher feed rate, figure 6b, the reinforce particles are more uniformly distributed inside the matrix material, thus reducing the abrasion.

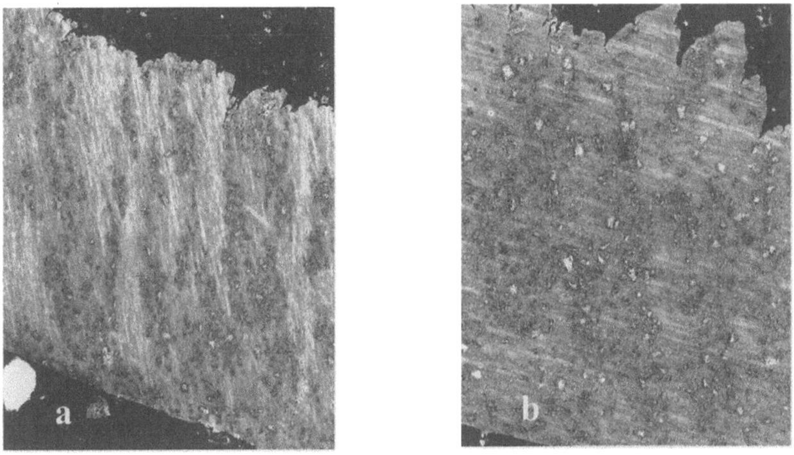

FIGURE 6. SEM image of MMCs chip: a) f = 0,14 mm/rev, b) f = 0,28 mm/rev.

The effect of such distribution is than to increase the lifetime of the PCD tools at increasing the feed even if the quantity of material removed is bigger at higher feed. The minor wear on the PCD insert can be attributable to the better abrasion resistance of such material [11]. For the other tool materials (A e B), their reduced abrasion resistance involve that, the increase of removed material, for higher feed, results in a shorter lifetime, even if the particles distribution is more uniform.

Moreover, the above reported results (see figure 5 and table 2) show that the finishing obtained with the A inserts is worst, for the same cutting parameter, than the one obtained by employing the CBN and PCD tools. Nevertheless it's important to highlight that the difference among the values of Ra reached at the end of the inserts tool life is always less than 25-30%.

4 CONCLUSION

On the basis of the obtained results we can draw that:

- The MMCs can be machined not only with PCD tools but also with coated Cermet ones.
- For the investigated cutting parameters the lifetime of the Cermet tools is about half the PCDs one.
- The CBN inserts doesn't show to be competitive with the other tools examined both as regard the lifetime and the surface quality.
- The most suitable tools appear to be the PCDs but, taking into account the costs of the tools, we suggest that it's profitable from the economic point of view to employ the Cermet tools (A), because these last have shown lifetime of about 50% of PCDs one, when the edge cost can be evaluated to be about 1/50.
- The use of PCDs can be limited for the cases when very low values of Ra are required.

REFERENCES

1. Mattheus, F.L., Rawlings, R.D., Composite materials: engineering and science — ed. Chapman & Hall.
2. Schwartz, Mel M., Composite materials properties, non-destructive testing and repair -polymers, ceramics, metal matrices.
3. Yoshikawa, H., Nishiyama, A., (1999), CVD diamond coated insert for machining high silicon aluminium alloys. Diamond and Related Materials 8, 1527-1530.
4. Chambers, A. R., (1995), The machinability of light alloy MMCs, Composites: Part A 27A, 143-147.
5. Tomac, N., Tonnesen K., (1992), Machinability of Particulate Aluminium Matrix Composites. Annals of the CIRP. Vol 41/1/92. 55-58.
6. Chandraserakaran, H., Johansson, J. O., (1997), Influence of Processing conditions and reinforcement on the surface quality of finish machined aluminium alloy matrix composites. Annals of the CIRP Vol. 46/1/1997. 493-496.
7. Polini, R., D'Antonio, P., Lo Casto, S., Ruisi, V. F., Traversa, E., (2000), Cutting performance and indentation behaviour of diamond films on Co-cemented tungsten carbide. Surface & Coatings Technology 123, 78-83.
8. Iuliano, L., Settineri, L., Gatto, A., (1998), High-speed turning experiments on metal matrix composites. Composites Part A 29A, 1501-1509.
9. Cronjiager, L., Meister, D., (1991), Drilling of fibre and particle reinforced aluminum. Composite Material Technology. ASME 1991, Vol. 37, 185-189.
10. Monaghan, J., O'Reilly, P., (1992), The drilling of an Al/SiC metal-matrix composite. Journal of Materials Processing Technology, Vol. 33. 469-480.
11. Weinert, K., (1993), A consideration of tool wear mechanism when machining metal matrix composites (MMC). Annals of the CIRP Vol.42/1/1993.

APPLICATION OF A ROTATING DYNAMOMETER FOR CUTTING FORCE MEASUREMENT IN MILLING

E. Kuljanic, M. Sortino, F. Miani

Department of Electrical, Management and Mechanical Engineering, University of Udine, Italy

KEYWORDS: Milling, Cutting Forces, Sensors.

ABSTRACT. The measurement of the cutting force is important from the theoretical and practical point of view. An investigation of cutting forces and torque in milling with a rotating dynamometer is described. Rotating dynamometer is more practical for milling, drilling and taping than the plate dynamometer, since it is mounted on the spindle of the machine tool independently of the size of the workpiece. The experimental results have been compared to estimated instantaneous forces computed by using Kronenberg's approach. The measurement system has been tested. Also, the dynamic and static characteristics of the rotating dynamometer are presented.

INTRODUCTION

The knowledge of cutting forces gives an important information which can be used both for theoretical and practical purposes [1]. Cutting forces are strongly correlated to chip formation mechanism, therefore, the measure of the cutting force can help understand the basic principles of metal cutting. In the past, the measurement of cutting forces allowed the researchers to develop the first mathematical models of cutting processes [2].

The measurement of cutting forces allows to investigate the influence of tool geometry on cutting process thus enabling the producer to develop better tools. Also, the cutting force data are indispensable in dimensioning parts and subsystems of the machine tool, and in optimizing the machining processes.

The dynamometers are instruments able to measure the force. Different dynamometers have been used in metal cutting to measure the cutting forces. More than hundred years ago the specific cutting force k_s N/mm^2 as a ratio of the main cutting force and F_c and uncut chip cross-section area A had been introduced.

$$k_s = \frac{F_c}{A} \tag{1}$$

Specific cutting force data are given for different workpiece material and different shapes of the uncup chip cross-section area in order to be able to determine the main cutting force without making experiments.

It has to be pointed out that the dynamometer has to be enough sensitive to register small force variations and to be stiff enough to avoid larger deflections and vibrations. It is difficult to make a good dynamometer, also, with or without small cross-sensitivity [3].

Published in: E.Kuljanic (Ed.) *Advanced Manufacturing Systems and Technology*,
CISM Courses and Lectures No. 437, Springer Wien New York, 2002.

In the latest years, the interest of dynamometers has moved to their application in the machining systems and tool condition monitoring system as machining sensors [4]. In particular, the application of dynamometers together with other sensors for tool condition monitoring seems to be very promising.

Several kinds of dynamometers can be found on the market, such as plates, rings, strain gages. A rule of thumb for cutting force measurement is to put the sensitive element very near to the tool tip because the interposed parts interfere with the measure. This statement explains why reliable measurement of forces, during milling, is difficult. Recently, special dynamometers for milling, boring and tapping have been developed. These dynamometers can be mounted directly on the spindle of the machine, i.e. very near to the tool tip.

The aim of this work is to investigate the application of a rotating dynamometer for cutting force measurement in milling as well as to investigate the characteristics of the rotating dynamometer.

1 DYNAMOMETERS

Basically, by using strain gages or piezoelectric sensors, it is possible to measure the forces. Strain gages must be attached to the parts where the highest deformation occurs in order to obtain maximum sensitivity. Usually, deformations are very low (10^{-3}-10^{-5} mm), thus the sensor sensitivity is highly dependant on the amplifiers used to condition the signal.

Piezoelectric sensors must be subjected to the forces requested to measure, hence they are inserted in the structure. The sensitive element releases electrical charges during deformation. By integrating the released electrical charges in time domain, the stress can be computed. Usually, the piezoelectric sensitive elements are inserted between pre-loaded plates in order to increase stiffness of the sensor and for linearity between deformation and output signal.

At present, two different kinds of dynamometers can be found on market: plate dynamometers and rotating dynamometers.

An example of plate and rotating dynamometers can be seen respectively in Figure 1 and Figure 2.

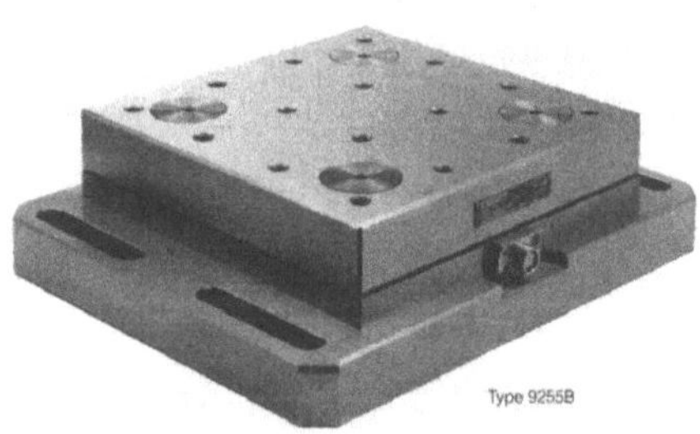

FIGURE 1. Plate dynamometer - Kistler

FIGURE 2. Rotating dynamometer - Kistler

Plate dynamometers are placed between the work piece and the working table of the machine tool. Both dynamometers can be used for tool condition monitoring in milling, drilling and threading.

Figure 3, shows different reference systems that are adopted by measuring forces with plate and rotating dynamometers: F_c and F_{cN} are the forces in the dynamometer reference system, F_f and F_{fN} are the forces in the fixed reference system. The axial force F_a is the same in both reference systems. Obviously, the dynamometer reference system rotates together with the dynamometer because the sensing elements are fixed on the dynamometer.

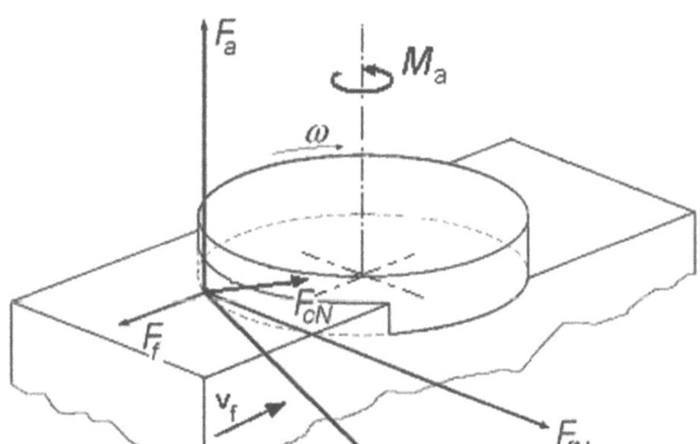

FIGURE 3. Rotating (F_c, F_{cN}, F_a) and fixed (F_f, F_{fN}, F_a) reference system

As it can be seen in Figure 3, the forces measured in the rotating reference system are always orthogonal to the cutting speed and they are very similar to theoretical forces in orthogonal cutting. Each reference system can be switched to the other one by using roto-translation algebra.

The forces measured in the fixed reference system could be affected by the position of the tool, which is not the case when the rotating dynamometer is applied.

The rotating dynamometer is a complex device composed of two parts the rotor, clamped in the spindle, and the stator. The rotor must communicate to the stator without contact by using telemetry. Therefore, the rotor must be coupled to the stator placed very close to the rotating tool-holder and fixed to the spindle housing. The stator transmits the power by mean of electromagnetic waves.

The rotating dynamometer contains the sensitive elements, charge amplifiers, digital signal converters for every channel and the telemetry electronics. It is very difficult to integrate all these instrumentation in a very narrow place such as the rotor itself. Also, the rotor must be enough stiff and light. To decrease the system complexity, the telemetry has a limited bandwidth, which limits the amount of information carried by every single channel and a number of channels.

Recently, a special tool-holder dynamometer for high-speed milling monitoring has been developed in the Department PTW in Darmstadt [5]. This device has enhanced bandwidth and is capable to measure forces and accelerations in milling. It is suitable for high-speed dies and moulds milling.

2 MILLING DYNAMOMETER SET-UP

As stated before, the dynamic and static characteristics of the dynamometer depend on the mechanical parts interposed between the tool tip and the sensing element.

Figure 4a) shows a very simple model of the rotating dynamometer. The rotating dynamometer can be modeled as a mass element m (the tool), connected by a spring element k and a damping element c, in the three main directions x, y and z to the dynamometer. In the z direction the system is very stiff. For the sake of simplicity, let consider the system independent of the x and y direction. The model in the x direction is given in Figure 4b).

As it can be seen, the model found in Figure 4b, the forces measured by the dynamometer are the structural reactions of the spindle to tool vibrations and forces applied to the tip.

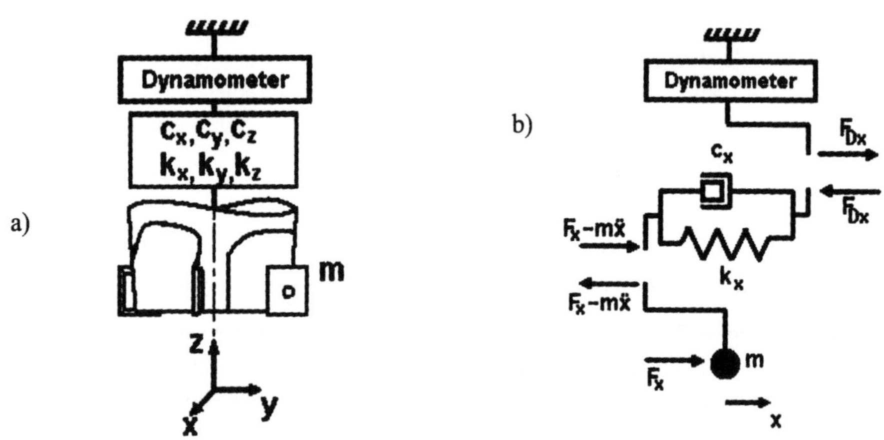

FIGURE 4. Spring-damping-mass model of tool holder dynamometer: a) 3 dimensional; b) one direction

From a mathematical point of view, dynamics of the tool can be written as follow:

$$\begin{cases} m\ddot{x}(t) = F_x(t) - k_x x(t) - c_x \dot{x}(t) \\ F_{Dx}(t) = k_x x(t) + c_x \dot{x}(t) \end{cases} \tag{2}$$

In frequency domain:

$$\begin{cases} \dfrac{x(s)}{F_x(s)} = \dfrac{1}{ms^2 + c_x s + k_x} \\ F_{Dx}(s) = (c_x s + k_x) x(s) \end{cases} \tag{3}$$

The frequency transfer functions for forces becomes:

$$\frac{F_{Dx}(s)}{F_x(s)} = \frac{(c_x s + k_x)}{ms^2 + c_x s + k_x} \tag{4}$$

The characteristic of the transfer function is neither constant nor linear in the frequency domain, thus requiring compensation. Indeed, the transfer function of the tool-dynamometer system is more complicated compared to this simple model. It is clear that the frequency characteristic of the tool-dynamometer system must be identified before using the dynamometer. To determine off-line the dynamic and static characteristic of the tool-dynamometer system, the pulse test can be applied [6]. The pulse test consists in hammering several times the tool tip and comparing acquired force signal from dynamometer to hammer-induced force signal. The frequency-transfer function of the dynamometer used in this study is given in Figure 5a. The theoretical transfer function plot of the mono-directional mass-clamping-mass system is shown in Figure 4b (see Figure 3). There is a good concordance between the experimental and the theoretical transfer function of the measuring system.

By using the transfer function, the system can be identified. It is possible to build a signal conditioner, which compensates the effects on measured force. However, it is not possible to obtain the original signal by inverse filtering due to the low-pass effect that eliminates higher frequencies. Moreover, the system contains some retarding elements that prevent compensation [7], thus only an approximated reconstruction is possible. In Figure 6, the difference between theoretical cutting force (a) and measured force by the dynamometer (b) is shown. It can be seen that the measured cutting force signal contains elements of the transfer function of the clamping system.

The identification of the system can also be accomplished during machining by comparing recursively the measured cutting force and the theoretical cutting force [8].

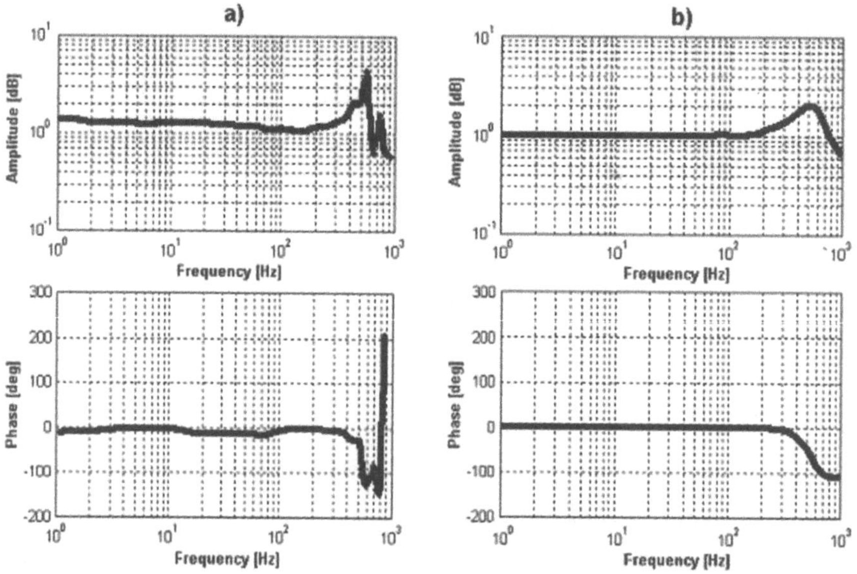

FIGURE 5. a) Experimental transfer function of the dynamometer and b) theoretical transfer function of the dynamometer

3 CUTTING FORCES IN MILLING

Face milling tests have been carried out in order to compare the relationship between theoretical cutting forces obtained by using M.Kronenberg's model and the cutting forces obtained experimentally. M.Kronenberg's equation [1] used in this research is as follows:

$$F_c = k_{s0} a_p h_x^{1-k_{hx}} \tag{5}$$

Where F_c is the cutting force, N, k_s is the specific cutting force (1), a_p is the depth of cut, mm, h_x is the uncut chip thickness, mm, and k_{hx} is the corresponding exponent of h_x.

The tests were performed on an Okuma MA 550-VB milling machine. The mill was a Sandvick-Coromant Coromill R290-050A32-15M, diameter 50 mm, inserts R290.90-1504M-WM 4040, on C60 workpiece for a face milling operation where a_p = 0.8 mm, a_e = 25 mm, f_z = 0.1 mm/tooth, z=2.

There is a significant similarity between the experimental cutting force curve and the theoretical cutting force curve. This proves the applicability of the M.Kronenberg's force model for turning in milling. The effect of the entrance of the cutter into the workpiece is more evident in the torque signal.

Figure 6 shows the experimental cutting force after inverse filtering where the theoretical cutting force. Figure 7 shows the experimental torque and the theoretical torque.

Both the eccentricity of the cutter and the eccentricity due to incorrect mounting of the insert have had a strong effect on cutting force, torque, vibrations and tool wear.

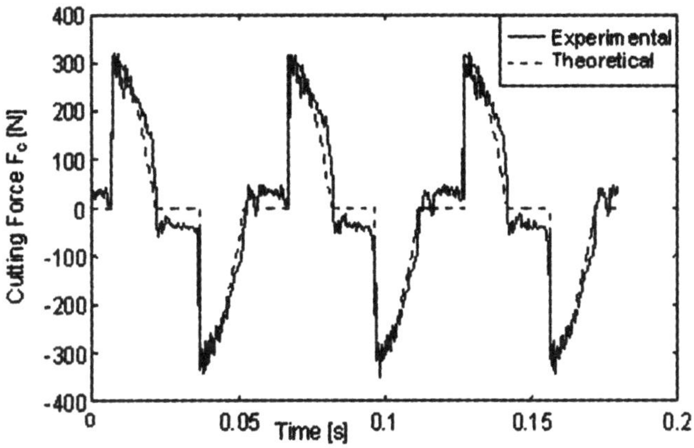

FIGURE 6. Acquired and theoretical cutting force

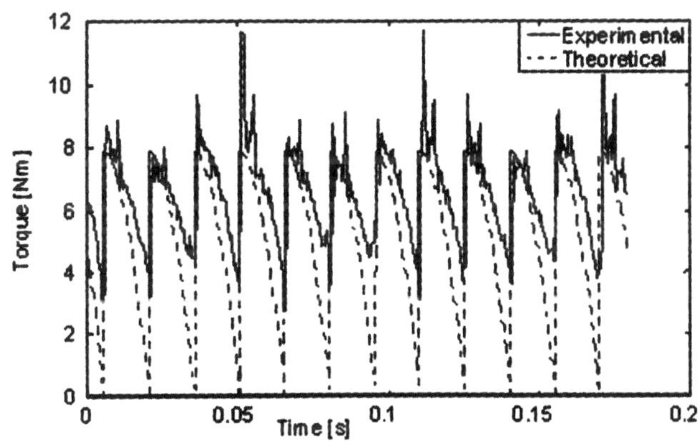

FIGURE 7. Acquired and theoretical torque

4 CONCLUSION

After many years of research in cutting force, using plate dynamometers in milling and drilling, there are available rotating dynamometers which are mounted on the spindle of the machine tool independently of the size of the workpiece. In accordance with considerations presented in this paper, we may draw the following conclusions.

The mathematical model (2), (3) and (4) of the dynamic characteristic of the measurement system is fitting well the experimental behavior of the system.

It has been proved that the tested rotating dynamometer gives satisfactory results of the cutting force and torque in milling.

There is a significant similarity between the experimental cutting force curve and the theoretical cutting force curve when applying M.Kronenberg equation. This is a proof of the applicability of M.Kronenberg's force equation in milling.

The rotating dynamometer can be used for tool condition monitoring and/or for process condition monitoring, and in industrial conditions.

It will be of further interest to compare the results obtained with a rotating dynamometer and the plate dynamometer, and to investigate the application of the rotating dynamometer for tool condition monitoring and process condition monitoring.

5 ACKNOWLEDGEMENT

The authors would like to thank Professor E.Schulz, Professor E.Abele, U.Fiedler, A.Versch and the staff of the PTW Institute of the Darmstadt Technische Universität (Germany) for their contribution to this work.

REFERENCES

1. G.F.Micheletti, (1977), Tecnologia Meccanica, UTET
2. M.C.Shaw, (1984), Metal Cutting Principles, Oxford Science Publications
3. E.Kuljanic, (1972), Study of Wear in Single-Tooth and Multi-Tooth Milling, Doctor of Philosophy Disertation, University of Cincinnati, Ohio, Congress Library, USA
4. G.Byrne, D.Dornfeld, I.Inasaki, G.Ketteler, W.Koenig, R.Teti, (1995), Tool Condition Monitoring (TCM) – Status of Research and Industrial Application, Annals of the CIRP, Vol. 44/2/1995, 541-567
5. H.Schulz, A.Versch, U.Fiedler, (2001), Process Monitoring with Mechatronic Tool Holders, Production Engineering, Vol. VIII/2, 115-118
6. I.E.Morse, W.R.Shapton, D.L.Brown, E.Kuljanic, (1972), Application of Pulse Testing for Determining Dynamic Characteristic of Machine Tools, 13th International Machine Tool Design and Research Conference, Birmingham, 1-13
7. H.Schulz, T.Herget, (1994), Simulation and Measurement of Transient Cutting Force Signal in High-Speed Milling, Production Engineering, Vol I/2, 19-22
8. J.Altintas, (2000), Manufacturing Automation, Cambridge University Press

SIMULATION OF CHIP FORMATION IN AN ORTHOGONAL CUTTING PROCESS USING FEM

G. Giorleo, R. Teti, A. Langella, D. D'Addona, U. Prisco

Department of Materials and Production Engineering, University of Naples Federico II, Italy

KEYWORDS: Orthogonal cutting; finite element method simulation; cutting modeling.

ABSTRACT - The objectives of this work are to develop a plain strain simulation model of orthogonal cutting with predetermined process conditions and to analyze the orthogonal cutting events with respect to plastic strain, equivalent stress and applied external force, with the goal to insure, as far as possible, the quality of the simulations. This work is based on a large deformation-large strain finite elements theory using an incremental approach in a 2D elastic-plastic model of an adiabatic steady state cutting, using the Marc AutoForge software package. Please, send any comments or questions by e-mail to: tetiro@unina.it.

1 INTRODUCTION

Cutting is one of the manufacturing processes more frequently used to produce parts of desired dimension and shape through the removal of unwanted material, in the form of chips, with a wedge-shaped tool. Due to the workmaterial deformation with large plastic strain, friction, high temperature, high strain rate, and their coupling effects, metal cutting is a highly non linear and coupled thermomechanical process [1]. An understanding of the physical phenomena involved in metal cutting is essential in order to predict cutting forces and temperatures, to select the tool material, to size machines and to choose the cutting parameters (cutting velocity, rake angle, feed velocity, etc.) for the economical operation of the process or the safety and performance of the machined product. The earliest analytical models of orthogonal cutting were proposed by Merchant [2], Lee and Shaffer [3] and Palmer and Oxley [4]. However, these analytical models show a significant divergence with experimental results due to the simplifying assumptions about the friction at the tool-chip interface, work hardening, strain rate and temperature dependencies of the work material properties. To take into account more realistically such process variables, the application of non-linear finite element techniques to the simulation of metal cutting becomes necessary. Over the last two decades, two kinds of mechanical formulation have been used for the finite element modeling of metal cutting. The first is the Eulerian-based approach [5, 6] in which the grid is not attached to the material and the cutting is simulated during the steady state, avoiding the need for a chip-separation criterion but requiring that the boundaries of the chip free surface must be known in advance. The second is the Lagrangian approach [5, 7, 8] in which the grid is attached to the workpiece and the cutting can be simulated from the incipient to the steady state, allowing for the prediction of chip geometry and residual stresses in the workpiece. This approach requires longer computational times to reach the steady state conditions, a remeshing algorithm to update the mesh, and a chip separation criterion to allow for the chip to separate from the workpiece [9]. Many separation criteria have been proposed in the last years and all of them are based on a certain characteristic quantity. When this quantity achieves a threshold value, the separation of the chip from the workpiece occurs. In the so-called geometric separation criteria [10], when the

Published in: E.Kuljanic (Ed.) *Advanced Manufacturing Systems and Technology*, CISM Courses and Lectures No. 437, Springer Wien New York, 2002.

tool tip approaches a node within a small critical distance, that node separates from the workpiece and becomes part of the chip. The physical separation criteria are based on the comparison of a "failure" variable with a critical value. The most common failure variables are effective plastic strain, strain energy density, and stress (normal, shear, von Mises equivalent stress) [8, 9, 11]. The difficulty to be overcome in the utilization of physical criteria is the determination of the critical value of the failure variables. The material flow behavior at high temperature, strain rate and strain encountered in the cutting process cannot be provided by the normal σ-ε test or impact tests. Thus, extrapolations are necessary.
In this paper, an original numerical model of continuous chip formation was developed using the Marc AutoForge FEM software package for AISI 304 stainless steel. The cutting process is carried out by a rigid tool and the fracture propagation model is realized by a stress separation criterion. The obtained results show the influence of the rake angle and friction coefficient upon the cutting force, chip geometry and the stress distribution in the chip and the workpiece.

2 FINITE ELEMENT MODEL

2.1 WORKPIECE AND TOOL MATERIAL MODELLING

WORKPIECE MATERIAL

The work material used for the plane-strain orthogonal cutting simulation was an AISI 304 stainless steel (0.08% C, 18-20% Cr, 8-10.5% Ni). This material is coded in the FEM Marc AutoForge as X5CrNi18_9. The physical and elastic properties of this steel at 25 °C are in Table 1.

TABLE 1. Physical and elastic properties of the X5CrNi18_9 steel

Mass Density	7850 (kg / m^{3})
Melting Temperature	1790 °C
Specific Heat	500 (J/kg °C)
Thermal Conductivity	16.2 (W/m-K)
Young's modulus	193 (GPA)
Poisson's ratio	0.28
Hardness	88 (HRB)

CONSTITUTIVE EQUATION

During cutting, the material undergoes high deformation ($\varepsilon >> 1$), high strain rate ($\dot{\varepsilon} = 10^4 \div 10^6$ s^{-1}) and high temperature (T > 300 °C). To account for such effects on the material properties, a constitutive equation describing the flow stress as a function of strain, strain rate and temperature is needed. In the FEM simulation, the material was modeled as isotropic elastic-plastic with isotropic strain-hardening through the Johnson-Cook formulation, eq. (1), where A, B, C, m and n are material parameters (Table 2), $\dot{\varepsilon}_0$ the equivalent strain rate at which the material starts to be sensitive to strain rate, T_{room} the room temperature and T_{melt} the melting temperature. The simulations were conducted at constant temperature $T = T_{room} = 25$ °C, which is a reasonable hypothesis in the case of a large flow rate of cooling fluid, so the thermal softening factor in eq. (1) is zero.

$$\sigma_{eq} = (A + B\varepsilon_{eq}^{n})\left[1 + C\ln\left(\frac{\dot{\varepsilon}_{eq}}{\dot{\varepsilon}_{0}}\right)\right]\left[1 + \left(\frac{T - T_{room}}{T_{melt} - T_{room}}\right)^{m}\right] \tag{1}$$

TABLE 2. Johnson-Cook parameters

Material	A (MPa	B (MPa)	n	C	M
X5CrNi18_9	154	198	0.0541	0.0035	0.207

TOOL

The tool is modeled as tungsten carbide and is considered a perfect rigid body.

2.2 FRICTION CONDITIONS

A critical aspect of cutting process numeric modeling is the characterization of contact and friction conditions at the tool/chip and tool/ workpiece interfaces. Modeling of these conditions has a fundamental role in the simulation since it directly ensures the agreement of the numerical analysis with the physical reality of the interaction between workpiece, tool and chip. Experimental observations revealed that the tool/chip interface can be divided into a sticking and a sliding region and the force distribution on the tool rake face cannot be considered linear [1]. If the total force is decomposed into two components, the component normal to the rake face varies along the rake face in an almost linear way. The tangential component evolves in a non linear way within the sliding zone and reaches a value of saturation in sticking zone, close to the tool tip. Friction conditions modeling in metal cutting should account for both behaviors. For simplicity, it may be hypothesized [7, 8] that the whole face of the tool is characterized by a sliding zone and the Coulomb friction model with a constant friction coefficient μ can be used. In this work, the friction is assumed to follow a Coulomb law, with a constant friction coefficient μ.

2.3 NUMERICAL SIMULATION

NUMERICAL MODEL AND LAGRANGIAN FORMULATION

The FEM used for the plane-strain metal cutting simulation is based on the Lagrangian formulation provided by the Marc AutoForge code. As is well known [1], the plane-strain assumption is a reasonable approximation of the orthogonal cutting process if the width of cut is at least five times greater than the uncut chip thickness. The elements used for the FEM simulations are four node quadrilateral elements (QUAD). The workpiece was modeled as composed of two bodies attached along a node line; these two bodies will separate as the tool advances according to a predetermined separation criterion (Figure 1). The upper part of the mesh represents the undeformed chip layer and the lower part the rest of the workpiece. The vertical and horizontal node displacements at the lower and the right workpiece boundaries are set equal to zero (boundary conditions). The chip layer consists of 80 QUAD elements with 84 nodes; its length and thickness are 50 mm and 0.4 mm. The rest of the workpiece has 160 QUAD elements with 189 nodes and thickness 20 mm. The tool moves from left to right with a constant cutting velocity of 150 mm/s.

CHIP SEPARATION CRITERION

The chip separation criterion is based on the node separation method [8, 10]. In this model, cutting is supposed to take place along the line dividing the undeformed chip thickness from the rest of the workpiece. Therefore, separation is assumed to occur only at the nodes lying on this line. The chip layer and the rest of the workpiece are initially in contact node to node along a line, but the two nodal lines (one on the chip layer and one on the rest of the workpiece) are distinct. When the equivalent von Mises stress reaches a threshold value, the separation of chip from workpiece occurs, eq. (2). The threshold value is set equal to the yield stress of the work material: 500 MPa.

$$\left[(\sigma_x - \sigma_y)^2 + \sigma_x^2 + \sigma_y^2 + 6\tau_{xy}^2\right]^{0.5} / \sqrt{2} = 500 \text{ MPa} \qquad (2)$$

3 RESULTS AND DISCUSSION

The first phase of chip formation (Figure 2) is characterized by a high compression at the tool /chip interface. The chip deforms progressively against the rake face and takes on a free shape. The distribution of the equivalent plastic strain rate is shown in Figure 2 where the primary and secondary deformation zones are also visible. The focus of this study is on the dependence of chip form and cutting force on the tool rake angle and the friction conditions at the tool/chip interface.

3.1 INFLUENCE OF THE TOOL RAKE ANGLE ON CHIP FORM

In this section, the effects of the tool rake angle will be considered. A set of four cutting simulations were carried out with identical cutting parameters except for the rake angle that varied between 5° and 20° by steps of 5°. The depth of cut is 0.4 mm, the cutting velocity is 0.150 m/s and the friction coefficient at the tool/chip interface is 0.1. The results of the analyses are presented after a tool path of 1.295 mm. Figure 3 shows the chip geometry at the end of the four simulations. The cutting ratio, $r = t_0 / t_1$, is seen to increase with increasing rake angle, in agreement with the physical phenomenon. It can be noted that the contour line of the forming chip free side presents a fairly regular shape for high rake angles. The main reason for these phenomena is that for small rake angles (5°), the cutting force can be assimilated to a compression force. The effect of compression yields the formation of the barreled chip shape and the considerable increase of chip thickness after cutting. By increasing the rake angle, the compression decreases and the metal cutting proceeds in a less coercive manner. This effect is confirmed also by the trend of the horizontal cutting force obtained by the simulation for the four rake angle values (Figure 4).

3.2 INFLUENCE OF THE FRICTION COEFFICIENT ON CHIP FORM

In this section, the effects of the friction coefficient will be considered. In a set of four numeric simulations with tool rake angle 10°, cutting velocity 0.150 m/s and depth of cut 0.4 mm, the friction coefficient varied from 0 and 0.3 by steps of 0.1. The results of the analyses are presented after a tool path of 1.295 mm. For high friction coefficients, the chip remains in contact with the tool rake face for a longer length than in the cases of low friction coefficients (Figure 5). The friction Coulomb law states that the slip between two contact surfaces occurs only when $F_t/F_n > \mu$. Therefore, for high friction coefficients, the chip in contact with the tool rake face has difficulties to flow as easily as in the case of weak coefficients (Figure 5).

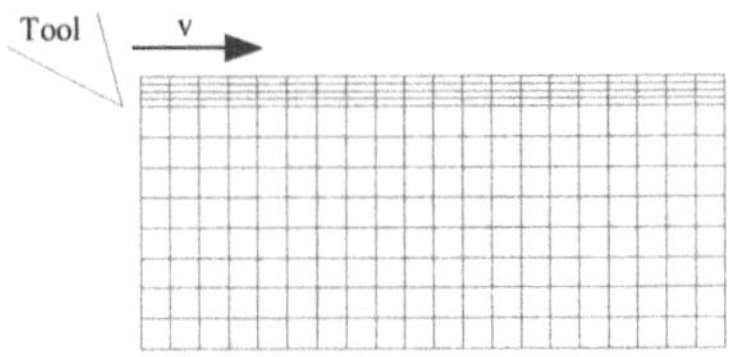

FIGURE 1. FEM mesh used in the orthogonal cutting simulations

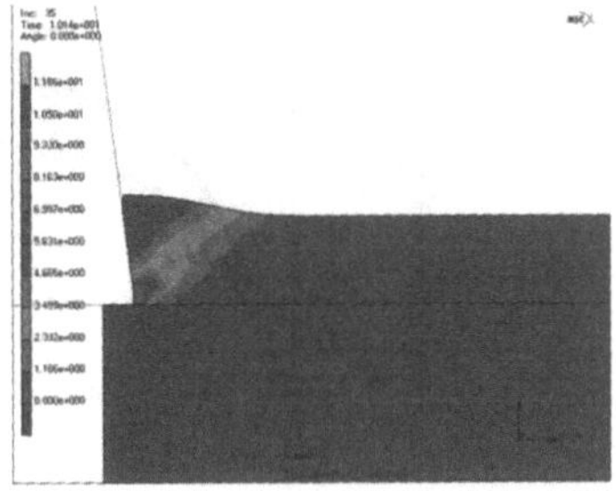

FIGURE 2. Equivalent plastic strain rate at incipient chip formation

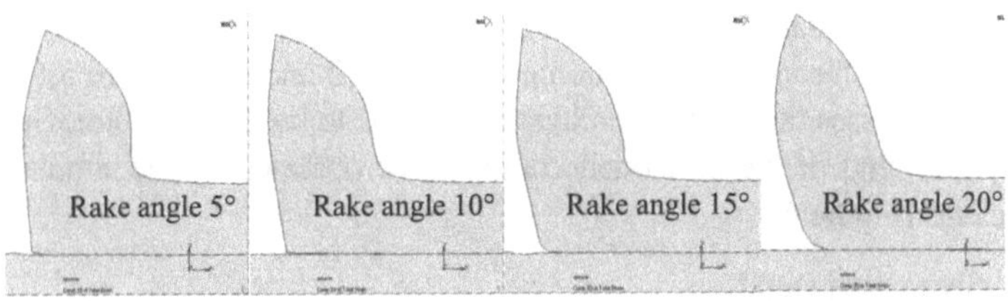

FIGURE 3. Chip geometry as a function of the rake angle

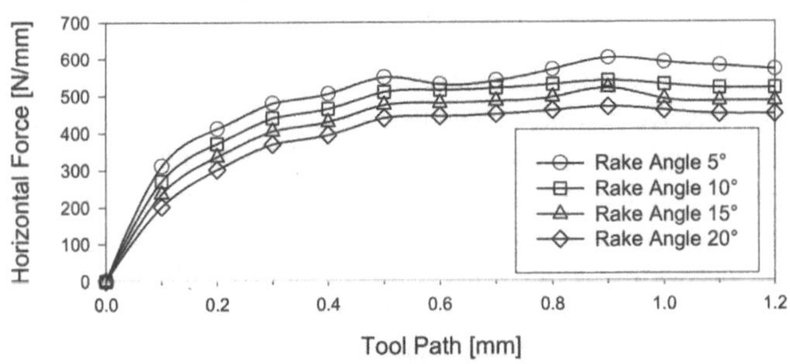

FIGURE 4. Horizontal force vs. tool path for the 4 rake angles

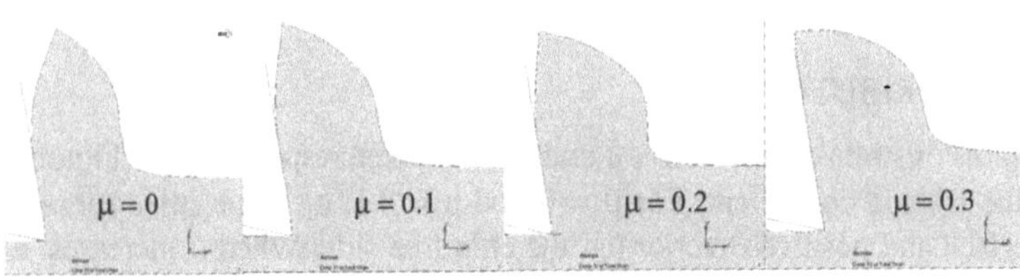

FIGURE 5. Chip geometry as function of friction coefficient μ

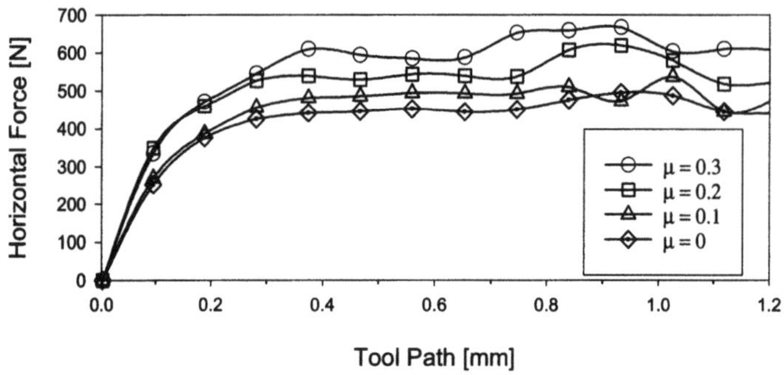

FIGURE 6. Horizontal cutting force vs. tool path for the 4 friction coefficients

Thus, the material accumulates and crams on the tool rake face. Very rounded shapes of the chip free sides can be noted for values of friction coefficient 0.2 and 0.3 (Figure 5); this is due to the effects of the high stresses that make the chip to accumulate on the rake face. Such effect is confirmed also by the trend of the horizontal cutting force obtained by the simulation for the four friction coefficients (Figure 6).

3.3 STRESS DISTRIBUTION

Numerical tests were carried out to study the influence of the friction coefficient on the stress distributions in the chip and work material. For all simulations, depth of cut was 0.4 mm, cutting velocity 150 mm/s, tool rake angle 10°. The results for σ_{11}, σ_{22} and the von Mises equivalent stress are reported in Figures 7, 8, 9, respectively, after a tool path of 1.5 mm.

Figure 7 shows that σ_{11} is negative (compression) in the chip and in the workpiece ahead of the tool tip, whereas it is positive (tensile) behind the tool tip and at the chip extremity. Figure 8 shows that σ_{22} is positive in a large zone ahead of the tool tip which during cutting is under traction. This is necessary for chip separation to occur in agreement with the separation criterion described in section 2.3. It can be observed that the σ_{22} tensile stress within the same zone decreases when the friction coefficient increases making chip separation harder. The plastic flow behavior can be examined from the von Mises equivalent stress distribution (Figure 9). The stress contours in the chip, ahead of the tool tip, are parallel and connect the tool tip with the turning point on the chip free side, thus identifying the shear angle. For $\mu = 0.3$, σ_{11} and σ_{22} ahead of the tool tip are very high at the surface to be manufactured. This can explain the observation of a very high Von Mises equivalent stress. Therefore, for this value of μ, the surface to be manufactured is subjected to higher stresses than for lower μ values.

3.4 STRAIN DISTRIBUTION

The contour plots of strain components ε_{11} and ε_{22} are given, respectively, in Figures 10 and 11 for 4 values of the friction coefficient. For $\mu = 0$ and $\mu = 0.1$, ε_{11} is negative (compression) at the tool/chip interface and positive (traction) at the chip free side. When μ increases, a zone defined by positive ε_{11} can be observed near the tool tip at the tool/chip interface. For ε_{22}, the observations are opposite to those for ε_{11}: for $\mu = 0$ and $\mu = 0.1$, ε_{22} is positive at the tool/chip interface and negative at the chip free side. When μ increases, a zone of compression ($\varepsilon_{22} < 0$) develops before the tool tip, at the tool/chip interface, determined by the "dry" contact (sticking zone) of the rake face and the chip, near the tool tip. The opposite behaviors observed in the chip for ε_{11} and ε_{22} can be interpreted in the following manner. The distortion in the chip is plastic and thus incompressible, which brings to the well-known relation $\varepsilon_{11} + \varepsilon_{22} + \varepsilon_{33} = 0$. The plane distortion condition implies $\varepsilon_{33} = 0$, and thus $\varepsilon_{11} = -\varepsilon_{22}$. The distribution of strain component ε_{12}, representative of the shearing, is shown in Figure 12 for 4 values of μ. The maximum shearing, naturally, takes place in the chip. The shearing in the workpiece is relatively weak in comparison. The existence of a large shearing gradient can also be noticed along the line that joins the tool tip with the turning point on the chip free side. This zone of strong gradient is the primary shearing zone. It can be noticed that, in each simulation, the gradient increases from right to left and, as μ increases from 0 to 0.3, the value of shearing becomes larger.

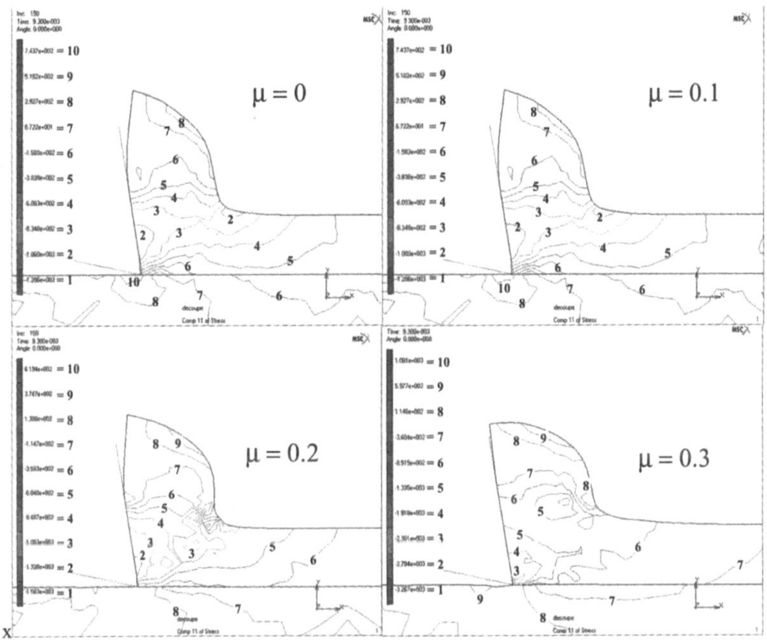

FIGURE 7. Distribution of σ_{11} for 4 different μ values; stresses are in MPa

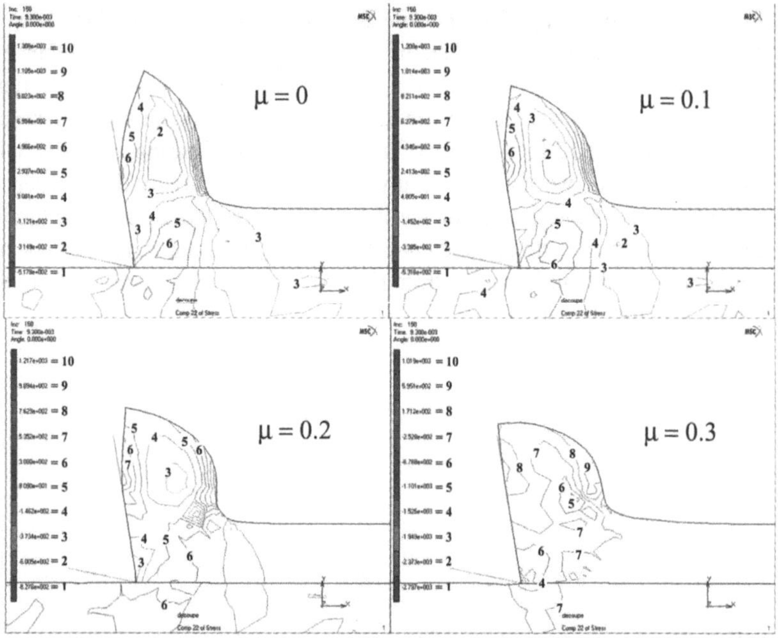

FIGURE 8. Distribution of σ_{22} for 4 different μ values; stresses are in MPa

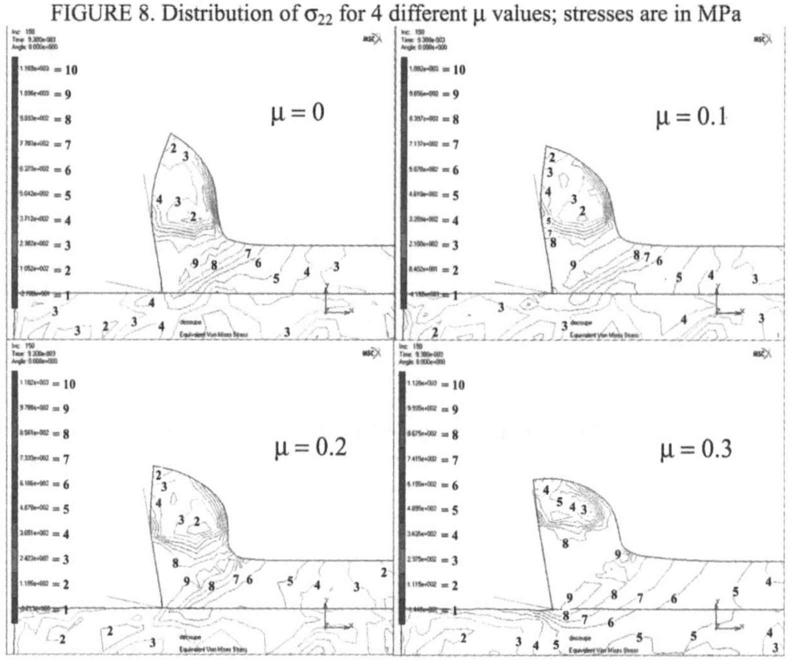

FIGURE 9. von Mises equivalent stress distribution; stresses are in MPa

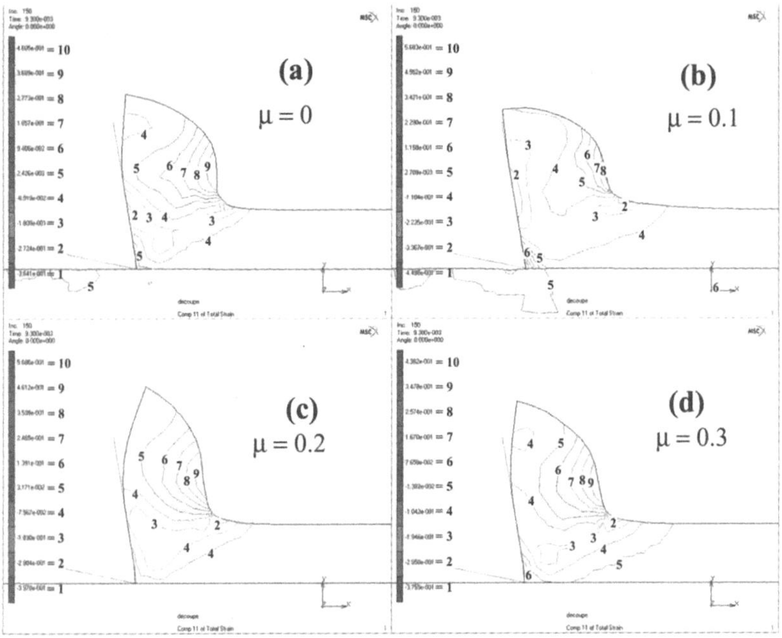

FIGURE 10. Distribution of ε_{11} for 4 different μ values

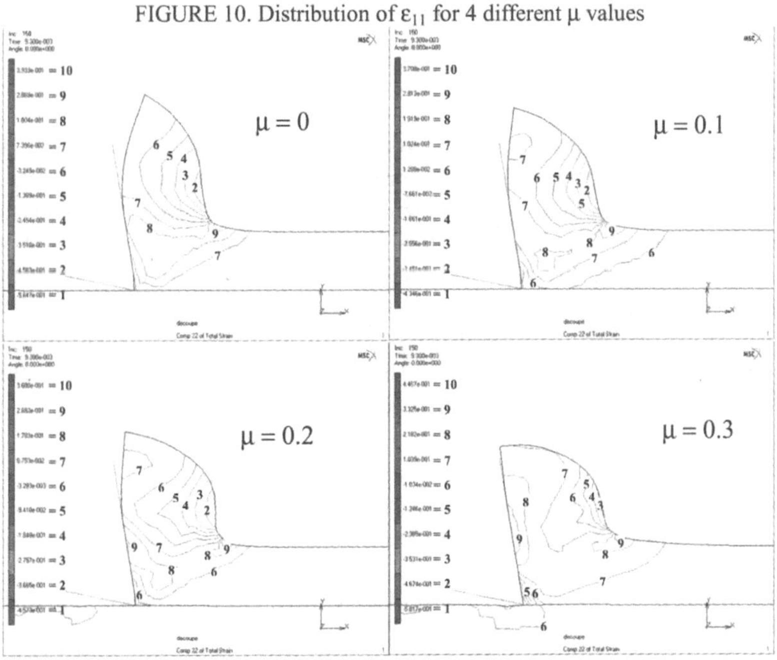

FIGURE 11. Distribution of ε_{22} for 4 different μ values

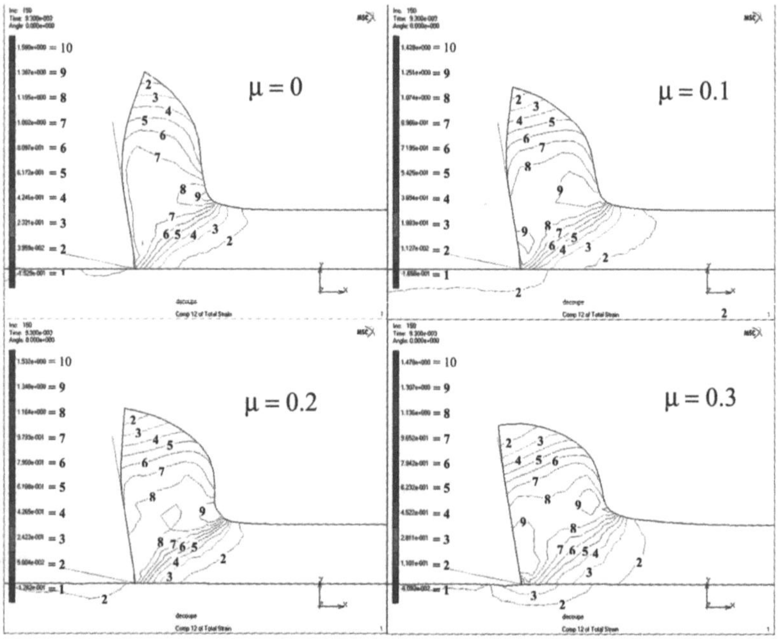

FIGURE 12. Distribution of ε_{12} for 4 different μ values

4 CONCLUSIONS

In this paper, FEM simulations of the orthogonal metal cutting process of an AISI 304 stainless steel were carried out using the Marc AutoForge code and a Lagrangian approach. A physical separation criterion for chip formation was applied utilizing a von Mises equivalent stress critical value and a nodal release technique. Chip/tool contact was modeled using the friction Coulomb law and a constant m value. The numerical simulations performed covered the incipient as well as the steady-state of the cutting process for a range of tool rake angle and friction coefficient values. The obtained results showed the dependence of chip morphology, stress and strain distributions and the cutting force values and trend on the rake angle and friction coefficient variations, in good agreement with the physical reality of the orthogonal cutting process.

REFERENCES

1. Shaw M.C., (1984), Metal Cutting Principles, Clarendon Press, Oxford.
2. Merchant M.E., (1945), Mechanics of the Metal Cutting Process, J. Applied Physics Vol 16, 267-318.
3. Lee E.H., Shaffer B.W., (1951), The Theory of Plasticity Applied To a Problem of Machining, ASME J. Applied Mechanics, Vol 18, 405-413.
4. Palmer W.B., Oxley P.L.B., (1959), Mechanics of Orthogonal Cutting, Proc. Inst. Mech. Engrs., Vol 173, 623-638.
5. Shih A.J., (1996), Finite Element Analysis of Orthogonal Metal Cutting, Int. J. Machine Tools & Manufacture, Vol 36/2, 255-273.

6. Kim K.W., Sin H.C., (1996), Development of a Thermo-Viscoplastic Cutting Model Using Finite Element Method, Int. J. Machine Tools and Manufacture, Vol 36/3, 379-397.
7. Ceretti E., Fallboehmer P., Wu W.T., Altan T., (1996), Application of 2D FEM to the Chip Formation in Orthogonal Cutting, J. Materials Processing Technology, Vol 59, 169-180.
8. Shet C., Deng X., (2000), Finite Element Analysis of Orthogonal Metal Cutting Process, J. Materials Processing Technology, Vol 105, 95-109.
9. Zhang L., (1999), On the Separation Criteria in the Simulation of Orthogonal Metal Cutting Using The Finite Element Method, J. Materials Processing Technology, Vol 89/90, 273-278.
10. Lin Z.C., Lo S.P., (1997), Ultra Precision Orthogonal Cutting Simulation for Oxygen-Free High-Conductivity Cooper, J. Materials Processing Technology, Vol 65, 281-291.

STUDY OF MACRO AND MICRO QUALITY DETERMINANTS OF PIECES OBTAINED FROM OPTIMIZED DRILLING PROCESS

A. Lima[1], M.Vieira Jr.[1]

[1] Post-graduate Program in Production Engineering – FEMP/UNIMEP – Brazil

KEYWORDS: Optimized drilling process; Dry cutting; Surface integrity.

ABSTRACT: The high necessity for productive quickness required nowadays, mainly due to globalization process, the research centers, industries and all social segments related to the development of productive activities are suffering a strong pressure to act drastically in order to avoid lost of time during products manufacturing. Among machining processes which have passed through this kind of requirement, the drilling process, subject of this research, has been suffering enormous changes in order to become better and optimized. These changes are the results of researches with the purpose of balancing the knowledge and the technological development concerned to drilling process in relation to other conventional machining processes with geometric defined edge cut. In such case, this issue aims a brief evaluation of macro and micro quality determinants (such as roughness and integrity of the inner surface) of parts obtained from optimized drilling process. It also aims to analyze which influences these determinants can have on the failure occurrence on the parts.

1 INTRODUCTION

Competitiveness is leading the companies to study various possibilities in reducing costs, especially in those sections directly related to the production. This is the main reason why new manufacturing technologies are being developed intensively, like the machining process without coolant or with a very high cutting speed.

However, these developments not always are followed closely by experts. In most of the cases they were made upon empirical decisions by professionals who were not qualified for that, and did not have enough knowledge of the requirements to which the manufactured pieces would be submitted in field [1].

Among machining processes which have passed through this kind of requirement, the drilling process, subject of this research, has been suffering enormous changes in order to become better and optimized. These changes are the result of researches with the purpose of balancing the knowledge and the technological development concerned to drilling process in relation to other conventional machining processes with geometric defined edge cut.

However, the researches for improving the process (usually through new materials and technologies in drill manufacturing, elimination of cut fluids or machining condition optimization) have very little stuck to macro and mainly to micro quality determinants of the machining products. Such a reality caused an interest in studying with more attention the existent relationship among the improvements made to optimize, specially the increasing of cut

Published in: E.Kuljanic (Ed.) *Advanced Manufacturing Systems and Technology*, CISM Courses and Lectures No. 437, Springer Wien New York, 2002.

speed and the quality determinants, more specifically the roughness and the integrity of the sub-superficial layer.

In the present research it is intended to analyze, from the results obtained in an optimized drilling process, the behavior of macro and micro quality determinants due to this optimization.

2 GENERAL FEATURES OF THE DRILLING PROCESS

The drilling process is the most frequently process used in general manufacturing industries, and it is largely employed in the automotive industrialization. This can be very easily observed, taking into account that almost all kinds of pieces has at least one hole, and that only a very small quantity of these pieces comes with the holes already done or almost done [2]. The pieces, generally, have to be full drilled or have their holes enlarged through drilling process, and this fact makes the study of the optimization of this process very important.

The drilling process is behind handed in relation to other machining processes for many reasons, but the main one is the drill diameter. A 10mm diameter drill (which is very common), to be used with speeds compatible to the hard metal (speed around 200 m/min), would need a rotation of 6400 rpm, something very far from the limits of the conventional drillers. For this reason, the number of use of CNC machining drilling centers for drilling process has increased [2].

Considering the importance of drilling in machining process, new materials and processes for manufacturing of pieces and drills has been emerged, producing tools with a large range of hardness and tenacity [3]. However, the most used drills are still made of HSS drills. This occurs in part because of a lack of information from the people who make decisions in companies.

On the other hand, inside machining world there are research groups working with the purpose of overcoming the limits inside of the set tool/machine/piece, and they always (or almost always) find new resources to brake the barriers of process optimization. This is the case of the elimination of cut fluid, as exposed below.

2.1 CUTTING FLUID ELIMINATION (DRY CUTTING)

The main functions of refrigerant/lubricant fluids are: to reduce the generation of heat by reducing the friction; to eliminate the influence in raising temperature in the material structure; and, transporting / removing turnings generated during the process. In machining with non-fluid cut the friction and adherence between tool and piece raises due to raising of the thermal charge present in the process. This propitiate significant raises in tool wear, in forming craters and long turnings that cause problems, specially in drilling processes. These difficulties can be minimized nowadays with the use of covered and geometric changed tools. [4].

In the last years, concerns with the environment have been pointed as a market request consequence (ISO14000 and "green point"). Works related to the banishment of the use of products which generate chemical residue (chorine, brimstone and phosphorus), harmful to human being and nature, have been developed, among them the lubricating and refrigerating fluids used in machining processes (specially in drilling) [3].

The global tendencies to environmental control are becoming rigorous and are focusing the manufacturing, watching that it does not cause damages to the environment [5]. The requests are present in rules of high importance for that companies which claim for acknowledgment as "Word Class Enterprise" or for those that are looking for partnership with these ones.

From these tendencies come, originally, the researches about the non-fluid cut drilling. Beyond the try of working with optimized parameters, people started working with non-fluid cut, looking for the banishment of residue and the reduction of production costs. The prices of lubricants and refrigerating, and of the equipment involved with them, for example the attention to avoid the development of bacteria inside the warehousing tanks, may reach approximately 15% of the production costs [5].

3 EXPERIMENTAL PROCEDURES

To make the analyses proposed in this paper, a set of holes drilling in the specimen material, steel 4340, in a vertical machining center Mori-Seiki (model SV-40 CNC), was made, without the use of lubricant / refrigerant, in optimized operational conditions as described in the method of COPPINI ET AL, 1988, and of PALLEROSI ET AL, 1991[6]. In this method, proved in laboratory and production environment, from the coefficients *x* e *K* of Taylor's equation, obtained from preliminary essays, the speed of reference from the Interval of Maximum Efficiency (IME) - V_{mxp} (velocity of maximum production) and V_o (velocity of minimum cost). In general, in cases of process optimization, people try to work with V_{mxp}, what make it possible to accelerate the production rhythm.

In the already made essays, from de figure of IME to optimize the process, the follow value was reached: V_{mxp} = 163m/min. The advance used was f = 0,18 mm/rev (data obtained from the drill manufacturer catalog).

The used tool was a one piece hard metal drill P40, with 10 mm of diameter, kind ALPHA 2, right version, with covering TiN-Al (model A3265 TFL - Titex), according to the Norm DIN 6537K for drill and DIN 6535HA for screw. The holes made were 32 (mm) depth. 384 drills were made (12.288 mm of total drilled length).

The analysis was done in two distinct moments: evaluation of macro quality determinants behavior (roughness, diameter variation and cilidricity variation), and evaluation of micro quality determinants behavior (sub-superficial layer integrity of the drilled piece and residual stress in the hole). To evaluate the roughness measurements with the portable Mitutoyo Surftest-211, with "cut-off" of 0,8 mm was used. For the diameter and cilidricity variations, a Starett 3D Machine was used. For the analyses of sub-superficial layer a high-resolution camera CCD Hitachi KP-M coupled to a metal graphic optical microscope Nikon was used to obtain the images of the macrostructure of the layer. The obtained images were analyzed with the help of "Global Lab Image View" software, which has screw and measurement recourses, due to the microscope obtained image enlargement. And to the residual stress, it was used a Residual Stress Analyzer (Digital Fastress).

4 RESULTS AND DISCUSSIONS

The analysis of the results is based on the quality of parameters that allow predicting the product behavior during its use; these parameters are known as micro and macro-determiners of quality. As a result of the experimentation described above, they had the behavior shown in the graphics of the figures 1 to 4, where the roundness and the cilindricity deviation of the holes, the surface stress and the roughness are represented. The HAZ formation is also shown at the figures 5 and 6.

It can be noticed that, except to some values, it does not exist significant increment in holes roughness value. This can be observed more clearly on the graphic on Figure 1, which shows a tendency to diminish the roughness as the number of holes increase.

On figure 2 it can be observed the behavior of the cilindricity deviation of the holes, which has the same tendency of roughness: diminishing during the drill life.

These behaviors can be the result of the wear at the main edge, that increases the role of the secondary edge during drilling due to the drill geometry, that assures its stiffness, and a consequent process stabilization, reducing the cilindricity deviation. This kind of stabilization is not so good, because it also brings a reduction on the hole diameter (fig. 4).

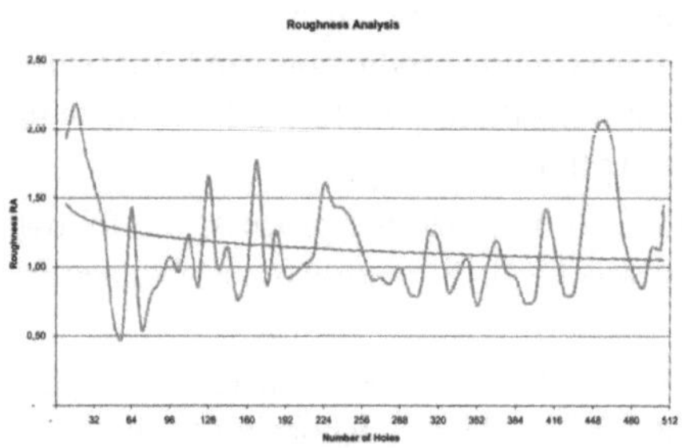

FIGURE 1. Roughness Analysis

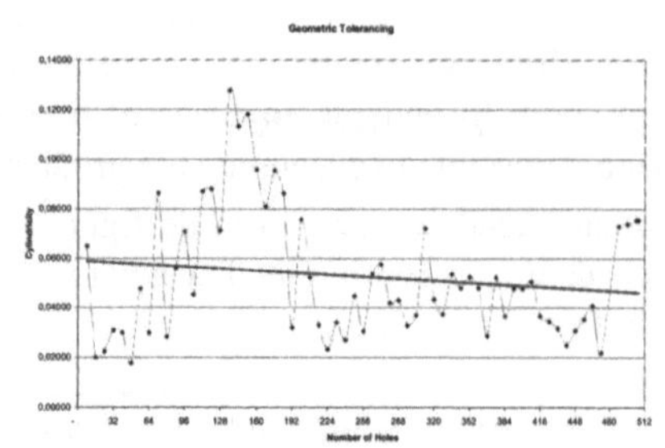

FIGURE 2. Cilindricity Analysis

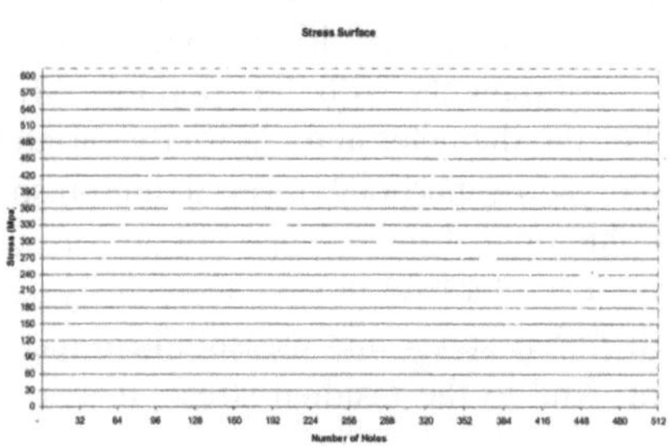

FIGURE 3. Stress Surface

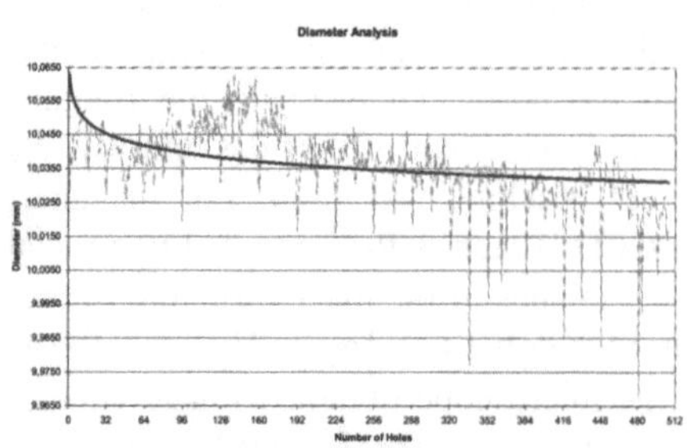

FIGURE 4. Diameter Analysis

The surface stress behavior shown at the figure 3 refers to the conditions found along the surface of the specimen at the holes nearby. Among the holes were chosen the points that represents regions that go from the beginning of the drilling until the region closely near to the last hole drilled, where are the presumed extreme conditions of the experimentation.

The behavior of the surface stress can be the result of a normalizing on the microstructure of the drilled material. Temperatures of the set drill/material during the process can reach levels between 800° to 1000°C (where the steel changes its phases) and assume a slow cooling velocity that causes this presumed normalizing; it is a no desired treatment because it occurs with no control and results on changes of the material structure that is different from the beginning of the experimentation.

The occurrence of this affected layer (as shown in the figures 5 and 6) can also be explained by the possibility of existence of friction between the drew and the drill wall, what causes a crush of the machining surface and the formation of a misshapen layer on the sub-superficial structure of the drill. This, however, must be the least prevailing factor for the formation of the affected layer, but it must be one of the reasons for the tendency of roughness diminishment.

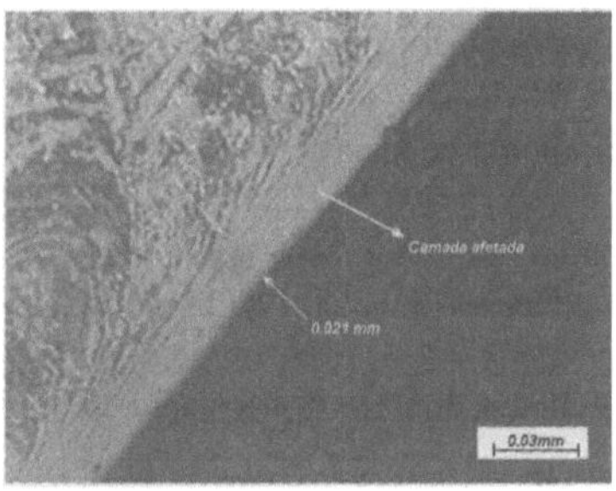

FIGURE 5. Typical HAZ

formation.

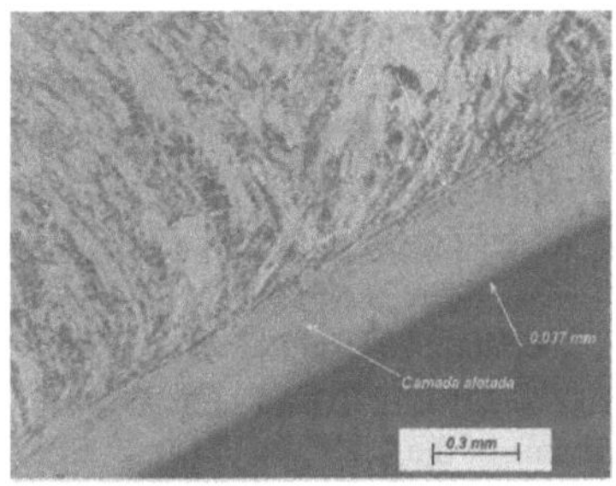

FIGURE 6. Typical HAZ

formation.

It is important to outstand that, usually, the drill elements contained in the pieces are used to fixation and so, they are submitted to propension forces. In this case, the formation of *HAZ* (or rather of a damaged layer on the drill wall) can be the point from where failures can start during the use of the pieces (trenches, tenseness accumulation, etc.).

It is also important to outstand that in the conditions the operation was made, it must be still another phase of manufacturing which correct the detected problems (an operation of conclusion for the removal of *HAZ*, for example).

5 CONCLUSION

According to the results obtained in this essay, it remains clear the need of a previous study on the application of the piece in the working area for the development of an optimized process of the directed machining parameters, with the objective of not harming its performance.

The banishment of use of cut fluid is possible due to innovations developed in drilling process. However, it is important to outstand that the possibility of HAZ formation is higher in this case, what can lead to the occurrence of failures in the pieces, originated from the drilling process.

It is also valid to bold that even with the roughness diminishment on the holes wall, the *HAZ* tends to appear, pointing to a possible need in realizing a new phase of the process to relief the tensions and to correct the dimensions.

The analysis of the results shows that it is important to focuses more than one characteristic of geometric deviation to help on the results interpretation. For instance, the roundness or the cilindricity all by themselves are not enough and can bring mistakes on the interpretation of the results.

In the stress analysis it is advisable to make at least one evaluation of the temperature gradient to reach the real condition of the material after machining, because of the different conditions of heat transfer found in the piece.

REFERENCES

1. MULLER, P. (2000)Ferramentas para Furar e Rosquear com HSC e sem Refrigeração. In: O Mundo da Usinagem, São Paulo/SP, Ed. Sandvik. p. 13 a 17.
2. COPPINI, N. L, DINIZ, A. E., MARCONDES, F. C. (2000) Tecnologia de Usinagem dos Materiais. ED. Art Liber Editora, SP, 2° ed.244 p.
3. SALES, W. F., Machado, A. R., Mello, J. D. B. (1998) Influência do Fluído de Corte no Desgaste de Brocas de Aço Rápido. In: IV Seminário de Desgaste. Associação Brasileira de Metalurgia e Materiais, São Paulo/SP, Ed: Édile,. p. 641 a 658.
4. KLOCKE, F. & EISENBLATTER, G. (1997)Lecture Presented at the Opening Session of Dry Cutting. Annals of the CIRP, v. 46 (2), p. 519-526.
5. NAKAGAWA, H. Mínima (2000)Quantidade de Lubrificação não Agride o Meio Ambiente. Revista Máquinas e Metais, ED. Aranda, , p 40-49.
6. PALLEROSI, C. A et al. (1991)Durability of Cutting Tools Under True Conditions. Proceedings of CANCAM 91 -Canadian Congress in Applied Mechanics, Canada, p. 173-175.
7. LIMA, A., Vieira, M. J., Libardi, R., Cancilieri, H. A. (1998) Análise da Camada Afetada pelo Calor na Superfície de Peças Retificadas. In: IV Seminário de Desgaste. Associação Brasileira de Metalurgia e Materiais, São Paulo/SP, Ed: Édile, p. 183 a 206

EXPERIMENTAL TEST FOR TOOL-LIFE PREDICTION IN TURNING

M. Nicolich

Università di Trieste. Dipartimento di Energetica – Italy

KEYWORDS: tool-life, cutting speed, turning

ABSTRACT. A comparison between two different experimental methods for predicting the cutting-edge tool life in turning is presented: the Standard ISO 3685 and the here proposed one, that make free the variation of cutting parameters, in particular the cutting speed.
The methodology here presented take into account the cumulative damage or the residual life of the cutting edge. On the contrary to the Standard, the cutting speed is not constant up to the complete wear of the edge, but different cutting speeds are to be employed. This brings to perform a test with a minimum use of two cutting edges and to evaluate the Taylor exponents analytically.
The comparison is based on experimental tests for both methods and the results in this previous work are encouraging.

1 INTRODUCTION

In chip removal, the cutting parameters are well correlated by the Taylor equation, which actually remain the most convenient despite the approach by more complicated models with not appreciable improvements [1]. For determining the unknowns of the Taylor expression, the exponents and the constant, an experimental test is mandatory because of the influence of uncontrolled factors which leads high uncertainty in computation if model is adopted [2].
Experimental tests are produced and often offered by manufacturers of cutting tool inserts, but data must be carefully used because of different tool machines and materials employed. Hence in factory it is significant to run tests proving the best behaviour of the cutting tool, therefore tests must be rapid and give as many information as possible. Several techniques have been proposed like RMS [3] and Factorial design[4] for choosing the number of tests to perform and to cover an assigned range of values, whereby a linear regression for Taylor relation would be determined.
The method presented in this paper relates to the cumulative damage or the residual life of the cutting edge and the base hypothesis is first posed, followed by the analytical exposition which leads to the formulation of the Taylor equation.

2 THE RESIDUAL TOOL LIFE

In considering valid the empirical Taylor equation $vT^n=C$ at the cutting conditions expressed by C, one can asks which amount of residual life L_1 should have the cutting edge after a cutting time $t_{01}<T_1$ at cutting speed v_1. With t_{01} the working time from the initial state 0 to 1 at the cutting condition expressed by 1: the residual life is $L_1= T_1- t_{01}$ and the rate of consumed life is $d_l= t_{01}/ T_l$, remaining the residual rate $l_1=(T_1- t_{01})/ T_1=1- d_1$.

Published in: E.Kuljanic (Ed.) *Advanced Manufacturing Systems and Technology*,
CISM Courses and Lectures No. 437, Springer Wien New York, 2002.

Under the hypothesis that the residual tool life rate l_1 is constant varying the operating conditions, for example the cutting speed v, it is defined:
L_{ij} : residual life at time t_i , with cutting speed v_j ;
L_{ii} : residual life at time t_i , after processing at cutting speed v_i ;

$$d_i=(T_i-L_{ii})/\, T_i = 1- L_{ii}/T_i\,.$$

Can be generalised as follows:

$$L_{ii} = T_i\,(1- d_{(i-1)}) - t_{(i-1)}\,;$$
$$L_{(i-1)i} = T_i\,(1- d_{(i-1)})\,;$$
$$d_i = \sum_{j=1}^{i} \frac{t_{(j-1)j}}{T_j} \quad ; \quad d_0 = 0\,.$$

Using the Taylor equation, as logarithmic transformation :
$Y = a + bX$, with $X = \ln v$, $Y = \ln T$, $c = \ln C$ and $a=c/n$, $b=-1/n$,
in fig. 1 the progression of the tool life is represented.

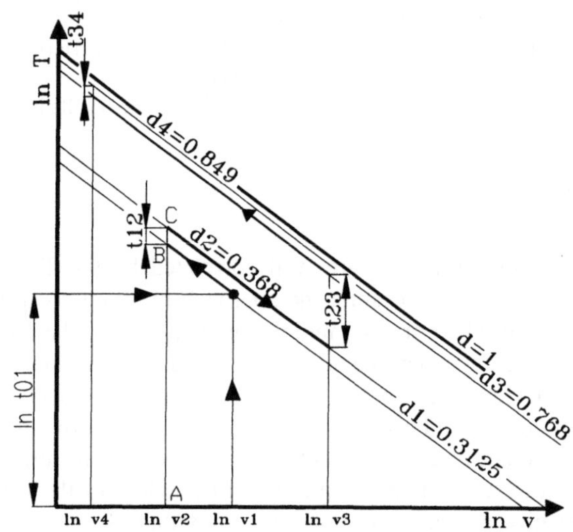

FIGURE 1. Graphic representation of the progression of the tool life.

By eliminating the unknown T_i with the Taylor equation, the above three equations can be expressed with the unknowns n and C.
If the cutting speed v_j , j=1,2,...i, will only vary during the test, other cutting parameters remaining constant in C, the cutting edge will be completely worn at time $\tau_i = \Sigma_i\, t_{(i-1)i}$ which correspond to L_{ii}=0 or d_i=1 :

$$d_i = \sum_{j=1}^{i} \frac{t_{(j-1)j}}{T_j} = \sum_{j=1}^{i} t_{(j-1)j} \cdot C^{-1/n} \cdot v_j^{1/n} = 1$$

For solving such an equation, which has the two unknowns C and n, it is necessary another one of the same type. This is obtained by using two cutting edges which are tested A and B times respectively up to their complete worn out:

$$\sum_{ja=1}^{A} t_{(j-1)ja} \cdot C^{-1/n} \cdot v_{ja}^{1/n} = 1$$

$$\sum_{jb=1}^{B} t_{(j-1)jb} \cdot C^{-1/n} \cdot v_{jb}^{1/n} = 1$$

with

$$t_{(j-1)ja} \cdot v_{ja}^{1/n} \neq t_{(j-1)jb} \cdot v_{jb}^{1/n} \; .$$

This means that two series of measures, two cutting edges, are needed for finding the constant terms of the Taylor equation in the form $vT^n=C$.
The problem can be extended to two independent variables for solving the Taylor equation in the form $vT^n x^m=C'$, with x the feed rate f or the depth of cut p.
Assuming the feed rate f as second variable and using the logarithmic transformation of the Taylor equation, the equation of a plane must be found:
Z=a X + b Y + c' , with $a=-1/n$, $b=-m/n$, $c'=\ln C'/n$.
The system is based on three equations where the unknowns are C', n, m as follows:

$$\sum_{ja=1}^{A} t_{(j-1)ja} \cdot C'^{-1/n} \cdot v_{ja}^{1/n} \cdot f_{ja}^{m/n} = 1$$

$$\sum_{jb=1}^{B} t_{(j-1)jb} \cdot C'^{-1/n} \cdot v_{jb}^{1/n} \cdot f_{jb}^{m/n} = 1$$

$$\sum_{jc=1}^{C} t_{(j-1)jc} \cdot C'^{-1/n} \cdot v_{jc}^{1/n} \cdot f_{jc}^{m/n} = 1 \; .$$

with

$$t_{(j-1)ja} \cdot v_{ja}^{1/n} \cdot f_{ja}^{m/n} \neq t_{(j-1)jb} \cdot v_{jb}^{1/n} \cdot f_{jb}^{m/n} \neq t_{(j-1)jc} \cdot v_{jc}^{1/n} \cdot f_{jc}^{m/n}$$

This means that three series of measures, three cutting edges, are needed for finding the constant terms of the Taylor equation in the form $vT^n f^m=C'$.
The extension to the third variable, depth of cut, is immediate.

3 THE FLANK WEAR CRITERION

For tool life experimental determination in turning the flank wear criterion has been adopted. This is a common industrial procedure because of the correlation with the surface roughness being obtained in turning chip removal. The test case refers to an industrial working condition: insert TiCN coated type ISO HC-P25, nose radius r=0.8 mm; steel material 34CrNiMo6; requested roughness R_a=3.2 μm. The cutting edge has been registered with an angle χ=93° and rake angle γ=-6°, β= 90°, λ=-7°.

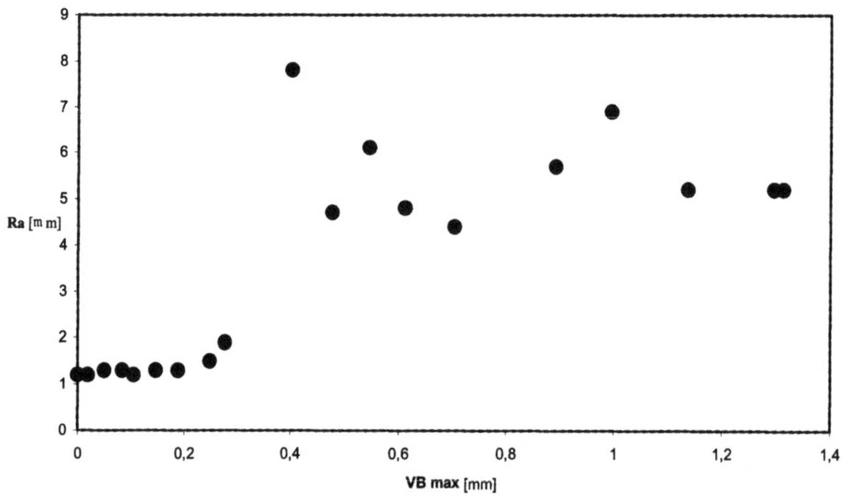

FIGURE 2. $R_a - VB$ experimental data.

The roughness has been considered as a constraint for the maximum flank wear assumption and the first measures have been addressed for finding the progression of the flank wear V_B versus the average roughness R_a , as in fig. 2, which bring to adopt a maximum V_B=3.2 mm. The working conditions of this test were: v=258.26 m/min, p=0.45 mm and f=0.182 mm/rev. Two series of tests has been performed because of comparing the here proposed procedure to the ISO 3685 [5]. A Taylor plane, taking the logarithmic representation (T, v, f), must be found so that the regression will be calculate for ISO 3685 procedure and the equation by solving the system in section 2. In following the ISO procedure will be indicated by constant cutting speed test v_c and the here discussed one by variable cutting speed test v_v.

A series of five values for feed rate has been chosen f=0.16, 0.182, 0.252, 0.308, 0.364 .

The cutting speed has been randomly selected in the range from v=150 m/min to v=258 m/min.

4 TEST WITH CONSTANT CUTTING SPEED.

Ten cutting edges have been employed obtaining N1=10 experimental data in the form (vi, fi, Ti) with i=1,...N1. The regression plane, using the least squares principles, for Taylor equation $Zc = acX + bcY + cc$, is

$$lnT=-8.4008\ ln\ v + -0.1533\ ln\ f + 46.8577$$

with a correlation coefficient R^2=0.9419.
If the test case is taken with f=0.182 mm/rev a line regression is obtained in the plane (T,v) as represented in fig.3:
lnT=-8.4008 ln v + 47.1206
In the following a bidimensional plot is used as intersection of the Taylor plane equation with the plane f=0.182.
The 95% confidence interval for the predicted value of T is also plotted in fig.3 which gives an error of 34.4% at v=190.56 m/min.

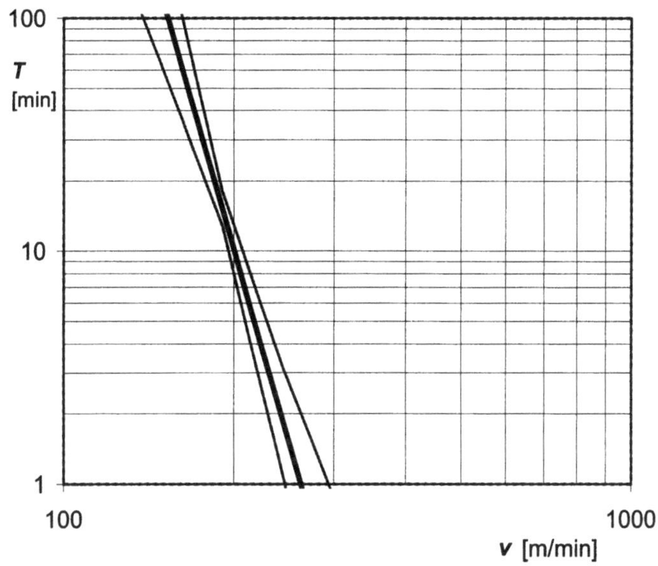

FIGURE 3. The regression curve and the confidence interval.

The number of test points will obviously decrease the error, but it will also be not lower of a minimum level [2] , in fact after 5-6 points it is drastically diminished and then it continues to decrease more lower.
The F test provide the statistics validation which gives F=17.22 that is greater if compared with F=3.79 given for a 95% probability and 7 degrees of freedom [6].

5 TEST WITH VARIABLE CUTTING SPEED

Three cutting edges must be used for solving the three equation system as before seen in sec. 2. for finding the Taylor's exponents. The experimental test has been repeated five times so that N2=5 systems have been solved finding N2 terns of (a_i, b_i, c_i) with i=1,...,N2 using an amount up to fifteen cutting edges:

$$Z_{iv} = a_{iv} X + b_{iv} Y + c_{iv} .$$

Each cutting edge was worn up to VB=0.4 mm with different cutting speed and different feed rate as for example C5 in tab.1 and with reference to fig.1.: it begins with v=156.1 m/min,

f=0.252 mm/rev for t=5.6 min, then with v=177.9 m/min, f=0252 mm/rev for t=7.3 min, again with v=167.8 m/min, f=0.252 mm/rev for t=6.7 min, and more with v=216.7 m/min, f=0.308 mm/rev for t=1.76 min, finally worn out with v=213.4 m/min, f=0.308 mm/rev for t=1.76 min.

TABLE 1. Test series for cutting edge C5.

n	v [m/min]	f [mm/rev]	t [min]
1	156.1	0.252	5.6
2	177.9	0.252	7.3
3	167.8	0.252	6.71
4	216.7	0.308	1.76
5	213.4	0.308	1.76

Solving the N2 systems, five terns of coefficients are obtained as follows in tab. 2.

TABLE 2. Coefficients of the plane equations.

	1	2	3	4	5
a_v	-8.37115	-8.32247	-8.26739	-7.51414	-7.67469
b_v	-0.12211	-0.1413	-0.1413	-0.52398	-0.4439
c_v	46.69237	46.3973	46.1437	41.91215	42.80765

Taking the mean value of the five, the plane equation can be written:
Z_v=-8.02997 X – 0.30296 Y + 44.79063
and $T= e^{Zv}$.
If the five equations are plotted (fig.4) in the (T,v) plane, at f=0.18 mm/rev, and compared with the regression before obtained from the standard ISO 3685, one can observe that they lay very close each other and with good agreement with the standard one.

6 VALIDATION OF THE RESULTS

If the agreement is good enough a mix F test provide it by taking the standard method measures from one side and the equation obtained in sec.5 on the other.
The test verify the agreement because F=11.37 is greater than F=3.79 requested for 95% of probability.
Focusing now the analysis to the bidimensional representation of the equation, taking f=0.18 mm/rev, the distribution of the coefficients can be shown in a scatter diagram containing the plot of the range of 95% probability.
Such a visualisation method [6] (fig.5) indicates on y axis a value A proportional to a_v and on y axis a value C proportional to c_v:

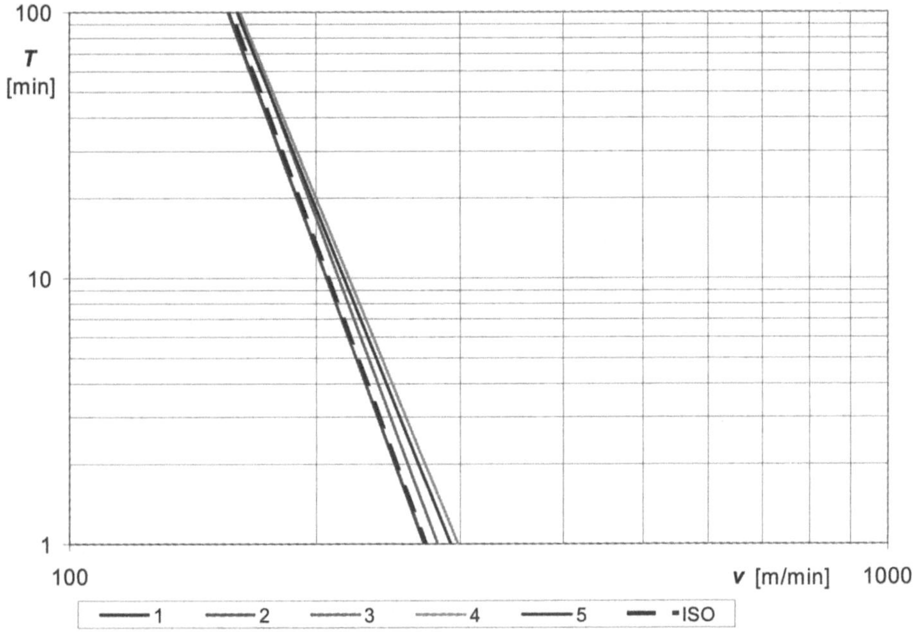

FIGURE 4. The comparison between the two methods

The range is bounded by an ellipse

$$N\cdot(C-c)^2+2\cdot(C-c)\cdot(A-a)\sum_{i=1}^{N}x_i+(A-a)^2\sum_{i=1}^{n}x_i^2=K$$

$$K=2\cdot s_y^2\cdot F_{2,N-2,1-a}$$

which become an ellipsoid if coefficients of a plane are discussed.

7 CONCLUSION

Two experimental test methods have been compared: the ISO 3685 and the here proposed one. An advantage is put on light, that is a minimum number of cutting edges is enough to evaluate the Taylor exponents and the constant. The equation is very close to the regression which need to wear a large number of cutting edges and a more complicate procedure requested from the constant cutting speed. From the industrial point of view is more important to know a reliable average performance of the insert, cutting waste time from experimental tests, instead of knowing the associated error. The variable cutting speed method presents a lower dispersion of data which can be considered an effect of the average of the various working conditions as a memory of its own story. Hypothesis on the constant residual tool-life will certainly simplify the approach and reliability of such an assumption will be discussed in an analysis in progress.

If that, the error will not be sensible, the control on the use of the tool could be well monitored in production.

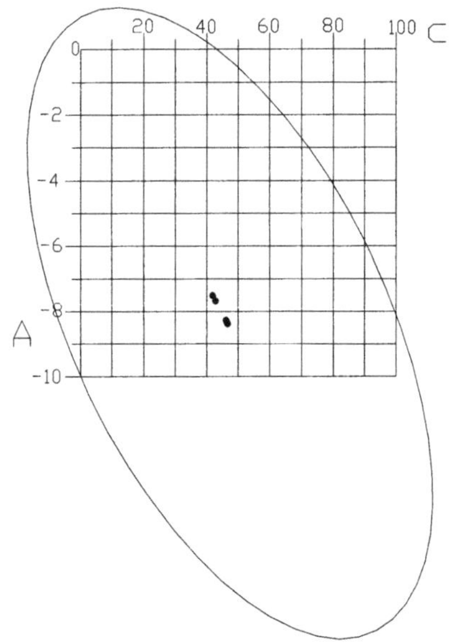

FIGURE 5. Scatter diagram of the coefficients

ACKNOWLEDGMENT

The Author will thank prof. R. Levi for precious suggestions and Wartsila Italia S.p.A.for material furnishing.

REFERENCES

1. Choudhury, L.A., El-Baradie, M.A., (1998), Tool-life Prediction Model by Design of Experiments for Turning High Strength Steel (290BHN). Jour. Of Material Processing Technology, Elsevier, 319-326.
2. Levi, R., Rossetto, S., (1975), Some Considerations on Tool-life Scatter and its Implications. Jour. of Engineering for Industry.
3. Kuljanic, E., Random Strategy Method for Determining Tool-life Equations. Annals of the CIRP, vol 29/1.
4. Montgomery, D.C., (1984) Design and Analysis of Experiments. 2nd ed., John Wiley and Sons, New York.
5. ISO 3685, Tool-life Testing with Single Point Turning Tools.
6. Dixon, W. J., Massey, F. J. Jr, (1969), Introduction to Statistical Analysis, 3rd ed., Mc Grow Hill, New York.

TECHNOLOGICAL MODELS BASED ON GRINDING WHEEL AND WORKPIECE CHARACTERISTICS IN THE GRINDING PROCESS OF METAL MATRIX COMPOSITES

A. Di Ilio[1], A. Paoletti[1]

1 Department of Energetics, Faculty of Engineering, University of L'Aquila, Italy

KEYWORDS: Metal Matrix Composites, Grinding, Modeling.

ABSTRACT. In this paper approaches to the modelling of grinding process of Metal Matrix Composites, based on empirical relationships, are proposed and discussed. To this purpose, experimental data obtained from tests carried out on a horizontal surface grinder have been employed. The simple relationships obtained could be utilised to predict how the grinding wheel type influences the grinding process and the machining quality of these non-traditional materials. The influence of shape, orientation and content of the reinforcement on the grindability of metal matrix composites is analysed. Investigations deal with grinding forces and degradation of the grinding wheel surface, acquired during the machining process and ground surface roughness. The effects of workpiece material composition on grinding wheel wear and ground surface quality is described considering grindability indices.

1 INTRODUCTION

The interest in machining of Metal Matrix Composites (MMCs) is rapidly growing due to the increased performances and requirements of high-technology and consuption industries. MMCs components need to be formed into the desiderate shapes and finished to the required dimension and tolerance. Metal Matrix Composites are given their required shape by brazing, bonding, powder metallurgy techniques, casting, metal spraying and by forming operations such as bending, swaging, drawing and extrusion [1], [2]. Although advances have been made in near-net-shape technology, finishing operations are often required to obtain dimensional tolerance as well as good finish.

Machining of these new materials requires tool materials of very high wear resistance because the reinforcement is extremely abrasive [3], [4]. Among traditional machining processes, grinding operation is important for MMCs, since it could be applied also in heavy-duty machining, in addition for obtaining desired dimensional tolerances and surface quality.

In recent years, considerable work has been carried out to understand the mechanisms of grinding conventional materials by regarding the grinding process as an interaction system between the surface of the wheel and the workpiece [5]. On the contrary, information about the grindability of MMCs is still insufficient [6], [7].

Since grinding is a complex manufacturing process with a lot of quantities which influence each other, modelling can be an useful tool to the comprehension and the simulation of the process itself [8]. Tönshoff et al. have described the state of the art in the modeling and simulation of grinding processes of traditional materials [9].

Published in: E.Kuljanic (Ed.) *Advanced Manufacturing Systems and Technology*,
CISM Courses and Lectures No. 437, Springer Wien New York, 2002.

In this work some empirical relationships for predicting how the grinding wheel and the workpiece material affect the grinding forces, the flat area percentage on the active surface of the wheel and the surface roughness are presented.

2 EXPERIMENTAL TESTS

Experimental tests have been carried out on a horizontal surface grinder, utilising grinding wheels based upon conventional abrasives and superabrasives. Grinding wheels, based upon alumina (Al_2O_3), silicon carbide (SiC), cubic boron nitride (CBN) and diamond (ASD) abrasives, have been used (Table 1).

TABLE 1. Characteristics of grinding wheels employed

Grinding wheel Identification	Diameter / Width (mm)	Abrasive	Grit size	Bond type
32A 46-IV ·	200 / 25	Al_2O_3	46	Vitrified
39C 60-KVS	200 / 25	SiC	60	Vitrified
CBN 126QB99	200 / 15	CBN (75% concentration)	126	Resinoid
ASD 126R75B99	200 / 15	Diamond (75% concentration)	126	Resinoid

Hereafter, the above four types of abrasive wheels are referred to as A46, C60, CBN and ASD, respectively. The materials employed for tests are aluminium alloys reinforced with silicon carbide.

Two types of MMCs, both made with an aluminium alloy matrix and SiC reinforcement in different shapes, have been investigated. The former, which is referred to as 15P-p in the course of the article, had a reinforcement in form of powder, the latter, 15W-n, had the reinforcement in form of whiskers. Specimens with dimensions 13 x 23 x 23 mm^3 have been cut from extruded rods. MMC samples reinforced with powders have been ground in direction parallel to the extrusion one, while MMC specimens reinforced with whiskers have been ground in the direction perpendicular to that of extrusion, i.e. perpendicularly to direction to which the fibres are preferentially oriented. Subsequently, in order to study the influence of workpiece material characteristics, different kinds of MMCs have been grounded, utilizing grinding wheel made by alumina (Al_2O_3) abrasive (Table 2).

TABLE 2. Workpiece materials employed for tests

Workpiece Material	Denomination	Hardness (HRB)	Grinding Wheel
Al-2009 / SiC-15P Parallel to extrusion direction	15P-p	83.4 ± 1.0	A46
Al-2009 / SiC-20P Parallel to extrusion direction	20P-p	62.4 ± 1.7	A46 C60 CBN ASD
Al-2009 / SiC-20P Normal to extrusion direction	20P-n	67.6 ± 1.5	A46
Al-2009 / SiC-25P Parallel to extrusion direction	25P-p	72.6 ± 1.0	A46
Al-2009 / SiC-15W Parallel to extrusion direction	15W-p	62.1 ± 1.4	A46
Al-2009 / SiC-15W Normal to extrusion direction	15W-n	70.5 ± 1.1	A46 C60 CBN ASD
Al-2009 / SiC-20W Normal to extrusion direction	20W-n	95.9 ± 0.7	A46
Al-6061 / SiC-25P Normal to extrusion direction	6061_25P	52.4 ± 0.6	A46
Al-7075 Normal to extrusion direction	7075	47.4 ± 0.6	A46

Dry conditions and constant cutting parameters have been adopted for all tests, as reported in Table 3.

TABLE 3. Grinding parameters

Depth of cut, a (mm)	0.01
Work velocity, V_W (mm/s)	300
Wheel peripheral speed, V_S (m/s)	20

For each test, 100 plunge cut grinding passes with 0.01 mm depth of cut have been performed. Force components, flat area percentage and surface roughness have been measured every 5 single up-grinding passes. The images have been acquired by means a CCD camera and a digital acquisition image card, illuminating the wheel with perpendicular light [10].

3 MODELLING RESULTS AND DISCUSSION

Empirical models are used in all fields of machining technology because their development requires less effort than physical models. However, every empirical model can be employed for the description of one machining application only. Material grindability is not dependent on a specific technological property, but is a function of a set of different characteristics which cannot be correlated each other. In order to compare the behaviour of different cases under test, the weighted average values of normal and tangential components of grinding forces, flat area, and roughness, have been calculated defining the following *grindability indices* [11], [12]:

$$I(x) = \frac{\sum_i (x_i \cdot V'_{wi})}{\sum_i V'_{wi}} \qquad (1)$$

where x is F'_n, F'_t, A and R_a respectively, V'_w is the specific material removal and index "i" refers to pass number. The results have been reported as a function of the ground material type and grinding wheel abrasive.

3.1 GRINDING FORCE

Figure 1a, b shows the weighted average values of normal and tangential components of specific grinding force for different workpiece materials.

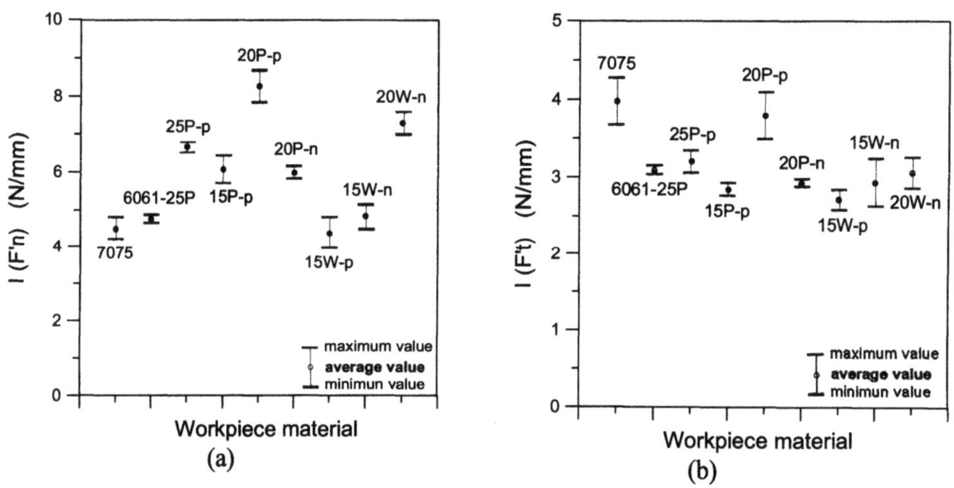

FIGURE 1. Weighted average values of normal (a) and tangential (b) components of specific cutting force for different workpiece materials

The normal and tangential components of the grinding forces are affected by orientation, shape and volume fraction of the reinforcement and by the type of matrix. For particulate composites having the same matrix, namely Al-2009/SiC-15P, Al-2009/SiC-20P, Al-2009/SiC-25P, both the components of the grinding forces are inversely related to the material hardness. On the contrary, materials reinforced with whiskers, namely Al-2009/SiC-15W and Al-2009/SiC-20W, show opposite tendency. This different behaviour is also found if it is considered the reinforcement orientation. In fact, while for particulate composite Al-2009/SiC-20P, the lowest force values have been obtained perpendicularly to the extrusion direction, for whiskers composite Al-2009/SiC-15W, the lowest force values have been reached parallel to the extrusion direction. However, the reinforcement, both powder or whiskers, contributes to reduce the tangential component of force. As can be seen in Figure 1b, the highest value of the tangential force has been obtained for non reinforced aluminium. The influence of the grinding wheel type on the forces are shown in Figure 2.

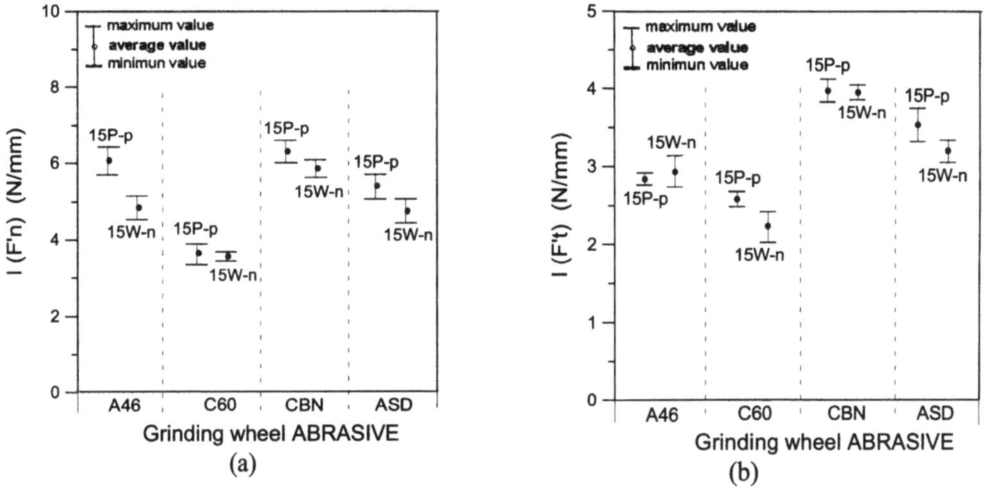

FIGURE 2. Weighted average values of normal (a) and tangential (b) components of specific grinding force for different grinding wheel abrasives

Diagrams in Figure 2, evidence that the lowest values of tangential and normal components are obtained with the wheel C60. Therefore, if the criterion of selection is based upon the grinding force, the best choice falls on the wheel C60. In this case, no significant differences between the two ground material exist. For other types of wheels, the composite 15P-p tends to give force values higher than the composite 15W-n, which exhibits a lower hardness (Table 2).

3.2 FLAT AREA

Figure 3a shows the weighted average values of flat area percentage for different workpiece materials. The influence of the grinding wheel type on the flat area is shown in Figure 3b.
Grindability indices concerning flat area percentage of grinding wheel for different workpiece materials are fairly close to those concerning specific tangential force. The highest flat area percentage has been revealed during grinding of non reinforced aluminium, while the composite Al-6061/SiC-25P seems to show the lowest tendency to clog the wheel.
As far as the flat area is concerned, the minimum values are obtained with A46 and C60 wheels. However, there exists a certain interaction between the type of wheel and the material being ground. The minimum clogging occurs for the composite 15W-n when ground with the wheel A46; this one exhibits the most open structure among the employed wheels. The contrary occurs when grinding with wheel C60, where the minimum clogging is obtained for the composite 15P-p. As aforesaid, in general, the composite 15W-n exhibits the lowest tendency to clogging phenomenon (Figure 3b), probably because of a higher notch effect associated with the fibrous reinforcement. From the above results, it descends that, if the criterion for wheel selection is the minimum tendency to clogging, the best choice will fall within conventional abrasives and, among these, it depends on the material being ground.

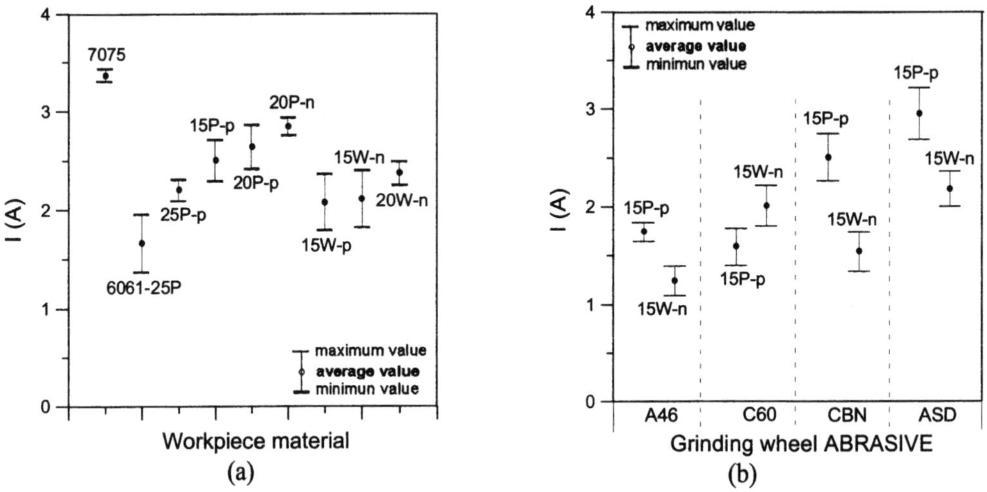

FIGURE 3. Weighted average values of flat area percentage for different workpiece materials (a), and weighted average values of flat area for different grinding wheel abrasives (b)

Grindability indices concerning flat area percentage of grinding wheel for different workpiece materials are fairly close to those concerning specific tangential force. The highest flat area percentage has been revealed during grinding of non reinforced aluminium, while the composite Al-6061/SiC-25P seems to show the lowest tendency to clog the wheel.

As far as the flat area is concerned, the minimum values are obtained with A46 and C60 wheels. However, there exists a certain interaction between the type of wheel and the material being ground. The minimum clogging occurs for the composite 15W-n when ground with the wheel A46; this one exhibits the most open structure among the employed wheels. The contrary occurs when grinding with wheel C60, where the minimum clogging is obtained for the composite 15P-p. As aforesaid, in general, the composite 15W-n exhibits the lowest tendency to clogging phenomenon (Figure 3b), probably because of a higher notch effect associated with the fibrous reinforcement. From the above results, it descends that, if the criterion for wheel selection is the minimum tendency to clogging, the best choice will fall within conventional abrasives and, among these, it depends on the material being ground.

3.3 SURFACE ROUGHNESS

The surface roughness is a characteristic quantity which determines the quality of the workpiece and refers to the microgeometry or topography of machined surfaces. Figure 4a shows the grindability index for surface roughness for different workpiece materials. The influence of the grinding wheel type on the surface roughness is reported in Figure 4b.

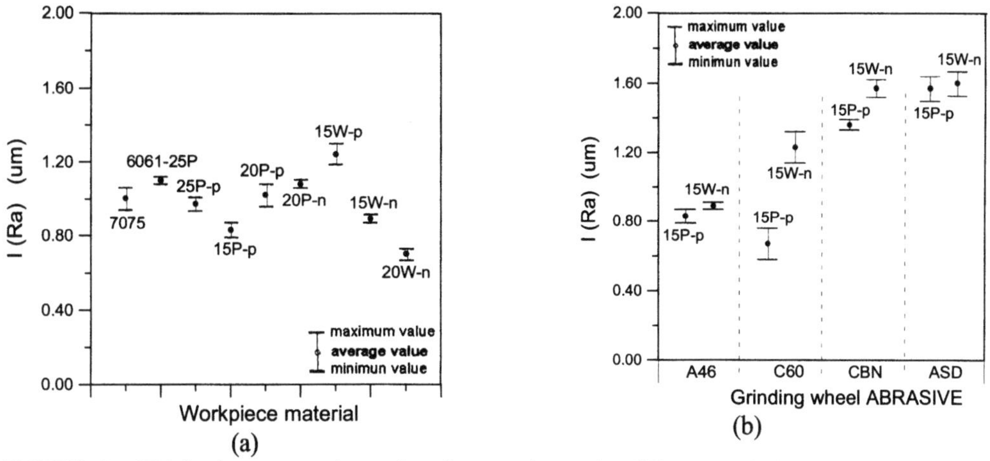

FIGURE 4. Weighted average values of surface roughness for different workpiece materials (a), and weighted average values of surface roughness for different grinding wheel abrasives (b)

Both for particulate and whiskers composites having the same matrix, the surface roughness follows a decreasing trend as a function of material hardness. However, the above mentioned correlation falls into failure when it takes into account the constituents of composite material, matrix and reinforcement, as well as reinforcement orientation. In fact, the highest roughness value has been obtained for composite Al-2009/SiC-15W, ground along the extrusion direction, notwithstanding its higher hardness with respect to both composite Al-6061/SiC-25P and non reinforced aluminium Al-7075.

The mean roughness of the ground surface is generally lower for the material 15P-p. However, also in this case, the characteristics of the material and the type of wheel show a mutual influence. For the composite 15W-n, the best surface finish occurs when using the coarsest grit wheel, i.e. A46, while for the composite 15P-p, which is harder than 15W-n, the best surface finish is obtained with the wheel C60. It may be noted that, for the conventional wheel C60, the highest value of surface roughness has been obtained for the material which exhibits the highest value of flat area, while the contrary occurs for superabrasive CBN wheel (Figure 4b).

3.4 TOTAL GRINDABILITY INDEX

Grindability indices considered are representative of a single characteristic, because take into account a single characteristic. It is possible to introduce a Total Grindability Index (TGI) by a linear combination of the singles indices:

$$TGI = a \cdot I(F'n) + b \cdot I(F't) + c \cdot I(A) + d \cdot I(Ra) \tag{2}$$

where the coefficients *a, b, c, d* (selected adopting the bond: $a + b + c + d = 1$), are dependent on the characteristics we want to underline and consequently on the grinding process. As far as the TGI has been concerned, better grindability is typical of a material exhibiting a lower value of the index. Figure 5 shows the TGI values for the different materials grounded by employing alumina grinding wheel, assuming equal values for the coefficients ($a = b = c = d = 0.25$).

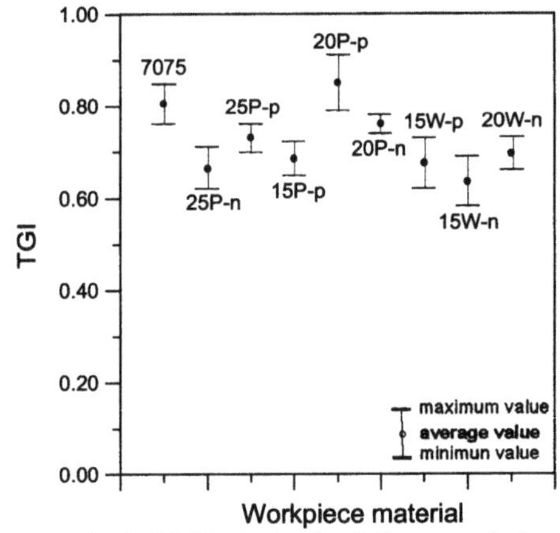

FIGURE 5. Total grindability index for different workpiece materials

Figure 6 depicts the influence of workpiece hardness and grinding direction with respect to extrusion direction on Total Grindability Index.

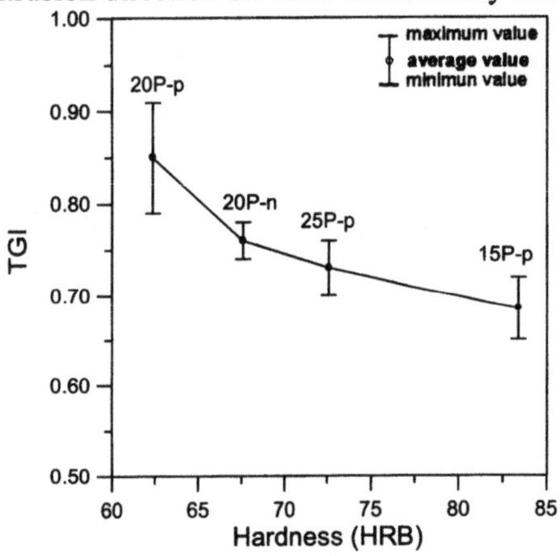

(a)

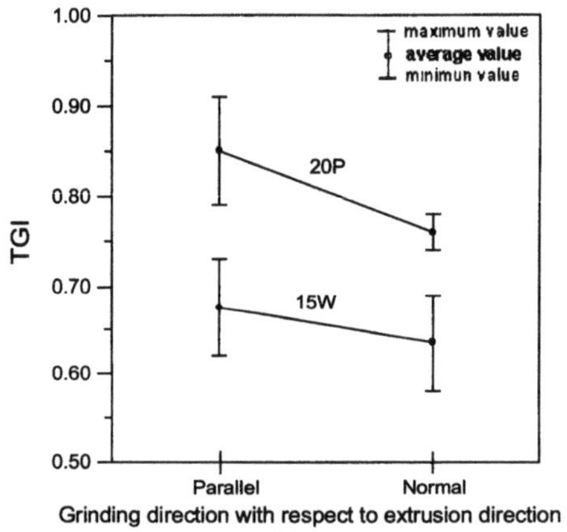

(b)

FIGURE 6. Total Grindability Index as a function of workpiece hardness for powder reinforced composites (a) and of grinding direction (b)

Both particulate and wiskers composites exhibit lower values of the index when they have been machined in direction orthogonal to the extrusion one. The TGI shows a decreasing trend as workpiece hardness increases.

As far as the TGI has been concerned, it takes into account the main three aspects of grindability that are tool life, surface finish and power required to cut. Giving different weights to coefficients a, b, c, d, it is possible to emphasize the influence of one or more aspects which characterize the machinability.

4 CONCLUSIONS

Metal Matrix Composites generally exhibit a certain free-cutting tendency which make them easier to be ground with respect to non-reinforced light alloys. Grindability indices have been settled taking into account grinding forces, wheel wear and workpiece surface roughness in order to compare the behaviour of different grinding wheel abrasives and workpiece materials. Among the grinding wheels employed in the tests, the silicon carbide one represents the best solution if the main concern is to hold low the grinding forces.

Flat area and roughness exhibit a more complicated behaviour. For the particulate composite the best results, in terms of both flat area and roughness, occur when using the silicon carbide wheel. The lowest tendency to clogging occurs with the aluminium oxide wheel that is characterised by the most opened structure among the wheels used.

The grindability of MMCs does not depend on a specific technological property, but is a function of a set of different characteristics which are not simply related each other. In fact, geometry, orientation and volume fraction of the hard reinforcement greatly affect the cutting ability of the grinding wheel and the surface finish of workpieces.

REFERENCES

1. Chandler, H.E., (1989), Machining of Metal Matrix Composites and Honeycomb Structures, Metal Handbook, Ninth Edition, Vol. 16, Machining, 892-901.
2. Srivatsan, T.S., Lavernia, E.J., (1991), Synthesis of Particulate Reinforced Metal Matrix Composites using Spray Techniques, Processing and Manufacturing of Composite Materials, ASME, Vol. 49, 179-221.
3. Clark, I.E., Cook M.W., (1994), Machining Highly Abrasive Composite Materials with Polycrystalline Diamond, Proceedings of Advancing with Composites, Milan, Italy.
4. Cronjager, L., Meister, D., (1992), Machining of Fibre and Particle-Reinforced Aluminium, Annals of the CIRP, Vol. 41/1, 63-66.
5. Zhang, L.C., (1994), Grindability of Some Metallic and Ceramic Materials in CFG Regimes, International Journal of Machine Tools & Manufacture, Vol 34, No. 8, 1045-1057.
6. Chandrasekaran, H., Johansson, J. O., (1993), The Role of Material and Grinding Parameters in Grinding of Alumina Fibre-Reinforced Aluminium Alloys Proceedings of Machining of Advanced Materials, Gaithersburg, USA, 20-22 July.
7. Di Ilio, A., Paoletti, A., Tagliaferri, V. Veniali, F., (1996), An Experimental Study on Grinding of Silicon Carbide Reinforced Aluminium Alloys, International Journal of Machine Tools & Manufacture, Vol 36, No. 6, 673-685.
8. Malkin, S., (1989), Grinding Technology - Theory Applications of Machining with Abrasives, Ellis Horwood Limited, West Sussex.
9. Tönshoff, H. K., Peters, J., Inasaki, I., Paul, T., (1992), Modelling and Simulation of Grinding Processes, Annals of the CIRP, Vol. 41/2, 677-688.
10. Di Ilio, A., Paoletti, A., (1995), Wheel Topography Evaluation During the Grinding of SiC Reinforced Aluminium Alloy, Proceedings of the First World Conference on Integrated Design & Process Technology, Austin, Texas, 6-9 December, 1995, 70-76.
11. Di Ilio, A., Paoletti, A., (1998), The Influence of Shape and Content of Reinforcement on Grindability of Metal Matrix Composites, Proceedings of European Conference on Composite Materials, Vol 2, Naples, 3-6 June, 541-549.

12. Di Ilio, A., Paoletti, A., (2000), A Comparison Between Conventional Abrasives and Superabrasives in Grinding of SiC-Aluminium Composites, International Journal of Machine Tools & Manufacture, Vol. 40, No. 2, 173-184.

OPTIMISATION OF PROCESS CONDITIONS USING "EXPERIMENTS WITH MIXTURES"

N. Šakić[1], V. Panić[1], I. Dubravec[1]

[1] Faculty of Mechanical Engineering and Naval Architecture, University of Zagreb, Croatia

KEYWORDS: Mixture experiment, MAG welding, Optimisation of process conditions.

ABSTRACT. In production the appearance of processes development in mixtures conditions is often present. For example, it almost regularly happens in chemical and process industry, but also in industry of polymer, rubber manufacturing and in welding and allied processes. For settlements of respond functions, to achieve optimal conditions of process, it is necessarily to use a special kind of experiments: experiments with mixtures. In this paper the basic thesis of such experiment on the example of welding (MAG treatment) under protective gas (gas mixture of Argon, Carbon dioxide, Oxygen) will be presented together with the procedure of numerical and graphical optimisation of process conditions, with previously defined optimisation's criterions: the mechanical properties of weld, the geometry of weld and loss of alloy elements.

1. INTRODUCTION

In production and technological processes a condition of mixture which is created by two or more components is often present, but in a such way that properties of the product or the result of the process depends on proportions of components. Generally, there is an assumption that the mixture is consisted from q components, where is x_i a value of the individually component In that case, the mixture can be defined by the next definitions (1) and (2):

$$x_i \geq 0 \; ; i = 1, 2, \ldots, q \tag{1}$$

$$\sum_{i=1}^{q} x_i = x_1 + x_2 + \ldots + x_q = 1 \tag{2}$$

The analysis of the design results, especially when there is a need for determing of influence of components on the design result, is based on determing the response surface:

$$E(y) = \Phi(x_1, x_2, \ldots, x_q) \tag{3}$$

The response surface can be linear:

$$E(y) = \beta_0 + \sum_{i=1}^{q} \beta_i \cdot x_i \tag{4}$$

Published in: E.Kuljanic (Ed.) *Advanced Manufacturing Systems and Technology*, CISM Courses and Lectures No. 437, Springer Wien New York, 2002.

The response surface can be higher-order:

$$E(y) = \beta_0 + \sum_{i=1}^{q} \beta_i \cdot x_i + \sum_{i \le j}^{q} \sum \beta_{ij} \cdot x_i \cdot x_j \tag{5}$$

In practical realisation of design with mixtures there is often an appearance of lower and/or upper boundary for proportions of individually components, i.e.:

$$0 \le a_i \le x_i \le b_i \le 1 , \qquad 1 \le q \tag{6}$$

Specific feature of such design models is limitation defined in (2), what is, for different values of q, realised as it is illustrate in Figure 1. The degree of response polynomial is defined by mark m, so the common mark of this models is: $\{q, m\}$. Figure 1. illustrates cases: $\{2,2\}$, $\{3,2\}$, $\{4,2\}$, $\{3,3\}$ and $\{4,3\}$.

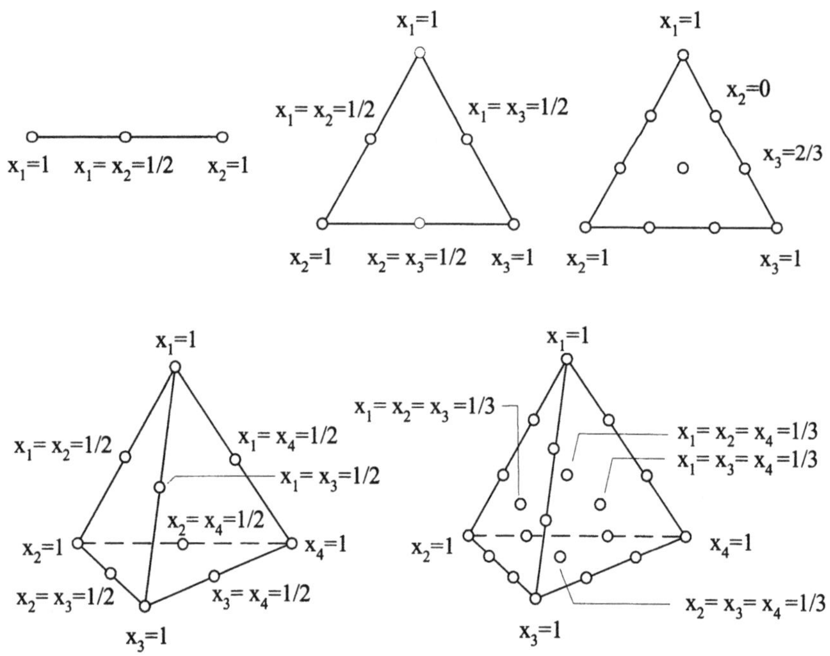

FIGURE 1. Some $\{2, m\}$, $\{3, m\}$ and $\{4, m\}$ simplex – lattice arrangements.

Table 1. illustrates polynomial $\{3,m\}$ and related simplex – lattice arrangements which are common forms for those models.

TABLE 1. $\{3,m\}$ polynomial and $\{3,m\}$ simplex–lattice.

Types of polynomial	Polynomial $\{3,m\}$	Forms of simplex-lattice
Linear	$E(y) = \beta_1 \cdot x_1 + \beta_2 \cdot x_2 + \beta_3 \cdot x_3$	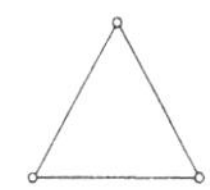
Quadratic	$E(y) = \beta_1 \cdot x_1 + \beta_2 \cdot x_2 + \beta_3 \cdot x_3 +$ $+ \beta_{12} \cdot x_1 \cdot x_2 + \beta_{13} \cdot x_1 \cdot x_3 + \beta_{23} \cdot x_2 \cdot x_3 +$ $+ \beta_{12} \cdot x_1 \cdot x_2 \cdot (x_1 - x_2) + \beta_{13} \cdot x_1 \cdot x_3 \cdot (x_1 - x_3) +$ $+ \beta_{23} \cdot x_2 \cdot x_3 \cdot (x_2 - x_3) + \beta_{123} \cdot x_1 \cdot x_2 \cdot x_3$	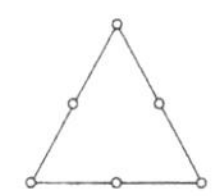
Full Cubic	$E(y) = \beta_1 \cdot x_1 + \beta_2 \cdot x_2 +$ $+ \beta_{12} \cdot x_1 \cdot x_2 + \beta_{13} \cdot x_1 \cdot x_3 + \beta_{23} \cdot x_2 \cdot x_3$	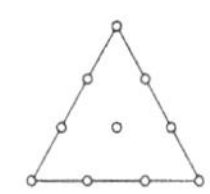
Special Cubic	$E(y) = \beta_1 \cdot x_1 + \beta_2 \cdot x_2 + \beta_3 \cdot x_3 +$ $+ \beta_{12} \cdot x_1 \cdot x_2 + \beta_{13} \cdot x_1 \cdot x_3 + \beta_{23} \cdot x_2 \cdot x_3 +$ $+ \beta_{123} \cdot x_1 \cdot x_2 \cdot x_3$	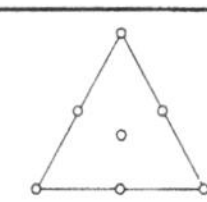

2. SIMPLEX AND PSEUDOSIMPLEX-LATTICE

Because of already described conditions in (6), complete simplex-lattice is often not possible for design model. Therefor, the princip of "pseudosimplex-lattice" is using, as it is illustrated in Figure 2.

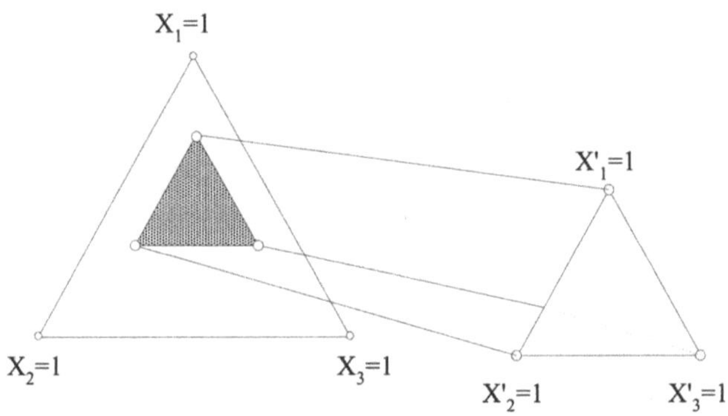

FIGURE 2. The full simplex and pseudosimplex-lattice.

Recaculating the original to pseudosimplex is represent by the next equation:

$$x'_i = \frac{x_i - a_i}{1 - L_d} \tag{7}$$

The lower boundary for proportions of individually component x_i is marked by a_i. Ld marks the summary of all lower proportions of components.

3. OPTIMISATION OF MAG WELDING PARAMETERS USING "EXPERIMENTS WITH MIXTURES"

It will be illustrate partial research results of mixture composition influence of protective gas on the geometrical characteristics of weld join. Entirely results are in reference [4]. The design is performed with constant weld speed, by two ways of metal transmissions. In this work are represented results for spray arc with speed v=8.1 meter per minute and power stream i = 284 amperes. Welding electrode was Bohler MK-8, φ = 1.2 mm, tin thickness was t = 10 mm, with arc length l = 4 mm. The experiment was performed on steel St 52-3 with composition: 0.150 % C, 0.480 % Si, 1.030 % Mn, 0.011 % P, 0.002 % S, 0.030 % Al, 0.0057 % N and 98.2913 % Fe. Composition of gas mixture was varies as Table 2. illustrates.

TABLE 2. Composition varieties of gas mixture

Design points	Pseudocomponents			Original proportions of components		
	x'_1	x'_2	x'_3	x_1 (Ar)	x_2 (CO_2)	x_3 (O_2)
1	1	0	0	100	0	0
2	2/3	1/3	0	94	6	0
3	1/2	1/2	0	91	9	0
4	1/3	2/3	0	91	9	0
5	0	1	0	82	18	0
6	2/3	0	1/3	94	0	6
7	1/2	0	1/2	91	0	9
8	1/3	0	2/3	88	0	12
9	0	0	1	82	0	18
10	0	1/3	2/3	82	6	12
11	0	1/2	1/2	82	9	9
12	0	2/3	1/3	82	12	6
13	1/3	1/3	1/3	88	6	6
14	2/3	1/6	1/6	94	3	3
15	1/6	2/3	1/6	85	12	3
16	1/6	1/6	2/3	85	3	12

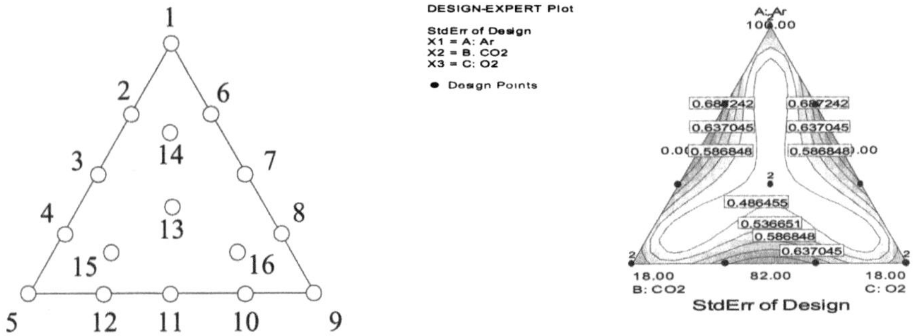

FIGURE 3. Points variates of gas mixture.

Beside geometrical characteristics and mechanical properties of weld join it was measured and analysed chemical elements ignition losses during the welding in dependency of gas mixture composition a welding conditions. All results are in Figure 3. Results of geometrical characteristics'are illustrated in Figure 4a., 4b., 4c. Results of experiment and 3D presentation of response surface are illustrated in Figure 5a., 5b., 5c. The figures are showing the Mixture Quadratic Model graphs.

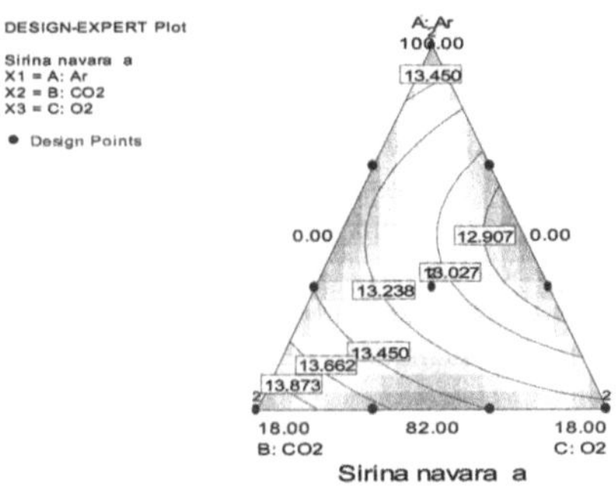

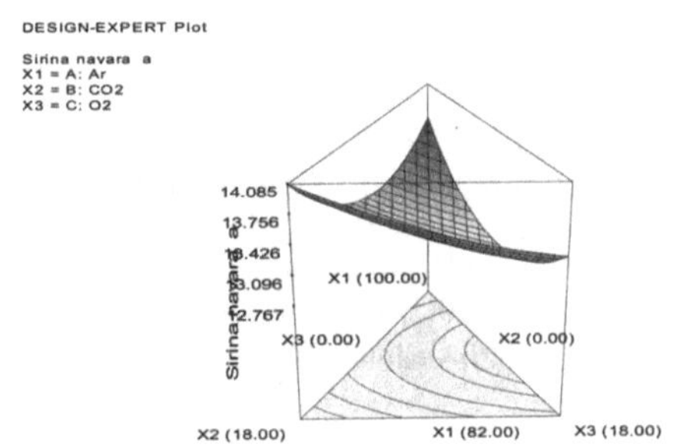

FIGURE 4a. Weld width FIGURE 5a.

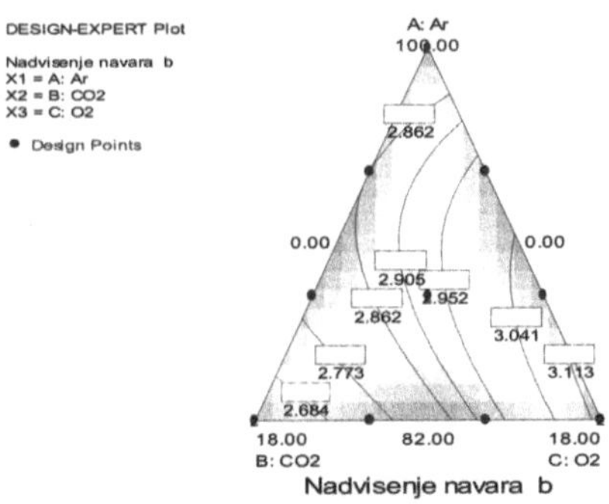

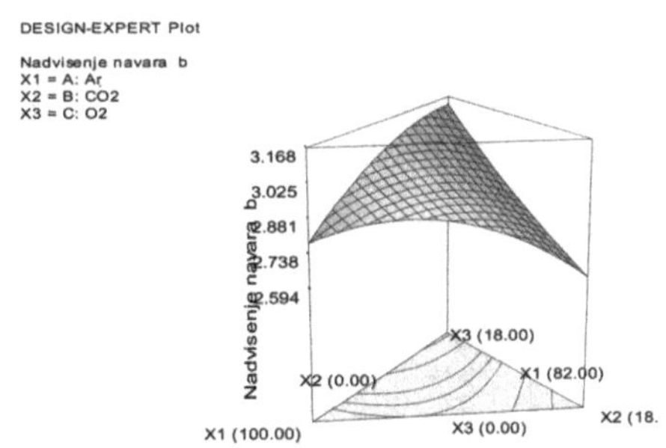

FIGURE 4b. Weld reinforcement FIGURE 5b.

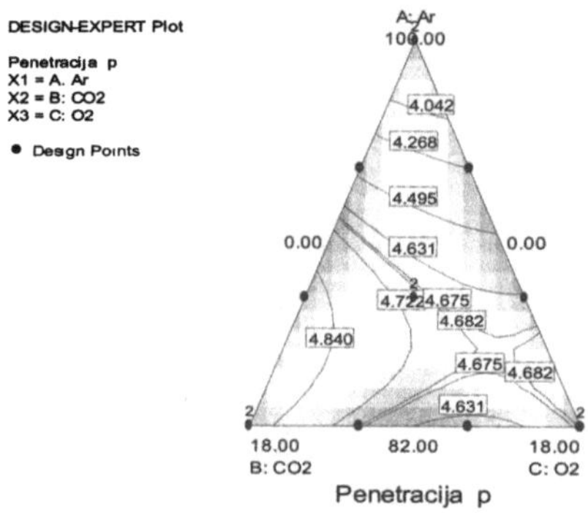

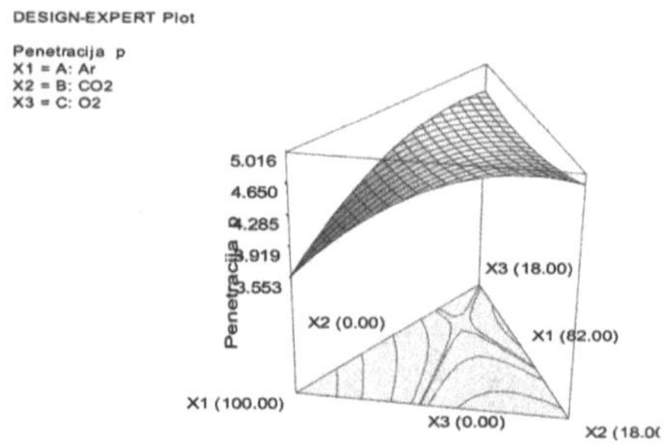

FIGURE 4c. Penetration FIGURE 5c.

4. CONCLUSION

With such model of design it is possible to get insight in effects of mixture compositions on responsive data. Technologists are able to directly optimise processes with graphical presentation of responsive surfaces. Numerical methods of optimisation are also available.

REFERENCES

1. Cornell, J.A., (1981), Experiments with mixtures, *John Wiley & Sons*, New York, USA.
2. Cornell, J.A., (1973), Experiments with mixtures: A Review, *Technometrics*, vol.15,No 3.
3. Myers, R.H., Montgomery, D.C., (1995), Response Surface Methodology (Process and Product
4. Optimization Using Designed Experiments), *John Wiley & Sons*, New York, USA.
5. Panić, V., (2001), Welding gasses influence on properties of welded joint, Master thesis,
6. Faculty of Mechanical Engineering and Naval Architecture, University of Zagreb, Croatia.

SELECTION OF OPTIMAL CUTTING CONDITION IN COMPUTER AIDED MACHINING BY GENETIC ALGORITHMS

V. Gecevska[1], V. Pavlovski[1]

[1] Faculty of Mechanical Engineering, University "St.Cyril and Methodius", Skopje, Macedonia

KEYWORDS: optimization, NC machining, cutting conditions

ABSTRACT: The productivity and the efficiency of machining centers are influenced inherently by the cutting conditions and the quality of NC programs. This paper proposes a methodology for incorporating classic and heuristic technique to analyze cutting conditions during the optimization process. The problem of selecting optimal cutting conditions, where the formulation involves the use of empirical relations, is considered. Optimal cutting conditions are determined by machining parameters during the optimization processes. Classic (non-linear) and heuristic (genetic and fuzzy) technique associated with empirical relations make the optimization.

1. INTRODUCTION

Flexibility, quality of the machining, productivity and efficiency are the tendencies of the development of manufacturing. They are the principles of the concurrent engineering. Development of the computer's technology enables these tendencies in the manufacturing industry by applying of high-speed machining systems with automatically and flexibility, which made possible performing optimal technological processes of machining.

Machining centers, as NC machines, are presented as flexible technological structures with: high level of automatisation, integration of many machining operations with high speed of cutting, manufacturing of complex mechanical parts, computer numerical control, automatic change of tools and depot for tools etc.

The questions of selecting optimal machining conditions and machinability development of machining centers are performed by research of the producers and the laboratories [1,3,4]. The productivity can by evaluated by productive and nonproductive time. In NC machining, the machining time consumed in running a NC program to machine a workpiece takes up almost all the productive time. In this sense, the productivity of the machining center depends directly on the quality of NC programs. For effectivity production, in the laboratory researches, optimization methods for analysis of machining parameters, as indicators of machinability are developed.

For increasing of the efficiency and the productivity of machining process, in this research, the methodology for determines the optimal parameters of machinability is proposed. The research is realized for the manufacturing of prismatic parts of the machining centers. The maximizing of the technological and the economical effects in manufacturing is used for the optimization of machinability at the machining centers.

Published in: E.Kuljanic (Ed.) *Advanced Manufacturing Systems and Technology*, CISM Courses and Lectures No. 437, Springer Wien New York, 2002.

2. THE OPTIMIZATION OF CUTTING CONDITIONS

A selection of the optimal machining conditions is done by optimization of machining parameters for NC production of machining centers. The optimization methods have been performed for the machining process as object of optimization, Figure 1.

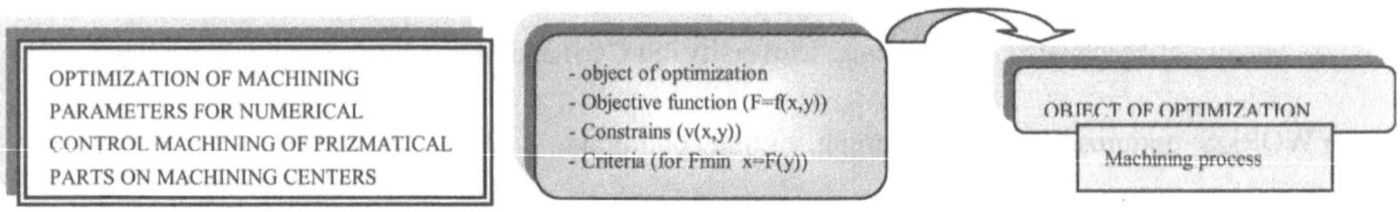

FIGURE 1. Defining the phases of optimization

In this research, the optimization process for NC machining with mathematical modeling of objective function, constraints and criteria for optimization is proposed. The mathematical model for objective function is defined to describe the object of optimization process and to determine the dependence between cutting condition, Figure 2. The equitation (1) is the objective function for machining time, for milling (1).

$$t_z = \frac{\pi \cdot D_{gl}}{1000} \cdot \sum_{i=1}^{n_p} \frac{a_i \cdot L_i}{V_i \cdot s_i \cdot \delta_i} + t_{pa} \cdot \frac{\pi \cdot D_{gl}}{1000} \sum_{i=1}^{n_p} \frac{a_i \cdot L_i \cdot V^{1/m^{-1}}{}_i \cdot s^{y/m^{-1}}{}_i \cdot b^{x/m^{-1}}{}_i}{(C \cdot K_v)^{1/m}} + B \tag{1}$$

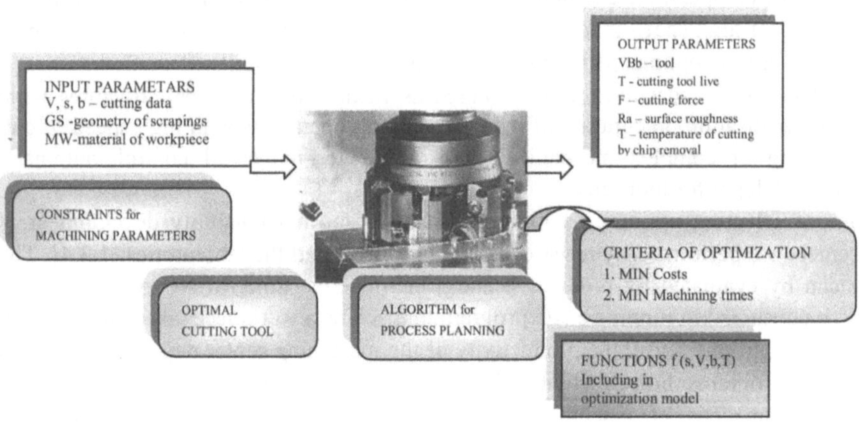

FIGURE 2. Mathematical model of objective function

Mathematical models, as equitation of constraints functions, are derived for the purpose in establishment of the interrelation between the machining parameters. Mathematical equitation are determined to use empirical and analytical relations for machining process and to involve experimental data [2].

The function of constrains are formulated from:

- Cutting tools characteristics and tool wear (2),
- Cutting tool life in different machining conditions,
- Quality and accuracy of the machining
- Properties of tool and workpiece materials,
- Geometry of the machining workpiece, characteristics of the main and idle movements (3) etc.

$$Vi \cdot s_{zi}^{\ y} \cdot \delta_i^{\ x} \leq \frac{C_v \cdot k_v \cdot D^i}{T_v^{\ m} \cdot b^q \cdot z^u} \qquad (2)$$

$$Vi \cdot s_{zi}^{\ y_1} \cdot \delta_i^{\ x_1} \leq \frac{60 \cdot 10^3 \cdot P \cdot n \cdot D^{i_1}}{C_F \cdot B^q \cdot z \cdot k_F} \qquad (3)$$

In this research, mathematical model of the function for cutting tool life in different machining conditions is proposed.

3. OBJECT-ORIENTED PROGRAMMING OF PROCESS PLANNING

In this research, it is created object-oriented algorithm for computer aided process planning for NC machining of prismatic parts of machining centers. A cutting operation (one cutting tool passes), as part from machining process is analyzed. Efficiency of the process has increased by the optimal order of the cutting operations.
The order of the operations influences economic effects of the machining. From this reason, in the research, project algorithm for the process planning of the optimal order of the cutting operations is defined, Figure 3, whiteout additional movement in machining process. Additional movement could not be involved in the algorithm, because of their constant machining values. Thus, they are defined constructively in the machining centers (time for turn off/on of the machine, time for change of tool, travel of workbench) etc.

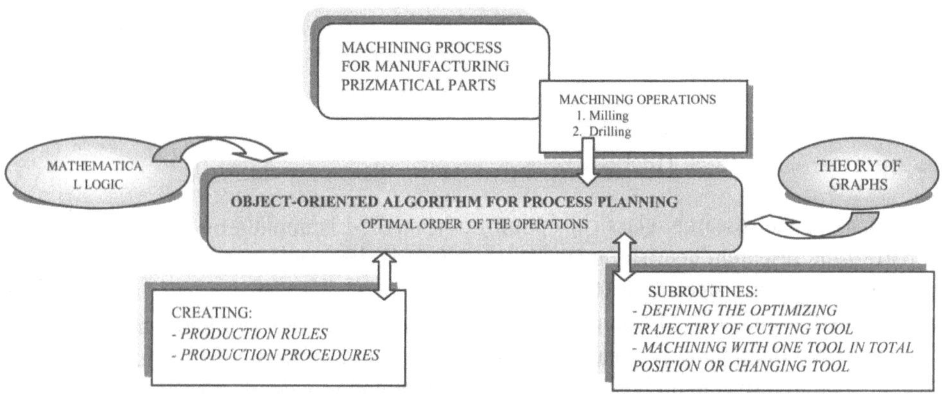

FIGURE 3. Projecting the object-oriented algorithm for process planning

Object-oriented algorithm for the process planning of the order of the cutting operation is modeled with C++. Using a theory of graphs and mathematical logic, production rules and procedures are defined. Their structure is based on the interaction between technological order of the machining (construction of workpiece, necessary machining operations, priorities of the cutting operations etc.).
The algorithm contents limits for optimized trajectory of tool movement among position points of machining of the prismatic part. In the algorithm, the several subroutines are created: the subroutine for defining the optimizing trajectory of cutting tool, the subroutine for evaluation of justification for machining with one tool in all positions where it is predicted, or changing tool in certain position.

4. HEURISTIC AND NUMERICAL OPTIMIZATION METHODS

The optimization of mathematical model in obtaining optimized machining parameters is made, using numerical and heuristic methods, Figure 4. Optimizing machining parameters are obtained as results (output) of process optimization of the mathematical model and included: cutting speed feed depth of cut and tool life.
The process optimization is made with criteria from technological and economical aspects. First criteria are expressed by process time structure. Second criteria are expressed by minimum costs of machining process. In this research, it is decided to create two different methods of optimization and comparative analysis of the results will by done.

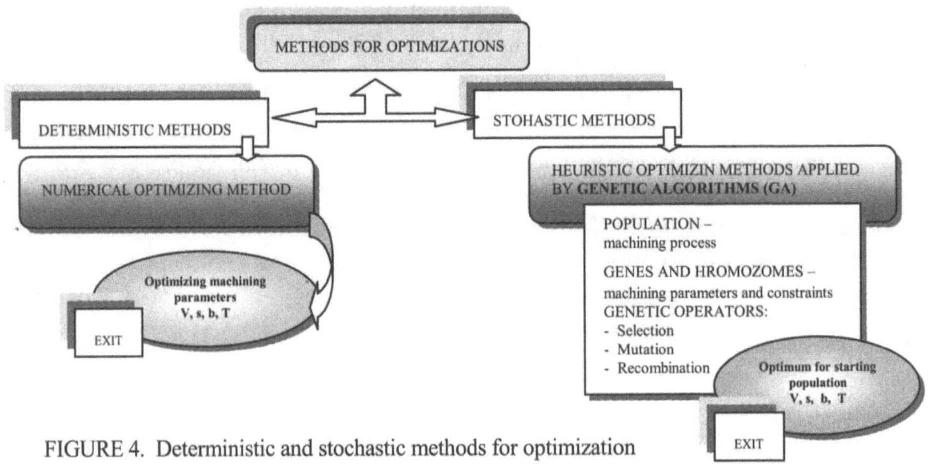

FIGURE 4. Deterministic and stochastic methods for optimization

In the first part of the research, classical optimization method is applied by Mat LAB. In the second part of the research, heuristic methodology of numerical optimization method are used by application the genetic algorithms (GA).
GA is the evaluative methods of optimization with heuristic approach [5], know-how methodology to decide the optimizing problems. GA, as heuristic optimization method, is applied for computer simulation of machining process. For GA, machining process represents a population, machining parameters with constraints are genes and chromosomes, and there must be defined genetic operators for evaluation, selection, mutation and recombination.

GA, with property for convergence after several populations, approaches an optimum for initial population. In the next phase of the research, it will be comparing between optimal results from classical optimization method and relative optimum from GA.
Verification and simulation of the results in experimental conditions are made. It is created computer program in C++, for connecting the optimizing results with NC program of machining center. In this mode, software has made adaptation of output parameters of NC program by GA optimizing parameters, for improvement of efficiency at machining process of machining centers, Figure 5.

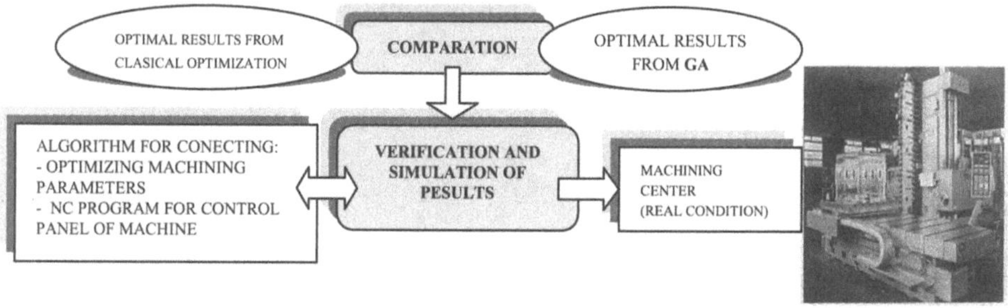

FIGURE 5. Verification and simulation of results on machining center

5. CONCLUSIONS

Methodology for receiving optimal machining parameters supplies a maximum of technical and economical effects in machining process for manufacturing the prismatic parts of machining centers. The contribution of this research is improvement of machining process and machinability. The directions of improvement are based at the aspect of performing the machining in conditions, resulted from optimizing algorithms, developed in this research: machining with optimized parameters, using cutting tools in optimal periods of tool life, optimal order of operations in process planning etc.
The intelligent methods of optimization, for machining process are applicable from many aspects of investigation. The intelligent methods produce interest for implementation in research of manufacturing.

REFERENCES

1. Carpenter I., Maropoulos P. (2000), Automatic tool selection for milling operations Part1. Cutting data generation; Journal of Engineering Manufacture, V214, 271-282
2. Rao S., Chen L. (1999), Determination of optimal machining conditions: A coupled uncertainty model; Journal of Manufacturing Science and Engineering, V121, 306-314
3. Stori J., Wright P. (1999), Integration of process simulation in machining parameter optimization, Journal of Manufacturing Science and Engineering, V121, 134-143
4. Wang W., Brunn P. (2000), An effective genetic algorithm for job shop scheduling; JEng. Manufacture, V214, 293-300
5. Goldberg D. (1989), Genetic algorithms in Search, Optimization & Machine learning; Addition - Wesley, USA, 407

AN ALTERNATIVE APPROACH TO ELASTO-KINEMATIC MODELLING OF MACHINE TOOL STRUCTURES

H. K. Tönshoff, S. Rehling, K. Tracht

Institute of Production Engineering and Machine Tools, University of Hannover, Germany

KEYWORDS: Hybrid multibody system, elasticity modelling, machine tool simulation.

ABSTRACT. When simulating manufacturing processes both, the kinematic behaviour of the machine tool structure and its elastic behaviour have to be taken into account.
The elasto-kinematic behaviour of machine tools can be calculated using the finite elements method. Alternatively, the machine tool structure can be modelled by a multibody system with extensions for elasticity modelling (hybrid multibody system).
In this paper a new approach for modelling elasticity within multibody systems is introduced – the Subelement Model. The calculation results of this model are compared to the calculation results of a finite element analysis on the basis of a simple structure with respect to accuracy. A further subject of research is the ability of the Subelement Model to model non-linearities (e.g. friction and contacts).

1 INTRODUCTION

Manufacturing accuracy of metal cutting machine tools is influenced by both the dynamical behaviour and the elastic deformations of the machine tool components. Especially at slim, heavily stressed components (e.g. slim cutting tools), the elastic deformations must not be neglected [1]. Consequently the modelling of these components has to take their elastic characteristics into account.

Currently, the mostly used method for modelling to elasto-kinematic behaviour of machine tool structures is the application of a dynamic finite element analysis (FEA). In that case the modelling of non-linearities like friction and component changing component contacts is very difficult. A dynamic FEA also needs a lot of calculation time depending on the complexity of the structure and the density of the mesh.

Within the scope of the development of simulation systems to simulate manufacturing processes, which demands easy modelling capabilities and moderate calculation times, this paper presents an alternative approach to model the elasto-kinematic behaviour of machine tools. This approach is based on multibody systems and extends them to hybrid multibody systems [4][5][6][7].

Hybrid multibody systems represent an extension to conventional multibody systems, in which structures are formed only by a number of rigid bodies which are linked with springs and dampers. Compared to normal multibody systems, hybrid multibody systems are extended by the ability to take the elastic behaviour of a structural component into account.

To model only the elasticity of heavily stressed components of the machine tool structure is usually sufficient. Thus all other components of the machine tool structure can simply be modelled as

Published in: E.Kuljanic (Ed.) *Advanced Manufacturing Systems and Technology*,
CISM Courses and Lectures No. 437, Springer Wien New York, 2002.

rigid bodies to imply their dynamical properties. In this case a multibody system with extensions for elasticity modelling can be deployed. An approach for such an extension on the basis of the multibody system ADAMS is introduced in this paper. It is called Subelement model and is exemplarily applied to the simple structure of a beam. The solutions are compared with the results of a finite element analysis using the FEA-system ANSYS.

2 THE SUBELEMENT MODEL

For the modelling of elasticity in any multibody simulation system, the approach of the Superelement-technique [6] and the approach of Wasmann [2] have been utilised and advanced to the Subelement Model. For this purpose, the modelled component is divided into Subelements. The following sections describe the structure of those Subelements.

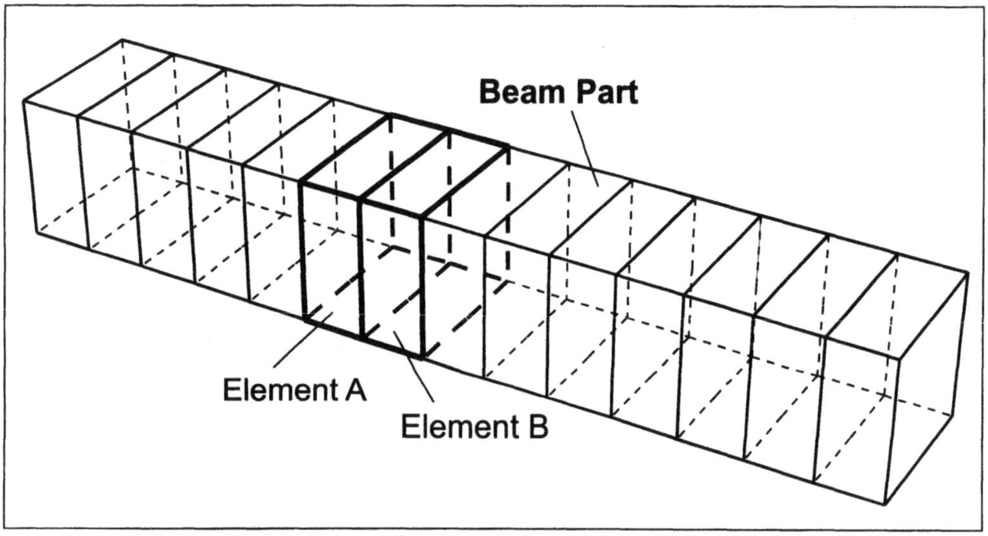

FIGURE 1. Beam Part divided into multiple rigid elements

2.1 SUBDIVISION OF COMPONENTS INTO SUBELEMENTS

As shown in figure 1, a machine tool component is subdivided into multiple Subelements along its predicted main deformation direction. Figure 1 symbolically illustrates this by an example of a beam, which is lengthwise subdivided into Subelements.

For the general case of load a two or three dimensional subdivision of the component into a planar or spatial grid is also possible. Size and physical properties of the Subelements may also differ. The variation of size and physical properties of the single Subelements allows the modelling of complex component geometries. Currently the shape of the Subelements is limited to cuboids. The development of shapes which are bounded by general polygons is in progress.

2.2 DEFINITION OF THE SUBELEMENT

As described in the previous section, elastic components of machine tools are assembled from Subelements. Those elements are linked with transmission objects, by which the elasticity of the component is modelled. Figure 2 shows the linkage between two adjacent Subelements using springs and dampers.

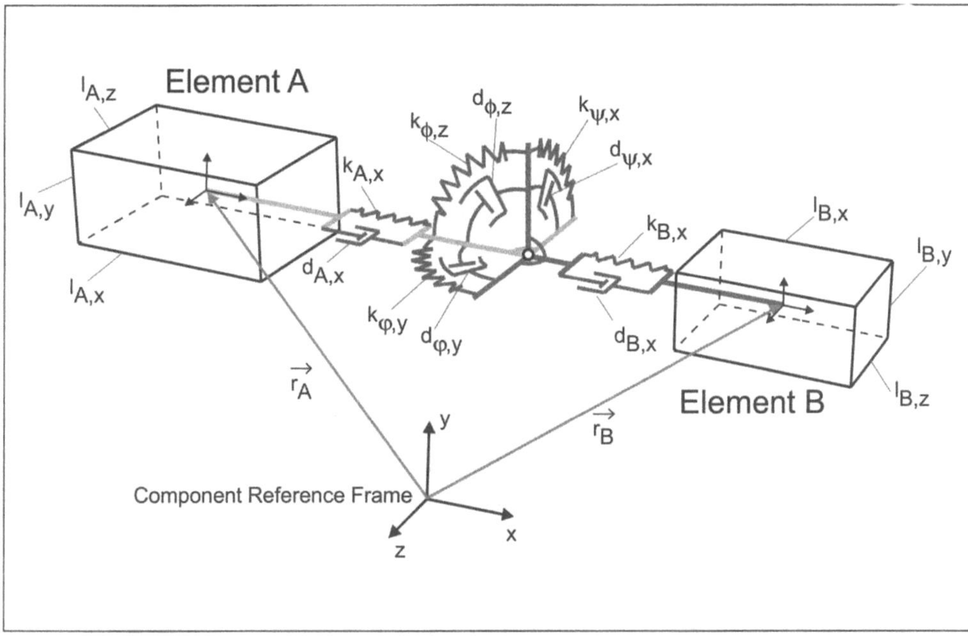

FIGURE 2. Linkage between adjacent Subelements

As shown in Figure 2 adjacent Subelements have four degrees of freedom (DOF) for deformation. These are bending around the y-axis ($k_{\varphi,y}$, $d_{\varphi,y}$) and around the z-axis ($k_{\phi,z}$, $d_{\phi,z}$), as well as torsion around the x-axis ($k_{\psi,x}$, $d_{\psi,x}$). The fourth DOF models the strain of the two Subelements due to axial forces along the x-axis ($k_{A,x}$, $d_{A,x}$, $k_{B,x}$, $d_{B,x}$). As suggested in [4], both DOF for shearing in direction of the y-axis and the z-axis are neglected.

The following section describes the calculation of the stiffness of the springs between the Subelements. With the modulus of elasticity of both elements (E_A, E_B), the stiffnesses $k_{A,x}$ and $k_{B,x}$ are obtained as follows:

$$k_{A,x} = \frac{2 \cdot E_A \cdot l_{A,y} \cdot l_{A,z}}{l_{A,x}} \qquad k_{B,x} = \frac{2 \cdot E_B \cdot l_{B,y} \cdot l_{B,z}}{l_{B,x}} \tag{1}$$

The stiffness of the bending spring $k_{\varphi,y}$ substitutes a series connection of two single bending springs ($k_{A,\varphi,y}$, $k_{B,\varphi,y}$) of the elements A and B:

$$k_{\varphi,y} = \frac{k_{A,\varphi,y} \cdot k_{B,\varphi,y}}{k_{A,\varphi,y} + k_{B,\varphi,y}} \tag{2}$$

The stiffnesses $k_{A,\varphi,y}$ and $k_{B,\varphi,y}$ are obtained by applying the principles of the calculation of bending beams [8][9]:

$$k_{A,\varphi,y} = \frac{24 \cdot E_A \cdot I_{A,yy}}{l_{A,z}^3} \qquad k_{B,\varphi,y} = \frac{24 \cdot E_B \cdot I_{B,yy}}{l_{B,z}^3} \tag{3}$$

The position of the Subelements with respect to the reference frame of the modelled machine tool component has to be taken into account when calculating the moments of inertia $I_{A,yy}$ and $I_{B,yy}$. With the displacement vectors $\vec{r}_A = \begin{pmatrix} r_{A,x} & r_{A,y} & r_{A,z} \end{pmatrix}^T$ and $\vec{r}_B = \begin{pmatrix} r_{B,x} & r_{B,y} & r_{B,z} \end{pmatrix}^T$ follows:

$$I_{A,yy} = \frac{l_{A,z}^3 \cdot l_{A,y}}{12} + r_{A,z}^2 \cdot l_{A,y} \cdot l_{A,z} \qquad I_{B,yy} = \frac{l_{B,z}^3 \cdot l_{B,y}}{12} + r_{B,z}^2 \cdot l_{B,y} \cdot l_{B,z} \tag{4}$$

The stiffness of the bending spring $k_{\phi,z}$ can be obtained the same way by applying equations (2) to (4). The stiffness of the torsion spring $k_{\psi,x}$ substitutes a series connection of two single torsion springs ($k_{A,\psi,x}$, $k_{B,\psi,x}$) of the elements A and B:

$$k_{\psi,x} = \frac{k_{A,\psi,x} \cdot k_{B,\psi,x}}{k_{A,\psi,x} + k_{B,\psi,x}} \tag{5}$$

The stiffnesses $k_{A,\psi,x}$ and $k_{B,\psi,x}$ are obtained by applying the principles of the calculation of torsion beams [8][9]. With the Poisson's ratios υ_A and υ_B of the two elements follows:

$$k_{A,\psi,x} = \frac{2 \cdot G_A \cdot I_{A,p}}{l_{A,x}} \qquad k_{B,\psi,x} = \frac{2 \cdot G_B \cdot I_{B,p}}{l_{B,x}} \tag{6}$$

with $G_A = E_A / (2 \cdot (1 + \upsilon_A))$ and $G_B = E_B / (2 \cdot (1 + \upsilon_B))$.

Once again, the position of the Subelements with respect to the reference frame of the modelled machine tool component has to be taken into account when calculating the polar moments of inertia $I_{A,p}$ and $I_{B,p}$:

$$I_{A,p} = \frac{l_{A,z} \cdot l_{A,y}}{12} \cdot \left(l_{A,z}^2 + l_{A,y}^2\right) + \left(r_{A,z}^2 + r_{A,y}^2\right) \cdot l_{A,y} \cdot l_{A,z}$$

$$I_{B,p} = \frac{l_{B,z} \cdot l_{B,y}}{12} \cdot \left(l_{B,z}^2 + l_{B,y}^2\right) + \left(r_{B,z}^2 + r_{B,y}^2\right) \cdot l_{B,y} \cdot l_{B,z} \tag{7}$$

Inside the modelled component only material damping occurs. As usual in structural dynamics, this is implemented by a stiffness proportional damping ratio ξ:

$$d_{i,j} = \xi \cdot k_{i,j} \tag{8}$$

2.3 APPROXIMATION OF ELASTICITY THEORY

The Subelement Model introduced in section 2.2 only approximates the elasticity of the modelled machine tool component. Beside of modelling only four elastic DOF the calculated stiffnesses are not constant values in reality. They vary with the deformation of the modelled component.

In case of modelling machine tool components, which have very little deformations relative to their dimensions, the arising error is negligible. Section 4 describes this error in detail on the basis of a simple example.

More critical are the analytically calculated modal parameters as described in [3]. These parameters have to be adjusted by applying a modal analysis on the machine tool component.

3 IMPLEMENTATION OF THE SUBELEMENT

The Subelements were implemented using an object-oriented architecture in C++. By integrating the application programming interface (API) of the multibody simulation system ADAMS (ADAMS/SDK), the connection to a simulation kernel was established. The implementation of the Subelements not only covers mechanical properties but also abilities which extend the standard functionality of multibody simulation systems. These extensions are described in the following sections.

3.1 MODELLING OF COMPONENT CONTACTS

In order to model contacts between components of the machine tool structure, a communication system was implemented into the Subelements. Via this system, forces between the elements, like friction or contact forces, can be added if necessary.

The boundary elements of two adjacent components, which can get into contact with each other are put together to a special group. Inside this group, the communication between the Subelements occurs autonomously without a controlling instance. By this technique, contacts between elements can be triggered. By variation of the simulation time step of the solver by the relating Subelement, the exact contact time can be determined.

3.2 DEFORMABLE GUIDANCES

Modelling of guidances is realised by restricting the relative DOF between the guidance component and the guided component. In case of a rigid linear guidance, this is implemented by fixing the orientation of the guided component's reference frame relative to the guiding component's reference frame. In addition, a guidance vector from the origin of the guided component is defined, which always has to have the same normal distance to the origin of the guiding component.

In case of a deformable guidance, the constraints between the two components are more difficult. This arises from the fact that the path of movement, on which the guided components moves, varies dynamically. It follows the deformation of the guiding component.

Thus the path of movement describes a spatial curve. This curve is not analytically describable, so it is described by discrete guidance points through which a spatial spline curve (NURBS) is interpolated. This spline curve is continuous in its second time derivative, so no steps in the acceleration function of the guided component occur.

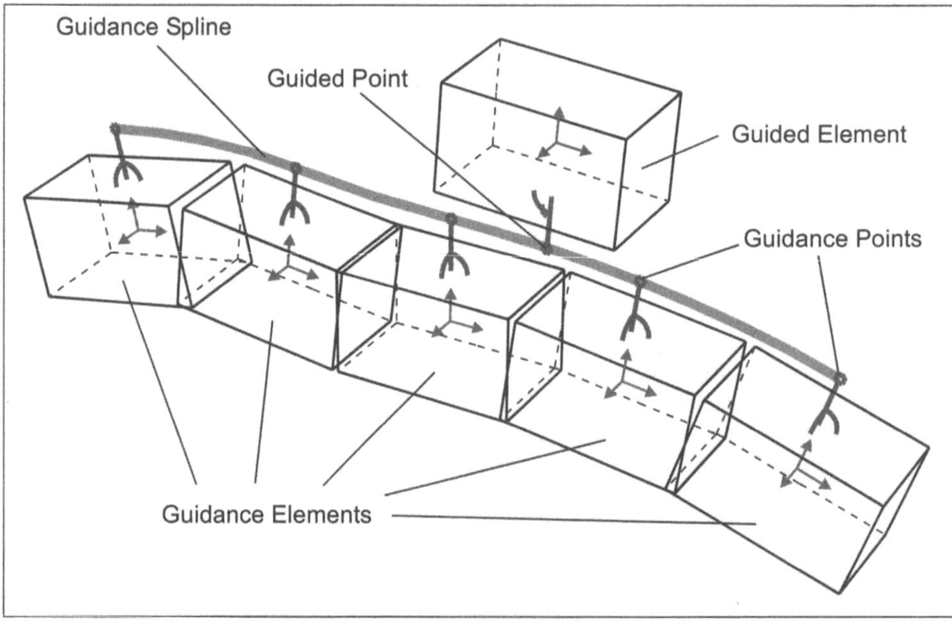

FIGURE 3. Deformable Guidance

Figure 3 shows the principle of such a guidance realised by the Subelement model. In each Subelement's reference frame one point is defined, which is part of the guidance – a single guidance point. The direction of the spline curve in the guidance points is parallel to one axis of the reference frame of the particular Subelement.

The guided Subelement also has a point, which is defined in the element's reference frame – the guided point. The guided point is also part of the guidance spline. The guided element's reference frame is oriented in such a way that the vectors of direction and curvature in the guided point are parallel to its axes. Consequently the guided Subelement follows the guidance spline.

4 COMPARISON TO FEA

This section compares the solutions of the Subelement model with the solutions of a FEA with ANSYS. This is exemplarily performed with the simple structure of a beam, which is fixed to ground at one end and deforms under its own weight.

Like shown in figure 4, the Subelement model of the beam consists of a spatial grid of connected Subelements. The number of Subelements in axis-directions of the co-ordinate system is given by the values n_x, n_y and n_z. The value for n_x is varied in steps from 5 to 100 to show the effect of the grid density on the solution.

The same beam was modelled and calculated in the FEA-system ANSYS with a total of 20 elements (SOLID95) and 248 nodes.

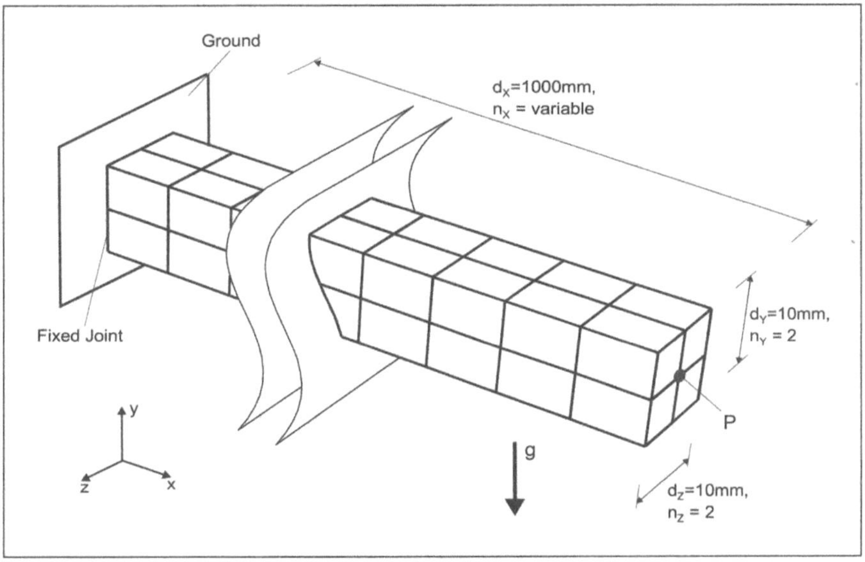

FIGURE 4. Beam modelled with Subelements

4.1 RESULTS

The solution of the FEA was taken as a reference for the displacement of the point P (shown in figure 4) and the first eigenfrequency around the z-axis. The displacement of the point P over the number of Subelements in direction of the x-axis n_x is shown in figure 5.

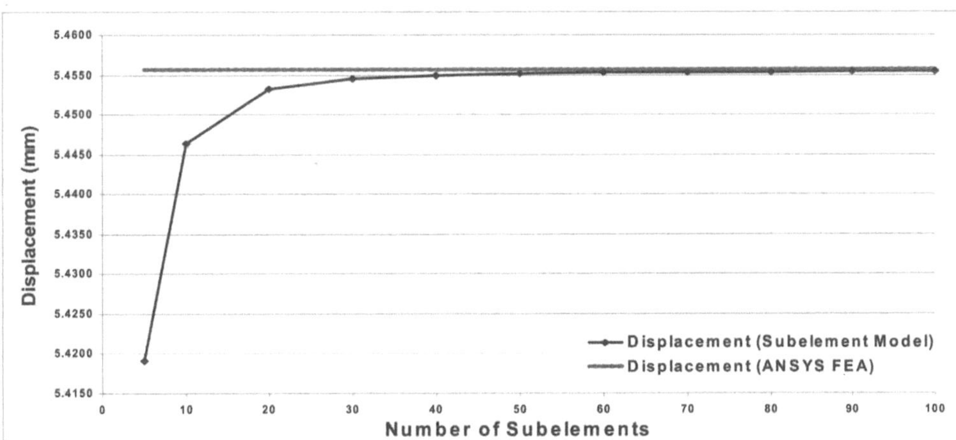

FIGURE 5. Comparison of displacement results

As shown in figure 5 the displacement value of the Subelement model converges to the displacement value calculated by the FEA for higher values for n_x. This shows that a sufficient approximation result for most applications can be achieved with a relative coarse grid.

The value of the first eigenfrequency around the z-axis was calculated by a modal analysis in ANSYS with the result of 8,3214 Hz. A Fast Fourier Transform (FFT) over the displacement function of the point P after applying an impulse force (Dirac delta-function) in this point gives a value of 8,3008 Hz for the first eigenfrequency around the z-axis. Here also is a good match between the solutions.

5 SUMMARY AND OUTLOOK

This paper described an alternative approach for elasto-kinematic modelling of machine tool structures, the Subelement model. The structure and implementation of this model was shown and the solutions were compared to the solutions of a finite element analysis.

The result is a simple calculation model based on multibody systems which can be used for machine tool structures in which the occurring deformations are small but not negligible. The simplicity and lesser calculation effort makes this model more applicable than complex FEA systems. Furthermore, the Subelement model can be implemented with any multibody simulation system.

The Subelement model will be further developed with respect to the element shapes. Cylindrical elements are currently under development to model round geometries like shafts. First approaches to shapes which are bounded by general polygons already exist. Furthermore an automated system for the segmentation of components into Subelements on basis of their CAD data is planned. Referring to this first approaches based on the meshing algorithms of FEA-systems have also been developed.

REFERENCES

1. Hann, V., (1983), Kinetik des Schaftfräsers, Fortschr.-Ber. VDI-Z Reihe 2 Nr. 66, VDI-Verlag Düsseldorf
2. Wasmann, U., (1995), Versuchsgestützte Modellierung von Werkzeugmaschinen auf der Grundlage statischer Nachgiebigkeiten, Fortschr.-Ber. VDI-Z Reihe 1 Nr. 263, VDI-Verlag Düsseldorf
3. Ahlborn, D., (1992) Meß- und Paramerterfehler bei der Modalanalyse von Maschinen, , Fortschr.-Ber. VDI-Z Reihe 11 Nr. 163, VDI-Verlag Düsseldorf
4. Hughes, P. C., Sincarsin, G .B., (1989), Dynamics of an elastic multibody chain, Part A: Body motion equations, Part B: Global dynamics, Dynamics and Stability of Systems, vol. 4, S. 209-244
5. Huston, R. L, Passerlello, C. E., Harlow, M. W., (1978), Dynamics of mulitrigid-body systems, Journal of Applied Mechanics, vol 45, S 889-894
6. Maißer, P., Tuan, P. A., Härtel, T. Freudenberg, H., Asch, M., Borgwardt, P., Kotzias, B., (1993), Modellierung und Simulation hybrider Mehrkörpersysteme mittels Superelementtechnik, Technischer Bericht im Auftrag der deutschen Agentur für Raumfahrtangelegenheiten Förderprojekt-Nr. 50 IB 9206, Institut für Mechatronik Chemnitz
7. Wie, H., (1992), Dynamische Simulation hybrider Mehrkörpersysteme mit beliebiger Systemstruktur, Fortschr.-Ber. VDI-Z Reihe 18 Nr. 108, VDI-Verlag Düsseldorf

8. Szabó, I., (1984), Einführung in die technische Mechanik, 8. Auflage, Springer-Verlag Berlin, Heidelberg, New York, Tokyo
9. Timoshenko, S. P., Goodier, J. N., (1982), Theory of Elasticity, Third Edition, McGraw-Hill Book Company Auckland, Hamburg, Paris

RELIABILITY AND AVAILABILITY OF COMPLEX PRODUCTION SYSTEMS

H.-P. Wiendahl , M. Hegenscheidt

Institute of Production Systems and Logistics (IFA), University of Hannover, Germany

KEYWORDS: complex productions systems, optimization procedures, operating curves

ABSTRACT: The increasing automation of production processes has given rise to a vast number of complex interlinked Production systems. According to a survey, the greatest hindrance to the use of complex assembly systems is a lack of efficiency caused by insufficient availability. Most complex assembly systems are linked in an elastic way, which means that short disturbances can be compensated by buffers. This sort of interlinkage of different systems is characterized as a highly dynamic process in which it is difficult to attribute the loss of availability to any particular production station. In order to get the best out of the investment, new approaches to analyzing availability need to be found. The paper gives an overview about existing procedures of optimization in the planning and operation period and presents a new optimization approach with operation curves.

1 INTRODUCTION

Numerous investigations by the Institute of Production Systems and Logistics have shown that especially the level of utilization and availability of interlinked production systems during operation frequently falls short of expectation [1]. This is mostly due to the configuration of the system structure (buffers, settings of cycle times, machine availability, number of stations, etc.) and other, indirect influences like personnel capacity and their qualifications. In connection with the disturbances occurring in the operating phase with different durations and probabilities highly dynamic systems arise whose yield performance is hardly to be prognosticated. Even today the so called "digital factories" do not cover this challenge in a proper way. The still unsatisfactory evaluation of the efficiency of production systems is rooted in the dynamic operational behavior, which predominantly depends on the breakdown characteristics and the type of interlinkage. It is of main interest to find out what influence the individual stations have on the entire system and how a harmonization in the sense of a "line balancing" can be achieved.

1.1 FAILURE PROBABILITY, RELIABILITY AND AVAILABILITY

The failure probability F (t) defines the probability of a failure or break down of a production station, a process or a component up to a point in time t. The complement of the failure probability is the reliability R (t) of a given station. R(t) defines the probability that a given station has not failed up to a point in time t [2]. According to example [3] the following applies:

$$F(t) + R(t) = 1 \tag{1}$$

With regard to several stations which are structured in a line or in parallel the entire reliability can be calculated as a function of the structure depicted in Fig. 1. This only applies when the events, e.g. failures at stations, occur independently. On the level of the production stations the number of "probable" failure cycles can be easily calculated, e.g. with the help of the average performance (*items per time*) and the failure probability.

Published in: E.Kuljanic (Ed.) *Advanced Manufacturing Systems and Technology*,
CISM Courses and Lectures No. 437, Springer Wien New York, 2002.

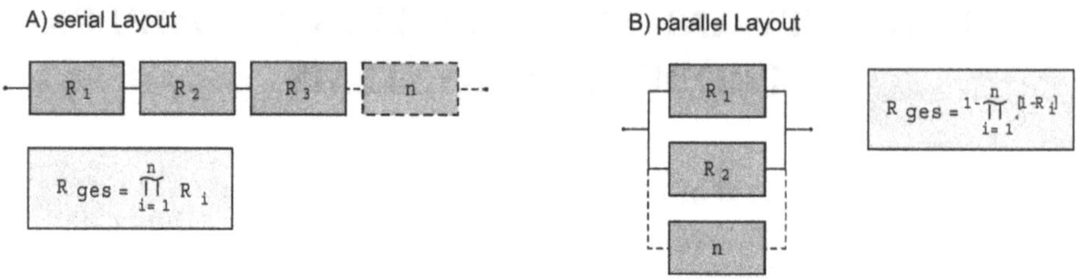

FIGURE 1. Reliability of stations located in a series or in parallel

By multiplying with the average standstill duration (mean time to repair, MTTR) one receives the summed breakdown time (TTR) for this period. The total of the productive time periods between the deadlocks (time between failure, TBF) results from the temporal difference from the runtime period (T_{run}) and TTR. The result is the technical availability:

$$A_s = \frac{\sum TBF}{T_{run}} \text{ with } T_{run} = \sum TBF + \sum TTR \tag{2}$$

A_s: technical availability

While the reliability exclusively represents the probability of an event, the availability is calculated by using the time periods derived thereof. Disrespective of the time periods considered, different types of availability exists. In the following, this term is understood without exception as a criterion for the evaluation of the technically caused standstills. A further evaluation criterion is the *utilization*. In addition to the time periods used for the calculation of the availability (T_{run}), the entire time period in use (T_{use}) is considered. This includes also time periods for preparation and maintenance procedures, pauses and the frequently neglected interlinkage losses (waiting and blocking of single stations) which are summarized in T_{org}. Hence, the utilization is much smaller than the availability:

$$U_{station} = \frac{\sum TBF}{T_{use}} \tag{3}$$

$$\text{with } T_{use} = \sum TBF + \sum TTR + \sum T_{org} \tag{4}$$

In particular for linked production systems the analysis of the entire utilization is crucial for the total yield performance. When considering the whole system the kind of interlinkage used is of essential importance.

1.2 TYPES OF INTERLINKAGE

In order to assess the effect a single station, especially the bottleneck, has on the level of system utilization, the type of interlinkage is of importance. In order to do so, two extreme cases are regarded, namely the ideally decoupled system (infinitely large buffers between the stations) and the rigid system (no buffers between the stations). In between these two extreme cases you will find the elastic coupling. The size of the decoupling is called *system elasticity*. It characterizes the ability of a station to compensate the influences caused by other stations. Figure 2 shows these three cases with the respective calculation of the system utilization. For the ideally decoupled system with unlimited buffers the level of utilization $U_{sys,decoupled}$ strictly depends on the bottleneck e.g. with the longest Cycle times within the production system. All standstills at every station can be compensated by the buffers. Therewith, the bottleneck determines the performance of the system in total. The level of utilization in a system with rigid interlinkage ($U_{sys,rigid}$) can be determined on the basis of the individual technical availabilities (A_i) of all stations.

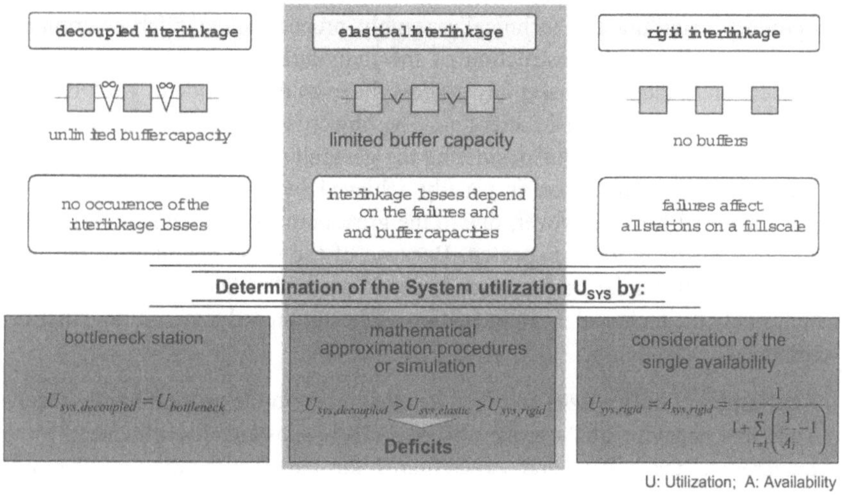

FIGURE 2. System Utilization in dependency of the type of interlinkage

The comparison to the sequential structure in Figure 1 is obvious. The substantial - and often disregarded difference – is, that in case of the rigid interlinkage large dependencies between stations occur. In the great majority of applications a station break down causes more or less immediately a waiting or blocking of all other stations due to the induced interlinkage losses. This reduces their runtime period which is used as the basis for the calculation of the failure probability. Following [2] and following tests carried out with real data this interdependence was studied at the IFA and lead to the formula depicted in figure 2. In summary, one can say, that in production systems with many stations we find a great amount of momentum, caused by varying cycle times, standstills due to breakdowns, routine maintenance, setup operations as well as by waiting and blocking periods of single stations. Hence, "line balancing" can only be achieved by taking the entire system performance into account – thus, focusing merely on cycle times will not do at all.

Main participants during the planning and configuration phase are the design engineers. Those who design the individual assembly operations usually are detail planners focusing on technological aspects. However, they generally have little knowledge of the dynamic behavior of the system as a whole. This shortcoming has to be eliminated with the help of a comprehensive technical design process.

2 THE CURRENT PLANNING PROCEDURE

2.1 DESIGN OF PRODUCTION LINES

As a rule, the starting point of the planning of assembly and production systems usually is a nearly finished product which has to be assembled as cost-effectively as possible. After analyzing the product regarding the technical assembly processes and after determining each working procedure, a detailed construction of the individual assembly stations takes place. During the allocation of the entire assembly process the even distribution of work contents onto the individual stations is the center of attention. The objective is to achieve a harmonized time tuning of the system to be installed. Considering the generally short cycle times a high level of utilization of each individual station is usually achieved by simultaneously automating and interlinking the material flow. However, due to the great number of automatic functions a high rate of short breakdowns is to be expected. Because of the close coupling of the individual stations this leads to substantial performance losses in particular. During the planning phase it is almost exclusively the technical implementation of the automated material flow that is placed into the foreground.

At first, the definition of buffer sizes and feed lengths linked to the process hardly receives any attention. Thus, the behavior of the system to be built is substantially influenced by this approach. Compared to the efforts and expenditures for the configuration of the individual stations, this step of designing the system as a whole receives relatively too little consideration.

3 EXISTING APPROACHES

The tools used in the first design phase for line balancing range from simple spread sheet analyses to analytic calculation models, based on the queuing theory (Markov chains, including the Simple Assembly Line Balancing Problem (SALBP) and Mixed Model Assembly Line Balancing Problem (MALBP)) [4], to the discrete, event-oriented simulation with partly specific analysis and interpretation modules.

3.1 STATIC CALCULATIONS

Firstly, technical designers execute simple calculations to determine the rate of utilization on the basis of forecast performance parameters. When following this approach the largest unknown quantity is the used characteristic, which is the availability in our case. If only those technical standstills are included which can be prognosticated by experience, losses from other unexpected events will be disregarded. In Addition, the dynamics of the standstills with regard to the point in time and the duration remains unconsidered. In case of an unfavorable constellation, e.g. in case of empty buffers, these standstills have an immediate effect on the entire system. If there are buffers with large capacities - and thus a high elasticity - between the individual operating systems and if standstills caused by technical reasons, maintenance and set up operations occur very rarely, a comparatively good prognosis can be obtained even with this static calculation.

3.2 QUEUING MODELS

The queuing theory is used for the planning of the configuration of chained production systems as well. The basic model of this theory consists of the actual queue, which characterizes an incoming stream of jobs or parts, and the work station (e.g. station 1), which processes these jobs. The arrival of the parts is calculated with the help of probability distributions (so-called Markov-chains use exponential distributions). The operating time, including standstills, is calculated accordingly. When using Markow chains the output-rate is an exponential distribution as well. This means it can be used as the input rate for the subsequent queue (e.g. station 2). Lager systems with plenty of stations can be mapped with the use of so called networks of queues, which are able to determine the results within a very short response time. An essential drawback is, that the input-process has no effect on other queuing elements (lacking an aftereffect). Thus, the recording of interlinkage losses is not possible. Another impediment is the lack of acceptance, which is not least caused by the very theoretic nature of the literature and an immense degree of abstraction which is necessary but hinders the direct use in practical applications. The accuracy of queuing models is clearly greater than the one obtained from the static calculation, it however can not reach the high level of simulation.

3.3 SYSTEM SIMULATION

Simulation is the most widespread technique for the study, interpretation and optimization of complex systems. Simulation offers the possibility of mapping real production systems with the use of computer programs. This model allows you to study and describe the behavior under modified simulation conditions. This works for systems, which can not be tackled in an analytical way. The discrete events of individual items occur along a global time calculation. That means that each new event can only be calculated and executed on the basis of the previous one. Thus the simulation technique can be applied to any type of problem. Despite its undisputed usefulness, simulation technology has not yet become a routine application in industrial planning departments. Small and medium sized enterprises in particular tend not to implement this technology because of expenditures still being high and a lack of necessary expertise and skills. Therefore, a mixed deductive-empirical approach is to be introduced in the following, which will empower the user to assess the performance yield of an production line much faster.

4 NEW APPROACH: OPERATION CURVES

At this point we would like to sum up the pros and cons of the above mentioned procedures. Our aim is to analyze the impact of the availability of the bottleneck station on the system utilization in a time- and cost-efficient way. A short response time - similar to the queuing models - with a high acceptance of the planners is to be achieved. The underlying assumption is that in a flexible linearly chained production system a bottleneck can be recognized on the basis of static observations. If one considers not only the cycle time but also the availabilities of the single stations in order to determine the system bottleneck, one can, according to Krüger [4], define in the case of a sufficient decoupling an efficiency factor. Depending upon the capacity of the buffers, stations with shorter cycle times than the bottleneck station are able to compensate pre- and postliminary standstills for a limited amount of time. This "supply of temporal work" can be evaluated by the inclusion of a performance component linked to cycle time in relation to the current bottleneck. The outcoming static bottleneck-linked efficiency E of a station i is:

$$E_i = A_i \cdot \frac{T_{c,bn}}{T_{c,i}} \text{ with } E_{bn} = A_{bn} \cdot \frac{T_{c,bn}}{T_{c,bn}} = A_{bn} \quad (5)$$

A_i : *technical availability of the station i*
$T_{c,bn}$: *Cycle time of the bottleneck station of the system*
$T_{c,i}$: *Cycle time of the station I*
bn :*bottleneck*

A performance profile can be generated from these calculations. The bottleneck does not depend on the longest cycle time. Instead it is characterized by the minimum value of the efficiency defined in (4). An optimization of this station appears most plausible and raises the question to what degree the total system is improved as a result.

By means of a characteristic curve the gradient of the system utilization increase is to be represented as a function of the availability improvement of the respective bottleneck. The basic consideration for the development of the function is the fact that the gradient of the curve for a real system, depending upon buffer capacity, approximates the rigid respectively the ideal decoupled interlinkage to a higher or lesser extent. The effects of modifying the availability at the bottleneck A_{bn} for these two types of interlinkage can easily be calculated in accordance to equations in figure 2. Related to the respective availability parameter they state the upper and lower limit of the gain to the system utilization (fig. 3). The above made considerations show that improving the weakest station with an ideal decoupling (infinite buffer) leads to a coefficiency from 1 to the value to which corresponds the level of efficiency (E) of the second weakest station (limited by A_{bp}). The gradient of this upper limit then no longer runs rising linearly, but horizontally without further utilization gain. Due to the reality the availability of the bottleneck can only moduled until 100%.

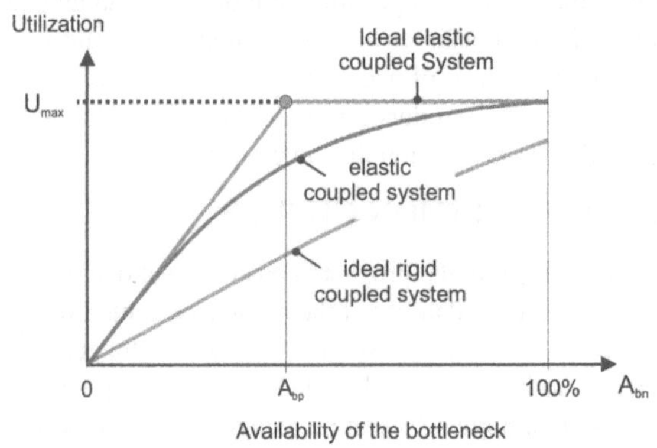

SEQARABICFIGURE 3. Operation curve model

The lower limit of the rigid interlinkage runs almost linearly rising over the entire area at a far lower level. At the bottleneck availability of 100% the rigid coupled system cannot reach the maximum possible utilization. This is caused by standstills of non-bottleneck stations. The behavior of elastic coupled systems is situated between the depicted borderlines. They strongly depend on the actual system elasticity. If the buffer capacity in the system rises, the course of the curve changes from the lower to the upper limit.

A describing function for the operation curve of the elastic coupled system was developed on the basis of the C-standard-function [5], [6] and generalizes the circle. Every point on a circle with an radius of 1 fulfil the equation:

$X^2 + Y^2 = 1$ In general the C-standard function is defined as $|X|^c + |Y|^c = 1$. (6)

After a few steps of transformation (switching, stretching and shearing), the result is an approximation to the real operating performance curve through the vectorial representation of the system utilization level. The running parameter is t with a value between 0 and 1.

$$\vec{U}_{sys} = \begin{bmatrix} t \cdot A_{bp} \cdot \alpha_1 - A_{bp} \cdot \left(1 - \sqrt[c]{1 - t^c} \right) \\ U_{\max} \cdot \left(1 - \sqrt[c]{1 - t^c} \right) \end{bmatrix}; t \in [0,1] \quad (7)$$

After elimination of the *parameter t* the equation becomes:

$$A_{bn} = U_{sys} \cdot \frac{A_{bp}}{U_{\max}} + A_{bp} \cdot \alpha_1 \cdot \sqrt[C]{1 - \left(1 - \frac{U_{sys}}{U_{\max}} \right)^C} \quad (8)$$

U_{max} : *max. reachable level of utilization*
U_{sys} : *utilization of the entire assembly system*
A_{bp} : *breakpoint of availability*
C : *form parameter*
α_l : *stretching modification factor*

With an adjusted parametric description the shape of the curve can be depicted between the extreme cases of an ideally elastic and an ideally rigid interlinkage.

With the easily established parameters U_{max} and A_{bp}, the development of the utilization level can be described in the form of the operating curve. A basic requirement is the experimental ascertainment of the form factor C and the stretching modification factor α_l which are determined, among other things, by the elasticity of the system. These two parameters are sufficient to approximate the specific system situation. They cannot be simply derived from generalized judgments. It is necessary to perform simulation studies to extract them.

Equation (7) calculates the bottleneck availability in dependency of the utilization. For a better understanding of the general statement the axes in the diagram are mirrored. The algorithm calculated as the best values c of 4,83 and an α of 908. The calculated function parameters are then compared afterwards to the data record of the simulation experiment already available in the data base. The first results of selected experiments are promising and show an average deviation between the calculated curve and the simulation results of approx. 2% within a range from -3,5% to +0.5 % (fig. 4).

This specific example shows that the system elasticity is by far insufficient to move close to the max. possible utilization. The cycle time-referred "supplies" are also not sufficient in order to adjust the interlinkage losses caused by a multiplicity of standstills.

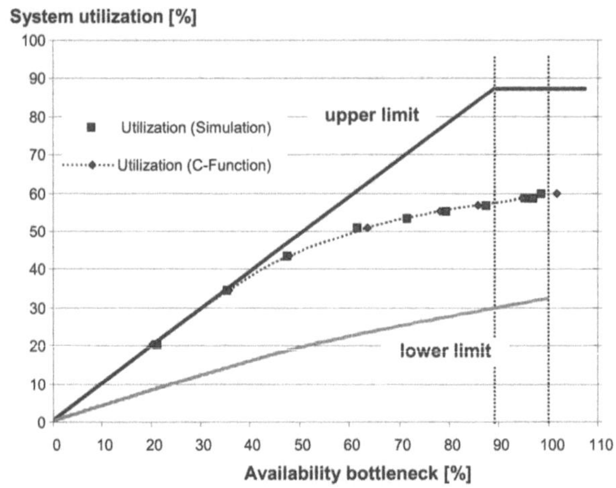

FIGURE 4. comparison of calculated and simulated utilization values

5 CONCLUSION

At present, the correlation between the optimization of the bottleneck behavior and the actual gain of utilization in an elastic coupled production system can only be determined sufficiently with the help of case-specific simulation experiments. According to many technical designers, however, the necessary efforts and expenditures for carrying out simulation projects are too numerous and too high. The objective must be to ensure a proper estimation of a case-specific situation without direct use of simulation. Therefore the developed approach is a first step. At present, the necessary function parameters must still be determined with mathematical algorithms. After detailed investigations into the most frequent basic patterns a further objective is to calculate these factors starting from purely static situation parameters also - and thus without any simulations. Based on these findings this methodology can be extended to further planning fields, e.g. the visualization of an optimal number of stations.

REFERENCES

1. Wiendahl, H.-P., Köhrmann, C.: International Strategies Used for Availability Optimisation of Assembly Systems. The International Journal of Advanced Manufacturing Technology, Volume 14, Number 9, Springer Verlag, London 1998
2. Bullinger, H.-J., Rieth, D., Euler, H. P., (1993), Planung entkoppelter Montagesysteme: Puffer in der Montage, Teubner Verlag, Stuttgart
3. Hans-Jürgen Warnecke (Hrsg.): Handbuch Instandhaltung (Maintenance Guide) Band 1: Instandhaltungsmanagement (Management of Maintenance), Köln, Verlag TÜV Rheinland, 1992
4. Krüger, Th. (2000): Nutzungssteigerung verketteter Produktionssysteme, Dissertation Universität Hannover, Fortschritts-Berichte Reihe 2: Fertigungstechnik, Nr.549, VDI-Verlag, Düsseldorf
5. Wiendahl, H.-P., Hegenscheidt, M.: Operating Curve Model For Describing The System Behaviour Of Complex Production Systems, International Conference of Engineering, Design and Automation (EDA) 2001, Las Vegas/USA 5.-8. August 2000
6. Nyhius, P., Wiendahl, H.-P., (1999), Logistische Kennlinien – Grundlagen, Werkzeuge, Anwendungen, Springer-Verlag Berlin Heidelberg New York

MODERN TENDENCIES IN ROTARY TRANSFER MACHINES

E. Gentili[1], C. Giardini[1], R. Pasquino[2], P. Quinzani[1]
[1] Department of Mechanical Engineering, University of Brescia, Italy
[2] Department of Mechanical Engineering, University of Salerno, Italy

KEYWORDS: RTM, parallelism, flexibility

ABSTRACT. Some important technologies changed Rotary Transfer Machines (RTMs): now all the productive system is governed by a computer, CNC reduced set-up time and improved precision. Today modular flexible transfer with 3-8 units (equipped with 3 CNC axes) perform sequential machining and can work infinite shapes and dimensions within a 300mm cube. Static turning using 3 axes units is also allowed. We want to relate the production costs to the new kind of transfer machines. The investment is bigger than that of traditional RTMs, but the production costs can be reduced. The paper presents an algorithm allowing the choice of the correct level of automation able to ensure the smallest production costs.

1 INTRODUCTION

A recent investigation, about Italian machine tools, shows that 32% of rotary transfers work more than 3 faces. This is possible with the new generation of rotary transfer machines. They are:

- transfers that execute all the machining in parallel. These machines are installed with a fixed operating unit, and 3,4 axes operating units. This kind of architecture gives high flexibility and productivity. In Italy some recent transfers can machine hundreds (more than 500) of different pieces with low set-up time (less than one hour) and high productivity: up to 10000 pcs/h (doubling productivity).
- modular transfer with 3-8 units 3 axes CNC perform sequential machining. The chief characteristic is the reconversion flexibility, but the productivity is lower than the precedent architecture. Today a modular transfer can machine infinite shapes and configurations in a 300mm cube.

The biggest result of the last few years of research is the possibility to machine (with RTM) complex pieces of brass, aluminium, steel, iron, as brass body of hydro-cleaners and complex pieces for the automotive sector.

For each situation there is the best kind of architecture and the right level of flexibility.

Flexibility is getting more and more important in rotary transfer machines. The most important dimensions of flexibility are:

- mix flexibility, measured by change-over time/saturation time;
- product flexibility, measured by time and costs to reach the markets, or measured by the number of kinds of articles that the RTM can machine;
- reconversion flexibility, measured by time and costs of reconversion;
- volume flexibility, measured by the maximum output obtainable.

A recent investigation about Italian machine tools shows as mix, product and reconversion flexibility are important for Italian use of RTM, while volume flexibility isn't so important

Published in: E.Kuljanic (Ed.) *Advanced Manufacturing Systems and Technology*,
CISM Courses and Lectures No. 437, Springer Wien New York, 2002.

Usually RTM machines work all day and they are never turned off. The only way to obtain the volume is to work a different number of shifts: two for low production, three for high production.
Some recent transfers are able to machine hundreds of different and complex kinds of pieces with low set-up times and good productivity.
Automatic load and unload are carried out using a robot with artificial vision. This means a unmanned production with a computerised surveillance system and tool control. Tool or insert change using economical tool life are allowed in order to prevent tool breakdown and scrap and rework costs.
The computer gives user-friendly and self-learning programming and gives the possibility to prepare part programs off-line; all working parameters are stored in the PC supervisor.
Production data and machine failure statistics, availability of hardware and software for DNC, connections, for modem tele-assistance are now at hand.
The software includes self-diagnostic functions and is able to display alarm conditions in real time.

2 FUNCTIONING OF ROTARY TRANSFER MACHINES

In a rotary transfer machine the transfer of pieces is made by a rotary cylindrical table. Rotary transfer machines can make complex cutting operations. To explain the functioning it is necessary to examine an elementary case [1, 2]:

- the workpiece has a shape of a cube and the faces are called C1..C6,
- the clamping devices are self centring jaws,
- the machining axes are perpendicular to the faces (e.g. drilling, vertical milling, threading, etc).

When a workpiece is locked two faces (for example C_1 and C_6) are parallel to the lateral area of the table. Face C_1 is in front of the table and is near the jaws, so it is impossible to machine this face. Faces C_4 and C_5 are partially encumbered by the collet of the clamping devices. Faces C_2 and C_3 are completely free, and it is possible to machine all the area utilising the lateral units. Face C_1 is completely free, and it is possible to machine all the areas utilising a radial unit. In this elementary case it is possible to machine simultaneously 3 faces (or 3 ways). Rotary transfer machines that machine simultaneously 3 faces are called 3-way.
To build a 3-way is easy and every constructor of rotary transfer machines can produce a good 3-way. The difficulty is to produce a good transfer that machines 4,5 faces of the cube in a single clamping. New technologies give the chance to machine 4,5 faces in a single clamping. With the rotary jaws we can work C_1,C_2,C_3 completely and C_4 and C_5 partially: five faces in a single clamping. With the rotary clamping devices we can work C_1,C_2,C_3,C_6 completely: four faces in a single clamping.

3 PARALLEL AND SERIES MACHINING

In a RTM we can have four levels of parallelism :

- The first level allows us to use more stations in a RTM. There are 3 kinds of stations: charge and discharge stations, auxiliary stations, and working stations. A RTM usually has 1-2 charge and discharge stations: in some RTMs there is just one station for charge and discharge, in other ones there are 2 different stations (S_c = 1-2). The working stations can

vary from 2 to 15 (S_w = 2-15). Usually RTMs do not have auxiliary stations (S_a). This kind of station is necessary only in particular situations. During the machining, the number of the works in progress is the same as the number of stations (WIP = S = S_c+S_w+S_a) and in every working station there is a different working stage.
- The second level of parallelism makes possible the use of more units in every station. Usually traditional transfers utilise up to 3 operating units per station. In particular situations it is possible to have 4 operating units per station. Modular transfers utilise one or two operating units per station: the limit is given by the dimensions of revolver heads.
- The third level of parallelism allows the use of multi-spindle heads on the operating units. In this way it is possible to use an operating unit to impart a spinning motion to many spindles: many parallel operations can be made simultaneously by a unit.
- The fourth level utilises special tools called forming or profile tools. These tools permit to execute more different and coaxial machining operations, such as drilling, milling and static turning. The important constraint is the rate of the spindle that involves different cutting speeds with different diameters of machining: for example if a profile tool executes simultaneously three coaxial holes: Φ5, Φ15, Φ25, the smallest one will have the lowest cutting speed and the largest one will have the highest cutting speed.

Now we can look at two new generations of RTMs:
- NC transfers that execute all the machining in parallel (as traditional transfers). These machines are installed with a fixed operating unit, and 3,4 axes operating units. They have similar machining centres. This kind of architecture gives high flexibility and productivity. In Italy some recent transfers can machine more than five hundred different pieces with a low set-up time (less than one hour). Other transfers allow to obtain high productivity, up to 10000 pcs/h, doubling the productivity. The most important constraint in this kind of machining is the dimension of the parts (less than 200x200x200mm), and the longer machining time that the piece needs. A recent investigation, about Italian machine tools shows that 58% of RTMs machine brass. There is correlation between brass and short time machining. This is the good workability of the brass.
- a modular transfer with 3-8 operating units, 3 axes CNC, performs sequential machining. Each unit has a revolver head with 4-8 spindles. The chief characteristic is the good performance in reconversion flexibility, but productivity is lower than in the precedent architecture. Today a modular transfer can machine infinite shapes and configurations in a 300mm cube. The chief constraints of modular RTMs are productivity and cost (more than 1.000.000$).

For every level of the parallelism the limits of the two classes are observed:
- first level: in the first class it is possible to have more than 16 stations whereas in the modular RTMs there are just 8 (the half).
- second level: in the first class it is possible to have 1-4 units per stations (usually 3) whereas in the modular RTMs there are just 1-2 units.
- third level: multi-spindle heads are often utilised in the first class, but seldom in the modular RTMs because of the encumberance of the system of revolver heads and multi-spindles.
- fourth level: in both these cases profile tools are used.

The capability of a traditional rotary transfer machine to carry out simultaneously more machining operations is better than the one of modular RTM (and this is the biggest

drawback), but modular transfers offer the opportunity to change tools using revolver heads (up to 8 tools per operating unit) carrying out serial machining.
Today many transfer machines can work different kinds of pieces. The numerical control improves set-up time because the time necessary to change the travels of the CNC units is zero. One way to shorten change-over times for rotary transfer machines is to get more use from each of the tool stations. The idea here is to keep the drive unit in place and only change the head and cutters. Then by adjusting the offset, feed rate and speed, the station is ready for the next workpiece. Specific tool stations are assigned a standardised operation, within certain parameters.
Revolver heads improve the times to change the tools because there are 6-8 tools on a head and the change is quick, only few seconds. Some important problems of revolver heads are: encumbrance, weight, inertia because the head is moving along X,Y,Z axis, eight is the maximum number of tools.
Constructors are searching for a new technology able to change tools with devices similar to the ones of machining centres (for example chain devices). This technology can solve the problem of 8 tools and the inertia, but can't solve the problem of the encumbrance which is perhaps more serious. Some clamping devices have special patented systems to quickly unlock. These devices shorten change-over times. Today the challenge is to develop a palletised system of clamping devices. The idea is like the palletised system in modern linear transfer machines, but they have a high level of encumbrance, so it is difficult to use this device in rotary transfer machine. This technology minimises the change-over time. In the last EMO (2001) a robotised system for changing jaws and clamping devices was presented by an Italian firm.
Another important innovation is a robot that charges and slots pieces automatically. The study is about recognising the kind of piece. Often the robot is too expensive or too slow, so another kind of charge and slot device is preferred.
The static turning units turn as a CNC lathe, so it is possible to carry out conical and profiled turning, single point treading. For the static turning is possible to have two different tools, and it is possible, for example to carry out sequentially turning rough and turning finish with one unit in short times.

4 ADVANTAGE OF CN AND COMPUTER IN RTM

Numerical control gives various benefits [3]. The most important are the opportunity to obtain lower set-up time and the opportunity to have great precision during the machining.
The number of axis controlled by numerical control can reach 60-70 axes per rotary transfer machine. Numerical control doesn't synchronise all the axes of the RTM, but there are groups of axes that need to be synchronised. The criteria used to create the groups are:

- The axis of a specific operating unit belongs to the same group, so they must be synchronised.
- The axes of a station (a station may have 1,2 or 3 operating units) belongs to the same group, so they mast be synchronised .

The reasons of for such a great number of controlled axes in RTMs are the great number of operating units in a transfer and the fact that the number of controlled axes per each operating unit is continuously increasing. While the traditional operating units have only the Z axis, or the spindle's axis synchronised with Z axis (max 2 axes), the recent operating units may have 5 or 6 axes that need to be synchronised by the numerical control.

A modern architecture utilises only modular operating units with 3 CNC axes (these modular operating units are like machining centres) installed on rotary transfer machine [4 - 11]. This architecture permits to work infinite shapes and configurations of workpieces. The numerical control is connected with a personal computer. The last one gives many benefits:

- Gives the possibility to insert the NC program, or simply to realise the program by means of user friendly graphic interfaces.
- Computer is a graphic interface and allows the operator to make sure that the machine functions correctly. Instead there are transducers placed in different parts of the machine connected with PC. In this way it is easy and instantaneous to see in witch zone of the machine the failure or problem has occurred.
- The network or modem connection gives the possibility to tele-monitor the right functioning of the machine (in the case of unmanned production). The PC identifies automatically the kind of failure, or the failures and if the failure requires the participation of the man. Software gives help on line to solve the failure, and eventually requires the participation, if needed, of a technician who can often modify software via Internet.
- Gives the opportunity to realise efficiently in process control and to correct automatically the dimensions of the workpiece.
- The PC supervisor stores working parameters and calculates efficiency measures, and stores the information coming from transducers and sensors and makes automatically control charts.
- The use of software is necessary to govern the robot. One of the robots (two robots) is /are utilised for the automatic charge and discharge of the workpieces. The robots can take the rough workpieces from palletised chains or conveyors. In the first case (palletised chains), the pieces have a specific position. In the second case (conveyors) the workpieces have random positions, so it is necessary to have an artificial vision system that identifies the orientation of the rough workpiece and allows orientation of the collet of robot in the right direction.

To choose a new RTM machine is a difficult task. The modern numerical control RTMs with computer and in process control needs great investment, but this gives important advantages that guarantee lower costs during the production. These advantages are:

- The production is more reliable (the number of failure decreases), CNC and computer permit a greater efficiency for the productive system (productivity increases).
- Better cycle times, thanks to the efficient synchronisation of the axes given by CN. It is possible to optimise the little passive time as rapid approach, time to rotate and translate (along the rotation axis to unlock the Hirth system) the table or the RTM.
- More flexibility in the machine, so it is easy reconvert the RTM to new production. This flexibility allows to lengthen the working life of machine tool. Often to change the kind of parts machined by traditional it is necessary reconvert the transfer, and it is necessary high reconversion costs. Besides the production system has a high lost production time.
- Time and Costs to Market is lower.
- Set-up time is strongly decreased.

The economic parameters are grouped in the following classes:

- Investment (cost to buy the RTM, and eventual future reconversions to change the production). The concept of working life of transfer is tightly correlated to the possibility to reconvert the machine to produce new kinds of parts. In the past there have been frequent cases of dismantling perfectly functioning transfer machines, but useless for the market (in facts they couldn't machine the new pieces that the market needed).

- Maintenance costs
- Variable costs directly or indirectly chargeable to the batch.

Tables 1,2,3 summarise costs regarding a new RTM.

TABLE 1. Investments costs

Investment	Opportunity to know the right value	Expensive charges
Costs of the productive system(RTM, robot for charge and discharge, etc..)	Known	Working life
Reconversion costs	Estimable	Working life
Costs of overhead exchanges of the lubricating oil treatment.	Known	Working life
Other starting investments	Known	Working life

TABLE 2. Maintenance costs

Cost items	Opportunity to know the right value
Routing maintenance	Known
Non-routing maintenance, in case of mechanic, electronic, CN, failure.	Estimable

TABLE 3. Variable costs

Costs items	Opportunity to know the right value	Expensive charges
Costs of pre-setting and balancing	Calculable	Directly to the batch
Manpower for set-up	Calculable	Directly to the batch
Other costs for set-up	Calculable	Directly to the batch
Tools cost	Calculable	Machining time
Direct manpower (during the machining)	Calculable	Machining time
Direct material (during the machining).	Calculable	Machining time
Electric power during the machining	Calculable	Machining time
Cooling by lubricating oil during the machining	Calculable	Machining time

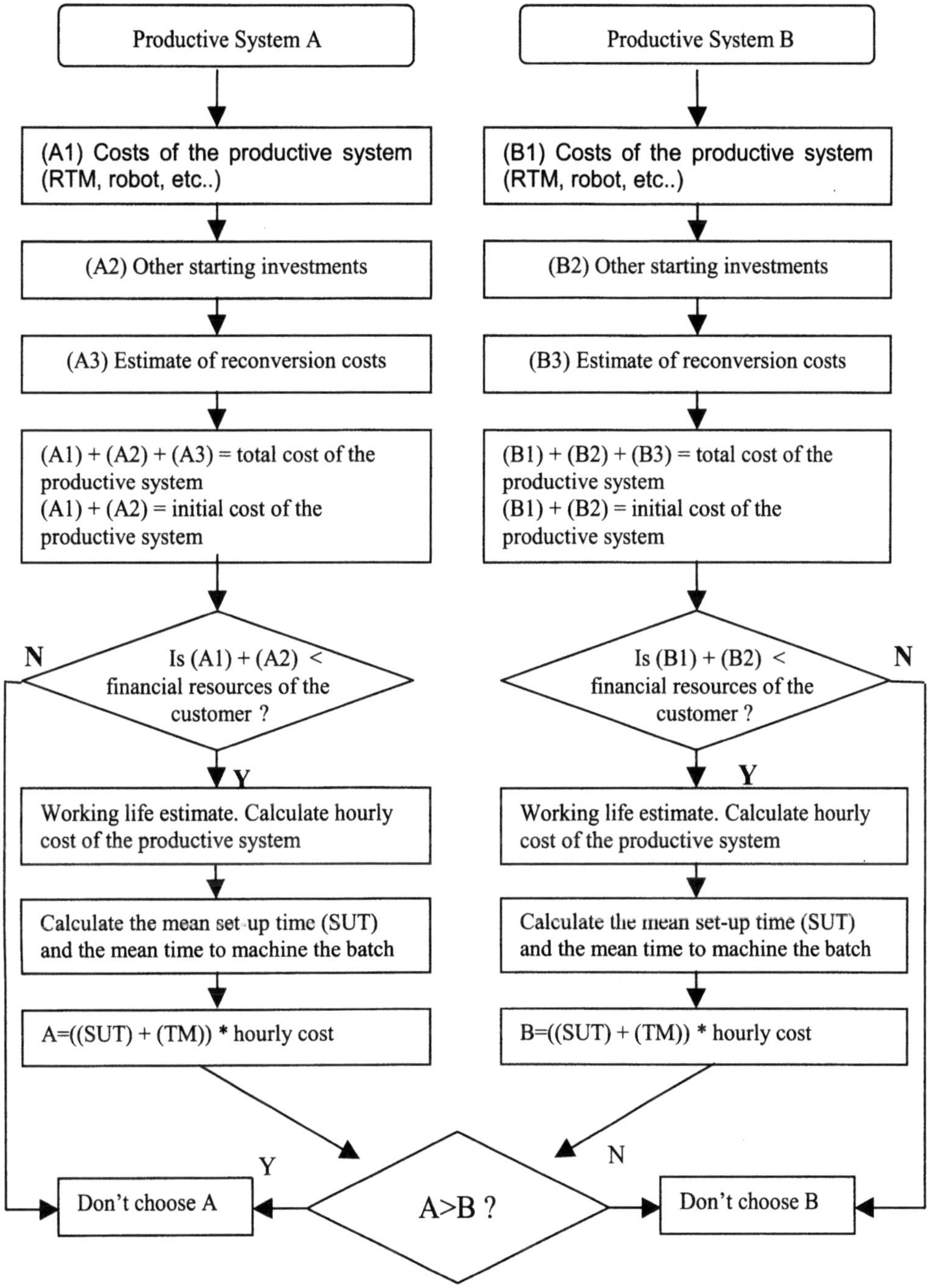

FIGURE 1. Choice between two productive systems: the proposed algorithm.

5 NEW TENDECIES IN ROTARY TRANSFER MACHINES (RTMs)

Modern NCs for RTMs allow to describe profiles using curves instead of points. One of the advantages of this methodology is the better surface finishing that is achievable. Another advantage is the reduction of the vibrations caused by the change of the cutting direction and this is very important considering that all the cutting operations are made simultaneously.
Another need of RTMs is the simplification of the communication systems [3]. The control for up to 60-70 axes simultaneously requires a great amount of wires. The solution, in the near future, is surely the innovation of the field bus.
As stated above another need is the possibility of small set up times, also when all the tools must be changed. The area where the worker would have to operate needs many hours to reduce the temperature and to be cleaned from the lubro-refrigerants and chips. The robot is the solution able to reduce to a hour the time for a complete set up of the machine.
The last improvement in cycle times will made possible by the use of linear engines, and this is a research field where the group of Mechanical Technologists of the University of Brescia is working.

ACKNOWLEDGEMENTS

The Authors wish to acknowledge the generous support offered by MURST in the realisation of this work. Special thanks to Mrs. Mary Flynn who checked the manuscript.

REFERENCES

1. Molinari S., Gentili E., (1999), New Technology in Rotary Transfer Machine, Proc. of the IV AITEM Conference, Brescia, 207-213.
2. Molinari S., Gentili E., (2000), New Ways to Manufacture Using Rotary Transfert Machine, Proc. of the Thirty-Third Int. MATADOR Conf., Manchester, 513-518.
3. Quinzani P. e D., Zapponi A., Gentili E., (2001), Nuove tipologie di controlli numerici per le macchine transfer a tavola girevole, Brescia Ricerche n.34, 21-27.
4. Giuliani http://www.giulianico.com
5. Gnutti transfer http://www.gnutti.com
6. Goss Deleeuw-Porta http://www.goss-deleeuw.com
7. Riello macchine utensili http://www.riellomu.it
8. Sinico S.p.a. http://www.sinico.com
9. Buffoli Transfer S.p.a. http:/www.buffoli.com
10. BTB Transfer http://www.btb.com
11. Porta Srl catalogue

THE MACHINE TOOL'S NEXT STEP: STEP-NC

M. Weck, J. Wolf

Laboratory for Machine Tools and Production Engineering, WZL RWTH Aachen, Germany

KEYWORDS: STEP-NC, ISO 14649, object oriented NC part programming

ABSTRACT. STEP-NC as a STEP compliant NC programming interface supports a bi-directional data exchange between CNC and CAM. Further more it helps to integrate the CNC into CAx world and supports high level geometry data formats.

1 INTRODUCTION

In modern industrial production the exchange of product data assumes ever-greater significance. Companies that wish to react flexibly, quickly and inexpensively to a new market situation must make information available as soon as possible in all areas, and must ensure that it can also be processed. Data conversions, costly follow-up processing of incomplete data and misinterpretations mean loss of time and consequently of money. Furthermore product quality and productivity also depend on an optimised production. It is thus a common interest of all large companies who are manufacturing and developing on a global scale at various locations to have unified powerful data formats and data interfaces all the way down from the planning to the production.

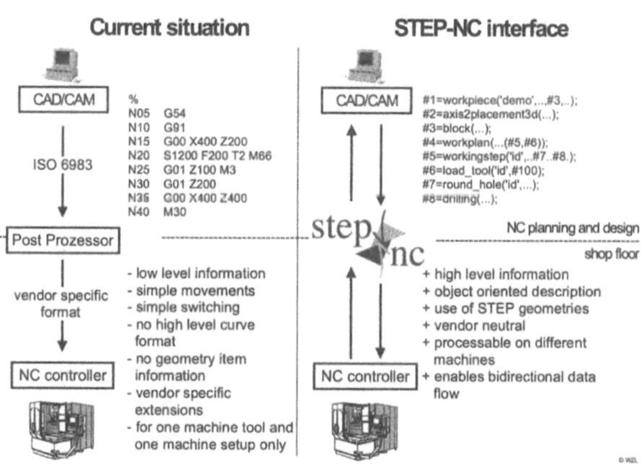

FIGURE 1. Deficits of current NC programming interfaces

While there are several new formats available for the exchange of geometry (IGES, STEP, etc.), the shop floor still deals with many paper drawings and poor NC programming interfaces, designed more than 40 years ago. Especially the part programming based on ISO 6983, also known as G and M Codes, is a bottleneck and forces the operator to often redo work already done in the planning department.

Published in: E.Kuljanic (Ed.) *Advanced Manufacturing Systems and Technology*,
CISM Courses and Lectures No. 437, Springer Wien New York, 2002.

A part programme compliant to ISO 6983 does only contain a small subset of the information available in the CAD/CAM system. Its sequence of simple linear and circular movements and instructions is not sufficient to support more complex machine functionalities, a modern graphical work piece visualisation or the modification of a complex NC programme at shop floor level.

By improvements during the last years, a large number of extensions were generated to enable today's numeric controls (NC) to realise powerful advanced functionalities like tilted tools on freeform surfaces or online process control mechanisms. Some of these extensions have been standardised, others are control-, machine tool vendor or end-user specific. As a result the NC programme is not useful for information exchange between the process planning department and the shop floor, but a list of instructions specific to produce one part on one machine tool.

Today the control as well as the CAM system has to be adapted to use the same specific codes and extensions. Otherwise the machine tool's functions cannot be addressed properly. Industries, using several different numerical controls, therefore are forced to use expensive post-processors, which generate control specific NC programmes that can only be used on particular machine tools. If the same NC programme should be executed on several basically identical machine tools, a common subset of programming codes and data formats having the same effect on each of these machines has to be used. Thus tool breakage control, collision avoidance or different kinematical abilities of the machine cannot be addressed in an optimised way.

Besides the limitations in part programming ISO 6983 and its extensions do not support bi-directional one-to-one exchange of NC programmes between different machine tools or computer aided systems (CAx). This means that any changes made directly in the NC code cannot automatically be fed back to upstream processes such as NC planning or design.

2 NEED FOR COMPREHENSIVE DATA IN THE NC

To realise complex functionalities like collision avoidance or a tool path generation depending on the tools available at a machine tool comprehensive information is needed. First the geometries of the work piece, the machine tool and the tools are needed. In addition a set of technological information is necessary to decide which overlap, cutting depth, feed etc. can be chosen. As in today's production only CAD/CAM systems and not the machine's numerical control have access to this information, these functionalities can only be realised offline during the NC planning and not at runtime. Variations in the clamping position, the tolerances of the raw piece or the actual tool's shape cannot be considered directly. First the changes have to be noted, passed back to the planning, where they are inputted into the CAD/CAM system and then the tool paths and machining sequence have to be calculated once again. The new data then is reduced to simple linear and circular movements and switch operations, as only they can be stored in common part programmes based on ISO 6983. In a last step the part programme is adapted to a specific machine tool's control by a postprocessor. Because of all these data manipulations and data reductions, conventional part programmes can hardly be changed at the shop floor. Furthermore they are bound to single machines, and they cannot be exchanged between different machines nor be used for the bi-directional exchange of information within planning, work preparation and production.

Thus the NC programme – the interface between planning and numerically controlled production – on the other hand represents a very weak link in a company's information flow towards production.

This weak link in data exchange is now being stabilised by a new, modern programming interface called ISO 14649. The standard has been developed in close collaboration with international CAD/CAM manufacturers, end users, renowned control manufacturers such as Siemens and the chair for machine tools of WZL at the University of Technology Aachen. In contrast to conventional NC programmes it is based on exchanging and processing characteristics – so-called features (e.g. pockets, borings, grooves), technological data sets like operations, tools and the processing sequence. Based on a feature the machining process is defined step by step with processing tasks. Roughing and finishing operations are part of this. Through the combination of several features and their processing steps, all operations necessary for the manufacturing of a part, starting from the raw material, can be described. The feature description is based on a data model containing geometrical, technological and organisational data. Since the data model is compatible with ISO 10303 (STEP), which is considered to become one of the major data models for geometry data exchange between CAD/CAM systems, the new standard offers an interface for constant forwarding of product information to a machine tool's NC control, which was hitherto only present in planning [1].

3 STEP-NC DATA MODEL

To receive high level information in the NC a suitable data model for information exchange between CAM and NC is required. This demand is the basis for the work of ISO TC 184 / SC1 / WG7, the working group that is responsible for ISO 14649.

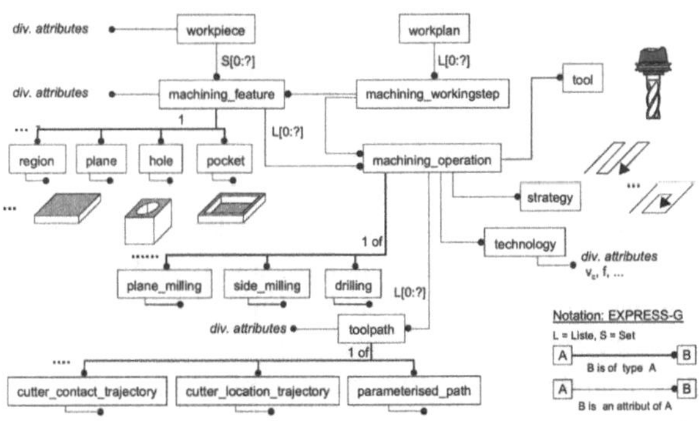

FIGURE 2. Simplified data model of ISO 14649 Part 10

In order not to lose the information's context ISO 14649 is designed in an object-oriented way. Each item can thus be defined as an individual object containing not only descriptive attributes, but also an implicit meaning. For instance the geometric data element "SLOT" might geometrical be considered as a subtype of "POCKET", but in terms of mechanical engineering there are different requirements concerning manufacturing and functionality related to these items [1].

By strictly separating geometrical, operational and process sequence data the access and the storage of the information is simplified and exchange between highly specialised modules becomes possible. A system which cannot process technological information but can display or manipulate geometry, as for example a computer aided design system (CAD), thus might easily use the STEP-NC programme file for data in- or output. At the machine tool there might be data masks to adapt the technological parameters without displaying all complex 3D surfaces.

Figure 2 contains a simplified abstract of the core data-elements as defined in Part 10 – General Process Data – of ISO 14649. It shows how geometry, represented by features, and the operations are defined separately and then combined in a working step, which can be processed.

The logical structure and sequence of the NC programme is defined by a so-called EXECUTABLE. An executable can be a simple WORKING_STEP, a list of actions (WORKPLAN) or elements for logical structuring: IF_STATEMENT or WHILE_STATEMENT.

The manufacturing of a part is defined by a sequence of several working steps. A working step defines which operation has to be executed based on a geometrical item, such as a feature (e.g. hole, pocket, slot), a region or a freeform surface. The operation itself is one more object-oriented element of ISO 14649, which contains technological information, tool data, security planes or distances and the process strategy.

The control is expected to interpret this data and to generate the required switching and movement operations to machine the part. If the functionality of the NC control is not sufficient or a very specific tool path is required to influence or optimise the process, STEP-NC here also offers solutions, ensuring the benefits of the data model's object-oriented and context based structure. For instance the attribute its_tool_path of ISO 14649 allows an operation to overload the path generation based on the feature's geometry and explicitly make use of predefined cutter contact, cutter location and tool orientation curves. Still the geometry information is available for display purposes, process simulation etc.

4 INTELLIGENT ANDS FLEXIBLE MANUFACTURING

In contrast to conventional NC programmes, STEP-NC for the first time provides a model for the complete description of the processing task. Information formerly limited to CAD/CAM systems becomes available to the shop floor and at the machine tool. Any loss of information to abstract movements and switching information, as in the case of conventional programming interfaces, is avoided. Instead, comprehensible units like features and operations are comfortably and clearly programmed. At the shop floor during the set-up of the machine tool and partly even at runtime the individual paths and switching commands are generated by the NC control. Since control is optimally adapted to the machine in question, not only the NC programming is simplified but complex processing tasks are also solved more efficiently than with CAD/CAM systems working offline and separately from machines. Controlled process data can be fed back into the control and there be used online for flexible path planning adapted to the processing situation. An example therefore is the active chatter avoidance. Whilst on the basis of conventional data interfaces such as ISO 6983 it was only possible to vary the technological parameters feed rate and rotational speed, the process can now be more efficiently regulated for instance by reducing the depth of cut. With conventional programming interfaces a change in the depth of cut could only have been performed offline in planning, since the necessary information on the volume to be machined has only been available at the CAD/CAM system. The data model of ISO 14649, on the other hand, embraces both raw-material and finished-part geometries, as in a CAD/CAM system. Volumes, tool paths or identations can be calculated by the numerical control.

One can imagine, that the machine tool is no longer only executing instructions, but also providing information and CAM functionalities. Production becomes more efficient and flexible.

5 FEEDBACK OF CHANGES AND KNOW-HOW

Changes in the NC programme are often necessary to improve the finished part and to optimise the production process. Today these changes cannot easily be fed back to the CAM system, as the context of single movement or switching instructions normally gets lost while generating conventional ISO 6983 NC programmes. That is why experiences of the shop floor are often not returned to the planning department and will not be available for the planning of future work pieces. The iterative improvement of the NC programme at the machine tool therefore becomes necessary even for similar parts.

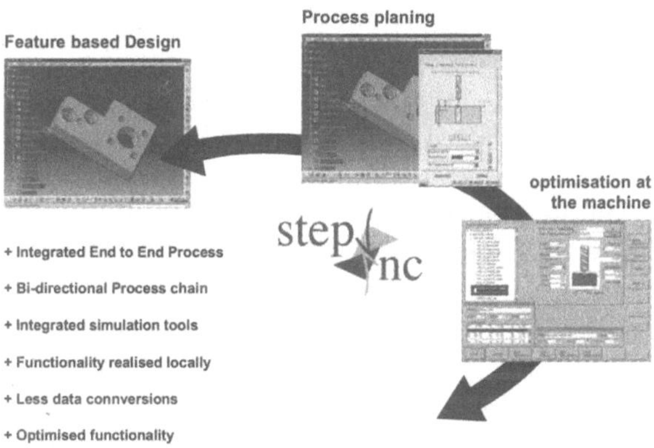

FIGURE 3. STEP-NC enables a bi-directional information exchange [Dassault Systèmes, OpenMind, Siemens]

At present sophisticated NC cycles or comments inside the part programme minimize the loss-of-context-problem. CAM and NC systems, which are able to re-interpret these user or vendor specific code extensions, can restore information and provide it through a data base to avoid redundant testing in the shop floor. But as the cycles are not standardised and as the included geometry model may vary, they are not suited to exchange information between different systems or companies.

Comparable to the NC cycles STEP-NC provides data elements for object oriented NC programming. But instead of defining single task, like a pocket cycle does, STEP-NC passes structured logical units (features, tools, working steps) and their internal relationships to the machine tool's control. Optimisations do not only result in changing single code lines, but in modifying the relevant attributes of an object. For example augmenting the depth of a drill hole does not simply mean to change "G01 Z100" to "G01 Z120" but the value of the attribute "its_depth" which is part of the feature's (object) "round_hole" description.

The modifications in an object oriented data model can be recognised and interpreted in NC planning or design. No information gets lost. Standardisation guarantees furthermore that this exchange is not limited to specific systems like the ones using vendor specific data element definition [2].

6 STEP-NC CONSIDERS VARIABLE COMPONENTS

A NC programme is used to transfer the movement and switching instructions from the CAM system to the machine tool in order to achieve one specific result, the part. Whilst the geometry of this work piece can be considered as a static data element, the tool paths followed to generate the work piece geometry, operation data and the hardware are variable and might depend on each other:

As Volvo Cars (Sweden) stated, one single NC programme executed on two identical machine tools with equal control- and tool configuration produces two different results. This difference is caused by the machine tools' specific tolerances and abilities, the tools and the part set-up.

Common NC programmes cannot consider all possible variations of these parameters. Therefore the programme always has to be tested and optimised at the machine tool it will be used on. The object oriented data model of STEP-NC treats geometry, technology or operations as single data. Only through a logical structure, the so-called MACHINING_WORKINGSTEP, the control gets the information which operation has to be processed on which geometrical item, e.g. a feature or a region. This allows the use of identical geometric descriptions on various machines while the more variable operation data is linked depending on the machine's characteristics and the actual machine set-up.

If a slightly modified work piece has to be produced (e.g. a pocket's depth is reduced or the position of a hole has to be changed) only the description of the geometry is changed. The already optimised technological and operational elements remain unchanged. Production on all machines can directly restart without the conventional need of programme generation and iterative parameter optimisation for every machine tool.

7 PRACTICAL RESULTS

Both, the European STEP-NC and the American Super Model project set up scenarios for milling. These first prototypes proved the benefits and potential of STEP-NC [3].

The prototypes realised by the European STEP-NC consortium for milling are based on commercial systems from Siemens, Open Mind and Dassault Systèmes. CAM modules from Open Mind (HyperFact, SoloCAM) and Dassault Systèmes (Catia V5) generate the STEP-NC programmes including high level CAD geometry data, operation data and sequencing information. This programme file then is passed to the shop floor, where a Siemens Sinumerik 840D control and ShopMill, Siemens' shop floor oriented NC programming tool, were adapted to process ISO 14649 encoded NC programmes.

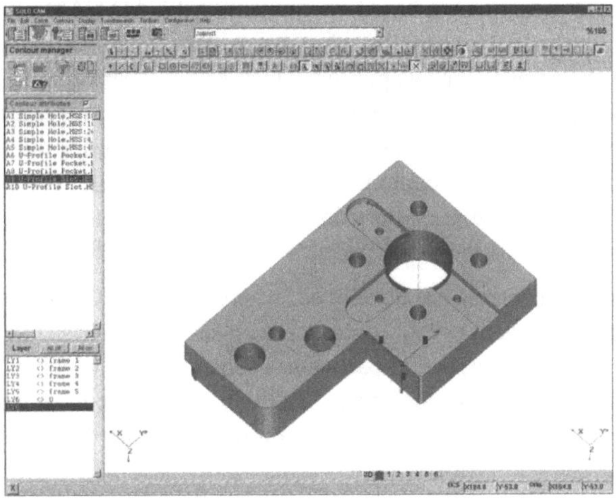

FIGURE 4. First part manufactured by STEP-NC

DaimlerChrysler and Volvo, participating as end-users in the STEP-NC project, use this prototype to gain experience with the new NC programming interface. As there are no more simple G- and M-codes passed to the machine, late changes in technology, programme sequence or even geometry can easily be realised and processed without the need of time consuming post processor runs and iterations with the planning department. Especially during the first test period some late changes in programme sequence and optimisation of technology had to be done. All these changes were visualised and simulated in ShopMill directly at the machine tool and stored back into the STEP-NC file to ensure, that no information got lost.

The next step within the European project will be the installation of a milling-scenario on machine tools with different kinematics and abilities (3 axis and 5 axis machines) in order to prove that STEP-NC supports exchangeability of programme files.

8 SUMMARY

Machine tools, their controls and additional components will always be installed depending on the end-users needs. Therefore every NC programming interface will be either too complex or much too simple depending on the required numerical control's functionalities.

While STEP-NC passes an object-oriented model to the control in order to enable NC controls to realise intelligent functionality in real-time and to integrate the shop floor not only as information recipient but also as an information supplier into modern industry's data exchange, there are other approaches that simply use the numerical control as a driver similar to a printer for a PC system. Here all information processing for axis-paths or PLC switch operations is generated offline directly by the CAM system. Real-time interaction as needed e.g. for process control cannot be realised.

The worldwide activities concerning STEP-NC are joined by the IMS STEP-NC project. The project focuses on several technologies in order to realise a complete and uniform up to date NC programming interface, ISO 14649. Further on this interface is tested in several different scenarios. On one side these scenarios are based on conventional CAD/CAM systems and NC controllers. They mainly evaluate the benefits of a more easy modification and exchange of NC programmes in the shop floor. For this purpose Open Mind integrated STEP-NC into their CAM module SoloCAM and Siemens implemented STEP-NC into their programming system ShopMill, which is running on a Sinumerik 840D. Volvo and DaimlerChrysler did already machine parts with drill holes, slots, pockets and contours, using this first industrial prototypes. During the next steps it is planned to extend the scenario using tool paths to manufacture and test more complex geometries. At the end of 2001 the Swiss partners will test the first STEP-NC prototype for EDM.

Besides more scientific scenarios are developed, where all CAD, CAM and NC data is stored in one database. Each CAx system or NC controller accesses the database to get or add information. There is no longer the need for several particular exchange files, which have to be archived and updated. Instead there is only one consistent STEP-NC data set, containing all relevant information to realise an intelligent and high productive production at the level of planning as well as in the shop floor.

REFERENCES

1. Brouer, N., (2000), NC-Steuerungskern mit Datenschnittstelle für eine Autonome Produktionszelle. Shaker Verlag Aachen, Germany.
2. Glockner, C., (1999), Integration von Facharbeiter-Erfahrungswissen auf Basis von Fertigungsfeatures. Shaker Verlag Darmstadt, Germany.
3. Hardwick, M., (2000), Art to Part in one STEP. Modern Machine Shop, Volume 73, No 2, USA.

AN AUTOMATED SYSTEM FOR DIMENSIONAL CONTROL BASED ON COMPUTER VISION

C. Borsellino[1], E. Lo Valvo[2], V. F. Ruisi[2]

[1] Department of Industrial Chemistry and Materials Engineering, University of Messina, Italy
[2] Department of Technology and Mechanical Production, University of Palermo, Italy

KEYWORDS: Measure, Computer vision, Automated visual inspection.

ABSTRACT. A totally automated control method for the dimensional control of pieces has been developed by Computer Vision techniques, combined with a movement equipment managed by the control system. By employing this technique it's possible to perform the dimensional control on manufactured articles having geometric characteristics that don't allow a trouble-free image acquisition maintaining a good resolution. By employing the same equipment an automated three-dimensional control system, based on stereoscopic vision has been developed.

1 INTRODUCTION

Automated visual inspection determines properties of products using visual information and it is very often automated by employing computer vision techniques. The computer vision field is growing rapidly, thanks to research efforts in the machine vision community and advancements in computer and vision technology. Over the years vision technology has matured; it has become more powerful and is now readily accessible to many more users, simply because it costs less and because it is easier to use. Nowadays automated visual inspection can provide cost-effective solutions to a number of "standard" visual inspection tasks, demanding accurate and reliable operation in real time [1-7].

The employ of computer vision increases the level of automation of several manufacturing processes, allowing substituting the human operator in such operations that, for their high recurrence, often give rise to errors or can be dangerous or difficult to realize. Moreover its use is necessary for measurement operations when the contact between the measure instrument and the workpiece can compromise the measure or the workpiece characteristics. This kind of problems is usual, for example, in controlling of workpiece for electronic industry. Actually these components are characterised by high precision and very small tolerance. Moreover, very often, the ratio between length and width of the final workpiece is very high; this characteristic doesn't allow an easy control by computer vision. In fact, in television images, the length/height ratio is 4/3, so if the whole piece is framed, there will be a loss of resolution because of the relevant length of the piece; if instead only a part of the piece is framed, the entire vision of the piece is loss.

In the present work a technique to solve this problem is presented: in completely automatic way, a series of images is acquired and then arranged in a unique image that has the resolution of the single frame. For this kind of operation a particular hardware system is needed and a software procedure that perform the acquisition and composition of the images with the maximum precision has to be developed, thus to obtain the necessary information for the workpiece dimensional

Published in: E.Kuljanic (Ed.) *Advanced Manufacturing Systems and Technology*,
CISM Courses and Lectures No. 437, Springer Wien New York, 2002.

control. In this paper a particular example is proposed to show the capability of the computer vision system, the case of a workpiece obtained by blanking has been examined; it is composed by twenty identical elements, each of them is a heat scatterer for electronic components (Figure 1).

FIGURE 1. The analysed element.

2 THE MACHINE VISION SYSTEM

2.1 SET-UP

The employed system (see Figure 2) is composed by a Personal Computer (Pentium III processor, 700Mhz) completed by a Matrox G400 Video Card and a frame-grabber card DIGITAL SE by CORECO, connected to a High resolution, Color Digital Camera CA-085A by SVS-VISTEK, equipped of a CCD sensor by SONY of 2/3" with a resolution of 1300x1300 pixel and 8 bit planes/color.

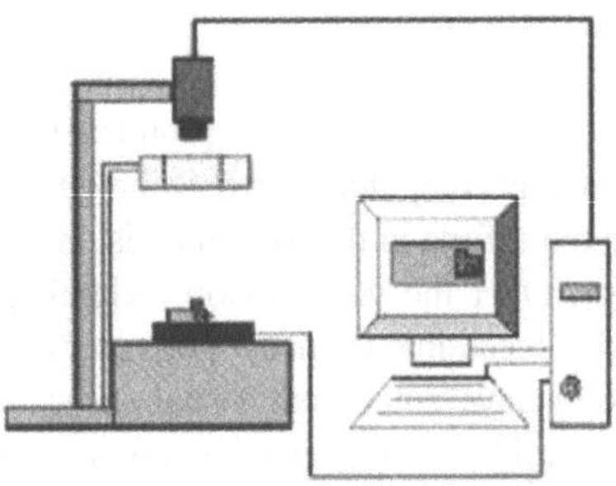

FIGURE 2. The set-up of computer vision system employed

To reduce some errors due to the image acquisition, the camera is equipped of a telecentric lens TEC M55 by Computar [3,8]. This lens allows a vision that strongly reduces the pillow and barrel effects that are typical effects in the traditional lens. In the case studied, after a calibration process, realized by using a sample, the obtained dimension of the pixel is 18.25 x 18.25 μm. A high precision motorized slide (IntelliStage by PI Instrument) has been employed for the motion of the workpieces. It has a stroke of 300 mm with a precision of 0.2 μm. A proper light system has been employed to reduce shadows and distortion in the acquired image.

2.2 THE EMPLOYED SOFTWARE

The software employed for the images acquisition and elaboration is WiT 7.1 by Logical Vision [9,10]. By its use it has been possible to perform the motion of the slide using proper inputs sent to the serial port RS-232. This software, that is extremely easy to program by means of block instructions (see Figure 3), is equipped with several libraries that allow a very high variety of functions such as filtering, visualization in several modes, geometrical and morphological transformations.

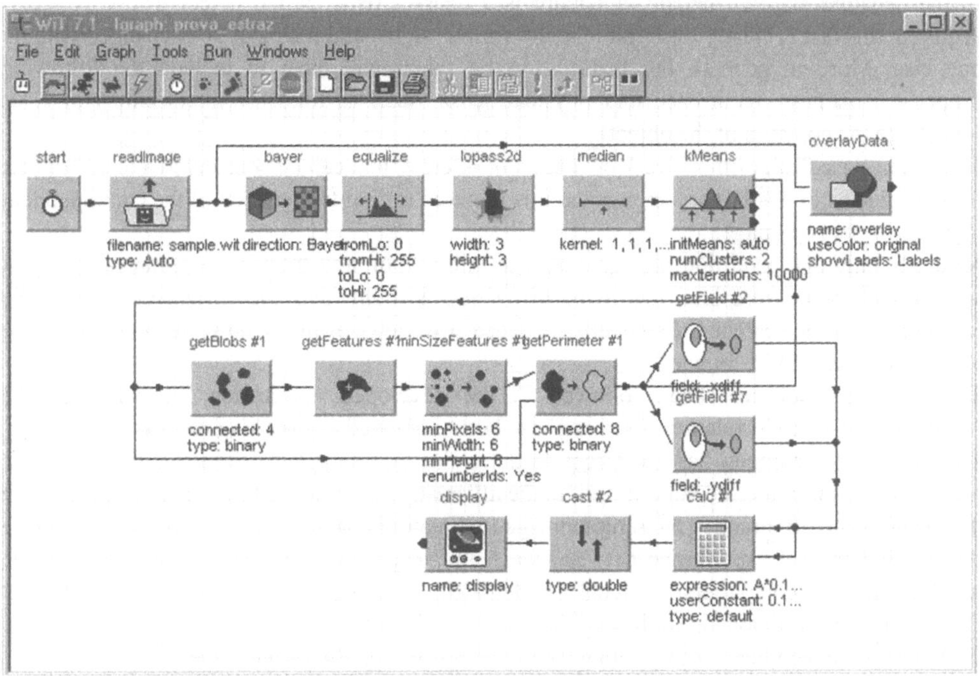

FIGURE 3. Example of program in WiT.

For our purposes some libraries have been particularly useful, they are the Segmentation, SerialIO e FastAlign libraries. The functions inside the Segmentation library allow the division of the image in groups of pixels with similar tonality, and calculate their areas, perimeters, barycentre and other geometrical characteristics. The function in the SerialIO library permit the envoy at the serial port of proper input strings, able to pilot the micrometric slide. The FastAlign functions allow the creation of models extracted by an image and the subsequent research of such models inside a second image; they finally find the displacement, the scale and the rotation to be applied to the model to let it match with the second image, with an approximation level defined by the user. Such operations are optimised to work in very short time and have shown to be very effective in the developing of our application.

3 THE PROBLEM OF THE MATCHING

The main topic for develop the dimensional control system is the overlapping of two associated images (matching). The problem is to locate a defined model with reference characteristics inside an image. As known, such problem has been faced defining several algorithms that can be summarised in two different approaches: the Area Based Stereo Algorithm and the Feature Based Stereo Algorithm. The algorithms related to the first approach generally perform the analysis of all the pixels present in the framed area, while the ones related to the second approach are based mainly on geometrical characteristics of the entities present in the acquired images.

In the "Feature Based" method, a feature is a characteristic of the object whose image is acquired, this last is clearly discernible in the various sights and it cannot be exchanged with no one else. More in general, the features, to be functional, have to include the following properties: oneness, recurrence and physical sense, such as lines, or intersections of lines (which are the contours of the object).
The "Area Based" algorithms work on a point of interest, that can be considered a local feature; the coordinates of this last are known and it is required to find the position of the correspondent point in the second image.
Naturally only one pixel does not give enough information and therefore a rectangular region around such point (template) has to be taken into account, thus the correspondence of the template has to be searched. Practically, between two views of the same scene a certain grade of similarity have to be founded.
This last approach has been followed in the developed application; in particular the "Normalized Cross Correlation Matching" has been employed as matching method [11].
This matching technique has the target to determine if a template, extracted from the first image, is present in the second one and, in the affirmative case, it has to estimate its position in the image plane. In practice the sample image (template) is passed trough all the second image (target) and the correspondence between the template data and the part of the target image involved in the comparison is locally calculated, locating the area for which the best correspondence is obtained. Supposing that an object image is described by the function $f(x-\xi, y-\eta)$, where the parameters (ξ,η) define the object position in the image plane.
If with $g(x, y)$ is indicated the observed image, this last can be written as follows:

$$g(x,y) = f(x-\xi, y-\eta) + n(x,y) \tag{1}$$

where $n(x,y)$ is a function that describes the background.
A measure of the local correspondence between template f and target g is the Euclidean distance that in the discrete case assumes the form:

$$d_{n,m}^2 = \sum_{k=-K}^{K} \sum_{l=-L}^{L} (g_{n+k,m+l} - f_{k,l})^2 \tag{2}$$

where $f_{k,l}$ is the light intensity of the point having coordinates (k, l) in the template and $g_{n+k,m+l}$ is the light intensity of the point of coordinates (n+k, m+l) in the target. Developing the square of the previous equation we obtain:

$$d_{n,m}^2 = \sum_{k=-K}^{K} \sum_{l=-L}^{L} f_{k,l}^2 - 2\sum_{k=-K}^{K} \sum_{l=-L}^{L} g_{n+k,m+l} f_{k,l} + \sum_{k=-K}^{K} \sum_{l=-L}^{L} g_{n+k,m+l}^2 \tag{3}$$

The first addend of the second term does not depend on the target and is defined norm of the template, and because this last is not function of (n,m) it does not influence on the research of the best correspondence, that is, in this case, the minimum of the defined function.
The last term instead depends only on the target and represents the "local energy" of the target, calculate in a region around the generic point (n,m) of (2K+1)x(2L+1) pixel.
Under the hypothesis that the local energy is constant, also the third addend can be considered as constant and than negligible.

The last term in the second addend is just the cross correlation that is the mutual correlation between the image and the template. Unfortunately, often there are uncertainty sources that cause the loss of validity of the described model; for example the lack of knowledge about the brightness of the target background. Normalizing the local brightness levels both inside the template and the target can solve the problem.
Thus the following entity are defined:

$$m_f = \frac{1}{(2K+1)(2L+1)} \sum_{k=-K}^{K} \sum_{l=-L}^{L} f_{k,l} \tag{4}$$

$$m_{g_{n,m}} = \frac{1}{(2K+1)(2L+1)} \sum_{k=-K}^{K} \sum_{l=-L}^{L} g_{n+k,m+l} \tag{5}$$

The so defined mean values have to be subtracted to the local value of f and g in the Cross correlation formula, this last will assume the following form:

$$\sum_{k=-K}^{K} \sum_{l=-L}^{L} (g_{n+k,m+l} - m_{g_{n,m}})(f_{k,l-m_f}) \tag{6}$$

Another parameter that can lead uncertainty is the contrast; also this problem can is solved by a normalization operation. So, defining:

$$\sigma_f^2 = \frac{1}{(2K+1)(2L+1)} \sum_{k=-K}^{K} \sum_{l=-L}^{L} f_{k,l}^2 \tag{7}$$

$$\sigma_{g_{n,m}}^2 = \frac{1}{(2K+1)(2L+1)} \sum_{k=-K}^{K} \sum_{l=-L}^{L} g_{n+k,m+l}^2 \tag{8}$$

we obtain the correspondence measure as follows:

$$\frac{\sum_{k=-K}^{K} \sum_{l=-L}^{L} g_{n+k,m+l} f_{k,l}}{\sigma_{g_{n,m}} \sigma_f} \tag{9}$$

If contrast and brightness are known, than it's possible to perform normalization with respect to both parameters thus obtaining the cross correlation coefficient:

$$\rho_{n,m} = \frac{\sum_{k=-K}^{K} \sum_{l=-L}^{L} (g_{n+k,m+l} - m_{g_{n,m}})(f_{k,l} - m_f)}{\sigma_{(g_{n,m} - m_{g_{n,m}})} \sigma_{(f-m_f)}} \tag{10}$$

This coefficient can vary between the value –1, for totally different images, and the value 1, for identical images. By means of the WiT the template recognizing procedure has been developed based on this parameter and employing in effective way the library called FastAlign.

4 THE CASE STUDY

As above mentioned, the case of the qualitative and quantitative control of a copper stripe for electronic component has been examined, the stripe is composed by twenty heat scatterer and it is obtained by blanking. The first step is the acquisition of ten images of length equal to 23.36 mm (1280 pixel of 18.25 μm each) that in total could contain the entire stripe. This operation is automatically performed by the software developed in WiT, by controlling the slide motion by proper commands on the serial port; a cycle of ten motion iteration is performed and then the image is composed, this last will be formed by 12800x1024 pixel, a dimension that is still manageable by the WiT. The second step was to highlight the main morphologic characteristics of the examined object.

A first comparison was done by recognizing the presence of the holes and the measure of their average diameter was performed. The measure of the same objects was obtained by a micrometric instrument to evaluate the performances of the system. In table 1 the values obtained with the two techniques are reported.

TABLE 1. Values of the holes diameter.

n. hole	D_C	D	n. hole	D_C	D
1	3.72	3.74	11	3.76	3.76
2	3.75	3.76	12	3.68	3.72
3	3.76	3.76	13	3.79	3.78
4	3.82	3.76	14	3.71	3.74
5	3.78	3.76	15	3.65	3.74
6	3.71	3.74	16	3.77	3.76
7	3.77	3.76	17	3.73	3.76
8	3.74	3.76	18	3.72	3.76
9	3.74	3.76	19	3.70	3.76
10	3.68	3.74	20	3.79	3.76

Figure 4 shows the trend of their difference:

$$E = | D - D_c | \quad (11)$$

Where D_c the average diameter obtained by Computer Vision and D the one measured. It can be observed that the difference is very small and always less than 0,1 mm.

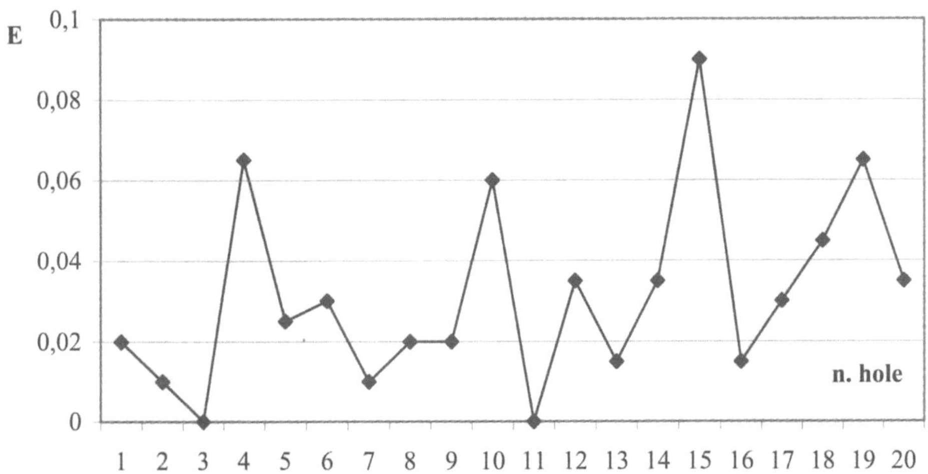

FIGURE 4. The E values obtained.

Subsequently the correct positioning of the stripe on the motion system has been evaluated. To solve the alignment problem, it is necessary to search on the total image the recurrence of a single element image (the heat scatterer) acquired in exactly horizontal position.

By employing the above-described technique of the *template matching,* the rotations of the single scatterer are evaluated in order to calculate the value of the most recurrent rotation angle. This information can be employed in two ways:

- It's possible to rotate the total image of the value of the calculated angle in such way to perform measures on the whole stripe (length of the stripe, distance between elements, center to center distances, between holes, etc...);
- It's possible to rotate the image of each template that has to be recognized inside the total image to extract information about a single element.

An example of qualitative control using a mask has been realized by creating some geometric entities (segments, rectangles, circles, polygons, etc...), by overlapping them to some details extracted from the image and measuring the distance between the intersections of the geometric entities with the profile of the detail.

To perform such measurement it is better to work on a binarized image, the threshold level is properly chosen by the system after an elaboration aimed to increase the contrast of the total image, with the aim to reduce the possibility of lost of detail in the image of the piece. On the obtained image a blob analysis phase is performed, this last consists in the extraction of entities with common characteristics and in the performing some measurements. Figure 5 shows an example of the measure performed.

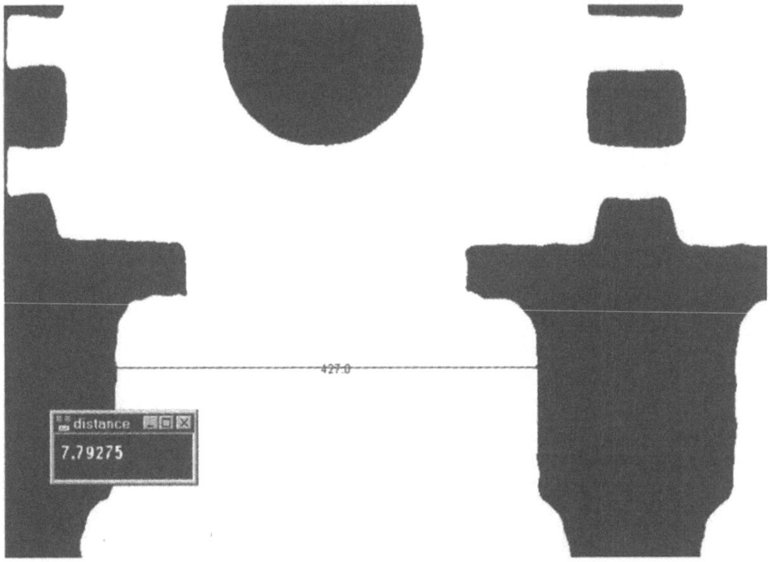

FIGURE 5. A measure example for the case study.

5 STEREOSCOPIC VISION

The developed system for the automatic measurement of plane workpieces can be employed also to perform three-dimensional measures. In fact the stereoscopic systems perform the three-dimensional assembly of the observed object starting from a couple of images of the same object acquired by two cameras with parallel axes.

The studies and the applications that utilize such techniques are numerous [11-16]. In our case the two images are not acquired from two cameras; they are acquired, instead from the same camera that frames the images of the same object in two different positions while the object to be measured undergoes a known displacement. However, with respect to the previous technique, a more accurate and complex calibration of the system is required. The calibration of the system consists in the determination of some intrinsic and extrinsic parameters, the first are the internal geometry and the optic characteristics of the camera, and the second is the spatial position of the camera reference system with respect to the piece coordinate system.

The measurement process is based on the differences between the two images obtained as a result of a workpiece displacement, and therefore the measure is founded on the location on the two images of two correspondent points. Also in this case the matching functions of the FastAlign library of the WiT were very useful both for the location of the characteristic elements of the framed object and in terms of rapidity of execution.

To increase the performances of the system, a proper dimension of the template has been chosen (in the examined case the best results have been obtained with a 64x64 pixels window), and by employing the epi-polar constraint, this constraint shows that, moving from one image to the other one, a characteristic point moves along a line called epipolar, that in our case results to be straight. This fact brings that the research of the extracted template do not have to be performed on the

whole second image, but only on a stripe of this last, having a height equal to the one of the template, locate at the same y coordinate, and as wide as the second image [12].
An example of three-dimensional measure, performed employing such methodology has been the evaluation of the height of a cylindrical element of the workpiece reported in Figure 6.

FIGURE 6. The workpiece employed for the 3D measure.

The comparison between the measure obtained by the stereoscopic system and the one performed with traditional systems has shown differences lower that 0,1 mm over a length of the part of 4 mm.

6 CONCLUSION

The proposed measure system allows, in a totally automated way, to perform measurement on workpiece of large dimensions, maintaining the level of approximation at very low values.
The operations are realized in very short times (normally lower than 2 seconds) and without any contact with the piece that has to be measured.
Moreover has been verified the possibility to utilize the same system to perform three-dimensional measures without any contact, based on the stereoscopic vision.
Therefore the use of such procedure will be employed for the measure of three-dimensional components and for automatic techniques for the wear evaluation of cutting tools.

ACKNOWLEDGEMENTS

This work has been supported by MURST found.

REFERENCES

1. Newmann, T.S., Jain, A.K., (1995), A survey of automated visual inspection, Computer Vision and Image Understanding, 61, 231-262.
2. Zang, J. B., (1996), Computer Aided Visual Inspection for Integrated Quality Control, Computer on Industry, 30, 139-158.
3. Lahajnar, F., Bernard, R., Pernus, F., Kovacic, S.; (2002), Machine vision system for inspecting electric plates, Computers in Industry, 47, 113-122.
4. Lanzetta, M.; Tantussi, G., (1999), Vision System Calibration and Sub-Pixel Measurement of Mechanical Parts, AMST'99, Proceedings of the 5th International Conference on «Advanced Manufacturing Systems and Technology», CISM Courses and Lectures N. 406, Edited by E. Kuljanic, Springer-Verlag, Udine, Italy, June 3rd-4th, 695-702.
5. Lo Casto, S., Lombardo, A., Lo Valvo, E., Ruisi, V. F., (1991), Computer Vision Profilometer Equipment and Evaluation of Measurements, WEAR, 146, 53-61.
6. Borsellino, C., Lo Valvo, E., Piacentini, M., Ruisi, V. F., (1996), A New On-Line Roughness Control in Finish Turning Operation, 4th International Conference on Advanced Manufacturing Systems and Technology (AMST'96), 661-668.
7. Galante, G., Piacentini, M., Ruisi, V. F., (1991), Surface Roughness detection by Tool Image Processing, WEAR, 148, 211-220.
8. Computar, 55 Telecentric, Accurate Machine Vision.
9. Logicalvision, WiT User's manual.
10. Logicalvision, WiT Operators reference manual.
11. van der Heijden, (1994), Image based measurement systems: object recognition and parameter estimation, J. Wiley & Sons.
12. Lanzetta, M., (February 1998), 3-D Vision in Production Processes, Part I - State of the Art (in Italian), Automazione e Strumentazione, n. 2, Year XLVI, 155-164.
13. Motta, J.M.S.T., De Carvalho, G. C., McMaster, R.S.; (2001), Robot calibration using a 3D vision-based measurement system with a single camera, Robotics and Computer Integrated Manufacturing, 17, 487–497.
14. Yang, C., Marefat, M., Ciarallo, F., (1999), Modelling errors for dimensional inspection using active vision, Robotics and Computer-Integrated Manufacturing, 15, 23– 37.
15. Karthik, A., Chandra, S., Ramamoorthy, B., Das, S., (1997), 3D tool wear measurement and visualisation using stereo imaging, International Journal Machine Tools & Manufacturing, Vol. 37, No. 11, 1573-1581.
16. Niranjan Prasad, K., Ramamurty, B., (2001), Tool wear evaluation by stereovision and prediction by artificial neural network, Journal of Materials Processing Technology, 112, 43-52.

STIFFNESS OF CNC MACHINE TOOL FEED DRIVES

Z. Pandilov[1], V. Dukovski[1]

[1] Department of Production Engineering, Faculty of Mechanical Engineering,
University "Sv. Kiril i Metodij", Skopje, Republic of Macedonia

KEYWORDS: stiffness, CNC machine tool, feed drives.

ABSTRACT. In this paper a model for the feed drive system with disturbance forces is given. Static and dynamic stiffness for the proposed model is analyzed. An equation for analytical calculation of the static stiffness is given. Correctness of the proposed equation is experimentally verified. Simulation of the influence of some parameters on the static and dynamic feed drive stiffness is performed with simulation program MATLAB & SIMULINK.

1 INTRODUCTION

Feed drives are widely applied to CNC machine tools, robots, manipulators, assembly machines etc. The feed drive stiffness may be defined as an influence of the disturbance force (torque) on the position (angular position) deviation.

In the theory of the automatic control stiffness can be defined as reciprocal value of the stationary error of the position (angular position) caused by the disturbance force (torque) [6].

Investigations about feed drives are very seldom presented in the literature. The results of Nieniewski [5], Nieniewski and Bollinger [4], Losic [2,3], Kakino et al. [1], are of particular interest. The most of the previous articles don't take into the consideration influence of the mechanical transmission elements on the feed drive stiffness. The research [1] is more complete, but still has one imperfection, absence of analytical equation for estimation the feed drive static stiffness.

2 A MODEL OF THE FEED DRIVE SYSTEM WITH DISTURBANCE FORCES

Fig.1 and fig.2 show an original model of the feed drive with disturbance forces.

All the relevant parameters in fig.1 are given bellow: Kv-position loop gain 1/s, T-sampling period s, s-Laplace operator, kg-coefficient of transformation of rotation in translation m/rad, D-damping of the electrical parts, ω-nominal angular frequency of the electrical parts 1/s, kvk-total stiffness of the mechanical transmission elements N/m, F-disturbance force N, kf-disturbance force gain 1/s, kb-gain of the mechanical transmission damping m/Ns, m-mass of the mechanical transmission elements kg, b-damping quotient of the mechanical transmission elements Ns/m, ki-integrator gain s, Xi-input position, Xo-output position.

Factors kf, kb and ki are always 1. They are taken into the consideration, only to have dimensional correctness of the models, given on fig.1 and fig.2.

Published in: E.Kuljanic (Ed.) *Advanced Manufacturing Systems and Technology*,
CISM Courses and Lectures No. 437, Springer Wien New York, 2002.

Transfer function between output position and disturbance force is given with equation (1).

$$\frac{Xo(s)}{F(s)} = \frac{k_f \cdot G_3(s) \cdot G_4(s)}{1+G_2(s)\cdot G_3(s)+G_1(s)\cdot G_2(s)\cdot G_3(s)\cdot G_4(s)} \tag{1}$$

With substituting the transfer functions G1(s), G2(s), G3(s) and G4(s) from fig.2 in equation (1) we obtain

$$\frac{Xo(s)}{F(s)} = \frac{b_3 s^3 + b_2 s^2 + b_1 s + b_0}{a_6 s^6 + a_5 s^5 + a_4 s^4 + a_3 s^3 + a_2 s^2 + a_1 s + a_0} \tag{2}$$

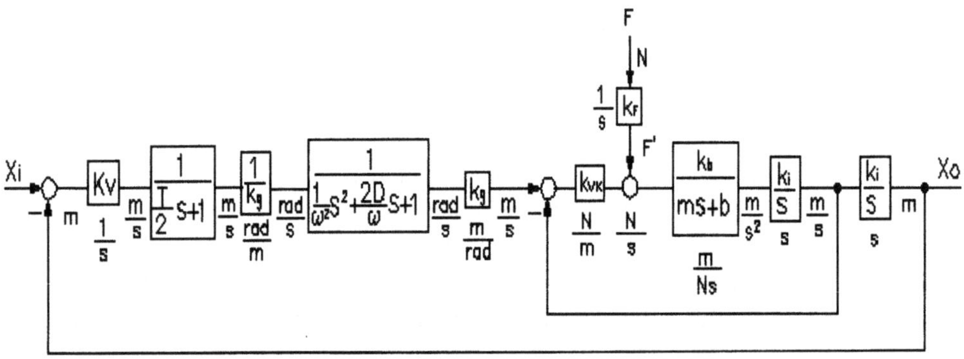

FIGURE 1. Model of the feed drive system with disturbance forces

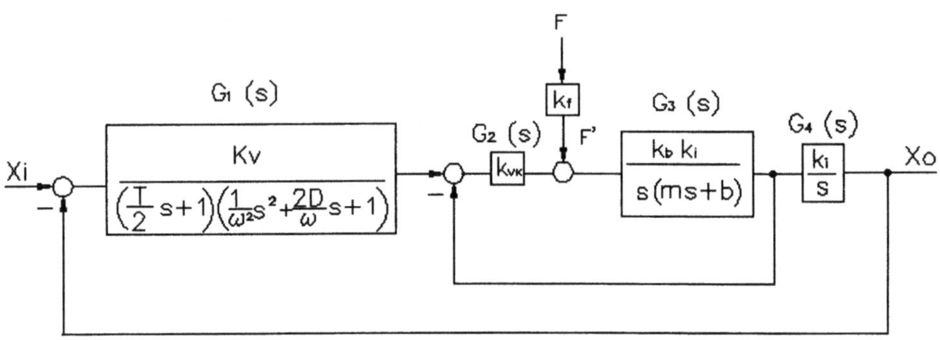

FIGURE 2. The model from fig. 1 after arranging

Coefficients in the equation (2) are:

$$b_3 = k_f k_b k_i^2 \frac{T}{2\omega^2},\; b_2 = k_f k_b k_i^2\left(\frac{1}{\omega^2}+\frac{DT}{\omega}\right),\; b_1 = k_f k_b k_i^2\left(\frac{2D}{\omega}+\frac{T}{2}\right),\; a_6 = m\frac{T}{2\omega^2},$$

$$a_5 = \left[m\left(\frac{1}{\omega^2}+\frac{DT}{\omega}\right)+b\frac{T}{2\omega^2}\right],\ a_4=\left[m\left(\frac{2D}{\omega}+\frac{T}{2}\right)+b\left(\frac{1}{\omega^2}+\frac{DT}{\omega}\right)+k_{vk}k_bk_i\frac{T}{2\omega^2}\right],$$

$$a_3 = \left[m+b\left(\frac{2D}{\omega}+\frac{T}{2}\right)+k_{vk}k_bk_i\left(\frac{1}{\omega^2}+\frac{DT}{\omega}\right)\right],\ a_2 = \left[b+k_{vk}k_bk_i\left(\frac{2D}{\omega}+\frac{T}{2}\right)\right],$$

$a_1 = k_{vk}k_bk_i$ and $a_0 = Kv\cdot k_{vk}k_bk_i^2$.

3 STATIC AND DYNAMIC STIFFNESS OF THE FEED DRIVE SYSTEM

One of the most important requirements with regard to feed drive system concerns its sensitivity to load disturbances.

The qualitative measure of this sensitivity is the feed drive system stiffness.

Dynamic feed drive system stiffness can be defined as a measure of influence of disturbance force F (torque T) on the output position Xo (angular position θo) deviation in the transient period.

$$Sd(s)=\frac{F(s)}{Xo(s)}=\frac{T(s)}{\theta o(s)} \tag{3}$$

For the model from fig.1 and fig.2 the equation for dynamic stiffness for feed drive systems become:

$$Sd(s)=\frac{a_6s^6+a_5s^5+a_4s^4+a_3s^3+a_2s^2+a_1s+a_0}{b_3s^3+b_2s^2+b_1s+b_0} \tag{4}$$

Where the coefficients $a_6, a_5, a_4, a_3, a_2, a_1, a_0$, b_3, b_2, b_1 and b_0 are equal to the coefficients in the equation (2).

The static stiffness Sst of the feed drive system in [5,4,3,2] is defined as:

$$Sst=\lim_{s\to 0} Sd(s) \tag{5}$$

The static stiffness of the feed drive system model on fig.1 and fig.2 can be defined as:

$$Sst=\lim_{s\to 0}\frac{a_6s^6+a_5s^5+a_4s^4+a_3s^3+a_2s^2+a_1s+a_0}{b_3s^3+b_2s^2+b_1s+b_0}=\frac{a_0}{b_0} \tag{6}$$

With substitution $a_0 = Kv\cdot k_{vk}k_bk_i^2$ and $b_0 = k_fk_bk_i^2$ in equation (6) we obtain

$$Sst=\frac{Kv\cdot k_{vk}k_b k_i^2}{k_f k_b k_i^2}=\frac{Kv\cdot k_{vk}}{k_f}\ \text{N/m} \tag{7}$$

As we mentioned always kf=1 1/s. In that case equation (7) becomes

$$Sst=\frac{Kv[1/s]\cdot k_{vk}[N/m]}{k_f[1/s]}=Kv\cdot k_{vk}\ \text{N/m} \tag{8}$$

From equation (8) it is obvious that the static feed drive system stiffness is defined as a product between position loop gain Kv and total stiffness of the mechanical transmission elements kvk.

4 SIMULATION OF THE FEED DRIVE SYSTEM WITH DISTURBANCE FORCE

Influence of the sampling period T, nominal angular frequency of the electrical parts ω, damping of the electrical parts D, damping gradient of the mechanical transmission elements b, mass of the mechanical transmission elements m, total stiffness of the mechanical transmission elements kvk and position loop gain Kv, on the dynamic and static stiffness has been investigated with simulation program MATLAB & SIMULINK.

In the simulations one parameter has been changed, and the other parameters were held constant.

In fact with the simulation, a position deviation in time domain Xo (t) caused by step disturbance force F=1 N is shown.

Simulation objects are models shown on fig.1 and fig.2.

$$Xo(t)=L^{-1}\left[\frac{Xo(s)}{F(s)}\cdot\frac{1}{s}\right]\ \text{m} \tag{9}$$

To estimate dynamic stiffness Sd we can use following equation,

$$Sd=\frac{F}{\max Xo(t)}\ \text{N/m} \tag{10}$$

where F is disturbance force and maxXo(t) is maximal position deviation.

In all simulation examples F was 1N.

We can calculate static stiffness Sst when the transient process will disappear and Xo(t) become constant.

Theoretically

$$Sst=\frac{F}{\lim\limits_{t\to\infty} Xo(t)}\ \text{N/m} \tag{11}$$

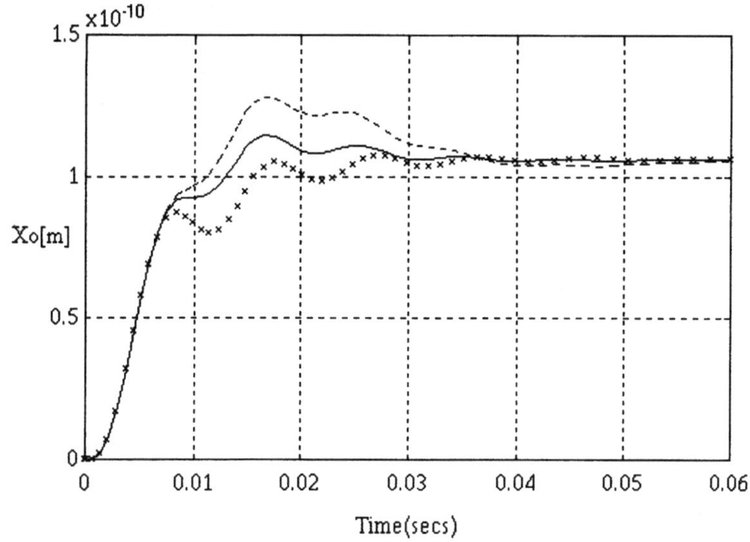

FIGURE 3. Influence of the sampling period T on the position deviation caused by disturbance force F=1 N

(D=0.7, ω=1000 s^{-1}, m=180 kg, b=25220 Ns/m, kvk=$88.34 \cdot 10^6$ N/m, Kv=106.34 s^{-1})

(——) T=0.006 s (- - -) T=0.010 s (x x x) T=0.002 s

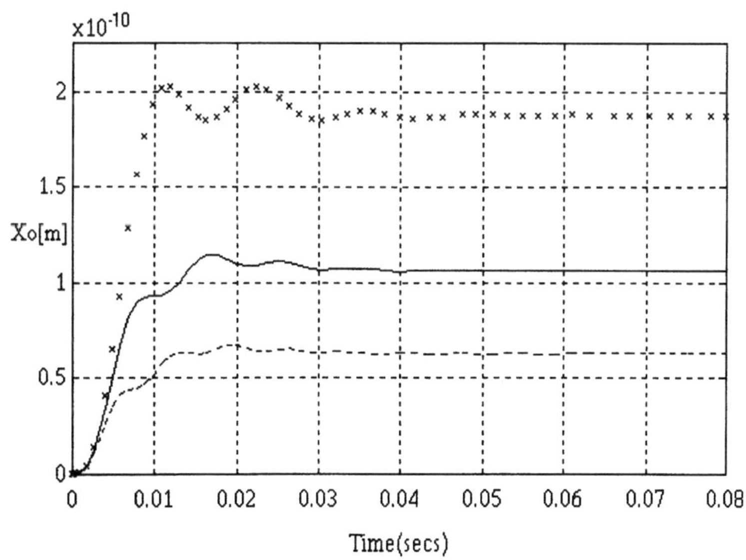

FIGURE 4. Influence of the total stiffness of the mechanical transmission elements kvk on the position deviation caused by disturbance force F=1N

(D=0.7, ω=1000 s^{-1}, T=0.006s, m=180 kg, b=25220 Ns/m, Kv=106.34 s^{-1})

(——) kvk=$88.34 \cdot 10^6$ N/m (- - -) kvk=$150 \cdot 10^6$ N/m (x x x) kvk=$50 \cdot 10^6$ N/m

Simulations have shown that increasing sampling period T, damping of the electrical parts D and mass of the mechanical transmission elements m, decrease dynamic stiffness. On the other hand increasing nominal angular frequency of the electrical parts ω, damping gradient of the mechanical transmission elements b, total stiffness of the mechanical transmission elements k_{vk} and position loop gain Kv, increase dynamic stiffness. With increasing total stiffness of the mechanical transmission elements k_{vk} and position loop gain Kv, increase static stiffness too.

Some examples of the simulation are shown on figures 3 and 4.

5 EXPERIMENTAL VERIFICATION OF THE EQUATION FOR CALCULATING THE STATIC STIFFNESS FOR THE FEED DRIVE SYSTEM

In order to verify the validity of the equation (8), an experiment on concrete feed drive system of CNC milling machine was performed.

The experimental installation is given on fig.5.

The results of the experiment are shown on fig.6.

Table 1 gives survey between experimentally obtained and analytically calculated with equation (8) static stiffness of the concrete feed drive system.

TABLE 1.

	Kv=28.33 s^{-1}	Kv=100 s^{-1}
Sst (exper.) N/μm	2650	8025
Sst (analytic.) N/μm	2503	8834
difference %	-5.55	+10.08

difference%=[Sst(analytic.)-Sst(exper.)] /Sst(exper.)·100%

It is obvious that with the appropriate changing of the position loop gain Kv, the elasticity of the system can be controlled.

From Table 1 it is evident that the difference between experimentally obtained and analytically calculated value of the static stiffness of the feed drive system is ±10% which is acceptable and sufficient for practice.

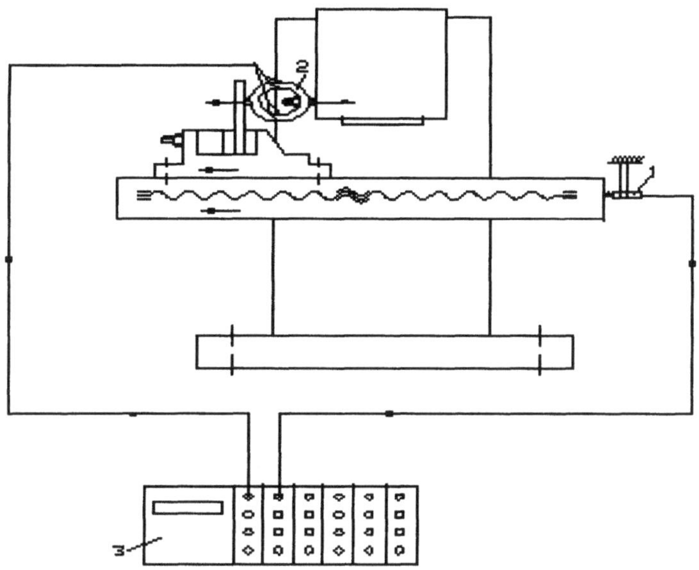

FIGURE 5. Experimental installation for determining static stiffness of the feed drive system for X-axes of CNC milling machine: 1. inductive transducer for displacement (HBM W1T/2), 2. device for generating disturbance force F, 3. device for registration the values of the displacement (position deviation) Xo and disturbance force F (HBM DA24 KWS3073)

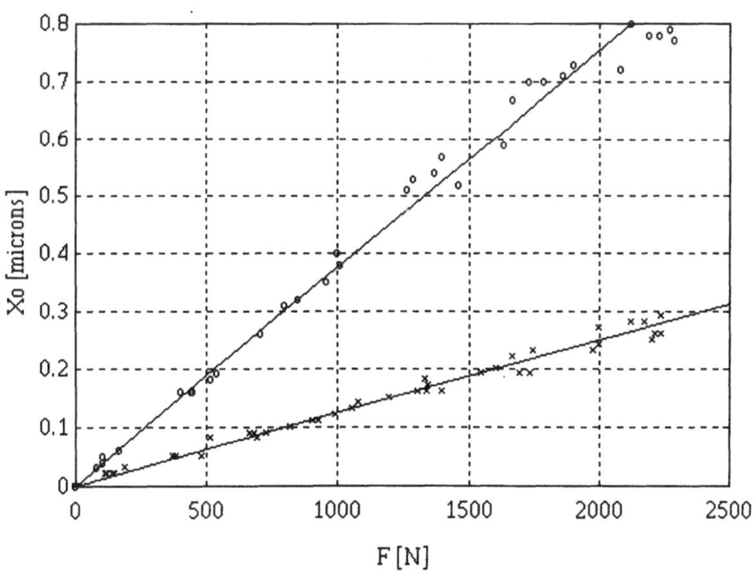

FIGURE 6. Experimentally obtained dependency of the position deviation Xo [microns] and disturbance force F [N] for feed drive system of CNC milling machine

(o - Kv=28.33 s^{-1}) (x - Kv=100 s^{-1})

6 CONCLUSION

A model of feed drive system with disturbance force was proposed. It has been shown by simulation that bigger values for sampling period T, damping of the electrical parts D and mass of the mechanical transmission elements m, decrease dynamic stiffness.

On the other hand bigger values for nominal angular frequency of the electrical parts ω, damping gradient of the mechanical transmission elements b, total stiffness of the mechanical transmission elements k_{vk} and position loop gain Kv, increase dynamic stiffness. Increasing total stiffness of the mechanical transmission elements k_{vk} and position loop gain Kv, increase static stiffness too.

An equation for analytically calculation of the feed drive systems static stiffness as a product of the total stiffness of mechanical transmission elements k_{vk} and position loop gain Kv, was proposed.

Correctness of the equation was experimentally verified. The difference was in ±10% limits, which is acceptable for practical use.

REFERENCES

1. Kakino Y., Matsubara A., Li Z., Ueda D., Nakagawa H., Takeshita T., Maruyama H., (1995), A study of the Total Tuning of the Feed Drive System in NC Machine Tools (2nd Report)-Single Axis Servo Parameter Tuning. Journal of the Japan Society of Precision Engineering, Vol.61, No.2, 268-272.
2. Losic A. N., (1985), Dynamic stiffness analysis of a PWM MOSFET DC feed drive for NC machine tools. Proceedings of the IECON'85 (International conference on industrial electronic, control and instrumentation), Vol.2, 761-767.
3. Losic A. N., (1983), Static stiffness and efficiency of a PWM MOSFET DC feed drive for NC machine tools. Proceedings of the industry applications society annual meeting,, pp.755-759.
4. Nieniewski J. M., Bollinger G. J., (1979), Computer Analysis of the Static Stiffness of SCR DC Feed Drives Used on NC Machine Tools. IEEE Transactions on Industry Applications, Vol.IA-15, No.5, September/October, 494-500.
5. Nieniewski M., (1979), Dynamic stiffness of SCR dc servo drives used on NC machine tools. Archiwum elektrotechniki, Vol.28, No.3, 645-647.
6. Netushil A., (1978), Theory of automatic control. Mir Publishers, Moscow.

A SET OF CRITERIA FOR SUPPLY CHAIN DESIGN

A. Villa

Department of Production Systems and Industrial Management
Polytechnic University of Torino, Italy

KEYWORDS: Supply Chains and Supply Chain Management, Large-Scale Business Lines, Integration

ABSTRACT. A "supply chain" is a network of production and distribution services which, connected through cooperation agreements, perform the functions of procurement of materials, transformation of them into intermediate and finished products, and finally distribution of final products to customers. "Integration" seems to be the new paradigm in organizing business lines, and "supply chain management" to be the related organization approach which could help in managing interactions among concurrent firms as well as marketing. But supply chains reveal to be complex systems, and their management a really complex task in which cooperation of several services together and their finalization to common industrial goals need methods and procedures still to be either developed or, at least, widely validated. The aim of this contribution is to propose some suggestions for the design and the organization of a new supply chain, such to make several SMEs able to operate together after a very short transient. To this aim a set of simple criteria for driving the supply chain design has to be made available for SME managers: their main scope will be to give simply-computable a-priori estimation of the convenience, for an enterprise, to stipulate an agreement for being included into a supply chain.

1 INTRODUCTION

A "supply chain" has been intended as a network of production and distribution enterprises which, connected through cooperation agreements, perform the functions of procurement of materials, transformation of them into intermediate and finished products, and finally distribution of final products to customers. Accordingly, the "supply chain management (SCM)" system is the decision-making structure to be developed such to make possible a real cooperation among these enterprises operating concurrently [1].

The crucial problem is how responsibility can be distributed to the number of managers of the different enterprises (often denoted "local agents"), owing to the more and more increasing dimension of the production and distribution networks: the so-called problem of decentralizing responsibility and autonomy, the solution of which will give directions in organizing a large-scale supply chain and its management.

This paper will approach this problem, with special attention to the definition of design criteria for management structures composed by a number of "agents" to whom a good autonomy is allowed [2]. The goal is to organize the supply chain as a network of cooperating intelligent agents, each one performing some supply chain functions and each one coordinating its own actions with the other agents.

2 BASIC CONCEPTS FOR SUPPLY CHAIN MODELING

A "supply chain" can be viewed as a temporary alliance among enterprises which share resources and competencies to give a better reaction to market requests, and the respective

Published in: E.Kuljanic (Ed.) *Advanced Manufacturing Systems and Technology*,
CISM Courses and Lectures No. 437, Springer Wien New York, 2002.

collaborations of which are supported by ICT instruments. Sometime, such an organized network of enterprises is also denoted "virtual enterprise" [3] or "strategic alliance" [4]. It means that a network of independent companies – suppliers, customers, etc. – linked by information technologies to share skills, costs and access to one another's markets, can be arranged through an agreement signed by the component firms just for a limited time horizon, depending either on expected future market conditions or on desired future strategies of each firm. Each enterprise, which agrees to be included into a "temporary network" and then be active inside the same supply chain, will state how to co-operating with the other components of the same supply chain in defining common production plans for specific products. Obviously, each enterprise could also maintain his own independence and autonomy for any other production.
In terms of conceptual model, a "network of enterprises" can be represented as a "graph of services and markets, alternatively connected". It means that each production service (i.e., a component firm) is an "intelligent agent" with proper autonomy and it can be only connected with nodes "market" both upstream and downstream [5]. On the other hand, each inter-stage market is a virtual model of a market place, i.e. a representation of a virtual space in which interactions among connected firms can occur.
Even if interested to preserve a portion of its own original autonomy, each firm, which aims to be included into a supply chain, surely aims to negotiate a cooperation agreement with its own candidate suppliers (at upstream stage along the chain) and buyers (downstream). Then the design of a new supply chain (which will be composed by a network of autonomous agents) must be based on the following necessary condition:

> "A multi-agents system must be cooperative, i.e. such that each local agent aims to evaluate its own convenience to contribute in improving the global system objective".

This intuitive concept of cooperation has to be specified more deeply [6].

The cooperation condition: Considering a set of agents into a supply chain, each one characterized by a proper autonomy, two types of cooperation among agents can be stated. A weak cooperation among agents can be established if agents have proper objectives non-conflicting each other and with the overall system objective, and each agent can measure its own convenience to contribute in improving the overall system objective. A strong cooperation among agents can be established if a weak cooperation among agents exists and, between each pair of agents directly related together, the information exchange can generate a precedence relation in the joint decision-making process.

The "cooperation condition" is the practical formulation of formal conditions usually assumed in multi-agents system control optimisation, as in applications of Large Scale System – LSS control theory as well as of game theory in manufacturing systems management problems (see [7]). Said conditions have to assure that an optimal control strategy, composed by non-conflicting strategies designed and applied by individual agents, can exist.

Accordingly, the supply chain design problem can be stated as follows:

> "Given a set of agents to be included into a new supply chain, for each pair of consecutive production stages, each one composed by either an agent (a firm) or a set of parallel agents, define under which conditions the "client-server" interactions between the agents belonging to

the two stages in terms of negotiations concerning the requests for purchasing items from downstream (the client) and the declaration of prices from upstream (the supplier), can be operated in such a way to generate either weak or strong cooperative strategies.

3 CRITERIA FOR DESIGNING A SUPPLY CHAIN

Let us refer to a supply chain organized in the form of a multi-stage production process, in which each stage is composed by a set of parallel firms, all connected to a common inter-stage market, both upstream and downstream.

The approach to the supply chain design consists of the following design steps, for each pair of consecutive stages in the chain:

1) (preliminary) specify the types of business line (i.e., products and industrial sector);
2) recognise the inter-stage information pattern, i.e. the set of connections linking together the different agents belonging to two consecutive stages, such as to recognize if a negotiation can be activated;
3) describe the different "strength" of each agent and of each stage, e.g. in terms of turnover or sales, such as to recognize potential dominance;
4) specify the types of agents' objectives (performance measures), with emphasis either to opportunity terms promoting formation of coalitions, or to cost terms forcing individual strategies;
5) describe the market exchange, such as to make evidence of the different types of demands and then of negotiations;

and, at each design step, verify the existence of conditions for a "strong cooperation" among the candidate firms to be included in the two consecutive stages of the supply chain.

The above stated design steps are now rewritten in order to describe how the investigation of a weak or strong cooperation situation could be performed by an industrial designer.

1) Characterizing the type of business line in terms of type(s) of items involved in the inter-stage business transactions, and of level (degree) of standardization.

Note that different business situations occur in case of either consumer goods or commodities, or components for a production process. In this last case, similar approaches to standardization can offer a relevant condition for good cooperation.

2) Characterizing the inter-stage "information pattern", by detailing how large the information pattern between two consecutive stages is; how the dominant structure of the information pattern is organized; how evident the "relative concentration" of a stage w.r.t. the other looks.

The wideness of the information exchange between two consecutive stages is given by the rate of the number of active connections among the complementary agents (i.e. included in the two stages), with respect to the number of connections which could be potentially activated. Referring to the information pattern organization, two main figures can be recognized: either "connecting the two stages" or "connecting at most firms within the same stage". In the former case, a "many-to-many" situation shows potential competition among firms (e.g. many suppliers for a same set of clients), whilst a "one-to-many" situation suggests existence of a driving agent in the connections. In the latter case, a "many-to-many" situation shows existence of a cartel, whilst a "one-to-many" situation declares existence of a hierarchy among firms in a same stage. Finally, a measure of the relative concentration of the upstream stage with respect to the downstream one is given by the rates of the respective Herfindhal index values (each Herfindhal index being the sum of squared values of market rates covered by the enterprises belonging to the considered stage of the chain). Depending on the value of the concentration measure, three different types of connecting networks can be envisaged: "one-to-one", "one-to-many" (or "many-to-one"); "many-to-many". In practice, the "one-to-one" chain occurs in case of reduced numbers of both suppliers and buyers; the "one-to-many" chain is characterized by either one supplier for a number of buyers or reverse; and the "many-to-many" situation can occur in large competitive situations.

3) Characterizing the "strength" of agents, in terms of existence (or not) of a "structural dominance" at the considered stage; and/or a "dominance in productivity", i.e. based on different production capacity at some stage; and/or a "dominance in the proceeds of sales".

Considering two consecutive stages, a measure of the structural dominance of a set of firms over the other is the rate between the number of upstream firms w.r.t. the number of downstream ones. Depending on the values of this measure, three different types of exchange can be recognized when, as assumed, suppliers operate upstream with respect to buyers: either "buyer-managed", or "supplier-managed", or "open market". The dominance in productivity is measured by the rate between the global production capacity values from up- and downstream stages. Depending on this rate, the "bottleneck" stage can be identified. Finally, the measure of the dominance in the proceeds of sales is stated by the rate between the global sales of the upstream stage w.r.t. the global sales of the downstream one. Depending on this rate, one can identify the "driving stage", i.e. the stronger stage in terms of sales volume.

4) Evaluating the "likeness" of agents and stages, depending on the types of agents' objectives.

Given two complementary agents with proper performance measures including several elemental costs of different types (e.g., production costs, distribution costs, inventory costs), the likeness of the two respective performance measures is given by the relative number of common elemental costs. Then, considering two consecutive stages in a supply chain, said stages only include either pairs of two complementary agents or terminal agents (i.e. agents with none inter-stage connection). Then, the likeness of the consecutive stages is measured by the average likeness of the pairs of complementary agents there included.

5) Specifying the market exchange, either in terms of type of market dynamics; and/or of type of demand periodicity; and/or of type of demand volatility (i.e., uncertainty).

The type of order matching in an intermediate market stage is suggested by a measure of the market dynamics, which specifies how interactions between the two sets of upstream (suppliers) and downstream (buyers) firms occur. This measure is stated by the rate between the pricing frequency (i.e. the rate of change of the price value during negotiations) and the market operation frequency (i.e. the frequency at which market negotiations are performed). Depending on values of the measure of the market dynamics, three different types of negotiation dynamics can be recognized: either "catalogue ordering", or "dynamic pricing", or "auction". These measures can respectively characterize the following situations: standard products at fixed prices for recurrent periodic (or steady-state) negotiations; complex products and services with dominant supplier (monopolistic situation); auction for products, which values can rapidly be modified (volatility). The demand periodicity should in principle be evaluated through a Fourier analysis of the past demand data: obviously, this is often practically impossible. But the Fourier analysis concept must still be used: it means that a "basic frequency" of the demand evolution (if anyone can be recognized over a short term history) must be identified, as well as the residual percentage of demand variations. Finally, the demand volatility has to be carefully estimated, mainly because it is an evident measure of the uncertainty about future demand values. A good estimation of it should be based on the evaluation of the incidence of high frequency components on the demand evolution (again through a Fourier analysis, or at least a visual recognition of them.

4 APPLYING CRITERIA FOR SUPPLY CHAIN DESIGN TO A SPECIFIC BUSINESS LINE

The above summarized design criteria have been applied to a specific business line with the aim of verifying the convenience of organizing supply chains: the Italian business line including the production of paper and its utilization both in printing companies and in publishing ones. According to available data, the inter-stage intermediate market includes negotiations between the paper production firms and the printing companies, on one hand (market A), and between the same paper producers and the publishing companies, on the other (market B).
For each production phase, the existing distribution of firms in Italy is described by Eurostat data (referred to 1998), summarized in the enclosed tables. These data allows obtaining estimated values for the Herfindhal index at each phase, the total number of firms and the number of firms covering a significant percentage of sales. In addition, further information has been obtained by direct contacts with the related industrial/trade organizations, such as to be able to answer almost all design questions (further investigation is still necessary, mainly for what concerns the evaluation of the firms individual objectives as well as the demand evolution).

T ABLE 1. Data concerning the business line phase of paper production
(*) Estimated over dates of average market rates per firm.

Range of work force [personnel units]	Number of firms	Market rates [%]
1 - 9	83	0.6
10 - 49	102	12.1
50 - 99	33	13
100 - 499	19	23.8
> 500	8	50.5
Herfindhal index = 0.0041 (*)		
Number of firms = 245 ; 27 firms cover cumulative market rate ≈2/3		
Annual sales [million] = 4146		

TABLE 2. Data concerning the business line phase of printing companies.

Range of work force [personnel units]	Number of firms	Market rates [%]
1 - 9	16485	28.3
10 - 49	2667	40.1
50 - 99	131	9.7
100 - 499	66	15.2
> 500	5	6.7
Herfindhal index = 0.000185 (*)		
Number of firms = 19354; 2869 firms cover cumulative market rate ≈ 2/3		
Annual sales [million] = 9771		

TABLE 3. Data concerning the business line phase of publishing companies.

Range of work force [personnel units]	Number of firms	Market rates [%]
1 - 9	5916	11.9
10 - 49	418	17.4
50 - 99	56	9.0
100 - 499	61	30.8
> 500	13	30.9
Herfindhal index = 0.00059 (*)		
Number of firms = 6464; 130 firms cover cumulative market rate ≈ 2/3		
Annual sales [million] = 9072		

The types of products there negotiated are "commodities" (design step 1) and negotiations based on catalogue look the best (design step 5). The three estimated Herfindhal index values suggest a "one-to-many" connection for both markets (design step 2). This pattern could motivate a prominent position of the paper production stage, but this situation is practically impossible because of the almost balanced values of annual sales of the three stages (design step 3). As previously mentioned, estimation of the agents' likeness is more difficult, but some significant evaluation in the average can be obtained by officials of industrial & trade organizations. In this case, information obtained from said organizations suggests that a low average likeness exists among the individual objectives at the three stages because of an evident competition (design step 4). This situation is evident at the paper production stage, including only 8 firms which cover half the market values. And a same competitive situation exists at the printing stage, owing to the extremely large number of very small companies. Then, organization of a supply chain seems to be convenient in case a strong cooperation agreement could be imposed: otherwise, the organization of a B2B e-market seems more profitable for all agents involved in the business line.

A further intermediate market exists between the printing companies and the publishing ones (i.e., an intra-stage market). Here the situation looks quite different because of the disequilibrium between a relatively small number of large publishing enterprises and the large number of printing SMEs: thus, hierarchies internal to this stage can be generated.

5 CONCLUSION

The analysis of the supply chain design steps, above presented, are giving some challenging suggestions for the supply chain organization as well as for the evaluation of the convenience, for an individual enterprise, to agree in being included into a supply chain. The five steps for the supply chain conceptual design, applied to an Italian business line (including paper production, printing and publishing companies) are revealing to be tools of industrial interest. Their application seems to be useful also in the conceptual design of B2B internet-based negotiation systems and market places. Further research is mandatory in order to completing a set of cooperation measures sufficiently wide to offer more accurate design criteria.

REFERENCES

1. Tayur, S., Caneshan, R., Magazine, M., eds. (1999), Quantitative Models for Supply Chain Management, Kluwer Academic, Boston.
2. Huang, C-H., Nof, S.Y. (2000), Autonomy and viability – measures for agent-based manufacturing systems, Int. J. Production Research, 38, 4129-4148.
3. Byrne, J.A. ed. (1993), The virtual corporation, Business Week, 8, 98-103..
4. Cantamessa, M., Villa, A. (2001), Negotiation models and production planning for virtual enterprises, Proc. MIM 2001 – IFAC Workshop on Manufacturing Modeling, Management and Control, Prague, 1-6.
5. Villa, A. (2000), Organizational structures for distributed networks of enterprises, Preprints MIM 2000 – IFAC Symposium on Manufacturing Modeling, Management and Control, Patras, 74-78.
6. Villa, A. (2001), Emerging trends in large-scale supply chain management, Plenary Lecture at Int. Conference on Production Research – ICPR-16, Prague (c-disk proc.)
7. Villa, A. (1991), Hybrid Control Systems in Manufacturing, Gordon & Breach Science Publ., New York.

APPLYING OF THE TYPICAL TECHNOLOGICAL OPERATIONS IN THE SINGLE PRODUCTION

G. Simunovic[1], N. Majdandzic[1], K. Simunovic[1], R. Lujic[1]

[1] Department of Organisation and Informatics, Mechanical Engineering Faculty, Slavonski Brod, Croatia

KEYWORDS: process planning, typical technological operation, single production

ABSTRACT. The article shows structure of Management Integrated Information System with a module, which enables automatic process planning for specific product group in the single production. The module enables creating of technological groups that are described with sequence of typical technological operations where single products are separated according to technological features. These operations are parametric defined and according to real features of product (diameter, width, length, and segments' number, piece's number etc.), process plans will be generated automatically.

In the single production, locksmith's works, welding and forming technologies are common, so, for such kinds of technologies experiential formulas have been developed. These formulas serve to calculate necessary manufacturing time for typical technological operations, requirements of auxiliary and consumable materials. According to previous sentences the manufacturing time and whole material costs (approximately) that are necessary for offer creating will be calculated.

1 INTRODUCTION

Basic market and production demands such as product quality, low production costs, fulfilled of due dates, rationalisation of material and information management and market competitiveness can be only realised by the application of new scientific approach in production preparation and in production itself. It is important not just for batch and low production but also for the single production which has not achieved significant improvement in designing and implementation.

The article shows structure of Management Integrated Information System with a module, which enables automatic process planning for specific product group in the single production.

Experiential developed models for manufacturing time calculating, material (auxiliary and consumable) requirements and whole material costs have been implemented in module [1].

For productivity and flexibility increasing significant influence has been achieved through computer implementation in production, process planning and management. Flexibility has become the most significant feature in CIM realisation where interaction and integration between phase of product design, materials selection, technological process and production activities are necessary.

2 MANAGEMENT INTEGRATED INFORMATION SYSTEM (MIIS)

Today's Management Integrated Information Systems have been designed on modular principle. It means that system consists of modules - program parts (organisational and business), which have been managed by the main program. Such a kind of structure can be acceptable and the reasons are: some modules are completed, modules can be implemented separately in the work and

Published in: E.Kuljanic (Ed.) *Advanced Manufacturing Systems and Technology*, CISM Courses and Lectures No. 437, Springer Wien New York, 2002.

every module can be built after. These Management Integrated Information Systems have not got process planning modules, so support to production preparation has not been given. Mostly, the technological parameters have been taken from other programs or have been inputted manually in Integrated Information Systems.

The figure 1 shows structure of Management Integrated Information System that has been developed for energy equipment enterprises [2,3]. Company Informaticki Inzenjering – ININ from Slavonski Brod, has made MIIS according to demands that user has set.

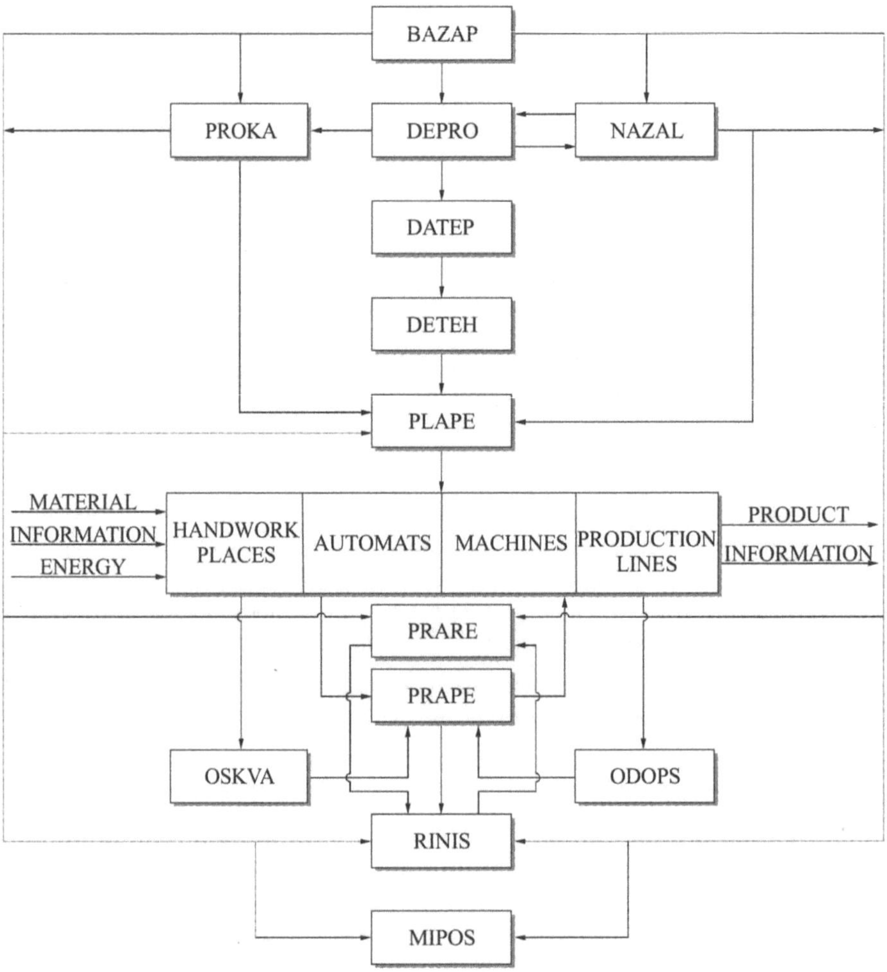

FIGURE 1. Structure of Management Integrated Information System (MIIS)

MIIS consists of:

- BAZAP – Common database,
- PROKA– Sales and calculation,
- DEPRO – Product definition,
- DATEP – Definition of technological demands and typical technological operations and tools,
- DETEH – Process planning,
- NAZAL– Spare parts and material purchasing and store subsystem,,
- PRARE – Realisation managing,
- PLAPE – Production and production preparation planning (MS-Project is involved),
- PRAPE – Production monitoring,
- OSKVA – Quality assurance,
- ODOPS – Maintenance of capacities,
- RINIS – Accounting and bookkeeping,
- MIPOS – Management information subsystem (Controlling and management).

Program support to technological works in MIIS has been given through module DETEH. Model and principles that have been implemented in DETEH will be explained in the next chapter.

3 PROCESS PLANNING MODEL AND ITS INTEGRATION IN MIIS

Figure 2 shows developmental diagram for process planning model [4,5,6,7]. Designing of subsystem DETEH has been based on the following according to our opinion.
Technological operations that are typical for production program in the enterprises have been described parametrically according to its position of manufacturing (machining work, welding, locksmith's works, bending of pipes etc.). These parameters are diameter, width, thickness, length, segment's number, piece's number, bending number etc. Before that, products have to be sorted in production groups, which have been defined according to products technological features. Every technological group has been characterised by the specific sequence of operations among the whole number of operations.

Process planning has been based on the typical technological procedures and has been developed on group technology principles. For selected group of work pieces (technological group) the typical technological procedure with specific features has been made. All specific segments of work pieces with calculation of cutting and machine parameters, requirement of auxiliary and consumable materials will be done through typical technological procedures. Parameters (cutting and work) and requirement of materials (auxiliary and consumable) have been calculated towards experiential set of formulas that were based on long time researching of influence parameters and relations between them.

Such a kind of process planning can be possible in production of energy equipment where products are very similar according to structure (described in subsystem DEPRO) but different towards power. Before the structure has been defined, the work pieces that are part of product have to be described. Work pieces are similar according to shape but can be made from different materials. If the sign has been set beside the appearance of operation and technological information has been enrolled (figure 3) according to typical technological procedures, than the real technological procedures will be done easily.

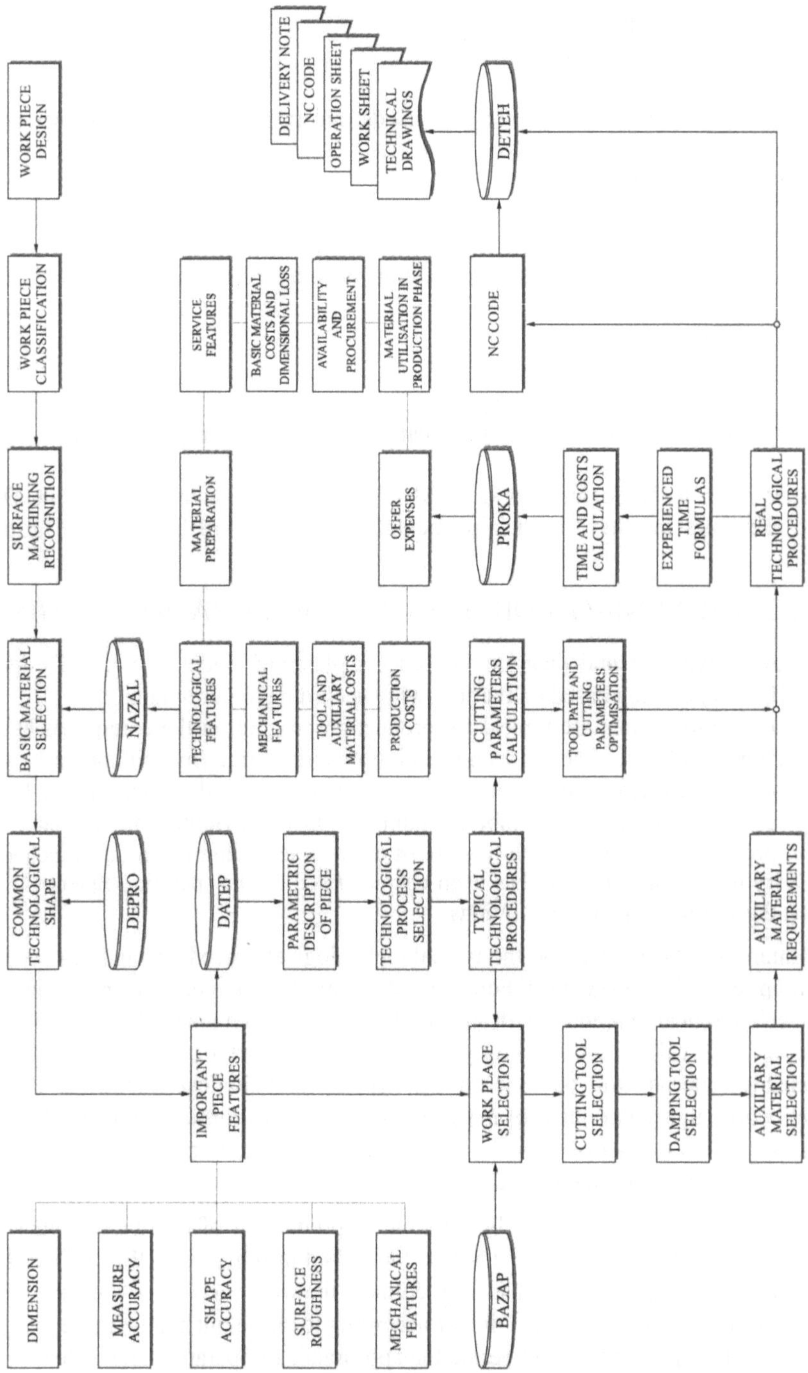

FIGURE 2. Process planning model integrated in MIIS

MECHANICAL ENGINEERING FACULTY SLAVONSKI BROD		Technological information									MANUFACTURING TIME PER OPERATION							Additional materials				
Questioning no.: 129/2000 Part or assemby name SPECIMEN														Welding								
Number	Technological procedures	Appearance of operation (1-yes, 0-no)	Diameter/width, mm	Wall thickness s, mm	Lenght, m/piece	Number, piece	Bending/piece	Piece	m²	ADDITION	Minute/measure unit	Locksmith's work	Pipe bending	Electric arc	Electric arc+TIG arc	Machining	Corrosion protection, min/m	Corrosion protection, kg/m²	Electrode, wire, kg	Gas, kg	Wire for welding, kg	Powder for welding, kg
1	Dimensional pipe cutting	1	80	6	0.03	4000		4300	33		7.6	32680										
2	Pipe grinding	0																				
3	Pipe welding	1						4300			2.6	8600		2521					10			
4	Frame manufacture	1		15		3803		3.8	65		6000	12550				4564	1260	16				
5	Holder manufacture	0																				
6	Frame welding	1						3.8						5705					67			
7	Chamber cutting	1	140	13	3.3	4		4	6		31	123					112	1				
8	Machining of chamber ends	1				8		8			38					307						
9	Holder welding	0																				
10	Lid manufacture	1	140	20				2	0.1		46					92	2					
11	Hole drilling	1						48			8.2					392						
12	Lids welding on chamber	1						2			160	4080			319				1	2		
13	...																					
14	...																					
OVERALL						Per machining type [Nh]						967		137	5.3	89.3	23	17	78	2		
						For work piece [Nh]						1221.6										

FIGURE 3. Product manufacturing plan

4 CONCLUSIONS

Metal Matrix Composites generally exhibit a certain free-cutting tendency which make them easier to be ground with respect to non-reinforced light alloys. Grindability indices have been settled taking into account grinding forces, wheel wear and workpiece surface roughness in order to compare the behaviour of different grinding wheel abrasives and workpiece materials. Among the grinding wheels employed in the tests, the silicon carbide one represents the best solution if the main concern is to hold low the grinding forces.

Flat area and roughness exhibit a more complicated behaviour. For the particulate composite the best results, in terms of both flat area and roughness, occur when using the silicon carbide wheel. The lowest tendency to clogging occurs with the aluminium oxide wheel that is characterised by the most opened structure among the wheels used.

The grindability of MMCs does not depend on a specific technological property, but is a function of a set of different characteristics which are not simply related each other. In fact, geometry, orientation and volume fraction of the hard reinforcement greatly affect the cutting ability of the grinding wheel and the surface finish of workpieces.

REFERENCES

1. Chandler, H.E., (1989), Machining of Metal Matrix Composites and Honeycomb Structures, Metal Handbook, Ninth Edition, Vol. 16, Machining, 892-901.
2. Srivatsan, T.S., Lavernia, E.J., (1991), Synthesis of Particulate Reinforced Metal Matrix Composites using Spray Techniques, Processing and Manufacturing of Composite Materials, ASME, Vol. 49, 179-221.
3. Clark, I.E., Cook M.W., (1994), Machining Highly Abrasive Composite Materials with Polycrystalline Diamond, Proceedings of Advancing with Composites, Milan, Italy.
4. Cronjager, L., Meister, D., (1992), Machining of Fibre and Particle-Reinforced Aluminium, Annals of the CIRP, Vol. 41/1, 63-66.
5. Zhang, L.C., (1994), Grindability of Some Metallic and Ceramic Materials in CFG Regimes, International Journal of Machine Tools & Manufacture, Vol 34, No. 8, 1045-1057.
6. Chandrasekaran, H., Johansson, J. O., (1993), The Role of Material and Grinding Parameters in Grinding of Alumina Fibre-Reinforced Aluminium Alloys Proceedings of Machining of Advanced Materials, Gaithersburg, USA, 20-22 July.
7. Di Ilio, A., Paoletti, A., Tagliaferri, V. Veniali, F., (1996), An Experimental Study on Grinding of Silicon Carbide Reinforced Aluminium Alloys, International Journal of Machine Tools & Manufacture, Vol 36, No. 6, 673-685.
8. Malkin, S., (1989), Grinding Technology - Theory Applications of Machining with Abrasives, Ellis Horwood Limited, West Sussex.
9. Tönshoff, H. K., Peters, J., Inasaki, I., Paul, T., (1992), Modelling and Simulation of Grinding Processes, Annals of the CIRP, Vol. 41/2, 677-688.
10. Di Ilio, A., Paoletti, A., (1995), Wheel Topography Evaluation During the Grinding of SiC Reinforced Aluminium Alloy, Proceedings of the First World Conference on Integrated Design & Process Technology, Austin, Texas, 6-9 December, 1995, 70-76.
11. Di Ilio, A., Paoletti, A., (1998), The Influence of Shape and Content of Reinforcement on Grindability of Metal Matrix Composites, Proceedings of European Conference on Composite Materials, Vol 2, Naples, 3-6 June, 541-549.

creasing of time for technology preparation, and it is also possible to calculate the manufacturing time quickly as well as total (approximately) material costs. According to previous sentences the better offer can be given.

REFERENCES

1. Majdandzic, N., Despotovic, B., Mileusnić, N., Simunovic, G., (2001), Welding Technology Program Support, Economical and qualitative aspects of highly effective welding processes, Slavonski Brod, 109-118.
2. Majdandzic, N., (1998), Management Integrated Information System Project.
3. Majdandzic, N., (1994), Enterprises computing, University in Osijek, Mechanical Engineering Faculty, Slavonski Brod.
4. Chang, P-T., Chang, C-H., (2000), An integrated artificial intelligent computer aided process planning system, International journal of Computer Integrated Manufacturing, Vol. 13, 483-497.
5. Ming, X.G., Mak, K.L., Yan, J.Q., (1999), A hibrid intelligent inference model for computer aided process planning, Integrated Manufacturing System, 10/6, 343-353.
6. Schey, J. A., (1997), Manufacturing Processes and Their Selection, ASM Handbook Volume 20: Materials Selection and Design, Hardbound Publisher.
7. CEEPUS project No PL-013/9697 "Development, testing and processing of contemporary functional, constructional and tool materials", Gliwice-Wisla, 1996.

PROCESS PLANNING APPLICATION AND INTERNET CONNECTION

P. Cosic[1] , D. Sever[2] , S. Milinovic[2]

[1] Department of Industrial Engineering, Faculty of Mechanical Engineering and Naval Architecture, University of Zagreb, Croatia
[2] Faculty of Mechanical Engineering and Naval Architecture, Aerospace Engineering, University of Zagreb, Croatia

KEYWORDS: Process Planning, Database, Internet.

ABSTRACT. This work is a part of the project creating the technological database necessary for improvement of process planning and connection database through SQL server by Internet environment. Developed web site for Database System and Process Planning is useful especially during development student's project's of technological process inside fields Process Planning, Production and Operations Management. Here we have elaborated the first phase of selecting tool machines, tools, operations and initial material on the basis of the chosen criteria.We have indicated the problems of selection and the possibility of working out a solution. The Internet technology gives us the possibility to share our data with many users and a chance to improve the elements of Virtual Manufacturing

1 INTRODUCTION

In the world of today different trends such as: individualism, multicultural way of life, globalization, shift of the workplace, virtual economy and manufacturing, explosion of communication, exchange of information, and the like predominate. Process planning determines how a product is to be manufactured and is therefore a key element in the manufacturing process. In spite of the importance of process planning in the manufacturing cycle, there is no formal methodology which can be used, or can help to train personnel for this job. Process planning activities are predominantly labor intensive, depending on experience and the skill and intuition of production labor. Industry has also observed in the recent years a decrease in the number of process planners. They are retiring from the field at a faster rate than new ones are being trained. The above situation, coupled with the objective to increase quality and reduce lead time and cost or alternatively improve productivity, has led to the widespread interest in CAPP – computer –aided process planning.

2 PROCESS PLANNING IN THE PRODUCTION STRATEGIES ENVIRONMENT

In the eightieth of last century the USA Vision Group gave answers to the following questions: *what to produce, what to buy, where to locate production* and *into which technological systems to invest.* The imperative of selecting an adequate production strategy is quite obvious because it is reflected in the price of the product, production cycle, possibility of delivery, etc. The idea of this project, in its initial phase, is to define selection criteria [1][2] of the initial materials,

Published in: E.Kuljanic (Ed.) *Advanced Manufacturing Systems and Technology*,
CISM Courses and Lectures No. 437, Springer Wien New York, 2002.

tool machines, tools, technological processes, sequencing of operations, mathematical terms for calculating cutting data, machining time, downtime/operation and the HELP with multimedia opportunities. In this manner, through elaboration of a variant of production, it is possible to come at a short notice to the most suitable way of production, production cost and the spatial and temporal distribution of the product.

3 TECHNOLOGICAL DATABASE FOR THE PROCESS PLANNING

To modernise process planning, to make it interesting to new generations of engineers, to solve the problem of "brain drain" (war, transition), to accept the laws of the global economy stands for connecting it to the systematised databases, CAM, CAPP, CAD, to the methods of simulation of the technological and production processes, to the management of manufacturing, were resulted with main criteria, relationships and databases. The presented concept of the systematised technological database (Fig. 1), rests on the necessary dynamic balance of the fast data updating because of the ever more present innovations in the work regulations, materials, tools and characteristics of the tools and the ever more stringent demands of a large range of users. The above mentioned diagram enables the analysis of the selection technology of tools, tool machines, production time, work regulations, and the economic analysis of the technological process adequacy. The general organization of the programming solution implies the following criteria for selecting the technological process:

- CRITERIA FOR SELECTING OPERATIONS

Product configuration, demanded precision of product shape and sizing, sequencing of operations, material quality and state of surface, heat treatment of material, batch size, availability of production equipment and technology,

- CRITERIA FOR SELECTING TOOLS

Material of workpiece, tool material, tool design, cooling and lubrication, tool life, cutting date, type of operations, sequencing of operations, application and machining conditions, selecting of tool holders and inserts,, tool price,

- CRITERIA FOR SELECTING MACHINE TOOLS

Way of tightening, quality of the treated surface, machining due to shape workpiece, technological parameters, machine cost, product complexity, batch size, .

- CRITERIA FOR SELECTING INITIAL MATERIAL

Product shape and sizing, entire workpiece quality, batch size, stress state, heat treatment of product, anisotropy of material, geometrical similarity between initial form and workpiece, type of process, material price.

4. EXPLANATION OF THE DEVELOPED SOFTWARE

The ultimate goal is to connect the recognition of the geometrical characteristics of the product to the database and to apply the methods of artificial intelligence in defining optimal technological solutions. The technological database is a part of the future virtual and intelligent manufacturing system [3] [4].

4.1 INITIAL MATERIAL SELECTION

The main idea was to research the basic elements for connecting the criteria of initial material selection with the information necessary for making a decision concerning.

TABLE 1. The main criteria for selection of initial material

Microsoft Excel - KRITERIJI ODABIRA ENG 4

B19 = good until very good depend of hardiness degree of material

CRITERIA SELECTION	METAL FORMING			
	ROLLING	DRAWING	EXTRUSION	FORGING
PRODUCT DIMENSIONS	standard dimensions		section until 160 mm, length until 1600 mm. wall thickness(for hollow workpieces) 0.5-15 mm	depend of dimensions, mass i.e. forging force of forging machine
PRODUCT SHAPE	standard rolling sheets and profiles	standard drawing profiles, wires and tubes	technological rotational symmetric hollow and solid shapes; tehnological form; geometrcal circumscription of the form	the most different elongated, rotational symmetrical and combined shapes
SURFACE QUALITY Ra (mm)	0.2-100	0.2-6.3	0.2-1.6	6.3-100
PRODUCTIVITY	mass production standardized pieces		40-100 pices/min, until 250 pieces/min for diminutive workpieces	from 100-6,000 kg/h depend of forging dimensions and kind of forging machine
GRAIN SIZE AND ORIENTATION IN BLANK	elongate structure and texture of material, characteristic for cold metal forming			fine grain size with good control of forging process
PRICE OF THE BLANK	according producer catalog		follow calculation	
RATIOS OF MATERIAL PORTION TO PRICE OF THE BLANK	near 100 %		after volume calculation of initial material, use catalogue of rolling and drawing profiles	after volume calculation of initial material, use catalogue of rolling and drawing profiles
RESISTANT STRESS IN BLANK	increase strength, yield limit and ultimate strength; decrease yield limit of material, resistance of corrosion, and electrical conductivity			depend of complexity of shapes, heterogeneous forging feature through the volume material
SCHEME OF STRESS STATE			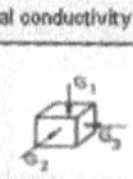	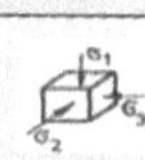
MACHINABILITY	good until very good depend of hardiness degree of material		bad if is not done normalization with annealing after forging	feature of cutting depend of complexity of impressions ; bad if is not done normalization with annealing after forging

Kriterji odabira sirovca

Ready

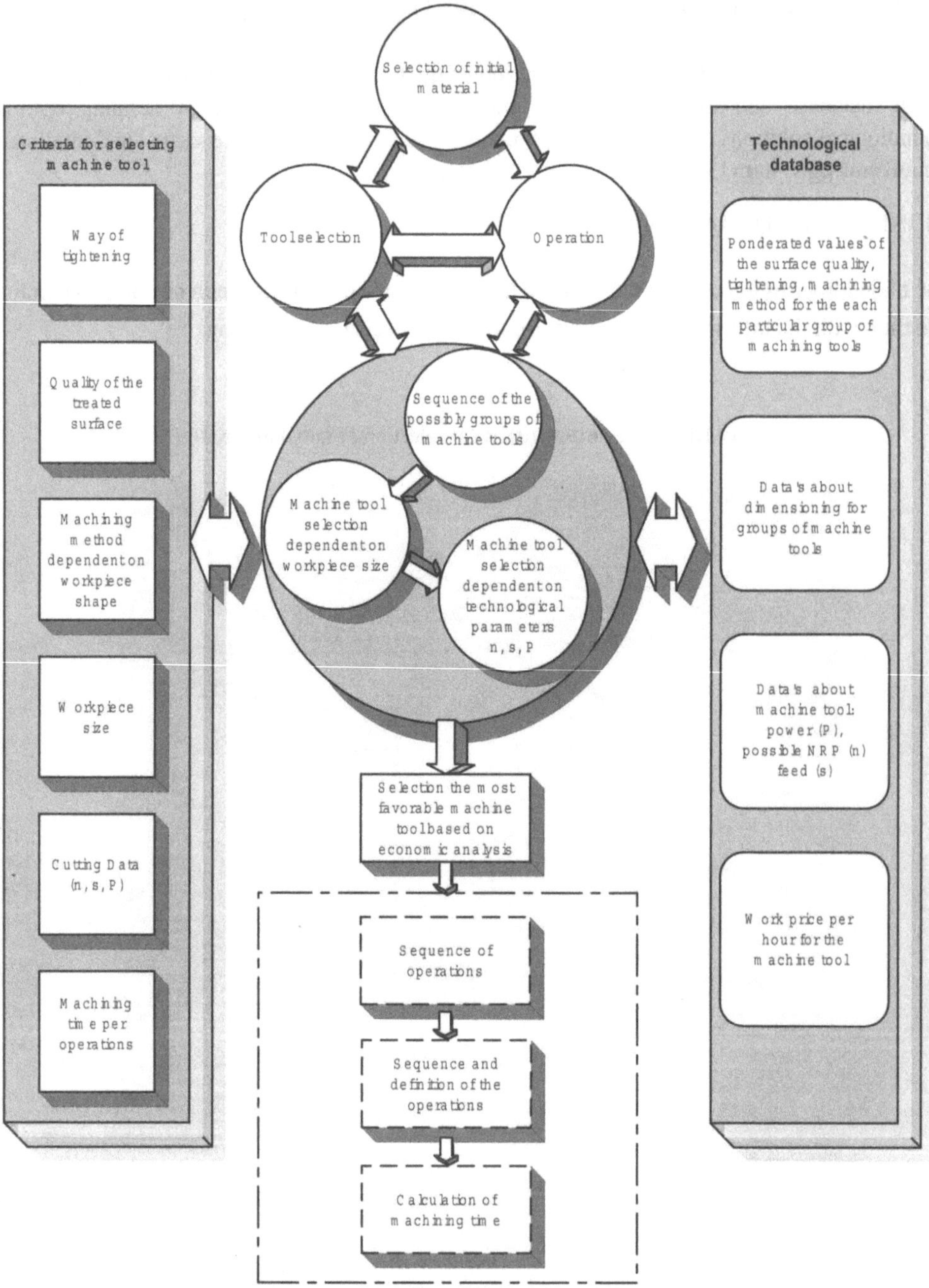

FIGURE 1. Flow diagram of the development of a technological process variant

Furthermore, in the technological database the data about the cutting data of the machine (rotations per minute (RPM), feed (s), power (P), its dimensions, and other important data can be found. The result of that research were defining of criteria for selection of material, methods for calculating initial material for the case of *the forging,* and *HELP text.* The possible criteria (Table 1) for initial material selection (initial product is a result of rolling, drawing, extrusion) are: *dimensions, product shape, surface quality, productivity, grain size and orientation, price of the blank, ratios of material portion to the price of the blank, residual stress in the blank, scheme of stress state, machinability* (considered in Table 1.), entire workpiece quantity, batch size, anisotropy of material, geometrical similarity between initial form and workpiece, heat treatment of product, etc. The end user can use the result of initial material selection only as a suggestion for planning a variant of technological process.

4.2 TOOL MACHINE SELECTION

From the flow diagram (Fig 1) we can see that the selection of the tool machine depends on the three logical units: the *criteria of selection*, the *mechanisms of data processing based on the criteria* and *the database.* For the criteria of selection we have chosen: in the first phase the *way of tightening and processing*, the *required quality of the treated surface*, *the dimensions of the workpiece* and the *technological data*, including performance (n, s, P), which together form the basis for further elaboration in the following phases of the database development. For solving the problem of the selection of the tool machine, the *database* contains the data about the groups of machines and about each machine separately. On the example of the *selection of the milling tool machine*, outlined in the (Fig. 2, 3), we are introduced and acquainted with the way of functioning of the application, with the input parameters and the output data, or with the responses to the user's demands. Inside the main window of the application we can see the possibility of access to the selection of the machine either from the menu bar Machine selection/ Milling or from the tool bar. By choosing a way of milling the window based on the multiplc choicc questions opens. The questions referred to the way of tightening, to the required quality of the treated surface relative moving of workpiece-tool, width of machining surface, depth of cutting. It is important to stress that it is possible to choose only one from the nine given options about the way of *tightening (tightening universal chuck, segmented head, flanged, universal, swival rise, magnetic chuck, dividing head, etc.*). We choose between the facemilling, squareshoulder milling, multipurpose milling, endmilling, side and facemilling, etc on the basis of the geometrical shape of the product. This application displays the descriptive and the pictorial illustration of the processing so that the user decides for the shape best corresponding to his own product.. In brief, it is possible to choose the desired combination of the already pond rated values of *tightening and processing, relative moving between workpiece and tool*, depth of cutting, the *quality of surface processing* dependent on the *geometry of the workpiece*, which together serve for selecting the *group of machines* (Fig. 2).

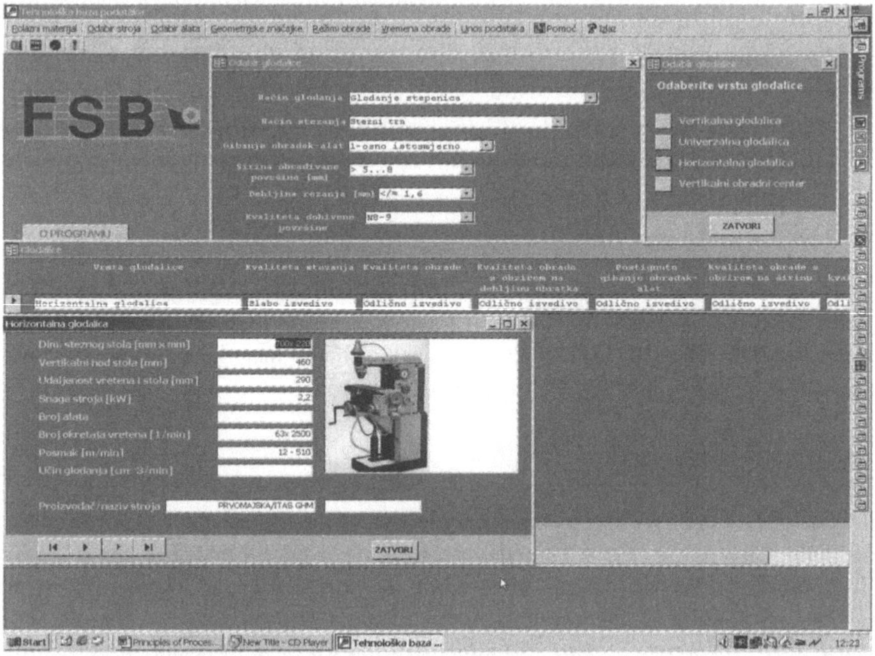

FIGURE 2. Procedure for selection possible groups of milling tool machines

4.3 TOOL SELECTION

The selection of tools involves many parameters, for instance, material of workpiece, geometrical features of workpiece, level of product complexity, number of phases, tool machine, defination of application and machining conditions, type of machining, insert shape, milling cutter, insert grade, tool holder type, method of holding the insert in the holder, direction of cut, coolant supply, diameter and depth of hole, type of hole, etc [2] [5].

During the process of tool selection (for example milling) a user has to:

- Define material and type of operation,

- Define material according to ISO (or other) standards and identify own operation from the table of contents,

- Select milling cuter (close pitch, coarse pitch, extra close pitch cutter),

- Define application and machining conditions (application: light, medium; conditions: good, average, difficult),

-Select the insert and see the feed per tooth and cutting speed,

At the end of the search proce dure, user can get tools recommendation and cutting data like speed, RPM, feed, coolant, tool life, number of holes per drill tool etc.

4.4 INTEGRATED SELECTION MACHINE TOOLS AND TOOLS

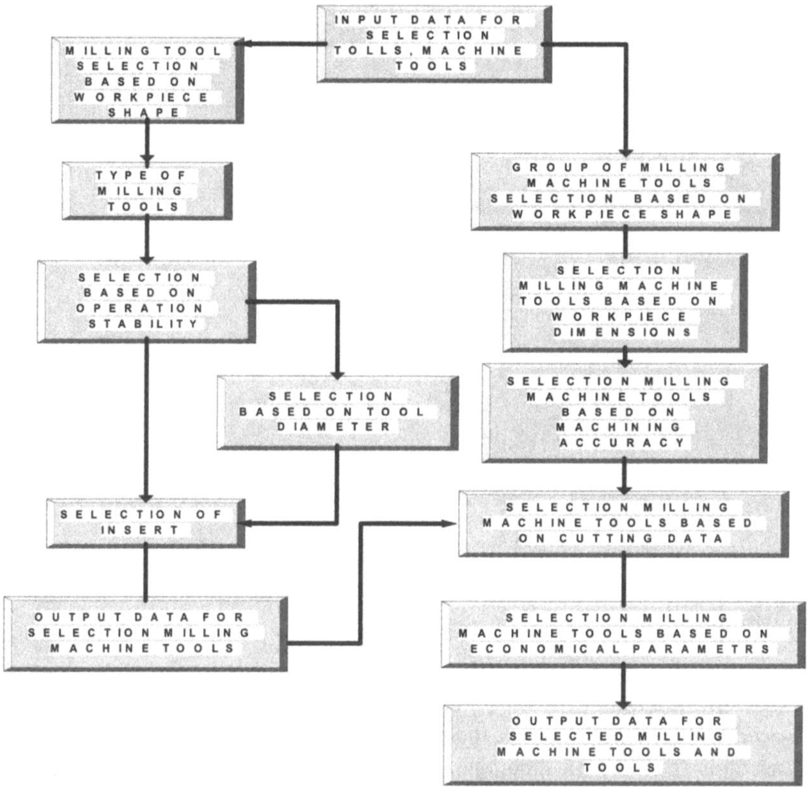

FIGURE 3. Interaction of selection milling machine tools and milling tools

The main difficulty in selection procedure of machine tools and tools is the strong dependence between workpiece size and shape, machine tool, tool, sequencing of operations. So, developers have to make decision: what would be first step in solution search and what interactions would be negligible in this phase of development. Our selection are tools selection as the first step and sequence of operations neglect.

5. PROCESS PLANNING IN DATABASE AND INTERNET ENVIRONMENT

Using the web site (Fig. 4) is just like using the application itself. The user must enter the basic data (starting from the selection of the operation, type of material ASP script communicates with database via SQL demands, which are sent to database server (SQL Server 7.0) that returns the desired data. It has also a second benefit because it gives us an easy way to incorporate data from other researchers or developers and build a distributed network. [6]. In a first period, connection between database and Internet was realized by ODBC. In the late phase of project, during some short period of time, we experimented with MSDE engine. But we have observed some instability accidents. So, as a good solution we change ODBC type of connection by newer OLE DB (Microsoft Jet OLE DB Provider) connection. As the result, our search and response speed became faster.

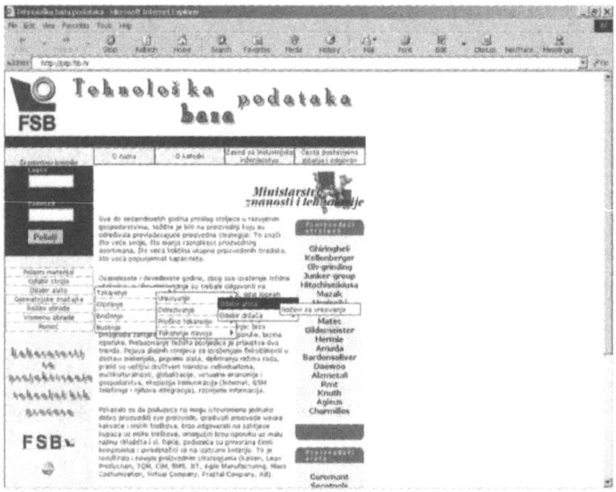

FIGURE 4. Web page for Database System for Process Planning (*http://ptp.fsb.hr/)*

6. CONCLUSION

The considered work presents the development of the variants of the technological process through insight in the criteria of selection, the mechanism of processing and database. The accent is placed on the selection of the tool machines and tools with the use of the ponderated criteria of selection (the way of tightening and processing, relative moving workpiece-milling tool, the quality of the treated surface, depth of cutting, conditioned by the geometry of the workpiece, the dimensions of the part, the cutting data, the times of the operations processing). The next phase would probably use fuzzy logics with the ponderated multiple solutions. The programming solution is realised for the client/server milieu and the Internet environment.

REFERENCES

1. Gu P.; Norrie D.H., (1995), Intelligent Manufacturing Planning, Chapman & Hall, London,
2. Buchmeister, B., Polajnar, (2000), Priprava proizvodnje, Maribor, Fakulteta za strojništvo.
3. Lin, E; Minis, I.; Nau, D.S.; Regli, W.C. *Virtual Manufacturing*,
4. Available from: http://www.isr.umd.edu/Labs/CIM/vm.html Accessed:1997-03-25.
5. Intelligent Manufacturing Systems (IMS), http://ksi.cpsc.ucalgary.ca/IMS/IMS.html. *Accessed from:* 1996-06-15.
6. CoroKey, Your Guide to productivity, Sandvik, (2000).
7. Chasqueira H; Santos L., Fonte A. J. P.D. ; Metrôlho, J.C. Development of Internet Toolkit for Computer Engineering Courses, Proc. of 17th CARS & FOF 2001, Durban, South Africa, Vol. 1, July 2001, pp. 365-370.

E-BUSINESS REQUIRES THE SIMULATION OF LOGISTICS PROCESSES

W. Sihn, T.-D. Graupner, U. Mussbach-Winter, H. Wiedenmann

Fraunhofer-Institute for Manufacturing Engineering and Automation (IPA), Stuttgart, Germany

KEYWORDS: E-Business, e-Services, Web-Services, Simulation, production, logistic management

ABSTRACT: E-business offers the chance to open new markets for selling and buying. E-business networks include both in-plant and cross-company processes. Therefore they are marked by a high complexity and strong demands regarding the management of the network. The Fraunhofer-Institute for Manufacturing Engineering and Automation developed several high-performance tools for designing and managing e-business networks. Three of these tools namely E-Simulation, E-Optimisation and the Internet-based Logistic Analysis. are described in this paper. E-Simulation allows users to simulate their own scenarios via Internet and benefit from the advantages of a dynamic simulation. If required, models can be adapted to changing conditions of the production environment. E-Optimisation is an Internet service for the determination of optimised production schedules based on flexible optimisation algorithms. The third tool described in the paper, the Internet-based logistics analysis is an electronic service that identifies rationalisation potentials in production logistics in a fast and cost-effective way.

1 INTRODUCTION

E-business opens up new opportunities to businesses while at the same time raising the demands on the structure and operation of new, more complex supply chains and networks. The spread of e-business and the related production networks have, without doubt, had a considerable impact on the organisation of order management processes throughout the supply chain. They include both in-plant and cross-company processes. In order to realise the benefits gained from e-business, a well functioning logistics chain is an absolute necessity. To enable order management in production networks it is no longer sufficient to consider the production capacities only. Existing or expected restrictions in supply, storage, and transport must also be included in planning. Therefore, when it comes to production and distribution, it is a must to use a high-performance tool to handle both E-business and Supply Chain Management. At our institute, several high-performance tools for designing and managing e-business networks have been developed. Figure 1 shows a selection of these tools, which will be described in the following [1], [2], [5].

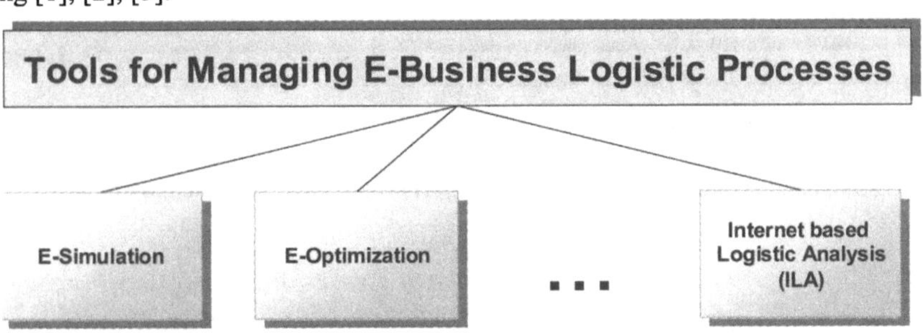

FIGURE 1. Tools for Managing E-Business Processes

Published in: E.Kuljanic (Ed.) *Advanced Manufacturing Systems and Technology*, CISM Courses and Lectures No. 437, Springer Wien New York, 2002.

2 E-SIMULATION

2.1 MEANING OF E-SIMULATION

Today, simulation systems are deployed in the planning of production and logistics systems. By visualising processes and evaluating simulation results, production planners are in a position to better understand complex systems. Difficult decisions are made more easily when a solid basis for decision-making is provided.

Simulation projects create customised simulation models for production and logistics systems. These models perfectly match organisational and technical processes with the required resources. Simulation models can thus be referred to as information stores. After completion of the simulation project, these information stores are usually no longer available to the user, unless the simulation systems are maintained and advanced by the enterprise. This requires that, apart from license fees and high-performance hardware, the company also has to provide skilled personnel [3], [4].

The aim of the e-Simulation service is to use and enlarge the information store for as long as the production and logistics system exists (see Figure 2). Therefore, it allows clients to use the generated models even after a project has been completed. By coupling simulation models to the Internet, users get constant access to the simulation tool. To make use of e-Simulation only an Internet connection is required.

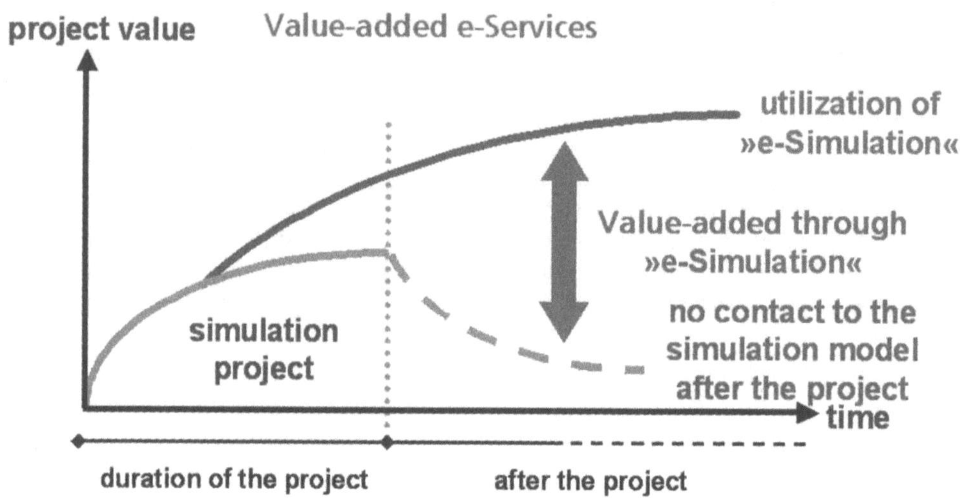

FIGURE 2. E-Simulation continues to increase the project benefit after the simulation project

2.2 BENEFITS OF E-SIMULATION

E-Simulation allows users to be actively involved in the modelling of processes during the early stages of the project. Thus, they are in a position to influence the model design at an early stage. They can simulate their own scenarios and benefit from the advantages of a dynamic simulation.
If required, models can be adapted to changing conditions of the production environment. This kind of customised support ensures the safeguarding of a company's investment. Adjustments can be made in a time and cost-saving manner to create the highest possible benefit for the entire life cycle of a production or logistics systems.

2.3 FUNCTIONING

When creating a simulation model, the required data are specified in close coordination with the user. These data include cycle times, set-up times or capacities needed for the simulation process. They are stored externally, e.g. in MS-Excel or MS-Access tables. The simulation models themselves are filed at the Fraunhofer Institute.

The user can now employ these data to create individual scenarios and upload or send them to the e-Simulation service via e-mail. On arrival, the transferred data are automatically checked for inconsistencies and transmitted to the simulation model for processing. Then, the simulation is automatically started. After the conclusion and evaluation of the simulation, the user receives an email with the results in tabular and graphic form. These evaluations consider the previously analysed company-specific parameters.

2.4 PRACTICAL EXAMPLE

E-Simulation has been successfully used for production optimisation in an automotive supply chain. During the project, simulation was employed to determine the necessary logistical areas and interpret the number and size of the cage boxes. Moreover, the future machine utilisation and throughput had to be established. Within a few days, a highly detailed simulation model for the production process was generated. The simulation model mapped 15 different machines and 40 different articles with specific cycle and set-up times.

Both the process and the control parameters of the simulation models were embedded in a MS-Excel table, ensuring the simple manipulation of production data. By applying the e-Simulation service, users at different sites could access the constantly updated models. Once the MS-Excel table was sent to a specified e-mail address, it took only 30 minutes until the user received the evaluation of the simulated scenarios, edited as graphics and nominal parameters. The simulation helped to carry out detailed analyses, identify bottlenecks and weaknesses before serial production started, and find appropriate solutions.

3 E-OPTIMISATION

3.1 MEANING OF E-OPTIMISATION

As mentioned above, e-business significantly increases the need for co-ordination in the production and logistics processes. Often, the problems are so complex that modern planning tools are required to facilitate decision-making in order to create and implement proposals in a reasonable period of time.

E-Optimisation is an Internet service for the determination of optimised production schedules based on flexible optimisation algorithms. E-Optimisation solves varied problems. The optimisation considers all the relevant constraints, such as stock costs, delivery performance, lead times and workload, and provides an optimal solution regarding the desired criteria.

3.2 BENEFIT OF E-OPTIMISATION

By applying e-Optimisation, the user saves time and money. Time is saved because a quick and uncomplicated utilisation of the planning tool is possible via Internet. The tool is based on flexible algorithms and runs on the service provider's own high-performance computers. Using the planning tool via Internet also helps to save costs. The only thing required to get access to e-Optimisation is an Internet connection. As the applications and hardware belong to the service provider, they do not have to be purchased. Thus, the user keeps both investment and operating costs for the maintenance of infrastructure at a low level. Only the service provider's performance has to be paid for.

3.3 FUNCTIONING

Using e-Optimisation is very simple: Either the user is directly connected to the e-Optimisation service via data link, or an encoded channel is used to transfer the production data at regular intervals.

The scheduler himself sets the operational planning frameworks through a specific graphic user interface, so that he can easily adjust them. Special attention has been paid to the design of the graphic user interface, as the man-machine interface is decisive for the correct and efficient use of a system.

E-Optimisation performs the optimisation tasks with the help of several evolutionary algorithms running in parallel on a computer network. The system is learning with every optimisation run and autonomously changing the parameterisation of existing algorithms. Thus, the system adapts to changing tasks, reducing maintenance costs and improving the system's performance.
After the optimisation, the results are shown on the desktop as loading plans or directly imported into an ERP system (see Figure 3).

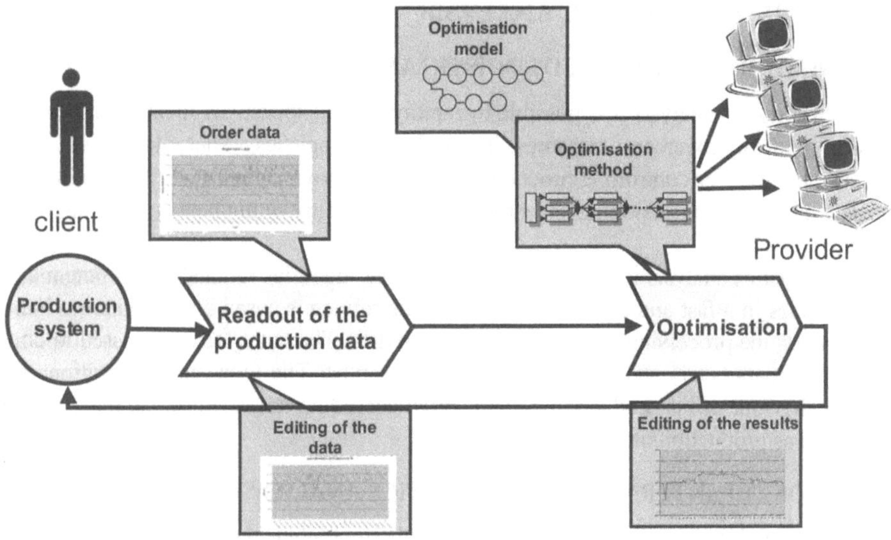

FIGURE 3. The E-Optimisation process

3.4 PRACTICAL EXAMPLE

E-Optimisation serves a wide range of applications mainly in the fields of automotive and steel industry, but also in the aircraft industry.

Taking an example from the steel industry, e-Optimisation was used to work out optimal loading plans for a cold rolling mill. A cold rolling mill produces extremely precise steel strips to be used, for example, in the automotive industry. It is the task of the optimisation planner to align the processing sequence of the steel strips in such a way as to achieve maximum throughput while letting the steel strips pass through the mill without delay. The system includes a number of restrictions and dependencies calling for the global optimisation of the overall process.

4 INTERNET-BASED LOGISTIC ANALYSIS

4.1 MEANING OF INTERNET BASED LOGISTIC ANALYSIS

With the aid of logistics analysis it is possible to rapidly and inexpensively identify and evaluate rationalisation and improvement measures in production logistics. Thus, logistics analysis becomes a suitable tool for controlling production planning and management (PPM); but it is also an appropriate means for dimensioning and structuring assignments within factory planning.

Internet-based logistics analysis is an electronic service that identifies rationalisation potentials in production logistics in a fast and cost-effective way. This service is based on the Hanover Funnel Model, automating the processing of existing production data. The latter are compressed to obtain relevant logistical parameters and be appropriately represented. The user gets information about the logistical behaviour of his production system sorted according to resources, i.e. machines and facilities, as well as production orders.

4.2 BENEFIT OF THE INTERNET-BASED LOGISTICS ANALYSIS

This service is available to every production plant with single-item production. Being Internet-based, the service avoids costly consulting visits and is available independent of time and location. For easy usage, the Internet-based logistics analysis provides a brief manual to interpret the obtained analysis results. The clarity of presentation and the tight construction of the underlying Funnel Model contribute decisively to the comprehensibility of the service.

The origin of the logistics analysis is the Funnel Model of the Institute of Production Systems (University of Hanover, IFA Institut für Fabrikanlagen). It links the production logistical parameters of stock, performance and turnover time, offering a valuable descriptive model for production logistics. The defined parameters of stock, performance, order structure, lead times, utilisation and schedule deviations help to rapidly identify those machines and facilities that significantly influence the logistical performance. This allows taking appropriate actions focusing on [6].

- lead time reduction,
- inventory reduction and
- improved adherence to schedules.

Most companies possess the necessary database by way of production data collection (PDC). Existing systems, however, create primarily parameters relating to monetary terms, from which production logistics improvements cannot be deduced directly.

The Internet-based logistics analysis is distinguished by an excellent cost-benefit ratio. By avoiding costly external consulting and giving simple access via World Wide Web, by exploiting available corporate data and making the modification of existing information systems superfluous, the potential benefits can be unlocked for production logistics.

4.3 FUNCTIONING

The Internet-based logistics analysis is divided into two parts, i.e. user registration and analysis. During registration the user enters his password at the Internet service's homepage. This initiates the creation of a "personal area" on the web pages from which the user can carry out the logistics analyses.

When performing the logistics analyses, the first step is to fill in a text file according to a pre-designed format. To this end, a template is available to be downloaded from the homepage (Figure 4). Next, the file including the company data is uploaded to the Internet service, which automatically carries out the analysis. The results are edited as web pages and stored in the personal area of the user, who is informed by e-mail. Thus, accessing results is possible independent of time and location.

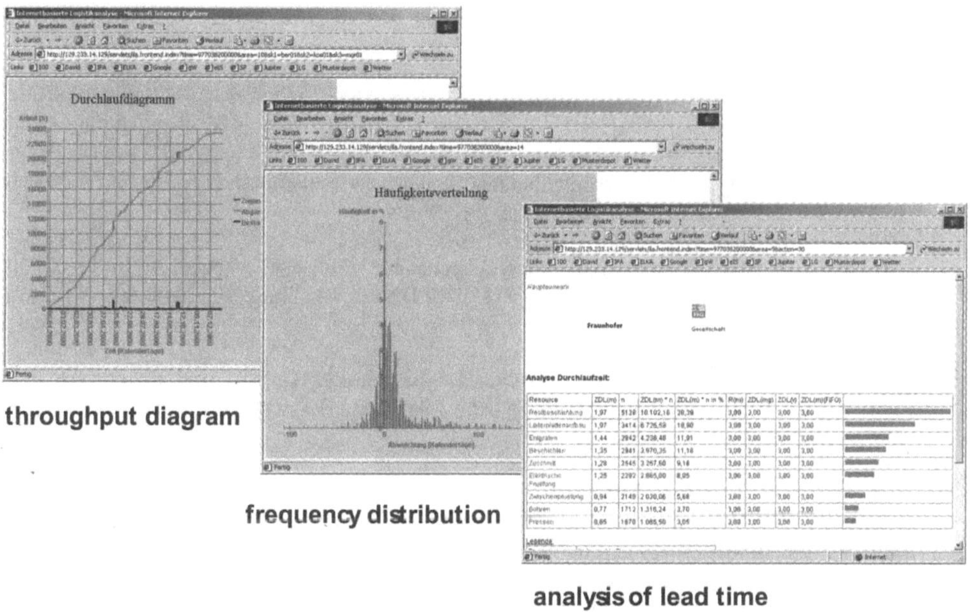

FIGURE 4. Analysis Results (Example)

The user may perform an unlimited number of analyses. The results remain available, but may be deleted by the user.

4.4 PRACTICAL EXAMPLE

The Internet-based logistics analysis is a brand-new tool. The development of the tool was finished a short time ago. Therefore a practical example is not available up to now.

5 CONCLUSION

The paper describes a selection of tools to support the production and logistical management processes in e-business networks. Our service range is constantly expanded thanks to several research and development projects being carried out in this field. Presently, we intend to develop new modules for an easier set-up of e-business simulation models.

REFERENCES

[1] Sihn, W., Halmosi, H., Mandel, J., (2001), Komplexe, simulationsgestützte Internetdienstleistungen – die Beispiele Produktionssimulation und Logistikanalyse. In: Schulze, Th., Schlechtweg, S., Hinz, V. (Hrsg.), Proceedings zur Tagung „Simulation und Visualisierung 2001“ am Institut für Simulation und Graphik der Otto-on-Guericke-Universität Magdeburg, March 22nd / 23rd, 2001.

[2] Sihn, W.; Graupner, T.-D., (2001), e-Industrial Services: Value-Added Services for the Producing Sector. In: Chryssolouris, George (Hrsg.); University of Patras / Dept. of Mechanical Engineering and Aeronautics / Lab. for Manufacturing Systems and Automation: Technology and Challenges for the 21st Century : CIRP 34th International Seminar on Manufacturing Systems, 16-18 May, 2001, Athens, Greece. Athens, Greece, 2001, S. 413-424

[3] J. Kulis & R.P. Paul. A review of web based simulation: whither we wander? In: In: Proc. 2000 Winter Simulation Conference (Eds.: Joines, J.A., Barton, R., Kang, K., Fishwick, P.), 1872-1881.

[4] E.H. Page. The rise of web-based simulation: Implications for the high-level architecture. Proc. 1998 Winter Simulation Conference, (Hrsg.: Medeiros, D.J., Watson, E.F., Carson, J.S., Manivannan, M.S.), 1663-1668.

[5] Sihn, Wilfried; Graupner, Tom-David: E-Industrial Services Value-Added Services for the Producing Sector Using the Example of Simulation. In: Jiang, Pingyu (Hrsg.) u.a.: ICECE 2001 : Abstracts for Proceedings of Internaional Conference on e-Commerce Engineering: New Challenges for Global Manufacturing in the 21st Century, Sept. 16-18, 2001, Xi'an, China. Beijing, China : China Machine Press, 2001, S. 40

[6] Nyhuis, P.; Wiendahl, H.-P.:
Logistische Kennlinien.
Berlin et al: Springer, 1999

TOOL PATH OPTIMIZATION BY HS MACHINING

J. Kopac, T.Roblek

Department of Machining Technology Management
Faculty of mechanical engineering,University of Ljubljana, Slovenia

KEYWORDS: high speed machining (HSM), tool path optimization, CAD/CAM

ABSTRACT. There are many components to an effective high speed (HS) machining process for die and mould manufacturing. Modern HS milling machine tools allow high spindle and feed rate speeds. Often forgotten is HS machining's impact on tool path programming techniques. Programming of NC tool path for HS machining takes more than just plugging in different numbers of feed, speed and radial and axial depth of cut. Therefore special techniques are needed to assure good surface quality, short machining time and low costs by machining. Efficient use of modern HS machines demands also modern CAD/CAM system.

1 INTRODUCTION

HSC technology has some basic demands by machining:

- maintain a constant chip load
- minimize feed rate losses
- maximize program processing speed

Maintain a constant chip load is important to achieve long tool life, good surface quality and high geometrical accuracy. To assure constant chip load is necessary to reduce air cutting, what cause also constant cutting force and less vibrations by cutting. Gentle entry cuts are advisable. Also try to minimize as much as possible the need for the tool to exit and re-enter the material. Ramping or helical entry from one level to another is better then plunging.

Minimize feed rate losses. The challenge to the CAM system is to make tool path with small stepowers at very high federate speed. This must be accomplished without forcing the cutting tool to make sharp turns. Modern HS milling machine tools have look-ahead feature to reduce the cutting feed rate when they detect a corner approaching. At sharp corner 90° is the cutting speed at the corner $v_f = 0$ mm/min.

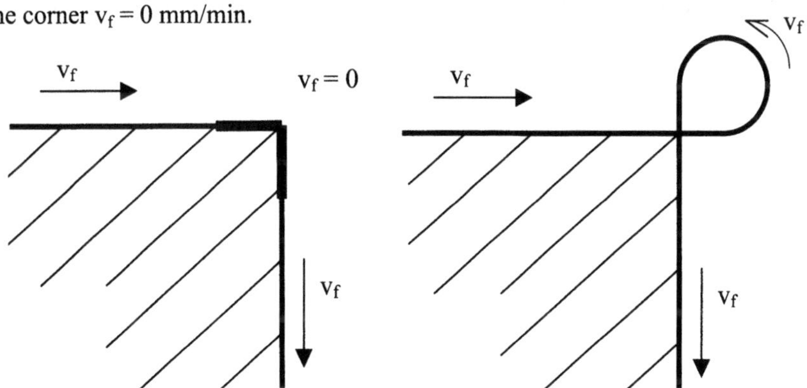

FIGURE 1. Cutting speed at sharp corner by conventional (left) and HSC tool path (right)

Published in: E.Kuljanic (Ed.) *Advanced Manufacturing Systems and Technology*,
CISM Courses and Lectures No. 437, Springer Wien New York, 2002.

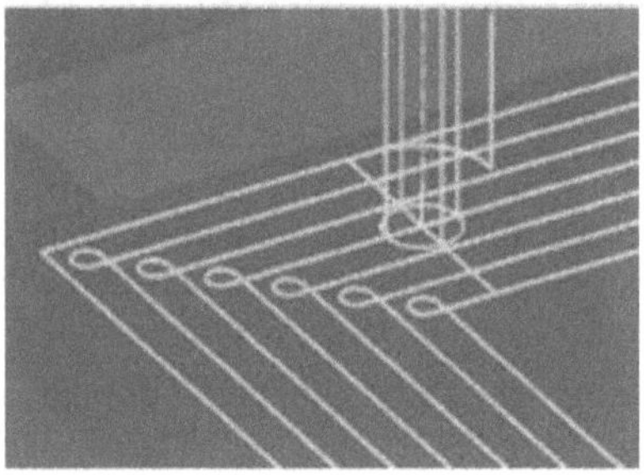

FIGURE 2. Generated fluid tool motions is important part of HS toolpath planning

A corner treatment function for HS machining rounds the sharp motions out of the path. If allowed to remain, this sharp motion would be seen by look-ahead function, which would reduce the feed rate accordingly. A CAM system that can generate fluid tool motions during corner machining can maintain more consistent high feed rates.
Maximize program processing speed is also very important for effective NC code producing. For complex parts (for example die and moulds) NC code generating time can be very long. In order to overcome data starvation – which will also impair the feed rate – the CAD/CAM system may be required to output tool paths appropriate to machine tool controls capable of running NURBS-based G-code. NURBS spline based G-code is called new G-code "G6.2". This command expands the choices from traditional linear and circular interpolation to interpolation along a spline represented by control points and knot points. By consolidating a complex, curving tool path into a single line of the program, this function saves on NC data, potentially resulting in more fluid HS machining.
The simpler the cut, the better, because that give opportunity to the process to operate at maximum feed rates and not to slow down for dense of clusters of data points or abrupt changes in direction. In zigzag cut connecting each successive pass with a looping motion likewise will minimize the sharp changes in acceleration and deacceleration required to execute a square end.
Stock remaining – local machining. In order to machine a part with different size of cutting tools is important that is the knowledge of stock remaining built in CAD/CAM system.

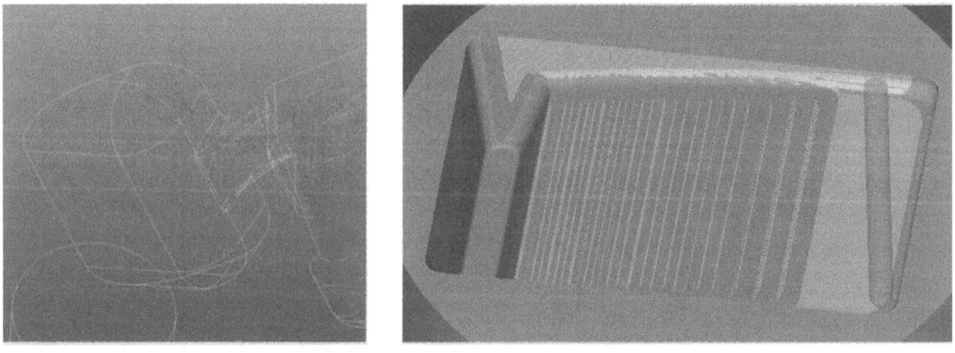

FIGURE 3. Stock remaining – local machining

Pencil milling automatically cleans up corners and concave fillets so that there is uniform stock left at surface intersections. Reliving the larger volume of material allows for less tool deflection and noise when cutting corners.
Rest milling is similar to pencil milling in that it defines all areas that were not cut based on one tool size, and automatically cuts only those areas with a smaller tool. The difference is that rest milling can cut anywhere, not just corners.
Side steps are the connections that crate effective transitions between adjacent tool paths when feed rates are particularly high. Parallel scan-line surface machining is the type of machining that has been used for the last teen years to finish multi-surface models. This type produce sharp stepover moves at the end of every pass. For feed rate speeds up to 1000 mm/min is advisable use of simple "looping" toolpath, for higher speeds we use a "golf club" stepover between passes.

2 OPTIMIZATION ANALYSES – MACHINING TYPE AND CUTTING TOOL DIAMETER

In the continuation an example of optimizing is shown. The aim of this was to produce a model with different machining types (methods) and different sequence of cutting tools.

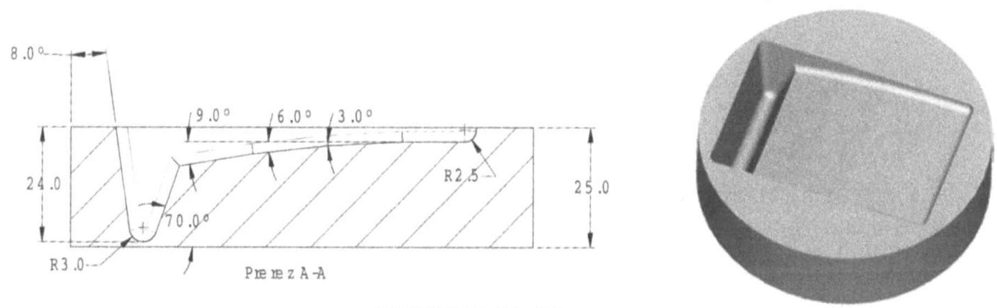

FIGURE 4. Model

In the optimization the CAD/CAM program called Pro/ENGINEER was used. 12 simulations were madc with different technological processes. The basic difference among the technological processes was in the type of machining the model and in type and number of used tools. Machining was limited to 3-axis milling with 6 different tools. In the selection of tools and cutting parameters we considered the type and hardness of working material (Table 1), the type of machining and also maximal spindle speed and feed rate of the MORI SEIKI FRONTIER-MI vertical milling machine, on which the machining took place. The machine reaches spindle speeds up to 20.000 min^{-1} and feed rates up to 5.000 mm/min. For machining we used four sharp endmills (z = 6) of 10, 8, 6 and 5 mm diameters and two ball nose endmills (z = 2) of 6 and 4 mm diameters. All endmills were made of hardmetal, coated with TiAlCN and suitable for rough and fine HS machining.

TABLE 1. Working material

Steel [DIN]	Hardness [HRC]	Structure [%]					
		C	Si	Mn	Cr	Mo	V
X40 CrMoV51	53	0.4	1.05	0.40	5.15	1.35	1.00

The results of this optimization are shown on figure 5. The differences in machining times are very high, so is important to produce in right way. In time, knowledge based systems may well automate virtually all of the process planning and programming for high speed machining. For now, there is no substitute for programmers understanding the high speed machining process.

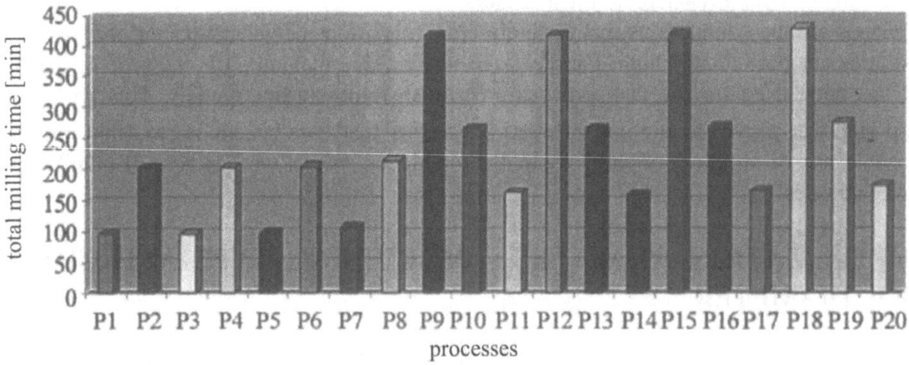

FIGURE 5. Total milling times by different ways of model machining

3 COMPARISON OF CONVENTIONAL AND HS TOOLPATH

We made a comparison between conventional milling toolpath method and HS milling toolpath method. The experiments were made on ProEngineer CAD/CAM system. The only difference between those methods is in type toolpath creation. All others parameters as cutting depth, spindle speed, etc. were the same. By conventional milling method we used spiral type of cutting tool movements. The tool machined all working area in 5.5 min.

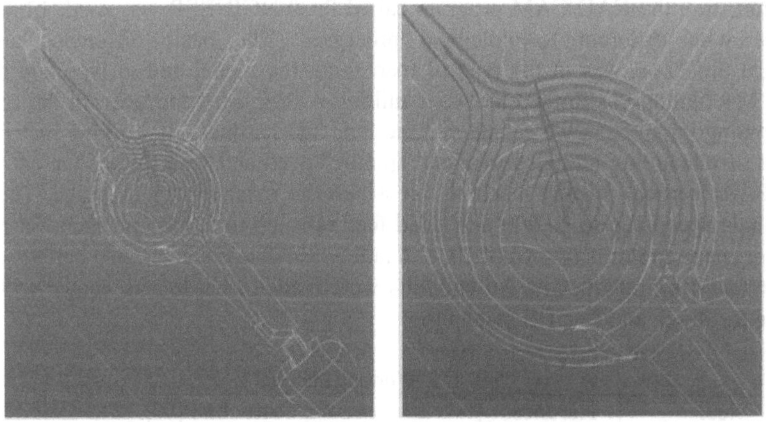

FIGURE 6. Conventional milling toolpath, time (first tool) = 5.5 min

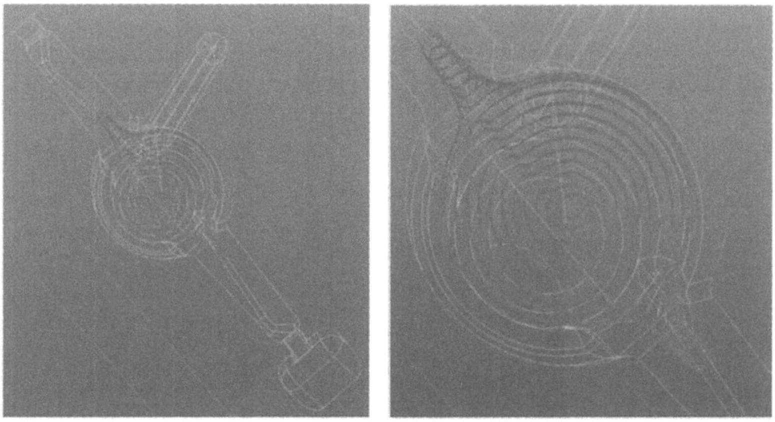

FIGURE 7. Conventional milling toolpath, time (first tool) = 5.7 min *(*some area in not machined)*

By HS milling method we used constant load type of cutting tool movements. Generated path was so produced without abrupt changes in directions. Calculating time increased approximately 3 times. Changes in directions were made with corner round radius 1 mm. The area of model where program has no possibility to make the path with corner round is not machined. By local milling is from that reason necessary to remove more material with smaller cutting tool, what increase total machining time.

4 CUTTING FEED RATE OPTIMIZATION

Some CAD/CAM systems have additional module for cutting feed optimization to optimize a milling operations. One of those is MasterCAM high feed optimization module which optimizes feed rates for 2.5-axis and 3-axis mill toolpaths. It computes feed rates that reduce machining time without sacrificing accuracy. Instead of using the same feed rate for the entire part, a different feed rate is computed for each tool position based on the volume of material currently being removed, resulting in a faster machining cycle and control over the machining forces. The high feed option optimizes both roughing and finishing toolpaths. Roughing toolpaths are optimized by using high feed rates in areas where a low volume of material is removed and by using low feed rates in areas where a high volume of material is removed. Finishing toolpaths are further optimized by volume calculation and by adjusting feed rates on arcs and at sharp corners, thereby reducing servo lag error (axis over-travel).

4.1 MACHINE DYNAMICS

The Machine Dynamics portion of the high feed option defines the machine tool's feed rate capabilities - how fast can the machine feed without over travelling. Typically, you define machine dynamics before entering other highfeed parameters.

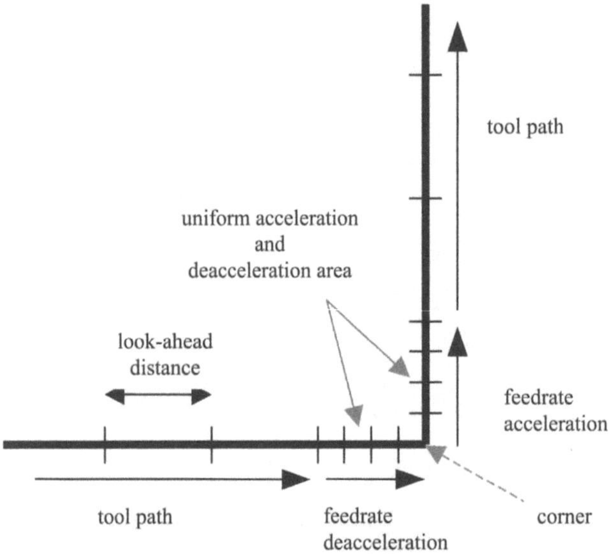

FIGURE 8. Machine tool dynamics influence on tool path

The cornering values are the most important machine dynamics parameters. Cornering parameters limit feed rates at sharp corners and arcs, resulting in increased accuracy and a better finish. At arcs and sharp corners machine tool needs to slow down to prevent over-travel. The minimum cornering feed rate is used when tool motion vectors change more then a specific angular amount.

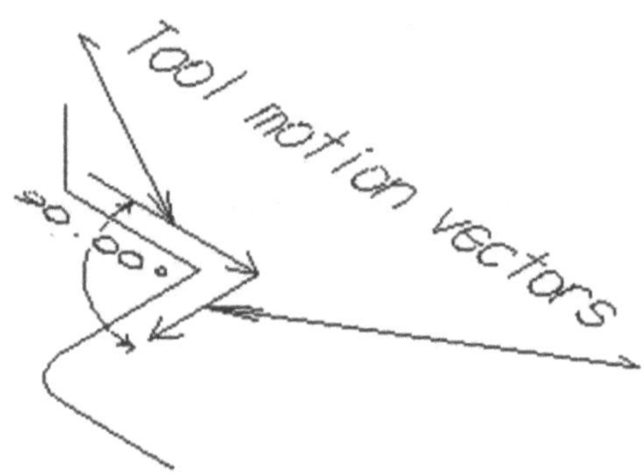

FIGURE 9. Tool motion vector

4.2 MATERIAL REMOVAL RATE

The material removal rate controls feed rates, and is especially applicable to roughing toolpaths. High feed function defines the material removal rate of each tool used in the toolpath. The reference tool table stores the tools and is generated automatically when selecting the operation for optimization. For material removal rate fhe following factors need to be described: cutting tool, maximum feed rate, minimum feed rate, feed rate, plunge rate, material removal rate (volume/min).

4.3 STOCK SETUP
Stock must be exactly defined, because the stock boundary divide tool path on air cutting and material cutting. Optimization module all toolpaths outside the stock boundaries make at higher feed rate.

4.5 PRACTICAL EXAMPLE
In laboratory we made an experiment and machined one part with and without high feed optimization module.

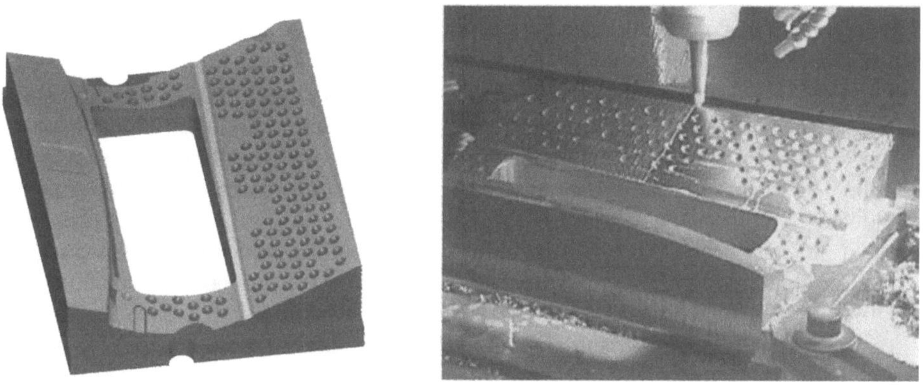

FIGURE 10. 3D model and manufacturing of bending insert

TABLE 2. Machining times of bending insert

Operation	without high feed function	with high feed function
	[min]	[min]
rough	210	171
semi finish	60	46
finish	220	172
additional finish	240	165
TOTAL	**730**	**554**

From the results represented in Table 2 is obvious that the total machining time with use of high feed function decrease for 31%.

5 CONCLUSIONS

Tool path optimization is part of complex process. Today's CAD/CAM systems allow handling with routine processes, but engineers' knowledge is a key to effective manufacturing. CAD/CAM system generated NC program usually base on accurate tool path (geometry), so special modules or program are needed to adjust NC program to cutting tool and machine tool properties.

Optimization module of specific CAD/CAM system can't be adapted to different CAD/CAM system. The costs for module are usually high. In our laboratory we develop a CAD/CAM system independent feed rate optimization module, which will handle with standard G-code.

Advantage of software look-ahead function regard to machine tool look-ahead feature is in better feed rate profile. Machine tool look-ahead function analyses a few hundreds of NC blocks ahead and calculates the maximum attainable feed rate. The contour error is defined in machine tool settings and is equal for rough and finish machining.

Software look-ahead function use recursive technique based on mathematical model. The generated feed rate profile is from that reason generated more accurate. The optimized feed rate for rough machining can be higher as for finish machining, because the contour error can be higher. Software optimized G-code is also possible to use on machine tool without look-ahead feature.

The next generation of CAD/CAM systems will combine manufacturing feature recognition and knowledge-based machining strategies to automate the complex machining process. Optimization module will become a part of CAD/CAM generated toolpath. The resulting system will provide complete automation and yet still allow experienced users to tailor the system.

REFERENCES

1. Marinc D., Tool path strategies for high speed machining, MMS Online, 2001
2. Beard T., Programming for high speed machining, MMS Online, 2001
3. Pegan P., High speed milling optimization, Diploma thesis, Ljubljana, 2000
4. Furlani M., HSC with application of modern CAM programming, Diploma thesis, Ljubljana, 2001
5. Kopac J., Sokovic M., Dolinsek S., Tribology of coated tools in conventional and HSC machining. J. mater. process. technol.. [Print ed.], 2001, vol. 118, no. 1/3, special issue "AMPT'99", part 1, str. 377-384
6. Kopac J., Roblek T., Problems solving at high speed cutting, Orodjarstvo '99, GZS, Združenje kovinske industrije, Odbor za orodjarstvo, 1999, str. 43-50
6. Balic J., A new NC machine tool controller for step-by-step milling. Int. j. adv. manuf. technol., str. 399-403
7. Cus F., The inclusion of the geometrical shape of the cutter into the optimization of the milling process. Int. j. adv. manuf. technol., 2000, vol. 16, no. 6, str. 392-403
8. Dolinsek S., Interrelation between cutting mechanics and machinability parameters; Drilling of Austenitic Stainless Steels. Metall (Berl. West), 2000, jg. 54, nr. 6, str. 184-189
9. Vajde Horvat R., Welzer Druzovec T., Rozman I., Sokovic M., An evaluation of process complexity. Stroj. vestn., 2001, letn. 47, št. 1, str. 15-27

UNCERTAINTY ANALYSIS AND OPTIMISATION OF A FMS BY EXPERIMENTAL DESIGN APPLIED TO SIMULATION

B. Palumbo[1], F. Caiazzo[2], V. Sergi[2]

[1] Department of Materials and Production Engineering, University of Naples "Federico II", Italy
[2] Department of Mechanical Engineering, University of Salerno, Italy

KEYWORDS: Flexible Manufacturing Systems; Computer Experiments; Design of Experiment.

ABSTRACT. The aim of this paper is to apply, in the simulation of a Flexible Manufacturing System (FMS), a statistical methodology that enables to evaluate the level of control factors, at low economical impact, that yields an optimal response. The methodology adopted to realise this objective is the Design of Experiment (DOE) applied to discrete event simulation. The analysis of the result is carried out by using the Analysis of Variance (ANOVA).
The first part of the paper describes, in general, the above-mentioned methodology. In particular, it takes in consideration the technological model that determines uncertainty (noise factor) and the choice of the stochastic model that interprets, in probabilistic terms, the same phenomenon.
The second part shows how it is possible to apply the aforesaid methodology to a realistic industrial example of FMS. The proposed FMS consists of six machining centres, an assembly and disassembly station of parts and pallets, two storage-shipping stations and three AGV.

1 INTRODUCTION

In the last few decades the manufacturing industry has progressed through the phases of industrialisation, mechanisation and automation. The concept of flexible manufacturing forms a part of global strategy of Computer Integrated Manufacturing. An FMS is an integrated, computer controlled complex of numerically controlled machine tools linked by an automated material handling system.
FMS are often characterised by a large part mix, alternate part routing, negligible tool changeover times, limited buffer storage between machines and faster processing times. These features make FMS capable of rapidly adjusting to product mix and product design changes. FMS have the efficiency and low-cost benefits of a well balanced transfer line, with the flexibility of the job-shop. These features make them particularly suitable for medium-volume or batch production.
The large number of variables that characterise a decision process in an FMS environment and the intricate nature of the mutual effects of decisions require hierarchical levels of decision making and the use of computer-based support tools.
Static allocation, queuing network and simulation are the most used techniques which, with increasing levels of detail and accuracy of the results, model the allocation process of the FMS resources to the processed part. Evaluating candidate solutions with these techniques allows the selection of the most promising ones, [1].
The simulation model requires, as input, a proposed production schedule as well as rules for dynamically adapting the schedules in consequence of stochastic machine down-time and throughput times. A high level of detail can be incorporated in the simulation model and, as result, we may

Published in: E.Kuljanic (Ed.) *Advanced Manufacturing Systems and Technology*,
CISM Courses and Lectures No. 437, Springer Wien New York, 2002.

obtain good point and interval estimates regarding, for example, the expected production costs and the likelihood and magnitude of up-side deviation from the target budget.

In the simulation field, practitioners often perform one-factor-at-time (OFAT) experiments, which vary only one-factor-at-time while keeping others fixed. However, statistically designed experiment that vary several factors simultaneously are more efficient when studying two or more factors, [2] [3]. In fact, statistical DOE refers to the process of planning the experiment so that appropriate data, that can be analysed by statistical methods, will be collected, resulting in valid and objective conclusions, [4].

Using a terminology usually adopted in the statistical field of experimental design, [4] [5], the input parameters and the structural assumption composing a simulation model of a FMS are called factors, and the output performance measures are called responses; each possible value of a factor is called a level of the factor and a combination of factors all at a specified level is called a treatment.

Factors may be classified as controllable or uncontrollable, depending on whether they represent action options to managers of the corresponding real-world system (policy variables). Reference [6] gives many practical examples of controllable and uncontrollable factors and shows how the factor's controllability may depend on the particulars of the situation.

If all the factors are deterministic, replicate observations from running the simulation model with the same factors will be identical. It is this lack of random error, due to variability in the experimental unit, that makes these deterministic computer experiments different from physical experiments, [7].

In literature there are two main statistical approaches to computer experiment, one based on Bayesian statistics and a frequentist one based on sampling techniques, [8]. In one or the other of these ways, it seems to be essential to introduce randomness to generate probability or confidence intervals.

This paper is based on a frequentist approach. In fact, process variability is incorporated in the simulation model by Monte Carlo sampling from statistical distribution used to model the uncontrollable factors. It has been natural, therefore, to design and analyse such stochastic simulation experiments using standard techniques for real experiments, [9]. Not considering these uncontrollable factors, named also noise factors, the simulation would be deterministic.

This paper takes in consideration only those control factors at low economical impact. The factors that don't change during the simulation are defined constant factors.

Looking at the typical phases in the simulation study — validation, screening, sensitivity analysis, uncertainty analysis and optimisation, [10] — this paper deals with the last two. In particular, the uncertainty analysis is relative to factors that are held intrinsically stochastic (noise factors) and not to modelling control factors by statistical distributions.

Section 2 discusses the FMS model and the control, noise and constant factors taken in consideration. Section 3 deals with the simulation model and DOE. Section 4 presents the results of the simulation.

2 FMS MODEL

2.1 LAYOUT AND FLOW OF PARTS INSIDE THE MANUFACTURING SYSTEM

The considered FMS reassumes some typical characteristics of a real production system. It is constituted by an entry station, an assembly station for mounting a part on the pallet, a station for loading the pallet in the system, 6 multipurpose machining centres, a transportation net with AGV, a station for the AGV battery change, a station for unloading the pallet from the system, a disassembly station for unmounting the part from the pallet and an exit station. The assembly/disassembly stations are connected, through conveyor belt, to the load/unloading stations. The schematic FMS layout is shown in Figure 1..

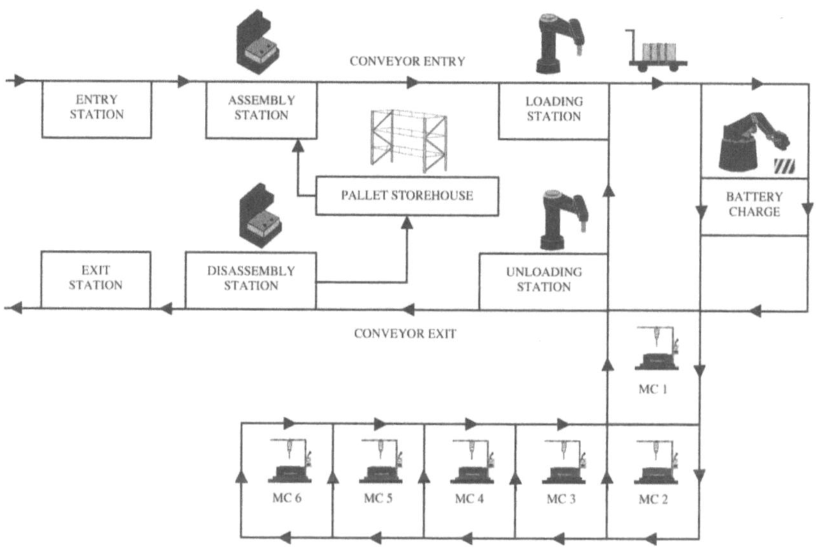

FIGURE 1. Schematic FMS layout

The system operates on a product mix of 5 different parts typology. The parts enter the system and they reach the entry station. If a pallet is available in the pallet storehouse, the pallet and the part are assembled and dispatched, through conveyor belt, to the loading station; otherwise, the part is queued to the entry station.

Arrived at the loading station, the part asks for the transport through AGV; the AGV collects the part and transfers it to the first machining centre scheduled by the working cycle. Alternative working cycles are foreseen, depending on the operational conditions of the system.

After having completed the technological cycle, the part is transferred to the unloading station. By a conveyor belt the part, still mounted on the pallet, is sent to the disassembly station; the empty pallet is sent to the pallet storehouse and the machined part is sent to the exit station.

2.2 CONTROL FACTORS

The control factors, at low economical impact, considered are the following: "Machine Scheduling Rules" (MSR), "AGV Scheduling Rules" (ASR), "AGV Speed" (AS) and maximum "Number of Pallets" (NP) inside the production system.

The MSR factor assumes three different qualitative levels: "First In First Out" (FIFO), "Short Process Time" (SPT) and "Long Process Time" (LPT). In the first case, FIFO, the parts will be worked in the same order with which they arrive to the machining centre. In the second case, SPT, the parts with a smaller residual process time will be worked first. In the third, and last case, LPT, the parts with a greater residual process time will be worked first.

The ASR factor assumes four different qualitative levels: "Fixing Routing" (FR), "Priority" (P), "Smallest Queue" (SQ) and "Short Travelling Distance" (STD). In the first case, FR, the parts follow the basic working cycles. In the P case, the parts follow, preferentially, the basic working cycles; if the machining centre to which the parts are destined reaches conditions of saturation, the latter will follow the first, the second and, sometimes, the alternative third working cycle. In the SQ case, the parts are always sent to the machining centre with the smaller queue among all those foreseen by the working cycles. In the STD case, the parts follow the cycles that minimise the distances for the transport among the various machining centres.

The factor AS assumes two quantitative levels: 30 and 70 m/min.

The factor NP assumes two quantitative levels: 35 and 50.

Table 1. summarises control factors and the coded levels taken in consideration during the simulation.

TABLE 1. Control factors

Control factors	Label	Levels			
		1	2	3	4
Machine Scheduling Rules	MSR	FIFO	SPT	LPT	
AGV Scheduling Rules	ASR	FR	P	SQ	STD
AGV Speed	AS	30	70		
Number of Pallet	NP	35	50		

2.3 NOISE FACTORS

The noise factors considered are the following: time between arrivals, time between failures, repairing time, working time, time between two battery charges, assembling and disassembling pallet time.

Table 2. shows the statistical distribution adopted to model the noise factors and the relative parameters; these values result from structural assumptions and data available by practitioners.

If we assume that the arrival, failure and repairing events recur at random in time, the time between these events can be usefully represented by an Exponential distribution, [11].

The working time is modelled by a Gamma distribution. The random nature results from the hypothesis that this time includes the set-up time that, for each machining centre, differs from part to part. The mean value, μ, is deducible from Table 3. for the different parts and working cycles. For example, type A part is assumed to use three working cycles. The first one is the basic working cycle, while the second and third one, are considered as alternative working cycles. The number in brackets shows the processing sequence number. The variance value is assumed equal to $0.0025\mu^2$, [12].

The time between battery charges is susceptible of small variations owed to the speed, to the transported load and to the effected run. A triangular distribution with parameters 14 and 18 minutes is assumed.

The assembling and disassembling pallet times are modelled by a Uniform distribution with parameters 1.5 and 2.5. The random nature results from the hypothesis that same pallets are employed for different parts, and so it is necessary to bring small changes to the action of the assembly and disassembly.

TABLE 2. Noise factors

Noise factors	Label	Statistical distribution	Parameters
Time between arrivals	NF1	Exponential, EXP(λ_1)	$\mu_1=1/\lambda_1=6$
Time between failures	NF2	Exponential, EXP(λ_2)	$\mu_2=1/\lambda_2=18$
Time between repairs	NF3	Exponential, EXP(λ_3)	$\mu_3=1/\lambda_3=25$
Working time	NF4	Gamma, $\Gamma(\alpha,\beta)$	$\alpha=400$ $\beta=\mu/400$
Time between battery charges	NF5	Triangular, TR(a, b)	a=14, b=18
Assembly and disassembly pallet time	NF6	Uniform, UN(a, b)	a=1.5, b=2.5

TABLE 3. Working cycles

		Mean working time, μ (minute)					
		MC 1	MC 2	MC 3	MC 4	MC 5	MC 6
Part A	Basic cycle	(1) 2	(2) 3	(3) 12			
	I alternative cycle	(1) 2			(3) 4		(2) 7
	II alternative cycle				(2) 4	(1) 9.5	
Part B	Basic cycle	(1) 3	(4) 38			(2) 4	(3) 18
	I alternative cycle			(1) 29	(2) 40		
	II alternative cycle		(2) 29	(1) 38			
Part C	Basic cycle	(1) 26			(2) 35		
	I alternative cycle	(1) 26		(2) 39			
	II alternative cycle			(2) 39		(1) 30	
Part D	Basic cycle				(2) 20	(1) 16	
	I alternative cycle	(1) 18		(2) 23			
	II alternative cycle	(1) 18	(2) 27				
Part E	Basic cycle			(1) 42			
	I alternative cycle	(2) 22				(1) 26	
	II alternative cycle					(1) 26	(2) 30
	III alternative cycle				(1) 29		(2) 30

2.4 CONSTANT FACTORS

The principal constant factors considered are the following: number of the machining centre, number of AGV, manufacturing layout, battery change time, conveyor speed, assembling and disassembling pallet time and productive mix.
A warm-up period sufficient to reach a steady-state has been adopted. Typically, data generated in the warm-up period instil bias in the analysis, [13]. Therefore, the output data in this first period have been truncated from the final analysis and only the output data relative to a following period have been considered.

3 SIMULATION MODEL AND DESIGN OF EXPERIMENT

The simulation of the production system has been implemented with Arena and its library "Advanced Manufacturing Build", [14].
This software allows a rapid implementation of the system's model used as case study; the only difficulty has been found in the insertion of the alternative working cycles. The solution to this problem has been that of employ the "Expression module", even if it is necessary to modify the "Parts module", assigning an increasing attribute index to every part typology. In the "Expression module" we define each service time of the basic working cycle and alternate ones for all the part typologies. Therefore the indexed "Process Time" is introduced in the "WorkCenter" module.
In Arena it is possible to define and control the random number generation for the principal discrete and continuous random variables.
A first technique could be that of using, for every source of uncertainty (noise factors) and for all the treatments, one random number stream. Such technique is known in literature as synchronised Common Random Number (CRN). The drawback is that in this way the replications don't constitute s-independent samples.
An alternative could be that of using, for every treatment, a fixed random number stream, usable by all the sources of variability of the system. For every treatment a different sequence would be used. In such way the replication would constitute s-independent samples.
The model of simulation introduced in this paper uses the second technique. For further deepening on the advantages and disadvantages of such techniques, see [6] and [14].
A complete factorial plan has been conducted for the management of the experimental tests. The number of treatments is equal to the possible number of combinations of the control factors, for a total of 48 treatments (3x4x2x2). The number of replications, randomly sampling from random variables, can be as great as wished. Twenty replications have been effected for each treatment.
The throughput time, W, has been considered as index of performance of the production system. In every replication, the simulator calculates the number of parts of ith type produced, n_i, and the throughput time of the jth single part, w_{ij}.
Results:

$$W = \frac{\sum_{i=1}^{5}\sum_{j=1}^{n_i} w_{ij}}{\sum_{i=1}^{5} n_i}$$

4 RESULTS

The observations in a factorial experiment can be described by the following effect model:

$$
\begin{aligned}
w_{ijkl} &= \mu + \tau_i + \beta_j + \gamma_k + \delta_l \\
&+ (\tau\beta)_{ij} + (\tau\gamma)_{ik} + (\tau\delta)_{il} + (\beta\gamma)_{jk} + (\beta\delta)_{jl} + (\gamma\delta)_{kl} \\
&+ (\tau\beta\gamma)_{ijk} + (\tau\beta\delta)_{ijl} + (\beta\gamma\delta)_{jkl} + (\tau\gamma\delta)_{ikl} \\
&+ (\tau\beta\gamma\delta)_{ijkl} \\
&+ \varepsilon_{ijkl} \\
& i = 1,2,3 \quad j = 1,2,3,4 \quad k = 1,2 \quad l = 1,2
\end{aligned}
\tag{1}
$$

where μ is the overall mean effect, τ_i is the effect of the ith level of MSR factor, β_i is the effect of the jth level of ASR factor, γ_k is the effect of the kth level of AS factor and δ_l is the effect of the lth level of NP factor; $(\tau\beta)_{ij}$ is the effect of the interaction between MSR and ASR factors at level, respectively, ith and jth (two factors interactions); $(\tau\beta\gamma)_{ijk}$ is the effect of the interaction between MSR, ASR and AS factors at level, respectively, ith, jth and kth (three factors interactions); $(\tau\beta\gamma\delta)_{ijkl}$ is the effect of the interaction between MSR, ASR, AS and NP factors at level, respectively, ith, jth, kth and lth (four factors interactions); ε_{ijkl} is a random error component. Obviously, an analogous definition is worth also for the other two and three factors interactions introduced in (1).

The null hypothesis, H_0, tested in this design, can be stated as follows:

$$H_0 \ : \ \{\text{All main effects and interactions} = 0\}$$

The alternative hypothesis, H_1, states that at least one of the main effects or interactions in non-zero:

$$H_1 \ : \ \{\text{At least one main effect or interaction } \neq 0\}$$

To test the hypothesis we use a multifactor ANOVA, [4]; the results are shown in Table 4..

The model diagnostic checking, by graphical analysis of residuals, has been successfully effected, [4].

The ANOVA table decomposes the variability of W into contributions due to various factors. The contribution of each factor is measured having removed the effects of all other factors. The p-values test the statistical significance of each of the factors. Since seven p-values, displayed in Table 4., are less than 0.01, these factors have a statistically significant effect on W at the 99% confidence level.

To assist in the practical interpretation of this computer experiment, Figure 2. and Figure 3., present, respectively, plots of main effects with the relative confidence interval at 99% confidence level, and the significant two factors interactions.

From Figure 2. it is presumable to believe there is a significant difference between the different levels of each factor. In fact, for each factor, there is always at least a value of the average, relative to a level, that falls out of confidence intervals related to the other levels. This approximate conclusion finds a rigorous confirmation in the execution of Duncan's multiple range test, [4].

In terms of main effects, the optimum combination of the control factors results:

MSR=FIFO, ASR=SQ, AS=70 m/min e NP=35.

It is also the best in terms of interactions. In fact, the interaction plots, Figure 3., among the qualitative factors, MSR and ASR, show that for ASR=SQ the throughput time, W, assumes the minimum values for all the levels of MSR. The discourse is analogous for the interactions among NP and the qualitative factors, MSR and ASR.

TABLE 4. ANOVA for throughput time, W

	Source	Sum of squares	Degree of freedom	Mean square	F-Ratio	p-value
Main effects	MSR	139695.00	2	69847.50	76.12	0.0000
	ASR	1100250.00	3	366750.00	399.67	0.0000
	AS	9751.84	1	9751.84	10.63	0.0011
	NP	342653.00	1	342653.00	373.41	0.0000
Interactions	MSR*ASR	45196.90	6	7532.82	8.21	0.0000
	MSR*AS	669.69	2	334.85	0.36	0.6948
	MSR*NP	9623.80	2	4811.90	5.24	0.0054
	ASR*AS	1175.88	3	391.96	0.43	0.7337
	ASR*NP	123402.00	3	41134.00	44.83	0.0000
	AS*NP	299.60	1	299.60	0.33	0.5669
	MSR*ASR*AS	5737.65	6	956.28	1.04	0.3963
	MSR*ASR*NP	13527.30	6	2254.55	2.46	0.0231
	MSR*AS*NP	2039.39	2	1019.70	1.11	0.3291
	ASR*AS*NP	3051.32	3	1017.11	1.11	0.3451
	MSR*ASR*AS*NP	6768.58	6	1128.10	1.23	0.2879
	RESIDUAL	836888.00	912	917.64		
	TOTAL	2640730.00	959			

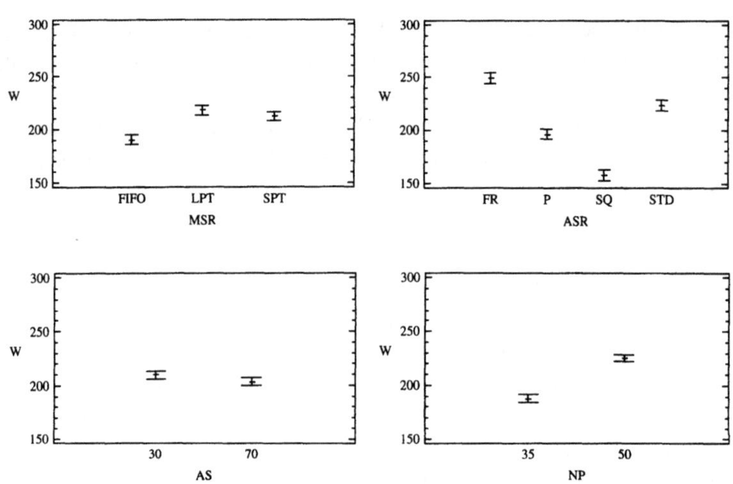

FIGURE 2. Main effects plots

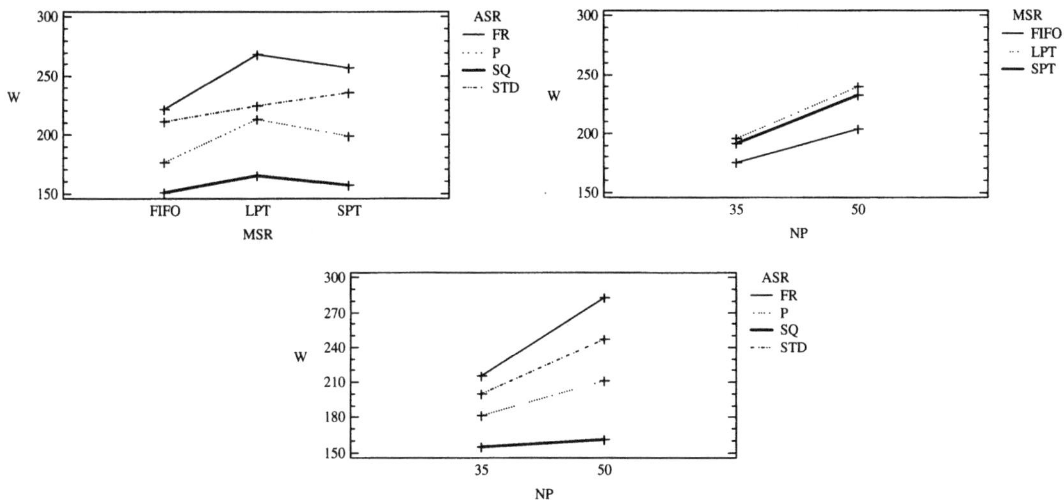

FIGURE 3. Interactions plots

5 CONCLUSION

The aim of this paper has been that of investigating the optimal choices of the variables, at low economic impact, that are under the control of the manufacturing system manager.
The manufacturing system employed as case study was simple enough, even if it contained all the peculiarities of the real systems. In a future paper we intend to examine the effect of the variability of the product mix, and of the number of machined parts typology, and to take in consideration, as performance indexes, not only the throughput time, but also the working in progress and the throughput, using a multivariate analysis of output variables or with the construction of a synthetic objective function. Other condition that will be explored will be that of using techniques able to obtain a response surface of immediate interpretation and able to help the manufacturing system manager to achieve an optimal management policy.

REFERENCES

1. Banks, J., Carson II, J. S., Nelson, B. L., Nicol, D. M., (2000), Discrete-Event System Simulation. Prentice-Hall, Upper Saddle River, N.J..
2. Czitrom, V., (1999), One-Factor-at-a-Time Versus Designed Experiments. The American Statistician, Vol 53/2, 126-131.
3. Prakash, A., Chen, M., (1993), Performance evaluation of flexible manufacturing systems using factorial design techniques. Proceedings of WESCANEX 93 "Communications, Computers and Power in the Modern Environment", *IEEE*, 407 –415.
4. Montgomery, D.C., (2001), Design and Analysis of Experiments. John Wiley & Sons, New York.
5. Box, G.E.P., Hunter, J.S., Hunter, W.G., (1978), Statistics for experimenters. John Wiley & Sons, New York.
6. Law, A.M., Kelton, W.D., (1998), Simulation modeling and analysis. McGraw-Hill, New York.

7. Sacks, J., Welch, W.J., Mitchell, T.J., Wynn, H.P., (1989), Design and Analysis of Computer Experiments. Statistical Science, Vol 4/4, 409-435.
8. Koehler, J.R., Owen, A.B., (1996), Computer Experiments. Handbook of Statistics, Vol 13, Ghosh, S., and Rao, C.. Elsevier Science, 261-308.
9. Kleijnen, J.P.C., (1987), Statistical Tools for Simulation Practitioners. Dekker, New York, 259-293.
10. Kleijnen, J.P.C., (1997), Sensitivity analysis and related analyses: a review of some statistical techniques. Journal Statistical Computation and Simulation, Vol. 57/1-4, 111-142.
11. Johnson, N.L., Kotz, S., Balakrishnan, N., (1994), Continuous Univariate Distribution, Vol. 1. Wiley, NewYork.
12. Yan, H.S., Wang, N.S., Cui, X.Y, Zhang, J.G., (1997), Modeling, scheduling and control of flexible manufacturing systems by extended high-level evaluation Petri nets. *IIE* Transaction, Vol 29, 147-158.
13. Centeno, M. A., Reyes, M. F., (1998), So you have your model: what to do next. A tutorial on simulation output analysis. Proceedings of the 1998 Winter Simulation Conference, December 13-16, 1998, Washington, DC, 23-29.
14. Kelton, W.D., Sadowski, R.P., Sadowski, D.A., (1998), Simulation with Arena. McGraw-Hill, New York.

COMPUTER AIDED DETERMINATION OF FEASIBILITY OF WORK OPERATION ON MACHINES IN FLEXIBLE MANUFACTURING SYSTEMS

I. Pahole[1]

[1] Laboratory for Automation and Optimization of Production, Faculty of Mechanical Engineering, Maribor, Slowenia,

KEY WORDS: machine tools, replacement machine tools, standard operation, CAPP, feasibility.

ABSTRACT, Nowadays, various methods of work can be used for planning the technology. Planning of technology can range from conventional manual method to use of various more or less capable programme tools. With respect to the price level the programme tools ensure different quality of results. This paper gives an example how to select a machine tool on the basis of known technology. Standard operations in accordance with DIN standards are used. The action required for planning in the CAPP module is described. Such a manner of determining feasibility of the operation can be used for the first selection of the machine tool as will as for searching for a replacement machine tool. The working operation on the machine tool is of assistance:

- to the technologist at the stage when he selects the machine tool for execution of order and
- during controlling the production it helps search for the replacement machine tool.

1. HOW TO SEARCH FOR A MACHINE TOOL IN CASE OF KNOWN TECHNOLOGICAL PROCESS

During planning of technology the technological process is determined, the work operations required by the shape and geometry of the product are determined. The operations are mainly defined by describing (to be face milled, to be milled up to depth etc.). Such type of describing however makes difficult the processing by known programme tools on computers. From the known standards it is possible to conclude that thc area of describing the work operations has been studied very precisely, see Table 1.

Such manner of designating the work operations maker describing of operations considerably easier.

Thus the technologist must use the standard designations. Graphic devices on computer can be of great assistance to him. We can obtain the following data resulting from the CAPP module:

- designation (serial number) of order for a certain period of planning,
- unambiguous designation of order,
- code or description of workpiece (s) in order,
- number of workpieces,
- required NC-programmes for certain operations on certain units,
- description of standard operations according to DIN standard,
- anticipated start time,
- anticipated calculated time for order, defined in advance,

Published in: E.Kuljanic (Ed.) *Advanced Manufacturing Systems and Technology*, CISM Courses and Lectures No. 437, Springer Wien New York, 2002.

2. DETERMINATION OF SPECTRUM OF FEASIBLE OPERATIONS ON MACHINE TOOL

For some known machine tool it is possible to determine the standard operations which can be effected on a machine tool. The standard operations are meant to be operations defined in DIN standard [1]. The table 1 shows the visible shape of data file for describing feasible standard work operations on some machine tool.

3. DETERMINATION OF REPLACEMENT MACHINE TOOL

When an order has already been introduced into production, it is often necessary to search for an alternative machine tool for certain operation or group of operations. Topical examples are:

- of stoppage, defect occur on the planned machine tool,
- of for some other reasons there are free working capacities on similar machine tools and reduction on even removal of kettle-neck are desired,
- of for any reasons the priority of orders (urgent order) has been changed,
- if proper working devices or tools, specifically connected with a certain machine tool, are not available.

4. REALIZATION AND SOLUTIONS

Automated searching for an alternative machine tool during preformeny of the production management is certainly the user's wish. We can come quickly to the conclusion that complete automation is not possible since it is almost impossible to describe workpiece by a suitable method [2], [3].

TABLE 1. Example of surrey of feasible work operations for machine tool

Surrey possibilities of execution of standard operations on a multiple – operation drilling and milling machine (example)		
Ser. No.	Code of operation according to DIN 8589	Description of operation for easier understanding
1	3.2.2.1.	Group of operations of face sinking
2.	3.2.2.2.	Group of drilling operations
3.	.....	......
4.	.....	......
5.	.....	......
6.	3.2.2.3.	Group of thread-cutting operations (is drilled holer)
7.	.....	......
8.		
9.	.....	......
10.	3.2.3.1.	Plans milling
11.	3.2.3.2.	Milling of round shapes

When searching for solutions of the problem presented across several factors by which the final solution - of existing at all – is conditioned. During the stage of planning it is possible to work out several alternative working procedures, which we optimise with regard:
- most favourable version with regard to time or,
- economically most favourable version.

If the number of these versions of working processes increases, such manner of planning

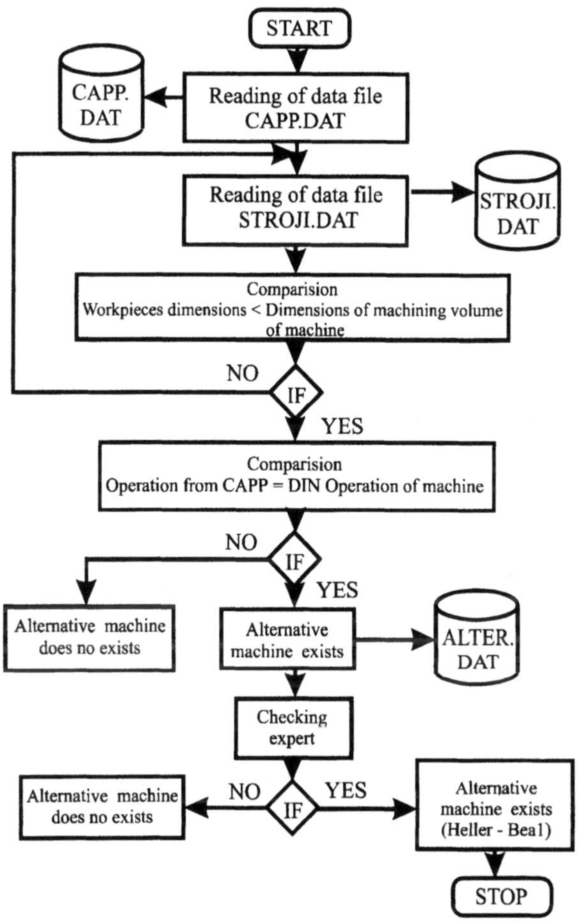

FIGURE. 1. Diagram of flow of programme module for determination of replacement machine tool

usually becomes economically questionable. It can also happen that such manner cannot be applied because it is time-consuming.

In parallel it is necessary to prepare the conditions for execution of certain order on several units of the flexible manufacturing system [4]. This applies particularly for orders of high priority. The programme equipment can be of great assistance to the technologist/operator for determination and/or selection of the alternative machine tool. The worked out programme package "ALTER" ensures that suitable machine tool can be found. The programme module o functions so that it compares the operations required for execution of the order with the spectrum of operations which the machine tools can perform. Even if the solution is favourable finally the interfere of the expert is anticipated who gives the final solution. Figure 1 shows a diagram of flow of the programme module.

5. FEASIBILITY OF ORDERS WITH RESPECT TO TOOLS AND WORKING DEVICES

Output data from such selection are:
- order selected,
- machine tool on which the selected order is fesiability,
- working devices which must be provided on the machine tool for selected order,
- tools which must be a available on the machine tool for selected order,
- start time of equipped the machine.

6. CONCLUSION

The presented manner of examining feasibility of the working operation on the machine tool is of assistance:
- to the technologist at the stage when he selects the machine tool for execution of order and
- during controlling the production it helps search for the replacement machine tool.

It is still man who maker the final decision. In practice it is impossible to reach complete automation for well-known reasons. In actual production unforeseeable events occur. The frequency of repeating is not known for unforeseeable able events. Automation is usually controlled by software on computers. This software is programmed for a certain area and does not control all events.

REFERENCES:

1. DIN 8589, (1981), Deutsche Normen, Beuth Verlag Gmbh, Berlin
2. Katalinić, B., (1990), Industrieroboter und Flexible Fertugugssysteme fürDrehteile, VDI Verlag, Düsseldorf.
3. Pahole, I.; (1996), Using data flow matrix for small and medium-sized enterprises. V: Proceedings of DAAAM- 95, Katalinic, B.(ed.), Wienna, 041-042.
4. Pahole, I., (1997), Planiranje vodenje in optimiranje inteligentnih fleksibilnih obdelovalnih sistemov z uporabo koncepta pretočne matrike, Fakulteta za strojništvo Maribor, Maribor, 35-46.

AN INTEGRATED TOOL FOR GEOMETRIC RECONSTRUCTION OF FUNCTIONAL SURFACES

D. Comelli, G. Concheri, R. Meneghello

Dipartimento di Architettura, Urbanistica e Rilevamento, Laboratorio di Disegno e Metodi dell'Ingegneria Industriale, Università di Padova, Italia

KEYWORDS: Reverse Engineering, software & hardware development, laser scanning.

ABSTRACT: At Laboratory of Design Tools and Methods in Industrial Engineering, University of Padova, the FIRE project "Functional Integrated Reverse Engineering" was started, which focuses in the realisation of a novel technological environment for performing the reverse engineering of functional prototypes or products. This environment consists of hardware technologies, a multi-probes system integrated in a 5-axis machine, and of software technologies, a software module aimed at phases of pre-processing, correction and analysis of the acquired data. In the present work, an introduction to the project is given together with the description of the equipment and the software module characteristic already set up, with some example of reconstruction of functional surfaces.

1 INTRODUCTION

The application of designing methods assisted by computerized tools is a practice influencing every aspect of the modern engineering processes. From the introduction of CAD systems, powerful tools have been developed, capable to assist the designer in every phase of his/her job. Today solid modelling plays a fundamental role in every productive phase. Beginning with the virtual model developed in the CAD environment the numerous techniques now available support the preliminary design, the engineering and finally manufacturing phase of the product fabrication.

Nevertheless in a lot of industrial applications the use of these engineering techniques is not yet extensive or even possible: as examples, the processes of industrialization of style concepts and the construction of medical prosthesis in which it is necessary to build virtual models starting from the physical prototypes. This process of copy of a real object in an ideal and immaterial model takes the name of Reverse Engineering (RE). The problem is often faced focusing on the selection of the instrumentation both hardware and software, while neglecting the functional and technological aspects relevant to the object under analysis. The usability of virtual model in the following phases of fabrication is considerably limited or favoured by the methodology and, the choices adopted on operative parameters. In this work it will be introduced a methodological approach based on the preliminary analysis functional and technological aspects which effectively condition the result of the reconstruction.

2 THE FIRE PROJECT

In literature, many contributions are finalized to the improvement of specific aspects of the reverse engineering process (e.g. [2], [3], [4], [5], [6]) or to the improvement of process integration (e.g. [7]). In general, it may be noted that there is not a unique best method for reverse engineering of real parts. The most suitable approach depends on the scope intended

Published in: E.Kuljanic (Ed.) *Advanced Manufacturing Systems and Technology*, CISM Courses and Lectures No. 437, Springer Wien New York, 2002.

for the model to be reconstructed and therefore such information should condition tool selection, operating methodology and model creation.

In order to identify the correct methodology for model reconstruction, it may be useful to consider the typical aspects on which the direct 3D modelling process of an industrial product is commonly based: part functionality, technological aspects, 3D model purpose(s).

The quality and the significance of a 3D model strictly depend on the level of "*knowledge*" implemented (e.g. the feature modeling sequence and the related parametrization, the manufacturing constraints taken into account and satisfied, etc.) and on how it fits the intended purpose(s) (e.g. computation of geometric parameters, carrying out engineering analyses and simulations, generation of NC paths for the manufacturing and inspection processes, etc.).

The reverse engineering process should be driven, since the initial stages, in order to generate "optimal" models for the intended purpose, at the same time synthesizing the relevant functional and technological aspects and requirements.

In the proposed reverse engineering process, the four classical phases, pre-processing, segmentation, fitting of the surfaces and creation of the model, are anticipated by a preliminary stage intended to plan the subsequent activities, applying a "*Design for X*" (X = Assembly, Manufacturing, ...) methodology. It is beyond the scope of the present paper to present, for instance, the Design for Assembly (DfA) or the Design for Serviceability (DfS) methodology. Referring to the wide literature on the subject (e.g. [8]), it is sufficient here to point out that DfX defines an analysis technique applied to models or objects in order to identify the characteristics that may result critical for the anticipated "X" application, and assess, according to a predefined set of rules and criteria, the most viable solution. In the preliminary stage the different "points of view" of the object considered and the related requirements on the subsequent phases can be taken into account. In this phase the physical object is analyzed considering:

- functionalities desired of the model;
- functionalities of the object;
- technological aspects;

fundamental characteristics of the object and of the technologies employed in its realization. From them fundamental information can be extracted for the improvement of the following phases of the reverse engineering.

In addition to the classical approach, a "*model abstraction*" phase is also integrated in order to explicitly extract the abstract level information from the reconstructed B-Rep model or even from the raw data cloud. As an example, if the object to be reconstructed is a cylinder, according to the functional requirements, the designer may be interested in the assessment of its nominal parameters (diameter, height), but also in the estimate of the associated errors either dimensional or geometrical (e.g. axis deviation or effective virtual size). Moreover, technological aspects could be investigated in the model, as e.g. the effective surface texture parameters. Depending on the desired information, data acquisition, model reconstruction and parameter extraction stages could be performed in a totally different manner.

3 INTEGRATED REVERSE ENGINEERING SYSTEM SPECIFICATIONS

At Laboratory of Design Tools and Methods in Industrial Engineering, University of Padova, the *FIRE* project "*Functional Integrated Reverse Engineering*" was started recently, which focuses in the realisation of a novel technological environment for performing the reverse engineering of functional prototypes or products

The proposed approach to the reverse engineering process, therefore, must be implemented into an integrated Computer Aided Reverse Engineering (*CARE*) environment, strictly interacting with a multi-probe 3D scanning system, that drives either the acquisition or the reconstruction stage, in order to finalize the various activities according to the function-oriented approach. In fig. 1, the CARE environment is described.

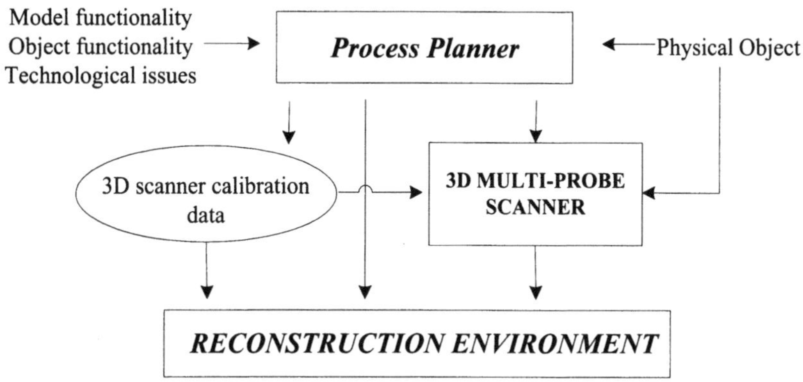

FIGURE 1. The CARE environment

The process planner module is responsible for the preliminary phase already described. It can be envisaged as an expert system, based on the acquired process knowledge and continuously developing, aimed at assisting the operator in detailing, according to the driving criteria mentioned in the previous section, the critical aspects of the whole process:

1. identification of the functional surfaces/features,
2. selection and set-up of the acquisition technology (tactile, laser or vision),
3. selection of viable object orientation(s) and related datums,
4. set-up of suitable scanning parameters (number of scan points and acquisition grid/path, clearance paths, etc.),
5. surfaces/features/model reconstruction strategy.

4 DEVELOPMENT OF THE FIRE PROJECT

The FIRE project is under development, but preliminary results, already available, are the set-up of a prototype 4-axes 3D scanner based on a laser probe, depicted in fig. 2, and a software module, denominated *Pascal* (Post-processing and Analysis of Scanning data from Laser equipped CMM) aimed at correcting and analysing the acquired data. The scanner is built up by three main components:

1. an orthogonal table (X and Y axes),
2. a vertical column (Z axis)
3. a rotary table whose axis can be oriented parallel to any of the three axes.

If planar surfaces are to be investigated, the rotary table can be removed. The axis motion and position control is performed by four stepper motor (one for each axis) controlled by an integrated module implemented in LabView that synchronizes axis motion and laser acquisition.

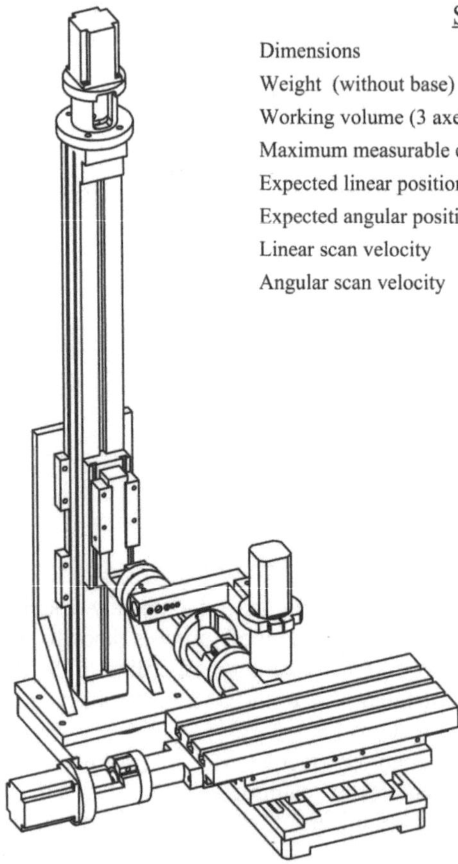

Scanner characteristics

Dimensions	950 x 770 x 1070 mm
Weight (without base)	95 kg
Working volume (3 axes)	250 x 310 x 500 mm
Maximum measurable cylinder	ø700 mm x 500 mm
Expected linear position accuracy	±0.05 mm
Expected angular position accuracy	± 25"
Linear scan velocity	8 - 25 mm/s
Angular scan velocity	0.24 rad/s

Laser characteristics

Producer	Wolf&Beck
Model	OTM3A-50/h/PC
Resolution	5 μm
Repeat accuracy	25 μm
Measuring frequency	1 kHz
Laser Class	3B
Dimension	ø58 x 180 mm

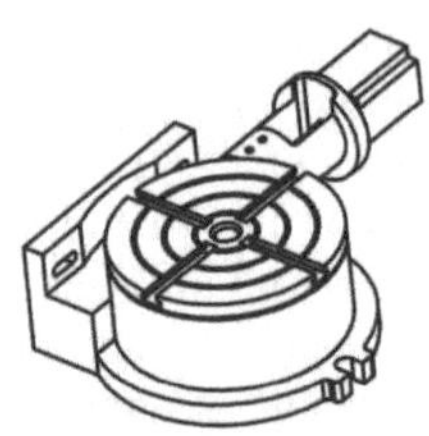

FIGURE 2. The CMM Laser and its characteristics

After the acquisition of the data-set, the phase of real reconstruction, where the choices and the acquired information converge in the reconstruction of the model, starts.

The data acquired have to be corrected through a 3D model of geometrical error compensation with the purpose to eliminate the inevitable construction errors of the CMM. It is currently in phase of development the mathematical model and the necessary calibration procedures in order to complete this fundamental phase.

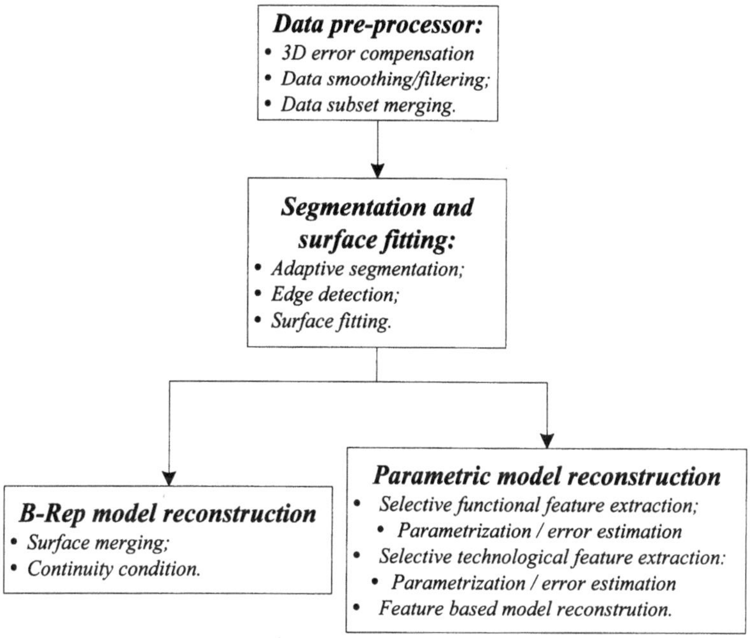

FIGURE 3. Specifications of the environment of reconstruction

After the correction, data require a phase of filtration, where the errors of reading of the laser and the background noise due to the physical principle of acquisition are corrected or at least attenuated. For this purpose they have been implemented a set of filters specifically studied for the correction of different effects, classified in three categories according to the mathematical principle and they can be listed to the type of used algorithm:

- Convolution through Fast Fourier Transform ,
- Minimization on average local value,
- Mixed Systems that use both principles.

The available filters are also classifiable in operation of the specific abilities, as illustrated in the following tables.

TABLE 1. Based filters on the Convolution

PillBox	Smoothing of the edges
Raised again cosine	Removal of the spikes
Exponential	Exaltation of the edges
SINC	Indicated in the zones with to constant bending

TABLE 2. Based filters on the Minimization

Averages	Rapid attenuation of the spikes

TABLE 3. Mixed algorithms

Gauss Edge Preserving	Indicated for noisy scannings

The software environment assists the user in previewing the result of the correction in order to facilitate the choice of the functional parameters.

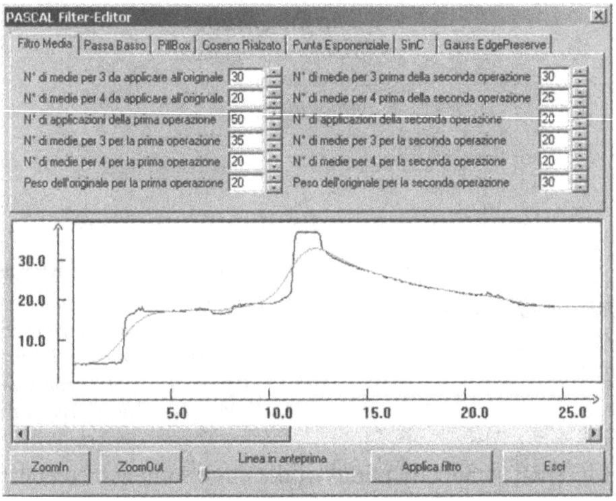

FIGURE 4. Pascal filter editor

The phase of merging of several not aligned scannings or different faces of the same object needs a more elaborate procedure, not free from the risk of systematic errors and uncertainty. They are currently under development methodologies to analyze and resolve this delicate phase.

After the pre-processing phase, is necessary to divide the data-set in homogeneous portions to analyze the single functional surfaces measured. This phase is based on the detection of the edges. In this case the grid structure of the scanning, used as a bi-dimensional image where the colour of the pixel/point is proportional to the height measured by the laser, allows to develop the procedure of edge detection using image processing methodologies. The system of elaboration gives the maps of the edges for the segmentation phase and the management of the single surfaces which are structured in a hierarchical tree.

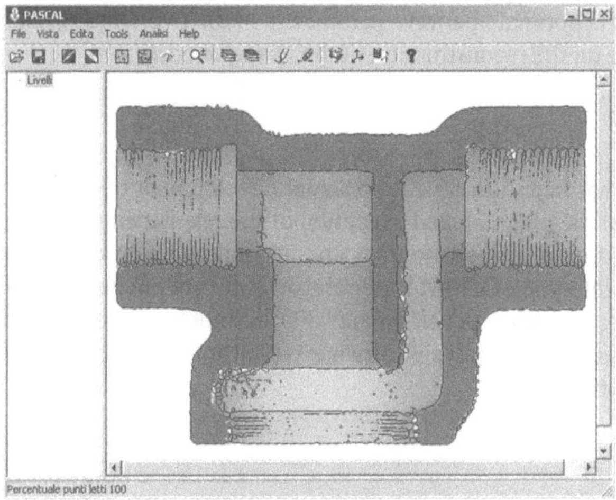

FIGURE 5. Results of edges of the functional surfaces

Thus it is possible to begin the phase of approximation of the surfaces, determining the geometric and "metrologic" characteristics of each surface. By the knowledge of these parameters the construction of the model takes place. At the present state of development, the reconstruction of surfaces is supplied by commercial software (as PTC Pro/ScanTools, Imageware Surfacer, Zeiss Holos and Dimension, Gestel SolidThinking).

At the end of the phase of segmentation and fitting of the surfaces it is possible to complete the process with the reconstruction of the model and the extraction of geometric parameters or the reconstruction of parametric features or the exporting of a B-Rep representation. The figure 6 illustrates two examples of reconstruction: the first one is a B-Rep reconstruction of an orthesis, where an excessive sampling would make inefficient the operations of reconstruction; on the contrary the second example depict the extraction of geometric parameters of functional surfaces of a mechanical component, where the abundance of information would allow a more accurate evaluation of dimensions and form errors.

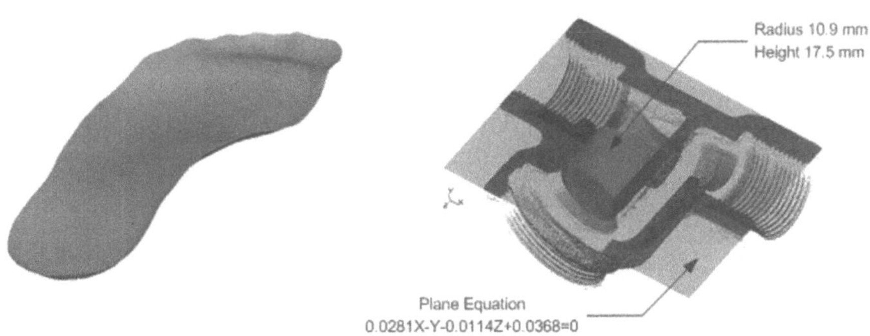

FIGURE 6. Examples of reconstruction B-Rep and extraction of geometric parameters

The choice of the optimal operative parameters in the two cases results to be fundamental in efficiently carrying on the reconstruction of the virtual model.

CONCLUSIONS

In this work it has been illustrated the actual development of the FIRE project, finalized to the improvement of the planning and execution of the reverse engineering process, focused to a more efficient utilization of the resources and an improvement of reconstructed models. For the time being, a prototype of CMM Laser and an software environment for pre-processing and analysis of data are under development. Furthermore the implementation of a contact measuring probe and of the analysis software within into OpenCASCADE CAD environment are under investigation.

REFERENCES

1. T. Varady , R.R. Martin, J. Cox , (1997), Reverse engineering of geometric models - an introduction, Computer-Aided Design, vol. 29, no.4, 255-268
2. P.N. Chivate., A.G. Jablokow, (1995), Review of surface representations and fitting for reverse engineering, Computer Integrated Manufacturing Systems, vol. 8 no. 3, 193-204
3. C.-C. Tai, M.-C. Huang, (2000), The processing of data points on design intent in reverse engineering, Int. J. of Machine Tools & Manufacture, vol. 40, 1913-1927
4. L.A. Piegl, W. Tiller, (2001), Parametrization for surface fitting in reverse engineering, Computer-Aided Design, vol. 33, no. 8, 593-603
5. K.H. Lee, H.-P. Park, (2000) Automated inspection planning of free-form shape parts by laser scanning, Robotics and Computer Integrated Manufacturing, vol. 16, 201-210
6. V.H. Chan, C. Bradley, G.W. Vickers, (2001), A multi-sensor approach to automating coordinate measuring machine-based reverse engineering, Computers in Industry, vol. 44, 105-115
7. K.H. Lee, H. Woo, (2000), Direct integration of reverse engineering and rapid prototyping, Computers & Industrial Engineering, vol. 38, 21-38
8. G. Boothroyd, P. Dewhurst, (1983) , Design for Assembly - A Designer's Handbook, Department of Mechanical Engineering, University of Massachusset , Amherst

COST-EFFICIENT MANUFACTURING BY EMPLOYING INCREMENTAL FORMING PROCESSES

P. Groche, M. Jöckel, C. Rachor , T. Rathmann

Institute For Production Engineering And Forming Machines, Darmstadt University Of Technology, Germany

KEYWORDS: Incremental Forming

ABSTRACT. Today's market is characterized by a high diversification and complexity of demanded products. With the flexibility of incremental forming processes these requirements can be achieved well. This paper focusses on the large varieties of geometry, prevention of material waste and possibilities to integrate parts by the application of processes based on kinematic shaping. Innovative investigations point out that incremental forming techniques offer great potentials to realize cost-efficient process chains.

1 INTRODUCTION

Companies within the field of production engineering are faced by an increasing demand of individual high quality products. As a result, the variants of parts that have to be manufactured have raised continuously in the last years. Saturated markets, the rising exhaustion of market niches and country-specific modifications as a consequence of increased globalisation are the basic reasons for decreasing lot sizes in plants. At the same time periods of product industrialization have to be reduced significantly [6,7]. These trends lead to increasing costs, caused by complexity of products and disposition. In this context studies by McKinsey indicate, that 15-20% of total losses of car manufacturers are based on complex product mixes (Figure 1) [9].

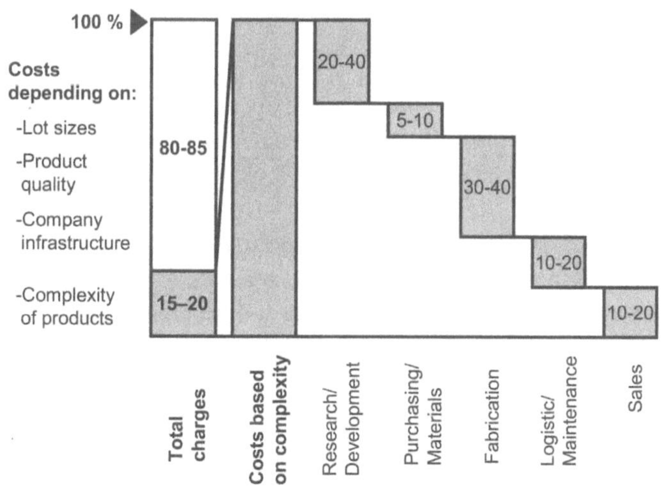

FIGURE 1. Costs depending on complexity of products [9]

Predominantly requirements of customers according to tailor-made products, high grade, low costs and short delivery times imply continuous optimizations of the factors time, quality and costs for the contractors. To raise efficiency, methods based on Kaizen Engineering, Lean Production and Just

Published in: E.Kuljanic (Ed.) *Advanced Manufacturing Systems and Technology*, CISM Courses and Lectures No. 437, Springer Wien New York, 2002.

in Time can be employed. In addition the selection of optimal manufacturing processes for defined product grades has to be performed [13].

2 REQUIREMENTS FOR FORMING PROCESSES

By the use of forming technologies short cycle-times, a high utilization of materials and low volumes of waste are feasible. Based on strain hardening of the manufactured parts lightweight design and savings of raw materials can be realized [2].

New targets arise by the increasing demand of complex products, which often require the employment of integral design principles. The resulting reduction of components to be processed affect cost-savings in multiple departments of companies, caused by the minimization of technical and logistical interfaces. Moreover, the number of component suppliers can be reduced, leading to a higher profitability of plants (Figure 2) [9].

According to integral construction new approaches in forming techniques are realized by Hydro Forming and Closed Die Forging. By the use of these processes multifunctional parts can be manufactured without additional assembling work. Due to high capital investments for tools and machines, their range of application is often limited to mass-production. For small and medium series of parts, combinations of metal-cutting, joining and casting are often employed [2].
Further attempts to increase the economic efficiency of process chains are based on the reduction of necessary metal finishing processes. In this connection cold extrusion can be applied to manufacture net shape components with high surface qualities. Since the range of geometries is limited, complex parts have to be fabricated by hot shaping. This induces additional production steps to finish tolerated areas [2].

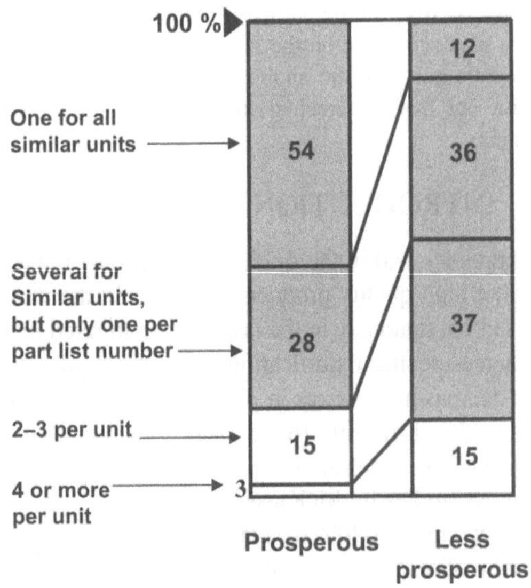

FIGURE 2. Numbers of component suppliers in prosperous and less prosperous companies [9]

With regard to the diversification of market needs the mentioned examples point out general disadvantages of conventional forming processes. Since the work pieces are mapped three-dimensionally according to the contour of the dies employed, the requirements according to short process chains and flexible production methods can only be partially fulfilled. This implies high costs for production facilities and long periods of industrialization. For axially symmetric components in particular, incremental forming processes can be a cost-efficient alternative.

3 COST-EFFICIENCY BY EMPLOYING INCREMENTAL FORMING PROCESSES

3.1 DEFINITION AND INDUSTRIAL APPLICATION OF INCREMENTAL FORMING PROCESSES

In contrast to conventional forming techniques, incremental forming processes use tools with rather simple geometries. Therefore the shaping of the end product is primarily determined by the kinematics of the tools.

Incremental forming processes are defined by the following characteristics:

- There are multiple steps to produce the targeted product
- It exists a sequential similarity of the steps
- There are intermittent boundary conditions
- Loading and unloading periodically take place at the same location of the work piece
- The tools are identical for more than one production step
- The deformation of the work piece occurs locally [1]

Figure 3 shows examples of already industrialized incremental forming processes.

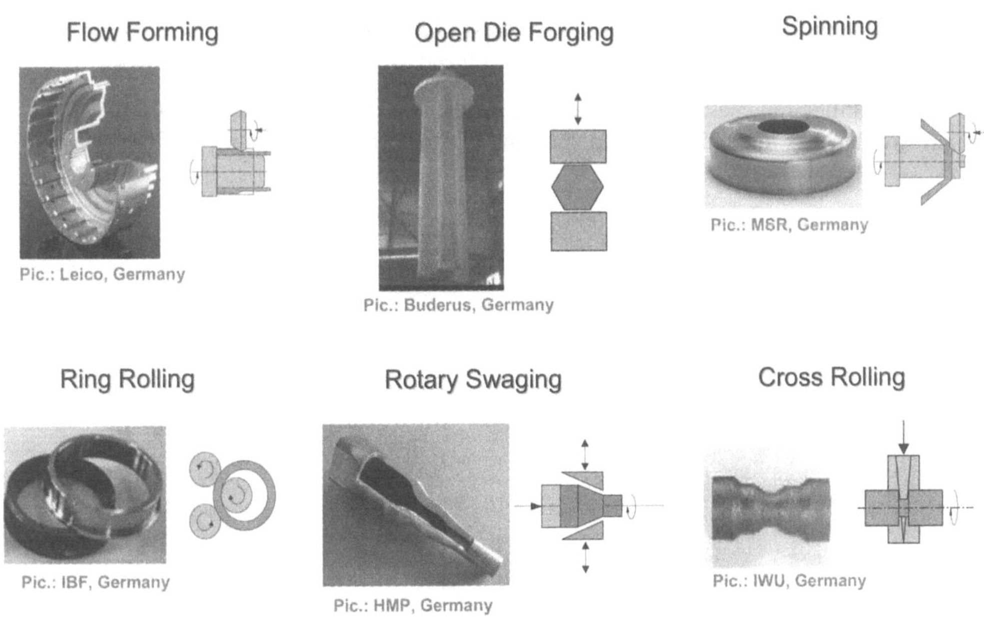

FIGURE 3. Industrialized incremental forming processes

In the following sections new process developments will point out, that replacing traditional process chains by incremental forming techniques offers significant improvements in terms of quality, time and costs (comp. [10]).

3.2 NEW APPROACHES IN KINEMATIC SHAPING

FLOW-SPLITTING OF DISK BLANKS

To manufacture V-pulleys, dynamic shapers and automobile wheels blanks having cup shape on both sides are required. Depending on order quantities these parts are fabricated by casting, sintering, forging or alternatively by welding several components.

A forming alternative for the described shapes is the so-called splitting-process. This is performed by feeding a tapered roll radially into a rotating disk blank. When manufacturing components by the conventional splitting process, several problems can be observed. On the one hand the splitting roll induces tensile stresses at the bottom of the split, which frequently lead to the formation of cracks. On the other hand, positioning the splitting roll and the disk blank relative to each other is problematic so that splitting of the workpiece will not mandatorily take place parallel to the disk blank. The consequences are different material thicknesses in the flange areas [3].

To eliminate the disadvantages of conventional splitting an innovative incremental forming process called Flow-Splitting has been developed. Therefore the disk blank is fixed into a modified spinning-lathe.

While rotating, two diametrically opposed tolling systems are moved radially onto the disk blank's edge and separate it into two flange areas. In contrast to conventional splitting each tooling system consists of three rolls, one obtuse-angled splitting roll and two supporting rolls (Figure 4).

While the splitting roll causes the material to flow into two flange areas, the support rolls induce compressive stresses in the deformation zone and thus increase the formability of the blank. Ex-

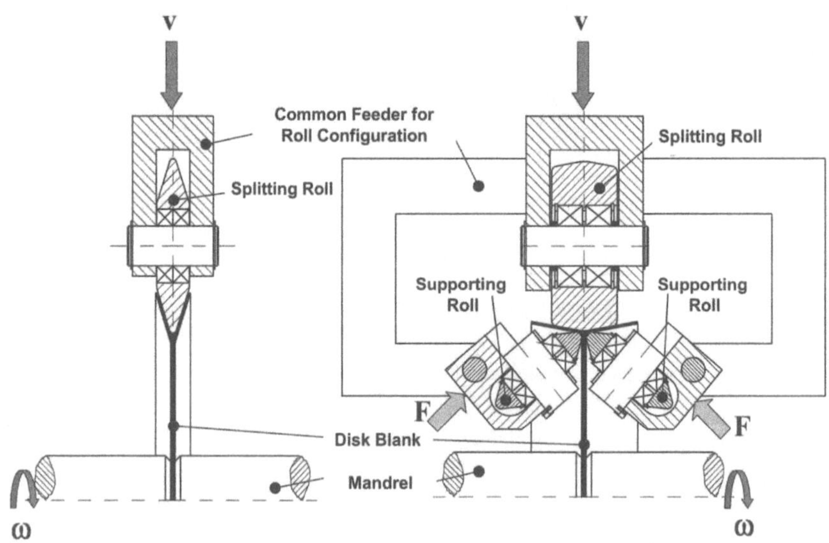

FIGURE 4. Conventional Splitting (left) and Flow-Splitting

periments were realized by a rotational speed of 1,5 rev/sec of the disk blank. The maximum radial velocity of the tool system was set to 1 mm/s [4].
The Flow-Splitting process shows basic advantages of incremental forming techniques. Based on local deformation zones reduced tool forces can be obtained. Due to the high amount of hydrostatic pressure, an extended formability of materials is attainable. Therefore the cold forming of even high-strength materials can be realized with in addition low tool costs. In most cases, end products are featured by high strain hardening in the formed areas and net-shape surfaces. This implies savings of raw materials and offers potentials for lightweight design. Figure 5 shows numerical calculated equivalent plastic strains and the distribution of micro-hardness of work pieces processed by Flow-Splitting.

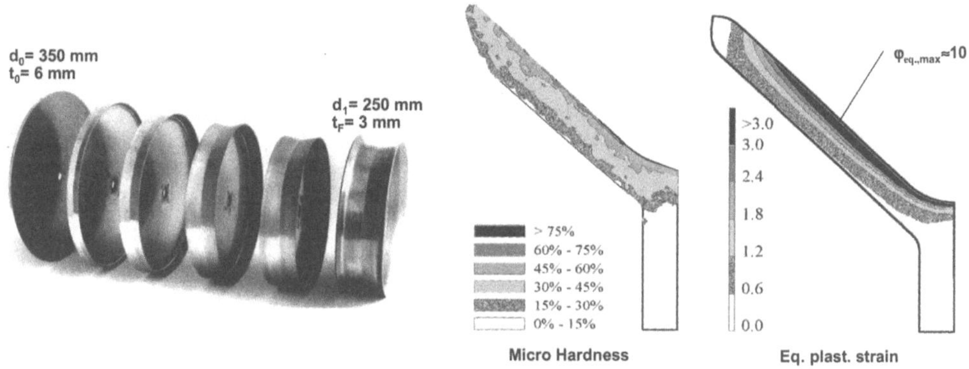

FIGURE 5. Manufactured parts (left) and comparison of micro hardness and calculated equivalent plastic strains (d: Diameter, t_0: Blank thickness, t_f: Thickness of flanges) [1]

Internal Gear Rolling

In planetary gears internal geared wheels are necessary to transmit turning moments. Conventional production methods mostly include a two-piece design of these wheels. Therefore a gear rim has to be manufactured by broaching and subsequent welded with a base plate [8]. Due to requirements according to shape tolerances and concentricity of end product, a large number of metal cutting processes have to be performed. This leads to a high complexity of technical and logistical efforts. Further on capital intensive investments concerning broaching tools and welding machines are necessary.

An innovative approach to reduce the described process chain is Internal-Gear-Rolling. It belongs to the group of Flow-turning methods and usually requires cup-like preforms, e.g. manufactured by forging and turning. An externally geared wheel fitted into the cup is used as a form-giving mandril. Work piece and mandril are set in rotation and rolls, which are arranged around the work piece, are translatorically moved along the cup. By the reduction of the cups outside diameter a material flow into the mandrils teeth is enforced. After the process has finished, an internally geared wheel with a net-shape gearing is formed (Figure 6) [2].

Due to the described process, the manufacturing of integral constructed, internal geared wheels by employing only one setting is feasible. To reduce costs, the forming rolls are designed as a tapered roller bearing, disassembling the inner raceway. As a result, kinematics of the rolls are easy to handle, which leads to decreased costs for machines and control techniques. To finish the formed wheel, only the outside surface has to be machined by metal cutting.

Internal gear rolling points out further advantages of incremental forming processes. Due to high plastic strains, integral construction of near-net-shape parts and thus the reduction of process chains are reachable.

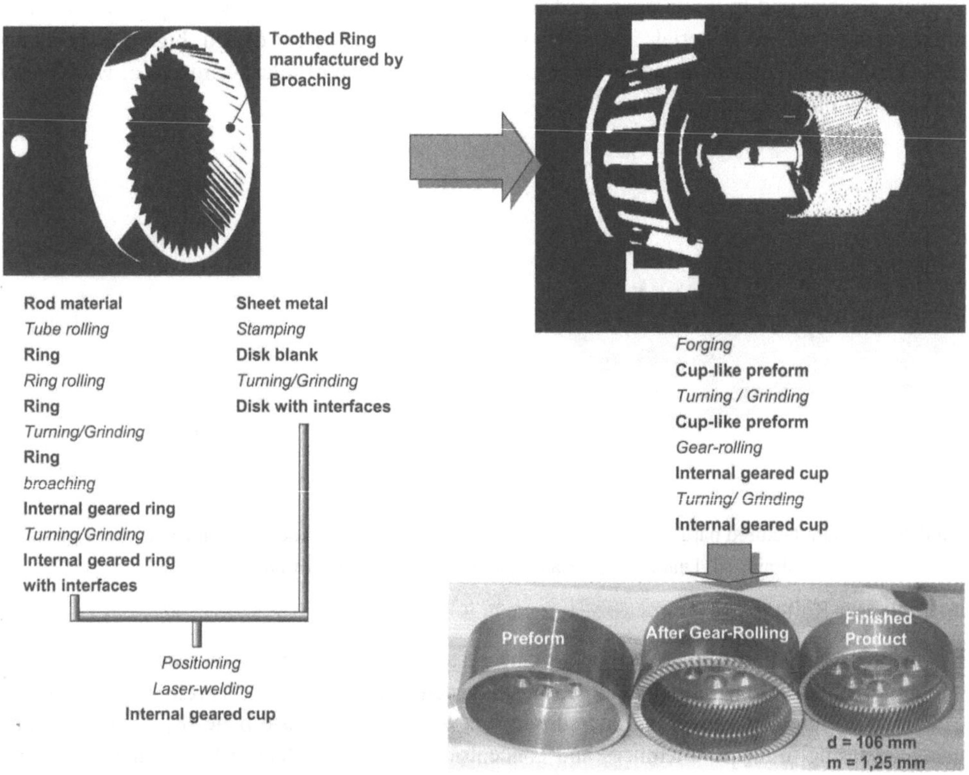

FIGURE 6. Conventional process chain to manufacture internal geared wheels (left) in comparison to internal gear rolling, manufactured parts (right, bottom)

Axial-Radial-Forming

To transmit torque in gear units and power trains, hollow shafts are often used. In industrial applications, the manufacturing of these shafts is realized by hot shaping (e.g. cross rolling) or by the joining of cold extruded component parts, which have to be finished by machining. In this connection, deep hole drilling processes particularly lead to a high amount of refuse and effect high production costs [5,11,12].

To reduce losses of material, Axial-Radial-Forming has been investigated. This process is a combination of Rotary Swaging and lateral extrusion. Rotary swaging as a non-cutting process is used for the reduction of outside diameters in the case of solid rods or tubes. Therefore, four die segments arranged opposite to each other apply radial forces directly to the workpiece. In order to avoid material penetration into the gap between the die segments, a relative rotation between work-piece and die segments has to take place. For Axial-Radial-Forming it is necessary to superimpose the Rotary Swaging process with a lateral extrusion. Thereby the flow behaviour in the deformation zone is improved by partial inductive heating. The contour of shaft collars to be formed are determined by the geometry of channels in the die segments. Production times for a shaft collar as shown in figure 7 are about 3-4 seconds [5].

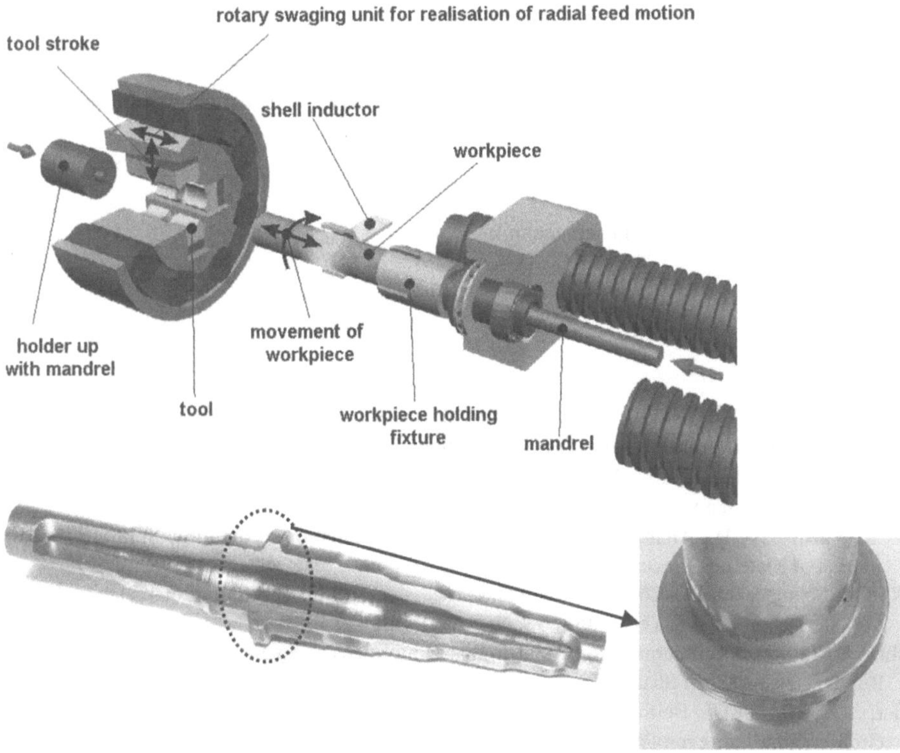

FIGURE 7. Schematic tool set-up for the Axial-Radial-Forming process (top), hollow shaft (middle) and manufactured shaft collar

Due to the significant reduction of metal finishing operations, process chains including Radial-Axial-Forming and Rotary Swaging offer possibilities to manufacture near-net-shape hollow shafts with minimized material costs.

4 DISCUSSION AND FUTURE OUTLOOK

New approaches in incremental forming techniques have shown, that these processes offer great potentials in respect to cost efficiency and product quality. Due to still unsatisfying methods of the numerical prediction of process parameters, the integration in industrial applications is inhibited. In many cases, the complex three-dimensional material flow has to be investigated by simplified numerical modeling or expensive experiments. To improve the precision of numerical calculated results and to reduce the still immense computing times, improvements in the fields of solvers and remeshing algorithms have to be made [1].

Further on, incremental forming processes require higher cycle times than processes like forging or deep drawing. Indeed, always complete process chains have to be analyzed. Direct and indirect cost savings due to reduced number of parts, low tool costs and savings of materials should be important reasons to consider incremental forming techniques during production planning of new components.

ACKNOWLEDGMENTS

The authors thank the following organizations for the promotion of investigations in the field of incremental forming:

- Deutsche Forschungsgemeinschaft
- Forschungsvereinigung Antriebstechnik
- Stiftung Industrieforschung

REFERENCES

1. Groche, P., et al., (2000), Modelling of Incremental Forming Processes. Proceedings of NAFEMS World Congress, Volume One in Lake Como, Italy
2. Groche, P. , Jöckel, M., (2001), Integralbauweise durch inkrementelle Umformtechnik. Festschrift zum 60. Geburtstag von Prof. Geiger, Bamberg Meisenbach
3. Hauk, S., Vazquez, V.H., Altan, T., (2000), Finite element simulation of the Flow-Splitting-Process. Journal of Materials Processing Technology 98, New York Elsevier
4. Hauk, S. & Schmoeckel, D., (2000), Tooling and process control for splitting of disk blanks. Journal of Materials processing Technology 98, New York Elsevier
5. Heislitz, F., (2001), Optimierung des Axial-Radial-Umformens – eine Verfahrenserweiterung des Rundknetens. Aachen Shaker
6. Köhler, A., (1988), Beitrag zur Verbesserung der Fertigungskostentransparenz bei Großserienfertigung mit Produktvielfalt. Berlin Springer
7. Müller, M., (1998), Qualitätscontrolling komplexer Serienprodukte. Aachen Shaker
8. Rachor, C., Köstermeier, K.H., (2000), Innenverzahnungswalzen – Ein innovatives Umformverfahren für Komponenten des Getriebes. 7. Umformtechnisches Kolloquium Darmstadt, Bamberg Meisenbach
9. Rommel, G., et al., (1993), Einfach überlegen – Das Unternehmenskonzept, das die Schlanken schlank und die Schnellen schnell macht. Stuttgart Schäffer-Poeschel
10. Schmoeckel, D.& Hauk, S., (1995), Flexible Production of Gear Molds Manufactured out of round blanks in a combined Chase-Roll process. Production Engineering Vol. II/2. München Hanser
11. Schmoeckel, D. & Gärtner, R., (1997), Manufacturing Strategy for Rotary Swaging. Proceedings of the 30th Plenary Meeting of the international Cold Forging Group ICFG, Den Bosch/ Holland
12. Siegert, K. & Krüßmann, M., (1997), Rotary swaging analysis. Wire Nr. 1/1997
13. Stricker, N., Jöckel, M., (2000), Optimierte Prozesskette im Schmiedebetrieb. 7. Umformtechnisches Kolloquium Darmstadt, Bamberg Meisenbach

THE USE OF A NEURAL NETWORK IN PREDICTING THE MECHANICAL PROPERTIES OF LOW CARBON STEELS AFTER HOT ROLLING

F. Caiazzo[1], F. Memola Capece Minutolo[2], R. Pasquino[1], V.Sergi[1]

[1] Department of Mechanical Engineering, University of Salerno, Italy
[2] Department of Material and Production Engineering, University of Naples, "Federico II", Italy

KEYWORDS: Concrete rebar, Hot rolling, Neural networks

ABSTRACT: A neural network using a learning procedure based on error backpropagation algorithm was employed to identify groups of hot-rolled low carbon steels with homogeneous mechanical characteristics. The mechanical properties of hot-rolled steel bars depend on two categories of variables: technological and structural. Very often, these interact with each other, and the resulting synergic effects are not always easy to evaluate. Moreover, as it is impossible to plan test sequences in which a single parameter at a time is made to vary within a sufficiently wide range, there is no way of designing a theoretical experimental model based on partial multiple regression.
As a matter of fact, as the technical standards for concrete bars currently in force only provide for compliance with the threshold values of certain parameters (yield tensile stress, ultimate tensile stress, ultimate elongation) and with other technological and geometrical specifications, operators in the building constructions industry are not always required to assess the exact mechanical characteristics of the reinforcing rods they produce. Nonetheless, it remains that the use of methods based only on the process manager's practical experience may result in serious technical drawbacks in terms of insufficiently constant quality standards, as well as financial difficulties associated with an inefficient use of available resources.

1. INTRODUCTION

This research project was carried out by the Department of Materials and Production Engineering of the University of Naples, "Federico II" and the Department of Mechanical Engineering of the University of Salerno, in collaboration with, and at the instigation of, a Rumanian steel plant which supplies a variety of products to Italian firms and is consequently interested in elaborating a method to predict the mechanical characteristics of concrete rebar processed from hot-rolled low carbon steels. Faced with growing demand for first-rate high-bond rebar with enhanced mechanical and technological characteristics, European operators must adopt innovative technological processes or upgrade conventional ones in order to secure higher quality standards in terms of rod diameters and constancy of the rebar characteristics over time.
As non-experts hold concrete rebar production to be a fairly uncomplicated technological process, numerous small-scale mills equipped with a low-capacity electric-arc furnace process steel directly from randomly selected scrap iron supplies, while others produce concrete rebar directly from billets purchased in their local markets or imported from newly industrialised countries at comparatively advantageous prices, with little guarantee concerning the quality

Published in: E.Kuljanic (Ed.) *Advanced Manufacturing Systems and Technology*,
CISM Courses and Lectures No. 437, Springer Wien New York, 2002.

levels of the steel they use. Others still employ various cast-offs sold at prices comparable to those charged for scrap iron, [1].

The mechanical specifications that conventional concrete rebar is required to meet in the country of our project partner only include minimum yield tensile stress, ultimate tensile stress, ultimate elongation and bendability. Additional geometrical standards covering the ribs processed in the last rolling stand are only applicable to high-bond concrete rebar. The most important of these mechanical characteristics is yield tensile stress, which can be scaled by suitable process parameter adjustments and whose value must be high without reducing ultimate tensile stress or resulting in flex cracks during bend stress tests.

Conventional methods used to guarantee the requisite strength characteristics include, in decreasing order of frequency:

- cold forming, a process which is carried out after hot rolling,
- adding of microalloying elements,
- heat treatment following completion of the hot rolling process,
- use of the raw material as processed in the rolling mill following still air cooling.

The final characteristics of the rods produced with the last method depend on a large number of technological variables. Indeed, it is assumed that yield tensile stress, ultimate tensile stress and ultimate elongation are mainly influenced by several factors including the original size of the ingots, rolling process temperature, final rod diameter and the number of passes in the stands, percent reduction, the shapes into which the ingots are made in the rolling stand sequence, as well as the chemical composition of the material used.

Generally speaking, if the variables affecting a measurable parameter are unrelated, i.e. independent of each other, the desired results can be achieved by combining an experimental factorial scheme with conventional multiple or partial regression analysis, [2,3]. In our case, this approach is not applicable because of strong interactions between the variables concerned. Consequently, to adjust individual parameters, it will be convenient to test out different technological processes instead of taking over data sets used in the industry based on operator experience. Years ago, numerous attempts were made to predict the mechanical characteristics of rolled rods based on empirical correlations between the chemical composition of the steel used and the technological parameters of the rolling process. As these attempts were not always grounded on satisfactory theoretical assumptions, the resulting correlation formulas are seldom used within the industry because they are only applicable in a few, very special situations. Although we were aware that working with fairly homogeneous product categories would be the best solution, we decided to do without conventional *cluster analysis* because at times this requires complex and extremely time-consuming combinatorial calculations whose results are not always reliable, [4]. In place of this, we turned to neural networks, which autonomously identify interactions between process variables. In point of fact, neural networks can perform adaptive recognition and classification tasks because they are not subject to strict rules and can flexibly respond to outer stimuli. One additional advantage of these networks is fast learning, though the quality levels of the results obtained are largely dependent on the system architecture and the learning algorithm used to establish internal correlations between data sets. The purpose of this research project is to establish if, and to what extent, neural networks can help predict the mechanical characteristics of hot-rolled rods processed from low carbon steel for use in reinforcing concrete constructions, [5].

2. BASIC NOTIONS CONCERNING NEURAL NETWORKS

A neural network can be described as a very complex and highly parallel processing system composed of numerous simple computing modules known as nodes or neurons, which autonomously identify existing correlations between sets of data.

Typically, its nonlinear analogue computing modules are interconnected by means of weights or synapses which can be gradually adjusted in order to enhance overall network capability. As there are no fixed rules applicable in all circumstances, the system's architecture is devised by reference to available process data and information based on previous experience. Each neuron simply determines the weighted sum of the input data and changes it into an output signal which is a nonlinear function of such sum:

$$out_j = f\left(\sum_{p=1}^{N} w_p - \theta\right) \tag{1}$$

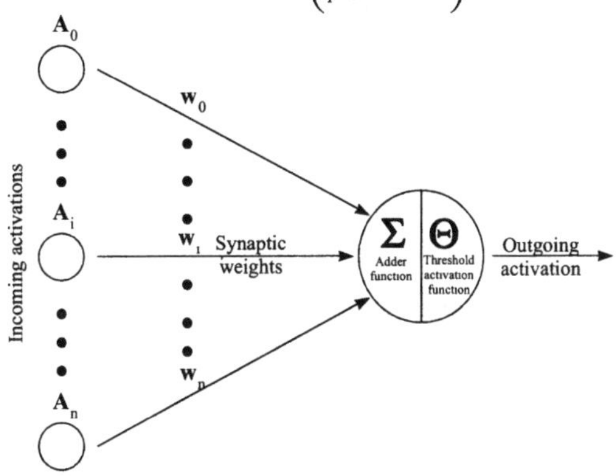

FIGURE 1. Diagram of abstract neuron model

The simplest and most widely adopted neural network is the perceptron, in which weights are either updated by means of a linear algorithm based on the values obtained when individual neurons are activated, or an iterative version devised by Widrow-Hoff in such a way as to guarantee stability of results. Both of these procedures work based on the following rule:

$$\mathbf{w}_{p+1} = \mathbf{w}_k + \eta\left(\mathbf{t}^p - \mathbf{w}_p^T \cdot \mathbf{x}^p\right) \cdot \mathbf{x}^p \tag{2}$$

In the above equation, **w** is the column vector of the weights or synapses, **x** is the column vector of the characteristics, or patterns, **t** is the vector of the expected correct values drawn from a set of examples used in the training process, and η is the learning rate. The weight vector is devised based on values that are initially chosen at random in the interval [0, 1] and subsequently updated through the use of (2), which may also be re-written in its more compact form:

$$\Delta\mathbf{w} = \eta\,\boldsymbol{\delta} \cdot \mathbf{x} \tag{3}$$

This is known as the Delta Rule, in which vector $\boldsymbol{\delta}$ is the difference between the expected result and the actual value as generated by the network.

The weak point of this model is that its simple and direct algorithm is not suitable for classes that are not linearly separable within the parameters space, for in this case the system will not converge towards a solution. Unfortunately, as most industrial applications subsume a large number of classes seldom separable in a linear fashion, it takes a multi-layered network in association with a nonlinear transfer function of the weighted sum of the inputs to generate the nonlinear decision boundary. And as higher-level correlations are extracted by the intermediate layers, this means that arbitrary decision surfaces may be generated.

The proposed multi-layer network has three unidirectional layers: the one with m input units is only a buffer and performs no computing work; the hidden layer with j units is connected both to the input layer and to the k unit output layer. Each of these layers is completed by a dummy neurone (whose input is always +1) simulating the threshold value, which in turn is determined in an adaptive way.

Using a superscript included in a square bracket to indicate which network layer is being considered (i = *input layer*, h = *hidden layer*, o = *output layer*), the input value for the jth unit of the hidden layer is determined based on the relation:

$$in_j^{[h]} = \sum_m w_{jm}^{[h]} \cdot x_m \tag{4}$$

while the output value for the same neuron is calculated as follows:

$$out_j^{[h]} = f\left(in_j^{[h]}\right) \tag{5}$$

where $f(\cdot)$, namely the activation function, is usually a sigmoid, but can be replaced by any other differentiable function.

The input value for the kth neuron in the output layer is:

$$in_k^{[o]} = \sum_j w_{kj}^{[o]} \cdot out_j^{[h]} = \sum_j w_{kj}^{[o]} \cdot f\left(in_j^{[h]}\right) \tag{6}$$

and, lastly, the output value of this same neuron is:

$$out_k^{[o]} = f\left(in_k^{[o]}\right) = f\left(\sum_j w_{kj}^{[o]} \cdot f\left(in_j^{[h]}\right)\right) \tag{7}$$

As the computed output value will normally differ from both the actual and desired values, a square error can be defined for each example and a mean value for the system as a whole:

$$E_p = \frac{1}{2}\sum_k \left(t_{pk} - u_{pk}\right)^2 \text{ and } E = \frac{1}{2P}\sum_p \sum_k \left(t_{pk} - u_{pk}\right)^2 \tag{8}$$

It is worth noting that the ½ factor in (8) only serves purposes of mathematical convenience.

The decisive turning point that led to the use of neural networks for a large number of optimization and classification problems occurred years after the perceptron had been formalised in mathematical terms, i.e. following the introduction of the error backpropagation algorithm which uses a recursive procedure to estimate the weights based on the number of errors in each layer. At this point, the weights can be updated using the steepest descent gradient method to minimise global error. The relevant calculations can be made using the chain derivation rule to express the partial derivatives in terms of two factors, one referred to the output value error variation rate and one referred to the variation rate of the output values of the kth neurone with respect to the input values of the same neuron. Formally speaking, the weight update algorithm reflects the Delta Rule, where the components of vector $\boldsymbol{\delta}$ are determined in different ways for the hidden layers and for the output layer:

$$\text{output layer } \delta_{pk} = \left(t_{pk} - out_{pk}\right) out_{pk}\left(1 - out_{pk}\right) \tag{9}$$

$$\text{hidden layer } \delta_{pj} = out_{pj}\left(1 - out_{pj}\right)\sum_k \delta_{pk} w_{kj} \tag{10}$$

3. TESTS AND ANALYSIS OF RESULTS

The high-bond concrete rebar types used for our research project are roughly equivalent to those marketed in Italy under the product codes FeB38K and FeB44K, have different strength characteristics and were produced in a variety of diameters ranging between 6 and 30 mm. The steel employed had been mainly processed from scrap iron purchased from maritime wrecking corporations and was mixed with rejects of the rolling process itself before being made to melt in an arc furnace. The chemical composition of the product is monitored during the refining process and adjusted by adding ferro-alloys on operator experience and/or empirical formulas of dubious validity. The chemical composition of the material is checked again at the last process step, just before tapping. Subsequently, a continuous casting process is used to make 130×130 mm square section billets which are then cut to a length of 10 m each and stored. Immediate transfer to the rolling plant is not possible because in spite of allowed deviations from standard characteristics the composition of the tapped material may nevertheless not be compatible with the diameter of the rods to be rolled.

Before the hot-rolling process is started, the billets are pre-heated in a gas furnace at a temperature of 1150°C. The temperature varies throughout the process due to heat increases (at the plastic forming step) and decreases (during transfers from one rolling stand to the next) whose values depend both on the area/volume ratio of the semifinished product at the step concerned and on the cooling system adopted from time to time. The rolling sequence includes a number of steps between 18 (for minimum diameter rods) and 8 (for maximum diameter rods) and is performed using close-groove rolls whose sizes and cross sections can be varied according to the final diameter to be obtained, [6,7]. The blooming mill input flow rate varies between 0.12 and 0.23 m s^{-1} for all formats, which means that output flow rates and total processing times will also vary accordingly. By way of example, the rolling sequence of a 12 mm diameter rod is illustrated in tab. I.

TABLE 1. Rolling sequence of 12 mm diameter rod

Step	Roll position	Roll diameter mm	Coss section shape	Cross section reduction %	Rod speed rate, m s^{-1}
1	H	550	S	29.0	0.15
2	V	550	S	25.5	0.20
3	H	550	S	28.2	0.29
4	V	550	S	26.0	0.39
5	H	550	O	33.8	0.58
6	V	550	R	24.1	0.77
7	H	425	O	33.9	1.16
8	V	425	R	25.1	1.55
9	H	425	O	33.3	2.33
10	V	425	R	23.7	3.05
11	H	425	O	30.6	4.40
12	V	425	R	19.3	5.44
13	H	335	O	19.1	6.73
14	V	335	R	13.3	7.76
15	H	335	O	18.7	9.55
16	V	335	R	7.9	10.36
17	H	280	O	19.3	12.85
18	V	280	R	19.7	16.00

Legend: H = horizontal, V = vertical, S = rectangular, O = oval, R = round

It is worth noting that at steps 13-14 of the processing cycle for up to 14 mm diameter rods the semi-finished product is brought to an 8-shaped cross section and is then split into two parallel rods; for technical reasons (length of the bar to be rolled) and in order to prevent excessive cooling, these receive passes through separate grooves until the final rod section is obtained. Thanks to this arrangement, all the rods leave the last rolling stand at a temperature exceeding 850°C, which is high enough to secure that the material remains in the austenitic state but not high enough to cause recrystallization. For this reason, the size of the austenitic grain, which determines the fineness of the final ferrite-pearlite structure, is greatly affected by the length of the forming process and especially by the last passes in the rolling stands.

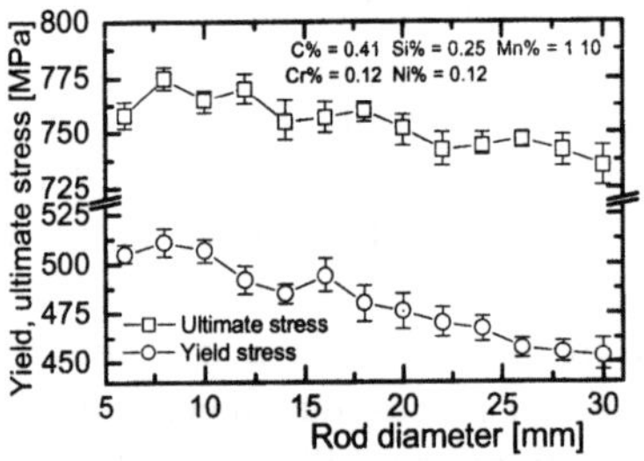

FIGURE 2.Yield and ultimate tensile stress values as a function of rod diameter, for a fixed material

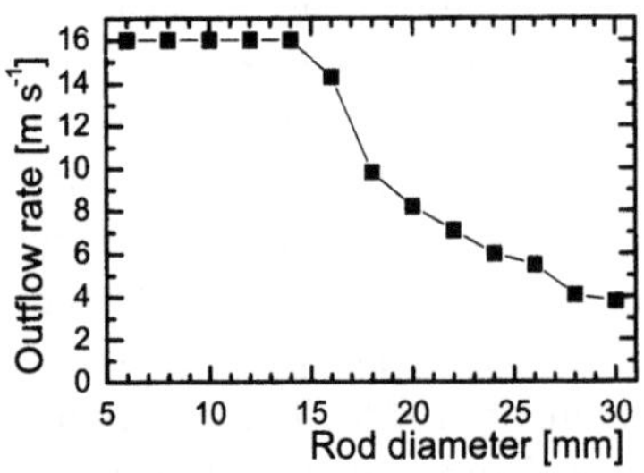

FIGURE 3. Outflow rates of different diameter rods on leaving the last rolling stand

At the initial stage of the project, different diameter rods were rolled using steel from a single production batch, since we had resolved to ascertain the effect of the rolling cycle on the strength characteristics of concrete rebar irrespective of the chemical composition of the steel used. The findings in fig. 2 suggest that based on their diameter, the rod products can be grouped into three categories with homogeneous yield tensile stress properties, whereas ultimate tensile stress is also dependent on factors such as inclusions or surface protrusions, etc., which might cause misinterpretations. The findings in fig. 2 show that 6 to 10 mm diameter rods on the one hand and 12 to 16 mm diameter rods on the other have fairly similar mechanical characteristics, while the rods with diameters exceeding 18 mm that form the third group have varying strength characteristics probably due to the different cooling rates of the core and external layers. Our test results can also be discussed in association with fig. 3, which reports the speeds at which the rods leave the last rolling stand in consideration of the synergic effects of the cooling and forming rates on such mechanical characteristics.

During this experimental application of a neural network in the identification of different categories of concrete rebar we resolved to restrict our analysis to maximum 16 mm diameter rods in order to avoid interferences from an additional variable, namely different cooling rates.

Consequently we analysed 450 production batches of the two types mentioned above, which were rolled to different diameters of up to 16 mm by reference to criteria mainly based on operator experience. From each of the batches rolled from one and the same billet we picked three bars and subjected them to chemical analysis and tests intended to determine characteristic unit loads. Subsequently we analysed our experimental results and extracted a set of 120 data items evenly distributed among the six project diameters and two product groups with very similar mechanical characteristics in order to find correlations between chemical composition and maximum rolled rod diameter. Our data items were divided into two sets, to be respectively used in the learning phase and in the process for assessing the network's overall generalisation capability, [8]. The results of this operation are reported in tabb. II and III.

TABLE 2. Experimental data sets used for connection learning

Sample	Typology	Carbon %	Manganese %	Chromium %	Nickel %	Yeld tensile stress. (MPa)	Ultimate tensile stress (MPa)
1	6_38	0.30	0.77	0.12	0.14	422.21	597.43
2	6_38	0.30	0.86	0.11	0.16	427.72	606.26
3	6_38	0.31	0.81	0.10	0.10	435.56	633.73
4	6_38	0.32	0.77	0.14	0.15	428.70	619.99
5	6_38	0.28	0.78	0.10	0.10	427.72	595.47
6	8_38	0.31	0.81	0.16	0.17	423.79	623.92
7	8_38	0.31	0.78	0.10	0.12	422.21	597.43
8	8_38	0.33	0.74	0.12	0.12	419.87	614.11
9	8_38	0.30	0.80	0.23	0.16	416.32	569.96
10	8_38	0.33	0.80	0.13	0.15	430.66	597.43
11	10_38	0.31	0.84	0.10	0.12	423.19	600.37
12	10_38	0.35	0.65	0.10	0.10	428.10	611.16
13	10_38	0.32	0.81	0.10	0.12	417.31	610.18
14	10_38	0.33	0.81	0.10	0.10	416.32	596.45
15	10_38	0.31	0.80	0.18	0.16	418.10	615.09
16	12_38	0.33	0.90	0.14	0.14	421.04	613.13
17	12_38	0.32	0.89	0.16	0.19	432.62	629.80
18	12_38	0.35	0.87	0.14	0.12	421.83	619.01
19	12_38	0.35	0.86	0.15	0.14	433.60	649.42
20	12_38	0.35	0.92	0.11	0.13	418.89	623.92
21	14_38	0.33	0.93	0.14	0.15	420.06	621.95
22	14_38	0.35	0.85	0.13	0.18	429.68	625.88
23	14_38	0.34	0.90	0.14	0.13	430.66	659.23
24	14_38	0.36	0.82	0.14	0.14	430.66	643.54
25	14_38	0.36	0.83	0.11	0.13	427.72	643.54
26	16_38	0.33	0.91	0.21	0.14	428.70	651.38
27	16_38	0.33	0.98	0.11	0.14	433.60	648.44
28	16_38	0.37	0.75	0.11	0.13	422.81	627.84
29	16_38	0.34	0.95	0.17	0.15	417.91	648.44
30	16_38	0.36	0.84	0.13	0.16	436.55	646.48
31	6_44	0.37	1.00	0.20	0.14	483.82	724.96
32	6_44	0.37	0.98	0.19	0.18	481.67	722.02
33	6_44	0.36	1.00	0.23	0.17	490.50	731.83
34	6_44	0.36	0.93	0.29	0.25	486.58	714.17
35	6_44	0.42	0.83	0.12	0.14	474.80	718.85
36	8_44	0.38	1.00	0.14	0.14	484.80	723.00
37	8_44	0.40	0.96	0.16	0.13	488.35	744.58
38	8_44	0.41	0.88	0.14	0.13	477.75	710.24
39	8_44	0.42	0.87	0.13	0.13	480.88	738.69
40	8_44	0.40	0.93	0.18	0.22	468.92	709.26
41	10_44	0.41	0.89	0.10	0.13	487.75	746.54
42	10_44	0.39	0.98	0.12	0.13	477.75	741.64
43	10_44	0.42	0.90	0.11	0.16	472.84	715.34
44	10_44	0.43	0.81	0.10	0.10	482.05	716.13
45	10_44	0.42	0.88	0.12	0.13	489.33	725.18
46	12_44	0.40	1.00	0.16	0.15	480.69	721.04
47	12_44	0.41	1.07	0.15	0.14	497.37	774.61
48	12_44	0.42	1.00	0.15	0.13	490.31	751.45
49	12_44	0.45	0.87	0.12	0.12	475.98	735.75
50	12_44	0.44	1.04	0.10	0.10	479.90	751.45
51	14_44	0.44	0.93	0.13	0.14	478.92	723.00
52	14_44	0.44	0.89	0.16	0.13	478.73	748.50
53	14_44	0.43	0.97	0.17	0.15	485.60	760.28
54	14_44	0.41	0.96	0.18	0.18	473.03	712.21
55	14_44	0.43	0.92	0.18	0.19	480.69	720.62
56	16_44	0.45	0.98	0.12	0.13	493.44	762.24
57	16_44	0.43	1.02	0.14	0.13	493.44	764.20
58	16_44	0.44	0.94	0.15	0.14	478.92	735.75
59	16_44	0.45	0.89	0.15	0.12	499.33	772.05
60	16_44	0.44	0.98	0.15	0.13	476.96	723.98

To avoid scaling problems, we resolved to normalise our input data in such a way as to have values between 0 and 1, i.e. compatible with those obtained by means of the activation transfer function:

$$x_{norm} = \frac{1}{2}\frac{x-\bar{x}}{3\sigma} + 0.5 \tag{11}$$

This data normalisation procedure was chosen because we assumed that in this way the network could also be used to process experiential data sets included in the statistical population adopted for the learning phase.

TABLE 3. Experimental data sets used for network control

Sample	Typology	Carbon %	Manganese %	Chromium %	Nickel %	Yeld tensile stress. (MPa)	Ultimate tensile stress (MPa)
1	6_38	0.29	0.88	0.12	0.13	419.87	600.37
2	6_38	0.30	0.80	0.19	0.18	425.75	593.51
3	6_38	0.31	0.73	0.12	0.14	417.12	569.96
4	6_38	0.30	0.91	0.12	0.16	425.75	596.45
5	6_38	0.31	0.87	0.11	0.14	421.83	591.54
6	8_38	0.31	0.88	0.12	0.13	424.77	622.94
7	8_38	0.32	0.78	0.12	0.12	429.68	650.40
8	8_38	0.31	0.75	0.15	0.17	418.29	573.89
9	8_38	0.32	0.82	0.14	0.13	420.25	587.62
10	8_38	0.34	0.70	0.13	0.13	422.81	627.84
11	10_38	0.31	0.76	0.11	0.11	415.15	561.13
12	10_38	0.32	0.73	0.11	0.15	424.96	565.06
13	10_38	0.32	0.82	0.12	0.12	424.96	602.33
14	10_38	0.32	0.82	0.13	0.18	415.94	635.69
15	10_38	0.31	0.88	0.14	0.13	415.94	623.92
16	12_38	0.30	0.93	0.25	0.23	430.66	645.50
17	12_38	0.33	0.98	0.14	0.14	428.70	627.84
18	12_38	0.37	0.76	0.10	0.12	426.74	640.59
19	12_38	0.35	0.90	0.14	0.14	424.77	623.92
20	12_38	0.35	0.86	0.16	0.16	437.53	632.75
21	14_38	0.36	0.75	0.12	0.13	429.68	640.59
22	14_38	0.34	0.88	0.13	0.14	415.15	593.51
23	14_38	0.33	0.90	0.14	0.14	430.66	647.46
24	14_38	0.36	0.87	0.10	0.12	419.87	603.32
25	14_38	0.36	0.88	0.10	0.10	418.89	643.54
26	16_38	0.32	0.95	0.12	0.14	417.91	633.73
27	16_38	0.35	0.90	0.11	0.13	423.79	643.54
28	16_38	0.34	0.89	0.16	0.17	421.83	641.57
29	16_38	0.35	0.80	0.12	0.15	418.89	630.78
30	16_38	0.36	0.80	0.16	0.15	430.66	660.21
31	6_44	0.38	0.95	0.17	0.17	473.82	703.38
32	6_44	0.40	0.91	0.14	0.13	476.77	687.68
33	6_44	0.40	0.93	0.13	0.13	498.35	718.09
34	6_44	0.40	0.84	0.24	0.17	490.50	728.88
35	6_44	0.40	0.90	0.17	0.14	498.35	715.15
36	8_44	0.40	0.83	0.16	0.13	501.29	737.71
37	8_44	0.40	0.90	0.16	0.13	474.80	733.79
38	8_44	0.39	0.96	0.17	0.17	506.20	741.64
39	8_44	0.39	0.94	0.15	0.14	488.54	741.64
40	8_44	0.40	0.86	0.21	0.15	478.73	730.85
41	10_44	0.40	0.88	0.16	0.21	464.01	699.45
42	10_44	0.41	0.83	0.17	0.12	482.65	729.86
43	10_44	0.40	0.97	0.15	0.12	497.37	765.18
44	10_44	0.40	0.99	0.15	0.18	483.63	728.88
45	10_44	0.41	0.92	0.15	0.17	489.52	726.92
46	12_44	0.40	1.00	0.19	0.19	463.03	700.43
47	12_44	0.42	0.97	0.17	0.16	486.58	752.43
48	12_44	0.39	0.98	0.36	0.23	495.41	754.39
49	12_44	0.39	1.00	0.12	0.15	498.35	781.86
50	12_44	0.44	0.99	0.15	0.15	465.98	743.60
51	14_44	0.42	0.93	0.16	0.16	482.65	715.15
52	14_44	0.45	0.92	0.13	0.12	478.92	720.05
53	14_44	0.43	0.91	0.18	0.18	480.69	754.39
54	14_44	0.42	0.95	0.18	0.16	471.86	712.21
55	14_44	0.41	1.00	0.17	0.18	474.80	714.17
56	16_44	0.41	1.07	0.12	0.11	485.60	746.54
57	16_44	0.42	1.08	0.14	0.13	476.96	731.83
58	16_44	0.42	0.93	0.23	0.14	484.61	746.54
59	16_44	0.42	1.03	0.16	0.15	497.37	772.05
60	16_44	0.44	0.89	0.18	0.14	485.60	737.71

Using conventional cluster analysis techniques, we checked for boundaries between convex classes and tried out different network configurations by means of a learning algorithm with error backpropagation. Our analysis was restricted to a single hidden layer with a varying number of neurones, as well as different learning rates and momentum factor values. The network which afforded optimal results in terms of training time and error minimisation had 12 hidden units, each one with a learning rate of 0.4 and a momentum factor value of 0.6. However, its weak points were limited flexibility and low generalization capability.

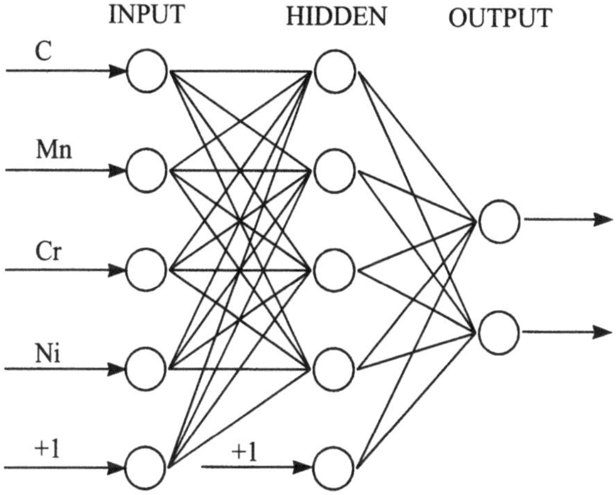

FIGURE 4. Diagram of the neural network used

As far as generalisation capability is concerned, optimal results were afforded by a fairly simple network, illustrated in fig. 5, with only four units in the hidden layer, a learning rate of 0.7 and a momentum factor value of 0.9. This network converges toward a solution after 200 complete iterations with a less than 10^{-4} global error level and affords the correct classification of 95% of the data items used in the testing phase. The system's errors invariably concern steel types comparable to the Italian variety FeB44K; in particular, samples 36 and 38 were included in the next higher category and sample 46 in the next lower category. However, based on the yield tensile stress and ultimate tensile stress values of the samples concerned, one would be inclined to accept the classification suggested by the system and conclude that it would be a mistake to use the corresponding production batches in rolling rods of the diameters concerned.

As our results proved fairly encouraging, we are planning to use the same classification procedure also for greater diameter rods and, at the same time, to increase the size of the input data vector to cover the magnitude of the deformation and the cooling rates through the insertion of appropriate parameters.

4. CONCLUSION

Simple neural network configurations can be effectively and rapidly used to subdivide different high-bond reinforced concrete rods into two categories with homogeneous strength characteristics (FeB38K and FeB44K), each of which is in turn split into two subcategories based on the diameter of the rolled bar (6÷10 mm and 12÷16 mm). On analysing our experimental data and the results of the rolling process we found that carbon, manganese, chromium and nickel levels in the steel used are major variables not subject to targeted adjustments which impact both the structural characteristics of the rolled rods and their response to heat treatment during still air cooling. Although technological variables also play a major role in determining the mechanical characteristics of the rolled bars, these variables were kept constant throughout our project work, regardless of the type of rods rolled from time to time. This means they will have to be taken into account when devising the weights for the synaptic connections.

The most efficient network is composed of two layers preceded by an input layer with four continuous-value input units feeding data into the network without performing any operations, one hidden layer with four computing units and, lastly, one output layer with two binary units capable of identifying four categories of products. The activation transfer functions of the individual neurones are of the logistic or sigmoid type and their threshold values are directly calculated by the network.

Thanks to the stability of the configuration adopted, no fluctuations were observed despite the fairly high learning rate ($\eta = 0.7$) adopted. This result was also due to the momentum factor that had been built into the system in order to continue the search for the minimum-energy value along the path previously followed; consequently, the error backpropagation learning algorithm converges towards a solution following several hundred complete loops on a group of 60 samples evenly distributed among the four categories. The main merits of the system include its high-level generalization capability and a classification error rate below 5%. As the results obtained were fairly satisfactory, it will be interesting to introduce penalty functions for every instance of wrong classification, thereby reducing the risk of producing rods with poor product characteristics and, conversely, guaranteeing first-rate products with specifications well above those currently required by applicable standards. The only weak point of the system is its unidirectional architecture, namely the fact that it only affords identifying the category of one production batch and thus can neither be used to specify compositions that will lead to a certain product category, nor suggest adjustments to the technical rolling process parameters.

REFERENCES

1. Korchynsky, M., Wille, P., (1984), Acciai saldabili di elevata resistenza per tondo per cemento armato con aggiunta di vanadio e azoto. La Metallurgia Italiana, No. 11, 484-491
2. Czitrom, V., (1999), One-Factor-at-a-Time Versus Designed Experiments. The American Statistician, Vol 53/2, 126-131.
3. Draper, N.R., Smith, H., (1998), Applied regression analysis. John Wiley & Sons, New York.
4. Everitt, B.S., (2001), Cluster Analysis. Edward Arnolds, London.
5. Haykin, S.S., (1998), Neural Networks: A Comprehensive Foundation. Prentice Hall, New York.
6. Johnson, W., Mellor, P.B., (1962), Plasticity for mechanical engineers. Van Nostrand, London.
7. Altan, T., Oh, S., Gegel, N., (1983), Metal forming: Fundamentals and applications. ASM Intl, New York.
8. Master, T., (1993), Practical Neural Network Recipes in C++. Morgan Kaufmann Pub., New York.

FEM MODEL COMPUTING LARGE DEFORMATION IN METAL SHEET FORMING BY HIGH POWER BEAM

G. Casalino[1], G. De Astis[1], A. D. Ludovico[1]

[1] Dipartimento di Ingegneria Meccanica e Gestionale, Politecnico di Bari, Italy

KEYWORDS: High Power Beam Forming, Large Deformation, FEM.

ABSTRACT. The use of high power beam technologies like the laser and the plasma arc has provided a mould-less metal sheet rapid forming technique. This technique can have application either for rapid prototyping and rapid manufacturing of mechanical products. The virtual manufacturing can be a boost for a rapid diffusion of the mould-less sheet forming. Even if the calculations can be cumbersome, the Finite Element Analysis (FEM) is unique for providing geometrical and physical information on product development and manufacturing via computer. Herein, a simplified FEM model for both laser and plasma arch is proposed. In particular, the heat source has been modeled so the computation is easier and faster. The model was validated by comparison with the experimental results and it is able to predict the forming angle, the final shape, and the residual stress of metal sheet induced by high power beam irradiation.

1 INTRODUCTION

High power beam is a promising technology for metal sheet forming. Using the heat induced by a high power beam a metal sheet sample can be either bent and formed without a hard tool.
Several authors have investigated the mechanism of deformation of laser forming and some important results have already been achieved. The energy intensity of the laser beam used has been about $6.0 \div 15.0 \times 10^3$ W/cm^2, the energy transfer ratio is 5÷10%. For enhanced absorption carbon-spray precoating of surface is necessary. The accuracy and the drawbacks of this technique have been evaluated [1]. As a result its applicability have been limited to the rapid prototyping and to the small batch manufacturing.
Other authors have proposed an alternative heat source that meets the basic requirements for beam for metal forming, which are high power density and adjustable power input and energy. That is the plasma arch in the low range for welding application (100 Ampere). It can obtain an energy intensity about $1.0 \div 4.0 \times 10^3$ W/cm^2. The energy transfer ratio of plasma can be 8 times higher then the laser's one. Qualification and safety protection equipment for personnel are less important and expensive then those for the laser [2].
The best results in tool-less forming study have been reached by experimental investigations, which are time and money consuming. The efforts to numerically model the forming process have also produced positive results even the complexity of the deformation mechanism and the number of the variables that affect the process [3]. The most interesting results have come from finite element method (FEM) and neural networks [4]. Nevertheless, none of these tools have been yet able to model satisfyingly the large deformation, which are typically of sheet forming. For FEM the computing time and memory allocation required for computation have been large. On the other hand, the artificial intelligence requires less time and memory but it does not provide the geomet-

Published in: E.Kuljanic (Ed.) *Advanced Manufacturing Systems and Technology*,
CISM Courses and Lectures No. 437, Springer Wien New York, 2002.

rical shape of the final product. As a result, either a personal computer and a small workstation have little ability to virtual manufacture complex shape obtained by high power beam. Since, computer is expected to greatly support assessing the manufacturability of a candidate design and to provide accurate estimates for processing times, cycle times and costs, as well as product quality, virtual manufacturing is a basic pre-requisite for the development of this technology. Therefore, numerical simulation of the process must be easily and rapidly performed.

This paper is dedicated to improve the ability of personal computer and small workstation to perform standard FEM thermo-mechanical analysis aimed to predict large deformation due to either laser or plasma arch beam irradiation. For this purpose, the heat source has been modeled by means of a temperature field, which has been determined and applied on the basis of the process features. During the training of the model the temperature field was estimated so the final bending angle of the sample was reproduced by computation. The thermal load was applied to the sample's surface.

In this fashion, several runs of the heat source on the irradiation path were simulated and large bending angle were calculated. The error to the experimental bending angles was as little as less then 2%. The effects of both the linear and curved irradiation have been simulated.

Further information were collected about the residual stress and the influence of different irradiation strategies on the final shape of the specimen.

Directly applying temperature field has proved to be a reasonable way for fast calculation of high power beam forming of steel plate. It has been demonstrated that is possible to save time and computer memory. In this way the manufacturing strategy of complex shape can be designed and tested by computer virtual manufacturing.

2 SETTING UP THE FEM MODEL

The finite element method is know as a powerful tool for studying the metal sheet forming process. The forming process is very sensitive to variations in parts geometry and treatment conditions, in this study an appropriate FEM model accounts for the temperature dependent behavior of the material (heating and cooling cycle) and the special geometrical conditions in die-less forming of complex shape.

The thermal-mechanic simulation can be uncoupled since the heat generated during the deformation work is much smaller than the laser induced heat and it may be neglected. The thermal analysis is performed to obtain the nodal temperature solution. For what concerns the formulation of the boundary conditions, the source is usually schematized as a flux of heat entering the workpiece surface which faces the laser head. Heat flux produced the nodal temperatures that are the loads applied to the structure in a subsequent mechanical analysis step (figure 1).

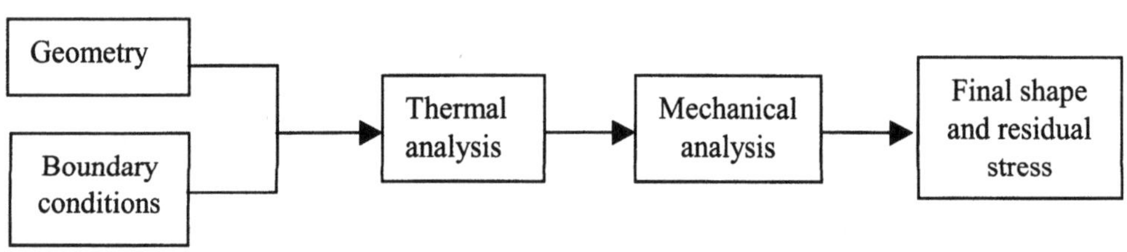

FIGURE 1. Uncoupled thermo-mechanical analysis.

This analysis has taken in account the thermal induced elastic-plastic with strain hardening behavior of the material with regard to the specific geometrical conditions. The work hardening between two successively runs can be neglected with no significant error since the high temperature balances that effect on material's strength.
The material was the AISI 304 austenitic stainless steel. The considered properties of material are the global coefficient of radiation, the convection, the specific heat, the thermal conductivity, the surface absorption coefficient, and the mechanical properties. All the material's properties varied with the temperature. The element utilized for the solution is a three-dimensional brick, with eight nodes.
Every heating-cooling pattern is divided into a finite number of steps. The movement of the thermal source was simulated loading a set of node in a moving local reference system, whose origin was positioned at the starting point of the feed path. According to the standard of the thermo-mechanical analysis the heat flux should be the thermal load. Figure 2 shows a close-up of the sheet surface related to the laser half spot area which is loaded with variable heat fluxes (gray scale) according to the gaussian distribution.
Figure 3 shows the different approach which has been used in this paper. A larger view of the metal sheet is provided in figure 3. In this case, the laser spot area is approximated by the square

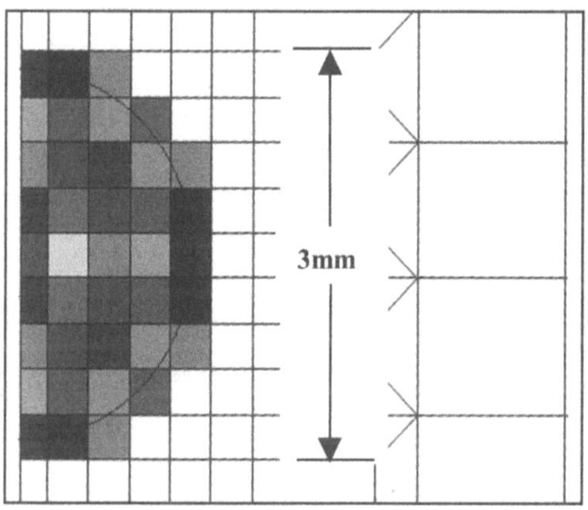

FIGURE 2. Close-up of heat flux distribution.

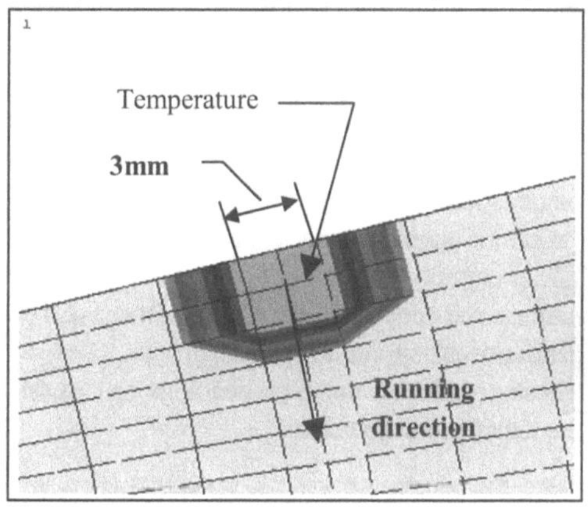

FIGURE 3. Approximated area of laser spot.

that circumscribes the actual spot. The square has a 3 mm side. All the nodes falling in this area have been loaded with an average process temperature which was determined by a trial-and-error technique.
In fact, a reasonable temperature of tentative was applied and the final bending angle was calculated after 16 runs. Thereafter, the numerical result was compared with the experimental one then the temperature was modified and the computation restarted. The initial temperature of the sheet for the first run was set to 20°C (room temperature). Figure 4 shows a flow chart of this trial-and-error process. The iterative analysis ends when the numerical versus experimental gap was less then 2% and the temperature was recorded as loading temperature for that case (T_L).
For the after-first runs, it has been necessary to consider the warm up of the sheet that increased the initial temperature of each run. These temperatures for the learning cases (see the tables in the "numerical results" paragraph) were calculated by the FEM method. The final temperature of the first pass was calculated and it was used as the initial temperature of the next step and so on (T_I).

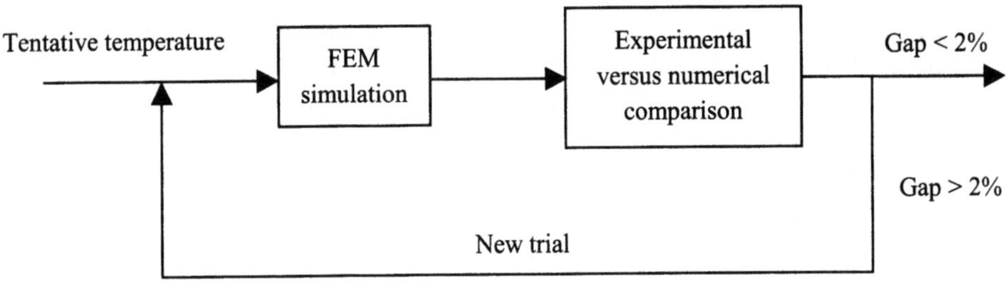

FIGURE 4. Trial-and-error approach.

Moreover, it came up that applying instantly the temperature to the node was incorrect and the convergence to 2% error was difficult. Therefore, the loading temperature of each step grew linearly from the initial to the final value according to the running speed of laser scansion (figure 5).
The experiments were performed by using a CO_2 laser. A Rofin Sinar CO_2 laser, mod. DC 025, operating in continuous wave regime was used. The exit 25 mm diameter beam is a TEM_{00} mode with a small TEM_{01} component and it has a 0.5 mrad divergence. Laser beam was focused on the work-piece by means of a 5-inch focal length ZnSe lens. The shielding gas nitrogen; it was ejected coaxially to the laser beam from a nozzle of 1.5 mm exit diameter with a pressure of 4 bar.
The sample was moved in a plane orthogonal to the laser beam by means of a four axis motorized table, controlled by a CNC program.
The geometry of specimen has varied from rectangular to circular sector shape. The thickness was constant and it was 2 mm. Otherwise, The dimensions of samples varied. For the rectangular shape, some length and width were tested. For the circular sector shape (see figure 6 for the mesh), the sector angle and the irradiation radius were considered [5].
The specimen was initially coated with carbon spray, which increases energy absorption.
Among the technological parameters, a particular attention was paid to the effects of the laser beam power, spot, and speed.
It was possible to vary the beam spot sizes on the work-piece surface by adjusting the distance of the lens from the sample.

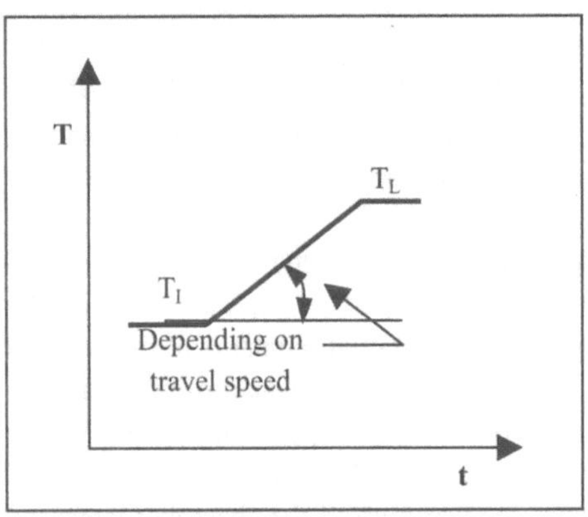

FIGURE 5. Temperature Loading Step.

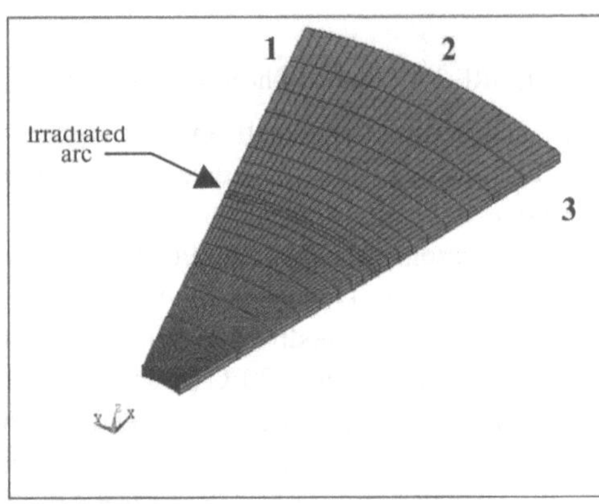

FIGURE 6. Circular sector mesh.

3 NUMERICAL RESULTS

A number of simulation were performed in order to validate the above-depicted numerical model. Table 1 and 2 show the parameters selection, the experimental and numerical angles, their percentage, and the loading temperatures for every simulation. For a matter of room the initial temperature for every step do not compare.

TABLE 1. Circular sector learning cases

N°	Radius mm	Sector °	Power W	Speed mm/min	Spot mm	Experimental angle °	Numerical angle °	Error %	T_L °C
1	67	30	500	2500	2	7.34	7.47	1.77	690
2	67	30	1500	2500	6	26.9	27.30	1.48	860
3	67	30	500	3375	2	7.18	7.29	1.53	680
4	67	30	1500	3375	6	19.08	18.90	-0.94	730
5	96	30	1500	3375	6	16.47	16.76	1.76	755
6	64	45	2000	2642	8	28.96	28.70	-0.89	880
7	45	45	1000	2500	4	16.34	16.57	1.40	670
8	48	60	500	3375	2	6.67	6.67	0	676
9	45	45	1000	3375	4	12.01	11.78	-1.9	600
10	64	45	2000	2850	8	17.21	17.67	2.06	700

TABLE 2. Rectangular-shaped learning cases

N°	Length mm	Path's posi-tion mm	Power W	Speed mm/min	Spot mm	Experimental angle °	Numerical angle °	Error %	T_L °C
11	64	48.9	1500	3375	6	18.57	18.13	-2.20	755
12	128	25.4	500	3375	2	8.42	9.77	13.80	676

Both the counterbending, the little elastic deformation away from the laser source typical of the laser bending, and the edge buckling of the sheet were correctly simulated. The experimental measurement of these phenomena were exactly reproduced by calculations.

Figure 7 shows the comparison between the numerical and experimental measurement of the edge's buckling, which is due to the inhomogeneous thermal cycle. In particular, the chart contains the measurement at the beginning, the middle, and the end of the plate's edge (see figure 6) of the sample n. 1

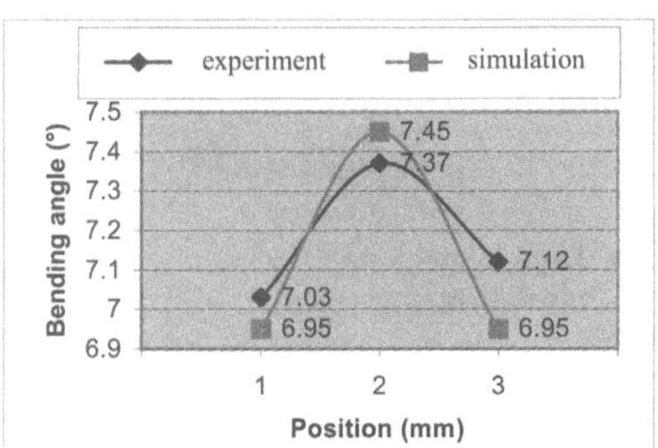

FIGURE 7. Measurement of edge's buckling.

referred in table 1. It is evident the very good agreement between the two sets of data.

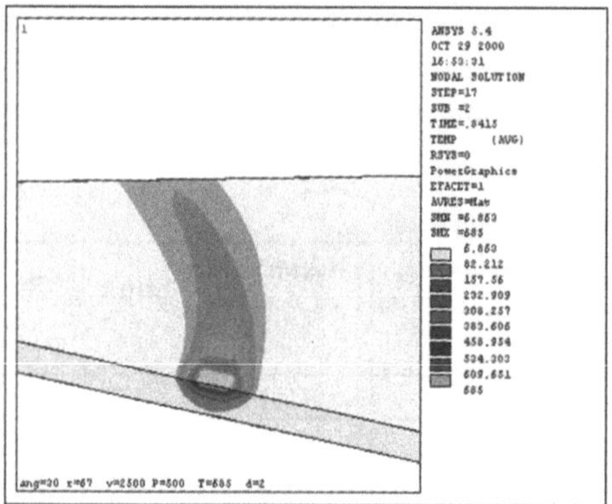

FIGURE 8. Thermal 1th step – upper side.

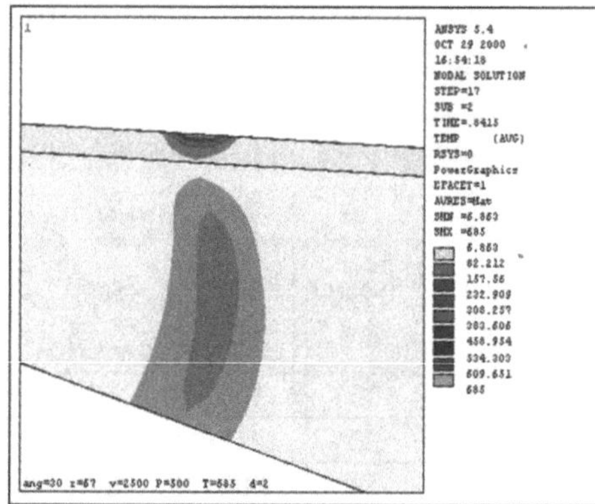

FIGURE 9. Thermal 1th step – lower side.

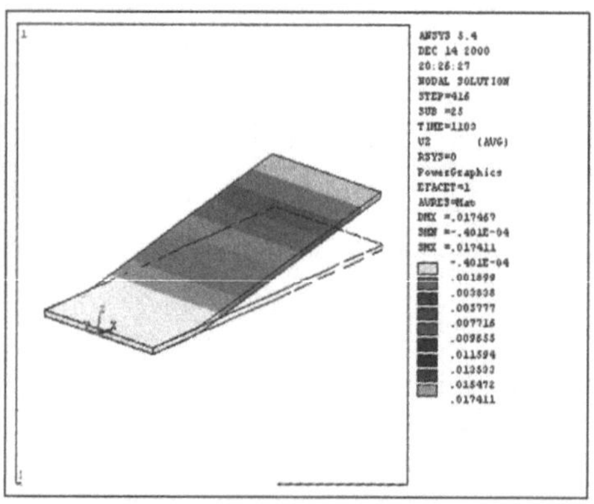

FIGURE 10. Deformed rectangular sample.

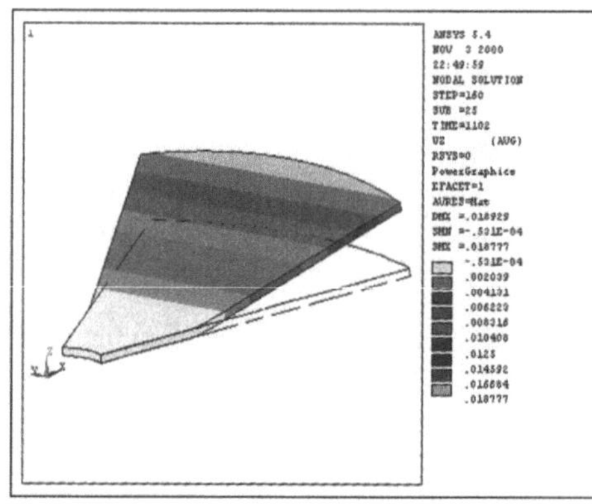

FIGURE 11. Deformed circular sector sample.

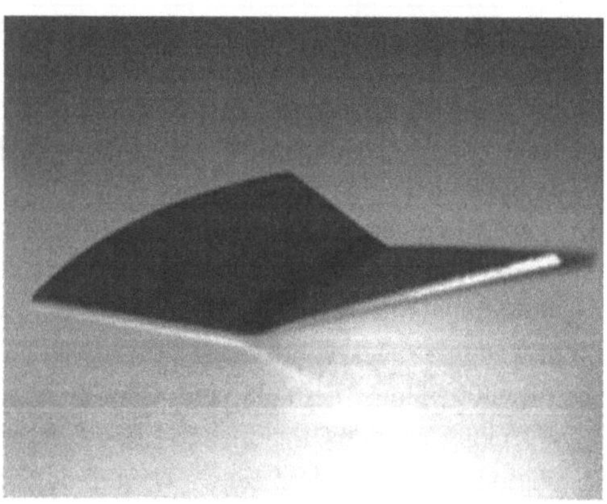

FIGURE 12. Experimental deformed sample.

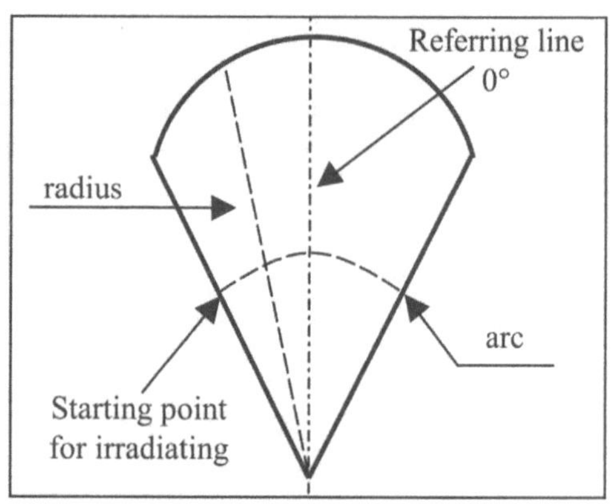

FIGURE 13. Residual stress measure.

Figures 8 and 9 show the thermal field of the upper and bottom side of the sample 1, respectively. The thermal deformation for both geometries, samples 11 and 6, are reported in figures 10 and 11 after 16 runs. The agreement between the experimental and numerical deformed shape is demonstrated by comparing figure 11 and figure 12.

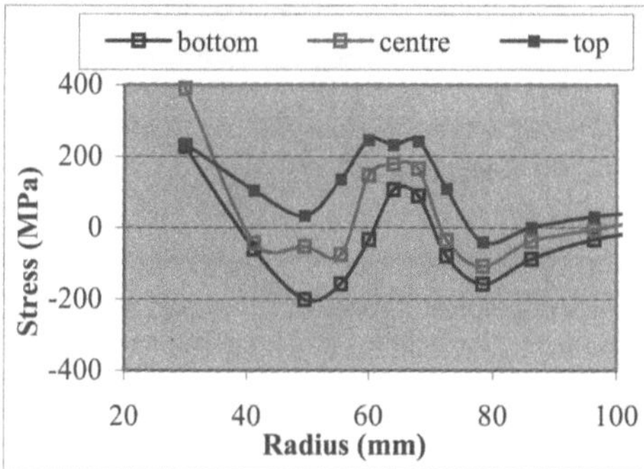

FIGURE 14. Tangential stress at 12.7°.

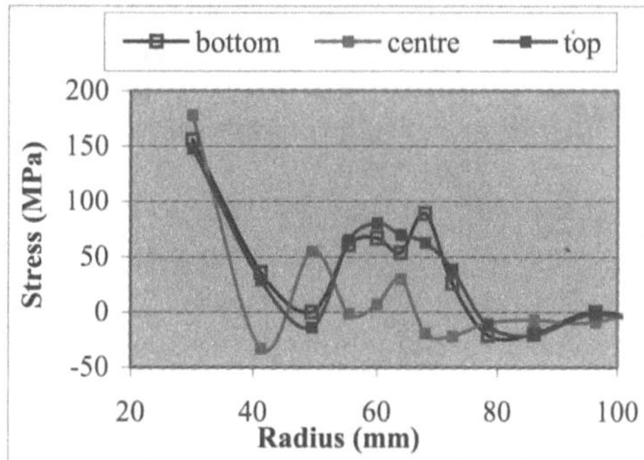

FIGURE 15. Radial stress at 12.7°.

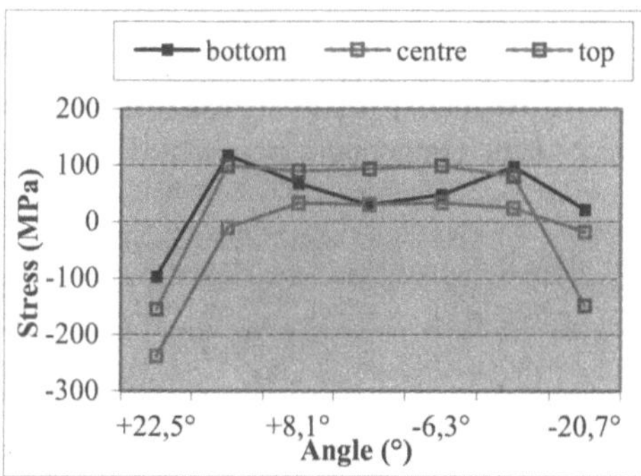

FIGURE 16. Radial stress at 60 mm arch.

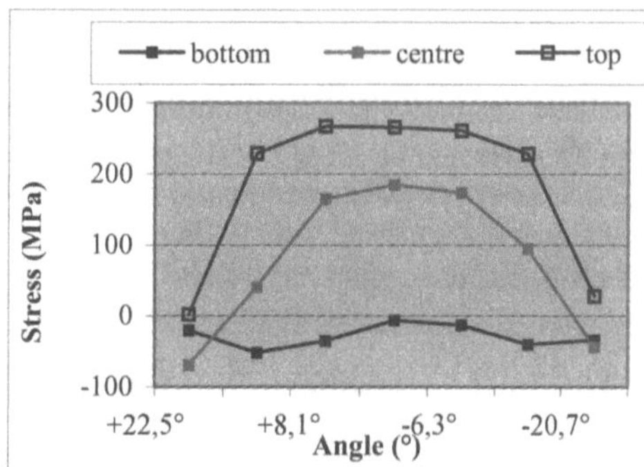

FIGURE 17. Tangential stress at 60 mm arch.

Observing the figures from 14 to 17, it is possible to analyze the residual stress of the sample 12 after 16 runs of laser. Figures 14 and 15 report the tangential and radial stress along a radius that was 12.7° to the right of the referring line (see figure 13). The tangential stress is tensile and it shows peaks at the beginning (in correspondence of the blocking device) and at the center of curves (the irradiated arch). Otherwise, the radial stress is compressive between the beginning and the center of curves. The three curves refer to the top, middle, and bottom of sample.

Figures 16 and 17 show the residual stress along an arc whose radius was 60 mm (lower then irradiated arch). In these cases, the stress are tensile in the middle of the arch and compressive at the edges of the sample. It is noticeable the dissymmetry of stress field around the referring line.

4 THE MULTIPLE REGRESSION MODEL

Multiple regression analysis is one of the most widely used of all statistical tool. In this paper a curvilinear model with one independent variable has been used. It was aimed to calculate the loading temperature and the initial temperatures of each run.

The variables were selected by means of the analysis of variance. It was demonstrated that the power (P), the speed (v), and their first order interaction have principal effects on the calculation of final bending angle [6]. The Pareto diagram in figure 17 show the variance analysis results. The

dotted line show the limit of significance of variables. The symbol θ (A) and l (B) are the sector's angle and the arch's length respectively.
Therefore, the model for temperature calculation initially contained those variables. Then, on the basis of the results of the analysis of variance (see table 3), the number of variables has raised to 7.

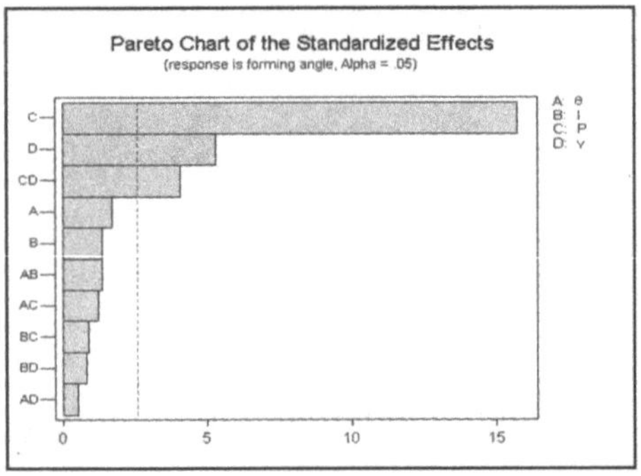

FIGURE 18. Effects on the bending angle.

TABLE 3. Variance analysis

	gdl	*SQ*	*MQ*	*F*	*p-value*
Regression	7	61542.15	8791.736	2.6841	0.2981
Residual	3	6550.751	3275.375		
Total	10	68092.9			

The table 4 contains the list of those variables and the corresponding coefficients. Using the regression model it is possible to calculate the initial and loading temperatures for every laser's run in the range of the investigated parameters.

TABLE 4. Regression coefficient.

power	speed	length	angle	$Power^2$	$Speed^2$	Power*Speed	intercept
21.669	-771.93	3.661	-3.042	1.411	13.205	-1.409	11730.37

5 CONCLUSION

In this paper the numerical calculation of large deformation of stainless steel sheet laser forming has been performed by FEM analysis. The FEM model has permitted to calculate the forming angle with 2% error, the related counterbending and edge's buckling. The computational time has significantly reduced. Useful information have been obtained about the residuals stress. This model can be applied to other power source like the plasma arch sheet forming.
On the basis of the results achieved and with further effort, it will be possible to perform the virtual prototyping and manufacturing of complex shape by high power beam using the computer.

REFERENCES

1. Hennige, T. (2000). Development of Irradiation Strategies for 3D-Laser Forming. Journal of Mat. Proc. Tech., 103, 2000, 102-108, S0924-0136(00)00392-7.
2. Li P.J., Chen Y.W., Male A.T., ZhangY.M., (2000). Flexible Forming of Sheet Metal Using a Plasma Arc. Journal of Eng. Manufac. Proc. Part B, 214, B2, 2000, 0954-4054.
3. Hoving W. (1997). Accurate Manipulation Using Laser Technology, Europto Laser Conf..
4. Casalino G., Ludovico A.D. (1999) "Parametric investigation of sheet laser bending by experimental and numerical analysis", Advanced Manufacturing Systems and Technology, CISM Courses and Lectures No. 406, Kuljanic, E. (Ed.), Springer Wien N.Y.

5. Casalino G., Ludovico A.D., Ancona A., Lugarà P.M.. (2001) "AISI 304 3-D Laser Forming for Rapid Prototyping". ICALEO 2001 – 20th International Congress on Application of Lasers & Electro-Optics , Jacksonville, Florida, (USA), LIA (Ed.).
6. Montgomery D. C. (1997). Design and Analysis of Experiments 4th ed. Wiley and Sons. Inc. (Ed.)

INVESTIGATION OF THE EFFECTS OF THERMAL AND MECHANICAL FATIGUE OF DIES IN HOT FORGING THROUGH A NEW SIMULATIVE LABORATORY TEST

G. Berti, S. Bruschi, S. Masiero, M. Monti, M. Zimmitti

DIMEG, University of Padova, Padova, Italy

KEYWORDS: Thermo-mechanical fatigue, Die, Forging.

ABSTRACT. The paper presents the design, setting up and conduction of a new simulative test to study the effects of thermo-mechanical fatigue of dies in hot forging. The aim of the test is to replicate the effective operative conditions to which a die is subjected during forging. Accordingly, the test has to reproduce coupled thermal and mechanical cycles as well as high thermal gradients in the central cross section of the specimen. Results in terms of number of cycles to crack initiation are presented and discussed.

1 INTRODUCTION

During service life, hot and warm forging dies are usually subjected to coupled thermal and mechanical cycles which deeply influences their thermo-mechanical fatigue life; both the thermal gradients in the layer near the contact surface with the deforming material and the mechanical cycles can produce initiation and then propagation of cracks [1-2]. The effects of thermo-mechanical cyclic loading can be studied through physical simulation on laboratory specimens, with the aim of evaluating the process parameters influence on thermo-mechanical fatigue life. From the scientific and technical literature, many attempts have been made in studying the fatigue behavior of die materials at high temperature, both at constant temperature (the so-called LCF tests) [3] and cyclically varying temperature (TMF tests) [4-6]. These TMF tests, in particular, introduce thermal cycles of at least 60 s and thermal gradients that are far from those occurring during automatic hot forging of small and medium components. Thermal conditions of hot forging dies during industrial operations consist of fast thermal cycles (10 s including the lubrication phase) and high and not negligible thermal gradients. From this standpoint, a proper test to study the effects of thermo-mechanical fatigue on dies has to be a simulative one, capable to replicate same thermal gradients, same thermal cycles and same effective stress field as in the forging process.

The objective of the paper is to present the design of a new thermo-mechanical fatigue simulative test and to discuss the effects on fatigue life of different thermal cycles. The designed experiment has to reproduce the same heating and cooling phases of the forging cycle, together with the thermal gradient in the central cross section of the specimen, reproducing the effective thermal field inside the die [7]. In the first part of the paper, the designed specimen will be presented, then the applied thermal and mechanical cycles will be introduced and finally the results in terms of number of cycles to crack initiation will be discussed accordingly to sensitivity to thermal parameters.

Published in: E.Kuljanic (Ed.) *Advanced Manufacturing Systems and Technology*, CISM Courses and Lectures No. 437, Springer Wien New York, 2002.

2 THE SPECIMEN

The non-standard geometry of the specimen is represented in Figure 1: it is based on a 10 mm square section bar provided with two threaded ends for the application of the mechanical cycle.

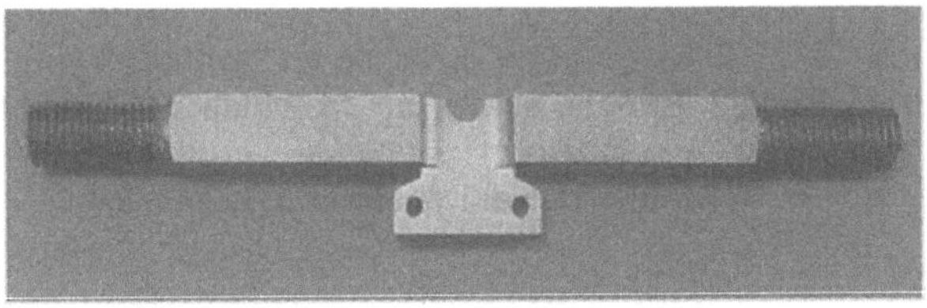

FIGURE 1. Specimen for the simulative TMF test

In the central zone, a 7 mm relief is introduced to increase the thermal gradient, for a more efficient cooling phase, and to obtain stress concentration. The lower part of the specimen, in its central cross section, is provided with an appendix, whose function is to allow the installation of a heat exchanger to obtain the thermal gradient. The copper heat exchanger, shown in Figure 2, is continuously cooled by 15°C water to maintain the bottom of the specimen at lower temperature.

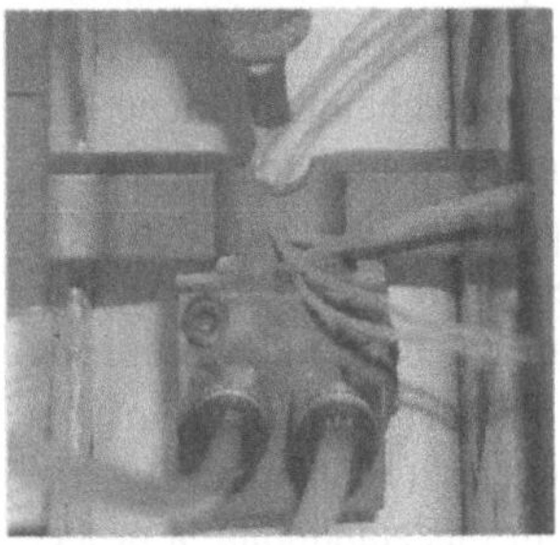

FIGURE 2. Specimen with heat exchanger mounted on Gleeble 3800™ simulator

With this configuration a thermal gradient is obtained from the top to the bottom of the specimen, replicating the thermal conditions of the some millimetre thick layer near the die surface. The material used for the specimens is a hot working tool steel (DIN X38CrMoV51) with the same heat treatment of hot forging dies; its chemical composition is presented in Table 1 while Table 2 reports the material yield stress at different temperatures.

All the experiments have been carried out on the thermo-mechanical simulator Gleeble 3800™, capable to accurately reproduce thermal and mechanical cycles in control of temperature, force, strain and strain rate.

TABLE 1. Chemical composition of X38CrMoV51

C	Si	Mn	Cr	Mo	Ni	V	W
0.36	1.10	0.40	5.0	1.30	-	0.40	-

TABLE 2. Yield strength of X38CrMoV51 at different temperatures

Temperature [°C]	Yield strength [MPa]
400	1100
500	900
600	600
650	400

3 DESIGN OF THERMAL CYCLES

Thermal cycles are designed in order to identify and represent some typical working conditions of the hot forging process, as determined by temperature acquisition during an automatic forging sequence carried out on a pilot plant located at DIMEG laboratories. The investigation of the thermal field, based on an integrated numerical and experimental approach described in details in [7], identifies a maximum temperature of 700°C on the die surface immediately after forging and a minimum temperature of 300°C during the lubrication phase. The Δ temperature on the die surface during forging cycle is then about 400°C. The thermal cycle imposed to the designed specimen consists of a heating phase followed by a cooling one. Heating is obtained by Joule effect in temperature control conditions. Cooling is obtained through compressed fluid spraying (this phase is not properly in temperature control but depends on the fluid characteristics and on pressure). Specimen temperature is controlled by a K-type thermocouple spot welded on the central top surface in the middle of the notch (TC1). A second thermocouple acquires temperature evolution at the bottom of the cross section near the heat exchanger (TC3). Four experimental conditions with different thermal cycles have been carried out in the range of temperatures between 700°C and 230°C. Table 3 summarises thermal cycle parameters of experimental campaign.

TABLE 3. Thermal cycle parameters

	Experimental conditions			
	1	2	3	4
Imposed Tc1 max	700°C	700°C	700°C	700°C

Real Tc1 max	700°C	700°C	700°C	700°C
Imposed Tc1 min	300°C	230°C	300°C	300°C
Real Tc1 min	350°C	230°C	400°C	400°C
Heating time	3 s	3 s	1 s	1 s
Delay time at T max	0 s	0 s	0 s	2 s
Cooling time	2 s	5 s	2 s	2 s

The first experiment consists of a thermal cycle with 3 s heating phase up to 700°C followed by a 2 s cooling phase down to 300°C. The second experiment differs from the first one for the duration of the cooling phase which is of 5 s down to 230°C. In the third experiment, the thermal cycle is faster, being characterized by a 1 s heating phase followed by a 2 s cooling one. The fourth experiment is characterized by a 1 s heating phase followed by a 2 s cooling one with a 2 s dwell time at the maximum temperature: during the soaking time, the average thermal gradient decreases due to TC3 temperature increase, as it can be observed in Figure 3, which reports both imposed and measured thermal cycles.

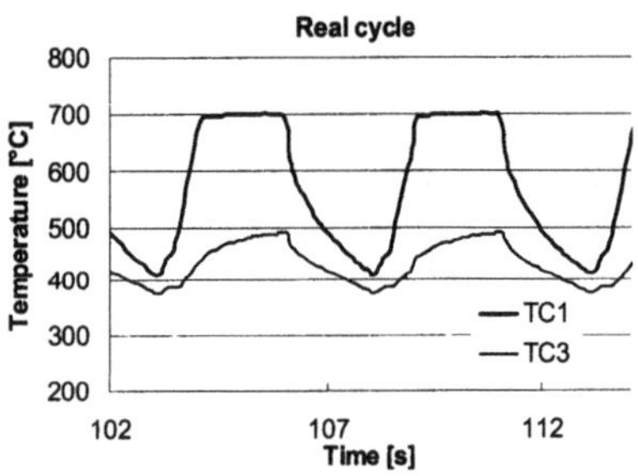

FIGURE 3. Imposed and measured thermal cycles in the fourth experimental condition

4 DESIGN OF MECHANICAL CYCLE

An out-of-phase mechanical cycle is over imposed to the previously defined thermal cycles. Minimum force is applied when temperature at the specimen notch is 700°C, while the maximum value is imposed at the minimum temperature. To identify the maximum load to be applied, the stress field at the specimen notch has to be quantified, as the equivalent stress has not to exceed the material yield strength, to avoid plasticization in the critical zone. The ELFEN™ code has been utilized to perform a fully coupled thermal and mechanical numerical simulation. This analysis allows to evaluate the stress intensification factor due to the notch and, therefore, the parameters of the mechanical cycle. In the FEM analysis, the specimen has been discretized with four-nodes tetrahedral elements and divided into four different volumes, each characterized by different boundary conditions. The load applied to the specimen consists of an uniform stress to surface A (Figure 4), being the mechanical cycle characterized by a triangular shape (max 136MPa for $T=T_{max}$; min 34MPa for $T=T_{min}$). From numerical simulations, the maximum value of the equivalent stress at the notch, accordingly to Von Mises criterion, doesn't exceed 1000MPa at minimum temperature and 250MPa at maximum temperature [8].

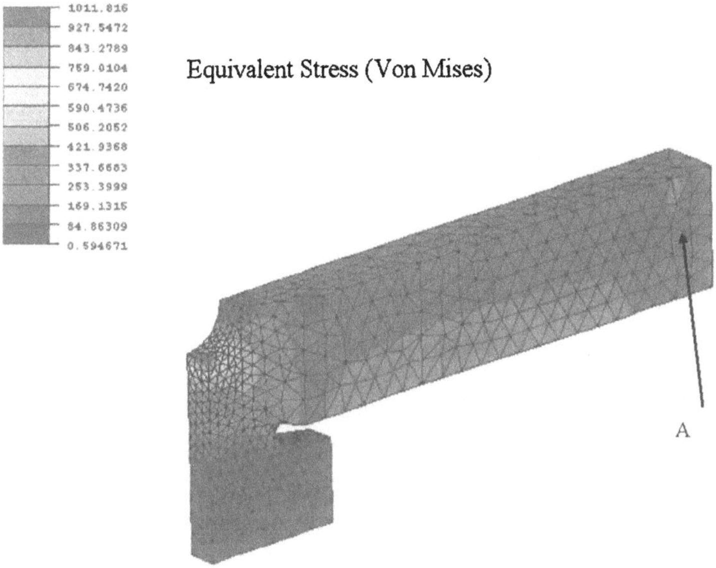

FIGURE 4. Equivalent Von Mises stress field

Comparing with the yield stress values of Table 2, the stress field at the notch can be always considered an elastic one. The main parameters of the mechanical cycle are finally summarized in Table 4.

TABLE 4. Mechanical cycle parameters

Parameter	Max Load	Min Load
σ (surface A)	136MPa	34MPa
σ eq VM (notch)	1000MPa	250MPa
Temperature (notch)	300°C	700°C
Force	13.6kN	3.4kN
$R=\sigma_{min}/\sigma_{max}$	0.25	

5 RESULTS

The experimental procedure consists in the execution of the thermo-mechanical cycles (described in previous sections) until an evident crack is detected at the top of the notch [9]. The number of replications of each experiment is equal to four. Figure 5 represents two examples of the over imposition of thermal and mechanical cycles referred to experimental conditions 2 (a) and 4 (b).

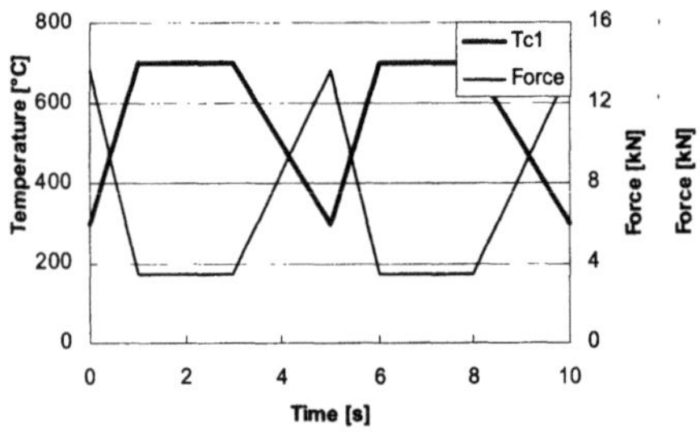

(a) (b)

FIGURE 5. Over imposition of thermal and mechanical cycles referred to experimental conditions 2 (a) and 4 (b)

Every 100 cycles the specimen is removed and inspected with an optical microscope to observe the presence of cracks. The test is stopped as a visible crack is identified. Micro structural analysis at the end of the test is carried out: the specimen is sectioned and chemically attached by NITAL 3 solution to evaluate the type of fracture it has been subjected to.

TABLE 5. Results

Experimental condition	Sample	N. cycles	Tc1 [°C] (range)	Tc3 [°C] (range)	Tc1 mean [°C]
1	1a	5500	680-340	480-370	490
	1b	4400			
	1c	4000			
	1d	4600			
	Mean	4625			
2	2a	4800	690-230	400-260	375
	2b	5200			
	2c	5800			
	2d	5300			
	Mean	5275			
3	3a	6000	700-390	500-380	490
	3b	5800			
	3c	5400			
	3d	6100			
	Mean	5825			

4	4a	3000	700-390	430-360	577
	4b	3100			
	4c	3800			
	4d	2800			
	Mean	3175			

Results in terms of number of cycles to crack initiation are reported in Table 5 and summarized in Figure 6. Accordingly to literature results, it's observed that fatigue life is influenced both by cycle mean temperature and by the soaking time at the maximum temperature. In fact the introduction of the dwell time (fourth experimental condition) drastically increases mean TC1 cycle temperature and reduces fatigue life (Table 5).

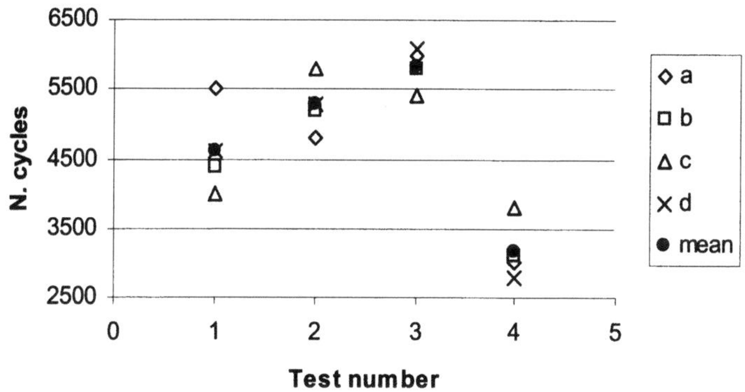

FIGURE 6. Results of the experiments

As regards repeatability of the test, all test configurations lead to final maximum scatter limited to 1000 cycles as shown in Figure 6.

Micro structural analysis shows that in all cases a main crack starts at the top of the notch and grows perpendicular to the notch surface (Figure 7 and 8). Other smaller cracks are observed in the area of the notch as shown in Figure 9. Results of micro structural analysis also confirm that in all cases the fracture is transgranular, proving that the main damage mechanism is mechanical fatigue, being negligible creep phenomena (Figure 10).

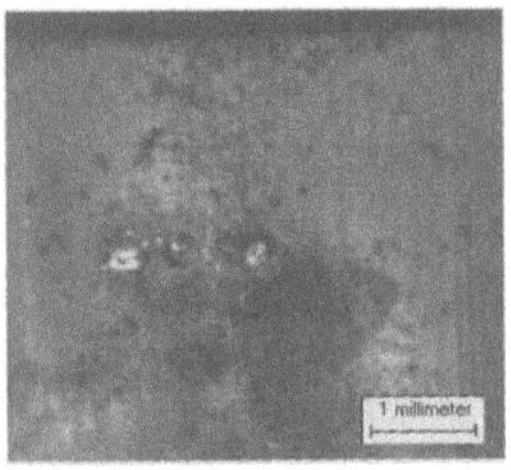

FIGURE 7. Main crack at the notch surface

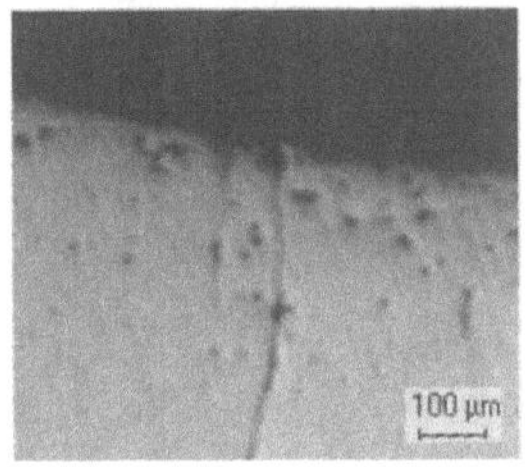

FIGURE 8. Direction of the crack

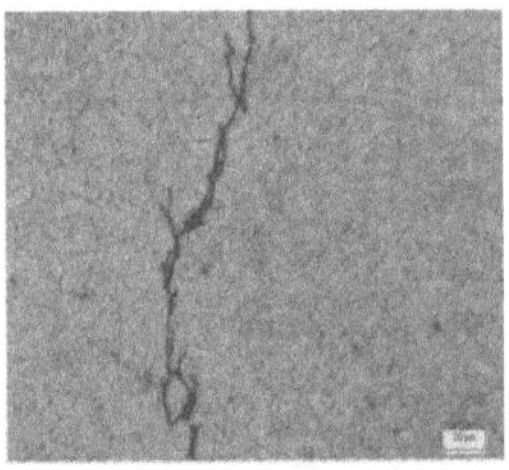

FIGURE 9. Small crack starting from the main crack

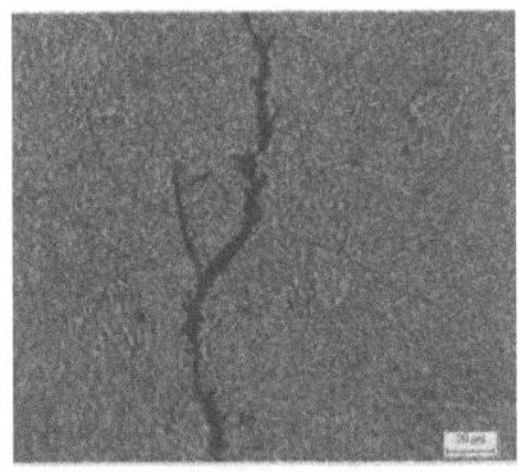

FIGURE 10. Transgranular crack

6 CONCLUSIONS

The test allows controlled replication of the industrial thermal and mechanical cycles, to which the die is subjected, together with the thermal gradients inside the die during forging operations.

The main parameters of the thermal cycle has been varied, showing that an increase of the cycle mean temperature due to a soaking time at the maximum temperature drastically decreases the number of cycles to crack initiation.

These preliminary results are encouraging in future developments of the proposed procedure which is a promising tool for gathering information on thermo-mechanical fatigue behaviour of die material as well as for comparing different operative conditions with the aim to optimise tool life.

REFERENCES

[1] Lange, K., Cser, L., Geiger, M., Kals, J.A.G., (1992), Tool Life and Tool Quality in Bulk Metal Forming. Annals of the CIRP, 41/2, 667-675.

[2] Vasquez, V., Khoerr, M., Altan T., Shivpuri, R., (1996), Determination of Fatigue Properties of Dies Steels for Hot Forging, Trans. of NAMRI/SME, 24/155-160.

[3] Sehitoglu, H., Karasek, M., (1986), Observations of Material Behaviour Under Isothermal and Thermo-mechanical Loading, J. of Eng. Mat. and Tech., 108/192-198.

[4] Boismier, D.A., Sehitoglu, H., (1990), Thermo-Mechanical Fatigue of MAR-M247 : Part I – Experiments, J. of Eng. Mat. and Tech., 112/68-79.

[5] Samrout, H., El Abdi, R., (1998), Fatigue Behaviour of 24CrMo5V-08 Steel Under Thermomechanical Loading, Int. J. of Fatigue, 20/555-563.

[6] Kadioglu, Y., Sehitoglu, H., (1995) Thermomechanical and Isothermal Fatigue Behaviour of Bare and Coated Superalloys, J. of Eng. Mat. and Tech., 117/95-102.

[7] Bariani, P.F., Berti G., Zimmitti, M., (2001), Integrated Approach to the Analysis of Thermo-Mechanical Fatigue of Dies in Hot and Warm Forgino, Proc. of V AITEM, Bari, Italy.

[8] Berti, G., Bruschi, S., Masiero, M., Zimmitti, M., (2001), Thermal and Mechanical Fatigue of Dies in Hot and Warm Forgino: a New Simulative Laboratory Test, Proc. of I PRIME, Sestri Levante, Italy.

[9] Mall, S., Nicholas, T., Pernot J.J., Burgess, D.G., (1991), Crack Growth in a Titanium Aluminide Alloy Under Thermomechanical Cycling, Fat. Fract. Eng. Mat. Struc., 14/79-87.

FLOW MODELLING OF AA 6082 ALUMINIUM ALLOY

C. Bruni, A. Forcellese, F. Gabrielli

Department of Mechanics, University of Ancona, Italy

KEYWORDS: Aluminium alloy, Constitutive equations, High strain rates.

ABSTRACT. The flow behaviour of AA 6082 aluminium alloy has been studied by means of torsion testing carried out at temperatures ranging from 425 to 500°C, with strain rates varying from 1 to 20 s^{-1}. For a given temperature and strain rate, flow curves exhibit a peak followed by flow softening up to fracture. Moreover, for a constant strain, flow stress increases with increasing strain rate and decreasing temperature. In general, the hot ductility shows an increase with increasing temperature up to a maximum, occurring between 450 and 475°C, followed by a decrease to the lowest value experienced at 500°C. An equation relating the hyperbolic sine of flow stress to the temperature modified strain rate was used to describe the flow behaviour versus the working parameters. The dependence of constitutive parameters on strain was also analysed. A very good agreement was found between predicted and experimental flow curves.

1 INTRODUCTION

The deformation processing technology depends on both advanced process modeling techniques, such as the Finite Element Method, and precise characterization of the plastic flow behaviour of the material under conditions of strain, strain rate and temperature, representative of the process. Equations describing the flow stress dependence on strain, strain rate and temperature are required as input to computer process models to predict the deformation history of any material element during forming [1]. Additionally, there are many practical hot forming operations where materials are subjected to high strain rates, so that the knowledge of a high strain rate response of the material is also required.

The aim of the present paper is to analyze the hot forming behaviour of AA 6082 aluminium alloy, by means of torsion tests performed at high temperatures (425 to 500°C) and strain rates (1 to 20 s^{-1}), in order to obtain the material parameters in the constitutive equation describing the dependence of flow stress on strain, strain rate and temperature. The predictions, in terms of equivalent stress-strain curves, are analyzed and compared with experimental data.

2 THE CONSTITUTIVE MODEL

The most used constitutive equation to model the hot working behaviour of metallic materials, proposed by Sellars and Tegart [2], correlates the equivalent flow stress σ to the equivalent strain rate $\dot{\varepsilon}$ and to temperature T as follow:

$$Z = \dot{\varepsilon} \cdot exp\left(\frac{Q}{RT}\right) = A \cdot \left[sinh(\alpha\sigma)\right]^{n'} \tag{1}$$

Published in: E.Kuljanic (Ed.) *Advanced Manufacturing Systems and Technology*, CISM Courses and Lectures No. 437, Springer Wien New York, 2002.

where Z is the Zener-Hollomon parameter representing the temperature modified strain rate, Q is the activation energy related to the deformation mechanisms taking place during the process, R is the universal gas constant, A, n', and α are material parameters. Such relationship applies across a broad range in hot working reducing to a power law at low stresses and to an exponential law at high stresses. Equation (1), that is generally used only for particular points of the flow curves, for example at the peak or at the steady-state regime of the σ-ε curves, has the advantage of simplicity. However, since equation (1) neglects the strain effects, it results in a very coarse approximation of the true behaviour of materials. An alternative method to obtain constitutive equations in hot working conditions is based on a phenomenological approach and consists in determining the relationships giving Q, A and n' as a function of strain. By substituting such relationships into equation (1), the flow stress is obtained:

$$\sigma(\varepsilon) = \frac{1}{\alpha} sinh^{-1}\left[\frac{\dot{\varepsilon}\, exp\left(\frac{Q(\varepsilon)}{RT}\right)}{A(\varepsilon)}\right]^{\frac{1}{n'(\varepsilon)}} = \frac{1}{\alpha} sinh^{-1}\left[\frac{Z(\varepsilon)}{A(\varepsilon)}\right]^{\frac{1}{n'(\varepsilon)}} \qquad (2)$$

Such procedure was utilized in the present work in order to supply a flow stress data bank for the wrought aluminium alloy under investigation. Equation (2) accounts for strain rate and temperature sensitivity by means of Zener-Hollomon parameter. Moreover, the sensitivity of A, Q and n' values to the various restoration mechanisms taking place during deformation processes, permits to develop a more general procedure for calculating flow stress data without choosing a priori which mechanisms may be operating during the deformation process.

3 MATERIAL AND EXPERIMETAL PROCEDURES

3.1 MATERIAL

The material investigated was AA 6082 aluminium alloy, in T6 condition, supplied in form of bars with a diameter of 25 mm; the nominal chemical composition is the following (wt %): 0.99 Si, 0.24 Fe, 0.04 Cu, 0.76 Mn, 0.92 Mg, 0.13 Cr, 0.03 Zn, 0.02 Ti, Al balance.

3.2 EXPERIMENTAL PROCEDURES

The hot formability studies used for determining the constitutive equations were performed by means of torsion tests. Such tests, in comparison with tension and compression, allow an analysis of the flow behaviour in extended ranges of strain and strain rate that are comparable with values of most metalworking operations [3,4]. Torsion tests were carried out until rupture of the samples on a hydraulically powered servo-controlled torsion machine, at temperature ranging between 425 and 500°C, with equivalent strain rates $\dot{\varepsilon}$ at the specimen surface varying from 1 to 20 s^{-1}. The samples were quickly heated by high frequency induction coil with a heating rate of 1°C/s. Temperature control was performed by a K-type thermocouple inserted inside the sample through a special blind hole machined, along the axis direction, at one side of the specimen shoulder. The gauge section of the sample was a solid cylinder with a length L of 10 mm and a radius R equal to 6 mm. The fillet radius between the gauge section and the shoulders was 3 mm. Surface shear stress τ and surface shear strain γ were calculated using the Field and Backofen relationships [5]. Equivalent stresses σ and strains ε were derived form surface shear stresses and strains by the von Mises criterion which has been generally accepted and gives reasonable agreement between results obtained in torsion and in tension or compression in hot working conditions [6]:

$$\sigma = \sqrt{3}\tau = \sqrt{3}\frac{\Gamma}{2\pi R^3}(3+n+m') \tag{3}$$

$$\varepsilon = \frac{\gamma}{\sqrt{3}} = \frac{1}{\sqrt{3}}\frac{2\pi RN}{L} \tag{4}$$

where Γ is the torque, N the revolution number, n $(= d\log\Gamma / d\log N|_{\dot{N},T})$ the work hardening rate and $m'(= d\log\Gamma / d\log\dot{N}|_{N,T})$ the revolution rate sensitivity of torque.

The value of n was assumed to be equal to zero: such hypothesis is true at the peak of the Γ-N curves whilst, in the other parts of curves, the use of value zero for n engenders negligible errors. Conversely, m′ values were calculated plotting torque versus revolution rate $\dot{N}$ at constant temperatures and numbers of revolution N.

The equivalent stresses were corrected for the increase in temperature ΔT due to the deformation heating. The ΔT values were used to offset flow softening effects correcting the equivalent stress by the following equation:

$$\sigma_c = \sigma + \Delta\sigma = \sigma + \Delta T \left.\frac{d\sigma}{dT}\right|_{\varepsilon,\dot{\varepsilon}} \tag{5}$$

where σ_c was the equivalent stress corrected for the deformation heating.
The values of $d\sigma/dT$ were determined from the uncorrected experimental data at low strain ($\varepsilon=0.3$) where the effect of the deformation heating was negligible [7]. In the following paragraphs the corrected equivalent stresses are considered.

4 EXPERIMETAL RESULTS AND DISCUSSION

4.1 HOT FORMABILITY

Flow curves at strain rate higher than 0.01 s^{-1} are not obtained in isothermal conditions. Since, due to high strain rates, adiabatic heating is not negligible [1]. As a consequence, flow curves of AA 6082 aluminium alloy were corrected to take into account of deformation heating. The effect of such correction is shown in Figure 1 where flow curves obtained at 450°C with strain rates of 3 and 10 s^{-1} are plotted both in corrected and uncorrected form.

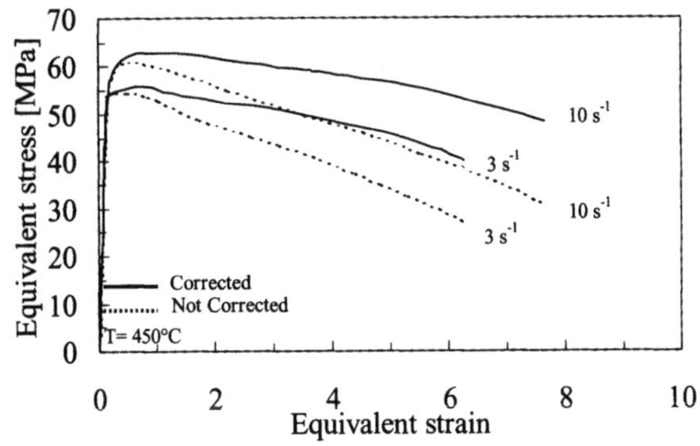

FIGURE 1. Influence of deformation heating on flow curves of AA 6082.

Flow curves confirms that the effect of deformation heating, at given temperature and strain, increases with increasing strain rate; moreover, Figure 1 also shows that, for the aluminum alloy under investigation, the effect of adiabatic heating can be considered negligible at low strains [7]. After the deformation heating correction, an increase up to a maximum value, followed by a limited flow softening up to fracture is apparent (Figure 2).

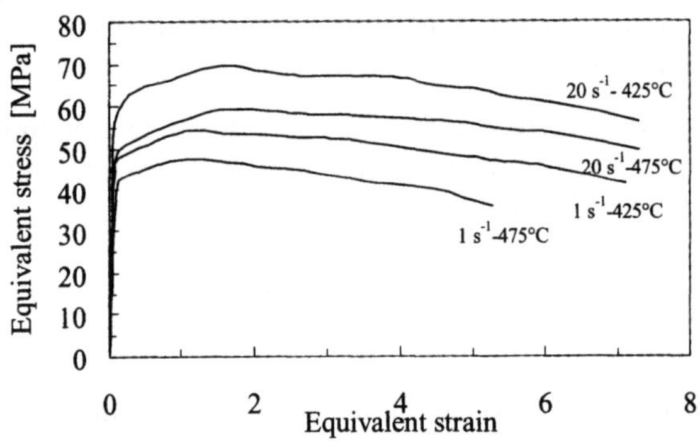

FIGURE 2. Effect of temperature and strain rate on flow curves of AA6082 corrected for the deformation heating.

Furthermore, it is shown by Figure 2 that, at constant strain, flow stress increases when strain rate increases and temperature decreases; such effects can be attributed to the increase of the mobile dislocation velocity with increasing strain rate and to the lower effectiveness of thermally activated processes with decreasing temperature. The presence of the peak in flow stress observed in σ-ε curves can be associated to the occurrence of restoration mechanisms such as dynamic recovery, solute depletion, and particle coalescence [8].

Hot ductility, measured as equivalent strain to failure, at different strain rates, is plotted versus temperature in Figure 3. For $\dot{\varepsilon} \geq 3$ s^{-1}, the equivalent strain to failure increases with increasing temperature up to a maximum, occurring in the temperature range varying form 450 to 475°C, followed by a decrease in ductility that is more marked at higher strain rates. It can be also seen that the ductility versus temperature curves for 10 and 20 s^{-1} are almost coincident.

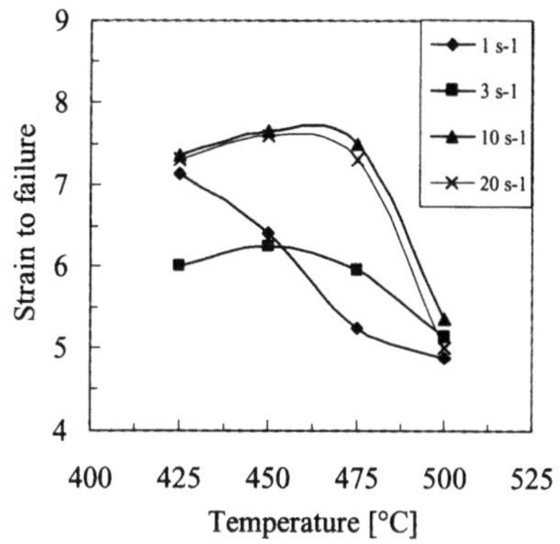

FIGURE 3. Temperature and strain rate effect on the hot ductility of AA6082.

Conversely, a monotonic decrease in ductility, with increasing temperature, at the lowest strain rate (1 s^{-1}), was observed. The behaviour of equivalent strain to failure versus strain rate and temperature, which can be probably attributed to high level of dynamic recovery occurring after certain values of both strain rates and temperatures, is consistent with that experienced by other aluminium alloys [9].

4.2 CONSTITUTIVE EQUATION AND VALIDATION

The flow curves of AA 6082 aluminium alloy were predicted using equation (2). The utilization of such equation requires the determination of the dependence of Q, n' and A on strain.

The activation energy of the deformation process was calculated by means of the following relationship [10]:

$$Q = -R \cdot \left.\frac{\partial \ln \dot{e}}{\partial (1/T)}\right|_{s} = 2.3 \cdot R \cdot \left.\frac{\partial \log \dot{e}}{\partial \log \sinh(\alpha\sigma)}\right|_{e,T} \cdot \left.\frac{\partial \log \sinh(\alpha\sigma)}{\partial (1000/T)}\right|_{e,\dot{e}} \qquad (6)$$

No significant influence of strain on Q was observed. Therefore, it is reasonable to assume, according to the other results [11,12,13], that the activation energy was independent of strain. The plots used to calculate the Q value are shown in Figure 4: a value of about 173 kJ/mol, larger than self-diffusion of aluminium (about145 kJ/mol), was found. Such effect was probably due to the pinning effect of fine particles on moving dislocations and subboundaries [9]. The value of stress multiplier α used in equation (6) was assumed equal to 0.052 MPa^{-1} [14]. However, the activation energy was not significantly influenced by changing the α value.

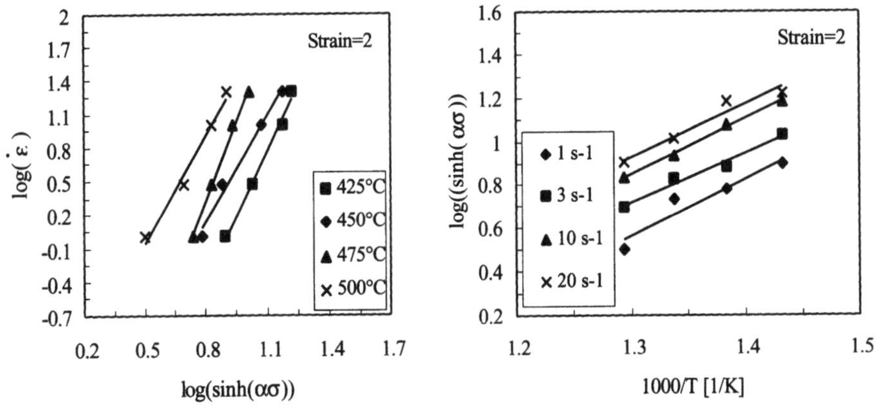

FIGURE 4. Plots used in the calculation of the activation energy of AA 6082.

Contrary to the Q behaviour, in the temperature range under investigation, the material parameters n' and A, representing the slope and intercept of *log(Z)-log(sinh(ασ))* plots (Figure 5), show a strong dependence on strain. Figure 6 shows that n' increases with strain up to a maximum and then decreases; instead, A decreases up to a minimum and then increases with strain. Such behaviours were also observed in other aluminium alloys [15].

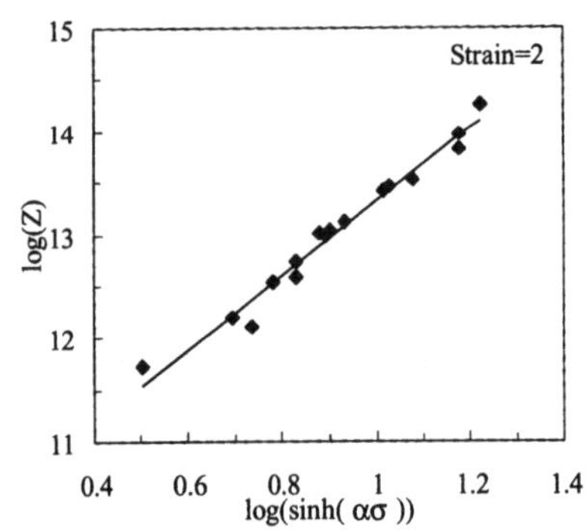

FIGURE 5. Typical correlation between flow stress and temperature compensated strain rate.

Both n' and A versus ε curves were described by an equation given by an exponential association with the material constants obtained by means of the least square fitting. The resulting relationships were the following:

$$A = 3.50\times10^{14}\exp(-7.50(2.80\varepsilon+0.78)) + 1.00\times10^{8}\exp(1.25(\varepsilon+1)) + 2.40\times10^{9}\exp(-3\varepsilon) \tag{7}$$

$$n' = 4.78\exp(-0.12\varepsilon) - 17.38\exp(-15.96\varepsilon) - 0.14 \tag{8}$$

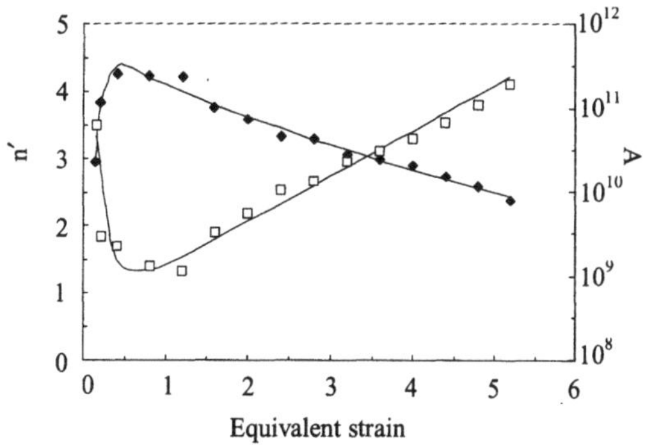

FIGURE 6. Typical plot used to calculate n' and A parameters.

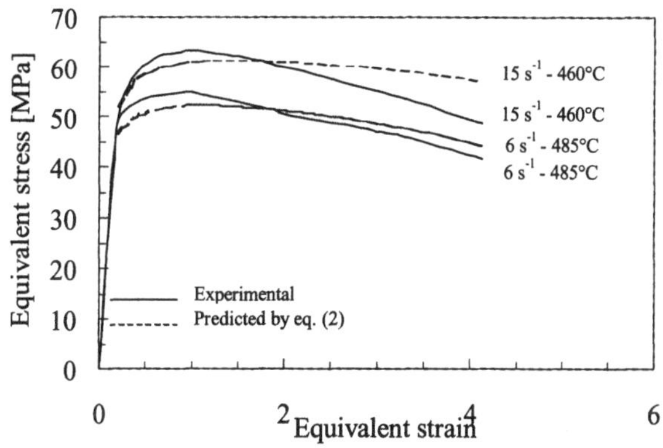

FIGURE 7. Comparison between predicted and experimental flow curves.

The validity of the constitutive model has been proven by comparing predicted flow curves with experimental ones obtained under testing conditions not used to develop the model itself. To this purpose, further torsion tests were carried out at 435, 460 and 485°C, with strain rates of 2, 6 and 15 s^{-1}. Two predicted flow curves, obtained by means of equation (2), using the activation energy derived by equation (6), and the values of n' and A calculated by equations (7-8), are shown in Figure 7, where they are compared with experimental ones. In general, an excellent agreement was found, as shown by the comparison between flow curves at 485°C - 6 s^{-1} even if, in few conditions, such as at 460°C-15 s^{-1} the maximum discrepancy between predicted and measured flow stress values was about 10%.

5 CONCLUSIONS

The hot formability studies of AA 6082 aluminium alloy have shown that the flow behaviour is well described by a constitutive equation of the hyperbolic sine of flow stress versus temperature modified strain rate. The material parameters of such equation are determined versus strain up to an equivalent strain of 5.2; in particular: i) the activation energy is independent of strain and equal to about 173 kJ/mol, ii) the n' exponent increases with strain up to a maximum and than decreases whilst the coefficient *A* decreases with strain to the minimum and then increases. The comparison between predicted and experimental flow curves shows an excellent agreement.

ACKNOWLEDGEMENTS

This paper has been performed in the context of the Research Project "Improving the reliability of FEM simulations of metal working operations through accurate modeling of material rheology, separation and fracture". The financial support of Italian MIUR is acknowledged. The authors also thank Dr. F. Tassi for his help during experimental work.

REFERENCES

1. Charpentier, P.L., Stone, B.C., Ernst, S.C. and Thomas, J.F., Jr, (1986), Characterization and Modelling of the High Temperature Flow Behavior of Aluminium Alloy 2024. *Metallurgical Transactions A.*, Vol. 17/A, 2227-2237.
2. Sellars, C.M. and Tegart, W.J. McG., (1966), La Relation entre la Résistance et la Structure dans la Déformation à Chaud. *Mém. Sci. Rev. Met.*, Vol 63, 731-746.
3. Moore, P., (1968), Methods for Studying Hot Workability: a Critical Assessment. Deformation Under Hot Working Conditions, The Iron and Steel Institute, London, 103-106.
4. Cotner, J. and Tegart, W.J. McG., (1969), High-Temperature Deformation of Aluminium-Magnesium Alloys at High Strain Rates. *J. Inst. Met.*, Vol. 97, 73-79.
5. Fields, D.S. and Backofen, W.A., (1957), Determination of Strain Hardening Characteristics by Torsion Testing. Proc. ASTM, Vol. 57, 1259-1272.
6. Jonas, J.J., Sellars, C.M. and Tegart, W.J. McG., (1969), Strength and Structure under Hot-Working Conditions. *Metall. Rev.*, Vol. 14, 1.
7. De Sanctis, A.M., Evangelista, E. and Forcellese, A., (1997), Assessment of the Forging Conditions of 6061/Al_2O_3/10p using Processing Maps and Stability Criteria . *Key Engineering Materials*, Vols. 127-131, 525-532.
8. McQueen, H.J., Evangelista, E., Forcellese, A., Smith, I.C. and Di Russo, E., (1991), Effects of Solution Treatment or Precipitation on the Hot Workability of 7012 and 7075 Alloys. Proc. Conf. Modelling the Deformation of Crystalline Solids, The Minerals, Metals & Material Society, Lowe, T.C. et al. (Eds.), New Orleans, 281-292.
9. Evangelista, E., Fiorini, P., Forcellese, A. and Gabrielli, F., (1992), Studies of Hot Workability of AA7012 PM Aluminium Alloy. Proc. Int. Conf. Materials Development in Rail Tire, Wing, Hull Transportation, Genoa, Vol. 1, 103-112.
10. Roberts, W., (1984), Dynamic Changes That Occur During Hot Working and Their Significance Regarding Microstructural Development and Hot Workability. Deformation, Processing and Structure. American Society for Metals, Metals Park, 109-184.
11. Sellars, C.M. and Tegart, W.J. McG., (1972), Hot Workability. *Int. Metall. Reviews*, Vol. 17, 1-24.
12. Ashby, M.F. and Frost, H.J., (1975), The Kinetics of Inelastic Deformation above 0 K. Constitutive Equations in Plasticity, A.S. Argon (Ed.), The MIT Press, 117-147.
13. Sherby, O.D. and Burke, P.M., (1966), Mechanical Behaviour of Crystalline Solids at Elevated Temperatures. *Progress in Mat. Sci.*, Vol. 13, 324-390.
14. McQueen, H.J. and Jonas, J.J., (1985), Role of Dynamic and Static Mechanisms in Multistage Hot Working. *J. Appl. Metalw.*, 410-420.
15. Forcellese, A. and Gabrielli, F. (1993), Modelling of the Hot Working Flow Behaviour of some Powder Metallurgy Aluminium Alloys. Proc. 3rd Int. Conf. on Advanced Manufacturing Systems and Technology, E. Kuljanic (Ed.), Udine, 65-72.

SIMULATION OF WELDING CONDITIONS IN PORTHOLE DIE EXTRUSION

L. Donati[1], L. Tomesani[1]

[1] Department of Mechanical Construction Engineering (D.I.E.M.), University of Bologna, Italy

KEYWORDS: extrusion, aluminum alloy, welding conditions, FEM, die stress.

ABSTRACT: In aluminum extrusion with porthole dies, the welding of material seams represents a major manufacturing problem inasmuch it affects the integrity of the final product. The way the problem is treated is yet not consistent in literature, and different quality parameters for welding are still used in practice.
In this work a new parameter for evaluating the welding quality is proposed. Its effectiveness is tested in comparison with other parameters by means of a parametric FEM analysis in which the distribution of pressure and velocity along the welding line is considered. The investigated geometry is a simple two-seams plane-strain die. Many die shapes have been considered, by varying the position and the shape of the leg, in order to understand the sensitivity of the parameters in a wide range of operative conditions. The solutions are also significant for evaluating process load, productivity and die-stresses.

1 INTRODUCTION

In the extrusion of hollow sections of aluminum, the billet is pressed through a die that determines the outer shape and over a mandrel that determines the inner shape. The material must, therefore, flow through a narrow gap. In porthole dies, as well as in bridge and spider dies, the material divides around the core supports which holds the stub mandrel, and then welds together in the welding chamber before passing through the gap. In this way, hollow or semi-hollow sections of great complexity can be produced. The process conditions in the welding chamber must be set at a level to produce perfect welds of the material fluxes and final products free of defects.
The production of good seam joints depends on many process parameters: pressure, temperature, residence time in the welding chamber, surface conditions and presence of contaminants. The complexity of the matter led historically to different ways of evaluating the welding conditions and different guidelines are still used in practice. Meier [1] suggests that both the width

Published in: E.Kuljanic (Ed.) *Advanced Manufacturing Systems and Technology*, CISM Courses and Lectures No. 437, Springer Wien New York, 2002.

and height of the welding chambers should be equal to 6 to 8 times the wall thickness of the hollow shape in the area under the support. Akeret [2] recommends that the local extrusion ratio in the final step has to be on the order of at least three for any part of the shape. More recently, Plata and Piwnik [3] proposed a quality factor as the integral of hydrostatic pressure for the time of contact on a definite path. Finally, the authors [4] stated that a modified parameter with material flow could also give interesting informations about the welding quality.
In particular the last two methods can be successfully used as a guideline in die design, when finite element simulation (FEM) of the process is performed. In fact, the knowledge of pressure and velocity distribution in the welding chamber can be utilized in order to evaluate a global factor related to the welding quality. On the other hand, these factors can be used, at the present time, just to simplify the comparison between different die geometries, and no links exist between these factors and the resistance of the weld itself. Only appropriate calibration tests on specific materials can give this knowledge and much work still needs to be done in this field.
In order to confront the validity and the effectiveness of the welding quality factors, the simulation of a hollow section extrusion process was carried out by means of the DEFORM finite element code; different welding conditions were produced in the chamber, by varying the shape and position of the legs.
In order to simplify the analysis, the complex 3D geometry of a commercial hollow product was replaced by a simple 2D plain strain geometry, where a single leg holds no mandrel and thus no hollow section is produced. Nevertheless, this condition is quite close to the real one and can be conveniently used in a confrontation analysis. All simulations were run considering heat generation and transfer.
The adoption of a FEM code with a lagrangian approach allowed to investigate the non-stationary part of the process. Here, much problems arise from the overloading of the legs, which is a major cause of die rupture. The evaluation of the solutions emerging from different angles on the legs allows to optimize their geometry with respect both on the process load and on the die stresses.

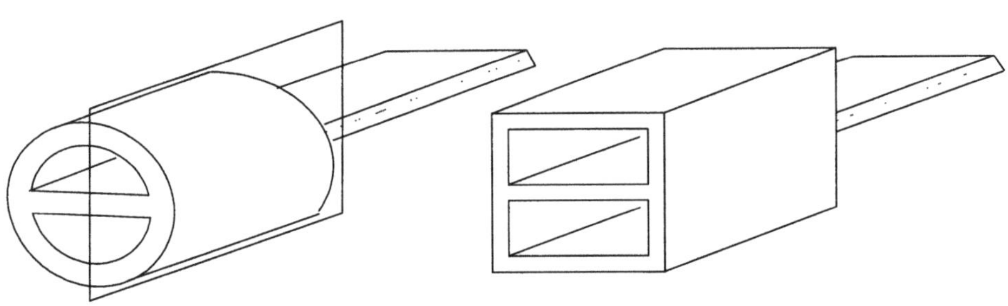

FIGURE 1. Schematized die (left), plain-strain die (right).

2 THE INVESTIGATED PROBLEM

The extrusion of AA 6060 aluminum alloy for hot deformation, by means of a simple die with one leg and a rectangular non-hollow bearing, was considered. This die (Fig.1, left) determines a weld plane in the middle section and has been chosen as base reference for the work, due to its great simplicity, with a reduction ratio similar to the usually ones. This allows to simulate real extrusion conditions, together with all the problems correlated with the formation of welding, in a low computational time. The same die is represented in a plain-strain version for FEM simulation on the right of fig.1. In fig. 2 the die section is represented.

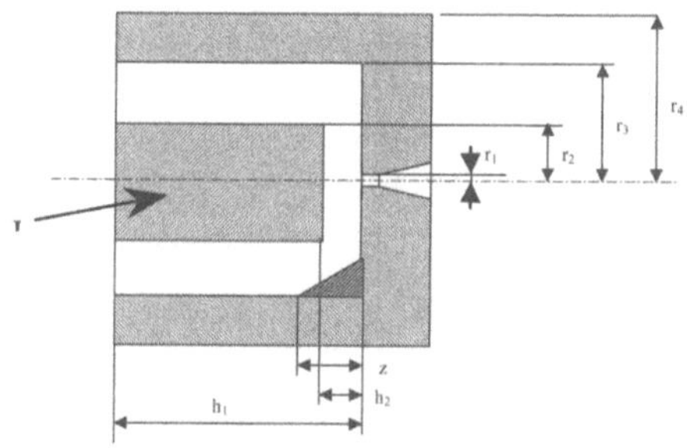

FIGURE 2. die

The influence of the principal geometrical variable were investigated: the leg shape and the chamber height. The leg shape (and in particular the leg main angle) is of primary importance owing to the fact it affects the shape of the filling channel within the die. In fact, a diverging leg determines converging flow channels for the material (fig.3-left) and, at the opposite, converging legs determine diverging channels (fig.3-center). All the three considered sections have the same area.

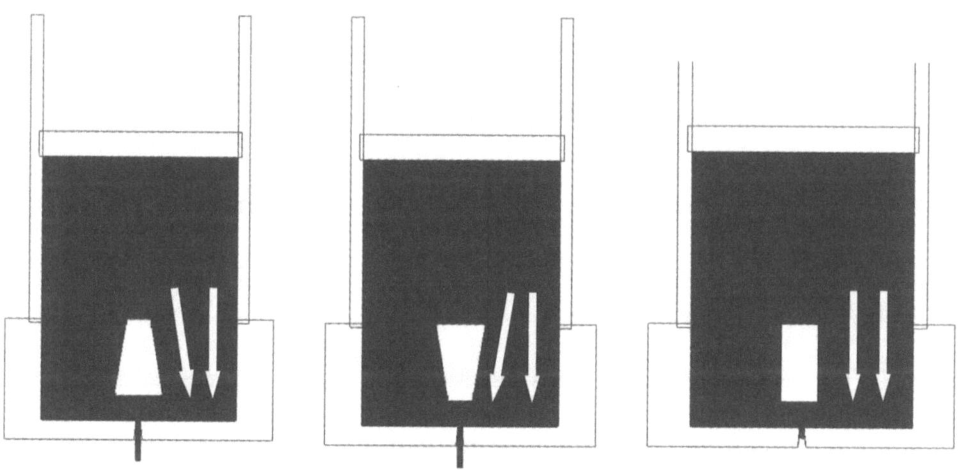

FIGURE 3. Investigated leg shapes (convergent, divergent and square channels).

It is interesting to understand the stress state on the leg, both in the non-steady die filling stage and in the steady state, when the material begins to flow in the final gap.
The height of the welding chamber (h_2 in Fig.2) was also considered because it affects the contact time of the two seams and the pressure conditions before the material is extruded. Clearly, the smaller the chamber height, the smaller is the whole die and its costs. The process is characterized by a typical area reduction r_3/r_1 (97%) that is hold constant throughout the simulations. The other geometrical parameters were also considered as constants.

3 THE FEM SIMULATIONS

The DEFORM code in Lagrangian formulation was used. Half geometry was considered due to symmetry and the portion of the axis after the leg was the considered line of welding. The deforming material (an AA 6060 aluminum alloy for hot deformation) constitutive equation was considered as follows [5]:

$$\dot{\bar{\varepsilon}} = A\left[senh\left(\alpha\bar{\sigma}\right)\right]^{n} \exp[-Q/RT]$$

$$\alpha = (0.03840 + 0.00318\ln\bar{\varepsilon})[MPa^{-1}]$$

The preheating temperature of the billet was 450°C. Heat generation and transfer toward the die system leaded to an exit temperature of nearly 520°C, as evidenced in the production of similar profiles. The friction between the die and deforming material at these process conditions can be considered as adhesive, so that a unitary friction factor was used. The ram speed was 8 mm/sec. The die material, an AISI H-13 hot work steel (X40CrMoV5), was considered for die stress analysis after simulations were run in a perfectly rigid tool condition.
As output parameters of the FEM simulation, the following were chosen:

1. For the process: maximum load, load-stroke diagram in non-steady state conditions;
2. For the die: maximum load, steady-state load, maximum tensile stress on the leg;
3. For the welding quality: normal pressure (P), shear stress (σ) and velocity (V) of the deforming material were used to evaluate two different quality factors, the first proposed by Plata and Piwnik [3], the second by the authors.

$$Q = \int q = \int \frac{P}{\bar{\sigma}} dt = \int \frac{P}{\bar{\sigma}} \frac{dl}{V} \quad (1) \qquad K = \int k = \int \frac{P}{\bar{\sigma}} V dt = \int \frac{P}{\bar{\sigma}} dl \quad (2)$$

4 RESULTS AND DISCUSSION

Except the leg stress analysis, all the values have been taken in the steady state conditions, with the same profile length. The plotted results are referred to the shape of the channel emerging from the leg angle (in the same sequence formerly proposed: convergent, divergent and square channels).

Dead zones

The importance of dead zones on the product quality is well explained in literature [6]. In fact, impurities are first captured in this zone and then they tend to converge to the butt. Thus, the migration of impurities is one of the major cause of defects in the product. A minimization of dead zones could be, then, used for improving profile quality.

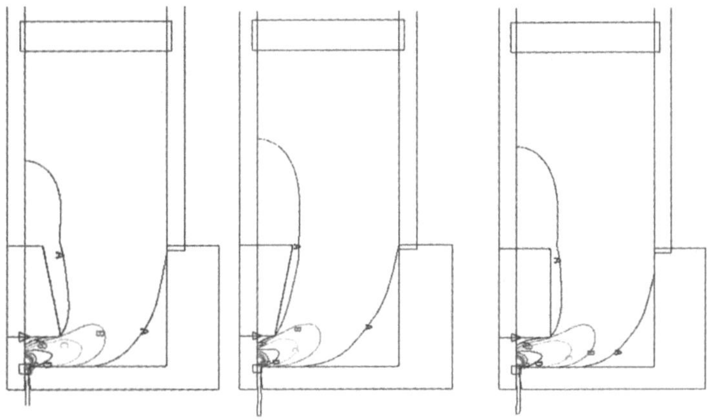

FIGURE 4. Dead zones formation

Three dead zones were evidenced by the simulations (Fig.4): upstream of the leg, downstream of the leg and at the die corner, as usual.
The first kind of dead zone is smaller in convergent channels, due to the smaller front area of the leg. It can also be noticed that thicker dead zones develop on the leg side in this case. The optimization of the leg shape can be used in order to reduce these phenomena.
The second dead zone extends on the back of the leg, with a peak dimension of 4 mm on the die axis when the chamber height is 20 mm. This dead zone is very important for welding quality evaluation, as will be explained later.
Concerning the third kind of dead zone, it can be stated that, in general, channels of constant grade of deformation will create in the material flow. Thus, minimal dead zone dimensions are obtained at the die corner when convergent channel are used; this zones can be further reduced if dies with bottom angles are utilized.

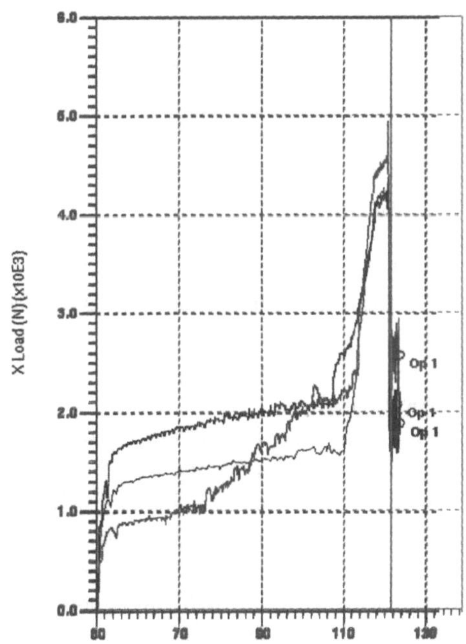

FIGURE 5. loads on the leg in the first stroke

Die stress

In fig.5 the load on the leg is plotted as a function of the ram stroke for the first billet of the production. It emerges that:

1) the maximum load in the die filling stage is around twice the steady state load; tensile stresses on the back side of the leg must be evaluated in this condition. The load decreases to the steady state value when the material fills the welding chamber and supports the leg from the back;
2) greater maximum load are produced in the converging channel solution; in the case studied, the maximum load with converging channels is 10% greater than the square and 7% greater than the divergent.
3) greater steady state loads are also produced with converging channel, being 44% greater than the diverging and 30% greater than the square.
4) the greater the front leg area, the greater the yielding load and the energy requirements.
5) lower maximum tensile stresses are produced with square legs (12% higher with diverging channels and 5% higher with converging ones
6) lower steady state tensile stresses are produced with divergent legs (2% higher with square channels and 24% higher with converging ones.

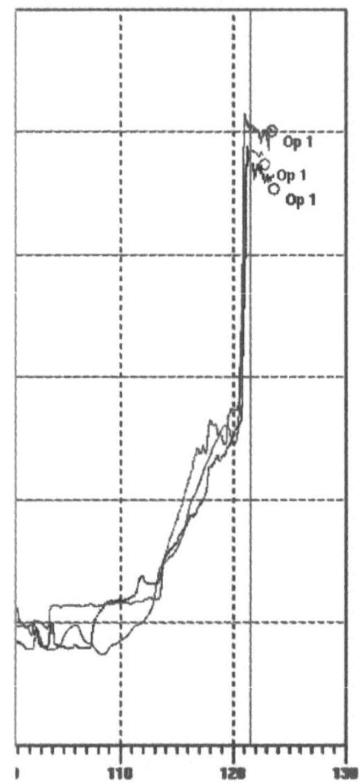

FIGURE 6. Load-stroke diagram

The differencees in tensile stresses on the leg produce, in the steady state condition, much greater differences in die life, since creep phenomena and crack grow are very sensitive to stress reduction.

Process load

In fig. 6 the three load-stroke diagrams are represented. It can be pointed out that the diverging channel solution is the one with the lowest maximum steady state load, nearly 10% lower than the convergent channel solution. Square channel produce a load intermediate between the others. The reason for this difference arise from the different material conditions in the channel (see also fig.3). In the converging solution, the material is forced to continuously deform: so, its flow stress is quite high throughout the channel (and consequently the shear and friction stresses). At the opposite, in the diverging solution the deformation is given at a single step, because the leg has the maximum section at the top. The material is consequently harder in a

much narrower zone (at the beginning of the channel), then it softens throughout the rest of the path before the welding chamber is reached. This considerations apply also to the above mentioned loads on the legs, where the diverging solution gives the lowest loads.
The energy required for the die filling stage is nearly the same in the three cases examinated; in the production stage the differences in energy are proportional to the differences in load, as explained earlier.

Welding quality

The pressure distribution on the welding line is evidenced in fig.7 for different leg angles in a 20 mm chamber height. The velocity fields are represented in fig.8. It can be seen that a peak of pressure (somhow influenced by the leg angle) is present near the leg contact. In this zone the material velocity is almost zero, indicating a dead zone behind the leg. The q parameter distribution (fig.9) is non-zero just in this narrow section, where times of contact tend to infinity (see eq.1).

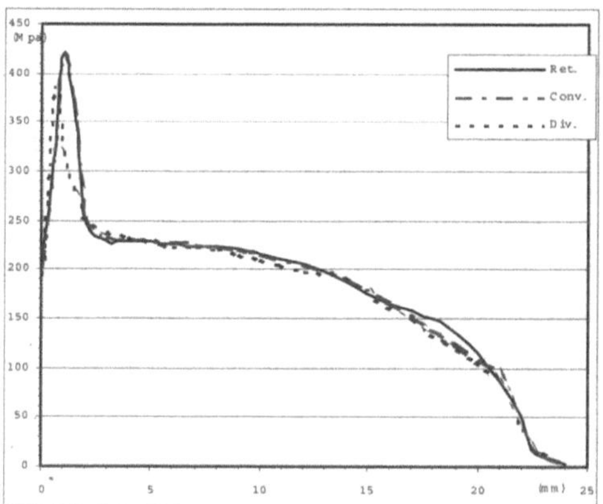

FIGURE 7. Pressure in welding

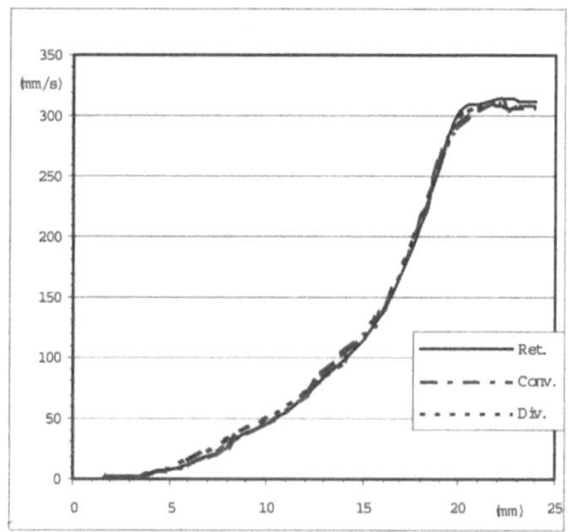

FIGURE. 8. Velocity in welding chamber

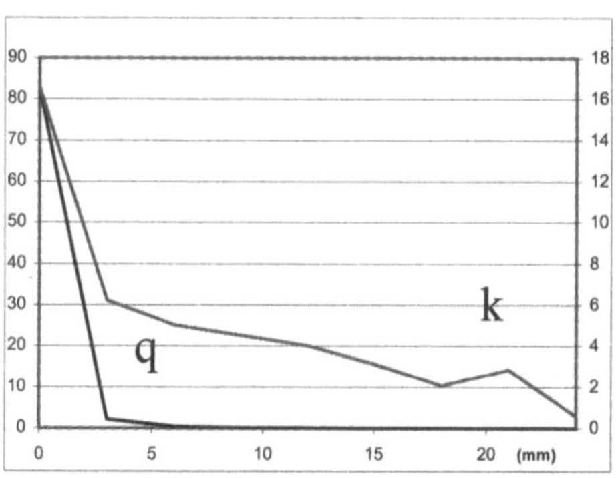

FIGURE 9. Distribution of q and k parameters

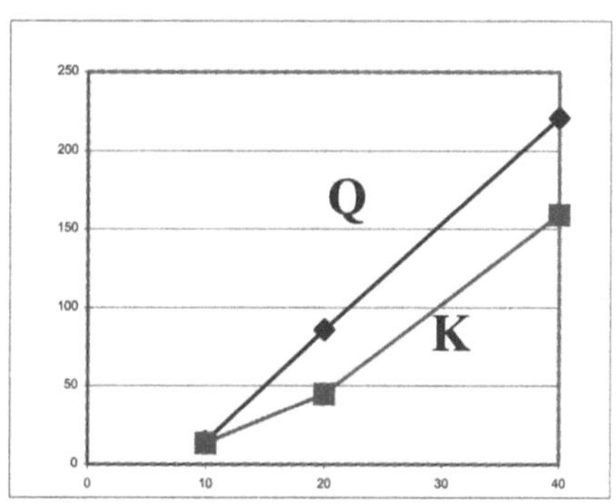

FIGURE 10. K and Q parameter changing h_2

In order to correct this unrealistic behaviour, the k parameter (eq.2) also considers the speed (the material flow) as a factor for judging the quality of the weld. The K distribution is then less steep along the welding line. In other words, the Q parameter quantifies just the dead zones, while the K parameter informs about the whole welding path.
In fig. 10, Q and K parameters are plotted with respect to the chamber height, which is known as a main factor of welding quality. Both parameters are well suited to inform about welding quality, although K evidences a better approach to zero for very short chamber heights.
The effect of leg angle on welding quality was also studied; the two investigated quality parameters gave diverging informations. The Plata criterion stated that converging channels improve welding quality, while the other predicts no practical changes.

5 CONCLUSIONS

The extrusion of a simplified hollow section has been investigated by means of a 2D parametric FEM analysis; the formation of dead zones, the stresses on the legs, the process loads and the welding conditions have been evaluated with respect to the position and shape of the legs.

- Dead zones are quite deeply influenced by the leg shape, both on the top of the leg and on the bottom of the die;
- Maximum loads on the legs are about twice the steady state loads; the lowest tensile stresses in the steady state are obtained with square legs.
- Lower process loads (up to –8%) are obtained with diverging flow channels. Corresponding differences are expected for energy.
- Welding conditions are deeply affected by chamber height (welding quality increases with chamber height) and much less by leg angle. A modified parameter for welding quality has been proposed, more sensitive to material fluxes in the whole chamber

ACKNOWLEDGEMENTS

This work has been performed with funding from MURST (Italian Ministry for University and Scientific Research), and with the collaboration of Compes S.p.A (die manufacturer). Thanks also to ing. Luca Guerra for the help in process simulations.

REFERENCES

1. Meier B., “Die design and material flow”, Baumjarten Journal, DGM informationsgesellschaft, Oberursel, Germany (1989), pp.171-178 (in german)
2. Akeret "Extrusion welds - quality aspects are now center stage" Proceedings fifth Int. Aluminum Extrusion Technology Seminar, ET 1992, vol.2, pp.319-336
3. Plata M. and Piwnik J., “Theoretical and experimental analysis of seam weld formation in hot extrusion of aluminum alloys”, Proc. seventh international aluminum extrusion technology seminar ET 2000, 1, pp.205-211
4. Donati L. and Tomesani L., Analysis of material flow and welding in aluminum extrusion of hollow sections, V AITEM Conference, Bari, September 2001, pp. 678-688
5. Verlinden B., Suhadi A., Delaey L., "A generalized constitutive equation for an AA6060 Aluminium Alloy", Scripta Metallurgica e Materialia, Vol.28, 1993, 1441-1446
6. Lefsta M., Reiso O., Johnsen V., "Flow of the billet surface in aluminum extrusion" Proceedings fifth Int. Aluminum Extrusion Technology Seminar, ET 1992, vol.2, pp.503-517

TOWARDS A GENERAL QUALITY INDICATOR FOR ADAPTIVE FEM SIMULATION OF SHEET FORMING PROCESSES

M. Strano, L. Carrino

Department of Industrial Engineering, University of Cassino, Italy

KEYWORDS: Sheet Metal Forming, FEM, Adaptive Simulation.

ABSTRACT. The concept of adaptive loading by FEM simulation (briefly referred to as Adaptive Simulation, AS) has been recently introduced. The AS method adaptively adjusts the loading paths as the simulation proceeds, reacting to the instantaneous results of the simulation itself, i.e. by monitoring the values of one or more state variables. A state variable is an indicator of the quality of the part undergoing the deformation. The formulation of state variables should be independent on the specific forming process. In this paper, a general quality indicator is proposed, which can be used as a state variable for the AS of different sheet metal forming processes. The paper includes two main sections. The first presents a brief review on possible state variables suitable for the AS and provides a justification of the proposed state variable. In the second section, the proposed approach is described and evaluated with two numerical examples.

1 INTRODUCTION

DESIGN OF SHEET FORMING PROCESSES BY FEM. The Finite Element Method is today a well-established technique for numerical simulation of sheet forming processes. The industrial success of the method is manly due the rapidity and reliability of formulations that use shell elements and explicit time integration schemes. In fact, numerous software houses (especially French and North American) offer commercial packages, suited for sheet forming applications. Several software houses are also developing and distributing optimization packages, to be used in combination with the FEM code, able to help and guide the user towards the best process design.

Optimization techniques are usually sequential, i.e. based on the interpolation of results obtained by several simulation runs. Generally, sequential optimization techniques are best suited for the geometric design of tools and dies. However, they can also be applied for the selection of the loading paths, i.e. for the optimization of time-dependent process control variables (velocities, forces, pressures, etc.). One of the most recent and convincing examples is given in [1]. However, sequential optimization methods show some limitations when used for the determination of loading paths in forming processes where control is a critical issue and where the process window is not very large. In fact, the main problems of sequential optimization techniques when dealing with control parameters are the following:

- an initial guess for the loading curves is required, which influence the final result;
- usually a high number of simulations is necessary before convergence is reached.

THE ADAPTIVE SIMULATION APPROACH. For the reasons mentioned above about the problems involved in sequential optimization, some authors have recently introduced the concept

Published in: E.Kuljanic (Ed.) *Advanced Manufacturing Systems and Technology*, CISM Courses and Lectures No. 437, Springer Wien New York, 2002.

of adaptive loading by FEM simulation, which is especially useful when the process control is a critical issue, as it happens in the Tube Hydroforming process [2], or in the deep drawing process with blankholder force control, or in the superplastic forming process.

Adaptive loading by FEM simulation (sometimes briefly referred to as Adaptive Simulation, AS) is a procedure designed to adjust the loading paths (e.g., the internal pressure vs. time in tube hydroforming) as the simulation proceeds, according to the instantaneous results of the simulation itself. The main goal of AS is not to optimize the loading paths, but rather to quickly determine a "good" feasible solution for difficult to form parts, where the process control is a critical issue.

The instantaneous selection of process parameters can be done either by artificial intelligence tools (as in [2]) or by the adaptive control theory (as in [3]). In the latter approach, the adaptive simulation method is designed to select the values of a vector of control variables $\underline{u}(t)$, by detecting and controlling the values of a vector of state variables $\underline{x}(t)$. A linear dynamic model (Eq. 1) is necessary to estimate the relation between process parameters $\underline{u}(t)$ and state variables $\underline{x}(t)$. The coefficients of the transformation matrix $\underline{\underline{\Gamma}}(t)$ are adaptively estimated at each control step as the simulation proceeds (1-step ahead) thanks to a least square method with a forgetting factor. The problem is solved by optimizing the quadratic cost function given in Eq. 2, where the weights of the matrix $\underline{\underline{W}}(t)$ are increasing with the time t, in order to give higher importance to the state variables as the simulation gets closer to the end of the process. In Eq. 1 and 2, t is the time at the beginning of the control step and Δt is the duration of the control step.

$$\underline{x}(t+\Delta t)=\underline{x}(t)+\underline{\underline{\Gamma}}(t)\,\underline{u}(t) \tag{1}$$

$$c(t)=\underline{u}(t)'\,\underline{u}(t)+\underline{x}(t+\Delta t)'\,\underline{\underline{W}}(t)\,\underline{x}(t+\Delta t) \tag{2}$$

This control problem can be defined as a deterministic LQ optimization problem, with terminal state costing and a 1-step ahead adaptive estimation of coefficients. At each time control step, the optimal solution of the stated problem is:

$$\underline{u}(t)=-[\underline{\underline{I}}+\underline{\underline{\Gamma}}(t)'\,\underline{\underline{W}}(t)\,\underline{\underline{\Gamma}}(t)]^{-1}\,\underline{\underline{\Gamma}}(t)\,\underline{\underline{W}}(t)\,\underline{x}(t) \tag{3}$$

Depending on the forming process being controlled, the user might decide that some of the coefficients Γ_{ij} of the dynamic linear model in Eq. 1 are null; the others are self-estimated by the model. The most critical point of the proposed procedure is the selection of the weights W_{ij} in matrix $\underline{\underline{W}}(t)$. In the following Section 3, some practical considerations on the selection of W_{ij} will be given. Both state and control variables should be non-dimensional and, where possible, normalized, in order to facilitate the selection of weights.

Analysis of robustness and stability of the proposed system depends on W_{ij} values and it is quite difficult. Since in practical cases it is necessary to add technological lower $\underline{u}_{low}(t)$ and upper $\underline{u}_{upp}(t)$ bounds on the $\underline{u}(t)$ values (Eq. 4), the analysis of the system becomes virtually impossible in a formal way, and it must be conducted empirically.

$$\underline{u}_{low}(t)\leq\underline{u}(t)\leq\underline{u}_{upp}(t) \tag{4}$$

STATE VARIABLES IN ADAPTIVE SIMULATION. A state variable is usually an indicator of the quality of the part undergoing the deformation. There are two different kinds of state variables:
- positive quality indicators, whose value secures a correct material flow (e.g. the thickness uniformity);

- risk indicators, that indicate the proximity to a condition of risk (e.g. a damage function or a wrinkle indicator [4]).

Besides, state variables can be direct or indirect. For instance, the thickness uniformity is a direct indication of quality, while the nodal velocity or force norm can indirectly serve as positive quality indicator. State variables can also be classified as local or global. As an example, the thickness uniformity can be expressed both as a local variable, when it is evaluated by small clusters of elements, or as a global variable, when it is evaluated considering all the elements in the work-piece. Local state variables are obviously more accurate, but they have to be evaluated element-by-element or node-by-node, and they are not suitable for adaptive loading, since they would generate too long a vector $\underline{x}(t)$. Finally, state variables can be dependent or independent on the specific forming process under FEM analysis. It is straightforward that, in order to favor a full success of the AS methodology, state variables should be as general and reliable as possible.

In this paper, a general approach is proposed which can be used for the adaptive loading by FEM simulation, by means of an adaptive controller, as the one described above. The method is based on a state variable or, more precisely, on a vector of state variables, suitable for the adaptive simulation of different sheet and tube metal forming processes.

The paper includes two further main sections. The next section (Section 2) briefly presents a review of possible state variables suitable for the AS and provides a justification for potential state variables suited for Adaptive Simulation. In Section 3, the proposed approach is described and evaluated with two numerical examples.

2 STATE VARIABLES FOR SHEET FORMING PROCESSES

There is available a very large number of quality indicators, to be possibly used as state variables. A complete literature review is beyond the scope of the present paper, and only a brief summary of the main possible indicators will be given, with a few representative citations. As already seen, state variables can be either positive quality indicators or risk indicators.

POSITIVE QUALITY INDICATORS. The only direct positive quality indicator, non dependent on the specific process, is the uniformity of thickness distribution. Indirect quality indicators can be obtained when controlling nodal or elemental variables (either geometric, kinetic or dynamic). Examples of process dependent indirect quality indicators are the nodal velocity in tube hydro-forming [5] and the flange thickness strain or the punch force in deep drawing. An indirect indicator, non dependent on the specific process, could also be built using some function of the elemental energies (total, internal, kinetic).

RISK INDICATORS. Risk indicators (or defect detectors) relate to possible defects. The risk of a defect is signaled quantitatively, and usually before a failure occurs; is some cases defect detectors only indicate the occurrence of a failure. The most common possible defects or failures in sheet metal forming processes are:

- defects caused by elastic phenomena after forming (springback, stress cracking);
- deviating shapes caused by insufficient calibration in case of force controlled processes or by an incorrect design of blank dimensions;
- deviating shapes due to excessive compressive stresses (wrinkling, puckering, buckling, etc.);
- deviating shapes due to directional differences in the plastic properties of rolled metals (earing);

- necking and fracture;
- surface defects (scratching and tool marks, orange peel, galling, Luders lines).

Only a subset of the listed defects can be considered for use in AS, i.e. only those defects that can be predicted by explicit FEM analysis during the process (although approximately or indirectly): deviating shapes, necking and fracture.

During the process, the shape can deviate for several different reasons (plastic instability, anisotropy, incorrect design of process parameters, etc.). In most practical cases, commercial FEM codes are able to reproduce these geometrical deviations, although with a certain degree on inaccuracy, especially in the case of plastic instabilities. Prediction of springback is obviously not possible during the simulation of the process (and somehow inaccurate also after the process). Surface defects cannot be estimated, unless special purpose numerical models or empirical indicators are used.

For all these reasons, it appears that a suitable vector of state variables could be one that incorporates variables indicating the amount of geometrical deviation (due to any possible cause: wrinkling, buckling, earing, etc.) and variables indicating the risk of necking or fracture.

STATE VARIABLES SELECTED FOR THE PROPOSED MODEL. The deviation of the forming blank from the desired shape can be effectively expressed as the averaged distance d between the die and the blank itself [1]. Another way of qualitatively indicating the risk of a deviating shape, especially due to wrinkling, is to control the increase of the workpiece surface area Sa [6].

On the other end, there are several different ways of predicting the risk of necking in sheet metal forming. However, these methods (FLDs, FLSDs, microstructural or and macroscopic damage functions, etc.) all involve costly experiments in order to determine the forming limits.

For these reason, in the proposed approach, no state variable for necking is considered, whereas the risk of fracture is reduced by designing the control variables $\underline{u}(t)$ in such a way that the controller tries to keep them at a minimum, as it tries to minimize Eq. 2. In other words, for each process, control variables must be selected so that their minimization reduces the risk of fracture. For instance, in deep drawing with blankholder force control, the control variable involving the blankholder force (BHF) should be built in such a way that the BHF is kept to the minimum possible level. The examples given in the next section 3 will further explain the proposed model.

With the given considerations, Eq. 1 can now be transformed into Eq. 5:

$$\begin{cases} x_1(t+\Delta t) = x_1(t) + \Gamma_{11}u_1(t) + \Gamma_{12}u_2(t) \\ x_2(t+\Delta t) = x_2(t) + \Gamma_{21}u_1(t) + \Gamma_{22}u_2(t) \end{cases} \tag{5}$$

Where: $x_1=(Sa-S_i)/(S_f-S_i)$, $x_2=(d-th/2)/(d_i-th/2)$, S_i is the initial workpiece surface area, S_f the final workpiece surface area, d_i the initial distance die-workpiece, th the initial sheet thickness. Consequently, Eq. 2 is transformed into Eq. 6:

$$c(t) = u_1(t)^2 + u_2(t)^2 + w_{11}u_1(t)^2 + w_{22}u_2(t)^2 + w_{12}x_1(t)x_2(t) + w_{21}x_1(t)x_2(t) \tag{6}$$

In the following, the method is tested by choosing $w_{12}=w_{21}=0$, $w_{11}=w(1-t/T)$, $w_{22}=w\ t/T$, where T is the final time. By this way, the importance of x_1 (the distance die-workpiece) is minimum at the beginning of the process and maximum at the end of the process, whereas the importance of x_2 (the surface area) is maximum at the beginning of the process and minimum at the end of the process. Given this cost structure, the only parameter yet to be selected is the coefficient w.

3 EVALUATION OF THE PROPOSED STATE VARIABLES

TUBE HYDROFORMING. As already previously stated, the role assigned to the first term of the cost function Eq. (6) is to ensure thickness uniformity and to prevent fracture. This goal can be achieved by minimizing pressure increments ΔP while maximizing axial feed increments Δf. The control system can therefore be designed with the following control variables:

$$\begin{cases} u_1(t) = \dfrac{\Delta P(t) - P_{\max}/n_c}{P_{\max}/n_c}; & P(t) \geq 0 \\ u_2(t) = \dfrac{\Delta f(t) - f_{\max}/n_c}{f_{\max}/n_c}; & \Delta f(t) \geq 0 \end{cases} \tag{7}$$

where n_c is the number of control intervals, P_{max} is the calibration pressure at the end of the process (estimated as a lower limit), f_{max} is the maximum possible axial feed increment (estimated as an upper limit). While Eq. 6 forces u_1 and u_2 to be minimized, the structure given by Eq. 7 ensures that P and f are kept as close as possible to, respectively, their lower and upper limits.

Several tests of the method on different tube hydroforming (THF) examples showed that a reasonable value for the weight coefficient w is 100. As an example, in Fig. 1, 2 and 3 the data and the results of an adaptive simulation run are shown. In Fig. 2b, it is clear that slight wrinkling occurs during the process, but wrinkles are flattened out at the end of the process (Fig. 2c). Final minimum thickness of the part is 1.69 mm (final maximum thinning 15.5%) which is a very small value for this kind of square bulged part.

In Fig. 3, the resulting pressure and feed vs. time curves are given. The plots show that the system rapidly finds an optimum slope for the feeding curve, which is kept approximately constant for the whole simulation. The pressure curve starts with a low slope (until t=3 sec), then little wrinkling starts occurring and pressure increases more rapidly until t=8 sec and it finally keeps a slightly oscillating behavior until the end of the process.

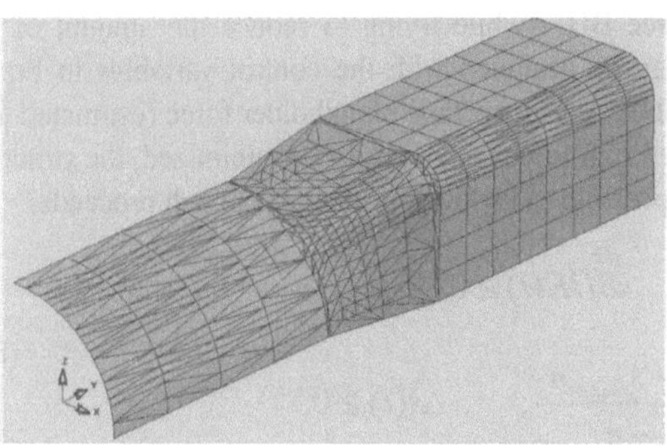

Data used for simulations
Isotropic hardening law:
$\sigma = K(\varepsilon + \varepsilon_0)^n$
K= 562 Mpa, n=0.22, ε_0=0.01.
Friction coefficient: 0.05
Initial thickness:
th = 2 mm.

FIGURE 1. Die (upper-right quarter) used for tube hydroforming simulations.

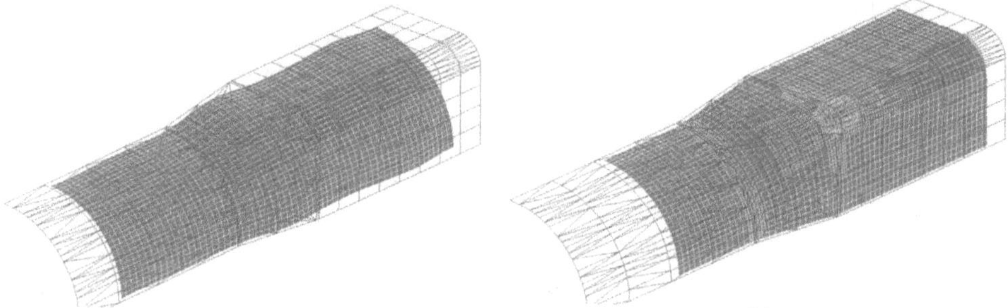

(a) (b)

FIGURE 2. Adaptive simulation of a THF process: (a) intermediate and (b) final state plots.

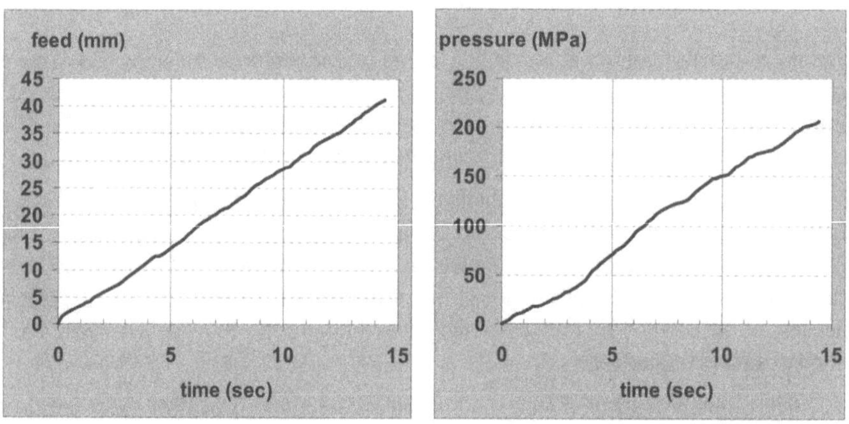

FIGURE 3. Pressure and axial feed curves obtained by adaptive simulation of a THF process.

DEEP DRAWING. The goal of ensuring thickness uniformity and preventing fracture can be achieved by minimizing the blankholder force BHF, while trying to reduce the amount of the stroke increments Δs. The control system can be designed with the control variables in Eq. 8, where n_c is the number of control intervals, BHF_{med} is an average blankholder force (estimated as a lower limit) and s_{max} is the total stroke. While Eq. 6 forces u_1 and u_2 to be minimized, the structure of Eq. 8 ensures that BHF is kept as close as possible to its lower limit as the punch proceeds.

$$\begin{cases} u_1(t) = \dfrac{BHF(t)}{BHF_{med}} - 1; \qquad BHF(t) \geq 0 \\ u_2(t) = u_2(t) = \dfrac{\Delta s(t) - s_{\max}/n_c}{s_{\max}/n_c}; \qquad \Delta s(t) \geq 0; \end{cases} \tag{8}$$

In the example described in the following Fig. 4, 5 and 6, a weight coefficient w =100 is used, as in the previous case. Maximum thinning can be expected either at the bottom corner of the pan

(point A in Fig. 4) or at the top of the dome in the center of the pan (point B). Distribution of thickness on the dome is controlled only by material properties and friction, but it cannot be influenced by the blankholder control. For this reason, given that the part is feasible, any loading paths that provide smaller thinning at point A can be considered as feasible.
Indeed, the loading paths obtained by FEM with adaptive loading do provide a thinning distribution with a maximum at point B.
In Fig. 6, the resulting blankholder force is plotted vs. the stroke. The BHF initially increases when there is practically no change in the state variable x_1 (surface area). As x_1 starts increasing the BHF initially decreases and it finally tends to stabilize (even though with strong oscillations) around the central value BHF_{med}.

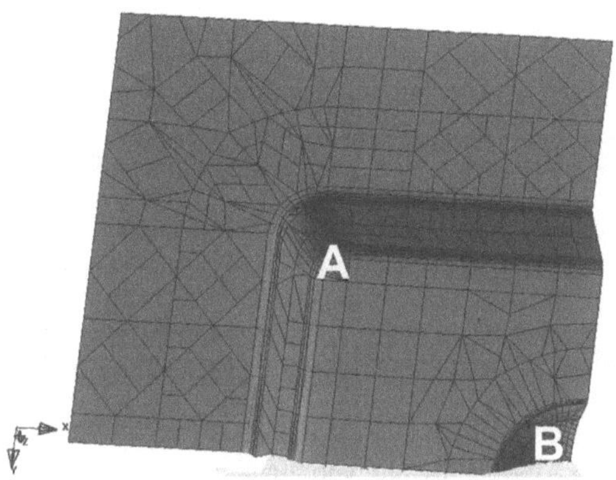

Data used for simulations
Barlat anisotropic hardening law:
K= 549 Mpa,
n=0.22,
ε_0=0.007.
$R_{00}=R_{45}=R_{90}=1.6$
Friction coefficient: 0.05
Initial thickness:
th = 2 mm.

FIGURE 4. Die (upper-left quarter) used for deep drawing simulations.

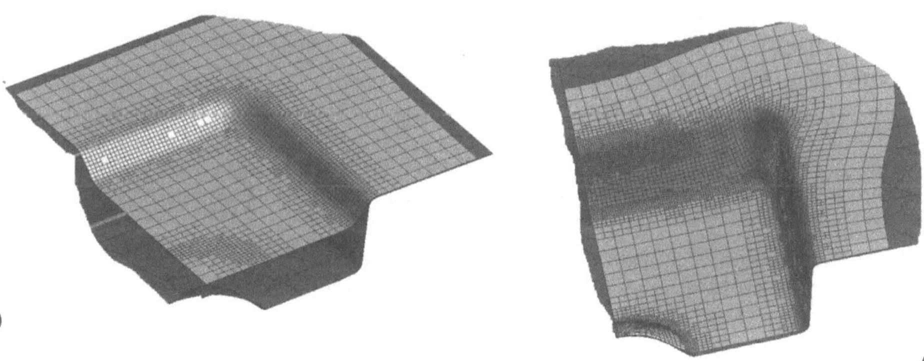

(a) (b)

FIGURE 5. Adaptive simulation of a deep drawing process: (a) intermediate and (b) final state plots.

4 CONCLUSION

The proposed Adaptive Simulation approach, based on predictive control theory, can be used for the simulation of sheet forming processes, in combination with a vector of two state that includes

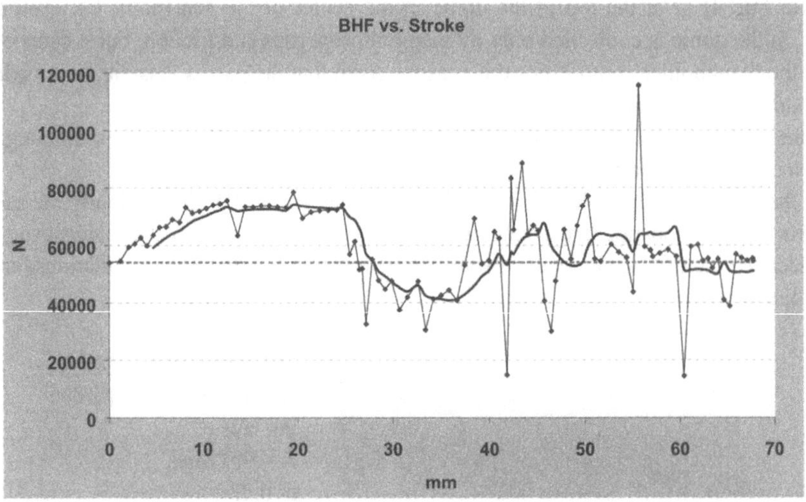

FIGURE 6. BHF vs. stroke curve, obtained by AS of deep drawing; a moving average trendline is added.

the distance between the workpiece and the die and the surface area of the deforming workpiece. Prevention of excessive thinning or fracture can be ensured by a proper selection of control variable to be used in the controller. Two examples of the model have been given with two different sheet forming processes: tube hydroforming and deep drawing with blankholder force control.

ACKNOWLEDGEMENTS

The authors gratefully acknowledge the Engineering Research Center at the Ohio State University (Columbus, Ohio), where this research line started under the guide of the director of the Center, Dr. Taylan Altan.

REFERENCES

1. Yang, J. B. Jeon, B. H. Oh, S. I., (2001). Design sensitivity analysis and optimization of the hydroforming process. J. Of Materials Processing Technology, 113, 1-3, 666-672.
2. Doege E., Kosters R., Ropers C, (1998). Determination of Optimised control Parameters for Internal High Pressure Forming Processes with the FEM. Proc. of the Int. Conf. on Sheet Metal '98, ed. by Kals, Geiger, et al.
3. Strano, M., Altan, T., (2001). FEA Simulation Strategies for Tube Hydroforming. V convegno AITEM, Bari (Italy), September 2001.
4. Nordlund P., Haggblad B., 1997. Prediction of wrinkle tendencies in explicit sheet metal-forming simulations. International journal for numerical methods in engineering, 40, 127-143.
5. Lundqvist, J., (1998), Numerical simulation of tubular hydroforming with an adaptive loading procedure in DYNA 3D. 11th Nordic Seminar on Computational Method, Stockholm, 39-42.

6. Strano, M., Jirathearanat, S., Altan,T, (2001). Adaptive FEM Simulation For Tube Hydroforming: a Geometry-Based Approach for Wrinkle Detection. CIRP Annals, 51.

THE MAIN PRINCIPLES OF FORMING INTEGRAL PANELS USED IN MODERN SHIPBUILDING INDUSTRY

R. Markovina[1]

[1] Department of Electrical Engineering, Mechanical Engineering and Naval Architecture, University of Split, Croatia

KEYWORDS: Shipbuilding industry, Integral structures, forming

ABSTRACT: In modern shipbuilding of high - speed vessels, like in the aircraft industry, the high requirements for quality, dynamic and static characteristics are placed, either in the phase of project or in the phase of production process. The corresponding speed in the light high-speed and super-high-speed structures can be achieved in various ways, all consideration however boil down to the reduction in weight of vessel proper, saving of structural material, rational increase of propulsion power, and modern form, modeling to reduce total resistance, and eliminating all needless lies elements. One of possible way for eliminating needless lies elements is using the integral panel (incorporated skins and longitudinals, produced by milling process). In this paper the main principles of the special technology of forming are done, which could be applicate in modern shipbuilding production process, especially in the production of super-high speed vessels.

1 THE PRINCIPAL METHODS OF INTEGRAL STRUCTURES FORMING

In forming integral structures, the following methods are usually used:

- forming by shot - peening (for thin structures) [4] using a steel shots by special machine system, on the outside of integral structure, with strongly controlled parameters, as shown in Fig. 2.
- forming by press "in three points" (English method) [1] with striker and dies special profiled, using elastics steel plates, about 250- 300 mm long. Thick structural parts are formed, firstly, to preliminary curve and then, by shot -peening, at he final curve. This is, practically, an experience method as shown in Fig.3
- forming by press "in three lines"(French method) [1] with very strong and very long press machines in two phases. Firstly, by pressing with system "striker-dies" along the integral structure, and then, by calibration with one of shot – peening systems, up to the final shape of curve as shown in Fig.4. The characteristics of this method are: decrease of working time, smallest working area, no outdoor transport operations, the experience of the operators is needed, small investments and small numbers of workers.

By author's experience, the combination of all three methods produce the best results, depending of curves complexity.

Published in: E.Kuljanic (Ed.) *Advanced Manufacturing Systems and Technology*, CISM Courses and Lectures No. 437, Springer Wien New York, 2002.

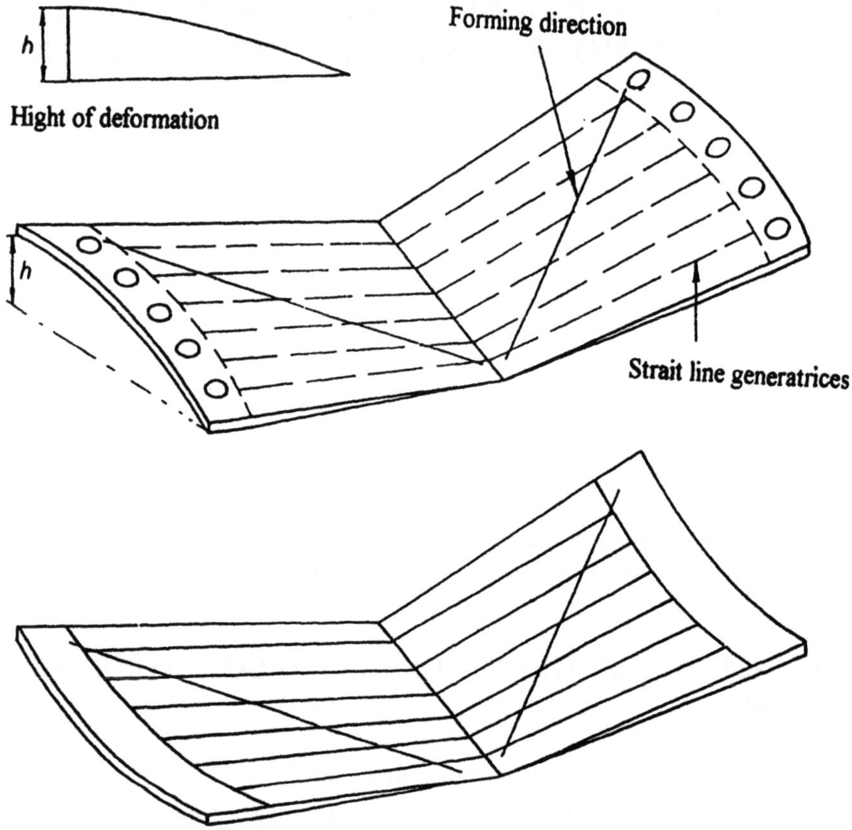

FIGURE 1. Formed complex integral structure

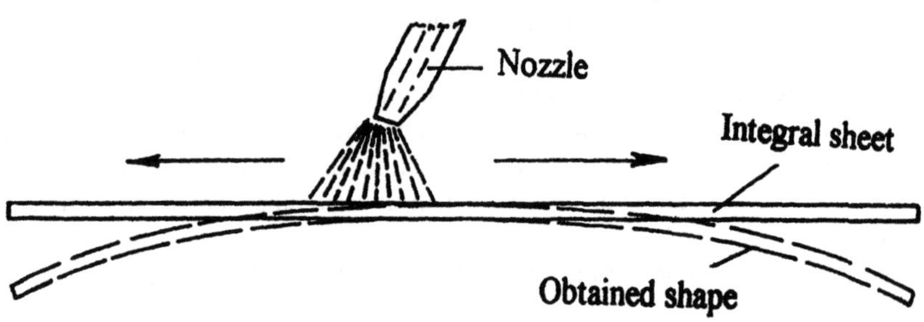

FIGURE 2. Forming by shot-peening method

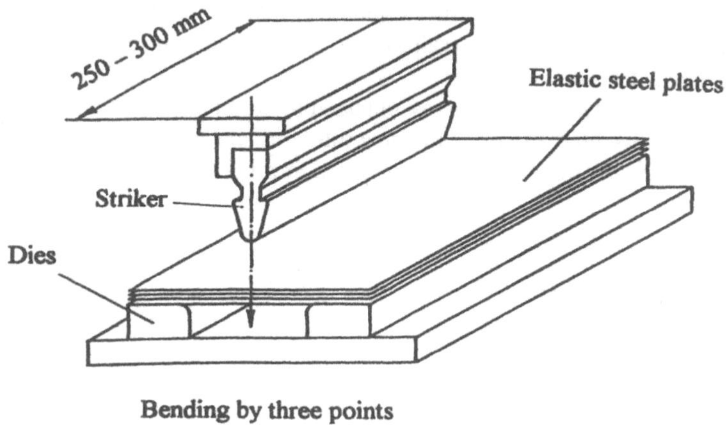

FIGURE 3. Forming by striker-dies method (English method)

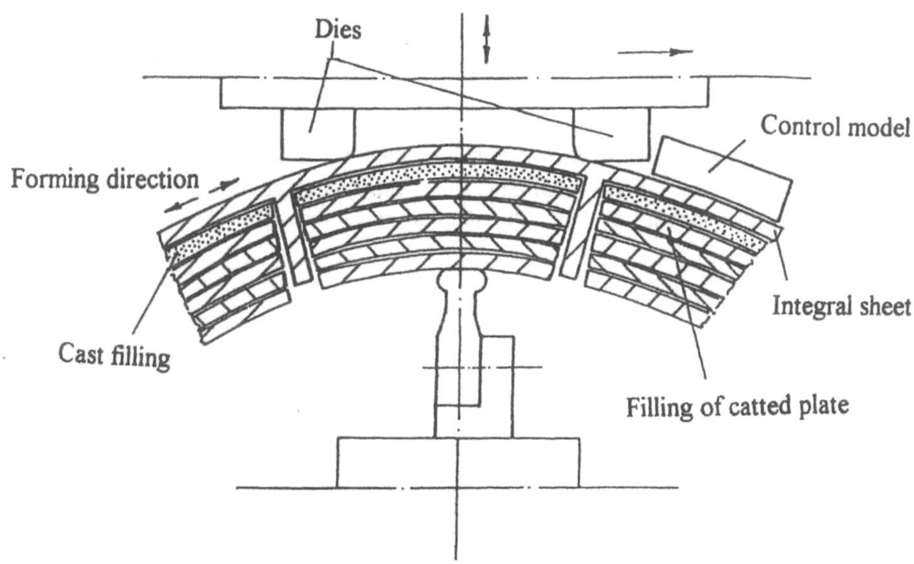

FIGURE 4. Forming by "striker-dies" (French method)

2 THE BASIC COMPONENTS OF INTEGRAL STRUCTURE FORMING

The main goal of the integral structure forming is to obtain regular, or irregular three-dimensional shape, depending on the position of integral structure on the vessel's hull.

According to the complexity of structure's geometry and final form of the vessels piece, the different profiles of curves can be realized by different forming technologies and can be divided in several categories [1]:

- one-direction (cylindrical) curves, obtained by forming in only one direction, by using the rolling machine, usually with three rollers as shown in Fig 5a. Generatrixes remain strait lines in the direction perpendicular to forming direction.

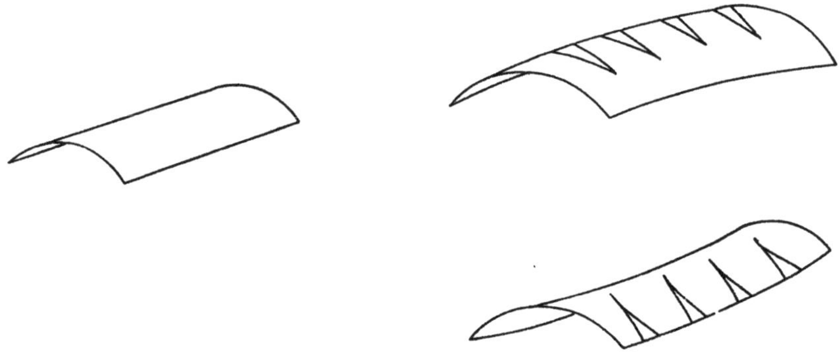

FIGURE 5. One- direction curve integral structure (a) and two- direction curves integral structure (b)

- two- direction curves obtained by forming in two directions, by using firstly the roller system for obtaining cylindrical curve, and then by using a special press or ironing machine system, for obtain final curvature with rectangular starting structural part's dimensions, as shown in Fig. 5b. Generatrixes changes theirs start geometrical characteristics.

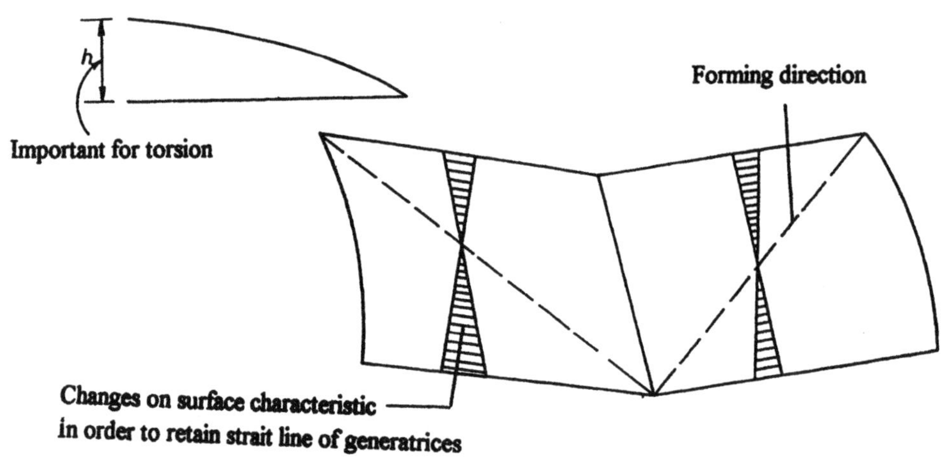

FIGURE 6. Complex geometrical shape forming

- complex geometrical shape, obtained by forming in three directions, by using several modes of forming and several production resources as: rolling machine, press – system by "three points" or "three lines" with and without special tools, depending on work piece geometry,
- very complex geometrical shape in three dimensions, obtained by forming in more directions, by using several modes of forming, and several production resources, and due to the requirement for final rectangular bordure in final shape, usually the "ironing machine" is used, for pressing and elongating integral structured parts at the same time.

3 THE PRACTICAL CALCULATION OF MAIN FORMING PARAMETRES

For well forming of integral structures it is very important to know the material characteristics as well as the forming parameters, like: the "elastic return"(spring back) zone, the forming force, the strain at the end of strike, the step and stroke of strike etc.

3.1 THE CALCULATION OF "ELASTIC RETURN" (SPRING BACK) ZONE

The calculation of "elastic return" means the knowing of the angle of "elastic returning", i.e. the R-ε diagram zone, where the "elastic return" will take place and the start and end of elastic deformation. That is an experimental procedure, when it is necessary to obtain the relation between the real radius of forming curve of the structural part R_1 (the final curve) and radius R_2 on which the integral structure has to be formed in order to obtain the real radius R_1, after "elastic return".
For Al-alloys this relation is:

$$R_1 : R_2 = 3 : 1 \qquad (1)$$

On the Fig. 7 the zone of "elastic return" is shown.

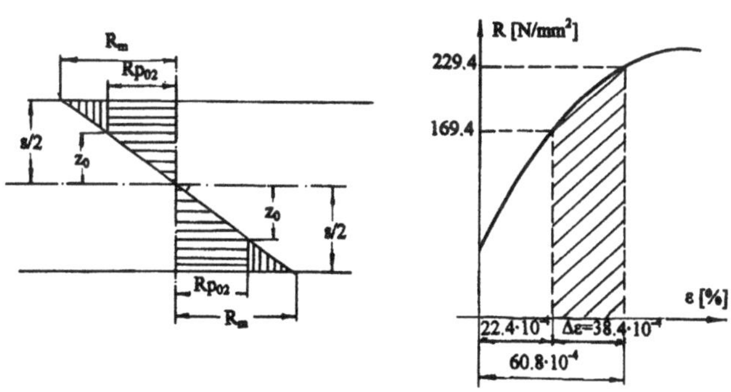

FIGURE 7. Strain diagram and "elastic return" zone

3.2 THE CALCULATION OF FORMING FORCE (F)

If the forming resistance is equal to product of forming modulus and strain R, than:

- bending moment is: $\frac{F \cdot l}{4}$, (2)

- bending modulus is: $\frac{I}{V} = \frac{b \cdot h^2}{6}$, (3)

and than:

$$\frac{F \cdot l}{4} = \frac{I}{V} \cdot Rp_{0,2} \quad (4)$$

The bending force can be obtained by equation:

$$F = \frac{\frac{b \cdot h^2}{6} \cdot Rp_{0,2} \cdot 4}{l} \; [\mathrm{N}] \quad (5)$$

3.3 THE STRESS DETERMINATION AT THE END OF THE STRIKE

The stress at the end of the strike can be determined by equation (5) as:

$$F \cdot l = \frac{I}{V} \cdot Rp_{0,2}, \text{ or } \; Rp_{0,2} = \frac{F \cdot l \cdot V}{I} \quad (6)$$

Compared to equation (5) the pressing force for forming can be calculated, for given material (filled with the filling) and than:

- for Al-alloy (400[N/mm^2])..........s = 6 mm...............................F_1= 288 [N],
- for the cast filling(35[N/mm^2].......s = 8 mm...............................F_2= 59 [N],
- for the cutting filling(40[N/mm^2]....s = 30 mm.............................F_3= 960 [N].

$\sum$ F =1307 [N]

The pressing force for forming of 130 [N] is the minimal necessary force for forming the present integral structure, and for calculated the strike characteristics.

3.4 THE PRACTICAL DETERMINATION OF THE STRIKE'S STEP

The strike's step depends of curvature radius, geometry of the integral structure and required tolerances in control of the final form. For the forms expected on super – high-speed vessels this step could be about s = 10 – 15 mm.

3.5 THE PRACTICAL DETERMINATION OF THE STRIKE'S STROKE

The determination of the strike's stroke will be realized by experience, and at every new pass of the strike the height of stroke has to be lower for h = 0,2 – 0,5 mm.

4 CONCLUSION

The forming of integral plates (skins) have to be a very important part of the modern production process in the production of super-high speed vessels. This process has to be strictly controlled, undivided and homogenous process in which the line – production of hull's and superstructure's modulus will exist with integral structures, like its fundamental part. Then it can realize the main goals in the new production as:

- increasing production quality,
- simplifying production,
- decreasing final production costs,
- decreasing human influence in the production, and
- best aesthetic view.

REFERENCES

1. Markovina, R., (1977), The basis of integral plates forming process, textbook, Mostar, Bosnia and Herzegovina.
2. Meguid, S.A., (1978), A study in the peen forming of metals, *Aluminium*, No. 54,Jahrg.1978- 3, 203-206.
3. Gold, R., (1979), The evolutionary design of Boeing's 767,*Precision metal*, Oct.1979, Canada, 37-44.
4. Kellock, B., (1982), Peen forming shapes up and extends life, *Machinery and production engineering*, No.4, 20-21.

THE MAIN PRINCIPLES OF USING INTEGRAL PANELS (SKINS) IN MODERN SHIPBUILDING INDUSTRY

R. Markovina[1]

[1] Department of Electrical Engineering, Mechanical Engineering and Naval Architecture, University of Split, Croatia

KEYWORDS: Shipbuilding industry, Integral structures, forming

ABSTRACT: In modern shipbuilding of high - speed vessels, like in the aircraft industry, the high requirements for quality, dynamic and static characteristics are placed, either in the phase of project or in the phase of production process. The corresponding speed in the light high-speed and super-high-speed structures can be achieved in various ways, all consideration however boil down to the reduction in weight of vessel proper, saving of structural material, rational increase of propulsion power, and modern form, modeling to reduce total resistance, and eliminating all needless lies elements. One of possible way for eliminating needless lies elements is using the integral panel (incorporated skins and longitudinals, produced by milling process). In this paper the main principles of using integral panels are done, like the application in modern shipbuilding production process of super-high speed vessels.

1 INTRODUCTION

In modern industry, as the shipbuilding industry is, the high requirements are placed for quality, dynamic and static characteristics for structural parts, weight diminution of production parts, increasing of speed, security in all regimes of using and good quality of products. The same requirements are demanded, either in the design phase or in the production phase, in building of super – high -speed vessels.

The main goal of this proposal for using *integral structure* in modern shipbuilding production is the simplification of production, particularly in the preassembly and assembly phases, as the positive influence on the technical characteristics of the vessels (low weight, high speed, security and stability in all regimes of navigation, satisfying stress and strain of construction, etc.). As the main material will be Al – alloys, one way to meet these demands is using the *integral structures* (integral panels and integral skins).

Though at the first moment, the production of integral structures seems very expensive and uneconomic, the final production costs of high – speed vessels will show its rent ability at the vital constructive parts like:

- the side plates of hull structures,
- the inside and outside plates of superstructure,
- the decks on the hulls and superstructure,
- the frames girders (bar – girders) of hull construction,
- the zones of side and deck openings (for cargo and passengers) etc.

The Fig. 1 shows the general view of one typical *integral structure.*

Published in: E.Kuljanic (Ed.) *Advanced Manufacturing Systems and Technology,*
CISM Courses and Lectures No. 437, Springer Wien New York, 2002.

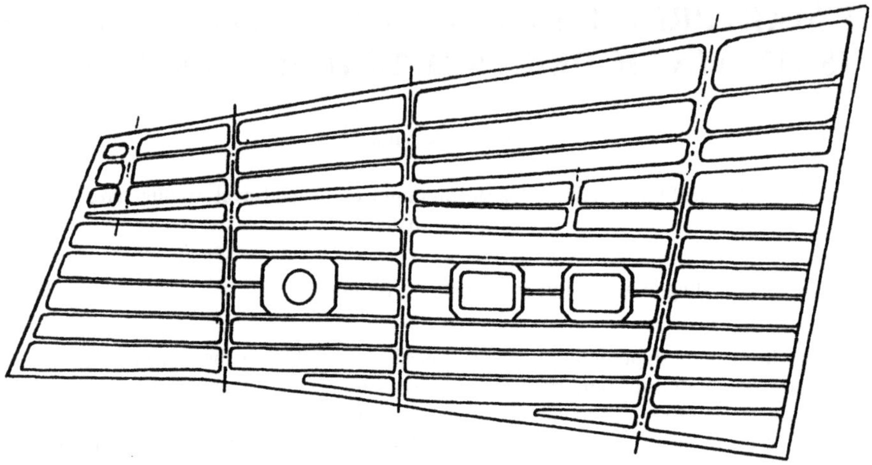

FIGURE 1. General view of one typical integral structure

2 DEFINITION OF INTEGRAL STRUCTURES

The hull parts produced like the elements of integral structure are the elements produced in one - piece, either by milling machine process (using NC – milling machine) or by chemical milling, made from thick Al – alloy plates in the integral form, that means homogenous structure with all necessary incorporated structural parts and without any kinds of joins between constructive elements in it.

That production of structural parts needs the special production resources and a severe quality control over the possible deformation during its production, caused by:

- releasing the internal strain of material,
- unusual heating of material during production,
- rate between thickness of plate and minimal thickness of machine structure (usually s=h/10),
- good flatness of the plate before integral structure machining, especially at the thin parts.

Depending on the geometrical shape and characteristic of integral structure, particularly the integral panels and integral skins, we can divide these structures [1] in the several categories with:

- "pockets" different depths,
- mutually parallel longitudinals, perpendicular at the forming direction ,
- mutually parallel longitudinals, along the structural part,
- combination of parallel and perpendicular longitudinals, at the direction of forming,
- intercostals longitudinals along the structural part,
- longitudinals and incorporated bottom of the frames and floors,
- special shapes and deformation profiles.

For each of these categories the production methods and resources have to be strictly determined, as well as the production technology of structural part with all special tools and outfits. From that reason, the following notices have to be kept in mind, particularly in the design phase:

- integral structures must have an adequate stress and strain,
- integral structures must have the most simple shape,
- integral structures must to be produced without steps of different thickness against side thicknesses,
- integral structures must to be produced with the same thickness along the alveoli between longitudinals,
- the ends of integral structures have to be finished as a thick part, and
- the production methods and resources, as well as the qualifications of operators, have to be counted in the design of the structural part,.

The advantages of the production high – speed vessels using integral structures are polyvalent:

- production at one – piece system, at one working place, without of any methods of joins on it, without the risks of cracks, no X –ray controls, no X –ray films, proper strain and stress, quality control consists of only dimensional control and fluorescent – penetrant method for cracks detection,
- weight of structural parts is decreased as well as the number of working operations,
- production time of final part and transport operations are decreased,
- quality control time is decreased,
- general security of construction is increased,
- good esthetical view is assured.

The defaults of the production high – speed vessels using the integral structures are:

- providing NC- milling machine with the big working table,
- providing the resources for forming integral structures with specific technology,
- making programs for NC – milling machine,
- operators must have a high qualification level.

3 THE PRINCIPAL INTEGRAL STRUCTURE CLASSIFICATION (BY FORMING MODE)

By the mode of forming, [1] the integral structures can be divided in:

- the integral structures without longitudinal profiles (thin structures named "pocket") and
- the integral structures with longitudinals, different directions.

For each of numbered integral structures there exists a "raison d'etre" for its using in construction parts, in the design phase, either for its specific geometry or for the technology of forming.

Integral structures without longitudinals (with the "pocket") are produced of thin plates, with the depth up to 8 mm, and forming is realized by rolling on rollers machine for cylindrical curvature, with the method "by down", because of an easier control of the curve during forming, and in order to eliminate the influence of its weights on the final curve. In some cases, forming can be realized by combination of rolling and pressing, using special tools and counting on "elastic return". The Fig. 2 shows that possibility.

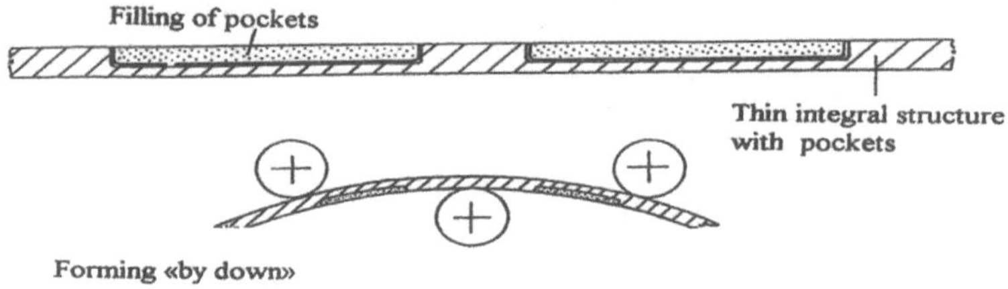

FIGURE 2. Forming of integral structure with the pockets with method "by down"

Integral structures with mutually parallel longitudinals (in the forming direction) are forming by pressing on the press- machine, type "striker – dies", either the striker or dies is moved, and is positioned on the upper side of the machine. In this case, firstly, it is necessary to fill up the "pockets" by prepared filling inside of the integral structure geometry, between longitudinals. This filling is made of specific material (both of LUCOFLEX and UR-325) with mechanical characteristics like the constructive material. The high of the fillings must be 3 - 5 mm over the highest part of integral structure, as presented in Fig. 3.

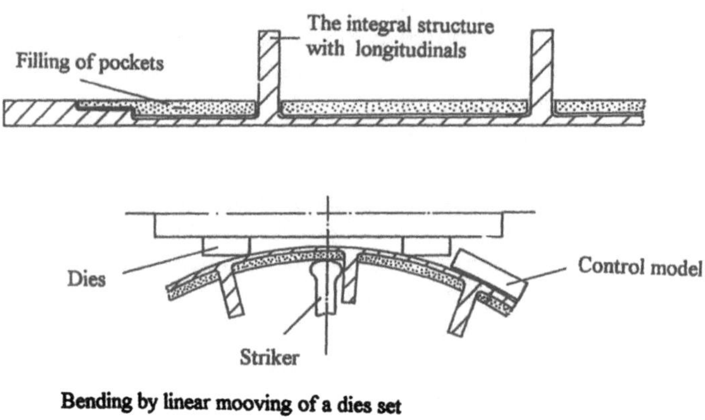

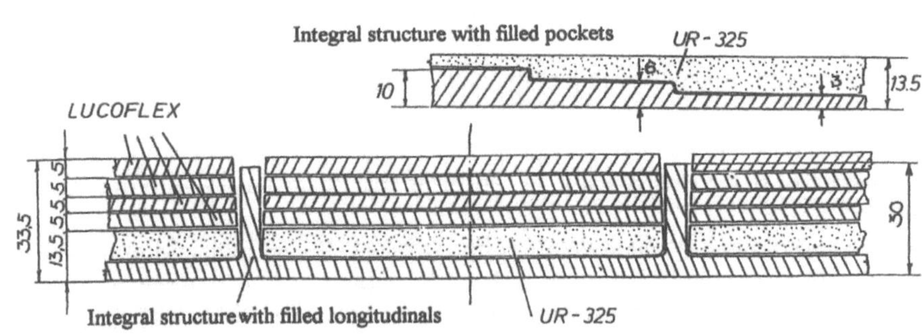

FIGURE 3. Forming of integral structure with mutually parallel longitudinals with method "by down"

Integral structure with longitudinals in both directions (in the direction of forming and perpendicular on the direction of forming) are formed "strike by strike" either " in three points" or "in three lines" at the special press with especially profiled striker. Of course, in this case the areas between longitudinals have to be filled up by previous prepared filling. For thin integral structures forming can be realized on the strong rolling machine. Forming has to be realized with the longitudinals "by up" because filling, in forming "by down", could fall down. Forming of simple regular forms by universal tools counting on "elastic return"(spring back) could be realized. For Al – alloys the "elastic return" is about 1/3 of real curvature radius, as presented in Fig.4.

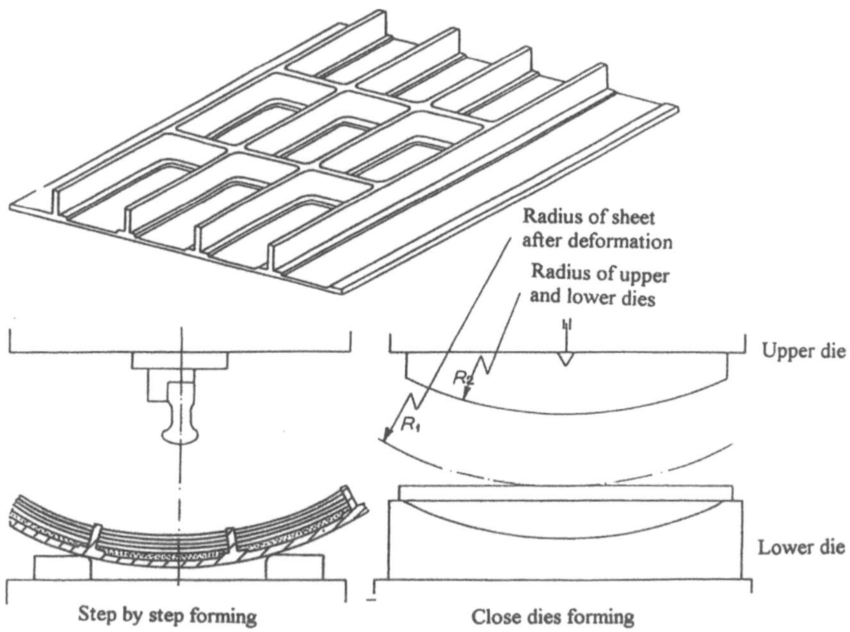

FIGURE 4. Forming of integral structure with longitudinals in both directions (a) and simple regular form (b)

The view of the strains for this case is presented in Fig.5.

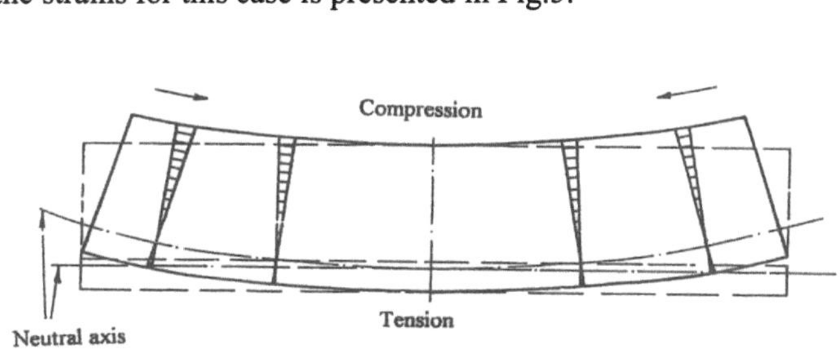

FIGURE 5. Strains of formed integral structure

In the case of the *integral structures with longitudinals around the bordures*, even in the machining phase, it is necessary to produce some additional surface around the bordure of final shape of the structural part with the same geometry, which is, in the phase of forming also filled with filling; and forming is effected by the method "strike by strike" ("in three points" or "in three lines"). In same cases, forming can be realized by the strong system of rollers. These structures are shown in Fig. 6. and are used for side openings or decks.

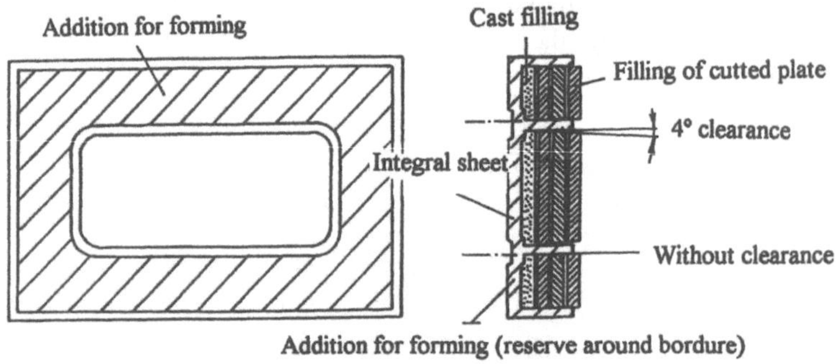

FIGURE 6. Filling the integral structure with longitudinals around the bordures

The integral structures with special shape and deforming profile are used for decks and side plates, at the shear strake and bilge strake. Due to different depths of the plates and used forces for forming, filling is done completely in two phases:

- with filling realized by polymerization of two components (UR-325), due to the fact that filling obtains the correct form of "pocket", and
- with positioning of filling produced of thin plate (LUCOFLEX), level by level, up to the high 3 – 5 mm over the highest part of integral structure.

The filling connecting between each level is done by the double side adhesive tape.

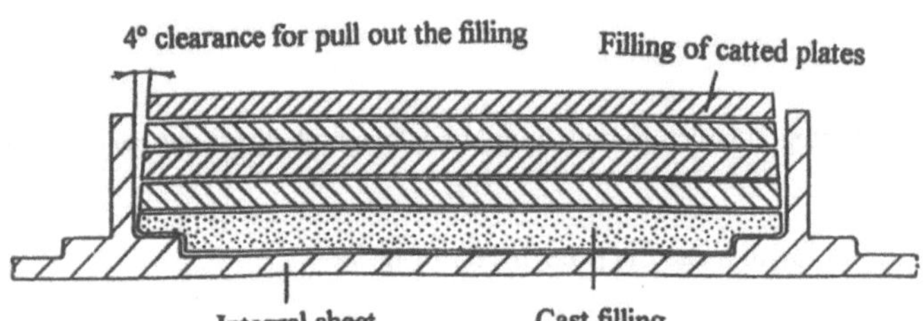

FIGURE 7. Filling positioning realized by polymerization and level by level

Forming is effected by "striker-dies" press – system ("in three lines") with calibration of skin thin parts, either by shot-peening or by pressing with special tools, calculating the effect of "elastic return.

The integral structures of decks, side plates, bilge and shear strake, as the bottoms sides of hulls have to be very precise and of excellent form. Their dimensions are very big (6000x2000mm), masses relatively small (40-50 kg) and with the small forming tolerances (± 0,2-0,5 mm).

4 CONCLUSION

Taking into consideration these facts the integral structures in modern shipbuilding industry will be very important and necessary, particularly in the serial production of the super – high-speed vessels. It will be necessary to eliminate all technical, technological and organization problems, and invest in indispensable machines and new knowledge of operators at the new production lines. Then the production with minimal working forces, material and production times can be obtained, and it can be possible to ensure the best quality.

These have to be the permanent tasks of all production participants in modern shipbuilding production, particularly in the production of the new generation's vessels.

REFERENCES

1. Markovina, R., (1977), The basis of integral plates forming process, textbook, Mostar, Bosnia and Herzegovina.
3. Gold, R., (1979), The evolutionary design of Boeing's 767,*Precision metal*, Oct.1979, Canada, 37-44.

DESIGN OF THE TUBE ROLL FORMING PROCESS THROUGH AN HEURISTIC ALGORITHM

G. Celano[1], A. Costa[2], S. Fichera[2], L. Fratini[1]

[1] Dipartimento di Tecnologia e Produzione Meccanica, University of Palermo, Italy
[2] Istituto di Macchine, University of Catania, Italy

KEYWORDS: Roll forming, Process design, Simulated Annealing.

ABSTRACT. In the paper the design of an industrial tube roll forming process is developed utilizing an heuristic algorithm, namely the simulated annealing (SA). In particular both the number and the shape of the deforming rolls are determined in order to define the forming sequence. The effectiveness of the results supplied by the SA algorithm is verified both by means of some experiments tests carried out on an industrial tube roll forming equipment and through a set of numerical simulations of the process.

1 INTRODUCTION

The tube roll forming is a continuous process for forming metal from sheet, strip or coiled stock into tubes of uniform cross section by feeding the stock through a series of roll stations equipped with contoured rolls [1]. The process is particularly suited to the production of large quantities and long lengths to close tolerances and involves a minimum of handling.

In this process the rolls are characterized by a progressive curvature enabling the desired final shape to be obtained; furthermore properly shaped blocks are located to favor the sheet movement and to control the elastic springback. At the end of the forming sequence a submerged arc welding operation is carried out to join the sheet borders.

The design of the process consists in the most suitable selection of the basic operating parameters: among them the number of passes and the geometry of the rolls utilized at each station are probably the most relevant [2]. In particular the geometry of the rolls is determined by the position of the center of curvature and by the curvature radius.

The above considerations show that the "flower" design (flower is the name generally given to the progressive section contours, starting with the flat sheet and ending with the desired section profile) is a very critical task. Actually a finite element approach can be very useful to evaluate an existing flower design, but it cannot be considered as a starting design method [3][4][5]. According to the above considerations in a previous paper the authors proposed the application of an Artificial Intelligence technique, namely Simulated Annealing, to the tube roll forming process [6]. The SA algorithm considered the roll forming sequences as vectors, the elements of which were the bending angles corresponding to each pass. The sequences had a variable size: the optimal number of passes was not fixed a priori and was determined through the evolution of the algorithm. A proper objective functions was utilized, either aimed to the reduction of the number of passes of the roll forming sequence or to the improvement of the quality of the final product avoiding shape defects on the strip edges.

Published in: E.Kuljanic (Ed.) *Advanced Manufacturing Systems and Technology*, CISM Courses and Lectures No. 437, Springer Wien New York, 2002.

In this paper the effectiveness of the results supplied by the SA algorithm is verified, both by means of some experiments tests, carried out on an industrial tube roll forming equipment, and through a set of numerical simulations of the process. Furthermore a sort of constrained optimization procedure was carried out in which the rolls to be utilized in the optimal sequence were chosen just among those in use in the actual industrial sequence. Again the obtained optimal sequences were tested both by experiments and FEM simulations.

2 THE INDUSTRIAL PROCESS

In the following the industrial experimental equipment utilized for the investigated process is described. It consists in a roll forming line, designed for artesian well tube production. The raw product is constituted by a coil which is unwind through an appropriate decoiler (Figure 1.a). Then the strip is regularized across a flatter and trimmer station (Figure 1.b) where it is both vertically and transversally positioned. In order to avoid discontinuity in the forming process a welding station is inserted along the line; in this way the conjunction between the final and the initial edges of two consecutive strips may be carried out (Figure 1.c). After that, the strip undergoes the bending process through the roll forming machine (Figure 1.d) made of 10 roll forming stations and 8 "controlling" stations in which properly shaped blocks are placed to avoid the insurgence of shape defects and to favor the progressive forming of the sheet metal. In each forming station a couple of properly shaped rolls is mounted and in this way at the end of the forming line a tube shaped final product is obtained which is longitudinally welded and then cut on the cutter station (Figure 1.e).

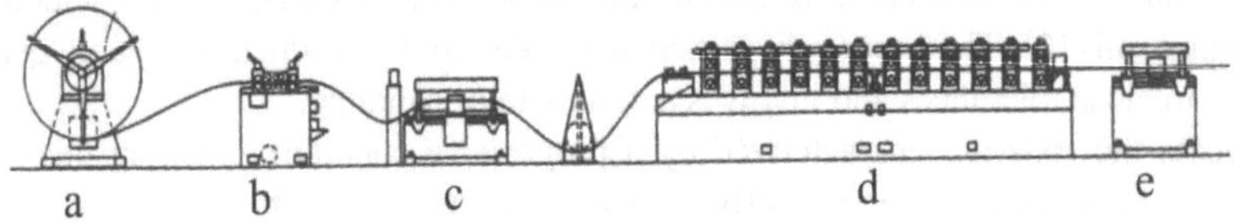

FIGURE 1. The forming sequence

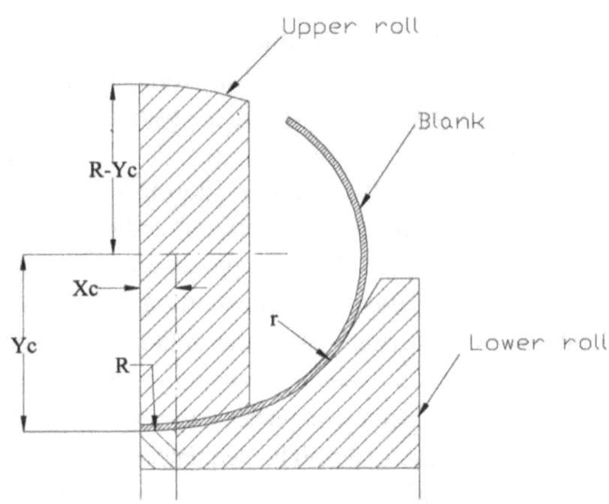

FIGURE 2. The roll geometry

In Figure 2 the main geometrical features defining the geometry of each single roll is reported. It should be observed that at least Xc, R and r have to be indicated in order to determine the geometry of the roll [6].

3 DESIGN AND OPTIMISATION PROCEDURE

In order to obtain an optimal design of the roll forming process, the following objectives have to be reached [7]:

- minimization of the number of roll passes in order to reduce the number of required rolls and the time spent for the machine setup when a production batch change occurs;
- maintenance of the longitudinal strain on the strip edges below the maximum value ε_a, to avoid the wavy edges phenomena and thus to assure a good quality of the welded tube.

On the basis of the above mentioned objectives, two optimizations procedures applied to the design of the roll forming process have been developed. The former aims to optimize the process considering the whole available set of rolls ("free optimization") so that, for example, a roll usually utilized for the production of a smaller diameter tube can be included in the sequence instead of a larger diameter one and vice versa. The latter procedure, on the contrary, takes the sequence pertaining to the factory as a starting point and then evolves reducing single roll passes according to the already analyzed objectives ("constrained optimization").

The optimization approach consists on the generation of an optimal roll sequence through a heuristic algorithm, namely simulated annealing, which minimizes the total number of roll passes and at the same time smoothes the average longitudinal strain at each roll pass.

In the last decade Simulated Annealing have demonstrated, their flexibility for solving many different classes of optimization problems, and their efficiency in respect to the traditional techniques of local optimization or iterative improvement [8].

In fact, SA combines to the structure of the traditional steepest descent method the possibility to probabilistically accept "retrogressive" movements towards worst solutions. This feature allows new possible optimal solutions to be found out and reduces the probability to accept poor local optimal solutions. SA operates through an analogy with statistical mechanics of condensed matter physics. Each stage of the algorithm represents a cooling "temperature" level which is maintained till an "equilibrium condition" is reached; then the temperature is cooled and the local search goes on. Because of its large flexibility, SA can be applied with success to the roll forming problem [6][7].

The input data elaborated by the proposed simulated annealing, are constituted by a detailed database containing the following rolls geometric features: Xc, R and r (see Figure 2). Whichever the type, single or double curvature, each roll is identified through a natural number. Therefore, the generic sequence to be optimized by the SA algorithm is represented by a variable size vector S whose elements are natural numbers: $\boldsymbol{S} = \{n_1 \quad n_2 \quad n_3 \quad \ldots \quad n_n\}$, where $n_1<n_2<\ldots< n_n$ and n_n is equal to the code of the roll that corresponds to the final diameter of the profile. It is relevant to point out that since the vector S has a variable size, the optimal number of passes is not fixed a priori but is determined during the evolution of the algorithm.

On the basis of such a feasible sequence, the optimization algorithm calculate the objective function based on the differences between the strains corresponding to each pass and the maximum allowable material elongation, and on a term depending on the number of passes.

$$OBJ = \frac{\sum_{i=1}^{n} \left|\varepsilon_a - \varepsilon(k)\right|}{n} + n^r \qquad (1)$$

where n is the number of passes, ε_a is the maximum allowable elongation on the strip, $\varepsilon(k)$ is the reached level of longitudinal strain derived by means of simple geometrical considerations [9] [10] and r is a constant (r=142). Of course, the objective function OBJ must be minimized.

In order to assure the feasibility of the sequences elaborated by the optimization algorithm, a set of constraints has been considered. In particular penalty weights acting on the OBJ function of each sequence have been introduced, essentially due to the following causes:

- since the objective function works on the average of the differences between strains, it is necessary to avoid unacceptable elongation values at a single pass, which are symptoms of too large values of conferred deformation at a single forming station;
- in order to facilitate the introduction of the strip into the roll forming machine, sequences characterised by too high values of the bending angles in correspondence of the first pass must be discarded;
- because of the work hardening of the material and of the final elastic springback, it is preferable to design roll forming processes characterized by decreasing values of the strains as the total bending angle increases.

It is worth to point out that, all these aspects have been taken into account in the algorithm evolution by selecting penalty weights whose importance is calibrated with respect to their effects on the final quality of the products.

In order to analyze a wide domain of feasible solution, a double stage neighbourhood search operator has been implemented. A proper description of the operators operating can be found in reference [6]. Furthermore, the above mentioned operators are randomly chosen with an equal probability to be selected.

Once a new forming sequence through the neighbourhood search method has been generated, a simple cooling schedule based on the variation of the objective function OBJ is carried out. It aims to accept or refuse the new sequence as perturbation of the original one, through a typical exponential probabilistic criterion. If the new sequence is refused, the actual temperature value is reduced by a multiplicative factor equal to the fixed cooling constant [6].

4 THE NUMERICAL MODEL

The roll forming process was simulated utilizing an explicit dynamic model [11]. A full three-dimensional analysis was developed taking into account the vertical symmetry plane.

The sheet was meshed with about 2,000 4-node shell elements characterized by 5 integration point through the thickness; moreover a geometrical remeshing algorithm was utilized during the numerical analyses in order to avoid a too relevant distortion level of the elements. During the strip movement several elements were in direct contact with the rolls and with the shaped blocks which were both modeled as rigid geometrical surfaces. Actually the use of no-meshed bodies improves the effectiveness of the contact algorithm since no inter-action between two different meshes has to be taken into account. Furthermore the utilization of small time increments, which characterizes the explicit models, permitted to properly follow the process mechanics and in particular the contact conditions at the rolls-sheet metal interface. In Figure 3 the deformed model utilized for the simulation of the 300 mm diameter (Φ) tube roll forming process is shown with reference to the industrial forming sequence.

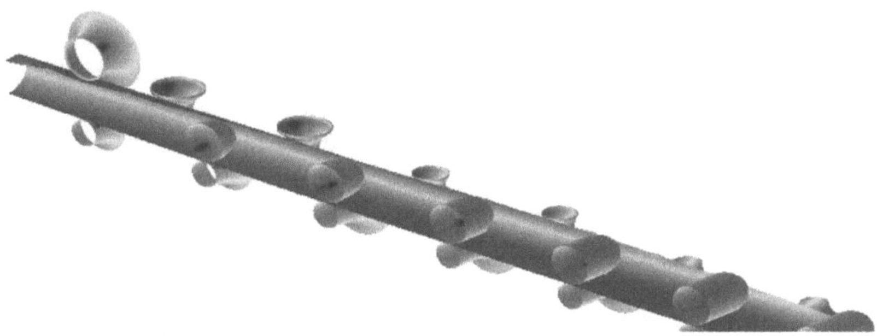

FIGURE 3. The numerical model (Φ = 300 mm, ten forming stations)

The forming processes were simulated applying to the lower forming rolls an angular velocity equal to 25 rad/s, 5 times greater than the actual one in order to reduce the CPU times. Nevertheless, no inertia effects were observed through a continuous check of the ratio of the kinetic vs. deformation energy, which has been limited to 0.1.

A Coulomb friction model was utilized with a friction coefficient equal to 0.15. Finally the sheet material was a 2.5 mm thick Fe 360 steel, whose plastic behavior is represented by the following power law:

$$\sigma=648*\varepsilon^{0.23} \quad (2)$$

obtained through a set of experimental tests.

5 THE OBTAINED RESULTS

In order to test the effectiveness of the proposed design procedure both numerical simulations and experimental tests were carried out. In particular the Φ = 300 mm tube, characterized by a sheet thickness of 2.5 mm, was taken into account. For such configuration the industrial practice utilizes a sequence of ten forming stations. First of all a "free" optimization procedure was developed and the SA algorithm designed an eight stations sequence [6]. In the next Figure 4 the Flower design corresponding to the "free" optimal sequence for Φ=300mm profile is reported.

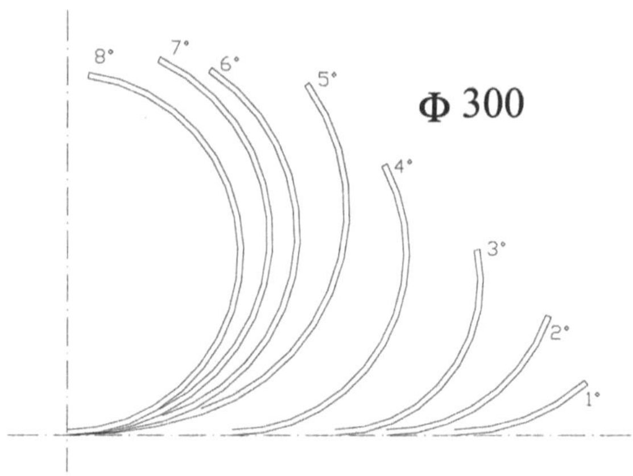

FIGURE 4. Free optimal Flower design (Φ=300mm)

Both the industrial sequence and the optimal one, proposed by the "free" design procedure, were numerically simulated in order to compare the obtained results in terms of reached levels of strain (Figures 5a and 5b). Actually a slight increase in the level of longitudinal strain was observed due to the reduced number of forming stations and the consequent uneven deformation mechanics along the production line. Nevertheless, as can be observed in the next figure 5b, the free optimisation procedure allowed to obtain a sound final product but the former part of the sheet which undergoes an initial transitory characterized by a stronger curvature than the required one.
The "free" optimisation sequence was also experimentally tested in order to verify the effectiveness of the design procedure: during the experiments sound components were obtained avoiding the insurgence of shape defects in the rolled tube. In figure 6 a final step of the forming process is reported and the joining edges of the bent sheet are shown highlighting a correct positioning for the incoming wending step.

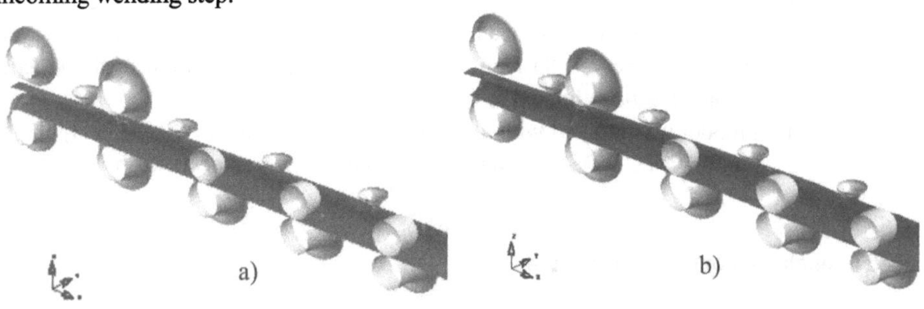

FIGURE 5. The deformed model: a) the industrial sequence; b) the "free" optimisation sequence

FIGURE 6. The experimental validation of the free optimisation sequence

Subsequently the verification of the "constrained" optimization sequence was developed; in other words the sequence was obtained limiting the domain of the objective function to be minimized in the design procedure, to the forming rolls utilized in the industrial sequence. In this way an optimal sequence was determined differing from the industrial one since the 3rd the 5th and the 7th upper rolls were released and did not produce any deformation on the sheet metal. The FEM simulation (Figure 7) allowed to verify the soundness of the final obtained tube apart from the former part of the sheet in which strong local deformations were observed due to the shoots which were impressed at each forming station to the sheet metal as it entered the rolls. Such defect, which

is typical of the initial transitory of the roll forming processes, in this case appears more relevant since an uneven progressive deformation is given to the blank because of the absence of three deforming rolls with respect to the industrial sequence.
Again the experimental verification of the optimized sequence was developed in order to test the effectiveness of the design procedure and of the FEM results. In this case a very simple set-up of the industrial plant was required since the optimized sequence was directly derived from the industrial one simply releasing three forming rolls. Again a sound product was obtained and in figure 8 the final stage of the process just after the welding process is shown.

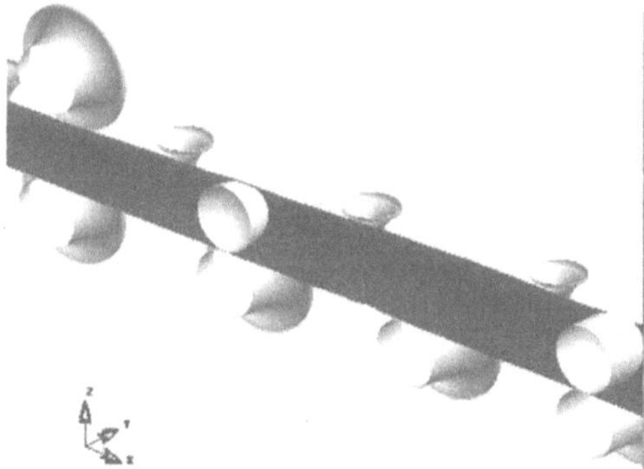

FIGURE 7. Constrained optimisation sequence: the deformed FEM model

FIGURE 8. Constrained optimisation sequence: the sound component after the welding step

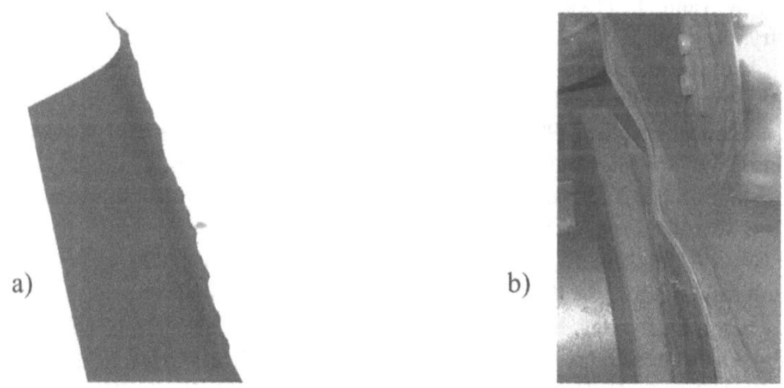

FIGURE 9. The shape defect: a) FEM simulation; b) experiment

Finally subsequent trials were developed in order to further reduce the number of forming steps; in particular both numerical simulations and experiments were developed releasing also the 1st upper roll. Actually such final tests demonstrated that shape defects are induced in the final product as a too high level of deformation is given at the first forming station and/or during the deformation process since in these cases the sheet metal is strongly beaten by the lateral surfaces of the rolls. In this way in the next Figure 9 (a and b) the typical observed shape defect is reported and furthermore in Figure 10 the consequences as far as the welding joint is regarded are highlighted.

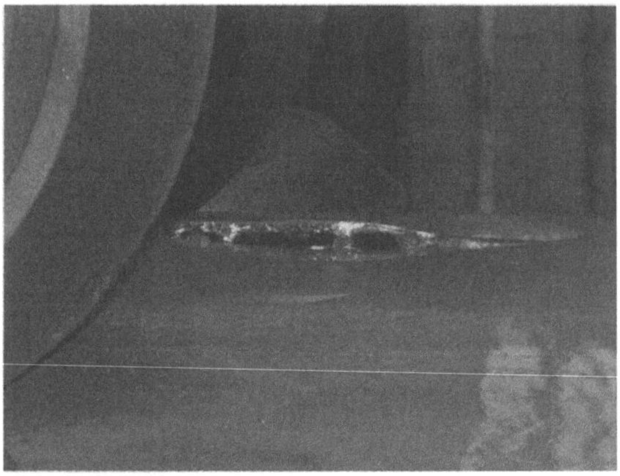

FIGURE 10. The ineffective welding due to the presence of the shape defect

6 ACKNOWLEDGEMENTS

This work has been made within a research project sponsored by the MURST (Ministry of the University and of the Scientific Research).

REFERENCES

1. ASM Metals Handbook Ninth Edition (1988), Vol.14, Forming and Forging.
2. Seldmaier, A. & Tann, J., (1999), Process simulation and quality management for the roll forming industry, Proc. 6th ICTP, 2381-2386.
3. Kiuchi, M., Abe, K., and Onodera, T., (1995), Computerized numerical simulation of roll-forming process, Annals of the CIRP, Vol. 44/1, 239-243.
4. Kiuchi, M. & Koudabashi T., (1985), Automated design system of optimal roll profiles for cold roll forming, Proc. 3rd Int. Conf. on Rotary Metal Working Processes, 423-436.
5. Brunet, M., Mguil, S., and Idrissi-Ronel, S., (1998), Simulation of roll-forming process with a specific F.E.M. code. Proc. of NUMIFORM '98, 941-946.
6. Celano, G., Costa A., Fichera, S., Fratini, L., Ricca A. and Micari, F., (2001), Design of an industrial tube roll forming process, Proc. of the V Convegno AITEM, 453-462.
7. Celano, G., Fratini, L., Fichera, S., Micari, F., (2001), Design optimisation of an industrial roll forming process through an heuristic algorithm, Proc. of the NUMIFORM 2001, pp. 589-594.
8. Kirkpatrick S., Gelatt Jr. C. D., and Vecchi M. P., (1983), Optimization by Simulated Annealing, Science, Vol. 220, 671-680.
9. Walker, T.R. & Pick, R.J., (1990), Approximation of the axial strains developed during the roll forming of ERW pipes, Journal of Mat. Proc. Tech., Vol.22, 29-44.
10. Bhattacharyya, D., and Smith, P.D., (1984), The development of longitudinal strain in cold roll forming and its influence on product straightness. Proc. 1st Int. Conf. on Tech. of Plasticity, 422-427.
11. Rebelo, N., Nagtegaal, J.C., Taylor, L.M., Passmann, R., (1992), Comparison of implicit and explicit finite element methods in the simulation of metal forming processes, Proc. of NUMIFORM '92, 99-108.

DESIGN ISSUES FOR MICROFORMED SHEET METAL COMPONENTS

D. Antonelli, P. Chiabert

Department of Production Systems and Economics, Politecnico of Torino, Italy

KEYWORDS: Sheet metal forming; miniaturization; size effects.

ABSTRACT. With the miniaturization of sheet metal working, a new method for designing the component and the associated manufacturing process must be sought. Standard methods for tolerances synthesis, formability study and part inspection are inadequate when applied to under millimeter components. The dimensional and geometrical tolerances are often of the same order of magnitude as the part dimensions, making harder the realization of functional assemblies. Micro-formed parts exhibit a peculiar behavior during the forming steps due to the so-called size effects. Furthermore, the inspection of micro-parts requires statistical approach and special instrumentation.

The paper analyses the performance of the microforming process by means of Finite Element simulation of the influence of size effects on the expected results. This is the basis for the setting or framing of design methodologies able to control surface topography errors by means of Geometrical Dimensioning and Tolerancing and statistical analysis tools. The paper highlights that simultaneous engineering techniques, already suggested for the development of standard components, are indispensable with micro-dimensional forming applications.

1 INTRODUCTION

Today industrial applications of thin metal sheet components regard specifically the electronic sector, where the miniaturization of circuitry compelled an analogous trend in the metal devices. A typical micro-forming process involves blanking and bending of a narrow metal strip on progressive dies to produce connectors, contact springs and leadframes. The goal is supporting the creation of new industrial fields with high added value, which are expected to steeply grow in the mechanical aspects, as much as in the electronic one.

Other forming processes of thin metal sheets are not yet investigated thoroughly [1]. Present study investigates the issues induced by transferring sheet metal forming technologies from the usual dimensions and thickness to the microscopic world. In this framework, investigations (mainly in Germany and in The Netherlands) focus on the scientific characterization of the rheological characteristics of thin foils and on the modelling of the plastic flow behavior of the material. Vice versa, present research deals with the industrial assessment of the forming processes, therefore analyzing the design of the process more than at its implementation.

Miniaturization complicates the control of geometrical variability of mechanical components. Tolerances assure functionality, economic manufacturing and quality of macro-mechanical products but in the conversion to micro-mechanical applications, the lack of empirical background and physical intuition of the process require a coherent, complete and unambiguous tolerancing methodology that is still to be developed.

Published in: E.Kuljanic (Ed.) *Advanced Manufacturing Systems and Technology*,
CISM Courses and Lectures No. 437, Springer Wien New York, 2002.

2 DESIGNING THE PRODUCT

Assuming to work in a Concurrent Engineering framework, designers have to assure the functionality of products as well as their economical manufacturing. The design methodology consists of the analysis of product functionality, its decomposition in functional requirements of components and, finally their translation into geometrical specifications. Designers establish a control on the geometrical variability of mechanical features, specifying size and dimensions, geometrical tolerances and geometrical properties of the surfaces according to Geometric Dimensioning and Tolerancing (GD&T) standards: ANSI Y14.5, DIN 7186-1, BSI 8888, ISO 286-1, ISO 1101, etc.

In spite of the enormous effort devoted to the formalization of tolerances definition there isn't a complete, coherent and unique tolerancing methodology. This situation is rapidly changing thanks to the Geometrical Product Specification (GPS) project driven by the ISO/TC213. The goal of the GPS project is the standardization in the field of surface properties, macro/micro geometry specifications, dimensional and geometrical tolerancing, verification principles, measuring equipment, calibration requirements. GPS standards could be profitably applied to micro-mechanical components, but some considerations are required.

The first issue concerns limits in the scaling process. GPS project classifies irregularities of surface or profile according to the ratio of the distance between irregularities (deviations, waves, cracks, etc.) to their depth. Form deviation has a ratio greater than 1000, waviness 100, roughness 5 and crack less than 5. It is obvious that overcoming some miniaturization level the spread among different types of irregularities keep on reducing until it disappears. In such a situation GPS standards couldn't apply and other methodologies should be developed.

A second issue is the type of control applied to geometrical variability: when the rate of miniaturization exceed the rate of improvement of the manufacturing process, geometrical errors cannot anymore be evaluated on the single component but have to be controlled over the entire manufacturing process.

According to an established practice in micro-electronic manufacturing, the worst case tolerancing is replaced by statistical tolerancing. The parallelism between electronic and mechanical microtechnologies is not so strict: large dimensional and geometrical errors in microelectronic components do not affect product's functionality because it is possible to stabilizes the nominal work point by applying an electrical retroaction, harder to implement in a mechanical device.

Statistical tolerancing is not fully developed in GPS or GD&T standards [2, 3]. The most advanced published standard in statistical tolerancing is the withdrawn German Standard DIN 7186 [4], which represented the first serious attempt in the direction of codifying statistical tolerancing by modifying the specification of a worst case tolerance interval [5].

Statistical tolerancing specifies the acceptability of a population of parts, without placing limits on individual parts. Some pieces, in a population not satisfying tolerances, possess the correct geometrical characteristics while some others, in a population satisfying tolerances, are so different from the nominal values as to be unusable. Any geometrical characteristic can have a statistical specification, indicated by the letters ST enclosed by a hexagon followed by a series of one or more frames which define the limits of acceptable workpiece distributions. There exist a number of ways in which these limits can be specified.

A first approach places limits on the population parameters (mean μ and standard deviation σ) using the Process Capability Indices (PCIs). A second approach defines a containment zone for the cumulative distribution function of the workpiece population (Distribution Function Zone - DFZ). Examples of these approaches are depicted in Fig. 1.

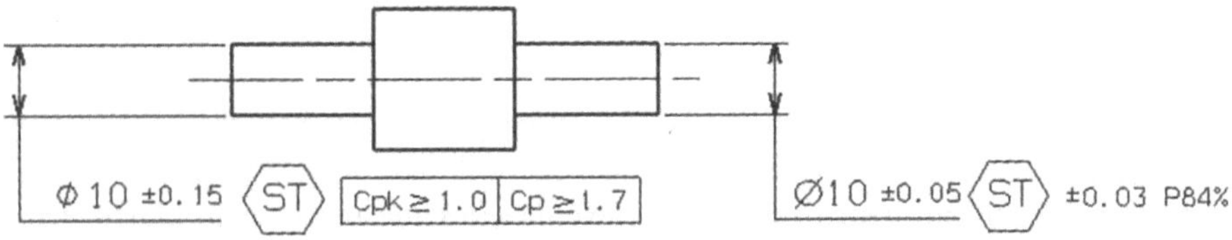

FIGURE 1. Specification of statistical tolerances with two different approaches

On the left side, PCIs-based specification controls diameter of cylindrical feature defining a finite region in the μ-σ plane describing possible workpiece distributions. Defined the Upper and Lower Specification Limit U and L of the worst-case tolerance and the target value τ of the toleranced characteristic, PCIs are computed from the mean μ and the standard deviation σ of manufacturing process according to (1-4). The trapezoidal shaded zone in Fig. 2a, based on (1-4), satisfies both conditions of the statistical reference frame of Fig. 1.

$$C_p = \frac{U-L}{6\sigma} \quad (1) \qquad C_{pk} = \min\left\{\frac{\mu-L}{3\sigma}, \frac{U-\mu}{3\sigma}\right\} \quad (2)$$

$$C_c = \max\left\{\frac{\tau-\mu}{\tau-L}, \frac{\mu-\tau}{U-\tau}\right\} \quad (3) \qquad C_{pm} = \frac{U-L}{6\sqrt{\sigma^2+(\mu-\tau)^2}} \quad (4)$$

On the right side of Fig. 1 a DFZ-based specification states that in any acceptable population of parts at least 86% of the diameters shall be within the interval [9.97, 10.3], at most 7% shall be within both the interval [9.95, 9.97] and [10.03, 10.05]. The resulting admissible DFZ is represented in Fig. 2b: any population of diameters whose distribution falls within the shaded zone is acceptable.

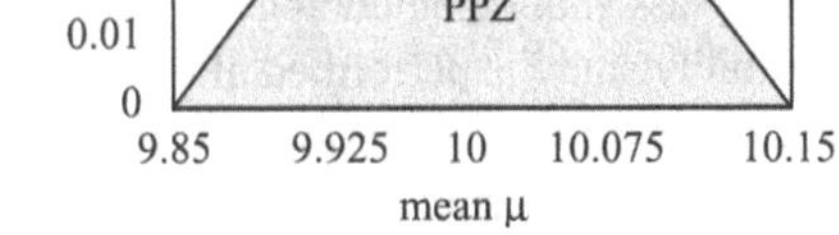

FIGURE 2a. Admissible region PCIs-based

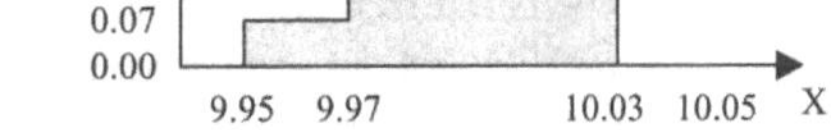

FIGURE 2b. Admissible region DFZ-based

As previously mentioned, statistical tolerancing does not assure a complete conformance with geometrical specification but it allows larger tolerances and improves the application of Statistical Process Control techniques.

3 ENGINEERING THE PRODUCT

3.1 PROCESS ASSESSMENT

Materials and shapes used in micro-forming are strongly different from the ones employed in conventional processes. The micro-formed components will have a wide usage that till now it is conceited to predict and to attempt classifying. Nevertheless some topics can be highlighted:

1. Micro-formed sheet components will be so small that material waste won't be considered a main issue, therefore the progressive dies with some strip traction mechanism could be the best solution to the material handling difficulties.
2. As assembly of many small parts is a challenging task, micro-formed techniques avoiding the movements of the parts from the initial fixture will be rewarding. Examples are multi-process integration of different functions, mainly forming and welding.
3. Esthetic will be scarcely considered a problem because also visual defects will be reduced to a barely visible scale.
4. As it will be impossible to detect flaws and defects inside the component, some ingenuity will be made necessary to implement a robust machines, i.e. a mechanical device able to withstand even some structural failure.

When engineering a micro-formed part, some precautions must be adopted. The manufacturing engineer must be always conscious of the poor precision - relative to the part dimensions - that he can achieve. Assembly issues become the actual technological challenge, while feasibility is nearly always guaranteed and the fabrication study is somewhat simplified. In order to better discuss the problems, we recall shortly the design process of a new sheet metal product in a modern facility [6].

The design begins with the development of the product concept. Then the conceptual design is converted into a draft in the feasibility step. At this stage materials, manufacturing sequence, machines are chosen, the process feasibility is tested by means of inverse FE codes and the overall economics is assessed. The production method consists in the design of the final shape and dimensions of the product, in the detailed process planning. To obtain these outcomes the forming process is analyzed using incremental FE codes and the process parameters are optimized. Eventually a prototype is realized and tested. In the same time the assembly methodology is assessed by choosing the method and the sequence. Usually assemblability study is performed after the manufacturability study. In the sheet metal forming field it is supposed that assembly problems can be solved by interventions on the process details without requiring an overall redefinition of the production process.

If we try to design a micro-formed product following the conventional sequence we face a number of contradictions. The product feasibility changes in every scale: bending limits, drawability, thickness reductions. Both the draft design and the formability study are realized before deciding the tolerances, therefore, following the discussion in the preceding chapter, their applicability is questionable. The process is simulated by means of FE codes using nominal values for dimensions and a standard flow curve for the material. Simulation is used in order to assess and optimize process parameters, blank size and shape and to define the exact shape of the dies for the intermediate forming steps whose number has been already defined in the feasibility stage. The only way to overcome the uncertainties of micro-forming processes is to execute different simulations covering the whole range of values present in the productive environment and to substitute the deterministic results that are expected from the simulation with a sensitivity study. In other words it will no more be possible to find the optimal blank profile to form a given component, but only a compromise one.

3.2 ACCOUNTING FOR SIZE EFFECTS

The miniaturization of sheet metal forming processes must cope with size effects [7, 8]. It is possible to make a distinction between first and second order size effects. First order size effects are caused directly by the reduction of dimensions and show themselves as a change in the relative importance of manufacturing parameters. Second order effects are indirectly linked to the sheet thickness. They are caused by changes in the material or process behavior during the scaling down.

What is important from our point of view is that first order effects can be accounted on by common engineering tools like CAD-CAM and FEM, second order effects ask for a new characterization of the materials together with different constitutive laws because of the non negligible importance of grain size and shape changes.

Between the first order effects, in order to assure industrial applicability, the tolerance design is the main issue. The list of second order effects is quite long and the way of action of every effect is not yet completely understood [9, 10, 11]: flow stress, anisotropy, ductility, friction, die deformation, bending behavior. In a thin metal sheet, the grains that are on the surface have a large part of their boundary free from contact with adjacent grains, therefore constraint forces on the surface are less. This explains many of the behaviors exhibited by micro-sheets.

4 DEEP DRAWING OF A SQUARE CUP

The proposed case study consists of a square cup having dimensions 20x20x8 mm. According to the GD&T approach, shape and dimensions of the cup are controlled by geometrical tolerances as seen in Fig.3. The bottom of the cup (datum A), shall be put on a perfect plane. Planarity control is required in order to avoid errors in the contact between datum feature and datum simulator. The position of three holes is verified with Maximum Material Condition. The pattern of the three holes also constitutes the secondary datum feature B, which locates the cup when inspectors verify the geometrical errors on the cup walls. It is possible to apply statistical tolerances to transfer the conformance verification from the single manufactured part to the manufacturing process.

Four different scaling levels (called length scales λ) are chosen to study the production method highlighting the differences among scales. Each level corresponds to a different sheet thickness. All the geometrical dimensions are scaled down following the geometric similarity. The four levels correspond to a thickness of 1, 0.5, 0.1, 0.01mm.

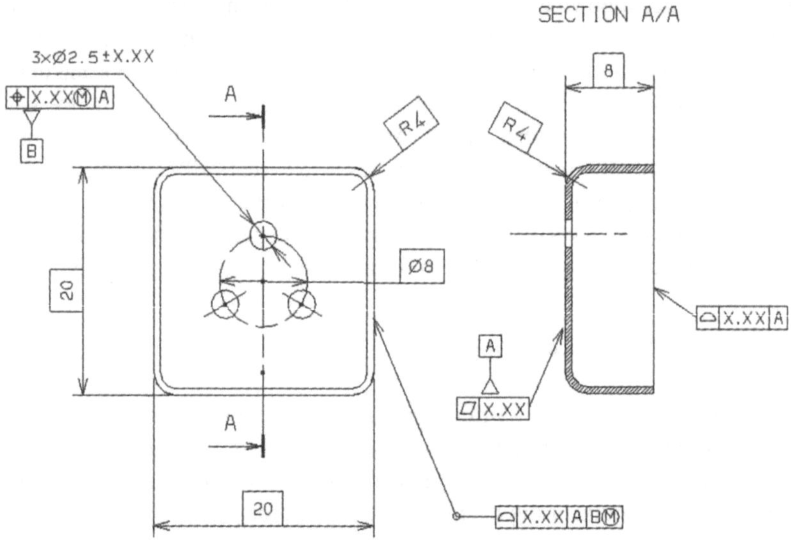

FIGURE 3. Square box

The FEM simulations are performed on a one-quarter symmetrical area for the CuNi18Zn20 alloy. The rigid square punch is a 20 x 20mm^2 flat surface rounded at the edges with a 4mm radius. The impossibility to guarantee a reasonable clearance between die and punch force the choice of the hydroforming process [12]. A liquid substitutes the die applying a uniform pressure all over the upper face of the sheet. In the real process a binder is provided opposite to the blankholder. Here it is not used to amplify the effect of different process parameters. Furthermore, the very small dimensions of the part makes harder the use of a binder.

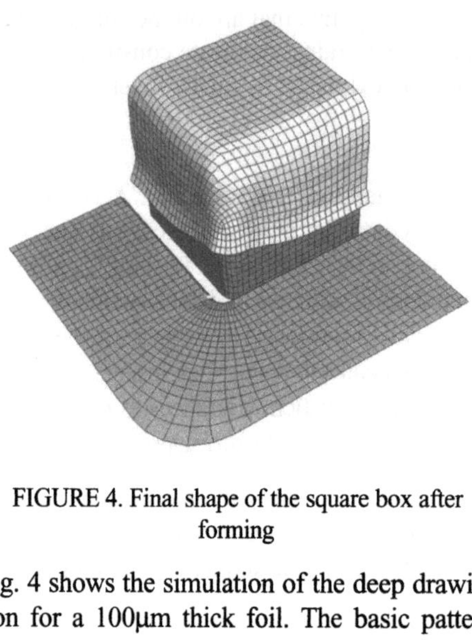

FIGURE 4. Final shape of the square box after forming

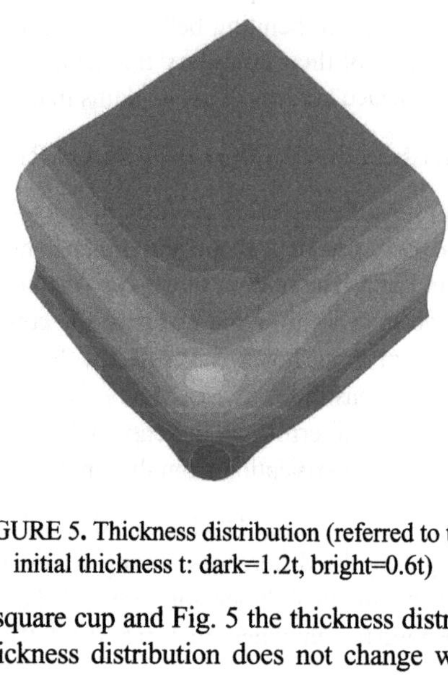

FIGURE 5. Thickness distribution (referred to the initial thickness t: dark=1.2t, bright=0.6t)

Fig. 4 shows the simulation of the deep drawing of a square cup and Fig. 5 the thickness distribution for a 100μm thick foil. The basic pattern of thickness distribution does not change when scaling down the dimensions: the upper corner shows the maximum thinning and the bottom corner has the maximum increase in thickness. The profile of the box edge is irregular with a visible tendency to form waves. The waviness is greater for large thickness and reduces to its minimum in the 10μm thick foil.

Three second order effects are considered [6,7]: the deformability of the die, the increase in the friction factor and the change in the material characteristic and anisotropy. The results of the simulation show a reduction of the minimum thickness for all the three factors (Fig. 6).

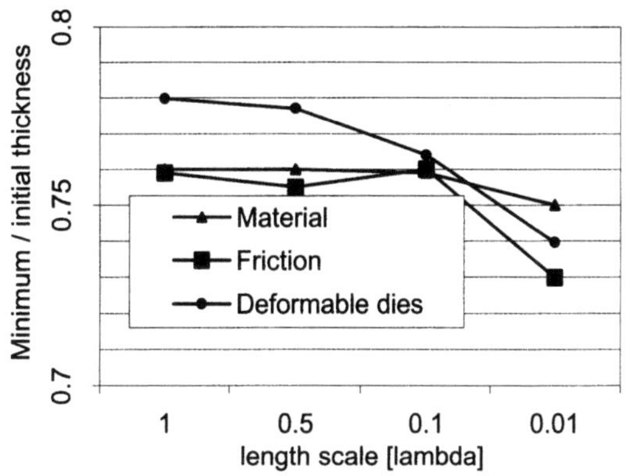

FIGURE 6. Minimum thickness after forming for different second order effects

The values of the principal indicators of process quality, thickness and principal strains, do not change very much among the different λ .

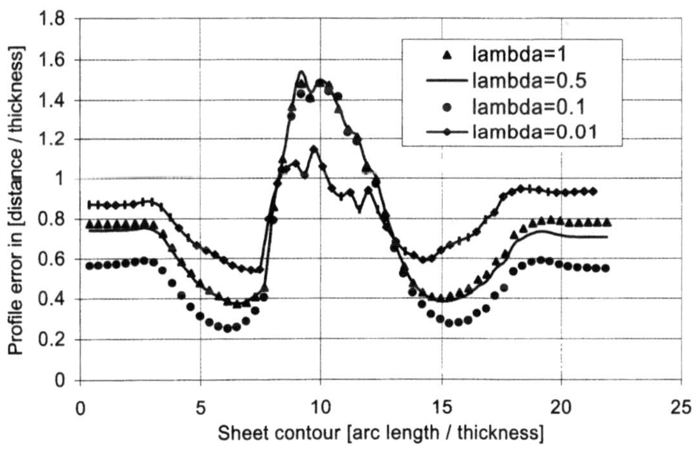

FIGURE 7. Profile error considering only the effect of die deformation

Table 1 shows that the ratio between standard deviation, computed for every effect and for every indicator, and average value is always under 10%. The greatest change is found in the bottom profile of the box, as shown in Fig. 7. Here the figure of the error with respect to the punch nominal profile is plotted as a fraction of the initial thickness. What is apparent is the fact that the minimum sheet thickness is always characterized by the smoother behavior of the profile (less waviness), despite a greater average error.

TABLE 1. Standard deviation of process variables as a percentage of the average values

Stand dev / average	ε_{max}	ε_{min}	t_{max}	t_{min}
Material	2%	-4%	1%	4%
Friction	2%	-3%	0%	2%
Deformable dies	3%	-8%	1%	2%

5 CONCLUSION

The paper analyses some methodologies applicable in the design of micro-formed mechanical components as well as in the choice of their manufacturing process. The investigation on the deep drawing process of a small square cup confirms that a simple translation of design methods from the macro to the micro-world does not provide good results. It also highlights that a full control on the design, manufacturing and inspection phase, requires a deeper comprehension and a better mathematical description of engineering activities involved in the production of micro-devices.

REFERENCES

1. Geiger, M., Kleiner, M., Eckstein, R., Tiesler, N., Engel, U., (2001), Microforming. Annals of the CIRP, vol.50, pp. 1-18.
2. Srinivasan V., O'Connor M.A., (1994), On interpreting statistical Tolerancing. Manufacturing Review, 7 (4), pp.304-311
3. Srinivasan V., O'Connor M.A., (1996), Towards an ISO Standard for Statistical Tolerancing. In: F. Kimura (ed.), Computer-aided Tolerancing. Chapman & Hall, pp.159-172
4. DIN 7186 – 1, (1974), Statistiche Tolerierung: Begriffe, Anwendungsrichtlinien und Zeichnungsangaben. Deutsche Normen, Berlin 30, BRD
5. Henzold G., (1995), Handbook of Geometrical Tolerancing. Design, Manufacturing and In-spection. John Wiley & Sons Ltd, London.
6. Geiger, M., Vollertsen, F., Kals, R., (1996), Fondamentals on the Manufacturing of Sheet Metal Microparts. Fertigungstechnologie, Erlangen, Germany.
7. Kals, T.A., Eckstein, R., (2000), Miniaturization in sheet metal working. J. Of Mat. Proc. Tech., Elsevier, vol.103, pp. 95-101.
8. Eckstein, R., Geiger, Engel, M., (1999), Specific Characteristics of Micro Sheet Metal Working. Proceedings of the SheMet'99, Erlangen, Germany, pp. 529-536.
9. Picart, P., Michel, J.F., (1999), Effects of size and texture on the constitutive behaviour for very small components in sheet metal forming. Proceedings of the 6th ICTP, vol. II, pp. 895-900.
10. Tiesler, N., Engel, U., (2000), Microforming - effects of miniaturization, Proceedings of the 8th Metal Forming, Rotterdam, A.A. balkema, pp. 355-360.
11. Raulea, L.V., Goijaerts, A.M., Govaert, L.E., Baaijens, F., Size effects in the processing of thin metal sheets. Proceedings of the SheMet'99, Erlangen, Germany, pp. 521-528
12. Siegert, K., Haussermann, M., Losch, B., Rieger, R., (2000), Recent developments in hydroforming technology. J. Of Mat. Proc. Tech., Elsevier, vol.98, pp. 251-258.

FINITE ELASTOPLASTICITY IN PLANE STRAIN COLD ROLLING PROBLEM

J. Brnic[1], M. Canadija[1], G. Turkalj[1]

[1] Department of Engineering Mechanics, Faculty of Engineering, University of Rijeka, Croatia

KEYWORDS: Cold rolling, Elastoplasticity, Continuum mechanics.

ABSTRACT. A detailed continuum mechanics formulation as a threshold for application of the Updated Lagrangian Hencky formulation to plane strain rolling problem is presented. The main part of this paper consists of detailed derivation of finite strain elastoplasticity constitutive equations. This formulation is especially suitable for analysis of metal forming processes since they generally involve large deformations. Based on this formulation, efficient numerical algorithm is developed and briefly described. In addition, several other aspects of numerical analysis are briefly addressed: friction, incompressibility and finite element model. The frictional conditions at contact interface are simulated by the friction layer technique. Incompressibility constraint is imposed by the mixed displacement-pressure FEM formulation. Efficiency is demonstrated on the numerical analysis of cold rolling of copper workpiece.

1 INTRODUCTION

Rolling is the process of reducing the cross section of the workpiece by a squeezing action as it passes between two rotating rolls. If the workpiece is a thin strip, with the thickness/width ratio of at least ten, then process is essentially plane strain problem. Only exceptions are narrow zones near the edges of the workpiece.

In the case of heavy reductions in cross sections, the process should be preformed at elevated temperatures. This is the case of hot rolling. Otherwise, if the process is carried out at room temperature and one can speak about cold rolling.

Accurate analytical apparatus is still not known. Therefore, only approximate solutions can be obtained with modest insight into process mechanics. At this pace of rolling production, this is not acceptable for modern industry. Open market competition demands constant increase of product quality with lower price. Consequently, engineers have to provide a wide palette of detailed results. The most suitable method to accomplish this is the numerical approach, such as finite element method or finite volume method. Due to its wide popularity, finite element method is probably the first choice.

2 REVIEW OF SOME BASIC RESULTS IN CONTINUUM MECHANICS WITH APPLICATION TO PROBLEMS IN MODELLING OF COLD ROLLING PROCESS

Numerical simulation of rolling involves numerous problems. Deformation process is non-steady when workpiece enters into the roll-gap. However, it becomes steady after the workpiece has passed fully into and through the roll gap. Model should provide good description of finite plastic strains, enforce isochoric character of plastic deformation and simulate friction at work-piece – die interface. Furthermore, contemporary finite strain theories [1][2][3] are based on

Published in: E.Kuljanic (Ed.) *Advanced Manufacturing Systems and Technology*,
CISM Courses and Lectures No. 437, Springer Wien New York, 2002.

multiplicative decomposition of deformation gradient and should be incorporated into rolling models. This was done recently in [4][5].

2.1 NEUTRAL POINT POSITION

Well-known principle from mechanics of continuum [6], continuity equation, in its local form using spatial coordinates states:

$$\frac{d\rho}{dt} = \rho \operatorname{div} \mathbf{v} . \qquad (1)$$

This applies to the workpiece material as well. Therefore, if the material is incompressible so that the density in the neighborhood of each material particle remains constant as it moves, continuity equation takes simpler form:

$$\operatorname{div} \mathbf{v} = 0 . \qquad (2)$$

The equation (2) is generally known as condition of incompressibility. In material coordinates, equation (1) takes form:

$$\frac{d}{dt}(\rho J) = 0 , \qquad (3)$$

where J is Jacobian determinant of deformation gradient and can be expressed as

$$J = \frac{dv}{dV} . \qquad (4)$$

Introducing $\rho = \text{const.}$ in (3) gives $J = 1$.

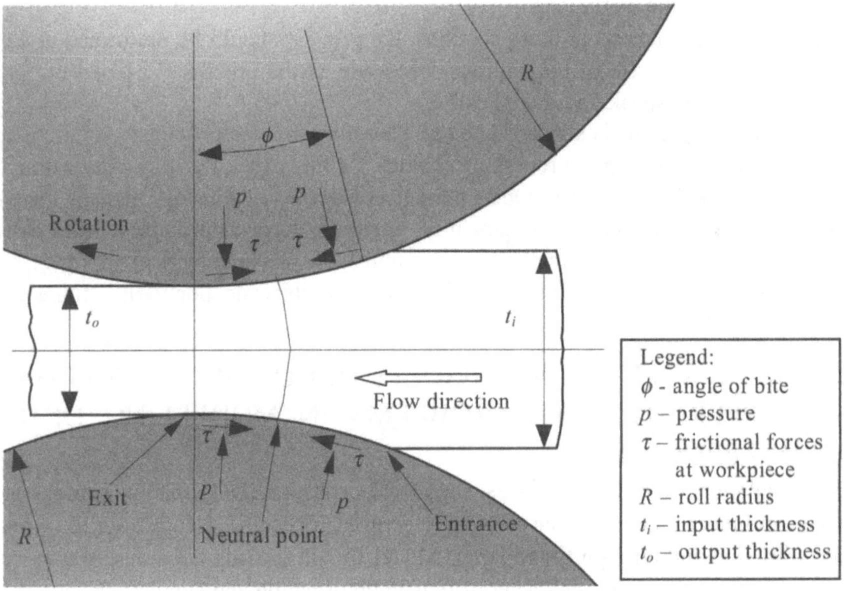

FIGURE 1. The definition of geometry in rolling

At the entrance to the roll gap, material is pulled between rolls by the frictional forces, Fig. 1. Direction of frictional forces obviously implies that velocity of workpiece material is slower than velocity of rolls at the interface. If we at this point neglect volume change caused by elastic deformation and treat the whole process as incompressible, then due to (3, 4) $J = 1$ and volume change should not take place. This means that as material flows to the exit, velocity must continuously grow and the end material velocity has to be faster than velocity of roll. As a consequence, direction of frictional forces at the exit is opposite to the frictional forces at the entrance. Since material flows slower at the entrance and faster at the exit, somewhere between these two points velocities of material and roll are equal. In hot rolling of metals this is a zone called neutral zone, while during cold rolling it is assumed that velocities are equal only in one point – neutral point [7]. This is the point where frictional forces change direction of action. Position of this point in not known in advance, and this presents difficulty in formulation of boundary conditions [8]. To circumvent these problems simple and efficient procedures have been developed [9][10].

2.2 MULTIPLICATIVE FINITE STRAIN ELASTOPLASTICITY

Finite strain elasto-plastic deformation is described by a multiplicative decomposition of deformation gradient into elastic and plastic part [3]:

$$\mathbf{F} = \mathbf{F}^e \mathbf{F}^p . \tag{5}$$

Velocity gradient is:

$$\mathbf{L} = \dot{\mathbf{F}} \mathbf{F}^{-1} . \tag{6}$$

It can be separated into elastic and plastic velocity gradient:

$$\mathbf{L} = \mathbf{L}^e + \mathbf{L}^p , \tag{7}$$

where are:

$$\mathbf{L}^e = \dot{\mathbf{F}}^e \left(\mathbf{F}^e\right)^{-1}, \qquad \mathbf{L}^p = \mathbf{F}^e \dot{\mathbf{F}}^p \left(\mathbf{F}^p\right)^{-1} \left(\mathbf{F}^e\right)^{-1} . \tag{8}$$

As a strain measure several tensors can be used. We select Hencky or logarithmic strain tensor defined as:

$$\mathbf{E}^e = \ln \mathbf{U}^e , \tag{9}$$

where $\mathbf{U}^e$ is elastic right stretch tensor obtained from polar decomposition theorem:

$$\mathbf{F}^e = \mathbf{R}^e \mathbf{U}^e . \tag{10}$$

$\mathbf{R}^e$ is elastic rotation tensor. Stress conjugate to Hencky strain tensor is rotated stress tensor $\bar{\boldsymbol{\tau}}$:

$$\bar{\boldsymbol{\tau}} = \left(\mathbf{R}^e\right)^T \boldsymbol{\tau} \mathbf{R}^e . \tag{11}$$

Therefore, variables that describe elasto-plastic process are $\{\mathbf{F}, \mathbf{F}^p, \boldsymbol{\tau}, \sigma\}$. Evolution of these variables completely defines whole elastoplastic deformation process. Last variable σ is isotropic hardening and defined to be a scalar function of equivalent plastic strain, i.e. $\sigma = \sigma(e^p)$. Due to simplicity, kinematic hardening is not included in this model. In order to define these state variables, constitutive equation is necessary for $\boldsymbol{\tau}$, while for $\mathbf{F}^p$ and σ evolution equations need to be defined. Considerations that follow solve these problems.
Helmholtz free energy per unit mass is defined by [6]:

$$\psi = u - s\theta , \tag{12}$$

where s is entropy, θ is temperature and u is internal energy. Time differentiation gives:

$$\dot{\psi} = \dot{u} - \dot{s}\theta - s\dot{\theta} . \tag{13}$$

From the first law of thermodynamics follows:

$$\rho \frac{du}{dt} = \boldsymbol{\tau} : \mathbf{D} + \rho r - \nabla \cdot \mathbf{q} , \tag{14}$$

Where $\boldsymbol{\tau}$ is Cauchy stress tensor, $\mathbf{D}$ = sym ($\mathbf{L}$) is deformation rate tensor, r is internal heat source and $\mathbf{q}$ is heat flux. Skew-symmetric part of $\mathbf{L}$ is assumed to be zero within the context of present theory. Also, local form of Clausius – Duhem inequality is:

$$\frac{ds}{dt} \geq \frac{r}{\theta} - \frac{1}{\rho} \nabla \cdot \left(\frac{\mathbf{q}}{\theta} \right), \tag{15}$$

This equation can be transformed to:

$$\rho\left(-\dot{\theta}\, s - \dot{\psi}\right) + \boldsymbol{\tau} : \mathbf{D} - \frac{1}{\theta} \mathbf{q} \cdot \operatorname{grad}\theta \geq 0 , \tag{16}$$

Introducing free energy and entropy per unit volume:

$$W = \rho \psi, \qquad S = \rho s . \tag{17}$$

gives:

$$-\left(\dot{\theta}\, S + \dot{W}\right) + \boldsymbol{\tau} : \mathbf{D} - \frac{1}{\theta} \mathbf{q} \cdot \operatorname{grad}\theta \geq 0 . \tag{18}$$

In isothermal case, i.e. $\dot{\theta} = 0$:

$$-\dot{W} + J\boldsymbol{\tau} : \mathbf{D} \geq 0 . \tag{19}$$

Separation of $\mathbf{D}$ into elastic and plastic parts gives further:

$$-\dot{W} + \boldsymbol{\tau} : \left(\mathbf{D}^e + \mathbf{D}^p\right) \geq 0 . \tag{20}$$

Using elastic work conjugacy [10]:

$$\boldsymbol{\tau} : \left(\mathbf{D}^e + \mathbf{D}^p\right) = \overline{\boldsymbol{\tau}} : \dot{\mathbf{E}}^e + \overline{\boldsymbol{\tau}} : \overline{\mathbf{D}}^p \tag{21}$$

where

$$\overline{\mathbf{D}}^p = \operatorname{sym}\left(\overline{\mathbf{L}}^p\right) = \operatorname{sym}\left(\dot{\mathbf{F}}^p \left(\mathbf{F}^p\right)^{-1} \right) \tag{22}$$

we obtain

$$-\dot{W} + \overline{\boldsymbol{\tau}} : \dot{\mathbf{E}}^e + \overline{\boldsymbol{\tau}} : \overline{\mathbf{D}}^p \geq 0 . \tag{23}$$

Furthermore, separating the free energy into elastic and plastic parts:

$$W\left(\mathbf{E}^e, e^p\right) = W^e\left(\mathbf{E}^e\right) + W^p\left(e^p\right). \tag{24}$$

gives:

$$\left(\overline{\boldsymbol{\tau}}-\frac{\partial W^{e}}{\partial \mathbf{E}^{e}}\right):\dot{\mathbf{E}}^{e}+\overline{\boldsymbol{\tau}}:\overline{\mathbf{D}}^{p}-\dot{W}^{p}\geq 0. \tag{25}$$

Standard argument that (25) must hold for all admissible processes state:

$$\overline{\boldsymbol{\tau}}=\frac{\partial W^{e}}{\partial \mathbf{E}^{e}},\quad \mathsf{D}=\overline{\boldsymbol{\tau}}:\overline{\mathbf{D}}^{p}-\dot{W}^{p}\geq 0. \tag{26}$$

Since $\overline{\mathbf{D}}^{p}=\mathrm{sym}\left(\overline{\mathbf{L}}^{p}\right)$ we use (22) and (26_2) to define evolution equation for the $\mathbf{F}^p$. Actual stress state $\overline{\boldsymbol{\tau}}$ maximizes dissipation function D and is subjected to constraints

$$\Phi\leq 0,\qquad \dot{e}^{p}\Phi=0,\qquad \dot{e}^{p}\geq 0, \tag{27}$$

where Φ defines yield surface. It is represented by:

$$\Phi=\sqrt{\frac{3}{2}}\left\|\mathrm{dev}\,\overline{\boldsymbol{\tau}}\right\|-\sigma, \tag{28}$$

Necessary condition for such maximum of D is:

$$\frac{\partial \mathsf{D}}{\partial \overline{\boldsymbol{\tau}}}-\dot{e}^{p}\frac{\partial \Phi}{\partial \overline{\boldsymbol{\tau}}}=0. \tag{29}$$

These conditions are classical in the convex mathematical programming literature [11] and are known as Kuhn – Tucker conditions. Following equations are easily established from (26) and (28):

$$\frac{\partial \mathsf{D}}{\partial \overline{\boldsymbol{\tau}}}=\overline{\mathbf{D}}^{p},\qquad \frac{\partial \Phi}{\partial \overline{\boldsymbol{\tau}}}=\sqrt{\frac{3}{2}}\frac{\mathrm{dev}\,\overline{\boldsymbol{\tau}}}{\left\|\mathrm{dev}\,\overline{\boldsymbol{\tau}}\right\|}. \tag{30}$$

Using equations (29) and (30) we obtain:

$$\overline{\mathbf{D}}^{p}=\sqrt{\frac{3}{2}}\dot{e}^{p}\overline{\mathbf{N}},\qquad \dot{e}^{p}=\sqrt{\frac{2}{3}\overline{\mathbf{D}}^{p}:\overline{\mathbf{D}}^{p}}. \tag{31}$$

Here $\mathbf{N}$ is normal to the yield surface in rotated stress space. Using (22) and (31) we obtain evolution equation for $\mathbf{F}^p$. Finally, evolution equation for σ is defined in common form:

$$\dot{\sigma}=H\cdot\dot{e}^{p}, \tag{32}$$

where $H=H\left(e^{p}\right)$ is hardening modulus.

2.3 NUMERICAL IMPLEMENTATION

Presented theory is implemented into finite element model. Both material and geometric non-linearity characterize nonlinearity of this model. Material nonlinearity is based on above description of elastoplastic strain evolution and computational procedure is basically a predictor – corrector method. Correction is performed by the standard radial return method. For details, see [3] and references therein.

As it is generally known, displacement based finite elements are not efficient in incompressible or nearly incompressible problems. Basic flaw is so-called locking behavior that results is

problems with calculation of spherical part of stress tensor. To deal with this obstacle, i.e. to enforce isochoric material behavior, mixed displacement – pressure FEM is used. In this approach, in addition to displacement variables pressure variables were also used. This method exhibit superior performance to standard displacement method [12].
To solve contact problems we used the friction layer technique [9][10]. This method is relatively simple to use and gives satisfactory results. Friction layer technique is applicable not only to rolling problems, but also in all metal forming problems.

3 EXAMPLE

To evaluate efficiency of presented procedure, cold rolling of copper is analysed. For this problem, a well – documented experimental results are available in [13]. This reference is a used as a standard verification for rolling models.
Geometrical and material properties are summarised in Table 1 and are selected according to mentioned experiment [13].

TABLE 1. Geometrical and material properties

Property	Symbol	Value
Initial thickness	t_i	6.350 mm
Output thickness	t_o	5.791 mm
Roll radius	R	79.375 mm
Initial workpiece length	ℓ_0	16 mm
Elasticity modulus	E	110 GPa
Friction factor	m	0.15
Yield stress	σ_{y0}	70.5 MPa
Poisson's ratio	ν	0.30
Hardening curve	$\sigma = 70.5\left(1+\frac{1}{0.022}e_p\right)^{0.49}$	

Thickness of the workpiece is considered small compared to the width and the process will be analysed as a plane strain problem. Obvious symmetry in geometry and loads implies that only one roll and half of workpiece need to be discretised. Rolls are assumed to be rigid.
This data is input into in-house developed software and analysis was carried out. Some results are presented below. Obtained pressure distribution at the roll – workpiece interface is presented in Fig. 2 and compared with experimental results [13].
To emphasise neutral point occurrence, field of relative velocities were also calculated, Fig. 3. Position of neutral point is determined by investigation of sign change of shearing stresses at roll – workpiece interface. Velocities are then calculated with respect to the neutral point velocity, i.e. horizontal component of velocity of the neutral point is subtracted from every other horizontal component of velocity. It is clearly seen that material is moving slower than roll before neutral point, while after neutral point its velocity increases and is greater than roll velocity. Material is therefore behaving as it was predicted by continuity equation (1).

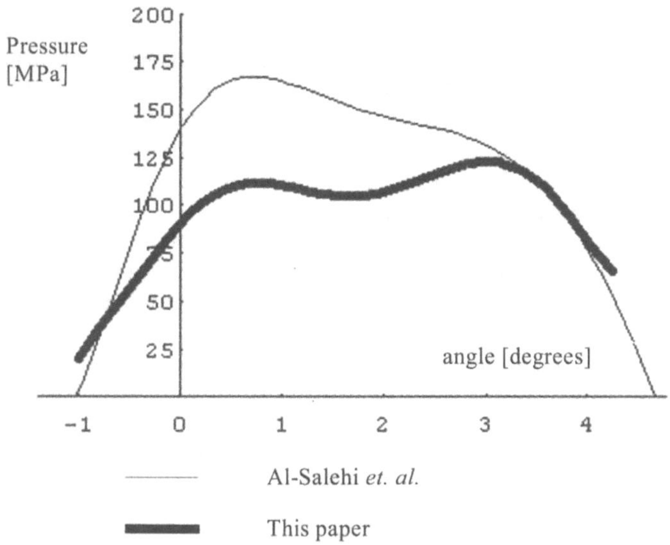

FIGURE 2. Pressure distribution at roll – workpiece interface

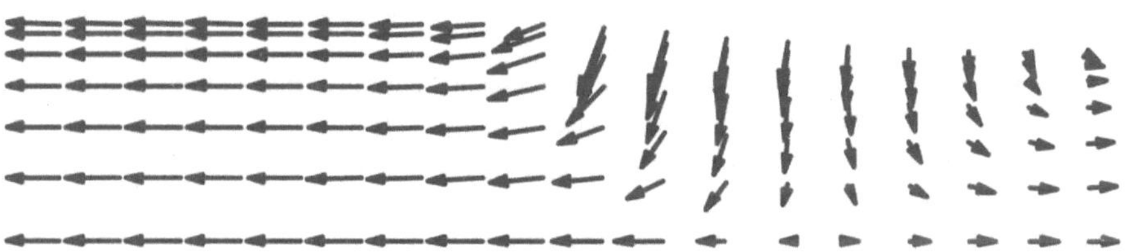

FIGURE 3. Relative velocities field

4 CONCLUSION

The main goal of this paper was formulation of procedure suitable for simulation of cold rolling process. Intention of authors was to include finite strain plasticity model that is in accordance with laws of finite strain continuum mechanics. Consequently, this approach circumvents problems and inaccuracies arising in application of additive decomposition of strain tensor model and other infinitesimal techniques. Simulation of frictional forces was performed by friction layer technique. Although today exist other perhaps more accurate contact schemes, we find this technique relatively simple to implement. As it can be noticed from example, it provides satisfactory results.

REFERENCES

1. Simo, J.C., (1988), A Framework for Finite Strain Elastoplasticity Based on Maximum Plastic Dissipation and the Multiplicative Decomposition: Part I. Continuum Formulation. Computer Methods in Applied Mechanics and Engineering, Vol 66, 199-219.
2. Simo, J.C., (1988), A Framework for Finite Strain Elastoplasticity Based on Maximum Plastic Dissipation and the Multiplicative Decomposition: Part II. Computational Aspects. Computer Methods in Applied Mechanics and Engineering, Vol 68, 1-31.
3. Eterovic, A.L., Bathe, K.J.,(1990), A Hyperelastic-Based Large Strain Elasto-plastic Constitutive Formulation with Combined Isotropic-Kinematic Hardening Using the Logarithmic Stress and Strain Measures. International Journal for Numerical Methods in Engineering, Vol 30, 1099-1114.
4. Maniatty A.M., Dawson P.R., Weber, G.G., (1991), An Eulerian Elasto-Viscoplastic Formulation for Steady-State Forming Processes. International Journal for Mechanical Sciences, Vol 33, 361-377.
5. Dixit, U.S., Dixit, P.M. (1997), A Study on Residual Stresses in Rolling. International Journal of Machine Tools & Manufacturing, Vol 37, 837-853.
6. Malvern, L.E., (1969), Introduction to the Mechanics of a Continuos Medium. Prentice-Hall, Englewood Cliffs, New Jersey.
7. Wusatowski, Z., (1969), Fundamentals of Rolling. Pergamon Press, Oxford.
8. Zienkiewicz, O.C., Jain, P.C., Onate, E., (1978), Flow of Solids During Forming and Extrusion: Some Aspects of Numerical Solutions. International Journal of Solids and Structures, Vol 14, 15-38.
9. Li, G.J., Kobayashi, S., (1982), Rigid-Plastic Finite-Element Analysis of Plane Strain Rolling. Journal of Engineering for Industry – Transactions of the ASME, Vol 40, 55-64.
10. Liu, C., Hartley, P., Sturgess, C.E.N., Rowe, G.W., (1985), Simulation of the Cold Rolling of Strip Using an Elastic-Plastic Finite Element Technique. International Journal for Mechanical Sciences, Vol 27, 829-839.
11. Liu, C., Hartley, P., Sturgess, C.E.N., Rowe, G.W., (1985), Elastic-Plastic Finite-Element Modelling of Cold Rolling of Strip. International Journal for Mechanical Sciences, Vol 27, 531-541.
12. Bathe, K.J., (1996), Finite Element Procedures. Prentice-Hall, Englewood Cliffs, New Jersey.
13. Luenberger, D.G., (1984), Linear and Nonlinear Programming, Addison – Wesley Publishing Company, Reading, Massachusetts.
14. Brnic, J., Canadija, M., Turkalj, G., (1999), Comparison of Measured and Computed Contact Pressure Distribution in Cold Sheet Rolling Process. Advanced Manufacturing Systems and Technology, CISM Courses and Lectures No. 372, Kuljanic, E. (Ed.), Springer Wien New York, 337-344.
15. Al-Salehi, F.A.R., Firbank, T.C., Lancaster, P.R. (1973), An Experimental Determination of the Roll Pressure Distribution in Cold Rolling. International Journal for Mechanical Sciences, Vol 15, 693-710.

BURNISHING VERSUS GRINDING FOR AUTOMOTIVE PARTS

L. Luca[1], I. Marinescu[1], S. Neagu-Ventzel[1]

[1] Department of Mechanical, Industrial and Manufacturing Engineering, University of Toledo, USA

KEYWORDS: Burnishing, Surface finish, Grinding

ABSTRACT. Burnishing is a chipless finishing method, which employs a rolling tool in order to achieve a plastic deformation of the surface layer of metallic parts. Some particular features of this process make it a good cheap alternative for other finishing methods in certain applications. Burnishing of soft metals (up to 45 HRC) has been a field extensively studied, unlike burnishing of hard heat-treated steels. This article describes a method of burnishing of hard steel surfaces (about 60 HRC) by means of a hard ball. The related tooling, based on a hydrostatic principle, is also presented. Experiments are exposed, together with the results and conclusions. In this stage, only the roughness was tracked, as a measure of burnishing performance. Influence of main burnishing parameters against final surface roughness is shown in diagrams. As a general conclusion, burnishing of hardened metallic surfaces proved to be a viable alternative for finishing.

1 INTRODUCTION

The principle of the burnishing process (figure 1) consists of the movement of a burnishing tool on the workpiece's surface, in the presence of a normal load. The burnishing tool can be a rolling tool (a roller or a ball) or a sliding tool (a diamond, carbide or ceramic tip). The initial surface has a microrelief made of a series of peaks and valleys of irregular height. When a normal load is applied on the ball, a compressive stress state occurs in the surface layer of the workpiece, resulting in plastic deformation as soon as the yield point of the workpiece's material is exceeded. These stresses are distributed asymmetrically and their magnitude is approximately uniform along the burnished workpiece. While the ball and the workpiece are rotating, this deformation progresses continuously over the entire surface.[1][2]

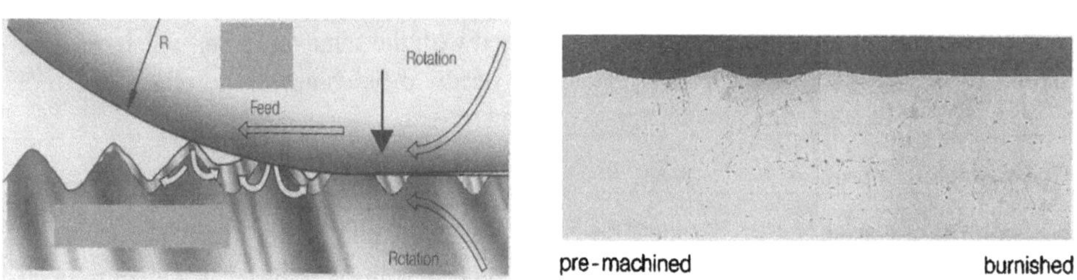

FIGURE 1. Schematics of the burnishing principle [3]

Published in: E.Kuljanic (Ed.) *Advanced Manufacturing Systems and Technology*, CISM Courses and Lectures No. 437, Springer Wien New York, 2002.

This technique allows obtaining a smooth surface, with roughness comparable with grinding, by plastic deformation of the surface layer. The peaks are pressed down to the surface, while the valleys are filled up from below. Flattening of asperities is not due to the bending or beading of the peaks, as it had falsely been believed, but to the fact that the entire material near the surface flows. The arrows in figure 1 (left side) shows this; the dash-and-dot-line represents the original surface geometry, at a reduced axial scale. Figure 1 (right side) shows a true scale undistorted section of the surface profile. At the same time, the compressive stresses induced within the surface layer lead to its work hardening and have a beneficial effect on mechanical properties of the layer. The compressive stresses in axial direction are of great importance for improving the fatigue strength.[4]

From the economic point of view, burnishing is a lot less expensive than grinding; table 1 presents a summarized comparison between these two machining processes from the effectiveness point of view:

TABLE 1. Burnishing versus grinding [5]

Grinding	Burnishing
Employs sophisticated and expensive machine tools.	Can be used on any precision lathe.
It cannot be coupled with other operations on the same machine tool. Increased lead and inter-operational times. Additional fixtures required.	Can be performed on the same lathe with rough and fine turning.
High energy consumption.	Very low energy consumption.
Very expensive tools.	Relatively inexpensive tools.
Auxiliary operations: truing and dressing, which are time consuming and require skilled operators.	No auxiliary operation needed.
Hundreds/thousands of parts machined with the same tool after which the tool is scrapped.	Thousands/tens of thousands of parts machined with the same tool after which only the roll needs to be changed, the rest of the tool being reused.
Chips generating process.	Chipless process.
No significant variations of hardness induced in the superficial layer; possible distortions and occurrence of microcracks.	Surface hardening with a 5-10% increase. Compressive stress induced in the superficial layer.

Burnishing of soft materials (up to 45 HRC) has been extensively studied, while burnishing of hard materials (over 45 HRC) was considered inefficient. Lately, due to development of new materials for the tool, burnishing could be extended to heat-treated metal components with hardness around 60 HRC. This type of operation can be used as a subsequent operation after hard turning. The usage of hard turning as a finishing operation is often restricted when rolling stress resistance is required. Moreover, wear of the turning insert leads to deterioration of surface. Burnishing can improve both the surface strength and roughness. The increase of surface strength serves mainly to improve fatigue resistance under dynamic loads. A big field of application is in the area of axles and slideways.[6]

2 EXPERIMENTAL SETUP AND METHODOLOGY

2.1 THE BURNISHING TOOL

In order to exceed the yield stress of the burnished material, a sufficiently high pressure must be developed at the contact area between the indenting tool and the workpiece. For burnishing hardened steel, with a very high hardness, this pressure can be achieved efficiently by means of a ball, due to the point contact. Moreover, given the tough circumstances of operation, any sliding motion of the ball onto the workpiece's surface must be avoided.

A common tool setup for burnishing is shown in figure 2 (left side). The burnishing force is ensured by the spring, while a retainer made out of low friction coefficient material (*e.g.* bronze) keeps the ball. This setup works well for burnishing of soft steel, but the contact between the ball and the retainer impeded periodically the free rotation of the ball in any direction; additionally, bronze from the retainer adhered on the ball's surface.[7]

The aforementioned disadvantages are eliminated through a recent hydrostatic design. The principle of this tool is based on what is shown in figure 2 (right side). The key element is a ceramic ball, which is the indenter. A hydrostatic pressure is applied behind the ball. This pressure generates the force, which presses the ball against the workpiece. The ball is supported entirely by the fluid, so it can freely rotate in any direction. The hard ceramic ball and the possibility of free rotation of the ball makes this setup effective for burnishing of hardened materials.

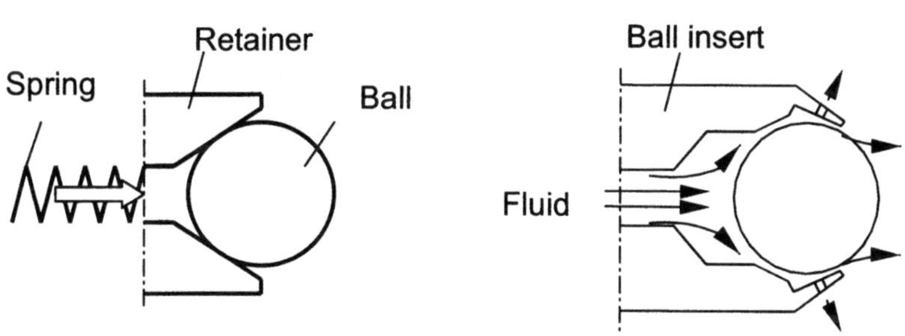

FIGURE 2. Setups of the burnishing tool [4][7]

The hydrostatic concept of this tool ensures a constant gap between the ball and its insert when the fluid pressure is applied. The normal force depends only on the fluid pressure and it can be kept constant when the tool follows a contour. The ball insert can also perform a compensation stroke, which achieves the contact of the ball with the workpiece after the pressure is switched on. After switching off the pressure line, the insert retracts in its initial position.

The hydrostatic burnishing tool offers several advantages. First of all, the hard ceramic ball makes possible that hardened steel be burnished (up to HRC 65). Aditionally, ball's ceramic has a low adhesion to the workpiece's material. Then, the normal force can be controlled and reproduced easily through the pressure at the pump; there is a proportionality relationship between the force and the fluid pressure. The normal force can be kept constant on a contour, due to stroke compensation of the ball. The coolant is the fluid in the system itself. Only a few drops of coolant are consumed during the process. The ball can freely rotate, without contacting the retainer; this prevents any sliding between the tool and the workpiece's surface. The tool can be clamped on any CNC or regular lathe. Maintenance and replacement of worn parts are very easy.[3][4]

2.2 THE HYDRAULIC UNIT

On purpose to create a hydrostatic pressure at the burnishing ball, a pump able to supply the required pressure is necessary. The hydraulic unit has the role to provide the required pressure to the hydrostatic tool. The working medium can be the coolant-lubricant already present in the machine, provided it respects certain specifications. The pressure can be adjusted manually or automatically.

A fluid having the following characteristics was employed: high levels of both mechanical and lubrication properties at 7% concentration; leaves a soft fluid residue to protect the machine tool; no foam in high-pressure operations; chlorine free.

2.3 FACTORS OF INFLUENCE

Many factors can influence the results of hard ball burnishing process. The factors influencing the ball-burnishing process can be divided into two groups: constant and variable factors. In our particular study only a few variables will be chosen; the others will be held constant at certain values.

The constant factors were: the workpiece material, the tool, the system stiffness, the lubricant is a constant, the number of passes (one pass). The variable factors were: the normal force, the feed, the speed, the initial roughness. Three experimental groups were employed, based on the initial (hard turned) roughness, as follows.

The first group has rather small differences of the initial roughness; therefore, this parameter was considered non-variable, and the normal burnishing load was varied. The second group shows large differences of the initial roughness, which was considered a variable parameter. The third group shows differences in initial roughness; since those differences are not very large, another parameter was varied, namely the feed.

The parameters tracked are some of the basic parameters of a surface, both for turning and for burnishing.

R_a – arithmetic mean roughness value (ISO), mostly utilized to characterize the surface roughness; R_z – Ten-point height (ISO), which in this case was considered more relevant for the burnished profile; R_{pm} – Mean peak height value above the mean line (DIN); this value was chosen because a large effect of burnishing upon the peak height of asperities was expected; T_p – profile bearing ratio at differend profile depths; flattening of asperities is expected to cause an improved bearing ratio of the burnished profile.

2.4 RESULTS

1. All ball-burnishing experiments lead to a reduction of roughness (figure 3). That means that burnishing is a good method to smoothen a turned surface. The ratio of reduction was in the range 1.76-3.53. A larger load determined a larger flattening ratio, due to higher pressure developed at the contact location between ball and workpiece. Also, the higher the initial roughness, the higher the reduction ratio. That means that plastic deformation was more intense for higher asperities. A lower feed is assumed to lead to a partial overlapping of ball's trace, so that asperities are better flattened and therefore the roughness reduction is higher.

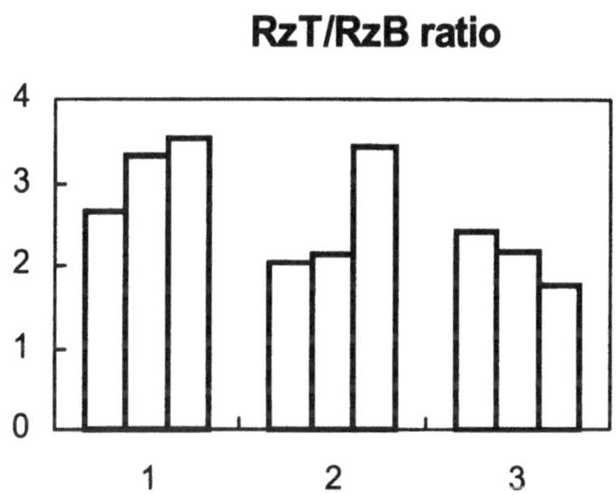

FIGURE 3. Roughness change ratio for hard turning, as compared with hard burnishing

2. Increase of the normal force leads to better surface roughness (figure 4). However, a very large force is expected to deteriorate the surface. The equipment used for present experiments didn't allow development of higher forces.

3. The lower the feed, the better the roughness (figure 5). This result is also in accordance to previous research on soft materials. Ball's traces are closer when feed is lower, so the final roughness should be also lower.

4. The initial roughness (after hard turning) is an important factor of influence on the final roughness (after burnishing). A low original roughness was expected to lead to a better final roughness. Figure 6 does not confirm entirely this expectation. A critical initial roughness might exist. Experiments should be extended in order find a deeper interdependence.

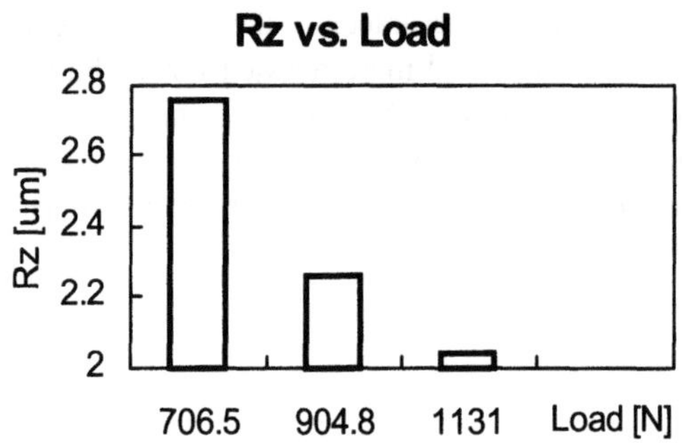

FIGURE 4. Roughness variation in terms of normal force

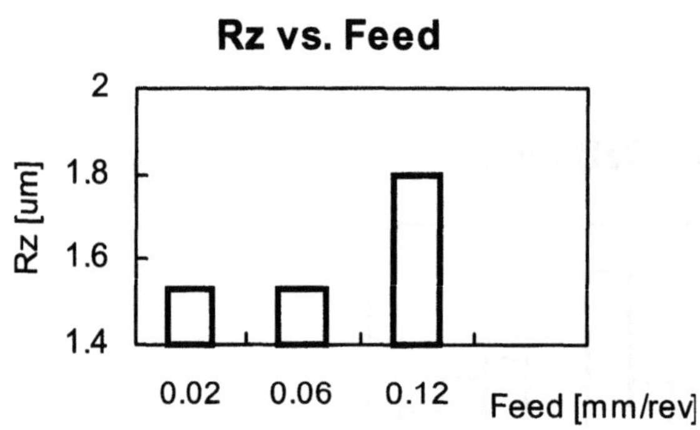

FIGURE 5. Roughness variation in terms of burnishing feed

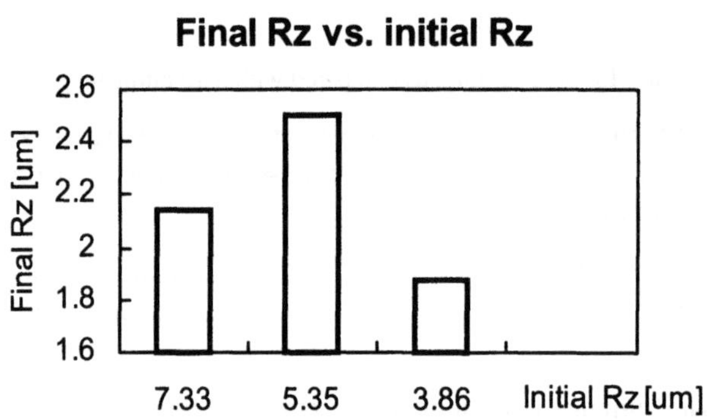

FIGURE 6. Initial hard turning roughness in terms of final burnishing roughness

5. The bearing ratio length goes to 100% much sooner for burnished surfaces (figure 7). That means better bearing properties occur on a burnished surface.

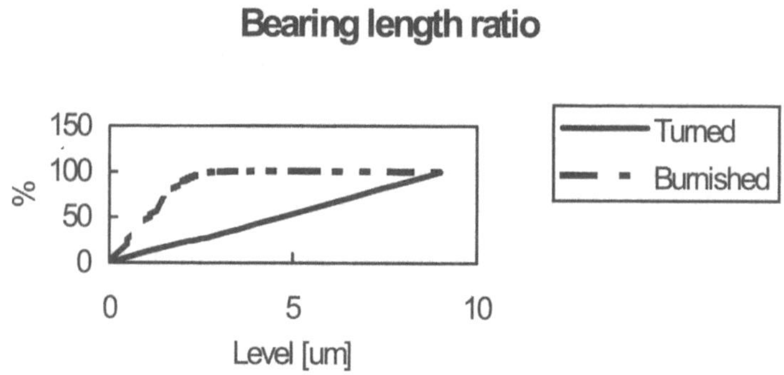

FIGURE 7. Bearing length ratio curves for hard turning, as compared with burnishing

6. Figure 8 shows the profiles of a hard turned surface (top) in comparison to a burnished surface (bottom).

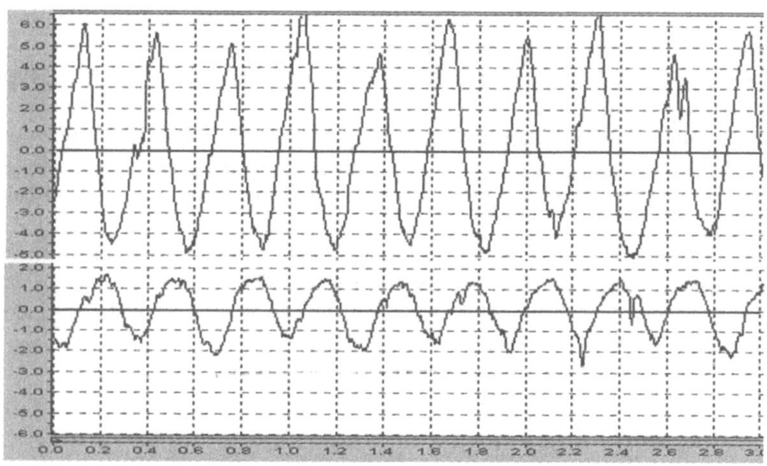

FIGURE 8. Profiles of a hard turned surface (top), as compared with a burnished surface (bottom)

3 CONCLUSIONS

As a preliminary investigation, the tests described in this paper proved the possibility of employing single-point burnishing of hard steels (about 60 HRC) as a finishing process. Previous hard turning was necessary in order to set accurate premises for subsequent burnishing.

The equipment necessary for ball burnishing was tested. The principle of hydrostatic tool, which prevents any sliding motion at the contact area, was utilized. The fluid had also the role of load generator and coolant as well. The tool worked smoothly and proved to be reliable and efficient. The punctual contact ensured a high local pressure for a relatively small load. Moreover, this load is easy to control by variation of the fluid pressure at the pump. A small amount of coolant was involved.

Influence of process parameters was also studied. The initial roughness from hard turning, the normal load applied onto the surface by a hard ball and the feed rate of the tool were considered the main working parameters, which determine the surface quality after burnishing. In general, the higher the initial roughness, the higher the final roughness. In addition, the reduction ratio of asperity height was higher for higher initial roughness. The normal load determines the deformation degree of asperities, therefore a bigger load lead to a smoother surface. The feed rate influenced burnished surface in the sense of better roughness for lower feed.

All in all, a considerably smoother surface could be obtained through burnishing of hard steel parts. Bearing ratio of the burnished surface was also significantly improved. The roughness range was comparable to the grinding process. The method described uses the same clamping of the workpiece for both hard turning and burnishing. Adjustments of the operation parameters and tools were very easy and didn't need much time. For all these reasons, burnishing proved that it could be a viable alternative to grinding.

REFERENCES

1. Loh, N.H., Tam, S.C., (1988), Effects of ball burnishing parameters on surface finish - A literature survey and discussion, Precision Engineering, Vol. 10, No. 4, 215-220.
2. Murthy, R.L., Kotiveerachari, B., (1981) Burnishing of metallic surfaces – a review, Precision Engineering, Vol. 3, No. 3, 172-179.
3. *** Tools for roller burnishing, deep rolling, forming, Ecoroll Prospectus, Germany.
4. *** Hydrostatic roller burnishing tool, Ecoroll, Germany, Operating instructions manual.
5. Neagu-Ventzel, S., (1999) Burnishing versus grinding, Research report, University of Toledo, U.S.A., 99-105.
6. Klocke, F., Liermann, J., (1998), Roller burnishing of hard turned surfaces, International Journal of Machine Tools Manufacturing, Vol. 38, No. 5-6, 419-423.
7. Morimoto, T., (1986), Examination of the burnishing process using a newly designed tool, Journal of Mechanical Working Technology, No. 13, 257-272.

EVALUATION OF WARPAGE OF INJECTION MOULDED PARTS USING RE AND FEM APPROACHES

L.M. Galantucci, G. Percoco, R. Spina and L. Tricarico

Dipartimento di Ingegneria Meccanica e Gestionale (DIMeG) - Politecnico di Bari – Bari - Italy

KEYWORDS: Injection molding, Reverse Engineering, Finite Element Method

ABSTRACT: The authors propose an integrated methodology to evaluate the warpage effects of plastic injection molding parts. Numerical simulations of the injection molding process allow the evaluation of the component manufacturability at an early stage in the development cycle, without fabricating prototypes and minimizing experimental tests. A detailed investigation of warpage causes is performed using a Finite Element system. To validate FE results, part surfaces subjected to warpage are reconstructed and measured using a Reverse Engineering system. Through the integration between Finite Element and Reverse Engineering systems, the warpage of parts is analyzed and a study of the influence of process parameters on product quality is performed. The integration of all data represents a powerful approach to critical aspects of part design and/or molding processing.

1. INTRODUCTION

Injection moulding is the main manufacturing process for the fabrication of plastic products Through this process, components with complex geometries are obtained with high production rates, good surface finishing and restricted dimensional tolerances [1][2]. The best practice of injection moulding requires full integration between design and manufacturing areas. The part and mould have to be simultaneously designed and their mouldability evaluated [3]. The product manufacturability has to be estimated using direct experimental validations from physical prototypes [2]. In this way, the product development cycle is continuously refined. This cycle becomes particularly difficult at the end of the design stage when part and mould geometries are fully defined and the process has to be implemented. The parameter choicc requires a complete understanding of the injection moulding process to obtain high-quality components. The presence of warped sections and defects has to be avoided, especially for products with ergonomic and aesthetic shapes [3][4]. The concurrent use of Finite Element (FE) and Reverse Engineering (RE) systems is a powerful approach to overcome these limitations. Through the development of numerical models, the part warpage can be predicted and then validated by direct measurements on components.

2. PARAMETER INFLUENCE ON WARPAGE

Thermoplastic injection moulding consists of three main phases: (i) filling, (ii) packing and (iii) cooling. During the first phase, a melt polymer fills the part cavity, moving through the sprue, runners and gates. In the packing phase, additional melt polymer enters the cavity to balance the part shrinkage caused by cooling [1][2][5]. Mould opening and part ejection complete the process. For an efficient implementation of injection moulding, the main parameters to control are associated to the process and mould design. The process parameters are: the temperature

Published in: E.Kuljanic (Ed.) *Advanced Manufacturing Systems and Technology*, CISM Courses and Lectures No. 437, Springer Wien New York, 2002.

and pressure profiles; ram speed; mould-wall temperature; injection time and mould opening time [6]. The design parameters are: part thickness, flow length, position and number of injection points, channel shape and dimension. All these parameters influence the part properties as thermal degradation effects, orientation and shrinkage. In particular, the part warpage depends on the residual stresses, mainly divided into flow-induced and thermal-induced [1][2][5]. The flow-induced stresses are determined by average flow length and fibre orientation [7]. The average flow length is the average distance covered by the polymer melt front during its advancement in order to completely fill the part cavity. This parameter is directly linked to the pressure profile because a great flow length is associated to a high pressure value. To reduce the pressure value, a possible solution is to increase the number of injection points. The flow advancement of the melt polymer is also important for the fibre orientation. The velocity of the melt polymer depends on the part frontal area. This velocity has to remain as constant as possible or an undesired polymer orientation will be obtained along the part section. An appropriate control of the ram speed reduces this effect. Particular attention should be paid to part thickness variations because the melt front velocity rapidly change between sections with different thicknesses. The thermal-induced stresses depend on volumetric shrinkage and unbalanced cooling. Volumetric shrinkage appears when temperature is decreased. The pressure-volume-Temperature (pvT) curve of each plastic polymer shows the reduction of specific volume between processing and room temperatures, reflecting material transitions from one physical state to another. The processing temperature is a function of melt and mould temperatures. The shrinkage is particularly high for a plastic material, reaching as much as 20 per cent by volume. Volumetric shrinkage can be limited choosing by appropriate packing conditions (pressure and time) after the filling phase and/or controlling the processing temperature. Non-uniform temperature distribution through part thickness and/or sections causes unbalanced cooling. This effect can be controlled by: designing an efficient cooling channel system; estimating temperature differences between mould cavity and cores; controlling temperature gradients on mould surfaces.

3. APPROACH

For complex interactions between process parameters of injection moulding, analytical studies are possible only for simple products [5][7]. A numerical methodology is very useful to evaluate components with complex geometries [8]. The aim of the approach is the development of an integrated environment for the evaluation of part warpage by coupling Finite Element and Reverse Engineering methods. This environment involves a continuous data flow between design, simulation and validation phases to analyse part manufacturability at the early stage of production (Figure 1). The product and mould are designed as 3D solid models using a general-purpose

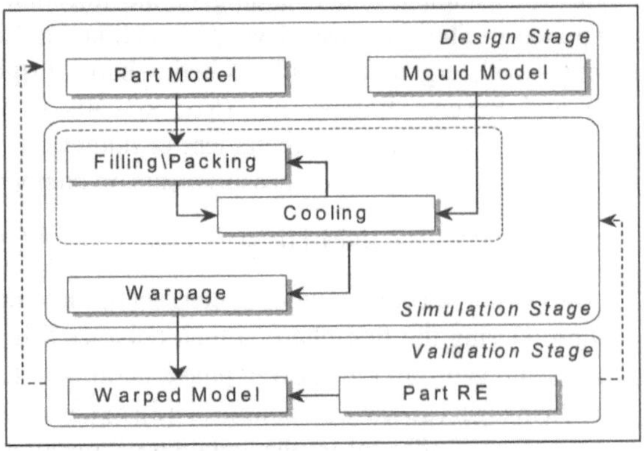

Figure 1: Diagram of the proposed approach

CAD system and then converted into mesh models for FE analyses. FE mesh models are designed using a double-skin surface representation. This representation leads to several advantages such as: the direct generation of meshes from 3D CAD solid models; the reduction of the computational effort due to the mid-plane generation [9]; the improvement of speed and accuracy for all double-skin model analyses. These FE analyses are performed using a step-by-step procedure where each step is related to a specific phase of the injection moulding process. During the simulation stage, several evaluations are made to identify possible modifications for improving the design of the part. The flow-induced and/or thermal-induced residual stresses are calculated at each step. In the first step of the simulation stage, the filling/packing of the part is evaluated under specific constraints on material properties, gating system and moulding machine. This step is important to identify the process parameters and the final part characteristics; estimate the flow-induced stresses. In the cooling step, the meshes of mould and cooling channel system are joined to the part to perform the thermal analysis. The simultaneous investigation of the part and mould during this step allows the estimation of the effects of each process parameter on the global performance of the final product. This permits the appraisal of the thermal residual-stresses of the injection moulding process. The filling/packing and cooling steps are reiterated until thermal flow stability between cavity, mould and cooling system is reached. At the end of this simulation stage, a model containing the warped surfaces is obtained.

To confirm the predicted results of the numerical model, a direct measurement of the product surfaces is crucial. Thus in the validation stage, the RE methodology is applied to reconstruct a computerized model from acquired product data points [10]. This phase is divided into two main steps: data acquisition and surface identification. The data acquisition is performed on the component using a 3D scanning system to acquire point clouds and/or scan curves representing the part. The second step is the recognition of surfaces from these unstructured acquired data, rebuilding the part model. Curvature and point deviation analyses are performed on this model to estimate the warpage of the real moulded part and to verify the correct response of the simulated warped model. As a result of the validation stage, important information is obtained to modify the mould model and/or the process parameter sets. The proposed integrated approach is therefore focused on the implementation of a FE model to predict the part warpage. This model will be used to realize products with a very limited warpage, decrease the experimental effort and reduce time-to-market and development cycle costs. The authors report the analyses performed for implement the FE model of a simple component to demonstrate the approach capabilities.

4. EXPERIMENTS

The part chosen to assess the proposed approach is a plastic casing for a home consumer electronic product. This casing consists of two parts to be assembled to form the final product. The functional requirements imposed on each part are respect of specific tolerances and a minimal warpage to guarantee correct assembly and prevent aesthetic imperfections in the final product. Only analyses on the upper component of the product are reported in this paper. The results of the proposed approach can be similarly used for the lower component. The part dimensions are 86x49.5x34 mm with a mean thickness of 1.5 mm. The experimental observations on the injection-moulded component were divided into two phases. In the first phase, data on current part production were collected to create a proper numerical model and some specimen parts were realized. In the second phase, the RE activity is performed on these specimens to measure their warpage and validate the designed FE numerical models. The component production was realized in a plastic manufacturing unit of an industrial firm while the RE activity was carried out in DIMeG department laboratory.

1.1 INJECTION MOLDING DATA

The experimental tests to realize injection mould parts were carried out on a hydraulic injection machine, the main technical specifications of which are summarized in Table 1. The mould, coolant and plastic materials were respectively P-20 steel, water and Enichem Edistir SR850 (Polystyrene).

The machine set-up was mainly realized by controlling: the ram velocity, the switch-over between high and low pressures, the packing and cooling times. During the velocity phase, the screw moves forward pushing polymer into the cavity until 99.5% by volume of the part is filled. Then the process is controlled by the packing pressure profile until the sprue is completely solidified and the cooling phase begins. The cycle is completed in roughly 20 s. Table 2 reports a summary of the process parameters for the current production. The part is produced using a double-cavity mould to improve manufacturing throughput. The parts are coupled to mould and cooling system. The meshes are realized from the CAD models of the complete assembly (Figure 2).

Table 1: Moulding machine characteristics

Parameters	Value	
clamping force	1000	KN
maximum injection pressure	170	MPa
maximum injection flow rate	122	cm^3/s
screw rotation speed	320	RPM
plasticating screw diameter	40	mm
screw length-to-diameter ratio	20	

Table 2: Moulding Parameters

Parameters	Value	
injection time	1.6	s
packing time	8	s
cooling time	8	s
mould opening time	2	s
polymer melt temperature	220	°C
mould temperature	45	°C
cooling entrance temperature	11	°C
hydraulic packing pressure	4.5	MPa
hydraulic-to-cavity pressure ratio	10	
packing/filling switch over	99.5%	
coolant flow rate	5	Lit/min

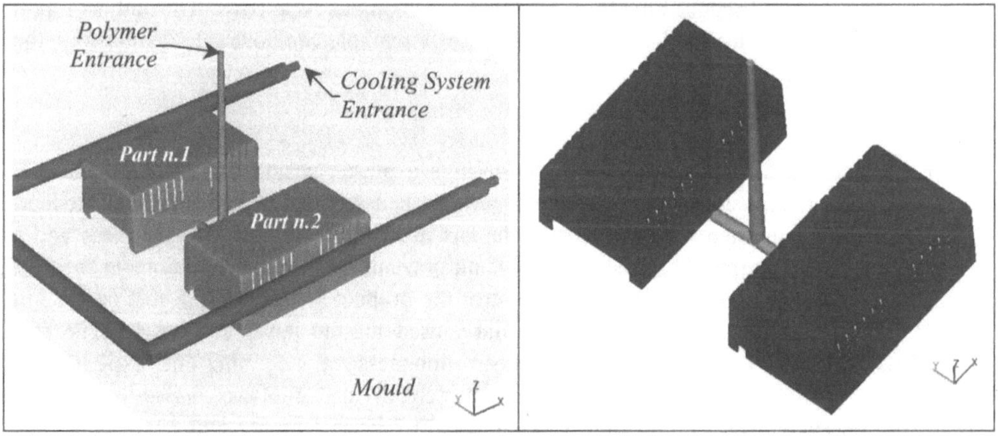

Figure 2: Model representation

The mesh density of each part and mould varies from a few elements along the largest dimension to a finer mesh for critical regions (e.g. rounded elements). The meshes of gating and cooling systems are represented using 1D elements with an associated thickness.

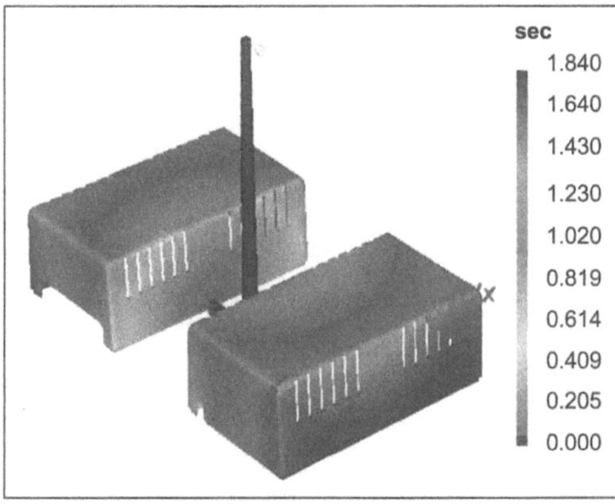

Figure 3: Fill Time

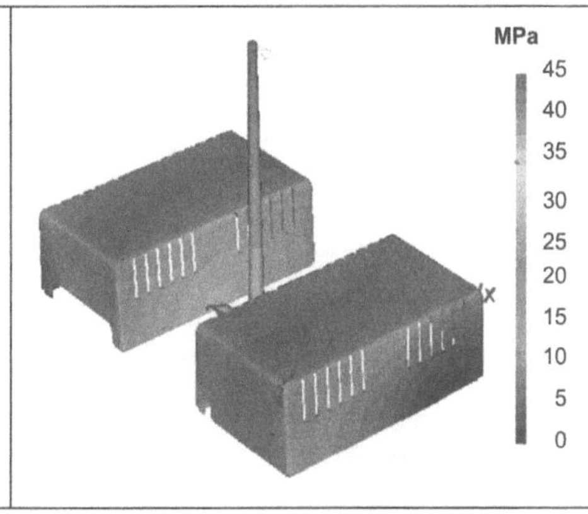

Figure 4: Pressure at switch over

The time necessary to fill parts is about 1.84 s and represents the sum of the imposed fill time (1.6 s) plus the delay time of the machine control and the additional time due to the polymer melt compressibility (0.24s). It is important to point-out that the moulder is not able to identify the exact instant of the end of filling due to the effects of the polymer melt compressibility which extends the filling time after the switch-over is applied. On the contrary, the FE simulation software does determine this additional time from pvT data, modifying the real filling time to compensate the polymer compressibility. This correction ensures a more accurate prediction of the material flow rate and other factors of interest when simulating the process. For example the subsequent pressure value at switch over between the filling and packing phase is calculated for a filling time equal to 1.84s. The filling time and the pressure results (Figure 3-Figure 4) report that the melt flow front does not present regions where the polymer stops or pauses, highlighting a good part processability.

1.2 REVERSE ENGINEERING OF THE PART

The RE methodology was applied on the part to reconstruct its 3D warped model. Using a contact-type 3D scanner (Roland Pizca PIX-30) based on a special piezo sensor technology (Figure 5), point clouds of different part surfaces were acquired and then joined in a single Stereolitography (STL) file. The scanning probe presents a declared minimum pitch equal to 0.05 mm along the X-Y axes and to 0.025 mm along the Z-axis. The maximum part dimensions are equal to 304.8x203.2x60.5 mm. The choice of this scanning system for small parts is justified by its lower cost than a traditional Coordinate Measuring Machine (CMM). The precision of this scanning system is obviously lower than a CMM, but it is adequate for the measurement errors expected for this application [11]. The model reconstruction step was carried out using a novel algorithm working on the STL file developed by the authors. This algorithm operates on the high-density data points of this file to eliminate redundant triangles without affecting the surface integrity and level of details of the part [12]. Surface detection is thus quickly achieved to reconstruct the final model. The reconstructed model is used to highlight the part warpage effects by measuring dimension deviations of the part from the original designed shape. Each surface of this part was reconstructed using the RE system, considering that the precision on the Z-axis is higher than those on the X- and Y-axis. Figure 6 show the top and side surfaces, acquired using the RE system.

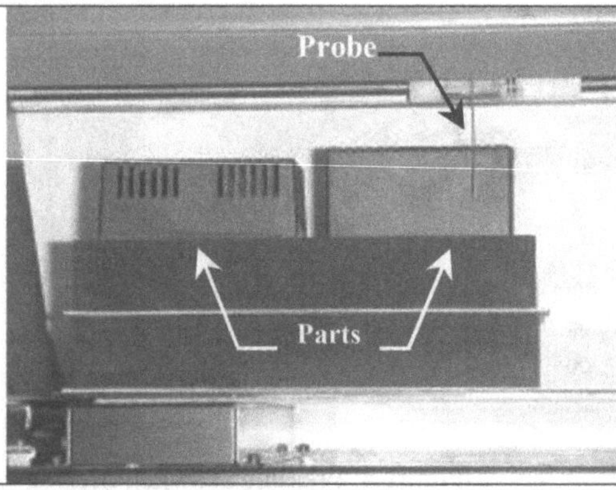

Figure 5: 3D scanning system

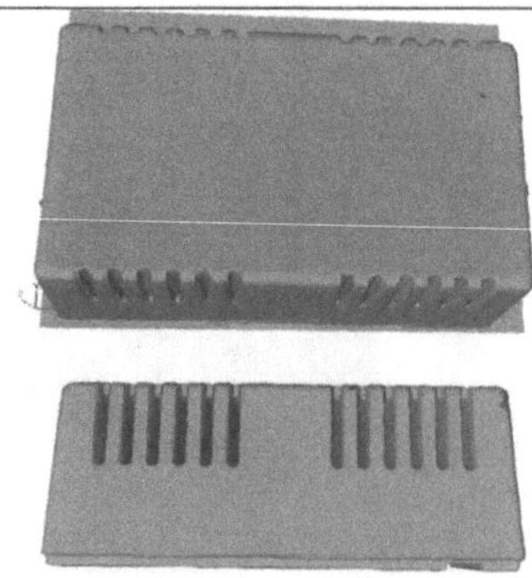

Figure 6: Top and Side surface

5. NUMERICAL/EXPERIMENTAL RESULTS

The numerical simulations of the moulded part concerned warpage analyses. Under the small deflection hypothesis, the first analysis was focused on the evaluation of total part deflection, defined as the vector of the point displacements in the X-,Y- and Z-directions from an origin.

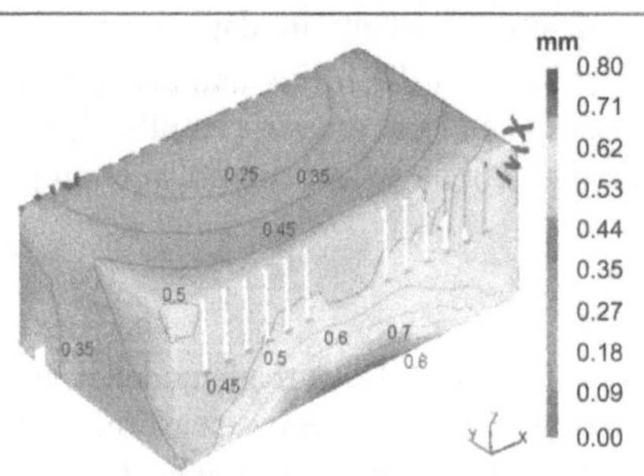

Origin is positioned on the gating system entry

Figure 7: Total deflections of part n.2

The total deflections were calculated for all surfaces using the FE system (Figure 7). To validate the predicted results of the FE simulation, a point deviation analysis was performed on surface point clouds acquired by the 3D-scanning system. Each point deviation is calculated from the original CAD surface, after compensating the positioning errors due to the experimental specimen set-up. The total deflections of the surfaces measured by RE system were in accordance with those predicted by the FEM analysis of the part. Figure 8 shows the point cloud of a side surface and its corresponding FE representation, considering only a half of it and positioning the origin W0 of FE analysis on the surface itself. However, the results of these analyses highlight the effects of warpage on the part deflections but do not give sufficient information to identify its dominant causes. The contributions of the flow-induced and thermal-induced stresses have to be isolated and their related deflections computed. In this way, possible adjustments to mould design and/or process parameters can be defined. The residual-stresses are primarily caused by thermal effects.

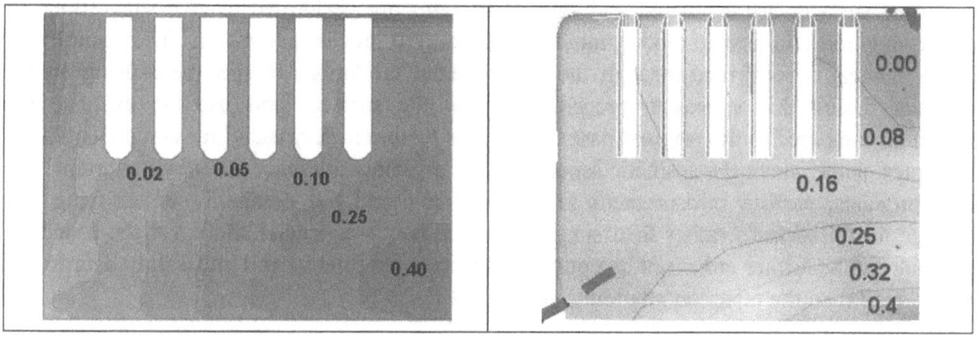

Figure 8: Point deviation analysis (RE analysis & FE analysis)

The appropriate design of the gating system and of the part guaranteed that the filling melt front through the component generates very low flow-induced stresses. In addition, the fibre orientation was absent due to the material characteristics of Polystyrene. Consequently, the warpage analysis required the separation of the thermal effects into the principal factors: cooling and shrinkage. The cooling effect is connected to the heat flux between part, mould and cooling system. The heat exchange was mainly localized on the side surfaces of the part near the cooling channels. As Figure 9 shows, the deflections on this side surface are more closely packed than those of the other surfaces of the part and oriented according to the coolant direction. The asymmetry of the contour lines along the large dimension of the part point-outs this behaviour. On the contrary, the shrinkage effect is more evident on the borders of the part due to the higher convex heat flux with respect to the central zones of the surfaces. Figure 10 shows the contour lines due to shrinkage and the symmetry of deflections.

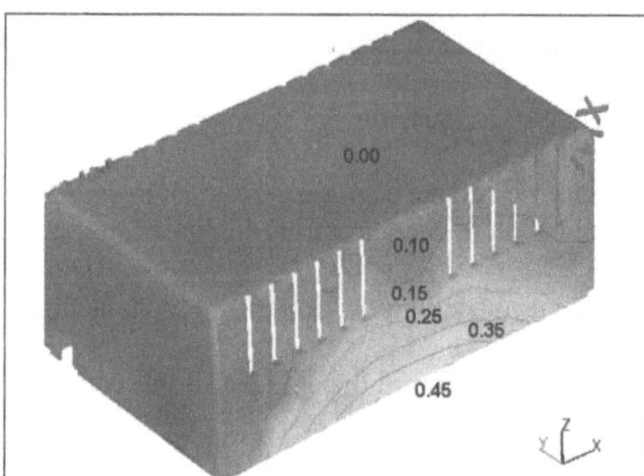

Origin is positioned on the gating system entry

Figure 9: Cooling

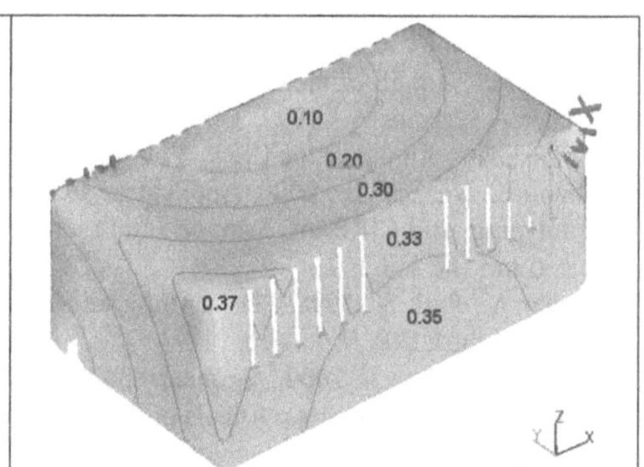

Origin is positioned on the gating system entry

Figure 10: Shrinkage

6. CONCLUSIONS

An integrated methodology to evaluate warpage for plastic components has been implemented. The proposed methodology can predict defects before production set -up is realized, giving the possibility to improve manufacturing efficiency. Through the FE numerical models, deflections of warped surfaces are estimated and several analyses are performed to give suggestions to process engineers on how to modify mould design and/or process parameters. The accuracy of the FE models is confirmed through the experimental validations of specimens using an RE approach. Using this approach, reconstruction of the surfaces was used to evaluate the deviations compared to the original part shape. This feature is very useful for components with a complex shape where the aesthetic aspect and the assembly tolerance are very important. For these products, surface measurement is very difficult and the possibility of verifying the warpage for all surfaces rather then single reference points is crucial. Moreover the proposed approach can be further enhanced by optimising process parameters and integrating data from a control system on the injection moulding machine [13].

7. REFERENCES

1. Potsch G, Michaeli W, 1994, Injection moulding: an introduction, Hanser Publishers, Munich (D)
2. Malloy R.A., 1994, Plastic part design for injection moulding: an introduction, Hanser Publishers, Munich (D)
3. Bernard A., Rapid product development case studies and data integration analysis, Computers in Industry, Vol. 43, 161-172
4. Varady T., Martin R.R., Cox J., 1997, Reverse engineering of geometric models: an introduction, Computer-Aided Design, Vol. 29, 255-268
5. Choi D.S., Im Y.T., 1999, Prediction of shrinkage and warpage in consideration of residual stress in integrated simulation of injection moulding, Composite Structures, Vol. 47, 655-665
6. Nguyen K.T., Prystay M., 1999, An inverse method for estimation of the initial temperature profile and its evolution in polymer processing, International Journal of Heat and Mass Transfer, Vol. 42, 1969-1978
7. Vincent M., Devilers E., Agassant J.F., 1997, Fibre orientation calculation in injection moulding of reinforced thermoplastics, J. Non-Newtonian Fluid, Vol. 73, 317-326
8. Choi G.H., Lee K.D., Kim S.G., 1994, Optimization of process parameters of injection moulding with neural network application in a process simulation environment, Annals of CIRP, Vol. 43/1, 449-452
9. Fischer A., Smolin A., 1997, Modelling in reverse engineering for injection moulding analysis of 3D thin objects, Annals of the CIRP, Vol. 46/1, 97-100
10. Fischer A., Manor A., 1999, Utilizing image processing techniques for 3D reconstruction of laser-scanned data, Annals of the CIRP, Vol. 48/1, 99-102
11. Iuliano L., Vezzetti E., 2000, Application of a piezo-electric reverse engineering system to jewellery prototyping, 2nd CIRP Int. Sem. on Intelligence Computation in Manufacturing Engineering, eds. R. Teti, Capri (I), 253-262.
12. Fischer A., Park S., 1998, 3D scanning and level of detail modelling for design and manufacturing, Annals of the CIRP, Vol. 47/1, 91-94
13. Wang K-K., Zhou J., 2000, A concurrent-engineering approach toward the online adaptive control of injection moulding process, Annals of the CIRP, Vol 49/1, 379-382

ANALYZING THE WARPAGE IN THE INJECTION MOLDING USING SLS TOOL INSERTS

J. Gabor, J. Kovacs

Institute of Machine Design, Budapest University of Technology and Economics, Hungary

KEYWORDS: Injection Molding, Warpage, Shrinkage, Deformation, Injection Molding Simulation, Simulation, Rapid Tooling, Rapid Tool Insert, SLS.

ABSTRACT. Around one third of all plastics are processed by injection molding, therefore economy and manufacture have fundamental importance. The major issues in the development of injection molding tools include the progress in material technology and the developments in tool design methodology, such as developing rapid tool inserts. It is essential to develop fast methods to manufacture tools for injection-molded-like prototype parts or mass-produced parts. Due to these high requirements, it is more and more significant to use simulation methods to optimize the part before manufacturing. This paper shows the differences and the advantages of these tool inserts in injection molding. The main difference between the conventional tools and SLS tool insert is also described in this paper. Beyond all question, the rapid tool inserts cause different properties such as different warpage of the polymeric part.

1 INTRODUCTION

Simultaneous engineering requires prototypes so the design errors can be detected in the earliest possible stage of product design, cutting back the cost and time involved in the modifications.

Injection molding produces highly accurate products in a very short cycle time. It is one of the most important polymer-processing operations in plastics industry. Even in modeling of injection molded parts one should be careful to fulfil the highly accurate product requirements. The rapid prototyping technologies usually produce inadequate models to fulfill these requirements. Rapid Tooling processes complement the Rapid Prototyping options by being able to provide higher quantities of parts in a wider variety of materials, even short-run injection molded parts in the intended production material.

2 INJECTION MOLDING TOOLS

Injection molding is a process by which hot polymer melt is forced into an empty, cold cavity of the desired shape and is allowed to solidify under high pressure and controlled cooling.

2.1 CONVENTIONAL TOOLS

The conventional mold material for injection molding is usually tool steel. The mold basically consists of two parts: the stationary part called cavity plate, and the moving part called core plate. Both halves are equipped with straight cooling channels in which cooled fluid is circulated to absorb the heat delivered to the mold by the hot thermoplastic melt.

Published in: E.Kuljanic (Ed.) *Advanced Manufacturing Systems and Technology*,
CISM Courses and Lectures No. 437, Springer Wien New York, 2002.

2.2 RAPID TOOL INSERTS

Rapid tooling is the term for either indirectly utilizing a rapid prototype as a tooling pattern for the purposes of molding production materials, or directly producing a tool with a rapid prototyping system.

Manufacturing of epoxy or aluminum- epoxy molds are reasonably quicker in comparison to machined molds. It is a relatively inexpensive way to create prototype and production tools. New epoxy resins offer much higher compression strength and heat resistance. If the molds are designed properly they can withstand the injection or the compression pressures with the use of aluminum standoffs or mold boxes. However the cycle time is between 5 to 15 minutes because of the poor thermal conductivity. Life time of the tool is a function of the thermoplastic material, fillers and part complexity. Some molds can create as few as 50 parts, while others can exceed 5.000.

The selective laser sintering (known as SLS) technique enables even highly complicated molds, dies and inserts to be built directly in metal from CAD data. In this case there is no geometrical restriction associated with the manufacturing tools. The laser-sintered parts can be post-processed, for example polished to produce tools suitable for injection molding. As with all layer-manufacturing techniques, even complex shapes can be built easily, including geometries, which are impossible to cut with conventional tools. One of the most important applications of this technique is to manufacture curved internal cooling channels. Depending on the type of plastic and the force and temperature of injection molding, core and cavity sets created through the SLS process can produce up to 50,000 parts.

3 SHRINKAGE AND WARPAGE

Shrinkage is inherent in injection molding process. It comes from the density difference between melt and the final product, see Figure 1.

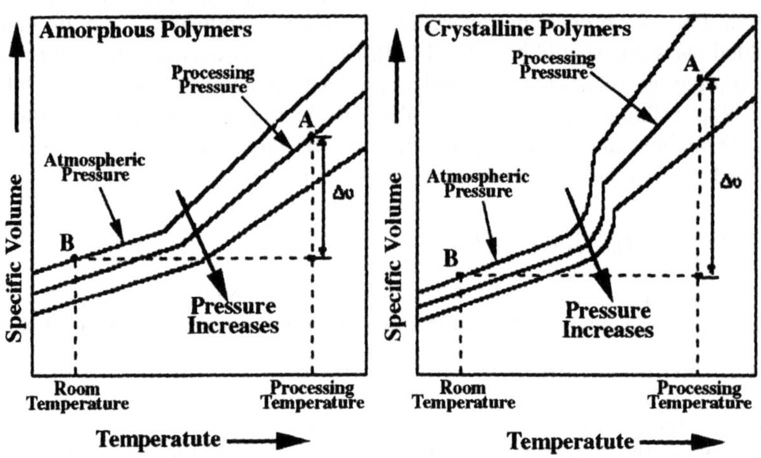

FIGURE 1. The PVT curves for amorphous and crystalline polymers; point A is the processing state and point B is the final state

Warpage is a distortion where the surfaces of the molded part do not follow the intended shape of the design. If the shrinkage throughout the part is uniform, the product will not deform or warp, it simply becomes smaller. However, achieving low and uniform shrinkage is a complicated task due to the presence and interaction of many factors such as molecular and/or fiber orientations, mold cooling, part and mold designs, and process conditions [1].

Using the rapid tool inserts may cause warpage of the part because of the difference thermal conductivity and other properties compared with conventional injection molds. Non-uniform cooling in the part, i.e. asymmetric cooling across the part thickness from the skin to the core, can also induce different shrinkage.

4 INJECTION MOLDING SIMULATION

For analyzing the injection molding the programs use the equations of continuity, momentum and energy. The shear viscosity is the most important material property in numerical simulation of the filling stage. It depends on the shear rate and the polymer temperature [2]. The shear thinning behavior of the viscosity is characterized either by the Power-law polymer viscosity equation or Cross-WLF equation. The Power-law polymer viscosity model characterizes the flow behavior of the material but it only works when shear rates are relatively high. This model does not account the effect of pressure and it might lead to inaccuracies. This model can be written as follows (Figure 2.):

$$\eta(T,\dot{\gamma}) = A \cdot \exp\left(\frac{T_a}{T}\right) \cdot \dot{\gamma}^{(n-1)}$$

where T is the actual temperature, T_a is the ambient temperature, $\dot{\gamma}$ is the shear rate, n and A are constants.

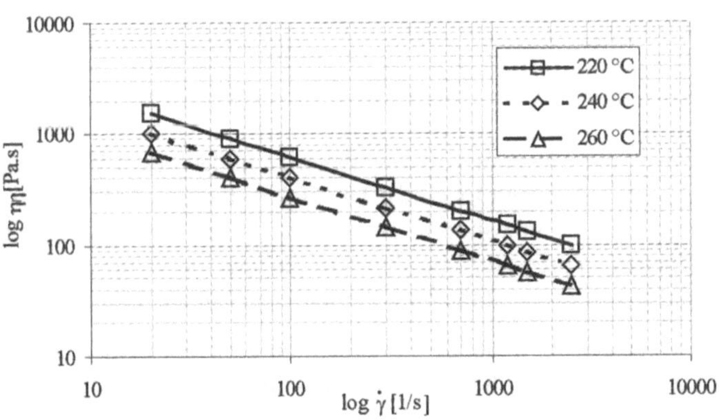

FIGURE 2. Viscosity – shear rate function described by equation 1.

The Cross-WLF model is more appropriate for injection molding simulations, as the temperature and pressure sensitivities of the zero-shear viscosity are better represented. The shear thinning behavior of the viscosity is characterized by the Cross-WLF (Figure 3.) equation:

$$\eta(T,\dot{\gamma},p)=\frac{\eta_0(T,p)}{1+\left[\frac{\eta_0(T)\cdot\dot{\gamma}}{\tau^*}\right]^{(1-n)}}$$

where $\dot{\gamma}$ is the shear rate, p is the pressure, T is the temperature, τ^* is constant and η_0 is the zero-shear viscosity which can be represented by the WLF form as follows:

$$T>T_{trans}\rightarrow\eta_0(T,p)=B\cdot\exp\left(\frac{T_b}{T}\right)\cdot\exp(\beta\cdot p)$$
$$T<T_{trans}\rightarrow\eta_0(T,p)=\infty$$

where T_{trans} is a reference temperature and is typically taken as the glass transition temperature of the material, B, T_b and β are constants.

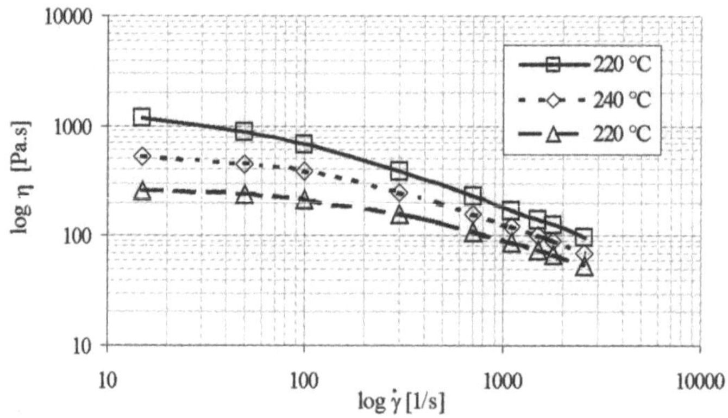

FIGURE 3. Representation of the Cross-WLF viscosity – shear rate dependence

The PVT data reflect the transitions as the material undergoes a phase change from one physical state to another (from melt to solid). The kink of the curve of PVT data of amorphous thermoplastics (Figure 1.) at atmospheric pressure is the glass transition temperature of the material (T_g). It depends on pressure. The slopes of the specific volume vs. temperature curves in the melt and solid states represent the bulk thermal expansion coefficients in the given states.

Modified Tait polymer density equation describes the variation of density (specific volume) with temperature and pressure in the melt and the solid states, between room and processing temperature over a wide pressure range:

$$\rho(T,p)=\left[v_0(T)\times\left(1-C\times\ln\left(1+\frac{p}{B(T)}\right)\right)+v_1(T,p)\right]^{-1} \qquad (4)$$

where ρ is the polymer density, C is a universal constant which equal to 0,0894, $v_0(T)$ is a temperature dependant specific volume, $v_1(T,p)$ is a temperature and pressure dependant specific volume. Above the transition temperature $v_1(T,p)$ is equal to zero.

Finite element methods are used to solve the coupled equations of continuity, momentum and energy. Three-noded triangular elements are used to describe the cavity and two-noded tube elements for the runners, connectors and channels. The melt front advancements are calculated by the control volume method [3]. The pressure, the temperature and the velocity field can be obtained in each time step. These results constitute the basis of the stress and deformation analysis.

5 RESULTS AND DISCUSSION

A number of simulation packages are commercially available for the simulation of the injection molding process. The basic idea is to create a model of the geometry or mold to be analyzed as the Figure 4 shows.

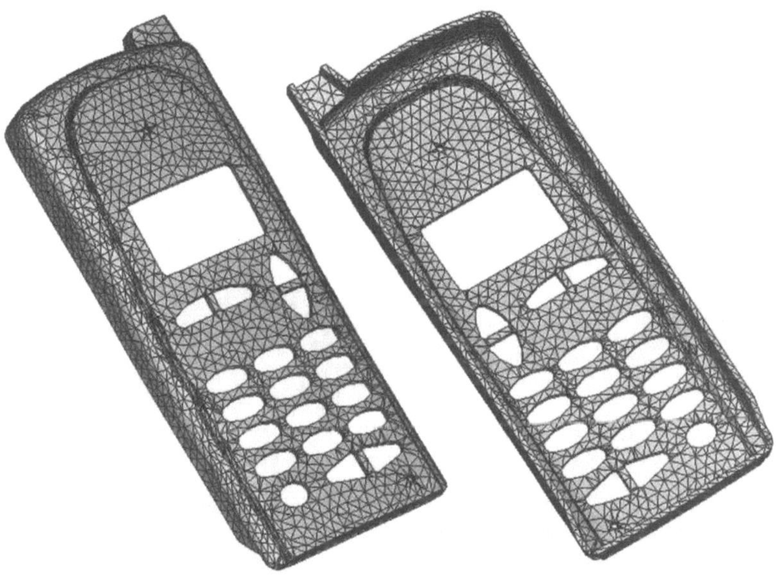

FIGURE 4. Front and back view of the FEM model of a cell phone

In this work the conventional injection molds was compared to the rapid tool inserts. There are significant differences between these two techniques. First of all the mold materials are different which cause different warpage. Secondly, the rapid tooling techniques allow to create curved internal cooling channels. Because of these differences, the efficiency of cooling could be better in the case of rapid tool inserts. Using these curved internal cooling channels the warpage could be decreased in addition manufacture could be cheaper and faster. Naturally the durability and abrasion of the rapid tool inserts are worse compared to the conventional molds, but it is worth to use to produce a couple of thousand parts [4].

In practice, rapid tooling mold material can be added to the simulation program (material database), and the irregular cooling channels can be designed. The different cooling channels are shown in Figure 5.

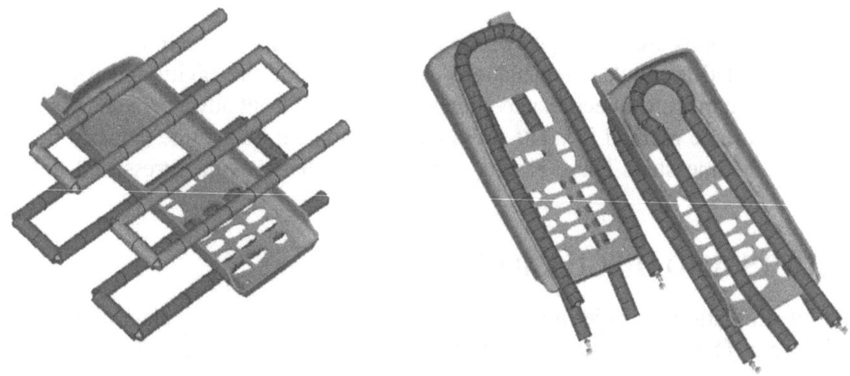

FIGURE 5. Conventional cooling channels (left)
and curved cooling channels in rapid tool inserts (right)

The result of the analysis is shown in Figure 6. The maximum deformation of the part caused by different cooling is 0.0285mm using conventional cooling channels. In contradiction to this result the maximum deformation is decreased to 0.0059mm using curved internal cooling channels.

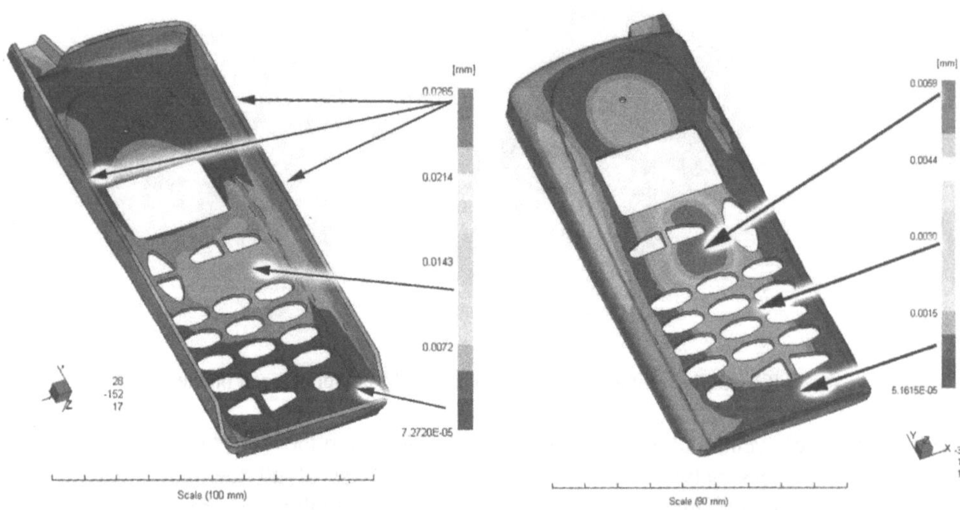

FIGURE 6. Deformation of the examined part caused by different cooling cannels,
differences between conventional mold (left) and rapid tooling (right)

As this result shows, the differences between the results of these methods could be considerable. The warpage could be decreased about 10 times if the cooling channels designed properly.

The influence of the mold thermal conductivity on the warpage of the mold was also examined. Figure 7. shows that the warpage of the part is function of the mold thermal conductivity.

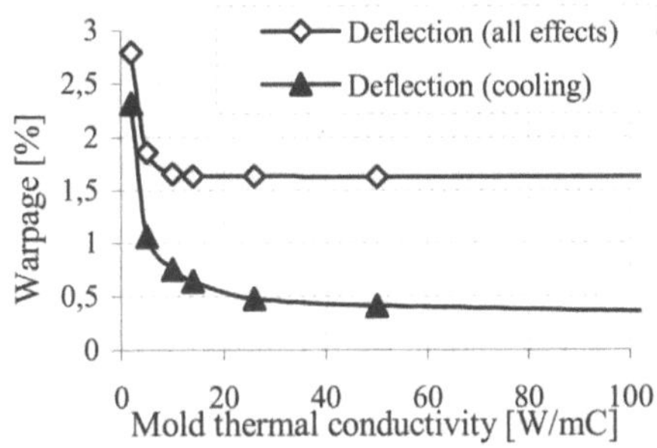

FIGURE 7. The warpage dependence of the mold thermal conductivity

The typical mold thermal conductivity is between 25 and 80 [W/mC] using conventional tool steel. The deformation dependence between these values is near constant, but using rapid tool inserts it could vary much more. The selective laser sintered tool insert's thermal conductivity is less than 15 [W/mC] and the unfilled epoxy resins thermal conductivity is around 2 [W/mC], which causes increased warpage of the part. Using rapid tool inserts it is essential to analyze the injection molding because of the above-mentioned reasons. The warpage of the part could be more significant but using simulation programs it can be decreased or it could be eliminated. The main advantage of the rapid tool inserts is that curved internal cooling channels could be used, but it must be optimized to achieve good quality part. These programs can analyze the cooling, and can optimize or minimize the warpage using these curved internal cooling channels.

6 CONCLUSION

It has been demonstrated that the rapid tool inserts are useful in the injection molding technology and the design could be optimized using simulation programs. The mass-production with these rapid tooling technologies will be available soon if the prototyping methods could produce accurate tools with better surface finish.

ACKNOLEDGEMENTS

The author would like to thank Fabio Miani of University of Udine and Károly Belina of Kecskemét College for assistance in this work and the ARBURG GmbH for the injection molding machine.

REFERENCES

1 C-Mold Reference Manual, (1998)

2 B. Abbes, R. Ayad, J.-C. Prudhomme, J.-P. Onteniente, Numerical Simulation of Thermoplastic Wheat Starch Injection Molding Process, Polymer Eng. And Sci, (1997)
3 C. A. Hieber and S. F. Shen, A Finite-Element/Finite-Difference Simulation of the Injection-Molding Residual Stress Process, Journal of Non-Newtonian Fluid Mechanics, Vol. 7 (1980).
4 E.Sachs, E. Wylonis, S. Allen, M. Cima, H. Guo, Production of Injection Molding Tooling With Conformal Cooling Channels Using the Three Dimensional Printing Process, Polymer Eng. And Sci, Vol. 40 (2000)

ORBITAL ELECTROCHEMICAL MACHINING OF ELECTRODISCHARGE MACHINED SURFACES

Z. Sadollah Bamerni, H. El-Hofy[1]

[1] Qatar University, Mechanical Engineering Department, Qatar

KEYWORDS : ECM, EDM, Orbital motion, Surface roughness

ABSTRACT. Electrochemical machining (ECM) provides one of the best alternatives for finishing parts machined by electro-discharge machining (EDM) in advanced materials used in die and mold industries. This paper presents experimental investigations that asses the impact of the orbital EC-finishing for the electro-discharge machined surfaces. Experiments were carried out using flat electrodes having single and multiple flow ports. During ECM, protrusions are normally formed under the flow ports. The use of orbital motion would lead to the reduction of the machining gap under these flow ports. The local current density, in that region, is thereby increased leading to a more uniform current density thus eliminating these protrusions. Effects of orbiting eccentricity, machining time on the protrusion shape and height, surface roughness and machining current are investigated. This work provides a method of producing surfaces which conformed better to the required shape especially in the production of dies and molds.

1 INTRODUCTION

Many ECM tools have slots or holes through which the electrolyte is supplied to the machining interface. These inlets have a detrimental effect as the machining does not proceed so rapidly in their vicinity and, as a result, protrusions are formed on the machined surfaces [1]. These protrusions have often been removed by time consuming mechanical means. The present work is carried out with the objective of removing these protrusions by relative orbital motion between the cathodic- tool and anodic-workpiece.

The use of orbital motion also improves the flow conditions since the gap size increases and decreases thus causing electrolyte agitation which, in turn, improves the current density and the resulting surface finish. During ECM without orbital motion the area under the flow port undergoes less metal removal due to the low current density which is caused by the large gap in this region. The height of these protrusions is known to be proportional to the square of the radius of the flow port, feed rate while it is inversely proportional to the applied voltage and electrolyte conductivity [2]. If orbital motion is applied, such area will be under the flow port for a fraction of time T [3]

Where

r The radius of the flow port.

R_E The radius of orbiting eccentricity.

$$T = \frac{2}{\pi}\sin^{-1}\frac{r}{2R_E} \tag{1}$$

Published in: E.Kuljanic (Ed.) *Advanced Manufacturing Systems and Technology*, CISM Courses and Lectures No. 437, Springer Wien New York, 2002.

Under such circumstances, the rate of metal removal could be increased to approach that of the rest of the surface by reducing the radius of flow port r and enlarging the radius of orbiting eccentricity R_E. However the decrease of r limits the flow of electrolyte which can be overcome by increasing the number of flow ports to ensure a homogeneous and a sufficient electrolyte supply to the inter electrode gap.

The reverse electrolyte flow and the presence of back pressure can, also, improve the surface quality and reduce the protrusion height. The surface quality could also be improved further if the sharp corners of the flow ports were smoothed to avoid cavitation and flow striation. In this work the effect of orbiting motion on the generated protrusions and surface quality are tested when finishing electrodischarge machined surfaces.

2 EXPERIMENTATION

The workpiece is 23.8 mm diameter and 40 mm length. Both ends were ground to an average roughness of 0.1 μm. The workpiece is then fixed to Eurospark 50 A EDM machine head. An EDM that removes 0.5 mm was carried out using no flushing conditions. The roughness was also measured using Talysurf 5 machine and was found to be 13 μm. The workpiece was then washed, dried and weighted before ECM tests.

Further tests to study the effect of radius of orbital eccentricity of the cathodic-tool electrode on the profile of the machined surfaces using multiple flow ports were also conducted using eccentricities ranging from 1 to 6 mm.

The orbital ECM tool electrode, was made from Duragraph 15 graphite having 2 mm central hole for electrolyte supply. The tool face was smoothed using fine emery paper. The inter-electrode gap was then measured using brass shims. For each experiment, the voltage (10 volt), the initial inter-electrode gap (0.2 mm), the orbiting speed (50 rpm) and; electrolyte (200 g/l NaCl) temperature (30 ^{0}C) and flow rate (24 l/min.) were kept constant. During machining, the current was recorded. The surface roughness and profiles were also measured and recorded. Different machining times, namely, 15, 25 and 35 seconds were experimented.

3 RESULTS AND DISCUSSION

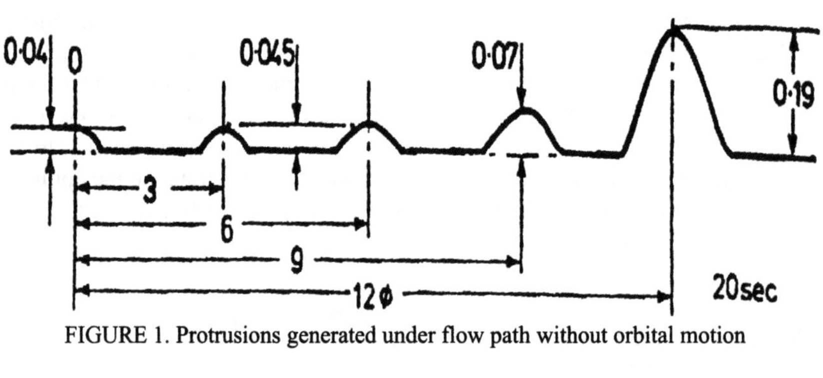

FIGURE 1. Protrusions generated under flow path without orbital motion

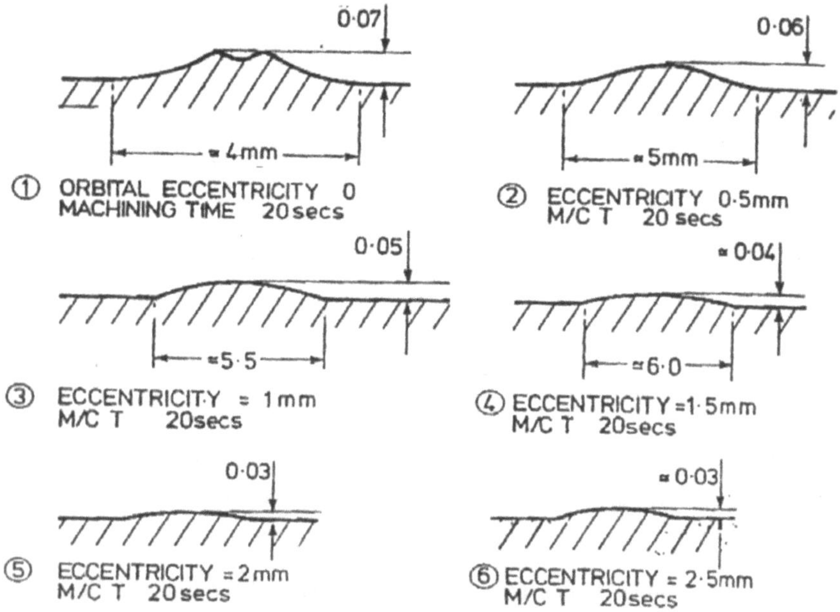

FIGURE 2. Protrusions formed at different orbiting eccentricities

A protrusion upon the profile surface was noted to occur in the center of the flow ports when the radius of eccentricity was zero. Figure 1 shows the profile of the surface obtained under flow path without orbital motion. Accordingly, the protrusion height becomes larger at greater workpiece radius. For a given radius of 9 mm and during 20 seconds of machining, the shape and height of protrusion were reduced as the radius of eccentricity was increased Figure 2. As the radius of orbit is increased the protrusion upon the surface took the form of circular ring. During 25 seconds of orbital ECM, the height of such protrusion was found to decrease from 179 μm to 90 μm for eccentricity values between 1 and 6 mm respectively, Figure 3.

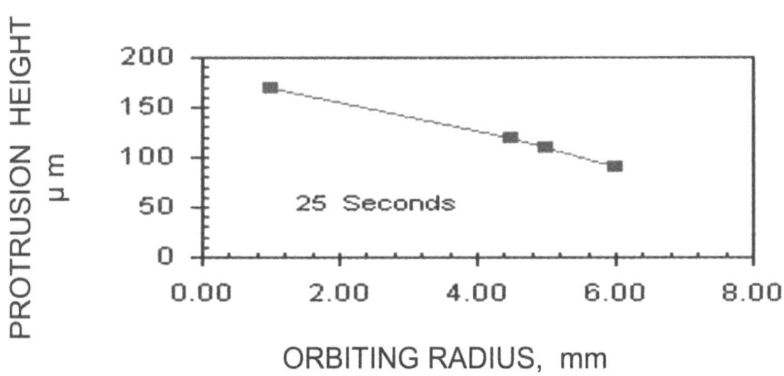

FIGURE 3. Variation of protrusion height with orbiting radius.

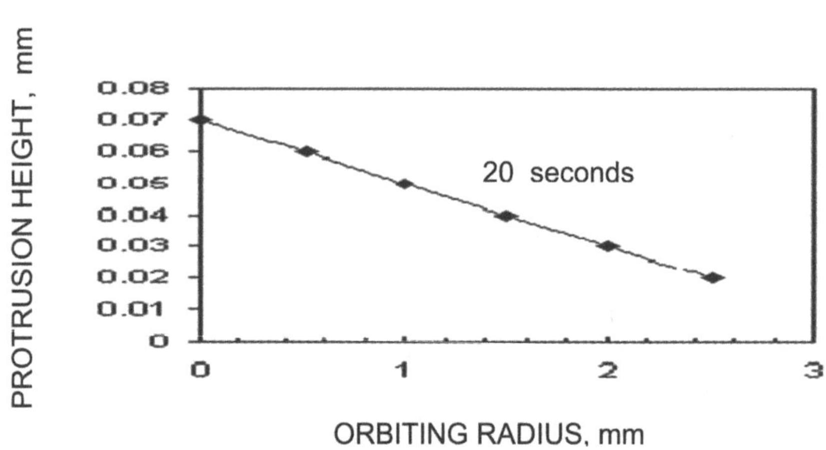

FIGURE 4. Protrusion height at different orbiting eccentricities.

Figure 4 shows similar trends for lower range of orbital motion and 20 seconds of orbital electrochemical machining time. These trends can be related to the increase of current density as the localized inter electrode gap decreases. Under such circumstances, the material removal rate is enhanced with consequent decrease in protrusion heights. For a given orbiting eccentricity of 2 mm, the longer the machining time the smaller would be the protrusion height, Figure 5.

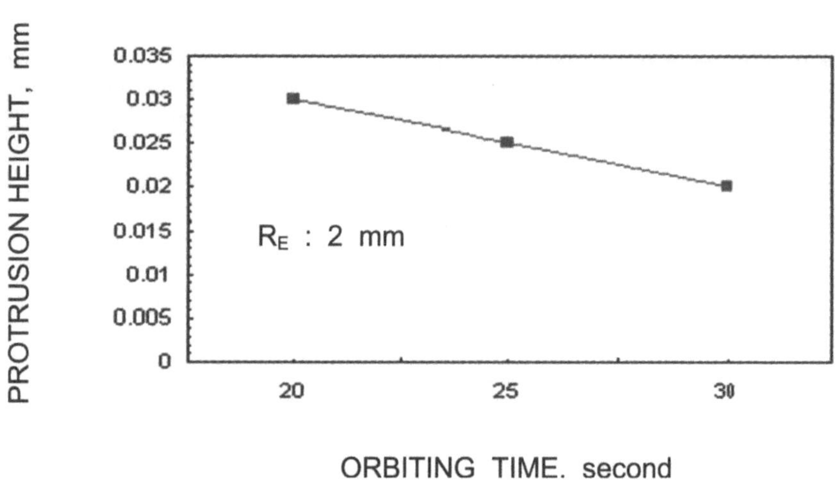

FIGURE 5. Effect of orbiting time on protrusion height.

Using single flow port experiments, the roughness measurements were taken at 6-8 mm from the center of the protrusion. Figure 6 shows the roughness profiles of the surface machined by EDM. However, the introduction of ultrasound in micro EDM enhanced the debris removal and improved the surface quality of micromolds [4]. Figure 7 shows the effect of orbital motion on EDM surface after orbital EC- machining time of 25 seconds and different orbiting eccentricities. It is evident that the produced surface roughness was found to be large for specimens machined at smaller eccentricities.

FIGURE 6. Surface profiles after EDM.

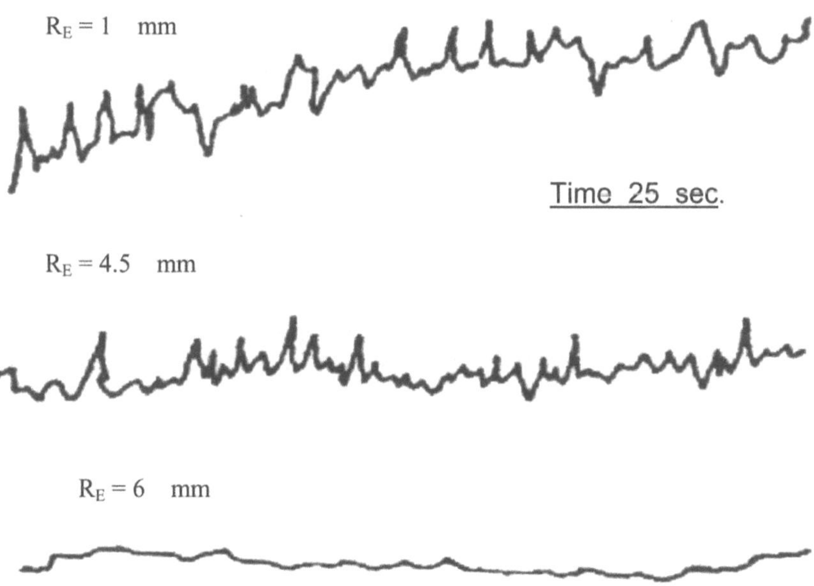

FIGURE 7. Surface profiles after orbital ECM at different eccentricities

The average and the maximum roughness values (Ra and Rt) for specimens subjected to EDM followed by ECM at different radii of eccentricity are shown in Figure 8. Accordingly, for machining time of 25 seconds the roughness values changed from 13 µm to 1.8, 1.6, 1.4 and 1.3 µm for eccentricity values of 1, 4.5, 5, and 6 mm while the maximum roughness Rt changed from 48 µm to 11, 10, 9 and 8 µm respectively. The decrease of surface roughness at greater eccentricities may be directly related to the rise in the current density and the improved flow conditions that flushes away the machining products as well as gas bubbles which leads to the improvement in the local current distribution and consequently the surface quality [5]. Under such conditions, the roughness of the machined surface was also greater at the outer perimeter than the center due to the reduction of the local current density as a result of the accumulation of the machining products that led to the higher protrusions as well as rougher surfaces.

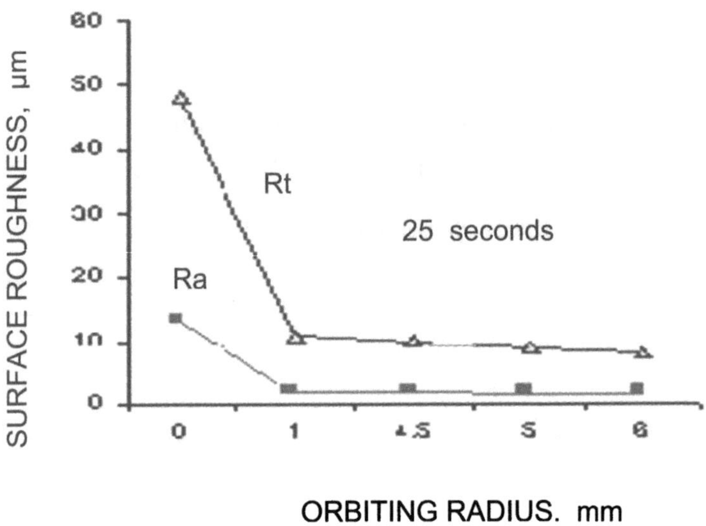

FIGURE 8. Variation of surface roughness with orbital eccentricities.

Figure 9 shows the current recordings at different machining times. Current oscillations occurs during the machining time due to changes in the inter-electrode gap associated with the orbital motion. The periodic fluctuations in the current recordings are related to the orbiting frequency. For the different machining times experimented, larger current was obtained at the start of machining as a consequence of the small inter-electrode gap at this stage of the machining cycle. The current subsequently decreased and the wave form adopted a similar pattern for each period of machining.

During ECM, the measured current indicated that passivation occurred at conditions of small radii of orbiting electrode. The experiments showed that passivation arose after 10 seconds of machining while normal active machining, at higher machining current, was possible after further 9 seconds [6]. No Passivation occurred when the radius of orbit was greater than 4 mm as the current remains steady at 190 – 240 Amperes.

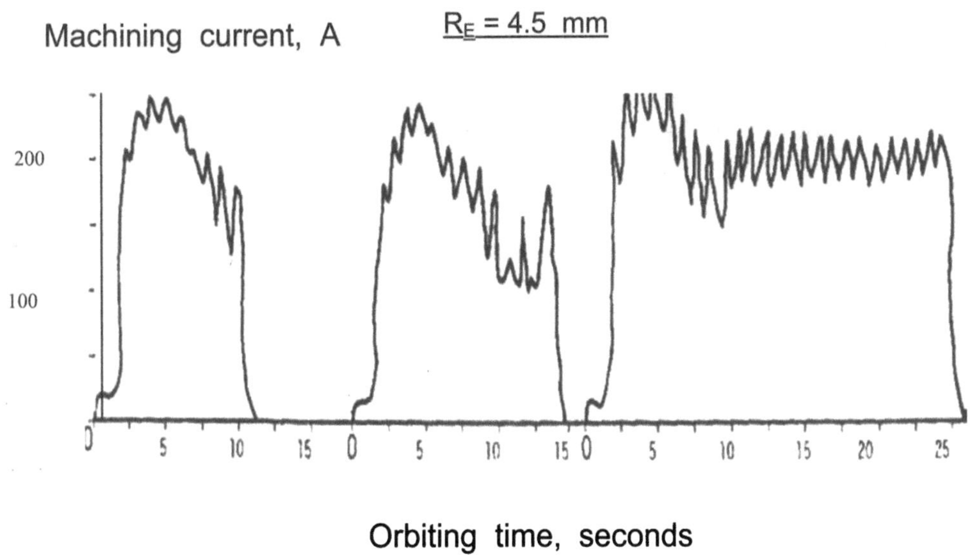

FIGURE 9. Current records for different orbital ECM periods

4 CONCLUSIONS

From the results obtained under the conditions descried in this work it can be concluded that;

1. Orbital motion of one electrode relative to the other, during ECM finishing reduces the height of the protrusions caused by the presence of the flow ports in the tool. This leads to an increase in dimensional accuracy.
2. The surface produced by EDM can bc smoothed by ECM in a very short time of machining. The average roughness, Ra, can be reduced from 13 μm to 1 μm, and Rt from 48 μm to 8 μm in 25 seconds.

REFERENCES

1. Tipton, ,H (1971), The Determination of the Shape of the Tool for Use in Electrochemical Machining , MTIRA report No 40.
2. Rajurkar, K. P; Zhu, D. and Wei, B, (1998). Minimization of Machining Allowance in Electrochemical Machining, CIRP Annals, Vol. 47/1, 165-168.
3. Sadollah, Z, (1984) Experimental and Theoretical Studies of the Electrochemical Machining of Electrodischarge Machined Surface Layer, Ph.D Theses, Aberdeen University.
4. Yeo, S. H. and Goh, K. M. (2001), The Effect of Ultrasound in Micro-EDM on Surface Roughness, Journal of Engineering Manufacture, Vol. 215 No B2,

271-276.

5. Rashed, A. F ; Youssef, H. A, and El-Hofy, H. A (1986), Effect of Some Process Parameters on the Side Machining During Electrolytic Sinking, 3rd PEDAC Conference, Alexandria, 733-746,
6. McGeough, J. A. , (1974), Principles of Electrochemical Machining , Chapman and Hall, London

DESIGN OF LOW EXPANSION COMPOSITES WITH REDUCED PROCESS INDUCED UNCERTAINTIES

F. De Bona[1], A. Gambitta [2], A. Somà [3]

[1] Dipartimento di Ingegneria Elettrica Gestionale e Meccanica, Università di Udine, Italy
[2] Accelerator Division, Sincrotrone Trieste S.C.p.A, Trieste, Italy
[3] Dipartimento di Meccanica, Politecnico di Torino, Italy

KEYWORDS: dimensional stability, composite laminate, coefficient of thermal expansion (CTE).

ABSTRACT. A procedure for the design of structural elements made of carbon-epoxy laminates with low thermal expansion in the main direction has been developed. By using an interferometric technique, the influence of material characteristics and process parameters on the thermal expansion behavior was assessed. In order to minimize the interval of uncertainty of the thermal expansion coefficient, a laminate with layers 0^0 and 30^0 coupled with layers with orientation between 45^0 and 60^0 was designed. The experimental results are in good agreement with those obtained numerically, confirming that the proposed procedure permits a low thermal expansion with high stiffness to be easily obtained.

1 INTRODUCTION

A growing request of structural elements characterized by low thermal expansion has recently been shown in several innovative industrial applications. In the field of optical communications, the possibility of using satellite networks for mobile communications and interactive multimedia services have been demonstrated and it seems very promising [1]. A space laser communication system generally consists of several hundreds of very compact and lightweight telescopes, whose structure is characterized by tight dimensional stability tolerances, that can be fulfilled only using materials with low thermal expansion coupled with high strength and stiffness. Even if high dimensional stability space telescopes have been already constructed for scientific purposes [2], in the case of a commercial communication satellite system the wide number of terminals required will call for a new production oriented method of design and production that have to replace the old manufacturing practices. A promising field of application of low-expansion structural elements is also industrial metrology. In this case it should be possible to have measurement robots characterized by high dimensional stability; consequently special temperature controlled environments could not be necessary, permitting the robots to operate directly in the manufacturing line.

Composite laminates are particularly suitable as structural material, when ultra-low thermal expansion properties are required. In fact some fiber materials (i.e. carbon, kevlar, silicon carbide, etc.) show a very low or even a negative coefficient of thermal expansion (CTE) [3]. Therefore, by embedding these fibers in epoxy matrices, having an high and positive CTE, it is possible to obtain a material with satisfactory mechanical characteristics and coefficient of thermal expansion very close to zero. The thermal expansion behavior of an angle-ply laminate can be evaluated following an approach based on the classical lamination theory [4] and a value of fiber orientation angle at which the CTE vanishes can be determined. Experience

Published in: E.Kuljanic (Ed.) *Advanced Manufacturing Systems and Technology*,
CISM Courses and Lectures No. 437, Springer Wien New York, 2002.

shows that this value has a low engineering interest as, due to the variability induced by the processing the actual CTE could differ significantly from zero. To overcome this problem a solution often followed is that of measuring the CTE of a large number of composite elements, selecting those with a low-expansion behavior and using the rejected ones for less demanding applications [2]. Another approach is that of making, after a thermal expansion measurement, a subsequent layup to compensate for the residual expansion [5]. Both the mentioned approaches involve very high processing costs.

The aim of this work is that of setting-up a procedure that, making use of standard processing, permits the thermal expansion behavior to be obtained with a low variability, in order to reduce the costs and thus make this material suitable for mass production industrial application. For this purpose firstly the theoretical aspect of the thermal expansion of composite laminate will be considered; than, following an experimental approach, the analysis of the variability induced by the main manufacturing parameters will be performed, with the aim of developing an design solution that allows to minimize their effect maintaining at the same time good strength and stiffness characteristics.

2 THEORY

The evaluation of the thermal expansion behavior of composite laminates is a topic that has received less attention in literature with respect to the evaluation of laminate stiffness and strength. In the case of angle-ply laminates the only complete theoretical treatise is reported in [4]; in their work the authors studied the deformations of a laminate induced by swelling, but suggested also an equivalent formulation to evaluate the effect of temperature changes.

In the case of pure thermal expansion it is possible to calculate the CTE of the single-ply versus any arbitrary orientation (x,y) in the plane of the layer using the following expressions:

$$\alpha_x = \alpha_1 \cos^2\theta + \alpha_2 \sin^2\theta \tag{1a}$$

$$\alpha_y = \alpha_1 \sin^2\theta + \alpha_2 \cos^2\theta \tag{1b}$$

$$\alpha_{xy} = 2(\alpha_1 - \alpha_2)\sin\theta\cos\theta \tag{1c}$$

where θ is the orientation angle of the fibers.

The expressions of α_1, α_2 can be obtained following the micro-mechanical approach suggested in [6]:

$$\alpha_1 = (\alpha_f E_f \eta + \alpha_m E_m (1-\eta))/(E_f \eta + E_m (1-\eta)) \tag{2a}$$

$$\alpha_2 = (1+\nu_m)\alpha_m (1-\eta) + (1+\nu_f)\alpha_f \eta - \alpha_1 (\nu_f \eta + \nu_m (1-\eta)) \tag{2b}$$

where E_f, ν_f, α_f and E_m, ν_m, α_m are respectively the fiber and matrix elastic constants; η is the fiber volume fraction.

Similar expressions for the lamina elastic constants E_1, E_2, ν_{12}, ν_{21} can be obtained [7].

The CTE of an angle-ply laminate in an arbitrary x direction is given by the following expression:

$$\alpha = [A_{11}^{-1} \sum (\overline{Q_{11}}\alpha_x + \overline{Q_{12}}\alpha_y + \overline{Q_{16}}\alpha_{xy})h_k] + [A_{12}^{-1} \sum (\overline{Q_{21}}\alpha_x + \overline{Q_{22}}\alpha_y + \overline{Q_{26}}\alpha_{xy})h_k] + [A_{16}^{-1} \sum (\overline{Q_{16}}\alpha_x + \overline{Q_{26}}\alpha_y + \overline{Q_{66}}\alpha_{xy})h_k] \tag{3}$$

In the case of angle-ply systems stacked symmetrically with respect to the middle plane, the laminate is defined to be symmetric and balanced because coupled bending effects are not allowed. In this case Eq. (3) becomes:

$$\alpha = \frac{A_{22}R_1 - A_{12}R_2}{A_{11}A_{22} - A_{12}^2} \tag{4}$$

where R_1 and R_2 coefficients are related to the lamina material constants and can be evaluated by means of the invariant method [8]; A_{ij} are the coefficients of the extensional stiffness matrix [7].

3 EXPERIMENTAL SENSITIVITY ANALYSIS

3.1 INTRODUCTION

As shown in the previous paragraph, the thermoelastic behavior of a laminate is affected by the properties of the fiber (E_f, ν_f, α_f) and of the matrix (E_m, ν_m, α_m) and by the process related parameters θ and η.

The elastic parameters (E_f, ν_f, E_m, ν_m) are always certificated by the supplier; generally their uncertainty is low and it does not affect significantly the laminate thermoelastic behavior; on the contrary, the value of the coefficient of thermal expansion of the fiber α_f and of the matrix α_m shows a high variability. In fact the thermoelastic behavior of low expansion fibers (i.e. carbon) can not be evaluated easily, as a standard measurement procedure is not available yet; in the case of the matrix material a standard dilatometer [9] can be generally used, but in this case an high uncertainty has to be expected, as the thermal expansion behavior is strongly affected by such process parameters, as curing time and temperature, resin aging, etc. In order to overcome the variability induced by the uncertainty on the thermal expansion of the basic materials, a possible approach is therefore that of introducing in Eq. (1) the values of α_1 and α_2 obtained from a direct measurement; this can be easily performed on unidirectional laminate samples obtained from the same prepreg tape that will be used for the final structure.

Among the so called process parameters, the fiber content η is generally certified by the supplier with an uncertainty below 1% , thus not affecting significantly the value of α. On the contrary, as shown in [10], fiber orientation angle θ can influence strongly the thermal expansion of the laminate. As laminates are generally obtained by manual layup, the most suitable approach to evaluate the variability of α induced by the fiber orientation angle errors seems that of following an experimental approach.

3.2 MATERIALS AND METHODS

A composite component manufacturer (Simbi S.p.A, Torino, Italy) was asked to supply the following types of laminates: $[0_{12}]$, $[90_{12}]$, $[\pm30]_{3S}$,$[\pm45]_{3S}$,$[\pm60]_{3S}$. Ten samples for each kind

of laminate were obtained separately by manual layup of prepreg tape (Texipreg FC12C HM120 REC) supplied by Seal S.p.A., Milano, Italy). Sample dimensions (60X160X1.5 mm^3) were chosen in order to limit non-homogeneity and border effects [11].
All the samples were cured at the same time according to the procedure suggested by the supplier. The linear thermal expansion in the temperature interval 293-303 K was obtained for each sample averaging 10 measurements.
Thermal expansion measurements were performed according to the method described in [11]. This procedure is based on a two-frequency laser interferometer system, that permits an accurate thermal expansion measurement of composite laminates to be easily obtained. In fact in the case of carbon epoxy laminates a resolution of 30 nm/m and an interval of uncertainty of 100 nm/m was obtained.

TABLE 1. Angle-ply CTEs

Laminate type	α_s [μm/mK]	σ_p [μm/mK]	σ_m [μm/mK]	α_t [μm/mK]
$[0_{12}]$	-0.28	0.03	0.05	-
$[\pm30]_{3S}$	-5.86	0.18	0.17	1.99
$[\pm45]_{3S}$	2.18	0.64	0.09	1.99
$[\pm60]_{3S}$	23.62	1.44	0.23	23.18
$[90_{12}]$	43.83	0.66	0.67	-

3.3 RESULTS AND DISCUSSION

Tab.1 shows the mean value of coefficient of linear thermal expansion α_s and the standard deviation σ_p, obtained performing measurements of different samples made of the same laminate type. The interval of uncertainty of the measurement technique σ_m was obtained from repeated measurements on the same sample. The theoretical value of coefficient of thermal expansion α_t was obtained introducing in Eq. (4) the values of α_1 and α_2 obtained from the measurement of the $[0_{12}]$ and $[90_{12}]$ laminates respectively.
It can be noticed that for the angle-ply laminates, the obtained values of α_t are always inside the uncertainty interval of the corresponding measured values α_s. The values of σ_p were obtained considering the thermal expansion of samples made of same pre-preg material and characterized by the same fiber orientation angle θ, but obtained separately by hand layup; it is therefore reasonable to conclude that this parameter is related to an error in orienting the prepreg layers during the laminate manufacturing.

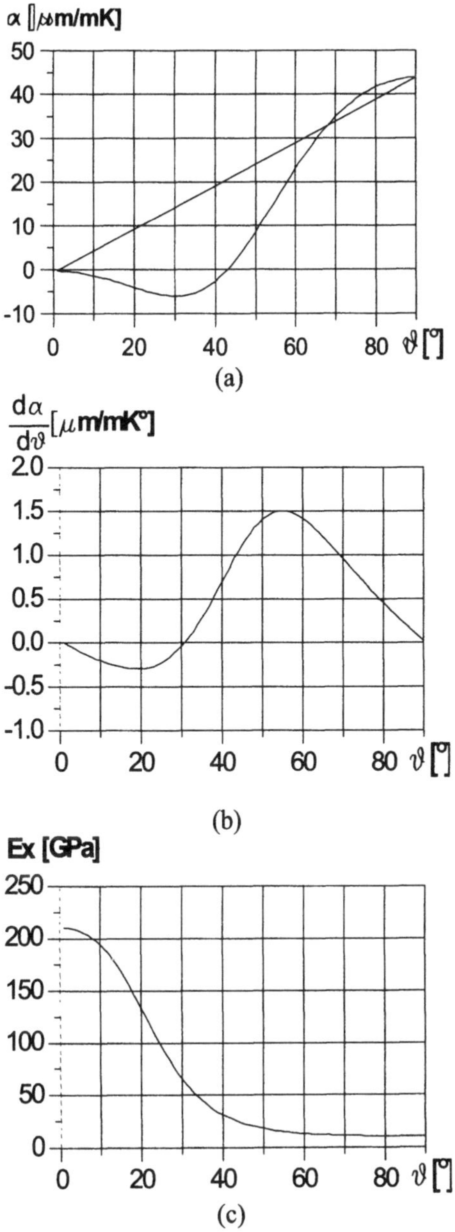

FIGURE1. Values of thermal expansion coefficient α (a), of its derivative $d\alpha/d\theta$ (b) and of the Young modulus E_x (c) versus fiber orientation angle θ for an angle-ply laminate.

In the case of the unidirectional laminates ($[0_{12}]$, $[90_{12}]$) and in the case of the angle-ply laminate with θ=30° ($[\pm 30]_{3S}$), the values of the process induces uncertainties σ_p show values similar to that of the corresponding measurement uncertainties σ_m. In order to explain this behavior, it could be helpful to consider the theoretical relationship between, $d\alpha/d\theta$ and θ (see Fig.1a,b). For θ=0°, θ=30° and θ=90° the values of the derivative is close to 0, therefore in these cases even significant orientation angle errors have limited effect on the thermal expansion of the laminate.

Angle-ply laminates $[\pm 45]_{3S}$ and $[\pm 60]_{3S}$ show a completely different behavior; in these cases the process induced uncertainty σ_s corresponds respectively to 28% and 6% of the value of α_s. In fact in the interval $40° \leq \theta \leq 70°$ the derivative $d\alpha(\theta)/d\theta$ is around its maximum value and therefore the linear coefficient of thermal expansion is strongly influenced by even small variations of θ.

Finally, it can be noticed that the procedure of evaluating α_1 and α_2 by direct measurements on unidirectional laminates seems correct, as in these cases (θ=0°, θ=90°) the derivative and consequently the uncertainty is low.

LOW EXPANSION DESIGN

3.4 PROPOSED PROCEDURE

As pointed out in [10], a 0 thermal expansion laminate can be designed simply considering an angle-ply laminate with θ corresponding to the point where $\alpha(\theta)$ vanishes. As in this case the derivative $d\alpha(\theta)/d\theta$ is high and therefore the process induced variability is significant, this solution has a limited practical interest.

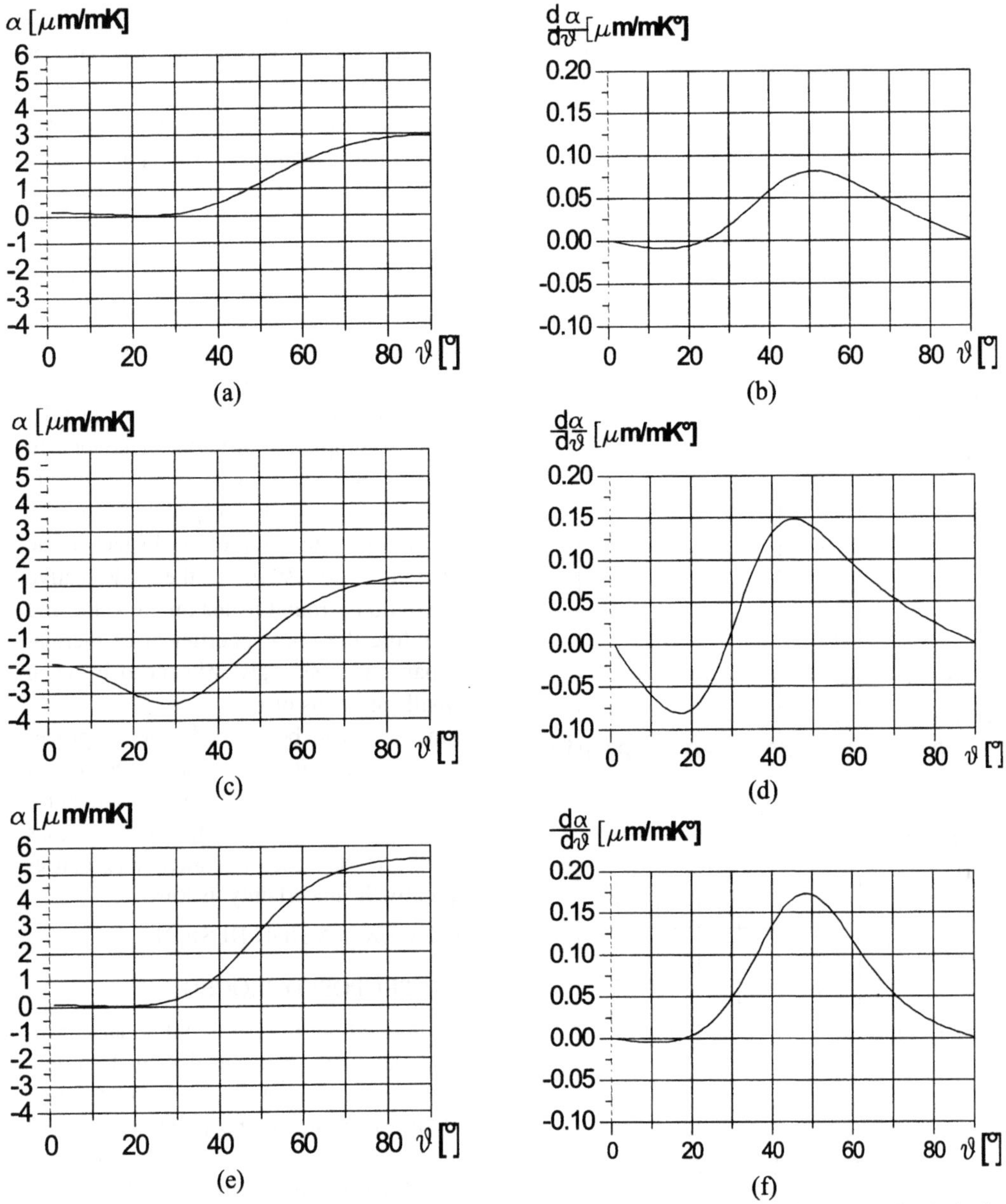

FIGURE 2. Values of α and of its derivative $d\alpha(\theta)/d\theta$ versus θ. a,b) [(±60°), (±θ) $(0°)_2$], c,d) [(±θ), (±30°), $(0°)_2$], e,f)[(±60°), (±30°), (±θ)].

A possible solution to obtain a low CTE with reduced process induced uncertainty could be that of designing a laminate with multiple layers at different fiber orientation angles. In order

to avoid post-curing bending the laminate has to be symmetric and balanced; moreover it would be preferable to avoid too different fiber orientation angles, in order to limits the lamina stresses that will be released after the curing process.
It has been shown that the minimum variability of α is achieved for θ= 0°, 30°, 90°. Among these fiber orientation angles, 0° e 30° have to be preferred; in fact, as shown in Fig. 1, in these cases a higher E can be achieved; moreover the CTE of the 90° laminate is strongly affected by the resin thermal expansion behavior and consequently by the curing process parameters, that generally can not be controlled accurately.
A possible design solution could be that of using a laminate mainly constituted by layers at ±30° e 0°, that show a slightly negative α; the unknown orientation angle θ^* of a third layer will be determined in such a way to set the overall laminate thermal expansion close to 0.
The proposed procedure to design a low expansion laminate can be summarized as follows:
-Firstly, the values of α_1 e α_2 are measured from unidirectional laminate samples obtained from the pre-preg tape lot that will be used for the final components; also E_1 and E_2 can be similarly obtained from tensile tests.
-Subsequently , according to Eq. (4), $\alpha(\theta)$ and $d\alpha(\theta)/d\theta$ curves are determined.
-Finally, following an iterative approach, the unknown orientation angle $\theta^*$of a symmetric and balanced laminate $[(\pm\theta^*),(\pm30),(0)_2]_S$ will be evaluated in order to counterbalance numerically the deviation from 0 of the $[(\pm30),(0)_2]_S$ laminate. Obviously the value of θ^* can not be evaluated "a priori", but strongly depends from the pre-preg lot properties and from the curing process followed.
For the prepreg material considered in this work a value of θ^* =60° has been chosen. In Fig.2 the values of α and of $d\alpha/d\theta$ versus θ values are reported; in this case the laminate must be considered as constituted by three parts: two of them have fixed fiber orientation angle, whereas the third part has a fiber orientation angle θ that varies from 0° to 90°. In all the three possible cases, low values of $d\alpha/d\theta$ around the chosen orientation angles (θ= 0° , 30° and 60°) are obtained; in the case $[(\pm\theta),(\pm30),(0)_2]_S$ (see Fig.2d) at θ=60° the proposed laminate exhibits a value of $d\alpha/d\theta$ one order of magnitude lower with respect to the 60° angle ply (see Fig.1b); in fact in this case the orientation induced errors are strongly limited by the presence of 0° and 30° layers. It must be also noticed that the proposed solution permits a value of Young modulus in the main direction E_x of 1.5 GPa to be obtained.

3.5 EXPERIMENTAL VERIFICATION

Ten samples for each of the following laminates were obtained: $[(\pm60),(\pm30),(0)_2]_S$, $[0_{12}]$, $[90_{12}]$ All samples were obtained individually from the same prepreg tape lot and underwent the same curing process. The CTEs at room temperature, obtained with the previously described procedure, are presented in Tab.2. The proposed solution ($[(\pm60),(\pm30),(0)_2]_S$) shows not only a very low mean value of coefficient of thermal expansion α_s, but, as expected, also a very low standard deviation σ_p. An even small, but significant difference between the measured and the computed values of CTE must be noticed. This can be easily justified, considering that α_t has been evaluated by using the values α_1 e α_2, obtained from the previous measurements (see Tab.1). Comparing Tab.1 and Tab.2 it is possible to notice that the thermal expansion coefficients of the two groups of unidirectional samples are slightly different; this is not due to errors in the orientation angle (the standard deviation in both cases is very low), but seems related to the non perfect repeatability of the curing process. If the values of α at 0° and 90° obtained from Tab.2 are introduced in Eq.(4) a value of α_t=-0.11 is obtained, which corresponds to the measured value.

TABLE 2. Low expansion laminate CTE

Laminate **type**	α_s [μm/mK]	σ_p [μm/mK]	σ_m [μm/mK]	α_t [μm/mK]
$([(\pm 60),(\pm 30),(0)_2]_S$	-0.10	0.04	0.04	0.08
$[0_{12}]$	-0.44	0.03	0.05	-
$[90_{12}]$	38.10	0.18	0.17	-

5 CONCLUSION

A procedure for the design of laminates for structural applications, characterized by low thermal expansion at room temperature have been developed and verified experimentally. The proposed method permits to design laminate components, whose thermal expansion behavior is characterized by limited variability, thus reducing the number of rejected components and consequently the costs.

REFERENCES

1. Lutz, H.P., (1997), Optical Communications in Space –Twenty Years of ESA Effort. ESA Bulletin, 91, 25-31.
2. Mc Carthy, D.J., Facey, T.A., (1982), Design and Fabrication of the NASA 2.4-meter Space Telescope. Optical Systems Engineering II, SPIE Vol.330, 139-143.
3. Stumm, J.E., Pynchon G.E., Kumweide, G.C., (1981), Graphite/epoxy Material Characteristics and Design Techniques for Airbone Instrument Application. SPIE Proceedings, Vol. 309, 188-198.
4. Halpin, J.C., Pagano, N.J.,(1969), Consequences of Environmentally Induced Dilatation in Solids. AFML-TR-68-395.
5. Krim, M.K., (1975), Design of Highly Stable Optical Support Structure, Optical Engineering, Vol. 16 (6), 552-558.
6. Shapery, R.E., (1968), Thermal Expansion Coefficients of Composite Materials Based on Energy Principles. J. Composite Materials, 2, 380-404.
7. Jones, R.M. (1975), Mechanics of Composite Materials. McGraw Hill Publishing.
8. Halpin, J.C. , (1984), Primer on Composite Materials. Technomic Pub.
9. ASTM E228-85 (1990), Standard Test Method for Linear Thermal Expansion of Solid Material with Vitreous Silica Dilatometer. Annual Book of ASTM Standard, 3.01, 378-382.
10. De Bona, F., Gambitta, A., Soma', A., (1992), Evaluation of Thermal Expansion Coefficient versus Fiber Orientation of Composite Laminate. Proc. of Experimental Techniques and Design in Composite Materials, Cagliari, 247-258.
11. De Bona, F., Somà, A., (1997), Thermal Expansion Measurement of Composites with Optical Heterodyne Interferometry. Experimental Mechanics, Vol. 37 (1), 21-25.

ANALYSIS OF HIGH POWER CO_2 DUAL BEAM LASER WELDING

E. Capello[1], P. Chiarello[1], E. Piccione[1], B. Previtali[1]

[1] Department of Mechanical Engineering, Politecnico of Milan, Italy

KEYWORDS: dual beam laser welding, analytical thermal model, high power CO_2 welding.

ABSTRACT. High power CO2 lasers are increasingly used in industrial welding, as this technology presents several advantages when compared to traditional welding processes. Nevertheless, the high power density of the laser beam can turn into a drawback when high carbon steels are welded. In the paper a high power (6 kW) CO2 laser source is used to perform a dual beam welding. The work was divided into three parts: the first is aimed to the modeling of the thermal field generated by the two portions of the laser beam. Then, based on this results, the optical equipment necessary to divide the beam into two portions with the desired characteristics was designed and developed.
In the third and final part welding experiments were carried out, and results have been compared to the traditional single beam welding, showing that the deal beam process can reduce the hardness of bead and HAZ, with beneficial effects on the toughness of the joint.

1 INTRODUCTION

The current large diffusion and success of the high power lasers in welding processes could be ascribed mainly to two features of the laser beam. The former is the high focalisation of the laser beam, which provides a tool with a small diameter. The latter is the high power of the beam which, together with the small spot area at focus, generates the power density needed to start the key hole phenomena. Thanks to the small beam diameter and the high power density, both the bead is narrow and deep penetrating, and the heat affected zone is small. Moreover, a high feed rate can be used since the power density is destined to the bead generation and not to the general heating of the workpiece.
However, in some cases these advantages could turn in drawbacks, which limit the laser beam welding application. Indeed, the high workpiece feed rate and the high energy density induce in the welding zone large temperature gradients and high cooling rates, which are particularly deleterious in the welding of high carbon steels, producing cracks in the bead and distortions in the workpiece.
Usually, in order to soundly weld these alloys the workpiece is globally pre-heated in oven to about 300-400° C. The pre-heating increases the workpiece temperature to a value below the transition temperature but sufficiently high to smooth the thermal cycle in the welding zone. Despite the pre-heating is nowadays a common industrial solution to welding of high carbon steels, the use of this device presents several drawbacks. In fact, the pre-heating is a time consuming and expensive operation. Moreover, the workpiece dimensions or location cold be a limit to this application.
An alternative solution, not yet deeply studied and applied, is the dual beam laser process. The dual beam laser process consists in the laser beam division into two beams: the former with a heating function, the later with a welding function. While the heating beam is not deeply

Published in: E.Kuljanic (Ed.) *Advanced Manufacturing Systems and Technology*,
CISM Courses and Lectures No. 437, Springer Wien New York, 2002.

focalised and pre-heats the welding zone, the welding beam is focalised and generates the bead in the keyhole condition.
The present work is a contribution to the study of the dual beam process. The paper deals with the application of the dual beam principle to high power CO_2 lasers . The main references to the dual beam welding are from Kannatey-Asibu and Liu. The Authors have investigated both the theoretic aspects (with analytical and numerical models) and the experimental aspects of the dual laser beam welding with medium power lasers (up to 3 kW) [1][2][3]. The medium power of the beam lasers does not allow high feed rate (and consequently high productivity) and limits the positive effects of the pre-heating beam.
The present work deals with the study and design of a dual beam system from a 6kW power CO_2 laser. First, the temperature field generated by the superimposition of the two beams has been investigated by a thermal model. The analytical thermal model was used to verify that, by the control of the process and design parameters, significant improvements in the cooling heat rate could be reached. Moreover, it was used in the design of the beam subdivision and focussing device. Then, a dual beam system has been built starting from a single high power CO_2 laser source. Experiments have been carried out, and the results of the dual beam laser welding have been compared to traditional single beam welding in terms of hardness of the welded zone.

2 THERMAL MODELS

The need for a thermal model arises from the fact that:

- the single beam heat cycle has to be compared to the dual beam heat cycle;
- the dual beam system has to be designed.

The temperature field in the workpiece arises from the simultaneous passage of the two beams, along the workpiece's surface. The two beams (Figure 1):

- are at a fixed distance *L*,
- move at the same feed rate *V* (the feed rate of the workpiece).

Since the two beams have different characteristics, two thermal models should be developed. Once developed, the two models can be composed together to describe the temperature field in the workpiece on the basis of the principle of superposition of effects.

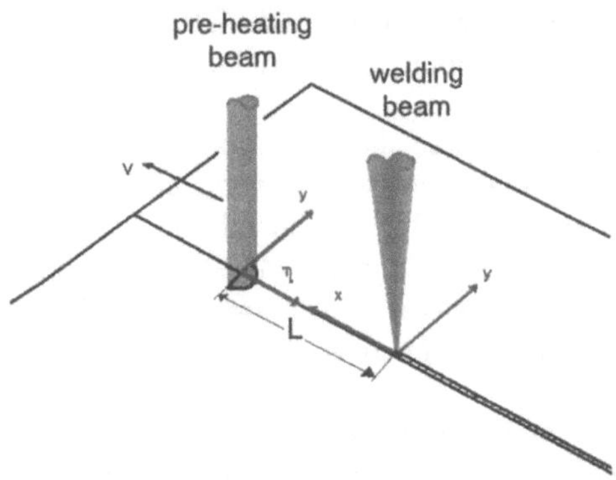

FIGURE 1. Dual beam laser system.

2.1 PRE-HEATING MODEL

The laser beam action could be described by a moving heat source travelling with constant feed rate V along the travelling direction. If the thermal properties of the material do not depend on temperature, the governing differential equation is:

$$\nabla^2 T - \frac{1}{\alpha}\frac{dT}{dt} - \frac{q_r}{\lambda} = 0 \tag{1}$$

where α is the thermal diffusivity, λ the conductivity and q_r the heat input rate per unit time. The pre-heating beam could be described by a cylindrical beam with a bi-gaussian power q_h:

$$q_h = \int_0^\infty\int_0^\infty q_{\max} e^{-\left(\frac{\eta^2+y^2}{r^2_B}\right)} dxdy \tag{2}$$

where q_{max} is the power density peak value and r_B is the distance from the beam centre at when the peak value has fallen to $1/e$.

An analytical solution for the temperature field $T(\eta,y,z)$, when a bi-gaussian laser beam of power q_h and radius r_B moves in the x direction, is available [4]:

$$T - T_0 = \frac{A_h q_h / V}{2\pi\lambda\sqrt{\left(t+\frac{\eta}{V}\right)\left(t+\frac{\eta}{V}+t_0\right)}} \exp - \left[\frac{1}{4\alpha}\left(\frac{(z+z_0)^2}{\left(t+\frac{\eta}{V}\right)} + \frac{y^2}{\left(t+\frac{\eta}{V}+t_0\right)}\right)\right] \tag{3}$$

where T_0 is the initial temperature, (η,y,z) are the coordinates fixed to the pre-heating source (reported in Figure 1), A_h is the absorbtion coefficient, t_0 the injection time and z_0 the apparent source origin. The injection time $t_0 = r_B^2/4\alpha$ is the time in which the heat diffuses inward when the pre-heating source covers a distance equivalent to r_B. The apparent source origin derives from the need to have a temperature of the surface limited to a finite value, and therefore the real source in the plane $z = 0$ is replaced by an apparent source in the plane $z = z_0$ above the surface. The actual value of z_0 is determined by imposing at time t_0 and at the surface $z = 0$ a temperature equivalent to the temperature reached in a solid heated for the time t_0 by a stationary heat source.

2.2 WELDING MODEL

Since the welding beam is focalised, a linear heat source, moving with constant velocity V along the workpiece surface, should yield a satisfactory picture of the key hole welding process. The analytical solution for the temperature field $T(x,y,z)$, when a linear source of constant strength q_w moves in the x direction, is [5]:

$$T - T_0 = \frac{A_w q_w}{2\pi\lambda} \exp\left(\frac{-Vx}{2\alpha}\right) K_0\left\{\frac{V\sqrt{x^2+y^2}}{2\alpha}\right\} \tag{4}$$

where A_w is the welding absorbtion coefficient, (x,y,z) are the coordinates system fixed to the welding heat source (see Figure 1), q_w is the heat rate per unit depth and K_0 is the modified Bessel function of the second kind and zero order.

2.3 DUAL BEAM MODEL

The thermal model of the dual beam laser process is the composition of the two afore - developed models: the pre-heating model and the welding model. Due to the linearity of the conduction equation, the superposition of the two effects, the pre heating and the key hole welding, leads to the global solution:

$$T-T_0=\frac{A_h q_h/V}{2\pi\lambda\sqrt{\left(t+\frac{L-x}{V}\right)\left(t+\frac{L-x}{V}+t_0\right)}}\exp\left[-\frac{1}{4\alpha}\left(\frac{(z+z_0)^2}{\left(t+\frac{L-x}{V}\right)}+\frac{y^2}{\left(t+\frac{L-x}{V}+t_0\right)}\right)\right]+\frac{A_w q_w}{2\pi\lambda}\exp\left(\frac{-Vx}{2\alpha}\right)K_0\left\{\frac{V\sqrt{x^2+y^2}}{2\alpha}\right\} \quad (5)$$

In the Equation 5 the two coordinates systems, the former relative to the pre-heating source, the later relative to the welding source, have been unified by the variable transformation $\eta = L - x$, where L is the fixed distance between the pre-heating and the welding sources (Figure 1).

Figure 2 shows the isotherms and the temperature field in the plane (x,y) of a carbon steels sheet subjected to the dual laser beam welding. In the example a power beam of 5.2 kW has been equally distributed: 2.6 kW to the pre-heating beam and 2.6 kW to the welding beam respectively. In order to evaluate the influence of the dual laser beam on the welding process two main thermal effects should be considered: the temperature field and the cooling heat rate.

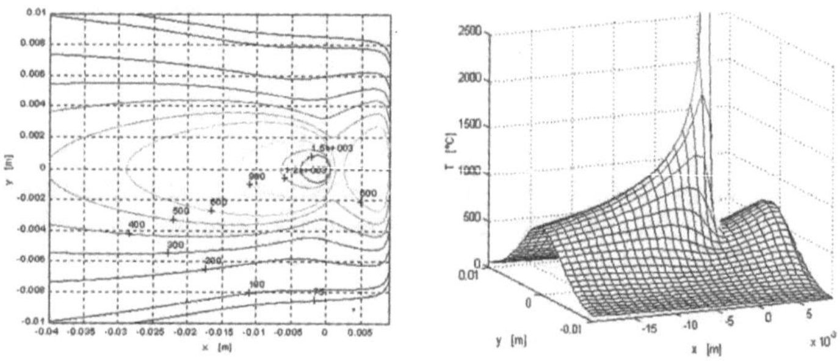

FIGURE 2. Isotherms and temperature field.

In the Figure 3 the comparison between a single beam laser welding (2.6 kW) and a dual beam laser welding (2.6 kW welding; 2.6 kW pre-heating) is shown. As could be noted, both the temperature and cooling time increase due to the pre heating action. The parameter $\Delta t_{800\text{-}Ms}$, defined as the cooling time from 800°C to the martensite start temperature, allows to compare the cooling rate curves.

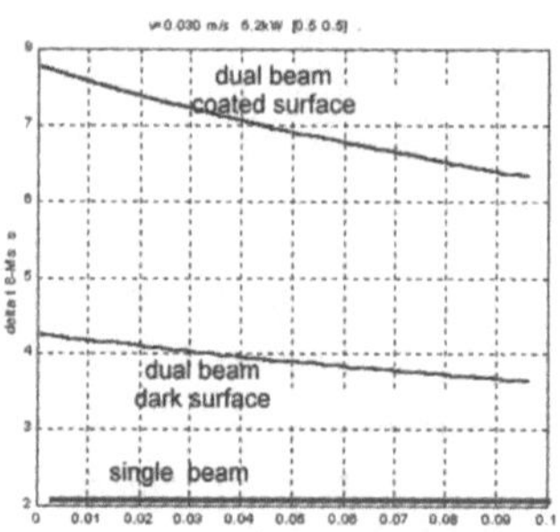

FIGURE 3. Single beam and dual beam welding comparison.

The $\Delta t_{800\text{-}Ms}$ is related to the cooling time of the carbon steel from the completely austenitesed state to the beginning of martensite solid state transformation. Since usually to a longer $\Delta t_{800\text{-}Ms}$ corresponds a less fragile bead, the response of the thermal model in terms of $\Delta t_{800\text{-}Ms}$ allows to compare different welding conditions. In order to obtain the $\Delta t_{800\text{-}Ms}$ values from the thermal model of the dual beam welding some parameters should be settled. The unknown parameters A_h, A_w, R_b of the Equations 2 and 5 have been calibrated against experimental data, collected by direct measurement of the temperature field with micro thermocouples, as reported in [6].

3 DUAL BEAM SYSTEM DESIGN

In the second part of the work the dual beam system has been designed and built. The dual beam system has been obtained by a conventional laser welding head. The source is a transverse flux 6 kW CO_2 laser. In order to obtain from a single beam two beams, the problem of the separation should be solved. Since the original beam is a high power CO_2 laser, the transmission by partially reflective lens is solution that should be rejected. The incident beam excessively heats the lens and may cause a severe damage to the beam splitting system. Therefore, if the dual beam system comes from a high power laser source, the reflection by a copper mirror should be the most suitable solution to separate the beam. Moreover, a moving mirror can be used to continuously modulate the repartition of the laser beam power in the two beams. The parameters of the dual beam system that should be controlled are:

- the relative distance between the two beam spots (L);
- the relative repartition of power between the two beams (P).

Both these characteristics can be obtained from the position of the copper mirror, which has two degrees of freedom (Figure 1). Therefore a device supporting the copper mirror has been designed in order to allow the accurately setting the following movements of the mirror:

- translation along the z axis,
- rotation around the y axis.

In the Figure 4, showing the dual beam system designed and realised, the single elements afore described could be observed: on the left the transmission path, in the middle the device supporting the copper mirror and on the right the welding head.

FIGURE 4. Dual beam laser welding system.

4 EXPERIMENT DESIGN

The dual system described in the previous paragraph allows to regulate two design parameters: the distance *L* and the percentage of the intercepted beam *P*. In order to limit the number of experimental conditions, this preliminary work deals only with the study of the effect of the distance *L* on the properties of the welds.
The percentage *P* has been fixed to 50 %: the copper mirror intercepts the 50% of the maximum power of the original beam (5.2 kW) and assigns 2.6 kW to both the beams.
Together with the percentage *P* the other design and process parameters which have been fixed are: the welding focus position (on the surface), the shielding gas (helium at 30 l/min), the travelling speed rate (30 mm/s).
In the experiments a high carbon steel (with a 0.55 % percent of carbon, C55) has been used. The C55 carbon steel has been furnished in oil quenched sheets of 1.25 mm thick with a dark surface. In the experiments two parameters have been varied:

- the distance *L*: 5 – 20 - 60 mm,
- the surface condition: as furnished dark surface and coated with graphite.

Each treatment has been repeated threes times. The response variable is the hardness of the bead and of the heat affected zone (HAZ). After the experiments the specimens have been sectioned, polished and etched in a 3% alcoholic solution of H_2NO_3 in order to reveal the grain structure. Vickers hardness measurements (2000 g load, UNI EN ISO 6507-1) have been made on the section normal to the welding direction to give the average hardness profile across the bead and the HAZ. Since the specimen's thickness is very small, according to the normative, only one single indentation line at half thickens has been measured.

5 RESULTS AND DISCUSSION

The dual beam laser welding of the C55 high carbon steel has been simulated by the thermal model for two different surface conditions: dark surface and graphite coated surface, changing the laser beam distance *L*. The single beam welding process in the same conditions has been simulated too. The outcome of the model has been the $\Delta t_{800\text{-}Ms}$, and is reported in Figure 5.

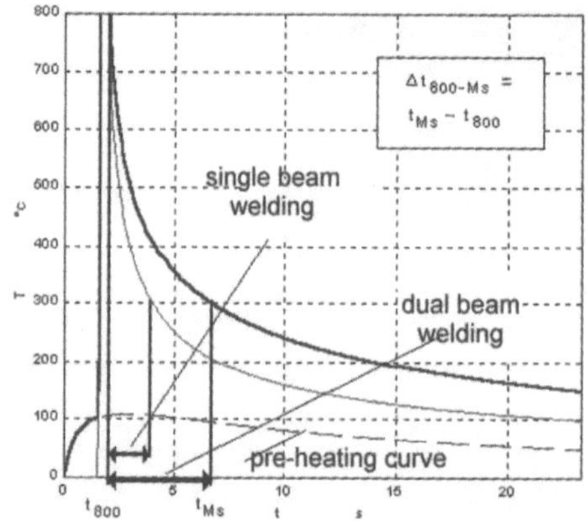

FIGURE 5. $\Delta t_{800\text{-}Ms}$ comparison.

As can be noted,

- the dual beam process generates in the workpiece a longer $\Delta t_{800\text{-}Ms}$ than the single beam welding process. Moreover, the cooling time $\Delta t_{800\text{-}Ms}$ increases when the distance L between the beams is reduced,
- the coated surface shows a longest $\Delta t_{800\text{-}Ms}$ due to the different absorption coefficient A_h.

In order to bear out the positive results predicted by the thermal model, the Vickers hardness have been measured on the surface normal to the welding direction. The data have been collected in Figure 6, which reports the HV values as a function of the distance from the centreline of the bead. In Figure 6 the term "reference" means the hardness of the specimen welded by the single beam process.

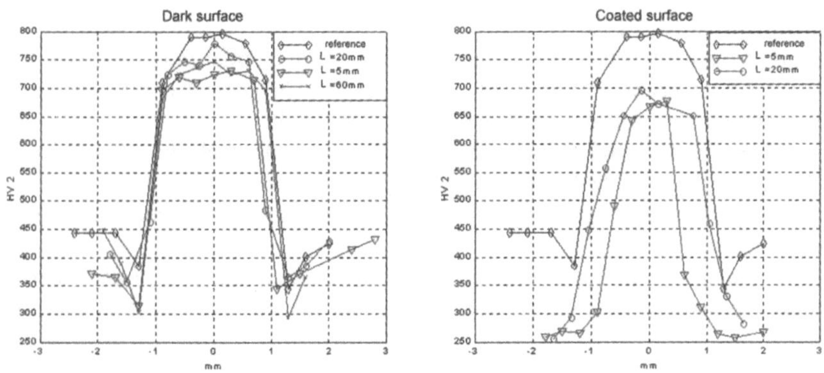

FIGURE 6. Hardness results.

The analysis of Figure 6 leads to the following considerations:

- the samples welded by the dual beam process have hardness values always lower than the samples welded by the single welding process,
- a remarkable difference could be pointed out in the case of the coated surface,
- the effectiveness of the pre heating treatment decreases when the distance between the two heat source becomes longer.

The theoretical and experimental results are correlated and compared in the Table 1 for the dark surface and the coated surface respectively.

TABLE 1. Results comparison.

	Dark surface		Coated surface	
Distance L (mm)	HV	$\Delta t_{800\text{-}Ms}$ (s)	HV	$\Delta t_{800\text{-}Ms}$ (s)
5	716	4.23	662	7.8
20	749	4.10	668	7.4
60	720	3.84		
Reference	789	2.04	789	2.04

As could be noted, the cooling time $\Delta t_{800\text{-}Ms}$ and the hardness values are correlated. Indeed, for both the surfaces in presence of the pre heating beam the toughness of the bead increases due to the longer cooling time. The increase of the toughness is more remarkable when the pre heating is performed on a coated surface. The absorption coefficient A_h for the steel surface (about 10%) increases up to the 80% for the coated surface [6]. As a consequence, for a distance L of 5 mm, when the $\Delta t_{800\text{-}Ms}$ is increased from 2.04 s (reference case) to 7.8 the hardness value decreases from 789 HV to 662 HV on average. Moreover, both the theoretical and the experimental results confirm that the distance L greatly affects the cooling time and the hardness of the bead: As the pre-heating beam is moved away from the welding beam, the cooling time is no longer affected by the presence of the dual beam system.

6 CONCLUSION

The present paper discusses the possibility of performing a dual beam laser welding as a solution to the negative effect of the large thermal gradient and high cooling rate on the toughness of high carbon steels.

The first part of the work was aimed to the modeling of the thermal field generated by the two portions of the laser beam. Then, based on this results, the optical equipment necessary to divide the beam into two portions has been designed and built.

Experiments have been carried out and, although limited to one steel, show that there is a fairly reduction in the hardness of the bead when a dual beam system is used.

Therefore, the preliminary conclusions pointed out by the thermal modelling and by the experimentation are:

- the smoothing of the heat cycle forecasted by the thermal model is in agreement with the hardness measurements, both of them confirming the positive effect of the pre heating treatment.
- if the surface of the specimen is opportunely treated in order to improve the absorption coefficient the influence of the preheating beam increases.
- reducing the distance between the two beams produces a positive influence on the thermal and mechanical behaviour of the specimen.

The afore mentioned considerations together with the easiness in the dual beam realisations gives the first encouraging results in the direction of improving the quality of difficult to weld materials.

ACKNOWLEDGEMENTS

The authors wish to thank all the people that have supported, and are supporting, the activities of the SITEC Lecco – Laboratorio per le Applicazioni Laser of the Politecnico di Milano.

REFERENCES

1. Kannatey-Asibu E.Jr., (1991),Thermal Aspects of the Split Beam Laser Welding Concept, ASME Journal of Engineering Materials and Technology, Vol. 113, 215-221.
2. Liu Y.N., Kannatey-Asibu E. Jr., (1993), Laser Beam Welding with Simultaneous Gaussian Laser Preheating, ASME Journal of Heat Transfer, Vol. 115 , 34-41.
3. Liu Y.N., Kannatey-Asibu E.Jr., (1997) Experimental Study of Dual Beam Welding of AISI 4140 Steel, Welding Research Supplement, N°.9, 342-348.
4. Ashby M.F., Easterling K.R., (1984), The transformation hardening of steel surfaces by laser beams – I.hipoeutectoid steels, Acta Metallurgica, Vol.32, 1935-1948.
5. Rosenthal D., (1946), The Theory of Moving Sources of Heat and its Applications to Metal Treatments, Transactions of the A.S.M.E, Vol 68, 849- 866.
6. Capello E., Chiarello P., Piccione E., Previtali B., (2001), Saldatura Laser Dual Beam, IIS Welding Congress, Milano.

FILAMENT WINDING: AN INTEGRATED SIMULATION ENVIRONMENT FOR AUTOMATED CELL PROGRAMMING

A. Anglani, F. Nucci, A. Spagnolo

Department of Innovation Engineering, University of Lecce, Italy

KEYWORDS: Filament winding, Automated process, Computer simulation.

ABSTRACT. Filament winding is a process in which tensioned resin-impregnated continuous fibers are placed on a specified path on a mandrel. Automated systems have been adopted to deliver the fiber with a fixed level of accuracy. Process simulation is a very important question in automated applications. Moreover, the necessity to process complex shape objects leads to consider different alternatives to deliver the fiber along the model. In this paper, we developed a methodology to obtain simulation models in an integrated software environment. The proposed solution, on the one hand, focuses on the exploitation of the degrees of freedom in the cell, on the other, it considers the part program generation when a rotating mandrel is adopted. A test case has been realized in order to check the feasibility of our approach. Two different filament winding cell models have been developed by following the proposed approach and an in-depth analysis is carried out in order to investigate on both solutions.

1 INTRODUCTION

Filament winding is an effective method to manufacture composite objects. In this process, composite layers are continuously wound on a mandrel. Previous studies suggested that product quality highly depends on process parameters. Since in such a kind of technology a wide range of variables is involved, it is very difficult to tune the system when a new product shape has to be manufactured. Simulation plays an important role in this context, hence innovative methodologies are continuously issued in order to produce high quality simulation models. A filament winding cell can consist in several components. Basically, the main ones are fiber supply spools, a resin impregnator (if needed), fiber delivery or delivery eye unit, and a mandrel. The delivery eye places fibers on a specified path on the mandrel. The strands can either be pre-impregnated with polymeric resin or impregnated during the process. Once the winding is completed, the resin is cured and, then, the mandrel is removed. Since the winding is performed under tension, depending on the surface geometry, fibers can slip from the chosen position. Hence, a stable winding path has to be carefully designed in order to avoid the slipping.

Major efforts have been made in the introduction of a robotic manipulator to perform the fiber placement activity. Indeed, in terms of shape, the high variability of such objects implies enormous difficulties in the design of automated machines. Existing flexible filament winding cells deal only with simple shape objects. Moreover, when machines work with complex geometry pieces, a low grade of flexibility is available. Therefore, only particular classes of pieces can be wound by using such machines. Recently, different filament winding cells have been designed in order to solve such a problem. Highly flexible cells have been issued and different complex shape objects have been obtained by filament winding technology. Depending on the manipulator workspace, using redundant degrees of freedom of a ma-

Published in: E.Kuljanic (Ed.) *Advanced Manufacturing Systems and Technology*, CISM Courses and Lectures No. 437, Springer Wien New York, 2002.

nipulator, it is possible to choose among multiple movement strategies in order to optimize several factors. Nevertheless, such robotic based filament winding cells entail several drawbacks: robots do not allow high production rates;different robot degrees of freedom are not exploited; the winding of particularly complex shape objects can lead robots to assume critical configurations.

Different studies have been made on this issue, [1][2][3][4]. Various theories suggest the way of obtaining the manipulator delivery eye point sequence once the path winding on the model is assigned, [5][6]. Recently, the determination of the corresponding robot part program has been investigated [7]. Additional studies have been made on this topic in order to find the solution of the manipulator inverse kinematic problem using a numerical algorithm, [8] [9] [4]. Other studies have investigated the possibility of evaluating, in qualitative and quantitative terms, the results of these inverse kinematic algorithms, [10] [11]. In this paper we have dealt with the evaluation of automated filament winding cells in which degrees of freedom are distributed between the mandrel and the delivery mechanism. The main advantage consists in the possibility of using - for a specific family of pieces – an overall number of degrees of freedom lower than the ones required by adopting a robotic cell with a fixed mandrel. Moreover, both high speed winding and low stress may be obtained in cell mechanisms. In the following section, the generation of the cell part program is presented. Then a methodology based on a simulation model for the study of such a cell is described. Finally, a test case is presented in order to show the potential of such a methodology and simulation results are reported.

2 DEGREES OF FREEDOM IN AN AUTOMATED CELL

Several approaches have been used to investigate on the possibility of adopting general-purpose machines for filament winding. Particularly, most of the studies focused on the application of anthropomorphic robots. The limited skill of such a kind of manipulator and the particular nature of the filament winding do not suggest the use of such a solution for a variety of products. The problem is of major importance for the complex shape objects. It is very important to note that a careful distribution – between delivery machine and mandrel – of the available degrees of freedom in such a cell, increases in the variety of obtainable pieces. An innovative approach to the classical robotic cell consists in the possibility of (1) moving the mandrel – in order to facilitate the machine work – and (2) using a simpler delivery machine (in terms of number of degrees of freedom). In this way, the main advantage consists in using a less complex delivery device and reducing stress factors during the working phase. It is important to note that the movement complexity reduction implies taking into account questions concerning automation science, such as minimization of joint wearing, working range for each joint and movement inversions. The main drawback consists in the need of controlling, simultaneously, different mobile systems. It is important to note that the distribution of available degrees of freedom can lead to the reduction of their overall number. For example a six-axes robot can be used to perform a winding operation for the pieces reported in FIGURE 1 – final piece (a) and mandrel (b) are represented. The piece is a wishbone shaped damper of a helicopter rotor. An alternative approach consists in rotating the mandrel (on z-axis) and using a delivery machine moving along a vertical plane. The combination of such movements leads to obtaining a feasible winding operation by using only 3 degrees of freedom. In order to guarantee a proper stress distribution and a right winding on the mandrel, the configuration has to take into account the fiber tension force. Indeed, the delivery eye point sequence is generated in order to maintain a particular mutual condition between mandrel and fiber – fiber has to be tangential to the mandrel in the delivery point. Such a condition must be ensured after the movements have been divided between mandrel and delivery device. In TABLE 1 different feasible alternatives are reported.

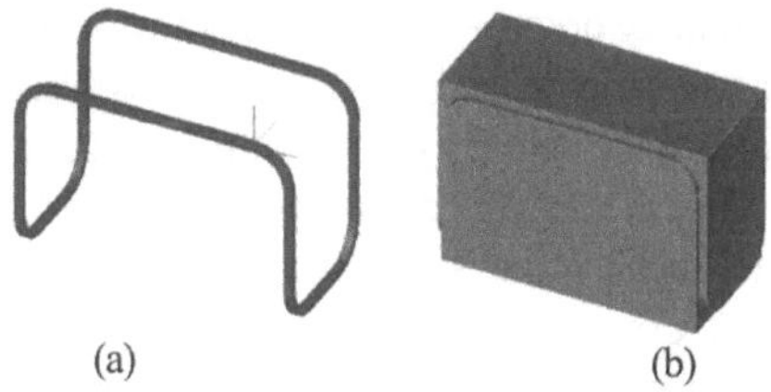

FIGURE 1. Wishbone shaped damper of a helicopter rotor

TABLE 1. Division of degrees of freedom for a winding operation

Degrees of freedom			
Delivery device	Mandrel	Total	Description
6	0	6	Six axes robot, difficulty to realize a part of complex shape objects
2	1	3	Efficiency in the division of movements. Wide range of complex shape objects can be obtained.
3	0	3	Robot with not enough skill to realize a complex shape object.
0	3	3	System having a fixed delivery device. It is difficult to control both tension and fiber delivering.

3 PART PROGRAM GENERATION

Part program generation basically consists in the determination of the exact list of movements for each component in time. Such an activity is very complex when the number of degrees of freedom is high. Indeed, the availability of different solutions to execute a particular movement leads to consider various alternatives. In such a context, complicated inverse kinematic algorithms have to be adopted. When the number of degrees of freedom is limited, this module can be significantly simplified by focusing on the particular division of movements. Once the point sequence – the delivery eye has to pass through – is determined, the methodology we adopted consists in expressing the coordinates in a way linked to the available degrees of freedom. In such a context, a coordinate transformation has to be performed in order to get the final variables. The above solution allows finding a solution to the problem expeditiously, with regard to the part program determination of the cell described in the previous section. The Cartesian coordinates (x,y,z) in the three-dimensional space are transformed in a cylindrical form (r,α,z), see FIGURE 2..

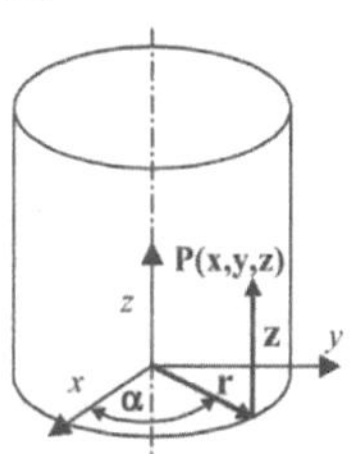

Variable	Association
r	horizontal movement of the delivery device
z	vertical movement of the delivery device
α	rotational movement of the mandrel

FIGURE 2. Cylindrical coordinate system

Both coordinate systems have the same origin – i.e. in the mandrel rotation axis, on the bottom part. The correspondence between coordinate variables and cell movements is described in FIGURE 2. Supposing the piece is fixed, the point sequence the delivery eye has to pass through – to lay the fiber on FIGURE 1a path – can be obtained, [6]. Since the delivery device has two degrees of freedom, it can lay down the fiber only on a YZ parallel plane. Such a plane is named *reachable plane* (RP from this point on) having a geometrical expression of $x=x_0$. Consequently, the piece has to be rotated so that the corresponding delivery eye point belongs to RP. Hence, a coordinate transformation is performed each time Referring to FIGURE 3a, $\{O_1, X_1, Y_1, Z_1\}$ is a referring frame rotated at angle α, with respect to $\{O, X, Y, Z\}$. Transformation formulas are reported in (1).

$$x = x_1 \cos\alpha - y_1 \sin\alpha, \quad y = x_1 \sin\alpha + y_1 \cos\alpha, \quad z = z_1 \tag{1}$$

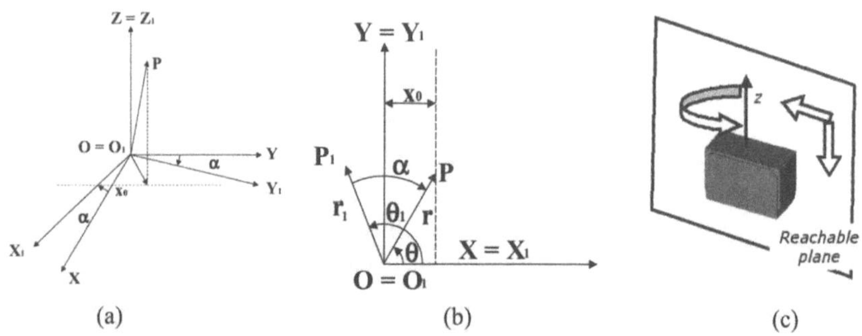

FIGURE 3. Rotation of the referring frame (a), geometrical description of the mandrel rotation (b) and reachable plane representation (c)

Supposing $\{O, X, Y, Z\}$ referring frame is fixed and $\{O_1, X_1, Y_1, Z_1\}$ rotates along with the mandrel. Delivery eye coordinate values are known in the $\{O_1, X_1, Y_1, Z_1\}$ because this referring frame is integral with the mandrel. Referring to the XY plane view, P_1 is the position of the delivery eye assuming piece is fixed, whereas P is the corresponding point on RP where delivery has to be actually carried out, by rotating the piece of α (see FIGURE 3b and FIGURE 3c). In order to determine α, the difference between θ e θ_1 has to be calculated, where θ (θ_1) is the angle between **OP** (**OP**$_1$) and X-axis. Since the winding path is available, it is possible to determine θ_1 by using (2). Supposing that $\rho=|\mathbf{OP}|=|\mathbf{OP}_1|=[(x_1)^2+(y_1)^2]^{1/2}$ and $\cos(\theta)=x_0/\rho$, it is possible to obtain θ by using (3). In this way, the angle α – the mandrel has to rotate in order to bring P_1 on RP – is equal to $(\theta-\theta_1)$. Once a rotation has been performed, the delivery eye can reach the final point and the next point in the sequence has to be considered – so a new rotation is made to bring it on the RP. Consequently, each point in the delivery-eye location list is subjected to consecutive coordinate transformations because of the repeated rotations.

$$\theta_1 = \operatorname{atan}(y_1/x_1) \tag{2}$$

$$\theta = \operatorname{acos}(x_0/\rho) \tag{3}$$

4 MODEL BUILDING

The proposed methodology enables the designing of a simulation model for an automated filament winding cell. A proper template has been produced for the selected software architecture, which is Visual Nastran 6.2 enhanced with Visual Basic 6 features under MS Windows O.S. The main advantage of the solution proposed consists in the possibility of straightforwardly producing a winding cell model once the main characteristics of the process are available. Indeed, the Windows-based interface and Visual Nastran features enable the simulation modeler to easily import cell components and describe how the whole system works. In order for this software package to be capable to meet the filament winding needs, Visual Basic integration has been developed. This makes it possible to read of a filament winding part point sequence – in terms of delivery eye locations and mandrel points – and reproducing it on the designed model. An additional advantage consists in the possibility of varying the process parameters during the simulation running. Indeed for actuators, it is possible to vary the following parameters: velocity, acceleration, force. Whereas for motors, the following quantities can be changed: angular velocity, angular acceleration, torque. For this reason, it is very simple to develop a proper Visual Basic module capable of controlling, over time, the value of such parameters in order to obtain a particular motion law [12]. Finally, it is possible to develop a generation part program module based on the characteristics of the designed model; in this context the availability of a programming language – such as Visual Basic – permits to reach an high level of flexibility and efficiency. Once the part program computation algorithm issues the list of movements required to lay the fiber down, the Visual Basic module sends the parameters - such as position, velocity, acceleration - to simulate the expected movement for actuators and motors modeled in Visual Nastran. Moreover, it is possible to reproduce the fiber tension – Visual Basic module can vary the module and the direction of such a parameter depending on the delivering path.

5 TEST CASE

In order to demonstrate the possibility of our approach to study a generic filament winding cell for the processing of complex shape objects, we designed two simple cells to lay down the fiber on the mandrel showed in FIGURE 1b. Simulation models have been compared in terms of performances by using features available in the developed template. Cell configurations are represented in FIGURE 4.

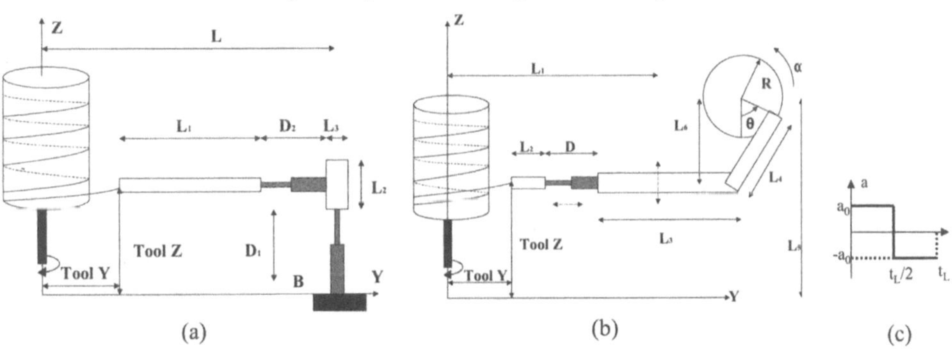

FIGURE 4. First (a) and second (b) robotic cell. Motion law(c)

Once the delivery device is considered, it is important to analyze the stress the delivery eye is subject to. The designed model provides force and torque data for each model component. In particular, reaction forces are very significant for the evaluation of each joint stress. The realized template allows running different simulations by using different limits for the simulation parameters (i.e. acceleration or torque) in order to study the system performances depending on such factors. For example, a constant acceleration motion law can be used to control both actuators and motors (FIGURE 4c). The winding path is sampled so that a point sequence is issued. The delivery device passes from one point to the next one by using the motion law. For each active element in the model, a maximum acceleration value and a specific motion length are available. First, the actuator/motor completing for last the own motion is detected – this element, at a given time t_L, completes the action by using the maximum available acceleration. The others can adopt an acceleration value lower than the maximum obtainable in order to complete the activity at the same time t_L. An interesting study can be performed by analyzing the connection between such acceleration limits and system performance, in terms of process completion time. Fiber tension is assumed constant. Two files provide the coordinates of the set point for delivery eye and the corresponding one on the mandrel. In the first model, an actuator is used for the delivery eye vertical movements. Whereas, in the second, a motor – controlling a connecting rod and a crank — is applied. In both models a motor controls the mandrel rotation and an actuator moves the delivery eye horizontally.

6 RESULTS

Several simulation runs have been performed in order to study the winding process by using the two models considered. For each actuator/motor, different values of maximum acceleration have been established. In other words, simulation runs use different values for a_{MAX} in order to determine which part of the model needs of performance improving. Threshold values are reported in TABLE 2 and refer to FIGURE 4a cell model. In the FIGURE 4b cell model mandrel and horizontal actuator thresholds are the same, whereas crank motor acceleration limit is calculated so that completion time is equal for both models.

TABLE 2. Maximum acceleration for the active elements in the two models

	Maximum acceleration		
	Mandrel	Horizontal Actuator	Vertical Actuator
Simulation run	[rad s^{-2}]	[cm s^{-2}]	[cm s^{-2}]
1	25	25	25
2	50	25	25
3	25	50	25
4	25	25	50

The first model analysis – concerning the mandrel motion – shows the selected threshold value is excessive for the kind of movement to be performed; hence processing time is not affected by mandrel rotation. Moving concerning the other active elements are crucial in both models. Indeed, completion time greatly depends on such movement parameters. Maximum value for the actual mandrel acceleration is reported in TABLE 3 – row 1. As it is possible to note no variation is detected when a_{MAX} changes. The maximum value of mandrel acceleration is 17 (upper bounds are reported in TABLE 2). Consequently, there is no need to improve performance in such a motor when a_{MAX} is greater than 17. For the first model, the maximum of delivery eye acceleration along the two directions is reported in TABLE 3 – row 3 and 4. When a_{MAX} is doubled (simulation run 3 and 4) it is possible to note that the actual acceleration maximum changes. It is possible to observe the decreasing of the processing time (see TABLE 3 – row 5) along with the different simulation runs. Passing from the simulation run 1 to 4, saved time is equal to 16.76%. Threshold acceleration values greatly affects reactions performed by the actuators and motors. In TABLE 3 – row 6 and 7 – these aspects are showed, differences in the two models are due to the unlike mass distribution in the models. It is important to note when horizontal (vertical) actuator acceleration increases, horizontal (vertical) constrain reaction gets higher; see TABLE 3 – row 6 (TABLE 3 – row 7).

TABLE 3. First model simulation results

Row	Simulation run	1	2	3	4
1	Mandrel Max Angular Acceleration [rad/ s^2]	17.0	17.0	17.0	17.0
2	Mandrel Max Torque [Ncm]	4.0	4.0	4.0	4.0
3	Max A_Y [cm/s^2]	22.4	22.4	47.3	22.4
4	Max A_Z [cm/s^2]	24.1	24.1	24.1	49.2
5	Processing time [s]	70.4	70.4	65.2	58.6
6	Max horizontal actuator reaction [N]	2.6	2.6	5.6	2.6
7	Max vertical actuator reaction [N]	177.4	177.4	177.4	181.9

On the basis of the results obtained, the second model behavior is quite similar to the other one in terms of mandrel rotation motor and horizontal movements. Instead, differences can be detected for the constraint reactions because of the variation in the mass distribution. Since different ways to obtain the vertical movements are, simulation results are significantly altered for such a question. In the first model, an actuator directly controls vertical movement accelerations. In the second model, the complexity of the relation standing between rotation motor and vertical displacement makes difficult the controlling of the vertical acceleration. The comparison of the maximum values for the vertical movements acceleration in both models is reported in TABLE 4 – rows 1-4. As it is possible to note in the second model, acceleration is greater than the one on the first. This behavior greatly affects the force values on the two models. Negative acceleration values are related to downward movements. Moreover, data reported in TABLE 4 – rows 5-8 – makes it easy to observe the second model is more stressed in the z-axis verse.

TABLE 4. Simulation result comparison

		Simulation run	1	2	3	4
Row	Model	Parameter				
1	1	Min vertical acceleration [cms^{-2}]	-24.1	-24.1	-24.1	-49.2
2	2	Min vertical acceleration [cms^{-2}]	-51.4	-51.4	-51.4	-115.6
3	1	Max vertical acceleration [cms^{-2}]	24.1	24.1	24.1	49.2
4	2	Max vertical acceleration [cms^{-2}]	54.6	54.6	54.6	79.3
5	1	Min vertical force on delivery arm [N]	112.6	112.6	112.6	109.7
6	2	Min vertical force on delivery arm [N]	109.4	109.4	109.5	101.8
7	1	Max vertical force on delivery arm [N]	118.3	118.3	118.3	121.3
8	2	Max vertical force on delivery arm [N]	121.9	121.9	121.9	124.8

The second model is subjected to wider variability. This is particularly clear in simulation run 4 – difference between maximum and minimum values is 22.9 N whereas in the first it is just 11.3 N. The analysis on the forces and torques in the other constraints leads to the conclusion that differences exist due to the distinct geometry. In

TABLE 5 variations of the crank gear angular acceleration and motor torque are shown. In particular, maximum values are reported. It is possible to note peak values increase from the first to the last simulation run, except for the second – only the mandrel motor acceleration is altered.

TABLE 5. Second model simulation results

Row	Simulation run	1	2	3	4
1	Max crank gear angular acceleration [rads^{-2}]	7.0	7.0	12.4	15.7
2	Max crank gear motor torque [Nm]	28.6	28.6	28.6	29.8

7 CONCLUSION

The proposed methodology permits to investigate on a given filament winding cell. In particular, the above mentioned analysis allows the selection of the appropriate winding parameters in order to maximize system performance taking into account the stress the different elements are subjected to. Moreover, the proposed approach exploits the possibility of using an integrated architecture to take advantage of the Visual Nastran and Visual Basic software packages. A theoretical approach is issued in order to manage the rotation of the mandrel and improve the utilization of the available degrees of freedom in a generic cell. The methodology enables the simulation model designer to quickly develop an accurate representation of a winding device. Moreover, a study is carried out in order to analyze in quantitative terms the results of such simulation models.

ACKNOWLEDGEMENTS

The work described in this paper has been funded by Project 488-92 "Sistema di produzione filament winding di manufatti a geometria complessa caratterizzati da alte prestazioni funzionali ed alta affidabilità", cluster C19, programma operativo del piano “Tecnologie Innovative per i Beni Strumentali”.

REFERENCES

1. Charoenseang, S., Srikaew, A., Wilkes, D., Kawamura, K., (1998), Integrating Visual Feedback and force feedback in 3-D Collision Avoidance for a Dual-Arm Humanoid Robot. Proceedings of 1998 International Conference on Systems, Man and Cybernetics, California, USA, October.
2. Imamura, T., Kuroiwa, T., Terashima, K., Takemoto, H., (1999), Design And Tension Control Of Filament Winding System. Proc. of IEEE Conference on System, Man and Cybernetics SMC'99, Tokyo, Japan, Oct 12-15, II, pp. 660/665.
3. Inoue, Y., Kitamura, S., Kidawara, Y., (1991), Force feedback control and collision avoidance of redundant manipulator. IEEE/RSJ Intern. Workshop on Intell. Robot and Systems IROS91, Nov. 3-5, Osaka, Japan.
4. Seereeram, S., Wen, J., (1991), An All-Geodesic Algorithm for Filament Winding of a T-Shaped Form, IEEE Transaction on industrial electronics, vol. 38, No. 6, December.
5. Scholliers, J., Van Brussel, H., (1996), Design and off-line programming of a robotic tape winding cell. Robotics and computer-integrated manufacturing, vol.12 n.1, pp. 93-98.
6. Carrino, L., Moroni, G., Turchetta, S., (1998), CAD/CAM for robotic filament winding process design, European Conference on Composite Materials, Naples, Italy, June 3-6.
7. Anglani, A., Manta, A., Nucci, F., (2001), Simulation for a Sound Analysis of Robot Part Program in Filament Winding Process, Proceedings of 15th European Simulation Multiconference, Prague, Czech Republic, June 6-9.
8. Eydgahi, A.M., Sheehan, J., (1991), A Computer Animation of Robotic Manipulators and Workcells, IEEE Control Systems Magazine, Vol.11, No.4, pp.56-59, June.

9. Mohseni, C., Willgoss, R., (1991), Offline Path And Motion Planning For Redundant Manipulators For Obstacle Avoidance, Proceeding of IECON'91, pp. 1005-1010.
10. Joshi, R., Sanderson, A., (1996), Application of feature-based multi-view servoing for lamp filament alignment, Proceeding of the 1996 IEEE International Conference on Robotics and Automation, Minneapolis, Minnesota, April.
11. Mitsi, S., Bouzakis, K., Mansour, G., (1995), Optimization of Robot Links Motion in Inverse Kinematics Solution considering Collision Avoidance and Joint Limits. Mech. Mach Theory, Vol. 30, No. 5, pp. 653-663.
12. Magnani, P.L., Ruggeri, G., (1986). Meccanismi per Macchine Automatiche, UTET, Turin, Italy.

METHODS FOR WELDING QUALITY IMPROVEMENTS

D. Pavletic[1]

[1] Faculty of Engineering, University of Rijeka, Croatia

KEYWORDS: Quality improvement methods, Welding parameters, Welded joints quality.

ABSTRACT. To establish welding parameters significance level and influence of welding parameters on occurrence of welded joints discontinuities and nonconformance it is necessary to perceive comprehensive welding parameters relationship as well as level of influence of each parameter. At present there are several tools for quality assurance and quality management of welded joints at disposal. Before all, there are Cause-and-Effect diagram, Cause-and-Effect matrix, Pareto diagram and Control charts. By appropriate application of this tools most influential parameters on quality of welded joints can be selected, which can be further, by application of Design of experiments methods as well as the other ways of parameters analysis, explored and optimal values of welding parameters obtained. Furthermore, paper provides an example of application of mentioned tools for obtaining optimal welding joint quality as a starting point for further numerical and other analysis and welding modeling.

1 INTRODUCTION

Constant improvement of welding processes, decrease or elimination of welds discontinuities and production of welded joints first time without failure is becoming a present necessity for successful production in highly competitive business environment. A good, in depth acquaintance with welding processes and an adequate process parameters management is essential to gain desired level of performance. To establish welding parameters significance level as well as influence of welding parameters on occurrence of welded joints discontinuities it is necessary to perceive comprehensive welding parameters relationship and level of influence of each parameter. At present, there are several tools for quality assurance and quality management of welded joints at disposal. Before all, there are Cause-and-Effect diagram, Cause-and-Effect matrix, Pareto diagram and Control charts. In the paper is presented an overview and an example of applications of methods for welding quality improvement by welding parameters management.

2 CAUSE AND EFFECT DIAGRAM

A cause-and-effect diagram is a graph that shows the relationships between a problem and its possible causes. These tree diagrams, originally developed by Kaoru Ishikawa in 1953, are used during brainstorming sessions where possible solutions for a problem are being sought. In this context, they provide a model for the connections between the problem, usually a quality characteristic that is not meeting customer needs, and factors that affect it. The problem is called effect, and the factors that affect it are called the causes. Cause-and-effect diagrams are helpful in eliminating problems by tackling their causes, and they are also useful in understanding the effects that several factors have on a process. They are also known as fishbone diagrams or Ishikawa diagrams.[1] When constructing a cause-and-effect diagram, it is often appropriate to consider six areas, causes, that can contribute to a characteristic response, effect: materials, machine, method,

Published in: E.Kuljanic (Ed.) *Advanced Manufacturing Systems and Technology*,
CISM Courses and Lectures No. 437, Springer Wien New York, 2002.

personnel, measurement, and environment. Each one of this characteristics is then investigated for subcauses. Subcauses are specific items or difficulties that are identified as a factual or potential cause to the problem, effect.[2] A cause-and-effect diagram is used for schematic presentation of arc welding parameters critical for quality of welded joint. Application of cause-and-effect diagram is effective in investigation of welding parameters influence on final quality of welded joints. Cause-and-effect diagram of arc welding parameters effect on welded joints quality is shown on figure 1 [3].

3 CAUSE AND EFFECT MATRIX

Cause-and-effect matrix is a simple QFD (Quality Function Deployment) matrix, used to ensure that the customer requirements are considered in determining the key characteristics of the product/process. Cause-and-effect matrix is used to relate the key inputs to the key outputs, while cause-and-effect diagram is used as primary sources of input information. Key outputs are rated in importance to external or internal customer, while key inputs are scored as to relationship to key outputs. There are two main approaches to the cause-and-effect matrix. First approach is a general approach that is used for relatively simple processes. Within this approach the outputs are placed across the top of the matrix and ranked, while inputs are placed down the side of the matrix starting with the first defined parameters and moving to the last one. Second approach is focused approach that is used for larger, more complex operations. Within this approach there are two phases. In the first phase the outputs are placed across the top of the matrix and ranked, the inputs are placed down the side of the matrix. The inputs are correlated to outputs and obtained results are used as input for Pareto analysis. In the second phase of this approach new cause-and-effect matrix is started with the inputs form the most critical three of four inputs defined in the first phase. In general, there are few steps in cause-and-effect matrix implementation. At the beginning it is necessary to identify key customer requirements (outputs) from cause-and-effect diagram. A priority factor should be rank ordered and assigned to each output. To be able to discriminate between levels at least are three to four levels needed. All inputs should be identified from cause-and-effect diagram. Correlation of each input to each output are then evaluated. Again, at least four levels are needed. If we obtain low score: changes in the input variable have small effect on output variable. In other case, if we obtain high score: changes in the input variable can greatly affect the output variable. To finish cause-and-effect matrix calculation correlation values should be cross multiply with priority factors and add across for each input. In the case of quality of welded joints influences of welding parameters on appearance of welded joint discontinuities are evaluated by cause-and-effect matrix, shown in table 1. In the matrix, for each type of discontinuities factor of importance is rank ordered and all listed parameters in matrix are correlated to every type of discontinuities. Finally, total value for each parameter is calculated by multiplying rating of importance with value given to parameters and adding across for each parameters. To be able to discriminate between levels, rating of importance for each parameter is divided in three groups, so for unacceptable imperfection rating of importance is 1,5, for critical discontinuities rating is 1,2 and for less critical imperfection rating is 1. Furthermore, parameters are also ranked in three groups. Critical influence of parameters is rated with 3 points, less critical with 2 points for and noncritical influence with 1 point.

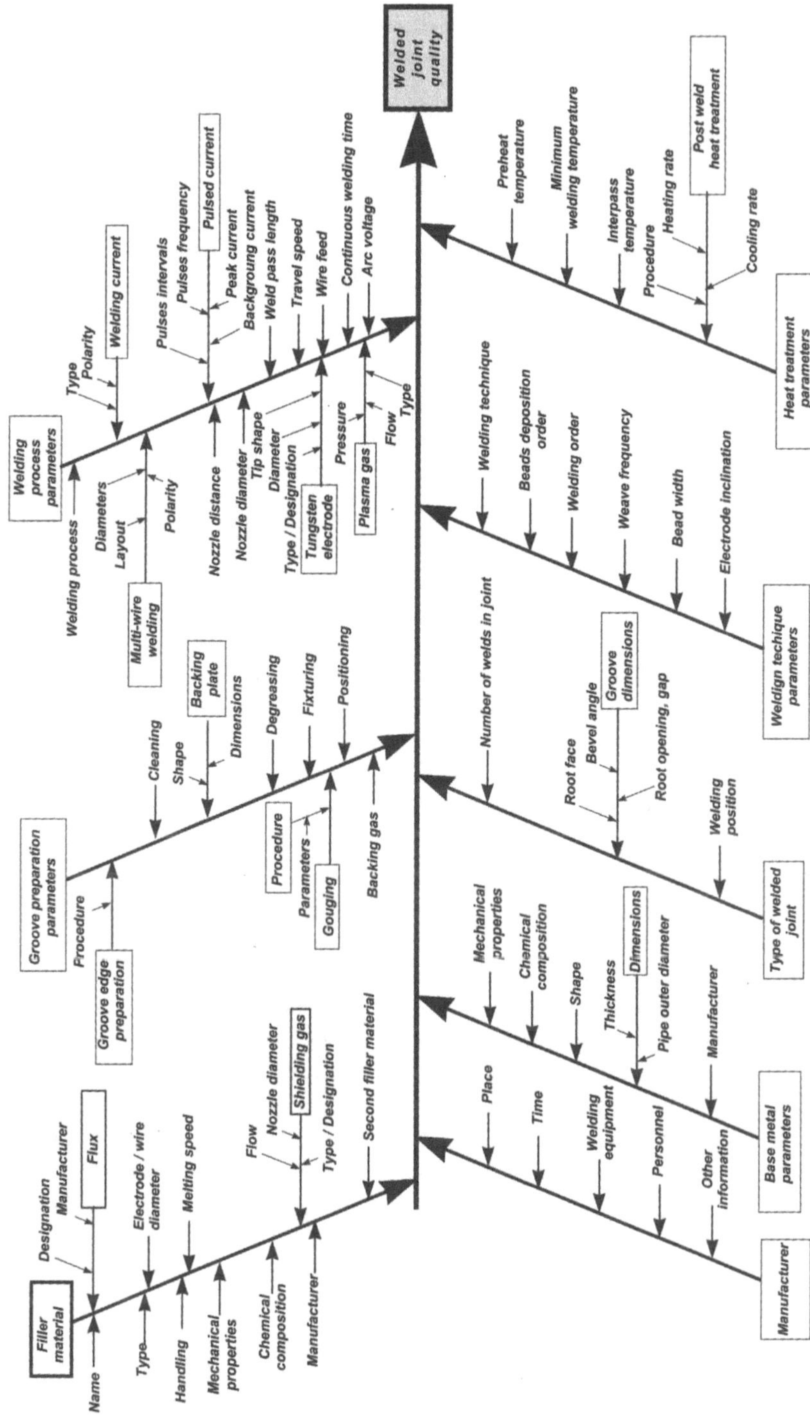

FIGURE 1. Cause-and-effect diagram for welded joint quality

TABLE 1. Influence of welding parameters on quality of welded joints

Welded joint discontinuities	**Cracks**		**Porosity**		**Inclusions**		**Incomplete penetration**		**Shape discontinuities**		**Other discontinuities**		**Total**
Rating of importance	**1,5**		**1**		**1,2**		**1,5**		**1,2**		**1,2**		
Base metal	3	4,5	1	1	1	1,2	1	1,5	1	1,2	1	1,2	**10,6**
Welding process	1	1,5	1	1	1	1,2	1	1,5	1	1,2	1	1,2	**7,6**
Type of welded joint	1	1,5	1	1	2	2,4	3	4,5	2	2,4	2	2,4	**14,2**
Welding order	3	4,5	2	2	2	2,4	2	3,0	1	1,2	1	1,2	**14,3**
Welding position	3	4,5	2	2	2	2,4	2	3,0	3	3,6	2	2,4	**17,9**
Groove preparation	2	3,0	3	3	3	3,6	3	4,5	1	1,2	1	1,2	**16,5**
Welding techniques	2	3,0	3	3	3	3,6	3	4,5	3	3,6	3	3,6	**21,3**
Filler material	3	4,5	3	3	3	3,6	2	3,0	3	3,6	2	2,4	**20,1**
Welding current	3	4,5	3	3	3	3,6	3	4,5	3	3,6	2	2,4	**21,6**
Arc voltage	3	4,5	3	3	2	2,4	3	4,5	3	3,6	2	2,4	**20,4**
Other welding parameters	2	3,0	3	3	2	2,4	1	1,5	2	2,4	2	2,4	**14,7**
Travel speed	3	4,5	2	2	2	2,4	3	4,5	3	3,6	1	1,2	**18,2**
Heat treatment parameters	3	4,5	2	2	1	1,2	1	1,5	1	1,2	1	1,2	**11,6**
Total		**48,0**		**29**		**32,4**		**42,0**		**32,4**		**25,2**	**209,0**

Results from the cause-and-effect matrix shows that welding parameters have different influence on appearance of welded joint imperfection. Pareto analysis is used to visualized and select significance level of welding parameters influence.

4 PARETO PRINCIPLE

Pareto diagram clearly exhibit information about relative importance of factors of certain problem. These information helps to identify the most important factors which will be firstly analyzed. With help of Pareto diagram the fields of possible improvements are clearly identified. In the case of welding parameters analysis, information obtained by the cause-and-effect matrix, specifically, total number of points for each parameter, are entered in table 2 where are made additional calculations necessary for Pareto analysis. In table 2 are then percentage of points calculated for each parameter. Cumulative sum as well as cumulative percentage of points are calculated and alphabetical identification of each parameter are added in the table. Resulting Pareto diagram is shown on figure 2. Drawn diagram shows cumulative percentage of points for each parameter. Pareto analysis shows that 50% of all points are distributed between five of 13 parameters. So, with adequate definition and selection of those parameters it is possible significantly to influence quality of welded joints. To establish adequate control over selected parameters application of appropriate control chart is proposed.

TABLE 2. Data for Pareto analysis

Parameter	Total points	% points	Cumulative	Cumulative %	Mark

Welding current	21,6	10,335	21,6	10,335	A
Welding techniques	21,3	10,191	42,9	20,526	B
Arc voltage	20,4	9,761	63,3	30,287	C
Filler material	20,1	9,617	83,4	39,904	D
Travel speed	18,2	8,708	101,6	48,613	E
Welding position	17,9	8,565	119,5	57,177	F
Groove preparation	16,5	7,895	136,0	65,072	G
Other welding parameters	14,7	7,034	150,7	72,105	H
Welding order	14,3	6,842	165,0	78,947	I
Type of welded joint	14,2	6,794	179,2	85,741	J
Heat treatment parameters	11,6	5,550	190,8	91,291	K
Base metal	10,6	5,072	201,4	96,363	L
Welding process	7,6	3,636	209,0	100,000	M
Total	209,0	100,000	-	-	-

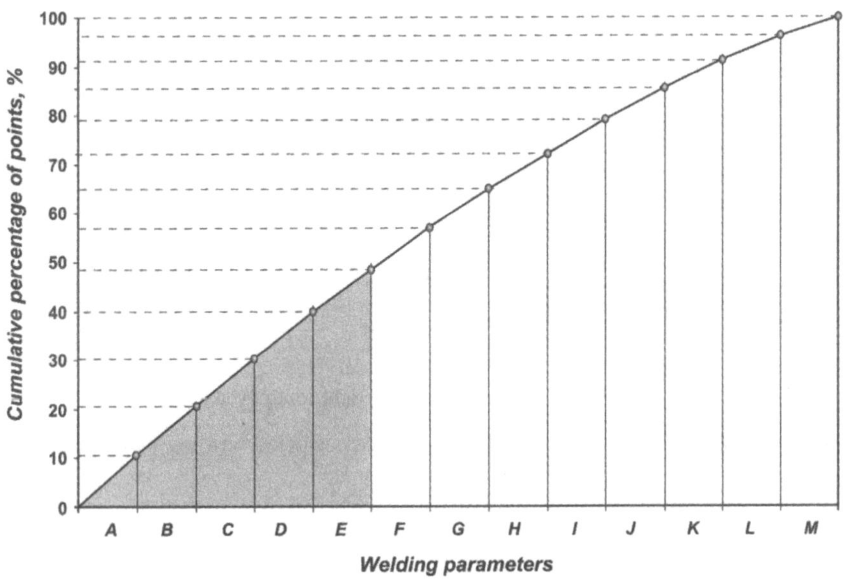

FIGURE 2. Pareto diagram

5 CONTROL CHARTS

A control chart is a chart used to monitor outputs or inputs of processes. The control chart is one of the most useful tools for studying variation. It employs some of the basic statistics and statistical theory to place limits of expected variation about a fluctuating pattern. Limits placed on the control chart give the control chart typical power to enable its user to determine if a process can be considered stable or *in control* and thus predictable, or unstable or *out of control* and thus unpredictable. Such information about an important quality characteristic can, therefore, enable one to continue to control, in the case of a stable process, or gain control, in the case of an unstable process. This is accomplished by repeated rational sampling of the process, analysis of the resultant data plotted on the control chart, and any necessary action on the process. Hence, the control chart shows the amount and nature of variation by time, indicates statistical control or lack of it, and enables pattern interpretation and detection of changes [4]. There are a number of different types of control charts to choose from, depending on the type of data involved. These fall into one of two categories: control charts for variables or control charts for attributes, figure 3.

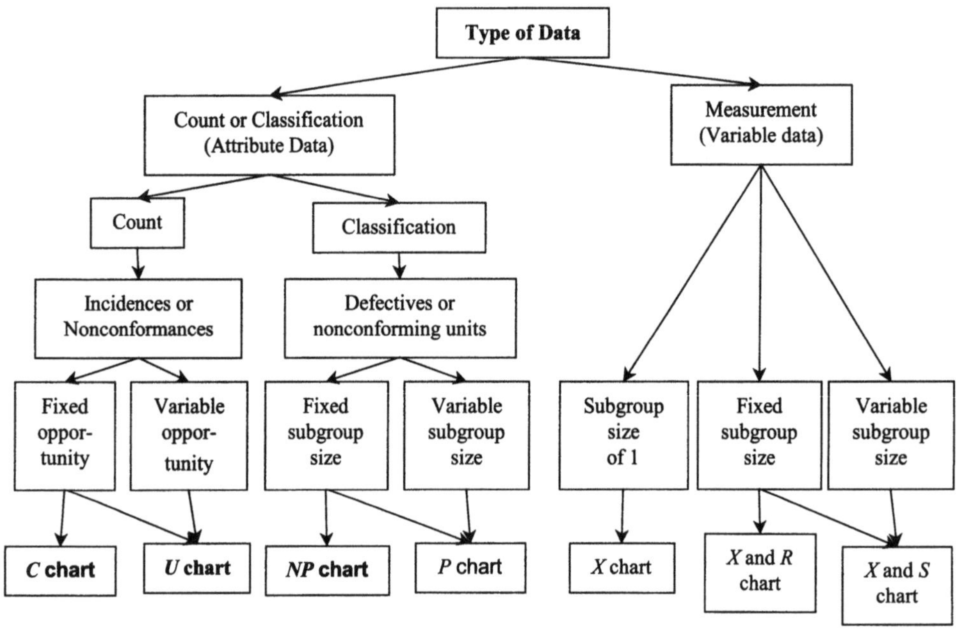

FIGURE 3: Guide for choosing right control chart type

An example of control chart application for previously selected welding parameters are provided on figure 4.

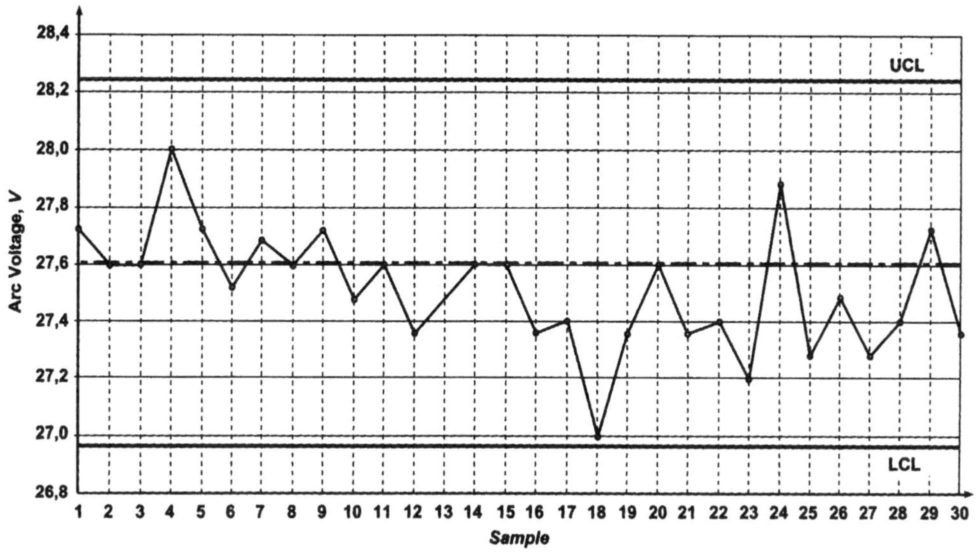

FIGURE 4. Control chart of arc voltage for gas metal-arc welding of carbon-steel but joint.

6 CONCLUSION

Application of presented method, as well as other methods for systematic and continuous quality improvement, should enable welding engineers and practitioner knowledge of level of importance of welding parameters and should provide adequate guiding tool to manage these parameters to eliminate or minimize occurrence of rework and scrap. Consequently, deployment and broader application of these methods will ultimately lead to higher level of welded joints quality and succeeding competitive advantage.

REFERENCES

1. Rao, A., et al., (1996.), Total Quality Management: A Cross Functional Perspective, John Wiley & Sons, New York
2. Breyfogle III, F. W., (1999.), Implementing Six Sigma: Smarter Solutions Using Statistical Methods, John Wiley & Sons, Inc., New York
3. Pavletić, D., (2000.), Computer Aided Quality Assurance of Arc Welding Parameters Selection, Master thesis, Faculty of Engineering, University of Rijeka, Rijeka
4. Wadsworth, H. M., et al., (1986), Modern Methods for Quality Control and Improvement, John Wiley & Sons, New York

RADIAL SEGREGATION OF WORKPIECE IN BARREL FINISHING

A. Boschetto, F. Veniali

Dipartimento di Meccanica e Aeronautica, Università di Roma "La Sapienza", Rome, Italy

KEYWORDS: Barrel finishing, Segregation.

ABSTRACT. Barrel finishing is a mass operation which is not yet completely known in its base mechanisms, even thought it is widely and effectively employed in the industry. As other operations which involve granular material movements, it is particularly complicated to be modeled. A laboratory apparatus has been contructed to experimentally analyze this operation. By using a technique based on digital image analysis it has been possible continuously catch the workpiece position within the barrel. So, it can be demonstrated that the entire finishing operation takes place in the so-called active layer. Yet, within this layer, the workpiece would be differently machined, due to the different relative working conditions at different radial position. In this work it is shown how the workpiece mean position within the active layer is influenced by workpiece density and media density.

1. INTRODUCTION

Barrel finishing is a mass operation which can attain very low surface roughness at very low costs even on very complicated mechanical parts, such as chains, eye glass frames, jewels [1]. Barrel finishing is controlled by a large number of parameters namely shape, density, dimension, surface roughness of the pellets, shape, density, dimension of the workpiece, shape, dimension and rotational speed of the barrel, surface roughness of the innerwalls of the barrel compound type. During the rolling phase [2, 3] the workpieces slide down along the active layer and are therefore machined. Due to the stochastic movements inside the barrel, the workpiece can segregate towards less effective zones of the barrel. Basic mechanisms of segregation are well known and widely investigated when powder mixtures are involved but are still relatively unknown when the purpose of the operation is the finishing of a workpiece rather than the mixing of a granular charge. The Oyama experiment [4] well depicts the segregation which occurs when mixing two granulars of the same type but of different sizes. Other typical examples are the large rocks which work their way up to the surface of tilled fields, and the Brazil nuts when transported in pick-up trucks over the rough back roads of South America [5]. A basic study of radial segregation is reported in [6] where the percolation, flow and vibration mechanisms are introduced. The segregations by flow and by percolation are the ones that characterize rotating barrel, but they were investigated only with the purpose of searching for pattern formation of the mixture rather than for machining.

Aim of this paper is to experimentally analyze the radial segregation of workpiece during barrel finishing in order to evaluate the efficiency of the process itself.

Published in: E.Kuljanic (Ed.) *Advanced Manufacturing Systems and Technology*, CISM Courses and Lectures No. 437, Springer Wien New York, 2002.

2. EXPERIMENTAL

2.1. EQUIPMENT

A horizontal axis barrel has been built up with a front surface made by transparent Pb-glass in order to limit the friction coefficient and therefore the wear of the windows itself. The barrel has an octagonal shape 400mm in diameter and 28mm depth, i.e. the barrel depth is marginally higher than the workpiece diameter of 24mm, so two dimensional motion is assured. It is mounted on a rig with speed range of 2-200 rpm adjusted by a digital speed controller. It is noteworthy that, with the used media, the rolling condition limits are within 10-20 rpm.

A digital videocamera 3COM HomeConnect was used to acquire the positions of the media and of the workpiece. The data acquired by the videocamera are directly sent into a PC via USB interface. After some preliminary experiments, the videocamera was focused on a $123x92mm^2$ area with a resolution of $160x120pixel^2$, thus leading to a linear resolution of 1,3pixel/mm. The frame rate was set at 50Hz. The area under analysis has been illuminated by two 5300lumen ligths. The video camera was positioned on a tripod to measure the absolute motion of the media and of the workpiece. Alternatively the videocamera was solidly connected to the rotating barrel via a protruding arm: this second layout allowed us to see the movement of the charge within the barrel regardless the rotation of the barrel itself.

2.2. PROCEDURES

Due to intrinsic stochastic behaviour of this process it was necessary to perform a statistical approach. Therefore, a custom software has been developed under Mathematica environment to automatically recognize the workpiece position. The acquired .AVI file has been exploded into a number of .BMP files. The duration of the animation has been set to 150s which, at the set frame rate led to 7500 .BMP files. These files were singularly analyzed by a suitable pattern recognition routine in order to catch the positions of the workpiece within the active layer.

2.3. DESIGN OF EXPERIMENTS

In this paper we present the main effects of the density of both media and workpiece and the rotational speed of the barrel. In order to do this, a nylon tube 10mm in diameter has been filled with different density materials and cut into cylinders 15mm in heigth. A hollow plastic sphere 24mm in diameter has been filled in turn with different density material. This procedure assures that both media and workpiece have, respectively, the same coefficient of friction,

TABLE 1. Design of experiments

Rotational speed	workpiece / media	Wood $760kg/m^3$	Nylon $1150kg/m^3$	Alluminum $2160kg/m^3$
11 rpm	Empty plastic ($1060kg/m^3$)	✓	✓	✓
	Plastic filled with iron ($2350kg/m^3$)	✓	✓	✓
	Plastic filled with lead ($3310kg/m^3$)	✓	✓	✓
18 rpm	Empty plastic ($1060kg/m^3$)	✓	✓	✓
	Plastic filled with iron ($2350kg/m^3$)	✓	✓	✓
	Plastic filled with lead ($3310kg/m^3$)	✓	✓	✓

the same shape, the same dimensions but have different density. As per the barrel parameters we have choosen two rotational speeds namely 11 rpm and 18 rpm which both assure rolling conditions of the media within the barrel [7]. The overall DOE is reported in Table 1. Three replicates had been performed for each set of parameters.

3. RESULTS AND DISCUSSION

In Figure 1 are reported three frames taken by the video camera fixed to the barrel and pointing at a zone far from the active layer, the so-called plug flow [8] or also passive region [9]. Each pellet surrounding the workpiece are evidenced by the drawn dotted lines. It is well evident from both the pictures and from the data reported in Table 2 that the pellets did not move at all around the workpiece for a relevant time (>1 s). This is a pretty interesting result, earlier found for mixtures [10,11] from wich we can say that the entire working action in barrel finishing takes place in the active layer. Matter of fact in this zone the whole media and the workpiece acts like a rigid solid material notwithstanding it is made by a granular material. We can therefore conclude that the whole working action takes place within the active layer since outside of it there is no relative motion between the workpiece and the neighbouring pellets.

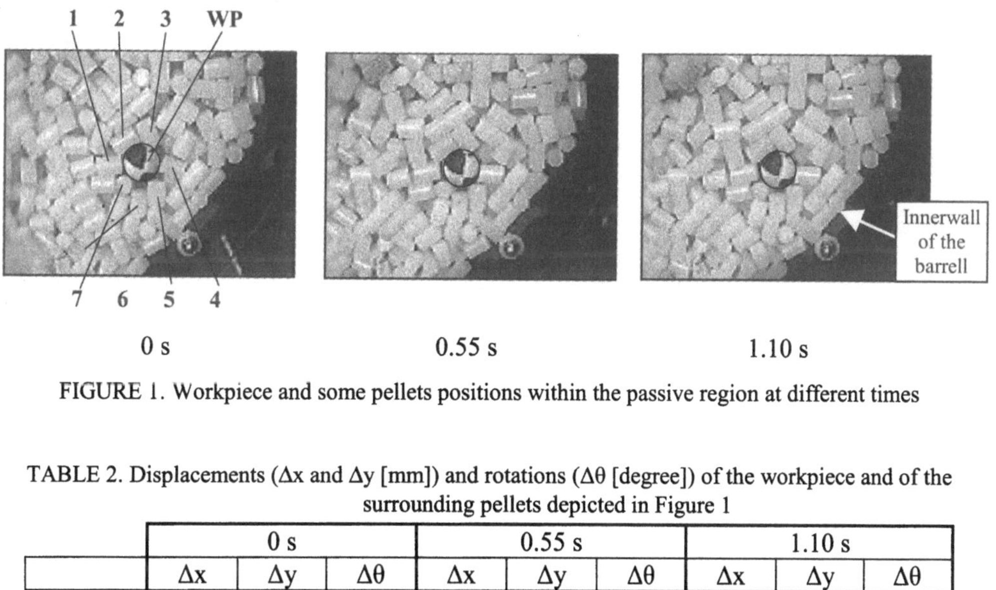

FIGURE 1. Workpiece and some pellets positions within the passive region at different times

TABLE 2. Displacements (Δx and Δy [mm]) and rotations (Δθ [degree]) of the workpiece and of the surrounding pellets depicted in Figure 1

	0 s			0.55 s			1.10 s		
	Δx	Δy	Δθ	Δx	Δy	Δθ	Δx	Δy	Δθ
Workpiece	0	0	0	0	0,1	2	0,1	0	5
Pellet 1	0	0	0	0	0	0	0	0	0
Pellet 2	0	0	0	0	0	0	0	0	0
Pellet 3	0	0	0	0,4	0,2	7	0,4	0,2	7
Pellet 4	0	0	0	0,2	0	22	0,2	0	22
Pellet 5	0	0	0	0	0	0	0	0	2
Pellet 6	0	0	0	0	0	0	0	0	0
Pellet 7	0	0	0	0	0	0	0	0	0

With the video camera positioned on the tripod and pointing at the active layer we did count the number of passages of the workpiece at different depths within the active layer itself. The corresponding results are reported in Figure 2, which shows the frequency distribution of the passages of the workpiece within the active layer for the eighteen working sets. It can be seen that the shape of the distribution markedly changes as a function of the different parameters. It is interesting to notice that the distribution moves towards the outer part of the layer as evidenced in the right top sketch of Figure 2 which is relative to the combination aluminum-plastic, i.e. heavy media and ligth workpiece. On the contrary, the distribution moves towards the inner part of the layer as evidenced in the lower left sketch of the same Figure 2 which is relative to the combination wood lead, i.e. heavy workpiece and light media. The other workpiece/media combinations show a mean value of the distribution positioned at about mid depth of the active layer. This behaviour can be explained in terms of different floating of the workpiece within pellets of different densities. This phenomenon is similar to the well-known "brazil nuts effect", which occurs when a workpiece different in size is submerged within pellets of a smaller size [5]. From the machining point of view this aspect will differently influence the properties of the workpiece when it is respectively in the depth or in the surface of the active layer.

In order to actually investigate the effects of the single machining parameters it is necessary to characterize the above distributions by means of simple and effective features. It is reasonable to think that the above distribution are of the Beta type as confirmed by the Chi-squared test performed on the residual distribution between the sperimental data and the beta distribution caracterized by the experimental mean and standard deviation.

The Beta distribution is defined by equation (1), and its mean and standard deviation are defined in the equations (2) and (3), respectively:

$$f(x) = \frac{\Gamma(\alpha+\beta+2)}{\Gamma(\alpha+1)\Gamma(\beta+1)} x^{\alpha}(1-x)^{\beta} \tag{1}$$

$$\bar{x} = \frac{\alpha+1}{\alpha+\beta+2} \tag{2}$$

$$\sigma = \left(\frac{(\alpha+1)(\beta+1)}{(\alpha+\beta+2)^2(\alpha+\beta+3)} \right)^{\frac{1}{2}} \tag{3}$$

We can now perform the ANOVA on these value to achieve informations on the effects of the parameters, and the results are reported in Table 3. It is well evident that both media and workpiece type influence the behaviour while the rotational speed doesn't. Also the shape of the distribution of the passages of the workpiece at different depth of the active layer, i.e. the standard deviation of the distribution, is influenced by both media and workpiece type but not by the barrel rotational speed. The effect of machining parameters on standard deviation should be intended as a broadening of the distribution when the densities of the workpiece and of the pellets are comparable.

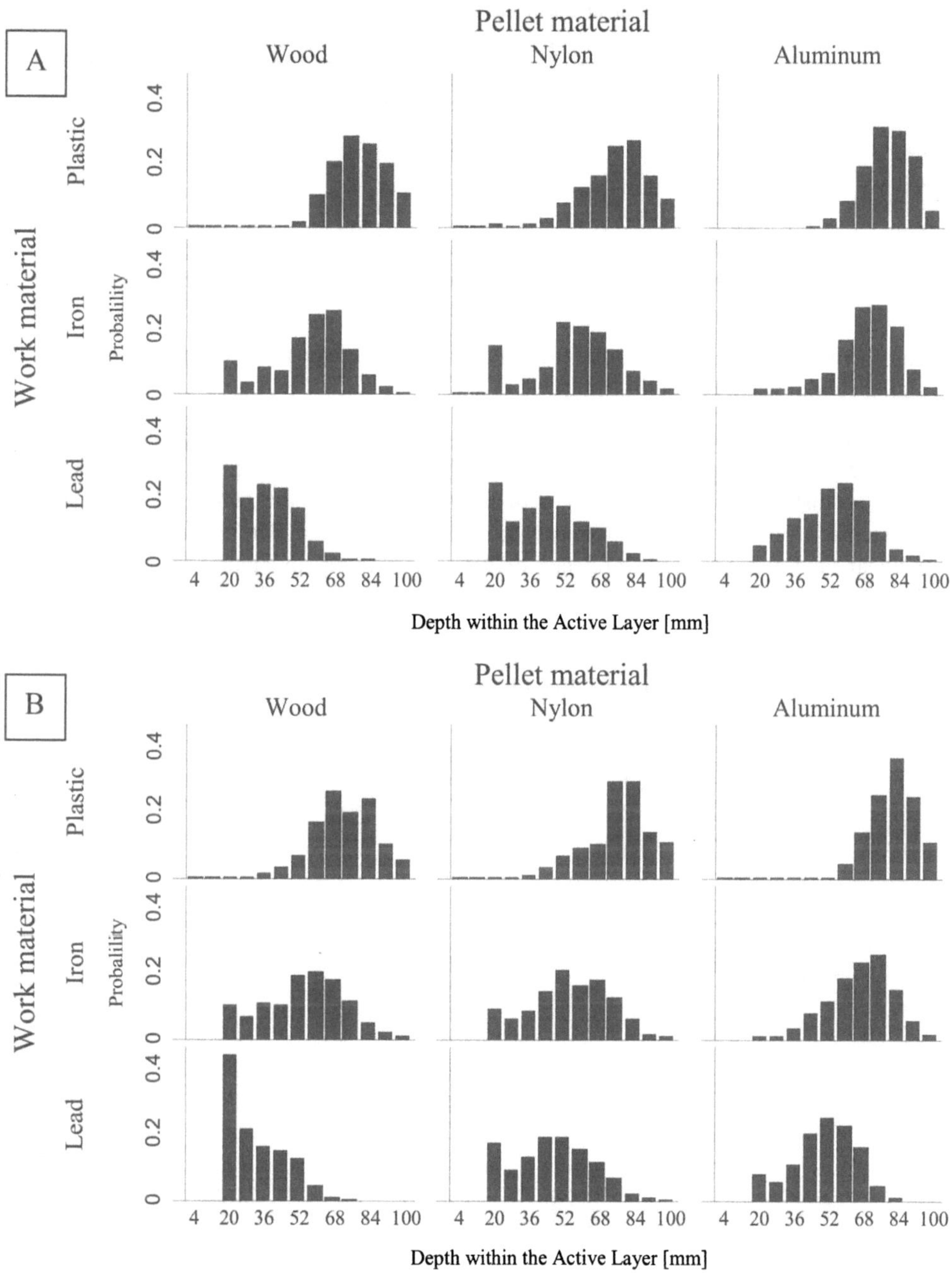

FIGURE 2. Frequency distribution of the workpiece position along the depth of the active layer for different media and workpiece density and for barrel rotational speed of 11 rpm (A) and 18 rpm (B)

TABLE 3. ANOVA applied to mean and standard deviation of the distribution of the position of the workpiece within the active layer

		mean				standard deviation			
Source	DF	SS	MS	F	P	SS	MS	F	P
media	2	1440.9	720.4	26.2	0.000	63.1	31.5	5.9	0.005
workpiece	2	10348.9	5174.5	188.7	0.000	141.8	70.9	13.2	0.000
rot.speed	1	28.6	28.6	1.0	0.312	0.2	0.2	0.0	0.825
Error	48	1316.0	27.4			256.8	5.3		
Total	53	13134.4				462.1			

We can therefore conclude that different material and different media will lead to different machining conditions in barrel finishing.

The above mentioned effect of the density of the media and the density of the workpiece can be summarized by taking into account the ratio of their densities. In Figure 3a it is shown the mean of the distribution of the passages of the workpiece for the entire set of parameters investigated as a function of the workpiece to media density ratio. The equation (4) is the best fit of linear type and gives a confidence of 95%. Conversely in Figure 3b it is shown the standard deviation of the distribution of the passages as a function of the same parameter with the parabolic best fit with a confidence of 95%, according to equation (5) :

$$\bar{x} = 85.40 - 11.33 \cdot \frac{\rho_{workpiece}}{\rho_{media}} \tag{4}$$

$$\sigma = 5.20 + 9.12 \cdot \frac{\rho_{workpiece}}{\rho_{media}} - 1.68 \cdot \left(\frac{\rho_{workpiece}}{\rho_{media}} \right)^2 \tag{5}$$

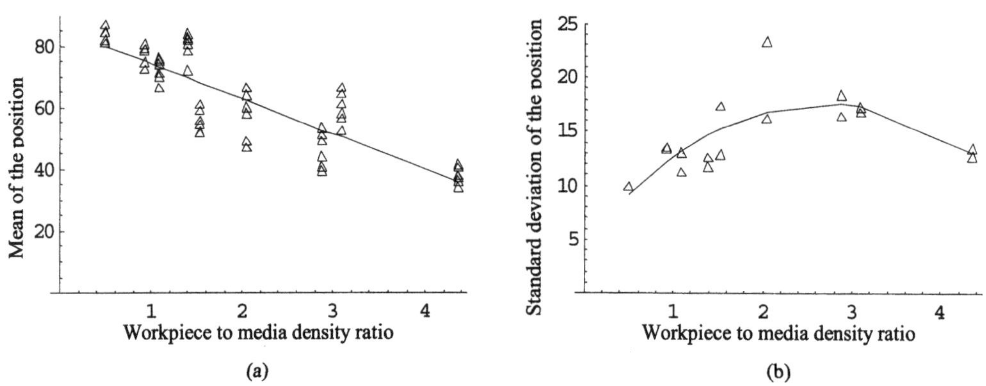

FIGURE 3. Mean and standard deviation of the distribution of the passages within the active layer as a function of workpiece to media density ratio

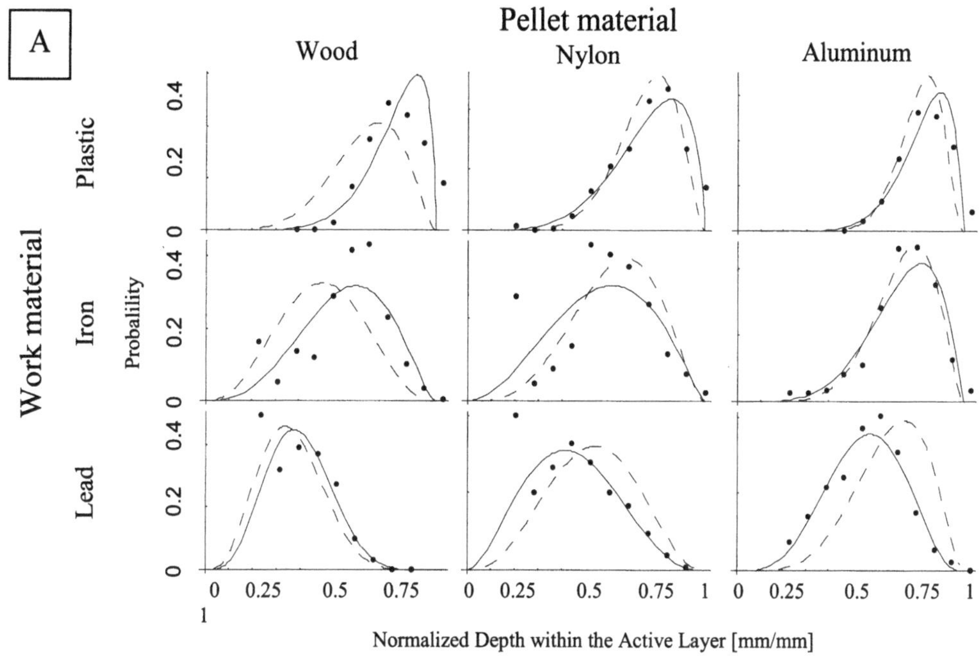

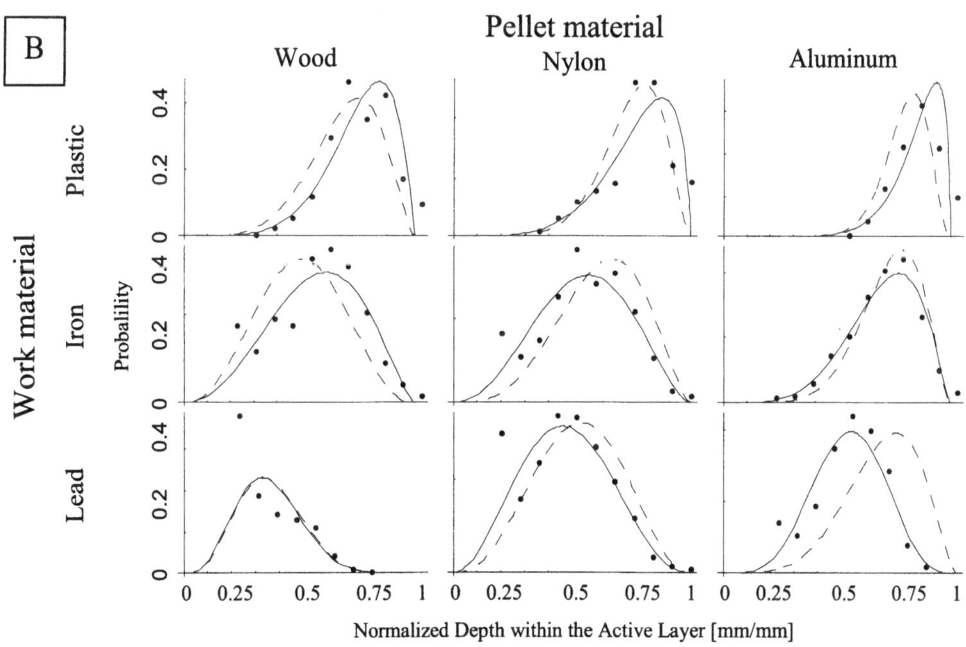

FIGURE 4. Distribution of passages of workpiece within the active layer at various depths within the active layer: experimental data (dots), best fit by beta distribution (hatched line), predicted values (continuous line), A = 11 rpm, B = 18 rpm

SHAPE-MACHINING OF AEROSPACE COMPOSITE COMPONENTS USING NON-TRADITIONAL ABRASIVE WATERJET CUTTING PROCESS

X. Zheng[1], L. Chen[1], E. Siores[1], P. Steele[2], P. Crothers[2]

[1] Industrial Research Institute Swinburne, Swinburne University of Technology, Australia
[2] Department of Engineering, Aerospace Technologies of Australia Ltd., Australia

KEYWORDS: Cutting, Two-Dimensional, Three-Dimensional.

ABSTRACT. The abrasive waterjet technology (AWJ) is a non-traditional method of machining which offers a productive alternative to conventional techniques. Aerospace composite materials although possessing special properties such as high strength to weight ratio and stability, are known to be difficult to machine, or even impossible to be machined by conventional methods. However, by using abrasive waterjet these composite materials can be efficiently machined. In this paper, two-dimensional and three-dimensional AWJ cuttings are addressed and a focus is placed on the challenging three-dimensional waterjet trial and due to its much more complexity involved, very little work has been done so far. The following subjects were studied: Off-line programming and simulation using IGRIP software, setup and control of the three-dimensional system etc. Finally, a few two-dimensional and three-dimensional abrasive waterjet machining conclusions and future research work are given that reflect our ongoing research.

1 HISTORY OF WATERJET CUTTING

In nature, the power of water in the form of erosion has already been active for millions of years. The extent of erosion through leaching and wearing of surface soil formations by water is well known with many examples. The advantageous use of water jet was already introduced more than 100 years ago for the mining of gravel and deposits. For example in ancient Egypt, river branches were diverted to wash out soil in search for gold and other minerals. Similar improved methods of mining reached New Zealand, South Africa, Russia, California and many other countries until the early 1900s [1]. Pumps were developed to obtain more energetic jets, for more profitable mining practice, soon in this century. In the 1930s, American and Soviet engineers made the first attempts to use high-speed water jets in mining, for the removal of rocks, minerals and coal. For this application they generated pulsating streams with pressures up to 1000 MPa [2]. Dr. Norman Franz conducted the first systematic study on ultrahigh-pressure (UHP) waterjet cutting in the late 1950's and early 1960's. Dr. Franz, a forestry engineer, wanted to find new ways to slice thick trees into lumber, therefore he dropped heavy weights onto columns of water, forcing that water through a tiny orifice. He obtained short bursts of very high pressures, and was able to cut wood and other materials [3]. The emergence of modern waterjet technology can largely be attributed to the development of high-pressure pumps, which can pressurise the water up to 400 MPa, with reliable performance for industrial use [2]. Today waterjets are capable to cut rocks and a wide range of nonmetals.

Published in: E.Kuljanic (Ed.) *Advanced Manufacturing Systems and Technology*,
CISM Courses and Lectures No. 437, Springer Wien New York, 2002.

A major boost to this technology was the introduction of abrasive waterjets to cut metals,composites and other hard metals. Abrasive waterjets use accelerated particles to cause erosion at the machining material. According to recent report from 'Frost And Sullivan', a market research company, abrasive waterjets are the fastest growing segment of the machine tool industry.

2 CLASSIFICATION AND PRINCIPLE OF WATERJETS

There are two different kinds of waterjet: the pure waterjet (WJ) and the abrasive waterjet (AWJ).

Pure waterjet is the original water cutting method. Even if water is the usual working fluid, there are other fluids like oil, liquid gases or emulsions, which can be used instead of water [2]. Pure waterjet can cut a lot of different materials. So it is possible to cut textiles, elastomer, thinner plastic, thermoplastics, fibrous materials, foodstuffs, paper, etc. with a velocity of up to 200 meters per minute [4].
Where the pure waterjet cuts soft materials, the abrasive waterjet cuts hard materials, such as metals, hard-rock, composites, ceramics and bullet-proof glass. Abrasive waterjets using standard parameters can cut materials with hardness up to and slightly beyond aluminium oxide ceramic (often-called Alumina, AD 99.9). Both, the pure waterjet and the abrasive waterjet, cause nearly no heat on the material. However, the water of a pure waterjet and abrasive waterjet are warmed as they are accelerated to high speed. Frictional forces and other factors warm the stream as it exits the orifice. In pure waterjet the water temperature is then raised 2 to 3 degrees for each 1,000 psi. Accordingly, if the stream is accelerated to Mach 3 the temperature is approximately 75° (170° F) to 85°(18 0° F) Celsius [3]. But compared with other conventional cutting processes, one can say that waterjet cutting is a cold cutting process. The advantage is that the machining with waterjets causes no hardening, wrapping, dripping slag, amalgamation, contamination pollutants (such as noxious gases, for example), or grinding dust. So the waterjet process is guaranteed to be particularly environmentally friendly. Furthermore waterjets fit in economical requests. Because of the very thin stream (diameter between 0.1 mm and 0.3 mm for WJ and 0.5 mm and 1.3 mm for AWJ) there is much more negligible loss of material than in case with conventional processes [3]. In respect to today's prices of raw materials this fact is also a welcome characteristic of waterjet technology. Moreover the cut edges are of good quality. There are no fraying or ridges. Correspondingly the trims do not require any costly post-cut treatments. Another big advantage of waterjets is that you can cut complex shapes. The technique is supremely suitable for the most multifaceted controlled cuts, sharp angles and bevelled edges. Minimal radii and extremely detailed geometries are easy to be manufactured as well. In addition, different from other machining processes, for waterjets it is possible to start the cutting process everywhere. It is not necessary to start at the edge of the material like you must do for example in band saw processes. Further advantages of waterjets are that the mechanical reactions take place on the microscope level [4] and that only low cutting forces (less 4.5 N while cutting) are inserted in the material [4]. Then the material's properties are not changed as much as in case of conventional machining.

2.1 PURE WATERJETS

Waterjets are based on the principle of pressurising water to extremely high pressures. After pressurising, the water stream is forced through a tiny hole, called orifice, in which the pressure energy is transformed to kinetic energy of the water (Figure 1). At approximately 276 MPa (40,000 psi) the resulting stream that passes out of the orifice is travelling at Mach 2. And at 413 MPa (60,000 psi) the speed is over Mach 3 [3].

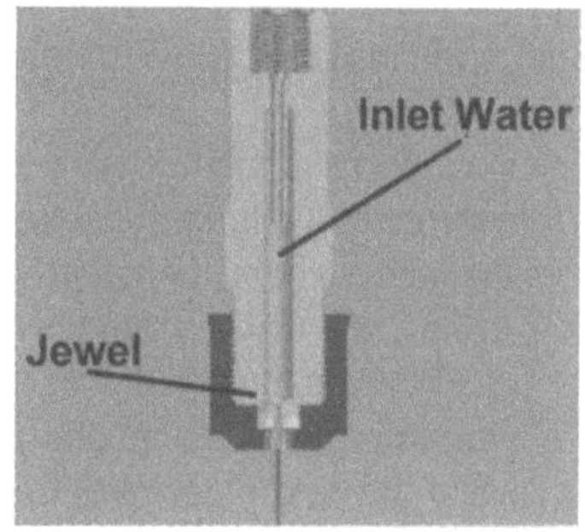

FIGURE 1. Pure waterjet

In case of pure waterjets, the material removal process can be described as a supersonic erosion process. The accelerated stream leaves the nozzle to impinge the material and causes stress at the surface of the material. Small cracks are then developed on the surface. Due to the fact that the water stream is very small, it hits only a small area of the material, which are removed. The machined chips are flushed away by the water.

2.2 ABRASIVE WATERJETS

The abrasive waterjet differs from the pure waterjet in just a few ways. In pure waterjet, the supersonic stream erodes the material. However, in the abrasive waterjet, the waterjet stream accelerates abrasive particles and those particles, not the water, cause the erosion of the material. After the pure waterjet stream is created, the abrasive is added (Figure 2). In general there are two possibilities to add the abrasive: the injector principle or the bypass principle.

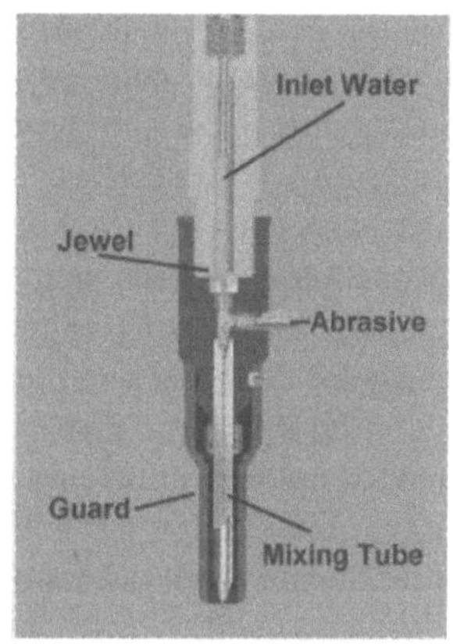

FIGURE 2. Abrasive waterjet

In case of the bypass principle the water stream is split into two parts, the main stream and the bypass stream. The bypass stream is led to a pressure vessel, in which an emulsion of water and high concentrated abrasive is situated. Through the volume of the bypass stream one can regulate the abrasive mass flow. Before passing the nozzle the main stream and the bypass stream are rejoined.

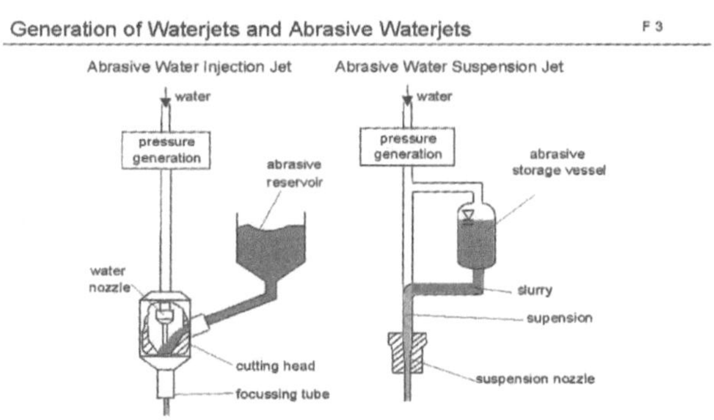

FIGURE 3. Injection and suspension jets

The abrasive injector system is based on the principle of a water jet pump. The water stream intakes air to a mixing chamber, in which the abrasive, the air and the water are merged. The task of the air is for the pneumatical transport of the abrasive material into the chamber. Because of the lower expenditure in case of the injector principle it is commonly used in industrial application [2]. After the mixing process the abrasive particles are accelerated, like a bullet in a rifle, down the mixing tube. Both principles are shown in Figure 3.

The abrasive used in abrasive waterjet cutting is hard sand that is specially screened and sized. Ordinary garnet, silicates or oxides are utilised as abrasives. The important parameters of an abrasive material are the material structure and hardness, the mechanical behaviour, the grain shape, the grain-size distribution and the average grain size [5]. The most common abrasive is garnet. Garnet is hard, tough and inexpensive. Different mesh sizes are used for different jobs. The abrasive waterjet is hundreds, if not thousands of times more powerful than a pure waterjet. Accordingly you can cut harder material than with pure waterjets.

3 APPLICATION OF AWJ IN AEROSPACE AND AIRCRAFT INDUSTRIES

Especial fibre-reinforced polymer-matrix composite is beginning to be used extensively in aerospace and automotive applications due to its low specific weight and high strength. The disadvantage of composite material is the machining, which is a challenging problem, because the mechanism of cutting is different from that of conventional structural materials. Traditional cutting processes, from welding to sawing, would destroy the structural integrity of such material. Moreover excessive degrees of cutting tool wear attributed to the abrasiveness and inhomogeneous nature of composites are common in conventional machining processes. Typical methods of machining include interlaminar and intralaminar delamination, fibre pullout, cracking and burning or matrix smearing of the composite material. These damage features, which are dependent on the fibre orientation, make most of the conventional machining methods uneconomical when applied to trimming of composite. Furthermore fibre-reinforced composites react unfavourably to non-traditional processes, like laser cutting, plasma processes, electrical discharge machining or ultrasonic machining, as well. Plasma processes often cause burns, fusion and gas formation along the cut edges. Most notably, when processing fibre-reinforced composite materials with laser technique, the coherent light beam of a laser fails completely. Electrical discharge and ultrasonic machining are two very precise machining methods that produce high quality surfaces, but the rate of machining is quite low. Waterjet and abrasive waterjet cutting methods have proven to be the perhaps most promising machining technique for composite material.

The high temperature composites for aircraft are used in turbine nozzle guide vanes and rotor blades for aero and land based gas turbine engines. At present they are being produced by conventional machining techniques like grinding, electro-discharge machining and vapour phase aluminising. But with AWJ cutting preliminary steps can be eliminated and only the final finishing steps have to be used by saving time and the cost of the components. State of the art CAD/CAM facilities can accept customer's aerofoil data for translation into robotics AWJ machining programs. Universal DXF and IGES are amongst the transfer systems that are available. Special programming techniques can be used to machine components in stainless steel, nickel based alloys, and titanium with product requirements from toilet bowls to engine cowls. A cost saving, single-piece, engine inlet cowl approximately 9 feet in diameter can be manufactured in a very less time [6].

Boeing, Textron, Lockheed, Martin Marietta, Northrop, Thiokol, Pratt & Whitney, Aerostructures, American Airlines, Northwest Airlines and Embraer (Brazil), aerospace and aircraft manufacturing industries with multi-process machine tools use waterjet cut because it can cut, drill and clean. Most of the composites used in U.S. commercial aircraft are cut with waterjets. A wide range of metals such as Inconel, titanium, aluminium, and high strength alloys are cut routinely today for aircraft use. With the help of AWJ machining dual mast low rail, mid rail, single mast low rail and bridge-type gantry robotic systems perform a variety of functions including abrasive waterjet cutting, parts inspection, routing, trimming, small hole drilling and milling. 6-axis composite cutting with touch probe can be carried out for most of the aerospace components, which are in curvature shapes. In aerospace spares supplies and maintenance ultrahigh-pressure waterjets remove thermal spray, abradable seals, acoustical composites, rubber, adhesives and other coatings from aircraft engine components such as stators, fan cases, combustion chambers, rotating components, spools and knife-edge seals.

In all, abrasive waterjet technology has a broad application in aerospace and aircraft industries. The acceptance of the aerospace industry to this technology has been a prime factor for its advancement over the last decade. A wide range of metals such as Inconel, titanium, aluminium, and high strength alloy are cut routinely today for aircraft use. The cutting of composites has been found ideal with abrasive waterjets. Other applications such as small hole drilling, and milling have been sparked by the needs of this industry. The job shop business has been growing to meet the demands of aerospace/aircraft needs. Aerospace industry first adopted three – dimensional abrasive waterjet cutting systems in the late 1980s which is ideally suited to cut the large, hard to handle, heat-sensitive parts endemic in the aerospace industry. AWJ machines were easily adapted to this application. Today waterjet plays a more and more important role in aerospace and aircraft industries.

4 TWO-DIMENSIONAL ABRASIVE WATERJET CUTTING

4.1 TEST EQUIPMENTS AND SETUP

Two-dimensional (2D) waterjet cutting is mainly concerned with the 2D shape machining. In two-dimensional research the following test specimen was cut (Figure 4) and different parameters influencing the shape machining were investigated that include cutting speed, pump pressure, abrasive flow rate, abrasive particle size etc. The test results are analysed and discussed.

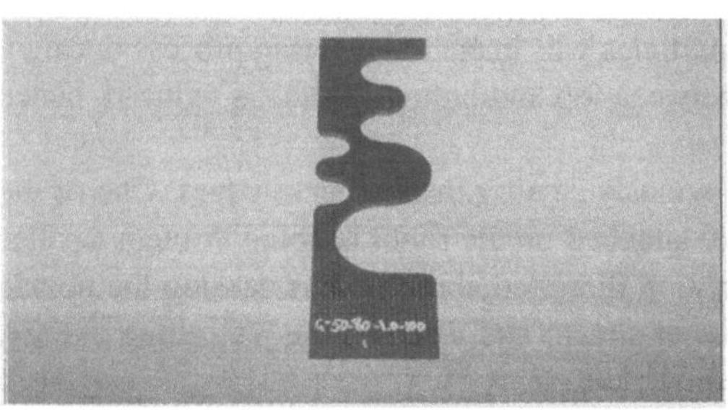

FIGURE 4. Test specimen cut by abrasive waterjet technology

Taylor-Hobson Surtronic 3 digital measurement machine with a cut-off of 0.76 mm is used to measure the surface roughness of the test specimens. In measuring kerf angles an OLYMPUS BX60 system microscope is used. The microscope has a scale overlayed over the eyepiece, which enables measurement of the kerf width directly without the need for a ruler.

In this two-dimensional research the test material is epoxy pre-impregnated graphite woven fabric and the standoff distance is set as 2mm which is based on our previous test results. Five cutting speeds are used namely 8.5 mm/s, 17 mm/s, 25.5 mm/s, 34 mm/s, 42.5 mm/s. Other machining parameters are set as the following: Pump pressure: 40 kpsi ≅ 275.6 MPa and 50 kpsi ≅ 344.5 MPa; mesh size: Mesh 80 and 120; Abrasive flow rate: 3.8 g/s, 5.3 g/s, 7.6 g/s.

4.2 TWO-DIMENSIONAL AWJ DISCUSSIONS AND CONCLUSIONS

- The surface roughness significantly increases as the cutting speed increases. The increase in the number of impacting particles at lower cutting speeds contributes to the improved surface finish. The additional particles serve to smoothen the surface that forgoing particles generated.
- Surface roughness changes dramatically at high cutting speeds. At comparatively low cutting speeds the surface roughness shows trivial changes.
- The bigger the difference between the cutting speeds the bigger surface roughness difference yields.
- An increase in the pressure improves the surface quality. These results are due to the increased fragmentation probability of the abrasive particles with an increase in their velocity. This fragmentation reduces the size of the impacting abrasive particles. Also, an increase in the pump pressure increases the abrasive waterjet kinetic energy. This process allows part of the excess energy to smoothen the surface.
- The influence of the pump pressure strongly depends on the depth of cut. The deeper the cut, the more pronounced the pump pressure influence is. The standard mesh size is mesh 80. Mesh 50 is larger yet coarser and the smallest is mesh 120. In the range of small abrasive diameters, the roughness almost linearly decreases as the particle diameter increases, except for very shallow cuts. This effect is very obvious in deeper cuts. The individual particle-impact characteristic is overcome by the general material removal characteristics as a certain depth of cut is obtained. Other effects, such as a decrease in the jet instability with a decrease in the abrasive diameter contributes to the surface structure. Nevertheless, as a certain abrasive diameter is exceeded, the roughness increases.
- The roughness decreases with an increase in the abrasive flow rate. A high number of abrasive particles involved in mixing increases the probability of particle collision that decreases the average diameter of the impacting particles thus improving the surface finish.
- It is apparent that the kerf angle decreases with a decrease in cutting speed because after reducing the cutting speed more abrasive particles will impact the cutting process during a certain period of time and the difference between top and bottom widths is reduced, hence minimising the kerf taper angle.
- The kerf angles at the straight edges are obviously smaller than at curve edges. One of the possible reasons for this is when the nozzle attached on the robot is going through smaller radius curves, the robot keeps changing motion direction abruptly thus causing the nozzle to vibrate. This vibration could disturb the jet stream and discount the jet cutting energy. As a result, bigger kerf angles are produced.
- Different pump pressures have a great impact on kerf angle. Increasing the pump pressure will reduce the kerf angle. However, this depends on the material being cut and for some materials the kerf angle will not linearly decrease by increasing the pump pressures. For this material the bigger the radius of the curve, the bigger the kerf angle produced on the corresponding curves.

- It is noticed that mesh 120 reduces the kerf angle. However, this is also related to the pump pressures. For a low pump pressures, a typical minimum in the taper appears at a certain abrasive particle diameter.
- The kerf angle decreases with an increase in abrasive flow rate. This is due to more abrasive particles act in the cutting process in a unit time and the difference between the top and bottom width is reduced. Thus the kerf taper is improved.

5 THREE-DIMENSIONAL ABRASIVE WATERJET CUTTING

Three-dimensional waterjet cutting is concerned with cutting three-dimensional profile parts and it's obviously much more complicated that two-dimensional cutting. Here we present this study on cutting of three-dimensional composite components using a stationary waterjet that reflects the ongoing research at Swinburne University of Technology. The aim of the trials is to develop the ideal methodology for cutting small three-dimensional parts to achieve best-cut quality and cycle times.

5.1 EQUIPMENT AND SETUP

STATIONARY WATERJET SETUP. The waterjet system at Swinburne University is used for these tests. The waterjet has been modified such that it can be removed from the robot and held by a stand. The robot can then be used to pick up and move the parts around underneath the waterjet.
PART NEST. To enable programming of the paths, the parts must be placed in a known location in an accurate and repeatable manner. This location will be known as the nest. To achieve this, AluFix will be used. AluFix is a modular fixturing system based on defined holes and accurately machined faces. The holes are spaced precisely and as such, an accurate fixture can be assembled / reassembled quickly and repeatably. The nesting of the parts is difficult due to the trim excess of the part not being accurate for use as reference. Therefore, the nest will be constructed such that the tool surfaces of the part are used for location.
FIXTURE. To hold the parts to the robot, a simple fixture was designed. This will enable a range of parts to be held without custom fixturing for each part. The support for the fixture attaches to the robot in place of the waterjet, attaching to the collision sensor. This support can be configured in a number of arrangements to enable a range of orientations of the fixture from the tool face.

5.2 OFFLINE PROGRAMMING

Delmia software IGRIP is to be used for the offline programming of the paths during the tests. The paths around three dimensional parts are complex and offline programming will be necessary to achieve accurate cut paths and ensure collisions do not occur. Also this is a good way to build confidence in performing real cuttings.

5.3 ADVANCED CUTTING TECHNIQUES

KERF ANGLE COMPENSION. Measure the affect of angling the jet into the part (perpendicular to motion) during cutting to compensate for the kerf angle. Discover the ideal angle for different cutting conditions – straight/curved and a range of thicknesses.

FORWARD ANGLE. Measure the affect of angling the jet forward in the direction of motion during cutting to compensate for the kerf angle. Discover the ideal angle for different cutting conditions – straight/curved and a range of thicknesses.
MAXIMUM ACCELERATION FOR CORNERS. Determine the maximum acceleration possible for cutting corners without accuracy problems.

5.4 SAFETY ANALYSIS

ROBOT COLLISION. There exists potential for collision of the robot into the tank or stationary jet. To avoid this, the following steps should be taken when running a new path and at the start of a testing session: (1) Check path for collisions with simulation and (2)Run robot at 10% of planned cutting speed with waterjet off. The use of world exclusion zones would not be practical due to the robot needing to move close to the tank and stationary jet.
WATERJET DAMAGE. Whilst the stationary jet is running, it presents the risk of damage to anything that passes underneath it. To limit the risk of damage, the following steps should be performed: (1) Simulation should check that nothing moves accidentally underneath the waterjet. (2) The waterjet should be turned on and off automatically in the program such that it remains on only when necessary. (3) The operator should be able to turn off the waterjet in case of an unexpected event.
LOSS OF PART. If the part becomes separated from the fixture, there exists a risk of danger if the part is moving quickly. To prevent the part from causing injury, the operator should not stand close to the robot when it is holding a part. When the robot is moving with a part at a high speed, the operator should not be inside the room.

5.5 FURTHER RESEARCH

In order to perform 3D cutting, fixtures and jigs need to be designed according to different components and this is a really challenging task. The component weight and cutting forces must be calculated and taken into account in designing the fixtures/jigs. Another challenging task in 3D cutting is how to protect the parallel surfaces of a component. Special clamping devices may be created to fix this problem. Also more studies should be given to orientation and collision prevention of the robot.

6 ACKNOWLEDGMENTS

The authors wish to thank the Industrial Research Institute Swinburne and Aerospace Technologies of Australia for their generous supports.

REFERENCES

1. Momber, A., Kovacevic, R., (1998), Principles of Abrasive Waterjet Machining, Springers, 1-76.
2. Louis, H., (2001), Grundlagen der Wasserstrahlanwendung, University of Hannover, 6-25.
3. Flow International Corporation, (2001), http://www.flowcorp.com/newsite/Faq.
4. Waterjet Sweden, (2001), http://www.waterjet-nw.de/start.htm.
5. Momber, A., (1998), Water Jet Applications in Construction Engineering, A.A.Balkema, 20-198.
6. Zheng, X., (2001), Study of Shape-Machining Aerospace Composite Components Using Abrasive Waterjet Technology, master thesis, Swinburne University of Technology, Australia, 35-92.

COMPARATIVE STUDIES OF BRUSH ELECTRODISCHARGE MACHINING WITH ELECTRODES OF ALLOY STEEL AND TUNGSTEN

S. Spadło

Department of Mechatronics and Machine Technology, Technical University in Kielce, Poland

KEYWORDS: electrodischarge machining, brush electrode, superficial layer.

ABSTRACT. A new metal removal process using a rotary brush as the hot electrode has been presented. The process of electrodischarge machining with brush electrodes in water-glass is a type of surface machining process of electrically conductive materials. The process is used to remove the excess material and to smooth the machined surface. The use of brush electrodes makes the process cost-effective because it reduce the time needed to machine large parts. Proper parameter settings, hot electrode material and the type of power supply generator help to control the intensity of phenomena occurring in the machining zone in order to achieve the desired properties. The process allows both the removal of burrs and flushes and the blunting of the sharp edges without damaging or undercutting the machined surface. An exceptional properly of brush electrodes is that they make it possible to produce a uniform layer which firmly adheres to the core and has a high content of alloy components.

1 INTRODUCTION

One of the basic problems of modern production processes is meeting the growing requirements concerning the resistance of machine parts to mechanical and chemical wear as well as heat and abrasive erosion. An important issue is also increasing the durability of technological tools such as blanking dies, plastic working tools, metal moulds, etc..

Improving the qualities of the above-mentioned tools by using materials engineering and proper superficial layer formation [1] [2] in the production process is one of the ways to meet the demands that have been set. The condition of the superficial layer is so essential because most of the wear and tear processes are effected by the environment. Almost all tribological and fatigue processes start or take place on the surface of the part.

The durability of the machine parts and the above-mentioned tools can be achieved as a result of a controlled technological process.

Searching for the possibility of improving the qualities of machine parts the author suggests changing the way of electrodischarge machining with a rotating brush electrode involving the use of electrodes made of alloys or metals such as tungsten (W), vanadium (V), chromium (Cr) or molybdenum (Mo) in the finishing process [3][4][5]. The author indicates that the physical phenomena accompanying electrical discharges in the machining process make it possible to modify the above-mentioned parts by enriching their surface layer with alloy components.

Published in: E.Kuljanic (Ed.) *Advanced Manufacturing Systems and Technology*, CISM Courses and Lectures No. 437, Springer Wien New York, 2002.

This variety of machining called brush electrodischarge mechanical machining (BEDMM) has the following features making it different from conventional EDM [6][7]:

- application of direct or pulse current generators,'
- random character of frequency, voltage and energy of the electric discharges,
- occurrence of periodic contact between the elements of the brush electrodes and the workpiece and the current flow through this contact,
- elastic fit between the brush and the shape of the machined surface.

Combining various physical and chemical processes acting on the workpiece material leads to achieving high machining efficiency. Such a method is often defined as the 'hybrid machining process' [8].

BEDMM uses a combination (Figure 1) of:

- electrochemical action,
- electroerosion,
- mechanical action,
- their interaction.

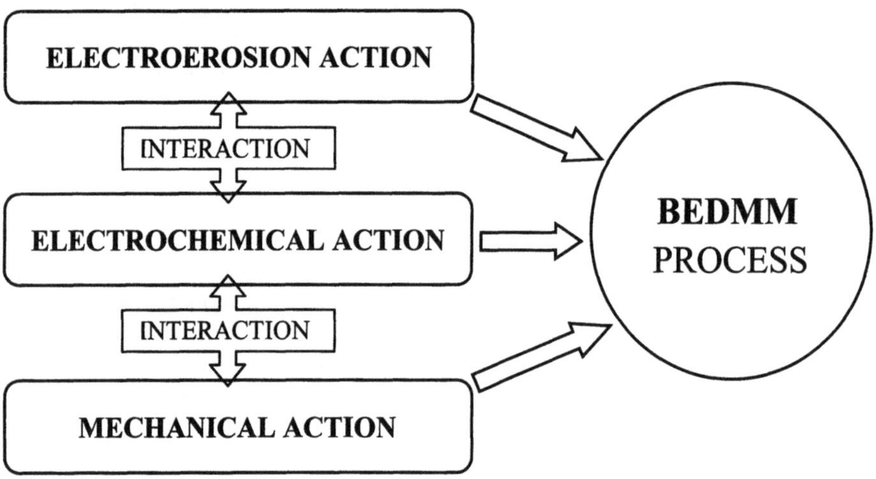

FIGURE 1. Process condition scheme in BEDMM

The main advantage of the method described is that both the alloying and the removing of the material from the surface can be performed while giving the part its final shape, smoothing its surface or regenerating it. The conditions of the process increase the costs of obtaining the final product only marginally because the process uses the same tools and equipment as brush electrodischarge mechanical machining.

The method suggested makes it possible to machine parts of complex geometrical shapes, e.g.: curvilinear shapes, without maintaining the strict kinematic dependencies in the tool setup.

2 PRINCIPLE OF BRUSH ELECTRODISCHARGE MACHINING (BEDMM)

The BEDMM process is based on cutting by electroerosion and friction interaction on metal. In this process a steel disk and the workpiece are parts of a direct current circuit. Machining takes place in the presence of water-glass solution in water. During machining electric arcs melt particles of the workpiece material which are subsequently removed by friction between the disk and the machined surface.

The tool in the form of a rotating brush is the cathode, the machined part being the anode. The hot electrode rotates while low concentration (< 10%) water solution of sodium silicate is being supplied into the machining area through a nozzle. Both the machined part and the tool can be powered by either direct or pulse current. The position of the brush should be properly chosen, as shown in (Figure 2.a) to ensure deflection (Δ) of the tool components big enough to allow mechanical rupture of the anodic layer on the surface of the workpiece and initiation of an electrical discharge [6].

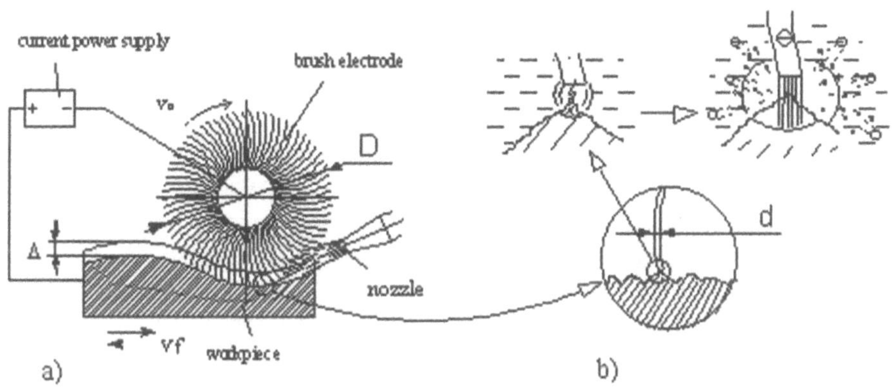

FIGURE 2. BEDMM process: a) schematic diagram and brush setting in relation to machined surface, b) occurrence of an electrical discharge

After the initiation is achieved and a plasma channel is built a rapid local increase in temperature occurs on the peak of roughness which causes melting, evaporation and metal removal. Figure 2b shows the principle of this process.

In typical machining conditions the mechanical contact between the tool and the workpiece occurs through the hard anodic-silicone layer. The movement of the elements of the brush along the surface and their pressure on the surface caused the removal of the anodic layer from peaks of roughness and initiation of electrical discharges. Results of the investigation show that the melted metal can be smeared on the workpiece surface by the brush elements. Under typical parameters of machining the mechanical contact of the brush with the melted metal can cause the liquid metal to spread over the machined surface (Figure 3.), as a result of which the peaks of roughness are flattened [9][10]. An increase in pressure, with low voltage applied, can cause the depassivated layer to be torn off. Eventually it leads to the direct contact of the electrodes and fading of the discharges.

The discrete structure of the brush induces sparking of high frequency (several kHz). The discharge duration is significantly shorter than in typical erosion mechanical cutting. Discharges cause melting and evaporation and metal removal from the workpiece. In a place where a discharge occurred the material is depassivated and some electrochemical reactions take place [10][11].

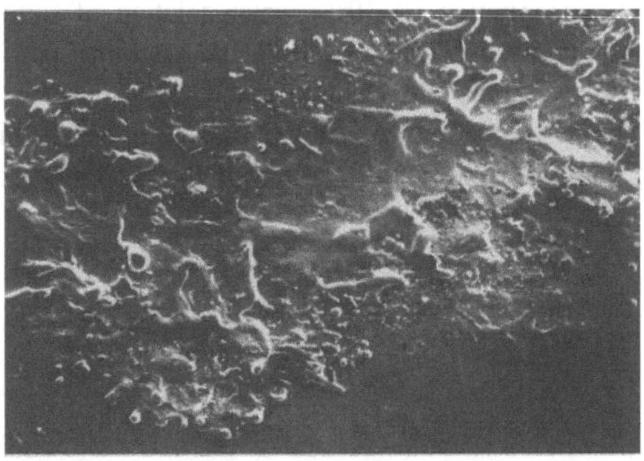

FIGURE 3. SEM photograph showing the effect of a single packed discharges, seen as a „path" of crater, $U = 8$ V, $d = 0.5$ mm (diameter of single wire), $v_o = 0.7$ m/s, magnification 75x.

In electrical discharge machining spark discharges are the main factor influencing the formation of the superficial layer. As a result of single discharges local melting and material vaporisation occur, resulting in roughening of the surface in the form of craters. They are determined, but as the craters overlap and are randomly distributed their structure must be random in nature.

The layer is shaped [12] by the energetic effect of the discharge on the electrodes and the following phenomena are caused by such a discharge:

- superficial layer melting of the anode and cathode,
- expulsion of the material into the inter-electrode area and its consequent re-solidification,
- transfer of the melted material from the tool electrode onto the part surface,
- mixing and diffusion of the particles of the transferred material into the workpiece material,
- temperature increase of the surrounding layers,
- very fast cooling of the superficial layer due to the heat transfer through the part core and quenched by the machining medium,

The above factors have influenced the following output parameters:

- productivity
- parameters describing the geometry of the surface layer (R_a, R_t, S_m, D_q, W_a, W_t, W_{sm}),
- surface texture,
- metallographic structure of the superficial layer,

- superficial layer chemical composition,
- microhardness distribution in the superficial layer.

3 THE INVESTIGATIONS CONDITIONS

Due to the complex space and time structure and random nature of the process it is not possible to foresee the effects of the machining with a brush electrode based on present-day mathematical models. Therefore, it was indispensable to conduct experiments to determine the parameters of the superficial layer.
The conditions the experiments were conducted under were typical of those of the material removal process. The experiments were conducted using brush electrodes made of materials of varied physical properties (melting-point, boiling-point, thermal conductivity, mechanical properties), e.g.: tungsten, molybdenum, chromium-nickel steel. Parts of the brushes (wires), used were made of stainless 1H18N9 steel (with the melting-point close to that of the machined part) and tungsten (with the melting-point substantially above that of the sample).

The initial investigation of the BEDMM modification SL process determined the following main factors [4], influencing the machining process.

- U – voltage - 8.0 [V],
- v_0 – tangential speed - 3.58 [m/s],
- vf - feed-rate - 12.85 [mm/min],
- deflection (Δ) - 0.4 [mm].

The following factors are assumed to be constant in the process:

- machining fluid, $Na_2SiO_3.nH_2O$ (water-glass) in water solution with the mass density 1,15 Mg/m^3 with temperature T = 293 K,
- the specimens used in the main experiment were made of construction 0.45C carbon steel
- characteristics of the brushes:
  - brush diameter D_z = 100 mm,
  - width of the brush W = 20 mm,
  - diameter of a single wire d = 0.3 mm,
  - material of the brush electrode carbon soft steel, tungsten, chromium-nickel steel (1H18N9T).

A scanning electron microscope JEOL JSM-5400 equipped with an X-ray microanalyses EDS ISIS-300 was used to examine the BEDMMed surfaces' structures and their chemical composition. The specimens were sectioned perpendicular to the machined surface and polished using standard metallographic techniques.

4 MICROSTRUCTURE AND X-RAY DIFFRACTION ANALYSIS

To evaluate the metallographic changes and the chemical composition of the superficial layer (SL) the high-grade construction 0.45C carbon steel was used. The choice of this particular type of steel made it possible to evaluate the structural changes caused by heat influence as well as the changes in the chemical composition of the superficial layer being the result of using electrodes made of alloy components, e.g.: chromium, tungsten, molybdenum, high alloy steels, etc.

During BEDMM there is high-energy density in those areas where the plasma channel strikes the workpiece, which results in a localized, and very high, temperature rise.
Subsequently the material is heated and evaporates and the liquid material is also flushed away. But some of the molten material remains and this is quenched at an extremely high rate by the machining medium and undergoes a change in physical and chemical properties. This process results in the production of a layer of re-solidified material called a recast layer [12][13].
To evaluate the influence of the material of the tool electrode on the superficial layer properties soft carbon steel, chromium nickel steel 1H18N9T and tungsten were chosen as the hot electrode and 0.4%C carbon steel as the workpiece.
The superficial layer in BEDMM consists of three layers. At the top, a molten and resolidified layer, called the recast or white layer, is observed. This layer is usually present because material removal in BEDMM is mainly based on the melting process of the workpiece material. In the recast layer mixing and diffusion of the material of the tool electrode and workpiece can occur due to a temporary direct contact of the electrodes. Below the recast layer a heat affected zone (HAZ) is present. This zone comprises the workpiece material, which has been affected by heat but has not been melted. In the case of a tool steel the layer is usually hardened. Below the hardened layer a temper zone is present.

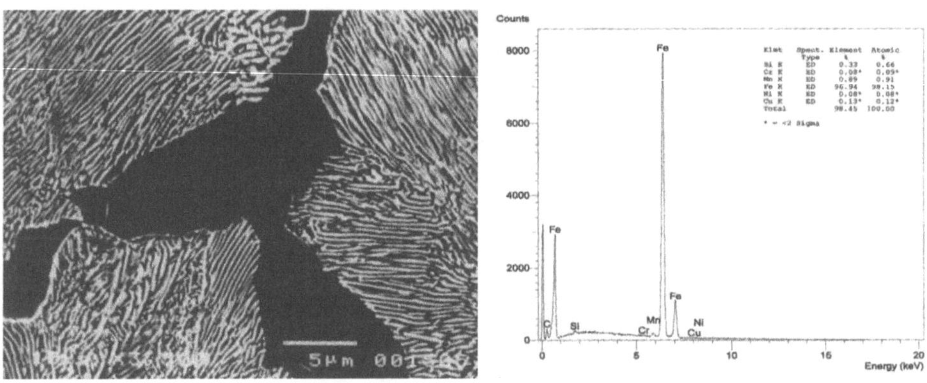

a) b)

FIGURE 4. Photograph of the metallographic microstructures of the workpiece - native material (steel 0.45 % C) magnification 3500x - a; and X-ray diffraction pattern obtained from the surface layer before machining - b.

Both the metallographic structure of the part core and its chemical composition were examined to compare the changes made to the superficial layer by the BEDMM process. SEM photograph (Figure 4a) shows the pearlitic-ferritic structure of the part core – 0.45C carbon steel. Figure 4b shows the results of an X-ray microanalysis of the chemical composition of the part core. These data were measured by an X-ray diffraction method using specimens, which were sectioned perpendicular to the BEDMMed surface.
To compare the changes, the experiments were conducted using an electrode made of the St5 steel of a similar chemical composition to that of the sample. Figure 5a shows the metallographic structure of the superficial layer of a sample made of the 0.45C steel after being BEDMMed at

a voltage of 8V. A zone of amorphous structure and a deeper-located heat-affected zone can be seen on the SL. Figure 5b shows the chemical composition of the superficial layer. An analysis of the distribution of the elements indicates that the layer is decarbonated.

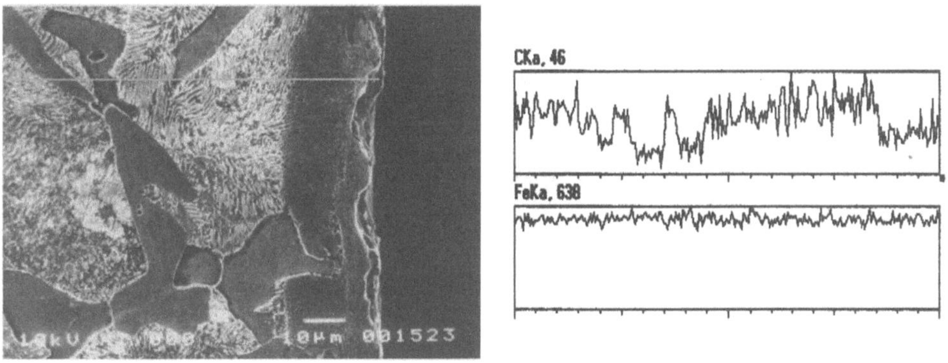

a) b)

FIGURE 5. Photograph of the metallografic microstructures and X-ray line scan of the superficial layer after the BEDMM process (U = 8 V) machined by soft carbon steel (0.35%C) electrode, magnification 3500x – a; X-ray analysis the recast layer- b

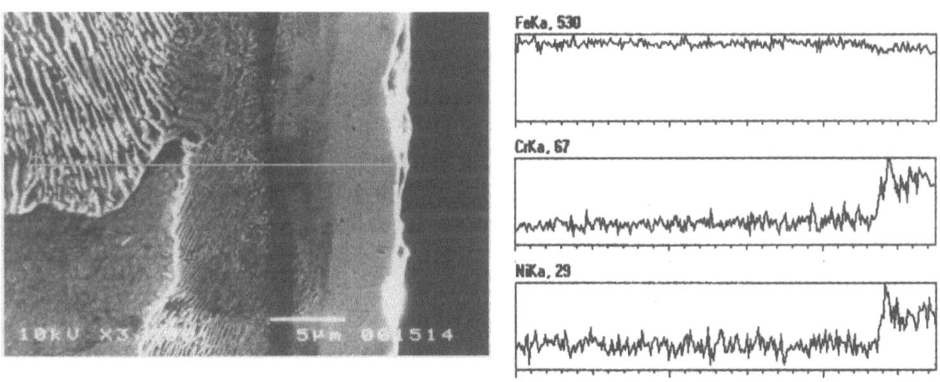

a) b)

FIGURE 6. Photograph of the metallografic microstructures and X-ray line scan of the superficial layer after the BEDMM process (U = 8 V) using a chromium nickel steel (1H18N9T) electrode, magnification 3500x – a; chemical elements (constituting the surface layer) distribution in the sub-surface layer - b

Figure 6a shows the metallographic structure of the superficial layer of a sample made of the 0.45C carbon steel after being BEDMMed with a chromium nickel steel (1H18N9T, 18% Cr, 9% Ni) electrode at a voltage of 8V. An outer layer can be seen. It has been altered but it adheres to the core quite well. The layer exhibits three zones: a zone of amorphous structure, a heat-affected zone, and a deeper-located transitory zone. Figure 6b shows the results of an X-ray microanalysis of the chemical composition of the superficial layer. The distribution of the

elements shows the presence of the elements the hot electrode was made of (chromium, nickel). The depth of the changes depends on the machining parameters and is 5 ÷ 8 µm.
An analysis of the chemical composition of micro-areas (Figure 7) of the altered layer provides a quantity assessment of the changes that have taken place. The recast layer exhibited diffraction peaks of chromium and nickel. It indicates that a large amount of chromium and nickel exists on the surface. Therefore, the machined layer may contain elements the hot electrode is of made of. The examined samples contain from 8% to 10 % of chromium and from 5% to 6 % of nickel.

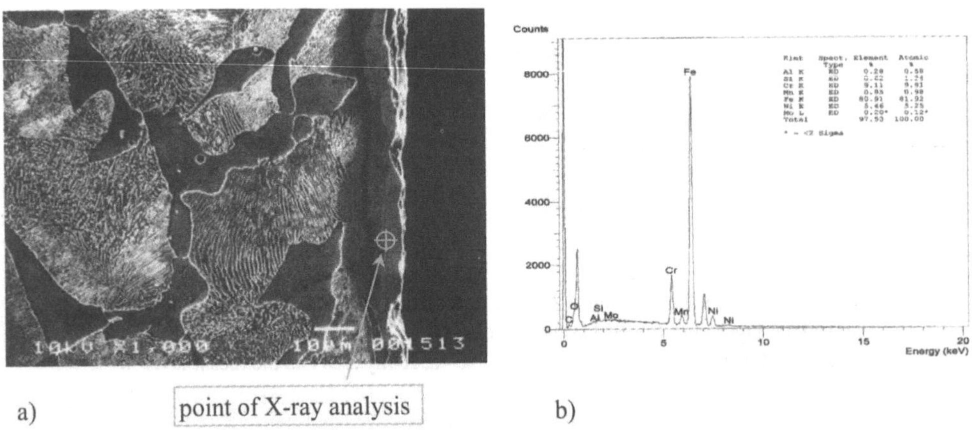

a) b)

FIGURE 7. Photographs of the metallographic microstructures of the surface machined by chromium nickel steel (1H18N9T) brush, U = 8 V, magnification 1000x - a; and X–ray diffraction pattern obtained from the surface layer - b

A microanalysis of the chemical composition of the superficial layer leads to the conclusion that the layer has been formed as a result of remelting the workpiece metal and the electrode particles. The results of the experiments show that it is possible to modify the core material with the elements the electrode is made of.

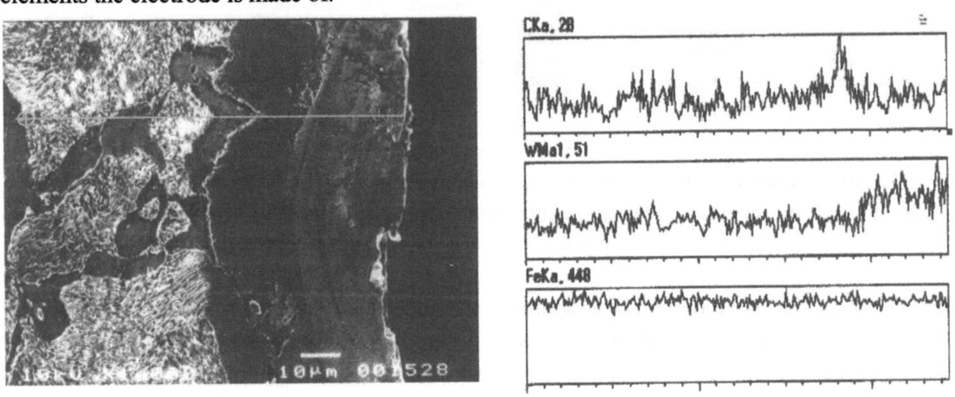

a) b)

FIGURE 8. Photograph of the metallografic microstructure and X-ray line scan of the superficial layer after the BEDMM process (U = 8 V) machined by tungsten electrode, magnification 1000x – a; X-ray analysis the recast layer- b

Figure 8 shows the results of an analysis of the chemical composition of the superficial layer. The results shows that the recast layer with an increased tungsten content has been formed as a result remelting and mixing the electrode and core material.

Figure 9a shows an SEM photograph of the metallographic structure of the superficial layer of samples made of the 0.45C steel after being BEDMMed with a tungsten electrode. The superficial layer exhibits local zones of amorphous structure and a deeper-located heat-affected zone.

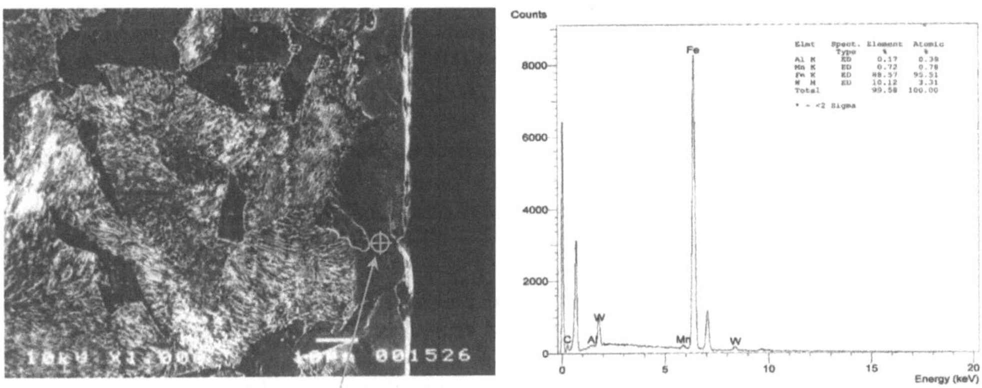

a) point of X-ray analysis b)

FIGURE 9. Photographs of the metallographic microstructures of the surface machined by tungsten brush, magnification 1000x - a; and X–ray diffraction pattern obtained from the recast layer - b

An analysis of the chemical composition (Figure 9b) of randomly chosen areas of the altered layer provides a quantity assessment of the changes that have taken place. The tungsten content of the examined samples ranges from 7% to 10 %.

5 CONCLUSION

Examination of brush electrodischarge machining shows that

- using brush electrodes makes the material/metal removal process cost-effective and allows the production of superior surface finishes,
- using appropriate metals or alloys to make the filaments of the brush makes it possible to introduce alloy components into the surface layer of the machined item, which consequently improves its quality,
- remelting (mixing) and diffusion are the basic mechanisms that allow the introduction of alloy components into machined objects,
- an X-ray microanalysis confirms the increased content of alloy components in the recast layer, with the level of chromium reaching 10% and nickel 5%.
- the results of the experiments show that it is possible to modify the core material with the elements the electrode is made of.

REFERENCES

1. Klocke, F., (1998), The Process Sequence in Tool – and Diemaking. Proceedings International Symposium for Electromachinning – ISEM-12, Aachen, Germany, 289÷298
2. Goto, A., et al., (1998), Formation of Hard Layer on Metallic Material by EDM. Proc. International Symposium for Electromachining ISEM-12, Aachen, 271 – 278,
3. Nowcki, B., Pierzynowski, R., Spadło, S. (1998), The Ectrodischarge - Based Surface Alloying Process Using Brush Electrode. Proceedings International Symposium for Electromachining – ISEM-12, Aachen, Germany, 289÷298.
4. Zołotych, B.,N., (1998), Ways of Solving the Problem of Controlling Composition and Properties of Machine Parts Surface Layer Formed under EDM Influence. Proc. International Symposium for Electromachining ISEM-12, Aachen, 253 ÷ 260.
5. Nowcki, B., Pierzynowski, R., Spadło, S. (1998), New Possibilities of Machining and Electrodischarge Alloying of Free-Form Surface. Proceedings of International Conference on Advances in Production Engineering. Part II. Warsaw, 310÷317.
6. Nowcki, B., Spadło, S., (1995), Brush Electro-Discharge Mechanical Machining BEDMM. Proceedings International Symposium for Electromachinning – ISEM-11, Lausanne, Switzerland, 191÷201.
7. Spadło, S., (1997), Complex Shape Surface Layer Finishing Process. Patent PL 172559 Departament Wydawnictw U P RP 10/97, Warszawa, 1÷4. (in Polish).
8. Kozak, J., Rajurkar, K., P., (2001), Abrasive electrical discharge grinding (AEDG). Proceedings of the Symposium Research on Clean Hybrid Micromachining (HMM) Processes Kraków, B5-1÷B-5-22.
9. Spadło, S., (1996), Formung der Oberschichteigenschaften durch erosiv-mechanische Bearbeitung der Materialien mit hilfe von Bürstenelektrode mit diskreter Struktur. Sammelheft des Projektes PL-1 des mitteleuropäischen Programms des Austausches der Universitätsstudien CEEPUS. Kielce, 157÷166.
10. Nowcki, B., Pierzynowski, R., Spadło, S. (2001), The Superficial Layer of Parts Machined by Brush Electro Discharge Mechanical Machining (BEDMM). Proceedings of II. International Conference on Advances in Production Engineering. Part II. Warsaw, 229-236.
11. Spadło, S., (2001), The Superficial Layer Forming by Anodic-Mechanical Machining with a Flexible Electrode. Proceedings of 7 th International Scientific Conference on Production Engineering. Computer Integrated Manufacturing and High Speed Machining - CIM'2001. Lumbarda, Croatia, 111-118.
12. Nowcki, B., Pierzynowski, R., Spadło, S. (2001), New Possibilities of Machining and Electrodischarge of Free-Form Surfaces. Journal of Materials Processing Technology Vol. 109, No. 3/2001, 371-376.
13. Nowcki, B., Pierzynowski, R., Spadło, S. (2000), Superficial layers of work-pieces subjected to electrodischarge – based alloying using brush electrode. Proceedings X. International Colloquium on Surfaces, Chemintz, 414÷421.

MEASUREMENT SYSTEM FOR mN FORCES

G. Manzoni[1,2], P. Miotti[2]

[1] DIEGM, University of Udine, Italy
[2] Mechatronic GmbH, Villach, Austria

KEYWORDS: balance, microthruster, microthrust.

ABSTRACT. Several recent experiments and demonstrations are carried out with very small satellites with mass in the range of 10 kg. One of the most important is the attitude control by means of microthruters which have to work in the rang of 100 - 1000 μN [1,2]. The aim of this work is the construction of a system for the measurement of the thrust produced by prototypes of micropropulsion system. A balance consisting of mainly two parts has been realised: a suspension system of the moving part; a position transducer. The suspension system is a flexural bearing and it behaves like a pendulum. In fact, two strips made of copper produce a restoring force, which counteracts the thrust generated by the propulsion system. The restoring force is due to two contributions: the elastic properties of the strips material and the gravity action on the moving mass. The position transducer is a laser system (LD 1605/Laser Micro-Epsilon) which allows to measure the displacement of the moving mass supported by the flexural bearing. The laser sensor is connected to a data acquisition board on a computer. The variation in the mean positions, obtained by a numerical filtering process, is proportional to the thrust.

1 INTRODUCTION

The current trend in space system design is to reduce mission costs. This can be done by reducing the mass of spacecraft and thus the launch costs. A great impulse to this direction is coming from the use of Microelectromechanical Systems (MEMS) which allow the miniaturization of spacecraft systems. With the reduction in scale for the next generation of satellite will be necessary to decrease the thrust level for the attitude control of the small spacecraft. For 10 kg satellite would be required a thrust level of about 25 μN – 1 mN to control accurately its position. Therefore it is necessary to measure very low thrust level for characterized the performances of new propulsion systems.

2 BALANCE CONFIGURATION

The balance has not only to move with a very small force, but it has also to carry the weight of the propulsion system which has a mass up to 0.45 kg. To reach this purposes a flexural bearing arrangement is used [3,4].

The propulsion system is fixed to the rigid beam and the thrust causes the structure to translate. Measuring the magnitude of the displacement "u" and knowing the stiffness of the whole system we can obtain the produced thrust. The beam not only translate, but also rotate about an axis going out from the plane of the drawing. This rotation can be neglected if the distance between the two strips is not too small.

Published in: E.Kuljanic (Ed.) *Advanced Manufacturing Systems and Technology*,
CISM Courses and Lectures No. 437, Springer Wien New York, 2002.

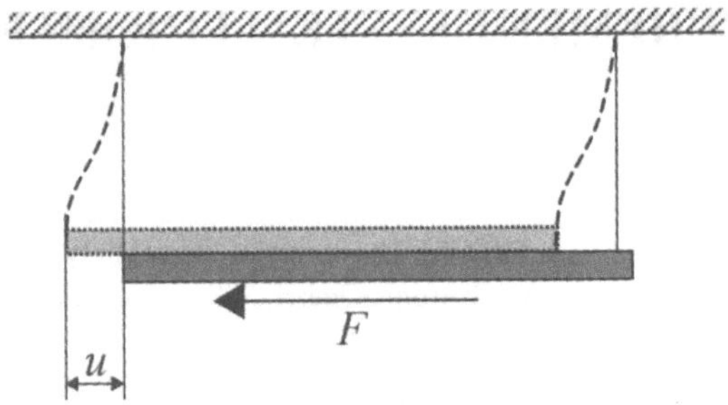

FIGURE 1. Flexural Bearing

For example, for 40 mm strips' distance the rotation is less than $5 \cdot 10^{-6}$ rad. Another feature of this device is the increasing of the stiffness with the weight carried by the strips. This is due to two effects: first is the direct action of the mass supported by the balance (the so called pendulum effect); second is the elastic restoring force of the two strips. The final stiffness is:

$$k = \frac{mg}{l}\left\{1 + \frac{2 \cdot [1 - \cosh(\lambda l)]}{2\cosh(\lambda l) - \lambda l \sinh(\lambda l) - 2}\right\} \tag{1}$$

$$\lambda = \sqrt{\frac{N_1}{EJ}} \cong \sqrt{\frac{N_2}{EJ}} \cong \sqrt{\frac{mg}{2EJ}} \tag{2}$$

The above relation is obtained considering a beam model with effects of the second order. In this way the effects of tension and bending of the beam are considered at the same time. In the figure 2 one can see the two components of the rigidity as function of the mass. It is drawn also the total stiffness. The plot refers to a couple of strips of this dimensions:
t = 30 μN thickness of strips
w = 15 mm width of strips
l = 100 mm length of strips
d = 40 mm distance between strips
strips made of copper

This type of balance was built and tested using a cold gas micro-thruster. The balance depicted in figure 3 was realized at Mechatronic Systemtechnik GmbH. To measure the displacement of the system from the initial position it is used a laser sensor (LD1605/Laser Micro-Epsilon) with a resolution of ±0.5 μm. The laser sensor allows to measure continuously the position of the moving mass without disturbing the measure itself. The laser sensor is connected to a personal computer by means of a data acquisition board (NI6070E National Instrument) and the final accuracy was ±1μm. This corresponds to a thrust error of 20 μN. This limit can be reduced using a position transducer with a better resolution, for example by means of an optic interferometer or a capacitor sensor.

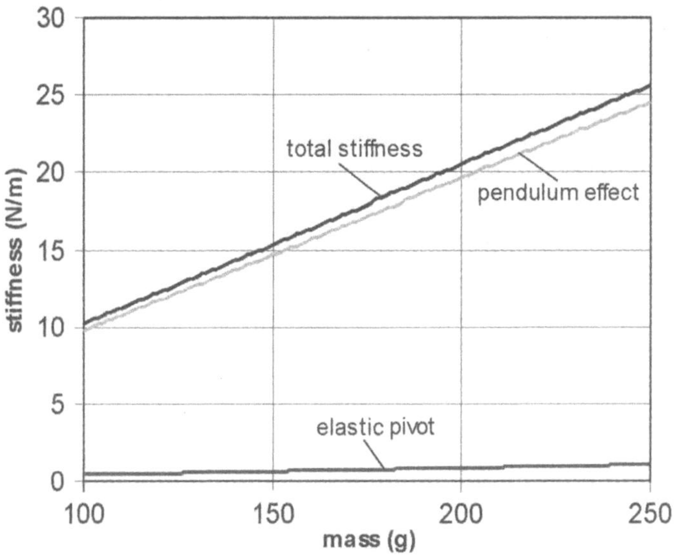

FIGURE 2. Stiffness as function of the mass

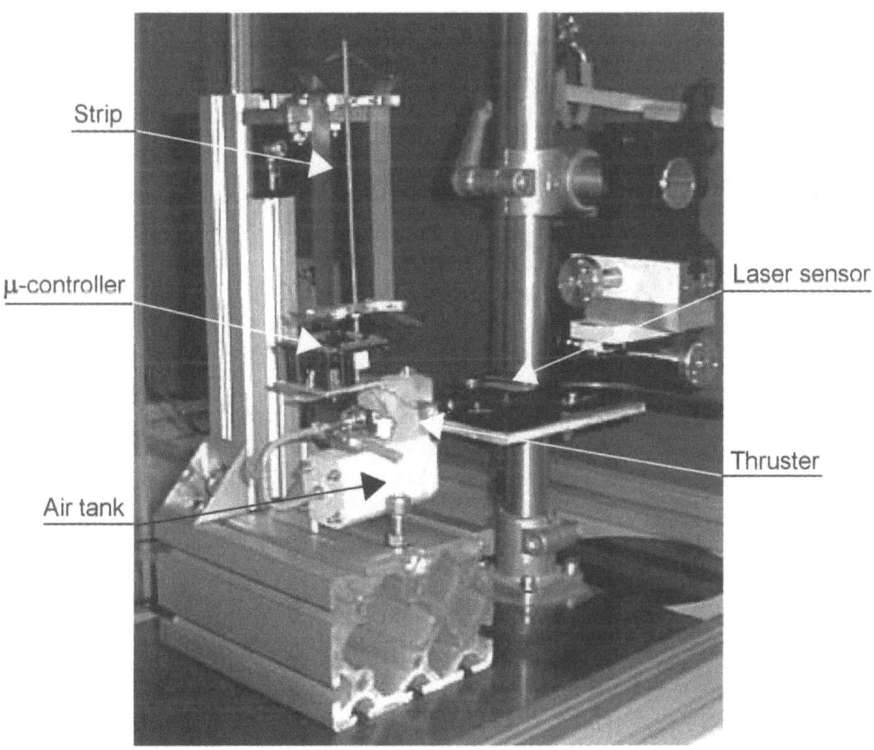

FIGURE 3. Balance

The strips are kept by the action of friction. In fact the strips are held in the middle of two blocks which are tightly fixed with screws. The friction generated between the blocks and the strips is enough to bear the system. In this way it is possible to reproduce a clamp constraint.

The moving mass supported by the strips is completely isolated from the surrounding. It comprises a gas tank and a micro-valve for feeding the micro-thruster with compressed gas; a pressure sensor also gives information of the gas before the thruster. The micro-valve and the pressure sensor are controlled by means of a micro-controller. The only connections with the external environment are the strips which allow the electrical supply (+15 V) to the system.

The signal obtained from the laser transducer is analyzed by a personal computer which permits to filter the data. In fact, the moving mass swings around a mean position with a sinusoidal behavior. In order to get only the mean position of the system eliminating the oscillations it is used a numerical filter which simply calculates the average value of the signal in one period.

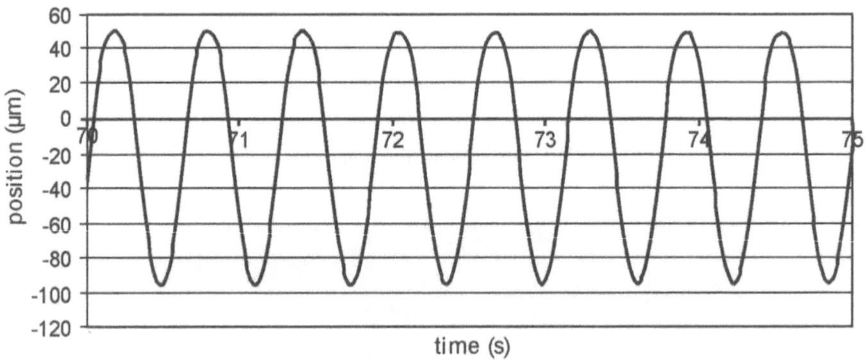

FIGURE 4. Oscillation

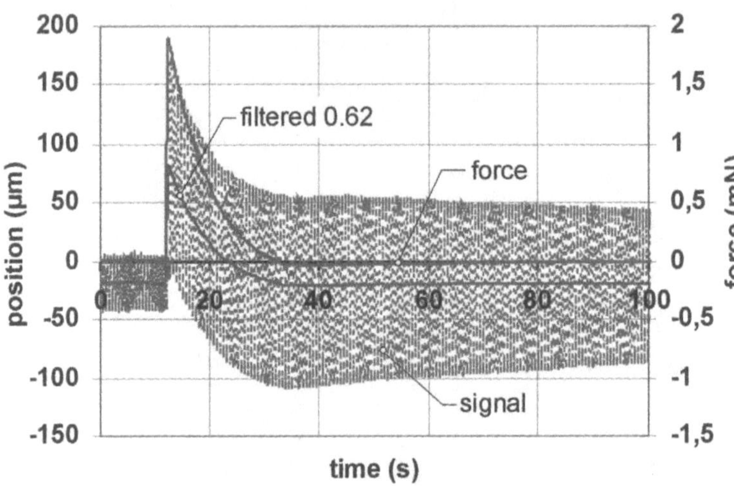

FIGURE 5. Thrust test

Some ways were studied for calibrating the system and to measure the real stiffness of the structure. The more precise is based on the oscillating period of the moving mass as described in the formula below:

$$\omega = \sqrt{\frac{k}{m}} = \frac{2\pi}{T} \tag{3}$$

From the period of oscillation T and the mass supported by the strips m, one can calculate the corresponding stiffness k.

Knowing the mean position and the total stiffness of the device it is easy to determine the thrust produced by the system. The picture 5 shows the produced thrust. Due to the low natural frequency of the device the rising part of the thrust is affected by greater errors than the other values. Thus it has not to be considered in the measuring range.

Not only it is measured the thrust, but also the impulse. This was simply obtained by a numerical integration of the laser sensor signal. In fact the impulse is:

$$I = m[v(t) - v(0)] + k\int_0^t x(t)dt \tag{4}$$

where $x(t)$ is the position of the moving mass at the instant t. The velocity $v(t)$ is also computed by a numerical derivation of the position $x(t)$. The picture below depicts the impulse measured with this device.

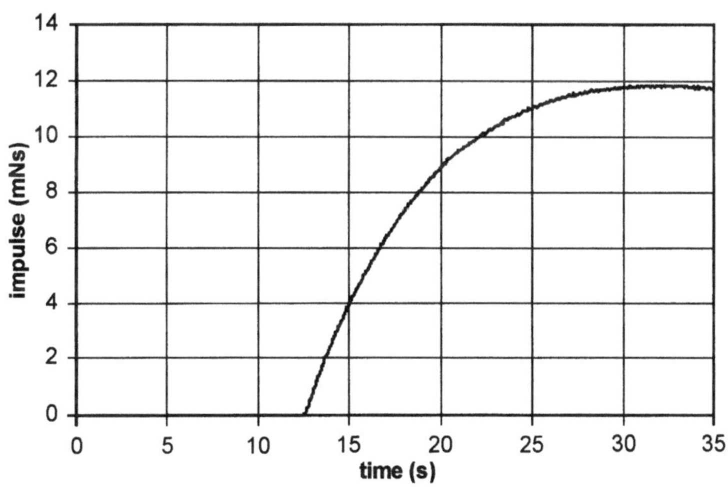

FIGURE 6. Impulse as function of time

3 COMPARISON

The measured thrust values can be compared with the theoretical ones based on the frictionless assumption of the gas expanding in the nozzle [5].

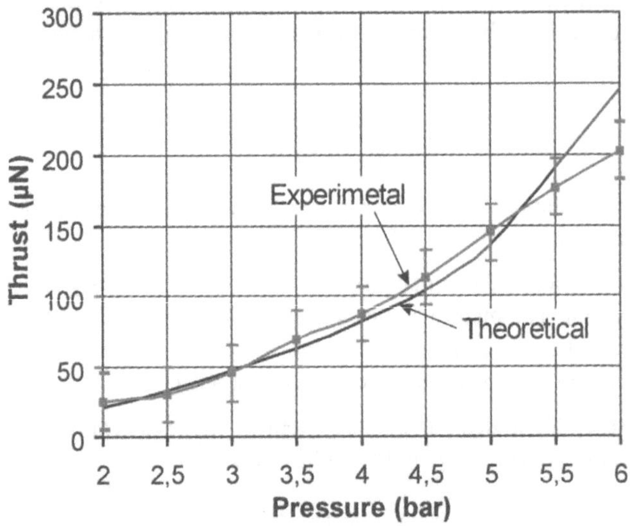

FIGURE 7. Thrust comparison

4 CONCLUSION

The resolution of the balance can be greatly increased by means of a position transducer with a higher precision. The most accurate sensors are those based on capacitive effects. From the PI catalogue it is possible to reach a resolution of 0.01 nm which allows measurement of forces with an accuracy of 0.01 μN. In fact assuming, as stated at the beginning of the chapter, a stiffness of 70 N/m a force resolution of 0.01 μN corresponds to a position resolution of 0.14 nm. Using a double set of capacitors in a push-pull configuration the electrostatic force on the moving mass is zero. This leads to a non intrusive system which does not modify the force measure.

REFERENCES

1. Janson, S., Helvajan, H., Breuer, K., (1997), Micropropulsion Systems for Aircraft and Spacecraft, 657-692
2. Manzoni, G., (2000), Design of a highly integrated micropropulsion system for microsatellites attitude control, AIAA-2000-3476
3. Eastman, F.S., (1937), The design of flexure pivots, Journal of the aeronautical sciences, vol.5
4. Jones, R.V., (1956), Some parasitic deflexions in parallel spring movements, Journal of Scientific Instruments, vol.33
5. Gordon, C., Aerothermodynamics of Gas Turbine and Rocket Propulsion, AIAA Education Series, ISBN 56347 241 4

APPLICATION OF STATISTICAL PROCESS CONTROL TO CONTINUOUS PROCESSES

G. Celano[1], A. Costa[2], S. Fichera[2]

[1] Dipartimento di Tecnologia e Produzione Meccanica, University of Palermo, Italy
[2] Dipartimento di Ingegneria Industriale e Meccanica, University of Catania, Italy

KEYWORDS Control charts, Autocorrelation, Average Run Length.

ABSTRACT. Control charts represent an efficient and easy tool to assure the state of statistical quality control in a manufacturing process. These tools are also implemented in continuous processes, where the critical parameters are often monitored by on line sensors measuring data with short time intervals. In this paper a continuous process is monitored by using control charts and its dynamic is modeled through linear time series that allow the effects of the autocorrelation to be eliminated. In this way, the control charts can operate on residuals that result identically and independently distributed. A statistical analysis on EWMA, CUSUM and control charts for individual measurements has been carried out to select the most performing tool for process monitoring.

1 INTRODUCTION

Statistical quality control tools are widespread in industry to achieve the control of critical process parameters. Nowadays, several industries implement these instruments to prevent sudden process failures, which can affect the quality of the output of their processes. In particular, control charts are a simply and powerful tool to plot the behaviour over time of some critical process parameters. Among them, the Shewhart control charts, the cumulative sum control charts (CUSUM) and the exponentially moving average (EWMA) control charts are easy to be implemented and interpreted. One of the basic assumptions for a correct application of these charts is that the collected data are independently and identically distributed (IID), [1]. But this hypothesis cannot be always verified. Unfortunately, for continuous processes, i.e. chemical processes and all manufacturing processes driven by inertial elements, the observations on the process are correlated over time. Therefore, the application of conventional control charts on these data can lead to misleading results in the form of excessive false alarms due to the data correlation. Alwan and Roberts [2] have developed an extension of traditional process control procedures that allows the effect of the systematic behaviour of the process to be separated from the variability depending on the special causes. The time series modelling of the collected data is based on the ARIMA models. Once a correct model has been implemented, it is possible to determine the residuals of the measurements, evaluated as difference between the real data and the fitted model. Then, a common cause chart on the fitted data is used to give a view of the level of the process and of its evolution through time. The state of statistical control of the process is evaluated by plotting the residuals on a Shewhart control chart or eventually on a chart for individual measurements. EWMA and CUSUM control charts applied on the residuals have also been proposed by [3,4,5,6] for the monitoring of a manufacturing process in a computer integrated environment.

Published in: E.Kuljanic (Ed.) *Advanced Manufacturing Systems and Technology*,
CISM Courses and Lectures No. 437, Springer Wien New York, 2002.

In the proposed paper a critical parameter representing the sulphur quantity included in the gas oil produced by a hydrofining chemical process is taken into account to evaluate the state of statistical control of the process. Due to the data autocorrelation, an approach based on time series analysis is proposed, that considers the application of control charts to the residuals resulting by the autocorrelation removal in the collected data. An autoregressive integrated model has been proven to be a good estimator for the controlled parameter; then, the state of statistical control of the residuals has been evaluated by an EWMA control chart. Moreover, a sensitivity analysis of a set of control charts for individuals measurements to small shifts in the mean of residuals has been carried out. The results show how EWMA and CUSUM control charts have equivalent statistical performances in revealing the occurrence of an out of control condition in the investigated process.

2 WORKING WITH AUTOCORRELATED DATA

A basic assumption for a correct implementation of control charts is represented by the independency of data and their distribution. When the process is in statistical control, an independent data having mean μ and variance σ^2 may be represented at time interval t as:

$$z_t = \mu + \varepsilon_t \tag{1}$$

where ε_t is independently and identically distributed with mean zero and variance σ^2. This is the so-called Shewhart model of the process, [1]. Unfortunately, this assumption often cannot hold because of the intrinsic correlation of data. In chemical processes and when the measurements are taken at time interval small with respect to the process inertia this problem becomes critical. In this condition it is necessary to implement a stochastic model based on time series that allows the autocorrelation to be eliminated from data. Time series modeling is very useful when a view of the level of the process and of its evolution through time is needed. The correlation between data can be evaluated qualitatively by means of the scatter plots or quantitatively through the autocorrelation function ACF, which is usually estimated through the sample autocorrelation function SACF:

$$r_k = \frac{\sum_{t=1}^{n-k}(x_t - \bar{x})(x_{t+k} - \bar{x})}{\sum_{t=1}^{n}(x_t - \bar{x})^2} \tag{2}$$

where k represents the time lag and n the number of observations.
The shape of the SACF and the values of the autocorrelation coefficients r_k allows the tentative time series model of the data to be formulated by the decision maker. The autocorrelated stochastic models can be modelled by considering that the time dependency among the values is generated by a series of independent "shocks" a_t [8]. These shocks are normally and independently distributed with mean equal to zero and variance σ^2_a. The sequence of the random variables a_t, a_{t-1}, a_{t-2},...are called white noise process. The white noise is transformed into the process variable x_t through a linear filter:

$$x_t = \mu + a_t + \Psi_1 a_{t-1} + ... + \Psi_p a_{t-p} = \mu + \Psi(B)a_t \tag{3}$$

where B is the backward shift operator which is defined as $Bx_t = x_{t-1}$.

Usually the $\widetilde{x}_t = x_t - \mu$ deviation from μ is considered; as a consequence, the equation (3) assumes the following expression:

$$\widetilde{x}_t = \Psi(B)a_t \tag{4}$$

The class of ARIMA (Autoregressive Integrated Moving Average) models is widely applied to model the correlated data. The general expressions for an ARIMA(p,d,q) model are:

$$\phi(B)(1-B)^d \widetilde{x}_t = \theta(B)a_t \tag{5}$$

$$\phi(B)w_t = \theta(B)a_t \tag{6}$$

where $w_t = (1-B)^d \widetilde{x}_t = \nabla^d \widetilde{x}_t$. These models are used to represent all the linear stochastic processes, both stationary and nonstationary. The equation (6) can be formulated as:

$$w_t = \phi_1 w_{t-1} + \ldots\ldots + \phi_p w_{t-p} + a_t - \theta_1 a_{t-1} - \ldots\ldots \theta_q a_{t-q} \tag{7}$$

The identification of the model to be tentatively entertained is carried out by considering the shape of the SACF and the values of the autocorrelation coefficients r_k. The parameters of the model that maximize a likelihood function are chosen as model parameters. Then, a diagnostic checking is developed to evaluate the effectiveness of the selected model. In [8] the portmanteau test is proposed to check the selected model through an analysis of the residuals. If the fitted ARIMA(p,d,q) model is appropriate, the following expression must be verified:

$$Q = n\sum_{k=1}^{K} r_k^2(\varepsilon_r) = \chi^2(K-p-q) < \chi^2_{df,cl} \tag{8}$$

where df represents the degrees of freedom K-p-q and cl the confidence level of the test. Once the model has been demonstrated to give a correct fit of the time series, the control charts can be applied on the residuals ε_r that represent a random variable identically and independently distributed. The charts to be applied on the process can be selected by considering a statistical criterion. In particular, an analysis on the ARL values of the charts with respect to different entity of the process shifts can be performed. The entity of the shift is expressed as a multiple δ of the residual standard deviation. The out-of-control ARL of a control chart represents the expected number of samples to be taken between the occurrence of a special cause and the out of control signal on the chart. The in-control ARL represents the expected number of samples to be taken between two successive false alarms. The statistical analysis can be performed by considering the same values of in-control ARL and comparing the out-of-control ARLs of the studied charts with respect to a wide range of possible shifts in the mean of process parameter.

3 EXPERIMENTAL RESULTS

A chemical process of gas oil desulphurisation is investigated. The gas oil coming out from the stripping column is characterised by a too high content of sulphur for an its correct functionality. The desulphurisation is carried out through an Hydrofiner where the gas oil reacts with hydrogen in presence of catalysts such as cobalt and molybdenum sulphides. The temperature of the reaction is fixed within the interval 320-380°C. The residual sulphur contained in the gas oil coming out from the Hydrofiner strongly depends on the reactor temperature, that must be kept within the mentioned interval in order to avoid an incorrect functionality of the desulphurisation process or a premature deactivation of the catalyst. The monitoring system of the desulphurisation process within the Hydrofiner consists of a sensor that detects the content of sulphur within the gas oil, expressed in ppm. A Courier 10-S on line sensor based on fluorescent x-rays has been used. A continue cathod beam directed within the gas oil energizes the sulphur atoms and the

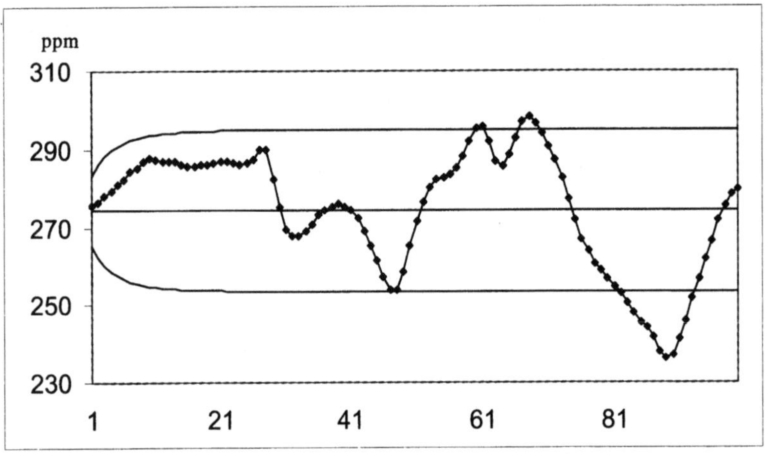

FIGURE 1. Preliminary EWMA chart on the residuals

corresponding fluorescence is measured and evaluated in terms of ppm. The sampling frequency is equal to one measure every hour. The specification limits USL and LSL of the process are respectively set equal to 320 ppm and 250 ppm. An EWMA control chart for individual measurements has been applied on 100 data collected through the on line sensor in order to investigate the state of statistical control of the process and to define a set of statistical control limits. The obtained EWMA chart is reported in Figure 1. The parameters of the charts are the same as reported in [7]. The EWMA chart signals an high number of out-of-control points. Such a situation is typical when dealing with strongly autocorrelated data: in fact, when the data independence assumption does not hold an high number of false alarms is expected.

In order to verify the possible data autocorrelation, an analysis of the SACF of the data has been carried out: the shape of the SACF and the value of the sample autocorrelation coefficients revealed an autocorrelated structure. Therefore, it has been necessary to use a time series linear model belonging to ARIMA class that allows the autocorrelation of data to be modeled and then the control chart to be applied on the uncorrelated residuals. An ARIMA(2,1,0) has been determined through the iterative approach of identification, estimation and diagnostic checking described in [8]. The initial values of the autoregressive parameters have been determined through the Yule-Walker equations [8] in particular $\hat{\phi}_1 = 0.72883 \quad \hat{\phi}_2 = -0.39612$. The eventual lack of fit of the model has also been evaluated by means of a statistical approach based on the portmanteau test. Table 1 shows the results obtained with the portmanteau test for group of $r_k(\varepsilon_r)$ having sizes K=10,15,18,22,25. A confidence level of 95% for the portmanteau test has been imposed.

TABLE 1. Portmanteau test for the ARIMA (2,1,0) model

K	**10**	**15**	**18**	**22**	**25**
Df=K-p-q: degrees of freedom	8	13	16	20	23
$Q_0=\chi^2(K\text{-}p\text{-}q)$	5.4	7.4	8.6	14.5	16.4
$\chi^2_{df,0.05}$	15.5	22.4	28.9	33.9	37.7

Because all of the observed values of Q are less than the respective critical values corresponding to a confidence level of 95%, there is no evidence to reject the fitted model. Furthermore, an analysis on normally distribution of the residuals has been carried out. Once the residuals have been proven to be normally and independently distributed, an EWMA chart has been applied on their values:

$$\begin{aligned} UCL &= \mu_r + 2.7\sigma_r\sqrt{\frac{\lambda}{2-\lambda}\left[1-(1-\lambda)^{2i}\right]} \\ CL &= \mu_r \cong 0 \\ LCL &= \mu_r - 2.7\sigma_r\sqrt{\frac{\lambda}{2-\lambda}\left[1-(1-\lambda)^{2i}\right]} \end{aligned} \tag{9}$$

where μ_r and σ_r represent respectively the mean and the standard deviation of the residuals.
The preliminary EWMA chart on residuals is reported in Figure 2. The residuals show a state of statistical control of the process. This means that the out-of-control signals reported on the data EWMA chart reported in Figure 1 depend on the autocorrelated structure of the measurements. Therefore, these out-of-control points can be considered as false alarms. As a consequence, the control limits of the EWMA chart are assumed as statistical control limits for the monitoring of the sulphur content in the gas oil.
An EWMA chart has been successfully applied on the on line process residuals thus allowing the occurrence of out of control conditions to be easily detected and eliminated before the misspecification of the USL or LSL, see Figure 3. The possibility of implementing other types of control charts to monitor the process has been taken into account; the charts have been evaluated by considering their statistical performances through an out-of-control ARLs comparison based on a range of shifts in the mean of the residuals. An IM control chart for individual measurements based on traditional X control scheme and a CUSUM control chart have been taken into account and compared with the adopted EWMA. The ARL evaluation for the IM control chart has been performed by considering the reciprocal of the power (1-β) of the chart, where β represents the

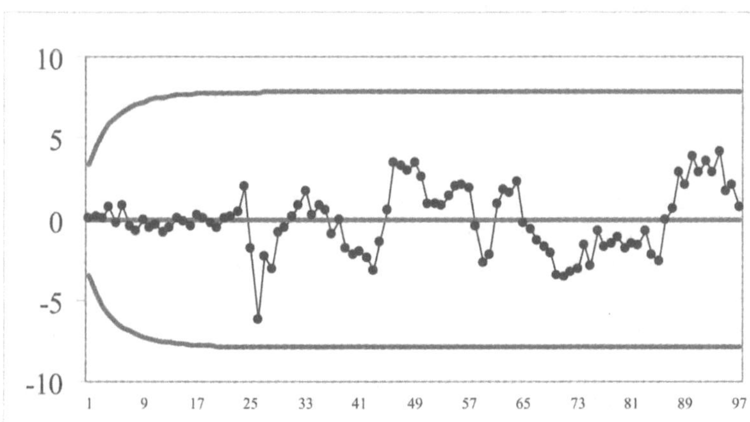

FIGURE 2. Preliminary EWMA chart on the residuals

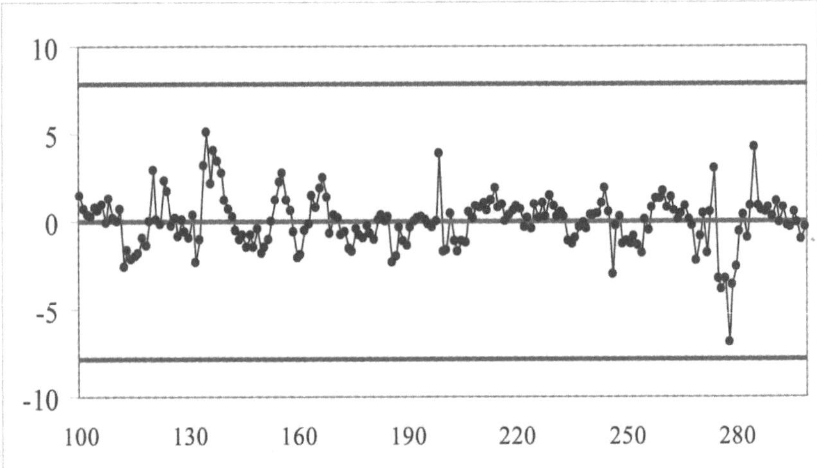

FIGURE 3. On line EWMA chart on the residuals

probability of not detecting an out-of-control condition when it effectively occurs. The ARLs for both CUSUM and EWMA charts are evaluated through a Markov chain approach [9,10]. In this approach the control interval of the charts is divided into equal subintervals. Therefore, the non absorbing states of the Markov chain are represented by a point falling within one of the subintervals, while the absorption condition of the chain coincides with a point falling outside the control limits. Once the transient states have been determined, the state transition matrix of the chain can be evaluated by considering the probabilities of each point falling within one of the subintervals. The ARL will be equal to the mean time to absorption corresponding to the Markov chain. In particular, when in-control ARL is evaluated, the in-control transient probabilities are considered; on the other hand, when the process is out-of-control, the determination of out-of-control ARL is performed by assuming the out-of-control transient probabilities. In order to compare the three different control charts, the same values of in-control ARLs must be assumed for each one of the selected scheme. This results in an equivalent performance of the charts with respect to the false alarms probability. In particular in-control ARL=370 and 500 and a range of δ values of the shift in the mean of the residuals between 0.2 and 2 have been selected, see Tables 2 and 3. The obtained results show how the IM control chart for individual measurements has poor statistical performances with respect the other two control charts. The out-of-control ARLs are very high especially when δ is small: this result is due to the intrinsic structure of the IM control charts that do not consider the history of the data. On the other hand, the EWMA and CUSUM charts are very sensitive to small and medium shifts. In particular, EWMA is the most performing chart when the shifts in the mean are $\delta \leq 1$ residual standard deviations. For example, when $\delta=0.2$ the reduction of ARL obtained with EWMA with respect to CUSUM is equal to 27%. When medium shifts are considered, i.e. when $1<\delta<2$ the CUSUM shows the better statistical performances with a maximum ARL reduction with respect to EWMA chart equal to 8%. Finally, it must be underlined how the value of the in-control ARL do not influence substantially the performance differences between EWMA and CUSUM charts.

TABLE 2. ARL comparison of IM, CUSUM and EWMA for ARL_0=370

Shift	IM L=3 n=1	CUSUM h=4.77 k=0.5	EWMA L=2.70 λ=0.1
0.2	308.41	169.57	123.22
0.4	200.07	54.42	41.20
0.6	119.66	24.59	20.89
0.8	71.55	14.39	13.38
1.0	43.89	9.90	9.73
1.2	27.82	7.50	7.65
1.4	18.25	6.02	6.60
1.6	12.38	5.04	5.38
1.8	8.69	4.34	4.70
2.0	6.30	3.81	4.18

TABLE 3. ARL comparison of IM, CUSUM and EWMA for ARL_0=500

Shift	IM L=3.09 n=1	CUSUM h=5.06 k=0.5	EWMA L=2.814 λ=0.1
0.2	411.99	205.92	150.17
0.4	262.18	61.35	46.70
0.6	153.86	26.65	22.81
0.8	90.41	15.34	14.33
1.0	54.55	10.49	10.33
1.2	34.03	7.92	8.06
1.4	21.97	6.36	6.62
1.6	14.68	5.33	5.63
1.8	10.15	4.58	4.91
2.0	7.25	4.02	4.36

Due to the reported statistical analysis, the use of an EWMA chart is suggested to monitor the process because the expected entity of the shift in the mean of the residuals is less than 1. When there is evidence of occurrence of special causes that give rise to larger shifts than 1 in the residuals mean, the CUSUM chart represents the best choice to monitor the process. When a CUSUM control chart is assumed to be the on line monitoring tool of the process residuals, its tabular form is suggested [7]; therefore, the two one sided CUSUMS are:

$$C_i^+ = \max[0; \varepsilon_r - (\mu_r + K) + C_{i-1}^+] \qquad (10)$$
$$C_i^+ = \max[0; (\mu_r - K) - \varepsilon_r + C_{i-1}^+]$$

The process is assumed to be out-of-control when the one of the CUSUMS falls outside the cotrol limits H or –H. The values of K and H are fixed as multiples of the standard deviation of the residuals σ_r. In general, K=0.5·σ_r and H=h· σ_r is selected to give a specified ARL performance, see Tables 2 and 3.

4 CONCLUSIONS

In this paper EWMA control charts have been applied to a continuous process. The sulphur content in the gas oil coming out from a Hydrofiner reactor has been considered as critical parameter to be maintained under control. In fact, the amount of sulphur is limited by the pollution caused by sulphur dioxide and trioxide emissions. A sensor based on fluorescent x-rays has been used to reveal the presence of sulphur. Due to the high level of data autocorrelation a time series approach was needed to fit the data and eliminate their autocorrelation. An ARIMA(2,1,0) has been proven to be a good estimator of the data and the corresponding residuals have been evaluated. Once the residuals have been demonstrated to be independently and normally distributed, an EWMA control chart has been implemented on their values to evaluate the state of statistical control of the process. The results show how this chart has been useful to define a set of statistical control limits for the process. Furthermore, the developed residuals EWMA charts can be easily implemented as on line tools to detect the occurrence of an out-of-control condition. A sensitivity analysis on the statistical performances of three different control charts for individuals has been carried out to select the best control chart for individual measurements to be used on the investigated process. The three compared charts are EWMA, CUSUM and control charts for individual measurements IM. The results show how the EWMA chart must be selected when small shifts in the mean of the residuals are expected; on the other hand, the use of CUSUM chart is preferable for medium shifts. Future work will be devoted to the use of run rules or adaptive schemes on the charts in order to improve their statistical properties.

REFERENCES

1. Montgomery, D.C., (1996), Introduction to Statistical Quality Control, 3rd edn. John Wiley & Sons, New York
2. Alwan, L.C., Roberts, H.V., (1989), Time series modelling for statistical process control, in: Statistical Process Control in Automated Manufacturing. Keats, J.B., and Hubele, N.F. (Ed.), Marcel Dekker Inc., New York, 45-65.
3. Yashchin, E., (1993), Performance of CUSUM control schemes for serially correlated observations. Technometrics, Vol 35, 37-52.
4. Timmer, D.H., Pignatiello, J.J., Longnecker, M.T., (1998), The development and evaluation of CUSUM-based control charts for an AR(1) process. IIE Transactions, Vol 30, 525-534.
5. Timmer, D.H., Pignatiello, J.J., Longnecker, M.T., (2000), Appling an AR(1) CUSUM control chart to data from a chemical process. Quality Engineering, Vol 13/2, 107-114.
6. English, J.R., Lee, S-C, Martin, T.W., Tilmon C., (2000), Detecting changes in autoregressive processes with $\overline{X}$ and EWMA charts. IIE Transactions., Vol 32, 1103-1113.
7. Montgomery, D.C., Friedman, D.J., (1989), Statistical process control in a computer integrated manufacturing environment, in: Statistical Process Control in Automated Manufacturing. Keats, J.B., and Hubele, N.F. (Ed.), Marcel Dekker Inc., New York, 67-87.
8. Box, G.E.P., Jenkins, G., (1976), Time Series Analysis: Forecast and Control, Holden-Day, Oakland (CA) (USA).
9. Lucas, J.M., Saccucci, M.S., (1990), Exponentially Weighted Moving Average Control Schemes: Properties and Enhancements. Technometrics, Vol 32, 1-12.
10. Gan, F.F., (1992), The Run Length Distribution of a Cumulative Sum Control Char. Journal of Quality Technology, Vol 35/3, 205-215.

DESIGN AND IMPLEMENTATION OF A CARTESIAN ROBOT

A. Gasparetto[1], G. Rosati[2]

[1] Department of Electrical, Management and Mechanical Engineering, University of Udine, Italy
[2] Department of Innovation in Mechanics and Management, University of Padova, Italy

KEYWORDS: Robotics, Robot Design, Cartesian Robot.

ABSTRACT. The paper presents the design and the implementation of a cartesian robot, carried out at the Department of Electrical, Management and Mechanical Engineering of the University of Udine, in cooperation with the Department of Innovation in Mechanics and Management of the University of Padova. The cartesian robot built is of the gantry type, i.e. it is intended to reach the object to manipulate from above. It will be described in the paper how the design requirements were set, with respect to the mechanical and the electronic part, as well as to the control system. The most important requirements to take into account were those regarding: accuracy, dimensions, velocity, weight of the object to be manipulated and the overall cost of the system. The robot will then be equipped with suitable tools, so as to be used for automation purposes in different fields (industrial, biomedical, etc.).

1 INTRODUCTION

The development of a manipulation system for different application purposes has been carried out at the Department of Electrical, Management and Mechanical Engineering of the University of Udine, with the cooperation of the Department of Innovation in Mechanics and Management of the University of Padova. Such a manipulation system was intended for the most general application scope, ranging from the generic "industrial" manipulation to more specific application such as medical tasks.

After an accurate analysis, it was chosen to build a cartesian robot with three degrees of freedom, made of three linear axes perpendicular to each other. This was due to the fact that the intended structure had to ensure the implementation of the first three degrees of freedom of the overall robot (namely, the positioning of the end-effector) in the simplest possible way. Moreover, the overall robot structure was to be of a modular type, so as to be able to add, remove and substitute the robot components without affecting the whole structure. For instance, the wrist with the last three degrees of freedom to be added to the robot could be a dedicated device for surgical operation (such as insertion of peduncular screws into human vertebrae), or a more generic tool for pick-and-place operation in an industrial environment. In any case, the supporting structure had to be simple, cost-effective, and easy to control. This led to the choice of a cartesian robot.

The paper is organized as follows. In Section 2, the design specifications for the robot are reported with many details. In Section 3, the implementation of the structure is described and discussed, and the effective technical data are reported as well. Moreover, the dedicated control system, that has also been realized in the laboratory, is described.

Published in: E.Kuljanic (Ed.) *Advanced Manufacturing Systems and Technology*,
CISM Courses and Lectures No. 437, Springer Wien New York, 2002.

2 SPECIFICATIONS AND DESIGN

Robot specifications are shown in Table 1.

TABLE 1. Robot specifications

Robot Type	Cartesian
Payload	10 kg
X-axis run	700 mm
Y-axis run	600 mm
Z-axis run	500 mm
Repeatability	0.5 mm
Max. Speed	1 m/s
Max. Acceleration	3 m/s^2

The payload quoted in Table 1 includes both the tool mass and the mass of the object to be manipulated. Moreover, all the requirements must be met within a limited budget.

The robot overall scheme is shown in Figure 1. The main structure is made of ten aluminum alloy bars, while the three Cartesian axes are provided by means of pre-assembled linear modules, that are a good compromise between costs and accuracy.

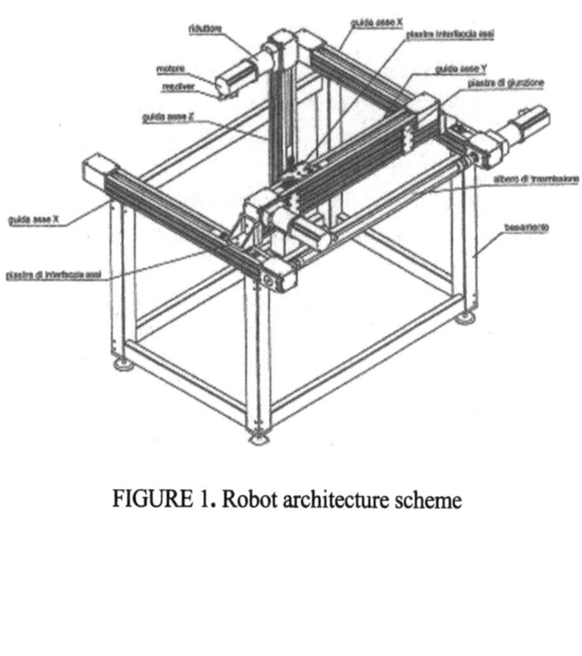

FIGURE 1. Robot architecture scheme

As it shows, the X-axis is made of two parallel linear modules, while the Y-axis is made of one aluminum bar plus a linear module mounted on the top of it. This way, we improved robot stiffness and reduced the loads acting on each module. The X-axis motor drives both X-axis linear modules by means of a rigid shaft permanent transmission, so as to reduce costs and avoid the driving control problems that would have risen if we had used one servomotor for each module. Moreover, the Z-axis linear module flange is mounted directly on the Y-axis flange, in order to let the robot tool approach the workspace from above, as it was set in the robot requirements.

Robot overall dimensions are 1600x1420x2100 mm^3, while the workspace measures 700x600x500 mm^3. The loss of workspace is the main counter-item of the choice of a Cartesian robot structure, whose major benefits are the increased robot stiffness and accuracy [1].

The structure and motors dimensioning took into account both static and dynamic loads [2]. No deformation analysis has been performed, as we chose pre-assembled linear bars and modules whose deformations are of a cent of a millimeter for static loads an order of magnitude greater than the robot operative loads. Moreover, in order to meet the repeatability requirement quoted in Table 1, we chose reduced play gear reducers and precision belt linear modules, gaining an overall positioning repeatability of nearly one tenth of a millimeter for each axis.

3 IMPLEMENTATION AND CONTROL

The linear modules chosen are the AS250D by "Automazione Macchine" (Amaro, Italy). Their main section measures 90mm by 110mm, while their overall length is of 667mm plus axis run. Linearity is ensured by means of a prismatic recirculating-balls linear guideway, while motion transmission is made through a toothed precision belt. Linear modules main specifications are shown in Table 2.

TABLE 2. Linear module "Automazione Macchine AS250D" specifications

Max. Speed	2.5 m/s
Max. Acceleration	10 m/s^2
Max. Torque	380 Nm
Repeatability	0.1 mm
Weight (run=0mm)	16 kg

The motors are three Siemens 1FK6042 AC Brushless servomotors, whose rated speed and torque are 3000rpm and 2.6Nm respectively. Each motor is provided with a built-in resolver and a Siemens Simodrive 611 driver, while the Z-axis motor has an automatic brake to compensate for gravity in static conditions. The motors are linked to the belt sprockets by means of three Tecnoingranaggi BGT1050 reduced play gear reducers, whose reduction ratio and efficiency are 1:6 and 95% respectively.

An overview of the robot is shown in Figure 2, while the picture on the right-hand side of Figure 3 depicts the Y-axis gearbox and motor. On the left hand-side of Figure 3, the electronic equipment is shown as well.

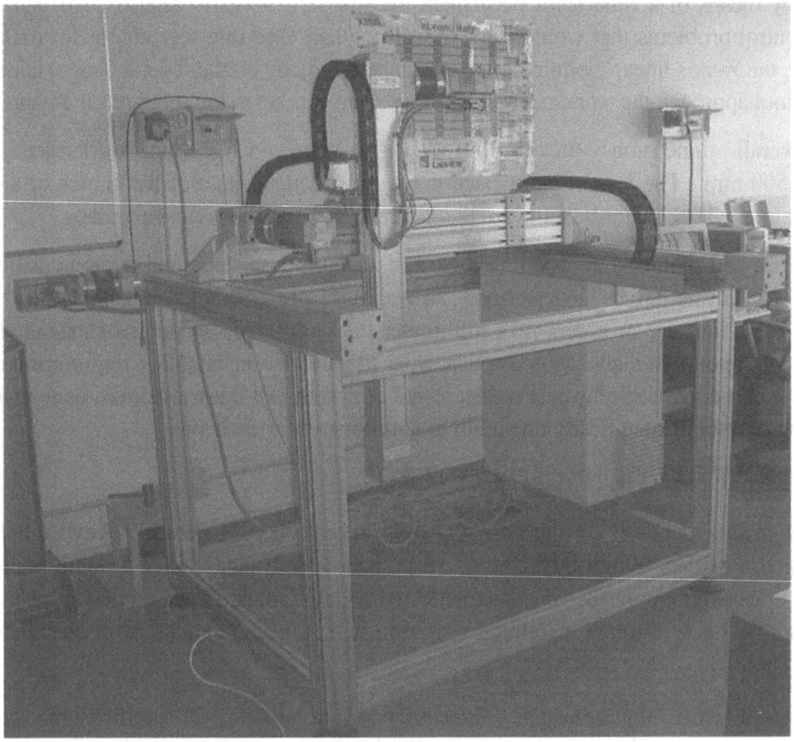

FIGURE 2. Robot overall view

FIGURE 3. Electrical equipment (left) and Y-axis gearbox and motor close-up (right)

Each axis of the robot is equipped with three inductive proximity sensors, that provide the zero position and the axis overrun signals. These sensors, together with the motor resolvers and drivers are connected to two National Instruments 7344 multiple-axis motion control PC boards, by means of two UMI 7764 interfaces. The PID control parameters for each axis were tuned using the Max software, provided together with the NI Motion Controllers.

A complete motion control user interface has been developed so far [3], and is currently undergoing an improvement process. The software has been implemented using Labview 6.0, so as to give the end-user the ability of easily driving the robot through simple 3D motion paths. The user can also modify all the control and trajectory parameters in a very simple way. Moreover, the software developed can be easily modified in order to meet the requirements of a specific application.

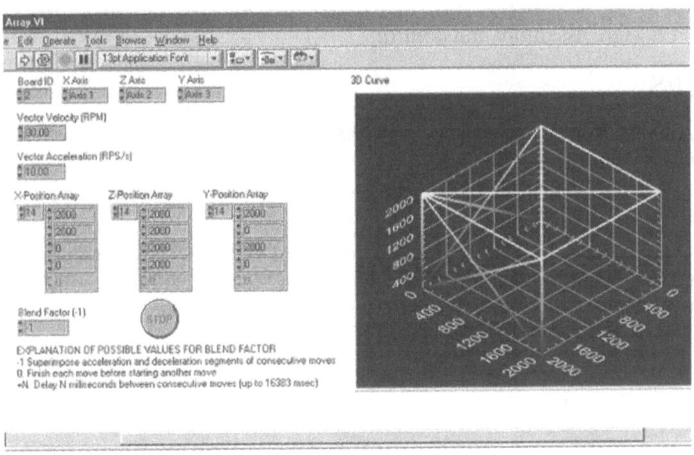

FIGURE 4. Control software main window

As far as it concerns further developments, a three degree of freedom tool will be designed and the robot, provided with a 6-axis force/torque sensor, will be tested in master-slave haptic teleoperation tasks. A wire-driven five degree of freedom haptic master is currently under development at the Department of Innovation in Mechanics and Management of University of Padova. The haptic master control will be interfaced to the Cartesian robot control in order to perform the teleoperation tests.

4 CONCLUSION

The development of a manipulation system for different application purposes has been described in this paper. This manipulation system was intended for the most general application scope, ranging from the generic "industrial" manipulation to more specific application such as medical tasks.

A cartesian robot with three degrees of freedom, made of three linear axes perpendicular to each other was built, so as to ensure the implementation of the first three degrees of freedom of the overall robot (namely, the positioning of the end-effector) in the simplest possible way. Moreover, the overall structure was built so as to implement a modular robot. In this way adding, removing and substituting the robot components without affecting the whole structure tuned out to be a relatively easy task.

Within the paper, the design specifications for the robot have been first reported with many details. Then, the implementation of the structure has been described and discussed, and the effective technical data have been reported as well. Finally, the dedicated control system, that has also been realized in the laboratory, has been described.

ACKNOWLEDGMENTS

We would like to thank the Italian Ministry of University and Scientific Research (MURST), that provided the funds within the two-year METAFORE Research Project. We would also like to acknowledge Mr. Roberto Camatta, Mr. Adriano Biason and Mr. Brecht Neirinck who helped us throughout robot design, assembly and control.

REFERENCES

1. Carlisle, B., (2000), Robot Mechanisms. *Proceedings of the 2000 IEEE International Conference on Robotics & Automation*, 701-708.
2. Camatta, R., Gasparetto, A., (1999), Progetto e realizzazione di un manipolatore cartesiano tridimensionale. University of Udine, Graduation Thesis.
3. Gasparetto, A., Neirinck, B., (2001), Motion control of a three-axis Cartesian robot. University of Udine, Department of Electrical, Management and Mechanical Engineering, Internal Report.

DESIGN OF A ROBOTIC WRIST FOR BIOMEDICAL APPLICATIONS

A. Gasparetto and A. Biason

Department of Electrical, Management and Mechanical Engineering, University of Udine, Italy

KEYWORDS: Robotic Wrist, Biomedical Applications, Design.

ABSTRACT. The paper presents the design of a robotic wrist for biomedical applications, carried out at the Department of Electrical, Management and Mechanical Engineering of the University of Udine. First, the intended use of the wrist, namely the insertion of peduncular screws into the vertebrae, together with the supporting structure (a cartesian robot) are described. Then, the design requirements for the wrist, especially with respect to the mechanical side, are set and discussed. The most important requirements that had to be taken into account were those regarding: accuracy, dimensions, velocity, and maximum payload. All the steps up to the final design of the wrist are described and discussed, with also the useful aid of a representation of the wrist in a CAD environment.

1 INTRODUCTION

The development of a manipulation system provided with three degrees of freedom (namely, a robotic wrist) has been carried out at the Department of Electrical, Management and Mechanical Engineering of the University of Udine, Italy. The robotic wrist is made of three rotation axes intersecting in a point, that is fixed with respect to the own reference system of the manipulator, so as to allow the orientation of a generic tool in the three-dimensional space. This manipulation system has been developed in the framework of the METAFORE project, funded by the MURST (Italian Ministry of the University and of the Scientific and Technological Research). The overall objective of the project is to realize a biomedical "slave" robot which is to assist the surgeon in piercing a vertebra, during the operation of insertion of peduncular screws (Figure 1), performed in order to increase the stability of the vertebral system. Up to date, such an operation is still performed manually.

The peduncular screws used in such operations have a maximum diameter of 5 mm, and a maximum length of 60 mm. Once they are inserted, a little stiffening board, chosen amongst various possible shapes, is fastened to any pair of screws, as shown in Figure 2. The tool used to pierce the vertebrae is a dedicated drill, having a maximum weight of 1 kg. This tool is to be connected to the robotic wrist described in this paper; moreover, the wrist is to be connected to a "father" manipulation system with three more degrees of freedom, consisting of a cartesian robot with three orthogonal axes. The cartesian robot has also been designed and built in the laboratories of the Department of Electrical, Management and Mechanical Engineering of the University of Udine, Italy. Moreover, a dedicated control system for the robot has been developed. In this paper, only a brief overview of the cartesian robot will be carried out in Section 2, whereas the design and the implementation of the robotic wrist will be carried out with more details in Section 3.

Published in: E.Kuljanic (Ed.) *Advanced Manufacturing Systems and Technology*, CISM Courses and Lectures No. 437, Springer Wien New York, 2002.

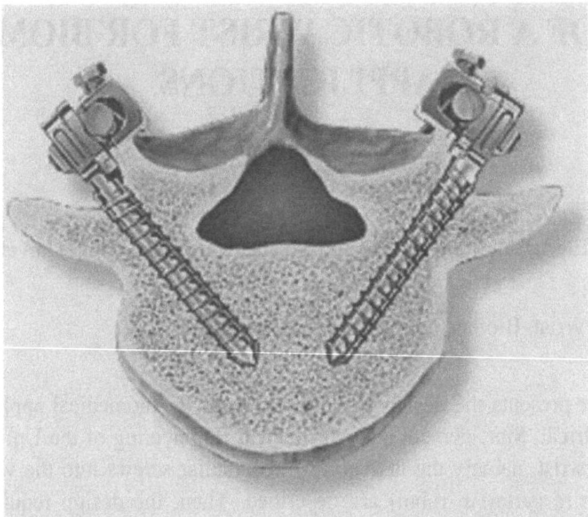

FIGURE 1. Section of a vertebra with a pair of peduncular screws inserted

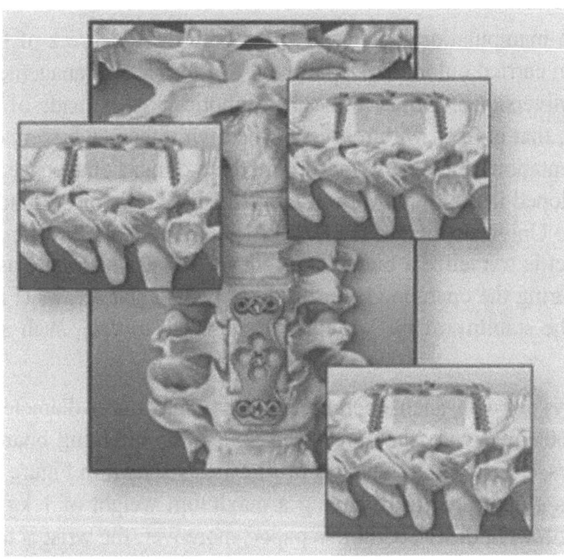

FIGURE 2. Stiffening boards for the peduncular screws

2 THE CARTESIAN ROBOT

The "father" robot to which the wrist is to be connected is a cartesian robot designed at the University of Udine [1]. The robot is provided with three cartesian axes (Figure 3), implemented by three linear pre-assembled units made of an aluminum alloy, and actuated by brushless motors. The motion transmission is implemented by means of precise toothed belts, so as to reach an accuracy of less than 1 mm. The vertical axis is self-braked in order to counteract the gravity force. Each motor is provided with a resolver, and is connected to the linear unit by means of a coaxial epicyclical gear train produced by Tecnoingranaggi. The maximum allowed excursion for each axis is:

- X axis: 700 mm
- Y axis: 600 mm
- Z axis: 500 mm

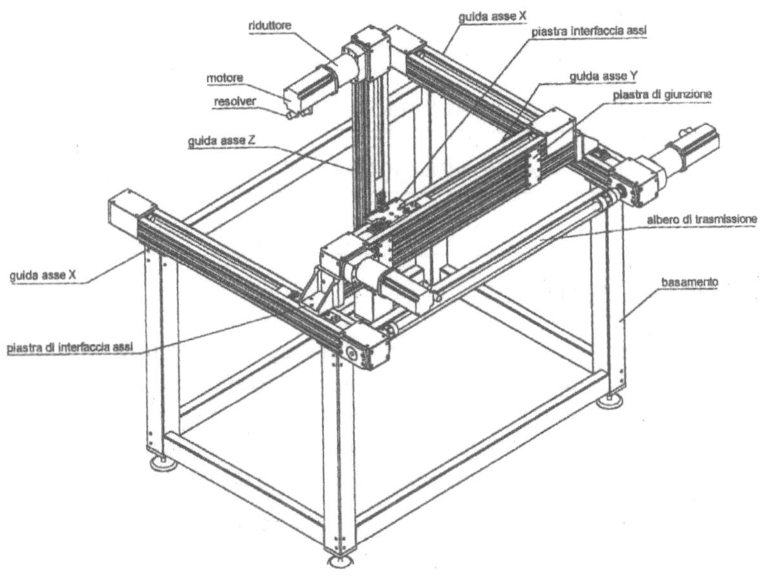

FIGURE 3. The designed cartesian robot

The dimensions of the cartesian robot are: 1600 x 1420 x 2100 mm^3, and its nominal payload is 10 kg. The robot is controlled by a PC, to which it is interfaced by means of dedicated National Instruments I/O boards. The high-level control system has been implemented by means of a software developed within the Labview environment.

3 DESIGN OF THE ROBOTIC WRIST

3.1 DESIGN REQUIREMENTS

The design requirements for the wrist specify that it must feature reduced weight and size. Namely, the wrist weight must be 3 kg maximum, and the drill connected to it must weigh 1 kg maximum. The maximum longitudinal and transverse loads acting on the wrist, and transmitted to the manipulator structure, have been estimated as 25 N for the longitudinal load, and 15 N for the transverse load. During the design phase, such values have been increased by 50% for safety reasons. Moreover, a torque sensor must be mounted on the wrist to measure the torque value. With respect to the positioning accuracy of the tool, it should be below 1 mm. The tool must be able to shift for 100 mm at once, so as to execute the piercing task. Table I reports the aforementioned design requirements.

TABLE 1. Design requirements

WRIST WEIGHT	≅3 kg
DRILL WEIGHT	≅1 kg
MAX LONGITUDINAL LOAD	25 N
MAX TRANSVERSE LOAD	15 N
POSITIONING ACCURACY	< 1 mm
DRILL SHIFT	100 mm

3.2 EVOLUTION OF THE DESIGN

The development of the robotic wrist followed a rather long evolution path. Many solutions were proposed: some were discarded because they were not feasible, some other were discarded because their efficiency/cost ratio was not satisfying. For instance, in the first proposed design, in order to make the three wrist axes intersecting into a single point, the choice was to set the brushless motors of the three axes (and the gear trains) within the wrist body.

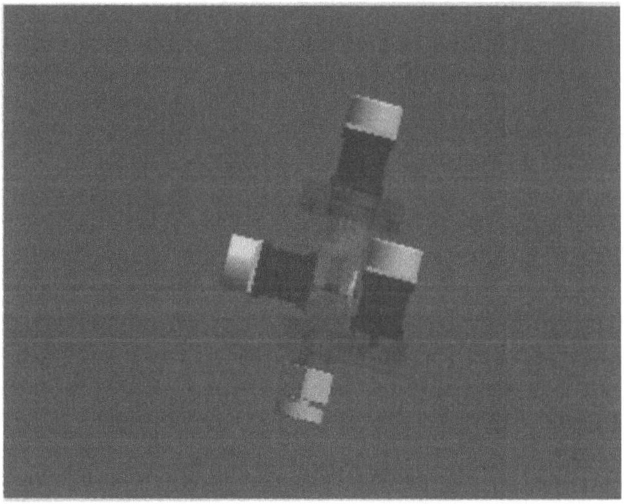

FIGURE 4. First version of the wrist

However, this caused a major increase of the wrist size as well as of the wrist weight; hence, more powerful motors were needed, and this in turn increased the weight of the structure, which could be avoided by using an extra transverse gear train. Such considerations led to the first version of the wrist, shown in Figure 4. However, the weight (6.5 kg) was still excessive, due to the high power required to the motors, that turned out to be too heavy. An important feature of this first version of the wrist lay in the fact that it was possible to connect to the third degree of freedom either a revolute joint or a sliding joint, in order to implement the translation of 100 mm set amongst the design requirements. This solution was possible because only two of the three Euler angle were necessary to define the end-effector orientation in this case.

A second version of the wrist contained a linear actuator (see Figure 5) which could implement the translational motion required. However, this solution also had several problems, due to the increase in the wrist weight as well as in the longitudinal dimension.

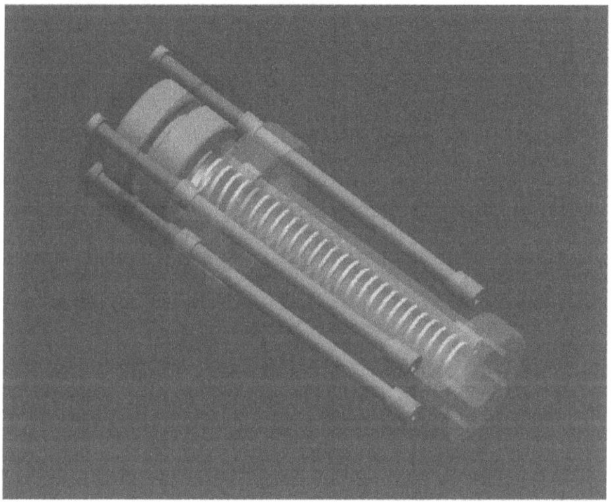

FIGURE 5. Linear actuator contained in one of the earliest versions of the wrist

Hence, this solution was discarded as well, and the choice was to definitely implement the translation by using the degrees of freedom of the cartesian robot to which the wrist is to be connected.

Several improvements to the design were then made, namely the dimensions were reduced as much as possible (without affecting the overall mechanical resistance of the wrist); moreover, the whole design was optimized by means of many simplifications and by suitably collocating the motors within the whole wrist structure. This allowed first to reduce the wrist dimensions, if not directly the weight; however, a reduction of the size caused a reduction of the nominal power required for the motors; hence, lighter motors could be used. In a word, the need for less powerful motors caused a reduction of the overall wrist weight.

3.3 THE CURRENTLY ADOPTED SOLUTION

Figure 6 shows the model of the wrist in a 3D CAD environment (Solid Edge 9), namely the currently adopted solution. Thanks to the optimization of the model, the designers managed to get an overall weight of the wrist that met the design requirements, still preserving a good mechanical resistance of the overall wrist structure. Namely, three brushless micromotors, with low weight and provided with light, high-precision gear trains were used in the definitive design. The search for such components was very accurate, and their most important features are reported in Table 2.

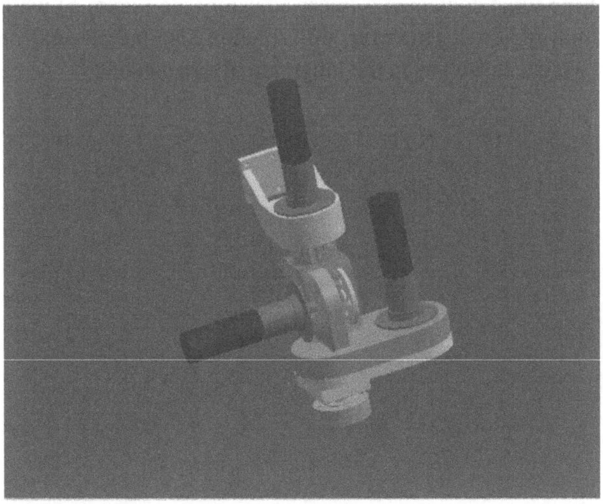

FIGURE 6. The currently adopted model of the wrist

TABLE 2. Main features of the brushless micromotors B1118

MASS	397 g
LENGTH	74.6 mm
MAX CONTINUOUS TORQUE	$86.1 \cdot 10^{-3}$ Nm
MAX RECOMMENDED SPEED	45000 rpm
EFFICIENCY	75-90%
AXIAL PLAY	114 μm
RADIAL PLAY	20 μm

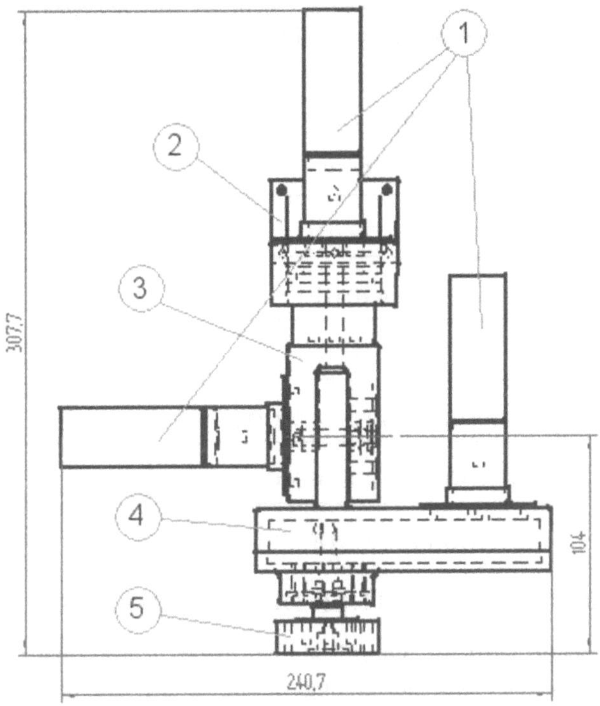

FIGURE 7. Design of the wrist

Referring to Figure 7, it is possible to observe the wrist design with more details. It is important to notice that the wrist has a modular structure, so that further developments will be possible with no need to completely redesign the system. For instance, the group "motor + gear + linkage", shown with the number "1" in Figure 7, can be substituted without modifying the connection between the linkage and the main wrist body. The wrist is to be connected to the cartesian robot by means of the linkage shown with the number "2" in Figure 7. The first motor (i.e. the first wrist axis) is connected to the former linkage and to the module "3", which in turn carries the second motor that actuates the module "4". The module "4" was designed in order to guarantee the intersection of the three rotation axes of the wrist. It is basically made of two shells making a little box which contains a gear for the transmission of the motion from the motor shaft to the shaft to which the tool, namely the drill, is to be connected through module "5". With respect to the module for tool connection, it should be noticed that it was not designed uniquely for connection of the drill, but it will be able to connect with many different types of tools, thus it can be used not only for biomedical purposes but also for several different tasks.

Referring again to Figure 7, it is also possible to notice that the size of the wrist was much reduced with respect to the previous versions. Table 3 reports the main technical features of the current version of the wrist.

TABLE 3. Main technical features of the current version of the wrist

MASS	≅3.2 kg
LENGTH	307.7 mm
WIDTH	240.7 mm
DEPTH	112.25
ROTATION RANGE OF MODULE "3"	360°
ROTATION RANGE OF MODULE "4"	180°
ROTATION RANGE OF MODULE "5"	360°

4 CONCLUSIONS AND FURTHER DEVELOPMENTS

The design of a robotic wrist for biomedical applications, carried out at the Department of Electrical, Management and Mechanical Engineering of the University of Udine, has been presented in this paper. First, the intended use of the wrist, namely the insertion of peduncular screws into the vertebrae, together with the supporting structure (a cartesian robot) has been described. Then, the design requirements for the wrist, especially with respect to the mechanical side, have been set and discussed. The most important requirements that had to be taken into account were those regarding: accuracy, dimensions, velocity, and maximum payload. All the steps up to the final design of the wrist have been described and discussed, with also the useful aid of a representation of the wrist in a CAD environment.

Further possible developments of the work are for instance those relative to the choice of the materials. The use of special materials featuring both low weight and high mechanical resistance can be studied. Moreover, a possible improvement of the design could be done by substituting the bearing currently used, namely axial bearings, with conical bearings, in order to increase the resistance of the structure to axial loads. The module "3" could also be improved, with respect to the connection between the cartesian robot and the wrist.

ACKNOWLEDGEMENTS

This work was supported by a grant of the MURST (Italian Ministry for the University and the Scientific and Technological Research), in the framework of the METAFORE project.

REFERENCES

1. Camatta R. (1999), Progetto e realizzazione di un manipolatore cartesiano tridimensionale, Master thesis, University of Udine

DEVELOPMENT AND PERFORMANCE ASSESSMENT OF A LASER TRIANGULATION SCANNER FOR DIGITIZING SHOE LASTS

N. De Rossi[1], A. Rossi[2], A. Trevisani[1], V. Zanotto[2]

[1] Department of Management and Engineering, University of Padova, Italy
[2] Department of Innovation in Mechanics and Management, University of Padova, Italy

KEYWORDS: 3D digitizer, laser triangulation, footwear industry.

ABSTRACT. This paper presents a non-contact laser triangulation 3D digitizer which offers a good combination of accuracy, efficiency, robustness and portability. The key component of the digitizer is an accurate laser triangulation rangefinder which moves on a vertical plane and continuously measures the distance of a last attached to a rotating shaft. A PD control action keeps the distance between the sensor and the last surface within the sensor measuring range, and close to the reference distance which minimizes measurement errors. One linear encoder, measuring the sensor horizontal displacements, and a shaft-position encoder, measuring the last angular displacements, provide the additional data needed for cross section reconstruction. The problem of digitizing a cross section which does not contain the center of rotation is faced using a photoelectric sensor detecting the cross section dimensions before the last reaches the rangefinder. The experimental results prove the effectiveness of the proposed system.

1. INTRODUCTION

The automation that is taking place in the footwear industry aims at producing high variety, low volume and low cost products. A phase of the traditional manufacturing process in which automation has help achieving better performances is the generation of a whole family of lasts for a shoe model. Lasts are initially made in wood by expert artisans, often in accordance with stylists' specifications. Generating scale copies of a wooden last into a resistant material is then necessary for producing a shoe model in several sizes and in different manufacturing sites simultaneously. Because of the continuous reduction of product development time, reverse engineering systems are generally used in this phase for rapid last modeling and manufacturing. The overall accuracy and efficiency of reverse engineering systems largely depends on the characteristics of the 3D digitizers employed to produce the 3D point clouds from which surface models are reconstructed [1,2].

In general, last digitizers can operate using either contact or non-contact methods. Contact digitizers use a touch probe which travels on the last surface to measure the position of a large number of points. In most recent contact digitizers, such as the Newlast® NL-DGT, lasts generally rotate with an angular velocity ranging from 10 to 15 rpm, about an axis that passes approximately through the toes and the heel, while the probe moves radially to gather cross sectional data (720 points per section) and along the last length (at 2 mm intervals). Contact digitizers are widely used in the footwear industry because of their reliability and accuracy. Yet they have some significant drawbacks, including large dimensions, high cost, and slow data acquisition. Moreover they do not

Published in: E.Kuljanic (Ed.) *Advanced Manufacturing Systems and Technology*, CISM Courses and Lectures No. 437, Springer Wien New York, 2002.

allow an accurate reconstruction of the last extremities, where lasts are clamped. To overcome such limitations, non-contact digitizers have been proposed recently.

A variety of non-contact rangefinding technologies can be employed to digitize the shape of an object, including photogrammetry, structured-light triangulation, time-of-flight, and interferometry. Most of the 3D scanners currently developed for the footwear industry make use of laser-stripe triangulation, and employ CCD cameras as measuring devices. For example, in [3] a system (FORMS) consisting of one CCD camera and two laser sources is presented. FORMS takes, on average, 50 s to digitize a foot to an accuracy of 0.5 mm. The 3D last scanner described in [4], which comprises one laser source and two CCD cameras, is much slower than FORMS (a scan takes 600 s) but has a higher accuracy (ranging from 0.05 to 0.2 mm). Commercially available 3D scanners are generally faster but their resolution is lower and often unsuitable for rapid prototyping. For example the Vorum Yeti®, which is employed in [5] and makes use of eight CCD cameras and four laser projectors, collects full shape information in approximately 3 s and scans to an accuracy of 0.5 mm. An identical number of cameras and laser sources is used in the INFOOT 3D scanner proposed in [6]. INFOOT scanning time is 10 s and its accuracy ranges from 0.7 to 2 mm. Two further 3D scanners commercially available are the Tecmath Pedus® (used in [7]) and the Knowledge & Information Technology® Model I. They both have a resolution of approximately 1 mm but the former takes 10 seconds to digitize a foot surface, while the latter takes 40 s.

In this paper a non-contact 3D digitizer for shoe lasts is proposed. The system does not employ CCD cameras, but a very accurate laser triangulation optical rangefinder, whose distance from the surface of the last rotating in front of it is continuously controlled and kept as close as possible to a reference value which minimizes measurement errors. Section 2 provides a schematic description of the 3D scanner and of its chief components. The procedure followed to calibrate the system is reported in Section 3 while Section 4 describes the tests performed to assess the system performances and the results achieved in digitizing three sample cross sections of a last.

2. SYSTEM DESCRIPTION

Figure 1 (a) shows a schematic representation of the proposed 3D last digitizer. A laser rangefinder (A) can be moved on a vertical plane by two orthogonal linear actuators (B-C). A last (D) is positioned in front of the sensor attached to a shaft driven by an electric motor (E). The shaft axis lies on the plane in which the sensor is moved. While the last rotates, the sensor projects a laser spot onto the last surface and measures its distance by optical triangulation. The 3D coordinates of the sensed points can then be computed, by simple matrix calculations, combining the positions of the sensor in the vertical plane, the angular positions of the last, and the measured distances.

While the last surface is digitized the sensor distance from the last must be continuously controlled so as to keep it within the sensor measuring range, and as close as possible to the reference distance which allows minimizing measurement errors. The horizontal linear actuator must therefore be fast, rigid and accurate to ensure an effective control. The vertical actuator must be rigid and accurate too, yet its control is simpler, since the sensor only has to be moved to predetermined vertical positions where cross section data are to be gathered. Hence in the prototype of the scanner used in this work to assess the performances of the proposed system, only the horizontal actuator is powered by a servomotor, while the vertical one is actuated manually (Figure 1 (b)).

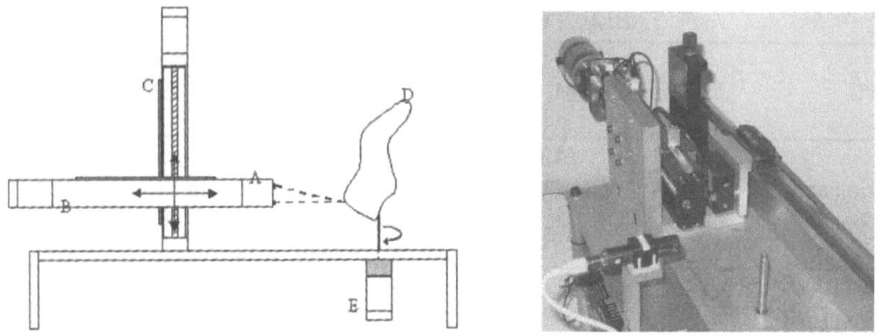

(a) (b)

FIGURE 1. (a) Schematic representation of the proposed 3D last digitizer. (b) Picture of the prototype

The optical rangefinder employed is the Micro-Epsilon optoNCDT 1605. This sensor permits high accuracy non-contact distance measurements and possesses good shock and vibration resistance, as well as low sensitivity to object surface properties and color. The linear actuator employed to move the sensor horizontally comprises a THK LM Guide Actuator KR3306A and a brush type DC motor MAE M540-0741. The motor amplifier is the AMC 12A8. Linear horizontal displacements are measured by the Heidenhain LS623 linear encoder rigidly connected to the linear actuator. An Owis VT 60 precision table is used to move the sensor vertically. The angular position of the rotating shaft is measured by the shaft-position encoder Eltra EH 63G. The shaft is supported by two ball bearings and is driven by a stepper motor MAE HY200 through a synchronous belt (velocity ratio: 2:1). The motor drive is the AEC SMD103. The chief technical features of the components employed in the prototype are reported in Table 1.

TABLE 1. Technical features of the prototype components

Micro-Epsilon optoNCDT 1605 (Light source: Red laser: 675 nm)	Measuring range [mm] (nominal)	± 6 (±10)
	Reference distance [mm]	65
	Resolution [μm]	6
	Measuring spot-diameter [mm]	0.9
THK LM KR3306A	Screw lead [mm]	6
	Rail length [mm]	300
	Positioning repeatability [μm]	3
	Positioning accuracy [μm]	20
	Backlash [μm]	3
MAE M540-0741	Maximum speed [rpm]	6000
	Torque constant [Nm/A]	0.071
AMC 12A8	Peak current [A]	± 12
	Maximum continuous current [A]	± 6
MAE HY200	Step angle [rad]	0.2865
	Step angle accuracy [%]	5
AEC SMD103	Drive	Bipolar
	Function	¼ step
Eltra EH 63G	Resolution [rad]	3.14E-4
Heidenhain LS623	Resolution [μm]	5
Datasensor S50-PL (Light source: Red laser: 650 nm)	Operating distance [mm]	0...350
	Resolution [mm]	0.5 (at 150 mm)

TABLE 2. CMZ Unimax technical features

CPU	Motorola MC68331, 32 bit, 16 MHz 512 Kb EPROM, 256 Kb Flash EPROM, 256 Kb SRAM, 512 Kb EEPROM Real-time clock
Acquisition and control boards	ACQSSG: analog input board, resolution: 12 bit CNV312: analog output board, resolution: ±5 mV DC MINENC: quadrature encoder board PIOD32: digital I/O board
Run time (1000 instructions) [ms]	3

The programmable logic controller (plc) CMZ Unimax (Table 2) allows performing real-time control of the whole scanning process, including the open-loop control of the last angular position, the closed-loop control of the relative position between the rangefinder and the last surface, and data acquisition from the transducers. In particular, the closed-loop control of the rangefinder position has been performed implementing a proportional-plus-derivative (PD) control. The controlled variable is the analog output of the rangefinder, the reference value is 0 V, and the control signal is the torque applied by the brush type DC motor. The task of the control action is keeping the rangefinder distance from the last within the measuring range, and as close as possible to the reference distance (65 mm) so as to minimize measurement errors. If the distance measured by the sensor goes beyond the measuring range (e.g. when the laser spot moves from the top side to the sole of the last) the controller stops the last rotation until the sensor reaches a correct position. As a consequence, scanning time depends on the last surface geometry: the smoother the changes in the surface curvature the lower the time required to digitize the last.

The second sensor which can be seen in Figure 1 (b), at the bottom-left corner, is a laser retroreflex photoelectric sensor Datasensor S50-PL. This sensor is employed to detect the angular positions where surface digitization is to be performed. As a matter of fact, when scanning a last surface, the rangefinder often has to execute measurements only over narrow angular intervals. This happens each time the last rotational axis does not pass through the cross section which is being digitized (e.g. near the last extremities). The photoelectric sensor allows determining the cross section angular dimensions and consequently setting the angular positions at which the control unit has to start and stop gathering data from the rangefinder. Only over these angular intervals are the output signals of the rangefinder meaningful and therefore useful to digitize the last surface and to control the motion of the sensor in the horizontal direction. The photoelectric sensor light beam and the rangefinder light beam form a neat right angle, which cannot change during and after the scans. Therefore after the photoelectric sensor has detected either the presence or the lack of the last reflecting surface, the plc can immediately compute the angular positions where the rangefinder has respectively to start and stop surface digitization.

3. SYSTEM CALIBRATION

One major drawback of the 3D non-contact digitizers employing CCD cameras is that they need to be accurately calibrated so as to deliver precise and reliable results. Mathematically it consists in determining a bijective function which identifies each pixel seen by the camera with its 3D coordinates. Although several effective and efficient calibration procedures have been proposed in literature (a comprehensive survey is given in [8]) calibrating a camera is still difficult if very accurate results are to be achieved: in fact a complete modelization of the camera is required, including all the geometric and intrinsic parameters. Calibrating the 3D scanner proposed in this paper is, on the contrary, an easy task. The procedure adopted here makes use of a precision-ground steel cylinder fixed to the rotating shaft. The cylinder diameter is 40.03 mm. In order to reduce measurement errors caused by the highly reflective surface of the cylinder, a Teflon film (thickness 50 μm) is laid on the cylinder surface.

First of all, the accuracy of the measurements made by the rangefinder over the measuring range is to be evaluated. The following procedure was adopted: by means of the horizontal linear actuator the sensor was moved to the position at the reference distance from the cylinder surface (sensor output: 0 V). Then the sensor was moved to several positions over the measuring range and the variations of the sensor output, which are proportional to displacements from the reference position, were compared with the displacements measured by the linear encoder. This procedure allows determining how the measurement error varies over the measuring range. Figure 2 shows the recorded errors, and the best fitting straight line for these results. It can be reasonably assumed that the error varies linearly with the distance, hence the following equation can be used to get a more accurate estimate of the actual distance D (measured in millimeters) of an object from the rangefinder value of the analog output U (measured in volts):

$$D = 65 - U/C \tag{1}$$

The adopted C value is 1.021 V/mm. This procedure allows a linear compensation of both the sensor and the acquisition board measurement errors over the measuring range.

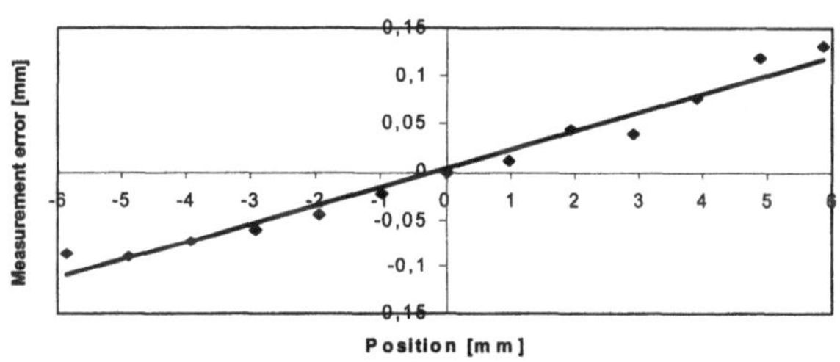

FIGURE 2. Measurement errors over the rangefinder measuring range

The steel cylinder was then used to compute the value of the constant K defined as follows:

$$K = R + D + r \cdot S = L - P \tag{2}$$

where R (mm) is the distance between the sensed point and the rotational axis, S is the distance of the encoder scanning unit from the encoder reference mark, measured in encoder steps, r (mm) is the encoder resolution, L (mm) is the distance, measured in the direction of the laser beam, between the rotational axis and the encoder reference mark, and P (mm) is the distance between the rangefinder and the encoder scanning unit. The K value employed is 45.90 mm.

Once the value of the constant K is known, it is possible to compute R on the basis of the values of D and S. Then simple calculations allow reconstructing the cross section of the object by combining the computed values of R and the angular measurements made by the encoder mounted on the rotating shaft.

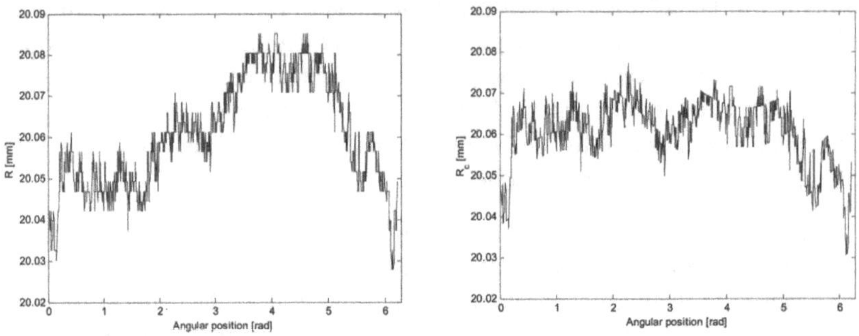

(a) (b)

FIGURE 3. (a) Recorded R values of the cylinder cross section. (b) Computed R_c values.

Although maximum care was taken in assembling the prototype, the measurements performed on the steel cylinder showed an oscillatory behavior which could be caused by a moderate deflection of the rotating shaft. Figure 3 (a) shows the R values computed through the data recorded on one sample scan of the cylinder cross section. The oscillation of the measured values can be clearly observed. The effect of such oscillation on the measured data was adjusted by subtracting to each value of R a contribution which depends on the angular position of the shaft, measured from a constant reference angular position. The following equation was adopted for each sensed point:

$$R_c = R - M \cdot sin(2\pi/n \cdot A + F) \tag{3}$$

where R_c is the adjusted R value, n is the total number of points measured per section, A is the current point number ($A = 0$ at the reference angular position). For the values shown in Figure 3 (a), and for all the measurements made on the same section, the values adopted were: $M = 0.014$ mm, $F = 3.4314$ rad. Figure 3 (b) shows the results attained after the compensation.

After calibration, the steel cylinder was measured ten times to evaluate measurement errors. The number of points measured on each scan was 1600 and the rotational speed 12 rpm. The mean error was 0.002 mm, the standard deviation 0.038 mm, and the maximum error 0.085 mm.

4. PERFORMANCE ASSESSMENT

When scanning a cylinder cross section the horizontal position of the rangefinder never changes being constant the relative distance between the cylinder surface and the sensor (the measurement oscillations cause very small changes of the controlled variable and consequently negligible variations of the control action). A proper assessment of the system performances should be made using an object whose cross sectional shape causes significant variations of the sensor output, and hence of the sensor position. Figure 4 (a) shows the cross section (quoted in millimeters) of the object employed. The object rotates about its axis of symmetry and allows evaluating the system performances in reproducing planar and round surfaces. The object cross section was scanned ten times, each time measuring 1600 points. Figure 4 (b) shows the R_c values computed for one of the tests, while Table 3 summarizes the results achieved by reporting the maximum and mean errors as well as the standard deviations.

TABLE 3. Measurement errors resulting from digitizing ten times the cross section shown in Figure 4 (a).

	Maximum error [mm]	Mean Error [mm]	Standard deviation [mm]
Planar surface	0.252	0.014	0.070
Round surface	0.099	0.012	0.041

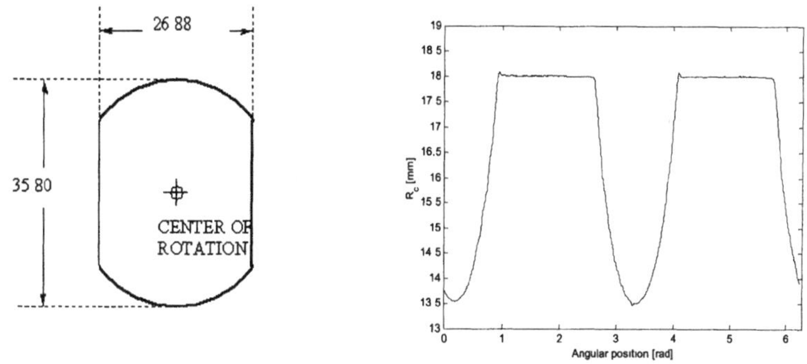

(a) (b)

FIGURE 4. (a) Cross section of the scanned object. (b) Computed R_c values

Figure 5 (b), (c), and (d) show the results achieved digitizing three sample cross sections of the last shown in Figure 5 (a). In the figures the symbol ⊕ shows the position of the center of rotation. Figure 5 (b) and (c) refer to a sections containing the center of rotation. Figure 5 (d) refers to a cross section that does not contain the center of rotation: the lack of points which can be noticed near the bottom-left corner of the plotted point cloud can be traced back to the decline in the photoelectric sensor performances when the last surface and the laser beam are nearly parallel.

5. CONCLUSIONS

Recently 3D object digitizers have begun playing an important role in the footwear industry, in particular for generating copies of shoe lasts. A non-contact digitizer for shoe lasts is presented in this paper. Compared with the contact digitizers used in the footwear industry the proposed scanner is faster (it can measure the position of up to 320 points per second while contact scanners usually do no go beyond 180 points per second), much lighter, has smaller dimensions, which makes it a portable system, and permits an accurate reconstruction of the last extremities. In comparison with the commercially available non-contact last digitizers, which make use of CCD cameras, the proposed scanner, though slower, is considerably more accurate and easier to be calibrated.

There are still some refinements that could be done to improve the system performances. In particular the maximum rotational speed at which scans are currently carried out is limited by the plc performances (clock frequency). Additionally by replacing the photoelectric sensor with a more accurate one, a higher resolution could be achieved when digitizing last cross sections which do not contain the center of rotation. Finally further study is needed to power and control the vertical linear actuator, and to perform an effective post-scanning data processing of the 3D point clouds.

ACKNOWLEDGMENTS. The authors wish to acknowledge Label Elettronica S.r.l. for supporting this research.

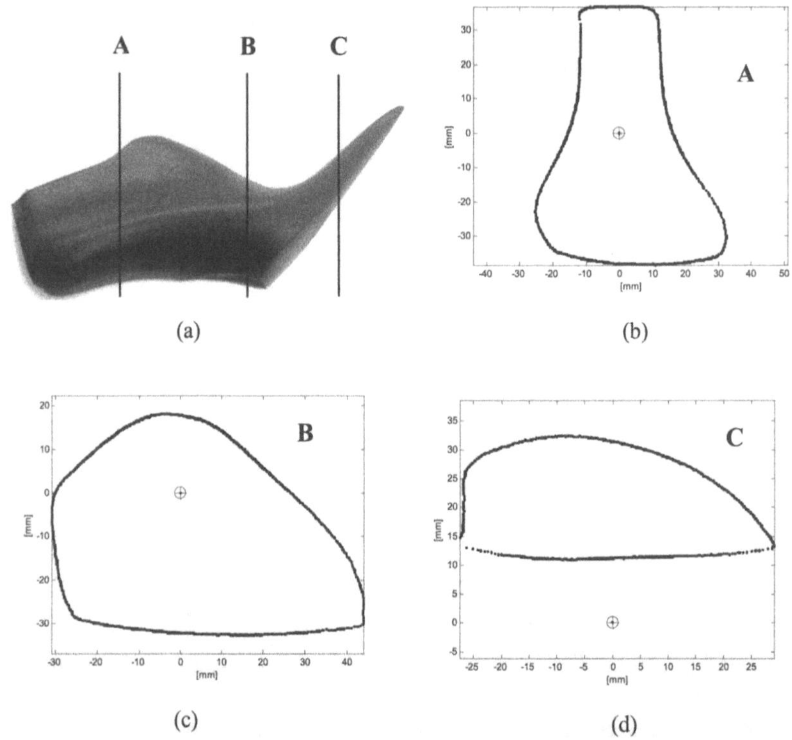

FIGURE 5. (a) The digitized last. (b) Cross section A. (c) Cross section B. (d) Cross section C.

REFERENCES

1. Curless, B, and Levoy, M., (1996), A Volumetric method for building complex models from range images. Proc. of the SIGGRAPH 1996, New Orleans, USA.
2. Li, L., Schemenauer, N., Peng, X., Zeng, Y., and Gu, P., (2002), A Reverse Engineering System for Rapid Manufacturing of Complex Objects, Robotics and Computer Integrated Manufacturing, Vol 18, 53-67.
3. Caracciolo R., Trevisani A., Rami G., Furegon R., and Toniato A., (2000), FORMS: a Mechatronic System for Rapid Foot Measurement and 3-D Reconstruction - Part I: Design and Machine Prototype". Proc. of the 7th Mechatronics Forum International Conference, Atlanta, USA.
4. Gonzáles F., Campoy P., Aracil R., Peñafiel F., and Sebastián J.M., (1993), Three-dimensional Digitizer for the Footwear Industry, SPIE Computer Vision for Industry, Vol 1989, 332-338.
5. Luximon, A., Goonetilleke, R., S., and Tsui, K., L., (2001), A Fit Metric for Footwear Customization, Proc. of the 2001 World Congress on Mass Customization and Personalization, Hong-Kong.
6. Kouchi, M., and Mochimaru, M., (2001), Development of a Low Cost Foot-Scanner for a Custom Shoe Making System, Proc. of the 5th Symposium on Footwear Biomechanics, Zurich, Switzerland.
7. Novotni, M., and Klein, R., (2001), Geometric 3D Comparison - an Application, ECDL WS Generalized Documents 2001.
8. Tsai, R.Y., (1987), A Versatile Camera Calibration Technique for High-Accuracy 3D Machine Vision Metrology Using Off-the-Shelf TV Cameras and Lenses, IEEE Journal of Robotics and Automation, RA-3, 323-344.

DESIGN AND DEVELOPMENT OF A VISION BASED LEATHER TRIMMING MACHINE

M. Lanzetta, G. Tantussi

Department of Mechanical, Nuclear and Production Engineering, University of Pisa, Italy

KEYWORDS: Artificial vision, Defect detection, Path planning.

ABSTRACT. The objective of the work described in this paper is to demonstrate a laboratory prototype for trimming the external part of a hide, assuming that the resulting machine would eventually form part of a completely automatic system in which hides are uploaded, inspected and parts for assembly are downloaded without manual intervention and prior sorting. Detailed literature and international standards are included. The expected advantages of integrating all vision based functions in a single machine, whose basic architecture is proposed in the paper, are also discussed.
The developed system is based on a monochrome camera following the leather contour. This work focuses on the image processing algorithms for defect detection on leather and the NC programming issues related to the path following optimization, which have been successfully tested with different leather types.

1 INTRODUCTION

Leather is animal hide or skin and it has been defined by the International Council of Tanners (Buenos Aires, 1978) and EEC directive (94/11, 23-04-1994). The most widespread commercial leather types are available from a variety of animals including cattle and calves, sheeps and lambs, goats, horses, deers and minor classes including reptiles, amphibians and aquatic animals tanned in producing countries usually for local consumption.
The trim of a hide is its shape having in mind its ultimate use (ISO 2820:1974). After preservation treatment and processing, trimming raw hides is the first step of the leather manufacturing process and is followed by nesting and cutting of components according to the size and shape of the final product, which is assembled by gluing and stitching: footwear, clothing, upholstery for automobiles and furniture, accessories (travelling and handbags, gloves, suitcases and briefcases, straps, handicraft, etc.) and technical goods (machinery belting, hydraulic packing and washers, etc.). The hide contours have a higher concentration of defects such as small holes and fringes which must be eliminated in the trimming operation.
The automation level in the leather industry is generally low, most operations are carried out manually by skilled craftsmen, so a lot can be done in this field to increase the automation level. Several attempts are available in the literature for different phases of production and most of them make extensive use of artificial vision techniques.

1.1 STATE OF THE ART

In [1], a vision based waterjet cutting machine is proposed, focusing on the image processing algorithms to measure the leather surface, using a fixed monochrome CCD camera with servo-controlled focal length and a backlighting system. This machine has a better resolution with

Published in: E.Kuljanic (Ed.) *Advanced Manufacturing Systems and Technology*,
CISM Courses and Lectures No. 437, Springer Wien New York, 2002.

respect to current methods based on photo-detectors. Regarding the machine evolution, examples of nesting taking into account the quality of the different hide areas and the leather homogeneity are provided. The waterjet cutting table is supposed to be separate from the inspection cell, which determines the cutting path. To fully exploit the potential of the waterjet technology several leather layers are supposed to be cut at the same time, but this increases the rejection of good material.

A similar approach has been followed by [2] which also describes a specific handling device mounted on a SCARA robot to pickup parts from a conveyor belt at variable speed up to 25 m/min and lay them flat on a pallet with a positional accuracy of 0.5 mm and 1°. Regarding handling, the gripper design is still an open problem and it is also under study at our department [3].

In [4], a patented machine prototype is described, which is able to stitch decorative patterns on a wide range of shoe leather and synthetic shoe materials. The system uses no part-dependent specialized fittings and accepts parts of predefined shape thrown onto a conveyor in any position and orientation whose silhouette is automatically identified. Patterns are stored in "teach mode" with a mouse as key points and are interpolated by cubic splines during the stitching operation.

The recognition and orientation problems in the manipulation of shoe uppers are also dealt with in [5], which concludes emphasizing the benefit of adapting the process to the workpiece by flexible computer control, e.g. a vision based approach, as opposed to manipulating the workpiece itself to achieve the desired orientation, because floppy materials are difficult to control with sufficient reliability and accuracy during robotic handling.

Other authors [6] extract geometric features (including the square root of its area and the length of 18 radii from its centroid to its boundary at equal distance from a datum) from binarized high-resolution images for component recognition purposes ranging between 25 and 400 mm^2 (most between 100 and 250) travelling at four per second.

Regarding cowhide defect detection, [7] and [8] propose a segmentation algorithm respectively using a fixed threshold and a new local contrast measure based on features extracted from the histogram of leather subimages.

Regarding the leather uniformity, international standards can be used as a reference, also for certification purposes into different classes. For instance it is required that in tests like rubbing, perspiration water and dry cleaning (ISO 11640 to 11643:1993), water spotting, migration into plasticized poly(vinylchloride), machine or mild washing (ISO 15700 to 15703:1998) or exposition to artificial light (ISO 105-B02), the color fastness be between 3 and 4 levels according to the international gray scale (ISO 105-A02), and only where the esthetic appearance is important; for instance, in the shoe industry, it is only required for uppers.

In [9], a vision system is used to correct the line drawings manually stamped on hides, by connecting the line interruptions, and to detect the quality of different areas, by recognizing the quality marks. By this, the system is able to provide a digital input to a Computer Aided Layout Design system (CALD) with a resolution of 1 mm/pixel in about 40 to 120 s per hide, depending on the drawing complexity. This operation, whose larger effort is the correction of errors coming from previous manual operations, due to the introduction of partial automation, also shows the usefulness of a system integrating all automatic functions.

The automation by a vision system of subsequent assembly operations in the shoe industry, such as skiving, which consists in reducing the thickness of the leather by the application of a

chamfer or scarf to edges that will overlap and which would otherwise be unacceptably bulky, is described in [5].
From the above examples, it can be noticed that many aspects of the leather manufacturing process have been approached by vision, but a complete system is not yet available. A possible architecture will be proposed in the first part of this paper and the main benefits discussed.

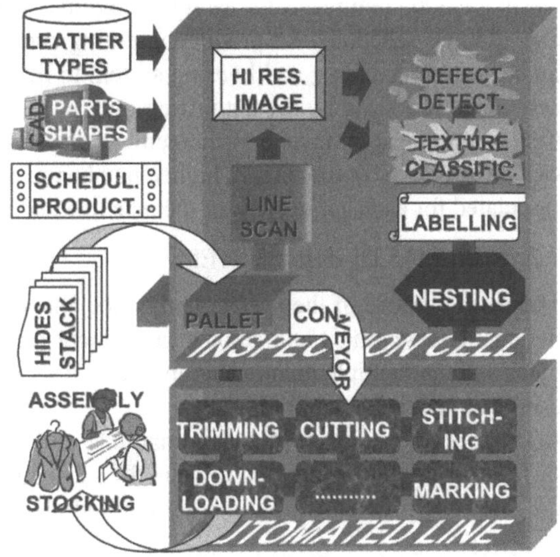

FIGURE 1. Architecture of the computer integrated automatic leather processing system. The information and material flows are respectively darker and brighter

2 THE PROPOSED SYSTEM

The computer integrated system described (Figure 1) can be installed at the end of a preservation processing line for trimming purposes or at the beginning of an assembly line to optimize cutting.
Hides are fed to the inspection cell where all vision based operations are carried out, followed by cells for other operations, such as cutting, stitching, etc., depending on the product type. Pallets on a conveyor are used to move the material through each operating cell in order to have an absolute reference with respect to all cells. All cells are connected to the inspection cell which provides the necessary data for processing.
The color texture classification is the most effective approach for defect detection, for homogeneity assessment and to label similar areas. The preferred acquisition device is a color line camera. The main benefits are:

- more repeatable lighting (on a single dimension);
- higher resolution with respect to matrix CCD.

The relative movement is achieved by the aforementioned conveyor. The required spatial resolution is 0.2 mm/pixel, corresponding to 41.6 cm with a 2048 pixel line camera [6]. For larger machines/hides, multiple cameras can be used to achieve the required resolution.
In addition to standard lighting, a structured lighting system is also used to discriminate between real defects and wrinkles due to insufficient hide stretching (especially for soft leather) after uploading. This method will also increase the system reliability for physical defect detection. A filter [7] is also used to reduce the shadows caused by the waviness.

2.1 NESTING

Nesting with hides has two special features: (i) the shape of the defect-free area is always different and (ii) the leather homogeneity and classification should also be taken into account.
An additional criterion is the radial distance of a part, because the leather is a fiber material. The preferential direction of fibers is radial, starting from the animal backbone, which is more homogeneous and resistant with respect to the neck, shank and belly, and which should be used for higher quality parts [10].

Nesting can be optimized for each hide by involving in this operation a large number of parts, e.g. all those scheduled for production during current shift. Special care will be required by the identification, handling and housing system to manage the high number of different parts to be assembled produced by the automated line. A compromise will be necessary depending on the nesting optimization level and the number of parts to be stocked. The amount of material that can be saved by optimizing a higher number of parts in the nesting operation can be easily calculated for each application of the machine and it depends on the product type.

2.2 BENEFITS OF THE SYSTEM

The main purpose of the system is the hide shape acquisition for the nesting optimization and the quality assessment and color measurement for leather sorting. The economic evaluation of hides and the multiple mentioned laboratory tests for color fastness (after the machine certification) would also be possible.

Additional operations required by man-made operations, like putting temporary marks on the surface of pre-cut or marking a locus for subsequent stitching (decorative in nature or part of the assembly process) would not be required anymore. Such operations were automated in [6] and [11].

The automation of leather trimming or cutting is very challenging because cycle times per component are very short (in the order of the minute) and it is important that they can be matched, but it is highly desirable for the following main benefits:

- improving the safety of the personnel working in a harmful environment and currently using sets of sharp knives at high speed;
- better material exploitation, through an optimization of the cutting path. The operator often follows shortcuts to speed up the trimming process thus increasing the rejection of good material (Figure 2);
- objective assessment of the various aspects of surface qualities;
- improving the process repeatability and productivity with a suitable hardware selection.

An advantage of a fixed reference, having hides in the same position on the pallet, is that parts of any shape and orientation can be extracted without additional recognition or localization algorithms for operations such as cutting, decorative stitching or handling, because these information are already stored in the system. From the literature it turns out that the resolution required for defect detection [8] [9] is 5 times higher than that required for recognition and orientation [6]. Additional benefits are:

- there is no need for training the system for recognition and orientation of each new shape processed as continuously required by the evolution of fashions;
- the elimination of recognition and orientation errors, estimated with an accuracy of 0.07 degrees and 0.2 mm/pixel in [6]; in particular, the difference in dimensions between adjacent sizes can be small and yet the process needs to be able to discriminate between them. It should be noticed that a typical manufacturing unit might have in process say 20-40 shoe styles, each one comprising left and right, each with ten sizes and perhaps as many as twenty components in the upper alone.

3 CURRENT WORK

Our work is focused on the trimming operation and it is based on a specific request from the industry. A laboratory prototype of the machine with its control software has been developed. It is intended for small size hides, while: cattle hides can fit a rectangle up to 3 × 4 m. The "home" position is in the center of the working space, represented by a rectangle of 321×575 ($\Delta X \times \Delta Y$) mm. After the hide has been positioned, the camera is moved from the home position towards the external border in the radial direction with angle Θ_0 (45° by default). To test the trimming algorithm, the camera moves around and grabs different portions of the whole hide. After acquisition each image is individually processed to determine the background and the defective areas to find the trimming path. This behavior simulates the analysis of sub-images extracted from the whole hide image scanned by a line camera.

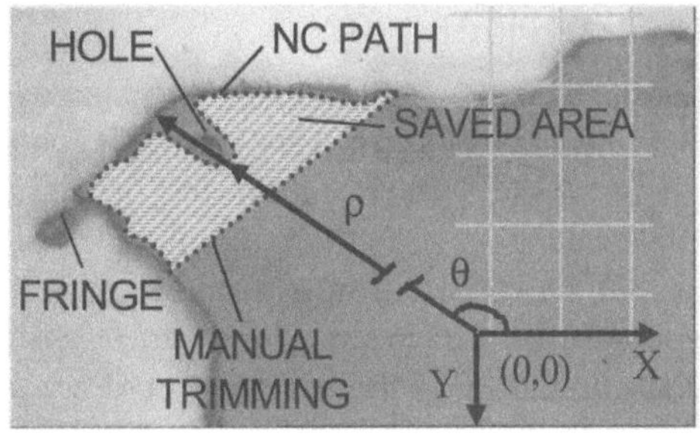

FIGURE 2. An image grabbed with leather defects, the optimal convex trimming path, the machine reference system and (part of) the control grid

No optical distortion (estimated below the order of the pixel) compensation algorithms have been used.

3.1 DEFECT DETECTION

Defects on hides can be divided into three categories: ante-, post-mortem and preservation defects (ISO 2822-1:1998). They include flaws, wrinkles, scars, mites (insect bites), brand and growth marks. Hides before trimming have a ragged border with fringes and small holes. To discriminate between leather and background, the mode of a 8 × 8 pixel control window in the image center is compared with the mode of the background (240). To give an order of magnitude, the size of a pixel with current magnification and camera resolution is about 0.4 mm/pixel. With current resolution this is the minimum statistically significant size.

To determine the defective areas, the current image is divided in a regular grid of 10 × 10 squares (Figure 2) and the mode is computed for each. A leather area is considered defective if its gray level differs significantly from the mean of the modes of all the observed squares, after excluding those containing the background (e.g. all squares whose mode is greater then 240). The mode of a square is considered instead of the mean because it is faster to calculate and it is not affected by the presence (up to about 50%) of pixels belonging to the background or to defective areas. In the rare event of a control window containing more than 50% of defective pixels, its weight in the calculation of the mean gray level of good pixels will be anyway low, because, in the worst case (only 1/4 of the image containing leather and the rest with the background), the mean is calculated on 25 squares. From the mean of modes computed, the acceptable range of "good pixels" is calculated. It depends on the leather uniformity, in our case (-15, +30).

After binarization, the leather appears as a black surface on white background and an opening (a 10 step erosion followed by a dilation) is also performed on the black area to eliminate small fringes from the contour. To determine the final contour of the good leather for trimming, the leather edge is detected by a convolution using a Laplacian mask. Considering that a binary image is input, the result is a white 1 pixel thick path on a black background.

3.2 PATH PLANNING

The function to optimize in the path planning strategy is the minimization of the scrap material. The software interface is very simple. It contains one main window and several control buttons. It displays the X and Y coordinates of the camera position in real time with respect to the absolute reference system joint to the machine. At each acquisition, the points belonging to the contour are converted from pixel to mm and from the image reference system to the machine absolute system, considering the camera position at the time of acquisition. The Cartesian coordinates returned by the controller card are also converted in polar coordinates R and Θ with respect to the machine center, which can be assumed as a center of rotation in the trimming operation. Hides should be positioned in order to be centered with respect to the machine center. Considering the radial distribution of fibers and the typical hide shape, this positioning allows optimal material exploitation and avoiding undercuts in path planning.

The use of polar coordinates is very effective in this application for the following reasons.

1. The initial angular position Θ_0 is recorded to determine the trimming end.
2. They are used to order in a sequence a set of points from a 2D matrix using the coordinate Θ.
3. To determine the optimal convex path which excludes a hole or a fringe (in the sense of Figure 2) the most internal points between those belonging to the hide contour and those belonging to the defect edge, if any, are selected considering the coordinate R. Only the radius of points with the same Θ approximated to the second decimal digit are compared to each other.

The path polar coordinates are then converted again in Cartesian coordinates that constitute the NC input. The unit of measure is the "count" which depends on the resolution of the stepper motor and the screw axial pitch. The output of the vision system is a sequence of 2D coordinates that are passed on to the axes NC. The path points are 1 pixel apart from each other, so the image resolution corresponds to the path resolution. The NC operates in vector mode and each new point is input as a difference from the current position.

The sequence of operations described is executed in real time, while the camera follows the profile determined after previous acquisition. The path is followed having the current point in the image center. The camera field of view has been selected taking into account the required resolution and the typical defect size to allow enough visibility for sudden changes of directions. Any difference in the overlapping of subsequent images is solved by linear interpolation by the NC low level routines.

The trimming path determined by the inspection cell also allows an accurate estimation of the defect-free leather area (ISO 11646:1993) that can be used to estimate its value and to schedule the amount of raw material necessary to fulfil a certain production request from the market.

3.3 FEATURES AND TESTING OF THE DEVELOPED SYSTEM

The inspection cell (Figure 3) is a gantry. The leather is positioned on a bright white panel to enhance the contrast with most leather types, which are dark. Two pentaphosphor Osram lamps model Lumilux Deluxe with high frequency power supply have been positioned on the cell ceiling symmetrically along the Y axis. The structure is isolated from the environmental lighting. Two recirculating ball-bearing slides carry a monochrome matrix CCD camera model M50 from Jai with a Computar Telecentric lens to reduce optical distortions.

A 4-axis numerical motion controller card from Galil model DMC-1740 controls two 2-phase 95 W hybrid stepper motor from Pacific Scientific with bipolar driver. The grabbing card is a Matrox Meteor-2MC with MIL. Both mentioned cards are installed on a PII MS Windows based PC. The vision and control program and interface have been programmed using MS Visual Basic. The geometric path computations are performed using the Mathworks Matlab 6.0 ActiveX.

FIGURE 3. Prototype of the inspection cell

The developed system has been used both to test the inspection and the trimming routines. By its motion the camera simulates the cutting head. It has been shown to be able to recognize most defects with different leather types and to simulate the trimming path. The surface finish and color has a strong influence on the outcome of inspection, yielding better results with brighter leather. Flaws on dark leather require higher light intensity. Highly reflective surface can create patchy saturated regions causing mis-detection.

The inspection and trimming time for a 320.000 mm^2 hide with a path of 2000 mm is about 2 minutes (the average speed with current hardware is upper limited). The time to process each image is less than a second.

4 FUTURE DEVELOPMENTS AND CONCLUSIONS

The simple defect detection algorithm developed is based on the average gray level of the examined leather. In the final system an additional routine will be necessary for the classification of the leather texture in different homogenous areas. The literature on this is very rich ([7] [8] [12] [13] and the bibliography included therein). This will also allow overcoming the hypothesis of a white background that could be too stringent with the same pallet used in the cutting cell. On the other side some defects are too subtle to robustly influence the parameters of the statistical model alone, so both detection approaches cooperating will be required.

The great potential of concentrating all vision based functions in a single machine has been shown from the analysis of literature and the description of the proposed system. The technology required for its implementation is state of the art, so the next step is their integration and engineering. The main drawback of this radical approach is the low technology level of the leather industry and consequently the small target market in the short term. So further applied research is required in order to increase the cost effectiveness of a completely automatic inspection and processing system. The defect detection and the path following algorithms implemented and described in this paper can be included in such system.

5 ACKNOWLEDGEMENTS

The authors would like to thank Ing. Paolo Bologni for his contribution to this study. Support from Mr. Flavio Antonelli, Mr. Enzo Peroni and Mr. Salvatore Balestrino of the Department of Mechanical, Nuclear and Production Engineering, University of Pisa, are acknowledged.

REFERENCES

1. Aranda Penaranda, J.D., Ramos Alcazar, J.A., Tomas Balibrea, L.M., Munoz Lozano, J.L., Torres Sanchez, R., (1994), Inspection and measurement of leather system based on artificial vision techniques applied to the automation and waterjet cut direct application. IEEE Int. Conf. on Systems, Man, and Cybernetics, Humans, Information and Technology, San Antonio, TX, USA, 2-5 Oct. 1994, Vol 1, 863-867, ISBN: 0-7803-2129-4.
2. Clot, R, Redarce, H.T., Betemps, M., Jutard, A., (1989), Sucker set configured by vision in real time: applicated grabbing flat and thin pieces. Proc. IASTED Int. Symp. Applied Informatics - AI '89, ACTA Press, Anaheim, CA, USA, 76-79, ISBN: 0-8898-6117-X.
3. Neri, F., (2001), Development of an automatic leather manipulation system for the tannery industry. Master Thesis (in Italian), University of Pisa, Italy.
4. Smith, D.L., Taylor, P.M., Taylor, G.E., Jolliffe, I., Reedman, D.C., (1991), Decorative stitching of randomly fed shoe parts. 91 ICAR, 5th Int. Conf. on Advanced Robotics, 'Robots in Unstructured Environments', Pisa, Italy 19-22 June 1991, Vol 1, 765-768, ISBN: 0-7803-0078-5.
5. Preece, C., Reedman, D.C., Simmons, J.E.L., Topis, S., (1994), Assembling non-rigid products in the shoe industry. IEE Computing and Control Division Colloquium on Intelligent Automation for Processing Non-rigid Products, London, UK, 19 Oct. 1994, 9/1-3, ISSN: 0963-3308.
6. Tout, N., Norton-Wayne, L., Reedman, D., (1991), Automated identification of shoe upper components. IEE Colloquium on Binary Image Processing - Techniques and Applications, London, UK, 10/1-10/5.
7. Hoang, K., Wen, W., Nachimuthu, A., Jiang, X.L., (1997), Achieving automation in leather surface inspection. Computers in Industry, Elsevier science, Vol 34, 43-54.
8. Azencott, R., Yao, J., (1994), Automated detection of cowhide defects using Markov random field techniques. Proc. 12th IAPR International Conference on Pattern Recognition, Jerusalem, Israel 9-13 Oct. 1994, Vol 1 Conf. A: Computer Vision & Image Processing, 791-793, ISBN: 0-8186-6265-4.
9. Lerch, A., Chetverikov, D., (1992), Correction of line drawings for image segmentation in leather industry. Proc. 11th IAPR Int. Conf. on Pattern Recognition, The Hague, Netherlands, 30 Aug.-3 Sept. 1992, Vol 1 A: Computer Vision and Applications, 45-48, ISBN: 0-8186-2910-X.
10. Manzo, G., (1999), Chimica e tecnologia del cuoio. Mediaservice, Legnano, Italy, 824 pp.
11. Preece, C., Simmons, J.E.L., Tout, N.R., Reedman, D.C., (1991), Marking and manipulation problems in the shoe industry. 91 ICAR (cited above), 786-789.
12. Lanzetta, M., Tantussi, G., Sviluppo e confronto di tecniche di controllo di qualita` di manufatti in granito (invited). Proc. "Studi e indagini sui materiali lapidei", Carrara 2000, XXI IMM, Carrara (Italy), June, 3rd 2000, 21-30.
13. Mirmehdi, M., Petrou, M., (2000), Segmentation of Color Textures. IEEE Trans. on Pattern Analysis and Machine Intelligence, Vol 22, 2, February 2000, 142-159.

ACCURACY OF A 3D VISION SYSTEM FOR INSPECTION OF COMPLEX GEOMETRY

S. Carmignato, E. Savio, L. De Chiffre

DIMEG, University of Padova, Italy

KEYWORDS: 3D vision, inspection, accuracy.

ABSTRACT. This paper illustrates an experimental method to assess the accuracy of a three-dimensional (3D) vision system for the inspection of complex geometry. The aim is to provide a procedure to evaluate task related measurement uncertainty for virtually any measurement task. The key element of the method is the use of a coordinate measuring machine (CMM) to supply reference measurements as basis for realistic statements of measurement uncertainty. Since robust techniques to establish traceability in CMM measurements of complex geometry are available, a CMM-based approach is suitable for the purpose to establish traceability. Accuracy performances of optical digitisation systems are assessed on the basis of deviations existing between acquired cloud points and the CMM measurements. To demonstrate the feasibility of the proposed method, the procedure is applied to an industrial case study.

1 INTRODUCTION

Traditionally, in manufacturing industry, geometrical inspection of items has relied strongly on techniques utilizing touch probe technologies. For regular geometry objects, such measuring techniques ensure both rapidity and accuracy. However, as components become free-form and more complex, non-contact methods become increasingly advantageous [1].

When inspecting complex free-form shapes with a touch probe method, only a small number of points are actually measured. Using non-contact methods, instead, a much denser point cloud can be acquired in a very short period of time [2]. This is just what is needed for the inspection of complex free-form surfaces, where the importance of the accuracy of a single point measurement is often irrelevant in comparison with the density of the inspected data and the coverage over critical areas.

In recent years, developments in non-contact metrology have increased dramatically. The increase of accuracy of optical systems has allowed the spread of such systems from reverse engineering applications to inspection of complex geometrical tolerances. However, in comparison with reverse engineering applications, the use of non-contact systems in dimensional and geometrical control requires an increased care to measurement quality assurance. The user, hence, needs to identify the metrological performances of such new measuring instruments.

Performance verification of measuring systems for complex measurements of free-form surfaces is not an easy task. In a digitisation made with 3D vision equipment, for example, it is not immediate to determine which is the degree of agreement between the result of measurement – a point cloud – and the true measured object. There is not a simple relationship, for instance,

Published in: E.Kuljanic (Ed.) *Advanced Manufacturing Systems and Technology*, CISM Courses and Lectures No. 437, Springer Wien New York, 2002.

between the accuracy of a single point measurement and the overall accuracy of the whole point cloud model.

Furthermore, there exists not an international standard on performance verification of optical systems. Thus, specifications provided by manufacturers on the accuracy of common commercial 3D vision systems are often unclear. The parameters with which such specifications are expressed and the way in which they are determined differ from a producer to another. This increases the difficulties that are met by users when comparing performances of different optical instruments.

2 APPROACH

In a quality management system, the measuring instruments shall be calibrated against certified equipment having a known and valid relationship to internationally or nationally recognised standards, in order to establish traceability [3]. According to the Guide to the Expression of Uncertainty in Measurement (GUM), a calibration includes the assessment of the uncertainties in the results of the measurement tasks. For a 3D vision system, however, an *a priori* assessment may be too complex because the influence quantities that can affect the result of an optical measurement are extremely numerous: they encompass surface characteristics of the measured object, lighting conditions, elaboration algorithms and many other factors [4].

The aim of this paper is to propose an experimental method to evaluate the accuracy of a 3D vision system for the inspection of complex geometry. The proposed methodology is intended to supply realistic statements of measurement uncertainties, with the purpose to establish traceability. The procedure is developed with a task-specific approach, in which the errors are evaluated for a specific measurement by comparing the results with a more accurate measuring system. The procedure allows the evaluation of task related measurement uncertainty for virtually any measurement task.

The key element of the method is the use of a coordinate measuring machine (CMM) as the link to the traceability chain. Such approach can be used in preference to the utilization of calibrated standards when inspecting free-form artefacts. The use of a state-of-the-art CMM, in fact, allows the inspection of a free-form object with the highest accuracy achievable nowadays. In general, CMM accuracy is often more than ten times higher than the accuracy reached with optical vision systems used for the same measuring task. Since techniques to establish traceability in CMM measurements of complex geometry are already consolidated, it appears clear that a CMM-based procedure is adequate to verify performances of optical digitisation systems in terms of deviation of measured points from the original physical object.

The proposed approach allows verifying the accuracy of measuring results independently from the specific non-contact system. The origin of the tested model, in other words, is not restrictive. Moreover, different kinds of model can be verified with the same methodology: besides point cloud models, triangulated and parametric surface models can be investigated with little adjustment to the procedure. Multiple views models, finally, are suitable for the verification as well as single view ones.

3 METHODOLOGY

The accuracy evaluation of non-contact systems for the inspection of complex geometry is achievable by implementing a CMM-based procedure. The outline of the method is shown in Figure 1. Essentially, it consists of the following phases.

3.1 NON-CONTACT MEASUREMENT

In the first phase, the digitisation with the vision system to be verified gives the non-contact model of the physical object. The object is scanned using one of the various non-contact techniques currently available. A typical 3D vision system projects light pattern on to the object and observes the intersection of the object surface and the light pattern through electronic cameras. A precise knowledge of the relative positioning of the light projector and the camera, calculated during calibration, is critical if the depth data are to be calculated accurately.

The resulting point cloud is generally extremely dense and is typically collected in a few seconds only. However, a single view is usually insufficient to fully measure an object in three dimensions. In this circumstance several digitisation from different orientation are required to fully cover the object. Each single view is assembled to the others by specific algorithms. The assembling accuracy can strongly influence the overall accuracy of the final model. The result of this phase, therefore, is a multiple view model of point clouds.

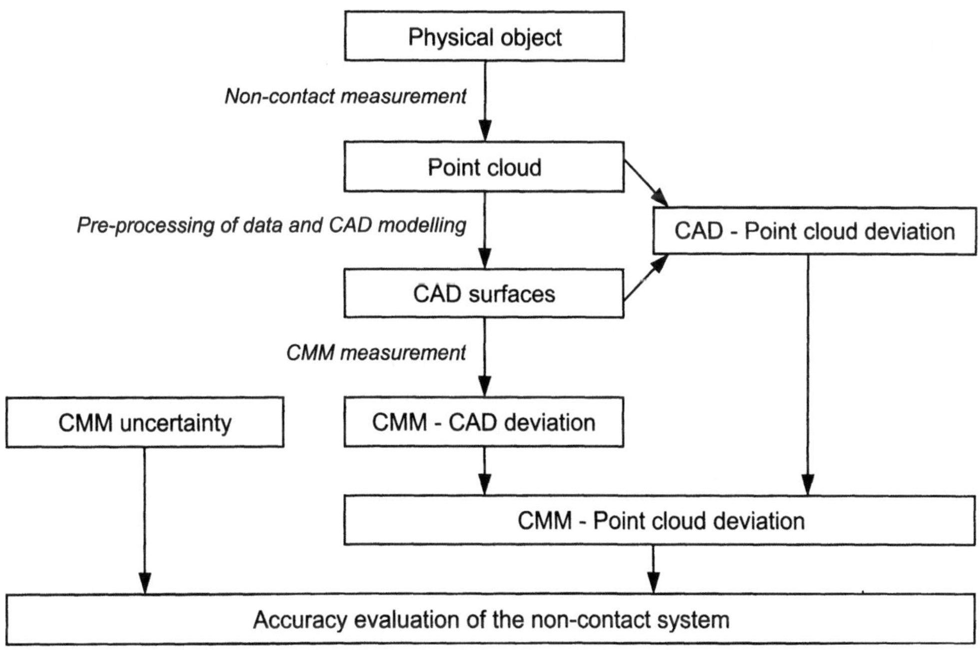

FIGURE 1. Outline of the method.

3.2 PRE-PROCESSING OF DATA AND CAD MODELLING

In the second phase, the acquired point clouds are pre-processed in order to prepare the data for the reconstruction of the CAD surfaces. Points can also be filtered in order to reduce the density of clouds and control the distribution over critical areas.

After pre-processing, parametric surfaces are defined over the point clouds. The strategy undertaken for surface reconstruction depends on the used CAD package. While some CAD packages can handle mass point clouds directly, others have limited point-handling capability or only can handle section line data.

In this phase it is also possible to evaluate the deviations existing between reconstructed surfaces and original point clouds. The assessment of such deviations is an essential step of the procedure for the estimation of the non-contact system accuracy. The surface can be analysed in terms of deviations against the cloud data (for example via colour plots). The result is a CAD model that represents the point cloud within a specified tolerance.

3.3 CMM MEASUREMENT

Previously generated CAD surfaces are imported in the CMM system. The CAD-based measuring software available on CMMs allows the programming of freeform CMM measurements and the evaluation of results with reference to CAD surfaces. The CMM can be programmed on the basis of paths drawn on the CAD surfaces, so that the CMM probing direction matches the normal direction to the CAD surfaces. The availability of the surface model provides the CMM with fundamental information for the measurement: dimension, shape and surface normal direction of the object. The touch probe measuring process is planned using predefined inspection plan functions available in the CMM software.

The coordinate system adopted in the optical digitisation is not used directly in the CMM system. Consequently, the errors caused by the 3D vision system have no influence in the coordinate system definition. The alignment is automatically improved using a 3D best-fit function available in the CMM software.

After the measurement, the deviations existing between the points acquired by the CMM (actual points) and those on the CAD surfaces (nominal points) are calculated. These deviations are defined as the shortest distance between a measured actual point and the point on the CAD surface calculated as the projection of the actual point.

3.4 ASSESSMENT OF DEVIATIONS

The local deviation between CMM measurement and non-contact measurement consists of two components:

- the deviation between optically measured point clouds and CAD surfaces (see 3.2) and
- the deviation between CAD surfaces and CMM measurements (see 3.3).

The final step of the procedure is the composition of such two components.

Both deviations are calculated as normal distances of points from CAD surfaces. The composition is made by an algebraic sum of the two distances. As a result, it gives back the total deviation between the non-contact point clouds and the CMM measurements. Since the measuring accuracy of the CMM is known, it is now possible to assess task-related accuracy.

4 CASE STUDY: THE TURBINE BLADE

The 3D-vision inspection of an industrial turbine blade has been used as example to demonstrate the feasibility of the proposed methodology.

4.1 INSTRUMENTS

The non-contact system used in the current work is a commercial whole field profilometer system based on active stereovision, where bi-dimensional patterns of light are projected on the part and acquired by a video camera. According to specifications given by the manufacturer, the system performs fast acquisition of surfaces with measuring accuracy of single-views within 120 μm (2σ). Such value is only a parameter given by the manufacturer for the used range of acquisition, without any mention about the specific measurement task.

For the accuracy verification, a coordinate measuring machine having scanning contact probe was used. The CMM system is equipped with dedicated software to measure free-form surfaces on the basis of CAD models. The mono-dimensional length measuring uncertainty of the specific CMM is $U_l = 2.2\ \mu m + L/300$ (where L is the measured length expressed in mm) [5].

4.2 PROCEDURE APPLICATION

The optical acquisition of the turbine blade surface was carried out by digitising eight partial views, corresponding to eight different measurements from different viewpoints. The system generates very dense point clouds: each range image has a resolution of 768 × 576 pixels (X × Y). Each view covers a measuring volume of 200 mm × 150 mm × 100 mm (X × Y × Z).

After that the eight distinct point clouds were aligned in a common reference system (see Fig. 2), the large set of points was pre-processed in order to produce regular clouds of points to facilitate the reconstruction of CAD surfaces. The undertaken strategy for the manipulation of points was to create section lines of the turbine blade surface as shown in Fig. 3. Points filtering and curve formation were performed by selecting points laying on parallel planes.

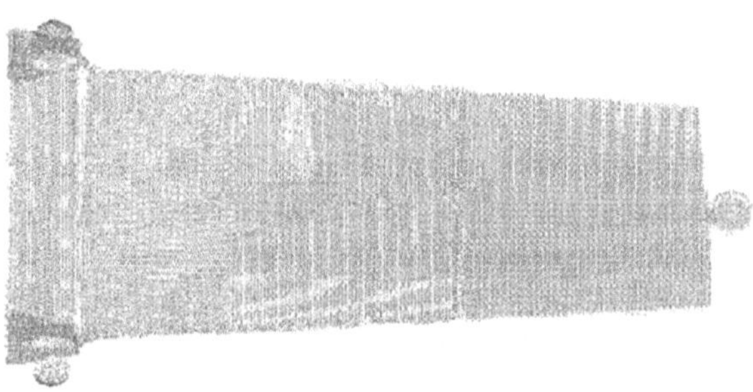

FIGURE 2. Acquired point clouds aligned in a common reference system.

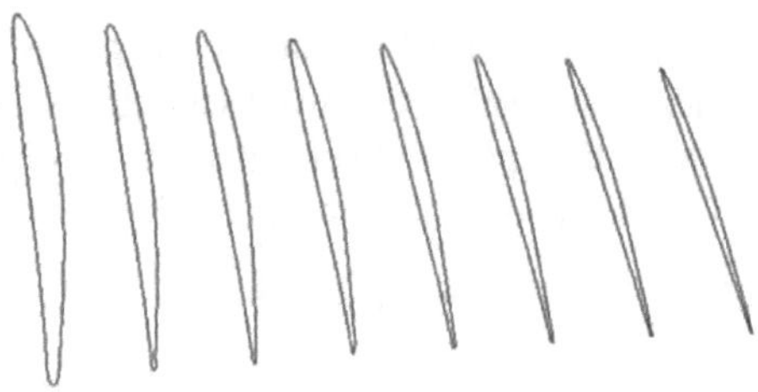

FIGURE 3. Curves generated for the reconstruction of CAD surfaces.

By means of the curves defined in the previous step, the CAD surfaces were automatically reconstructed. The final CAD model of functional surface is shown in Fig. 4. The maximum deviation of the parametric surfaces from point cloud was verified within a specified tolerance of 0.008 mm.

The CAD model was used to drive the contact probe during CMM measuring. Since the CMM software easily performs mathematical alignment and path planning definition, the turbine blade was fixed on the CMM table with no reference to the alignment in use for optical measurements. Then, using predefined functions available in the CMM environment, the off-line programming was carried out and, finally, scanning techniques were applied to measure the part.

The local deviations between optically measured point clouds and CAD surfaces were calculated by means of the CAD system, while the deviations between CAD surfaces and CMM measurements were given by the CMM software. The final stage of the process, the sum of the two deviation components, gave back an assessment of task-related accuracy, as exposed in next section.

FIGURE 4. Shading of the reconstructed CAD model.

4.3 EXPERIMENTAL RESULTS

In the specific application of the procedure to the turbine blade, the deviation between nominal points on the CAD surface and actual points from CMM measurements was found to be bigger than 0.1 mm for the majority of inspected areas. In comparison with such considerable deviation, the point clouds to CAD surface distance (less than 0.008 mm for every point) and the CMM uncertainty for scanning measurements over CAD model (less than 0.01 mm) [6] can be neglected. For this reason the accuracy assessment for the specific task can be directly estimated from the deviation existing between nominal points on the CAD surfaces and actual points measured with CMM.

Experimental results show a significant difference between the accuracy of a single view acquisition and the accuracy of the whole multiple-view model. In the following details on this are given.

SINGLE VIEW. At first, point clouds coming from different views were examined separately. The alignment of each examined point cloud was automatically improved using a 3D best-fit function available in the CMM software. This allowed to estimate the digitising error of a single view acquisition. For CMM measurements over nominal points belonging to a same view it was found that for 95% of data the deviation between nominal and actual points are within 0.15 mm.

MULTIPLE VIEWS. The whole model obtained from the alignment of distinct point clouds in a common reference system was verified by following scanning paths over the entire surface of the turbine blade. This allowed to estimate the digitising error of the multiple view acquisition. For CMM measurements over nominal points belonging to distinct views it was found that for the 95% of data the deviation between nominal and actual points are within 1 mm.

The considerable difference existing between the estimated accuracy of a single view acquisition (0.15 mm) and the estimated accuracy of the whole multiple-view model (1 mm) is largely imputable to the errors introduced during the alignment of distinct views. The alignment of views is a critical operation that has a strong influence on the overall accuracy of the reconstructed model. In the dimensional inspection with 3D vision systems, therefore, techniques for the assembling of views play a main role in the achievement of better metrological performances.

5 CONCLUSION

In this paper, an experimental method for performance verification of a 3D vision system has been described, with the specific purpose to assess the accuracy of the system when inspecting complex geometry. The procedure can be used to evaluate task related measurement uncertainty for virtually any measurement task. The key element of the procedure is the use of a coordinate measuring machine as the link to the traceability chain.

The accuracy estimation is achieved by determining the deviation existing between the points measured with the non-contact system and the true geometry of the original physical object. To identify the true geometry of the original physical object with a known degree of uncertainty, a CMM is used to measure the original physical object, following paths drawn on CAD surfaces reconstructed over the point clouds coming from the optical system.

A concrete example was also presented: the inspection of an industrial turbine blade by means of a commercial whole field profilometer system. Experimental results showed a significant difference between the accuracy of a single view acquisition and the accuracy of the whole multiple-view model, the first being within 0.15 mm (95% of points) and the second within 1 mm (95% of points). The difference is largely imputable to the errors introduced during the alignment of distinct views. Assembling of views, therefore, is a critical step for high accuracy measurements.

REFERENCES

1. Clark, J., (2001), Implementing Non-Contact Digitisation techniques within the mechanical design process, Sensor Review, Vol. 20, No. 3, 195-201.
2. Carbone, V., et al., (2001), Combination of a Vision System and a Coordinate Measuring Machine for the Reverse Engineering of Freeform Surfaces, Int J Adv Manuf Technol, Vol. 17, 263–271.
3. ISO 9000 series of standards – Quality systems, (1994).
4. De Chiffre, L., Hansen, H. N., (1995), Metrological limitations of optical probe techniques for dimensional measurements, Annals of CIRP, Vol. 44, No. 1, 501-504.
5. VDI/VDE 2617, (1986), Accuracy of coordinate measuring machines. Characteristic parameters and their checking. Parts 2.1, Measurement task specific measurement uncertainty. Length measurement uncertainty, Verein Deutscher Ingenieure, Düsseldorf, Germany.
6. Savio, E., De Chiffre, L., (2002), An Artefact for Traceable Free Form Measurements on Coordinate Measuring Machines, Precision Engineering, Vol. 26, 58-68.

AUTONOMOUS AGENT SYSTEM USING DISPATCHING RULES IN THE NEGOTIATION PROTOCOL

L. Castelli[1], A. Nicola[1], R. Pesenti[2], W. Ukovich[1]

[1] DEEI, University of Trieste, Italy
[2] DIINFO, University of Palermo, Italy

KEYWORDS: Autonomous Agents, Negotiation Protocols, Scheduling.

ABSTRACT. In this paper, the most important results obtained by the simulated application of autonomous agent paradigms to a real factory are presented. The classical rules of dispatching are compared with the autonomous agents approach. In particular, the possibility of redesigning the negotiation rules in terms of currency in order to take into account even non-time-related costs is considered. Finally, a new project on the effective application of the autonomous agent system to a test bed, modelling a simplified firm, is proposed.

1 INTRODUCTION

This work originates from a case study on an existing firm producing printed plastic items. A detailed simulation model has been developed for it, taking into account details sometimes disregarded in the literature, such as set-up times, transportation times, machine failures, etc. The model has been used as a test bed to compare the performance yielded by traditional dispatching rules versus autonomous agent paradigms as possible shop control schemes. Tardiness-based measures have been used for evaluating the relative performance of the dispatching rules based scheme. A thorough description of this work can be found in [1]. The most interesting findings have been developed in [2].

Based on these results, we propose a project for the effective application of these negotiation schemes to an ideal, simplified firm. A scheme of negotiation protocol based on currency functions is also considered.

The structure of the paper is as follows: in the next section some features of the real firm considered are described. Then, the autonomous agent system is briefly introduced. In section 4 the dispatching rules used for minimizing tardy jobs and the related versions of the negotiation protocol are presented. Performance evaluation results are summarized in section 5. A new negotiation protocol based on currency scheme is presented in section 6. The effective application of the autonomous agent system to a new test bed, is introduced in the last section.

2 CASE STUDY: THE REAL FIRM

The simulated test bed, based on an existing firm producing printed plastic items from simple plastic sheets, has been already presented in a previous work [2]. We summarize here its most important features:

Published in: E.Kuljanic (Ed.) *Advanced Manufacturing Systems and Technology*, CISM Courses and Lectures No. 437, Springer Wien New York, 2002.

- There are 21 machines (some machines can be clustered together when they perform the same action);
- Some machines work 24 hours per day, some others either 16 hours per day, or 8 hours per day; during the weekend the factory is closed;
- More than 1000 different items are produced;
- Each item has to be worked at least on 2 machines;
- There are not re-workings on the same article;
- If an item needs a particular action, it can choose between two or more machines of the relative machine cluster;
- The switching between different items on a same machine often requires a set-up time that we have considered;
- The simulated model includes the possibility of machine failures;
- Work orders can arrive at unforeseen times in the day;
- Each order is divided on pallets: each of them carries 400 plastic sheets;
- Preemption of an order is allowed between the processing of two different pallets, not between the sheets of the same pallet;
- "Just in time" approach is applied in the real factory production;
- The flow of work in progress is assured by a system of AGVs (automatically guided vehicles);
- Due dates are calculated using *TWK-method* presented in [3].

The simulation model has been implemented using eM-Plant® software (version 4.0), provided by Tecnomatix Technologies Ltd.

3 THE AUTONOMOUS AGENT SYSTEM

The basic entities of the manufacturing process modeled as autonomous agents are:

- The machines;
- The work orders;
- The pallets of each work order,

which interact dynamically to generate the production schedule. Two other types of agent are introduced:

- Transportation agents (AGVs are modeled as transportation agents);
- Database agent (as a simple *staff holon* [4]), with the function of collecting only immutable data, such as fixed sequences of machines, which have to be visited by each article.

The last two mentioned agents do not have an active part in the protocol negotiation and are usually called objects.

The interaction scheme between agents uses a bidding mechanism based on the contract net protocol idea. Agents have to know time dependent information only through mutual interaction. The negotiation protocol based on the dispatching rules based scheme has been developed in different versions, whose details can be found in [1]. Each version introduces new features to one or more agents of the system in order to increase the quality of agent interaction and system performance.

4 DISPATCHING RULES BASED SCHEME AND SCHEDULING EVOLUTION

In [1] and [2] two dispatching rules particularly suited for minimizing tardiness-based measures (see [5]) have been used: RR (Raghu and Rajendran) and PT+WINQ+SL (processing time plus work in next queue plus slack). These rules, using slack values, process preferably orders that can be completed sooner, thus trying to reduce the number of tardy orders.

The negotiation protocol includes only the PT+WINQ+SL rule for the priority evaluation, because it involves a smaller number of variables than RR.

PT+WINQ+SL works out the priority value Z_i^t of the order i at time t, using the processing time p_{ij} of the generic order operation j, the relative slack s_i^t and the work W_m^t in the queue of machine m at time t:

$$Z_i^t = \begin{cases} p_{ij} + \min(s_i^t, 0) & \text{if } j = k_i \\ p_{ij} + W_m^t + \min(s_i^t, 0) & \text{if } j < k_i, \end{cases} \tag{1}$$

where k_i is the number of operations needed to process the order i.

The negotiation protocol includes the PT+WINQ+SL rule for the priority evaluation. In [1] and [2] the use of the dispatching rule is different from what found in literature (see [6], [7], [8], [9]) and, in addition, the used protocol considers set-up times, that often are not considered for simplicity in other works.

As mentioned above, different versions of the scheduling mechanism have been proposed. The first, and most elementary one uses the PT+WINQ+SL rule implemented in a distributed architecture. In absence of a central coordination mechanism, work load is not decided by a central dispatcher as using simple dispatching schemes, but is derived on-line from the concurrent cooperation between agents.

The system operates as follows. Each pallet, once an operation has been completed or the first operation is going to begin, requests from the database agent the list of the machines able to perform the next needed action. Once this list has been acquired, the pallet agent notifies its presence to all these machines. Then machine agents record this notification of presence in a list. The list of each machine contains the information about the pallets that could be immediately processed by that machine. When the input buffer of a machine empties, the machine agent, if the machine is not failed, selects one of the pallets in its list.

Each pallet belongs to a work order, whose agent behaves as a supervisor for its pallet agents, so the machine agent sends a "request for information" to each order agent which has at least one pallet recorded in the machine list. Order data, which consist of the information necessary for priority value computation, are collected and recorded in proper fields of the different machine lists.

The pallet to be processed is then selected using the dispatching rule in order to evaluate the convenience of processing a particular order. In the negotiation phase, machine agents try to limit the number of setups processing without interruptions batches of pallets carrying the same article. In this first version of the system, pallet agents are passive: they have only to notify their presence to the machines and wait for a call for processing. Accordingly, control is performed mainly by machine agents.

The negotiation between agents has been improved in several ways. Each improvement to the interaction between pallet and machine agents has improved performance measures, in terms of experienced tardiness.

In the final version, the pallet has a more active role in the negotiation protocol. Each machine agent has two lists:

- *Ctrl_list,* in which pallet notifications of presence are recorded (as explained above);
- *Order_list,* which has the function of a reservation register for the machine agent: here pallets which are going to be processed on that machine are recorded.

Now the system operates as follows. Once the input buffer has become empty, if *order_list* is not empty, the machine "calls for processing" one of the pallets recorded there in FIFO order. If *order_list* is empty, the machine agent, by means of the dispatching rule, "elects" one of the pallets recorded in its *ctrl_list*, giving it the chance to begin a bargaining phase with all machine agents able to perform the needed action.

The elected pallet communicates with these machine agents (whose buffer or *order_list* may be not empty), in order to have both an evaluation of waiting time and a permission of being processed. Then the pallet agent chooses the machine with the lowest waiting time evaluation among the ones which have given positive permission of processing and reserves its next operation by means of a record in the *order_list* of that machine.

Whenever a machine failure occurs, the machine agent releases the pallets in its *order_list* (they have to begin a new phase communicating their notifications of presence to the machines and so on) and gives no more permissions of processing as far as failure persists. Pallet agents which can be processed by other machines, automatically avoid the failed machine without the necessity of special algorithms or behaviors to manage the problematic situation.

Beyond the *order_list* introduction, this version of the system shows other two important features:

- dispatching rule is slightly changed in order to give more importance to production order slack;
- possibility of accelerating production orders with work breaks among pallets over 24 hours is introduced. Order agent acts as a supervisor for its pallet agent: if it detects a pause longer than 24 hours, a boolean flag is set as true and the pallet will be processed with high priority.

According to [8], autonomous agent systems may be classified in:

- *part-centered bidding*: machines make bids and compete with each other to obtain a part to process. The part chooses the machine offering the best bid;
- *machine-centered bidding*: parts make bids to the machines; the machine accepts to process the more attractive part.

The scheme described above can be considered as a mixture of the previous ones: the negotiation protocol leaves to the pallet agent the final choice, but only between machines which have previously given their permission.

5 PERFORMANCE EVALUATION

Results are an average of 50 simulation runs, each one composed by 2.500 work orders.

In Table 1 experimental results about mean tardiness, max tardiness and percentage of tardy orders are summarized.

	Mean tardiness (sec)	Max tardiness (sec)	% Tardy orders
RR	12.176	---	---
PT+WINQ+SL	10.652	---	---
AA first version	5.274	107.795	34,51%
AA final version	2.917	61.049	30,13%

Table 1: Mean Tardiness, Max Tardiness, Percentage of Tardy Orders

In terms of mean tardiness, the autonomous agent versions obtain a strong improvement respect to RR and PT+WINQ+SL. In the autonomous agents last version, the mean tardiness is reduced to 45 % of the first one.

The final autonomous agents version obtains significantly the best performances even in terms of maximum tardiness, variance of tardiness and percentage of tardy orders.

ANOVA and Duncan's multiple range test [10] have been used with significance level $\alpha = 1\%$ in order to verify that performance measures are significantly different and normality assumption of data samples has been verified using a chi-square test again with significance level $\alpha = 1\%$.

Our test-bed and simulation experiments have generated interesting results in heterarchical production scheduling. The merits of the proposed bidding technique have been established and analysis of results have pointed out directions for extending our protocol.

6 CURRENCY SCHEME

On the developed models based on the PT+WINQ+SL dispatching priority rule, we are currently evaluating the introduction of a new negotiation scheme based on currency values [11].

The most important feature of this approach is the possibility to introduce a non linear dependence of the currency on the time before and after the due date, using shape factors in the conversion functions. This kind of negotiation model may include all costs that cannot be expressed in temporal terms, as order costs and penalty on tardy orders. The parameters in the shape functions may be used also to customize the protocol to firm requirements.

The negotiation protocol structure is the following. Production orders enter the system with some budget, depending on their priority evaluated, for instance, according to a non-linear function of the order slack. Order agents broadcast processing requests to all machines able to process them. Each machine has a limited buffer to store the parts to be processed. The machines that have not exceeded the buffer limit will reply with a bid, based on expected completion time and on the past bid success rate [11]. The order agent selects the lowest bid, and controls that it is not over its budget. The selected machine, after re-verifying its buffer limit, accepts the process and periodically confirms its availability. In any case in which the previous conditions are not verified, the order agent works out a new budget according to the new slack condition of the order, and broadcasts again a bid request.

7 APPLICATIVE PROJECT

As mentioned in the introduction, at present we have devised a project, which has, as final objective, the implementation of a general model, which can be applied with low cost to most manufacturing firms. The basic features of the model are:

- The flexibility toward the application to different structures. The model must not be developed *"ad hoc"* for the *case study*, but should be general enough to be customizable at a low cost for many sensibly but not radically different manufacturing firms.
- The interchange ability of different strategies that can be applied. We should easily substitute the algorithms of the agents without reprogramming the model.

Beyond all this, to improve and finalize the project, the project contemplates producing a new test bed, so we can gradually insert real hardware autonomous agents, starting with machine agents. In order to achieve it we are working with an engineering and research firm, specialized in using electronics and information technologies to globally manage and control industrial production processes.

Obviously, in order to manage the project, we intend to achieve it in several steps. The first one is the application of real autonomous agents to a simple and featurable firm, created with a laboratory simulator and composed of 4 different machines. Production orders should be composed of pallet and each item has to be processed by two or more machines, with a fixed sequence.

The engineering and research firm uses simple process computers connected to the machines, having two functions:

- Checking the machine status (on, off, out of order, etc.);
- Sending and recording information about the current operation (they are programmable by the machine operator).

The collected information is stored in a central database and it gives a complete description of production process and firm status. The model in which we are introducing hardware autonomous agents is proposed in figure 1. Presently, the process computers are used only to acquire and store data in a database. The database is connected to a scheduler that tries to improve the system performance and stores crucial data in the study of consequences of machine failures.

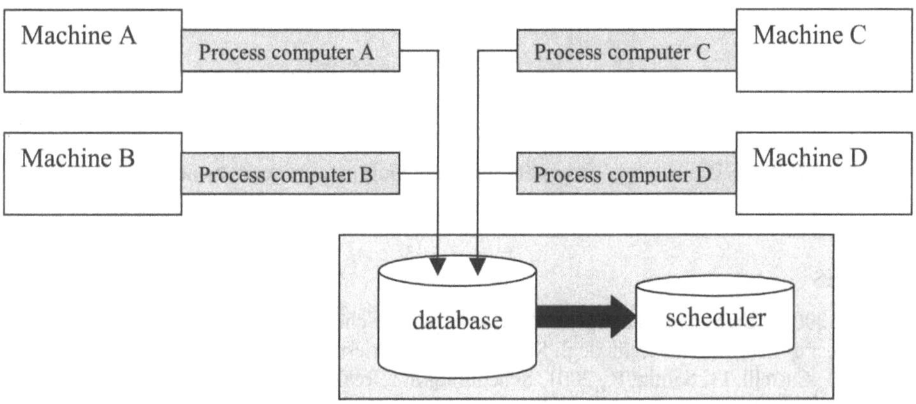

FIGURE 1. Acquiring and storing scheme, using process computers

We are working to improve this passive role of the process computers, in order to convert them in hardware machine agents, as proposed in figure 2. In the scheme of figure 2 all process computers are connected, and use XML standard for communications with TCP-IP protocol. Process computers, used as machine autonomous agents, will be programmed with UML standard, or similar and specifics languages as AUML. When the laboratory simulator is complete, we should test several negotiation protocols and their adaptability on different kind of production orders.

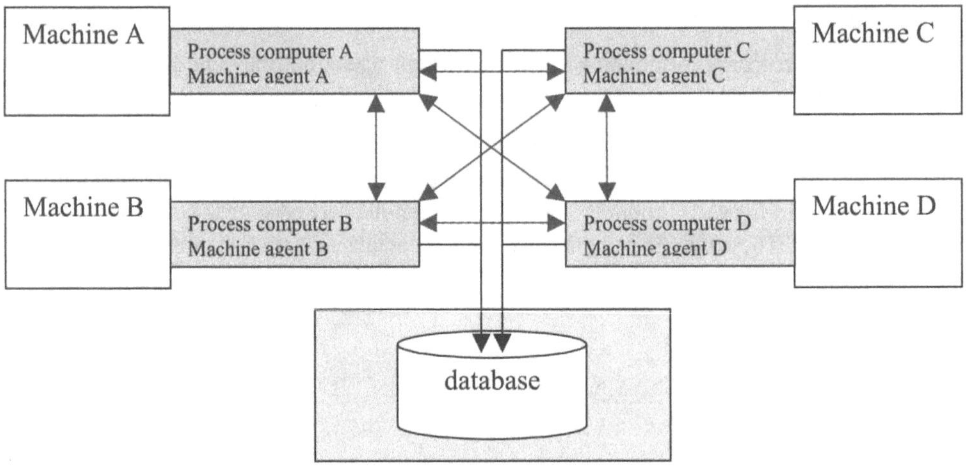

FIGURE 2. Process computers used as machine autonomous agents

REFERENCES

1. Santin, P. 2001. "Un Sistema ad Agenti Autonomi per la Schedulazione della Produzione in Tempo Reale". Tesi di Laurea. Università degli Studi di Trieste, Trieste, Italy – in italian.
2. Pesenti, R., Castelli, L., Santin, P., 2001, Scheduling in a Realistic Environment Using Autonomous Agents: A Simulation Study. Proceedings of Agent Based Simulation II Workshop, Passau, Germany, April 2-4, pp.149-154.
3. Blackstone, J.H.; D.T. Phillips, and G.L. Hogg. 1982, "A State-of-the-Art Survey of Dispatching Rules for Manufacturing Job Shop Operations". *International Journal of Production Research* 20, No.1, 27-45.
4. Wyns, J. 1999. "Reference Architecture for Holonic Manufacturing Systems" Phd Thesis, KULeuven, Katholieke Universiteit Leuven, Faculteit Toegepaste Wetenschappen, Departement Werktuigkunde Afdeling Productietechnieken, Machinebouw en Automatisering, Celestijnenlaan 300B-B-3001, Heverlee (Leuven), Belgium.
5. Holthaus, O. 1997. "Design of Efficient Job Shop Scheduling Rules". *Computers and Industrial Engineering* 33, No. 1, 245-252.
6. Baker, A.D. 1998. "A Survey of Factory Control Algorithms which can be implemented in a Multi Agent Heterarchy: Dispatching, Scheduling and Pull". *Journal of Manufacturing Systems* 37, No.4, 297-320.
7. Brennan, R.W. and W. O. 2000 "A Simulation Test-Bed to Evaluate Multi-Agent Control of Manufacturing Systems". In *Proceedings of the 2000 Winter Simulation Conference* (Orlando, FL, Dec. 10-13). J.A. Joines, R.R. Barton, K. Kang, and P.A. Fishwick, eds, 1747-1756.
8. Saad, A.; K. Kawamura; and G. Biswas. 1997. "Performance Evaluation of Contract Net-Based Heterarchical Scheduling for Flexible Manufacturing Systems." *Intelligent Automation and Soft Computing* 3, No. 3, 229-248.
9. Kouiss, K; H. Pierreval; and N. Mebarki. 1997. "Using multi-agent architecture in FMS for dynamic scheduling." *Journal of Intelligent Manufacturing* 8, No.1, 41-47.
10. Montgomery, D.C. 1997. *Design and Analysis of Experiments.* John Wiley & Sons, New York, N.J.
11. Krothapalli, N.K.C. and A.V. Deshmukh. 1999. "Design of Negotiation Protocols for Multi-Agent Manufacturing Systems". *International Journal of Production Research* 37, No.7, 1601-1624.

ROBUSTNESS ANALYSIS OF FREQUENCY-DOMAIN PI CONTROLLER DESIGN PROCEDURES

W. Krajewski[1] A. Lepschy[2] S. Miani[3] * U. Viaro[3]

[1] Systems Research Institute, Polish Academy of Sciences, Warsaw, Poland

[2] Dip.to di Elettronica e Informatica, University of Padova, Italy

[3] Dip.to di Ingegneria Elettrica, Gestionale e Meccanica, University of Udine, Italy

KEY WORDS: Process control, Robustness, PI controllers, Time-delayed processes, Stability margins

ABSTRACT: PI controllers for first-order time-delayed processes can be designed on the basis of a normalized model according to open-loop frequency-domain specifications. To this purpose, it is convenient to refer to the charts representing the loci of constant stability margins and cross-over frequencies in the plane of normalized parameters. It is shown how the designer can avail himself of these charts to easily evaluate the effects of changes in the process parameters and, thus, control system robustness. In particular, the situation corresponding to the least stability margin is identified together with the related cross-over frequency. The procedure is illustrated with the aid of representative examples.

1 INTRODUCTION AND PROBLEM STATEMENT

PI controller tuning can be based on a normalization of the open-loop transfer function and on frequency domain specifications along the following lines.

Let the actual process transfer function be approximated by the standard first-order lag plus delay (FOLPD) model [1], [2],[3]:

$$P(\hat{s}) = K\frac{e^{-L\hat{s}}}{1+T\hat{s}}, \tag{1}$$

where the independent variable has been denoted by a hat because the standard symbol s will be reserved to the normalized variable, and let the PI controller transfer function be represented by:

$$C(\hat{s}) = K_P + \frac{K_I}{\hat{s}} = K_P\left(1+\frac{1}{T_I\hat{s}}\right). \tag{2}$$

The above-mentioned normalization is based on the variable transformation:

$$s := T\hat{s}, \tag{3}$$

*Corresponding author. Email: miani.stefano@uniud.it

Published in: E.Kuljanic (Ed.) *Advanced Manufacturing Systems and Technology*, CISM Courses and Lectures No. 437, Springer Wien New York, 2002.

leading to the normalized open-loop transfer function:

$$G(s) = \frac{(a+bs)e^{-\tau s}}{s(1+s)}, \tag{4}$$

where:

$$a := K_I K T, \quad b := K_P K, \quad \tau = \frac{L}{T}. \tag{5}$$

Note that parameter τ depends on the process only, whereas a and b are also proportional to the controller parameters K_I and K_P.

Since (2) contains two "free" parameters, it makes sense to refer to a pair of independent performance specifications; here we consider the phase margin m_φ, accounting for stability robustness and transient precision, and the gain cross-over frequency ω_A, accounting for response promptness and filtering properties (alternatively, the pair m_φ, m_g could be considered).

As shown in [4], the points of the (a,b)-plane characterized by a constant value of m_φ form a curve (indicated in the sequel by m_φ=const) whose shape depends on the normalized delay τ, whereas the loci ω_A=const (ellipses) are independent of it. Therefore, from a practical point of view it is required: (i) to consider the diagram with the loci m_φ=const for a number of τ values, and (ii) to superimpose the curves of ω_A=const on these diagrams.

By way of example, the family of lines ω_A=const and m_φ=const for $\tau = 0.5$ is represented in Fig.1. The boundary of the stability region on the (a,b)-plane is formed by the right half-plane arc where $m_\varphi = 0^o$ and by the segment of the b-axis ($a = 0$) included between its intersections with such axis.

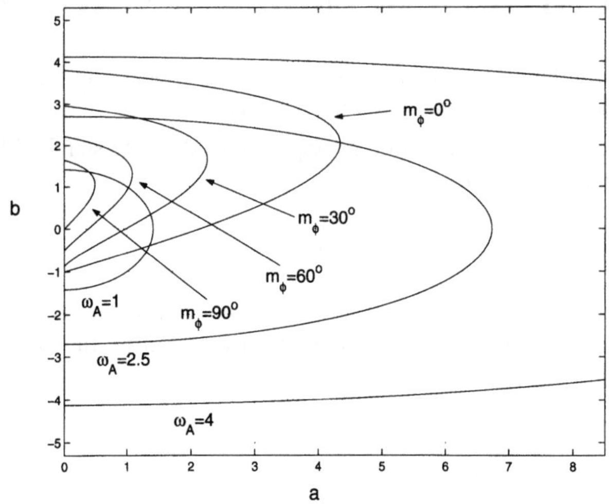

Figure 1: Chart for $\tau = 0.5$ with the loci m_φ =const and ω_A =const

After the process transfer function (1) has been identified [5] and the relevant value of $\tau = \frac{L}{T}$ has been determined, from the specification on the gain cross-over frequency, i.e., $\hat{\omega}_A = \hat{\omega}_A^*$, one immediately obtains the desired normalized cross-over frequency $\omega_A^* = T\hat{\omega}_A^*$. Then, the coordinates (a^*, b^*) of the intersection, if any, between the curve $\omega_A = \omega_A^*$ and the curve corresponding to the desired phase margin $m_\varphi = m_\varphi^*$ on the chart for τ are evaluated. Finally, the values K_P^* and K_I^* of the controller parameters ensuring that $C(\hat{s})P(\hat{s})$ exhibits the cross-over frequency $\hat{\omega}_A^* = \frac{1}{T}\omega_A^*$ and phase margin m_φ^* are computed as:

$$K_P^* = \frac{b^*}{K}, \quad K_I = \frac{a^*}{KT}. \tag{6}$$

The procedure for finding a^* and b^* is illustrated in Fig.2 with reference to the process transfer function:

$$P(\hat{s}) = \frac{e^{-0.5\hat{s}}}{1+\hat{s}}, \tag{7}$$

for which $K = T = 1$ and $L = 0.5$ (so that $\tau = 0.5$), and to the specifications: $m_\varphi = m_\varphi^* = 60^o$, $\hat{\omega}_A = \hat{\omega}_A^* = 1$ (so that $\omega_A^* = 1$). From (6) it follows that $K_P^* = b^* = 0.9761$ and $K_I^* = a^* = 1.0233$.

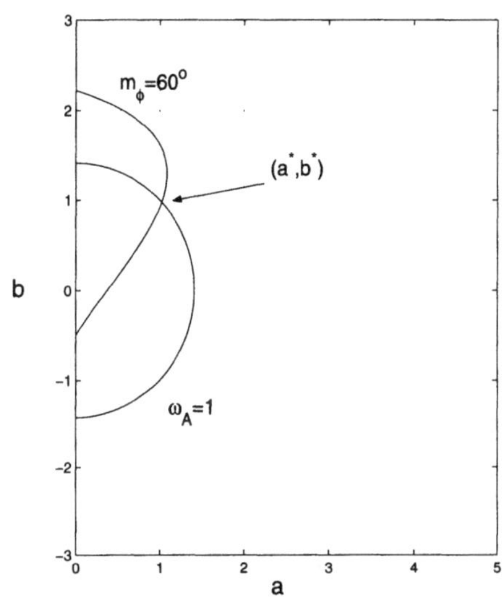

Figure 2: Determination of the parameter values a^* and b^* satisfying the specifications $\omega_A = 1$ and $m_\varphi = 60^o$

The next section analyses the robustness of the resulting design when the model parameters are allowed to vary over prescribed intervals.

2 ROBUSTNESS ANALYSIS

Let us assume that the uncertainty on the process parameters is described by the inequalities:

$$K_m \le K \le K_M, \quad T_m \le T \le T_M, \quad L_m \le L \le L_M, \tag{8}$$

and that the controller parameters are tuned to the values K_P and K_I computed according to the procedure outlined in Section 1 with reference to the nominal values K_n, T_n and L_n of the process parameters K, T and L.

Relations (8) define a parallelepiped in the (K,T,L)-space which is mapped by (5) into a 6-face solid $\mathcal{U}$ of the (a,b,τ)-space as exemplified in Fig.3 (eight edges of $\mathcal{U}$ are straight-line

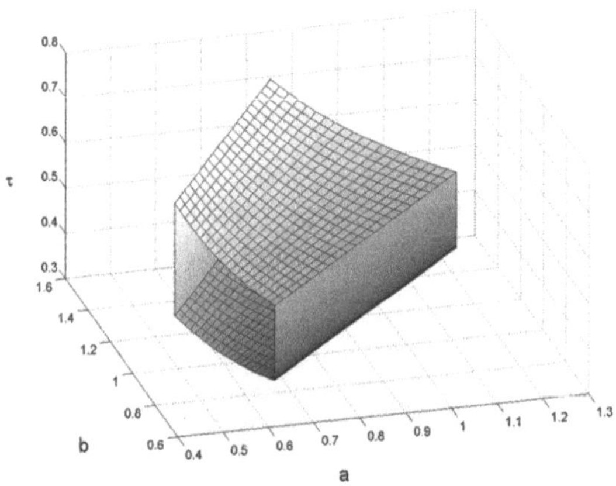

Figure 3: Representation of the uncertainty solid in the (K,T,L)-space

segments and four edges are curved arcs). The worst situation compatible with the given parameter ranges, i.e., the one leading to the least phase margin, can easily be evaluated using the above-mentioned charts as follows. First, the range $\mathcal{D}$ of τ-values corresponding to (8) is computed as:

$$\mathcal{D} := [\tau_m,\ \tau_M] := \left[\frac{L_m}{T_M},\ \frac{L_M}{T_m}\right]. \tag{9}$$

Then, the intersections of $\mathcal{U}$ with a suitable number of planes $\tau = \hat{\tau}$, $\hat{\tau} \in \mathcal{D}$, are determined. The resulting plane figures have the form of a trapezium, like the one in Fig.4, whose two parallel sides are always on the same (horizontal) straight lines:

$$b = b_m := K_P K_m, \quad b = b_M := K_P K_M. \tag{10}$$

The other (slanted) sides belong to the straight lines through the origin with slopes:

$$\gamma_{\hat{\tau},m} := \frac{K_P}{K_I}\frac{1}{T_{\hat{\tau},M}}, \quad \gamma_{\hat{\tau},M} := \frac{K_P}{K_I}\frac{1}{T_{\hat{\tau},m}}, \tag{11}$$

where:

$$T_{\hat{\tau},m} := \max\left\{T_m, \frac{L_m}{\hat{\tau}}\right\}, \quad T_{\hat{\tau},M} := \min\left\{T_M, \frac{L_M}{\hat{\tau}}\right\}. \tag{12}$$

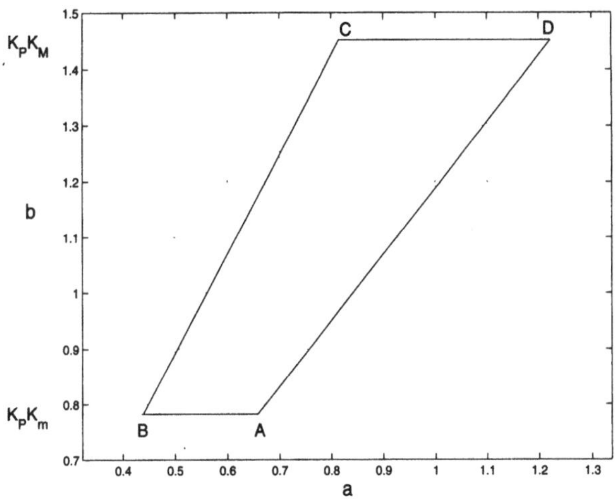

Figure 4: Intersection of the uncertainty solid with a plane $\tau = \hat{\tau}$ =const.

It follows that the coordinates of the four vertices of the trapezium of Fig.4 are:

$$\begin{array}{ll} \text{- point A:} & (a_A, b_A) = (\frac{1}{\gamma_{\hat{\tau},m}} b_m, b_m) = (T_{\hat{\tau},M} K_m K_I, K_m K_P), \\ \text{- point B:} & (a_B, b_B) = (\frac{1}{\gamma_{\hat{\tau},M}} b_m, b_m) = (T_{\hat{\tau},m} K_m K_I, K_m K_P), \\ \text{- point C:} & (a_C, b_C) = (\frac{1}{\gamma_{\hat{\tau},M}} b_M, b_M) = (T_{\hat{\tau},m} K_M K_I, K_M K_P), \\ \text{- point D:} & (a_D, b_D) = (\frac{1}{\gamma_{\hat{\tau},m}} b_M, b_M) = (T_{\hat{\tau},M} K_M K_I, K_M K_P). \end{array} \quad (13)$$

Both for $\hat{\tau} = \tau_m$ and for $\hat{\tau} = \tau_M$ the trapezium degenerates into a segment (so that $A = B$ and $C = D$) of slope $\frac{K_P}{K_I}\frac{1}{T_M}$ and $\frac{K_P}{K_I}\frac{1}{T_m}$, respectively. As $\hat{\tau}$ increases from τ_m to $\tau_1 := \frac{L_m}{T_m}$ the left side moves to the left, whereas for $\hat{\tau} \geq \tau_1$ this side remains unchanged; similarly, the right side remains unchanged for $\hat{\tau} \leq \tau_2 := \frac{L_M}{T_M}$ and, as $\hat{\tau}$ increases from τ_2 to τ_M, it moves to the left until it eventually overlaps the left side along the straight line with slope $\gamma_M := \gamma_{\tau_M,M} = \frac{K_P}{K_I}\frac{1}{T_m}$.

As clearly shown by Fig.1, on any horizontal segment (b =const) inside the stability region of the chart for $\hat{\tau}$, the smallest stability margin is achieved at the rightmost point. Also, since the region where the phase margin is less than a given value is convex in the area of interest, the worst case may only correspond to either point A or point D; usually, the least value $m^l_{\varphi,\hat{\tau}}$ of m_φ is the one corresponding to point D (even if it could correspond to point A, when the trapezium lies within the region where $m_\varphi > 90^o$).

In general, as $\hat{\tau}$ changes in $\mathcal{D}$, $m^l_{\varphi,\hat{\tau}}$ changes too, because the curves m_φ=const approach the b-axis for increasing $\hat{\tau}$. Usually the overall minimum:

$$m_{\varphi,m} = \min_{\hat{\tau} \in \mathcal{D}} \left\{ m^l_{\varphi,\hat{\tau}} \right\} \quad (14)$$

is achieved either for $\hat{\tau} = \tau_2$ (because the right side of the trapezium remains the same for $\tau_m \leq \hat{\tau} \leq \tau_2$, so that $m^l_{\varphi,\tau_2} \leq m^l_{\varphi,\hat{\tau}}$) or for $\hat{\tau} = \tau_M$, as shown by the examples in the next section.

Since the cross-over frequency ω_A of the normalized transfer function $G(s)$ depends on transformation (3), to obtain an estimate of the rise time $\hat{t}_r$ of the actual step response (and to evaluate its changes due to parameter variations), it is necessary to determine the value of the actual cross-over frequency $\hat{\omega}_A = \frac{1}{T}\omega_A$ beforehand. This, in turn, requires computing T. Now, the values K^* and T^* of the process parameters K and T corresponding to a point $P^* = (a^*, b^*)$ on the considered chart are given by (cf. (5)):

$$K^* = \frac{b^*}{K_P}, \tag{15}$$

$$T^* = \frac{a^*}{K_I K^*} = \frac{K_P}{K_I}\frac{a^*}{b^*}. \tag{16}$$

Therefore the relationship between the actual cross-over frequency $\hat{\omega}_A^*$ and the normalized cross-over frequency ω_A^* at P^* is:

$$\hat{\omega}_A^* = \frac{K_I}{K_P}\frac{b^*}{a^*}\omega_A^*. \tag{17}$$

3 EXAMPLES

Let us consider again a process whose nominal transfer function is given by (7) so that: $K_n = 1$, $T_n = 1$, $L_n = 0.5$, and the nominal value of the normalized delay is: $\tau_n = \frac{L_n}{T_n} = 0.5$, and assume that the process parameters may vary over the intervals: $0.7 \le K \le 1.3$, $0.8 \le T \le 1.2$, $0.4 \le L \le 0.6$ (so that $0.\bar{3} \le \tau \le 0.75$.) The system robustness will be evaluated with reference to each of the following specification pairs:

(i) $m_\varphi = 50^o$, $\hat{\omega}_A = 1$,

(ii) $m_\varphi = 60.05^o$, $\hat{\omega}_A = 1$,

(iii) $m_\varphi = 70^o$, $\hat{\omega}_A = 1$,

the second of which only slightly differs from that adopted at the end of Section 1 (where $m_\varphi = 60^o$). The above specifications give rise to three different situations (yet, not exhaustive of all possible cases): if the design is based on (i), the least phase margin occurs for $\hat{\tau} = \tau_2 = L_M/T_M = 0.5$; if it is based on (iii), it occurs for $\hat{\tau} = \tau_M = 0.75$; whereas specifications (ii) lead to the "intermediate" situation in which $m_{\varphi,m} = m^l_{\varphi,\tau_2} = m^l_{\varphi,\tau_M}$.

According to (5), the (constant) controller parameters K_p and K_I are computed as:

$$K_P = \frac{b_n}{K_n}, \quad K_I = \frac{a_n}{K_n T_n}, \tag{18}$$

where a_n and b_n denote the nominal values of the normalized model parameters, which could be determined either graphically (using the previous charts) or analytically from the "interpolation conditions":

$$a_n = \left[\sin(L_n\hat{\omega}_A + m_\varphi) + T_n\hat{\omega}_A\cos(L_n\hat{\omega}_A + m_\varphi)\right] T_n\hat{\omega}_A, \tag{19}$$

$$b_n = T_n\hat{\omega}_A\sin(L_n\hat{\omega}_A + m_\varphi) - \cos(L_n\hat{\omega}_A + m_\varphi), \tag{20}$$

immediately obtained from $C(j\hat{\omega}_A)P(j\hat{\omega}_A) = \cos(m_\varphi - \pi) + j\sin(m_\varphi - \pi)$ by separating real and imaginary parts. In all the above-mentioned cases, the overall minimum $m_{\varphi,m}$ corresponds to point $D = (a_D, b_D)$ of a trapezoidal section but, as already pointed out, does not occur for the same value of τ (and of T).

Specifications (i) lead to:

- controller parameters: $K_P = 0.7836$, $K_I = 1.1773$;
- minimal phase (worst case) margin: $m_{\varphi,m} = 34.33^o$;
- values of the process parameters at the minimum: $K = 1.3$, $T = 1.2$, $L = 0.6$ ($\tau = 0.5$);
- worst case gain cross-over frequency: $\hat{\omega}_A = 1.135$.

Fig.5 shows the control system step responses for the nominal plant (curve a), for the worst-case plant (curve b) and for the plant corresponding to $K = 1.3$, $T = 0.8$, $L = 0.6$ (curve c), i.e., to point D of the (degenerate) trapezium for $\tau = 0.75$.

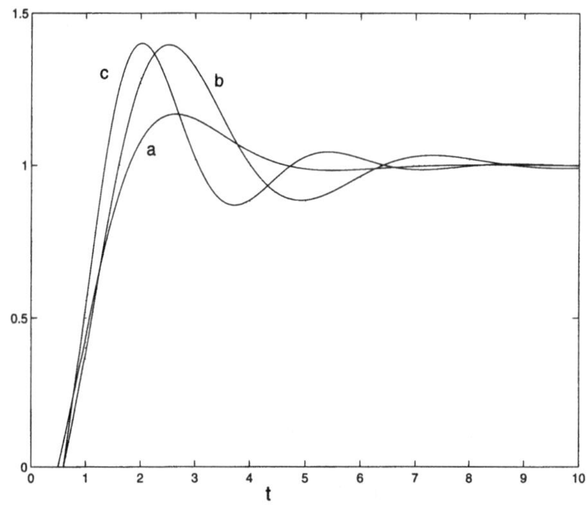

Figure 5: Step responses of the control system designed according to specifications (i)

Specifications (ii) lead to:

- controller parameters: $K_P = 0.9770$, $K_I = 1.0225$;
- minimal phase (worst case) margin: $m_{\varphi,m} = 43.83^o$;
- values of the process parameters at the minimum: $K = 1.3$, $T = 0.8, 1.2$, $L = 0.6$ ($\tau =$ 0.5, 0.75);
- worst case gain cross-over frequency: $\hat{\omega}_A = 1.158$.

Specifications (iii) lead to:

- controller parameters: $K_P = 1.139$, $K_I = 0.8383$;
- minimal phase (worst case) margin: $m_{\varphi,m} = 48.03^o$;
- values of the process parameters at the minimum: $K = 1.3$, $T = 0.8$, $L = 0.6$ $(\tau = 0.75)$;
- worst case gain cross-over frequency: $\hat{\omega}_A = 1.607$.

4 CONCLUSIONS

The robustness of the PI control of a FOLPD process has been analysed by referring to the parameter space of the normalized open-loop transfer function (4). Specifically, the solid representing the structured plant uncertainty in the (a,b,τ)-space has first been constructed. Then, its contacts with the loci characterized by constant stability margins and cross-over frequencies have been determined, which allows us to easily identify the situations leading to the minimal phase margins and the related cross-over frequencies. Usually ($m_\varphi < 90^o$), the worst case with respect to stability robustness corresponds to the upper right corner of the trapezoidal intersection of the above-mentioned solid with a plane τ=const, whose coordinates have been provided; however, as shown by the examples of Section 3, it can occur for different values of the plant model time constant.

References

[1] Ziegler, J. G., and Nichols, N. B. (1942). Optimum settings for automatic controllers. *Transactions of the ASME*, 64, 759–768.

[2] Åström, K. J., and Hägglund, T. (1995). PID controllers: Theory, design, and tuning. *Instrument society of America*, Research Triangle Park, NC, USA.

[3] O'Dwyer, A. (2000a). A summary of PI and PID controller tuning rules for processes with time delay. Part 1: PI controller tuning rules. *Proceedings of PID'00: IFAC workshop on digital control*, Terrassa, Spain (pp. 175–180) (published in Digital control: Past, present and future of PID control, Pergamon Press, ISBN: 0-08-043624-2).

[4] Krajewski, W., Lepschy, A., and Viaro, U. (2001). PI compensation of time-delayed processes based on frequency domain specifications. Internal Report, University of Udine, Department of Electrical, Mechanical and Management Engineering.

[5] O'Dwyer, A. (2000b). Time-delayed process model parameter estimation: A classification of techniques. *Proceedings of the international conference on control 2000*, Cambridge, England.

A NEW APPROACH TO THE USE OF VIBRATION SIGNALS IN TOOL-WEAR MONITORING IN HIGH SPEED METAL TURNING

Dimla E Dimla Snr.[1], M.Sortino[2]

[1] School of Engineering & Technology, De Montfort University, Leicester, United Kingdom
[2] Department of Electrical, Mechanical and Management Engineering, University of Udine, Italy

KEYWORDS: Tool Wear Monitoring, Vibrations, High Speed Metal Cutting

ABSTRACT: Tool wear monitoring plays a very important part in production quality, volume and downtime of a machine tool. In order to reduce scrap levels, tool wear must be monitored and cutting tools changed when they might adversely affect the product quality. The investigation reported in this paper was aimed at furthering our understanding of the relationship between cutting tool wear and vibration signatures. A series of turning tests were conducted using double-coated insets to machine alloy steel. The prevailing wear forms on the cutting tool were measured while the vibration signals were recorded. An analysis of the vibration signals indicated that there was significant correlation between some peak frequencies of the vibration signal and tool wear.

1. INTRODUCTION

A metal cutting tool is considered to be worn when the surface finish quality (i.e. roughness) on the workpiece significantly deteriorates or when the dimensional integrity, in terms of tolerances are grossly violated. Tool wear processes generally occur in combination with the predominant wear mode and are dependent upon the cutting conditions, workpiece, tooling material and the tool insert geometry. Clearly, dependency of the cutting tool wear on these factors makes tool wear monitoring an extremely daunting task.

During a typical metal cutting operation, a complex and dynamically varying system develops between the workpiece and the cutting tool. The complexity and stochastic nature of the cutting process mean that identification of the actual cutting tool condition is difficult, but not impossible. Measurement of the cutting forces, power requirements, vibrations, temperature, acoustic emission, etc. could be achieved through the application of sensors, and their behaviour used in predicting and monitoring tool wear. The aim of such investigations is often to establish distinctive sensor signal characteristics whose trend bears a meaningful relationship between signal features and measured tool wear.

The investigation described in this paper centred on a turning operation conducted to determine characteristic vibration signal features considered to be sensitive to tool wear. High speed turning of hardened carbon steel using carbide tools was conducted, and on-line vibration data (components of acceleration signals in the three principal directions) was collected, analysed and features correlated to tool wear. It was possible to identify distinct differences in signal characteristics when cutting was performed with a new insert compared to that carried out with a worn insert. Frequency based features correlated well to the tool wear, with the determined

Published in: E.Kuljanic (Ed.) *Advanced Manufacturing Systems and Technology*,
CISM Courses and Lectures No. 437, Springer Wien New York, 2002.

characteristics showing the sensor signals affected by the different wear modes.

2. BRIEF LITERATURE REVIEW

Over the years, attempts at tool condition monitoring (TCM) have utilised mechanical and physical sensors such as touch trigger probes and proximity sensors, but these require interruption of the cutting process for measurements to be made. Focus on indirectly detected parameters (e.g. cutting forces, vibration and chatter, acoustic emission, temperature, current, torque, etc.) whereby process changes resulting as a consequence of tool wear have been the way forward. Advances in technology permit these cutting tool variables to be measured through the application of sensors and their behaviour monitored to facilitate tool state prediction. After more than a decade of investigations, a generally applicable TCM system is still to be found (Lister, 1993), but TCM systems currently exist for particular applications such as acoustic emission (AE), vibrations and force measurement (Byrne et al, 1995). The vast amount of literature in this field suggests that a variety of process parameters in the metal cutting environment can be tapped and used to predict the cutting tool-state. Several literature surveys have been conducted; for example, see Dimla (2000) and Byrne et al (1995).

3. EXPERIMENTAL PROCEDURE

The instrumentation set-up is shown in Figure 1. The set-up consisted of a centre-lathe with a Kistler triaxial accelerometer (type 8730A) to measure mutually perpendicular components of vibration. A Kistler power supply coupler (type 5134) was used to decouple the acceleration signals, sampled using a 386 SX PC with a 12-bits Amplicon PC-30 data acquisition card at 4096 samples per channel at 30 kHz.

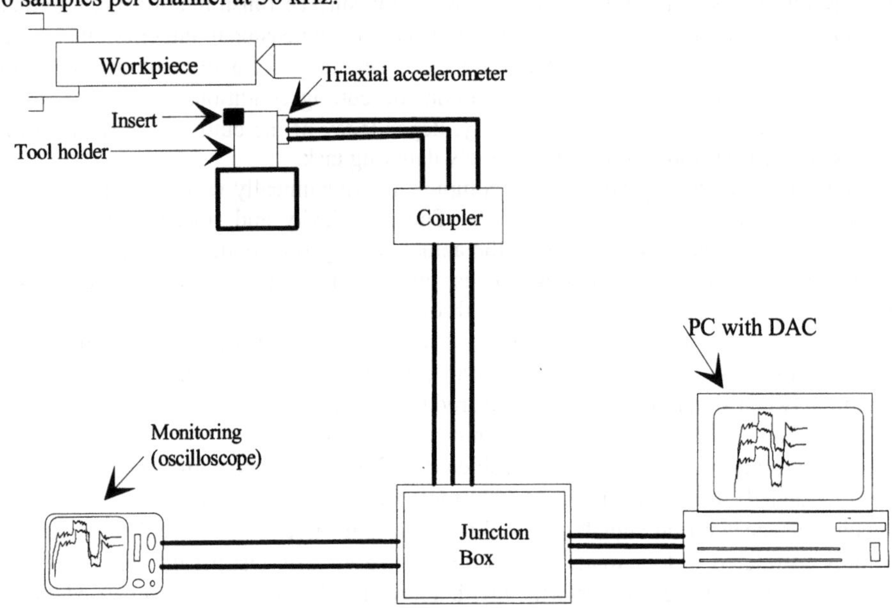

FIGURE 1. Experimental test rig

The cutting conditions were selected based on the toughness of the workpiece within the tool manufacturer's recommended range for the workpiece-tooling geometry configuration (Sandvik, 1995). The tests were conducted within the following cutting conditions:

- Cutting speeds: 275 m/min., 300 m/min., 350 m/min.,
- Feed-rates: 0.1 mm/rev, 0.2 mm/rev and 0.3 mm/rev., and
- Fixed depth of cut of 2 mm.

The tool inserts used (Sandvik P15 and P25) were coated via chemical vapour deposition with P15 having a thick wear resistant coating on a hard resistant substrate. Its inner coating was 8 μm thick comprising of 6 μm aluminium oxide (Al_2O_3) on top and a thinner outer layer of titanium nitride. The P25 insert on the other hand had a thick (≈ 10μm) layer of Al_2O_3 on top of a medium size titanium carbon nitride layer. The workpiece used was low carbon alloy steel, oil quenched, rolled and tempered of the EN24 type, with a Brinell hardness of 255.
Generally, interrupted cuts were conducted using a fresh tool at the beginning until the tool failed or the cutting edges had accumulated comprehensively high wear levels at which continued cutting risked catastrophic failure. The duration of each test was such that complete stabilisation of the cutting process would be attained while allowing significant tool wear accumulation. Measurements of flank, nose and notch wear lengths were made just after the vibration signals had been recorded. Tool wear measurements were made using a toolmaker's microscope and the observations supplemented by scanning electron microscope profiling. A tool life diagram is shown in Fig. 2. The chip type formed, presence of built-up edge, and tool features such as chipping or fracture were also recorded.

4. RESULTS & DISCUSSION

The raw experimental data was processed through dedicated software in the versatile Mathwork's Matlab coding programme on a PC environment. The acceleration data was transformed from time to the frequency domain through implementation of a fast Fourier transform algorithm, thus reducing the size of the sampled data and alleviating co-linearity amongst the principal axes components. Details of the algorithms and codes can be found in Dimla (1998). Results obtained were presented in both spectral (waterfall) and contour plots (Fig. 3-8).

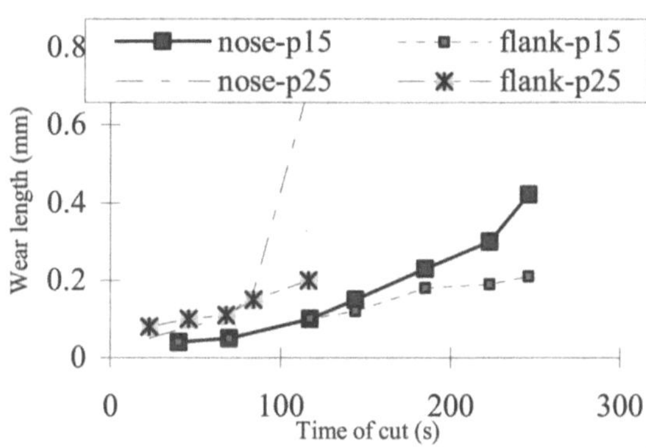

FIGURE 2. Wear length Vs. Time (V=300m/min & f=0.2mm/rev)

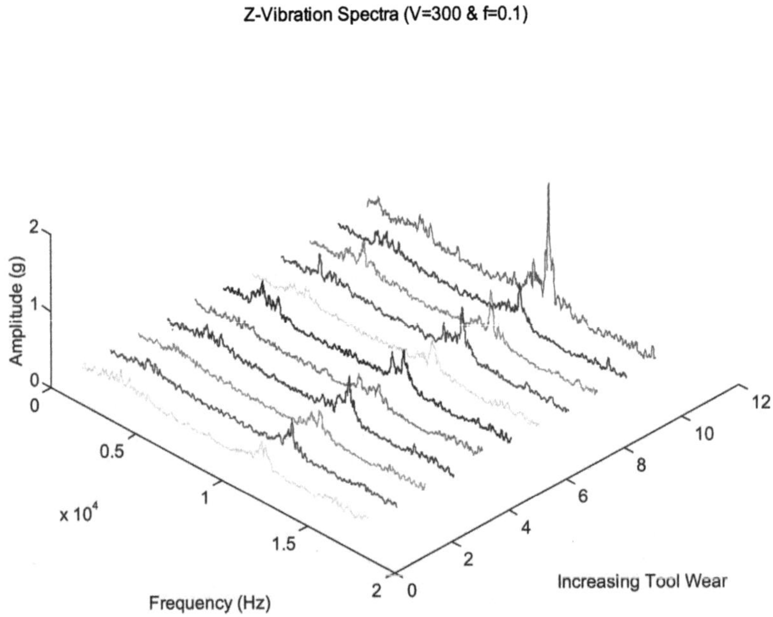

FIGURE 3. Z-acceleration spectra - flank wear (V=300m/min. & f=0.1mm/rev.)

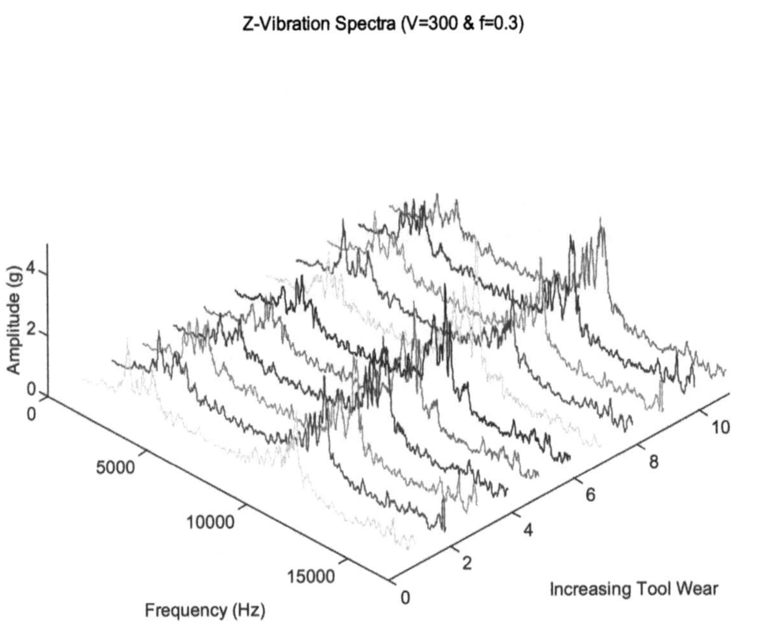

FIGURE 4. Z-acceleration spectra - flank wear (V=300m/min. & f=0.3mm/rev.)

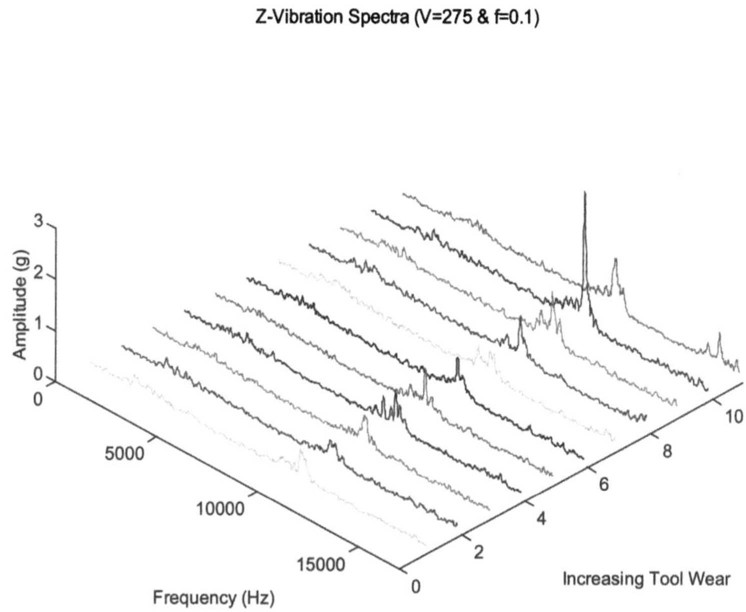

FIGURE 5. Z-acceleration spectra - flank wear (V=275m/min. & f=0.1mm/rev.)

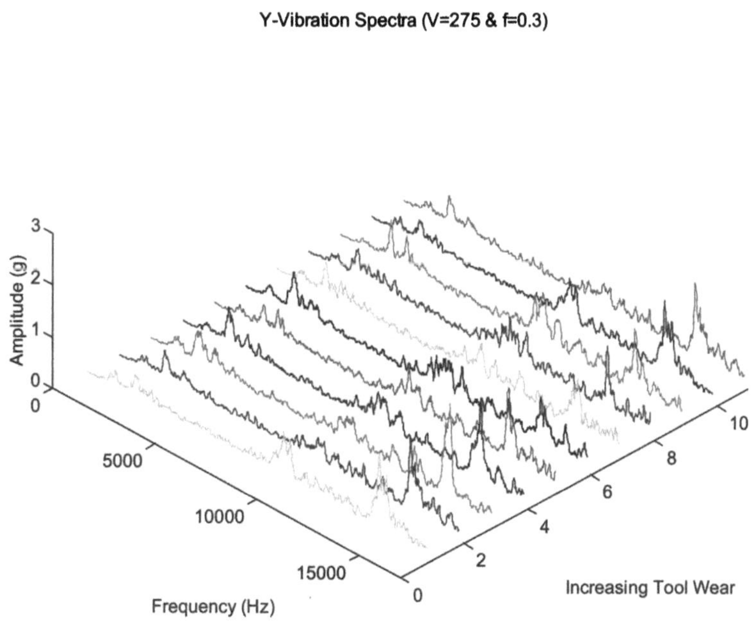

FIGURE 6. Y-acceleration spectra - flank wear (V=275m/min. & f=0.3mm/rev.)

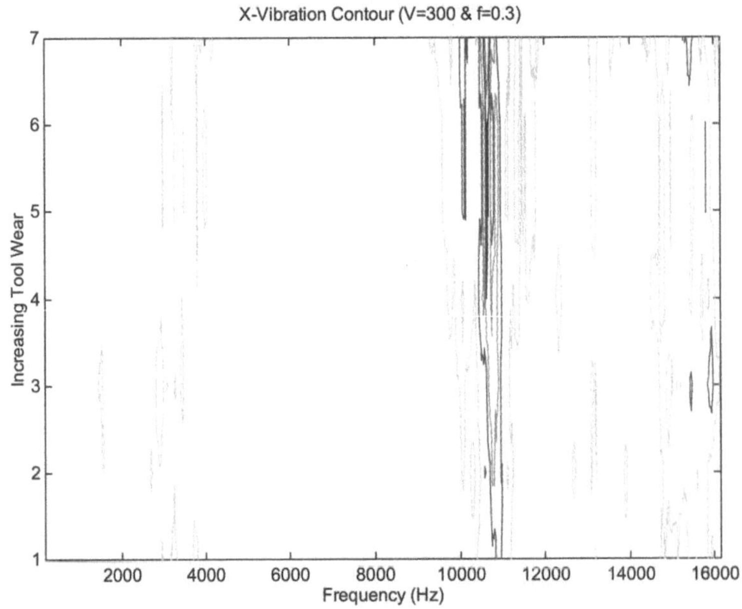

FIGURE 7. X-acceleration contour (V=300m/min. & f=0.3mm/rev.)

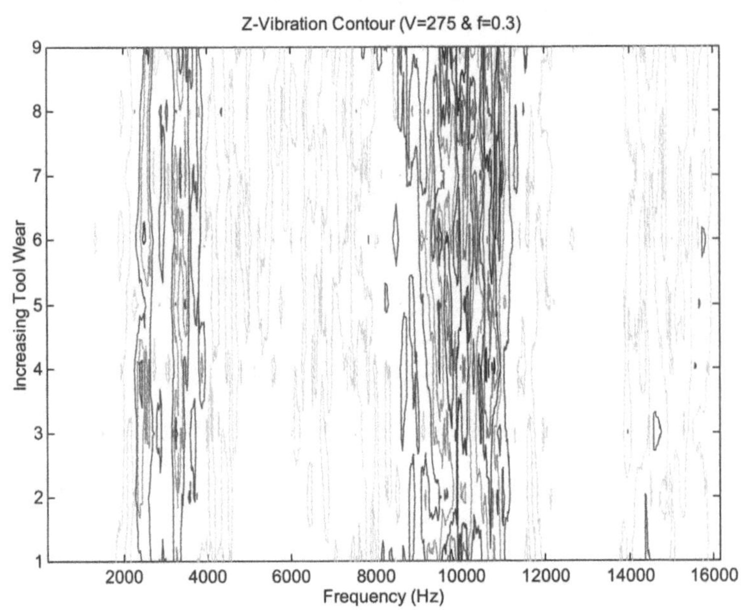

FIGURE 8. Z-acceleration contour (V=275m/min. & f=0.3mm/rev.)

The wear-time plot illustrates the effective wear rate at the prescribed cutting speed and feed-rate. At low feed-rates, the tool wore much gradually than at higher feed-rates. In terms of tool life (longevity), the tool lasted longer at low cutting speeds and the wear phases were more visible and pronounced. The most noticeable indicator of imminent tool failure was the sudden escalation of nose wear. Typically, when the nose wear was higher than 0.2 mm, one ran the risk of catastrophic failure. This level of wear was also found to be an applicable safety threshold for the flank wear length. Notch wear on the other hand increased steadily in the early part of the tests, and then remained practically the same for the remainder of the tool life. Because of the bulky nature of the results, only those for P25 tool insert are presented.

SPECTRA PLOTS

Two main frequency bands can be identified (Fig. 3-6): a 2-4 kHz (the fundamental resonant frequency of the tool holder) and a 9-11 kHz band (chip lamination frequency). The z-axis was the most sensitive to wear accruement as the 2-4 kHz band increased steadily peaking before falling until tool failure. Meanwhile the 9-11 kHz peak increased as wear accumulated with the last spectrum showing the largest increase. Increases in the feed-rate led to a proportional increase in the magnitude of both peaks, indicating the same trend as with low feeds. Comparison of spectra depicting cutting at different speeds showed that increasing the cutting speed actually decreased the amplitude of the resonant peak.

CONTOUR PLOTS

No effects of tool wear were visible, but changes in the cutting conditions led to a proportional change in the energy and spread nature of the contours (Fig. 7-8).
A more elaborate description of this work can be found in Dimla and Lister (2000) and Dimla (1998).

5. CONCLUSIONS

An experimental investigation programme was conducted on a turning centre from which the cutting tool wear and vibration signature signals were recorded. An analysis of the signals obtained was conducted to identify wear sensitive features in the frequency domain. At certain frequencies, there was some evidence of correlation between vibration signals and tool wear. This study has shed more insight into the complicated process of correlating tool wear to vibration signatures and one hopefully contributed towards the development of an effective universal tool wear monitoring system.

REFERENCES

1. AB Sandvik Coromant (1995) Turning Tools, Metal Working Products Catalogue, Automation Products
2. D E Dimla Snr. and P. M. Lister (2000) On-line metal cutting tool condition monitoring - Part I: Force and Vibration analyses. Int. J. Machine Tools & Manufacture, 40(5), 739-768.
3. D. E. Dimla Snr. (1998) Multivariate tool condition monitoring in a metal cutting operation using artificial neural networks. Ph.D. Thesis, School of Engineering, The University of Wolverhampton, UK
4. D. E. Dimla Snr. (2000) Sensor signals for tool wear monitoring in metal cutting operations - a

review of methods. International Journal of Machine Tools and Manufacture, 40(8), 1073-1098
5. G. Byrne, D. Dornfeld, I. Inasaki, G. Ketteler, W. König and R. Teti (1995) Tool Condition Monitoring (TCM) - The State of Research and Industrial Application, Annals of CIRP, 44(2), pp 541-567
6. P. M. Lister (1993) On-line measurement of tool wear. Ph.D. Thesis, Manufacturing and Machine Tools Division, Mechanical Engineering, UMIST, UK

MARKET VALUE OF RECYCLED MATERIAL MIXES: A KEY PARAMETER FOR DESIGNING FOR RECYCLING

G. Berti[1], W. A. Knight[2], G. Lucchetta[1]

[1] Department of Industrial and Manufacturing Engineering, University of Padova, Italy
[2] Department of Industrial and Manufacturing Engineering, University of Rhode Island, RI, USA

KEYWORDS: Design for recycling, Recycling industry, Price determination.

ABSTRACT. Research on Design for Recycling has long focused on disassembly sequence planning and ease-of-disassembly evaluation. Recently, a few studies have begun considering the material separation processes that characterize bulk recycling downstream in the recovery chain. However, the boundary conditions of bulk recycling are outside the scope of the above-mentioned investigations and therefore important economic parameters, such as process costs and the market value of recycled materials, are arbitrarily defined. This study presents a method for determining the revenue for different mixes of recycled metals in order to analyze how the value of the remaining mixed material changes when a part is disassembled or a separation process is carried out.

1 INTRODUCTION

Due to the wide diffusion of electrical and electronic products and their short useful lifetime, the amount of waste electrical and electronic equipment (WEEE) generated in the European Union (EU) is rapidly increasing. Consequently, public opinion and increased environmental legislation focused on product stewardship have led product manufacturers to undertake efforts to improve product design with a recycling-friendly approach [1]. In the last ten years, research on Design for Recycling (DFR) has placed more emphasis on the optimization of the disassembly process [2], addressing issues that can be grouped in two major areas: planning of disassembly sequences [3-4] and determination of disassembly depth [5-6]. However, recent studies have underlined that disassembly is only the first step of the current recycling practice and it is often followed by mechanical separation processes, better known as bulk recycling, and disposal by landfill or incineration [7].

Table 1 summarizes main cost drivers and relevant benefits for different end-of-life (EOL) process. Most successful DFR methods developed up to date try to perform a quantitative evaluation of product recyclability, opening the way to the economical optimization of costs and benefits of the recycling processes. The above-mentioned DFR methods usually neglect boundary conditions of bulk recycling and are based on a rough simplification, often assumption, of market value of recycled materials and processing costs.

A more accurate analysis of these end of life processes suggests that:

- Manual disassembly labor rate and landfill or incineration disposal costs depend only on regional factors and can be considered fixed during a DFR analysis. Furthermore, these data are easy to gather and are stable in the short period.

Published in: E.Kuljanic (Ed.) *Advanced Manufacturing Systems and Technology*,
CISM Courses and Lectures No. 437, Springer Wien New York, 2002.

- For a given electrical and electronic product, only a few components are taken apart to be sold as spare parts, because the value of the remainder is not sufficient to compensate for the labor cost of disassembly. The used components market is well established and their market prices are easy to determine.
- Bulk recycling costs depend only on the used technology and process parameters that can be considered fixed during a DFR analysis (e.g. throughput time or batch volume [8]).
- Revenue from secondary raw materials is strongly influenced by product design and previous disassembly decisions. The value of the obtained secondary raw materials mix can change due to:
    - a design decision (e.g. avoiding incompatible materials);
    - a component disassembly, which is a source of contamination;
    - a separation process, which increases the value of the remaining materials mix.

TABLE 1. Economical parameters of recycling processes

End-Of-Life Process	Costs	Benefits
Disassembly (manual)	Labor rate	Revenue of components
Bulk Recycling	Processing costs	**Revenue of materials**
Disposal	Disposal costs	-

Resulting from this analysis, determination of revenue of recycled material is a crucial point and a procedure to guide designers in the determination of the revenue for different mixes of recycled materials is needed. The procedure can be used to select product materials and assembly methods for the best combination of disassembly, bulk recycling and disposal for a given product.

2 SCOPE OF THIS STUDY

An extensive study of the emerging recycling industries in Italy has been carried out, as a basis for the following analysis [9]. Industrial waste has not been considered but the study has been limited to those industries dealing with materials recycled from post-use or post-consumption sources.

This analysis provided a clear view of the current end-of-life scenarios for the following categories of electrical and electronic products:

- white goods, e.g. refrigerators, washing machines, dishwashers and air conditioners;
- brown goods, e.g. TVs, VCRs and audio equipment;
- IT equipment, e.g. PCs, monitors and printers;
- small household appliances, e.g. vacuum cleaners and hairdryers.

A preliminary investigation has been performed aimed to the determination of factors influencing the market value of different mixes of recycled materials. The analysis has focused on the ferrous and non-ferrous recycled metals market for the following reasons:

- metals have a higher specific value than other recycled materials and are relatively easier to separate;
- the market is well regulated and a considerable amount of economical data is available, since these recycling industries are better established in the EU than the other ones.

The study was carried out through an extensive literature survey and numerous interviews with industry associations, recycling companies, metal smelters and other people involved in the end-of-life industry. In addition, many recycling plants have been visited.

3 OVERVIEW OF THE RECYCLING INDUSTRY

A recycled material supply chain begins with the end-of-life product take-back and ends up entering a production process in form of secondary raw material (Figure 1). When customers buy new electrical and electronic products, they can give back the used ones to the dealer. Otherwise, WEEE can be brought to special municipal areas where they are collected and temporarily stored. In both cases the take-back service is free of charge for the consumer. After the collection phase, WEEE are then transported to recycling firms and treated according to the environmental legislation. Recycling plants tend to specialize on few end-of-life product categories adopting the best available technology.

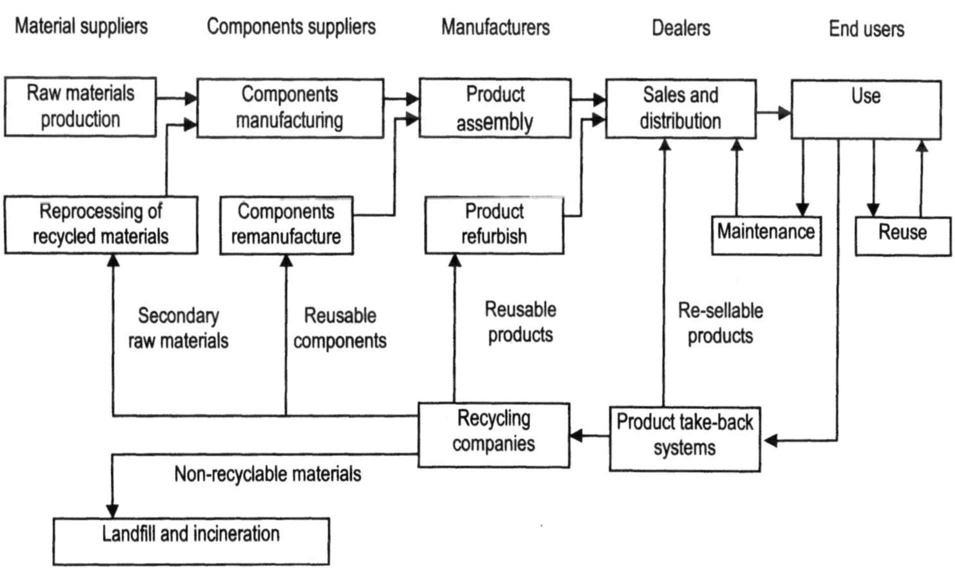

FIGURE 1. Recycling options for end-of-life products.

The revenue of the treatment, paid respectively by product manufacturers or municipalities, represents the major part of the recyclers' income. Its price varies between 15.50 € and 20.65 € per product. The process itself includes the following mandatory steps:

- Draining cooling circuits and insulation containing CFCs;
- CRTs disassembly and treatment, which contain toxic lead powder;
- Batteries disassembly;
- Hazardous components disassembly (e.g. capacitors containing PCBs and switches containing mercury).

Apart from harmful components and substances, only few other valuable components are taken apart manually, due to the high labor cost of disassembly (about 0.25 €/min). The rest of the product is then shredded to reduce its size by generating a mixture of flakes of uniform size that consume less space - this reduces transportation and landfill costs (0.13-0.16 €/kg) - and are easier to manage and separate. Successive separation processes divide the main material stream into distinct material types, such as ferrous metals, non-ferrous metals and plastics [10]. These mixes of materials are sold to a variety of traders on the market, followed by to metal smelters (e.g. for iron, steel, copper, aluminum, brass, lead, zinc, etc.) and to the reprocessing industry (for engineering plastics). The rest fraction that cannot be sold on the market is eventually disposed of by landfill or incineration.

4 CLASSIFICATION OF RECYCLED MATERIALS

Trading of different mixes of secondary raw materials is partially regulated by industry specifications, which classify them into categories according to parameters such as grade, source and shape. As a matter of fact, the chances to sell many recycled materials depend on the specification of standard commercial categories. If a mix of recycled materials belongs to any of these categories, its market price is not arbitrarily decided but it can only vary within a well defined price range, according to negotiation between buyer and seller. Price ranges for the most traded categories are tracked by industry associations – such as the Chamber of Commerce of Milan for the Italian recycling market – and are available to members and subscribers.

On the other hand, the existence of standard categories is a basic condition for the optimization of the recycling process. In fact, target parameters for the output of the separation processes can be defined through the identification of the requirements that distinguish each category from the others. From a DFR viewpoint, a modification in the product design - in terms of material selection or assembly methods – or the previous disassembly of a 'contaminating' component can make the output mix satisfy more stringent requirements. In turn, this can yield:

- a shift to a category with a higher price range or
- a price increase within the same category.

4.1 FERROUS METALS

Ferrous metals are contained mainly in white goods (e.g. refrigerators, washing machines and dishwashers), which can be sold to metal smelters after removal of the hazardous and valuable components (e.g. electrical motors and stainless steel). Otherwise, WEEE can be shredded to generate a mixture of chips ranging from 10 to 15 cm. Ferrous metals are then separated by means of strong magnets and sold to metal smelters. In this case, scrap market value is higher and varies according to the following parameters:

- fragment size (the smaller the better);
- metal source;
- concentration of inert materials (e.g. plastics, rubber, glass and concrete);
- concentration of non-ferrous metals (especially copper, zinc and tin).

This mixture of ferrous metal chips is known as Proler and it is the most valuable kind of scrap. In Europe, the trade of ferrous scrap is regulated by the CECA classification. Table 2 reports the price ranges for the categories of ferrous scrap recovered from WEEE [11].

TABLE 2. Ferrous scrap categories for WEEE (25 Jan 2002 prices)

Category	**Limitations**	**CECA**	**Price [€/ton]**	
			min	**max**
Old light ferrous scrap	Thickness ≤ 6mm, density ≥ 0.5, Cu ≤ 0.4%, Sn ≤ 0.02%,	51	21.50	27.00
Flakes (Proler)	Max dimension ≤ 200mm, density ≥ 0.9, Cu ≤ 0.25%, Sn ≤ 0.02%	33	113.50	119.00

4.2 NON-FERROUS METALS

The non-ferrous metal scrap categories take into consideration only copper and aluminum. The reason for this is twofold: these metals have a high specific market value and are largely utilized for the manufacture of electrical and electronic products. These categories are defined by the UNI (aluminum, nickel, lead, zinc and tin) and UNI EN (copper, brass and bronze) classifications. Each category has different requirements for the following parameters:

- chemical composition,
- grade,
- fragment size,
- type (source).

For these categories, price ranges are determined by means of data surveys conducted by the Chamber of Commerce of Milan. The UNI EN classification take into account only high grade copper scrap (grade > 90%). Except for covered cable fragments, low grade copper is not classified into any category and therefore has no price quotation. Table 3 reports the price ranges for some categories of copper scrap recovered from WEEE [12].

TABLE 3. Copper scrap categories for WEEE (25 Jan 2002 prices)

Category	Grade	UNI EN	Price [€/ton]	
			min	max
Covered cable fragments	30-70%		290.0	918.0
Recovered wires, slabs	92.0%	12861-S-Cu-9	1299.0	1351.0
Recovered end losses, burnt wires	96.0%	12861-S-Cu-8	1402.0	1454.0
Recovered end losses, burnt wires	98.5%	12861-S-Cu-6	1516.0	1568.0
Recovered pipes, tapes, etc.	98.0%	12861-S-Cu-7	1480.0	1532.0
Recovered pipes. tapes. etc	-	12861-S-Cu-5	1532.0	1584.0
Flakes	99.5%	12861 S-Cu-10-C	1480.0	1532.0
Flakes	99.8%	12861 S-Cu-10-B	1532.0	1584.0
Flakes	99.9%	12861 S-Cu-10-A	1583.0	1635.0
Recovered electrolytic copper	-	12861-S-Cu-1	1583.0	1635.0

Aluminum and copper scrap markets are very similar. Scrap categories are defined by the UNI classification and the Chamber of Commerce of Milan estimates their price range quotations. Table 4 reports the price ranges for some categories of aluminum scrap recovered from WEEE [12].

TABLE 4 Aluminum scrap categories for WEEE (25 Jan 2002 prices)

Category	Grade	UNI	Price [€/ton]	
			min	max
Covered cable fragments	-	10297/2	400.0	425.0
Floated flakes	>95%	10297/7	744.0	795.0
Old laminated aluminum scrap	>85%	10297/4	943.0	995.0
Cables and wires	99.5%	10297/2	1148.0	1200.00

5 MARKET PRICE DETERMINATION

The revenue for a mix of recycled metals results from a negotiation between buyer and seller. If the mix characteristics comply with the requirements of a standard category, its market price can not be arbitrarily decided but varies within the category price range. For some of the categories (e.g. copper scrap) ranges are defined for the concentration of the target metal. In this case, the higher the grade the closer the scrap value to the upper limit of the price range.

For mixes of recycled metals that are not included in a standard classification, the price negotiation depends on several more factors. However, the most important factors that mostly affect the revenue, in order of importance, are the following [9]:

- the concentration of the target metal, roughly estimated by the buyer (i.e. the smelter) usually through a visual inspection;

- impurities and contaminant elements which penalize the target metal refinery process;
- the dimensions of the scrap, which influences the melting process;
- the actual price of the virgin metal [13].

The buyer usually calculates the starting price for this kind of negotiation as follows:

$$P_r = (P_{LME} \times c_{Cu}) - C_r \quad (1)$$

where: P_r = the market price for the metal mix,

P_{LME} = the actual value of the London Metal Exchange price,

c_{Cu} = concentration of the target metal,

C_r = cost of the metal refinery process.

The processing costs may vary considerably depending on the concentration of the target metal and the contaminents present. For example for the processing of recycled copper, for a given concentration of copper, the cost of the refinery process depends on the kind and concentration of the following impurities and contaminant elements: plastics (PVC), varnish, oils, aluminum and nickel. In particular, when the concentration of aluminum and nickel exceeds 1% of the mixture, pure copper has to be added in an appropriate amount, thus increasing the process cost. On the other hand, iron, zinc and lead can be easily separated.

6 CONCLUSIONS

A procedure for determining market prices for mixes of recycled metals has been outlined. This procedure can be used for evaluating the benefits of the recycling processes and, therefore, optimizing the combination of disassembly, bulk recycling and disposal for a given product. The recycling market structure delineated in this paper gives a basis for further studies that should be focused on the variables that can explain price fluctuations within category ranges. For mixes of recycled materials that are not included in a standard classification, work is ongoing to determine the relationship between the processing cost and impurities present for a range of target materials.

REFERENCES

1. Commission of European Communities, (2000), Proposal for a Council Directive on Waste Electrical and Electronic Equipment. COM(2000) 347 def. 2000/0158 (COD).
2. Jovane, F., Alting, L., Armillotta, A., Eversheim, W., Feldmann, K., Seliger, G., (1993), A key issue in product life cycle: disassembly. *Annals of the CIRP*, Vol 42/1, 640-672.
3. Subramani, A. K., Dewhurst, P., (1991), Automatic generation of product disassembly sequences. *Annals of the CIRP,* Vol 40/1, 115-118.
4. Lambert, A. J. D., (1997), Optimal disassembly of complex products. *International Journal of Production Research,* Vol 35/9, 2509-2523.
5. Zussman, E., Kriwet, A., Seliger, G., (1994), Disassembly-oriented assessment methodology to support design for recycling. *Annals of the CIRP,* Vol 43/1, 9-14.

6. Harjula, T., Rapoza, B., Knight, W. A., Boothroyd, G., (1996), Design for disassembly and the environment. *Annals of CIRP,* Vol 45/1, 109-114.
7. Sodhi, M. S., Knight, W. A., (1998), Product design for disassembly and bulk recycling. *Annals of CIRP,* Vol 47/1, 115-118.
8. Sodhi, M. S., Young, J., Knight, W. A., (1999), Modeling material separation processes in bulk recycling. *International Journal of Production Research,* Vol 37/10, 2239-2252.
9. Borsetti, A., (2002), Design for Recycling. Sviluppo di un metodo di valutazione economica della riciclabilità di apparecchiature elettriche ed elettroniche. MS thesis, Dipartimento di Innovazione Meccanica e Gestionale, University of Padova.
10. Vesilind, P. A., Rimer, A. E., (1981), *Unit operations in resource recovery engineering.* Prentice-Hall, Inc., Englewood Cliffs, N.J.
11. Camera di Commercio di Milano, (2002), Prezzi dei rottami, *Il Mercato Metalsiderurgico,* 47/2, 10.
12. Camera di Commercio di Milano, (2002), Prezzi metalli non ferrosi, *Il Mercato Metalsiderurgico,* 47/2, 5-6.
13. Macavoy, P. W., (1988), *Explaining metal prices,* Rochester Studies in Economics and Policy Issues, Rochester.

IDENTIFICATION AND ASSESSMENT OF NON CONFORMITY PARAMETERS FOR QUALITY COST REDUCTION

M. Boscolo[1], E. Padoano[1]

[1] Department of Energetics, University of Trieste, Italy

KEYWORDS: quality assurance, non-conformity costs, 100% inspection.

ABSTRACT. The paper presents the main results of an empirical study carried out in a manufacturing firm that produces heat exchangers. The study showed that the indexes used in the ISO certified quality system were not able to correctly evaluate quality costs and that the introduction of new indicators could bring to a more effective quality improvement. The analysis deals with the internal failures which can be detected in the testing stage: a statistical analysis of malfunction detection and repair activities for the different products and failure types was performed. This made it possible the quantitative assessment of the parameters, taking into consideration three years in view of the definition of their trend which was compared with that of the indexes already in use. Eventually, the most critical areas of the products, in terms of costs, were identified in order to start an improvement program.

1 INTRODUCTION AND RESEARCH OBJECTIVES

In order to put into practice an improvement process a company must evaluate its production process effectiveness with respect to business objectives. This is particularly evident for a firm that wants to pursue strategic factors such as quality and costs: there is a need to control processes and therefore to measure their performance with respect to these two dimensions.

Such performance evaluation is based on a system of measures that should be an integral component of the company's quality system, it should be easily manageable and it should provide a way of communicating between different areas and functions. The quality system is the means of putting into effect company's objectives, with respect to quality, in terms of improvement process, correction and prevention strategies and actions: measurement activities are a necessary step of the whole process [1]. Performance measures should produce the information needed to drive the improvement process; they should therefore translate general quality and cost objectives, such as customer satisfaction or efficiency, into more structured indexes (e.g. change of orders due to errors, number of pieces reworked etc.) able to give immediate feedback on problem areas within production processes.

Several authors (see, as examples, [2] and [3]) emphasized the need of operability of performance measures. This could be met if such measures are objective, straightforward (easily understandable and useable) and quick to perform. These aspects make it possible to implement a system of measures that will be carried out and managed simply by operators and that will provide meaningful data for decision makers. The results of measures should be visible to each function that is involved in the improvement process, starting from top management that is critically responsible for

Published in: E.Kuljanic (Ed.) *Advanced Manufacturing Systems and Technology*, CISM Courses and Lectures No. 437, Springer Wien New York, 2002.

company's commitment to quality advance efforts. Of particular importance is the outcome offered by indicators which aim to identify production process areas that bear a greater impact on output quality or major profit improvement opportunities [4]. In particular, a central role can be played by the identification of costs related to internal and external quality [5] [6].

These issues are not always clearly understood when quality performance measures are projected and implemented, the major drawbacks may be an intervention limited to few less important areas of improvement or a lack of support by several enterprise's decision makers [7]. In summary, the above mentioned facets should be carefully taken into account to make a quality performance measurement system really effective in its contribution to the overall improvement policy.

The present research, started in 2000, has thoroughly explored such issues in a medium enterprise which produces heat exchangers. The company has implemented a quality system, ISO9000 certified, and is committed to continuous improvement. In order to meet this objective, a performance measurement system was developed which is mainly based on standard parameters such as number of defective units produced in a period, number of assistance interventions on units sold and so on. Data collection started in 1998.

One of the main results of the implementation of this system was the coming out of the importance of preventative actions during the production process aimed at reducing non-conformities detected in quality control activities. Notwithstanding, it was perceived that the measures adopted were not completely effective in evaluating the actual outcomes of preventative programs implemented. This could be, on the one hand, a serious setback to the whole company's perception of the benefits of such programs, in particular when compared to costs, and, on the other hand, it could make it difficult the identification of the areas of production process that constitute the most valuable candidates to quality enhancements. In brief, it was felt as necessary to introduce new measures, not strictly connected to ISO9000 norm; this confirms the findings of other researchers who pointed out some unsatisfactory aspects of the norm with respect to continuous improvement and total quality (see, as an example, [8]).

In summary, the objectives of the study were to carry out new indexes of performance directly related to company's quality goals and able to gain top management attention, and to evaluate production performance through new and old indexes in order to show major problem areas that were unnoticed. It was therefore necessary to develop measures that could use data already collected, in view of a comparison of the improvement process results by means of the old indexes and of the new ones.

2 THE GENERAL DEVELOPMENT OF THE STUDY

A first exploratory analysis of the company's production processes and quality system, and a thorough debate with quality managers and top management made it possible to precisely define the company's operational goals with respect to the overall "continuous improvement" objective. The issues identified were:

- to evaluate the benefits attained by preventative actions already put into effect with respect to their costs;
- to identify possible areas of improvement that could bring to better results in terms of non-conformity cost savings;

- to assess the magnitude of such savings compared with costs due to preventative actions;
- to carry out a performance evaluation system that could be easily managed by operators.

After the classification of all costs due to, on the one hand, the appraisal and prevention activities and, on the other hand, the internal and the external failures, an extensive analysis of the data already collected was performed to compare such costs to production and supply costs and to annual revenues. It was then possible to identify two major problem "macro-areas" that needed further analysis:

- a significant part of defects were due to external suppliers;
- the most important portion of internal non-conformities were detected in the hydraulic testing of products and were due to the welding stage of the process.

The following section presents the analysis of this second issue and the main results achieved, as it proved to bear the greatest impact on the preventative programs of the company's quality function and got the utmost attention of the top management.

3 THE ANALYSIS OF INTERNAL POOR-QUALITY COSTS

The hydraulic testing shop performs inspection and, when needed, fixing operations of the heat exchangers coming from the welding shop. The four types of products, vertical boilers (VB), low power heat exchangers (LPHE) medium power heat exchangers (MPHE) and boiler units for thermal sets (BU), are filled with water, at a testing pressure which is 1,5 times the working pressure, and, after a defined time interval (which is dependent on product type), all welds are checked. Water leaks or humidity are considered non-conformities and are subject to a fixing procedure and to a new test.

The test is performed on all units produced (100% detailed inspection) and each time the fixing procedure is activated, shop operators record on pre-formatted tables the type of defect and its position. This made it possible to identify the most frequent non-conformities with respect to the number of total defects detected; the evaluation is based on the following index:

$$fd_i^k[\%] = \frac{nd_i^k}{\sum_{i=1}^{p_k} nd_i^k} \times 100 \tag{1}$$

where:

- $k = 1, \ldots, 4$ indicates the product type;
- $i = 1, \ldots, p_k$ indicates the position of a weld (that depends on the product type);
- nd_i is the number of defective welds of position i;
- fd_i is the frequency of defective welds of position i.

Number and position of welds are directly related to product complexity, so more complex products (e.g. MPHE) need a more thorough inspection. Existing preventative procedures were mainly aimed at improve welding with respect to more frequent defective positions.

The following index was also computed, in order to identify the incidence of a defective weld in position i with respect to total inspected units of type k in a given year (N^k):

$$Fd_i^k[\%]=\frac{nd_i^k}{N^k}\times 100 \qquad (1')$$

A different information, that was collected by operators, was the duration of the repair interventions: unfortunately the reports gave an account only of the total time spent for each position in a year (Tr_i^k). The analysis of such data showed that the average time spent in the fixing activity of a weld in a position i of a product k is, with good approximation, constant in the three years. This fact was also confirmed by a six months monitoring of repairing activities and denotes that these operations are performed by shop personnel routinely.

The data above mentioned made it possible to compute an index that proved to be of particular interest in the context of the study, namely the incidence of total time spent in fixing defects in a position i with respect to total repairing time:

$$ftr_i^k[\%]=\frac{Tr_i^k}{\sum_{i=1}^{p_k} Tr_i^k}\times 100 \qquad (2)$$

This index identifies the defective welds that are more costly in terms of time spent in repairing operations. It was then made a comparison between the data obtained by means of equation (1) with those attained using equation (2). It was therefore possible to identify two extreme non-conformity typologies:

- defects which were frequent in a year but that were responsible only for a limited amount to total repairing time;
- defects less frequent but that had a major incidence on the total repairing time.

Few weld positions, for each product type, were markedly of the second typology: such welds were consequently considered *critical*. Figure 1 and Figure 2 show the charts that followed the data processing for two products (respectively MPHE and LPHE) and in relation to years 1999 and 2000. The labels of the x-axis display the weld positions: for each position the column *defects%* indicates the results of equation (1), while the column *time%* the results of equation (2). Critical defects are those of position 14 for MPHE whereas 8, 10 and 12 for LPHE. It is worth noting that while for LPHE the more frequent defective welds bear also the higher effects in terms of time, this is not true for MPHE: position 1, 7 and 8 have much higher percentages, with respect to total non-conformities, compared with the value of position 14. For the other two product types, VB and BU, the analysis made it possible to identify, respectively, two and five critical positions.

Other results, which proved to be of the greatest interest for company's management, are those summarized in Table 1: the improvements due to preventative programs, with respect to total number of defects and total repairing time, are compared. The values in the second and third columns report the diminution of the number of detected defects with respect to 1998 (year in which the quality system was implemented), while the values in the fifth and sixth columns report the reduction of total repairing time with respect to the same year.

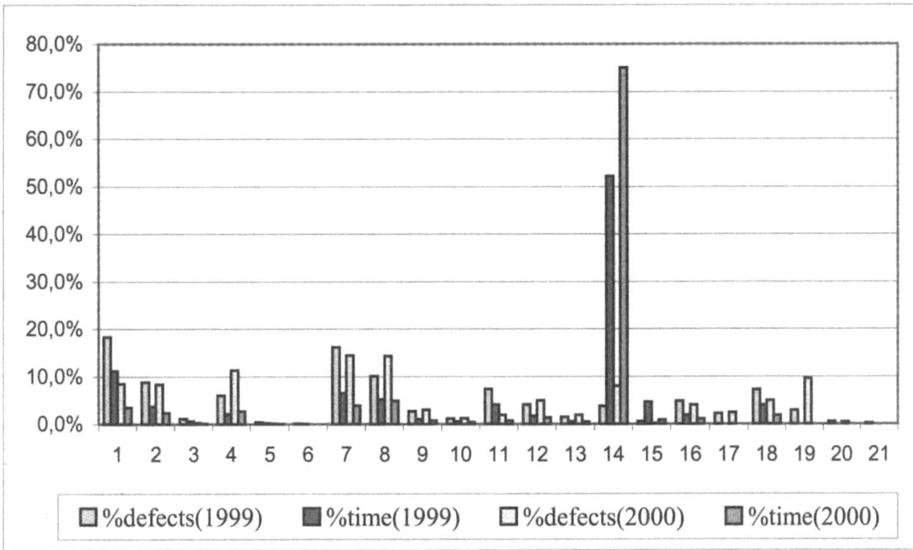

FIGURE 1. Comparison % of defects - % repairing time (MPHE)

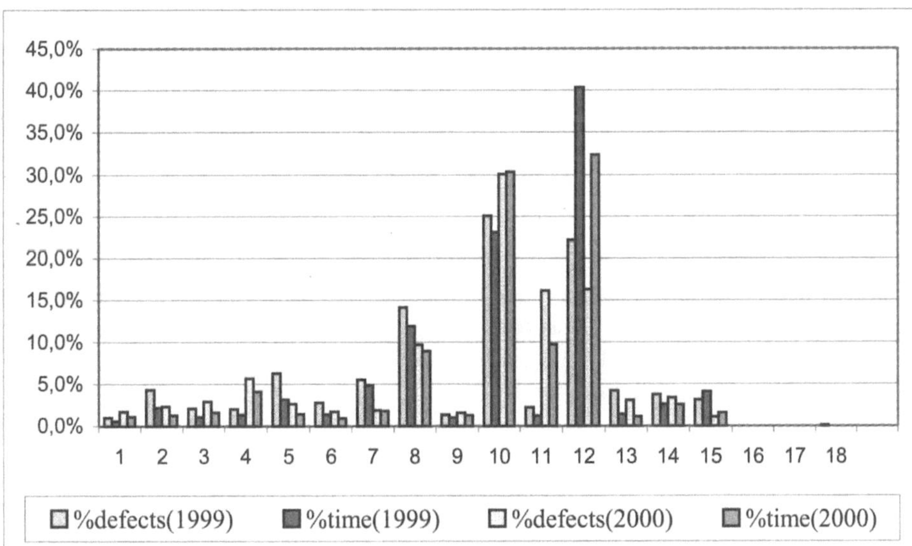

FIGURE 2. Comparison % of defects - % repairing time (LPHE)

TABLE 1. Effects of existing improvement program

	% of dimin. of defects			*% of dimin. of repairing time*		
	1998	1999	2000	1998	1999	2000
MPHE	100%	-43,3%	-75,0%	100%	-12,3%	-36,6%
LPHE	100%	-37.1%	-50,0%	100%	-23,5%	-45,0%
BU	100%	-43.5%	-35,9%	100%	-42,1%	-35,5%
VB	100%	-26,7%	-64,4%	100%	-22,6%	-62,0%
Totals	100%	-42,2%	-55.5%	100%	-23,5%	-51,9%

By means of Table 1 it is possible to deduce that:

- for MPHEs, the existing improvement program brought about a notable diminution of number of non-conformities, but they had less influence on repairing time; thi is due to the fact that the greatest effect of prevention was the reduction of more simple defects;
- for the other product types, the preventative actions had similar outcomes both on number of non-conformities decrease and on repairing time diminution.

In view of planning new preventative actions on welding operations, the research team considered of great importance to identify the cost related to internal non-conformities. The data and process analysis performed showed that the main components of cost related to weld defects are:

- repairing time, which included the time spent in actual repairing activities (major fraction) and other ancillary activities (e.g., emptying and re-loading of heat exchangers);
- operator cost (per hour).

This last component is due to the fact that shop operators are sufficiently flexible to be employed in other activities, so an hour spent in fixing a defect is an hour subtracted to other productive tasks (*opportunity cost*). Considering that each fixing can be performed by a single operator, even if for a short time the support of a second one may be advisable, the total cost of all non-conformity in position *i*, repaired in a year, is:

$$C_i^k[€] = c_h \times Tr_i^k \tag{3}$$

where:

- c_h is the cost of an operator per hour [€/hour];
- Tr_i^k must be expressed in hours.

From an analytical point of view, repairing cost is a function of Tr_i^k times a constant, so the same critical areas with respect to time are critical with respect to costs. Notwithstanding, the results reported in economic values, presented to the company's top management, had a significant impact on future decisions related to improvement programs.

In fact, non-conformities detected in the hydraulic test represented in 1998 a 0,4% and in 2000 a 0,27% of total production costs (in contrast, non-conformities detected externally were, in 1998, a 0,09% of total production costs). Taking into account the results of the analysis, top management has decided to support the new operational objective, suggested by the study, of planning preventative actions to reduce critical non-conformities, which were responsible of major repairing time and therefore of the greatest portion of internal poor-quality costs.

4 CONCLUDING REMARKS

The new measures of performance proposed broadly satisfy the goals of the research in terms of simplicity and of information provided. The company is now aware of the production issues that have the higher impact on internal non-conformities and that must be firstly considered in prevention programs.

A first program has been developed with respect MPHE and its critical areas. The results obtained to date show that with a reduction of 50% of the number of defects it is possible to reach a 40% of non-conformity cost savings for annual MPHE production. Improvement programs interesting the other products will follow shortly.

The parameters of performance carried out show their greatest potential if they are collected continuously by operators of the hydraulic testing shop, because this makes it possible a monitoring of poor-quality costs. The relative simplicity of the indexes proposed brought about the opportunity to screen the values related to each welding position for a product type, in order to control the outcomes of preventative actions. Then, a project for electronic collection and treatment of the data of internal non-conformities has been started.

Other issues that will be investigated in the near future include:

- the magnitude of cost savings with respect to prevention costs;
- the improvement of repairing processes with respect to critical positions;
- the outcomes that could be derived from critical positions analysis in terms of new product design or re-engineering.

It is, above all, felt as necessary to perform a thorough investigation of the first issue, for the period of monitoring suggested that a deeper improvement of weld related defects could be accomplished only at higher production costs.

REFERENCES

1. Ishikawa, K. (1985), What is total quality control? The Japanese way, Prentice-Hall, Englewood Cliffs.
2. Kumar, S., Gupta, Y. P. (1993), Statistical process control at Motorola's Austin assembly plant, Interfaces, 23, 2, 84-92.
3. Speirs, J., Nash, T. J. (1995), The use of standards in performance measurement, IEE Colloquium on Measures of Performance, 5/1-5/7.
4. Bain, W. (1989), The real value of the cost of quality, Proceedings of the IEEE Aerospace and Electronics Conference, 2076-2079.

5. Crosby, P. B. (1979), Quality is free: the art of making quality certain, Mentor, New York.
6. Juran, J. M. (1988), Quality costs. Quality Control Handbook, Juran, J. M., Gryna, F. M., Bingham, R. S. (Eds.), McGraw-Hill, New York, 4.2-4.30.
7. Rungtusanatham, M. (2000), The quality and motivational effects of Statistical Process Control, Journal of Quality Management, 4, 2, 243-264.
8. Curkovic, S., Pagell, M. (1999), A critical examination of the Ability of ISO9000 Certification to Lead to a Competitive Advantage, Journal of Quality Management, Vol. 4, No. 1, 51-67.

MINIMIZING THE COST OF QUALITY: AN INNOVATIVE SAMPLING PLAN

C. Giardini[1], G. Valentini[2]

[1] Department of Mechanical Engineering, University of Brescia, Italy
[2] IESE Business School, Barcelona, Spain

KEYWORDS: Quality assurance, Sampling plan, Quality costs.

ABSTRACT. This paper deals with an innovative procedure for quality sampling intended for the printed circuits industry. The Authors provide evidence of the failure of common sampling plans in situations like the analyzed one, describe the new plan developed, and give account of the empirical results obtained. The proposed sampling plan takes into account the peculiar nature of the possible nonconformities, the common features of the industry and of the customers, and the possible sources of costs, with the aim of minimizing the firm' s total costs of quality through the use of statistical and informatics tools. Developed and tested thanks to an in-depth case study in a leading Italian firm, the described method can also be applied to several other industrial situations.

1 INTRODUCTION

Over the last decades, quality has been progressively recognized as a fundamental success factor in pursuing sustainable competitive advantages. Yet, despite the diffuse espoused firms' acknowledgment of this fact, we are still far from a conscious and sound implementation of the available quality tools and principles. And this is also true for SMEs, which often limit themselves to a blind application of standards and practices on fashion, rather than focusing on their peculiar process, needs, and problems.

This paper deals with an innovative sampling plan, developed thanks to an in-depth case study in a leading Italian firm in the printed circuits industry, whose aim is that of minimizing the so called ' cost of quality' . Rsults are general and – as we will see – applicable to a wide variety of industrial situations. Still, the strict link with a real case assures the usefulness of the task and demonstrates the convenience of the application of the sampling model.

The paper is organized as follows. First, some of the characteristic attributes of the printed circuits industry will be presented, as well as a short discussion of the manufacturing process and of its main features. Second, the general problems will be analyzed through the result of the case study and a new alternative model will be presented. Third, the necessary preliminary steps for its implementation will be described. Results obtained and concluding remarks will follow.

2 THE SETTING

A printed circuit is a flat board made of plastic or fibreglass – typically 1.6 mm thick – on which electronic components such as resistors, diodes, CPUs etc. are attached on thin ships of

Published in: E.Kuljanic (Ed.) *Advanced Manufacturing Systems and Technology*, CISM Courses and Lectures No. 437, Springer Wien New York, 2002.

conductor [1]. Actually, there are a great many varieties of printed circuits that can be produced. The fabrication process is rather complex: for the purpose of this paper, it will suffice knowing that the process is constituted by a long series of phases, among which electroplating, dry film lamination, imaging, developing photopolymers and etching, and that at the end of the process two quality tests are carried out. A first one, the electrical test, assures the perfect conductivity of the circuit of all circuits. A second one, performed on the whole lot or on a sample depending on the contingent situation, looks for other possible kind of nonconformities.

The printed circuits industry is pretty developed in Italy. An autonomous survey of the Authors conducted among the largest fifteen firms of the sector in Northern Italy, highlighted the results presented in table 1. The percentage of scrap is highly dependent on firms' contingencies: the type of circuit (i.e. double-sided *vs.* multi layers) and the number of different items launched every day constitute by and large the most important and distinguishing characteristics.

The analysis also highlighted that the great majority of the companies adopted standard in process sampling plans, generally based on the UNI-ISO 4842-75, given a desired AQL. The need for in-process sampling, despite the final 100% inspection, is plainly motivated by the fact that in its absence items that are nonconforming from the very beginning may be worked through the whole chain, and finally discarded, with the two possible negative consequences of sustaining useless manufacturing costs, and being not able to meet the demand. Furthermore, while reworking is almost never possible at the end of the productive chain, it is usually allowed in process.

In summary, the relevant information collected from the external analysis is that:

- Not only does the quality of the output depend on firms' peculiar managerial and behavioral practices, but also on product specific characteristics.
- There are some process phases particularly critical as for quality output.
- Generally speaking, since the focus is on the final 100% inspection, no particular weight is given to in-process inspection, which is usually carried out with traditional techniques adopted from other industries.

The case study allowed to deepen these facts. The specific industrial situation faced was characterized by an average percentage of scrap of the 2.1% (85% of the produced circuits are double sided, the remaining 15% multi layers), the 24% of which was given by working phases given in outsource. Taking into account internal phases, it was found that three phases on their own were responsible of more than 60% of the nonconformities in accordance with the Pareto principle. Deepening the observation, it was found that not only is this result justified by the technical complexity of these phases, but also by the frequent event that a large amount of nonconformities was found in the inspected sample. This circumstance was due to the fact that the cause of nonconformities was generally a problem that would keep producing defects until the cause of the problem itself had been removed. This kind of defect is very different in nature from the pure casual one, that may appear with a certain probability, but that does not affect the subsequent result of the process. The completely different character of these defects led the Authors to think of the essential need of a new sampling plan.

TABLE 1. Data relative to the 15 most important firms in terms of sales in Northern Italy

FEATURE	AVERAGE VALUE
% of scrap	4.7%
% of double -sided circuits	63.0%
% of multi layers circuits	34.5%
% of flexible circuits	2.5%
Max number of layers	14
N. of different items launched every day	20

3 THE NEW MODEL

As above mentioned, the sampling plan derives from the distinction between two main possible kind of defects which may take place in production of printed circuits: we defined them as *continuous* and *casual* defects.

We define continuous defects those for which, given the event E_t "the tth piece is non-conforming", we have that:

$\Pr(E_t) = q$, with $0 \leq q < 1$ if the $(t\text{-}1)^{th}$ produced item was conforming;

$\Pr(E_t) = 1$ if the $(t\text{-}1)^{th}$ produced item was non-conforming.

In other words, there is a fixed probability q that during the process there appears a first defect. Once this defect appears, all the following items will be nonconforming, too. On the other hand, a casual nonconformity is such that $\Pr(E_t) = p$, with $0 \leq p < 1$, for any t. That is, $\Pr(E_t \mid E_{t-1}) = \Pr(E_t)$.

It is important to notice from now that the phases depicted as related to continuous defects are not excluded by the possibility of occurrence of casual defects [1]. In any case, this fact is so marginal -experimental data confine this contingency to about the 3% of detected nonconformities - that the final inspection suffices. Evidently, given these different families of defects, it is not wise using the same sampling procedures for every phase, since they clearly cannot be always effective. More specifically, a procedure based on the extraction of a random sample from a batch once it has finished a phase, is not successful if applied to phases characterized by continuous defects. The usual procedure applied for all the phases before our study was the following. A random sample of size n was selected from the batch. The number of nonconforming items in the sample was found and compared to the acceptance number c. If the observed number was less than or equal to the acceptance number, the lot was accepted. If more than c nonconforming items were found in the sample, the lot was "rejected" and a 100% batch inspection followed. However, think of applying this procedure after a phase characterized by continuous defects in the extreme case in which the very first item worked in a

[1] In the following, we will indifferently use the locutions "phases characterized by continuous/casual defects" and "continuous/casual defects", meaning the same concept.

batch is nonconforming: then the whole batch will be nonconforming. The lot will be surely rejected, but it will have to be entirely re-worked – in the best case, when re-working is always possible – with a consistent cost involved. Can' t we find a more efficient way to organize the in-process sampling? The idea behind our work is that of developing a sampling plan which is able to minimize the (expected) cost of quality in the case of both continuous nonconformities and casual defects.

3.1 CONTINUOUS NONCONFORMITIES

We explained how an ordinary single sampling plan is not efficient in the case of continuous defects. It appears more effective a sampling plan that is based on the inspection of items with a certain frequency during the production itself. In this way, possible nonconformities can be early identified, their cause removed, and, at the same time, costs related to re-working or scraps may be saved. In our approach, the optimal frequency of sampling is that which minimizes the total cost of the plan.

Let us analyze more in depth the situation, assuming that:

- We are dealing with batches of size N.
- The sampling procedure is the following: an item is extracted and inspected during the production every f items worked. If the item is nonconforming, then all the items are inspected backwards until the last nonconforming is found.
- The probability for any worked item to be the *first* nonconforming one is q.

What' s the expected number of nonconforming items? Given our assumptions, f items have been worked after the last inspection. All of them will be nonconforming if the first worked one is already nonconforming; by the same token, f-1 items will be nonconforming if the first nonconformity appeared with the second worked item, and so on.

Thus, the number of nonconforming items NCI among the f just finished are:

$NCI=f$ with probability q

$NCI=f$-1 with probability $q(1\text{-}q)$

$NCI=f$-2 with probability $q(1\text{-}q)^2$ etc.

In general, the expected number of nonconforming items since the last inspection is:

$$E(NCI)=\sum_{i=1}^{f}(f-i+1)\cdot q\cdot(1-q)^{i-1} \tag{1}$$

As a consequence, the expected number of items inspected backwards (NII) will be that indicated in (2), which derives directly from (1) remembering that the inspection will end when the first conforming item will be found backwards.

$$E(NII)=\sum_{i=1}^{f}(f-i+2)\cdot q\cdot(1-q)^{i-1} \tag{2}$$

3.2 CASUAL NONCONFORMITIES

In the case of casual nonconformities, a binomial distribution can be used to find the probability of observing x nonconforming items in a sample of size n. Assuming that the lot proportion of nonconforming is p, the probability of finding x nonconforming items is given by:

$$P(x)=\binom{n}{x}\cdot p^x\cdot(1-p)^{n-x} \tag{3}$$

As a consequence, given the acceptance number c, the lot acceptance probability is obtained calculating the probability P_a:

$$P_a=\sum_{x=0}^{c}P(x)=\sum_{x=0}^{c}\binom{n}{x}\cdot p^x\cdot(1-p)^{n-x} \tag{4}$$

3.3 QUALITY COSTS

Once p is given, the aim is that to develop an algorithm abŀe to calculate the optimal sample size and acceptance number so as to minimize the total cost of the sampling plan. The same must be done for continuous nonconformities. The general costs we considered, the ones we wanted to minimize, are: a) the cost of inspection, b) the cost of internal failure, and c) the opportunity cost related to inspection due to machine stop [2]. Let us describe how we calculated these voices in the two different cases, stressing from the beginning that, since we are dealing with random variables, we are looking for *expected* costs.

The cost of inspection, in the case of casual defects, is simply given by the cost of examining the given number of items prescribed by the sampling plan, i.e. the average inspection time of one item multiplied by the number of items times the cost of labor. In the case of continuous defects, not only is the inspection with a given frequency taken into consideration, but also the eventual backward inspection, whenever the examined item is nonconforming. To this end, *E(NII)* as calculated in (2) is fundamental. In both cases, trying to adapt to different contingent situations, the fact that the inspection could be carried out with the manufacturing process forcedly stopped or not, was considered. After the in spection, the detected nonconformities should be corrected: this leads to the costs of internal failures. Depending on the specific phase, a certain proportion of nonconformities could be reworked while the remaining part will have to be discarded[2]. In addition, in the case of casual defects, the eventual cost of rejection, i.e. of total inspection of the lot, will have to be considered. This is calculated weighting the probability of inspecting a whole lot of size N, by the probability of not accepting it (i.e. a number of nonconforming items greater than c is found in the sample of size n), given by the complement to one of (4). In both cases, finally, we considered the possible opportunity cost of not producing while inspecting items. Quality costs considered are summed up in table 2.

In summary, the following task is that of optimizing the usual trade-off between the cost of control and the cost of internal failure.

[2] The cost of internal failures is based on internal accounting data that provide the added value of an item after any phase.

TABLE 2. Quality costs

CONTINUOUS DEFECTS COSTS	CASUAL DEFECTS COSTS
INSPECTION = cost of inspection every *f* items + average backward inspection cost	INSPECTION = ordinarily inspected items x average unitary cost
INTERNAL FAILURE = reworking cost + scrap cost	INTERNAL FAILURE = *Pa* x expected scrap cost + (1-*Pa*) x (100% inspection cost + expected cost of reworking and scrap)
OPPORTUNITY COST = average inspection time x unitary opportunity cost	OPPORTUNITY COST = average inspection time x unitary opportunity cost

3.4 OPEN PROBLEMS

The analysis so far conducted presents two main problems remained open. First, how to calculate the optimal sampling frequency (in the case of continuous defects) and the optimal sample size and acceptance number (in the case of casual defects), given p and q? Second, how to obtain an unbiased estimator of p and q?[3]

We defined q as the probability that any worked item has to be the first nonconforming one in a production phase characterized by continuous defects. Let' s consider the random variable R which represents the progressive number of the first nonconforming item after the production has started, i.e. how many items have been produced to get to the first defective item, this one included. Then, $\Pr(R=k)=q\,(1-q)^{k-1}$ and $E(R)=q^{-1}$ being that:

$E(R)=\sum_{n=1}^{\infty} n(1-q)^{n-1}q = q\sum_{n=0}^{\infty}(n+1)\cdot(1-q)^{n} = q\cdot\frac{1}{q^{2}} = q^{-1}$. As a consequence, $\bar{q}$ a consistent and unbiased estimator of q, can be calculated as the inverse of the average number of items produced to get to the first nonconformity over a large enough sample of lots. The estimation is even simpler in the case of casual defects, when p is easily estimated with the average proportion of nonconforming items with reference to the inspected items.
The following paragraph describes a practical solution to the mentioned problems.

4 LOOKING FOR THE DATA AND FOR THE OPTIMUM

An intensive experimental campaign of three months was launched, after the design of a provisional sampling master plan, whose goal was that of getting the necessary data to estimate, as accurately as possible, p, q, and the proportion of nonconforming items that could be reworked after each phase. Having this aim, the participation of the whole firm and the development of on purpose data-sheets and data-management routines were required so as to collect the necessary information, that is, the number of inspected items, the number of nonconformities found, and the degree to which a reworking was possible.

Data collecting for continuous defects looks pretty obvious, whereas casual defects deserve a

[3] By the same token, the proportion of nonconforming items that could be reworked after each phase has to be estimated.

deeper look. This problems descends from the fact that while in theory continuos defects do not get to the final QC control (by definition of our sampling plan, which prescribes to inspect with a given frequency and eventually backward), it is still possible that many casual defects reach the final inspection. Thus, why confine to sampling data in order to estimate p, when data from electrical and QC tests are available? Conceptually, it has to be noted that the QC test and the electric test detect different kind of nonconformities (and this is obviously the reason why they are both carried out). Hence, data concerning detected defects at the final inspection were taken into account, too.

Given the gathered data, that referred to more than 1000 lots, the first step was that of calculating a suitable estimator and a confidence interval for p and q. Considering that: firstly, the proportion of nonconformities depends on the kind of printed circuit but it was too complex to get and analyze the data separately and the production mix changes frequently; and that, secondly, p and q are likely to change in any case over time, both for exogenous (e.g. raw material) and endogenous (e.g. learning) reasons, an Excel application was created so that it is easy to live-update the estimate thanks to new experimental data collected in process.

Finally, two specific application were created to design the best sampling plan for phases characterized by casual and continuos defects. The applications were simply designed with a recursive routine that, given the expected proportion of nonconforming items, the added value to the phase at stake, and the usual percentage of possible reworking, is able to calculate respectively the frequency f and the lot size n and the acceptance number c leading to the minimum expected cost sampling plan.

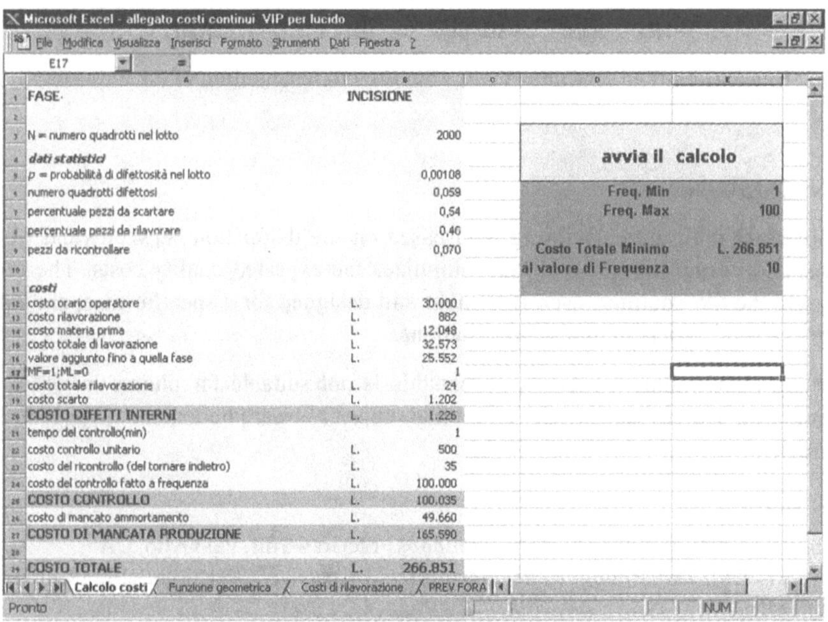

FIGURE 1. Example of the SW routine for continuous defects

5 IMPLEMENTATION

The new sampling plan has been implemented and tested over a three-month period. Results are significantly positive. General nonconformities decreased to less than one half, from 2 to 0.6% (see Figure 2), and nonconformities caused by internal phases decreased from the 76% to the 60% of the total. In particular, we noticed how the improvement is sustained by those phases that were first considered as the most problematic. True enough, it has also to be considered that the simple facts of a new attitude towards quality and the acquired habit of daily observation and interpretation of data may have brought an important contribution to the final result. Still, this does not diminish the value of the new sampling plan, but strengthens the idea that quality tools could not only have a direct, "technical" effect on firm' s performance, but may also contribute to reinforce and empower the consciousness and the disposition of the organization towards quality, in a real spirit of continuous improvement.

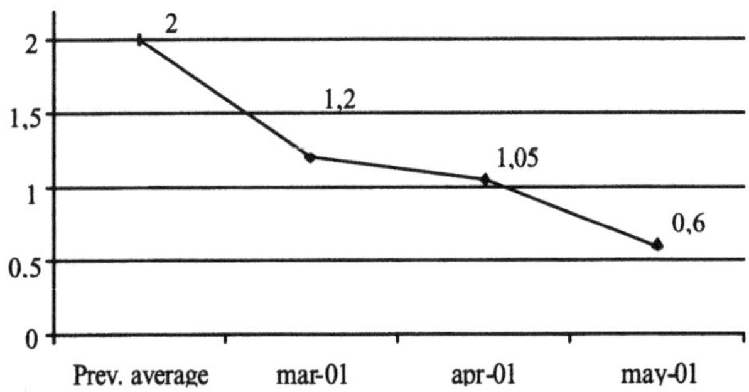

FIGURE 2. Nonconforming items percentage after the implementation of the new sampling plan

6 CONCLUSIONS

We developed an innovative sampling plan based on the distinction between what we called *continuous* and *casual* defects, and that minimizes the expected quality costs. The idea, the statistics, and the SW routines, even if thought and designed for a specific case, are valid and applicable to a wide variety of industrial situations.

The sampling plan' s main limitation is that this is not suitable for phases in which several operations are carried out sequentially with tools subject to wear phenomena [3] (e.g. milling for printed circuits).

REFERENCES

1. Coombs, C., (1998), *Printed Circuits Handbook*, McGraw Hill, Palo Alto, CA.
2. Evans, L., (1996), *The Management and Control of Quality*, West Publishing Company, St. Paul, MN.
3. Giardini, C., Ceretti, E., Maccarini, G., Bugini, A., (1999), Economic design of control charts considering the influence of the tool wear, *Proceedings of AMST 99.*

OPTIMIZATED DESIGN OF NOTCH SHAPES FOR MICROFABRICATED COMPLIANT MECHANISMS

E. Dalla Serra[1], F. De Bona[1], M. Munteanu[2]

[1] Department of Electrical, Management and Mechanical Engineering, University ofUdine, Italy
[2] Transilvania University, Brasov, Romania

KEYWORDS: Microsystem design, Structural optimization, Finite Element Method (FEM).

ABSTRACT. In this work a procedure for the optimal design of flexural hinges to be microfabricated with a lithographic process is proposed. The structural optimization problem is approached by coupling a parametric finite element model to an optimization algorithm. A computer code was developed to generate the mesh at each optimization step accordingly to the values of the design parameters provided by an optimization toolbox. The objective function is the maximum displacement of the mechanism, which must be maximized. The solution is constrained by strength andkinematic requirements. The notch shape is described by spline functions according to an original procedure developed by the authors. Different cases were proposed showing that the procedure is always convergent.

1 INTRODUCTION

Very often microsystems make use of design solutions based of compliant mechanisms [1], [2]. In fact microfabrication technologies permit monolithic structures to be easily obtained, thus reducing microassembly problems related to the use of microdevices composed of several parts; moreover the compliant approach allows backlash, stiction and wear to be eliminated [3]. There are two types of approaches to design compliant mechanisms: continuum synthesis approach and kinematic synthesis approach; in the first case a topology optimization technique is used to obtain a structure with the desired kinematics [4]; this approach, even if quite promising, is still in progress, as the formulation of the optimization problem is not easy, particularly concerning the definition of the objective functions. In the case of the kinematic synthesis approach, the compliant mechanism is obtained simply introducing lumped compliance in a traditional rigid link configuration; in [5] a method based on a so called pseudo-rigid body model is introduced to design compliant mechanisms; the flexible elements (flexible pivots) are obtained by means of leaf springs, whose length is quite smaller in comparison with the length of the rigid parts. It is therefore possible to model the flexural pivots as torsion springs attached to rigid body joints, whose stiffness is obtained from elementary beam theory. A different and more widely used design solution to obtain flexible pivots is suggested in [3], where the flexure is obtained machining a circular cutout on either side of a rectangular cross section blank; also in this case stiffness is evaluated by means of elementary beam theory.
It must be pointed out that in [3], in [5] and in [4] only the relationship between flexure geometry and stiffness is studied, even if following a strongly simplified approach. On the other hand the stress analysis of the pivots is completely missing. As pointed out in [1] the main drawbacks of flexural pivots include high stress concentration and therefore low fatigue strength; it seems

Published in: E.Kuljanic (Ed.) *Advanced Manufacturing Systems and Technology*,
CISM Courses and Lectures No. 437, Springer Wien New York, 2002.

therefore more correct to consider already, in the flexure design phase, not only stiffness, but also strength requirements.

For this purpose it is in general necessary to refer to a numerical model based on FEM. In fact analytical solutions for an accurate stress analysis are not available for several of the possible geometries of the flexural pivot. Moreover a numerical approach permits a more accurate estimation of the flexure stiffness to be achieved.

The necessity of referring to a computer aided method (i.e. FEM) to evaluate the mechanical behavior of the flexure suggests to use a computer aided method also to design the device. In fact, from the design point of view, in general it is necessary to find out the flexural pivot geometry that permits to obtain the required kinematics with very low stiffness and stresses. This is a typical structural optimization problem that can be therefore solved by coupling a parametric finite element model to an optimization algorithm. This approach seems quite helpful particularly in the case of microsystems based on compliant mechanism. In fact the most commonly used microfabrication processes (bulk and surface micromachining and LIGA) are based on lithography [2], a process that permits prismatic microstructures, characterized by even complex in-plane shapes, to be easily obtained. It follows that, if in the case of macro-devices obtained by metal cutting the flexure geometry is limited to very simple shapes, in the case of micro-devices obtained with lithography, solutions with no-circular shapes can also be foreseen.

In this work a computer aided procedure for the optimal design of flexible pivots to be microfabricated with a lithographic process is therefore proposed. In this way it becomes possible to obtain, following a kinematic synthesis approach, compliant micromechanisms with the required kinematics and strength.

2 BASIC PROBLEM

Fig. 1a shows the most common type of flexural hinge: a plate characterized by a thickness t, a width $2b$ and a height $2h$ with a notch on each side. The plate is therefore compliant in bending about the z-axis, but quasi-rigid about the other degrees of freedom.

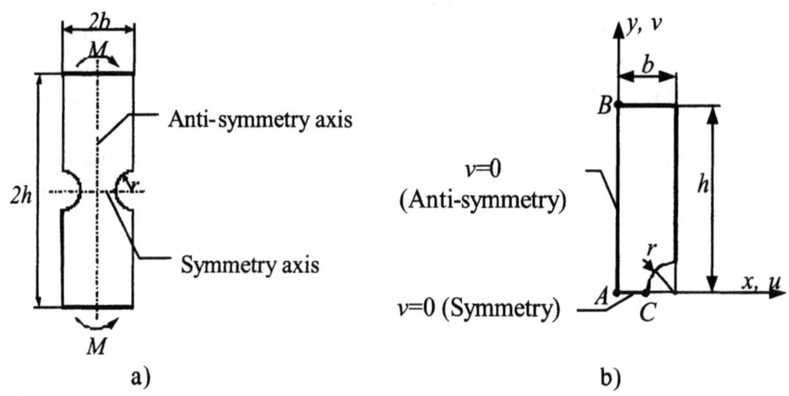

FIGURE 1. Flexural hinge considered (a) and details of its structural model (b)

In general different kinds of loads can be applied, but in this work only the basic problem of the plate loaded by a couple will be considered (Figure 1). Obviously the same approach could be applied also to different cases.

Figure 1b shows the model used for the structural analysis. As the horizontal axis is a symmetry axis and the vertical one is an anti-symmetry axis, only one quarter of the structure can be considered. In addition to the boundary conditions related to symmetry and anti-symmetry, the horizontal displacement u_A of point A was set to 0 in order to avoid a horizontal rigid body motion.

3 STRUCTURAL OPTIMIZATION PROCEDURE

3.1 CIRCULAR SHAPED NOTCH

In the case of flexural hinges that have to be obtained using a traditional mechanical machining process the possible configurations of the notch are in general limited to circular shapes. The case of a plate with opposite semi-circular notches is therefore firstly considered. This case will be used as a reference in the following treatise.

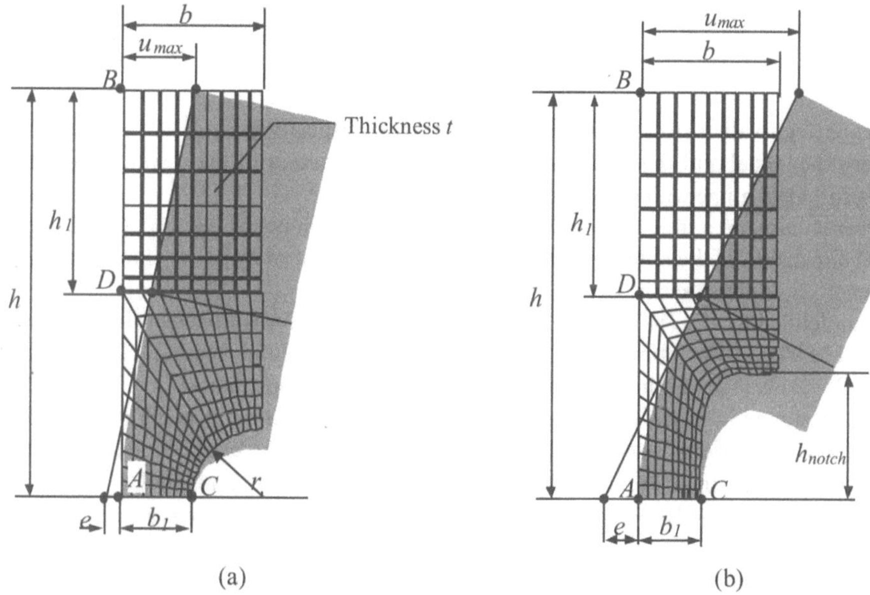

FIGURE 2. First iteration FEM model (a) and example of optimized solution (b)

The main requirement that has to be fulfilled in designing a flexural hinge is that of increasing the flexibility with respect to the so-called "sensitive" degree of freedom (that in this case is the rotation around the z axis). The design effort in this sense is obviously limited by strength considerations; moreover a well-defined kinematics has to be followed (in this case the rotation around point A). It follows that even in this simple case, the optimal design of the flexure can not be obtained easily. An analytical solution for any shape of the notch is difficult to be found in general [6] and a numerical approach is therefore proposed. The structural problem can be solved using a parametric Finite Element model and the optimal solution can be obtained by using a suitable optimization algorithm. In this work a dedicated FEM code was developed using MATLAB™ programming language. As optimization algorithm, MATLAB™ Optimization Toolbox was used [7] and the function *fmincon* provided by this toolbox was used.

Figure 2a shows the FEM model used at the first step of the iterative optimization technique. The dimensions h and t are generally imposed by overall dimension requirements. So, in the case of circular notches only two geometrical parameters characterize the structure: the notch radius r and the width b. On the other hand, it must be noticed that, for a given value of r, the value of b will be automatically limited by strength consideration. In fact the strength condition imposes uniquely the dimension $b_1=b-r$. In the following paragraph, for a given b, the shape of the notch will be modified in order to improve the compliance (see Figure 2) and the design variables of the problem will be those required to describe the notch geometry.

The maximum horizontal displacement u_{max} occurs in point B (Figure 2); u_{max} characterizes the flexibility of the structure and therefore the objective function that must be minimized is:

$$f = 1/u_{max} \tag{1}$$

Strength and kinematics considerations permit to define two constraints of the optimization problem. The following expressions are suggested:

$$g_1=\max(\sigma_{eq})-\sigma_{adm}<0 \tag{2}$$

$$g_2=e-e_{imposed}<0 \tag{3}$$

where:

σ_{eq}, σ_{adm} are the equivalent stress and the admissible stress respectively;

e, $e_{imposed}$ are the actual and the imposed eccentricity, i.e. deviation of the flexural kinematics from that of an ideal hinge with center in point A (see Fig. 2b).

In the general case of a free shaped notch the constraint imposed by Eq. (2) limits the transversal dimension of the notch b_1, while the constraint described by Eq. (3) limits the height of the notch h_{notch}; in the circular case only Eq. (2) is imposed.

The problem defined by Eq. (1) is a non-linear optimization problem with constraints. Function `fmincon` of MATLAB™ Optimization Toolbox follows the Sequential Quadratic Programming method (SQP) to solve the non-linear optimization problems with constraints [7].

The notch is designed in order to have a localized elasticity simulating as much as possible an ideal hinge. The eccentricity e represents the error of motion of the elastic notched beam in comparison to an ideal hinge. Figure 2 shows how the eccentricity e is evaluated: the positions of the nodes situated on the B-D segment in the deformed plate are considered and interpolated by means of a linear regression analysis. The value of e is obtained considering the intersection of the line of regression with the x-axis. In the case of an ideal hinge the segment B-D rotates around point A without any effort. In the case of the elastic notched plate loaded by the moment M, the motion of segment B-D generally is not a pure rotation around point A, moreover segments B-D bends. As a first approximation the value of e permits to estimate the deviation from the desired kinematics of the real hinge. Large plates ($b>>b_1$) with localized notches ($h_{notch}<<h$) obviously will exhibit lower values of e.

A MATLAB™ computer program was written to generate the finite element mesh at each optimization step accordingly to the values of the design variables provided by MATLAB™ Optimization Toolbox. Four-nodes isoparametric finite elements were used and selective integration over each finite element was applied.

At present several commercial FEM codes have structural optimization toolbox already implemented, in this work the structural optimization problem was reconsidered using the "first order" optimization method of FEM code ANSYS™ [8]. Eight-nodes isoparametric finite elements were used. The notch shape was described through the pre-processor tools of ANSYS™: an arc was defined by simply imposing center, radius (design variable) and arc length (90°). The maximum number of iterations was set to 50 and the tolerance of convergence to 10^{-5}.

3.2 FREE SHAPED NOTCH

Microfabrication techniques based on lithography permit non-circular geometry to be easily obtained; it is therefore interesting to examine, starting from the circular notch case, if even slight variations of shape can improve the flexure characteristics. For this purpose a procedure that permits a free shaped notch to be considered, was also implemented.

Figure 3 shows the procedure followed using MATLAB™; the notch shape is described by means of spline functions characterized by three points, called "spline points", which may not coincide with the finite element nodes (in Figure 3 spline end points are filled).

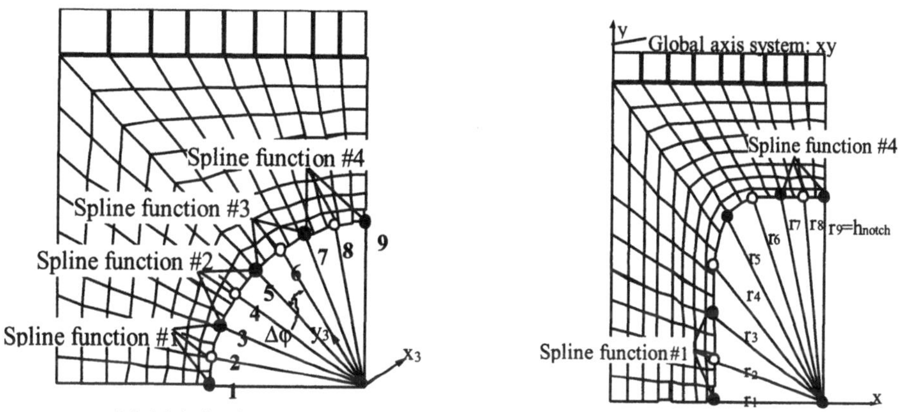

a) Initial circular notch b) Final "stretched" shape of the notch

FIGURE 3. Notch geometry definitions by means of spline functions

Spline functions are polynomial of 3rd degree in their own local system of reference (local y axis is always the radius corresponding to the middle point of the spline function). Each spline function introduces 4 unknown coefficients and thus n spline functions introduce $4n$ unknowns. The following conditions are imposed: each spline function has to pass through its three points ($3n$ conditions); in the end points the slope of the neighboring spline functions must be the same (n-1 conditions); in point 1 (see Figure 3a) the slope must be vertical as the symmetry condition holds (1 condition). It results that the total number of conditions is $4n$ too. As shown in Figure 3a, to define the notch geometry it is necessary to estimate the unknown values of the radii r_i of the $2n$+1 spline points, as their initial relative angular position $\Delta\varphi_i$ are set equal to $\pi/(2n-1)$. The angles are not design parameters, but during the optimization process the mesh around the notch is "stretched" only vertically, if h_{notch} increases, and therefore the angles are modified. Usually 3 or 4 spline functions permit the notch geometry to be described quite accurately.

The ANSYS™ model of the free notch was described by cubic spline functions. The number of interpolated points is equal to the number of design variables. Each point has been defined by its coordinates. The ordinates are function of the height of the notch h_{notch}, which is an optimization variable, the abscissas are the design variables. The maximum number of iterations was set to 150 and the tolerance of convergence to 10^{-6}.
In the case of the free shaped notch, the optimization process starts from the optimized circular notch for both MATLAB™ and ANSYS™ methods.

4 RESULTS AND DISCUSSION

The proposed procedure proved to be always convergent. The results are presented in a non-dimensional form, accordingly to the following notation:

$\beta=b/t$, $\lambda=h/b$ and $\gamma = r/b$ (geometrical parameters)

$\mu=M/(\sigma_{adm} b^2 t)$ (load parameter)

$\eta=e\ E\ b\ t/M$, where E is the Young modulus (kinematic parameter)

$\upsilon_{max}=u_{max}\ E\ b\ t/M$ (stiffness parameter)

Tab. 1 shows the results obtained in the case: $\beta=2$, $\lambda=3$ and $\mu=0.125$. To evaluate η, the value $h_1/h=0.5$ (it determines the length of segment BD) was always considered.

TABLE 1. Comparison between results given by MATLAB program and ANSYS

N	Notch type	$\eta=eEbt/M$	Optimization Function: $\upsilon_{max}=u_{max}\ E\ b\ t/M$		Eccentricity:
			MATLAB™ 3 spline functions	ANSYS™ 4 spline functions	
1.	Circular notch	4.3	20.63	20.85	It results from: $\max(\sigma_{eq})<\sigma_{adm}$
2.	Free shaped notch	5	27.48	26.01	It is imposed as a constraint (Eq. (3))
3.		8	39.47	38.57	
4.		10	45.01	44.79	

In the case of the circular notch, the optimization procedure gives $\gamma=0.5$. As pointed out in Paragraph 3.1 the circular notch case, which is the initial configuration for the optimization process, was taken as reference and the corresponding value of the objective function was denoted by υ_0. It is interesting to compare the free shaped notch with the circular notch; for this purpose the value of the ratio $\upsilon_{max}/\upsilon_0$ is reported in Figure 4, that shows the equivalent stress distribution around the notch obtained with the MATLAB™ program for 3 values of η. As it results from Table 1 and Figure 4, the allowable maximum eccentricity strongly influences the maximum displacement u_{max}. Obviously the compliance of the notch increases for higher values of the imposed maximum eccentricity. For instance, for $\eta=10$ it results $\upsilon_{max}/\upsilon_0=2.18$ and this improvement is quite remarkable. It results also that the height of the notch, h_{notch}, increases, as expected, with the eccentricity, consequently, also the stiffness parameter υ_{max} increases with the height of the notch, h_{notch}.

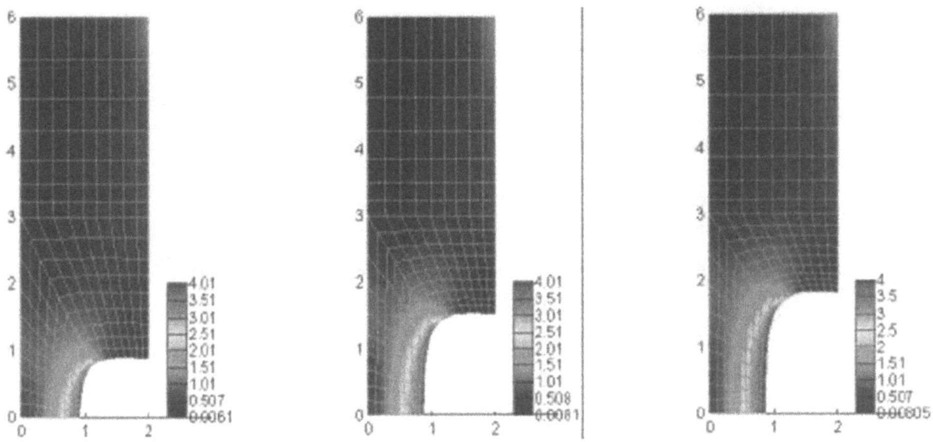

a) η=5, υ_{max} =27.48 b) η=8, υ_{max} = 39.47 c) η=10, υ_{max} =45.01

h_{notch}/r =0.881, $\upsilon_{max}/\upsilon_0$=1.33 h_{notch}/r =1.521, $\upsilon_{max}/\upsilon_0$=1.91 h_{notch}/r =1.820,$\upsilon_{max}/\upsilon_0$=2.18

FIGURE 4. Optimized notch shapes at different values of η (MATLAB™ program)

As shown in Figure 5 the results given by the procedure implemented in MATLAB™ and by ANSYS™ are very close, although the shape in the upper part of the notch is quite different. This might be explained mainly by the fact that in this region the stresses are very low (see Fig.4) and therefore it follows that this part of the plate does not influence significantly the objective function and the eccentricity. Obviously the shape of the upper part of the notch has a negligible influence on the stress distribution in the notch, too. In fact it was noticed that it exists several local optimum notch configurations that leads to very closed values for the objective function defined by Eq. (1). The cases presented in Figure 5a and Figure 5b may be considered as two extreme cases. MATLAB™ computer program makes use of function "`fminco`" of the Optimization Toolbox that allows the implicit values for tolerances and for the minimum change of the design parameter to be defined. In order to ensure a very good precision of the optimization process a value of 10^{-5} for objective function and a value of 10^{-8} for the minimum change of the design parameters were imposed. Therefore the solution was obtained in a rather large number of iterations (500-1000); nevertheless the solution converges to a quite regular notch shape (see Figure 5b). On a PENTIUM III computer, a single iteration (FEM model, mesh generation, displacement and stress solutions) takes approximately 1.2 s for a 600 DOF model with 7 design variables (free shaped notch). It follows that the procedure implemented in MATLAB™ requires computer times always lower than 15-20 minutes. In the case of the procedure implemented in ANSYS™ the free shaped notch (9 design variables) requires a computational time more than 10 times higher.

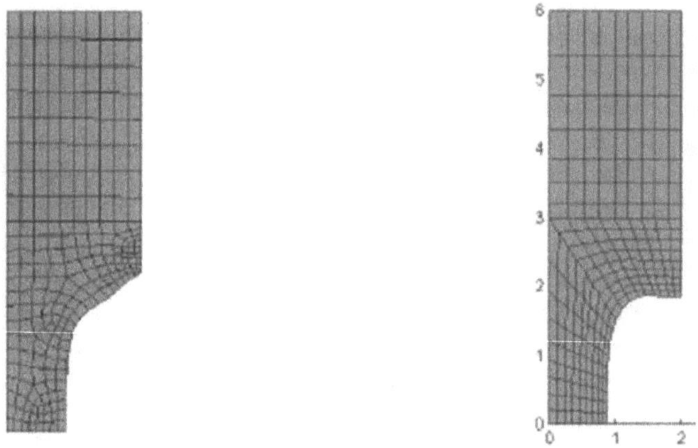

a) b)

FIGURE 5. Optimized solution obtained with ANSYS™ (a) and with MATLAB™ (b)

5 CONCLUSION

An original method to design flexible hinges microfabricated with lithography was developed; the proposed approach was implemented both using a program written in MATLAB™ programming language and also using the Design Optimization Toolbox of the FEM code ANSYS™. The results obtained following the two approaches are quite similar; they clearly show that, with respect to the usual design approach, the proposed method permits a significant improvement of the hinge characteristic to be obtained.

REFERENCES

1. Anathasuresh, G.K., Sridhar, K., (1995), Designing Compliant Mechanisms. Mechanical Engineering, November, 93-99.
2. Madou, M., (1997), Fundamentals of Microfabrication, CRC Press.
3. Paros, J. M., Weisbord, L., (1965), How to Design Flexures Hinges. Machine Design, November 15, 151-156.
4. Nishiwaki S., Min, S., Ejima, S., Kikuchi, N., (1998), Structural optimization Consideing Flexibility. JSME International Journal, Series C, Vol. 41, 476-484.
5. Howell, L.L., Midha, A., (1994), A Method for the Design of Compliant Mechanisms With Small-Length Flexural Pivots. Journal of Mechanical Design, Vol. 116, 280-290.
6. Peterson, R. E., (1974), Stress Concentration Factors, John Wiley and Sons.
7. Coleman, T., Branch, M. A., Grace, A., (1999), Optimization Toolbox For Use with MATLAB, The MathWorks Inc, Natick, MA, USA.
8. ANSYS Advanced Analysis Technique Guide, (1999), Forth Edition, Ansys Inc., PA, USA.

QUALITY FUNCTION DEPLOYMENT AND TAGUCHI METHODS: A PRAGMATICAL APPROACH

E. Gentili[1], F. Galetto[2]
[1] Department of Mechanical Engineering, University of Brescia, Italy.
[2] Department of Production Systems and Economics, Politecnico di Torino, Italy.

KEYWORDS: Quality Management, DOE, QFD.

ABSTRACT. The goal of "Quality Engineering" is to design Quality into every product and all the processes that build them. Design is the most important phase for Quality both of products and processes. Two methodologies, originated in Japan, are claimed as very important for Quality design: Quality Function Deployment (QFD) and Taguchi Methods (TM). Often they are called upon as the best synergy a Company can use during development of products and processes.

1 INTRODUCTION

In order to make Quality of products (processes, systems and services) a good knowledge of Quality ideas and a good experience about Quality tools for achieving Quality are absolutely needed. A person involved in Company management, especially Quality Managers need that. In order to find and use the Quality tools for Quality achievement, education of Managers on Quality is essential. Unfortunately too many managers know very little about Quality ideas and Methods.
Design of Experiments (DOE) is a major element of the design activity and one of the most important methodologies to achieve Quality through intelligent testing of factors that influence Quality, during the design stage of product and processes. For processes DOE allows us to find the optimum setting of factors that provide the best Quality and economic yield, that is minimization of "disquality cost". One of the most widely used methodologies is the Taguchi Method (Robust Design). Quality Function Deployment (QFD) is considered by many people as "a tool which is able to ensure that the voice of the customer is deployed throughout the product planning and design stages", via the use of various "houses of quality" that allow the collection and the organization of information about quality, features, characteristics, parts, processes and goals relative to a product or a service. The literature on these methods is rapidly expanding. Therefore it seems important to stand back a bit and meditate from a managerial point of view. The paper shows, using actual cases that the Scientific Approach is able to provide the right route towards the good methods for Quality.

2 QUALITY AND PREVENTION

One can see the level of "disquality" (the opposite of Quality) analyzing the new ISO documents on the Quality Systems:
ISO 9001:2000 "Quality Management Systems - Requirements"
ISO 9004:2000 "Quality Management Systems - Guidance for Performance Improvement"

Published in: E.Kuljanic (Ed.) *Advanced Manufacturing Systems and Technology*,
CISM Courses and Lectures No. 437, Springer Wien New York, 2002.

The new ISO standards on the "Quality Management Systems" still lack correct ideas of Quality and of Quality Management for Prevention: they use badly the ideas on Improvement and Prevention. Item 8.5 of the ISO 9004:2000 is worse than before. The Members of the Tech. Committee ISO/TC 176 fall short on "Prevention", as one can see reading the "new ISO Standards" and comparing them with this paper and others in the references; in ISO 9004:2000 the word "prevention" is very rare.
The standards ISO 9000:2000 "Quality Management Systems – Fundamentals and Vocabulary" and ISO 9004:2000 state that "eight quality management principles have been identified that can be used by top management in order to lead the organization towards improved performance."
The seventh principle “Factual approach to decision making” states: “Effective decisions are based on the analysis of data and information”. The subject is important and we think we need to remember Deming’s ideas [1, 2] such as the following, recurring in his books:

- Experience alone, without theory, teaches management nothing about what to do to improve quality and competitive position, nor how to do it.
- Understanding of quality requires education.
- There is no substitute for knowledge.
- It is a hazard to copy. It is necessary to understand the theory of what one wishes to do or to make.

ISO/TC 176 defined the term “quality” as the “degree to which a set of inherent characteristics fulfils requirements”. It is like confounding the concept of “temperature” with the concept of “heat” (which is energy). In the authors’ opinion, the first step to Quality achievement is to define logically and correctly what Quality is. In order to provide a practical and managerial definition F. Galetto in 1985 proposed the following one: Quality is the set of characteristics of a system that makes it able to satisfy the Needs of the Customer, of the User and of the Society. This definition highlights the importance of the needs of the three actors: the Customer, the User and the Society. Prevention is the fundamental idea present: you possibly satisfy the needs only by preventing the occurrence of any problem against the needs. [3, 4]
The second step is “Understand the word Prevention”. ISO 9004:2000 uses the three terms: needs, expectations and requirements quite a lot (Planning, § 5.4). Prevention is not considered fundamental. At § 8.5.3 one finds the statement "Loss prevention in the form of planning should be applied to realization and support processes, activities and products to ensure the satisfaction of interested parties”. Unfortunately planning is confounded with prevention. You can plan and not prevent [5]. Various practitioners suggest using a panacea for quality planning, too many times considered equivalent to "prevention”. Quality Function Deployment, like ISO 9004, is full of "needs and expectations" (and "customer satisfaction"), that are put in the "customer voice room" of the House of Quality. To hear that voice we must listen very carefully to the customer’s needs.

3 QFD AND DEMING’S SCIENTIFIC OPINION

Let us consider a few definitions of Quality Function Deployment [6, 7]: QFD is considered by many people as "a tool which is able to ensure that the voice of the customer is deployed throughout the product planning and design stages", via the use of various "houses of quality" that allow the collection and the organization of information about quality, features, characteristics, parts, processes, goals, ... relative to a product or a service. But sometimes the same people, say things to the contrary: that QFD is method (not a tool) for structured product planning and development to specify clearly the customer's wants and needs, and then evaluate each proposed product or service capability systematically in terms of its impact on meeting those needs. Again, QFD is a process (not a tool, or a technique) that puts companies in touch with their customers. QFD is a methodology that employs the customers' wants and needs as its basic input: QFD is a process or a methodology for planning product and services or a method. Also: QFD is not a tool: it is a planning process, as opposed to a tool for problem solving or analysis. Again, QFD is a process (not tool or a method, or a technique) that provides structure to the development cycle. Again, some other people say that QFD is a system (not a process or a method) for translating customer requirements into appropriate company requirements. "Quality Deployment is a major force in the Total Quality effort in the United States" [8]. Let us hope that in the future we will not have the Total Quality Improvement: it would be very hard! "We can define quality function deployment as converting the customers' demands into "quality characteristics" and developing a design quality for the finished product by systematically deploying the relationships between the demands and characteristics ... quality function deployment, broadly speaking is a general term that means deployment of quality through deployment of quality functions." [9]. "QFD is a structured process ... It establishes customer value using the voice of the customer and transforms that value to design, production, and manufacturing process characteristics. The result is a systems engineering process which prioritizes and links the product development process so that it assures product quality as defined by the customer/user" [10] ".

QFD “practitioners” misuse the term Quality. Actually the term “quality” in the acronym QFD stands for “qualities”. This is the fundamental reason for the different meanings given by different authors to the acronym QFD. Practitioners tell that QFD converts consumers' demands into quality characteristics: in effect they confuse quality with qualities. The quality characteristics become demanded qualities to convert into measurable and engineering elements. Bossert [11] admits: “Customer requirements are translated into manufacturer's terms”. The example in Fig.1, derived from ASI (American Suppliers Institute), does not help in making Quality: Taguchi Methods are used instead of the Design of Experiments.

It is clear from the above statements that they are important to define QFD. Sometimes reading QFD papers and books you find the same authors claiming that QFD is a Tool in some parts and a Process (not a tool) in other parts. QFD is said to be a planning process (methodology), but planning does not mean Prevention: you can plan and do not make any Prevention. We need to define QFD: Quality Function Deployment is the process setting the qualities able to satisfy the needs of the customers, the users and of the society. A set of qualities doesn't make necessarily Quality, but it can be a good base to reach this target.

The literature on QFD is rapidly expanding. Therefore it seems important that Managers stand back a bit and meditate from a managerial point of view. If you look at the plethora of terms used in the "customer voice room" (WHAT room) of the HoQ, you notice that Needs (in the definition of Quality given above) is a term quite different from any of the items mentioned in the HoQ. Moreover, it is important to notice that the Customer is quite different from the User. Looking at the plethora of terms used in the "technical room" of the HoQ (HOW room) we note explicitly that Prevention is always absent in QFD literature: this absence is evident from the first two rooms of the HoQ and this is a limitation that can be taken into account or, better, must be solved.

We think that the following statement of the Nobel winner M. Gell-Mann is very suitable for

describing the "reality of application of Quality Methods": "Once that such a misunderstanding has taken place in the publication, it tends to become perpetual, because the various authors simply copy one each other". QFD deals with the deployment of the qualities needed by the customer and this is the first important step to reach Quality.
These who demean Taguchi methods have missed the whole point. Taguchi methods are not just a statistical application of design of experiments. Taguchi methods include integration of statistical design of experiments into a powerful engineering process. The true power of Taguchi methods comes from their simplicity of implementation" [10]. "Traditional experimental designs and Taguchi techniques are tools utilized to understand the relationships in the body of the QFD matrix. There is much controversy among statisticians as to which tools do the job best. .... Use whatever works best for what you are trying to understand. ....." [11] "Dr. Taguchi ... allows us to evaluate at the design stage the cost of quality loss in order to strike the optimum balance between the manufacturing cost and the quality loss. It has been my privilege to work closely with Dr. Taguchi during the past four years, and to be a leader in the implementation of his methods in the United States. For those who wish to be technical leaders in this new industrial revolution, this book is essential." [12]. "I ask Don Clausing who provided several consultancies on Taguchi Methods to companies.... to answer some questions on ... ("The Professionals and Taguchi Methods") [13]. Taguchi Methods don't work and to advise what is proposed by Don Clausing is therefore dangerous.
Let's look at figure 1, made according with what suggtested by ASI ; you see two "rooms of the House of Quality". Unfortunately

- the "whats" don't represent the customer needs
- the "hows" are not the right response.

	HOWS		
	Implement simultaneous engineering partnership with suppliers	Apply Taguchi methods and QFD at design stage	Institute preventive maintenance training for customers
WHATS			
Improved product quality and reliability			
Reduced delivery times and costs			

FIGURE 1. A QFD matrix for "quality development", according to ASI [7]

The Deming replies can only be taken from his very good books *Out of Crisis* [1] and *The New Economics for Industry, Government, Education* [2]:
Experience alone, without theory, teaches management nothing about what to do to improve quality and competitive position, nor how to do it. ### Anyone that engages teaching

by hacks deserves to be rocked. ### ... Understanding of quality requires education. ### Teams of unqualified examiners visit suppliers to rate them. ### There is no substitute for knowledge. ### There is widespread resistance of knowledge. ###Everything best is not enough. ### It is a hazard to copy. It is necessary to understand the theory of what one wishes to do or to make. ### Without theory, experience has no meaning. ### The result is that hundreds of people are learning what is wrong. ### I make this statement on the basis of experience, seeing every day the devastating effects of incompetent teaching and faulty applications. ### Without theory, there is no learning. ### Again, teaching of beginners should be done by a master, not by a hack. ### Beware of common sense.

All these statements are useful for everybody.

The analysis of the "completed House of Quality" that is found in the book "The Management and Control of Quality" [14] gives the possibility to measure the distance of the "quality practitioners" from Quality. Fig.2 represents what reported in [14] in fig.2.3, where students are wrongly considered as customers. Students are the input, but they are not surely the customers. They are prepared for the society.

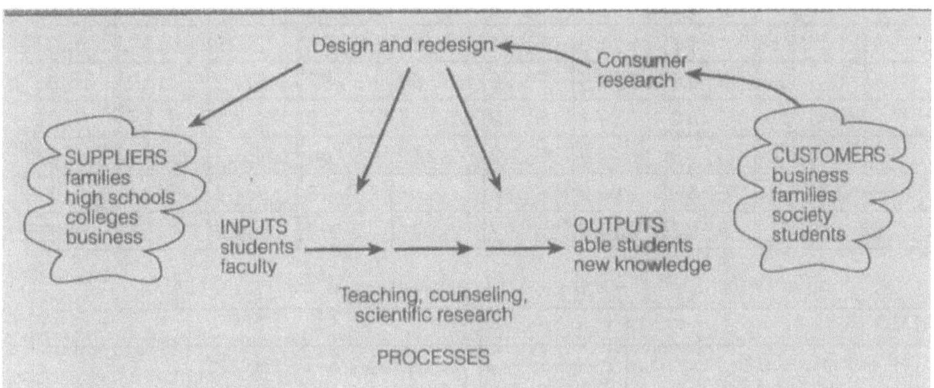

FIGURE 2. Students are not customers but the input and the output of the process

4 DESIGN OF EXPERIMENTS VERSUS TAGUCHI METHODS

Prevention is very important for Quality achievement; Quality comes forth during development. DOE helps a lot in preventing problems; therefore it is very important during product development (fig. 2): managers have to learn Design of Experiments ideas to draw good decisions during product/process/service development, and act according the "statistical thinking" [1, 2] and use the Scientific statistical approach to experiments which entails

a) statistical design b) correct statistical conduct c) scientific statistical analysis of the data.

The "Scientific Approach", is strongly required by GIQA (Golden Integral Quality Approach) [3, 4, 15, 16, 17]. The opposite of the "Scientific Approach" is the Taguchi Method. Mathematics, Logic and Physics can prove that "Robust Design" Methods using the Taguchi approach is wrong especially when G. Taguchi writes ".... when there is interaction, it is because insufficient research has been done on the characteristic values". There are many cases that can show the erroneous decisions taken using Taguchi Methods (suggested in connection

with the QFD, as we have seen before) and the benefits of managerial and scientific conduct of testing activities. Let's look at an application of "Robust Design Methodologies" [20]: the paper helicopter, one of the devices used to illustrate the application of Taguchi's statistical methodologies to experimentation.
The observations (the response) are the time (in seconds) between the release for the helicopter and its impact to the ground. The example considers three experimental phases and at the 3rd experimental phase the conclusion is that only 3 factors are considered important B and E at 3 levels and N at 2 levels. The design is a full factorial one, unbalanced, with replications. Table 1 shows the experiment data and table 2 the analysis of the data reported in [20].

TABLE 1 Experiment data.

Run	B	E	N=-1					N=1		**Mean**
1	-1	-1	5,82	5,78	5,68	6,14	5,96	5,69	6,02	*5,87*
2	-1	0	5,92	5,83	5,8	5,79	6,13	6,15	5,76	*5,91*
3	-1	1	5,35	5,28	5,48	5,48	5,31	6,62	6,93	*5,51*
4	0	-1	6,2	5,68	5,91	5,88	5,75	5,99	5,91	*5,90*
5	0	0	5,48	5,86	5,8	5,67	5,82	5,91	6,11	*5,81*
6	0	1	5,82	5,74	5,82	5,84	5,8	5,61	6,69	*5,81*
7	1	-1	6,53	6,29	6,67	6,82	6,64	5,87	5,82	*6,64*
8	1	0	6,56	6,24	6,28	6,8	6,58	5,53	6,16	*6,39*
9	1	1	5,97	6,17	5,97	6,24	5,97	5,62	4,38	*5,76*

TABLE 2 Analysis of data [19]

Source of variation	g.d.l.	SS	MS	F-ratio	P
Intercept	1	2236	2236	21826	0.000
B	1	2.640	2.640	25.77	0.000
E	1	2.059	2.059	20.10	0.000
BE	1	0.484	0.484	4.721	0.034
BB	1	0.417	0.417	4.072	0.048
EE	1	0.200	0.200	1.952	0.168
Error	57	5.839	0.102		
Corrected total	62	11.639			

First of all, the experimental design is unbalanced and therefore you need the G-Method [18] for a scientific analysis: you can not get the "residual error" as difference between the corrected total and the (intercept+B+E+BE+BB+EE). The "real" error is given by the total minus the "SS full model" (with 18 df) [18]. The SS provided in table 4 are all wrong as you can see from table 3. Beside that, with the G-Method [18] you can get the linear and the quadratic effect of the factors and of the interaction (table 3). With the scientific analysis you get quite a different picture regarding the significance of factors and interaction.

TABLE 3 G-Method Analysis of Variance [20]

Source	Df	SS	MS	Fc	F(5%)	significance
Total	63	2247,6				
Mean	1	2236				
Corrected total	62	11,639				
B	2	1,148	0,574	9,5716	2,4245	*
E	2	0,668	0,334	5,5698	2,4245	*
BE	4	1,038	0,2595	4,3273	2,0742	*
Residual error	45	2,6986	0,06			
Blinear	1	0,9123	0,9123	15,212	2,8205	*
Bquad	1	0,2357	0,2357	3,9309	2,8205	*
Elinear	1	0,5858	0,5858	9,7674	2,8205	*
Equad	1	0,0823	0,0823	1,3722	2,8205	
Bl*El	1	0,4837	0,4837	8,065	2,8205	*
Bl*Eq	1	0,0535	0,0535	0,8922	2,8205	
Bq*El	1	0,2929	0,2929	4,8837	2,8205	*
Bq*Eq	1	0,208	0,208	3,4685	2,8205	*

5 CONCLUSIONS

QFD pragmatic books for practitioners lack greatly a Quality and Scientific Approach. According to Deming "It is a hazard to copy. It is necessary to understand the theory of what one wishes to do or to make. Without theory, experience has no meaning."

Hence, it is much better to be both pragmatic and theoretic [4, 18, 19], instead of being practitioners. Managers must take care of needs more than they care for "wants": needs are more important.

This paper stresses again that Prevention with "future consequences of present decisions" (the futurity), in the Company space-time continuum, through holism, intellectual honesty and scientific approach, is essential in order to provide the Customers/Users/Society integrally the Quality they need. The Customers, the_Users and the Society need flexible managers who can think integratively, prevent and solve problems, be life-long learners using their intelligence and intellectual honesty. Decisions based solely on data analysis by applying the (ISO 9000:2000 and 9004:2000) seventh principle "Factual approach to decision making" which states: "Effective decisions are based on the analysis of data and information" are usually ineffective and wrong. DOE (and Reliability Tests) is actually very important both for product development and for improvement.

Quality of methods for making Quality is important to prepare managers and to achieve Quality everywhere in our Society [18, 19].

ACKNOWLEDGEMENTS

The Authors wish to acknowledge the generous support offered by MURST in the realisation of this work. Special thanks to Mrs. Mary Flynn who checked the manuscript.

REFERENCES

1. Deming, W.E. (1986), Out of the Crisis, Cambridge Press.
2. Deming, W.E. (1997), The New Economics for Industry, Government, Education, Cambridge Press.
3. Galetto, F., (1989), Quality of Methods for Quality is Important, EOQC Conference, Vienna
4. Galetto, F., (1999), GIQA, the Golden Integral Quality Approach: from Management of Quality to Quality of Management, Total Quality Management, Vol. 10, No. 1
5. Galetto, F., Gentili, E. (1999), The Need of Quality Methods Used for Quality, Proc. of CAPE '99, Durham, UK, 281-289.
6. Galetto, F. (1999), QFD: Some Managerial Concerns, AITEM 99, Brescia, 281-291.
7. Galetto F., Gentili E. (2000), In Search of Quality in QFD and Taguchi Methods, Proc. of CAPE 2000, Edimburgh, UK, 511-520.
8. King, R, (1987), Listening to the Voice of the Customer, National Productivity Review 6, no. 3, 277-281.
9. Akao, Y., 1990, Q.F.D. Integrating Customer Requirements into product Design, Productivity Press.
10. Dean ,E.B., 1999, Quality Function Deployment, Internet Paper.
11. Bossert, J.L., 1991, Quality Function Deployment, ASQC Quality Press, Milwaukee, USA.
12. Clausing, D.P.,The House of Quality, Harvard Business Review, May-June 1988, 63-73.
13. Sullivan, L., 1989, Quality Function Deployment, Quality Progress, Prentice Hall.
14. Evans, J.R., Lindsay, W.M., (1996), The Management and Control of Quality, West Publishing Company, St Paul, USA.
15. Galetto, F., LEVI, R., (1993) Planned Experiments: Key Factors for Product Quality, 3rd AMST 93, Udine, 39-49.
16. Galetto, F., (1996), Managerial Issues for Design of Experiments, 4th AMST 96, Udine, 839-848.
17. Galetto, F., (1999), Quality Methods for Design of Experiments, 5th AMST 99, Udine, 833-842.
18. Galetto. F., (2001). Qualità. Alcuni metodi statistici da Manager, CLUT, Torino.
19. Galetto, F., Gentili, E., (1999), Quality of the Quality Methods, AITEM 99 Conference, Brescia, 293-302.

DATUMS CONCEPTS BY ASME AND ISO STANDARDS

I. Cristofolini

Department of Mechanical and Structural Engineering, University of Trento, Italy

KEYWORDS: Geometric Dimensioning and Tolerancing (GD&T), Geometrical Product Specifications (GPS), Datums

ABSTRACT. A clear, concise and unequivocal communication of the design intents from engineering to manufacturing to verification is essential aiming at ensuring products quality. Datums and Datum systems, allowing defining the location and the geometrical characteristics of parts, play a central role in establishing a common language. By this way it is necessary analyzing Datums concepts as by the standards to which refer in the dimensioning step, being ASME and ISO approach to geometric dimensioning and tolerancing generally different. Datums concepts as by ASME and ISO standards are evaluated and discussed in this work, also considering related GPS ISO Drafts (ISO/TC 213 work in progress), aiming at evidencing the evolution in ISO standardization.

1 INTRODUCTION

In these last years the scenery of global market changed so deeply and quickly that industries had to front a real revolution. Sophisticated CA-x systems, advanced materials, new technologies lead to products showing even more complex geometries, revealing the need for standards, methodologies and procedures, allowing describing them properly. Geometrical characteristics deeply influence aspects as functionality, manufacturing, assembly and verification so that it is very important adequately considering them in the first steps of design, when dimensioning and tolerancing [1].

Analyzing Datums and Datum systems as a powerful tool in establishing a common language from design to manufacturing to verification it is very important evaluating the standards used for reference; it is necessary verifying if the same terms, as by different standards, have the same meaning or not.

ASME reference for Geometric Dimensioning and Tolerancing (GD&T) is actually ASME Y14.5M-1994 (next version is expected to be disposable 2005), defining the guidelines for dimensioning and tolerancing, also considering the geometrical characteristics [2, 3]. This standard also reports Datums concepts definitions, meaning and uses. On the other hand, a relevant number of standards describe the ISO guidelines for geometric dimensioning, now collected and harmonized in the Geometrical Product Specification matrix (GPS matrix) [4]. ISO 5459-1981 is actually the main reference for Datums [5].

It is however necessary underlining that the GPS matrix also identifies the beginning of an important work concerning the general revision and development of standards related to the Geometrical Product Specifications by the ISO Technical Committee 213 (ISO/TC 213) [6]. Many Work Groups belong to ISO/TC 213: Work Group 2 is explicitly dedicated to Datums and datum sys-

Published in: E.Kuljanic (Ed.) *Advanced Manufacturing Systems and Technology*,
CISM Courses and Lectures No. 437, Springer Wien New York, 2002.

tems. The revision of ISO 5459 concerns now four important Projects: ISO/CD 5459-1:1998(E) *Datums and datum systems for geometrical tolerancing – Part 1: General terms and definitions* [7], ISO/CD 5459-2 *Datums and datum systems for geometrical tolerancing – Part 2: Indication in technical product documentation* [8], ISO/WD 5459-3 *Datums for geometrical tolerancing – Part 3: Association methods for datums and datum systems for the assessment of geometrical tolerances* [9], ISO/WD 5459-4 *Datums for geometrical tolerancing – Part 4: Metrological establishment of datums and datum systems for the assessment of geometrical tolerances* [10].

Aforementioned standards are analyzed and compared in this work, also aiming at identifying if the evolution in standardization adequately reflects the need of the evolving industrial reality.

2 DATUMS CONCEPTS DEFINITIONS BY ASME AND ISO STANDARDS – COMPARISON AND DISCUSSION

Firstly, aiming at verifying the correspondence in meaning of the terms used by the different standardization, the related definitions are compared and discussed.

2.1 DATUMS DEFINITIONS

Datums definitions as by ASME Y14.5M-1994, ISO 5459-1981 and ISO/CD 5459-1:1998(E) are reported in the following table.

TABLE 1. Datum definitions

ASME Y14.5M-1994	ISO 5459-1981	ISO/CD 5459-1: 1998(E)
Datum: theoretically exact point, axis or plane derived from the true geometric counterpart of a specified Datum Feature. A Datum is the origin from which the location or geometric characteristics of features of a part is established. ***True geometric counterpart***: the theoretically perfect boundary (virtual condition or actual mating envelope) or best-fit (tangent) plane of a specified datum feature	**Datum**: a theoretically exact geometric reference (such as axes, planes, straight lines etc.) to which toleranced features are related. Datums may be based on one or more features of a part.	**Datum**: situation feature used to define the position and/or orientation of a tolerance zone or to define the position and/or the orientation of the virtual condition (in the case of complementary requirements, e.g. maximum material requirement) ***Situation feature***: ideal feature which is a point, a straight line, a plane or a helix, from which the position and/or the orientation of a feature can be defined

Firstly considering the two definitions reported in the published standards, we can observe that, although they can look similar, some important differences can be evidenced: referring to the true geometric counterpart of a specified Datum feature, ASME standard clearly states how the Datum relates to the part, while ISO standard generically establishes that Datums may be based on one or more datum features on the part.

In the ISO Draft the correlation between Datums and features on the part, although not explicitly established in the definitions, can be evidenced by further analysis of the Drafts contents, and it will be discussed following. Moreover, the introduction of the virtual condition concept looks more closely reminding to ASME.

2.2 DATUM SYSTEMS DEFINITIONS

Datum systems definitions as by ASME Y14.5M-1994, ISO 5459-1981 and ISO/CD 5459-1:1998(E) are reported in the following table.

TABLE 2. Datum systems definitions

ASME Y14.5M-1994	ISO 5459-1981	ISO/CD 5459-1:1998(E)
Datum Reference Frame: sufficient datum features, those most important to the design of a part, or designated portions of these features are chosen to position the part in relation to a set of three mutually perpendicular planes, jointly called a datum reference frame. This reference frame exists in theory only and not on the part. (…)	***Datum system***: a group of two or more separate datums used as a combined reference for a toleranced feature	***Datum system***: ordered list of at least two datums and most three datums, which may be single or common datums ***Single datum***: datum based on one surface considered alone ***Common datum***: datum based on two or more surfaces considered simultaneously
Datum target: a specified point, line, or area on a part used to establish a datum	**Datum target**: a point, line, or limited area on the workpiece to be used for contact with the manufacturing and inspection equipment, to define the required datums in order to satisfy the functional requirements	**Target**: portion of an integral feature (surface) which is an area and which is used for establishing a datum

By ASME Datums exist within a framework of three mutually perpendicular intersecting planes known as the Datum Reference Frame, as shown in figure 1. Positioning and orienting a part means fixing it in relation to the Datum Reference Frame: the part must contact the primary Datum plane by a minimum of three points, the secondary Datum plane by a minimum of two points and the tertiary by a minimum of one point. By this way, it is commonly said that the part is completely constrained (being the six degrees of freedom of the part completely removed); more exactly we should say that the method of measuring the part is constrained [11]. ISO, instead, both in the published standard and in the Drafts definition, does not establish a Datum Reference Frame, only referring to a Datum system, defined as a group of two or more separate Datums used as a combined reference for a toleranced feature. The definition reported in the Draft just underlines that the Datums have to be ordered in the Datum systems. The different approach between the two standards becomes now even more evident: the Datum system defined by ISO looks to be related to single toleranced features on the part, rather than giving a Frame where a complete definition of the geometrical characteristics of the part can be established.

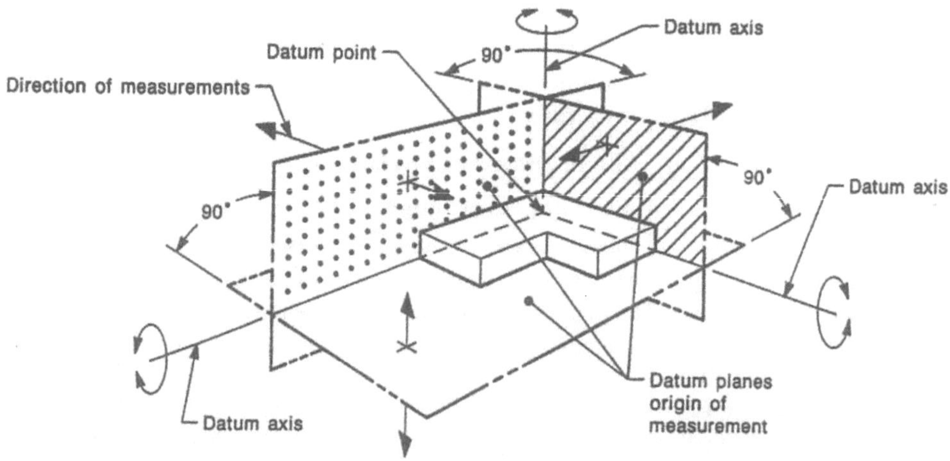

FIGURE 1. Datums and Datum Reference Frame representation as by ASME Y14.5M-1994 [2]

About Targets definitions, we can observe that ISO Drafts definition is quite similar to the ASME (it looks to be limited to areas, but the concept is further extended to lines and points [8]), reminding to a correlation between features on the parts and Datums; while ISO published standards definition relates to Datums verification and manufacturing equipments. Datums definition by means of manufacturing and inspection equipment is also recalled in the Simulated Datum feature definition and will be discussed following.

2.3 DATUM FEATURES DEFINITIONS

Datum features definitions as by ASME Y14.5M-1994, ISO 5459-1981 and ISO/CD 5459-1:1998(E) are reported in the following table.

TABLE 3. Datum features definitions

ASME Y14.5M-1994	ISO 5459-1981	ISO/CD 5459-1:1998(E)
Datum feature: an actual feature of a part that is used to establish a datum	**Datum feature**: a real feature of a part (such as an edge, a surface, or a hole etc.), which is used to establish the location of a datum	***Restricted surface:*** portion of an integral feature (surface) which is an area and which is used for establishing a datum

By ASME definition, where features of a part have been identified as Datum Features and used to establish Datums, the part is oriented and immobilised relative to the three mutually perpendicular planes of the related Datum Reference Frame in a selected order of precedence. By ISO 5459, a Datum feature is a real feature of a part, which is used to establish just the location of a Datum, not Datums selves. Again, it looks difficult univocally defining the correlation between

Datums and parts by ISO definitions, while Drafts definition of restricted surfaces (used for establishing Datums) looks more closely reminding to ASME Datum features.

2.4 SIMULATED DATUMS DEFINITIONS

Simulated Datums definitions as by ASME Y14.5M-1994, ISO 5459-1981 and ISO/CD 5459-1:1998(E) are reported in the following table.

TABLE 4. Simulated Datum definitions

ASME Y14.5M-1994	ISO 5459-1981	ISO/CD 5459-1:1998(E)
Simulated Datum: a point, axis, or plane established by processing or inspection equipment, such as the following simulators: a surface plate, a gage surface or a mandrel	**Simulated datum feature**: a real surface of adequately precise form (such as a surface plate, a bearing or a mandrel, etc.) contacting the datum feature(s) and used to establish the datum(s)	

Being the Datums theoretical entities, the concept of Simulated Datum is introduced. ASME and ISO different approach is again evidenced: while ASME standards defines the simulated Datums to represent a theoretical concept in the reality, ISO standard looks defining the theoretical entities by means of the real ones. The definition of simulated Datum is absent in ISO Draft, but the correlation between Datums and inspection equipment is concerned by ISO/WD 5459-4, *Datums for geometrical tolerancing – Part 4: Metrological establishment of datums and datum systems for the assessment of geometrical tolerances* [10] and will be discussed.

2.5 DISCUSSION

The analysis of the datum concepts definitions allows for some considerations.

ASME Datums concepts definitions evidently belong to a coherent and well defined system of rules for geometric dimensioning and tolerancing. Theoretical and real entities are clearly and univocally related: through the true geometric counterpart of the datum features, the part may be related to the Datum Reference Frame, whose Datums allow properly defining the geometrical characteristics of the whole part. The Datum Reference Frame concept is strictly related to the possibility of using functional gages for verification, according to the general ASME point of view, as synthesized by Rule 1 and virtual condition concept [2]. Anyway, the possibility of defining the whole geometrical characteristics of the parts referring to a Datum Reference Frame clearly and univocally related to the part is of course helpful for any verification method could be used. The Datum Reference Frame concept, basic in GD&T, can result as a key-point in establishing a common language for design, manufacturing, assembly and verification.

ISO 5459:1981 definitions, instead, lead to a more difficult correlation between theoretical and real entities. Datums are established by means of the processing or inspection equipment and located by means of the datum features. The datum system is defined as for tolerancing single features, rather than giving the possibility of completely defining the geometrical characteristics of the part. Some difficulties can be encountered, when trying to establish the aforementioned "common language" by means of such definitions.

ISO Drafts definitions, although reminding those reported in the ISO 5459:1981, evidence some important differences. The use of Datums again looks to be limited to the definition of the geometrical tolerances of single features on the part: the possibility of completely defining the geometrical characteristics of the whole part, strictly related to the aforementioned "Datum Reference Frame concept", is not well considered, as by the Datum system definition. On the other hand, the relations between features and Datums are more precisely defined, thus reflecting the need of establishing "links" between features "as designed" and the same features "as to be verified".

3 DATUMS CONCEPTS EVOLUTION IN GPS ISO DRAFTS - DISCUSSION

According to the general GPS main objective [12, 13], a strong effort is performed, aimed at establishing clear correlations between the specification and the verification steps [14]. Concerning Datums identification, some basic concepts are introduced, reminding to the interrelationships of the geometric features described in other standards [15, 16]; the main guidelines are shown in figure 2.

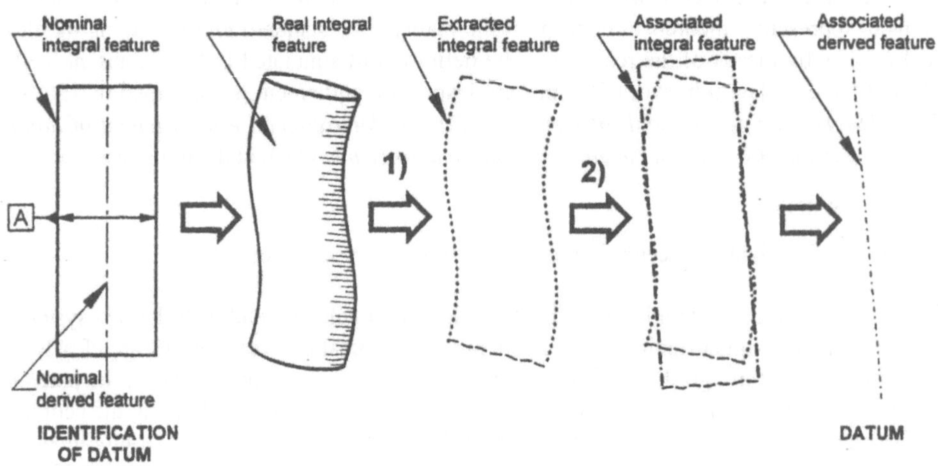

FIGURE 2. Basic concepts for Datums identification [7]

Datums are defined on the basis of surfaces, belonging to different classes characterised by the Degrees Of Freedom (DOF) for which the surface is unvarying: *complex surface* (DOF: none; situation features: plane, straight line, point), *prismatic surface* (DOF: 1 translation along a line of a plane; situation features: plane, straight line), *surface of revolution* (DOF: 1 rotation around a straight line; situation features: straight line, point), *cylindrical surface* (DOF: 1 translation and 1 rotation around a straight line; situation features: straight line), *planar surface* (DOF: 1 rotation perpendicular to the plane and 2 translations along 2 lines of the plane; Situation feature: plane), *spherical surface* (DOF: 3 rotations around a point; situation features: point).

The concept of "Feature with intrinsic characteristic" is introduced, reminding to the ASME "Feature of size" concept; thus confirming the tendency at minimising the differences between ISO and ASME point of view [17]. Such features are then distinguished between Feature with Variable Intrinsic Characteristic (FVIC) and Feature with Fixed Intrinsic Characteristic (FFIC), also specifying the cases whether Maximum (respectively Minimum) Material Condition is applied or not. A long series of Rules, precisely defined, covers the more various situations for Datums, common datums, and Datums systems [8].

The association methods allowing establishing Datums from nominal integral features are also well defined, by means of Constraints (describing the conditions that must be satisfied) and Objective Functions for Association (Gaussian, Least squares, maximum inscribed, minimum circumscribed…) [9].

Finally, relevant work is spent analysing the metrological problems related to the establishment of the associated Datum feature [10].

The analysis of the aforementioned concepts introduced by ISO Drafts leads to some interesting considerations, despite they are relating to a "work in progress", thus necessarily subject to revision. The basic evolution principles can anyway be individuated and discussed, especially with reference to the concerning industrial reality.

The intent of defining classes (of features, of situations…), as well as that of establishing a mathematical definition of concepts (also well defined by ASME [3]), is of course positive, but the need for completeness often lead to long series of Rules, which interpretation could result difficult to the users. The language self, describing principles and applications, may require a strong effort, considering the industrial reality, which can be normally expected. The risk could be to perform a perfectly precise set of rules, which are too difficult to understand and to follow, so that they could be not followed at all. It could be a big risk, considering how deeply the geometrical characteristics of the parts affect functionality, assembly, thus substantially the quality of products.

Probably the whole GPS philosophy will require for some time and particular attention to be diffused and understood, but special effort should be spent to describe at least the basic concepts simply and unambiguously (as more is possible). Although it is a very difficult task, because of the complexity and variety of situations to describe, it could be maybe worthy trying to develop a core of well defined basic concepts, to which all the particular cases could be clearly referred. The task should really be performing a set of rules describing a complex reality in a simple way.

4 CONCLUSION

Datums concepts as by ASME and ISO standards are analyzed in this work, evidencing analogies and discrepancies by means of the comparison of Datums concepts definitions. Particularly, while in ASME standard real and theoretical entities are clearly and univocally related, it is more difficult establishing such a univocal correlation by ISO 5459:1981 definitions. On the other hand, as recognized by related ISO Drafts analysis, a relevant effort is actually spent by ISO/TC 213 aiming at defining precise correlations between Datums and features on the parts, being the intent always establishing "links" between the specification and the verification steps. It is however necessary underlining how, establishing such correlations, a complex language often describes a long series of Rules, which understanding and application could result difficult in the industrial reality.

REFERENCES

1. Meadows, J. D., (1995), Geometric Dimensioning and Tolerancing, Marcel Dekker Inc., New York, NY.
2. ASME Y14.5M-1994, Dimensioning and Tolerancing, The American Society of American Engineers, New York, NY, 1994.
3. ASME Y14.5.1M-1994, Mathematical Definition of Dimensioning and Tolerancing Principles, The American Society of American Engineers, New York, NY, 1994
4. ISO/TR 14638: 1995(E), Geometrical Product Specification (GPS) – Masterplan, ISO, Switzerland
5. ISO 5459: 1981, Technical drawings – Geometrical Tolerancing – Datums and Datum Systems for geometrical tolerances
6. ISO/TC 213, http://129.142.8.149/isotc213/index.htm
7. ISO/CD 5459-1:1998(E) Datums and datum systems for geometrical tolerancing – Part 1: General terms and definitions (1998)
8. ISO/CD 5459-2 Datums and datum systems for geometrical tolerancing – Part 2: Indication in technical product documentation (1998)
9. ISO/WD 5459-3 Datums for geometrical tolerancing – Part 3: Association methods for datums and datum systems for the assessment of geometrical tolerances (2000)
10. ISO/WD 5459-4 Datums for geometrical tolerancing – Part 4: Metrological establishment of datums and datum systems for the assessment of geometrical tolerances (2000)
11. Podda, G.,(2001), A Geometric approach to the Datum Reference Frame, Proceedings XII ADM International Conference, Vol. W1, 12-25
12. The challenge of ISO/TC 213, (1996), http://129.142.8.149/isotc213/challeng.htm
13. Business Plan of ISO/TC 213 - Dimensional and geometrical specifications and verification, (2001), http://129.142.8.149/isotc213/213business%20plan.pdf
14. Srinivasan,V., (2001), An Integrated View of Geometrical Product Specification and Verification, 7th CIRP Seminar on Computer Aided Tolerancing
15. ISO14660-1:1999, Geometrical Product Specifications (GPS) – Geometrical features – Part 1: General terms and definitions, (1999)
16. ISO14660-2:1999, Geometrical Product Specifications (GPS) – Geometrical Features – Part 2: Extracted median line of a cylinder and a cone, extracted median surface, local size of an extracted feature, (1999)
17. Concheri, G., Cristofolini, I., Meneghello, R., Wolf, G., (2001), Geometric Dimensioning and Tolerancing (GD&T) versus Geometrical Product Specification (GPS), Proceedings XII ADM International Conference, Vol. D1, 37-50

QUALITY DECISIONS AND ISO 9000:2000 PRINCIPLES

F. Galetto

Department of Production Systems and Economics, Politecnico di Torino, Italy

KEYWORDS: Quality Management, DOE, QFD.

ABSTRACT: Quality Engineering looks at designing Quality into every product and all the processes. Design is surely the most important phase for Quality. To make Quality of products (processes, systems and services) good knowledge of Quality ideas and good experience about Quality tools for achieving Quality are absolutely needed. The new ISO Standards on the "Quality Management Systems" state that "eight quality management principles have been identified that can be used by top management in order to lead the organisation towards improved performance." The seventh principle "Factual approach to decision making" states: "Effective decisions are based on the analysis of data and information". Well-designed tests in industrial framework provide data for decisions. Using real applications we will see that Quality Managers need great care for the correct implementation of ISO principles. They need Quality Methods scientifically founded and applied to scientifically designed tests. That will help the implementation of another essential ISO 9004:2000 principle "Continuous Improvement".

1 INTRODUCTION

At various Conferences on Total Quality Management for Higher Education Institutions, from 1998 to 2001, the author presented many actual cases of wrong application of the concepts and methods related to Quality, done by professors. Deming (1986) would have said

- The result is that hundreds of people are learning what is wrong. I make this statement on the basis of experience, seeing every day the devastating effects of incompetent teaching and faulty applications.

The cases referred to Quality definition, Analysis Of Variance (ANOVA), Reliability, Control Charts, Reliability Tests, Taguchi Methods, "Robust Design" Methods, ...

One professor (TQMHEI 1999) told me he disagreed with me, because he knew hundreds of good applications of Taguchi and the so-called "Robust Design" methods. I asked him for just 2 out of those hundreds of good applications: after 3 years, I am still awaiting for one of them ...

In another Conference (AITEM 1999) a session chairman on Quality Methods stated that "in his experience, interaction are less probable than factors"; my (78, ..., 2000b) experience is opposite.

Another told me that he did not understand why I considered very bad the book of Montgomery (1996), saying "If you, via Internet, look at quality courses of many Universities, you will find many professors suggesting that book (Montgomery 1996) to students; therefore, you Galetto, you are in a minority. Hence, we many, we are right." Quality definition "Quality is inversely proportional to variability", given in Montgomery book is nonsense: according to this definition Ferrari, the winner of 2001 Formula One Championship, has the lowest Quality!!! Professors, teaching in quality courses, generally do not know reliability and then do not understand the consequences of Montgomery definition.

But there is something worse: Montgomery is in contradiction and his lovers, suggesting his book to students do not realise that; let's consider "example 6-6" (page 291, in the book of Prof. Montgomery). It is written there: "A chemical engineer wants to set up a control chart for monitoring the occurrence of failures of an important valve. She has decided to use the number of hours between failures as the

Published in: E.Kuljanic (Ed.) *Advanced Manufacturing Systems and Technology*, CISM Courses and Lectures No. 437, Springer Wien New York, 2002.

variable to monitor. ... Clearly, time between failures is not normally distributed. [... constructing a time-between-events control chart is essentially equivalent to control charting an exponentially distributed variable.] Figure 6-23 is a control chart for individuals and a moving range control chart for the transformed times between failures ... If a process change is made that improves the failure rate (such as a different type of maintenance), then we would expect to see the time between failures get longer. ...".
The data are the following

816	729	4	143	431	8	2837	596	81	227	286	948	536	124	603	492	1199	1214	2831	96

I agree with Montgomery about his statement "If a process change is made that improves the failure rate, then we would expect to see the time between failures get longer.". This statement proves that Prof. Montgomery (and his lovers, suggesting his book to students) does not realise that, according to his definition, <<quality gets worse, "If a process change is made that improves the failure rate", because "the time between failures gets longer.">>. One can easily prove the mistake of Prof. Montgomery, if he/she reads and understands the Galetto books and papers (1995, ..., 2000b). I had the chance to point out this to several professors, teaching Quality related matters: they did not understand. I asked them to analyse scientifically the data: they were incapable, because they did not know the theory (you can find it in Galetto 1995, ..., 2000b). Montgomery book is a bad book, and professors do not realise it.
The new ISO Standards on the "Quality Management Systems" state that "eight quality management principles have been identified that can be used by top management in order to lead the organisation towards improved performance." The seventh principle "Factual approach to decision making" states: "Effective decisions are based on the analysis of data and information".
Well-designed tests in industrial framework provide data for decisions.
Using real applications we will see that Quality Managers need great care for the correct implementation of ISO principles.
They need Quality Methods scientifically founded and applied to scientifically designed tests. That will help the implementation of another essential ISO 9004:2000 principle "Continuous Improvement".

2 ISO 9000:2000 "QUALITY PRINCIPLES", ISO 9001 "ANALYSIS OF DATA"

The new Standard ISO 9000:2000 and ISO9004:2000 suggest the use of eight quality management principles. Here they are:
4.3 Use of quality management principles (ISO 9004:2000).
To lead and operate an organisation successfully, it is necessary to manage it in a systematic and visible manner. The guidance to management offered in this International Standard is based on eight quality management principles. These principles have been developed for use by top management in order to lead the organisation toward improved performance. These quality management principles are integrated in the contents of this International Standard and are listed below.
a) Customer focus Organisations depend on their customers and therefore should understand current and future customer needs, should meet customer requirements and strive to exceed customer expectations.
b) Leadership Leaders establish unity of purpose and direction c>f the Organisation. They should create and maintain the internal environment in which people can become fully involved in achieving the organisation's objectives.
c) Involvement of people People at all levels are the essence of an organisation and their full involvement enables their abilities to be used for the organisation's benefit.
d) Process approach A desired result is achieved more efficiently when activities and related resources are managed as a process.

e) System approach to management Identifying, understanding and managing interrelated processes as a system contributes to the organisation's effectiveness and efficiency in achieving its objectives.
f) Continual improvement Continual improvement of the organisation's overall performance should be a permanent objective of the organisation.
g) Factual approach to decision making Effective decisions are based on the analysis of data and information.
h) Mutually beneficial supplier relationships Any organisation and its suppliers are interdependent and a mutually beneficial relationship enhances the ability of both to create value.
Successful use of the eight management principles by an organisation will result in benefits to interested parties, improved monetary returns, the creation of value and increased stability.
The new Standard ISO 9001:2000 states the following:
8.4 Analysis of data (ISO 9001:2000)
The organisation shall determine, collect and analyse appropriate data to demonstrate the suitability and, effectiveness of the quality management system and to evaluate where continual improvement of the effectiveness of the quality management system can be made. This shall include data generated as a result of monitoring and measurement and from other relevant sources.
The analysis of data shall provide information relating to

a) customer satisfaction (see 8.2.1),
b) conformity to product requirements (see 7.2.1),
c) characteristics and trends of processes and products including opportunities for preventive action, and
d) suppliers.

Particularly important for the sequel is noting that ISO Standards still do not care of Quality of Quality Methods. Professors do the same.

3 WRONG APPLICATIONS OF ISO 7TH PRINCIPLE

We are going to present some very few cases related to data analysis, taken from books. They show that books are in agreement with the new ISO Standards 9004:2000 and 9001:2000, notwithstanding that Quality is absent in them.
Design of Experiments (DOE) is a major element of the design activity and one of the most important methodologies to achieve Quality through intelligent testing. One of the most widely used methodologies is the Taguchi Method (in connection with "Robust Design").
Let's see the erroneous decisions taken using Taguchi Methods (suggested in connection with the QFD), using the data and the analysis of G. Taguchi (1987, Chap. 11). Neglecting the interactions, as it is usually done by "Taguchians", from these data Taguchi finds the optimum state $A_5B_{-1}E_1$, where he thinks there is the maximum.
Actually, as shown in table 2 (using the G-Method), interactions are generally more important than factors; hence it is not wise to neglect them. It is interesting to note the following statement of Taguchi: "In analysing, we calculate as though No. 1 to No. 6 had been experimented on twice each. We do this so that the factorial effects will be orthogonal.". Taguchi misses completely the scientific approach.
Actually, if one uses the scientific G-method, (Galetto 1978, ..., 2000b) he can really find that

1. the mean is "entangled" (& symbol used) with the interactions BC, CD, BE, BCDE: I&BC&CD&BE&BCDE
2. the 5-level factor A is "entangled" with the interaction BC, CD, BE, BCE, CDE, AB, AC, AD, AE: A&BC&CD&BE&BCE&CDE&AB&AC&AD&AE

3. the 2-level factor B is "entangled" with the interactions CDE, AB, AC, AD, AE: B&CDE&AB&AC&AD&AE
4. the 2-level factor C is "entangled" with the interactions …
5. the 2-level factor D is "entangled" with the interactions …
6. the 2-level factor E is "entangled" with the interactions …
7. the interaction DE is "entangled" with the interactions BC, CD, BE, BCE, CDE, AB, AC, AD, AE: CD&BC&CD&BE&BCE&CDE&AB&AC&AD&AE
8. the interaction BD is "entangled" with the interactions …

TABLE 1 Data and analysis of G. Taguchi [1987, chapter 11]

FACTORS					Data		
A	B	C	D	E	y1	y2	y3
1	-1	-1	-1	-1	6	4	5
1	1	1	1	1	2	0	1
2	-1	-1	1	1	9	8	11
2	1	1	-1	-1	1	0	-1
3	-1	1	-1	1	14	13	14
3	1	-1	1	-1	7	4	5
4	-1	1	1	-1	9	8	12
4	1	-1	-1	1	6	6	4
5	-1	1	1	-1	10	13	11
5	1	-1	-1	1	9	10	7

Taguchi Calculations				
Source	f	SS	V	KV
Total	29	967.98		
A	4	374.81	93.70	59.97**
B	1	475.02	475.02	271.44**
C	1	1.69	1.69	0.96
D	1	0.19	0.19	0.11
E	1	77.52	77.52	44.30**
e1	1	2.08	2.08	2.08
e2	20	36.67	1.84	
Wrong analysis Taguchi misses the scientific approach				

TABLE 2 Scientific Analysis (G-Method, with & ...) of the data of table 1

Source	df	SS	MS	Fc	F5%	sign	Source	df	SS	MS	Fc	F5%	sign
TOTAL	30	1998.00					BC & ...	1	192.20	192.20	104.84	4.35	*
Mean &	1	1442.13					D & ...	1	4.80	4.80	2.62	4.35	
Corr. Total	29	555.87					AD & ...	4	292.88	73.22	39.94	2.87	*
A & ...	4	221.53	55.38	30.21	2.87	*	E & ...	1	13.33	13.33	7.27	4.35	*
B & ...	1	249.69	249.69	137.19	4.35	*	AE & ...	4	284.32	71.08	38.77	2.87	*
AB & ...	4	51.13	12.78	6.97	2.87		DE & ...	1	192.20	192.20	104.84	4.35	*
C & ...	1	1.20	1.20	0.65	4.35		Residual	20	36.67	1.833			
AC & ..	4	296.48	74.12	40.43	2.87	*	Residual NOT obtained by difference						

Due to this important entanglement the optimum is different from Taguchi findings. He found as optimum state the following one, $A_5B_{-1}C_{indifferent}D_{indifferent}E_1$, while actually it is $A_5B_{-1}C_{-1}D_{-1}E_1$: the error is due to the interactions and to the fact that the factorial effects ARE NOT orthogonal!!!! [as it can be shown by the Normal Equations (G-Method, Galetto 1978, ..., 2000b)]!!! Besides the previous example, many other cases can show the pack of lies of the "Taguchians", using the G-method, as mentioned at the plenary session of EOQC Conference (F. Galetto, 1989) by J. Juran. Taguchi makes decisions based on data analysis and applies correctly the (ISO 9000:2000 and 9004:2000) seventh principle “Factual approach to decision making” which states: “Effective decisions are based on the analysis of data and information”. BUT his decisions are NOT effective: they are wrong! It is interesting to note that Taguchi missed many interactions that are more important than factors.!!!!

Let’s see now prof. Montgomery (1996) that lies on the Experimental Design in his book [the book

suggested to students]; he wrote (in the example 11-4, page 545):"An article in Solid State Technology ("Orthogonal Design for Process Optimisation and its Application in Plasma Etching", May 1987, pp. 127-132) describes the application of factorial design in developing a nitride etch process on a single-wafer plasma etcher. ... we will concentrate on etch rate for silicon nitride. We will use a single replicate of a 2^4 design to investigate this process. Since it is unlikely that the three-factor and four-factor interactions are significant, we will plan to combine them as an estimate of error." Neglecting the interactions, as it is usually done by "Taguchians" and "Robust Design practitioners", from HIS DATA (false data) Montgomery finds and writes: "clearly, the main effects of A and D and the AD interaction are significant ...". Besides lying on data, Montgomery makes a "strange" data analysis. Moreover it is easily seen he uses the same data for very different experiments: the actual experiment was a 3^{4-2} design (fractional, with 3 replicates of run 1); the actual experiment has a very complicate ALIAS Structure, forgotten by Yin and Jillie, and by Montgomery who invented the experiment as a 2^4 design and made wrong assessment of significance. Using the G-Method (F. Galetto 1978, …, 2000b) one can find the ANOVA table 4 [using 2 interactions and factor B (not significant) as the Residual Error estimate].
Montgomery makes decisions based on data analysis and applies correctly the (ISO 9000:2000 and 9004:2000) seventh principle "Factual approach to decision making" which states: "Effective decisions are based on the analysis of data and information". BUT his decisions are NOT effective: they are wrong! It is interesting to note that Prof. Montgomery missed many interactions that are more important than factors.!!!! (table 4) Why professors suggest his book to students ???????????
In table 3 the letters stand for: A=C'=Gap, B=B'=Pression, C=D'=Flow-Rate, D=A'=Power
The actual experimental design is a "fractional factorial 3^{4-2} design" in the controlled factors A', B', C', D'. The authors do not provide the "alias structure", as always do the "Taguchi lovers". The use of the G-method (Galetto 1978, ..., 2000b) prove that every factor is "entangled" with various interactions: A'&B'*C'&B'*D'&C'*D'&A'B'C'&...; B'&A'*C'&A'*D'&C'*D'&...; C'&A'*B'&A'*D'&B'*D'&...; D'&A'*B'&A'*C'&B'*C'&...
More precisely, there is also the ALIAS structure (the symbol @ stands for "equivalent to"), neglected by all professors ...:

(A'+B') @ C'*D'@...	(A'+C') @ B'*D'@...	(A'+D') @ B'*C'@...	(B'+C') @ A'*D'@...	(B'+D') @ A'*C'@...	(C'+D') @ A'*B'@...

This means that changing "additively" any two factors is exactly the same as changing "interactively" the other two factors and ....
Again, Montgomery, in his book, makes decisions based on data analysis and applies correctly the (ISO 9000:2000 and 9004:2000) seventh principle "Factual approach to decision making" which states: "Effective decisions are based on the analysis of data and information".
BUT his decisions are NOT effective: they are actually wrong! He is not scientific.
The "Scientific Approach", is strongly required by GIQA (Golden Integral Quality Approach) [Galetto 1999]. The opposite of the "Scientific Approach" are the Taguchi Method and "Robust Design" Methods. Mathematics, Logic and Physics prove that the Taguchi approach is wrong especially when G. Taguchi writes ".... when there is interaction, it is because insufficient research has been done on the characteristic values". There are many cases that can show the erroneous decisions taken using Taguchi Methods (suggested in connection with the QFD).

TABLE 3 Data in Solid State Technology (Yin and Jillie 1987)

False data of Montgomery [1996]											
Run	A	B	C	D		Run	A	B	C	D	
1	-1	-1	-1	-1	550	9	-1	-1	-1	1	1037
2	1	-1	-1	-1	669	10	1	-1	-1	1	749
3	-1	1	-1	-1	604	11	-1	1	-1	1	1052
4	1	1	-1	-1	650	12	1	1	-1	1	868
5	-1	-1	1	-1	633	13	-1	-1	1	1	1075
6	1	-1	1	-1	642	14	1	-1	1	1	860
7	-1	1	1	-1	601	15	-1	1	1	1	1063
8	1	1	1	-1	635	16	1	1	1	1	729
"...single replicate of a 2^4 design to investigate this Process. Since it is unlikely that ... ... plan to combine them as an estimate of error."											

Actual data [Yin et al. 1987]					
Run	C'	B'	D'	A'	
1	-1	-1	-1	-1	1075
2	0	0	0	-1	633
3	1	1	1	-1	406
4	0	-1	1	0	860
5	1	0	-1	0	561
6	-1	1	0	0	868
7	1	-1	0	1	669
8	-1	0	1	1	1138
9	0	1	-1	1	749
1'	-1	-1	-1	-1	1052
1"	-1	-1	-1	-1	1037

TABLE 4 Scientific Analysis of false Montgomery data

Source	df	SS	MS	Fcalc	Sign.	Source	df	SS	MS	Fcalc	Sign.
Tot Cor.	15	531420.9			10%						10%
A	1	41310.56	41310.56	4105.40	*	D	1	374850.1	374850.1	37252.18	*
B	1	10.56	10.56	1.05		AD	1	94402.56	94402.56	9381.62	*
AB	1	248.06	248.06	24.65	*	BD	1	1.56	1.56	0.15	
C	1	217.56	217.56	21.62	*	ABD	1	68.06	68.06	6.76	*
AC	1	2475.06	2475.06	245.97	*	CD	1	18.06	18.06	1.79	
BC	1	7700.06	7700.06	765.22	*	ACD	1	126.56	126.56	12.58	*
ABC	1	976.56	976.56	97.05	*	BCD	1	2575.56	2575.56	255.96	*
Error	3	30.19	10.06			ABCD	1	6440.06	6440.06	640.00	*

TABLE 5 Scientific [G-method] data analysis of article in Solid State Technology (actual data)

Source	df	SS	MS	Sign (10%)	Source	df	SS	MS	Sign (10%)
A' & ...	2	33030.89	16515.44	*	A'*B' & ...	4	362280.44	90570.11	*
B' & ...	2	56336.22	28168.11	*	A'*C' & ...	4	67596.44	16899.11	*
C' & ...	2	351020.22	175510.11	*	A'*D' & ...	4	407356.44	101839.11	*
D' & ...	2	11260.22	5630.11	*	B'*C' & ...	4	44291.11	11072.78	*
					B'*D' & ...	4	384051.11	96012.78	*
Residual	2	732.67	366.3333		C'*D' & ...	4	89367.11	22341.78	*

Let's look at another application of "Robust Design Methodologies": it is a tolerance design dealt in Park's book (1996), presented by Prof. Lanzotti (University of Naples) at a SIS (the Italian Statistical Society) Course "Sperimentare per la Qualità" (Turin, February 2001). The data are tensile strengths of items, dependent from 4 controlled factors (A, B, C, D) at 3 levels. The full factorial Experimental Design needs 81 test states. Since they are too many a fractional factorial with 9 states was carried on (inner array). Other 4 factors (A, B, C, D) at 3 levels were used as noise factors (outer noise array with 9 states); therefore 9 data were collected for each of the 9 inner array states. The four factors were A= temperature, B=heating time, C=additive and D=working speed. Prof. Lanzotti gave the students the following conclusion: "factors A and D are significant and therefore have to be selected for optimisation". At my statement that actually interactions BC, BD, AB were more significant, due to the ALIAS structure, and therefore factors B and C deserved, as well, consideration for optimisation, he replied that "there were not interactions": therefore he showed he did not know the importance of the ALIASES. Prof. Lanzotti made decisions based on data analysis and applied correctly the (ISO 9000:2000 and 9004:2000) seventh principle "Factual approach to decision making" which states: "Effective decisions are based on the analysis of data and information". BUT his decisions were NOT effective: they were wrong!

TABLE 6 Data and ANOVA (presented by Prof. Lanzotti) [Parks, 1996]

Experiment for tensile strength				
A	B	C	D	RESPONSE
-1	-1	-1	-1	9 data
-1	0	0	0	9 data
-1	1	1	1	9 data
0	-1	0	1	9 data
0	0	1	-1	9 data
0	1	-1	0	9 data
1	-1	1	0	9 data
1	0	-1	1	9 data
1	1	0	-1	9 data

"Neapolitan" ANOVA			
Source	Df	SS	MS
Total	8	59.73	
A	2	48.93	24.47
B	2	1.74	
C	2	0.71	
D	2	8.35	4.18
(e)	4	2.45	0.613

ALIAS structure missed
Wrong significance evaluation
Scientific approach lacking

Since the experimental design is a "fractional factorial 3^{4-2} design" in the controlled factors A, B, C, D, there are the "entanglement" and "alias" relationships as seen before.
If Prof. Lanzotti had used the G-method (Galetto 1978, ..., 2000b) he would had found that every factor is "entangled" with various interactions ("&" stands for the "entanglement relation" and "..." for the not shown higher order interactions): A&B*C&B*D&C*D&ABC&...; B&A*C&A*D&C*D&...; C&A*B&A*D&B*D&...; D&A*B&A*C&B*C&... Again, he made decisions based on data analysis and applied correctly the (ISO 9000:2000 and 9004:2000) seventh principle: he simply was wrong!

4 CONCLUSIONS

We have seen some cases that are given in books lacking Quality.
Unfortunately professors suggesting those bad books do not know the following Table of "Quality and Pragmatheory", which highlights that Management Must (and actually Need to) improve their Culture: without Culture methods are wrong and decisions risky ... For methods used in the Quality field, Deming would say: "Misleading, if not useless.", "... on the basis of experience, seeing every day the devastating effects of incompetent teaching and faulty applications.", "Beware of common sense.", ... Since Deming said (1997) "The purpose of a school should be to prepare students for the future ...", professors MUST teach, in a scientific way: we need "Quality Professors, EDUCATED on Quality.". One way is "The Profitable Route to Quality: the FAUSTA VIA [F. Galetto 1999, 2000].
There no space for giving more examples ...
We only have room for reminding professors the words of J. Juran at Vienna (1989) EOQC [European Organisation for Quality Control] Conference: "read F. Galetto papers", hoping that they do not forget the following

Table of QUALITY and PRAGMATHEORY (F. Galetto 2000)

F. Galetto (11/98) εQ^{IO}_{GE}	Concepts and ideas	
	SCIENTIFIC	Wrong
Teaching Method	⇓	⇓
EFFECTIVE ⇒	IDEAL εQ^{IO}_{GE}	VERY DANGEROUS
NOT effective ⇒	Almost Useless	Expensive

If I do not Know That	I DO NOT Know	⇒	I think I Know
If I do not Know That	I Know	⇒	I think I DO NOT Know

(F. Galetto, 2000)

♥ facts and figures are useless, if not dangerous, without a SOUND THEORY (F.G.98)

♥ The best practical thing is a sound theory (F. Galetto, 1970)

♥ Misleading, if not useless …. (W. E. Deming)

ACKNOWLEDGEMENTS

The author acknowledges the support offered by QuASAR (via A. Moro 8, Buccinasco, Italy).

REFERENCES

1. DEMING W.E. (1986), Out of the crisis, Cambridge Press
2. DEMING, W.E. (1997), The new economics for industry, government, education, Cambridge
3. GALETTO, F. (1978), An application of exp. design in the Automotive field, SIA Congress
4. GALETTO, F. (1984) Assessment of Product Reliability, World Quality Congress '84, Brighton
5. GALETTO, F. (1986) Quality/Reliability: How to get results, EOQC (Autom. Section), Madrid
6. GALETTO, F. (1988) Quality and reliability: a must for industry, ISATA, Montecarlo
7. GALETTO, F. (1989) Quality of methods for quality is important, EOQC Conference, Vienna
8. GALETTO, F. (1990) Basic and managerial concerns on Taguchi Methods, ISATA, Florence
9. GALETTO, F. (1993) DOE. Importanti idee sulla Qualità per i manager, DEINDE, Torino
10. GALETTO, F. (1995) Affidabilità, Volume 1: Teoria e Metodi di Calcolo, CLEUP, Padova.
11. GALETTO, F. (1995) Affidabilità, Volume 2: Prove di affidabilità, CLEUP, Padova.
12. GALETTO, F. (1996) Managerial Issues for Design of Experiments, 4th AMST 96, Udine
13. GALETTO, F. (1999) The Golden Integral Quality Approach, Total Quality Manag., Vol. 10, N 1
14. GALETTO, F. (1999) Quality Education on Quality Management 2nd TQM for HEI, Verona
15. GALETTO, F. (1999) Quality Methods for Design of Experiments, 5th AMST 99, Udine, 833-842
16. GALETTO, F. (2001) Looking for Quality in Quality books 4th Conf. on TQM for HEI, Mons, Belgium
17. GALETTO, F. (2000) Qualità. Alcuni metodi statistici da Manager, CLUT, Torino
18. GALETTO, F., GENTILI, E., (99), Quality of the Quality Methods, AITEM 99 Conf., Brescia, 293-302
19. MONTGOMERY, D.C. (1996), Introduction to Statistical Quality Control, Wiley & Sons
20. PARK, S. (1996), Robust Design and Analysis for Quality Engineering, Chapmann & Hall
21. TAGUCHI G. (1987), System of Experimental Design, vol. 1 & 2, UNIPUB/Kraus Publications
22. YIN, G., JILLIE, D. (1987), Orthogonal Design for Process Optimisation Solid State Technology

FAILURE SCENARIO FMEA: THEORETICAL AND APPLICATIVE ASPECTS

E. Locatelli[1], N. Valsecchi[1], G. Maccarini[1], A. Bugini[1]

[1] Department of Engineering, University of Bergamo, Italy

KEYWORDS: FMEA, Failure Scenario, Expected Costs.

ABSTRACT. This paper aims to give a new growth perspective of Failure Modes and Effects Analysis that can be called "Expected Costs of Failure Scenarios". Traditional FMEA discipline has been used to evaluate Failure Effects in aerospace and automotive industry since '60 [1, 2] and today is applied to anticipate Failure Modes of products and services already in design session concerning different industry contexts. The evaluation of Failure Effects perceived by the customer through the common Risk Priority Number does not offer certainty of success in giving priorities intervention on risk reduction due to its lack of perceiving the Failure Cost sustained by industry. If FMEA focuses on Failure Costs of Failure Scenarios we can derive new intervention priorities based on total Expected Costs of Cause & Effect chain of events. In these "scenarios" the priorities are based on the cost of the whole product life cycle and are connected with the corrective actions which are to be planned, even if they are often in conflict with the classic RPN ranking

1 PROBLEMS ABOUT FMEA USED IN NON-AUTOMOTIVE INDUSTRIES

In order to conform new regulation and quality standards (Vision 2000 – QS9000) many industries nowadays are looking for solutions and applicative methodologies suitable to their own manufacturing context [3, 4, 5]. Most producers want to get new decision-making parameters about Traditional FMEA such as a successful compromise among Reliability (the basic Traditional FMEA parameter), Availability and Costs. Traditional FMEA can give a great contribute to product's monitoring and reliability increase but yields under potential Failure Modes cost evaluation and under Cause & Effect chain of events definition.

The next steps consists of wondering if Risk Priority Number (RPN) parameters are consistent with identifying traditional FMEA limits as well as innovative elements to be introduced to fulfill FMEA economical evaluation approach (availability, costs). FMEA basic analysis is really successful in sorting out product's weak and strong points (from FMEA point of view) together with the aim of exalting both products reliability and efficiency (availability, costs). Many industries strongly claim to make risk reduction strategies (reliability, service and diagnostic increase) agree with production and service cost reduction goals and strategies, charging on them the management of the whole product life cycle. The purposes of a potential evolved FMEA will then: (i) assist design decisions through "design errors proofing", (ii) help to make suitable choices when prevention from process error is needed, (iii) support preventive measures, economically based, which emphasize a criticality level linked to Failure Mode Expected Costs (that means to reduce error occurrence, to increase product's reliability, to perform a "safe" product).

Published in: E.Kuljanic (Ed.) *Advanced Manufacturing Systems and Technology*, CISM Courses and Lectures No. 437, Springer Wien New York, 2002.

2 TRADITIONAL FMEA'S BASICS ANALISYS

2.1 BASIC WORDS

- *Failure Mode*: negation of a product own function.
- *Effect* (of the Failure Mode): failure occurrence, described as perceived by the customer.
- *Cause of Failure*: design/process errors, whose consequence is a Failure Mode.
- *Current Controls*: standard design/process control able to identify Failure Cause/Mode.
- *Severity* (S): severity is the seriousness evaluation of the potential Failure Mode's Effect.
- *Occurrence* (O): it is the probability of particular Cause to happen and evolve into the forecast Failure Modes (during expected life of the product). It is the product of particular Causes probability to happen (P1) and probability to evolve into the Failure Mode (P2).
- *Detection* (D): it is the measure of Current Control capability to identify Causes or Failure Modes (linked to design defects). It's the evaluation of the probability that process controls might intercept a Failure Mode before it determines internal or external customer effect.

S, O and D are classified into a range from 1 to 10, where 10 represents the highest risk value.

TABLE 1. Example of Severity, Occurrence and Detection classification

	Severity - S	Occurrence - O	Detection - D
1	Customer doesn't perceive failure.	REMOTE failure probabili-ty. Simple design solution, applied on similar products.	VERY HIGH PROBABILITY that design tests (test bed, calculations and simulations) will identify design weakness.
10	Serious safety risk. Immediate unpredictable failure.	VERY HIGH failure proba-bility. Complex solution never applied before.	Verification program is INSUFFICIENT. Failure surely reaches customer.

- *Risk Priority Number* (RPN). It's the mathematical product of Severity, Occurrence, Detection index. It helps give a priority to Corrective Actions development, introduced when RPN values exceed a pre-fixed threshold. RPN's rank starts from 1 up to 1000 (with 120 possible RPN values). The values of RPN threshold are mainly determined by means of fixed value (for example RPN threshold = 100) over which some corrective actions must be taken.

2.2 FMEA CHARACTERISTICS

FMEA methodology aims to identify and analyze potential product/process failures giving as output the RPN classification. The traditional FMEA approach shows some limits:

- RPN is inconsistent with priority risk definition.
- Detection (D) cannot measure risk contribution with accuracy.

These are the main reasons why it is necessary to introduce a different FMEA approach, in agreement with a multiplicity of Failure Modes, simply defined as "Failure Scenario". For example environmental conditions may have different influence on the resulting Failure Modes which derive from a particular Cause and Effects Failure chain of events. The aim of FMEA is to increase product reliability so that manufacturing costs can be enhanced. This methodology focuses on most risky failures, planning action rankings in order to reduce failure risk with corrective actions which can decrease Occurrence and Detection parameters.

RPN limits. RPN is a way to rank corrective action with the purpose to foresee products failure risk. RPN doesn't provide sufficient information by which developing cost variable-based decisions may be assumed. As a matter of fact RPN, in the best situation, provides a qualitative evaluation of the risk. Referring to RPN threshold, it is difficult to find justifications or motivations for the threshold choice: the commonly applied RPN threshold is fixed 100 one, which exclude from the rank 51,3% of the possible S, O, D combinations exceeding the threshold (too high risk).

D Limits. The definition of Detection may drive to think that a recurrent but easily detectable failure has to be considered not critical. Furthermore the possibility that a recurrent but easily detectable Failure Mode has to be associated **to high repair/reject cost is undervalued**.

Some questions can follow the above definitions:

- Does design or process control capability to detect a Failure Mode mean that such Failure Mode may be prevented?
- Does the identification of a potential Failure correspond to a real Failure symptom/detection?
- Which is contribution of Detection to a risk?

We must remember that Detection is neither a conditional probability nor an action correlated with the measurement associated to a Failure Risk. Detection represents the capability of a control to intercept a Failure but it does not mean that such control can prevent Failures or that this Failure will not reach the final customer.

3 "FAILURE SCENARIO" FMEA

As already said, this paper aims to find the relationship, if any, between RPN and Failure Cost. It's then necessary to analyze the connection between failure occurrence and its "First Cause" (in order of time). When using the traditional FMEA approach some inadequacy of space (columns of FMEA form) may create an obstacle to a complete analysis of Cause and Effects chains of events. Sometimes it is difficult to recognize and show the difference between Cause, Failure Modes and Effects. Then, Failure Modes coming from previous Cause and Effects chain, cannot be correctly linked to main Cause and Effects chain of events.

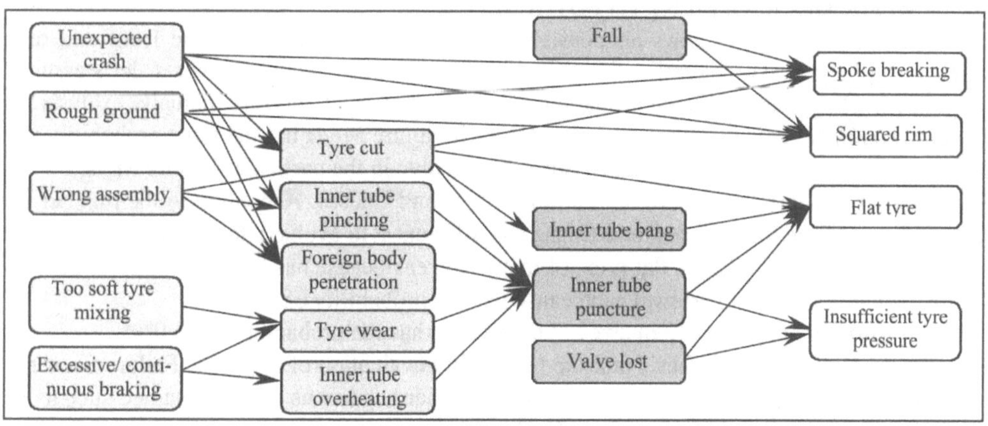

FIGURE 1. Example of "Failure Scenario"

In order to overcome this situation, we try to analyze, as an example, all the possible Cause and Effects chain concerning a collection of events linked to "Wheels failure of a racing bicycle". Let's assume the term **"Failure Scenario"** [6] to be representative of the Cause and

Effect chain of events, from the initial Cause (C_1) to the end Effect (E_N), including all intermediate Effects ($E_1 \div E_{N-1}$), driving to a Failure Mode; a certain Occurrence will be associated to each Failure Scenario. Fig. 1 shows 31 independent paths (Failure Scenarios), each describable within a line of traditional FMEA Form [7].

3.1 BASIC PARAMETERS OF "FAILURE SCENARIO" FMEA

- *Severity (S) and Occurrence (O):* defined as in section 2.1.
- *Detection (D):* assume Detection to be 1, to annul its contribution to RPN calculation.
- *Probability (P):* indicates the possibility of a certain event to happen.
- *Cost (C):* economical evaluation of the damage caused by a Failure Mode.

"Failure Scenario" FMEA can be used to evaluate the hypothetical Failure Cost and suggests a logically consistent parameter-based evaluation of intervention priorities such as Failure Probability (P) and Failure Costs (FC). It is important to underline that the development of a "Failure Scenario" FMEA allows the combination of both "Multiple Causes" driving to one Effect and "Single Cause" driving to multiple Effects.
The use of this procedure implies to answer the following main questions:
- what is the probability (P) of a "Failure Scenario" to happen?
- if such Scenario happens, what are the consequences (material/economical)?

TABLE 2. Comparison between risk reduction techniques

METHOD	STRATEGY	FAILURE PROBABILITY	FAILURE COST	PRODUCT COST	PRODUCT LIFE CYCLE COST
Traditional FMEA	IMPROVE reliability	↓	=		Unknown
"Failure Scenario" FMEA	REDUCE failure and product Cost	Cost based decision	Cost based decision	Cost based decision	↓

3.2 WORK DEVELOPMENT HYPOTHESIS

Let's assume that Fig. 1 shows all possible Failure Scenarios, linked to the front and rear wheels of a bicycle. To determine the probabilities of Failure Scenarios events, let's assume the following hypotheses: "Equal probability" or "Experience", which are mutually exclusive.
Equal probability. This hypothesis means that each Failure Mode has the same probability (P) to be induced from all the Failure end-Causes that evolve in the particular Failure Mode.
Fig. 1 shows the correlation between Failure Modes and Causes. According to the number of in-arrows to Failure Modes, each end-Cause has different P to evolve into Failure Mode:
- for "spokes breaking" and "flat tyre" (4 in arrows) each cause has the probability of 25%;
- for "squared rim" (3 in arrows) each cause has the probability of 33.3%:
- for "insuff. tyre pressure" (2 in arrows) each cause has the probability of 50,0%

Experience. Under the hypothesis that the probability evaluation of a specific Failure Scenario is the output of a "bicycle technician experience" further conditions will be assumed such as:
- Failure Scenario is independent from Cyclist experience.
- The distance covered per year is: 5.000 km.
- Bicycle apparatus and gear is medium quality (purchasing cost about 1.500÷1.750€).
- Tyres are changed 3 times per year and their failure probability is P^T = 3 failures / 5.000 km per year equivalent to 0.0006 (failures / km).

- Inner tube are changed 7 times per year and their failure probability P^{IT} is 7 failures / 5.000 km per year equivalent to 0.0014 (failures / km).
- Total failure probability is P^{tot} = (7+3) failures / 5.000 km per year = 0.002 (failures / km).

In Fig. 2, each Failure Scenario is related to a certain occurrence probability, obtained as conditioned probability of chain of events represented in Fig. 1. According to Bayes Theorem, the total occurrence probability ($P^{\#}$) of a Failure Scenario is given by the conditioned probability P(D)*P(C/D) of Causes-chain and the conditioned probability of the end-effect occurrence $P(E/D\cap C)$:

$$P^{\#} = P(D)*P(C/D)*P(E/D\cap C) \quad (1)$$

where: $P(Y/X) = P(Y\cap X)/P(X)$.

So the evaluation of a particular Failure Mode probability (E - deriving from an assigned chain of Causes A, B, C, D) or Scenario is given in Fig. 2:

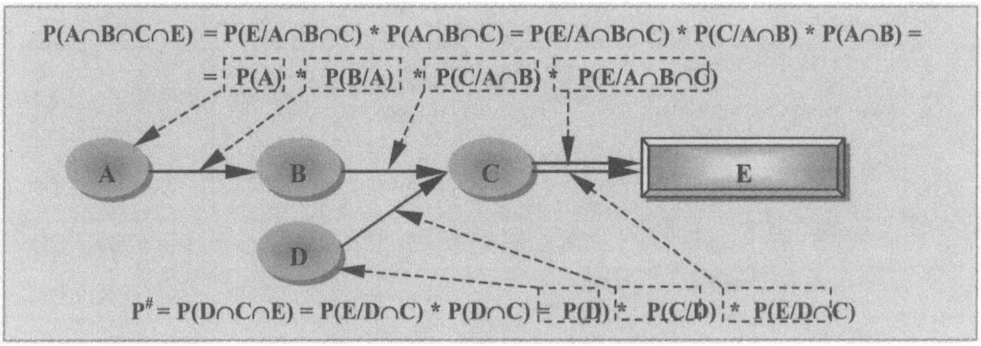

FIGURE 2. Example of Failure Scenario 2

Example: Referring to Fig. 1 the probability P^0 to get "flat tyre" due to "foreign body penetration", caused by "rough ground", is given by:

$$P^0 = P_{rought\ ground\ \cap\ foreign\ body\ penetration\ \cap\ inner\ tube\ puncture\ \cap\ flat\ tyre} = \ldots = P_{(rough\ ground)} * P_{(foreign\ body\ penetration\ /\ rought\ ground)} * P_{(inner\ tube\ puncture\ /\ rought\ ground\ \cap\ foreign\ body\ penetration)} * P_{(flat\ tyre\ /\ rought\ ground\ \cap\ foreign\ body\ penetration\ \cap\ inner\ tube\ puncture)}$$

Failures cost-based ranking. This paper aims to show a new FMEA method basing Failure risk ranking on Expected Cost evaluation. The index Failure Expected Cost (FEC) is the mathematical product of Failure Probability (P) and Failure Cost (FC). Since P and FC evaluation metrics are proportional, multiplication is an admissible operation, and the total FEC (as risk index) becomes:

$$\sum_{i=1}^{n} (P_i \cdot FC_i). \quad (2)$$

where: P_i = Failure Mode-i Probability and FC_i = Possible Cost of a Failure Mode-i.

3.3 RPN VS EXPECTED COSTS

Assuming that:

- the correspondence between Failure Probability and Occurrence (O), is reported in Tab. 3;
- Failure Modes interception capability, Detection (D), is normalized to 1;

- the seriousness of a potential "Failure Scenario" is based on Failure Cost (FC) factor; costs metrics will result linear respect to consequences, Severity (S) range is [1÷10], Tab. 3;

as a consequence we can calculate the FEC for each combination of P, FC and the correspondent RPN from the related O, S, (see Fig. 3).

TABLE 3. Severity-Cost and Occurrence-Probability correspondence

S/O	1	2	3	4	5	6	7	8	9	10
Cost	50 €	100€	150€	200€	250€	300€	350€	400€	450€	500€
Probability	6.7E-7	6.7E-6	6.7E-5	5.0E-4	2.5E-3	1.2E-2	5.0E-2	1.2E-1	3.3E-1	5.0E-1

Now taking into account, front or rear wheel of a common competition road bicycle, 5 different Failure Scenarios can be assumed according to Tab. 4.

TABLE 4. Comparison between RPM and "Failure Scenario FMEA" FEC

Scenario	Function	Failure Mode	Failure Effect	S	Failure Cause	O	P	C (ª)	RPN	FEC (ª)
A	Work in traction to keep the rim round	Spokes breaking, unsuccessful traction	Potential fall	7	Unexpected crash	3	0.000067	350	21	0.023
B	Running / rolling	Flat tyre	Remarkable increase of cyclist effort	3	Tyre cut	7	0.05	150	21	7.5
C	Running / rolling	Flat tyre	Fall	10	Inner tube bang	4	0.0005	500	40	0.25
D	Running / rolling	Insufficient tyre pressure	Increase of cyclist effort	2	Valve loss	5	0.0025	100	10	0.25
E	Running / rolling	Flat tyre	Fall	10	Inner tube puncture	5	0.0025	500	50	1.25

Two methods (RPN and "Failure Scenario FMEA" FEC) provide different outputs for risk intervention priorities (basing on the same "Failure Scenario"); in fact there are (see Fig. 3):

- scenarios A and B have the same RPN (21) but quite different Failure Expected Costs (FEC)
- scenarios C and D have the same FEC (0.25 ª) but quite different RPN
- scenarios B and E show a remarkable disagreement in terms of intervention priority.

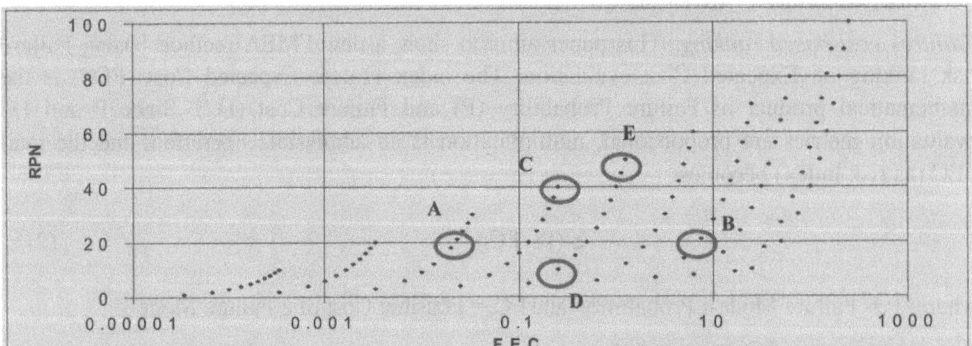

FIGURE 3. Relationship between Expected Costs and RPN, in logarithmic scale.

4 "COST BASED DECISION" ON INTERVENTION PRIORITIES

Now, consider the Failure Scenario "Flat Tyre" of a front or rear wheel of a common competition road bicycle that can happen both in racing or during training. Tab. 5 shows data for the two scenarios: Failure during racing and Failure during training. The Race Cost means: the registration, the discomfort, the eventual spare part, the participation, etc. costs

TABLE 5. Provided Failure Scenarios

	RACING (Failure Scenario 1)	TRAINING (Failure Scenario 2)
Failure scenario	Flat tyre	Flat tyre
Probability P of wheel failure (per year)	(7+3)/5.000 km per year = 0,002 (failures / km)	
Inner tube replacement cost	4[a]	4[a]
Tyre replacement cost	25 [a]	25 [a]
Probability of "flat tyre failure" due to inner tube bang or tyre cut	0,60	0,40
Inner tube Race Cost	16 [a]	-
Tyre Race Cost	100 [a]	-

Let's take into account the opportunity to implement some Corrective Action (CA), (the Extra Costs shown in Tab. 5 are hypothetic):

CA I: Do nothing (P=0.002), ExtraCost of 0%

CA II: Failure probability reduction (P=0.0015), Extra Cost of 15%

CA III: Reduction of replacement cost, Extra Cost of 30%

CA IV: Maintenance.50%potential failure in race detected (Pdet= 0.5), Extra Cost of 5%

In order to calculate the FEC the following expressions are used:

- for Potential Corrective Actions I, II, III:

$$\text{Total flat-tyre-Failure Expected Cost} = FEC_{race\ (scenario\ 1)} + FEC_{training\ (scenario\ 2)} = (P_{wheel\ failure})*[(P_{failure\ during\ race})*(C_{race}+C_{replacement})+(P_{training})*(C_{replacement\ during\ training})] \qquad (3)$$

- for Potential Corrective Action IV:

$$\text{Total Failure Expected Cost} = FEC_{race\ (scenario\ 1)} + FEC_{training\ (scenario\ 2)} = (P_{wheel\ failure})*[(P_{failure\ during\ race})*(1-P_{detected})*(C_{race}+C_{replacement}) + (P_{failure\ during\ race})*(P_{detected})*(C_{replacement\ during\ training})+(P_{failure\ during\ training})*(C_{replacement\ during\ training})] \qquad (4)$$

where:

- Failure during race reduction = $P_{failure\ during\ race}*(1-P_{detected})*(C_{race}+C_{replacement})$
- Increase of failure detection before race = $P_{failure\ during\ race}*P_{detected}*C_{replacement\ during\ training}$.
- Failure during training not influenced = $P_{training}*C_{replacement\ during\ training}$.

In conclusion, we must remark that extra costs strongly depend on what kind of Corrective Action is meant to be introduced since they modify the cost rank of the product life cycle. Let us remember that the first solution "Do nothing" sometimes is a better solution from the economical point of view.

Tab. 6 shows the comparison between costs, comparison that gives answer to the question put at the beginning of this paper.

TABLE 6 Comparison between product's life cycle costs and potential corrective actions.

Corrective Action		P	$P_{\text{failure during race}}$	$P_{\text{failure during training}}$	$C_{\text{replacement during training}}$	$C_{\text{race+}}C_{\text{replacement}}$	FEC	Extra Costs (Hypoth.)	FEC+ Extra Costs	Rank
Inner tube	I	0.0020	0.6	0.4	4 [a]	20 [a]	0.0272 [a]	+0%	0.0272 [a]	3^
	II	0.0015	0.6	0.4	4 [a]	20 [a]	0.0204 [a]	+15% = 0.0031 [a]	0.0235 [a]	2^
	III	0.0020	0.6	0.4	3 [a]	19 [a]	0.0252 [a]	+30% = 0.0076 [a]	0.0328 [a]	4^
	IV	0.0020	0.6*0.5 = 0.3	0.4+0.6* 0.5 = 0.7	4 [a]	20 [a]	0.0176 [a]	+5% = 0.0009 [a]	0.0185[a]	1^
Tyre	I	0.0020	0.6	0.4	25 [a]	100+25 [a]	0.1700 [a]	+0%	0.1700 [a]	3^
	II	0.0015	0.6	0.4	25 [a]	100+25 [a]	0.1275 [a]	+15% = 0.0191 [a]	0.1466 [a]	2^
	III	0.0020	0.6	0.4	20 [a]	100+20 [a]	0.1600 [a]	+30% = 0.0480 [a]	0.2080 [a]	4^
	IV	0.0020	0.6*0.5 = 0.3	0.4+0.6* 0.5 = 0.7	25 [a]	100+25 [a]	0.1100 [a]	+5% = 0.0055 [a]	0.1155 [a]	1^

5 FINAL CONSIDERATIONS

Today, FMEA methodology is spreading from the automotive and aerospace contexts, where it was generated, to other manufacturing industries. A new effort is required to evaluate interventions priorities of both reliability and cost reduction basing them on different "Failure Scenarios". The traditional RPN appears sometimes useless: it should be independent from Detection parameter, which is too subjective, arbitrary and incapable of preventing failure events. Moreover RPN has no correlation with Expected Cost increase, which has a great impact on industrial managers.

The main difficulties concerning Expected Costs are their Failure Cost (C) assumption and Probability (P) association. A Cost Based decision FMEA is still unexplored, little known and probably lacking of a systematic methodological approach suitable for machinery and manufacturing industrial sectors.

"Scenario based" FMEA may be considered as an opportunity to spare resources as well as to undertake methodological innovation for future market exploitation.

REFERENCES

1. Aerospace Recommended Practice ARP926A, 1979.
2. Automotive Industry Action Group (1995), Potential Failure Modes and Effects Analysis (FMEA) Reference Manual, Detroit, USA.
3. Kara-Zaitri, C., Keller, A., Barody, I., and Fleming, P. (1991), An Improved FMEA Methodology, Proceedings of the 1991 IEEE Annual Reliability and Maintainability Symposium, pp. 248-252.
4. Kmenta, S., and K. Ishii (1998), Advanced FMEA Using Meta Behaviour Modelling for Concurrent Design of Products and Controls, Proc. of the 1998 ASME Design Eng. Tech. Conf., Atlanta, USA.
5. Stamatis, D.H. (1995), Failure Mode and Effect Analysis: FMEA from Theory to Execution, ASQC Quality Press, Milwaukee, USA.
6. Kmenta, S., and Ishii, K., (2000), Scenario-based FMEA: a Life Cycle Cost Perspective, Proceedings of the 2000 ASME Reliability, Stress Analysis and Failure Prevention Conference, Baltimore, USA.
7. Locatelli, E., Maccarini, G. (2001), Gestire la Metodologia FMEA nell'ottica informatica, Brescia Ricerche, 34/01.

FROM PLASTIC WASTE TO RECYCLED-CONTENT END PRODUCTS: THE CHAIR INDUSTRY CASE

A. Meneghetti, G. Nardin, P. Simeoni

DiEM – Department of Energetics and Machines, University of Udine, Italy

KEY WORDS: recycling, plastic waste, MSW management optimisation

ABSTRACT. The paper describes a virtuous ecological cycle of recovering and recycling plastics from source separated collection that is realised in the chair industrial district in Friuli-Venezia Giulia. A process for converting the reclaimed plastic material into recycled-content end products has been running; in particular chairs with a 55% weight of recycled plastics are successfully manufactured by a local enterprise. This new product is able to close plastic waste management chain by recycling that type of recovered plastic without a defined market. Potential integration with local MSW system is proposed, in order to optimise waste management chain from both an environmental and economic point of view.

1. INTRODUCTION

The great production of municipal and industrial waste is one of the most critical environmental issues our society is facing nowadays.

In Italy, the main disposal methodology adopted has been direct landfilling, but this has turned environmentally unsustainable. The consciousness that waste can be a resource in terms of material supply and energy recovery has induced Italian authorities to affect waste system management deeply by the so-called Ronchi's Decree, proclaimed in 1997.

Thus, Italian waste management has been modified and updated basing on the European Community environmental policy. In particular, two main objectives are recognised as guidelines for the future:

- Improve source reduction of waste, re-use, materials recycling and energy recovery from waste;
- Optimisation of waste disposal, which should be performed by a proper set of facilities in the same region where waste is produced (*proximity and self-sufficiency principle*) and adopting the best available technologies in order to reduce any environmental impact. Also, waste can be sent to landfill only if downstream of a treatment process devolved to recovery most of its material or energy content.

Considering waste as a resource and not as a problem leads to change future scenarios for waste management and to identify new ways of valorisation for recovered materials. Only an

Published in: E.Kuljanic (Ed.) *Advanced Manufacturing Systems and Technology*, CISM Courses and Lectures No. 437, Springer Wien New York, 2002.

integrated approach to the whole waste management chain can lead, in fact, to both environmentally and economically sustainable systems [1].
In the paper attention is focussed on the plastics management chain. In section 2 it is described how plastic waste is processed in Italy and how its management chain is organised at the moment. In section 3, a virtuous case of a new recycled-content product developed in the local chair industry is analysed in order to identify an alternative solution to landfilling for the less valued part of collected plastic waste. In section 4 it is analysed a possible optimisation of the local plastics chain by affecting its organisation. How this new market opportunity can be interfaced with MSW (municipal solid waste) management system in the province of Udine is explained and advantages from both an economical and environmental point of view highlighted.

2. THE PLASTIC WASTE MANAGEMENT CHAIN IN ITALY

During the last few years the use of plastic materials was skyrocketing. This marked increase has emphasised the problem of disposal for the growing amount of plastic waste produced.
The incidence of various fields to the production of waste related to preconsumer and postconsumer plastics is shown in Figure 1. MSW has the highest incidence, equal to 67%, followed by distribution and industry, that are relatively less significant in plastic waste production [2]. As a result, the growing attention to stimulate separate collection of plastics in particular close to residential communities.

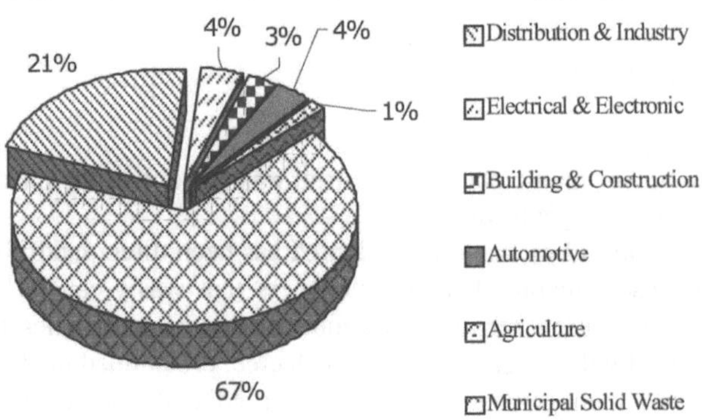

FIGURE 1. Origin of plastic waste

The collection of source separated materials is generally followed by the sorting phase. Separating different types of plastics, with homogeneous properties (see Figure 2), makes possible to send recovered materials to their own markets [3].

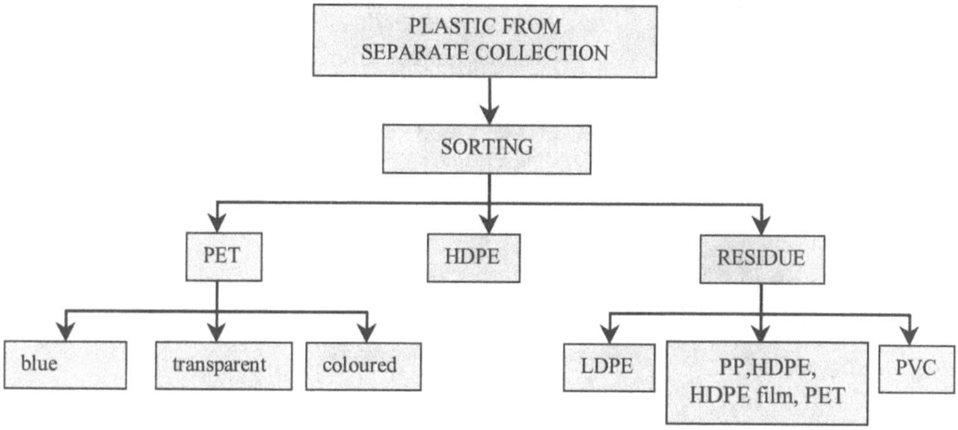

FIGURE 2. Types of plastics from source-separated collection obtained by local sorting facilities

PET is a very versatile material, using which are produced fiber and textile yarns, building materials and thermoformed packaging; it has a solid market and an high commercial value. The high density polyethylene (HDPE) has a steady market too: it's used to product bottles and others containers made by co-extrusion, boxes and packagings, and pipes used in the building trade. Residues are used to produce *mixed* or *heterogeneous* plastic; regarding the origin of these materials, is useful to pinpoint that, in general, HDPE comes from bottle's caps, while PP from their gaskets and HDPE-film from their labels. The content of PET in the mixture represents small objects that are not captured during the separation of PET from the other kinds of polymers. Mixed plastic, which counts for 26% of total weight of plastic waste on the basis of local sorting facilities' mass balance, has not a defined market, on the contrary of the other plastic types separated in the same facility. As a consequence it is generally sent to landfill. This material has performance characteristics, in terms of strength and duration, and a low cost, which should, however, make it competitive respect to many traditional materials. Beside the environmental impact related to landfilling materials with slow decomposition rate, the incapacity of recognising a virtuous market allocation for this part of plastic waste leads to administrative problems for subjects involved in MSW management. Ronchi's Decree establishes, in fact, that a 35% level in MSW source-separated collection must be reached by 2003. To achieve this target, only those materials, which are collected in separated way and effectively sent to recycling or energy recovering, can be considered. If mixed plastic is sent to landfill, it cannot contribute to target fulfilment, thus lowering performance indicators of public administrations. Therefore, the identification of a new market for mixed plastic could improve MSW management systems from environmental, economical and institutional point of view.
Organisation of the Italian plastic waste management chain is represented in FIGURE 3

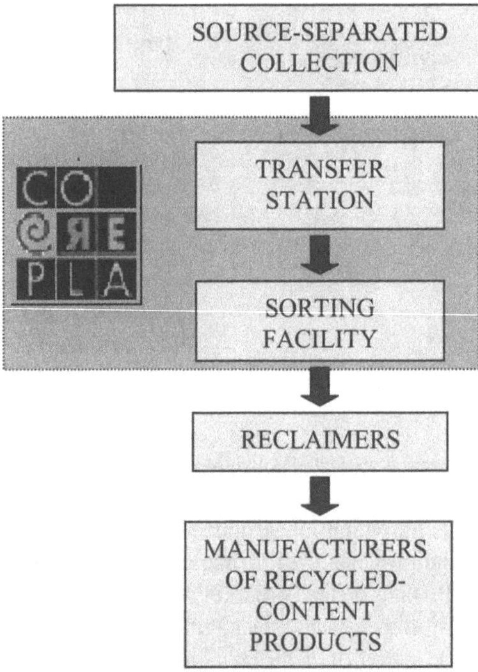

FIGURE 3. Organisation of Italian plastic waste management chain

The main role is played by CO.RE.PLA., the national agency for collection, recycling and recovering of plastic packaging, which is responsible of plastics management after source-separated collection performed by local operators. By stipulating specific agreements, CO.RE.PLA. assures plastic waste withdrawal from transfer stations to which operators must deliver their own amount of plastic waste: operators are paid by CO.RE.PLA. on the basis of the quality of collected waste. The Agency moves waste from transfer station to sorting facilities where the above described types of plastic materials are separated. The next step in the plastic chain involves reclaimers, who transform the recovered product bought from CO.RE.PLA. into feedstock materials that will be sold to manufacturers of recycled-content products.

3. A VIRTUOUS CASE OF PLASTIC RECYCLING IN THE CHAIR INDUSTRY

CHENNA S.p.A. has launched into the market panels for the unitized body of office chairs, realised with materials whose only alternative destination is landfill.

The panel is produced by hot pressing of a mixture consisting of ground wood and mixed plastic in the ratio of 45% and 55% by weight.

The wood comes totally from the production scraps of the chair industrial district in Friuli Venezia Giulia that have no alternative use. Plastic material is a mixture of 85 % of HDPE, 6,5 % of PET, 4,5% of PP and 4% of HDPE-film (all expressed in percentage by weight). Therefore, it can be derived from the residues of sorting facilities, the above-mentioned mixed plastic, in which all these types of polymers are present and from usual recycled HDPE to reach the required amount. Obviously, all the required types of polymers in mixed plastic from sorting facilities must be separated in order to dose them on the basis of percentages required in the panel.

The mechanical properties of the panel are summarised in Table2.; they are comparable to the performance of traditional wooden chairs.

TABLE 1 – Mechanical properties of a recycled-content panel.

	Measure unit	Value
Density	kg/m^3	1020-1060
Tensile strength	Mpa	10-15
Stretching	%	1.4-2.0
Young modulus	Mpa	2500-3200
Flexural strength	Mpa	18-28
Flexural modulus	Mpa	1000-1800
Resilience	kJ/m^2	5-9

The manufacturing process of this new panel doesn't require any kind of virgin material, but is able to close either the plastic and the wood waste management chain.

In Figure 4 we compare the CO_2 emissions related to the production of the amount of virgin materials associated with Chenna's present production volume and those related to recovering the same amount from waste. In Table 2 costs paid to suppliers for both virgin and recycled plastics are showed. It can be recognised how the use of recycled materials leads to savings in environmental and economical terms. In particular CO_2 reduction is coherent with Kyoto protocol on global warming prevention, to which Italy adheres.

In addition it has to be noted that further environmental improvement is gained due to the subtraction of the recycled amount of mixed plastic to landfilling.

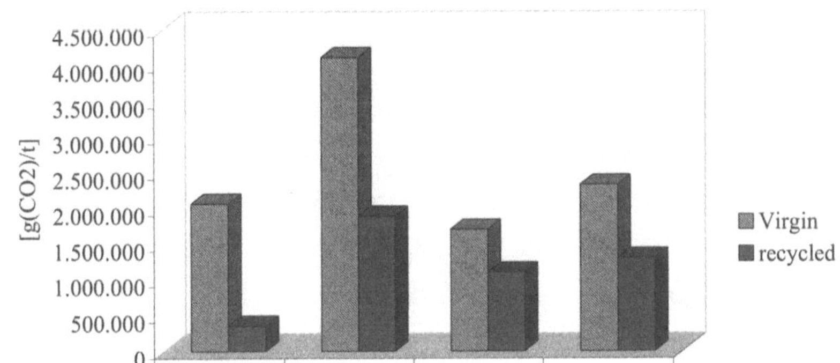

	HDPE	PET	PP	HDPE film
Virgin	2.060.000	4.100.000	1.700.000	2.320.000
recycled	353.325	1.885.000	1.100.000	1.299.900

FIGURE 4. CO_2 emissions related to the production of virgin polymers and recycled ones (Data from [4])

TABLE 2. Supply costs of virgin polymers and recycled plastics related to the present production volume

	Measure unit	Cost of virgin plastic supply	Cost of recycled plastic supply
HDPE	ton	174752	117810
PET	ton	22673	13213
PP	ton	9252	4782
HDPE film	ton	10072	5266,8
Total materials costs	€	216747	141072
Materials cost per panel	€	0.103	0.0672

4. OPTIMISING LOCAL PLASTIC WASTE MANAGEMENT CHAIN

The manufacturing process of recycled-content panels can be usefully interfaced with MSW management in the Province of Udine (520,451 people), where Chenna is placed. In Table 3 we compare total annual amount of plastic MSW coming from source separated collection in the Province, to annual requirements for Chenna's present production volume.

TABLE 3. Annual amount of plastic MSW from source separated collection in the Province of Udine and Chenna's requirements.

Material type	*Plastic waste collected in the Province of Udine [ton/year]*	*Materials requirements for panels [ton/year]*
PET	2,378	-
HDPE	1,529	1,963.5
Sorting Residue:		
LDPE	1,189	-
PET	170.4	150.1
PP	118.9	103.9
HDPE film	107.0	92.4
Others	169.9	

As it can be seen, the new product is able to recycle most of residues from sorting facilities and all the HDPE recovered in the Province: the lacking amount of HDPE can be reasonably considered to be achievable in the short term for both the marked increasing trend of plastic content in MSW and expected growing of source separated collection according to Ronchi's Decree.

At present, Chenna purchases a minimum amount of polymers separated from mixed plastic from reclaimers placed in Lombardia and the other amount is imported from Germany. Therefore, waste produced in the Province of Udine are collected by local operators, sent to stocking points in the Province since the lack of a real CO.RE.PLA.'s transfer station, and successively transported to CO.RE.PLA.'s sorting facility in the near Province of Pordenone. Then, separated plastic materials are moved towards reclaimers mostly placed in a non-adjacent region (see left side of FIGURE 5), from which a portion of them come back to Udine to be processed by Chenna. This "tourism of waste" causes environmental impact in terms of transports emissions as shown in FIGURE 6, in contrast to Ronchi's Decree principles.

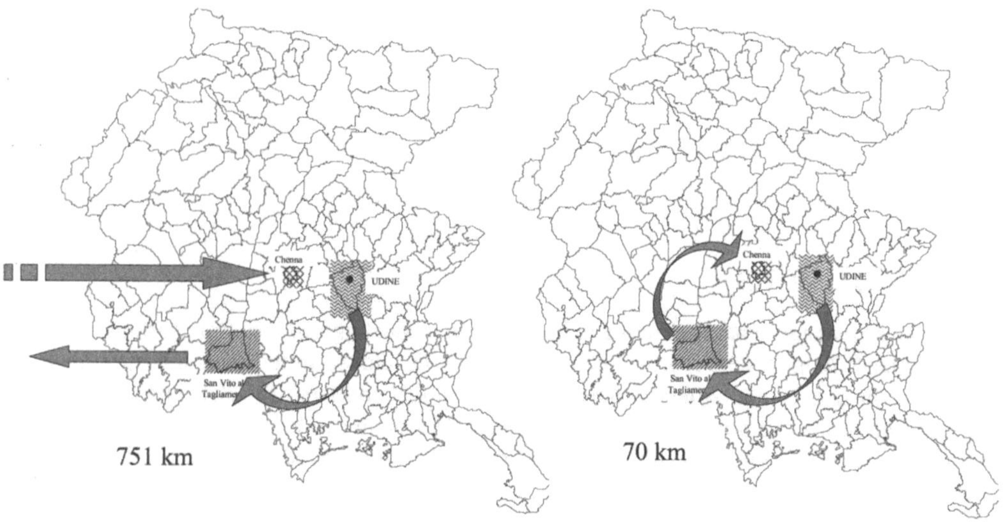

FIGURE 5. Present and virtuous organisation of local plastic waste management chain

To better intercept the plastic waste produced in the Province, in order to avoid waste import and to promote local MSW source separated collection so that a 35% level can be really obtained, a new CO.RE.PLA. facility could be built in the Province where plastic waste could be either transferred or sorted. Mixed plastic could be locally processed and sent to Chenna with conspicuous savings on transport costs and CO_2 emissions (see right side of Figure 5 and FIGURE 6). This solution would optimise the plastic waste management in the Province of Udine and also assure Chenna of proper supply.

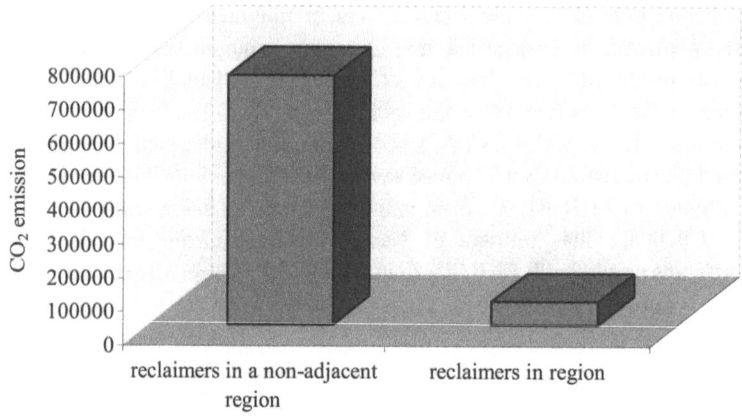

FIGURE 6. Environmental impact in terms of transports emissions

5. CONCLUSIONS

Plastic waste management chain needs to be closed by allocating residues from sorting facilities avoiding landfilling. A new recycled-content product, developed in the chair industrial district of Friuli-Venezia Giulia, offers a new solution for the problem. By changing the present organisation of the chain the local MSW management system can be optimised by nearing the source of waste to the end-user of the most difficult type of recovered plastic to be placed in the market.

The chair industry case can be considered as a virtuous example of recycling that can be further enhanced by a proper integration with MSW management system. In our opinion, this could be the new path towards a holistic management of waste, from cradle to grave, needed to support sustainable development.

REFERENCES

1. McDougall F., White P., Franke M., and P. Hindle, Integrated Solid Waste Management: a Life Cycle Inventory, Blackwell Science, 2001;

2. Eggels P.G., Ansems A.M.M., Van Der Ven B.L., "Eco-efficiency of recovery scenarios of plastic packaging", APME-TNO, Brussels, 2001;

3. Buwalda T., "Plastics", The McGraw-Hill Recycling Handbook, H. B. Lund ed., McGraw-Hill, 2001;

4. BUWAL 250/II, Life Cycle Inventories for Packagings, Vol. II Swiss Agency for the Environment, Forests and Landscape, CH, 1998.

QUALITY PRODUCT CONTROLS OF SILICON PIXEL SENSORS FOR A HIGH ENERGY PARTICLE PHYSICS EXPERIMENT

G. Cabras[1], D. Cauz[1], D. Cobai[1], M. Cobal-Grassmann[1], B. De Lotto[1a], C. Del Papa[1], H.Grassmann[1], L. Santi[2]

[1] Department of Physics and INFN, University of Udine, Italy ([a] corresponding author)
[2] Department of Physics and INFN, University of Trieste, Italy

KEYWORDS: Silicon Pixel Sensor, High Energy Physics, Detector.

ABSTRACT. This paper describes the quality control tests on the silicon pixel sensors for the ATLAS detector [1], performed at the Semiconductor Detectors Laboratory of the Physics Department of the Udine University. A silicon pixel detector system [1] for high multiplicity charged particle pattern recognition has been developed by the ATLAS collaboration to meet the CERN Large Hadron Collider [2] requirements. Every component is tested at each stage of the construction by the producers according to agreed standards, under the control of the collaborating institutes. Nevertheless, some of the quality controls must be performed in the institutes, since they require tools available only in specific laboratories of the collaboration. The Udine group of the Istituto Nazionale di Fisica Nucleare is among the groups which have the skills for such specialized controls.

1 INTRODUCTION

Semiconductor pixel detectors are currently used in fundamental high energy physics research [3], and important applications of this advanced technology are foreseen in various fields such as, for example, digital medical imaging [4].

A silicon pixel detector system for high multiplicity charged particle pattern recognition near the collision point has been developed by the ATLAS collaboration to meet the CERN Large Hadron Collider requirements. It is composed of 2228 modular units, each containing 328 x 144 pixel cells of 50 μm x 300 μm (250 μm thick). Each module is connected to 16 read-out chips, covering an active area of 2.2 m^2. Signals will be collected by about $1.4 \cdot 10^8$ electronic channels.

A climatized clean room has been installed at our Department, including a probe station equipped with a micro mechanical movement plate, a microscope, a video camera and a data acquisition system, to perform a series of specific quality controls on the sensors. Those sensors passing the requirements will be subsequently bump bonded to the electronics and mounted in the final detector.

In the following sections the electrical and mechanical control measurements performed at our laboratory will be briefly described, and first results will be presented.

Published in: E.Kuljanic (Ed.) *Advanced Manufacturing Systems and Technology*,
CISM Courses and Lectures No. 437, Springer Wien New York, 2002.

2 GENERAL PROCEDURES

The quality assurance plan for the ATLAS pixel sensors [5] consists of mechanical and electrical test measurements on pixel sensor wafers performed by the institutes under standard conditions: temperature between 20°C and 24°C, relative humidity < 50%, clean room environment with separated, filtered air supply and an environmental control capable of avoiding extremes in temperature and humidity. To assure comparable results over the whole period of quality measurements and between the different labs participating to the measurement process, calibration and cross calibration are also of primary importance.

3 MECHANICAL TESTS

3.1 VISUAL INSPECTION

Each wafer is inspected before any measure is performed, by checking mask alignment structures (visible misalignment < 2μm), scratch pattern correctness and visible irregularities or mechanical damages. The equipment consists of a microscope (at least 1:60 magnification), a microscope mounted camera, with the possibility to store pictures as computer image files.

3.2 PLANARITY MEASUREMENTS

One randomly chosen wafer from each batch is put into the measurement device and coordinates are measured across a matrix measurement points covering most of the wafer's surface; from these coordinates an average surface plane is calculated. The wafer planarity is defined as the distance between two planes parallel to the average plane enveloping all the matrix points, and is required to be < 40 μm.

3.3 THICKNESS MEASUREMENTS

Two opposite test sector pieces of one randomly chosen wafer from each batch are measured using a mechanical micrometer. The thickness is measured near two opposte planarity marks and compared to the wafer thickness data provided by the vendor. The two measured thickness values are required to be < 260 μm, and the approximate thickness uniformity, defined as the maximal difference between the two values, must be < 10μm.

4 ELECTRICAL TESTS

4.1 CURRENT-VOLTAGE MEASUREMENTS

Each wafer contains different sensor structures (tiles, diodes, single chips and mini chips) [1], which have to be tested. The equipment consists of a probe station with chuck and needles for two side contact points, a voltage source and a picoamperometer. High voltage is applied to the metallized contact pad on the p-side. The bias grid on the n-side is connected to the picoamperometer, which is in turn connected to the low side of the voltage source (ground). This can be done on an insulated chuck across the p-side edge. A ramping high voltage is stepped through from 0V down to –500 V.

RELEVANT QUANTITIES. Typical I vs. V curves for diode and tile sensors are shown in Figure1.

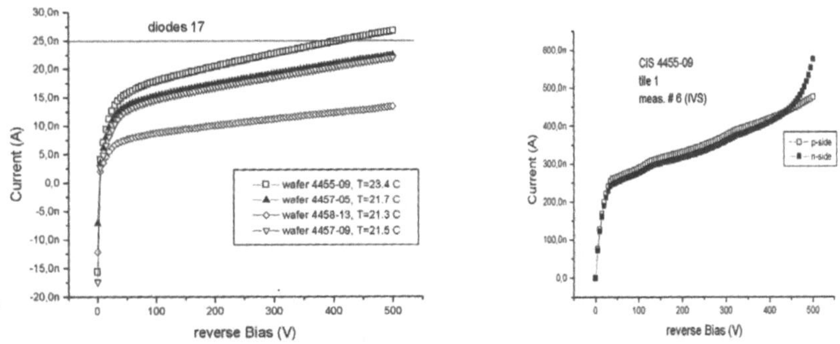

FIGURE 1. Current vs. voltage curves for diodes and tiles

The breakdown voltage V_{bd} is defined as the highest measured absolute V with absolute normalized I < 25nA, (I < 100nA for single chips and I < 2μA for tiles). Defining I_{op} and I_{50} as the absolute normalized currents at $-V_{op}$ and $-(V_{op} - 50V)$ (V_{op} is determined in CV measurements as explained in the following paragraph), a current slope is defined as I_{op}/I_{50}.

CRITERIA The quality assurance requirements are:

- $V_{bd} \geq V_{op}$;
- current slope < 2;
- I_{600} < 800μA (equivalent to 36μA at –10°C)(conformity relevant for microchips after irradiation).

4.2 CAPACITANCE-VOLTAGE MEASUREMENTS

The high voltage side of an LCR-meter is connected to the diode contact pad on the p-side. The metallized n-side ring of the diode is connected to the low voltage side (ground). The high voltage is stepped through from 0V up to -V_{bd} or –200V, whichever is higher in absolute value. In cases of high measurement noise resulting from this set-up, the measurement polarity can be inverted, with the high voltage side connected to the n-side and the voltage stepped through to V_{bd} or 200V.

RELEVANT QUANTITIES. A typical C vs. V curve for diodes is shown in Figure 2.

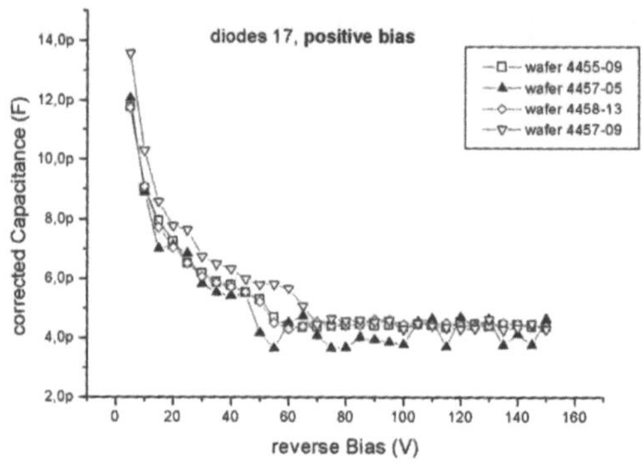

FIGURE 2. Capacitance vs. voltage curves for diodes

The depletion voltage V_{dep} is the absolute voltage at the sharp bend levelling into a plateau in the plot (usually between 50V and 120V). The depletion capacitance C_{dep} is the capacitance at an absolute $V = V_{dep}$. The operation voltage V_{op} is defined as the maximum between 150 V and (V_{dep} + 50 V). The resistivity is

$$\rho = \frac{d^2}{2 \cdot V_{dep} \cdot \mu_e \cdot \varepsilon_{Si} \cdot \varepsilon_0} \qquad (1)$$

where d is the sensor thickness, $\mu_e = 1427 cm^2/Vs$ the electron mobility, $\varepsilon_{Si} = 11.75$ the dielectric constant for Silicon, $\varepsilon_0 = 8.854187817 \cdot 10^{-6} pF/\mu m$ the permittivity constant.

CRITERIA

The quality assurance requirements are:

- $30\ V \leq V_{dep} \leq 120\ V$ before irradiation, ≤ 200 V after irradiation with a flux $\Phi_1 = 5 \cdot 10^{14} cm^{-2}$ of protons having an energy of 24 GeV ($3.1 \cdot 10^{14} cm^{-2}$ neutron equivalent);

- $V_{op} < V_{bd}$

- $2\ k\Omega cm \leq \rho \leq 5\ k\Omega cm$.

4.3 OTHER MEASUREMENTS

Additional electrical measurements are performed on one wafer randomly chosen from each batch:

- I vs. V and C vs. V curves on a MOS oxide test structure, where a $V_{bd} > 50$ V is required.
- I vs. V curve on one gate controlled diode. To test the bump bonding process and the effects of irradiation these structures will be measured after dicing and after being irradiated.
- V on the n-side pixel contact against common ground on one punch-through structure.

4.4 CURRENT STABILITY MEASUREMENTS

A good tile on one randomly chosen wafer from each batch is measured after the wafer has undergone all the other measurements. The high voltage is set to $-V_{op}$, and a current measurement is taken every 10s for at least 15 hours.

RELEVANT QUANTITIES. A typical I vs. t curve is shown in Figure 3.

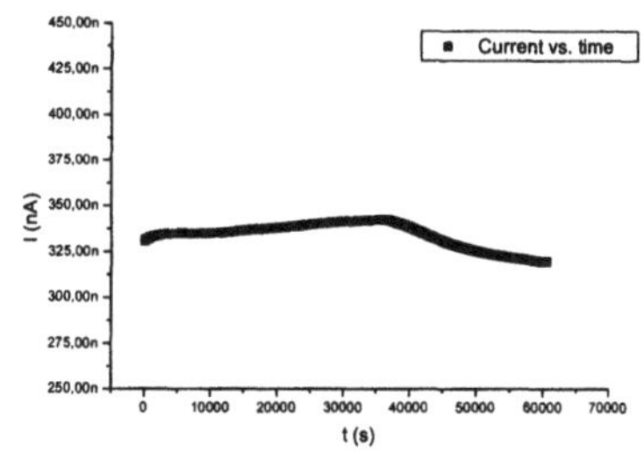

FIGURE 3. Current vs. time curves for a tile

- starting current I_{start} and final current I_{end}: absolute normalized currents at t = 0s and t = 54000s.
- time current slope = I_{end}/I_{start}.

CRITERIA - slope ≤ 1.3.

5 PRELIMINARY RESULTS

A series of pre-production measurements have been performed at our laboratory on 16 wafers from two different vendors. The quality checks previously described have given the following results:

TABLE 1. Quality controls on sensor wafers

			% good	
			Cis	Tesla
Planarity			100	100
I V	$V_{bd} > V_{op}$ & slope	Tile	100	95
		Single chip	78	25
		Mini chip	86	44
C V	$30\text{ V} < V_{dep} < 120\text{ V}$		100	100
	$2\text{ k}\Omega\text{cm} < \rho < 5\text{ k}\Omega\text{cm}$		100	100

6 CONCLUSIONS

The increasing complexity of future high energy particle physics experiments has led people to join their efforts in order to be able to afford the cost, in terms of financial and human resources, of the construction of such big experiments. This has also led to committing to industry for the implementation of parts of the apparatus, under the requirements imposed by experimental physicists and engineers. Nevertheless, experience has shown that the quality controls of such a production must be performed in the institutes, since they require not only specific tools, but also time to think about specific problems which could have escaped the industry production chain. This process is difficult to automatize, even if improvements in this sense can be imagined.

REFERENCES

1. ATLAS Inner Detector Technical Design Report, (1998), CERN/LHC/98-13.
2. http://lhc.web.cern.ch/lhc.
3. Rossi, L., (2001), Semiconductor pixel detectors for particles and X-rays. Nuclear Instruments and Methods A46, 1-265.
4. See for example Proceedings of IEEE 2000 on Medical Imaging, Lyon (2000).
5. ATLAS Project Documents N. ATL-IP-QA/1-16,(2001).

EVALUATION OF DFX TOOLS IN ECODESIGN

M. Kljajin[1]

[1] Faculty of Mechanical Engineering in Slavonski Brod, University of Osijek, Croatia

KEYWORDS: Ecodesign, DFX tools, LCA/LCI tools, DFA, DFD, EOL, DFL, Valuation.

ABSTRACT. This paper discusses the essence of ecodesign, respectively three key terms: sustainable development, cleaner production and life cycle management compares traditional product development with ecodesign. The term DFX tools is used to sum up all tools which provide a methodology that is focused on part of a product's life cycle or focused on one of a number of ways to improve an aspect of a product. In this paper it will be given an idea how it could be possible to valuate DFX tools which is using in ecodesign (the ECO Tools). This paper discusses some similarity of DFX tools with the LCA/LCI tools, as in providing an inventory of the inputs and outputs during part of a life cycle of a product, process or activity. An advice to future users of ECO tools will be also given in this paper.

1 INTRODUCTION

Ecodesign means that "the environment" helps to define the direction of design decisions. In other words, the environment becomes a co-pilot in product development. In this process the environment is given the same status as more traditional industrial values such as profit, functionality, aesthetics, ergonomics, image and overall quality. In some cases the environment can even enhance traditional business values.

The subject of ECO tools and the implementation of the tools into the actual design and production processes of companies. ECO tools are either software or non-software tools with the purpose of analysing and improving environmental performances of products, processes or whole design strategies. As there are large numbers of ECO tools available, the problem for future users is to find the tool for their specific problem from the lists of tools available. Usage of ECO tools at the moment and differentiation of the tools are also explained in the paper.

The term DFX Tools is used to sum up all tools, which provide a methodology that is focused on part of a product's life cycle or focused on one of a number of ways to improve an aspect of a product. This paper discusses some similarity of DFX tools with the LCA/LCI tools, as in providing an inventory of the inputs and outputs during part of a life cycle of a product, process or activity. However, the inventory is confined to the area the tool is specialised in and extra information is provided to help improve that aspect. Some tools are directed towards more than one area and can for instance improve a product's properties for the assembly stage as well as the EOL stage. In this case the tool has to be provided with different information, more often focused on its design in a contractual contents. The emphasis will be put on construction methods, fastening methods and the materials used. An advice to future users of ECO tools will be also given in this paper.

Published in: E.Kuljanic (Ed.) *Advanced Manufacturing Systems and Technology*, CISM Courses and Lectures No. 437, Springer Wien New York, 2002.

2 DEFINITION OF ECO TOOLS

As the term ECO tools is not used in any of the reports, there is not a clear description for it. The following definition has been made to explain the term: ECO tools are either software or non-software tools with the purpose of analysing and improving environmental performances of products, processes or whole design strategies. Most ECO tools available at this moment are aimed at analysing and improving the design of products, such as consumer goods and electronics.

Currently most tools are used in companies that are improving the environmental impact of their products and are require systems to assist. Research institutes collaborating with industry or government institutes are also using many tools. As there are large numbers of ECO tools available, the problem is to find the tool for a specific problem. As the difference between the ECO tools can often be found in the purpose they were developed for, the first step is to define the objective. This is also coherent with the phase of the design process where the tool is intended to be used. The usual objectives given by developers of ECO tools are the following:

- DFX tools - Analyse existing products and processes and use that information to improve certain aspects of a product;
- LCA/LCI tools - Analyse existing products and processes and use that information as feedback to improve the product's environmental behaviour;
- Pollution Prevention or Waste Prevention tools - Compare certain materials and processes to give options to different levels of impact on the environment;
- Improvement tools - Implement improvement throughout the design methodology in order to improve the whole design process.

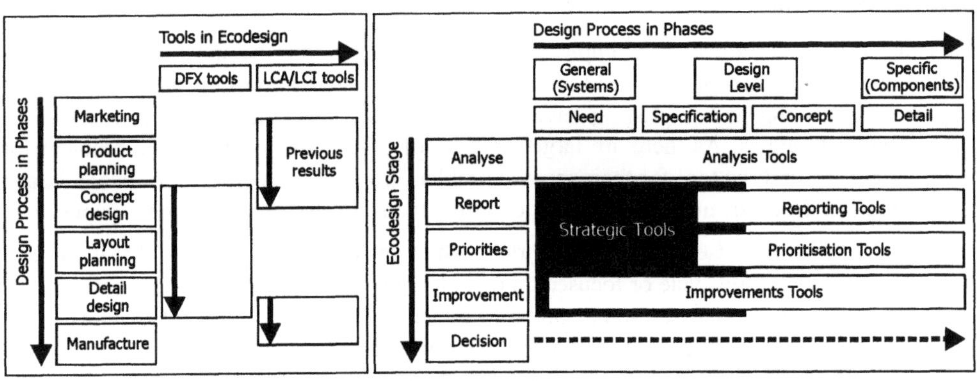

FIGURE 1. The placement of DFX and LCA/LCI tools in the design process

These are the objectives that were also used most by vendors and reports. Results from the research might indicate that some vendors have extended the objective of their tools, but in most cases it can be explained by a more free translation of the question.

In some cases the vendors have given improvement as the objective instead of just assessment or inventory. According the SETAC definition of LCA the tools should implement an improvement analysis as well, and as in most cases this not done, these tools will be placed in the group of LCA/LCI tools [1][2][3][4][5].

As tools only serve a particular objective, they are also only used in certain stages of the design process. To explain the placement of DFX and LCA/LCI tools in the design process, the Fig. 1 has been given.

The methodology behind LCA tools is to calculate the "outputs" as can be seen in Fig. 2 of a product's full life cycle, in other words "cradle to grave" and then to translate those effects into the "effects" and even into "damage" [6][7][8][9][10].

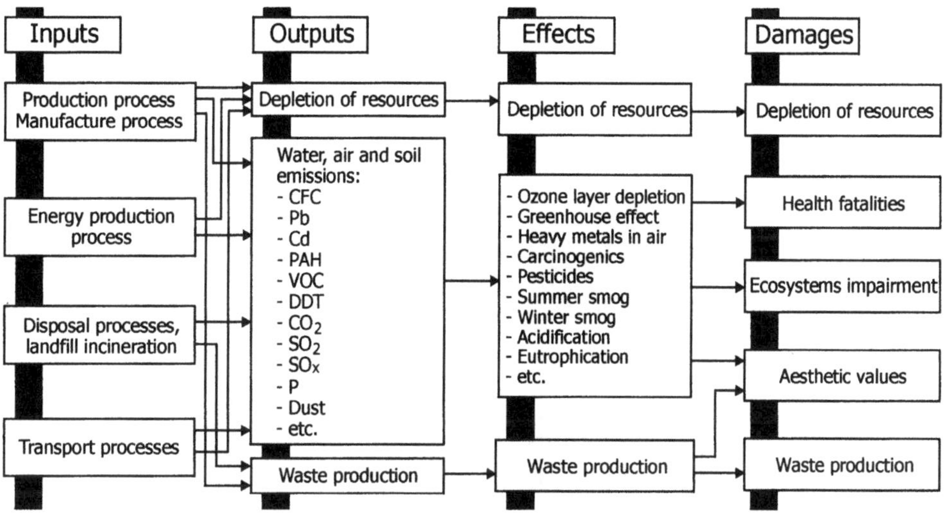

FIGURE 2: Life Cycle Analysis Structure

2.1 DEFINITION OF DESIGN FOR X TOOLS

The term DFX is used to sum up all tools which provide a methodology that is focused on part of a product's life cycle or focused on one of a number of ways to improve an aspect of a product. The tools known to us so far can be divided into the following three groups: DFA - Design For Assembly, DFL - Design For Life and DFD - Design For Disassembly/EOL - End Of Life.

DESIGN FOR ASSEMBLY. This methodology is mainly used to improve the design of a product in its manufacturing life phase [12][13]. The environmental aspect in these tools is quite minimal as they are more directed towards the manufacturing process, however, they do provide a possible comparison between financial and environmental impacts of changes in the design. More specific tools in this area are: DFM - Design For Manufacture and DFMA - Design for Manufacturing and Assembly. Two tools that have been reviewed cover this area and they are:

DFE, DFA AND DFM SOFTWARE TOOLS. The DFA software tool provides designers with information in the concept stage of the product development. DFA results provide estimated assembly times, assembly costs and operation times as well as suggestions for redesigns. According to Boothroyd Dewhurst, reductions can be achieved in assembly times (61 %), assembly operations (53 %), separate fasteners (69 %), assembly defects (68 %), service calls (57 %) and time to market (50 %). Fig. 3 shows the effect of DFA on part count reduction summarising published results of success stories resulting from the application of the Boothroyd-Dewhurst DFA software.

The DFM tool consists of five different early cost estimating modules: machining, injection moulding, sheet metal working, die casting and powder metal parts. Each of these modules is designed to provide the user with component cost information in the early design stage and all the software modules contain interactive material and equipment databases. Both tools can be combined with the DFE tool also provided by this company to allow users to make informed trade-off decisions regarding the value of environmentally friendly products versus their cost. No prices are available for the tools DFA and DFM, however when referring to the price of the DFE tool, it will probably be in the same area, somewhere around EUR 18.000 per version.

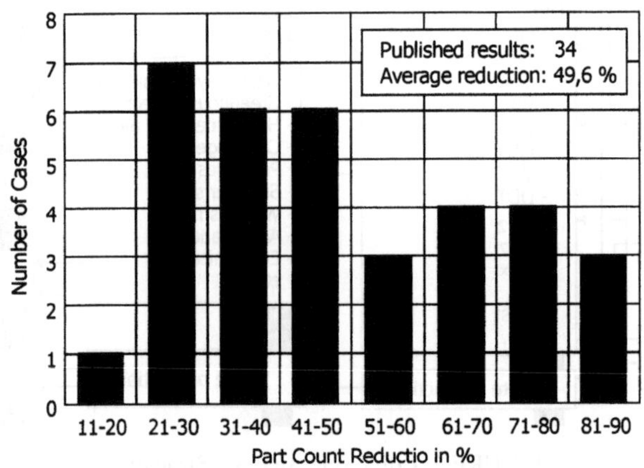

FIGURE 3. Summary of published data showing effect of Boothroyd-Dewhurst DFA on part count reduction

PRICE Systems HL and M. PRICE Systems have developed a number of tools, some of which can be used for Life Cycle Cost Analysis combined with either Design for Assembly or Design for Manufacture.

PRICE HL (Hardware Life Cycle) Model calculates cost effectiveness and operational availability at the system, subsystem, major assembly and sub-assembly levels. It provides cost outputs for all phases in the hardware life cycle and provides a system trade-off capability to determine hardware life cycle costs during early stages of hardware design.

PRICE M (Electronic Module and Microcircuit Model) Model produces estimates of development costs for electronic modules and Application Specific Integrated Circuits (ASIC). The model uses cost estimating relationships based on the amount of microcircuit components, percentage of new circuit cells, circuit cell design repeat, specification level, degree of computer aided design, products familiarity, engineering experts and calibration factors to estimate ASIC specifications and design costs.

The PRICE tools have primarily been sold to Aerospace and Defence organisations and are quite expensive (see Table 1). Although they have experience with a large number of product ranges, consumer goods are not one of them. However, PRICE M is specialised in electronic components, a growing market. The database supplied with the tools is not very large, however, new information from the customer can be included. PRICE Systems use a calibration process to provide the necessary information for the program.

TABLE 1. Merits, demerits and prices of some ECO tools

No	Name of ECO Tool	(+) Positive & (-) Negative	Price in EUR
1	AMETIDE	+ simple to operate	free downloadable and on-line
2	BDI Range	+ user friendly and applicable in early stage of design process - environmental data is limited so that a full life cycle approach is not taken - with no CAD link, entering data is slow	approx. 16.100 for each of these tools
3	DFR-Recy	+ simple Excel/Access based software	1.000
4	DIANA	+ extensive database of mechanical operations - only mechanical disassembly is incorporated	1.700
5	EUROMAT	+ useful the early stages of the design process	unknown
6	GE Plastics DFR Manual	+ easy to use - limited use - plastics only	available free from GE Plastics
7	LASeR	+ different analyses possible - extensive amount of information has to be inserted	available free by download
8	PRICE Systems	+ specific tools for each purpose + good support and software - expensive	24.200 - 33.900
9	ReStar	+ designed to work as simple as possible - easy to use + allows optimisation related to recovery costs	unknown
+ merits, - demerits			

DESIGN FOR LIFE. These tools are directed towards improving the products behaviour during its life phase, by either decreasing its impact or by increasing the length of its life. In that way, they sometimes work towards more sustainable products, fit for a different kind of use by consumers and producers/suppliers. These tools can be considered to improve a products impact during life: DFMAIN - Design For Maintainability and DFS - Design For Serviceability. Two tools have been suggested in this section:

DFS software tool allows designers and engineers to evaluate the Serviceability of a product when it is in the early design stage, where changes to the product can be made at minimal cost. It will provide reports with suggestions for redesign that highlight areas in the service task which should be examined for service improvement. No information was available on the price of this tool, but in the line of the other BDI tools, it can be estimated around EUR 18.000 per version.

LASeR, Life-cycle Assembly Serviceability and Recycling Prototype Program. This windows based program evaluates the serviceability and assembly of mechanical designs. The user has to define a structural design of a mechanical system along with cost, labour and material data. The program then offers the option to analyse the assembly or the serviceability after which a determination is made of the disassembly costs and reprocessing costs associated with the given strategy. The part of the assembly analysis is done by the GE-Hitachi Assembly Evaluation, which is used with permission of the General Electric Company.

DESIGN FOR DISASSEMBLY OR END OF LIFE TOOLS. As the two previous groups were directed towards the first two stages of a product's life, this group is focused on the end or disposal stage. This is where they try to improve the product's performance by stimulating reuse of certain components and/or materials and if possible further use of the complete products by giving them a second life [14]. Tools included in this section are: DFR - Design for Recycling and Design for Disposal. Some tools were surveyed in this group and they are also included into Table 1:

AMETIDE, A METhodology for Time Disassembly Estimation. This is mainly a Design for Disassembly tool with an easily updateable database from where all necessary disassembly techniques available corresponding with a fastener can be taken. The designer gets a panel of all possibilities for removing a part and can choose which is the best depending on what the part is assembled for. Parameters like frequency of the task and position can be given, materials can be selected and the mass of a part can be entered after which a theoretical maximum time for disassembly can be determined. Also comparisons between disassembly and machining operations can be made where for instance machining time is 90 % preparation and 10 % effective use.

DFR-Recy is a relatively simple Excel/Access based software for evaluation and documentation of product recyclability. Product structures such as parts, materials, and numbers of parts are given as input, after which the program calculates recyclability values as percentages. A reference Access database is given for the evaluations and reports are obtained in numerical and graphical format. The problem indicated by the vendor is the strong dependence on local conditions such as available materials, technical recycling methods and so on, that the user has to customise the database to the conditions in which the evaluations are to be made. As the author know the program and manual are only available in the Finnish language, but a possible translation would be fairly easy to make with moderate efforts.

EUROMAT is development-accompanying instrument for environ-mental conscious and recyclable materials. This tool provides comprehensive assistance for finding the best environmental conscious and recyclable materials concerning a certain profile of requirements. The principles of the instruments are top-down, integrated, interactive and iterative approaches. The following three characteristics are used to reach the objective:

- Identification of generally qualified composites and materials out of all theoretically existing combinations with a given profile of requirements.
- In the early stages of the design process of a new product not only the properties of usage and manufacturing are taken into account but also (dis)advantages regarding recyclability, working conditions, environmental impacts as well as economical indications.
- Research and development can be made more efficient because misleading directions of development can be recognised and corrected at an early stage utilising the interactive expanded profile of requirements.

ReStar is an AI-based software planning tool to design for disassembly, which uses the A* algorithm, a form of dynamic programming that guarantees an optimal solution over multiple objectives. The program also optimises the balance between the costs per hour of disassembly with what can be sold, which ensures that the disassembly plan makes more money than it costs. ReStar plots a cost curve of the effort required for disassembly, testing, repair, remanufacturing and product changes that enable recovery. It also plots a curve of the revenue from resale and reuse. The tool helps design engineers find the optimal point of the two curves. The program is being used by the Green Engineering Corporation that use it in engagements with companies.

3 CONCLUSION

The number of names of ECO tools is very large, even though a lot of them are not actually commercially available or fully developed yet. The number of tools is still growing and new names kept emerging. Therefore this paper will not be complete but will give an indication of what tools are available and what can be done with them. It is difficult to develop a tool for a specific product or a whole industry range and general tools can often not be used for all products either. This indicates a need for interactive tools so that the information specific for use can be put in by the users themselves. However this asks for a lot of specific information from the user and therefore a lot of time to put that information in as well.

Most of the tools were used by industry and research (84 %), it can be assumed that most of the use by research is done in co-operation with industrial partners as well. Most tools have been developed for general design or have gone from specialised to general tools (83 %). In case of specialisation, the areas of electronics, packaging and chemicals are most common. Most data has been obtained by interaction with industry (65,5 %) and in combination with other sources (61 %), which indicates fairly good source material. Also the scientific reports, from BUWAL and EPA are used often (39%). The updating of the data is something, which a lot of vendors are not really clear about. They either provide no information on the time schedules they use, which might indicate they do not have a system for updating.

DFX tools have some similarity with the LCA/LCI tools, as in providing an inventory of the inputs and outputs during part of a life cycle of a product, process or activity. However, the inventory is confined to the area the tool is specialised in and extra information is provided to help improve that aspect. Some tools are directed towards morc than onc arca and can for instance improve a product's properties for the assembly stage as well as the EOL stage. In this case the tool has to be provided with different information, more often focused on its design in a contractual contents. The emphasis will be put on construction methods, fastening methods and the materials used.

The databases of DFX tools are different from those of LCA or LCI tools although some similarities can be found [5]. There is more specialisation in certain areas, such as fasteners and production processes. In some cases also the aspect of time will play a large influence, and even more detailed knowledge of manufacture processes of the product are demanded from the user, as these are not considered to be available as general knowledge. Interaction between databases and users is a definite must in this case.

The number of DFX tools is not as extensive as the LCA or LCI tools [5], but a number of tools, which also specialise in a certain aspect of a product design, are not included in this list, such as tools for cost analyses. Most tools cover more areas and are commonly known as DFE tools.

An advice to future users of ECO tools is to apply the less expensive non-software tools to have a look at environmental impacts in general before starting the use of more expensive software tools. If there has been decided to use software tools to perform analyses or to achieve improvement, then the choice for the tool is still very much open to factors such as the budget available and the wanted precision.

REFERENCES

1. Sweatman, A.; Simon, M., (1996), Life Cycle Assessment of an Electrolux Vacuum Cleaner, an Evaluation of LCA Tools, DfE Manchester Metropolitan University, Manchester.
2. Billatos, S. B.; Basaly, N. A., (1997), Green Technology and Design for the Environment, University of Connecticut, Connecticut.
3. NNa, (1991), Packaging and the Environment (Life-cycle Analyses of Selected Packaging Materials, Quantification of Environmental Loadings), Tillman, Baumann, Eriksson and Rydberg.
4. NNb, (1991), A Technical Framework for Life Cycle Assessment, SETAC Society of Environmental Toxicology and Chemistry.
5. Kljajin, M., (2000), Ecodesign Approach and DFX Tools, Proceedings of the 5th Engineering Systems Design and Analysis Conference, ESDA 2000, G. K. Wolfe and O. Arnas (Ed.), ISBN 0-7918-1665-6, ASME International, New York, PD-Vol. 82, 51-58.
6. Rice, G., (1996), LCA software review, Center for Environment Strategy, University of Surrey, Surrey.
7. Pedersen, S.; Wilson, C.; Pitts, G.; Stotesbery, B., (1996), Electronics Industry Environmental Roadmap (Chapter 4: Design for Environment: Evaluation and Tool Needs), Microelectronics and Computertechnology Corporation, MCC Texas.
8. Menke, D. M.; Davis, G. A., (1996), Evaluation of Life Cycle Assessment Tools, University of Tennessee, Center for Clean Products and Clean Technologies, Tennessee.
9. Keoleian, G. A.; Menery, D., (1993), EPA Life Cycle Design Guidance Manual (Environmental Requirements and the Product System), National Pollution Prevention Centre, University of Michigan, Michigan.
10. NNc, (1996), UNEP-Life Cycle Assessment, What is it and how to do it?, CML, Novem, RIVM and UNEP, supported by the Ministry of Housing (the Netherlands) and Unilever.
11. NNd, (1993), The LCA Sourcebook, Society for the Promotion of LCA Development, SPOLD, Business in the Environment, BiE SustainAbility Ltd.
12. Boothroyd, G.; Dewhurst, P., (1979), Product Design for Assembly, Designers Handbook, Boothrayd Dewhurst, Inc., 138 Main Street, Wakefield, RI 02879.
13. Kljajin, M., (1998), Design for Assembly, Tehnički vjesnik/Technical Gazette, Vol. 5, No. 3/4, 37-44.
14. Kljajin, M., (1999), Design for Disassembly, Tehnički vjesnik/Technical Gazette, Vol. 6, No. 1/2, 41-48.

TIME LAG ERROR OF INVENTORY IN INFLATION

J. Mrša

Economic Faculty, University of Rijeka, Croatia

KEYWORDS: Inventory, Process Planning, Inflation, Time Lag Error

ABSTRACT. An algorithm of real value of inventory in inflation has been made. The model is made taking into consideration International Accounting Standards. In this point of view this model is based on maintenance of financial capital principle. The computer program for simulation of real value of inventory is made for modern process planning of engineering components manufacture. The program is an autonomous entity, but it can be part of one general program. Mathematical model of inventory was made from historical costs analyze. For the big inflation rate real value of inventory outflow has to be calculated by using precise methods.

1 INTRODUCTION

The effects of the rate of inflation on the calculated inventory depend on the rate of an increase in the general level of prices between the moment of input costs acquisition of inventory and output costs of sales that includes calculated inventory.

Inventory evaluation as current asset is usually based on historical costs. Moreover, estimation of historical costs of inventories is not unambiguous. Inventories are contemplated as one unit, but we know that they are consisted of large number of different units.

A recorded amount of inventory depends on accounting method of inventory flow, chronology of inventory flow and of used accountancy calculation method, i.e., of inventory costing methods.

Management decision-making cannot be efficiently without realistic financial statements, or without realistic view on a cost-benefit situation. A really cost-benefit situation cannot be made in satisfactory way if the value of real inventory is not known. Special difficulties in evaluation of real inventory are arising in condition of inflation.

Inventories could be evaluated by traditional system of historical costs or by IFO methods[1]. IAS 2 recommends using the FIFO method or by method of weighted average price.

Whole inventory values, from the beginning of accounting period and increasing by acquisition trough the actual period, could be divided on residual inventories and inventory costs.

---

[1] FIFO (first in, first out), LIFO (last in, first out), HIFO (highest in, first out)

Published in: E.Kuljanic (Ed.) *Advanced Manufacturing Systems and Technology*,
CISM Courses and Lectures No. 437, Springer Wien New York, 2002.

2 EXPENDITURE INCREASING DUE TO INVENTORY TIME LAG ERROR

Expenditure increasing due to inventory time lag error under the FIFO method, or method of weighted average price directly depends of day of inventory holds.

Expenditure increasing due to inventory time lag error is equal:

$$R_z = \sum_{m=1}^{M} Z_m [(1 + p)^{1-(t-t_m)} (1 + p)^{1-t}] \qquad (1)$$

$$R_z = \sum_{m=1}^{M} Z_m (1 + p)^{1-t} [(1 + p)^{t_m} - 1 \qquad (2)$$

where are:

R_z - expenditure increasing due to inventory time lag error

M - number of outflows

Z_m - quantity of particular outflows

t - moment of inventory outflow, $t=d/365$

t_m - inventory withholding time in days, $t_m = d_m/365$

d_m - inventory withholding time in years

Particular determining of time lag error of every inventory item is impossible practically. For this purpose, average time of inventory withholding could be used with satisfied accuracy. Average time of inventory withholding is equal [1]:

$$t_s = \frac{1}{k_o} \qquad (3)$$

$$k_o = \frac{Z_{iz}}{Z_s} \qquad (4)$$

where are:

t_s- average time of inventory withholding

k_0 - coefficient of inventory turnover

Z_{iz} -historical inventory outflow

Z_s - historical average quantity of inventory

An outflow of inventory during the time dt, for steady state condition, and by presumption that the inventory amount is constat, will be equal:

$$d\,Z_{iz0} = \frac{Z_{uk}}{t_{ob}}dt \tag{5}$$

where are:

dZ_{iz0} - value of an inventory outflow during the time dt

Z_{uk} - total inventory outflow

t_{ob} - accounting period, $t_{ob} = 1$

If the prices are arising by rate p, value of an inventory outflow vs. time is growing up by:

$$dZ_{iz} \approx dZ_{iz0}(1+p)^t \tag{6}$$

If the inventory costing method FIFO is applied, and if an inventory outflow was evaluated as historical costs the recorded value of entered inventory in accounting period is equal:

$$Z_{iz} = \int_0^1 \frac{Z_{uk}}{t_{ob}}(1+p)^t\,d\,t \tag{7}$$

$$Z_{iz} = \frac{Z_{uk}}{t_{ob}}\left(\frac{1+p}{\ln(1+p)} - \frac{1}{\ln(1+p)}\right) \tag{8}$$

Inventory time lag error is equal to total difference between the recorded and real value of inventory in the prices at the end of accounting time. A final amount of inventory that calculated at the end of accounting period is equal:

$$R_z = \int_0^1 \frac{Z_{uk}}{t_{ob}}(1+p)^t[(1+p)^{1-(t-t_s)}(1+p)^{1-t}]dt \tag{9}$$

$$R_z = \frac{Z_{uk}}{t_{ob}}(1+p)[(1+p)^{t_s}-1] \tag{10}$$

TABLE 1. Time lag error calculated by equations (11) and (13)

Time t_s ↓ / Rate p→		0.010	0.1000	0.3000	1.000	10.00
0.01	Eq. (13) Eq. (11) Error %	0.010 0.010 0.004	0.1000 0.1000 0.0489	0.3004 0.3000 0.1484	1.005 1.000 0.496	10.51 10.00 4.89
0.02	Eq. (13) Eq. (11) Error %	0.020 0.020 0.007	0.2002 0.2000 0.0957	0.6018 0.6000 0.2935	2.020 2.000 0.984	22.12 20.00 9.57
0.06	Eq. (13) Eq. (11) Error %	0.060 0.060 0.001	0.6016 0.6000 0.2628	1.8153 1.8000 0.8434	6.177 6.000 2.858	81.41 60.00 26.30
0.24	Eq. (13) Eq. (11) Error %	0.239 0.240 - 0.278	2.4167 2.4000 0.6896	7.4088 7.2000 2.8181	26.674 24.000 10.024	844.05 240.00 71.57
1.00	Eq. (13) Eq. (11) Error %	0.964 1.000 -3.708	9.9499 10.0000 -0.5034	32.0434 30.0000 6.3770	138.629 100.000 27.865	141817.90 1000.00 99.29
5.00	Eq. (13) Eq. (11) Error %	3.887 5.000 -28.627	42.1919 50.0000 -18.5062	153.0385 150.0000 1.9854	1075.056 500.000 53.491	13000900000.0 5000.00 100.00

Time lag error is equal:

$$R_z = Z_{iz}(1+p)\frac{(1+p)^{t_s}-1}{\dfrac{1+p}{\ln(1+p)}-\dfrac{1}{\ln(1+p)}} \tag{11}$$

For small values of inflation rate *p*, $\ln(1+p) \approx p$ and $(1+p)^{ts} \approx 1+p\ t_s$. There it could be written that time lag error of inventory withholding is equal:

$$R_z \approx Z_s\, p \tag{12}$$

$$R_z \approx Z_{iz}\, p\, t_s \tag{13}$$

If the inflation rate *p* is small, the equations (12 and (13) could be applied with just good accuracy. But if the inflation rate is not small the equation (11) have to be applied.

3 AN ANALYSIS OF APPLICABILITY OF TIME LAG ERROR ESTIMATION METHOD

The time lag error of inventory withholding that calculated by equations (11) and (13) and relative errors between them are shown in Table 1and Figure 1. Time t_s was relative value of accounting period, and time lag errors are presented in percents of inventory outflow values during the accounting period. Calculation was made by computers program MRSAACC.

Equation (13) gives a possibility for very simple calculating of time lag error of inventory withholding, but with very low accuracy in comparison with Equation (11). In the calculation of inventory time lag error different values of inflation and time of withholding have been applied.

If it is accepted that the tolerance limit of error is 5% percent, simplified Equation (13) is tolerable in a normal business circle with time of withholding, t_{ob}=0.24 or for one year accounting time t_{ob}=88 dais till to an inflation rate of 30%. The tolerable inflation rate is arising if the inventory withholding time is arising.

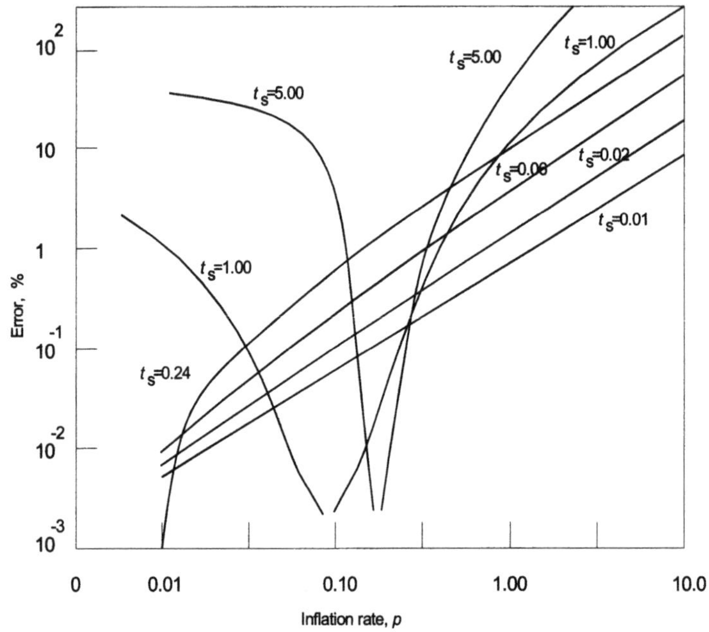

FIGURE 1. Time lag error

4 CONCLUSIONS

A really cost-benefit situation cannot be made in satisfactory way if the value of real value of inventory is not known. Mathematical modeling of inventory in inflation is efficiency way of improving the cost benefit analyze.

If the profit is calculated according to the system of historical costs, the effects of inflation on the calculated values of recorded inventory from the moment of its adding to other expenditures have to be determined by applying the same method as for the other costs.

If the inflation is emphasized, the real value of an inventory outflow should be calculated by using precise formulas.

REFERENCES

1. Popović Z . Economic analyze, Informator, Zagreb, 1983., p. 468. (In Croatian)

2. ASC, Accounting for the Effects of Changing Prices, Accounting Standards Committee, London, 1986.

3. Baxter, W. T., Inflation Accounting, Philip Allan, London, 1984.

4. Carsberg, B. V., Current Cost Accounting the Benefits and the Costs, Prentice Hall International, 1984.

5. Solomons, D., Making Accounting Policy, Oxford University Press, New York, 1986.

6. Tweedie, D., and Whittington, G., The Debate on Inflation Accounting, Cambridge University Press, Cambridge, 1987.

7. Whittington, G., Inflation Accounting an Introduction to the Debate, Cambridge University Press, Cambridge, 1989.

8. Mrsa, J., Mathematical Modeling of Inventory In Inflation, Proceedings of MicroCAD '98,, ed. Jula Patko Miskolc, 1998.

COLLABORATIVE DESIGN AND MANUFACTURING OF A CUSTOMIZABLE VOICE PROSTHESIS

G. Concheri[1], S. Filippi[2], R. Meneghello[1], C. Miani[3], D. Comelli [1]

[1] DAUR – Design Tools and Methods for Industrial Engineering Lab., University of Padova, Italy
[2] Department of Electrical, Management and Mechanical Engineering, University of Udine, Italy
[3] Department of Surgical Sciences, University of Udine, Italy

KEYWORDS: Collaborative development, Surgical devices, Rapid Prototyping, Rapid Tooling, Fits.

ABSTRACT. An innovative approach to the development of biomedical devices must be intrinsically collaborative, because it requires a strict correlation among medical specialists and engineers, and a careful calibration of Rapid Prototyping and Rapid Tooling techniques in order to account for the required prosthesis design optimisation, customisation and lead time. In this paper, besides a discussion on some critical aspects relevant to the collaborative development of a voice prosthesis, some preliminary results on the functional design and the analysis of the fitting conditions of a composite mold manufactured using Rapid Prototyping (RP) techniques will be presented.

1 INTRODUCTION

There are areas among human activities that heavily impact life quality and that could profitably take advantage of the most innovative methodologies and tools or techniques developed in the industrial engineering field. The biomedical area is undoubtedly one of these, and the possible contribution is particularly relevant in those fields that are less attractive for major companies or private research centres. Besides a strict collaboration among medical specialists and engineers, a key aspect for the successful development of medical devices and prostheses is the set up and verification of viable manufacturing processes assuring the required flexibility, lead time, etc. From this point of view, Rapid Prototyping (RP) and Rapid Tooling (RT) techniques present many favourable features, but some operating aspects influencing the global design and manufacturing process need to be investigated accurately. Therefore, besides a discussion on some critical aspects relevant to the collaborative development of a voice prosthesis, some preliminary results on the functional design of a composite mold manufactured using Rapid Prototyping (RP) techniques will be presented.

2 MEDICAL ASPECTS

Laryngeal cancer involves 8 people out of every 100,000 of the European Union population per year with a crude 10 year survival rate of 50%.
Every surgical patient require a post-operative rehabilitation procedure regardless of survival rate. Particular attention should be paid to those patients who undergo a total laryngectomy in order to restore their voice. For this reason about 150000-200000 European Union inhabitants need a periodical follow-up. Full rehabilitation enables the patient to resume a satisfactory social and economic quality of life.

Published in: E.Kuljanic (Ed.) *Advanced Manufacturing Systems and Technology*,
CISM Courses and Lectures No. 437, Springer Wien New York, 2002.

Post total laryngectomy can be managed both with surgical shunts or, more simply, with in-dwelling or non in-dwelling voice prostheses, as depicted in Figure 1. Nowadays there is a large consensus about it. The first option is harder to achieve but, when effective, allows the patient to avoid the periodical replacement of the valve that it is needed about twice per year (some recent papers suggest every 4-5 months; [1]) when a prosthesis is put in; reducing, therefore, the individual and social costs (hospitalisation, cost of the valve, etc...). On the other hand it is also true that many surgical shunts are quickly complicated by leakage of saliva and food, although you can easily solve the second problem by putting a valve into the surgical shunt.

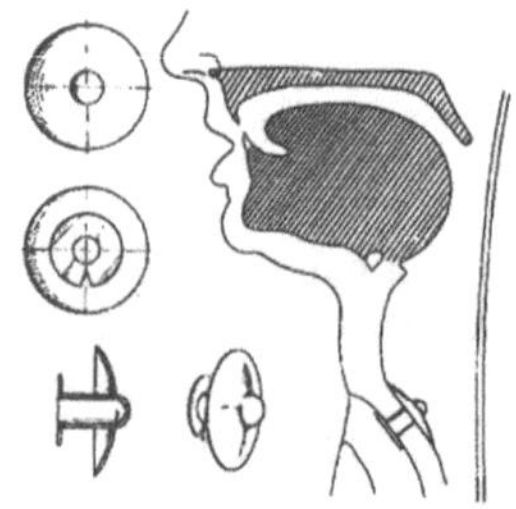

FIGURE 1. Staffieri's voice prosthesis and the indication of the anatomical site where the valve is placed

There are neither standardization about the various surgical procedures available nor evident data about their results; this is why it has been difficult to achieve repeatable results and to choose between so many different techniques up to now. The surgeon is requested to be particularly skilled in order to perform a surgical shunt without valve insertion and avoid common pitfalls and complications, otherwise it is better to opt for prosthetic rehabilitation. Sometime it is better to take advantage of the double option given by the use of the surgical shunt and, later, the surgical shunt + valve, if it proves to be necessary. This is maybe the best option for the patient and for the social costs.
Nowadays the lab "in vitro" prosthesis resistance levels for the new commercial devices have reached values very close to the resistance of the normal larynx during phonation [2].
A comparison of "in vitro" and " in vivo" patient's aerodynamics of the different voice buttons shows the failure of the previous correlation between low pressure and resistance on one hand and better clinical performance on the other one [3].
This is confirmed by the fact that for decades the old type of voice button has been, and it is yet, used with excellent results. It is, therefore, clear that prostheses with high resistance can only work well in certain patients. Indeed, the patient should be evaluated before as well as during insertion [4] of the phonatory prostheses to determine the intrinsic resistance to the new, post-surgical anatomy. Moreover, such an analysis should be independent from the voice button itself.
The lower pressure device seems to be more useful just in those patients who have high intensity phonation needs. Thus, the absolute best voice button has still to be found and it could never be. This is showed by the fact that clinical testing of devices developed many years ago keeps on giving nearly analogous results, with just a slight difference in the intensity range.
In some patients, the so-called low-(in vitro) resistance devices are unable to maintain their aerodynamic characteristics because of interference from anatomical resistance which accounts for as much as 80-85% of all resistance. This markedly increases the overall, subjective effort

(or intratracheal pressure) needed for phonation. On the other hand, patients with very low "anatomical" resistance may sometimes produce a voice of very poor quality without any effort. This is particularly true in patients who had, during surgical procedure, pharyngeal segment replaced by free flaps imported from other areas of the body.
However, it has still to be determined whether these different behaviours operate in a neurogenic manner (peripheral or central) or if they are solely based on a dynamic variation in muscle tone, scarring of the operated field, or if multiple factors are called into play. Hence, such sites have revealed a dynamic effect on the anatomical resistance exerted.
The research is now addressing this issue to the aim of finally providing a mean for effortless speech in "high resistance" patients and a better quality of voice in "extremely low resistance" patients.

Nowadays the choice of a phonatory prosthesis must be determined on a patient-by-patient basis and may be solved by a custom-made device for every single patient.

3 ENGINEERING ASPECTS

3.1 FORMALIZATION OF THE ENGINEERING PROBLEM

From an engineering point of view, the results described in the previous section can be summarized in the following considerations:

- The geometric and dimensional configuration of the voice prosthesis has not been fully optimised yet and therefore it could be further refined and developed. A deep comprehension of some subtle physical phenomena that influence the prosthesis functional behaviour is still lacking and should be addressed carefully [5]. Such investigations are primarily responsibility of the medical specialists, but they may be supported by a rigorous identification and verification of the functional parameters involved in the modelization of the prosthesis and by the adoption of virtual prototyping techniques in order to analyse different device configurations. Such a methodology, that may directly apply also to the development of other medical devices, and that should deserve attention because of the possible innovation that may induce, will not be discussed further in the present paper.
- In addition to the general activities of device optimisation stated above, a specific prosthesis customisation phase is needed in order to tailor the functional characteristics to the individual needs of each patient.
- The prosthesis must be replaced frequently and therefore the study of the manufacturing process of the device should address the feasibility of small batch production of custom devices accounting for economic issues, flexibility, lead time, etc.
- Another critical aspects is related to the identification of suitable materials for the voice prosthesis. For the scope of the present work a silicon semi-solid paste was identified, but a dedicated work should be addressed, in order to consider a wider range of possible materials and to assess the actual bio-compatibility of such materials.

The collaborative development of the voice prosthesis is undergoing [5], but the accurate definition of its functional and geometric characteristics cannot leave out of consideration the manufacturing issues imposed by customisation and small batch production. These aspects, which are fundamental in order to direct the design activities, deserve careful consideration and will be discussed below.

3.2 MANUFACTURING ISSUES

Assuming that the optimal set of parameters characterizing device operation has been identified, a parametric model of the prosthesis, implementing the desired functional aspects and accounting for the expected customisation, can be generated that could be used to instantiate a family of parts simply adapting the set of driving parameters. In the present work, the parametric solid model of the prosthesis depicted in fig.2 was generated using PTC Pro/ENGINEER [6].

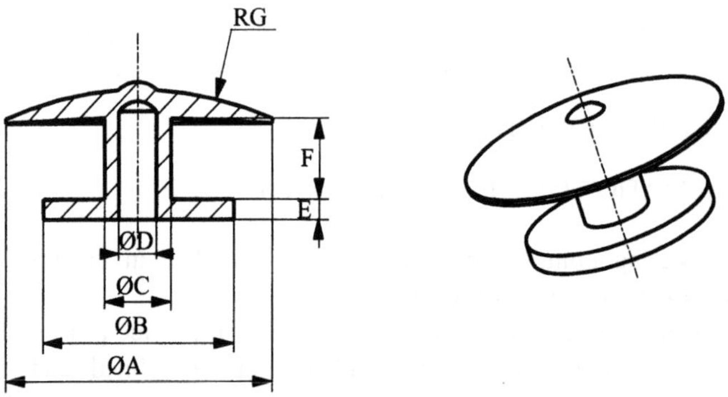

FIGURE 2. Parametric model of the voice prosthesis

In addition to the prosthesis development phase, the need for customized devices also suggests the implementation of a collaborative approach to device manufacturing based on Rapid Tooling and Rapid Prototyping techniques. Rapid Tooling can be defined as the rapid production of tools using RP systems either directly or indirectly.
Two different Rapid Prototyping (RP) systems are available in the FP@ laboratory of Agemont at Amaro (Udine - Italy) and have been considered in the present work:

- Stereolitography (SL): 3D System SLA 3500.
- Direct Metal Selective Laser Sintering (DMSLS): EOS M250X.

Direct SL manufacture of a custom device depends on the availability of suitable photo-polymers assuring either the needed mechanical characteristics or bio-compatibility. Moreover the short duration of the prosthesis requires the production of more than a single device. For these reasons, the direct SL manufacture was not considered.
Material properties and small batch production considerations therefore suggest molding as the most suitable method for the voice prosthesis manufacturing, using RT techniques for mold forming.
An RP model of the part to be produced can be used to form the injection mold, but in the present case the complexity of the device shape, that comprises thin walls, wide undercuts and a deep cavity, suggests the adoption of a direct RT approach in order to generate a composite mold, built-up by more parts intended to be assembled together.

3.3 COLLABORATIVE DEVELOPMENT OF VOICE PROSTHESIS

In the development of the voice prosthesis, various actors in three different locations are involved:

- the Medical Centre, where the medical specialists manage the post-laryngectomy rehabilitation of surgical patients;
- the Engineering Department, responsible for the design and customisation of the prosthesis, the relevant mold and its composing parts;
- the Manufacturing Facility, that is in charge for the rapid prototyping, post-processing and set-up of mold components, and for the production of the required batch of custom devices using injection molding.

In Figure 3, the proposed development process for the voice prosthesis is summarized.

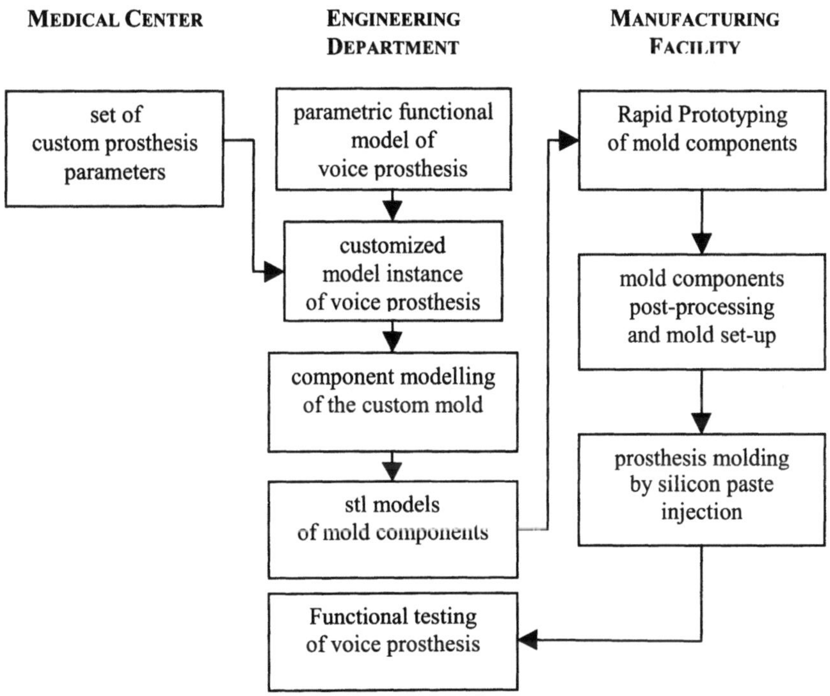

FIGURE 3. Collaborative prosthesis development process

4 PRELIMINARY RESULTS

4.1 MOLD ASSEMBLY DESIGN

The development process discussed above is still undergoing and further refinements should be implemented. However, in order to verify some critical phases, a tentative design was proposed for the mold components, based on the prosthesis parametric model in Figure 2. Two sets of mold components have been produced at FP@ laboratory by means of the DMSLS system, using proprietary metal powder with an average particle size of 80 μm, and the SL system, using Vantico SLA 5190 photopolymer. In Figure 4, the exploded view of the mold assembly CAD model and the relevant SL and DMSLS rapid prototypes are presented.

FIGURE 4. CAD model, SL and DMSLS Rapid Prototypes of mold components

Finally, for a preliminary evaluation of molding capabilities of the tools produced, a trial manual process of injecting a silicon semi-solid paste was employed using the SL generated mold assembly. The resulting raw part is shown in Figure 5.

FIGURE 5. Tentative prosthesis produced by manual mold injection using the SL mold assembly

4.2 ANALYSIS OF FITS IN DMSLS MOLD ASSEMBLY

A critical aspect of the proposed prosthesis development process is the production of directly mountable mold components by means of DMSLS techniques. In order to verify the direct assemblability of DMSLS generated parts, the fit conditions indicated in Fig. 6 have been measured by means of a CMM (a Zeiss Prismo Vast 7, courtesy of Unilab Laboratori Industriali S.r.l., Venezia – Italy) .

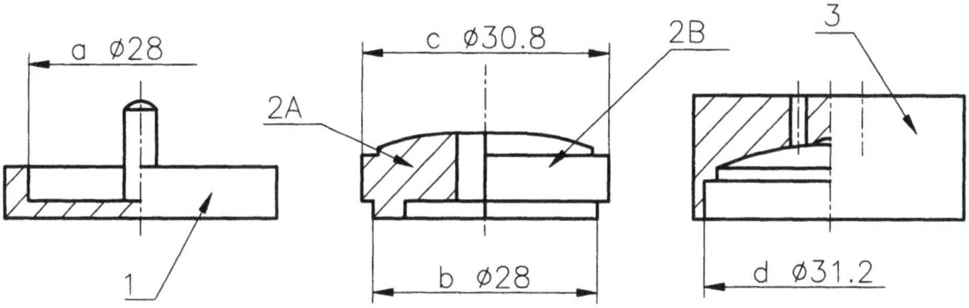

FIGURE 6. Verified fitting dimensions of mold components

TABLE 1. Experimental measures of fit sizes. (All values are in mm)

Part	No.	Dim.	Description	Nominal size	Actual size	Lower dev.	Upper dev.	Round-ness	Std.dev.	Virtual size
Base	1A	a	Inner Dia	28	27,908	-0,183	0,128	0,311	0,040	27,726
"	1B	a	Inner Dia	28	27,860	-0,218	0,126	0,344	0,045	27,642
Shell	2A	b	Lower Outer Dia	28	27,980	-0,124	0,118	0,243	0,041	28,098
"	"	c	Lateral Outer Dia	30.8	30,664	-0,210	0,242	0,452	0,052	30,905
"	2B	b	Lower Outer Dia	28	27,964	-0,143	0,168	0,311	0,050	28,132
"	"	c	Lateral Outer Dia	30.8	30,678	-0,445	0,138	0,583	0,060	30,816
Cover	3A	d	Lateral Inner Dia	31.2	31,025	-0,147	0,095	0,242	0,041	30,878
"	3B	d	Lateral Inner Dia	31.2	30,990	-0,251	0,105	0,355	0,040	30,739

From Table 1, it can be recognized that features modelled according to the same nominal size of Ø28 mm (1a-2b) cannot fit one into another: selecting the larger "hole" (feature 1Aa, virtual size 27.726 mm) and the smaller "shaft" (feature 2Ab, virtual size = 28.098mm) an interference fit results of -0.372 mm.

The features modelled imposing a gap (3d - 2c = 31.2 - 30.8 = 0.4 mm) can be hardly assembled: the larger "hole" (feature 3Ad, virtual size 30.878mm) can accommodate only the smaller "shaft" (feature 2Bc, virtual size = 30.816mm) resulting in a slight clearance fit (0.062 mm). The other possible choices always produce interference fits (from -0.027 to -0.166 mm).

The reasons for these results can be found in the technological aspects of the DMSLS process. Given the size of the sintered powder particles (about Ø 0.08mm), the resulting surface necessarily presents a texture characterized by consequent parameters (Rz ≅ 80, Ra ≅ 20-25). Surface roughness influences also size deviations and form error. It can be noted, by the way, that the sintering process tends to alter the "mating size" towards the maximum material condition MMC of either inner or outer features. This effect is probably emphasized by the sticking of partially sintered powder particles at the boundary surface of the prototype, thus worsening the size accuracy. It should be noted that the average roundness error is 0.355 mm, slightly more than 4 times the particles diameter, and, if it is allowed to extend the results obtained from such preliminary and incomplete tests, this value could be considered an estimate of the predictable reduction of the fit clearance or increment of the fit interference.

In order to verify the effect of the "last" layer of powder on prototypes dimensions, we tried to manually remove the "sticked" or the just partially sintered particles using a copper wire brush (more accurate techniques are described in [7]). This procedure is close to finishing operations (shot-peening or wire brushing) which is performed on the laser sintered moulds that are used in industrial conditions for the production of plastic parts by mean of injection moulding.

TABLE 2. Experimental measures of fit sizes (brushed features). (All values are in mm)

Part	No.	Dim.	Descript.	Nominal size	Actual size	Lower dev.	Upper dev.	Round-ness	Std.dev.	Virtual size
Base	1A	a	Inner Dia	28	28,147	-0,145	0,314	0,459	0,065	28,002
Shell	2A	b	Lower Outer Dia	28	27,697	-0,152	0,179	0,331	0,050	27,876
"	"	c	Lateral Outer Dia	30.8	30,492	-0,161	0,278	0,439	0,059	30,770
Cover	3A	d	Lateral Inner Dia	31.2	31,300	-0,127	0,150	0,277	0,475	31,172

Table 2 shows that a slight removal of the superficial partially adherent particles using a copper wire brush varied the actual sizes of the handled features of about 0.24 mm, obtaining mating sizes similar to the nominal values (therefore restoring the nominal fit conditions: 1Aa-2Ab = 0.127 mm; 3Ad-2Ac = 0.403 mm). The effect of such operation on the form errors cannot be estimated by such a preliminary test (average roundness error = 0.376 mm).

5 CONCLUSIONS

The implementation of suitable collaborative methodologies for the generation of a voice prosthesis requires also the verification and the set up of viable innovative manufacturing processes based on RP and RT techniques. Some preliminary results on the fit conditions of a composite mold manufactured using Rapid Prototyping (RP) techniques have been presented that tentatively indicate the suitability of the proposed process to the direct development of custom devices.

REFERENCES

1. De Carpentier JP, Ryder WD, Saeed SR, Woolford TJ. Survival times of Provox valves. J Laryngol Otol 110(1): 37-42, 1996.
2. Smitheran JR, Hixon TJ. A clinical method for estimating laryngeal airway resistance during vowel production. J Speech Hear Dis 46:138-146, 1981.
3. Miani C, Bellomo A, Bertino G, Staffieri A, Carello M, Belforte G. Dynamic behavior of Provox and Staffieri prostheses for voice rehabilitation following total laryngectomy. Eur Arch Otorhinolaryngol 255(3):143-148, 1998.
4. Callaway E, Truelson JM, Wolf GT, Thomas-Kincaid L, Cannon S. Predictive value of objective esophageal insufflation testing for acquisition of tracheoesophageal speech. Laryngoscope 102:704-708, 1992.

5. C. Miani, A. M. Bergamin, A. Staffieri, S. Filippi, F. Miani, M.Zanzero, Experiences in Rapid Prototyping: voice devices for patients who have undergone total laringectomy", AMST'99 Advanced Manufacturing System and Technology, E. Kuljanic ed., Springer, Wien, 573-582, 1999
6. PTC, Pro/ENGINEER 2000i2 Documentation, 2001
7. Tay F.E.H., Haider E.A.,Laser sintered rapid tools with improved surface finish and strength using plating technology, J. Mat Proc Tech, 121:318-322, 2002

ANALYSIS OF GEOMETRICAL DISTORTIONS OF SPUR SHAFTS DUE TO HEAT TREATMENTS

L. D'Angelo - Department of Engineering, University of Ferrara, Italy

KEYWORDS: Heat Treatments, Geometrical distortions, Design Of Experiments (DOE).

ABSTRACT. A multidisciplinary study is being conducted on noise emission of spur gear pumps. Primary noise sources, as for instance pump body vibrations, have been detected through measurements in a tailor-made anechoic chamber. Vibrations are in turn heavily influenced by geometrical defects of shafts, often arising during heat treatment and surface hardening.
After a short presentation of the industrial process to manufacture spur gears, the paper describes in more detail (i) the selection of hardening parameters to be monitored, and (ii) the experimental procedure adopted, the focus being on the search for correlations between the selected geometrical defects and process parameters. Due to the large number of process parameters taken into account, Design Of Experiments (DOE), together with available information and company know how, have been used to reduce the required number of tests. Some preliminary results are surveyed at last.

1 INTRODUCTION

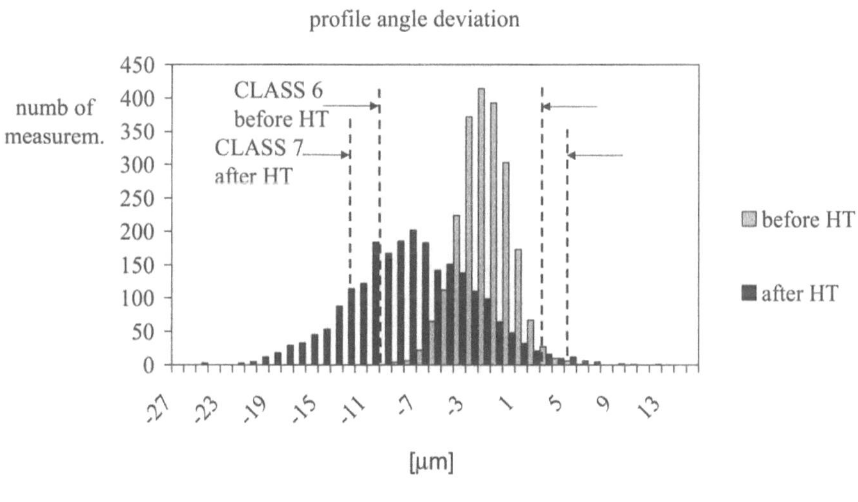

FIGURE 1. Deviation of the shaft profile angle due to the heat treatment

The work presented is derived from a degree thesis [1] developed in cooperation with TRW-Marzocchi, a company producing gear pumps for power steering equipments. In this category

Published in: E.Kuljanic (Ed.) *Advanced Manufacturing Systems and Technology*, CISM Courses and Lectures No. 437, Springer Wien New York, 2002.

of products, very tight tolerances are used, because dimensional deviations of gears dramatically reduce pump efficiency. In addition, experimental tests have shown that deviations also have direct influence on noise level of pumps which is one main problem in steering equipments. In a previous thesis on a more general subject [2], dimensions of 50 shafts are compared before heat treatment and at the end of the manufacturing process. It appears that final dimensions are more scattered and, in some cases, they are so shifted that parts must be discarded, being out of the specified allowances. One case when both these effects are evident is shown in Fig. 1. These results suggest that it could be useful to know in more detail the geometrical effects of heat treatment in order to reduce these distortions. If the geometrical distortions were repetitive, as confirmed in [3], it would also be possible to compensate for them in the preceding manufacturing operations.
Several studies can be found in the technical literature on residual stresses and distortions due to heat treatments (see, for example, refs. [4 – 9]), often presenting new material models. In [8] a list is presented of commercial codes suitable for such analyses and, for each code, main peculiarities are highlighted. It seems that studies are usually quite specific in terms of (i) material, (ii) geometry and (iii) treatment, and it is not clear at what degree the presented results can be extended to different situations. In addiction, "the accuracy of the results depends of course on the accuracy of the input data, the major difficulty seems to be the description of the heat treatment process; that is, the accurate prediction of the temperature distribution in a piece" [10].
The main objective of this work is the reduction of distortion in components like those in Fig. 2. Uncertainties related to (i) extendibility of material models found in the literature, together with (ii) the difficulty to evaluate the initial temperature distribution inside the furnace charge and inside the single parts have suggested to give up with the numerical approach. A strictly experimental procedure have been used to solve the problem, assisted by statistical considerations.

2 THE MANUFACTURING PROCESS

The shafts of the pump are shown in Fig. 2, the driven shaft on top, the driving at bottom; this work has been addressed to the study of geometrical distortions in the driving shaft because it comprises all the geometrical features of the driven. The material is UNI 18 Ni Cr Mo 5 Pb, a case-hardening steel in which lead is added to increase its machinability. The manufacturing process is made of the following steps:

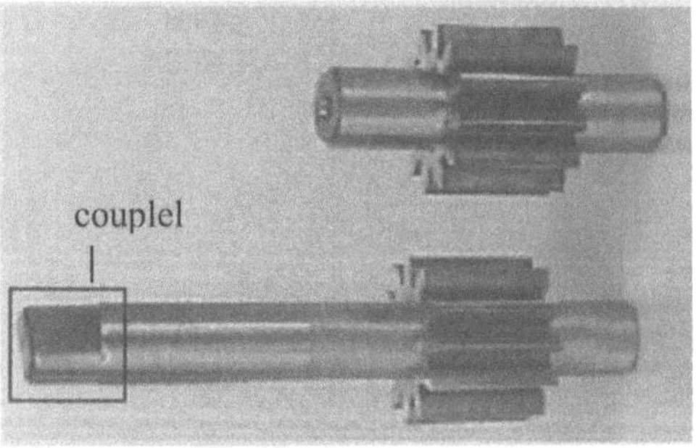

Figure 2. The pump shafts

- turning,
- teeth cutting,
- heat treatment and carburise case-hardening,
- grinding and superfinishing.

Turning. It is made on a six-mandrel GM 35 AC Gildemeister lathe. In the first part of the process the longer shank is rough turned, finished and dimpled. The shorter shank is also rough turned. The following step is milling one end of the shaft, to obtain parallel surfaces; this is necessary for coupling to the electrical engine. In the rest of the paper, this area will be called *couplel*, contraction of coupling element (see also Fig. 2). Then teeth flanks are machined and the shorter shank is finished, end cut and dimpled. After turning, dimples eccentricity is measured.

Teeth cutting. A CNC Pfauter Mikron hobber is used to cut shaft teeth. AAA class coated carbide hobs are used to this aim, up milling is preferred for better finishing teeth.

Heat treatment and case-hardening. Before heat treatment, shafts are washed and dried, so that hardening is more homogeneous during heat treatment. Shafts are then case-hardened and quenched. This process will be analysed in more detail in the next section. It is followed by hardness testing and metallographic analysis by means of optical microscopy. Afterwards, shafts runout is measured and reduced by an automatic PAS – E/6 straightener.

Grinding and superfinishing. Teeth flanks and shanks are grinded to Ra = 0.2 μm, to improve coupling with bushings and, therefore, to increase the volumetric efficiency of pumps. Areas where shanks are mated to bushings are superfinished to reach Ra = 0.16, in order to improve tightness.

3 HEAT TREATMENT AND CASE-HARDENING

Shafts are vertically positioned on grids called blisters which, in turn, are stacked to form pallets. A charge is made of two pallets of six blisters each, 1800 shafts in total. The treatment is made up of four steps (see Fig. 3): preheating, case-hardening, quenching and recovery. The steps 1 – 3 take place in the same furnace. The inner atmosphere is made up of liquid methanol, which determines the hardening environment, methane and nitrogen. Methane is a secondary hardener component. Methanol and methane decompose to CO and H_2.

Preheating. It takes place at about 850 °C for 30 minutes. The carbon content in the furnace atmosphere is kept low so that shafts heat more evenly.

Case hardening. This step is the longest, 100 minutes, and requires the highest temperature, 990 °C. The carbon diffusion inside austenite is also determined by the carbon content in the furnace atmosphere, which is kept at 0.84%. The aim is to obtain an hardened depth of 0.6 mm.

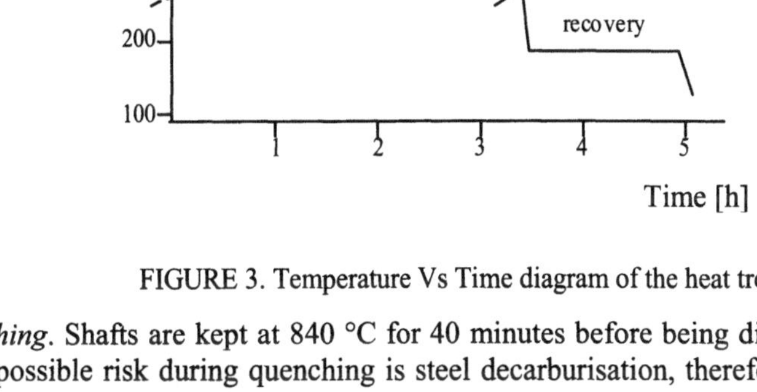

FIGURE 3. Temperature Vs Time diagram of the heat treatment

Quenching. Shafts are kept at 840 °C for 40 minutes before being dipped in a oil bath at 110 °C. A possible risk during quenching is steel decarburisation, therefore carbon content in the

furnace atmosphere must be kept under control. In this case a percentage of 0.78% is chosen. After quenching, shafts are again washed and dried, similarly at what was done before heat treatment, to remove dust residuals on the surface.
Recovery. This step is quite time consuming, shafts are kept at 180 °C for 90 minutes and it takes place in a different furnace.
In the sensitivity analysis explained at §5, 8 process parameters have been taken into account to study geometrical distortion of shafts. They are, as follows:

- preheating time (t_1) and temperature (T_1),
- hardening time (t_2), temperature (T_2) and carbon percentage in the furnace atmosphere (C_2), and,
- quenching time (t_3), temperature (T_3) and carbon percentage in the furnace atmosphere (C_3).

4 MEASUREMENT

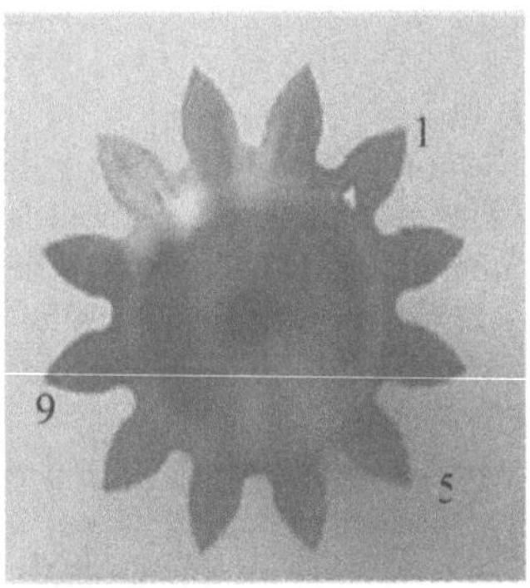

FIGURE 4. Cross section of the gear

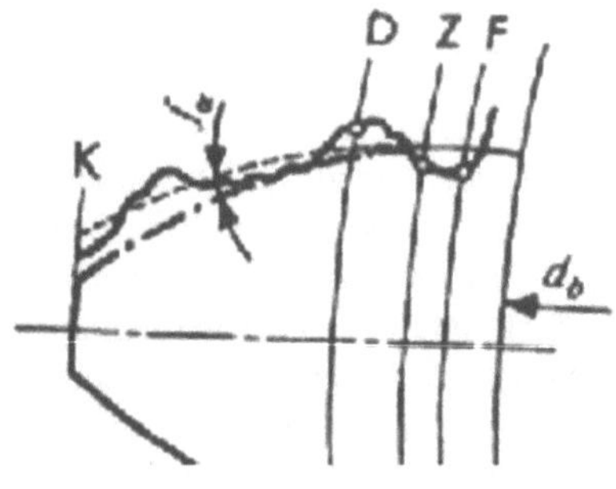

FIGURE 5a. Tooth profile

The (i) geometrical parameters taken into account for evaluating distortions as well as the (ii) measuring procedure and (iii) tolerances are based on DIN norm 3962 [10]. In particular, TRW gears must be in the 6th quality class before the heat treatment and in the 7th after the treatment (see also Fig. 1). The couplel thickness is measured on a digital micrometer; for all the other parameters it is used a Zeiss Prismo Vast 5 HGT Coordinate Measuring Machine (CMM) with 4 CNC axes, especially equipped for rotational parts. The software is GON-UX 8.0, specific for the measurement of spur gears. Parameters are both measured in the cross section of

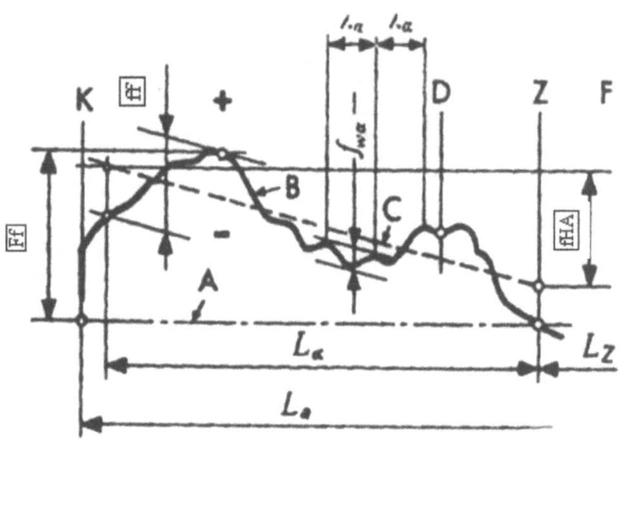

FIGURE 5b. The straightened tooth profile

the gear (Fig. 4) and in the longitudinal section of the shaft.
The measured parameters are, as follows:

- couplel thickness,
- shaft runout,
- step between each pair of adjacent teeth, both on left and right side, and,
- for 3 teeth (see Fig. 4) their
  1. thickness,
  2. left and right profile, and
  3. helix (alignment of tooth axis with shaft axis).

In particular, the measurement of both the tooth involute profile and helix requires the evaluation of three parameters. In Fig. 5a the nominal profile is the bold dash-dot line, the continuous irregular curve is the real profile and the thin dashed curve is the involute which best approximates the real profile. In Fig. 5b the nominal profile is straightened into an horizontal line and the other two lines are modified accordingly. Three parameters are inside rectangles for highlighting. ff is a measure of the shape deviation, fHA express the slope error of the real profile, Ff is the total deviation, it takes into account both errors but it is not the arithmetic sum of ff and fHA. Helix errors evaluated using the same procedure, the relevant parameters being fBf, fHB and FB.

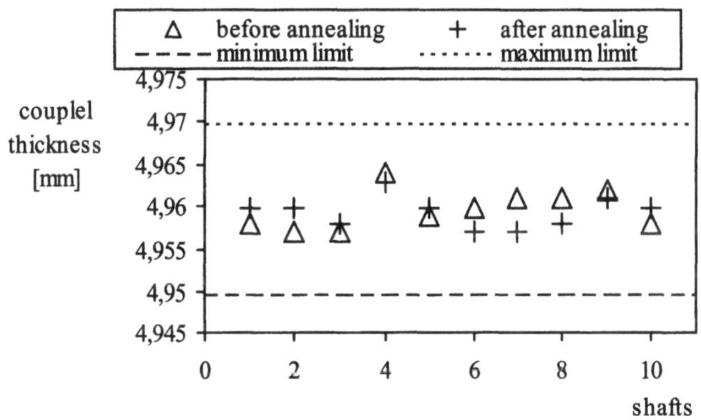

FIGURE 6. Comparison of dimensions before and after anneal-

5 PROCEDURE

As the first step of the procedure, it has been verified that geometrical distortion measured after the heat treatment is only due to that process. A doubt can arise, that machining operations induce distributions of residual stresses which only become apparent after stress recovery. For this reason, 10 shafts have been measured after gear cutting, annealed at 600 °C for 90 minutes and then measured again. Results are similar to that in Fig. 6, where no appreciable dimension changes can be seen. It is possible to conclude that, to study geometrical distortions, it is not necessary to anneal shafts before the heat treatment.
To tackle the study, due to the large number of process parameters involved, it has been decided to divide the analysis into two parts; the former (*screening analysis*) is addressed to determine the process parameters which mostly influence distortions, the latter (*optimisation*

STEP	T_{MAX} (°C)	T_{MIN} (°C)	T_{MAX} (min)	T_{MIN} (min)	$\%C_{MAX}$	$\%C_{MIN}$
preheating	T_{1A}= 870	T_{1B}= 850	t_{1A}=40	t_{1B}=20		
hardening	**T_{2A}=900**	**T_{2B}=870**	**t_{2A}=110**	**t_{2B}=90**	**C_{2A}=0.88**	**C_{2B}=0.8**
quencing	**T_{3A}=840**	**T_{3B}=820**	t_{3A}=40	t_{3B}=30	C_{3A}=0.8	C_{3B}=0.75

TABLE 1. Values assumed by process parameters in the test, bold is used for the parameters selected for the fractional analysis

analysis) is devoted to quantify relations (when existing) between main process parameters and geometrical distortions. This work presents some results of the screening analysis.
As anticipated at §3, eight process parameters have been taken into account. For this kind of study, the Design Of Experiments (DOE) theories [11] suggest a factorial analysis where, for each process parameter, two values are chosen at the extremes of the range to study, showed in Tab. 1. For a complete sensitivity analysis, to study also the influence of parameter combinations, 2^8 (=256) furnace charges would be necessary. To shorten the task, a fractional factorial 2^{8-4} (=16) design has been used. In this case it is necessary to choose four parameters, out of the eight previously mentioned, which are expected to have larger influence on distortions. The selection is usually based on a specific know how of the process to be studied. The selected parameters are the hardening temperature T_2, the hardening time t_2, the hardening carbon percentage C_2 and the quenching temperature T_3. Each one of the selected parameters is combined with all the others; this is not true for the remaining. When a fractional factorial 2^{8-4} design is developed, it is not possible to differentiate between some effects of the second order, which are said to be aliased.

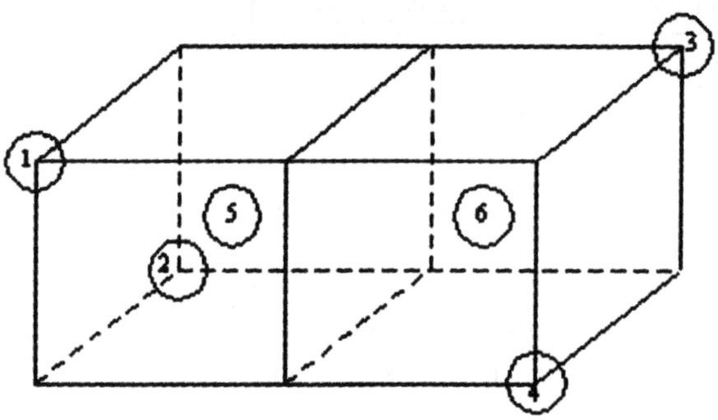

FIGURE 7. Positions of measured shafts

Another process parameter has been taken into account, which does not affect the test length: the position of shafts in the charge. To this aim, 6 positions have been chosen, 4 at charge vertices and 2 in the middle of the two pallets, as shown in Fig. 7. For each position 4 shafts have been measured and the mean values have been used in the analysis.
Another statistical method used in the course of the study is the *two-sample t-test*.
The two-sample t-test is useful when it must be decided if the differences between two samples are significant. It is based on the evaluation of the parameter given by the formula below:

$$Z_0 = \frac{\overline{X} - \overline{Y}}{\sqrt{\frac{\sigma_X^2}{n_x} + \frac{\sigma_Y^2}{n_y}}} \qquad (5.1)$$

where $\overline{X}$ and $\overline{Y}$ are the sample means, σ_x and σ_y are their variances, n_x and n_y are the sample sizes.
According to this method, one dimensional change is significant when, under the hypothesis of normal distributions, the Z_0 parameter is outside a predefined interval.
This method has been used, together with the comparison of dimensions with the relevant allowances, to evaluate dimension changes due to (i) the annealing (see also Fig. 6) and to (ii) the heat treatment.

charge	couplel thickness	runout	fHA	fHB
1	1	1,5	1	1,2,3,4
2	2	1,5,6	1,5	
3	5	3,5,6		6
4	5	1,2,6		1,2,6
5	1	1,2,5		1
6	1,3,5,6	5,6		
7				6
8	3,5	1,4		
9		1,3,5,6		6
10		1,3,6		
11	5	2,5		
12		1,4		
13		1,2		
14	5	1		
15	2	1,4,3	5	4,3
16		5		1

TABLE 2. Significant distortions related to charges and positions

6 RESULTS

The reported results are relevant to the influence of (i) the part position and (ii) the charge parameters on distortions. Results are summarized in Tab.2. Apart from the first one, each column is dedicated to present distortions of a specific parameter. The first important result is that only four dimensions out of the nine measured show significant distortions in some cases. The term "significant" must be intended in the sense given by the formula (5.1). The four dimensions are the couplel thickness, the shaft runout, and the slope error of both profile and helix. Tab. 2 shows the list of significant distortions in relation to charges and positions. It seems that the runout is by far the geometrical element most sensitive to thermal distortions, which are significant in 36 cases out of 96. It can also be observed that the positions in the charge have different influence on distortions. Position 1 appears 20 times, position 4 only 4 times. The attempt to understand the reasons of these results will be a matter for further investigations. In a similar way, the number of significant distortions is very different in the 16 charges, with a maximum in the charge nr. 1 (8 distortions) and a minimum in the charge nr. 7 (one significant distortion). Charges 7, 14 and 16 seems very promising because few significant distortions have been found. They are all possible candidates as starting points for the refining analysis. To select which one of the three is the best, a complementary investigation is being conducted to know the hardness profile of components.

7 CONCLUSIONS

The Department of Engineering of the University of Ferrara and TRW- Marzocchi of Ostellato (FE) are collaborating to reduce the distortions of shafts due to the thermochemical treatment. A strictly experimental procedure has been chosen, assisted by statistical methods to reduce the test length. The whole activity has been divided in a screening analysis and a refining analysis. The paper reports some results of the former, aimed at identifying the process parameters which most influence distortions. In particular, it has been shown that (i) only 4 out of 9 geometrical parameters are affected by significant distortions during the treatment, (ii) significant distortions take place more often in some positions within the furnace charge than in others, and (iii) the combination of treatment parameters heavily influence the number of significant distortions.

ACKNOWLEDGEMENTS

The whole management of Trw-Marzocchi is gratefully acknowledged for the financial support and for the active collaboration throughout this work. A special thank to Mr. Lomi for his competent suggestions on items relevant to the heat treatment and for his enthusiastic participation.

REFERENCES

[1] Balboni, S., (2002), Analysis of geometrical distorsions induced on spur gears during carburise case-hardening treatments. Degree Thesis dissertation at Univ. of Ferrara (in Italian).

[2] Castellani, D., (2000), Experimental investigation on effects of geometrical defects on noise of a gear pump. Degree Thesis dissertation at Univ. of Ferrara (in Italian).

[3] Clarke, P.C., (1998), Close-tolerance Heat Treatment of Gears, *Heat Treatments of Metals*, Vol. 1998, n. 3, p. 61-64.

[4] Liu, C.C., et al., (1998), Effect of stress on transformation and prediction of residual stresses, *Materials Sciences & Technology*, Vol. 14, p 747-750.

[5] Thuvander, A. et al., (1998), Simulation of Heat Treatment Response and Distortion of Bearing Steels, *Bearing Steels : Into the 21st Century*, ASTM STP 1327, J.J.C. Hoo and W.B. Green, Eds, p. 265-278.

[6] Ehlers, M., Müller, H., and Löhe, D., (1999), Simulation of stresses, residual stresses and distortion in stepped cylinders of AISI 4140 due to martensitic hardening by immersion cooling, Journal de Physique IV, Proc. of Euromech-Mecamat'98, p. 333-340.

[7] Ericsson, T., (2000) Residual Stresses Induced by Heat Treatment and Thermochemical Treatment, *20th ASM Heat Treating Society Conference Proceedings*, p. 392-401.

[8] Ferguson, B.L., Freborg, A.M. and Callabresi, M.L., (2000), *20th ASM Heat Treating Society Conference Proceedings*, p. 79-86.

[9] Denis, S and Simon, A., (1996), Modeling of Heat Treatment of Steels: From Concepts to Process Simulation, Proc of the 2nd Conference on Quencing and the Control of Distorsion, p. 239-245.

[10] AA. VV., (1987), Tolerances for Cylindrical Gear Teeth, DIN norm 3962.

[11] Montgomery, C.D., (1997), Design and Analysis of Experiments, J. Wiley & Sons.

BENCHMARKING OF ATTRIBUTE TWO-LEVEL SAMPLING SCHEMES WITH ZERO ACCEPTANCE NUMBER

M. Dassisti [1]

[1] Dipartimento di Ingegneria Meccanica e Gestionale , Politecnico di Bari - Bari, Italy

KEYWORDS: attribute sampling schemes; zero acceptance number; skip-lot.

ABSTRACT. The paper presents a benchmarking method for attribute sampling schemes, with specific reference to zero acceptance number. The method decision making in selection of appropriate sampling schemes, in the case of very high level of vendor's quality. Closed form expressions are provided for a single-level and a two-level sampling scheme. The aim of the paper is to provide an useful and easy-to use tool for benchmarking attribute sampling schemes, with zero acceptance number, to be adopted in the daily practice. The potentiality to infer conclusions up to more complex schemes is argued.

Notation

AQL = poorest level of quality for the vendor's process that the consumer would consider to be acceptable as a process average.

α = producer's risk

B(f) = benchmarking function

c = acceptance number, integer $c \geq 0$

C_{index} = cost of lot inspection in the indexed plan

f = sampling fraction , with $0<f<1$

i = clearance number

index (foot letter)= R=reduced, Sk = skip-lot, N=normal; LS=less-severe, N-R = normal-reduced scheme; SkSP = skip-lot sampling plan; N-Sk = normal - skip-lot sampling scheme

L = indirect cost for inspecting a whole sample

m = sample size for the reduced plan

n = sample size for the reference plan, integer $n>0$

n_0 = sample size for the normal plan, integer $n_0 > 0$

N = lot size

p = lot fraction defective, with $0 \leq p \leq 1$, characteristic of the incoming lot quality from vendor (supplier)

P = probability of lot acceptance of the reference plan for the skip-lot plan

P'_{index} = probability of lot acceptance of the indexed plan, with $c=0$

P_{index} = probability of lot acceptance of the indexed plan

Q_{index} = probability of rejecting a lot

U = direct cost per unit item inspected belonging to a sample

X_{index} = parameter representative of the average time spent on the indexed plan

Published in: E.Kuljanic (Ed.) *Advanced Manufacturing Systems and Technology*, CISM Courses and Lectures No. 437, Springer Wien New York, 2002.

1. INTRODUCTION

Despite the supplier integration strategy strongly promoted by the Total Quality Management approach, the use of acceptance inspection still remains an recurrent practice, commonly adopted to prevent non-quality costs downward the production process.

Main strategies for inspecting incoming lots of products, and thus vendor's lot sentencing, are: ST1) 100% inspection of lots (complete control of N items); ST2) acceptance sampling by inspecting a sample of n items, which can be performed by attributes or by variables; ST3) no inspection (free-pass). Strategies ST1 and ST3 represents the two extremes, being ST3 the ideal case; strategy ST2, the most adopted one, lies in between.

In the paper we will focus only on ST2 acceptance sampling by attributes, where a unit of product is classified as conform or non-conform. Acceptance sampling by attributes can be implemented using several different sampling schemes, which present their pro's and con's; amongst others [1]: lot-by-lot single or multiple sampling plans; lot-by-lot normal, tightened and reduced with appropriate switching rules; Dodge-Roming's sampling plans; lot-by-lot sampling plans with skipping (also called skip-lot SP or SkSP); continuous SP (CSP).

Deciding the most appropriate scheme to adopt is strongly related to the specific operating reality, being the number of boundary conditions to be considered high. Choices amongst different schemes is not, in fact, only a matter of statistical reasoning; it might depend, also on the cost of parts, vendor's quality level, suppliers relationship strategies; etc.
Quite often in TQM environments, depending also on decision of the authority responsible for sampling, many times a ST3 strategy is adopted. The skip-lot scheme responds to this tendency, being the inspection performed on a randomly selected fraction (f) of samples extracted from the incoming lots. In order to set a skip-lot sampling plan (SkSP) the confidence on the supplier's quality level should be high: usually this decision is taken after a clearance of i samples. Nevertheless, appropriate sensitivity of the inspection scheme adopted should be set, namely by appropriately selecting the value for the acceptance number (c). In the industrial practice, several conditions hold for employing acceptance sampling plans with zero acceptance number. Sampling plans with zero acceptance number is usually adopted in the area of safety-related items or in situation of costly nature of testing, or whenever there is the desire to maintain a steep operating characteristic (OC) curve [2]. A sampling plan with zero acceptance number has poor statistical performances; this is the reason why Norms strongly do not recommend their use [3]. Nevertheless, it is also possible to find the use of sampling plans with zero acceptance number still in the industrial practice, whenever very huge incoming lots are treated, as a compromise between sample size reduction versus non-conformance risks. The benchmarking method proposed in the paper will focus on sampling schemes with zero acceptance number.

2. THE BENCHMARKING METHOD PROPOSED

The paper aims at presenting a method to compare acceptance sampling schemes, with specific reference to zero acceptance sampling ($c=0$). The following two conditions are the key points

(KP) of the benchmarking method proposed, to provide a clear evidence in comparing the performances of two sampling schemes:

KP1) the same acceptance probability at the Acceptance Quality Level (AQL) should be imposed, i.e. the plans compared have the same acceptance probability (producer's risk, α) at $p=AQL$. This assumption make sense in the light of comparison two sampling schemes; it also cope with the primary focal point of design of many acceptance sampling systems [1].

KP2) the properties of a generic sampling scheme can be inferred from the benchmarks of less-severe individual sampling plan belonging to it, provided that the other plans of the scheme to be compared have the same features.

The method proposed focuses on the performance of the less-severe individual plans belongings to a generic sampling scheme, being oriented to the use in the case of in case of very high level of incoming lot quality. The less-severe (LS) plans here addressed, corresponding to a vendor's production being at steady rate, are: a reduced (R) plan, where a lower sample size (m) than the normal one (n_0); a skip-lot (Sk) plan, where only a fraction f of lots of the reference plan (n) are inspected [4]. Two acceptance schemes will be considered here for benchmarking purposes, for the sake of simplicity, namely: a Normal-Skip-lot scheme (corresponding to the well known SkSP-2 scheme [1]); a Normal-Reduced (NR) plan, as in Figure 1.

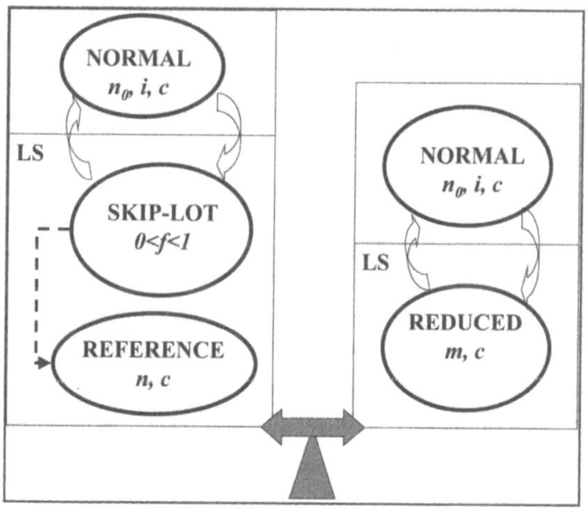

FIGURE 1. N-Sk sampling scheme vs. N-R scheme considered.

The following switching rules between plans for the generic acceptance scheme will be adopted:

R1) start with the normal inspection;

R2) According to the standard inspection schemes, such as the ISO 2859, the less-severe sampling inspection is instituted after a normal inspection whenever the preceding i lots passed a normal inspection, with a total number of defectives in the sample less or equal to the applicable limit number. Thus, if and only if i consecutive lots are accepted on normal inspection, switch to the less-severe inspection; otherwise restart counting accepted lots from zero in the normal inspection;

R3) remain into the skipping phase until a single lot is rejected, in this latter case switch to normal inspection;

R4) screen each rejected lot and correct or replace all defective units found

The following hypotheses will be adopted for the rest of the paper:

H1) Lot acceptance conditions assumed are that the number of nonconforming unit is below or equal to the acceptance number

H2) producer's risk α will be assumed constant; also the lot fraction defective (p) will be supposed to be constant for the benchmarking purposes.

H3) lots to be inspected are selected at random: results of lot inspection will be thus stochastically independent from each other. The same hypothesis will be made for the items within a sample.

3.1 PERFORMANCE OF INDIVIDUAL SAMPLING PLANS

In the present paragraph properties of individual sampling plans will be derived, in the case of zero acceptance number ($c=0$). These will be used later for benchmarking purposes of acceptance schemes. From a statistical point of view, several measures of performance can be adopted to compare different sampling plans; these are namely [5]: Operating Characteristic curves (OC); Acceptable Quality Level (AQL), defined as the maximum proportion of variant unit is a lot that, for purpose of acceptance sampling, can be considered satisfactory at process average; the Average Sample Number (ASN), which has an economical significance. Most of the formulae derived are well assessed ones, but will be presented in a simpler and more attractive form, revisited for benchmarking purposes.

3.1.1 THE OPERATING-CHARACTERISTIC CURVE

The discriminating power of a sampling plan is portrayed by its operating characteristic (OC) curve, which maps the acceptance probability over the percentage of defective (p). In this paragraph we derive expression for evaluating the probability that a lot will be accepted, given a generic fraction of nonconforming p. As concerns the skipping phase, only a fraction f of all the incoming lots are inspected; other plans will have $f=1$. According to H3, two generic events of selecting a lot for inspection (S) will result stochastically independent. The probability of rejecting a given inspected lot (say D the event of rejecting a lot) thus results from the definition of conditional probability:

$$Q_{plan} = P(D|S) = P(D) \cdot P(S). \quad (1)$$

According to the sampling rule before mentioned, P(S) = f holds in the case of skip-lot plan, while P(S)=1 otherwise (normal and reduced plans). The probability of rejecting an inspected lot, for a given probability of acceptance P_{plan} of a plan adopted is:

$$P(D) = (1- P_{plan}) \quad (2)$$

By using (1) the operating characteristic curve for the skip-lot plan can be easily derived, where $P=P(n,c)$ for the reference plan, as

$$P_{SK} = 1-Q_{SK}= 1 - (1-P) \cdot f \quad (3)$$

Formula (3) gives the first measure of performance for a given skipping phase, provided the P is defined for the reference plan. The operating characteristic curve, in fact, can be derived and studied mapping of P_{SK} versus the process fraction defective (p), which is a vendor's characteristic. In the case of reference plan with acceptance number set to zero ($c=0$), say n the sample size; the probability of acceptance thus descends from the binomial probability distribution as $P=(1-p)^n$.

In this case, it is thus easy to derive the OC curve for the skipping phase, for $c=0$ from (3) as:

$$(P_{sk})_{c=0} = P'_{sk} = 1-[(1-(1-p)^n)\cdot f] \quad (4)$$

while, for the reduced plan with $c=0$, descends:

$$P'_R = (1-p)^m \quad (5)$$

and, for the probability of acceptance for normal plan with c=0:

$$P'_N = (1-p)^{n0} \quad (6)$$

being, generally, $m<n_0$. The OC curves should be compared with the ideal characteristic curve, that is defined as [1]:

$$P_{id} = 1 \quad for \quad 0 \leq p < AQL; \qquad P_{id} = 0 \quad for \quad p \geq AQL \quad (7)$$

3.1.2 THE AVERAGE RUN LENGTH

The ARL is a measure of the sampling plan's responses to a change; it is defined as the average number of lots to the occurrence of a rejection [6]. It is easy to show, see also [7], that for the normal plan– where there is no switch to the less-severe plan as far as i consecutive lots are accepted – holds:

$$ARL_N = (1-P_N{}^i)/[(1-P_N)\cdot P_N{}^i] \quad (8)$$

In the less-severe plan, as far as a defective lot is found, a normal inspection scheme is instituted immediately. For this plan, being P_{LS} the probability of lot acceptance of the less-severe phase, it is:

$$ARL_{LS} = 1/[1-P_{LS}] \quad (9)$$

According to (9), the equation for the reduced plan with c=0 it is:

$$ARL_R = 1/(1-P_R) \quad (10)$$

and, for the skip-lot plan with c=0 it is:

$$ARL_{SK} = 1/[(1-P)f] \quad (11)$$

being P the probability of acceptance in the relative reference plan.

3.1.3 THE AVERAGE SAMPLE NUMBER

The Average Sample Number represents the expected total sample size; it is a indirect measure of the sampling costs. Given the notation above used, for the individual sampling plans it is easily to see that:

$$ASN_N = n_0$$

$$ASN_R = m \qquad (12)$$

$$ASN_{Sk} = n\,f$$

4. BENCHMARKING OF INDIVIDUAL LESS-SEVERE PLANS

The criteria here suggested for benchmarking sampling schemes bases on comparing their less-severe plans. Say U the cost per part sampled (we will refer to as the 'direct cost' of inspection, being proportional to the number of items inspected) and L the cost of inspection for the whole sample (we will refer to as the 'indirect cost' of inspection, being independent from the sample size), which are expressed into appropriate unit of measure.

Two cases will be here addressed, whether only the direct costs of sampling items are significant (§4.1) or the indirect costs become significant (§4.2).

4.1 INDIVIDUAL SAMPLING-PLAN PERFORMANCES: NEGLIGIBLE INDIRECT COSTS

According to the above, whether L is negligible the performance comparison reduces to comparing only the number of inspections to be performed, which easily descend from statistical measures derived in §3. The two generic acceptance plans here addressed are: a reduced lot-by-lot SP and skip-lot sampling plan.
Let's first analyze the OC curves of the two plans at $p=AQL$. Given the KP1 conditions in §2, we impose the two plan to have the same acceptance probability at $p=AQL$; thus from (4) and (5) it results:

$$P'_{sk} = 1- [1-(1-p)^n] \bullet f = 1-\alpha = P'_R = (1-p)^m = 1-\alpha \quad \text{for} \quad p=AQL \qquad (13)$$

where, again, n represents the sample size of the reference plan of the skipping phase, in the case of $c=0$, and m the sample size for the reduced plan to be compared.

It is possible to derive the following expression, by replacing p with AQL in (13):

$$1- [1-\alpha/f]^{(1/n)} = 1- [1-\alpha]^{(1/m)} \qquad (14)$$

From eq. (14), descends the expression of the average number of samples inspected for the skipping phase (in the case of $c=0$) :

$$ASN_{sk}=f\, n = f \bullet m \bullet [\, ln\,(1-\alpha/f) \,/\, ln(1-\alpha)] = f \bullet m \bullet A \qquad (15)$$

being the parameter A expressed as

$$A = ln\,(1-\alpha/f) \,/\, ln(1-\alpha) \qquad (16)$$

It is easy to see that A is always greater than 1, being always $0< \alpha/f = 1-(1-p)^n <1$ from eq. (13) imposed according to KP1 benchmarking condition. As a consequence, it is possible from (15) to derive m of the reduced plan. From this descends the first statement:

ST1) the reduced plan of inspection has always a lower sample size than a skip-lot plan, with the benchmarking assumptions made in §2.

This conclusion remains true independently of the specific *AQL* value, as far as a individual sampling plans are considered, provided also that $0<f<1$.
Let's now consider the behaviour of sampling plans for all the other points $p \neq AQL$, by considering the following function ($c=0$):

$$D(p) = P'_R - P'_{sk} = (1-p)^m - 1 + (1-(1-p)^n) \cdot f \tag{17}$$

$D(p)$ is calculated as the difference between the acceptance probabilities of reduced plan and skip-lot plan. It is easy to see that $D(p=0) = 0$, $D(p=1) = f-1$ and also that is $D(p=AQL)=0$.

By a simple analysis of the first derivative of $D(p)$, it can be seen that:

$$\begin{aligned} D(p) &> 0 \qquad \text{for } 0<p<AQL \\ D(p) &< 0 \qquad \text{for } AQL<p<1 \end{aligned} \tag{18}$$

This means that the OC curve of the reduced plan tends to be closer to the ideal OC curve of (6) than in the case of skip-lot plan.
As a conclusion, in the case of $c=0$, it can be stated that:

ST2) the individual reduced plan always outperforms in term of OC curve the individual skip-lot plan of inspection, under the benchmarking conditions put as in §2.

Again, statements ST1 and ST2 holds whenever the sampling cost per unit prevails over other cost figures (i.e. whenever the only benchmarking criteria is the sample size).

4.1 INDIVIDUAL SAMPLING-PLAN PERFORMANCES: NON-NEGLIGIBLE INDIRECT COSTS

Indirect cost might be , for instance, set-up costs for testing instruments, transportation costs; etc. Whenever these are no more negligible with respect to the direct costs the following reasoning should be made.
The total cost of inspection per each lot, in the case of the reduced plan, results (remember eq. (12)):

$$C_R = ASN_R \cdot U + L = m \cdot U + L \tag{18}$$

As concern the total cost per lot for the skip-lot plan, from (15) and remembering the meaning of *A* in (16), the following expression holds:

$$C_{sk} = ASN_{sk} \cdot U + f \cdot L = f \cdot m \cdot A \cdot U + f \cdot L \tag{19}$$

Remembering that, under the benchmarking conditions made, $f\,A$ is always greater than 1 (being $0<\alpha/f<1$), it results that the term ($f\,m\,A\,U$), corresponding to direct cost of inspection, is always greater than U. On the other hand, being always $0<f<1$, the term $f\,L$ is always lower than L. The skip-lot plan thus results to increase the direct cost of inspection (proportional to the number of items inspected) while reduces the indirect costs of inspection. Benchmarking the two acceptance plans means, easily, to compare C_R versus C_{Sk}. To this aim, the following difference function can be used:

$$B(f) = (C_R - C_{Sk}) / (m \cdot U) = (1 - f\,A) + (1 - f)\; L/(m \cdot U) \tag{20}$$

where, again, A is expressed as in (16). By studying the function in (20) for given α and n, at different levels of the ratio between indirect and direct costs L/U, it can be seen that skip-lot perform better as far as this ratio increases. Graphs will be omitted here; these can be easily derived to draw appropriate conclusions, depending on the specific values of L, U, f and n, being α derived from condition KP1 as in (13), and m from eq.(15), taking the first integer greater than the reckoned value. Remembering the properties of fA and that $m>1$, it can be stated that:

$$B(f)<0 \quad when \quad L/U < m \bullet (fA-1)/(1-f), \quad i.e.\ L<U$$
$$\text{and} \tag{21}$$
$$B(f)>0 \quad when \quad L/U > m \bullet (fA-1)/(1-f), \quad i.e\ L>U.$$

In general, it is possible to state, perfectly according to the results obtained in §4.1, that

ST3) the reduced plan outperforms the skip-lot whenever the indirect cost of inspection (*L*) is lower than the direct unit cost of inspection (*U*). Vice-versa is true for the skipping plan, again under the benchmarking conditions put in §2.

5. BENCHMARKING OF A TWO-LEVEL SAMPLING SCHEMES

5.1 DECOMPOSITION APPROACH

Expression derived in the present paragraph descend from the hypothesis made in §2, thus each plan meeting the requirements of a discrete-state Markov Process [8]. Formulae can be easily derived either using the normal transition matrix approach (see, e.g., [9]) or the simplified method (see, e.g., [7] or [10]). The Markov chain models has been proved to be a well assessed tool [9]. They allowed also to formulate very simple and attracting closed form expression here, having a decomposed structure, more manageable and intuitive.

We will refer to two parameters, which are representative of the average time spent on each inspection plan, namely

$$X_N = ARL_N \,/\, [ARL_N + ARL_{LS}] \tag{22}$$

and, being LS either R or Sk,

$$X_{LS} = ARL_{LS} \,/\, [ARL_N + ARL_{LS}] \tag{23}$$

Descending from these parameters, it is possible to benchmark two-level sampling schemes (refere to figure 1) according to the following general formulae (according to a simplified method similar to [7]):

$$P_{scheme} = P_N \bullet X_N + P_{LS} \bullet X_{LS} \tag{24}$$

as concern the the probability of lot acceptance, and, in the same way,

$$ASN_{scheme} = X_N \bullet ASN_N + X_{LS} \bullet ASN_{LS} \tag{25}$$

for the average sample number.

Each scheme can thus be compared by simply evaluating the respective ASNs, once the respective ARLs has been evaluated with (8), (10) and (11). Accordingly, the probability of acceptance can be derived for the N-R scheme:

$$P_{N\text{-}R} = P_N\,(1- P_N{}^{i})\,(1- P_R) + P_R\ P_N{}^{i}\,(1- P_N) \,/\, [\,(1- P_N{}^{i})\,(1- P_R) + P_N{}^{i}\,(1- P_N)] \tag{26}$$

and the probability of acceptance can be derived for the N-Sk scheme [6]:

$$P_{N\text{-}Sk}= f\,P_{Sk} + (1\text{-}f)\,P_{Sk}{}^{i} / [f + (1\text{-}f)\,P_{Sk}{}^{i}] \tag{27}$$

In the same way, it is easy to derive the expressions for the ASN in the case of N-R scheme as

$$ASN_{N\text{-}R}= X_N \bullet n_0 + X_R \bullet m \tag{28}$$

And in the case of N-Sk scheme [6]

$$ASN_{N\text{-}Sk}= f\,n / [f + (1\text{-}f)\,P_{Sk}{}^{i}] \tag{29}$$

Also, the economical benchmarking function can be derived, being

$$C_{N\text{-}R} = ASN_{N\text{-}R} \bullet U + (X_N + X_R) \bullet L = ASN_{N\text{-}R} \bullet U + L \tag{30}$$

and

$$C_{N\text{-}Sk} = ASN_{N\text{-}Sk} \bullet U + (X_N + f \bullet X_{Sk}) \bullet L = ASN_{N\text{-}Sk} \bullet U + L - (1 - f) \bullet X_{Sk} \bullet L \tag{31}$$

As

$$B(f)= (C_{N\text{-}R} - C_{N\text{-}Sk}) / U = (ASN_{N\text{-}R} - ASN_{N\text{-}Sk}) + (1 - f) \bullet X_{Sk} \bullet L/U \tag{32}$$

with the same meanings of symbols adopted before.

5.2 BENCHMARKING TWO-LEVEL SAMPLING SCHEMES: INFERENCE METHOD

As it can be seen from formulae in §5.1, the same general rules of performances are not so evident for the scheme, if it is $n \neq n_0$. The case where $n=n_0$ is here address, showing how conclusions drawn in §3 for the individual less-severe plans can be used to benchmark more complex schemes, provided all the other plans have the same characteristics.
This descends thanks to the simple decomposed formulation made in eq. (23) and (24).
It easy to derive the closed forms for OC curve and ASN, for $c=0$, by replacing P_R with P'_R of eq. (6) and P_{SK} with P'_{SK} of eq.(5) respectively in (24) and (25). By a simple analysis of the functions X_{SK} and X_N, under the KP1 benchmarking conditions, it is easy to show that it is always

$$X_{SK} > X_R \tag{33}$$

According to this, it is easy to see from (24) and (25) how conclusions ST1 and ST2 provided §4 still hold, as far as benchmarking conditions KP1 are respected. Therefore it results

$$P_{N\text{-}R} < P_{N\text{-}Sk} \quad \text{and} \quad ASN_{N\text{-}R} < ASN_{N\text{-}Sk} \tag{34}$$

It easy to derive how, form eq. (32) the same conclusion derived in ST3 still holds.
As a conclusion form (30), it is possible to state that

ST4. benchmarking of individual sampling plans can be useful to infer conclusions on two-level sampling schemes, under the benchmarking conditions put in §2.

6. CONCLUSION

The paper presents a practical benchmarking method, mainly oriented to compare acceptance sampling schemes, in case of very high level of vendor's quality, with specific reference to null acceptance sampling (c=0). Quite often adopted in practice, this latter condition has been usually neglected or avoided by norms.

The usefulness of the approach is the practical applicability to real industrial cases and the simplicity of the method, which might result very useful for decisional purposes.

It would be interesting to extend the application of the benchmarking method to several different schemes, with different design features; further research would be necessary in this direction for practical usability.

AKNOWLEDGEMENTS

The author wish to acknowledge dott. G. Di Santo for the fruitful discussion had.

REFERENCES

[1] Montgomery, D.C., (1997), Introduction to Statistical Quality Control. J. Wiley & Sons, ISBN 0-471-30353-4

[2] Soubdararajan, V., Vijayaraghavan, R., (1990), Construction and selection of tightened-normal-tightened (TNT) plans, Journal of Quality Technology, Vol.22, No.2, April, 146-153.

[3] UNI ISO 2859-0; (2001), Procedimenti di campionamento nel collaudo per attributi, UNI – Milano - Italy

[4] Perry, R.L., (1973), Two-level Skip-lot Sampling plans – operating characteristics properties, Int. J. of Quality Technology, Vol.5, No.4, October, 160-166

[5] KENNET, R.K.; ZACS, S.; (1998), Modern Industrial Statistics, Duxbury Press, ISBN 0-534-35370-3

[6] Perry, R.L., (1973), Skip-lot Sampling plans, Int. J. of Quality Technology, Vol.5, No.3, July, 123-130

[7] Brugger, R.M., (1975), A simplification of Skip-lot procedure formulation, Int. J. of Quality Technology, Vol.7, No.4, October, 165-167.

[8] Parzen, E., (1999), Stochastic Processes, SIAM, Oakland, ISBN 0-89871-441-9

[9] Grinde, R.B., McDowell, E.D., Randhawa, S.U., (1987), ANSI/ASQCZ1.4 Performance Without Limit numbers, Int. J. of Quality Technology, Vol.19, No.4, October, 204-215.

[10] Brugger, R.M., (1989), A simplified Markov Chain Analysis of ANSI/ASQC Z1.4 Used Without Limit numbers, Int. J. of Quality Technology, Vol.21, No.2, April, 97-102.

ECONOMIC EVALUATION OF ALTERNATIVES FOR MICROELECTRONICS LITHOGRAPHY PROCESSING

Z. Wang[1], W.A. Knight[1]

[1] Department of Industrial and Manufacturing Engineering, University of Rhode Island, USA

KEYWORDS: Microelectronics, Lithography processing, Economic modeling

ABSTRACT. Cost-of-Ownership models for microelectronics fabrication equipment have been adapted to the analysis of the economics of optical lithography and associated processes such as mask making and photoresist application. This requires modification of the cost-of-ownership approach to the series of processing steps for wafer fabrication. These models enable the economic impacts of microelectronics design and processing decisions on silicon wafer fabrication to be estimated. The models have been incorporated into a software tool that uses defaults and data based on current industry standards and practice. Using this program the sensitivity of wafer processing costs to basic integrated circuit design parameters can be evaluated. The basic assumptions and data used in the economic models are presented together with results from application to typical case studies. The availability of these models facilitates the design for manufacture of microelectronic integrated circuits.

1 INTRODUCTION

The availability of suitable cost models at the early stages of product design can lead to more competitive products because design teams can predictively assess alternative design concepts leading to more effective decisions. This approach to design for manufacture (DFM) has been applied effectively in many sectors of industry [1-4]. The development of appropriate cost models for microelectronics products fabrication is necessary and this paper outlines applications for to the critical lithography stages of microelectronics manufacture.

Integrated circuit (IC) fabrication costs can be modeled by an extension of Cost-of-Ownership (COO) models for fabrication equipment. SEMATECH began developing COO models in 1990 and since then a commercial COO model has been prepared through joint development projects [5]. Cost-of-Ownership modeling is an advanced form of activity-based cost (ABC) analysis that considers the costs of acquisition and operation, including the scrap value of work in progress. The lifetime cost-of-ownership per good device is generally sensitive to production throughput rates, overall equipment reliability, and yield, but is relatively insensitive to initial equipment purchase price. While initial COO models have been developed for wafer fabrication equipment, these methods can be extended to other applications [6,7].

2 COST MODELING BASED ON COST OF OWNERSHIP

A COO model is a tool to analyze the total cost of owning a piece of equipment to fabricate a product. Such models include both the cost of production preparations (fixed costs), the cost of production processes (variable costs), the important factors of productivity (throughput, utiliza-

Published in: E.Kuljanic (Ed.) *Advanced Manufacturing Systems and Technology*, CISM Courses and Lectures No. 437, Springer Wien New York, 2002.

tion, and yield) and production wastes (yield loss). The basic cost-of-ownership model for wafer fabrication is described by:

$$C_W = (C_F+C_V)/(TPT\ Y\ U)+C_Y \qquad (1)$$

where: C_W = Cost per wafer; C_F = Fixed cost; C_V = Variable cost; C_Y = Cost due to yield loss; TPT = Throughput; Y = Composite Yield and U = Utilization.

In practice IC fabrication consists of a large number separate processing steps (up to 200 or more) and the COO model needs to be modified as the cost-modeling framework of IC fabrication. The COO model takes into consideration two parts of cost: total add-on cost for fabrication at a particular process step/equipment, which is represented by the term $(C_F + C_V)/(TPT * Y * U)$ and the present value of the IC product which is lost due to the yield of the process step/equipment, which is represented by the term C_Y. The fixed cost and the variable cost in the first term are nothing but the add-on cost of fabrication investment and the add-on value of consumables amortized by the total number of good products. The second term is the value of a waste IC translated from the percentage of scrapped wafer. Current COO models are equipment centered rather than process centered, because they were initially developed to address the cost of owning individual equipment, but these models of IC manufacturing can be extended to address processes rather than equipment [6,7].

A number of basic assumptions need to be made. For example, overall manufacturing yield is a product of several factors, including wafer processing yield, die fabrication yield, and packaging yield. For most technologies, the die fabrication yield and packaging yield are so close to 1 that to a first approximation only the wafer process yield need be considered as the manufacturing yield, Y_{fab}. In addition, the total manufacturing costs include both wafer fabrication costs and chip packaging costs. Some studies [8] have shown that the wafer fabrication costs are the main cost driver to determine the total costs when the chip die area is reasonably large. Only when the die size is small do the packaging costs significantly affect the total manufacturing costs. Given the large die size of today's IC products the packaging costs can often be ignored. Thus a revised cost model of IC manufacturing is:

$$C_{fabi} = [\Sigma\ [1/\ (Y_{pari} * Y_{fi})](C_{fi}\ /\ (TPT_i * C_{vi}) + C_{yi}]/N_{IC} \qquad (2)$$

where: C_{fabi} = Manufacturing cost of ith process step per die
C_{fi} = Total fixed costs which include all nonrecurring costs associated with ith process step
C_{vi} = Consumable costs per wafer of ith process step
C_{yi} = Cost of yield loss
Y_{fi} = ith process step-generated functional yield
Y_{pari} = Parametric yield of ith process step
N_{IC} = number of chips (dies) on a wafer.

The cost of yield loss, C_{yi}, consists of two components:

$$C_{yi}=[V_{i-1}(1-Y_{pari})+V_{wafer}(1-\Pi C_i)(1-Y_f)e^{-(n_0-1)\Pi Ci}]/[Y_f+(1-\Pi C_i)*(1-Y_f)e^{-(n_0-1)\Pi Ci}]w_i \qquad (3)$$

where: V_{i-1} = Value of wafer after i-1th process step
V_{wafer} = Value of completely fabricated wafer at functional test stage
Y_f = Total functional yield

w_i = The percentage of functional yield loss due to broken down component of functional yield at ith process step
V_{i-1} (1 - Y_{pari}) = Value of scrapped wafer due to parametric yield loss at ith process step

DL (Defect Level), it is the percentage of bad wafer recognized by the final functional test. Detailed discussions on this can be found in [6].

$$DL= [(1-\prod C_i)(1-Y_f)\, e^{-(n_0-1)\prod C_i}] \;/\; [Y_f+(1-\prod C_i)(1-Y_f)\, e^{-(n_0-1)\prod C_i}] \tag{4}$$

The value of scrapped wafer at final functional test translated onto ith process step is:

$$V_{wafer} [(1-\prod C_i)*(1-Y_f)e^{-(n_0-1)\prod C_i}] \;/\; [Y_f+ (1-\prod C_i)*(1-Y_f)e^{-(n_0-1)\prod C_i}]\; w_i \tag{5}$$

All of the above terms require the development of specific detailed analyses for the individual processing steps in fabrication. These detailed analyses can be found in [6].

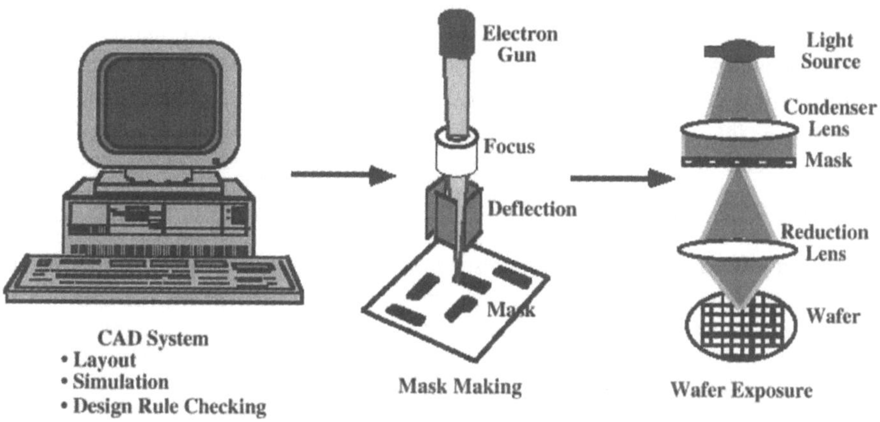

FIGURE 1. Lithography Process Flow form IC Design to Wafer Exposure

3 LITHOGRAPHY TECHNOLOGY

Modern IC technology is based on "planar technology" where devices, interconnects, and electrical isolation are fabricated on the surface of a wafer using successive deposition, implantation, lithography, and etching. Lithography is the most complicated, expensive, and critical process in microelectronics fabrication and is the key for further reduction of VLSI device feature size. Pattern transfer and, in particular, lithography steps present the toughest challenges from both technical and financial viewpoints. The lithography process consists of several steps (Figure 1). For each circuit layer a mask is made from a CAD system output. The silicon wafer is then coated with a layer of photoresist polymer. The mask and an optical system are then used to expose the photoresist. After development and fixing the wafer is etched to produce the circuit elements for the layer. This process is repeated for each layer of the device.

There are three basic measures of performance for a lithography system; resolution (minimum feature size that can be exposed); registration (overlay accuracy from layer to layer) and throughput (a key factor for economic impact). Although these three criteria are usually used to benchmark the technical performance of a lithography system, they are also the key parameters to the economic estimation. Both the parametric yield and functional yield of lithography are closely dependent on the line resolution and registration/overlay control. The throughput is also the key factor to determine the economic impact of a production system.

One approach to increased resolution is to use bigger lenses, but this also decreases depth of focus (DOF). During projection, maintaining the DOF across a 200 mm wafer is difficult since the wafer topology alone may be as much as 2 μm unless some planarization is done. Added to this are the wafer bow and flatness. Therefore, some compromise must be achieved between resolution and DOF.

The other avenue to reducing the minimum feature size is to use shorter wavelength light sources. Traditionally, Hg vapor lamps have been used which generate many spectral lines from a high intensity plasma inside a glass lamp. Modern lithography systems make use of g-line ($\lambda = 436$ nm) and i-line ($\lambda = 365$ nm, used for 0.5 μm, 0.35 μm) light sources. More recently the Deep Ultra-Violet (DUV) light source has been used to achieve finer resolution and higher light intensity, which in turn minimize the resist exposure time and achieve a higher throughput. The brighter sources in DUV are KrF ($\lambda = 248$ nm, used for 0.25 μm) and ArF ($\lambda = 193$ nm, used for 0.18 μm) excimer lasers. As the wavelength of the light sources get smaller and smaller, the problems become finding suitable resists and transparent optical components at these wavelengths.

4 COST MODELING OF OPTICAL LITHOGRAPHY TECHNOLOGY

The COO for a given semiconductor process step is dependent not only on fixed costs such as depreciation of equipment and amortized installation costs, but variable costs such as the cost of consumables, as well as the productivity factors of tool throughput, utilization and process yield. This approach to model development provides the versatility and flexibility to be tailored to any particular IC production environment, and to any particular wafer process step such as thin films/CVD, or ion implantation.

There are three major stages during one lithography process step: the mask making, the photoresist application and processing before or after the exposure, and the photoresist exposure. Therefore, when applying and modifying the SEMI COO standard model, the total optical lithography cost is given by as follows:

$$C_{photo} = [C_{lph} + C_{res} + C_{msk}/(N_{wafer}\ Y_{par}\ Y_f) + C_{yloss}] / N_{IC} \qquad (6)$$

where: C_{photo} = Total lithography cost per IC
C_{lph} = resist exposure related lithography system cost per wafer
C_{res} = photoresist and related process cost per wafer
C_{msk} = one mask making cost
N_{wafer} = total number of wafers can be exposed by the one mask, it is the mask lifetime
Y_{par} = Parametric yield of lithography process step, including exposure and photo-resist yield
Y_f = Functional yield of lithography process, including both exposure and photo- resist yield
C_{yloss} = average yield loss cost per wafer.
N_{IC} = number of chips (dies) on a wafer

To calculate the lithography system cost per wafer, the following model is developed:

$$C_{lph} = \Sigma[(C_{sysi}/D_{Yi} + \beta_i * C_{sysi}/T_y + C_f * A_{cri}/T_y + C_{labi})/(TPT_i * U_i * Y_{pari} * Y_{fi}) \qquad (7)$$

where: C_{sys} = cost of a lithography exposure machine ($)
D_Y = depreciation time of the machine (hrs)
β = O/M coefficient to describe the cost for the machine operation and maintenance
T_y = total operating time (hrs) per year
C_f= clean room facility cost ($)
A_{cr} = the footprint of the machine with respect to the total area of clean room
C_{lab} = lithography fab line labor costs ($/hr)
TPT = throughput of the optical lithography exposure system (wafers/hr)
Y_{par} = Parametric yield at present process step
Y_f = Functional yield of present process step
U = lithography exposure machine utilization

The model of the photoresist and the related process cost is developed as follows:

$$C_{res} = (G_{res}/W_g)/Y_{fi}+\Sigma[(C_{pri}/D_{Yi}+\beta_I C_{pri}/T_y+C_f\ A_{cri}/T_y + C_{labi})/(TPT_i\, U_i\ Y_{pari}\ Y_{fi}) \qquad (8)$$

Where: G_{res} = resist cost per gallon (G)
W_g = the number of wafers coated per gallon
C_{pr} = the cost of the machines for photoresist processes

Exposure system throughput can be rewritten in the following format:

$$TPT = (3600-T_{lot})/(T_{oh} + N_{exp}\,(T_{exp} + T_{step} + T_{settle}) \qquad (9)$$

where: T_{oh} = overhead time (sec.) associated with loading/unloading the wafer, prealigning the wafer, and re-calibrate the mask after a lot.
N_{exp} = number of exposure per wafer, equal to the number of exposure fields fitting in the wafer.
T_{exp} = exposure time per exposure field (sec.)
T_{step} = average stage movement time from one exposure field to another exposure field (sec.)
T_{settle} = stage settling time when moving from one field into another field (sec.)
T_{lot} = lot overhead time (sec./hr)

Usually the equipment throughput of photoresist coat/developer matches each lithography exposure system in the fab line. Thus, the throughput of the exposure tool is used as the throughput of the resist related processes.

As far as the equipment utilization is concerned this can be obtained by dividing the uptime by the total operating time Ty or by 168 hrs/week as provided by the SEMI E10-90 Guideline [6]. An equation for the calculation of utilization is:

$$U = 1 - (USD + SD + Eng + STBY +Test) / 168 \qquad (10)$$

where: USD = Unscheduled maintenance downtime; SD = Scheduled maintenance downtime; Eng = Engineering assist time; STBY = Standby time and Test = Nonscheduled time, mainly due to production test
A similar model to equation 6 is used to estimate the costs associated with the photo-resist handling, application, pre-baking, post-baking, and development processes. The parametric

yield of photo-resist spin is ignored because the process is very simple and well controlled. The process variation is disregarded.

5 CASE STUDIES IN APPLICATION

These models have been incorporated into a web based software tool that can be used to investigate the effect of alternative design configurations and processes. Continuous developments in lithography systems have taken place. The basic systems now use I-line projection (λ = 345nm), but more recently deep-ultra-violet (DUV) systems, KrF (λ = 248nm) and ArF (λ = 193nm) have become available, but at increased cost. In general it has become popular to adopt a mix and match strategy for the utilization of lithography systems. With this approach the more established and less costly technologies are used for the less critical layers.

TABLE 1: Lithography Case Study Results.

Key Results	ArF	KrF	i-line
Exposure Throughput	119.5 wafer/hr	90.4 wafer/hr	66.1 wafer/hr
Exposure Yield	98.59%	98.38%	99.62%
Resist Yield	85.7%	86.8%	92%
Exposure Cost	41.2 $pwle	3.27 $pwle	3.11 $pwle
Resist Related Cost	3.74 $pwle	4.36 $pwle	1.11 $pwle
Mask Related Cost	14.6 $pwle	8.84 $pwle	3.74 $pwle
Number of Layers	3	6	12
Total Cost	67.38 $/wafer	98.82 $/wafer	95.52 $/wafer
Chips per Wafer		61	
Cost per Chip		4.29 $/chip	

A case study based on a typical wafer fabrication requiring 21 lithography levels, with 3 ArF layers, 6 KrF layers and 12 non-critical i-line technology layers produces the results shown in Table 1. All data embedded in the tool are extracted from the latest SEMATECH studies of the lithography industry and the 1999 SIA Technology Roadmap for Semiconductors [9]. In general, DUV (ArF and KrF) technology has better equipment throughput than old i-line technology because a pulsed laser light source is used with much higher illumination intensity and also a photo-resist with less sensitivity can be used, but with significant increase of equipment costs. The results in Table 1 show that the cost of i-line technology is much lower than that of the more advanced DUV (ArF and KrF) technology. Thus it can be seen why the mix-and-match fabrication strategy has gained popularity. With technology advances to sub 0.18-micron levels, the main cost driver becomes the high cost of mask making rather than the lithography process itself.

From the case study conducted above, the resulting total lithography cost is 67.38 + 98.82 + 95.52 ≈ $260/wafer. For a fabrication study conducted by Hewlett Packard [10] in 1997, the photo-

3. Bariani, P.F., Berti, G., D'Angelo, L., (1993), Tool Cost Estimating at the Early Stages of Cold forging Process Design, *Annals of the CIRP*, 42/1:279-282.
4. Russel, G.A., (1985), Design for Manufacturability of Printed Circuit Board Assemblies, *Annals of the CIRP*, 34/1:37-40.
5. Anon, (1996), Cost of Ownership for Semiconductor Manufacturing, Semiconductor Equipment and Materials International (SEMI), California.
6. Wang, Z., (2000), Early Estimates of the Economic Impacts During Microelectronics Product Design, Ph.D. Thesis, University of Rhode Island.
7. Wang, Z., Knight, W.A., (2001), Economic Impacts During Microelectronics Product Design, *Annals of the CIRP*, 50/1:97-100.
8. Gray, P.R., Meyer, R.G., (1993), Analysis and Design of Analog Integrated Circuits, 3rd Edition, John Wiley, New York
9. Anon, (1999), International Technology Roadmap for Semiconductors –1999 Edition, Semiconductor industry Association.
10. Duley, J. R., Varma, V., Wood, S.C., (1997), Sense and Sensibility: The Scalable MiniFab, Proceedings of the Int. Symposium on Semiconductor Manufacturing, San Francisco, October: A45-A48

CERAMIC COMPONENTS BUILT BY STEREOLITHOGRAPHY

C. Esposito Corcione, A. Greco, A. Licciulli, M. Martena, A. Maffezzoli

Department of Innovation Engineering, University of Lecce, Italy

KEYWORDS: stereolitography, alumina.

ABSTRACT.Stererolithography (SL) is a rapid prototyping process traditionally used to produce polymeric parts for various applications in aerospace, biomedical, automotive and electronics fields. Though it is a promising process for production of advanced ceramics, a number of critical points need still a proper solution: develop materials for functional prototypes, avoid interlayer delamination, obtain mechanical performances. In this work the production of ceramic green starting from specially formulated photocurable suspensions is presented. In particular, Al_2O_3 ceramic precursor have been built with an experimental stereolithography system, and the ceramic objects have been obtained from pyrolisis of the organic binder and subsequent sintering. Finally, characterization of mechanical properties and microstructure of the samples is presented. In addition, a 3D System SLA250 machine has been modified in order to build ceramic objects. A new loading vat with varying capacity has been projected and built, in order to produce parts with lower material quantities. The building and recoating parameters of the machine have been studied and modified to work with high-viscosity formulations.

1 INTRODUCTION

Conventional shaping processes for ceramics are traditionally based on a powder molding process using a negative mold and subsequent thermal compaction. Especially for the production of complex shape prototypes the outlay of molds is the major costing factor. The use of Rapid Protoyping (RP) processes can reduce the costs and time for the production of single objects or of very limited series[1]. Since the mid 80's the need of fast automatic fabrication techniques lead to the development of many additive RP technologies [1]. The new fabrication concept allowed the construction of complex parts, starting from a 3D-CAD model, without using a mould. In RP, the object is obtained bonding, layer by layer, its sections by an additive process rather than cutting or «sculpturing» the parts from an homogeneous block. The key point of RP is given by the direct link between the CAD model and the solid object, generated in few hours suggesting that tailor made or single functional parts can be fabricated, for example, for prosthetic applications in very a short time. Traditionally, most of the RP processes are used to produce polymeric objects. However, over the last decade, several solid free-form fabrication (SFF) methods have been investigated as techniques to fabricate metal or ceramic parts[1] directly from computer 3D projects. In particular SFF of ceramics is aimed to produce high performance, near net-shape structural [3], [4] and/or functional, [5] [6] ceramic parts. The interest in these fabrication process arises

Published in: E.Kuljanic (Ed.) *Advanced Manufacturing Systems and Technology*,
CISM Courses and Lectures No. 437, Springer Wien New York, 2002.

from different application fields: ceramic materials are needed as prototype for functional tests and pre-series production tests, ceramic moulds in unique copy are needed in metal casting e.g. jewellery, ceramic cores are needed in investment casting of hollow objects. Since ceramics are good biomaterials, inert, bioactive and reabsorbable implants are made by alumina, zirconia, carbon, hydroxyapatite, bioglass. Many applications are possible with the convergence of three distinct technologies: medical imaging (computer assisted tomography CT, Magnetic Resonance Imaging MRI), CAD and computer graphics for image processing and edge detection, rapid prototyping of ceramic implant.

However, only recently laser sintering of metal powders has been commercially introduced. In the last years, the significant efforts devoted to the development of technologies of RP based on non-polymeric materials lead to the market equipments based on the selective laser sintering of metallic powders. On the other hand, the use of ceramic materials in RP is applied at laboratory level in different processes:

- LOM (Laminated Object Manufacturing). Object building is commonly performed by sequentially cutting stacking and bonding sheets of paper. Therefore, LOM technology can be exploited for building ceramic or ceramic matrix composites parts using sheets of ceramics or fiber reinforced ceramic prepregs [7].
- FDM (Fused Deposition Modelling). Usually a thermoplastic polymer is extruded and the object is built layer by layer according with the CAD data. The application to RP ceramic processing is obtained extruding a suspension of a ceramic powder and a thermoplastic binder [8]. The green parts must be heated in order to burn the binder and to sinter the ceramic [9].
- DCJP (Direct Ceramic Jet Printing). In this process a ceramic powder is dispersed in a liquid medium suitable for a modified ink jet system that is used for building the parts layer by layer. Multilayer objects can be obtained loading different ink jet heads with more than one type of ceramic powder [8].
- SLS (Selective Laser Sintering). Thin layers of ceramic, metallic or polymeric powders are selectively sintered using a laser beam (typically CO_2) of high power [10]. The laser beam moves on the surface of a vat filled with powder according with the desired section geometry and the objects are still built adding further layers. One advantage of this process is related with the direct production of ceramic parts, skipping the production of the intermediate green.
- STL (Laser Stereolithography) [2]. This represents the most assessed process for rapid prototyping and can be adapted to building of ceramic objects. The objects are obtained polymerising a low viscosity liquid resin layer by layer. The shape and the dimensions of the parts are directly transferred from a three dimensional CAD system to the stereolithography equipment where a laser beam (usually He-Cd or Ar) polymerises the different sections. Suspensions of ceramic powders in a photoreactive resin can be used in a standard STL equipment in order to build green parts [11]. A candidate ceramic stereolithography suspension must satisfy several requirements. Since a high-quality ceramic is the aim, the ceramic green body must have high density, in order to be readily

sinterable to form a dense ceramic. SLA suspensions. It also must have high reactivity in order to minimize the build time and finally its viscosity must be comparable with that of commercial SLA resins (usually less than 3 Pa*s).

In this work the production of ceramic objects using a stereolithographic apparatus is presented. UV curable pre-ceramic suspensions for alumino-silicate parts have been studied using photocalorimetric and rheological characterizations. Thermogravimetry and dilatometry have been used in order to analyse the behaviour of the preceramic green during thermal treatment. An experimental set-up has been used to build layer by layer green bars. Stereolithography apparatus (SLA), SLA-250 (3D System, Valencia, CA) has been modified in order to fabricate ceramic green. Ceramic objects have been obtained by pyrolisis of the organic binder and subsequent sintering of the green at high temperature. Finally, a characterization of mechanical properties and microstructure of the samples has been performed.

2 EXPERIMENTAL

2.1 MATERIALS

The adopted suspensions are prepared using commercially available acrylic and silicone acrylate monomers: SR350 (trimethylolpropane trimethacrylate , TMPTMA),SR203 (tetrahydrofufuryl methacrylate, THFMA), by Cray Valley, MEMO, by Huls. The resins are loaded with alumina powder AES 23 (alumina easy sintering) from Sumitomo, possessing a density after sintering of 3.77g/cm^3 (fired density at 1600 °C), mean particle size of 1.8 μm, surface area of 0.42 m^2/g and linear shrinkage on sintering of about 16%. In every formulation 3 phr of 1-hydroxy-cyclohexyl-phenyl-khetone Irgacure 184 from Ciba, as photoinitiator, is added..

2.2 CHARACTERISATION TECHNIQUES

Every formulation needs to be characterized in terms of reactivity, viscosity, behaviour during thermal treatment, and finally mechanical properties of ceramic parts have to be tested. The techniques used for materials characterization are the following:

- Differential Scanning Calorimetry (DSC). The cure of the resin was carried out in air at 25 °C in a DSC Perkin Elmer DSC-7. The DSC is modified for irradiation of the sample using transparent quartz windows. The light reaching the sample, produced by a 300 W Xenon lamp Cermax LX 300, is limited to a wavelength interval of 325$\pm$4 nm using a monocromator, in order to simulate the irradiation band of a He-Cd laser beam. The beam is focused on the sample using a system of lenses and a mirror.
- Rheology. The viscosity of suspensions at 25 °C as a function of the amount of filler loaded is measured using a stress controlled rheometer Bohlin CS.
- Thermogravimetry (TGA). The behaviour of the preceramic green samples during pyrolysis of the organic binder and sintering is studied using TGA, Netzsch STA 406. Dimensional changes associated with pyrolisis and sintering are measured by a TMA, Netzsch TMA 402.

- Mechanical testing. Three point bending test of sintered ceramic specimens are performed using a dynamometer Instron 4507.

2.3 EXPERIMENTAL SET-UP

The experimental set-up is composed of the following parts assembled according to the sketch of Fig. 1:

- A laser He-Cd, from Omnichrome series 3056, with a specific power of 17mW/mm^2, emitting at the wavelength of 325nm, and with a beam diameter of 0.2mm;
- A scan system, which drives the laser beam on the suspension surface;
- A software which controls the scan system allowing the construction of rectangular parts scanning the vat surface along parallel lines. The parameters which can be controlled are the scanning direction, the dimension of the part, the scan speed and number of scans per unit width. The energy per unit area was controlled by properly setting the last two parameters;
- A moving platform with manual control, which can move downward to build a new layer after suspension re-filling.

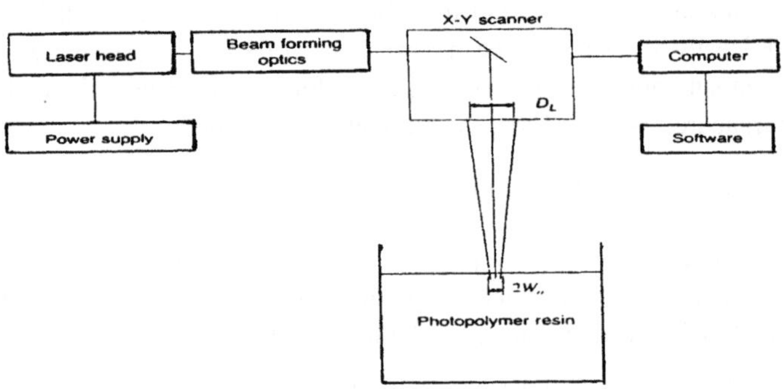

FIGURE 1: sketch of the experimental set-up

3 RESULT AND DISCUSSION

3.1 FORMULATION OF SUSPENSIONS

Different formulations of base resin for ceramic suspension have been prepared with the aim of maximising the reactivity and minimising the viscosity. The use of a silicone acrylate resin is considered a key factor in order to obtain a solid residue of silica after pyrolysis of the organic binder in the suspensions and in order to improve the dispersion of the allumina powder in an hydrophobic resin. This residual silica is thought to reduce shrinkage and improve sintering. The modified differential scanning calorimeter described above is used to compare the rate of reaction of the formulations with that of commercial resin from Ciba, SL5170 an epoxy base resin characterised by lower reactivity. The kinetic properties of SL5170 are considered as a lower limit of reactivity for effective use in STL.

The DSC measurements have been used for determination of the advancement of polymerisation by assuming that the heat evolved during polymerisation reaction is proportional to the overall extent of reaction given by the fraction of reactive groups consumed. Following this approach the degree of reaction, α, is defined as:

$$\alpha = Q(t)/Q_{tot} \qquad (1)$$

where Q(t) is the partial heat of reaction developed during a DSC experiment and Q_{tot} represents the total heat of reaction measured when the reaction is completed. The reaction rate, $d\alpha/dt$, is thus obtained from the heat flow dQ/dt as:

$$d\alpha/dt = 1/Q_{tot}\,(dQ/dt) \qquad (2)$$

The time dependence of the degree of reaction calculated by DSC for the base resin (45% TMPTMA, 45% THFMA, 10% MEMO) is compared in Figure 2 with that of the STL resins SL 5170. As shown in Figure 2, the time to reach full conversion for the base resin is lower than that of SL5170. This suggests that the same conversion can be obtained with lower energy dose.

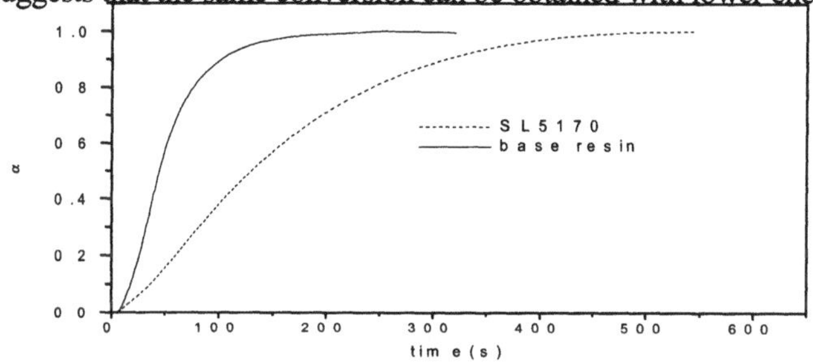

FIGURE 2: plot of the degree of reaction vs. time

The base resin has been filled with 45% by volume of alumina. The suspension has been further characterised with the experimental set-up on the basis of its working parameters. The reactivity of a STL resin is characterised by two parameters relating the cure depth with the energy dose at the vat surface (E_o) [2]:

$$C_d = D_p \ln \frac{E_o}{E_c} \qquad (3)$$

where the working parameters D_p and E_c, are the penetration depth and critical energy respectively and the cure depth (C_d) represents the thickness of gelled resin. E_c is the minimum value of energy needed to promote polymerisation on the suspension surface, while D_p is a characteristic parameter proportional to the gelled thickness.

The working parameters of suspensions are calculated measuring the hardened thickness at different values of E_o. A linear regression on experimental C_d values plotted as a function of $\ln(E_o)$ is used for the determination of D_p and E_c from the slope and the intercept respectively. The working curve is compared in Figure 3 with that of SL5170, obtained using literature values of D_p and E_c [12].

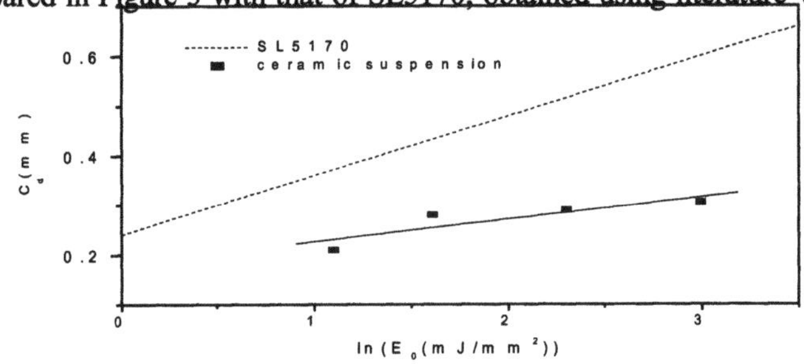

FIGURE 3: working curves of ceramic suspension compared with that of commercial SL resin

The calculated and reference values are listed in Table 2.

TABLE 1: Working parameters of adopted formulations and commercial resins.

Formulation	E_c(mJ/cm^2)	D_p(mm)
SL5170	13.5	0.12
Ceramic suspension	25.48	0.075

A good resin for the use in stereolithography should have low values of Ec, in order to start the reaction with low energy dose, and high values of Dp, in order to have higher cured thickness. From the results reported in Fig. 3 and Table 1, it can be observed that for the suspension B009 the slope is higher (lower D_p) than for commercial resin. Further, it presents high energy E_c, due to the presence of a monofunctional methacrylate component. The high loading of alumina powders, results in lower values of Dp, because of scattering of light from ceramic particles. Anyway, this formulation lead to sample parts without any curl distortions.

Another important parameter for STL formulations is the viscosity. The aim is to provide an low viscosity, auto leveling suspension. Such a requirement is difficult to be fulfilled with highly solid loaded ceramic suspensions that behave like non-newtonian fluid and display a yield point. In particular, the viscosity of the pure monomer resin is extremely important. A lower starting viscosity usually results in a lower final viscosity for the suspensions. In this situation at least good levelling and flatting by recoater blade should be obtained. Thus suspensions were prepared in low-viscosity UV-curable liquids and dispersants were carefully selected to control the colloidal chemistry, and therefore the rheology.

Viscosity vs. shear rate measurements are reported in Figure 4. Measurements were performed in the range between 0.1s^{-1} and 100s^{-1} .The values of shear rates acting on the resin during recoating are related to the speed of the recoating unit, which is quite high (maximum 30mm/s). Using the equation $\gamma = V / \Delta x$ (γ is the shear rate, V is the speed of the recoater unit and ΔX is the layer thickness) a rough indication of the maximum shear rate acting on the suspension is obtained to be $300s^{-1}$. In order to be used in an SLA, the ceramic suspension viscosity must be of the same order of that of conventional resins. The available materials for SLA are all chraterized by a viscosity lower than 3 Pa s. As it can be seen in Figure 4, this viscosity is obtained at about $12s^{-1}$, which is a value much lower than that one occurring during recoating. This suggests that the suspension can be used in the stereolithographic apparatus.

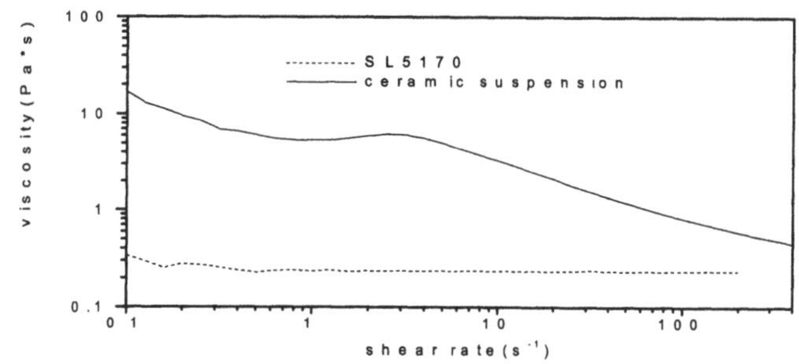

FIGURE 4: viscosity vs. shear rate

TGA has been used to identify the temperature intervals where pyrolysis takes place and evaluate the solid yield of the ceramic suspensions.The weight loss of the base resin is shown in Figure 5. In the case of ceramic suspension the resin burn out occurs up to 550 °C with a weight loss of about 19%. Two degradation steps are detected in Figure 5: the first can be attributed to the degradation of acrylates and the second to silicone acrylate monomers. At higher temperatures the weight of the sample keeps constant and no other degradation phenomena can be monitored. Alumina content of the suspension corresponds to about77% by weight. The difference between this value and the one experimentally obtained is attributed to silica residual after resin burn-out of organic binder.

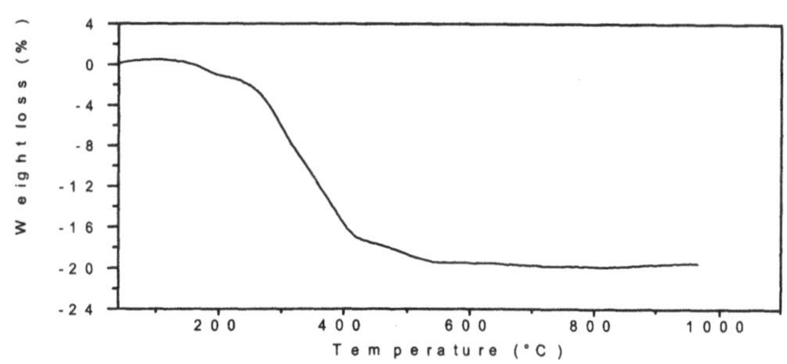

FIGURE 5: TGA analysis on ceramic suspension

Following the indications of TGA experiments the resin burn out between 200°C and 550°C must be very slow, because in this phase the strong shrinkage can determine thermal stresses. The overall thermal treatment has then been divided into three steps:

- Post-cure: the green ceramic part is held in a furnace at 160°C for 100min, in order to allow a complete reaction of the resin eventually not fully reacted at the end of the laser exposure. The resulting pre-ceramic green is characterized by poor mechanical properties because of the presence of the resin which is also responsible of low density values;
- pyrolysis: heating at about 60°C/h to 500°C. This is a very delicate phase where the resin burn-out is accompanied by an increase of density and a resulting shrinkage;
- sintering: heating from 500°C to 1550°C at 300°C/h. The fraction of voids in the ceramic is reduced with a significant increase of density.

3.2 MECHANICAL PROPERTIES

The ceramic parts have been obtained on the experimental set-up by laser exposure of each layer to an energy dose of 10mJ/mm^2. The thickness of each layer has been chosen to be 0.1mm, resulting a good compromise between cure depth, processing times and shape tolerances. In order to obtain a ceramic piece about 3 cm thick 30 layers are needed. Three points flexural test were carried out on a Lloyd LR5K instrument on a 35mm long sample made out with the chosen suspension. The stress-strain plots are shown in Figure. 6.

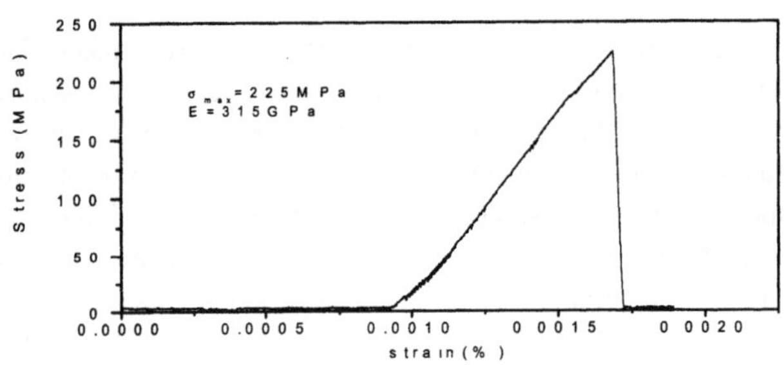

FIGURE 6: mechanical behaviour of sintered ceramic

Other samples obtained with different formulations went on delamination, because of the not very good adhesion between different layers during STL building. However, delamination did not occur for this suspension as a consequence of the low stiffness and high stress relaxation properties of the matrix. The value of the modulus is about half of that for traditionally sintered alumina powders. The density value is 2.9 g/cm^3, compared to the value of 3.72g/cm^3 of the fully dense alumina powder used.

4 CONCLUSIONS AND FUTURE WORK

Ceramic objects have been obtained from photoactivated ceramic suspensions using an experimental stereolithographic set-up. The ceramic suspensions have been studied. Their photoreactivity and viscosity are considered to be very important in the forming process efficiency, whereas their thermal behaviour and ceramic residual are important during sintering phase in determinig the structural properties of the final ceramic. Mechanical properties i.e. strength and modulus of sintered products are good enough, not so far from those of traditionally sintered ceramics. More likely the differences are due to the high residual porosity, not less than 10%. It is confirmed that ceramic structural component can be designed and fabricated by stereolitography from photoactivated ceramic suspensions. At present the ceramic suspension has been testing in a stereolithography apparatus, Model SLA-250 with a He/Cd laser source. Typically, the laser power at the surface of the vat is about 30mW. SLA 250 has been modified in order to build ceramic objects. A new loading vat with varying capacity (8-45Kg) has been designed and built, in order to produce parts with lower material quantities. "Windowpanes test parts" [2] are in construction, in order to determine on the apparatus the working parameters E_c and D_p. The values for the suspension are waited to be quite different from those of SL 5170, so building parameters of the machine have been modified in order to work with lower reactivity and higher viscosity suspensions.

Future investigations will be devoted to optimise the process and the mechanical performances, improving the rheological properties of suspensions and increasing alumina content. Furthermore a deeper analysis of the STL process parameters (linewidth compensation factor, shrinkage compensation factors in plane and cure depth) relating the final dimensions of the sintered parts to the CAD model is needed.

REFERENCES

1. R.Knitter, W. Bauer, D. Gohring, P. Risthaus, (2001), Fabrication of Ceramic Microcomponents by Process Chain, Euro RP 2001.
2. P.F. Jacobs, "Rapid prototyping and manufacturing", (SME, Dearborn, USA 1992)
3. A. Greco, A. Licciulli, A. Maffezzoli, (2001), Journal of Materials Science, 36, 99-105
4. G.A. Brady and J.W. Halloran, (1998), J. Mater. Sci. , 33, 4551
5. D. J. Waller, and A. Safari, (1992), Am. Ceram. Soc., 7 5 [6] 1648-55
6. H. L. Marcus, and D L. Bourell, (1993),Advanced Materials & Processing, 9, 28-35.
7. M. Burns, "Automated fabrication improving productivity in manufacturing", (Prentice Hall, 1993).
8. M.J. Edirisinghe, Mat. World, (1997), 5, 138.
9. C.I. Alwood, M.C. Maguire, B.T. Pardo and A.A. Bryce, (1996) SAMPE J., 32(1) 55.
10. M. Agarwala, D. Bourell, J. Beaman, H. Marcus and J. Barlow, (1995), Rapid Prototyping J. 1, 26
11. G.A. Brady and J.W. Halloran, (1998), J. Mater. Sci. 33, 4551
12. CIBA-GEIGY, Technical data sheet.

CT-BASED REVERSE ENGINEERING AND RAPID PROTOTYPING: EXPERIENCES IN DIFFERENT APPLICATION DOMAINS

C. Bandera, M. Felice, S. Filippi

DIEGM Department, University of Udine, Italy

KEYWORDS: Computer Tomography, Reverse Engineering, Rapid Prototyping.

ABSTRACT. Current Rapid Prototyping (RP) techniques are becoming available in many different application domains and hence for a growing number of users, even if some features make them still usable by trained personnel only. The same evolution is happening for what concerns Computer Tomography (CT) activities: moving from medical diagnostic issues, now CT is used in industrial environments (quality testing, etc.), in artistic ones (non-destructive acquisition of data from monuments and similar), and, always in the medical field, during the process planning for surgery activities and the development of surgical tools.

This paper summarizes our experiences related to these domains, where the acquisition of the reality, performed by means of the proper CT equipments, has been the first phase of the process that, through the recognition of the original structures, has finished with the generation of the physical prototypes using RP techniques.

1 INTRODUCTION

Among the Reverse Engineering (RE) techniques that are assuming a major role in various fields, the Computer Assisted Tomography (CT) is now widely used not only in medical but also in industrial and archeological contexts, mainly because of the non-destructive nature of the data acquisition practices.

Moreover, a strong link is becoming evident between CT and Rapid Prototyping (RP) technologies. Two main factors can explain this interaction. First, there is a conceptual link between the method used by CT to acquire the object geometry (slice by slice) and the method used by RP to build a new geometry (layer by layer). This similarity has the practical effect that a direct interface between these techniques improves the quality of the result in terms of accuracy and consistency of the replicated part with the original one.

Another reason that can explain the link between CT and RP is the possibility offered by RP techniques to produce a real model, directly from the virtual one, without any specific tool and independently from the geometric complexity of the object to be produced. This is particularly advantageous in the medical field where the geometrical complexity of the structures that have to be reproduced can be very high.

Published in: E.Kuljanic (Ed.) *Advanced Manufacturing Systems and Technology*, CISM Courses and Lectures No. 437, Springer Wien New York, 2002.

2 METHODS AND TOOLS

2.1 COMPUTER TOMOGRAPHY

As it is widely known, CT is a data acquisition technique that uses a flux of highly collimated X-rays to obtain information about a certain section of an object. A detector is placed behind the object and a multidirectional analysis allows reconstructing the density distribution which is illustrated by different levels of gray representing different densities.
This brings a notable advantage if compared with other Reverse Engineering (RE) techniques, that is the possibility of describing both the external and internal geometry of the object. Moreover, the images obtained allow the possibility of describing the object in terms of material density and, therefore, components weight.
In the meantime, this technique is expensive because of the cost of the tomographer itself and because of the computation and professional resources needed for 2D image processing and for 3D model reconstruction [1].

2.2 FROM COMPUTER TOMOGRAPHY TO RAPID PROTOTYPING

The images obtained from the CT allow the identification of the structures which are defined by different gray values. To obtain a replica of the original part through RP, these images need to be segmented, i.e. a region of interest is selected in every image on the basis of the gray level, and the 3D structure is rebuilt by piling up the successive regions of interest.
The profiles obtained from the segmentation can be used to define the bi-dimensional geometry of every slice, thus producing directly the SLI file for the RP machine [2]. Since the distance between the CT images is usually greater than the layer thickness for the RP machine, additional slices should be created and this is done by cubic interpolation.
From the same segmented images, a virtual mock-up can be obtained by reconstructing a polygonal mesh of triangles by linear interpolation from the region contours. In this way, the STL file is generated, which can be sliced and then used for RP.
The advantage of the first method stands in the higher accuracy that can be obtained, whereas the STL model can be useful for simulations or as a reference for designing new parts.

2.3 RAPID PROTOTYPING

Rapid Prototyping includes a large number of manufacturing technologies which build a part layer by layer, starting from a virtual model, thus generating a three dimensional model by material addition. One of the most well known and consolidated RP techniques is called Stereolithography (SLA), which uses a liquid resin which is polymerized, layer by layer, by a UV laser beam, which scans the region representing the area of each section. The part is built on a grid that is lowered after the polymerization of every layer; in this way, the solid part is built inside the liquid resin. When the building process is finished, the platform is lifted, the part is drained, the supports, which have allowed the production of hanging parts, are easily removed and the polymerization is completed using a UV oven [2].
Between the various advantages of this technology, three are the most relevant for our interests: the possibility of producing very complex geometries without splitting the part into

different pieces and without the need of specific tools; the direct interface with the CAD or the RE software; and the rapidity of the overall process and therefore the possibility of immediately checking any errors.

3 CASE STUDIES

In the following paragraphs, three case studies will be illustrated, the first two regarding medical applications, the third one related to the archeological field. The tools used are the same in both application domains, the only difference being in the tomographer: a medical one in the first two case studies and an industrial micro-tomographer in the last one.

3.1 MEDICAL APPLICATIONS

The use of CT in the medical field is widely known and appreciated, since this tool can describe the three dimensional anatomical space as a sequence of bi-dimensional images instead of a unique projection as in conventional X-ray images. Moreover, the most recent CT devices allow to get automatically three dimensional renderings of the anatomical structures. Nevertheless, these data are still used as virtual images or as a visual description of the three dimensional context, rather than as a real volume with proper dimensions, weights, densities, etc. Less common is the use of this data volume to build a solid virtual or real model, which could be tested, manipulated, and modified.

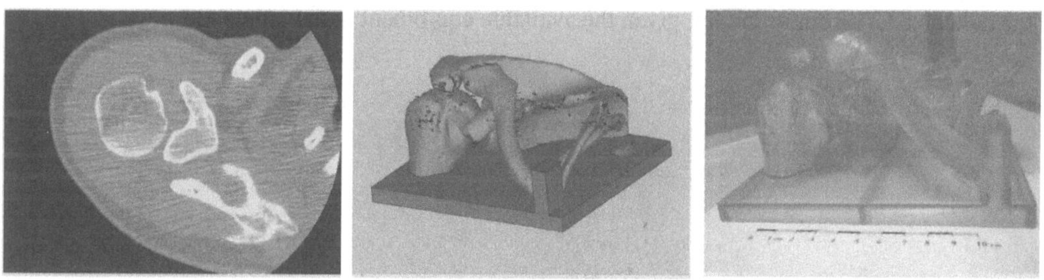

FIGURE 1. CT image, virtual model and RP replica of the shoulder joint.

SHOULDER PROSTHESIS

In many cases of joint prosthesis implants, the data acquisition is performed using CT, because it is much more effective than a radiography for understanding the disposition of the anatomical structures with special interest to the three dimensional relations of the joint bones. In this specific case study, data were obtained from the hospital on a magneto-optical disk and were then transformed in a stack of images, in the form of project of Mimics software, developed by Materialise, which is the package we have used for image processing and 3D reconstruction [3]. On the basis of the gray levels, the anatomical structures were identified and the three dimensional model was reconstructed as a mesh of triangles. In this case, even in presence of minimum artifacts and noise, the structure could be identified quite easily, since the contours of these bones (humerus, scapula, ribs, clavicle) were quite regular. Thanks to

that, automated tools could be used to eliminate contour irregularities due to artifacts (i.e. screws from previous implants): peaks or cracks have been eliminated by eroding and dilating the selected masks, that is adding or subtracting one or more pixels to the contours of the segmented area.
The mesh of triangles thus obtained has been used as a virtual model for simulating the implant of different prosthesis: qualitative but also quantitative evaluations have been performed on these models. For example, it has been evaluated the amount of material to be removed for generating the clearance for the prosthesis and the contact surface thus obtained between prosthesis and bone.
Finally, the virtual model has been used for building the RP model: first, a base platform has been added to preserve the relative position of the bones, then the complete model has been sliced in layers of thickness equal to 0.125 mm and the file thus obtained has been sent to the RP machine.
All these steps, from CT images to triangle mesh, and finally to RP model, can be seen in figure 1.
The major problems encountered in this case are related to the raw data format and to their physical support. This is a relevant problem because the output of many CT devices is written in proprietary format and the physical support is usually a magneto-optical disk or a data cartridge. This requires a specific hardware and software equipment to read the data and the finding of a proper equipment can require a time that is not compatible with the patient needs, especially in cases of trauma which usually need to be solved within days. Therefore, if the entire process, from RE to RP must take less than a week, it is absolutely necessary to obtain the data on a CD-Rom or via ftp, given the available equipment at the moment.

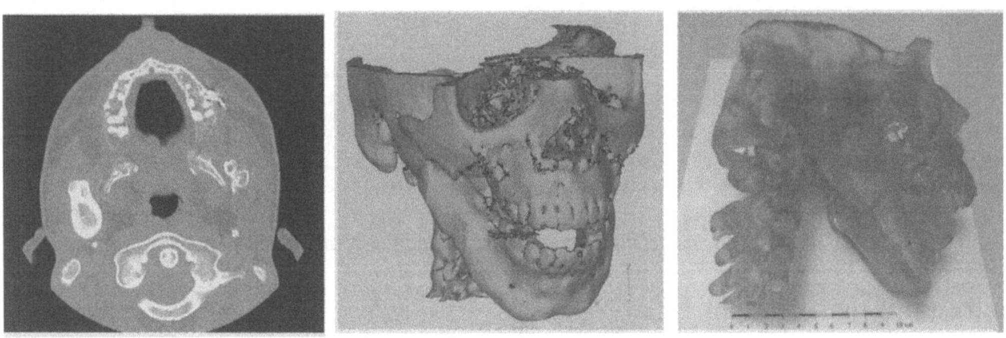

FIGURE 2. CT image, virtual model and RP replica of the skull.

MAXILLO-FACIAL SURGERY

Maxillo-facial surgery deals with the very complex geometry of the facial and skull bones; in this context, the availability of a three dimensional real model can optimize many stages of the surgical procedure, i.e. the understanding of the anatomical context and the specific pathology, the planning of the surgical cuts, the shaping of the required artifacts, etc.[4][5][6][7]. The most relevant advantages include the possibility of performing these operations before the intervention (thus reducing the surgical procedure duration), not onto the patient (thus reducing his/her risks), and onto a faithful replica of the patient morphology (so that errors are minimized).

The major problems in the RE activity are connected with the artifacts generated by dental closures and prostheses. In this case study (Fig.2), but, more generally, in maxillo-facial context, these problems could not be overcome with automated tools because of the high complexity of the bone structures, which are intrinsically irregular; the best practice resulted the manual editing and correcting of the segmented areas. This operation required the active involvement of the doctors, who have the knowledge of the anatomic context and the understanding of the accuracy which is effectively required in different anatomical areas [8].

3.2 ARCHEOLOGICAL APPLICATIONS

The application of CT in archeological field is due to its non destructive nature and to the possibility of analyzing both external and internal properties of the objects. This has appeared very useful mainly in paleontology and for fossil analysis [9][10][11].

CELTIC FIBULA

In this case study, a celtic fibula was scanned with an industrial micro-tomographer: 195 images were obtained, with isotropic voxels of 0.2 mm. A first analysis of the images highlighted the evidence that the fibula was constituted of at least five components, probably of different materials: the main body, a spring, a pin inside the spring and two caps at the end of the pin (figure 3).

Nevertheless, the separation of the models of these components and especially of the spring from its pin has been difficult and a combination of automated tools and manual editing was used to obtain the result. Two series of models of the fibula were obtained with this procedure: a unique STL model of the complete fibula and five STL models of its single components. All these models were replicated in epoxy resin, using stereolithography, with layers of 0.1 mm thickness value. The epoxy model of the fibula body has been used in investment casting to replace the wax pattern: it has been covered with ceramic material and air dried, the pattern material has been melted and evacuated leaving an empty ceramic shell, which has been filled with molten silver.

As it can be seen in the figure, the silver fibula presents some cracks in the frontal part; these cracks can be caused by solidification problems in this thin area or by the melted epoxy resin which can break the ceramic shell when it is melted.

4 CONCLUSIONS

The experiences presented above showed the advantages of the application of CT and RP in medical and archeological fields, not only as a research activity but also as a common practice. Moreover, new possibilities are offered in both domains by the combination of these tools and practices: in medicine, the three dimensional model of the patient anatomy can be used for designing custom prostheses or for simulating the structural interaction between prostheses and bones, in archeology, the models can be used not only for replicating the object with different materials but also for virtually or physically reconstructing missing parts of them.

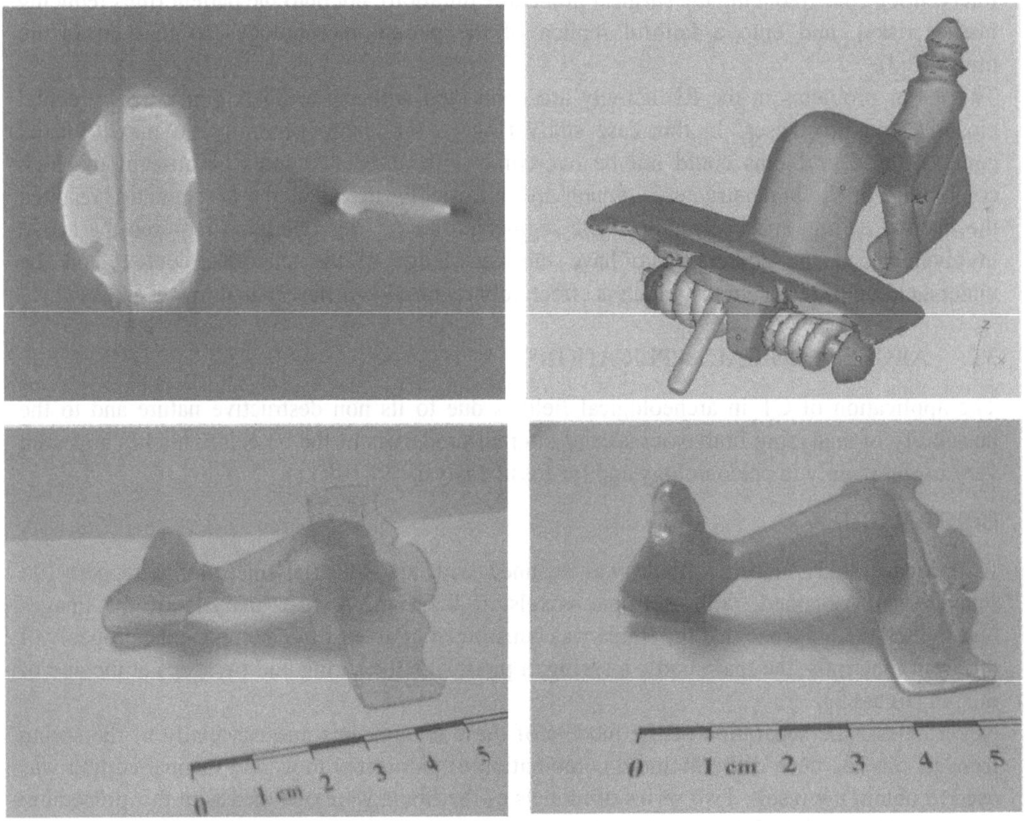

FIGURE 3. CT image, virtual model, RP replica and silver reproduction of the Celtic fibula.

ACKNOWLEDGEMENTS

The authors gratefully acknowledge the help of dr. Facchini and dr. Dalla Pria (Lima Group), prof. Bazzocchi (Radiology Institute, Polyclinic of the University of Udine), prof. Politi and prof. Robiony (Maxillo-facial Department, Polyclinic of the University of Udine) and dr. Buora (Civic Museums of Udine).

REFERENCES

1. Kalender, W.A., (2000), Computed Tomography: Fundamentals, System Technology, Image Quality, Applications. Publicis MCD Verlag, München.
2. Gatto, A., Iuliano, L., (1998), Prototipazione rapida: la tecnologia per la competizione globale. Milano (Italia), Tecniche nuove.
3. Mimics Reference Guide, (2001), Materialise Software.

4. Lightman, A.J., (1998), Image Realization - Physical anatomical models from scan data. SFF Proceedings, SPIE Symposium, Medical Imaging, San Diego, CA.
5. Swann, S., (1996), Integration of MRI and stereolithography to build medical models: a case study. Rapid Prototyping Journal, Vol. 2, No. 4, 41-46.
6. Kermer, C., (1999), Preoperative stereolithographic model planning in craniomaxillofacial surgery. Phidias Newsletter No. 2, Denmark.
7. Haex, J.K.Th., Poukens, J.M.N., (1999), Preoperative planning with the use of stereolithographic model. Phidias Newsletter No. 3, Denmark.
8. Filippi, S., Bandera C., Felice, (2001), Cooperative Work in medicine: linking together distributed expertise in hip-prostheses development and surgical planning. International conference ISPE-CE2001, Advances in CONCURRENT ENGINEERING, California (USA).
9. Zollikofer, C.P.E., Ponce de Leon, M.S., (1998), Tools for Rapid Prototyping in the Biosciences. IEEE Computer Graphics & Applications, Nov. 1995, 48-55.
10. Zhang, G., Tsou, Y., Rosenberg, A.L., (2000), Reconstruction of the Homunculus skull using a combined scanning and stereolithography process. Rapid Prototyping Journal, Vol. 6, No. 4, 267-275.
11. D'Urso, P.S., Thompson, R.G., Earwaker, W.J., (2000), Stereolithographic (SL) biomodelling in palaeontology: a technical note. Rapid Prototyping Journal, , Vol. 7, No. 4, 212-216.

EXPLOITING THE EVALUATION OF RP ARTIFACTS TO UPDATE KNOWLEDGE IN DESIGN GUIDELINES

C. Bandera, S. Filippi, F. Miani, G. Toneatto

DIEGM, Dept. University of Udine, Italy

KEYWORDS: Rapid Prototyping, Knowledge Based Systems

ABSTRACT. The DMSLS (Direct Metal Selective Laser Sintering) RP technique, even more than the other ones present on the market, shows features that make it usable only by well-trained personnel. To avoid this limitation, hence aiming at lowering the distance between designers and potentialities of this RP technique, our research group has developed a set of guidelines which may be used during the first phases of the product lifecycle. One of the main issues is, of course, the quality of the knowledge present inside these guidelines. Up to now, information has been gathered mainly by analyzing the user manuals of the RP machines and exploiting our own experiences in the field. The aim of this paper is to describe a more rigorous approach for collecting new information to be used to update the existing knowledge. DOE practices, metrological activities and more, have been used to generate a set of study cases, sufficient to highlight at most limits and potentialities of the DMSLS technique. Coming from the experimental data, after the required phases of analysis and synthesis, we have got the information that makes the guidelines day by day a more effective tool for helping designer during product definition and engineering.

1 INTRODUCTION

The Direct Metal Selective Laser Sintering (DMSLS) is one of the most promising Rapid Prototyping (RP) technologies, because it allows generating metal objects in a direct way [1]. This lets to perform a great number of tests about the prototypes but, even more, it makes possible the generation of industrial tools (rapid tooling [2]), mainly related to the plastic injection field. Both the fast procedures and the advantages related to the construction modalities (layer by layer), where very complex geometries and cooling channels can be easily built up, make the adoption of DMSLS really interesting. Nevertheless, DMSLS technology shows some limitations related to the process itself; i.e. scarce surface finishing, dimensional tolerances (around 0.1 mm), lower mechanical properties than the steel usually adopted in the field, frequent need for supports, etc.

Because of these potentialities and lacks of the DMSLS technology, designers and engineers must be given some help to operate the best choices related to a domain that usually they don't know at all. In particular, they should be able to evaluate the added value of the adoption of this technology if compared with the classic ones, and to modify the design and the engineering of the product to make it compatible with the requirements of DMSLS.

Starting from here, our research group is developing some guidelines to assist the user in modeling his/her product from the DMSLS point of view. Nonetheless, the results could be quite easily extended to other application contexts.

Published in: E.Kuljanic (Ed.) *Advanced Manufacturing Systems and Technology*, CISM Courses and Lectures No. 437, Springer Wien New York, 2002.

The first partial result of the research has been the definition of a homogeneous structure to collect the knowledge and to make it available maximizing the usability during the design phase.

At this time, the main aspect to investigate about is the content of the knowledge database inside the guidelines. The importance of this point is reinforced by the fact that some recent tests of the first version of the guidelines, implemented by means of the DBMS Microsoft Access, have validated the structure and the knowledge accessibility, but at the same time they have shown some lacks for what concerns the information content.

The main goal of the activities described in this paper has been the investigation about different ways to collect the information to update the knowledge inside the guidelines.

In the following, the guidelines will be shortly described, highlighting the importance of the knowledge database and the lacks of the information content of the classic sources (user manuals of RP machines, etc.). Then all the activities performed to collect new data will be described, starting from experimental ones and ending with rapid tooling practices. Some considerations about the results and potential future developments will close the paper.

2 THE GUIDELINES

Limits and potentialities of DMSLS reconfigure the modeling domain. For this reason it is necessary to make some important choices during the design phases when considering the opportunities of using this RP technology. Some help in this sense could come from the classic design rules presented by the user manuals of the RP machines, but they show deep lacks for what concerns structure, completeness and usability. The result is that they are scarcely adopted by the designer [3], remaining at disposal of RP technicians only.

The development of the guidelines has started form this consideration. They are a knowledge based tool [4] with a high level of usability, oriented to users not necessarily expert about the RP/DMSLS domain [5]. More closely, the guidelines have been thought to help the designer in developing a new product (design), so as so to guide him/her in modifying an existing one from the point of view of the RP generation (re-design). Guided by a system that is based on the evaluation of the costs of the modifications and of the costs and benefits in respecting or violating the suggested rules, the user could establish if the adoption of the DMSLS represents a valid alternative to the classic approaches and, in the same time, which one could be the best re-design configuration.

As a result of this phase of the research, a first version of the guideline has been implemented. Figure 1 shows a couple of pictures taken during a test session.

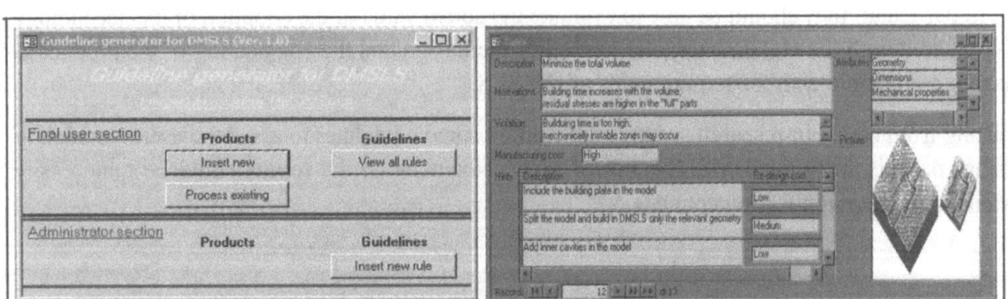

FIGURE 1. Some pictures of the guidelines taken during a test session.

At this stage, the information present inside the database has been collected from the user manuals of some DMSLS devices and from personal experiences using an EOS250X machine, placed inside the RP lab of the University of Udine. From the beginning, it has been quite clear the difficulty in finding information related to the meaningful aspects for the design and engineering phases. The lack of knowledge has risen during the first study case [6], where the guidelines have been used in designing a coffee machine. These lacks were mainly related to mechanical properties of thin elements and to the behavior of the prototype when filled with water. By the way this test has highlighted the validity of the structure of the guidelines; so to suggest once again focusing the attention about the information content.

The main lacks of the guidelines could be summarized as follows:

a) they don't address the physical and mechanical properties of the sintered materials;

b) there is scarce attention to the validity of the tools generated for rapid tooling purposes;

c) there is no quantification of some aspects related to the DMSMS process, i.e. the amount of residual stress, deformations, etc.

Then, in trying to improve the knowledge content of the guidelines, priority has been assigned to those activities able to collect data related to these aspects. After that, the effort has been to understand how these data could influence the design phase and how they had to be integrated inside the database.

The following paragraph describes the activities conducted to collect the data and the results coming from them.

3 ACTIVITIES

To fill the lack of data in literature about the properties of objects built by DMSLS technology a complete set of experiments has been designed and carried out to investigate the main physical, mechanical and dimensional characteristics of laser sintered pieces. At the same time sintered mould inserts have been tested in the field to evaluate the behaviour and the validity of these objects as industrial tools. These activities are described in the following.

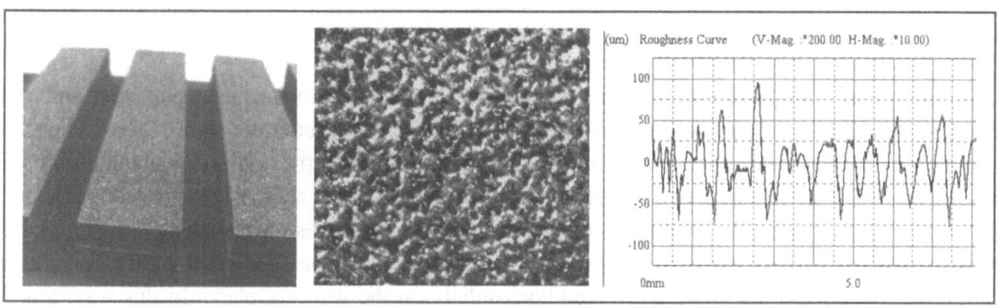

FIGURE 2. experimental measurements of surface roughness; from left to right: specimens used for the measures, an image of a DMSLS surface taken by a stereomicroscope and a roughness curve taken by a surface texture measuring instrument.

3.1 EXPERIMENTAL TESTS ON LASER SINTERED BODIES

First some measurements of the density have been done using the Archimedean method as suggested by the standard ISO 3369. The data give some initial information about the quality of the sintering process and the mechanical properties. Another characteristic investigated in the first stage is the surface finishing. To this purpose some measures of the roughness on the sides of sintered parallelepipeds have been done (figure 2). Also the improvements obtained by finishing processes as shot peening and brushing have been evaluated. Then metrological testing has been carried out to determinate the dimensional and geometrical accuracy of the pieces. In this case a CCM machine has been used to measure the tolerances on geometries studied on purpose. At last mechanical tests have been made as: tensile test, compression test, impact test, hardness test. Thanks to the experimental work done, it has been possible to collect clear data on the properties of the iron-based sintered material currently used by EOS in DMSLS process (Direct Steel 50 v2). The properties of a metal powder developed inside the department have been analysed too. Some results for the commercial powder are reported in Tab. 1. More details on the tests are presented in another paper [7].

Besides the problems in carrying out the tests in a competent way and in using proper instruments, which leaded us in setting up collaborations with other universities and laboratories, the most relevant difficulties have raised in the designing phase of experiments.

TABLE 1. summary of experimental tests on laser sintered objects made in commercial powder DirectSteel50.

Property	Experimental value	Notes
Hardness	230 HV10	Milled surface
Tensile strength	280 MPa	Load perpendicular to layers plane
Compression strength	800 MPa	Load perpendicular to layers plane
Roughness	15 – 20 μm	
Roughness after shot peening	10 – 15 μm	

The first question highlighted has been the absence of standards for testing metal laser sintered objects, standards to be followed to have results of real technical/scientific significance. Thus we decided to base the measurements on the ISO and MPIF standards for traditional sintered objects, similar to the laser sintered ones under many aspects. But this choice has proved to be not satisfying in some cases. An example is represented by the specimens for the tensile test (figure 3), characterised by a high ratio of the length on the transversal cross-section (ISO 2740). To reduce building time the specimens were oriented horizontally on the working plane. In this way, a non uniform distribution of the residual stresses occurred, causing bending lengthwise and, for some pieces, fracture during the removal of the supports. It seemed clear that the remaining specimens wouldn't have given meaningful results to evaluate the mechanical properties of DMSLS object built at the state of the art, that is respecting the rules suggested by the guidelines. This experience has highlighted the need for redesigning some DMSLS specimens in such a way that critical situations are avoided; the collected data are actually representative of the behaviour of the objects commonly built by this technology.

For the tensile test a better solution, yet not the best, has been to use the traditional axial-symmetric tensile specimen (UNI EN 10002), to be built vertically, so that the residual stresses are minimised. On the other way, a long building time and the risk of breaking caused by the recoater sweeping have to be accepted. The compression test specimen, a small cylinder, 10mm in height, 10mm in diameter, has proved to be the most suitable for mechanical testing as it can be built successfully in a very short time and gives good reproducibility of results. Moreover the data obtained by the compression test can be used to estimate, through proper equations, other mechanical properties, as hardness and tensile strength.

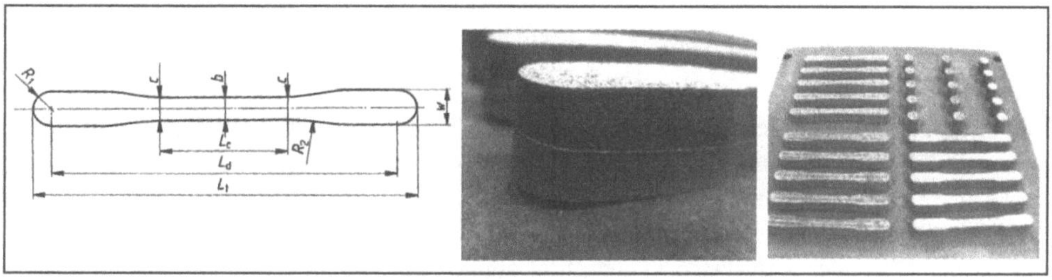

FIGURE 3. mechanical tests; from left to right: the geometry of the ISO specimen for tensile tests on traditional sintered objects, an image highlighting the bending of this specimen made by DMSLS processing, the plate with tensile and compression specimens.

Another problem arisen during the planning phase of experiments [8] has been in defining the type and the number of tests necessary to *investigate the effects of the processing parameters* on the properties of DMSLS objects. It has to be stated precisely that a careful analysis on the influence of the processing parameters is mostly of technological interest but it can also concern the field of design methods. In fact the choices taken during the design phase may affect and/or be affected by building parameters, considering their influence on the good course of the manufacturing process and the quality of results. At this time the guidelines assume implicitly standard machine settings but, thanks to the experimental tests, we are studying the possibility to integrate the technological variables into the database, in such a way that a product configuration, expressed by a set of accepted rules and changes, generates automatically the main processing parameters. Among these parameters it is possible to list: scanning speed, powder grain size, layer thickness and hatching type, this one defined by several factors like hatch spacing, beam compensation, building strategies as pre&post contouring and skin and core sintering. It can be easily noticed that the number of experiments needed to study the effects of all these variables and their combinations is too high by far. So we first have to renounce to study interactions between factors of high order (commonly not over the second). Moreover it is possible to apply statistical methods as DOE [9] to reduce the number of experiments further on. One last important help in planning the tests may come from the development of theoretical models about the process which can forecast the parameters that mainly affect the property investigated case by case, excluding the others.

3.2 RAPID TOOLING TESTS OF LASER SINTERED OBJECTS

Although experimental testing is of great importance in evaluating the properties of laser sintered objects properly and in understanding how they are affected by processing parameters, it's not enough to forecast the behaviour of DMSLS components in rapid tooling applications. As already mentioned, DMSLS technology allows building inserts or small moulds for plastic injection moulding in very short times, regardless of the geometrical complexity of the object and with the possibility to make conformal cooling channels (figure 4a), which can reduce the cycle time and improve the quality of the final parts. Nevertheless the properties of DMSLS moulds are worse than the ones of traditional moulds, so that is often necessary to resort to subsequent processing to partially improve the quality of rapid tools, considering at the same time that the gain in terms of time and cost is not lost.

From the point of view of design methods it becomes of great importance to collect information to define with accuracy the *domain of applicability* of DMSLS technology in rapid tooling, referring in particular to injection moulding. For this purpose the best solution is testing the behaviour of DMSLS tools directly in manufacturing processes, to have an evaluation, maybe scientifically less rigorous but of greater importance in industrial applications, of properties difficult to analyse through lab tests (finishing, machinability, weldability, coupling, water sealing and mechanical behaviour under the complex conditions of moulding process).

Thus we have set up collaborations with companies specialised in injection moulding, for the purpose of building and testing DMSLS tools and finding new solutions to improve their quality. These activities are still ongoing, firstly because a lot of work is required in such a recent field and still full of limitations as rapid tooling, secondly because it is been not so easy to find companies willing to invest time and money in this technology, which is very promising for the future but currently can't compete with traditional approaches. Nevertheless the first tests have given encouraging results and much useful information has been collected to improve the knowledge content of the guidelines.

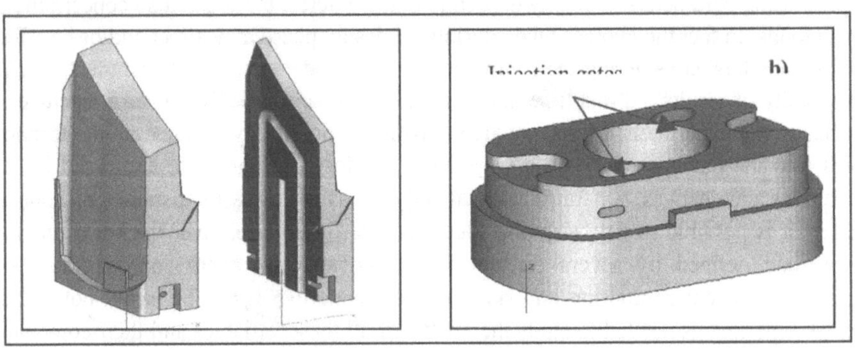

FIGURE 4. an example of mould insert with conformal cooling channel (a) and the model of the DMSLS insert for a car light mould used in rapid tooling tests (b).

As an example we report the experience of a DMSLS mould insert for a car light body (figure 4b). The object has a simple shape, with a central hole, two opposite injection gates and two holes to be threaded in the bottom. The polishing of the insert, necessary for esthetical and functional reasons (ejection of pieces can be compromised), has shown the impossibility to reach the finest finishing because of the micro porosity of the sintered material. Coupling dimensions were 0,1 – 0,15mm less (0,2mm after polishing) than the nominal ones. This has made necessary adding new material by cold welding operations to respect the tolerances, which has taken a remarkable amount of time. Moreover, the two injection gates were slightly different, with the consequence of an unbalanced flow and a bed filling of the cavities. For this reason, two new gates have been built beside the original ones using electrical discharge machining. The sintered tool, after the operations described, has been used for moulding tests: it has proved to bear well the injection pressure and furthermore the mould has worked automatically. Due to cost reasons it has not been possible to verify the maximum life of the insert, yet it hasn't shown any sign of deterioration after 50 cycles. This experience has given quantitative information to control better some well-know limitations of DMSLS technology (finishing, tolerances, etc.) and has contributed to demonstrate that mechanical properties of rapid tools, although worse than the ones of traditional moulds, can bear typical injection pressures in small series applications.

3.3 UPDATING THE KNOWLEDGE BASE OF THE GUIDELINES

While completing the data collection phase, it has been possible to start the analysis on how to integrate them inside the guidelines knowledge database. This operation could come at two levels:

1) information could be directly inserted inside the database, if it is simple to manage and use by the designers (i.e., surface roughness, tensile strength, etc.);

2) there could happen a *projection* of the technological information onto the design domain with subsequent modification of existing rules and generation of new ones.

The second level highlights the fact that the activities to collect information and to insert it inside the guidelines are not a simple refinement of the database content but, sometimes, a real upgrade of the database itself. For example, as following by the experiences in the field, some new rules are needed to address the elimination of some feature of the CAD model before to generate the sintered prototype; these features could be machined later, using classic procedures (threaded holes, multiple cooling channels, etc.). The information related to the removed features should be maintained inside the model description, because the RP operator must know the position of them to be able to set the job of the RP machine in the correct way.

4 CONCLUSIONS AND FUTURE DEVELOPMENTS

The activities related to the analysis of the properties of the sintered prototype and their usefulness in the rapid tooling case are still on the way, but they are allowing a meaningful improvement of the guidelines knowledge base. Up to now, the only aspect not covered by these investigations is related to the critical situation of the DMSLS, i.e. the presence of residual stresses, deformations and flaws. All of this would require the development of an analytic model of the whole sintering process. This model is under study by many researchers and some ideas are documented in [10],

[11], and [12]. Maybe in the future these models will allow the numerical simulation of the laser sintering process, giving some indication about the effects of the process parameters. By the way, the state of the art still shows some uncertainty and this produces limitations with no doubt on the design domain. For what concerns future developments, the guidelines show the possibility to automatically generate an output for the RP operator, containing precise indications on how to set up the model for the job (i.e. listing the features that will be machined afterwards) and the process parameters. Another important aspect will be the specialization of the guidelines in the field of rapid tooling, obtained by the development of a collection of new rules.

ACKNOWLEDGEMENTS

Authors of this paper would like to thank R. Meneghello, A. Zambon, S. Bruschi, and T. Del Negro for the help in performing the experimental activities. Nonetheless CSAT and ATS industries have given their best in using the DMSLS technology in industrial field; M. Zanzero helped us in generating the sintered prototypes using the RP lab of the University of Udine.

REFERENCES

1. Agarwala, M., Bourell, D., Beaman, J., Marcus, H., Barlow, J., (1995), Direct Selective Laser Sintering of Metals, Rapid Prototyping Journal, Vol. 1, No. 1, pp. 26-36.
2. Jacobs, P. F., (1995), Stereolithography & Other Rp&m Technologies: From Rapid Prototyping to Rapid Tooling, Society of Manufacturing Engineering.
3. Filippi, S., Bandera, C., Toneatto, G., (2001), Rapid Prototyping: Design Rules to Achieve Good Results from DMSLS Processes, IIEC 2001, 1st International Industrial Engineering Conference.
4. Haffey, M., K., D., and Duffy, A., H., B., (2000), Knowledge Discovery and Data Mining within a Design Environment, In Cugini, U., Wozny, M. (eds.): Knowledge Intensive CAD Volume 4, Proc. Fourth IFIP WG 5.2 Workshop on Knowledge Intensive CAD, 72-87.
5. Nielsen, J., (1993), Usability Engineering, Academic Press, Cambrige, MA.
6. Bandera, C., Filippi, S., Toneatto, G., (2001), Generation and Testing of Guidelines for Effective Rapid Prototyping Activities, ADM 2001.
7. Miani, F., Bruschi, S., Dal Negro, T., Meneghello, R., Zambon, A., (2000) New Iron Based Experimental Powders for the Direct Metal Selective Laser Sintering, Proceedings 9th European Conference on Rapid prototyping, Athens.
8. Box, G. E. P., Hunter, W. G., Hunter, J. S. (1978), Statistic for Experimenters: an Introduction to design, Data Analysis and Model Building, John Wiley & Sons, New York, NY.
9. Montgomery, D. C., (1996, 4th Edition), Design and Analysis of Experiments, John Wiley & Sons, New York, NY.
10. Zhang, Y., Faghri, A., Buckleym, C.W., Bergman, T.L., (2000) Three-Dimensional Sintering of Two-Component Metal Powders With Stationary and Moving Laser Beam, Journal of Heat Transfer, Vol. 122, 150-158.
11. Shiomi, M., Yoshindome, F., Abe, F., Osakada, K., (1998), Finite Element Analysis of Melting and Solidifying Processes in Laser Rapid Prototyping of Metallic Powders, International Journal of Machine Tools & Manufacture, Vol. 39, 237-252.
12. Matsumoto, M., Shiomi, M., Osakada, K., Abe, F., (2002) Finite Element Analysis of Single Layer Forming on Metallic Powder Bed in Rapid Prototyping by Selective Laser Processing, International Journal of Machine Tools & Manufacture, Vol. 42, 61-67.

A CONTRIBUTION TO THE ANALYSIS, CHARACTERIZATION AND IMPROVEMENT OF SURFACE PROPERTIES GENERATED BY DIRECT METAL LASER SINTERING (DMLS) FOR RAPID TOOLING

M. Vezzani[1], A. O. Andrisano[1], R. Groppetti[2], P. Onesti[3], A. Rossi[2], A. Scrivani[2]

[1] Department of Engineering Sciences, University of Modena and Reggio Emilia, Italy
[2] Department of Industrial Engineering, University of Parma, Italy
[3] Democenter, Modena, Italy

KEYWORDS: Direct Metal Laser Sintering (DMLS), Rapid Prototyping and Tooling, injection moulding dies.

ABSTRACT. Rapid Prototyping and Tooling (RP&T) technologies are applied extensively and intensively in the industry and their importance is increasing because it reduces the development time and cost required for product prototyping and injection moulding dies and EDM electrodes tooling. Namely additive technologies are getting interest due to the possibility to build a component by adding material. The paper addresses the study and application of Direct Metal Laser Sintering (DMLS) of copper alloys for Rapid Tooling of dies for thermoplastic polymer injection moulding. Specific samples have been designed by CAD, in order to be representative of the geometrical and morphological features and functional characteristics of the dies. The main limitations of DMLS prototypes are relevant to surface properties as porosity, roughness, hardness, corrosion, erosion and wear resistance. The study takes into consideration the characterization and improvement of DMLS surface properties in order to meet their functional requirements. To this purpose, different surface treatments have been studied and applied, as blasting and peening, resin infiltration, electroless nickel plating. Sample surfaces has been fully characterized and analysed, as obtained by DMLS, and after surface treatments. 3D surface roughness and microgeometrical morphology and porosity and microhardness properties have been systematically investigated.

1 INTRODUCTION

Rapid Prototyping and Tooling (RP&T) technologies are applied extensively and intensively in the industry and their importance is increasing in order to reduce the development time and cost required for product prototyping and injection moulding dies and EDM electrodes tooling [1-3]. Several RP&T technologies have been proposed and widely applied in the industry, but their application domains and capabilities have not yet always fully defined and their potential has not yet fully exploited. Particularly additive technologies are getting interest due to the possibility to build a component by adding material, rather than removing it, using different materials, as polymers, metal alloys and sand, etc. Different additive technologies can be considered, although they show common characteristics: it is possible to build the product layer by layer directly from CAD models, without necessity of dedicated tooling and fixtures. These technologies are really flexible and allow to get components with complex geometrical shape and from several materials. All these advantages brought a lot of interest to these technologies during last years and their use has been addressed mainly to the prototypes production. Never-

Published in: E.Kuljanic (Ed.) *Advanced Manufacturing Systems and Technology*, CISM Courses and Lectures No. 437, Springer Wien New York, 2002.

theless some aspects need to be improved: dimensional precision, surface finishing as well as the use of materials with mechanical properties comparable with the traditional ones [4]. This paper addresses the study and application of Direct Metal Laser Sintering (DMLS) of copper (Cu) alloys for Rapid Tooling of dies for thermoplastic polymer injection moulding. The principle of DMLS is the powder sintering by means of IR laser beam: a layer of powder is prepared in the machine and the laser beam draws the shape of the section to be produced. The powder is sintered and layer by layer the object is built. With this technology it is possible to obtain object from several materials: polymers, wax, metals. The main limitations of DMLS prototypes are relevant to surface properties as porosity, roughness, hardness, corrosion, erosion and wear resistance. This study is a contribute to the characterization and improvement of DMLS surface properties in order to meet their functional requirements. Different surface treatments have been studied and applied to tooling samples specifically designed by CAD, in order to be representative of the geometrical and morphological features and functional characteristics of the dies; their model has been processed and post-processed and they have been produced by DMLS. Their surface properties have been systematically investigated in order to compare the different surface treatments.

2 TECHNOLOGICAL CONSIDERATION

This paper addresses the study and application of DMLS of Cu alloys powder for the Rapid Tooling of dies for thermoplastic polymer injection moulding. The main advantages of DMLS technology is the time reduction between design and production of new products. The main characteristics of the dies must be: good mechanical properties, good thermal conductivity, resistance to corrosion and wear caused by melted materials, good surface properties in order to obtain product with high surface finishing.

The first properties could be reached through the development and use of metallic alloys materials which result in higher mechanical performances on respect to polymeric materials. Nevertheless the main limits of the DMLS technology from the point of view of the above mentioned properties are: surface roughness and porosity, that result in a difficult sliding of the thermoplastic material during the injection moulding; difficulties in the extraction of the product from the die after solidification because the thermoplastic material penetrates into the pores of the die surface; low product surface quality.

For this reason, it is necessary to improve die surface porosity and roughness by means of surface treatments like peening or blasting, infiltration and coatings with materials suitable to fill die surface pores and improve die surface roughness: in fact conventional surface finishing processes like machining are time consuming and costly on dies with complex shape.

Blasting and peening have been considered with the purpose to reduce surface roughness with a single or double stage process to be applied alone or in combination with other treatments. Impregnation has been carried out in order to improve mechanical as well as surface properties of dies by filling the pores with metacrylate based polymer. Care has been taken to remove any residual layer on the surface that would be obviously unacceptable during the dies functional life. Electroplating had purpose to close and fill up the surface pores by adding 100 – 200 μm thickness coating. It is important to consider that these coatings do not need any surface finishing after the deposition.

Specific tooling samples have been designed by CAD, in order to represent the geometrical and morphological features and functional characteristics of the dies. Surface treatments and char-

acterization have been performed on the samples while final in-service trial on the actual die is in progress.
Figure 1 shows the shape of the sample developed in this study with concave and convex surfaces as well as horizontal and vertical surfaces with different thickness, outer sharp corners and inner corners with different fillets (size 30 mm by 30 mm by 30 mm). These features have been designed in order to be representative of the main geometrical properties of the dies. Several surfaces have been considered as representative of the geometry and identified with a number as explained in Figure 1.

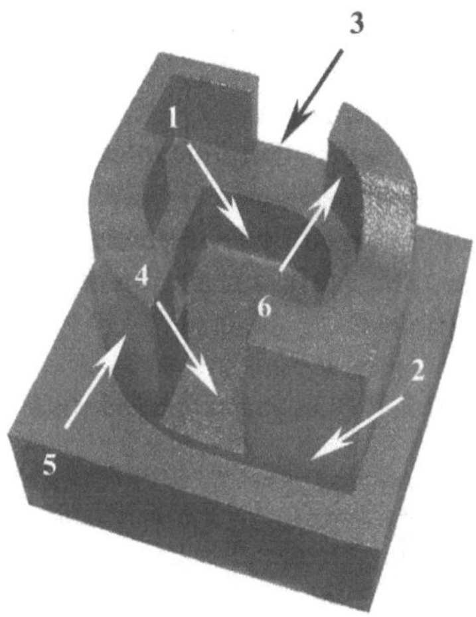

FIGURE 1. DMLS tooling sample with identification of the analysed surfaces (vertical flat surfaces: 1:V13; 2:V41; horizontal flat surfaces: 3:P23; 4: P42; cylindrical vertical surfaces: 5:O11; 6:O13)

The CAD model of the sample has been processed and post-processed in order to obtain the part program for the DMLS machine. The tooling samples as well as prototype dies have been produced by DMLS as described below.

3 EXPERIMENTAL

3.1 SAMPLES PRODUCTION

Tooling samples and dies have been produced with powder specially set up for rapid prototyping by DMLS. The powder contains brass, nickel and copper phosphide particles so the main elements are copper, tin, nickel and phosphor. Brass and copper phosphides have eutectic compound at low melting point: these particles melt when heated and wet the nickel powder that partially melts and in part remains solid. At the end of this process the particles are completely connected. Powder with two different grain size distribution has been used: the coarsest in a range of 40 – 80 μm with the average diameter of 50 μm, while the finest powder in a range of 20 – 50 μm with the average diameter of 30 μm.

Powder sintering is performed by means of infra-red CO_2 laser (λ 10.6 μm, power 200 W). Care should be taken to control the process temperature in order to optimise the physical and chemical phenomena that occur during sintering, namely the volume variation due to the state change. Surface properties have been investigated in order to state differences among the regions of the same sample and determine the limits of DMLS technology. Furthermore the properties of the samples treated in different way have been compared.
Different methodologies have been separately or simultaneously considered in order to improve the surface properties of the dies:

- blasting by corundum,
- glass beads and zirconium oxide (zirconia, ZrO_2) peening,
- infiltration by means of metacrylate based polymer,
- nickel and nickel-copper electroplating.

The treatments have been applied on raw samples or, in case of infiltration and electroplating, to samples already treated by blasting or peening.
A 2^3 factorial experiment has been designed by DOE in order to find the set of parameters for the blasting and peening processes. In the first stage three parameters have been chosen and varied on two levels as shown in Table 1.

TABLE 1. DOE parameters and levels

			Symbol	Measure unit	Level	
					Low level	High level
Variable	Bead/grain size	x_1	D	μm	100	250
	Blasting/peening material density	x_2	M	Kg/dm^3	2,5 (glass)	3,9 (corundum)
	Blasting/peening pressure	x_3	P	bar	2	4

Blasting material has been corundum (Al_2O_3 99,1%) with average grain size of 250 μm, while glass contains SiO_2 (72%), Na_2O (13%) and CaO (8%) with an average grain size of 105 μm. 3D roughness characterization has been carried out and SEM analysis has been performed on the lowest roughness samples. After the evaluation of the experimental results below described, a two stages process has been planned: a first abrasive phase carried out with corundum and a second phase carried out by glass or zirconia peening. The best conditions of the two steps have been selected using the materials and parameters shown in Table 2 (where p is the blasting/peening pressure).

TABLE 2. Parameters for the planning of corundum blasting phase and glass and zirconia peening step

		Process conditions				Sequence	Sample
Phase	T_i	D [μm]	Material	p [bar]	t [s]		
1	1	177	Corundum	3	120	1	6
	2	500		3		4	15
	3	177		6		2	36
	4	500		6		3	25
2	5	335	Zirconium oxide	4	90	1	10
	6			3		2	33
	7			2		3	28
	8	250	Glass	4	90		33b

Impregnation tests have been carried out on both raw sintered samples and blasted/peened samples, using a commercial metacrylate based polymer, that is an anaerobic sealing with good performance in penetrating not only the surface pores but also the open pores internal to the bulk material. The infiltrated samples have been characterized with 3D microgeometrical surface analysis by contact and SEM analysis.

Nickel and nickel-copper electroplating has been chosen because it can be performed at low temperatures and nickel is contained also in the substrate Cu alloy. This technology has also good performances in coating the parts with different shapes and hidden edges. Furthermore nickel plating is commonly used for antiwear and anticorrosion application.

After sample cleaning and surface activation, nickel plating has been performed by galvanic deposition in solution of nickel chloride, nickel sulphite and boric acid at 60 – 70 °C temperature and a current density of 2.5 A/dm^2 [5].

Table 3 shows the deposition parameters used for nickel and nickel-copper electroplating. Three stages have been performed: in the first stage Ni and Cu-Ni coatings have been deposited and tested. In the second stage the effect of the current density has been investigated while in the last stage the target has been to improve the exposition of the sample to the anode and avoid thickness disuniformity and the difficulty to coat internal surfaces (as resulted by the characterization below described).

TABLE 3. Nickel and nickel-copper electroplating process parameters

Stage	Sample identification	Initial condition	Material	Treatment condition
1	14N	Raw	Cu + Ni	Standard
	17N	Blasted	Cu + Ni	Standard
	16N	Blasted	Ni	Standard
2	36N	Blasted	Cu + Ni	Increased current density to 3 A/dm^2
	6N	Blasted	Cu + Ni	Increased current density to 5 A/dm^2
3	25N	Blasted	Cu + Ni	Improved disposition cathode/anode and higher anode area

3.2 CHARACTERIZATION TECNIQUES

In order to evaluate the surface properties of the DMLS tooling samples, 3D microgeometrical measurement and analysis were performed by means of a Three-Dimensional Digital Stylus Profilometer (3DSP) and a data acquisition module and an off-line analysis module, that receives digitized surface data from the acquisition module and applies a series of data processing techniques to provide information and evaluate attributes to characterize the surface properties [6]. Digitised surfaces have been flattened by least squares mean polynomial in order to remove form errors. Flattened surface digital models have been visually investigated using axonometric projection techniques and plan views. Axonometric view algorithms have generated lateral and aerial point plots, line plots and shading techniques, while orthogonal projection has been used to obtain contour plots and plan views with shading and height based colour assignment.
The 3D attributes obtainable by this method are based on the three-dimensional extension of 2D surface parameters defined in the ISO 4287 and DIN 4776 standards. The acquisition of the surface data has been carried out by the scanning of a square area and sampling a 128×128 matrix of point (0,0250 mm acquisition steps). The amplitude parameters, used in this study, that can be evaluated and extracted by this technique, are the following:

Arithmetic mean deviation: S_a
Ten point average height: S_z
Total Roughness : S_t.

Furthermore the analysis module is able to execute spectral and fractal analysis, basin analysis and surface porosity determination. 3D spectral analysis identifies the spatial frequency components of the surface. Some spectral functions can be evaluated in order to characterize surface isotropy and periodicity, and to discriminate the presence of high frequency constituents (directly related to surface roughness) from low frequency components (related to waviness). Functional parameters evaluated in this study are the following:

Surface bearing index: S_{bi}
Core oil retention index: S_{ci}
Valley oil retention index: S_{vi}

Basin analysis determines a set of planar domains characterised by the presence of a local pit. The basin distribution visualisation highlights morphological periodic features as scratches and valleys and can be used to characterise surface isotropy, periodicity and auto-correlation length.
Porosity analysis determines the percentage of total volume occupied by the void phase.
Metallographic analysis has been performed by means of Scanning Electron Microscopy (SEM). Vickers microhardness measurement has been carried out on cross section by means of a Vickers Microhardness Tester (50g load applied for 15 s). The characterization has been performed on the above mentioned surfaces (see Figure 1).

4 RESULTS AND DISCUSSION

4.1 CHARACTERIZATION OF THE RAW RAPID TOOLING SAMPLES AS OBTAINED BY DMLS

2D roughness measurements have been executed on 14 rapid tooling samples chosen random and do not show significant differences between the horizontal, vertical and curve surfaces as shown in Table 4.

3D roughness parameters determination has been carried out on the above considered surfaces of the same sample taken as reference. Table 5 shows the results that highlight little differences between the surfaces: vertical and curve surfaces show higher average roughness values than the horizontal ones. These surfaces are more difficult to be generated because they are obtained by superposition of finite dimension layers and curves that are obtained for each layer by interpolation. The analysed results show that this interpolation is appropriate because the roughness values are anyway very close.

TABLE 4. 2D roughness parameters values [μm] from raw samples as obtained by DMLS

Sample identification		P23	V41	O11	P42
R_a	Mean	14.3	13.7	14.0	14.9
	Min.	11.6	11.3	10.9	11.9
	Max.	16.5	15.5	15.4	17.7
R_z	Mean	75.0	71.2	73.1	78.3
	Min.	61.3	62.0	59.7	63.3
	Max.	85.8	79.6	88.5	93.9
R_t	Mean	91.5	83.0	87.1	93.6
	Min.	71.2	73.4	59.7	72.3
	Max.	114.2	96.6	107.3	124.8

TABLE 5. 3D roughness parameters values from raw samples as obtained by DMLS

		Parameters						
		Geometrical			Functional			Basin
		S_a [μm]	S_z [μm]	S_t [μm]	S_{bi}	S_{ci}	S_{vi}	
Raw samples	V13	17.3	137.7	173.4	0.61	1.54	0.11	222
	V41	17.1	113.8	130.8	0.72	1.50	0.10	245
	P23	17.1	112.0	128.7	0.62	1.48	0.10	204
	P42	15.9	114.3	141.4	0.60	1.57	0.11	231
	O11	15.2	118.9	143.4	0.60	1.56	0.12	227
	O13	17.1	128.3	142.2	0.61	1.55	0.11	234

Figures 2 shows digitalized vertical flat (a), horizontal flat (b) and cylindrical vertical (c) surfaces. In all the examined surfaces spectral analysis shows the periodicity due to the sintered particles size, while in the curve surfaces a periodicity can be noted that results from the border sharpening operation performed to get the desired curve. Vertical surfaces do not show any particular anisotropy.

Figure 3 shows the autocorrelation functions (ACF) of different surfaces respectively horizontal flat, vertical cylindrical and vertical flat: from the spectral and ACF analysis the periodicity due to the size of the sintered particles and the influence of powder feeding system of DMLS can be evaluated, that characterise the surfaces in different locations.

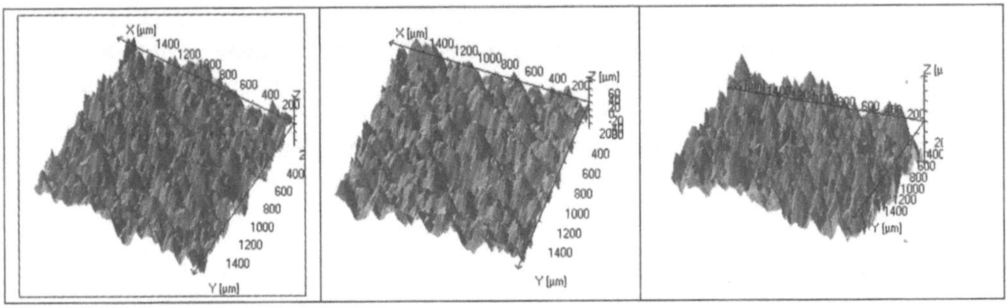

FIGURE 2. Digitalized vertical flat (V41, a), horizontal flat (P42, b) and vertical cylindrical (O11, c) surfaces

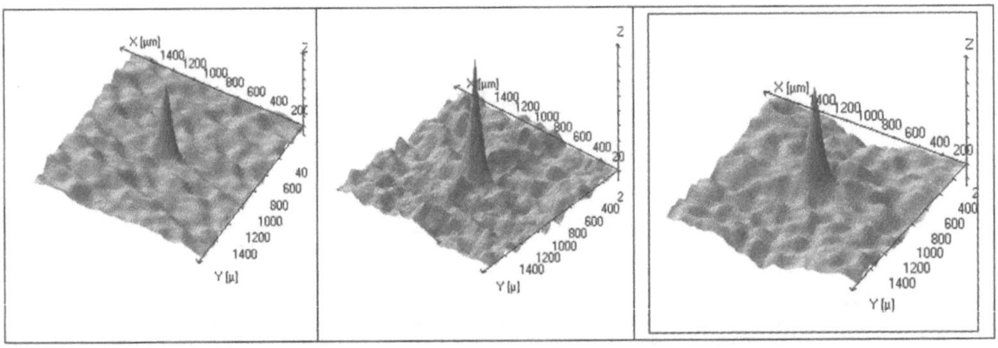

FIGURE 3. Autocorrelation function axonometric view of horizontal flat (P42, a), vertical cylindrical (O13, b) and vertical flat (V13,c) surfaces

Figure 4 shows the morphology of the DMLS sample obtained by SEM analysis. The porosity of DMLS parts is higher than the porosity obtained by conventional sintering, because of the uncompleted melting and cohesion of the particles during DMLS process.

FIGURE 4. Morphology of the DMLS sample obtained by SEM analysis

4.2 DOE RESULTS AND BLASTING AND PEENING OPTIMIZATION

The 2^3 factorial experiment results for the blasting and peening processes allow the evaluation of principal and interaction effects, of the variance and importance of an effect and the building of a process model, that can be used for process condition prediction in order to get a specific surface roughness Ra. While Ra can assume values included in the range from 3.9 µn to 13 µm, the lowest value is obtained by corundum blasting (250 µm grain size, 4 bar blasting pressure).

From the statistical analysis and the analysis of the residuals a predictive model has been defined as follows:

$$Ra = a_0 + a_1x_1 + a_3x_3 + a_{13}\,x_1x_3 \qquad (1)$$

where x_1 is the abrasive particle bead/grain size; x_3 is the blasting/peening pressure and a_0, a_1, a_3, a_{13} are coefficients evaluated by experiment results analysis. Figure 5 shows the representation of the model surface that allows the selection of the process conditions, depending on the specified roughness.

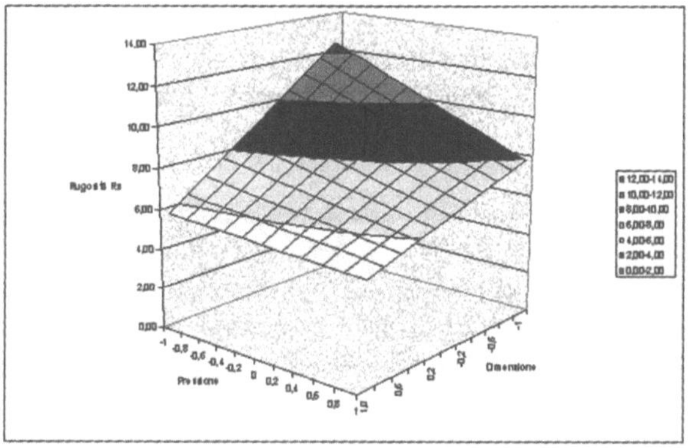

FIGURE 5 Surface obtained by roughness measurements processing.

Figure 6 shows glass peened and corundum blasted surfaces analysed by SEM. In both cases good improvement in the surface finishing has been noted on respect to the raw samples. Furthermore the peening glass beads produce a decrease of roughness by plastic deformation of the surface without any material removal while corundum blasting removes material. In both cases residuals of abrasive materials (glass or corundum) have been found near the surface blasted by abrasive particles. Corundum is able to decrease roughness removing the external peaks while it does not reach the internal surface pores: removal of material could some time discover new surface pores. Glass peening has a benefit action because crush the external peaks closing the internal pores. Nevertheless, the use of excessive blasting/peening pressure could produce damages on the surface.

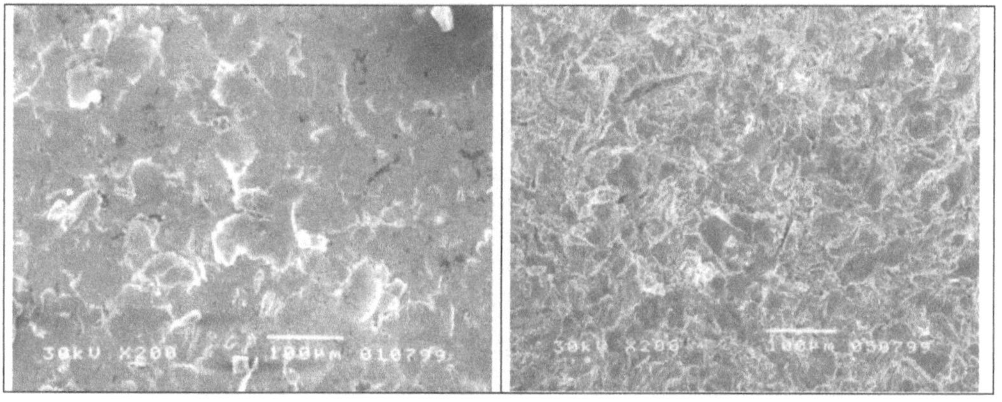

FIGURE 6. Glass peened surface (left) and corundum blasted surface (right).

According to these considerations it has been planned to execute both processes with the parameters reported in Table 2.

CORUNDUM BLASTING ANALYSIS

The analysis of the 2D roughness measurements gives Ra in a range of 3.1 – 8.3 μm. The results seem to be worst than in the first experiments with the exception of the case with 6 bar pressure and 500 μm particles. Nevertheless these parameters are suitable for the sample internal surfaces while the external ones are overblasted and damaged by the passes of blasting nozzle. These considerations are confirmed by the measurements of 3D roughness parameters. These tests allow to consider that higher abrasive particles size and pressure better surface finishing, but care should be taken in order not to damage the treated surfaces.

ZIRCONIA PEENING ANALYSIS

The best result obtained in the first stage (glass beads with 250 μm size, 4 bar pressure and application time of 90 s) has been taken as reference for the parameter selection as shown in Table 2. Ra values are function of the pressure; in this case best Ra values (2.4 μm) are obtained with the lowest pressure values. Also in these trials the internal surfaces are favourite with respect to the external ones. The trials resulted in a substantial equivalence of the zirconia and glass treatments.

The parameters have been selected in order to obtain final Ra value equal of 2.5 μm and surface porosity equal to 3-4%. The final parameters are: first stage by corundum blasting (250 μm size, 4 bar pressure and 90 s application time) and a second stage by zirconia peening (250 – 425 μm size, 2 bar pressure and 90 s application time).

4.3 INFILTRATION RESULTS

It is important to verify the following aspect of the infiltration process: internal pores infiltration, surface pores infiltration, film formation on the sample surfaces.
Table 6 shows the results of 3D roughness parameters. Comparing these results with the ones obtained on the raw not infiltrated sample, it seems that the surface roughness decrease only slightly. This is confirmed by a simple visual inspection of the samples.

TABLE 6. 3D parameters for infiltrated samples

		Parameters						
		Geometrical			Functional			Basin
		S_a [μm]	S_z [μm]	S_t [μm]	S_{bi}	S_{ci}	S_{vi}	
Infiltrated samples	V13	15.8	114.1	140.4	0.59	1.63	0.10	201
	V41	14.7	110.1	126.1	0.62	1.51	0.12	202
	P42	17.1	119.2	139.3	0.61	1.55	0.11	175
	O11	15.2	116.3	136.4	0.62	1.51	0.11	206
	O13	14.5	111.3	158.6	0.61	1.56	0.11	204

Basin analysis confirms that infiltration fill the surface pores so that the basin number decreases on the surface P23 about 25%. Infiltration produces a reduction of the range peaks-valleys by filling the surface pores.
Infiltration produce roughness decreasing from the raw condition but it is not yet enough in order to solve completely the problem. Infiltration could be used in combination with other treatments in order to improving the structural characteristics of the sintered material by improving the cohesion and filling the pores.

4.4 ELECTROPLATING RESULTS

Nickel coating thickness is higher on the external surface than in the internal ones due to the exposition to the anode. Thickness measured by SEM analysis is equal to 15 μm on the external surface, it is still higher (20 μm) on fillets and sharp corners, while decreases to 5 μm in the internal surfaces. SEM analysis showed some spallation point of nickel coating.
On these external surfaces (V41) roughness decreases from values of Sa equal to 17 μm on the raw sample to 4 μm after the electroplating while on the surfaces far away from the anode (V13, P23), where the thickness is lower, the roughness remains practically the same before and after plating. Figure 7 shows digitalised P23 surface before and after nickel plating on the sample plated after peening. The analysis confirms the trend. The better initial status allows to get better results also on the surface that are not directly in front of the anode.

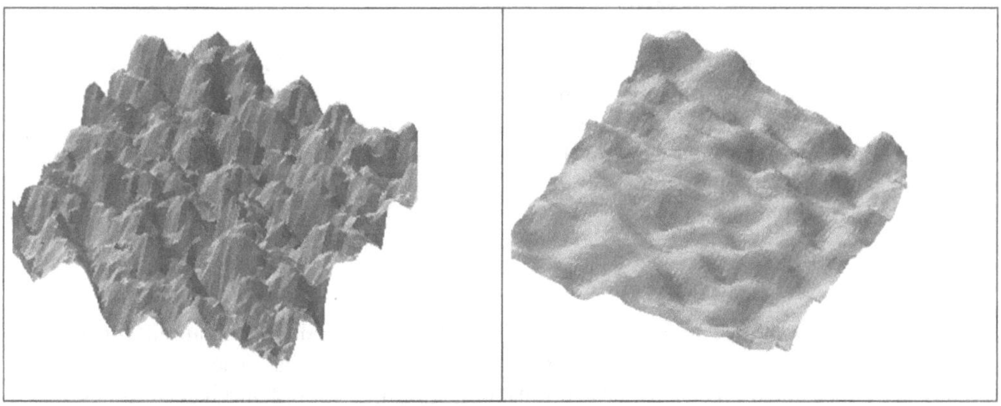

FIGURE 7. Digitalised surface P23 of peened sample before (left) and after nickel electroplating (right). (Nominal size is 3.175 mm by 3.175 mm)

The main problem of electroplating is thickness disuniformity that results in surface finishing disomogeinity. This problem has been considered in the samples treated in the further trials stages. In these cases the thickness distribution seems to be better due not only to the increased current density but also to a better disposition of the anode; nevertheless the problem is not totally solved. Table 7 shows the 3D roughness results before and after plating for sample 36 (2nd stage of trials) and sample 25 (3rd stage of trials).
In the sample where the current is the highest, the thickness is still more disuniform, namely on the edges high deposit is found.

From these trials it is possible to consider that the highest thickness nickel coating contributes to the reduction of surface roughness but, with a complex shape, it is difficult to reach the internal hidden corners. Nevertheless by current increasing, it has been noted high deposition on the exposed corner. The planning carried out in the third stage gives suitable results from the point of view of the thickness uniformity with high decreasing of roughness parameters in comparison with the same sample not plated. Vickers microhardness measurements executed on plated surface give average values of 600HV.

TABLE 7. 3D roughness parameters before and after plating for sample 36 (2nd stage of trial) and sample 25 (3rd stage of trials)

		Sample 36 (2nd stage)		Sample (3rd stage)	
Surface		Peened	Plated	Peened	Plated
P23	S_a	6.5	5.2	8.2	6.21
	S_z	45.3	30.8	5.6	39.7
	S_t	49.5	41.21	71.7	46.2
P42	S_a	6.2	6.3	7.4	6.1
	S_z	53.1	51.0	56.2	50.0
	S_t	63.7	68.1	66.8	61.8
V13	S_a	8.9	9.2	7.9	6.9
	S_z	65.9	59.3	57.0	52.7
	S_t	80.2	71.6	67.8	64.0

In conclusion nickel plating allows to get good surface finishing with high hardness, improving the surface properties. Nevertheless two problems need to be solved: thickness disuniformity on the surface far away or not directly exposed to the anode. From the examined results this aspect could be improved by the right value of the current density and a good disposition of the anode with respect to the part to be coated.

5 CONCLUSION

DMLS technology allows to reduce the development time and cost required for product prototyping and injection moulding dies and EDM electrodes tooling. The limits of this technology are dimensional precision, surface finishing as well as the low mechanical properties if compared with the traditional sintered materials. This study addressed the improving of surface properties of rapid tooling samples obtained by DMLS of Cu alloys. Three methodologies have been considered: peening and blasting, infiltration by metacrylate materials and nickel electroplating. Blasting and peening showed good performances but these technologies are not able alone to get the roughness good for the application on injection moulding dies. After a DOE and further planning, it has been selected that a suitable treatment requires two stages: the first phase by corundum blasting (250 μm size, 4 bar pressure and 90 s application time) and a second phase by zirconia peening (250 – 425 μm size, 2 bar pressure and 90 s application time).
Infiltration has showed good performances to close internal porosities but not to fill completely the surface pores. This treatment could contribute to increase cohesion of sintered material more than the surface properties. At this purpose infiltration could be used in combination with other treatments like blasting or coating. Suitable in service injection moulding tests should be carried out to verify thermal behaviour of the resin used for the infiltration.

Nickel electroplating showed good performances in surface roughness reduction, coupled to interesting improvement of hardness that should result in wear and corrosion resistance. Nickel electroplating could be used in combination with other treatments. A suitable combined treatment could be: production of DMLS sintered tooling, infiltration to improve mechanical properties, blasting and peening to reduce partially the surface roughness and nickel electroplating to improve wear and corrosion resistance and improve further the surface finishing.
Trials on actual dies to test the in-service behaviour of both DMLS and surface treatments are in progress.

ACKNOWLEDGMENTS

The authors acknowledge the support of the industrial companies: SM (Lombardore, Torino), CGM (Correggio, Reggio Emilia), Norblast (Bologna), Elsy Ersearch (Treviso), Loctite Italia (Mialno) for the surface treatments and the financial support of MIUR (Italian Ministry of University and Scientific and Technological Research).

REFERENCES

1. Jacobs, P.F., (1992), Rapid Prototyping and Manifacturing, First Edition, Society of Manufacturing Engineers, Dearbon, MI.
2. Printz, F.B., (1997), JTEC/WTEC Panel Report on Rapid Prototyping in Europe and Japan Vol. I. Analitical Chapters, Baltimore, MD.
3. Charney, C., (1990), Time to Market: Reducing Product Lead Time, Society of Manufacturing Engineers, Dearbon, MI.
4. Trimmer, D., (1991), The Exploitation of Rapid Prototyping, Second International Conference on rapid Prototyping, Conference Proceedings, University of Dayton, Dayton, OH.
5. Metals Handbook Volume 5 (1982), Surface cleaning, finishing and coating, ASM, OH.
6. Bracali, P., Groppetti R., Scrivani, A., (1999), On a micro-topographic three- dimensional acquisition and analysis system for engineering surface characterisation, Proc. Of AITEM99, Brescia, Italy.

LASER SINTERING OF SAND AND ITS APPLICATION IN RAPID TOOLING

X.H. Wang*, J.Y.H. Fuh, Y.S. Wong, Y.X. Tang

Department of Mechanical Engineering
National University of Singapore
10 Kent Ridge Crescent, Singapore 119260
(*email: mpewxh@nus.edu.sg)

KEYWORDS: Rapid tooling, Laser sintering, Sand casting.

ABSTRACT More and more materials are developed for rapid prototyping (RP) technology. Metal, polymer, fiber, paper, wax, etc. have been successfully processed by different RP processes. Silica sand is commonly used in the foundry industry. The process and parameters for directly laser sintering of the silica sand, without any binder, are studied by the authors. The development of direct sand laser sintering process could enhance the method of fabricating sand casting mould. On the one hand, with the RP technology, the fabrication of a complicated pattern has become easier with less manual intervention. On the other hand, the RP technology can be applied to directly fabricate sand casting mould, called rapid sand casting mold. Without the time-consuming procedure of the pattern fabrication, the rapid sand casting mould has the potential of further reducing the lead-time of producing a casting product. The process of the direct sand laser sintering and sand mold casting are presented in this paper.

1. INTRODUCTION

Casting is one of the oldest methods of manufacturing metals, in which sand casting is a widely used manufacturing process [1, 2]. The first requirement in making a sand casting mold is to design and make a pattern. The pattern is usually made of wood, wax, or metal. Sprue, runner, riser, draft and shrinkage allowance must be considered in the pattern fabrication[3]. A pattern is traditionally made by skilled workers or CNC machine. Since the emergence of RP technology, fabrication of a pattern has become easier with less manual intervention. After the pattern is completely built, a specially prepared sand that is mixed with different binders and additives is put on the pattern to make a mold. After the mold is dried, the pattern is removed from the mold. The mold is then ready for casting. A metal is poured into the mold to cast the metal part. The full process of the conventional sand casting is shown in Figure 1.

2. EXPERIMENTAL PROCEDURE

2.1 PROCESS PARAMETER OPTIMISATION ON SILICA SAND

Silica sand is commonly used in the foundry industry. In this experiment, the silica sand was used as the material of sand casting mold. The result of Energy Dispersive X-ray analysis (EDX) and X-Ray Diffractomter (XRD) showed that the main composition of the silica sand is

Published in: E.Kuljanic (Ed.) *Advanced Manufacturing Systems and Technology*,
CISM Courses and Lectures No. 437, Springer Wien New York, 2002.

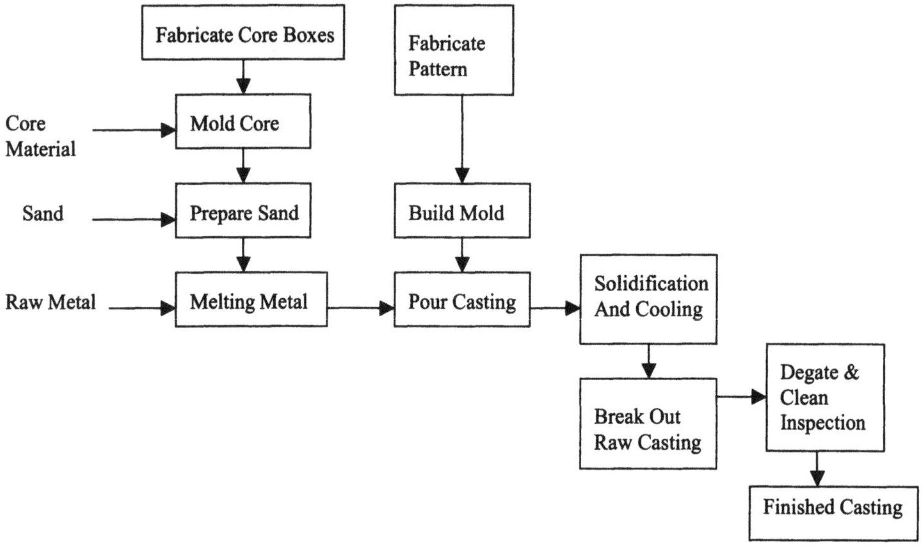

FIGURE 1. Production sequence in sand casting [4]

silicon dioxide (SiO_2). The particle size is 100 µm. The influence of the laser power and scan speed to a single sintered line and single layer was investigated.
Firstly, the scan speed was kept constant at 100 mm/s, and the only variable is laser power, which varied from 80 watts to 160 watts in increment of 20 watts. The width of the sintered line was measured. The relation between the laser power and sintered line width is shown in Figure 2. The following trends were observed. The sintering width increases with increase in laser power at a constant scan speed. When the laser power was less than 80 watts, the single line could not be fully sintered. When the laser power was higher than 160 watts, the sintered sand powder balled up, and many spheroids appeared.

Secondly, the laser power was kept constant at 120 watts. The only variable is scan speed, which was varied from 80 mm/s to 160 mm/s. The width of the sintered line was measured. The relation between the scan speed and sintered line width is shown in Figure 3. The sintered width decreases with increase in scan speed for a constant laser power. When the scan speed was higher than 160 mm/s, the single line could not be fully sintered. As a scan speed of less than 80 mm/s, the sintering efficiency has been found to be too low.

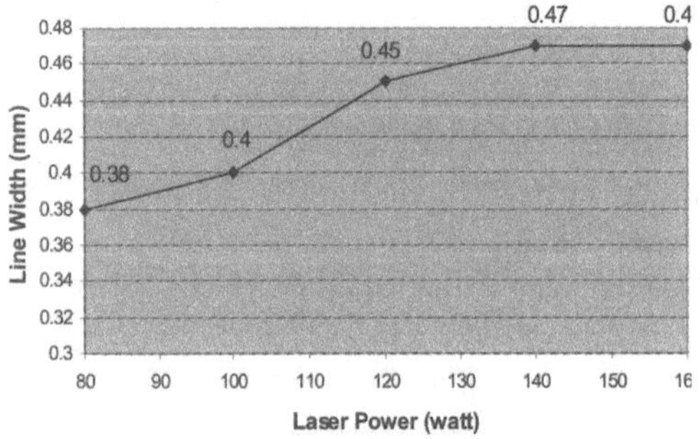

FIGURE 2. Influence of Laser Power on the Sintered Line Width (100 mm/s Scan Speed)

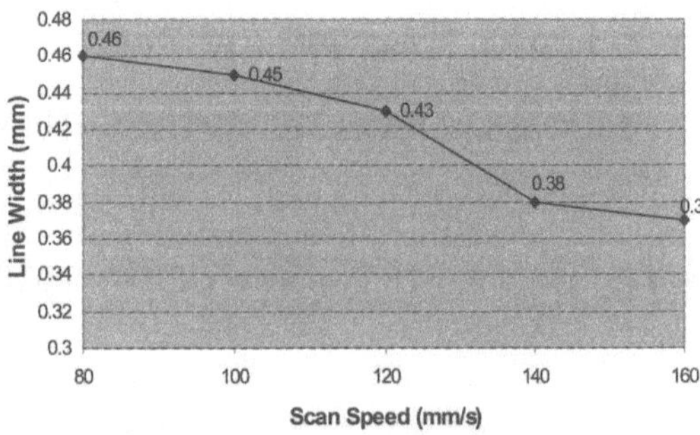

FIGURE 3. Influence of Scan Speed on the Sintered Line Width (120 watts laser power)

Thirdly, from the parametric study on the single-line sintering, we found that the line width was also about 0.4 mm, and the line width changed very little with the scan speed and laser power. A 50% overlap sintering between the adjacent lines was found to be able to guarantee tight adhesion. Therefore, the scan space was set to 0.2mm in this experiment. Firstly, the scan speed was set constant at 160 mm/s. The only variable, laser power, was varied from 40 watts to 140 watts. The results of the single layer sintering are shown as Table 1. From the results, we found that the layer thickness and strength increase with the increase in laser power. However, with a lower laser power, the surface finish is better.

Fourthly, the laser power was set constant at 80 watts, and the scan space was 0.2mm. The only variable was scan speed that was varied from 60 mm/s to 200 mm/s. The results of the single layer sintering are shown as Table 2. From the results, it has been found that the layer thickness and strength increase with the decrease in scan speed. However, with a faster scan speed, the surface shows a better finish.

TABLE 1. Influence of laser power on the single layer sintering

LASER POWER (WATT)	SCAN SPEED (MM/S)	SCAN SPACE (MM)	LAYER THICKNESS (MM)	REMARKS
40	160	0.2	0.5	Layer is weak; surface is smooth
60	160	0.2	0.6	Layer is weak; surface is smooth
80	160	0.2	0.7	Layer strength is improved, but surface is rougher than above
100	160	0.2	0.75	Layer strength is improved, but surface is rougher.
120	160	0.2	0.78	Layer strength is improved, but surface is rougher than above and distortion is obvious.
140	160	0.2	0.80	Layer strength is improved, distortion is obvious and large points appear on surface.

TABLE 2. Influence of Scan Speed on the single layer sintering

LASER POWER (WATT)	SCAN SPEED (MM/S)	SCAN SPACE (MM)	LAYER THICKNESS (MM)	REMARKS
80	60	0.2	1.25	Layer strength is good, distortion is obvious and large voids appear on surface.
80	80	0.2	0.95	Layer strength is good; large voids appear on surface.
80	100	0.2	0.85	Layer strength is good; surface is smoother than above
80	120	0.2	0.75	Layer strength is weaker than above; surface is smooth.
80	140	0.2	0.70	Layer strength is weak; surface is smooth.
80	160	0.2	0.65	Layer is weak; surface is smooth
80	200	0.2	0.60	Layer is very weak; surface is smooth

2.2 FABRICATION OF SAND MOLD

3D models of a gear sand mold were generated by Unigraphics 15.0 (as shown in Figure 4). The STL-format file was loaded into the control computer of a specially developed high-temperature laser manufacturing (HTLM) system. According to the parameters required, such as scan pattern, laser power, scan speed, and layer thickness, the software automatically generates the data for sintering.

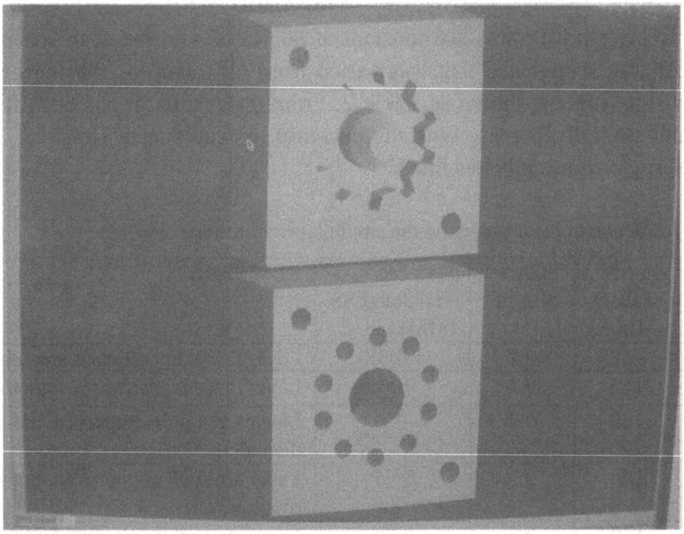

FIGURE 4. 3D Models of the Gear Sand Mold

2.2.1CREATING THE BASE LAYER

The first layer must be bonded to a base plate, otherwise, placing a new layer by the scraper may lead to shifting of the previous layers, and the building process could be a failure. The base plate can be made of either mild steel or aluminium. A mild steel plate was used as the base plate in this experiment. Silica sand is very difficult to be bonded on the metal plate directly. Therefore, a base layer between the sand layer and base plate is required. A mixture of silica sand and sodium silicate solution was used as the base layer material. Sodium silicate solution could be solidified when it is exposed to CO2 or is heated up. Silica sand was mixed with sodium silicate solution at the proportion of 2:1 by volume. The mixture is a kind of slurry. The slurry was evenly placed on the base plate at the thickness of 0.3 mm. The base plate with the slurry was exposed to the laser beam. The parameters used to the base layer were: laser power of 80 watt, scan speed of 100 mm/s, and scan space of 0.2 mm. Because sodium silicate solidifies more easily than silica sand, the energy intensity is less than that needed in silica sand sintering. The base layer serves as a medium between the metal base plate and the pure sand layer, which has a good adhesion characteristic with either metal plate or silica sand.

2.2.2 PART BUILDING

To ensure good adhesion between the part and the base layer, the contour of the first layer structure is exposed at a scan speed of 80 mm/s and laser power of 100 watt. Due to the spot size of the laser beam and the resultant heat effect zone, the compensation was considered into the contour scan. According to the result of the parameter study, the laser beam offset was set at -0.2mm. From the second layer, the parameters were set back to those listed in the following:

Scan pattern: raster
Laser power: 160 watts
Scan speed: 120 mm/s
Scan space: 0.20mm
Layer thickness: 0.20mm

The laser beam selectively sintered the sand powder according to the cross-sections of the mold. Layer by layer, the part was built up as shown in Figure 5.

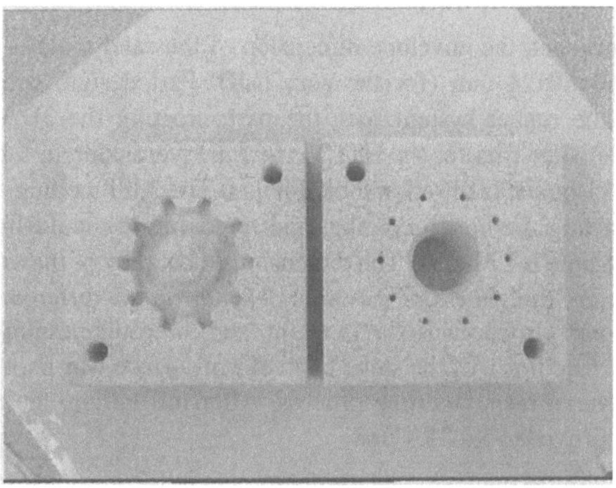

FIGURE 5. Sand Mold by HTLM System

2.2.3 SILICATE INFILTRATION

After removing the mold from the HTLMS, infiltration process was carried out so as to improve the strength, surface finish, and density of the mold. The infiltration process was very simple. Using a brush, sodium silicate solution was painted on the surface of the sand mold, and the liquid permeated into the mold. After several rounds of painting, the sand mold was put under the normal atmosphere for solidification, or the sand mold can also be put in an oven at 200 °C to speed up the solidification process. Thereafter a sand mold with a good surface finish and better strength was obtained.

2.2.4 CASTING

The sand mold was preheated for 2 hours to 200°C. The metal cast was AlSi with a melting point of 660°C. AlSi was put into a furnace to melt at 800°C for 3 hours and poured into the sand mold. The cavity was fully filled and all features were reproduced (see Figures 6 and 7).

FIGURE 6. AlSi Casting in Sand Mold

FIGURE 7. Casting part – Gear after removal

3. RESULTS AND DISCUSSIONS

In this experiment, the envelope dimension of the sand mold is 60x60x25 mm (for the cavity half) and 60x60x24 mm (for the core half). Part design, constructing the runner systems, combining the runner system with the part, creating the 3D model of the sand mold, and preparing the slice data for the HTLM machine were done in 4 hours. Sintering the sand mold took about 11 hours. Infiltration took 2 hours. The AlSi melting and sand mold preheating took 3 hours. Casting the metal, cooling, and removing the mold from casting took 2 hours. The overall lead-time is 22 hours. The dimensional accuracy of the sand mold built by the HTLM is within 0.4 mm. Roughness of the mold depends on the different sintering area. The roughness of sintered part surface was Ra 35.6 μm, and the roughness of the vertical surface was 26.2 μm. The surface finish of the slope or freeform surface was poor because of the staircase effect from the layered manufacturing process. After infiltration, the roughness of sintering surface could be improved to Ra 25.4 μm.

Compared with the conventional casting process, the presented rapid casting process (as shown in Figure 8) is much simple. In the rapid sand casting, there is no pattern fabrication . In the conventional casting process, much manual work is involved in the pattern fabrication that is very complicated and time-consuming. However, with the rapid sand casting process, the pattern making is not required. The whole mold building process can be implemented automatically by the HTLM system. The lead-time of a new sand mold was much reduced.

No single process has yet demonstrated a universal advantage without disadvantages, so the choice of a process still involves a compromise for the particular design requirements [5]. The reduction of the lead-time depends on the size of the sand mold. For the sand mold with a huge size, sintering the big volume of sand takes quite a long time. In that case, the rapid sand casting process is less advantageous in lead-time. The rapid tooling process is more suitable for fabricating complicated and relative small sand molds.

Owing to the layered manufacture nature of the RP process, the staircase surface feature is unavoidable, especially for an inclined plane. Therefore, the surface finish of the rapid sand casting mold is not as good as that of conventionally made sand casting mold. With finer silica sand and thinner sintering layer, the surface finish could be improved.

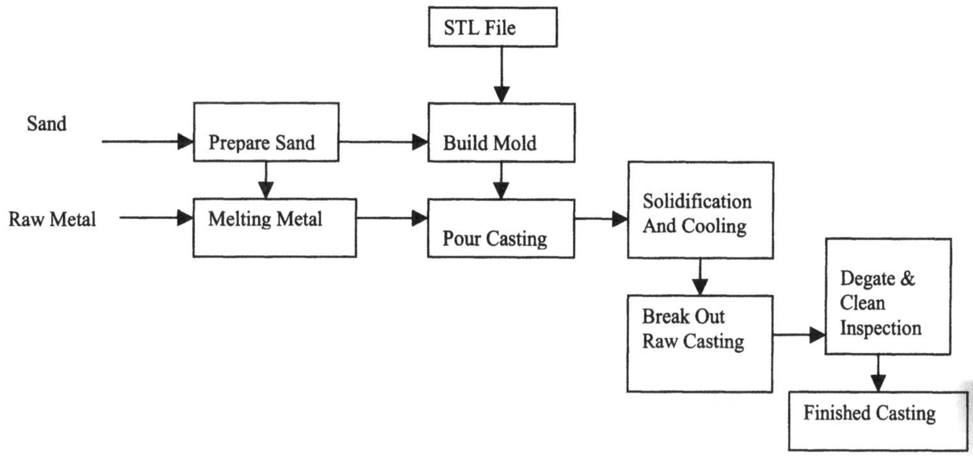

FIGURE 8. Production sequence in Rapid Casting Process

There are several error sources specific to the conventional fabrication process. First, in the fabrication of the pattern, there is an error between the real pattern and the designed model. Second, the error rises when duplicating the sand mold from the pattern. Most errors will accumulate and be propagated by subsequent processing steps. As the rapid sand casting mold is built directly by the HTLM according to the 3D file, a pattern is not needed. Therefore the two kinds of errors mentioned above are avoided. Shrinkage from sintering the silica sand is very little, and with further optimisation of the process parameters, accuracy can be controlled within the range accepted by industry.

4. CONCLUSIONS

In this paper, the conventional fabrication process of sand casting mold is reviewed. A novel fabrication process using the RP technology to directly build a sand casting mold is demonstrated. Several important issues relevant to the suitability of the process for making sand casting tooling are discussed. The following conclusions are drawn:

- When the part is relatively small and complicated, the rapid fabrication process is an effective way to reduce the tooling lead-time in sand casting process. Therefore, for an appropriate part geometry, the use of RP technology to directly produce tooling for casting processes offers significant time and cost savings.
- Owing to the layered manufacturing nature of the RP process, the staircase feature affects the surface finish of the rapid sand casting mold.
- Compared with the conventional fabrication process, the rapid sand casting mold process reduces the errors arising from the pattern fabrication. The error sources are

less in rapid casting mold process. Therefore the error can be controlled to an acceptable range.

ACKNOWLEDGEMENT

The authors would like to gratefully acknowledge the research grants provided by the National University of Singapore. The authors also thank Dr Hu Bang Hong, Mr Yang Yong and Mr Ning Yu for their helps in experiments.

REFERENCE

1.Wang, WL., Conley, J.G., Stoll, H.W., 1999, Rapid Tooling for Sand Casting, Using Laminated Object Manufacturing Process, Rapid Prototyping Journal, Vol 5., No. 3.

2.Gregg Bruce, R. , Mileta, M. Tomovic, John E. Neely, Pichard R. Kibbe (1998), Modern Materials and Manufacturing Processes.

3.White, Neely, Kibbe, 1977, Meyer, Machine Tools and Machining Practices, Vol. 1.

4.Groover, M.P., 1996, Fundamentals of Modern Manufacturing: Materials, Processes and Systems.

5.Johnson, J. L. 1994, Principles of Computer Automated Fabrication.

DIRECT LASER SINTERING OF CU-BASED METALLIC POWDER FOR INJECTION MOULDING

H. H. Zhu*, J. Y. H. Fuh and L. Lu,

Department of Mechanical Engineering, National University of Singapore, Singapore
(* correspondence: engp9861@nus.edu.sg)

KEYWORDS: Direct laser sintering, Metal powder, Injection mould

ABSTRACT. Direct laser sintering is an attractive Rapid Prototyping (RP) process for producing injection mould inserts, because of its capacity of fabricating any desired three-dimensional (3-D) parts directly from metal powder. This paper reports a research work on the fabricating Cu-based injection mould inserts using direct laser sintering. A pair of mobile phone cover mould inserts was successfully fabricated in a self-developed rapid prototyping machine in ambient atmosphere at room temperature. The total sintering time is 40 hours. 65% relative theoretical density and average surface roughness Ra 14-16μm were achieved with negligible distortion and curling. Infiltration epoxy was employed as post-process to improve the density and the strength of the sintered parts. The moulds were used to mould ABS-PC phone mobile cover after finishing. No defects were detected after moulding of 100 components.

1 INTRODUCTION

Direct laser sintering is a sequential layered approach to manufacture any desired three-dimensional parts with simple or complex shapes from metal powder directly. Direct laser sintering of metallic parts, similar to liquid-phase sintering, is a complex metallurgical process that can be influenced by several phenomena, such as wettability, viscosity and flow [1]. Therefore, till now, only several metal powders can be sintered directly by laser for industrial applications. The main drawbacks limiting the application of this technology are the " balling" phenomenon and low accuracy [2].

Cu-based material system is attractive because of its good thermal conductivity, high electrical conductivity and low cost. The feasibility of producing Cu-based metal parts directly by Selective Laser Sintering (SLS) has been demonstrated using various metal systems, such as Cu-Sn, bronze-Ni, Cu-Solder and Cu-Fe [3-4]. Although full density parts can be produced using direct laser sintering, the low accuracy caused by high shrinkage limits its industrial application. The sintered parts produced by direct laser sintering usually with 60~70% relative density. The porous parts need to be further densifed, such as by sintering or infiltration. The advantage of infiltration is its low shrinkage and high accuracy characteristics compared to sintering. Epoxy is often used as infiltrant to low melting material because of its high viscosity and wettability to metal materials [5].

The present work focus on the research of direct laser sintering of a Cu-based material system which mainly consists of Cu and Cu_3P. A pair of injection mould inserts for mobile phone

Published in: E.Kuljanic (Ed.) *Advanced Manufacturing Systems and Technology*, CISM Courses and Lectures No. 437, Springer Wien New York, 2002.

cover has been fabricated in a self-developed Rapid Prototyping machine in ambient atmosphere at room temperature.

2 EXPERIMENTAL SETUP AND PROCEDURE

2.1 SETUP

The experiments were carried out on the High Temperature Laser Manufacturing Systems (HTLMS) developed at the Department of Mechanical Engineering, The National University of Singapore. The basic setup for the HTLMS is shown in Figure 1. This system consists of a CW CO_2 (λ=10.6μm) laser capable of a maximum output power of 200W, a chamber with atmosphere control and powder delivery system.

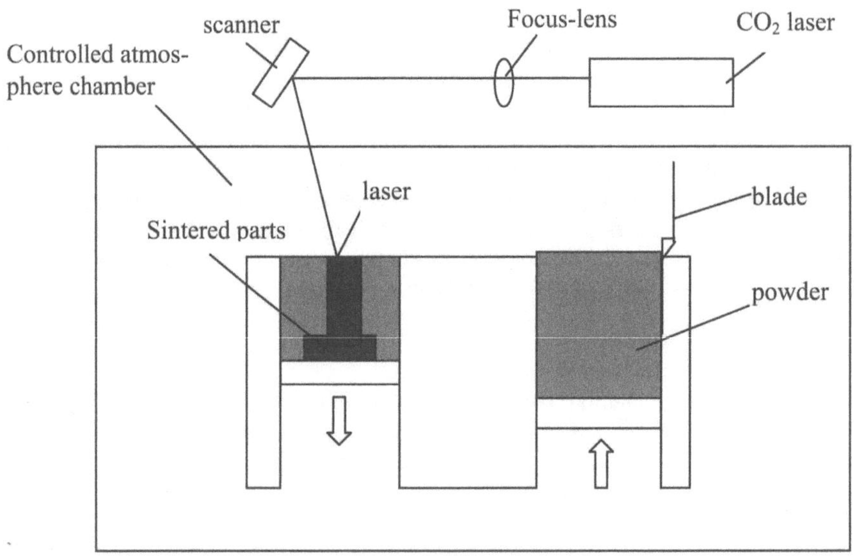

FIGURE 1. Schematic diagram of HTLMS

2.2 MATERIAL

Electrolysis 99% Cu powder with dendrite shape and mean particle size of 40μm, and pre-alloyed powder SCuP with spherical shape and particle size ranging from 5μm to 20μm were used in this experiment. The pre-alloyed powder SCuP mainly consists of Cu and P, other metal elements were added to improve its sinterability and mechanical properties. Two powders were mixed manually by the ratio of 60:40 by volume.

2.3 PROCESS PARAMETERS SETTING

The mixture of powder was first loaded into the process cylinder and levered by the blade to obtain a flat powder surface. All the inserts and the samples for testing were fabricated in air atmosphere at room temperature. The laser power used in this experiment was kept constant at 120W. The scan line spacing was controlled at 0.2mm. Scan speed was setting at 200mm/s and layer thickness was 75μm. After sintering, the inserts and a part of samples were infiltrated with epoxy. Injection moulding process was carried out using Arburg 221k 75-350 injection moulding machine. Samples for metallographic examination were prepared and etched with a mixture of $FeCl_3$, HCl and distilled water according to the standard metallographic procedure. Microstructure was observed using a JEOL JSM-5600LV scanning electron microscope (SEM). The density was measured using a micromeritics AccuPyc 1330 machine. The surface roughness was measured using a Surfcom 120A instrument. The value of the surface roughness was obtained from the average value of five measurement readings.

3 RESULTS AND DISCUSSIONS

3.1 SINTERED PART

The sintered injection mould inserts are shown in Figure 2. The parts show relative dense and smooth surface. 65% relative density and Ra 14-16μm average surface roughness have been achieved. No observable curling and distortion could be found. Dimensions of the insert are 150x100x20mm.

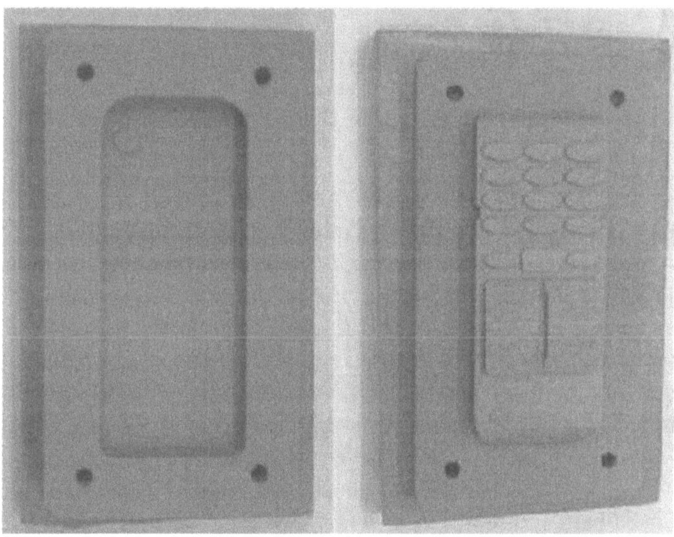

FIGURE 2. Mould inserts fabricated by direct laser sintering

The microstructure of the sintered part is shown in Figure 3. From the picture, it can be seen a network bonding the structural particles clearly. The EDX result shows that the white network is phosphor-rich P containing Cu, it indicating the pre-alloy SCuP was molten during sintering. The gray area is the Cu solid solution. Therefore, the mechanism of this process is liquid phase sintering. The X-ray diffraction analysis shows no CuO or Cu_2O oxide in the sintered parts, while instead of a new phase Cu (PO_3). This is because phosphor reduced the oxygen and formed Cu (PO_3).

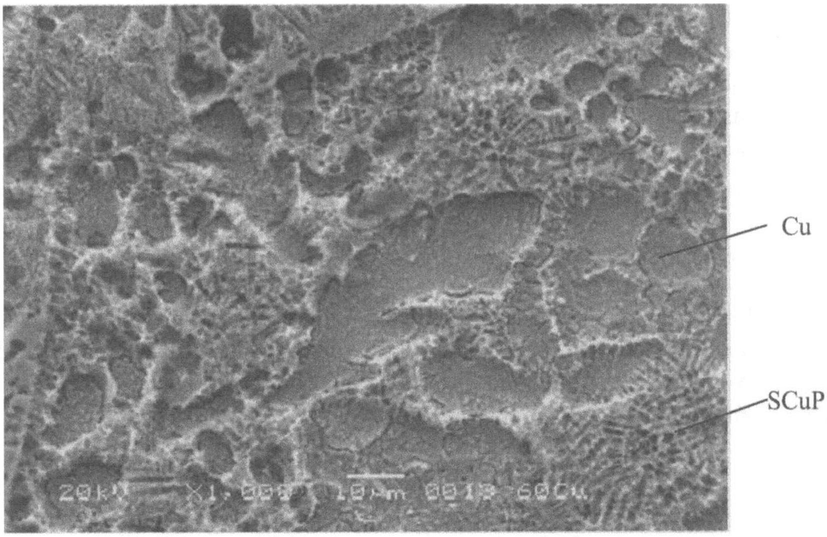

FIGURE 3. The microstructure of the sintered part shows network binder

3.2 INFILTRAED PARTS

Complete infiltration with epoxy was achieved after 4 hours of total infiltration time. Figure 4 shows the SEM image of the infiltrated sample. It can be seen from the image that almost completely infiltration was achieved. But the crevice of the interface predicts no good wetting between the epoxy and Cu-based material. The hardness of infiltrated part show a slight increase in the hardness compared to the sintered parts (Table 1). It can be predicted that the strength of the part improved by in filtration. It is noted that the hardness of infiltrated parts is still very low. This is mainly attributed the soft Cu matrix. Although the full density of sintered parts is obtained, it contains about 35% of soft epoxy which cannot provide strong restriction on the deformation. After infiltration, the inserts were finished to about 3μm. Table 1 gives some properties of the sintered and infiltrated part. 100 pieces of injection moulding were performed. No defects were detected after that. Figure 5 is the photo of one of the components.

TABLE 1. Properties of the sintered and infiltrated part

MATERIAL	HARDNESS	SURFACE ROUGHNESS	RELATIVE DENSITY
SINTERED PART	HR15-T 66	14-16μm	65%
INFILTRATED PART	HR15-T 40	2-3μm	100%

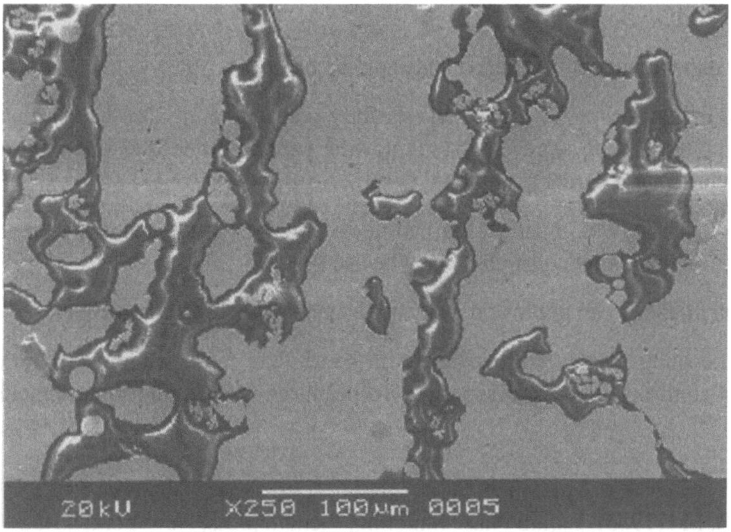

FIGURE 4. The SEM image of the infiltrated sample

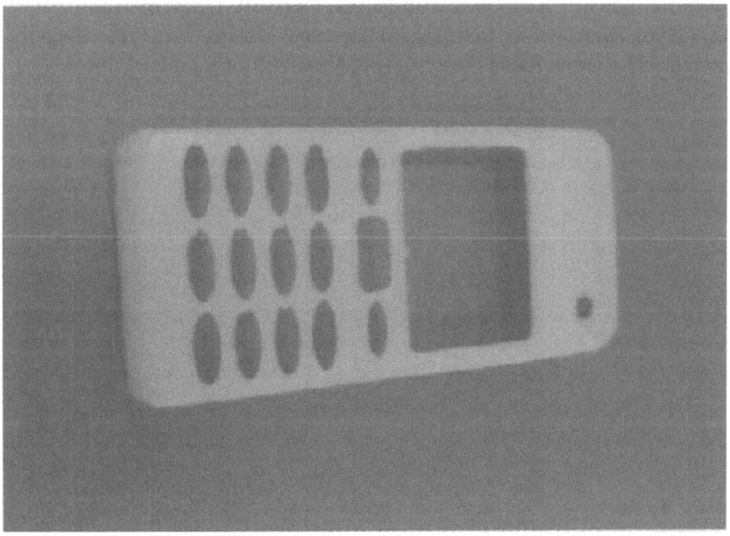

FIGURE 5. One of the moulded mobile phone cover

4 CONCLUSIONS

Based on the experiments, following conclusions can be drawn:

(1) Cu-based mould inserts have been successfully prepared using direct laser sintering. It was observed that phosphorous segregated along powder particles after laser sintering indicating formation of a liquid phase during sintering.

(2) About 65% density was obtained. Further increase in the density is possible but more percentage of the low melting point constituent must be used.

(3) Infiltration of epoxy can improve density but is unable to increase in hardness.

(4) Surface finishing of the infiltrated parts is better than those without infiltration. Injection moulding test confirms direct laser sintering is a promised processing for rapid moulds manufacturing.

REFERENCES

1. German, R. M., (1985), Liquid Phase Sintering, Plenum press, New York, 43-62
2. Agarwala, M., Bourell, D., Beaman, J., et al, (1995), Direct Selective Laser Sintering of Metals, Vol 1/1, 26-36.
3. Niu, H. J., Chang, I. T. , (1998), Liquid Phase Sintering of M3/2 High Speed Steel by Selective Laser Sintering, Scripta Materialia, Vol 39/1, 67-72.
4. Lu, L., Fuh, J., Wong, Y.-S., (2001), Laser-Induced Materials and Processes for Rapid Prototyping, Kluwer Academic Publishers, Norwell, USA, 89-138
5. Nyrhilä, O., (1996), Direct Laser Sintering of Injection Moulds. The Proceedings of the 5th European Conference on Rapid Prototyping & Manufacturing, Finland, 186-191.

MOLECULAR DYNAMIC SIMULATION OF NANOINDENTATION OF SILICON CARBIDE BY DIAMOND TIP

A. Noreyan[1], I. Marinescu[1], J. Amar[2]

[1] Mechanical, Industrial and Manufacturing Engineering Department, University of Toledo, USA
[2] Department of Physics and Astronomy, University of Toledo, USA

KEYWORDS: Molecular Dynamic Simulation, Leonard-Jones potential, FCC structure.

ABSTRACT. The interaction between silicon carbide (3C structure) and diamond indenter was investigated using ninety-thousand atom molecular dynamic simulation with Leonard-Jones potential. At the special value of applied strain to the surface, a crack forms on the silicon carbide surface and moves toward perpendicular direction of indentation. The changes of energies (kinetic potential and total), pressure and temperature are observed during the process. The depth of indentation, the speed of indentation and the initial temperature of workpiece were varied to study the phenomena.

1 INTRODUCTION

Current research efforts in fabrication are focused on achieving machining accuracy of less than 100 nm. To guarantee such accuracy, the surface roughness of the worked material should be an order of magnitude smaller, and to obtain such smooth surfaces the machining tools must have similar dimensions. This has led to the realization that progress in controlling/obtaining brittle-to-ductile transition for grinding of hard materials hinges on understanding the fundamental processes at the atomic level when a tool (such as a diamond grain) with an applied force (load) acts on a workpiece. Depending on the load, these processes include friction, elastic deformation, plastic deformation, and fracture, while the workpiece undergoes indentation, scratching, and eventual detachment of a chip of material.
Such ultraprecision machining and microcutting is not trivial to characterize in the laboratory because of increasing in-process measurement problems, inaccessible contact area of tool and workpiece, and the difficulty of surface analysis in this range. On the other hand, atomistic simulations of the workpiece material/machining-tool interface, coupled with an analysis of the surface/interface mechanics provide a powerful approach to understanding factors that govern nanoscale precision production. Therefore our purpose is to carry out systematic atomistic simulations along with the corresponding experiments in order to obtain a fundamental understanding of the microscopic factors affecting nanomachining. This understanding will enable the development of fabrication processes for nanomachining.

2 CONCEPT AND BASIC ELEMENTS OF MOLECULAR DYNAMICS

Molecular dynamics comprises macroscopic, irreversible thermodynamics and reversible micro mechanics. The thermodynamic equations form a link between the micro mechanical state, a set of atoms and molecules, and the macroscopic surroundings, the environment. The

Published in: E.Kuljanic (Ed.) *Advanced Manufacturing Systems and Technology*, CISM Courses and Lectures No. 437, Springer Wien New York, 2002.

thermodynamic equations yield the quantities *system temperature* and *hydrostatic pressure* into the model and allow to determine energy changes involving heat transfer. In mechanics it is usual to consider energy changes caused by displacement and deformation.

By the term "mechanical state" of a microscopic system, a list of present coordinates {r} and velocities {v(t)} of the constituents is meant [1]. For this information about the state of the system to be useful equations of motion, capable of predicting the future, must be available. As the governing equations of motion for a system of constant total energy, the well known Newton's equations of motion can be chosen.

Newton's equations of motion:

$$d(v_i(t)) / dt = 1 / m_i \sum_{i<j}(F_{ij}(R_{ij}, \alpha, ...)) \qquad (1)$$

$$d(R_i(t)) / dt = v_i(t) \qquad with\ i,j = 1\ to\ n \qquad (2)$$

The resulting force on an atom i is expressed by an integral over all force contributions F_{ij}. Numerically this is calculated as a sum over all forces acting on each atom i (1). Hence, two bodies at close distance interact through this sum of force contributions in the equation of motion.

The force contribution is calculated by employing partial derivatives of so-called potential functions, which describe the energetical relation between atoms with respect to the separating distance, bonding angle and bonding order, possibly. This is also where different materials can be considered as well.

To advance the atoms in space, the equation of motion has to be integrated with respect to time once to obtain the new velocity and twice for the new position of each atom. Numerically, this operation is more efficiently carried-out by approximation schemes, for instance using finite difference operators and the so called Verlet or Stoermer algorithm [1,2].

Verlet algorithm (velocity form):

$$R_i(t + \Delta t) = R_i(t) + \Delta t * v_i(t) + ½ * 1/m_i * \Delta t^2 * F_i(t) \qquad (3)$$

$$v_i(t + \Delta t) = v_i(t) + h/(2*m_i) * \{F_i(t + \Delta t) + F_i(t)\}, \quad with\ i = 1\ to\ n \qquad (4)$$

With the present positions ($R_i(t)$), velocities ($v_i(t)$) and forces ($F_i(t)$), first the new positions and forces at $t + \Delta t$ can be calculated and then the new velocity as well.

Given the equations of motion, forces and boundary conditions, i.e. knowing the current mechanical state, it is possible to simulate future behavior of a system. Mathematicians call this an initial value problem. A reasonable distribution of the initial velocities can be obtained from the Maxwell-Boltzmann distribution function.

2.1 MATERIAL REPRESENTATION: THE POTENTIAL FUNCTION

While the original molecular dynamics theory is well based within the physics, empirical elements were introduced from the material science area in order to match the results of experiments with the theoretical, and so far physical model. The key to computational efficiency of atomic-level simulations lies in the description of the interactions between the atoms at the atomistic instead of the electronic level. This reduces the task of calculating the complex many-body problem of interacting electrons and nuclei like in quantum mechanics to the solution of an energetic relation involving, basically, only atomic coordinates [1].

The central element of the MD-code is the force calculation. As it is the most time-consuming part in an MD computer program, it determines the whole structure of the program. Efficient algorithms for the force calculation are important for large-size system, i.e. for systems with a large number of atoms (see [2]).

PAIR POTENTIALS

First van der Waals described a model of a material which can form liquid and solid condensed phases at low temperatures and high pressures. Such condensed phases require both attractive and repulsive forces between atoms [1]. Since the simplest possible representation of many-body interactions is a sum of two-body interactions, the so-called pair potentials were the first potential descriptions of such type. A typical course of the functions is shown in figure 1.

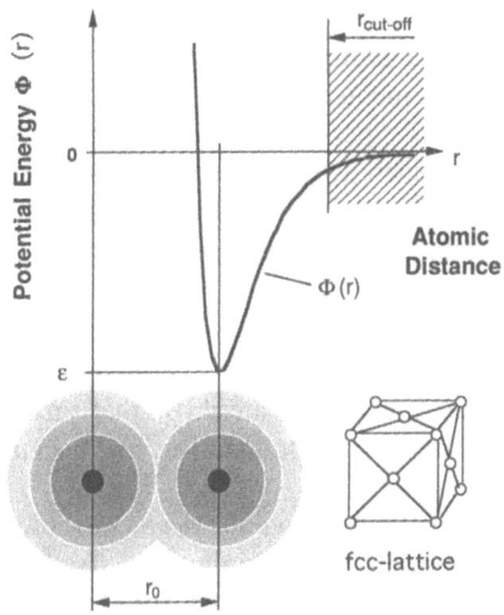

FIGURE 1. Course of a pair potential function

The best known pair potential functions are the Lennard-Jones and the Morse potential (eqs. 5 and 6). The well-depth of the functions are given by the parameter ε and D for the minimum potential energy or sublimation energy, while σ and r_o are constants that define the position of the energy minimum. These parameters are derived from fitting to experimental data like lattice constant, thermodynamic properties, defect energies and elastic moduli.

$$\textit{Lennard-Jones:} \qquad V_{LJ}(r) = 4\,\varepsilon\,[(\sigma/r)^{12} - (\sigma/r)^{6} \tag{5}$$

$$\textit{Morse:} \qquad V_M(r) = D\,[e^{-2\alpha\,(r-ro)} - 2e^{-\alpha\,(r-ro)}\,] \tag{6}$$

The potentials describe chemically active materials as bonds can be established or cut at the long-range part. They represent reasonable descriptions for two-body forces to the extent, that they account for the repulsion due to overlapping electron clouds at close distance and for attraction at large distances due to dispersion effects. Generally in solids, a shielding effect is expected to make interactions beyond the first few neighbors of limited physical interest. Hence potential functions are commonly truncated at a certain cutoff distance, preferably with a smooth transition to zero, and result in so-called short-range forces. Besides, the long-ranged Coulomb forces are usually beyond the reach of MD model sizes [1].

Tersoff [3] developed an interatomic potential for multicomponent systems and applied it very successfully to the SiC system. The Tersoff potential is a many-body potential that describes accurately the single elemental structures as well as the multicomponent ones.

$$E = \sum_i E_i = \frac{1}{2} \sum_{i \neq j} V_{ij}, \qquad (7)$$
$$V_{ij} = f_C(r_{ij})[f_R(r_{ij}) + b_{ij} f_A(r_{ij})]$$

In the case of SiC, the potential was first fitted to describe very accurately the elemental systems Si and C independently. This is an essential feature of the potential as the workpiece should mimic realistically crack propagations, fractures and other structural modification occurring during a micro-machining process. A single additional parameter is then added to up-grade the potential into treating mixtures of the two elements in a variety of structures. The potential was then tested for the SiC system and compared to experimental and theoretical available data. Tersoff found an excellent agreement for quantities such as lattice constants and elastic constants. This interaction potential is simple enough to be used in large-scale simulations and yet accurate enough to describe the interatomic force field during abrasive processes.

2.2 BOUNDARY CONDITIONS

Boundaries are an intrinsic, vital part of models. Thermodynamic properties are thought of as characterizing "bulk" matter, that is enough material so that surface effects and fluctuations can be ignored. To decrease the influences of boundaries, the system size need to be chosen "big enough" [1].

Besides the option of free surfaces, which would result in a cluster in free space, basically 2 types of boundaries are common in MD simulations: fixed and periodic boundaries. The simplest type, in terms of realization, is the fixed atom boundary which confines all freely propagating atoms inside or provides support for them at one or more sides. It is simply realized by taking away the dynamics of such atoms. The consequences of such infinitely hard boundaries for the simulation can be significant as no energy can be passed through the boundary and phonons will be reflected at it. The sole use of hard boundaries represents a poor representation of the surrounding environment / material.

Periodic boundary conditions (PBC) were introduced to avoid the hard boundary reflection and allow to study bulk and bulk/interface structures without the strong boundary influence in small models (see [1,2,4,5]). It is imagined, that the bulk of the material is made of many similar systems along the axis perpendicular to the periodic boundary plane, i.e. there are no surfaces along this axis. The system reacts as if there are identical systems at both sides of the PBC, exposed to the same conditions and changes. Practically the system is connected to itself and atoms at one side interact with atoms on the other side or transfer through the PBC from one to the other side.

A consequence of periodic boundaries is that energy and phonons are not reflected, but travel through the system by means of PBC's. Two-axis PBC is employed where symmetry axes are available and the lattice structure allows an undisturbed bonding through the PBC planes. Additionally deformation compatibility across the PBC is fulfilled by an appropriate alignment of preferred slip systems relative to the PBC's, in order to avoid artificial deformation patterns.

3. SIMULATION MODEL

In recent years atomistic studies involving molecular dynamics (MD) simulations of abrasive processes from the perspective of nanofabrication have taken a big leap, particularly with a group of researchers in Japan [6-12]. Initial MD studies on small systems based on simple pair potentials for metallic systems and those including three-body interactions and angular terms like Stillinger-Weber [13] and Tersoff [3] potentials for Si and C have been supplemented recently by large-scale simulations using thousands [14] of atoms. For metallic systems such atomistic studies have been carried out [15] using millions of atoms and more realistic interaction potentials like those based on the embedded atom method (EAM) [16]. Calculation of macroscopic stress and strain distributions [17], atomic potential energy profiles [17,15], and radial distribution function [18] have provided preliminary information on surface integrity, crack propagation and brittle-ductile transition. It has been found, for example, that dislocation loops are emitted from the crack front and the sequence of dislocation emission events depends on the crystallographic orientation of the crack front [15]. Related MD simulations [18] have also shown that most materials, in spite of their brittleness, can be machined in ductile mode under a sufficiently small scale of machining. A large scale simulation of the diamond turning of silicon surface [14] has addressed the issue of graphitization and carbide formation and also that of chip formation and surface damage. These initial studies for perfect single crystal surfaces and interfaces have established the importance and usefulness of MD simulations of abrasive processes. The stage is now set to carry out the much needed systematic study of micro-machining and cutting processes which extend from simulations at the fundamental atomistic level to experimental verification of such simulations, and to experimental investigation of processes with industrial applications.

The aim of the simulation is to examine the atomistic mechanisms taking place during brittle and ductile nano machining of single crystals and polycrystals of Si and SiC. A careful analysis of the structural and energetics changes of the individual parts (at the atomic level) of the system during simulated machining under a variety of initial conditions will allow us to rationalize the brittle to ductile transitions and to provide criteria for the occurrence of this transition from a knowledge of the intrinsic properties of the materials (work piece and indenter) that govern plasticity and fracture.

There are several stages in a theoretical study of the structure and energetics of a given system. The choice of atomic structure, i.e. crystallographic orientation, presence of defects, and voids etc., forms the first stage. Next the atomic positions are allowed to relax to their minimum energy configuration. Relaxation patterns are then analyzed in terms of atomic coordination. Finally, the energetics of different regions of the system are calculated.

In order to directly study the dynamics of nanoindentation, molecular dynamics simulations of these processes is carried out using large-scale molecular dynamics code - SPaSM (Scalable Parallel Short-range Molecular dynamics),which was developed by T.C. Germann and P. S. Lomdahl. The SPaSM code uses domain-decomposition techniques in order to efficiently carry out parallel molecular dynamics simulations. The SPaSM code has also previously been used to study fracture with Lennard-Jones and embedded-atom-method (EAM) potentials [15].

3.1 CHARACTERIZATION OF THE INDENTER AND THE WORKPIECE

There are several stages in a theoretical study of the structure and energetics of a given system. The choice of atomic structure, i.e. crystallographic orientation, presence of defects, and voids etc., forms the first stage. Next the atomic positions are allowed to relax to their minimum energy configuration. Relaxation patterns are then analyzed in terms of atomic coordination. Finally, the energetics of different regions of the system are calculated.

CHARACTERIZATION OF THE INDENTER

In all simulations, the indenter is made from a closed packed stacking of carbon atoms in the diamond structure. The indenter is constructed as a pyramid with the number of atoms varying from several hundred to few thousands. The base of the indenter varies from a few atoms, with a radius in the nm range, to over a hundred atoms, with a radius in the 10nm range. For this indenter, the minimum energy configuration is calculated by allowing the atomic positions to relax via the conjugated gradient scheme. The structural changes and stability of the indenter by itself will be studied using molecular dynamics studies at finite temperatures. This step is essential as the temperature near the indenter-workpiece interface may rise by several hundreds of degrees during machining and alter the structure of the indenter.

CHARACTERIZATION OF THE WORKPIECE

SiC material is used extensively in several industries including those involved in micro-electronic devices and memory components. The materials used in the industrial fabrication lines are polycrystalline made of hard grains, which make them both hard and brittle at the same time. Because of this brittleness, these materials fracture easily during grinding. Polishing, on the other hand, proceeds by mainly ductile plastic removal of matter with no fracture. The aim of nano-grinding is to operate in the ductile regime of the polishing while maintaining the ultra-high precision of the grinding. It is then the aim of this study to determine the conditions under which this ductile grinding maybe possible. Since plastic removal of matter proceeds in confined regions in which the local structure may be that of a perfect crystal, or contain a local defect (a void, a dislocation or a grain boundary) and since fracture essentially occurs in polycrystalline multigrained systems, we are planning to include all these factors in our simulations in three different stages.

The single crystal work piece will be made from a stacking of closed packed layers with a free surface. Periodic boundary condition is applied in the x and y directions (parallel to the planes) while the free surfaces will be allowed to relax. Since the SiC material that is most commonly used, and the one that will be used in our experiments is in the 3C structure, this is the structure that is used in our simulations.

The system is allowed to relax to its minimum energy configuration. A multi-grained polycrystal is made from several nanoclusters. These clusters contain several hundred to several thousands of atoms. The clusters are put together and then compressed in the simulations. The final product will be allowed to relax to a zero pressure configuration with periodic boundary conditions in x, y and z directions. As another method to create defects, repeated nanoindentation simulations will be carried out over different parts of the workpiece and then the workpiece will be relaxed.

3.2 NANOINDENTATION

The external forces are applied to the indenter. Since the system is not isolated (the indenter is kept under external forces) it is necessary to dispose of this extra energy realistically. In a real system, this energy will dissipate in the bulk. One way of simulating this disposal of the energy is by placing some thermostat layers above bottom clamped layer of the workpiece. The atoms in this thermostat region are kept at constant temperature all along the simulation. Energy will dissipate also into the indenter, and hence, we will keep the layer just below the static layers at constant temperature.

Currently SPaSM code is being modified for Tersoff potential, which, as it was stated above, is the best match for diamond and SiC materials.

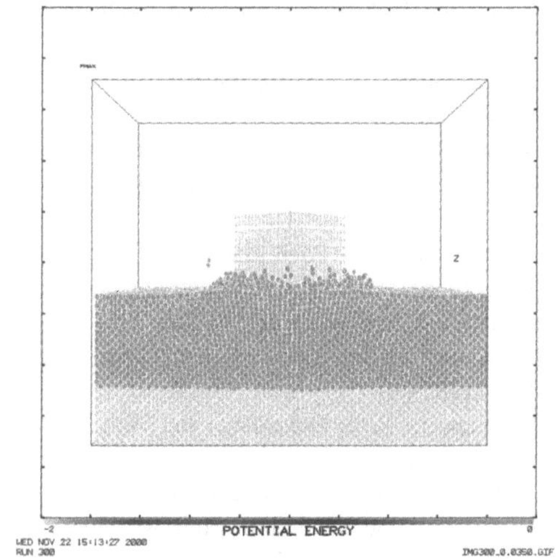

FIGURE 2. INDENTATION

The figure 2 shows a snapshot of the early stages of indentation using Leonard-Jones potential with fcc crystal structure. As can be seen, the indentation process leads to a localized heating of the substrate and indenter in the interaction region. Light gray substrate atoms undergo unconstrained molecular dynamics while the dark gray atoms at the bottom of the substrate undergo constant temperature (Langevin) dynamics and serve as a heat sink. Bottom layers correspond to fixed atoms.

REFERENCES

1. Hoover,W.G., 1991, Computational Statistical Mechanics, Studies in Modern Thermodynamics 11, Elsevier Science Publisher, Amsterdam-Oxford-New York-Tokyo, 313 pages.
2. Allen, M.P., Tildesley, D.J., 1987, Computer Simulation of Liquids, Clarendon Press, Oxford.
3. Tersoff, J., 1989, Modeling Solid-State Chemistry: Inter atomic Potentials for Multi-component Systems, Phy.Rev. B, Vol.39, No.8, 5566-5568, and 1990, Vol.41, No.5, 3248.
4. Rentsch, R., Process modeling by means of molecular dynamics (MD), 2nd. Int . Conf. on Machining of Advanced Materials, Aachen, Germany, 30.9.-1.10.1996, VDI - Berichte No.1276, pages 175-195.
5. Rentsch, R., Dissertation, Molecular Dynamics simulation for nanometer chip removal processes, Keio University, Yokohama, Japan, 1995, 150 pages
6. Shoichi Shimada, Naoya Ikawa, Hiroaki Tanaka, Junichi Uchikoshi, icStructure of Micromachined Surface Simulated by Molecular Dynamics Analysis,ln Annals of the CJRP, Vol. 43, 51(1994).

7. Shoichi Shimada, Naoya Ikawa, Toyoshiro Inamura, Nobunior Takezawa, Hitoshi Ohmori, Toshio Sata, iaBrittle-Ductile Transition Phenomena in Microindentation and Micromachining,le Annals of the CIRP, Vol 44 , 523 (1995).
8. Toyoshiro Inamura, Nobuhiro Takezawa, Yasuhiro Kumaki, iuMechanics and Energ,y Dissipation in Nanoscale Cutting,lo Annals of the CIRP, Vol.42, 79 (1993).
9. I. Inasaki, "Application of Simulation Technologies for Grinding Operations,l. VDI Berichte Nr. 1276, 197 (1996).
10. R. Rentsch, I. Inasaki, inInvestigations of Surface Integrity by Molecular Dynamics Simulation,li Annals of the CIRP, Vol.44/1, 295-298 (1995).
11. N. Ikawa, S~Shimada, H. Tanaka, G. Olimori, ilAn Atomistic Analysis of Nanometric Chip Removal as Affected by Tool-Work Interaction in Diamond Turning,lo Annals of the CIRP, Vol.40, 551-554 (1991).
12. R. Rentsch. I. Inaskai, inSimulation of Single Point Matching of Poly-Crystalline Metals bv Molecular Dvnamics (MD),lc in International Progress in Precision Engineering, Ed. M. Bonis et al., Proc. of the 8th, IPES, Elsevier, 1995, F. Compiegne), pp.347-350.
13. F.H. Stillinger, A.W. Weber, ieComputer Simulation of Local Order in Condensed Phases of Silicon,ln Phys. Rev. B31 5262-5271(1985).
14. D.B. Boercker, J. Belak, I.F. Stowers, R.R~ Donaldson, and W.J. Siekhaus; Simulation of Diamond Turning of Silicon Surfaces,ln unpublished.
15. S.J. Zhou, D.M. Beazley, P.S. Lomdalil, and B.L. Holian, ioLarge-Scale Molecular Dynamics Simulations of Three-Dimensional Ductile Failure, Phys. Rev. Lett. 78, 479 (1997).
16. M.S. Daw, M.1. Baskes, iaEmbedded-Atom Method: Derivation and Application to Impurities, Surfaces, and other Defects in Metals,l, Phys. Rev. B29, 6443-6453 (1984).
17. T. Inamura, S. Shimada, N. Takezawa, and N. Nakahara, iaBrittle/Ductile Transition Phenomena Observed in Computer Simulations of Machining Defect-Free Monocrystalline Silicon,la Annals of the CIRP, 46, 31 (1997).
18. Shoichi Shimada, Naoya Ikawa, Hiroaki Tanaka, Junichi Uchikoshi, icStructure of Micromachined Surface Simulated by Molecular Dynamics Analysis,ln Annals of the CJRP, Vol. 43, 51(1994).

INFLUENCE OF THE HOT ISOSTATIC PRESSING ON THE MECHANICAL BEHAVIOUR OF CVD-COATED HARDMETALS

M. Tolazzi[1], S. Kursawe[2], H.G. Sockel[3]

[1] Department of Mechanical Engineering, University of Erlangen-Nürnberg, Germany
[2] Bayerische Motorenwerke AG, Dingolfing, Germany
[3] Department of Materials Science and Engineering, University of Erlangen-Nürnberg, Germany

KEYWORDS: Hardmetals, Fatigue, Hot Isostatic Pressing, Coatings.

ABSTRACT. CVD-coated hardmetals came into market almost thirty years ago. These materials perform well, where extreme difficult cutting conditions, such as interrupted cutting and precision works, prevail. In this work, two different grades have been investigated under monotonically increasing and cyclic loadings. For each grade two sets of samples were available, produced by means of two different sintering processes: Vacuum Sintering and Hot Isostatic Pressing. The aim of the work was to relate the mechanical properties to the sintering strategy. The mechanical testing revealed that, under both montonically increasing and cyclic loading, there were no significant differences related to the sintering process. This makes the more convenient conventional vacuum sintering more suitable for the production of the substrate hardmetal for CVD-coatings.

1 INTRODUCTION

CVD-coated hardmetals have been employed as steel cutting tools since the early 1970s [1]. Their combination of hardness and toughness makes them suitable, especially under extreme difficult cutting conditions, such as interrupted metal cutting or precision works. The substrate material for those tools is industrially produced by two different technologies: the traditional Vacuum Sintering (VS) and the Hot Isostatic Pressing and Sintering (HIP). For this work, two CVD-coated hardmetal grades were available, the grade G8 and A7. For each of them, two sets of samples, produced respectively by means of VS and HIP, have been tested.

TABLE 1. Investigated materials

grade	wt.-% WC	wt.-% Co	wt.-% γ-phase	CVD-coating thickness (μm)
G8	86.0	6.0	8.0	12.5 ± 0.3
A7	90.5	6.0	3.5	18.8 ± 0.2

Published in: E.Kuljanic (Ed.) *Advanced Manufacturing Systems and Technology*,
CISM Courses and Lectures No. 437, Springer Wien New York, 2002.

The aim of the work was to investigate the mechanical behaviour of these grades under monotonically increasing and cyclic loadings, in order to relate their mechanical properties to the sintering process by which they were produced. The microstructures of the samples were characterised metallographically.

2 EXPERIMENTAL

2.1 INVESTIGATED MATERIALS

An overview of the investigated hardmetals is given in Table 1. The composition of the γ-Phase was (W,Ta,Ti,Nb)(C,N) for the grade G8 and (Ta,Nb)C for the grade A7. The two grades differ in the thickness and composition of the coating (TiC/Ti(C,N)/TiN for G8; TiN/Ti(C,N)/Al2O3/Ti(C,N)/TiN for A7) and in the presence of a functional gradient in the grade G8 (Fig. 1). The employment of G8 in interrupted cutting requires in fact improved toughness, while the thickness of the coating of A7 confers to the grade the high wear resistance needed in precision works [2]. The WC grain size of the hardmetals was in the range of 1.35 μm for G8 and 0.95 μm for A7.

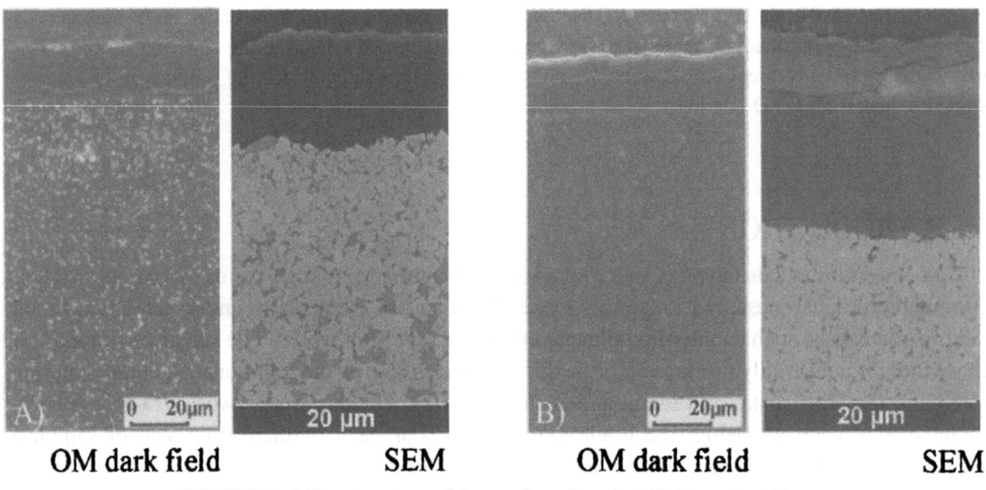

FIGURE 1. Microstructure of the grades: a) grade G8; b) grade A7.

2.2 SINTERING PROCESS

Vacuum sintering is the conventional sintering process. It consists of heating the previously formed part to 1350-1650 °C. At this temperature, a liquid phase is formed with solution of the carbides in the molten cobalt, while diffusion takes place in the hard phase, changing its original composition and properties [3].

The so-called HIP-treatment consists in applying a pressure of 100-150 MPa to the compact, during the conventional vacuum sintering, by means of an inert gas which presses the part from all directions, while being heated from an external source [1]. The benefits of this treatment on uncoated hardmetals is well known and can be summarised in a reduction of porosity which leads to an increase in transverse rupture strength and wear resistance [4]. In contrast, the effects of this combined and more expensive process on CVD-coated hardmetals are still not well known.

2.3 CVD COATINGS

Chemical vapour deposition is the conventional process used for the application of TiC coating layers. The $TiCl_4$ contained in a gaseous phase is converted, at high temperatures, into titanium carbide crystals, which grow from the substrate of the sintered carbide [5].

2.4 MECHANICAL TESTING

All tests were carried out in bending stress in an apparatus based on a principle of Fett et al. [6]. The apparatus allows testing under monotonically increasing loading and alternating cyclic bending with constant displacement of the cantilever. The inert strength σ_i was determined as bending strength without the influence of subcritical crack growth [7]. The frequency of the fatigue tests under cycling loading was 4 Hz, and all tests were carried out at room temperature until failure of the sample or 10^6 cycles. Fig. 2 shows a schematic representation of the testing apparatus in the case of alternating cyclic loading.

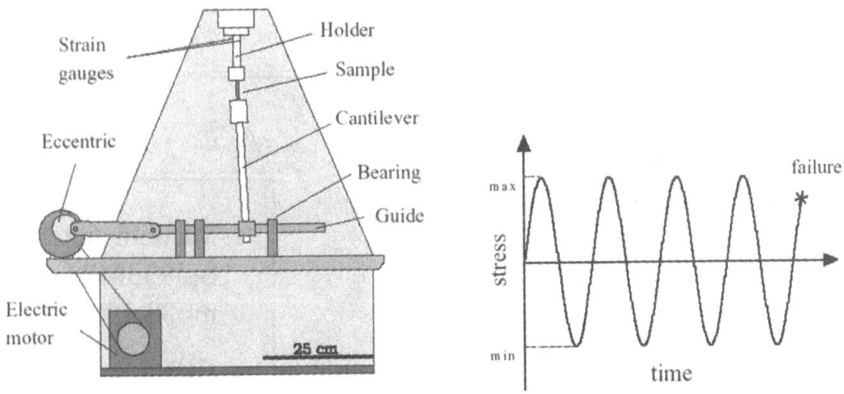

FIGURE 2. Apparatus for the alternating bending loading.

3 RESULTS AND DISCUSSION

3.1 MONOTONICALLY INCREASING LOADING

Figures 3 and 4 compare the results of the inert strength for the VS and the HIP hardmetals in a Weibull diagram. F in this figures is given by $F(\sigma_0) = 1\text{-}\exp(-(\sigma_i/\sigma_0)^m)$, m is the slope of the lines. In both materials, the mean inert strength σ_0 is slightly higher in the case of vacuum sintering, while the scattering of the results, measured by theWeibull modulus m, is smaller.

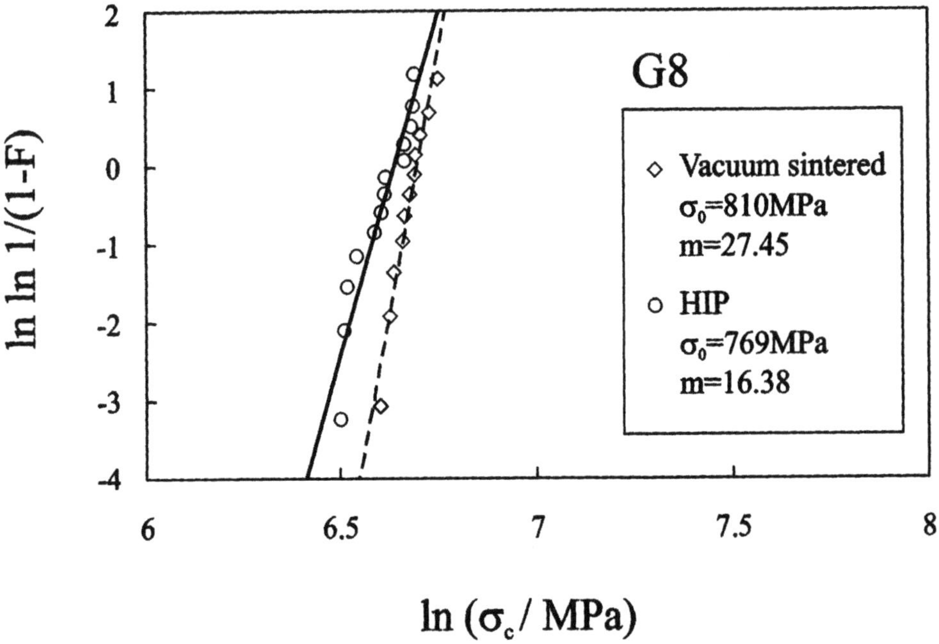

FIGURE 3. Weibull distribution for the inert strengths, hardmetal G8.

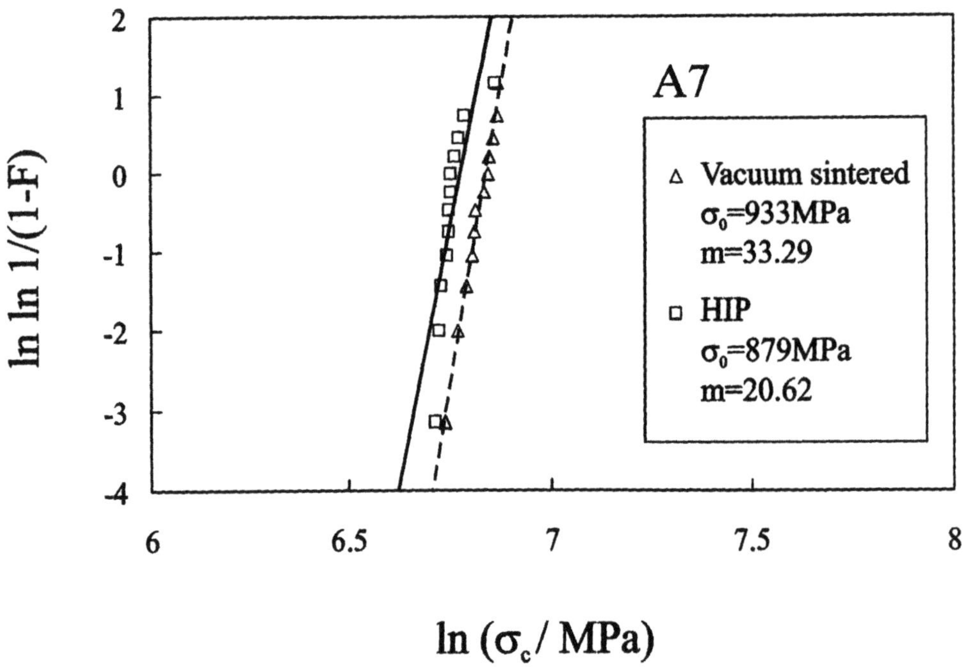

FIGURE 4. Weibull distribution for the inert strengths, hardmetal A7.

In contrast to HIP treatment of uncoated materials, the HIP does not lead to any improvement in the mechanical behaviour of the hardmetal in the present case of coated materials. The presence of a brittle coating, with a thickness in the same order of magnitude as the critical crack length, has probably a much stronger influence on the inert strength than the properties of the sintered matrix.

3.2 CYCLING LOADING

Figs. 5 and 6 show the Wöhler diagrams for the two investigated hardmetals. As observed in the results of the inert strength, the curves in both cases do not show a significant difference. This means that the sintering process does not influence the mechanical behaviour of CVD-coated hardmetals. What can be also noticed here is an anomaly in the relationship between inert strength and Wöhler diagram for the grade A7. It was shown in previous works [8-10], and it is partially confirmed in the present study (for the grade G8) that the Wöhler curves start from a value of $\Delta\sigma/2$ which coincides with the inert strength of the material. For the hardmetal A7, the range of the inert strength lies much higher (about 20%) than the starting point of the Wöhler curves. Such a behaviour has not been reported previously in the literature.

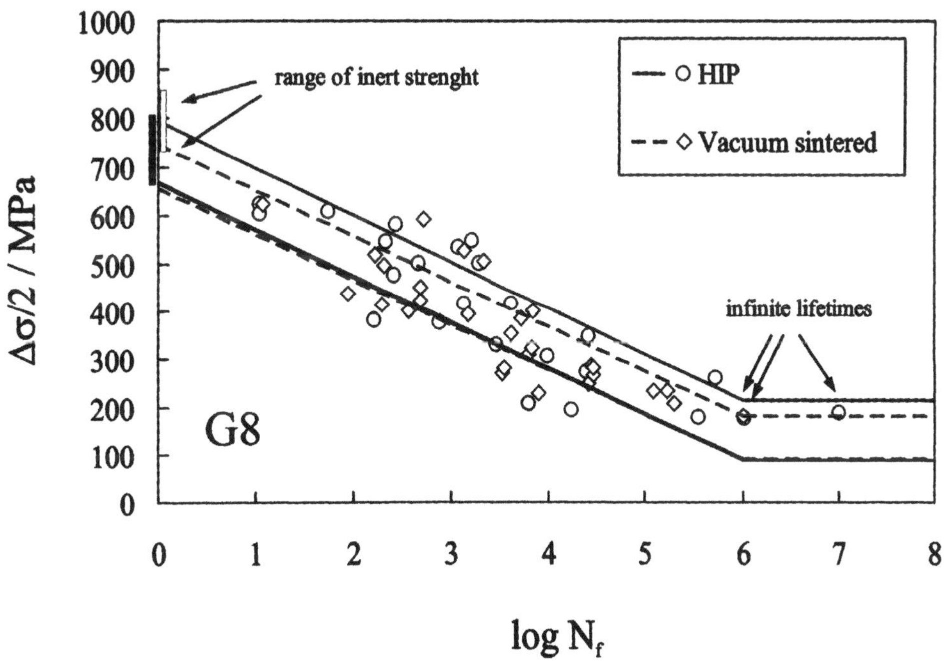

FIGURE 5. Wöhler diagram of the fatigue life results for the hardmetal G8. The endured stress amplitude $\Delta\sigma/2$ is plotted vs. the number of cycles until failure, N_f. The bars at $\log N_f = 0$ mark the range of the inert strength.

The reason for this is not completely clear but can be attributed to the extremely thick coating (Tab. 1), which can cause the premature failure of the samples, when subjected to high stresses and thus a decreasing in the slope of the curve. A crack can grow fast to about 20 μm through the coating and then, at stress values close to σ_0, needs to grow only about 10 μm more to become a critical crack and break the sample (the calculated value for a critical crack in this case is ca. 28 μm). At lower stress values, on the other hand, the critical crack length is bigger, and the influence of the coating decreases.

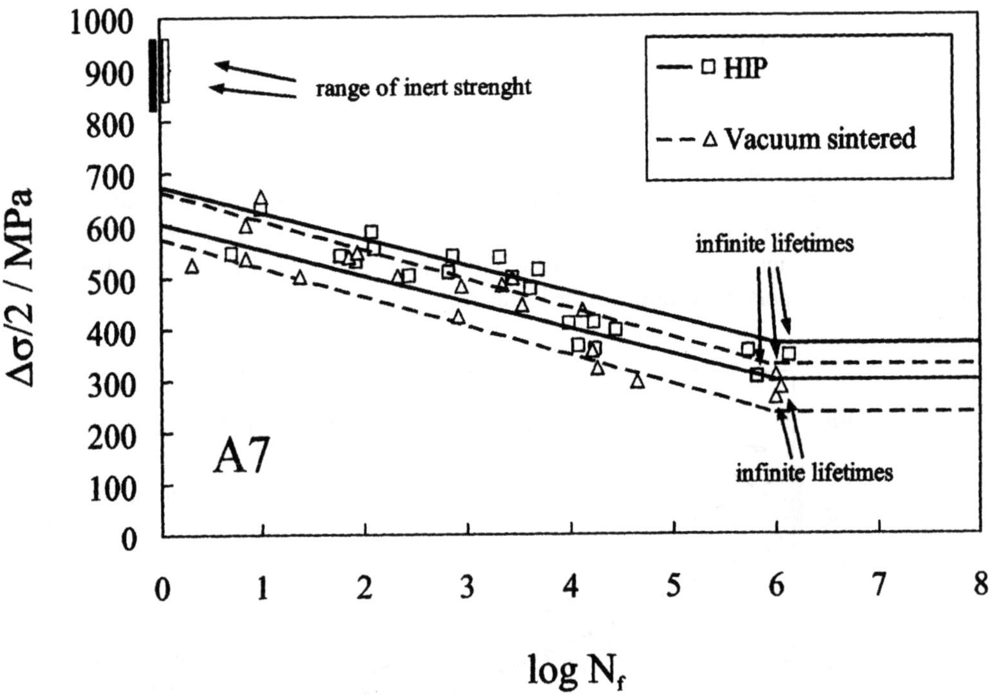

FIGURE 6. Wöhler diagram of the fatigue life results for the hardmetal A7. The endured stress amplitude $\sigma/2$ is plotted vs. the number of cycles until failure, N_f. The bars at log $N_f = 0$ mark the range of the inert strength.

4 CONCLUSION

The mechanical investigations of the grades led to the following results:

- For both grades, no significant differences in the mechanical behaviour under monotonically increasing loading were found. Moreover, the inert strength was found to be slightly higher in the case of vacuum sintering than after HIP treatment.
- For both grades, no significant differences in the mechanical behaviour under cyclic loading were found.

It is concluded that the presence of a CVD–coating has a much stronger influence than an improved sintering process (such as Hot Isostatic Pressing) on the mechanical behaviour of hardmetals. In the case of CVD–coated hardmetals, nearly the same mechanical behaviour can be obtained by employing the conventional and more convenient vacuum sintering.

REFERENCES

1. Brookes, K.J.A., (1996), World Directory and Handbook of Hardmetals and Hardmaterials, Sixth Edition. International Carbide Data.
2. Kennametal Hertel AG, Mistelgau. Personal notice.
3. Sandvik Coromant. Modern Metal Cutting, a pratical handbook.
4. Göransson, M., Odebo, U., (1981), Isostatic pressing of hardmetals. Procedings of the 10th plansee-seminar, 1:647-657.
5. Konyashin, I. Yu., (1995), PVD/CVD technology for coating cemented carbides. Surfaces and Coatings Technology, 71:277-283.
6. Fett. T., Martin, G., Munz, D., Thun, G., (1991), Determination of $da/dn - \delta k_I$ curves for small cracks in alumina in alternating bending tests. Journal of materials science, 26: 3320-3328.
7. Munz, D., Fett, T., (1989), Mechanisches Verhalten keramischer Wersktoffe. Springer-Verlag, Berlin.
8. Schlund, P., (1998), Das mechanische Verhalten von beschichteten Hartmetallen und Cermets unter monoton ansteigender und zyklisch wechselnder Beanspruchung. Doctorate Thesis, University of Erlangen-Nürnberg.
9. Kindermann, P., (1999) Ermüdungsverhalten von Hartmetallen und Cermets unter zyklisch wechselnder Beanspruchung zwischen 25 °C und 900 °C. Doctorate Thesis, University of Erlangen-Nürnberg.
10. Pott, Ph., (2000), Einfluss von Zusammensetzung der Bindephase auf die mechanischen Eigenschaften von ausgewählten Hartmetallen. Doctorate Thesis, University of Erlangen-Nürnberg.

EFFECT OF CRYSTAL ORIENTATION ON SURFACE INTEGRITY

J. Prohászka[1], J. Dobránszky, M. Horváth[2], A.G. Mamalis[3]

[1] Research Group for Metals Technology, Budapest Univ. of Technology and Econ., Budapest, Hungary
[2] Department of Manufacturing Engineering, Budapest Univ. of Technol. and Econ., Budapest, Hungary
[3] Manufacturing Technology Division, National Technical University of Athens, Greece

KEYWORDS: Crystal orientation, Roughness, Plastic deformation

ABSTRACT. Anisotropy of materials mechanical properties is one of the main factors, determining surface finish of machined surfaces. In the interaction between the cutting tool and the cut material both plastic and elastic deformations appear. Effect of crystallites in the machined surface on the workpiece surface finish depends on crystal direction perpendicular to the surface. Value of elastic deformation is determined by the ratio of applied stress and the determinant modulus of elasticity in given direction. Since E modulus depends on crystal direction, surface roughness changes from one crystallite to other one. The rates of these changes depend on the deviations of Young modulus in different crystal directions. In the paper the changes of surface finish in different materials are shown, as functions of cutting conditions and those of deviations in modulus of elasticity, in different crystallite orientations.

1 INTRODUCTION

In the conventional machining depth of cut and feedrate are big relatively to the sizes of different crystallites and the surface roughness, so it is not necessary to consider influence of the microstructure features and anisotropy of elastic properties of crystallites. When one makes mirror like surfaces using cutting, the size of chip is less than that of the crystallites of the machined material. So, the roughness of the mirror surfaces is determined not by the deformation features of polycrystalline materials, but by those of the single crystals, especially in the areas, far from the boundaries of crystallites [1-5]. In the following lines mirror making machining experiments and some conclusions are shown, based on metallographical measurings of microstructures concerning the elastic and plastic deformations of mono crystals.

2 THE SPECIMENS

The specimens were made of oxygen free (OF, Cu=99,99%) copper, because its module of elasticity shows very big dependence on the orientation. Considering the biggest differences of copper's Young modulus the ratio E_{111}/E_{110} is more then 3.0.

We made specimens of conventionally recrystallized material with small crystallites, than the specimens were sheeted (rolled, ε=90%) and by secondary recrystallization we got some cm^2 big crystallites to „amplify" the influence of cutting inside and on the boundaries of them, relatively to cut surfaces with small crystallites. After the rolling the specimen were heated

Published in: E.Kuljanic (Ed.) *Advanced Manufacturing Systems and Technology*, CISM Courses and Lectures No. 437, Springer Wien New York, 2002.

(T≅1350 K, t=6 h). Metallographical pictures of two specimens with 2.10 mm thickness are shown on Figure 1.

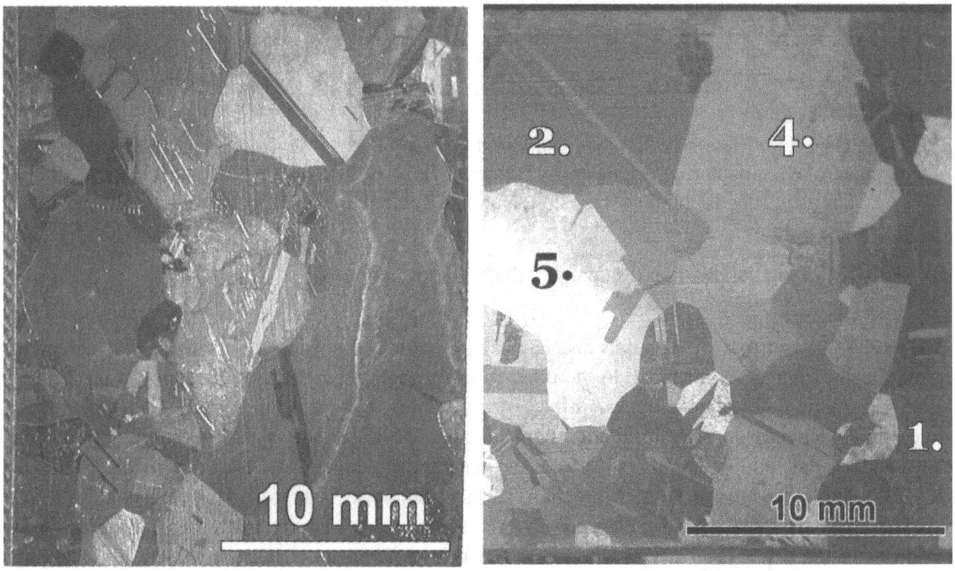

a) b)

FIGURE 1. Metallographic pictures of macroetched specimens

3 THE MACHINING AND MEASURING

Orientations of bigger crystallites were determined by Laue method before measuring for being able to consider features of active slip systems.

The specimens were turned using single crystal diamond tool with nose radius r_ε=2 mm, edge radius is some 20 nm. With cutting speed $v \leq 600$ m/min, feed rate f = 1,3 μm/rev and depth of cut a = 1.3 μm mirror like surfaces were made. Two versions of turning were applied:

– Face turning on small (20 mm) diameters, when the direction of cutting and deformations were continuously changing the active slip systems – where the gliding took place as a result of dislocations movements – drastic changes occurred.

– The specimen was fixed on a disc and face turning was made on big diameters (≥ 100 mm), so the changes in the active slip systems, especially inside of single crystallites were negligible (Figure 1b).

4 THE RESULTS OF MEASURING

On the Figure 2 a picture of mirror surface is shown, made using interference contrast. One can see, that the surface is formed by crystallites on different levels. The less is the size of crystal the bigger is the difference relatively to the adjacent crystallites. Figure 3 is also a picture of a mirror surface [6].

The roughness of different crystallites could be seen using the scale on the left side of the picture. Sudden level changes are seen on boundaries of crystallites. Arrows mark the exceptions when the roughness is showing large changing in the inner parts of the crystallites surface. In both crystallites twin areas took place and on their boundaries the cutting forces drastically change.

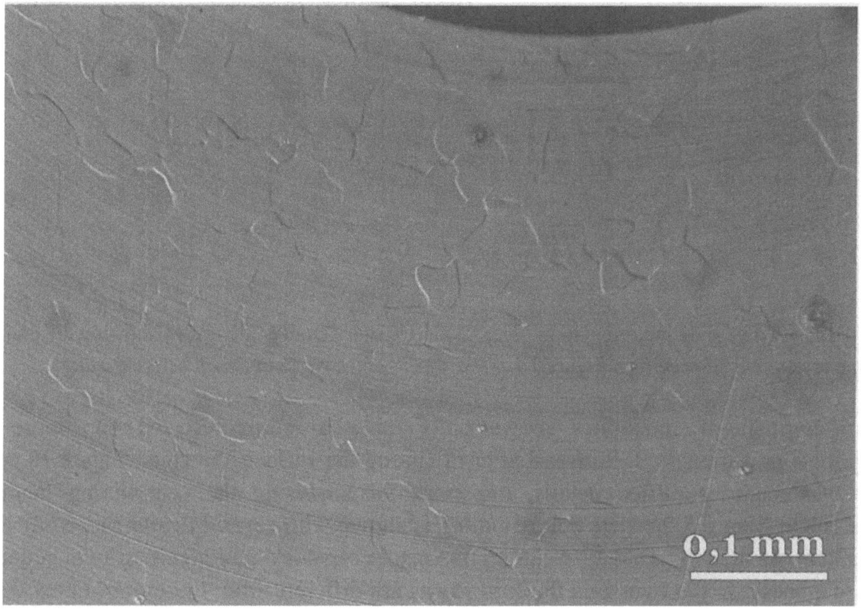

FIGURE 2. Nomarski interference contrast picture of a cut mirror surface

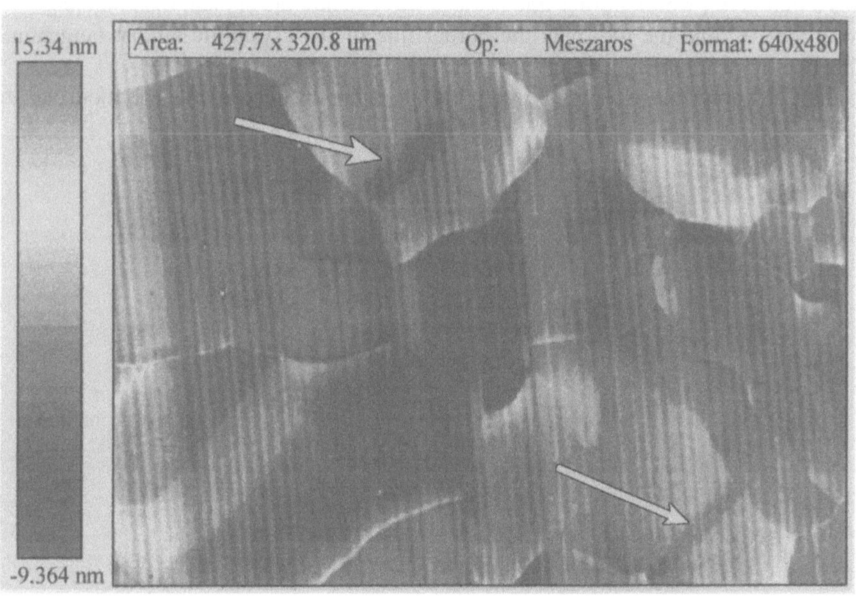

FIGURE 3. Mirror like surface showing roughness and crystal boundaries

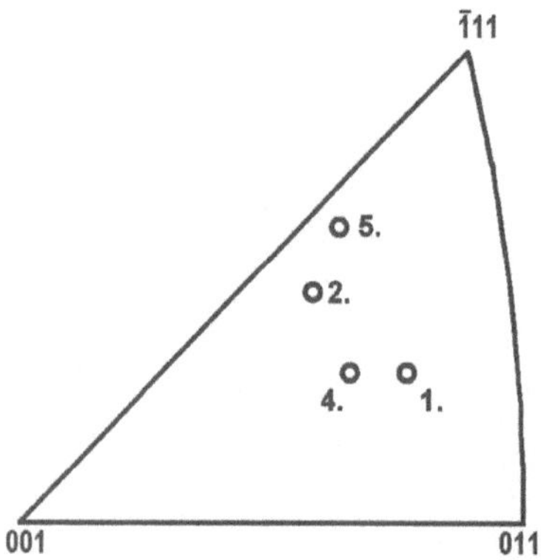

FIGURE 4. Poles showing orientation of some crystallites of specimen Figure 1b

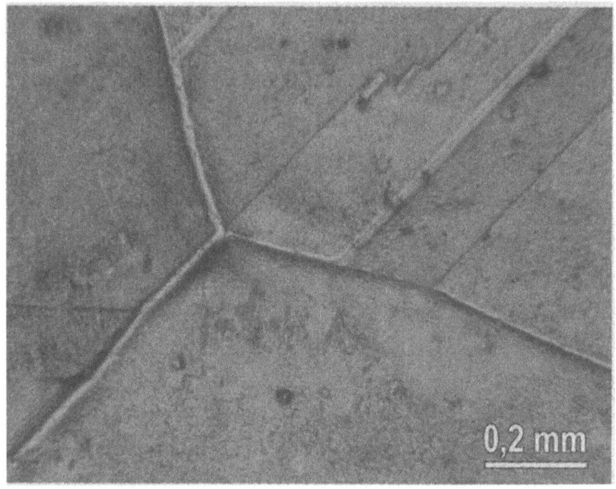

FIGURE 5. Detail of Figure 1b showing the grain and twin boundaries before cutting

The $\mathbf{n}_i$ normals of crystallites shown on Figure 1b are represented on Figure 4 in stereographic unit triangle. Before and after of cutting the surface shown on Figure 1b several picture was made amplified details, important for analysing the determining factors of roughness. On Figure 5 the state before cutting is shown with several twin areas, while a twin area could be seen on Figure 6 after cutting: the reader can see the influence of twin boundaries on the roughness. In the twin area the feed marks are different from those out of area. Picture on Figure 7 made on the same cut surface and shows the effect of crystallite on the feed marks:

- When the tool approaches perpendicularly the crystallite boundary along the AB line it generates a thin layer, which border line is more wavy inside the crystal, then outside.
- Along the AC line the tool cuts the border line of the crystallite almost tangentially, so along this line there are no significant changes.
- On the CD borderline we can see very big changes, the feed marks are modified in very wide band.

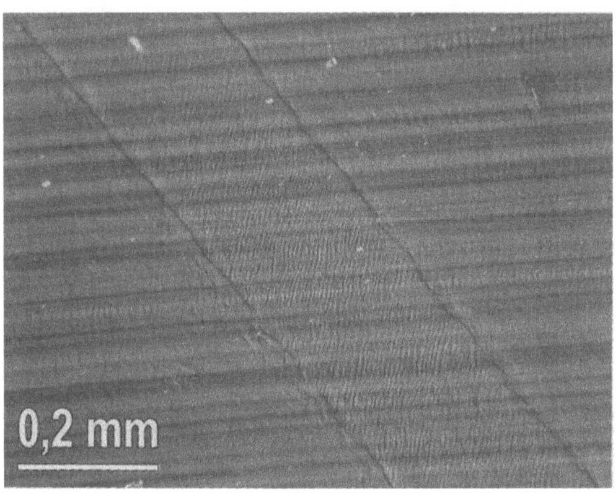

FIGURE 6. Detail of Figure 1b showing effect of cutting on twin boundary

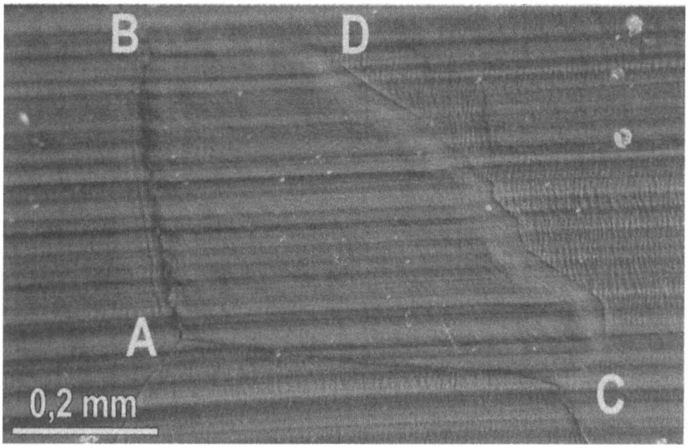

FIGURE 7. One crystallite of specimen 1b showing effect of crystallite boundaries

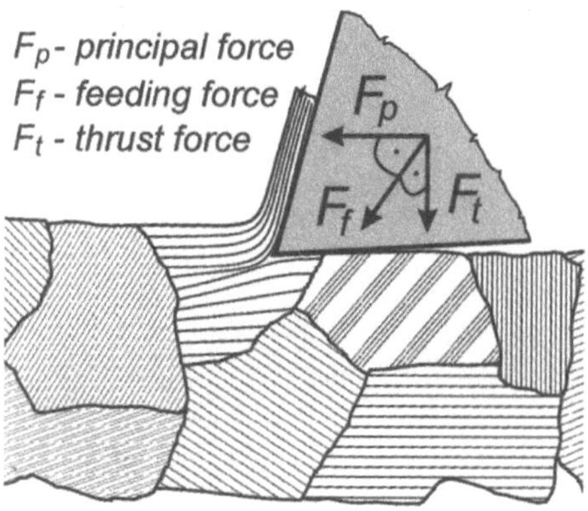

FIGURE 8. Active slip system in the area of chip and three components of cutting force

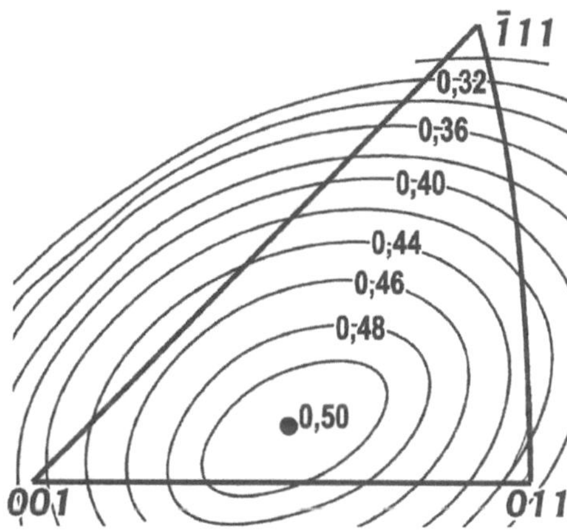

FIGURE 9. Constant lines of Schmid factor for surface centred metals considering active slip system (111) [101]

5 EVALUATION OF MEASURING RESULTS

On Figure 8 we represent the applied scheme of the cutting model [7, 8]. Thick lines outline the grain boundaries, while inside of them cross lines of the active slip system with the plane of picture are drawn by parallel lines. In all crystallites of the specimen (Figure 1b) maximum shear stress occurs only in one slip system, on the slip plane (111), and in slip direction [$\bar{1}$01]. We cannot exclude activity of other slip system(s), because poles of some crystallites are very close to pole 111, and the value of shear stress is

$$\tau = \frac{F_e}{A}\cos\alpha\cos\beta\,, \tag{1}$$

where F_e is the resulting cutting force, A is the unknown cross section, α is the angle between force and the normal to slip plane and β is the angle between force and slip direction.

Values of Schmid factor in base triangle are shown on Figure 9. Near the vertices the Schmid factors are changing only in very small way for more than one slip systems. The slip starts in the slip plane, in which to the activation of Frank-Read source minimum stress is needed:

$$\mathrm{T}_{F-R} = k\frac{Gb}{D}, \tag{2}$$

where k is a constant, close to 1 (depending on the orientation of the dislocations), G is the shear modulus of elasticity, b is the Burgers vector of dislocation and D is the distance between two fixed points of Frank-Read source.

The situation is much more sophisticated, when the curvature of tool path is big, e.g. when the tool is cutting the same crystal during one rotation (Figure 10).

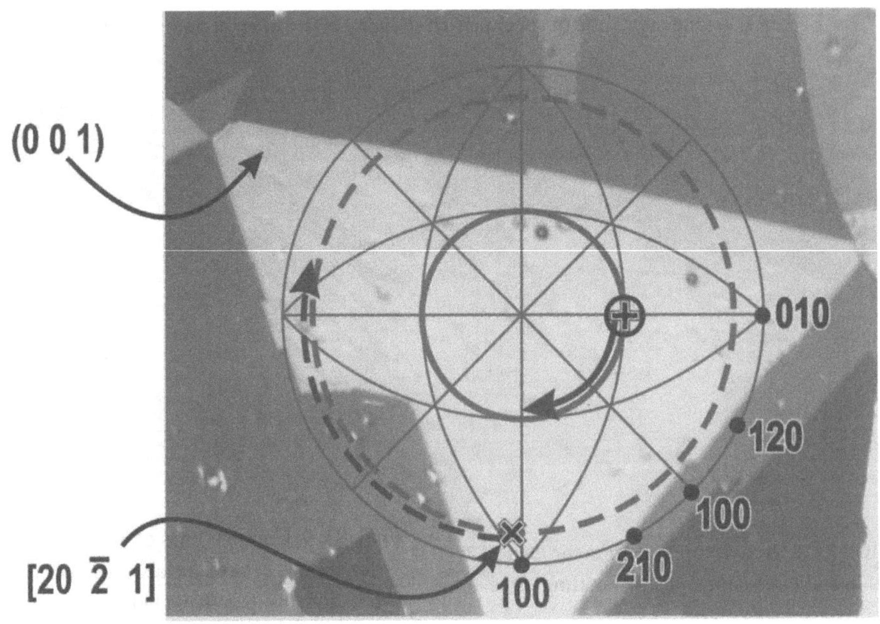

FIGURE 10. Trajectory of tool and the resulting cutting force when cutting one crystal [001] on axis of rotation

The path of the tool on the central crystallite is a closed circle arc. The orientation of this crystal is [001]. The thin line shows the tool path, the arrow – the direction of rotation.

On one of the quarter circle of the tool path arc several crystal direction poles and their indexes are marked, where the tool moves on a plane perpendicular to direction [001]. In this case shear stress is changing continuously and slip systems are changing suddenly, and in some tool positions number of active slip systems and Miller indexes, marking their orientations change, too. In the case, when tool position is marked with ⊕ the resulting force is parallel with direction [20$\bar{2}$1], its pole is marked ⊗ on the pole picture 001 connected with inner crystallite.

Change of this force is marked by dotted line. Angle between the actual position of the tool marked ⊕ and the poles of resulting force marked ⊗ is constant during machining.

In every point, where the dotted line crosses one of the line of any triangle of pole picture the active slip system is changing. It means, that during one revolution the active slip system changes 16 times. Of course, in crystallites far from the centre number of changes is less.

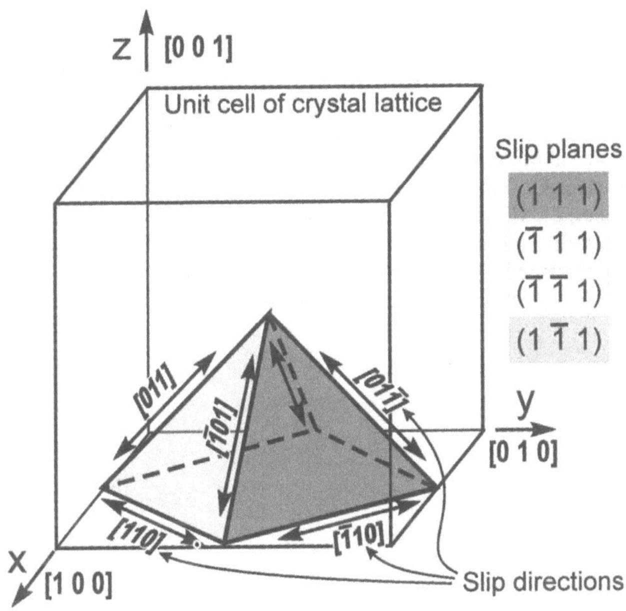

FIGURE 11. Slip planes and directions of FCC crystallites on a half octahedron

On Figure 11 on a half octahedron the possible slip planes and directions are shown.

When the cutting force is eliminated a mechanical state of equilibrium is recovered, resulting elastic of deformation. Due to elimination of thrust force the elastic deformations are perpendicular to the cut surface and they in high grade depend on the orientations of the cut crystals. This is the reason of impact of Young modulus on roughness. In the case of copper with the same value of thrust force or stress the elastic deformation in crystallites with [111] orientation is three times more, than that of in crystallites with [100] orientation.

6 CONCLUSIONS

Roughness of mirror-like cut surfaces with same cutting conditions is determined by microstructure of material, by its deformation features. As it is shown, the roughness is mostly influenced by grain and twin boundaries, by number and character of active slip systems, features of elastic and plastic deformation and by the orientations of crystallites in the surface layer. Our measurings were carried out on soft materials with 10^6 /cm^2 dislocation density.

Analogical experiments and measuring should be carried out of for other materials, e.g. soft and martensitic steels. We need much more knowledge on microstructure of materials and machining processes for controlling reliably the quality of mirror like surfaces during cutting.

REFERENCES

1. Grabchenko, A., Horváth, M., Mamalis, A.G., Mészáros, I., Paulmier, D., (1999), Ultraprecision Machining of Mirror-Surfaces. OSIN'99, Krakow, 189-206.
2. Moriwaki, T., (1989), Machinability of Copper in Ultraprecision Micro Diamond Cutting, Annals of the CIRP, Vol. 38/1, 115-118.
3. Spenrath, N.M., (1991), Technologische Aspekte zum Feinstdrehen von Kupferspiegeln, Dissertation, RWTH Aachen, 31-34.
4. König, W., (1981), Fertigungsverfahren Band 1, VDI Verlag GmbH, Düsseldorf 1981.
5. Mamalis, A.G., Prohászka, J., Mészáros, I., (2000), The Effect of the Anisotropy of the Material on the Surface Topography in Case of Ultraprecision Machining, 1st EUSPEN Topical Conference on Fabrication and Metrology in Nanotechnology, Copenhagen, May 28-30, 440-446.
6. Nyírő, J., Mamalis, A.G., Prohászka, J., Mészáros, I., (2001), Analysis of Ultraprecision Turned Mirror Surfaces, EUSPEN 2nd International Conference, in Turin, May 28-31, 718-721.
7. Nyírő, J., (2000), Mathematische Modell des Microdrehens, Periodica Politechnica'99, 107-110.
8. Mészáros, I., Nyírő, J., (2000), The Analysis of the Microcut Surface Topography, The 11th International DAAAM Symposium, Wien, October 19-21, 147-148.

SINTERED ALLOYS MECHANICAL PROPERTIES SIMULATION WITH CELL METHOD

F. Cosmi

Department of Energetics, University of Trieste, Italy

KEYWORDS: Sintered alloys, mechanical properties, computer simulation

ABSTRACT. Aim of this paper is to introduce a new numerical model that may be employed to assess sintered alloys mechanical properties, depending on residual porosity. The proposed methodology makes use of Cell Method (CM), a recently developed numerical method. CM is particularly suitable for problems in which discontinuities or heterogeneities are present. The model consists of a constituent matrix of cells in which randomly distributed void cells are present, accounting for residual porosity. Such a model can be solved with CM. In this paper, some aspects of CM, which are important for the specific application, will be briefly recalled. The proposed model will be illustrated and results from several simulations in both the elastic and plastic field will be discussed and compared with experimental data available in literature.

1 INTRODUCTION

It is well known in literature that mechanical properties of sintered alloys strongly depend on residual porosity, which affects the effective section and induces stress concentrations [1]. Therefore, the presence of pores results in a reduction of strength and ductility with respect to wrought materials.

This paper introduces a numerical model, which may be used in order to predict mechanical properties of sintered alloys, given the residual porosity. The model consists in a matrix of cells of the constituent in which randomly distributed void cells are present, accounting for residual porosity, as shown in Figure 1. A compression test is simulated on this model.

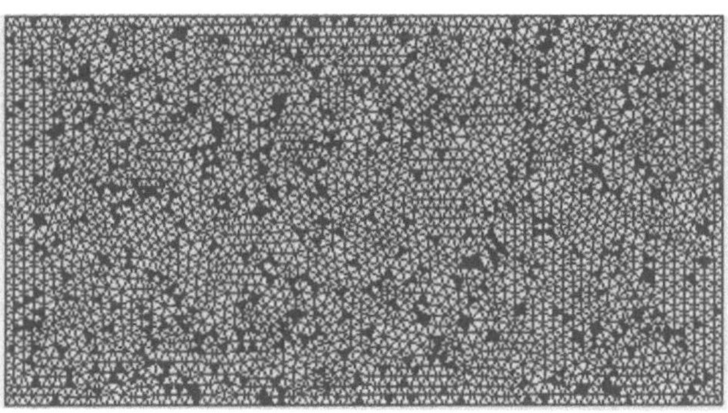

FIGURE 1. Randomly distributed void cells (black) among the matrix of constituent cells (white)

Published in: E.Kuljanic (Ed.) *Advanced Manufacturing Systems and Technology*, CISM Courses and Lectures No. 437, Springer Wien New York, 2002.

Cell Method (CM), a recently developed numerical method that is particularly suitable for problems involving heterogeneities and discontinuities, is used to solve the proposed model. Some aspects of CM, which are important for the specific application, will be briefly recalled in Section 2. The model used and simulations results will be presented for both elastic and plastic field in Sections 3 and 4 respectively. Conclusions will be drawn in Section 5.

2 NUMERICAL MODEL

As already introduced, the proposed numerical model consists of a homogeneous matrix in which pores are randomly distributed. This model can be solved with Cell Method. A few aspects of CM, which are important for the purpose of the paper, will be briefly recalled. A more detailed description and comparisons with FEM results for plane elasticity may be found in [2].

CM is a recently developed numerical method [3] that has some advantages over other previously introduced and widely used methods, such as Finite Element Method (FEM).

One important fact of FEM is that in order to write equilibrium equations, a differential formulation is used. But when discontinuities are present it is no longer possible to perform a differentiation and special elements must be developed in order to overcome this problem. As a general statement, it can be said that the use of FEM implies the size of the mesh to be smaller than any characteristic length involved in the problem [4].

On the contrary, CM directly writes equilibrium equation for an influence region of each node, using only global variables instead of "traditional" field functions.

The basis of the method comes [3] from a classification of global variables involved in field problems, which can be always organized as belonging to one of the following categories: *Configuration variables* – displacements, strain tensor, etc.; *source variables* – forces, stress tensor, etc. and *energy variables*. Different variables are associated to different spatial elements. Consistently with this classification, two staggered complexes of cells – meshes – are used:

- *Primal cells*, in these simulations 3 nodes triangles from a Delaunay mesh, to which configuration variables are to be associated;
- *Dual cells*, from the Voronoi tessellation associated with the previously defined triangles, to which source variables are to be linked (see Figure 2).

It should be noted that each node – and only one node – of the primal cell falls inside a dual cell.

Strain components in a point within the cell, $\{\varepsilon\}_c$, will be expressed as a function of nodal displacements $\{u\}_c$ that are the unknowns of the problem:

$$\{\varepsilon\}_c = [B]_c \{u\}_c \tag{1}$$

Using Hooke's law, stress components $\{\sigma\}_c$ will be given by

$$\{\sigma\}_c = [D]_c \{\varepsilon\}_c = [D]_c [B]_c \{u\}_c \tag{2}$$

where $[D]_c$ represents the constitutive matrix of the cell.

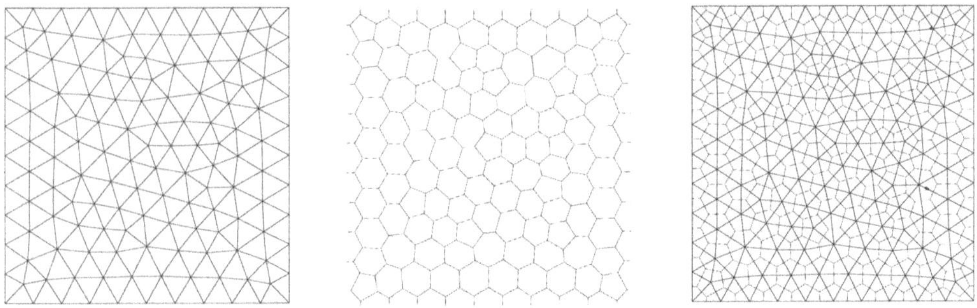

FIGURE 2. Primal, dual cells and their ensemble

In this work, displacements of a point inside the primal cell are expressed as a linear interpolation of displacements of the nodes of the cell. As a consequence, strain and stress components are uniform within the cell.
The previous steps are similar to what is done with FEM, but the methods differ deeply in how field equation is obtained.
In fact, CM considers the dual cell as an *influence region* for the inner node (see Figure 3).

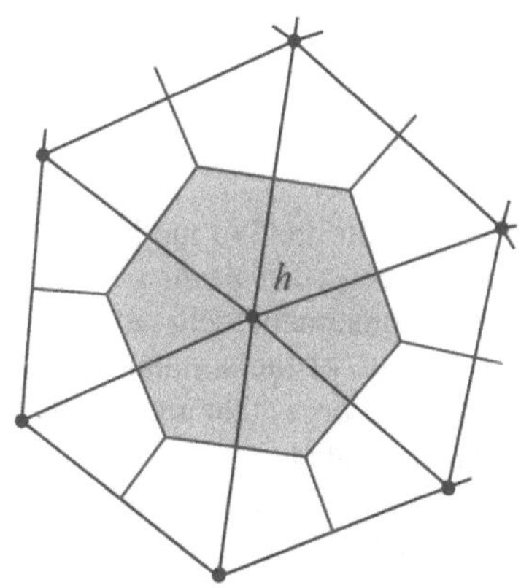

FIGURE 3. Dual cell of node *h*

From stress components $\{\sigma\}_c$ it is possible to express the forces acting on each side of the dual cell. A balance of all the forces acting on the dual cell is then possible, adding up contributions from all the sides surrounding the node and the external forces acting on the dual cell, if present (see Figure 4). Therefore, CM writes an equilibrium condition directly in a discrete form for this influence region.
A linear system in the usual form

$$[K]\{u\}=\{F\} \tag{3}$$

is then obtained, which can be solved with the usual methods.

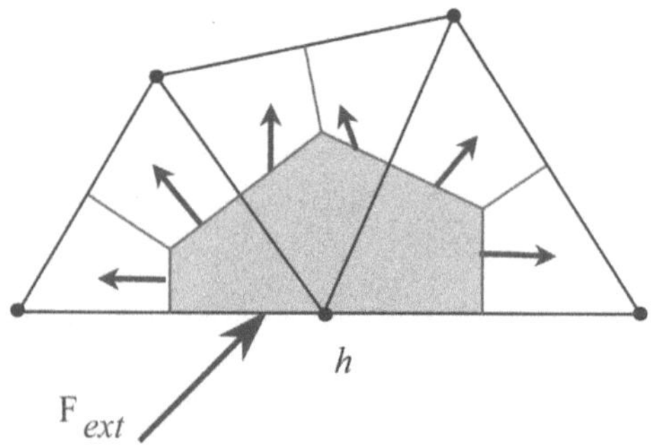

FIGURE 4. Forces on a boundary dual cell

It must be noted that the linear system has been written directly in a discrete form, as a sum of contributions. Therefore, no differentiation has been used and discontinuities can be easily dealt with.
A model where heterogeneities are the same size of the cell, such as the one shown in Figure 1, can be then solved with CM.

3 SIMULATIONS IN ELASTIC FIELD

In a first set of simulations in the elastic field, a model consisting of a matrix of 5340 primal cells has been used. In the model, cells can be of two kinds: ferrous and voids. Voids are randomly distributed among the constituent cells in order to obtain the desired residual porosity, as shown in Figure 1 for one of the simulations. Size of the cells in the model has been chosen as to compare with size of pores that are usually found in sintered materials.
Five compression tests have been simulated.
Two different sintered materials, A1 and A2, have been considered, obtained from the same commercial powder, namely NC100.24, with different porosity levels. Extensive experimental results for these materials can be found in [5].
For each alloy, porosity changed slightly from one simulation to the other, while porosity distribution was random. Compression Young modulus values have been computed and compared with those available in literature for the simulated alloys. Simulation results in the elastic field for a larger number of alloys may be found in [6].
The model at present is limited to compression. In fact, small non-linearities in strain/stress behavior below the macroscopic yield take place in experimental tension tests of sintered alloys. This progressive damage accumulation at the moment has not been taken into account by the model.
In the following:

E_0 = Young modulus of constituent material
E_S = experimental compressive Young modulus
E_{CM} = computed compressive Young modulus
Average results for the two considered alloys are shown in Table 1.

TABLE 1. Average results

Alloy	A1	A2
Residual porosity	13.4	9.8
E_0 (GPa)	207	207
E_S (GPa)	150	168
E_{CM} (GPa)	138	155
Deviation	**8 %**	**7.7 %**

Details of the simulations results are given in Table2. It can be seen that, although a general trend is present – increasing porosity will decrease apparent Young modulus – porosity alone is not sufficient to account for all variations. For instance, porosity being equal, different distributions of voids in the matrix lead to different "structures"– see results from simulations 1 and 4 for alloy A2. On the other side, different porosity distributions may lead to similar results, although porosity is changed.

TABLE 2. E_{CM} vs. *porosity*

Alloy	A1		A2	
Simulation	*porosity*	E_{CM} (GPa)	*porosity*	E_{CM} (GPa)
1	12.8	143	9.7	155
2	13.9	138	10.6	154
3	13.6	128	10	156
4	13.4	137	9.7	158
5	12.6	142	10.4	152

A graph showing the dependency of E_{CM} on porosity for the mentioned alloys is shown in Fig. 5. Again, the general trend is followed and the previously discussed scatter can be appreciated.

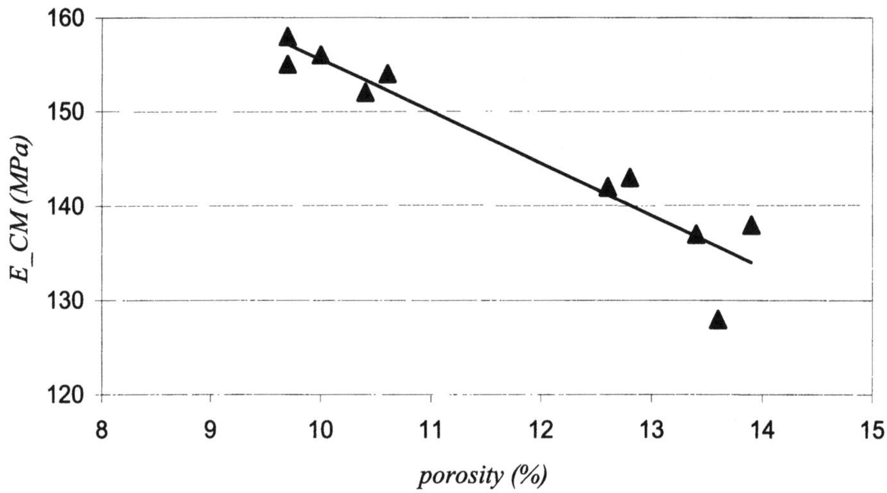

FIGURE 5. Dependency of E_{CM} on porosity of materials A1 and A2.

Although the model assumes a simple linear constitutive law for the constituent cells and linear interpolation functions for the displacement field, the obtained results are very promising.
Deviation between experimental and simulation results was less than 10%, which is within the usual range of variability for such materials.

4 SIMULATIONS IN PLASTIC FIELD

In order to introduce rheological behaviors other than the elastic one, an elastic-plastic incremental model for CM has been developed and discussed in [7]. A first attempt to apply this model to sintered materials mechanical behavior modeling in plastic field is discussed in the following.
Assuming an elastic-perfectly plastic constituent material, the model for monodimensional stress state is given by the well-known relationship:

$$\varepsilon = \frac{\sigma}{E},\ \lambda = 0 \quad \text{when} \quad 0 \le \sigma < \sigma_Y \qquad \text{and} \qquad \sigma = E(\varepsilon - \lambda) \quad \text{when} \quad \sigma = \sigma_Y \tag{4}$$

In a plane state, the expression for stress components now becomes

$$\{\sigma\}_c = [D]_c [\{\varepsilon\}_c - \{\lambda\}_c] \tag{5}$$

where $\{\lambda\}_c$ collects inelastic strain components.
Load history is divided in a convenient number of finite steps. Constituent material behavior is non-linear elastic within each step. Fundamental equation (3) in incremental terms now reads

$$\{\Delta F\} = -[K]\{\Delta u\} + [L]\{\Delta\lambda\} \tag{6}$$

where $[L]\{\Delta\lambda\}$ collects the plastic strain equivalent forces. At the beginning of each step displacements and internal stresses are known. Von Mises yield condition has been assumed and the radius of the yield locus used to update $\{\Delta\lambda\}$. The process is repeated until convergence is obtained, then a new step is considered.

A preliminary simulation has been performed to test if such a simple model could be suitable to model sintered alloys behavior. A model of 1498 cells as shown in Figure 6 was used.

FIGURE 6. Cells in plastic field (grey) in the matrix of 1498 cells during simulation. 216 void cells (black) are randomly distributed among constituent (white)

For constituent cells, E=207 GPa and Ry=120 MPa were assumed. Results are shown in Figure 7 as a stress/strain plot for both sintered alloy and assumed constituent without voids.

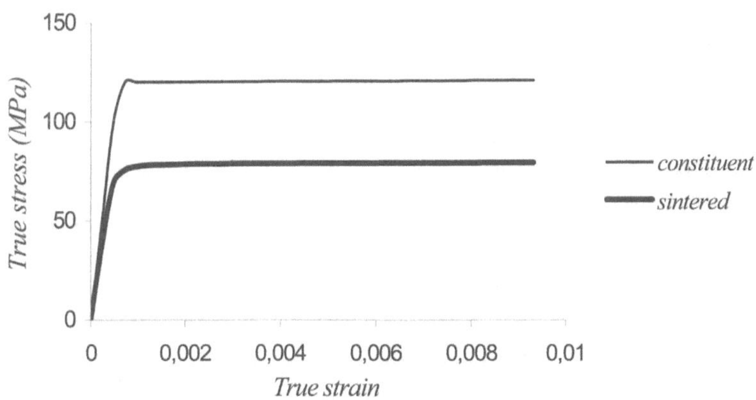

FIGURE 7. Results from simulation in elastic and plastic field

From simulation, Young modulus in the initial elastic field for the sintered alloy with 13.4 % residual porosity was 158 GPa, a result in good agreement with those of the previous simulation and experimental data (see Table 1)).

As reported in literature [5], experimental data for materials A1 and A2 show a discontinuity in the region of yielding in the stress/strain plot. It can be seen that in the simulation, slope changes at yield, although at the moment this may only be regarded as a qualitative result.
We shall recall also that Hollomon's equation is often used to represent mechanical behavior of sintered alloys in the plastic field, but is not suitable in the elastic region and is not able to take into account for the above-mentioned discontinuity.
On the other side, simulation results tend to be "flat" in the plastic region. This is due to the fact that an elastic-perfectly plastic material has been assumed as powder constituent. Better results should be achieved with a plasticity model including hardening, which is currently being implemented.

5 CONCLUSIONS

The present paper introduces a new numerical model that may be employed to assess sintered alloys mechanical properties, depending on residual porosity. The proposed methodology makes use of Cell Method, particularly suitable for problems involving heterogeneities. Simulation results are in good agreement with experimental data in the elastic field and seem promising also in the plastic field, although more simulations are needed to validate the model in plastic field and, moreover, other behaviors for the constituent should be tested, namely models including hardening.

AKNOWLEDGEMENTS

The author wishes to thank prof. Enzo Tonti, University of Trieste, for his precious advice on Cell Method. Thanks also to dr. ing. Fabio Miani, University of Udine, for sharing his experience on sintered materials.

REFERENCES

1. Danninger, H., Jangg G., Weiss, B., and Stickler, R., (1993), Microstructure and Mechanical Properties of Sintered Iron – Part I, PMI, Vol.25, N. 3, 111-117
2. Cosmi, F., (2001), Numerical solution of Plane Elasticity Problems with the Cell Method, CMES, Vol.2, No.3, 365, 372
3. Tonti, E., (2001), A Direct Discrete Formulation of Field Laws: The Cell Method, CMES, Vol. 2, N. 2, 237-258
4. Roux, S. (1990), Continuum and Discrete Description of Elasticity, Statistical Models for the Fracture of Disordered Media, H.J. Hermann and S. Roux (editors), Elsevier Science Publishers B.V. (North Holland), 248-255
5. Bertini,L., Fontanari,V., Straffellini,G.,(1998), Tensile and Bending Behavior of sintered alloys: Experimental results and Modelling, Jour. of Engineering Materials and Technology, vol.120, pp. 248-255
6. Cosmi F., Di Marino F., (2000), A New Approach to Sintered Alloys Mechanical Behavior Modeling, Proc. of 17th Danubia Adria Symposium on Experimental Methods in Solid Mechanics, Prague, Czech Republic, 75-78
7. Cosmi,F., (2001), Elasto-plasticità con il metodo delle celle, Atti del XXX Convegno Nazionale AIAS, Alghero, 1235-1244

SOME PROPERTIES OF NITRIDED TOOL STEELS

B. Smoljan

Chair of Materials Science and Engineering,
Faculty of Engineering, Rijeka, Croatia

KEYWORDS: Tools, Tool Life, Nitriding

ABSTRACT: Efficiency of tools and tool life are investigated. One of very effective way of steel tool quality improvement is nitriding and especially plasma nitriding. Plasma nitriding process is on the one hand very good adapted to the demands of practice and on the other hand is fit for automation and free from environmental disadvantages has gained good acceptance. Increasing of toll performance, diminishing of abrasive and cohesive wear and the improving corrosion resistance make the plasma nitriding a valuable process for surface treatment.

1 INTRODUCTION

The nitriding process is one of the more important surface treatments for improvement of tool properties and has been known for long time. Nitrogen is incorporated into the surface layers of engineering components during the nitriding and by that contributes to an increase of mechanical and chemical resistance.

For the nitriding of tool steels several methods are used which differ in the nitriding medium, nitriding duration, nitriding temperature, and kind of compound layer. They are powder nitriding, bath nitriding, nitriding in gas and plasma nitriding [1].

The powder nitriding method practically is no longer used today. The bath nitriding is mainly used for short time nitriding of bulk material. However, the cleaning of parts after nitriding from salt, especially in needs high expense of manpower. Finally, a disadvantage of this nitriding process is the high, process temperature that usually exceeds the tempering temperature of tool steels. There is an increasing problem with the disposal of the poisonous cyanide salts after use. The nitriding in gas can be provided at lower temperatures but the process takes considerably more time and can take several days. The plasma nitriding is the most recently developed process [2]. The advantages of nitriding are used for many tools made out of cold-work tool steel, hot-work tool steel, plastic mould steel or high speed steel. Important is, however, that the nitriding treatment is adapted to the tool material and the load of the tool, which can be done best by the plasma nitriding process.

Opposite to all other mentioned nitriding processes, in the plasma nitriding process nitrogen is incorporated into the surface layer not by a thermo-activation, but by an electric current, which is applied between the furnace and the nitriding material. The wall of the furnace is connected as an anode while the material acts as a cathode. The process is started by the migration of electrons from the cathode to the anode. On their way through the furnace, which is filled with a mixture of nitrogen and hydrogen gas with low pressure, the electrons hit the nitrogen or hydrogen molecules, and split them up into ions. The positive charged ions are now accelerated on to the cathode, where they hit the nitriding material with high speed. By this, the high kinetic energy of ions will

Published in: E.Kuljanic (Ed.) *Advanced Manufacturing Systems and Technology*,
CISM Courses and Lectures No. 437, Springer Wien New York, 2002.

be absorbed and transformed into heat, by which the nitriding material is heated up. By the hit of the ions on the surface of the tools to be nitrided, they knock out small particles such as oxides and other pollutions. This sputtering process cleans the surface and causes a depassivation, which is necessary for a uniform nitride layer. When nitrogen ions hit the tool surface, nitrides will be formed and nitrogen may be incorporated into the tool material by diffusion.

2 MICROSTRUCTURE OF NITRIDED LAYER

When nitrogen is penetrating into the surface of steel, in the beginning, it is fixed within the atomic lattice, i. e., it is kept in the spaces between the iron atoms of the lattice. At higher nitrogen contents, if the solubility limit is reached, nitrides are formed according to the phase diagram iron-nitrogen. The nitride compounds existing in tool steels are the fcc Fe_4N with the designation γ' and the hexagonal $Fe_{2-3}N$ with the designation ε. The diffusion of nitrogen into the surface of tools and the formation of nitride compounds can result in a strong increase of hardness. The depth of nitriding can be determined by hardness measurements. To ensure a uniform determination of the nitriding depth, the method of measurement was standardized .

The nitrogen diffusion layer and the nitride compounds that have been formed on the surface can be made visible by etching a metallographic cross-section. The diffusion layer can be recognized with thickness of several millimeters. The nitrides can be recognized as a separate layer with a thickness of about 5 to 10 microns. Sometimes, within this nitride layer or at the surface of it, porosity can be found. Content of the nitrogen and the carbon as a function of the distance from the surface for a bath nitrided and a plasma nitrided specimen are shown in Figure 1.

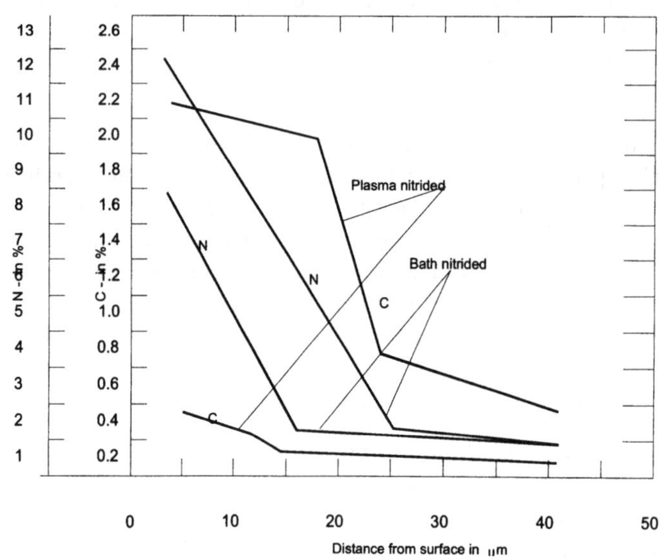

FIGURE 1. Carbon and nitrogen content in plasma and bath nitrided specimens

The bath nitrided specimen contains a nitrogen content of about 11 % at the surface while the plasma nitrided specimen contains about 6 % nitrogen at the surface. According to the phase diagram this reveals that the compound layer is ε-nitride for bath nitriding and γ'-nitride for plasma nitriding. Typical for the bath nitriding is the high carbon content of the compound layer, which has a level of up to 2 %. Contrary to that, γ' compound layers have a very low carbon content. The content of alloying elements such as Cr, Mo and Mn is lower in the compound layer than in the matrix. Only silicon seems to be enriched in the compound layer.

The plasma nitriding process enables to vary the kind of compound layer and the depth of nitriding according to the properties demanded of the tool. Changing the nitriding parameters, as well as, gas composition, nitriding time and nitriding temperature can make these variations [3]. By increasing the nitriding time, the depth of the nitrided layer will grow according to Figure 2.

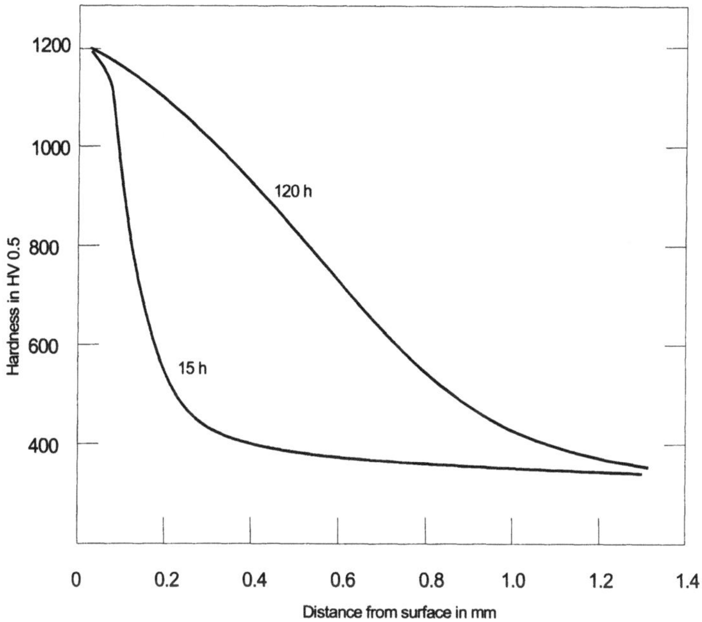

FIGURE 2. Surface hardness of the hot-work toll steel X 37 CrMoV 5 1 KV

A significant influence on the composition of the nitrided layer and on the surface hardness has the chemical composition of the steel. Increase of C, Cr, Mo, V and Al amounts cause an increase of surface hardness, while the depth of the nitriding layer is decreased in the same order.

Different tool steels have a very wide variation of alloying contents, and they have a different surface hardness pattern. Typical examples of the thickness and properties dependence of the composition of plasma nitrided tool steels are shown in Table 1.

TABLE 1. Nitrided layer

Steel	Compound layer, μm	Nitrided network layer, mm	Diffusion layer, mm
35 CrMo 8 KV	7	0.13	0.20
X 37 CrMoV5 1 KV	6	0.08	0.13
X 155 CrVMo 12 1 KV	-	-	0.08

Incorporation of nitrogen into the surface layer causes an increase of specimen volume. The dimensional variation of nitrided tool steels is less than 5 to 10 microns. At the edges, this increase of volume may affect oversizes. For sensitive cutting edges these oversizes increase the danger of spal damage of the tools. When using the plasma nitriding process, this influence is kept on a low level by the sputtering activity of the ions, which is especially strong at the edges. The sputtering activity of the ions furthermore avoids local defects on the tool surface, caused by insufficient cleaning. This is important for steels with high chromium contents, which tend to form passive layers.

3 AN ANALYZE OF THE NITRIDED TOOLS PROPERTIES

Improvements of the mechanical, tribological and chemical properties can be achieved by nitriding of the surface of tools. However, the toughness may be reduced and the probability of tool failure may be increased, especially at areas with high tensile stress or cutting edges [4].

The following circumstances for a good functionality of nitrided tools have to be taken in consideration: correct selection of the tool material; correct selection of the structure, strength, toughness and tempering properties of the tool steel; equal and homogeneous layer without defects; appropriate composition of the layer to the requirements; appropriate hardness surface of the layer to the requirements; appropriate nitriding process parameters to the tool material and the required layer [5].

The plastic stress relaxation is not possible into nitrided layer. Therefore, cracks may easily occur on the layer. Material with low toughness may crack, starting from surface defects such as notches caused by the some manufacture process. Even at elevated temperatures the toughness of nitrided layers will not be improved [1].

The depth of the layer and the composition of the tool steel have a remarkable influence on the toughness of nitrided tools (Figure 3).

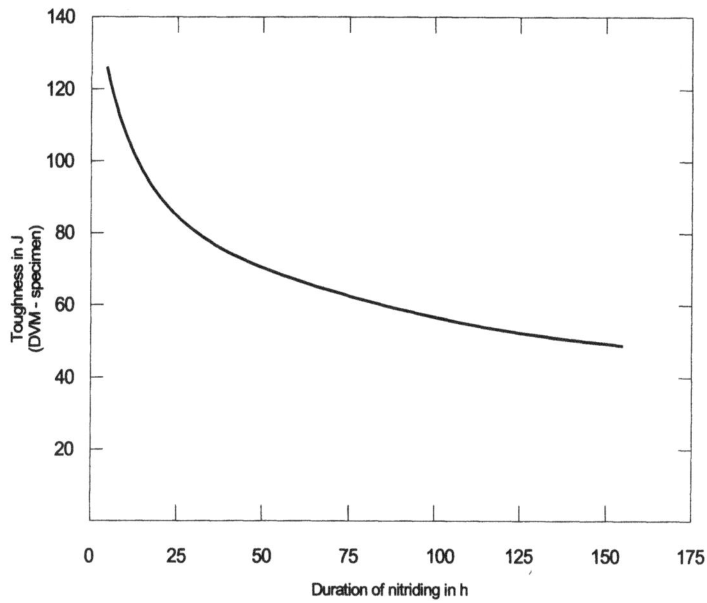

FIGURE 3. Toughness of nitrided tool steel 31 CrMo 12

The toughness furthermore is depending on the atmosphere and the temperature used during plasma nitriding. The microstructure of the tool can also, effects the surface toughness. Martensitic structure reduces the toughness by approx. 60 %, while with bainitic structure the situation is worse, Figure 4.

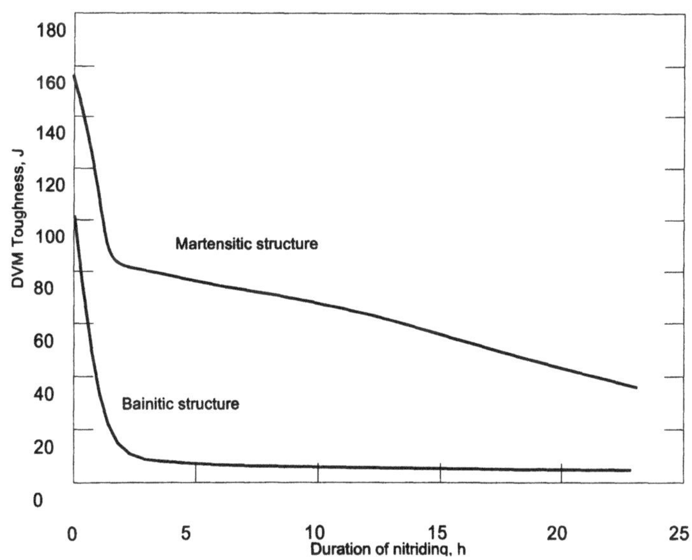

FIGURE 4. Toughness of nitrided tool steel with different microstructure

The corrosion resistance of tool steels generaly can be improved by nitriding. Besides the alloying content of the tool steel, the kind and thickness of the nitrided layer as well as the corrosive medium affect the corrosion properties. The corrosion resistance of the tool steel for instance can be improved by nitriding, if it is affected by hydrochloric acid. At the attack of acetic acid, however, its corrosion resistance was decreased. In the salt water spray test, nitration caused a small, and in the condensed water spray test it caused a high reduction of corrosion. Moreover, it could be proved that γ -compound layers are more corrosion resistant than ε-compound layers.

The major influence regarding the engineering and economical importance of nitriding is the improvement of the antifriction properties, the reduction of adherence and the increase of wear resistance. By the hard nitride layers the tool performance can be multiplied and in many cases an economic use of tools is possible only after nitriding. Generally, the γ-nitride compound layers are more wear resistant [4]. For instance, at the processing of plastics, nitrided tools with γ-compound layers revealed better wear resistance than bath nitrided tools with ε-compound layers, Figure 5.

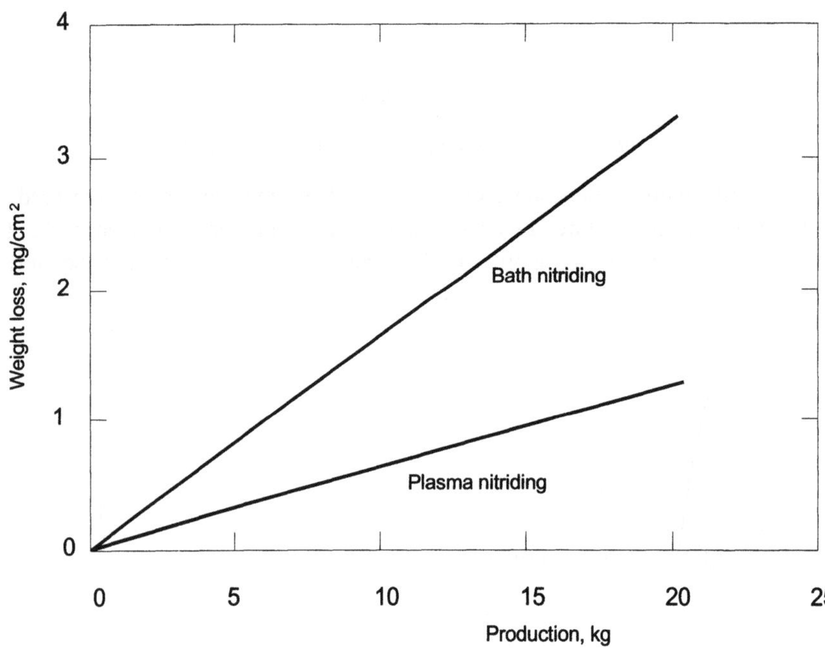

FIGURE 5. Tool wearing during processing of plastics

In many cases the γ-compound layers enable higher performance possibly due to their better toughness.

The increase of volume by the incorporation of nitrogen atoms causes compression stresses in the surface of the tools. Applied tensile stresses can be reduced by the residual compression stresses and the fatigue bending strength can increase.

4 CONCLUSION

The nitriding process and especially plasma nitriding is one of the more important surface treatments for improvement of tool properties.

Opposite to all other mentioned nitriding processes, in the plasma nitriding process nitrogen is incorporated into the surface layer not by a thermo-activation but by an electric current, which is applied between the furnace and the nitriding material.

Plasma nitriding process is on the one hand very good adapted to the demands of practice and on the other hand is fit for automation and free from environmental disadvantages has gained good acceptance. In future plasma nitriding will maintain its position within the variety of other new developed surface coating processes.

By the plasma nitriding process is easy to vary process parameters. There is good reproducibility, suppression of pores in the compound layer, possibility to influence to control the kind of layer, possibility to nitride steels with low tempering resistance by lowering the nitriding temperature, low dimensional change possibility of local nitriding no pollution of the environment, no use of poisonous chemicals.

Increasing of performance, diminishing of abrasive and cohesive wear and the improving corrosion resistance make the plasma nitriding a valuable process for surface treatment. Toughness, wear resistance, fatigue strength, corrosion resistance of tools can be effective improved by plasma nitriding, but for cutting and punching tools the surface hardness must not be too high as this can be the cause for damage of the cutting edges.

REFERENCES

1. Krauss, G., (1995), Steels: Heat Treatment and Processing Principles, ASM International, Materials Park.
2. Slycke, J., (1981), A Study of the Reactions Occurring During the Carbonitriding Process, Journal of Heat Treatment, Vol. 2, (No. 1), 3-19.
3. Pickering, F., (1987), The Properties of Tool Steel for Mould and Die Applications. Tool Materials for Mould and Dies, Ed. G. Krauss, Colorado School of Mines Press, Golden, 3-32.
4. Haberling. E and Rasche, K., (1992), Plasma Nitriding of Tool Steels. New Materials Processes Experiences for Tooling, Ed. G. Berns, Mat Search, Andelfingen, 369-393.
5. Berns, H., (1987), Strength and Toughness of Hot Working Tool Steel. Tool Materials for Mould and Dies, Ed. G. Krauss, Colorado School of Mines Press, Golden, 45-65.
6. Jurković, M., Smoljan, B., Jurković, Z., (2001), Mathematical Modelling and Benefits Analysis of Application Surface Engineered Tools, Proceedings of 8th of IFHTSE, Dubrovnik, 373-381.
7. Roberts; Krauss, G., Kemmedy, R., (1998), Tool Steel, ASM International, Materials Park.

FILLET STRESS RELATED TO NON-UNIFORMLY LOADED GEAR TOOTH GEOMETRY

G. Marunić

Department of Mechanical Engineering Design, University of Rijeka, Croatia

KEYWORDS: Spur gear, Fillet stress, Non-uniform load.

ABSTRACT. The influence of non-uniform load distribution along spur gear tooth facewidth, upon the fillet stresses, has been investigated by means of the Finite Element Method p-version. There have been taken into consideration different load distribution cases corresponding to the contact without errors, the possible alignment error in fitting the gear, and the effect of gear crowning. The results of the investigation performed by 3D numerical approach, have enabled the determination of spur tooth stress maximum value and its position both, along the tooth fillet and facewidth, for complex combined influence of gear geometrical parameters and non-uniform load distribution. The obtained maximum equivalent von Mises stress and axial stress that arises in the direction of gear axis, were compared for different load distribution acting along the tooth facewidth.

1 INTRODUCTION

In this paper spur tooth root stress state has been the subject of investigation considering the factors that certainly effects its complexity. First factor is non-uniform load distribution along the tooth facewidth, as ideal uniform load distribution is practically never achieved. The uniform load distribution along the facewidth has been taken as a reference. The second factor is a gear rim thickness, the subject of numerous investigations mainly based on 2D approach. The third factor of interest, is the actual spur tooth facewidth value, that has been not taken into account by 2D approach.

For the evaluation of load-carrying capacity of gear tooth root, the attention has been devoted to maximum tensile stress that occurs on the tooth side where the load acts.

By use of numerical methods, considerably more adequate tooth stress-strain state determination is enabled, related to the conventional ones. The investigations of spur gear stress field by means of 3D numerical methods are not so numerous in relation to 2D ones, being in the same time the Finite Element Method and its versions dominantly used.

An early experimental work [1] is worth to be mentioned as an effort made to point to the differences in 2 and 3D approach for the stress analysis of gear. The experiments were carried out on spur rack teeth with load applied along the tooth tip line, subjected to different load distribution and, with uniform load distribution, varying the tooth length. Besides, tests were done on helical rack teeth in several mesh conditions.

The stress field in the tooth fillet along the facewidth of spur gears were investigated by means of 3D hierarchical finite element models [2]. The parametric analysis was performed by utilising an integrated system of programs for automatic generation of 3D numerical models. A wide range of facewidth/modulus values was considered in order to show the actual complexity of stress field, also pointing out the gradual passage from plane stress to plane strain conditions, as the facewidth increases.

Published in: E.Kuljanic (Ed.) *Advanced Manufacturing Systems and Technology*,
CISM Courses and Lectures No. 437, Springer Wien New York, 2002.

As conventional standards generally (there is exception) do not take into account the peculiarities of thin-rimmed gear stress-strain state, there have been numerous studies on this subject. Related to the aim of this paper, 3D stress analysis [3] by means of 3D p-FEM approach was given valuable results about parallel-axis, external, spur gears (both standard and HRC-High Contact Ratio) and standard helical gears. In the analysis, the influence of both rim thickness and a gear geometry is investigated, the latter being taken into account by considering several values of the gear teeth number for each kind of gear. The computed tensile and compressive stress distributions were utilised to define the alternating stress in the tooth root area and the corresponding fatigue stress.

To take into account the influence of thin rim on the stress-strain state of tooth, in [4] two different approaches were given: the use of numerical methods of theory of elasticity when carrying out the check calculations, and combined theoretical and experimental methods in the case of design and approximate check calculations.

In [5] the influence of contact area on the stress field in gear tooth root using the 3D FEM for numerical simulation, was reported. Numerical results were compared with results from analytical methods provided by DIN 3990. Practical experience shows that the acting force on a gear tooth is almost never distributed uniformly along the tooth width, consequently the effects on the stress field were analysed and compared for different load distributions cases.

2 BASIC CONSIDERATIONS FOR 3D FEM CALCULATION

The single gear models were established consisted of the loaded tooth and its adjacent teeth, in the form that corresponds to the angular extension of 120°. The models of thin-rimmed gears were constrained only along both radial edges, for solid gear being additionally constrained inside of sufficiently thick rim. The tooth tip was loaded by the same total normal load, differing in the distribution along the tooth facewidth. There were simulated by uniformly distributed load the contact without errors (case 1), by asymmetrically distributed load in relation to the tooth facewidth, the possible alignment error in fitting the gear (case 2), and by symmetrically non–uniformly distributed load the effect of gear crowning (case 3). The models were meshed with parabolic tetrahedron elements and solved using 3D p-version FEM of the software package I-DEAS [6].

As mentioned before, the emphasis has been added to the selection of: - the tooth facewidth values b/m_n (expressed by the modulus m_n), in order to simulate narrow (b/m_n=2) and wide (b/m_n=4) facewidth tooth i.e. to investigate the influence of the actual tooth three-dimensionality on the stress state in the root area, - the rim thickness s_R/m_n that is considered as thin (s_R/m_n=1,5; 2) and thick (s_R/m_n=8) i.e. that corresponds, in relation to the tooth root stress behaviour, to a solid gear. Additionally, the stress analysis was performed for gears with fewer (z=20) and larger (z=100) number of teeth z, while the following gear parameters were the same for all models: modulus m_n=10 mm, pressure angle α_n=20°, dedendum h_f=12,5 mm, and addendum modification factor x=0.

3 ANALYSIS OF MAXIMUM TENSILE FILLET STRESSES

On the basis of the established stress state on tensile side of the loaded tooth area, the subject of analysis has been maximum values of equivalent von Mises stress σ_{eqmax}, as meritorious stress for tooth root load-capacity evaluation, and axial stress σ_{zmax} that obviously arises when the tooth facewidth increases. The distribution of these stresses along the tooth facewidth varies along with the change of the chosen gear geometrical parameters, depending upon the load distribution under consideration. Fig. 1 shows the distribution of these stresses in the tooth fillet area; the stress position on the fillet is expressed by the value of fillet tangent angle φ and the tooth centre line, while several positions along the tooth facewidth are fixed and

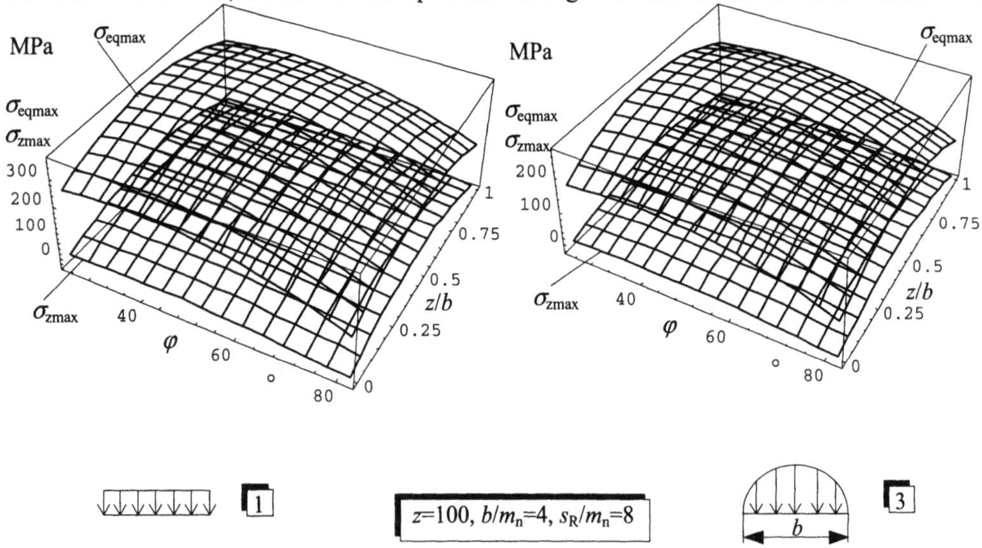

FIGURE 1. Stress distribution area on tensile side of a wide loaded tooth fillet for load distribution cases 1 and 3 (z/b – adimensional coordinate, φ - angle between fillet tangent and tooth centre line)

expressed by adimensional coordinate z/b. The analysis of results is dealing with maximum stress values achieved at certain position along the tooth fillet, as well as along the facewidth; the variation of these stress positions has not been the subject of the analysis.

In Fig. 2 and 3, maximum equivalent stress σ_{eqmax} and axial stress σ_{zmax} are presented in relation to the rim thickness s_R/m_n, for narrow and wide facewidth, in three cases of load distribution along it. The gears of different geometry were taken into consideration by fewer (Fig. 2) and larger (Fig. 3) number of teeth. In relation to the arise of axial stress value that has been neglected by analytical and numerical 2D investigations, the obtained results point to the influence of facewidth value upon the increase of axial stress magnitude, especially for the gear with fewer number of teeth. The rim thickness decrease causes axial stress slight decrease, with exception of the thinnest rim under consideration, regardless of load distribution case.

For ideal uniform load distribution case 1, the decrease of equivalent stress as the rim thickness decreases, is more expressed for the gear with larger number of teeth, regardless of the facewidth value. At the lower boundary of the chosen rim thickness range, in some cases (z=20) the trend of its increase can be expected. For the gear with larger number of teeth, axial stress decrease is more influenced by the rim thickness decrease, but this influence diminishes as the facewidth increases.

Related to non-uniform asymmetrical load distribution of case 2, equivalent stress is generally smaller for the thinnest rim under consideration, compared with the thick rim. This is more obvious for wide gear with larger number of teeth. Axial stress takes its peak value for wide gear with fewer number of teeth.

Equivalent fillet stress of gear tooth subjected to symmetrical non-uniform load distribution of case 3, is the mostly influenced by variation of rim thickness for narrow gear with larger number of teeth. As regards axial stress, it takes maximum values for wide gear with fewer number of teeth.

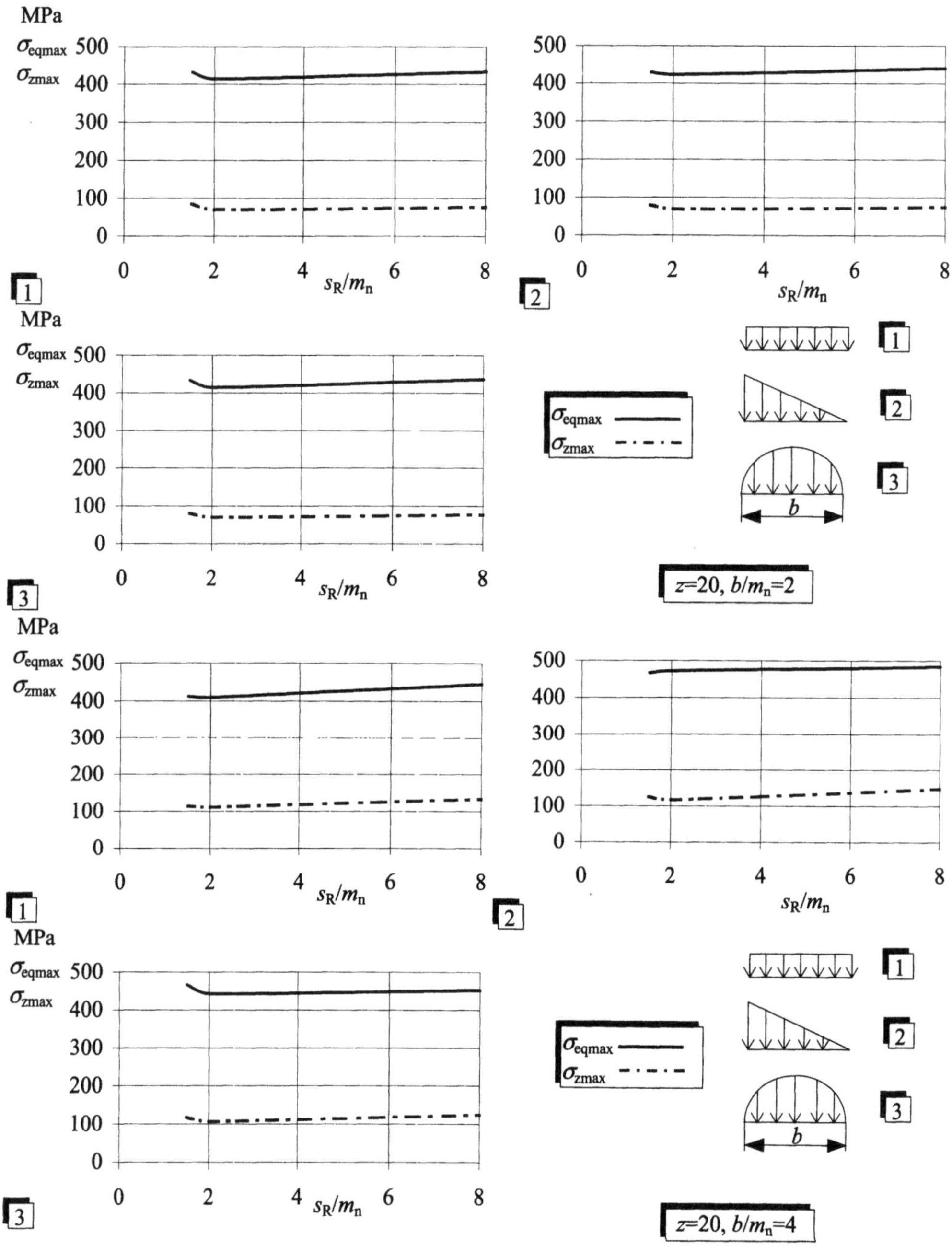

FIGURE 2. Relation between maximum tensile fillet stresses (equivalent and axial) and rim thickness, for narrow and wide gear with fewer number of teeth, subjected to uniform and non-uniform loading

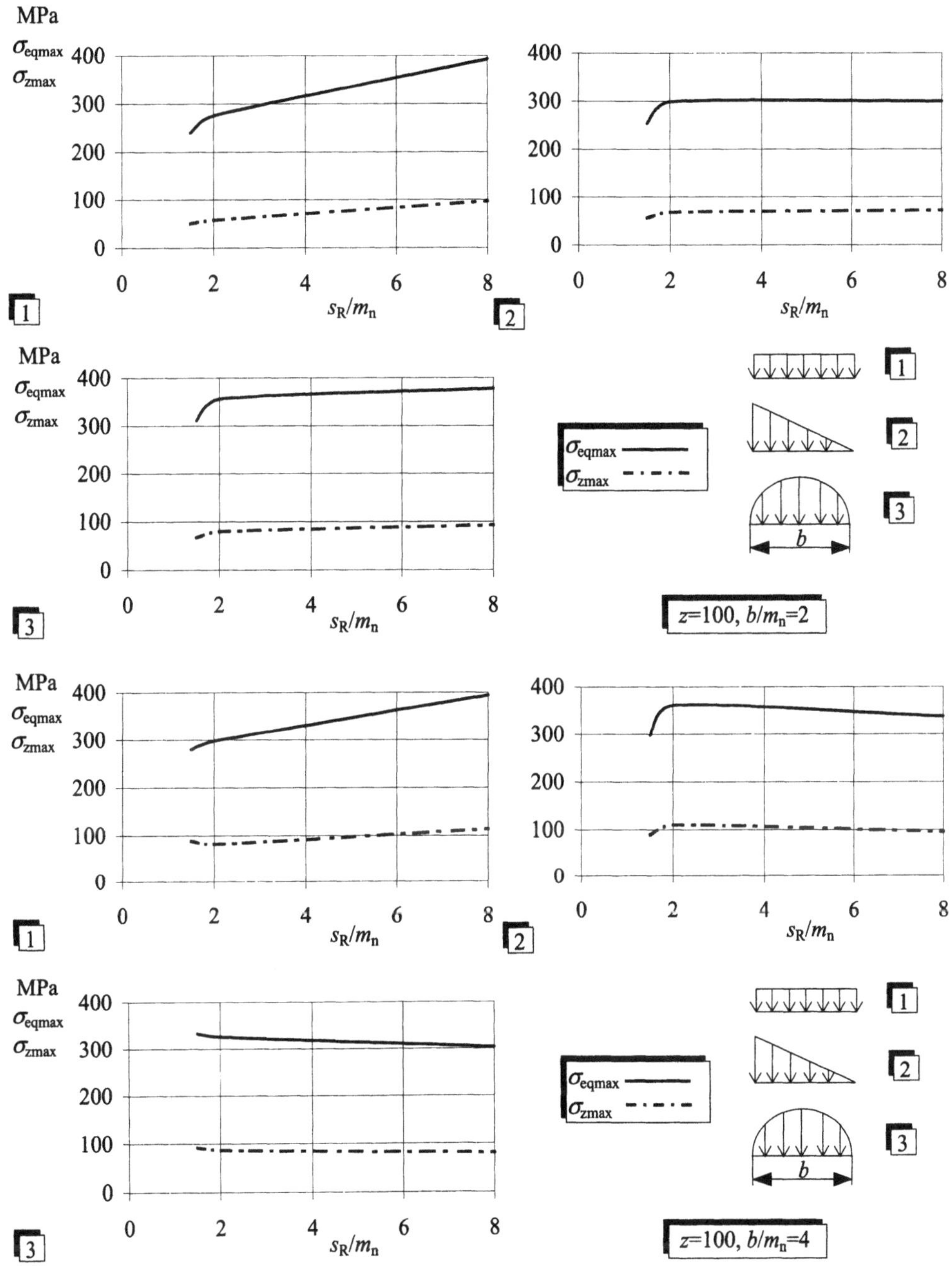

FIGURE 3. Relation between maximum tensile fillet stresses (equivalent and axial) and rim thickness, for narrow and wide gear with larger number of teeth, subjected to uniform and non-uniform loading

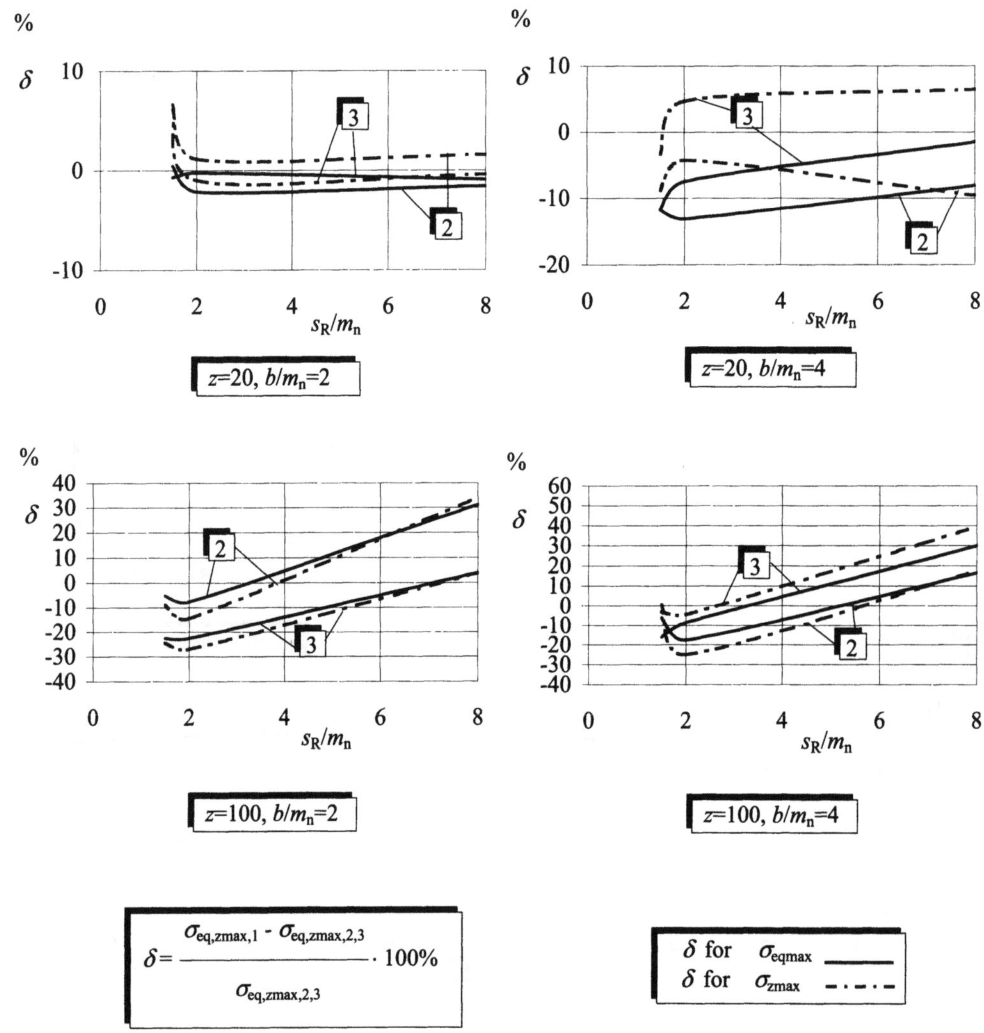

FIGURE 4. Deviation of maximum tooth fillet stresses (equivalent and axial) for non-uniform, from the corresponding ones for uniform load distribution, related to rim thickness, facewidth value and number of teeth

4 DEVIATION OF STRESSES FOR NON-UNIFORM TOOTH LOADING FROM IDEAL UNIFORM ONE

In Fig. 4 the deviation δ of maximum filletstresses for non-uniform load distribution (cases 2 and 3), from the corresponding ones for uniform load distribution (case 1), is shown. The severe increase of stress deviations occurs for wide gear with larger number of teeth, strongly depending upon the rim thickness. For the lowest value of rim thickness, the difference between the stress values for uniform and non-uniform load distribution diminishes, especially for gear with larger number of teeth, regardless of stresses and facewidth values under consideration.

5 CONCLUSIONS

The results of investigation obtained by 3D numerical approach, has enabled the determination of spur tooth stress maximum value and its position both, along the tooth fillet and facewidth, for complex combined influence of gear geometrical parameters and non-uniform load distribution.

The possibility is offered for the choice of gear geometrical parameters in relation to the values of maximum tooth fillet stresses, by taking into consideration two non-uniform load distributions along the tooth facewidth, that can appeare in practice instead of ideal uniform one.

REFERENCES

1. Parenti Castelli, V., Meneghetti, U., (1981), Three-Dimensional Stress Analysis in Spur and Helical Rack Gears, International Symposium on Gearing&Power Transmissions, Tokyo, 153-158.
2. Baret, C., Pidello, A., Raffa, F.A., Strona , P.P., (1989), Stress Path along the Facewidth in Spur Gears Fillet by 3D p-FEM Models, International Power Transmissions and Gearing Conference, Chicago, 173-179.
3. Curti, G., Raffa, F. A., Garavelli, D., Baret, C., (1991), Three - Dimensional Stress Analysis of Thin - Rimmed Gears by the p – FEM Approach, JSME International Conference on Motion and Power - transmissions, Hiroshima, 787-794.
4. Anoshin, G. V., Derzhavets, Yu. A., Pinezhaninov, F. P., Sholomov, N. M., (1990), Stress-Strain State of the Teeth of Gears with a Thin Rim, *Vestnik Mashinostroeniya*, Vol. 70, No.11, 20-23.
5. Flašker, J., Glodež, S., Pehan, S., (1993), Upliv funkcionalne nosilne ploskve na napetost v zobnem korenu zobnika, *Strojniški vestnik*, Ljubljana, Vol.39, No. 11-12, 299-308.
6. I-DEAS ™, (1996), Users` Guide, Milford.

USE OF ULTRASOUND TO DETECT INTERNAL DEFECTS OF SPHEROIDAL GRAPHITE CAST IRON PRODUCTS

E. Ceretti[1], A. Attanasio[1], C. Giardini[1]

[1] Department of Mechanical Engineering, University of Brescia, Italy

KEYWORDS: **Ultrasound technique, non-destructive analysis, graphite cast iron pieces.**

ABSTRACT. The use of non-destructive techniques, such as penetrant liquids, magnetic particles, ultrasonic methods or thermographies, to control the integrity of pieces it is becoming widely used by the companies. In fact, non-destructive methods can be applied to control the quality of the produced parts without destroying the part itself. In this study the attention will be focused on the analysis of surface and internal integrity of spheroidal graphite cast iron products. At present, these products are controlled by using X rays. The technology is very accurate but there are several drawbacks related to the fact that this exam it is very expensive and time consuming since it is, in general, performed by suppliers being the X-ray equipment too expensive for single company. With this research, the Authors aim to develop a procedure able to detect, with a good reliability, the internal defects of spheroidal graphite cast iron parts by using ultrasonic methods. These techniques do not need to destroy the piece, they are not expensive, very fast and can be easly used by the company.

1 INTRODUCTION

Spheroidal graphite cast iron parts are getting more and more diffuse in car industry, thanks to the high performance of this material. Frequently these pieces are critical and they need to guarantee people safety. Consequently, it is very important to apply a detailed control technique on each piece without destroying it.

Non-destructive methods can be classified under the detectable defects. The Ultrasonic testing and the X-ray technique are able to identify internal defects, the penetrant liquid and the magnetic particles are utilized for superficial or sub-superficial defects, the thermography test is applied to anomalous thermal differences [1].

The present research, developed with Streparava S.p.A., has defined a reliable procedure to detect the internal defects of the spheroidal graphite cast iron by using the ultrasonic technique.

2 THE ULTRASONIC TECHNIQUE

Non-destructive testing with ultrasonic waves is widely used for internal defect detection. This technique uses acoustic waves reflection which is generated by the interaction between the

Published in: E.Kuljanic (Ed.) *Advanced Manufacturing Systems and Technology*,
CISM Courses and Lectures No. 437, Springer Wien New York, 2002.

waves and the defects. Ultrasonic waves are sound waves with high frequency and this allows to obtain a defined geometry of the ultrasonic beam which detects the presence of small discontinuity in the metal matrix thanks to reflection [2].

To generate an ultrasonic wave into a workpiece, it is necessary to impose a periodic oscillating motion to the material of piece. The so obtained wave front, of particles that shift in phase, spreads inside the material.

The ultrasonic wave used to detect defects, can be classified with two parameters: the direction of the wave front propagation and the direction of particle vibration. As a consequence the wave can be longitudinal, shear, superficial or Lamb (Figure 1).

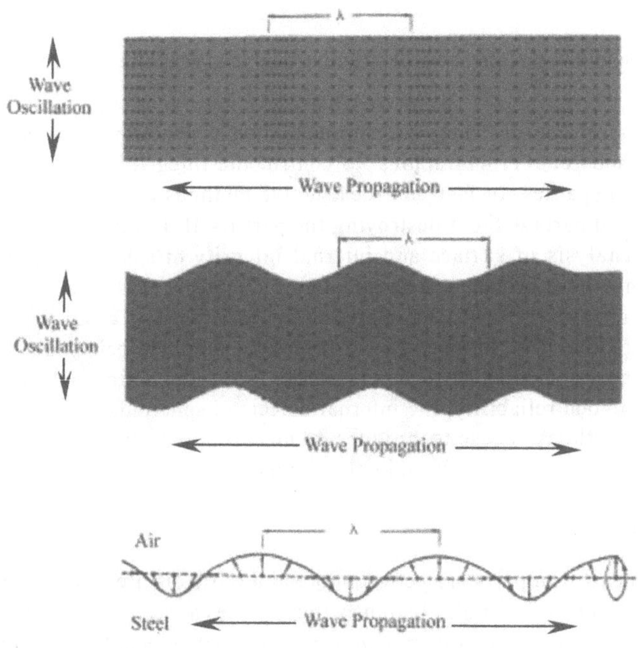

FIGURE 1. Longitudinal, shear and lamb waves.

The physical quantities that define the acoustic wave, are:

- Frequency (*f*), is the opposite of the time that a wave takes to go from a maximum (or minimum) in amplitude to the next one. The frequency range in ultrasonic testing varies from 0,5 up to 25 MHz;
- Wave length (λ), is the distance between two consecutive maximum (or minimum);
- Wave velocity (*v*) in the material, is the ratio between distance and time;
- Acoustic pressure (*P*), is the applied pressure in the unit surface;
- Wave intensity, is the energy that goes through the unit area in 1 second.

3 THE ULTRASOUND

The ultrasound wave generation is realized by probes [3]; these are special transducers (piezo-electric, electrostrictive or magnetostrictive) which convert electrical pulses into mechanical pulses.

The wave geometry is influenced by two parameters: the effective dimension of probe and the wave length.

As shown in Figure 2, two zone can be identified; the first, named near zone or Fresnell zone, is characterized by a plane geometry, the latter, named distant zone or Fraunöfer zone, has a spherical geometry.

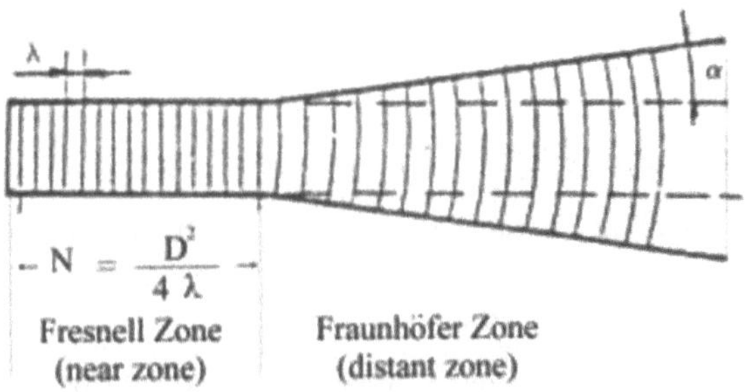

FIGURE 2. Geometry of the ultrasonic beam.

The equation (1) defines the extension of the near zone:

$$N = \frac{D^2}{4 \cdot \lambda} = \frac{D^2 \cdot f}{4 \cdot v} \tag{1}$$

where **D** is the effective diameter of probe;

λ is the wave length;

f is the frequency;

v is the wave velocity in the material.

From equation (1) it can be seen that the beam divergence decreases by increasing the probe diameter and wave frequency.

Another zone can be defined into the beam: the blind spot area, where the probe does not detect internal defects. This is because as long as the vibration generated from pulse is present, the probe is "deaf" to defects. The blind spot is the volume of the workpiece situated under the control surface where the presence of defects is not perceived by the probe. It is important of not misunderstanding the blind spot with the near zone, in fact the blind spot can extend further the near zone.

It is necessary to examine the dynamic of the ultrasound; if the material is the same, the wave velocity depends on the wave typology. The waves that must be applied are the longitudinal and shear waves. Equations (2) and (3) can be used to determine the value of the wave velocity for the two wave typologies.

$$v_l = \sqrt{\frac{E}{\rho} \cdot \frac{1-\nu}{(1+\nu)\cdot(1-2\nu)}} \tag{2}$$

$$v_t = \sqrt{\frac{G}{\rho}} = \sqrt{\frac{E}{\rho} \cdot \frac{1}{2\cdot(1+\nu)}} \tag{3}$$

where **E** is the Young' s modulus;

G is the shear modulus;

ν is the Poisson coefficient;

ρ is the density of the material.

These equations are very important, because the probes transfer into the workpiece longitudinal wave (straight probe) or shear waves (angular probe) and the value of the wave velocity is a basic parameter to compute the defect position.

Others phenomena related with the ultrasounds are reflection, refraction, diffusion and dffraction [3].

It is very important to consider the loss of the acoustic pressure due to the propagation through the material. This decreases with an exponential trend as shown by equation (4):

$$P = P_0 e^{-\alpha x} \tag{4}$$

where **P_0** is the initial acoustic pressure;

P is the actual acoustic pressure;

x is the distance covered by ultrasound;

e is the radix of the natural logarithm;

a is the loss coefficient.

The attenuation of the wave in the material is caused by two phenomena: the absorption of energy from the material itself and the diffusion of the beam in different directions.

It is important to underline that by increasing the frequency, the signal is attenuated; as a consequence, ultrasonic waves with high frequency result in better geometry signals, but with lower explorable depths.

The material studied in the present research is cast iron. In this material the attenuation changes with the graphite structure. If the graphite is lamellar the attenuation is very high; with spheroidal graphite the attenuation is equal to steel attenuation. So it is possible to apply non-destructive techniques with ultrasound on spheroidal cast iron [4].

4 INTERNAL DEFECTS CLASSIFICATION

Different typologies of defects can occur in cast iron pieces due to: wrong cooling, wrong dimension of the gate, wrong filters, ...The classification of cast iron defects is repo rted in UNI EN specifications [5]; only the defects detectable with the US testing will be examined in this research [6].

The first type of internal defects is related with the solid contraction during the cooling process, that is the shrinkage cavity (Figure 3), and the spongy contraction (a cloud of micro cavities) (Figure 4).

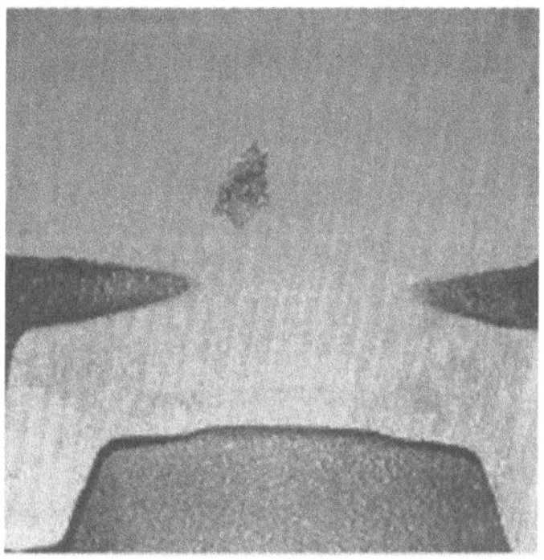

FIGURE 3. Shrinkage cavity.

FIGURE 4. Spongy contraction.

The second defect family is due to the presence of gas in the liquid metal.

When the gas is not removed during the solidification, a blowhole defect can be generated into the workpiece. This defect can reach the surface or can be trapped into the material, so generating cavity with a smooth surface (Figure 5). Figure 6 shows the pores or pinholes, which are similar to the spongy contraction but are caused by the gas emission. The diameter of these defects is less than 1.5 mm.

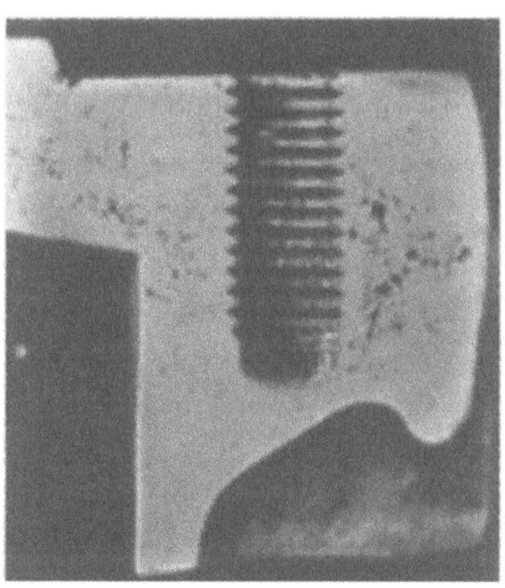

FIGURE 5. Blowhole.

FIGURE 6. Pinhole.

The last family of defects is related with the presence of slags in the casting, which generates inclusions into the workpiece [7,8]. When the slags are caused by the different components of the alloy are defined as endogenous inclusions; when they are due to external factors, are called exogenous inclusions. Often these imperfections are easily visible because they are superficial (Figure 7), except when the endogenous inclusions are caused by oxides; the US testing and the micrographies can detect them as shown in Figure 8.

FIGURE 7. Exogenous inclusion.

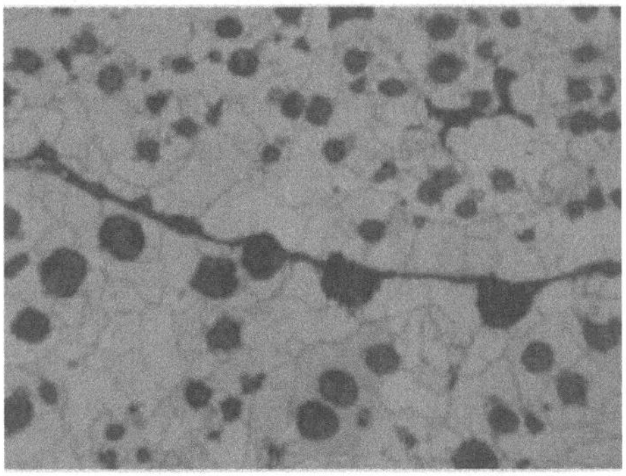

FIGURE 8. Endogenous inclusion.

5 CONTROL PROCEDURE

A control procedure for spheroidal cast iron parts based on [9,10,11], has been defined showing a good reliability of the obtained results.

5.1 CALIBRATION

First of all it is necessary to execute a precise calibration for probes (Figure 9) and materials parameters. To do this it is necessary to have a sound sample of the material under control. On this sample, at least two sets of flat holes, whose diameter ranges from 1 to 5 mm, and with different depths, have to be performed (Figure 10).

The reference sample, for each cast iron material, allows the determination of : signal propagation speed, acoustic delay, zero of instrument, gain, AVG, ...

FIGURE 9. Probe DDM 5/5.

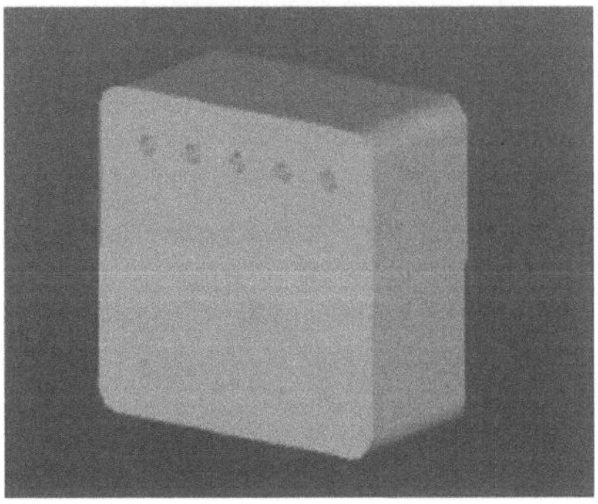

FIGURE 10. Reference sample.

5.2 CONTROL PROCEDURE

The identified control procedure can be described as follows:

1. visual inspection of the piece surface to detect superficial defects;
2. first fast scan of the part using a probe of 10 mm diameter and a frequency of 4Mhz (DDM 10/4) to identify the defect areas;
3. polishing of the critical surfaces;
4. detailed scan of the part using a probe of 5 mm diameter and a frequency of 5Mhz (DDM 5/5) to identify the defect extension;
5. if it is possible repeat the procedure on perpendicular surfaces.

The UT signal changes with the transversal section of the part. As a consequence, it is necessary to distinguish between a section parallel or not to the opposite surface, such as when the opposite surface is a fillet radius.

When the opposite surfaces are parallel, the UT signal furnishes the bottom echo. This element is very important, because if the bottom echo decreases and new echo is generated that means that a defect is present (Figure 11). When the opposite surfaces are not parallel, the bottom echo is absent. The presence of defects is related to several picks, generated by the signal reflection caused by imperfections (Figure 12).

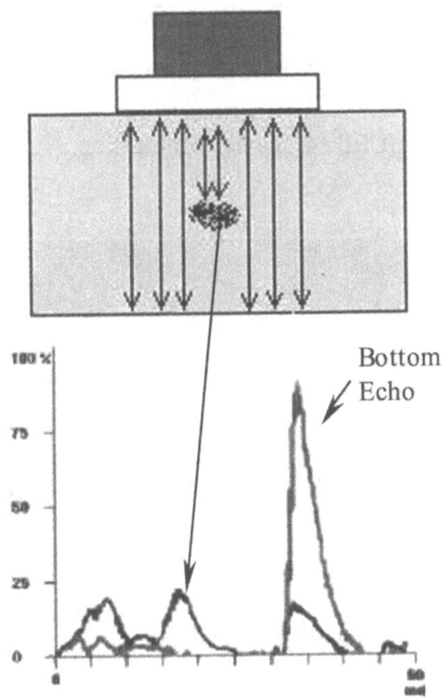

FIGURE 11. Parallel surfaces.

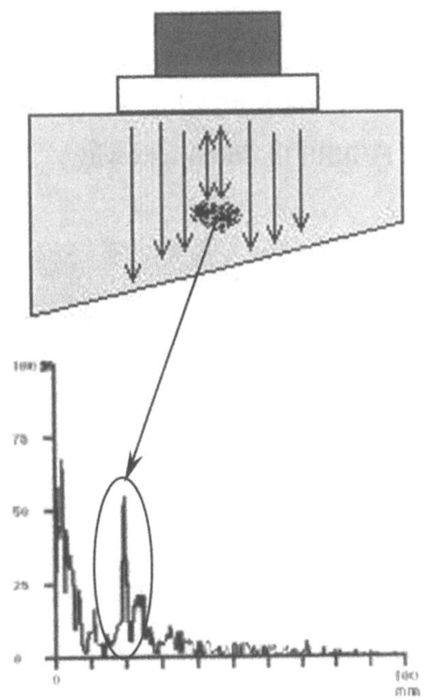

FIGURE 12. Not parallel surfaces

To design a good sampling campaign, it is necessary to identify, for each piece typology, the most critical areas and the minimum acceptance level for each detected defect.

Moreover, critical zones (where the working stresses are very high) have to be defect free, when an anomaly is detected and it is impossible to evaluate its criticality, the part can be destroyed to analyze the magnitude of the defect itself.

6 SIGNAL DETECTED

The developed research has identified the ultrasound signals related with all internal defect typologies. Figures 13 – 18 represent the classical signal mapping. From the images it is possible to see that when the defect typology is similar (spongy contraction and pinhole) the output signal is similar. As a consequence, a very good training is needed for technicians working with non-destructive techniques.

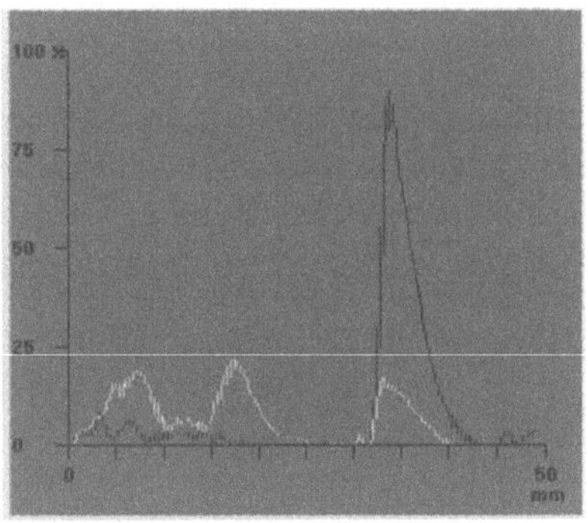

FIGURE 13. Shrinkage cavity.

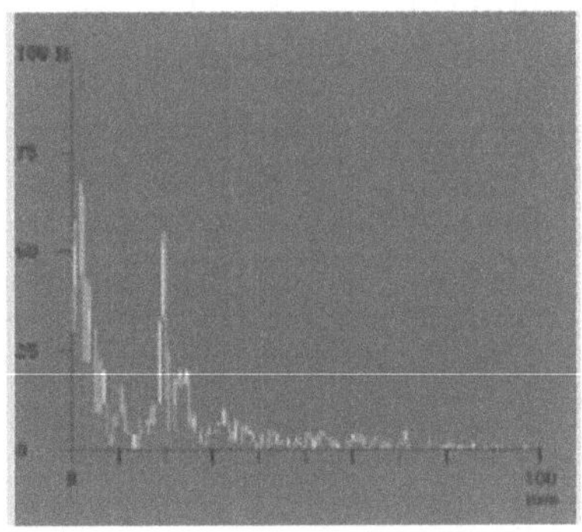

FIGURE 14. Spongy contraction.

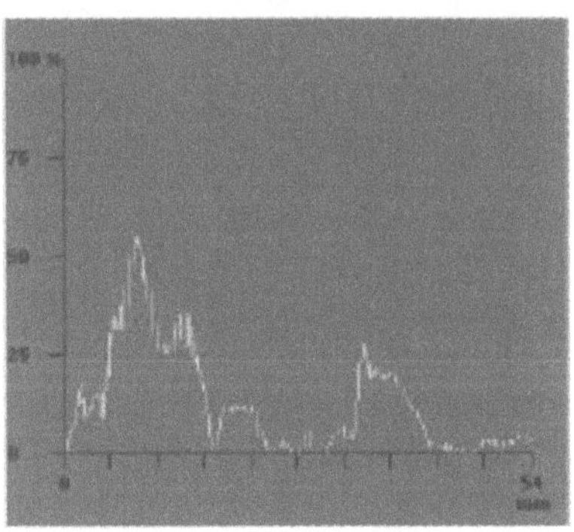

FIGURE 15. Blowhole.

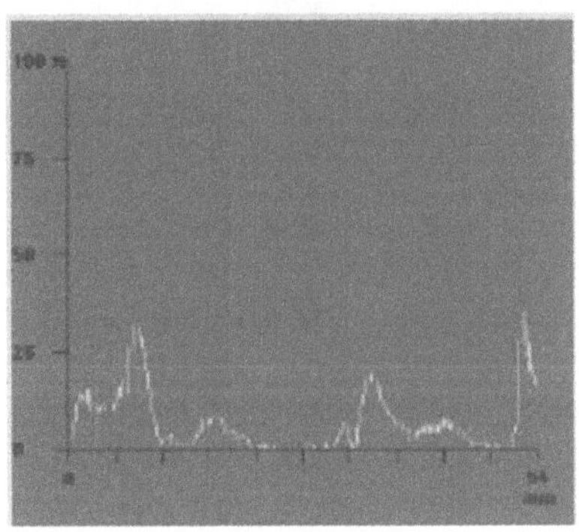

FIGURE 16. Pinhole.

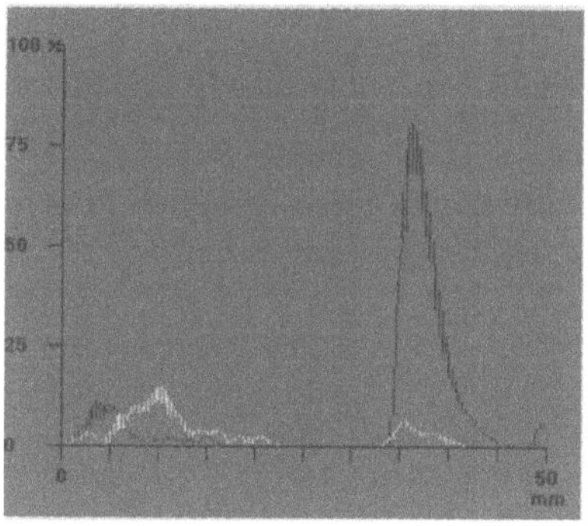

FIGURE 17. Exogenous inclusion.

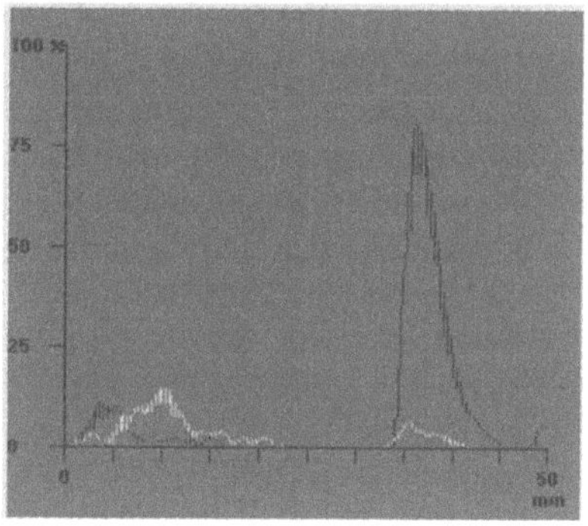

FIGURE 18. Endogenous inclusion.

7 CONCLUSIONS

To verify the reliability of the developed procedure, several US tests have been executed on different cast iron parts.

Afterwards, the obtained results have been compared with radiographies and with the sectioned parts showing a very good agreement as reported in Figure 19.

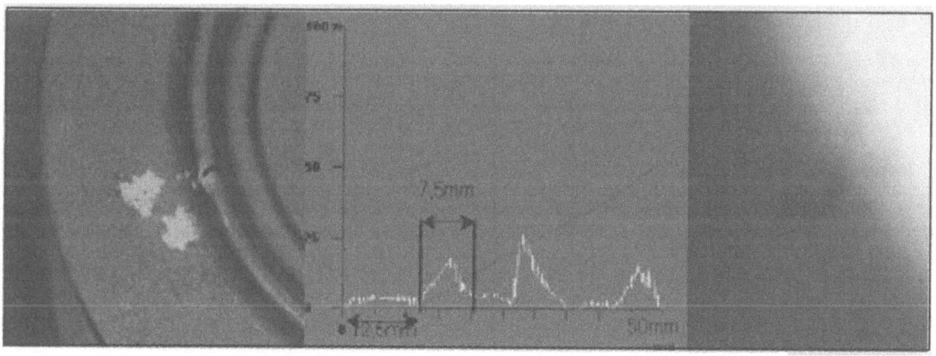

FIGURE 19. The same defect noting with ultrasonic and radiography testing.

8 ACKNOLEDGEMENTS

This work has been made possible thanks the co-operation of STREPARAVA S.p.A Company (Adro – Brescia – Italy).

REFERENCES

1. AA. VV., (1991), Nondestructive Testing, , ENEA Italian Commission for Nuclear and Alternative Sources.
2. AA. VV., Controllo non Distruttivo con Ultrasuoni, ENEA Italian Commission for Nuclear and Alternative Sources.
3. Calcagno G., La Torre C., Ansaldi E., Re Fiorentin S., Magistrali G., Torrida G., Corso Teorico sui Controlli non Distruttivi con Ultrasuoni. Fisica degli Ultrasuoni – Apparecchiature per Controlli con Ultrasuoni, C.R.F., Torino, Italy.

4. Magistrali G., (1996), Controllo delle Ghise Sferoidali con Tecniche Ultrasoniche.
5. UNI-EN 6047.
6. Remondino M. S., Corso per Specialisti in Tecnologia di Fonderia. Identificazione dei Difetti. Diagnosi e Terapia, Associazione Nazionale delle Fonderie.
7. Barton R., (1981), Nodular (SG) Iron: Possible Structural Defects and Their Prevention, Fourth International Conference of Licensees for the GF Converter Process, Schaffhausen, Switzerland, 340 – 353.
8. Goodrich G.M., (1997), Cast Iron Microstructure Anomalies and Their Causes, AFS Transaction, Vol.105, 669 – 683.
9. prEN 190/411-2: 1993.
10. prEN 12680-2: 1996.
11. Iveco Standard 18-5602: 1993.

MILLING TESTS OF AUSTENITIC STAINLESS STEEL WITHOUT NICKEL

L.Zaquini [1], P.Reusser [1]
[1] Machine Tool Laboratory, Department Mechanics, EICN, University of applied science, Canton of Neuchâtel (Switerland)

ABSTRACT

This paper presents an experimental approach aiming to find optimal cutting parameters on milling of a new material. The austenitic stainless steel without nickel is famous for its very bad machinability .This experiment has allowed determining optimal cutting parameters, behaviour of this material and tools to be used. The results are encouraging and show that it's possible to machine this material which is of interest for the watch industry.

1 INTRODUCTION

Most of the watchcases or are made of austenitic stainless steel (Example: 1.4435, 316L, 1.4539, 904L). These materials contain nickel and nickel rejection could cause allergies. The watch industry wants to replace these materials, for the production of bracelets and watch cases, and would replace them by using austenitic stainless steel without nickel. The hardness of this new material (>280 HV) generates a bad machinability compared with standard stainless steels (FIGURE 2, Mechanical and chemical properties of austenitic stainless steel. Some previous milling tests showed very bad results. In addition the question was the feasibility and the economical convenience.

The machine tools laboratory of the EICN was engaged in a project, aiming to verify the machinability of austenitic stainless steel without nickel, find suitable tools and define optimal strategies and cutting parameters on milling.

1.1 The austenitic stainless steels

In the austenitic stainless steels, there is quite a bit of nickel (>15%) which produces the austenitic structure. As we know, nickel rejection can generate health problem and, in particular, allergies. In the austenitic stainless steel without nickel the nickel is replaced by nitrogen and molybdenum which generate the austenitic structure. This substitution gives better chemical and mechanical properties and a worse machinability.

This new material is the result of research mainly done at ETHZ (Eidgenössische Technische Hochschule Zurich). The final result of the research is an alloy named PANACEA (Protected Against Nickel Allergies Corrosion Erosion Abrasion) which is the base of a set of new alloys which from today are available on the market.

	Cr	Mo	Si	V	Nb (+Ti)	Ni	Mn	P	S	N	C
PANACEA	17.1	3.25	1	0.04	0.105	0.026	12.3	0.01	0.007	0.089	0.017
1.4435	18	2.8	0.5			14	1	0.02	0.01	0.01	0.03
1.4539	20	4.5	0.4			25	1	0.01	0.01	0.01	0.02

FIGURE 1, Table of chemical composition

Published in: E.Kuljanic (Ed.) *Advanced Manufacturing Systems and Technology*, CISM Courses and Lectures No. 437, Springer Wien New York, 2002.

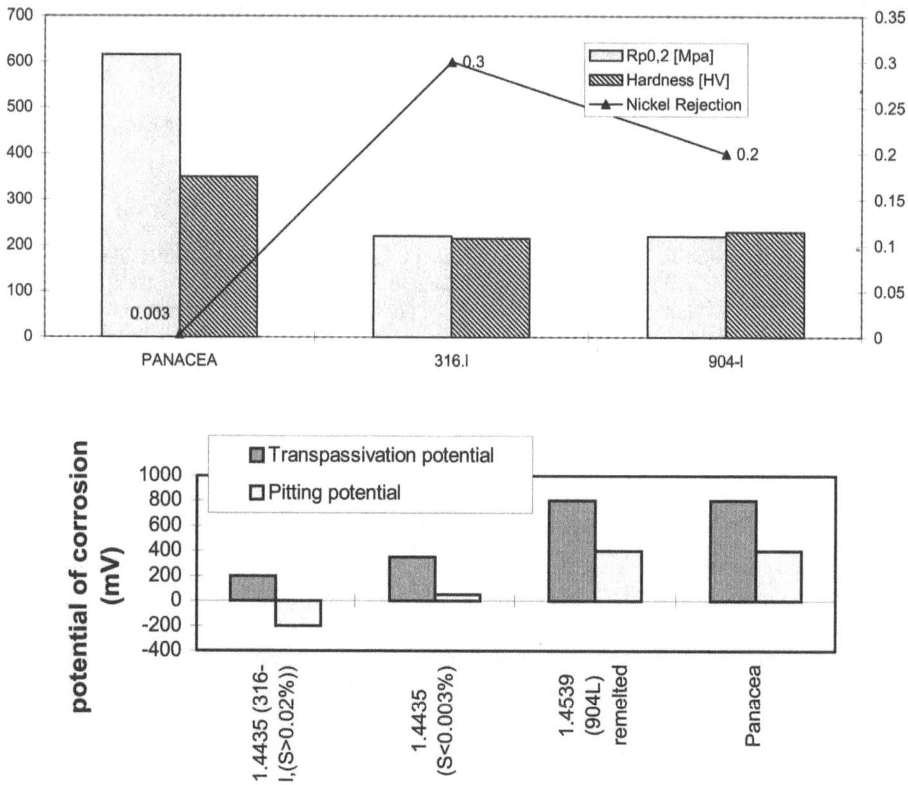

FIGURE 2, Mechanical and chemical properties of austenitic stainless steel

2 OPTIMISATION OF CUTTING PARAMETERS

Cutting parameters are considered optimised when the product of tool life and volume of the material removed per time is maximum. We decided to adopt the standard AFNOR: NF E 66-520 for finding optimal parameters [1]. It is based on the measurement of specific energy (power due to the cutting forces divided by the volume of the material removed per time). The method is based on the hypothesis that the optimal parameters correspond to the parameters, which produce a minimal specific energy during the machining.

We are aware of the fact that this standard neglects the temperature and other important factors (vibrations, absolute cutting forces etc.). In addition this method neglects the interactions of the parameters.

For these reasons, as in the common practice of our laboratory, we validated the parameters by carrying out tool life tests.

The standard recommends the following steps:

- Preliminary tests
- Determination of optimal cutting speed (V_c) and feed per tooth (f_z)
- Validation

2.1.1 Preliminary test

First of all, we have to check if the tool is compatible with the material to be machined. That means that we have to define the area where the machining is stable and no catastrophic events occur (no tool breakage, no formation of building up edge, satisfactory roughness of the surface machined ... etc).

2.1.2 Determination of optimal cutting speed (V_c) and feed per tooth (f_z)

We have to take a constant feed per tooth and we have to measure the specific energy for different cutting speeds.
In a second time, we have to take a constant cutting speed and we measure the specific energy for different feeds per tooth. The power is measured by a dynamometric table. The specific energy is given by the equation I

$$W_c = \frac{P_c}{Q} \quad \text{equation I}$$

where W_c is the specific energy [J/m^2], P_c is the cutting power [N m/s], Q is the volume of the material removed per time [m^3/s]
In order to measure the cutting forces we used a dynamometric table connected to an acquisition system. We developed a software package for the computation of the average specific energy during the milling. Explained in FIGURE 3, Measure chain of cutting force acquisition system.

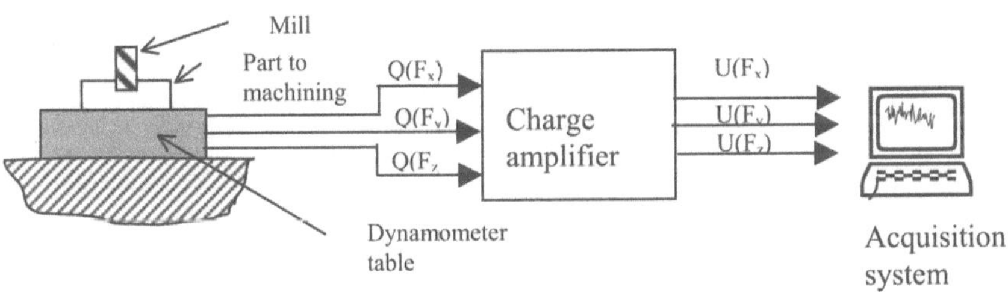

FIGURE 3, Measure chain of cutting force acquisition system

Facilities

Our tests were carried out using the following devices:

Designation	Manufacture	Characteristics
Milling center 3 axis	Heidenrich Harbeck	Milling center, with a HSC spindle 24'000 rpm
Tool	DIXI Tools	Reference: 7543, Geometry with high cut, for austenitic stainless steel. z = 3. Coating: Monolayer TiAlN from Balzers (Xidur®). Tool Material: WC helix angle ($\lambda = 45°$) Rake angle (γ), we can not declare it for confidentiality reasons.
Dynamometer Table	Kistler T5007	Range axis x and y:0...5000 N. Range axis z: 0...10000 N. Linearity ≤ %FSO : 0,3. Sensibility [Pc/N]: -7.92(x), -7.96(y),-3.71(z)
Acquisition card	NI	Measure voltage in function of time
Acquisition force software	Labwindow (NI)	Created by LMO of the EICN. For each axis, give the cutting force mean values for a given time.

Our milling tests were carried out using bars fixed on the dynamometer table as in the following picture (Fig 4).

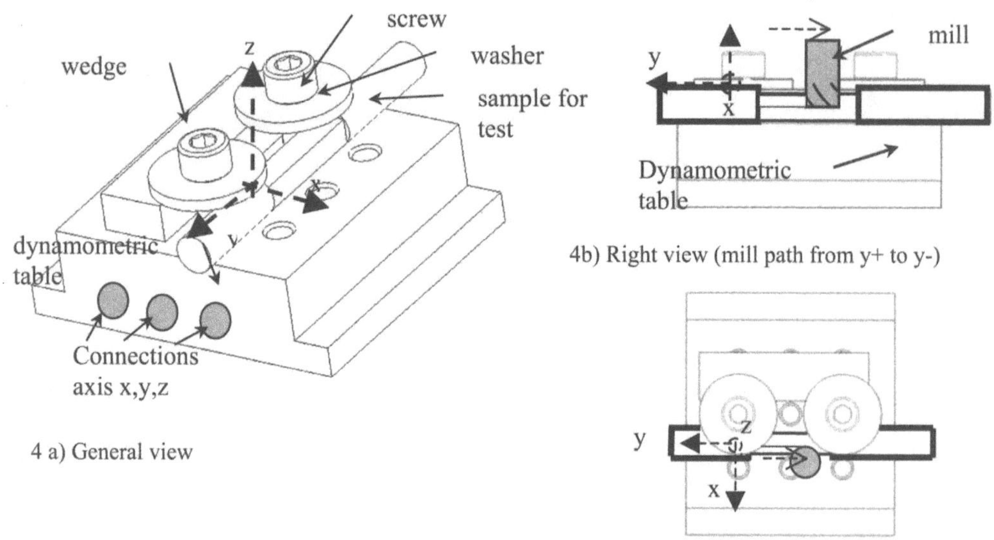

FIGURE 4 , Different views of experimental assembly

The results of the preliminary tests showed us that the feed per tooth is:
f_z from 0,005 mm to 0,05 mm.
Vc from 50 m/min to 350 m/min (limited by the spindle speed).

The axial and radial depth in our tests were: a_e= 1mm, a_p=3mm.

The results of the measurement of the specific energy were:

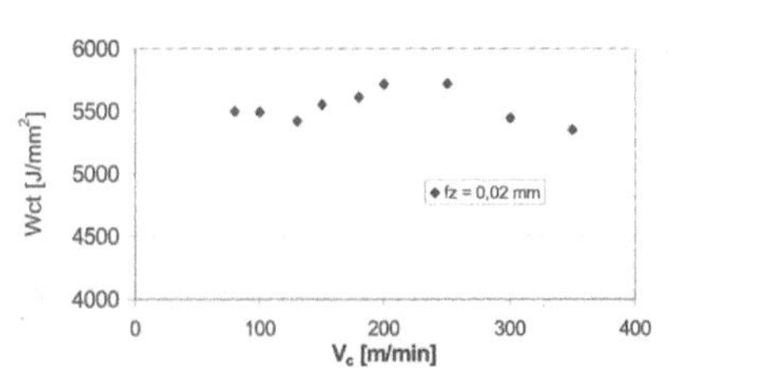

FIGURE 5, Specific energy = f(Cutting speed)

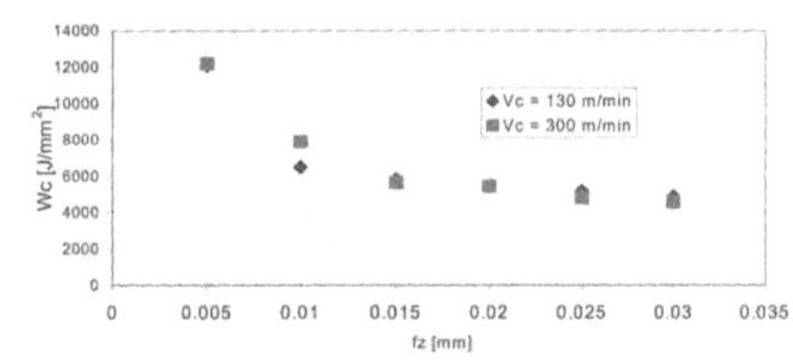

FIGURE 6, Specific energy = f(Feed per tooth)

We have to point out that there are two domains, related to the cutting speed, where the specific energy is minimal.

1- Between 80 m/min and 150 m/min.
2- Starting from 300 m/min. We assume that it corresponds to the beginning of HSC domain [2].

Between the two domains we recognize a zone where the specific energy is high. We assume that it corresponds to the "dead zone" [3] well known by the people used to adopt HSC.

When considering FIGURE 6, Specific energy = f(Feed per tooth), we see that the specific energy decreases if feed per tooth increases. For the choice of the optimal value of fz, we have to consider that the cutting forces increase when increasing the fz. Therefore it will not be convenient to adopt values of fz which do not produce important reductions of the specific energy.
The people who commissioned the work to us weren't interested in HSC.

We will adopt Vc = 130 m/min and f_z = 0,02 mm/dent.

2.2 Validation Test

The validation tests consist of carrying out the measurement of the tool life.
We mill with the parameters obtained in the previous tests and listed above. Every two minutes, we photograph the cutter and we measure the wear. For this test the milling centre was the HURON KX10. This machine is conceived for HSC (Spindle Step Tech 42000 rpm, Feed >30m/min axis x and y, 18m/min); we know that the rigidities of the two spindles are similar.

2.2.1 Results

Test n°1

The parameters are a_e= 1mm, a_e=3mm, V_c=150 and 130 m/min, f_z=0,02mm, mill DIXI 7543 (see facilities on chapter 2), ∅6mm, coating TiAlN monolayer. We saw that the tip of the tool was easily broken. Nevertheless the machining was possible. The measurement of the tool wear was based on the measurement of the V_B, neglecting the breakage of the tip of the cutter.

FIGURE 7 Wear n°1 = f(Machining time)

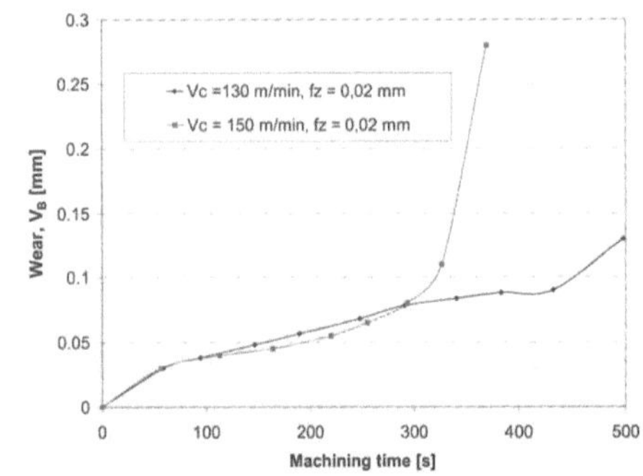

FIGURE 8, Photo of mill test n°1, machining duration 500 sec

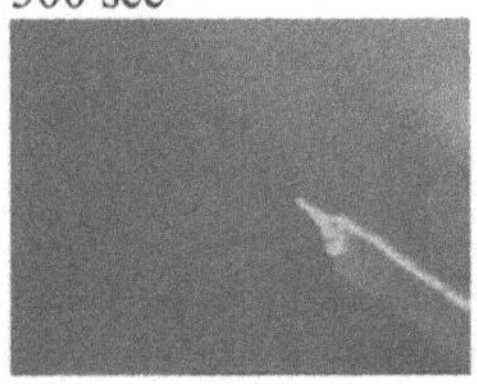

FIGURE 9, Photo of mill test n°1, machining duration 615 sec

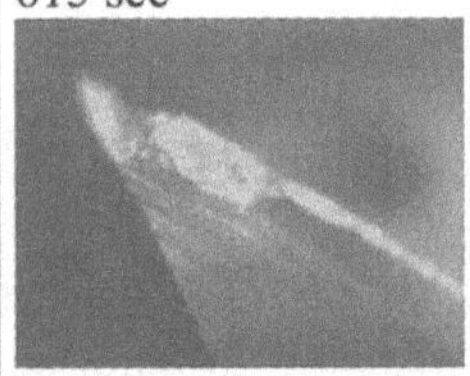

Test n° 2

We try autolubricant coating (MoS_2) on TiCN and TiAlN. The parameters are the same as that of the test n° 1.
In this test we were obliged to use ∅ tool = 5mm and a_e =0,5mm.

FIGURE 10, Wear test n°1 = f(Machining duration)

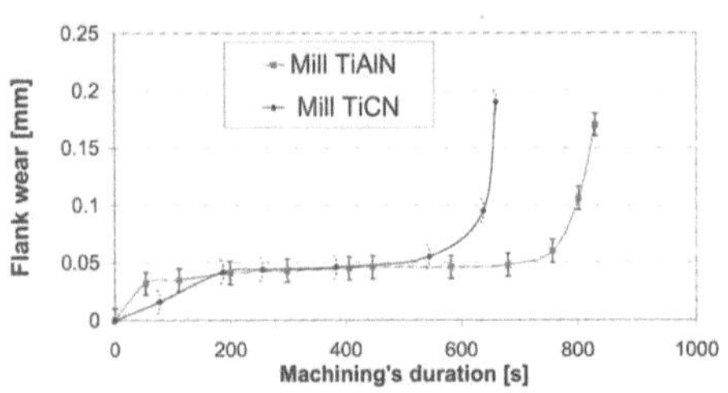

FIGURE 11,Photo mill test 2 (795 s)

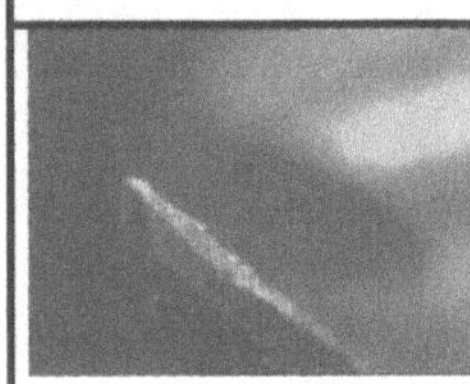

FIGURE 12, Photo mill test 2 (820 s)

The tool life is similar to that of the previous test. The radial pass was the half of that used in the test n°1.
Therefore the material removal during the tool life was reduced. The auto lubricant layer is not justified.

Test n°3

We adopted a tool with a different geometry with a reduced helix angle ($\lambda = 30°$).
The tool tip is more robust. This tool was originally conceived for carbon steels, brass and bronze. ∅tool = 5mm, z = 3, coating: TiAlN monolayer from Balzers (Xidur ®). The cutting parameters are the same as test n°2

FIGURE 13, Wear test n°3 = f(Machining duration)

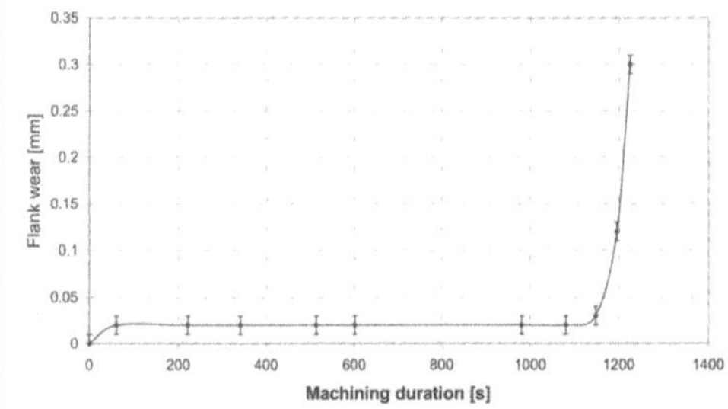

FIGURE 14, Photo mill test 3 (980s)

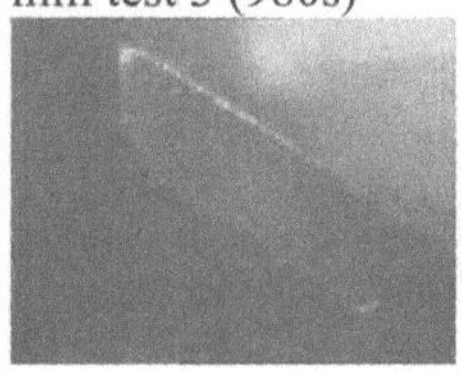

FIGURE 15 , Photo mill test 3 (1200s)

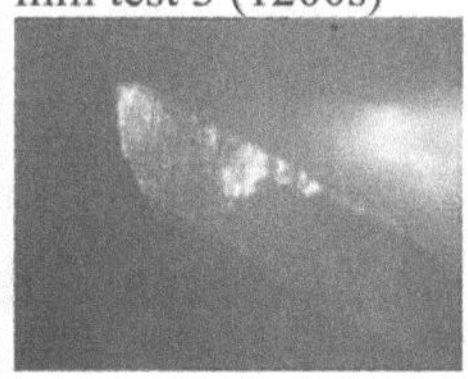

This test offer the best results. The geometry of the tool appear to be really more adapted for this material. The wear remain constant for a long time, till the coating layer is removed (1100s). The breakage is less marked.

3 ANALYSIS EDS

With our electronic microscope for EDS (Energy Dispersion Spectrometry) analysis, we examined the chemical composition of the wear. We have to point out that the results are less accurate on rough surfaces (as in the case of analysis of the tool wear) than on flat surfaces. Nevertheless the following indications are worth noting.

Results

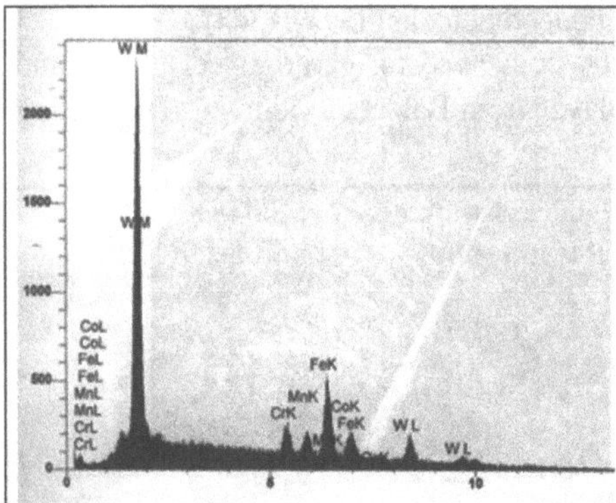

FIGURE 16, EDS Analysis of the wear, test n°1

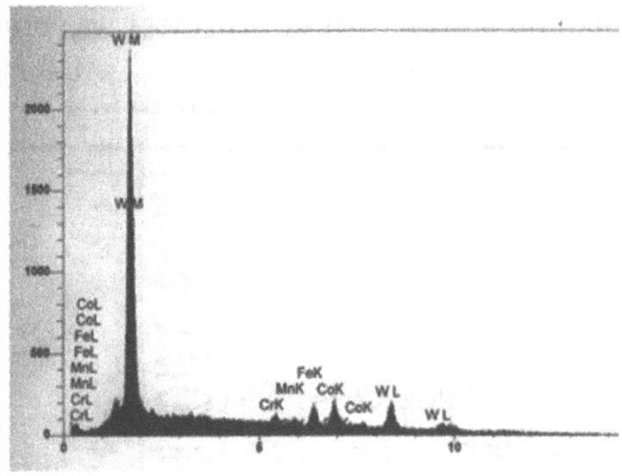

FIGURE 17, Analysis of the wear, test n°3

The pictures 16 and 17 show the results of the analysis at the end of the test. We have to point out that no aluminium, nor titan are present. Therefore the coating layer has totally disappeared.
We see some inclusions of Mn, Fe and Cr. They are the principal components of PANACEA. The peaks corresponding to the components of the stainless steel are higher in the case of test n°1. We think that the contact pressure and the temperature is weaker on the cutter of test n°3.

4 CONCLUSIONS

Austenitic stainless steel without nickel has a worse machinability than other austenitic stainless steels. Nevertheless it is machinable in industrial applications and environments. We researched the stable working area, we identified optimal cutting parameters for milling. We discovered that tools conceived for stainless steels are not suitable for our material. We must use more resistant tool geometry and resistant coating (monolayer TiAlN) on carbide tool.

5 BIBLIOGRAPHY AND REFERENCE

1. Couple outil-matière, partie 5, Norme française NF E 66- 520-5, septembre 99 Editions AFNOR
2. Herbert Schulz, (1996), Hochgeschwindigkeits Bearbeitung, Hanser Verlag, 24
3. Herbert Schulz, (1997), Translation Serge Torbaty, Fraisage Grande Vitesse, editions Technologies d'aujourd'hui, 3

UNCERTAINTY MANAGEMENT FOR PRODUCTIVE PROCESS CONTROL

M.Russo, D.Freguia

Danieli & C., Buttrio (UD), Italy

KEYWORDS: uncertainty management, process control, measure conformance assessment

ABSTRACT: Today, there is a need for measuring and granting critical measures of final products. It is important to assess the adequacy of the measuring instrumentation in order to give to the customer an objective reference. A reference guide for the evaluation of the adequacy of the instrumentation is given in the ISO protocol 14253-1. In this paper, the method developed at Danieli & C. for the evaluation of the measure uncertainty during the measure of mechanical parts is illustrated. An example of application of the method applied on a micrometer is presented.

1 INTRODUCTION

The need for measuring and granting critical measures of the final products, together with the research of methodologies for the increment the global efficiency of production processes, led Danieli & C. to verify the adequacy of its measuring system to these purposes. This task has been accomplished by considering a perspective of progressive narrowing tolerance intervals in the future. In order to obtain the international comparability of a measure, the "metrologic traceability chain" has been achieved by Danieli & C. in two steps: first, all measures had been referred to national or international Standards in agreement with the prescriptions of the General Conference on Weights and Measures; second, it has been established a measure uncertainty management along the traceability chain.

As known, measure uncertainty is the parameter, associated with the result of a measurement, that characterizes the scattering of values that could reasonably be attributed to the measurand. This parameter provides information on the reliability of the measuring process. Uncertainty is the combined result of multiple sources of variability associated to the measuring process, such as instrumentation, environment or human interaction with the process.

Published in: E.Kuljanic (Ed.) *Advanced Manufacturing Systems and Technology*, CISM Courses and Lectures No. 437, Springer Wien New York, 2002.

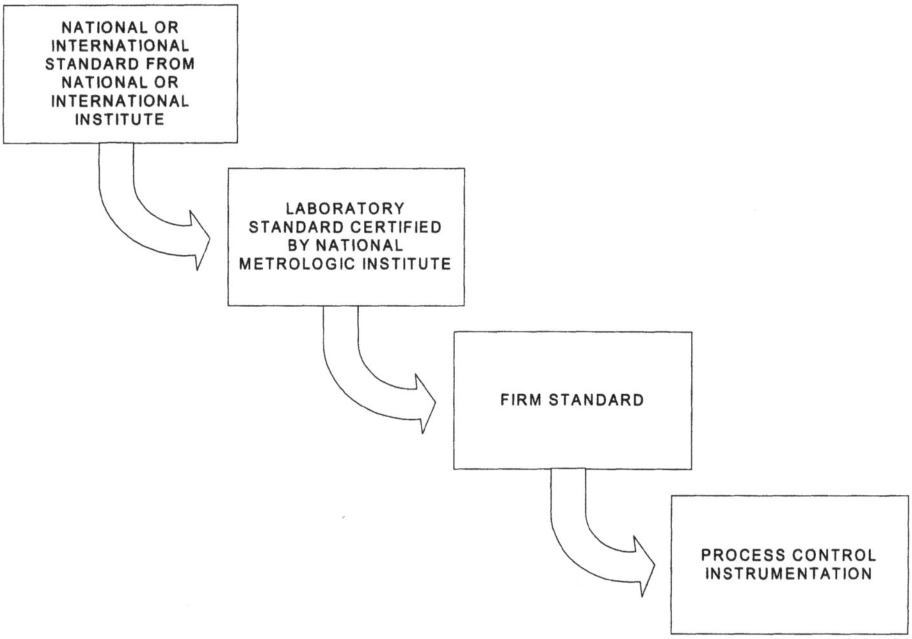

FIGURE 1. Traceability Chain

2 UNCERTAINTY MANAGEMENT SYSTEM DESIGN

Up to this introductory consideration, the first step has been the evaluation of uncertainty in order to develop the adequacy analysis of the systems used for in-process measurement.
A reference guide for this task is the ISO protocol 14253-1, "Decision rules for proving conformance or non-conformance with specifications". The protocol specifies that the conformity of a measure of a product to its design tolerance must be evaluated by considering the uncertainty of the measure.

For every measure, a "conformance zone" can be defined as the "tolerance zone", defined during design, minus the "uncertainty range". As can be seen in Figure 2, it is obvious that if the measured value is within the "uncertainty range", not necessarily it is CONFORMANT if it is within the zone U-, and not necessarily it is NOT CONFORMANT if it is within the U+ zone.

It has been established that the conformance zone must be at least the 75% of the tolerance zone defined during design; if this target is met the measure is considered reliable. If the ratio between conformance zone and tolerance zone is between 0.60 and 0.75, it is necessary to consider that the measurement result can be within the uncertainty range. In this case, the conformance of the measurand to the design must be carefully evaluated. A ratio below 0.60 is not considered acceptable.

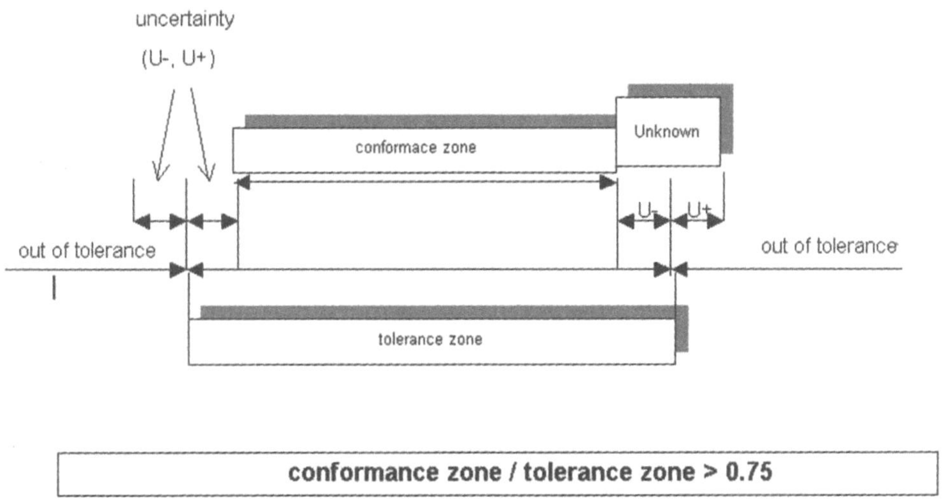

FIGURE 2. "Conformance zone" definition

3 MEASUREMENT UNCERTAINTY ANALYSIS

Evaluation of measurement uncertainty is referred to all tasks involved in the measurement method, taking into account instrumentation calibration, instrumentation uncertainty and uncertainty derived from the environment.
Measurement can be classified as prescribed into UNI EN 20286, as follows:

- Hole measurement (internal parts measurement)
- Shaft measurement (external parts measurement)

First, for every type of instrumentation, variability components must be identified. The reference scheme used for mathematic model of variability is:

Measure = Instrumentation measure + error component "a"+....... + error component "n"

All significant systematic errors has been corrected during the warm-up phase of the production system thus they are not considered.
In order to determine the uncertainty, variances must be evaluated. All constant and proportional components of the variances must be added. Then, the expression must be linearized by superimposing that linear and quadratic values coincide with extremes of the measuring range of the considered instrument. The result is the measure uncertainty. By multiplying the result for a cover factor of 1 a confidence interval of 68.26% is obtained that is considered sufficient in the steel industry.

4 UNCERTAINTY EVALUATION

4.1 ESTIMATION OF UNCERTAINTY COMPONENTS

An example of uncertainty evaluation for an external dimension measurement instrument is reported. The instrument is a bi-millesimal arc micrometer, measuring range 175-200 mm. This kind of instrumentation is used very often, prior or after machining, or for intermediate and final tests with medium-high sampling frequency.
In the first part of this study all uncertainty components had been evaluated.
The uncertainty components can be classified as follows:

u_1. Instrument repeatability
u_2. Human operator
u_3. Temperature gap between the instrument and the workpiece
u_4. Calibration uncertainty of the micrometer

The uncertainty on thermal dilation coefficients has not been taken into account because its contribute is not significant compared to thermal gradient contribute.
Moreover, the variability component connected to the difference of the thermal dilation coefficients is not considered because workpiece and instrument are both made of steel: computer simulations demonstrated that this contribute is not significant. Normally, it is necessary to take into account the difference of the thermal dilation coefficients: operators and process designers must be trained to consider this contribute. Variances has been considered not correlated.

The component u_1 can be estimated by repeatability tests. The relative square variance can be calculated by using the formula 1:

$$s^2 = \frac{\frac{1}{n-1}\sum_{k=1}^{n}(x_k - \bar{x})^2}{n} \tag{1}$$

Where x_k are the single measures, $\bar{x}$ is the mean measure and n is the number of measures.
The experimental test consisted in repeated measures taken on a standard adopting in repeatable conditions: measures have been taken as quickly as possible, by the same operator, in a controlled environment. By operating in this way it is possible to obtain results affected only by intrinsic variability of the measuring method.
At least 20 measures must be taken in order to suppose normal the distribution in repeatable conditions on a standard of 200 mm made of rectified steel, between plane surfaces (measurement of plane surfaces is the worst condition).
From the measures the following values has been obtained:

$$\text{Mean: } \bar{x} = 0.200002 \text{ [m]}$$

$$\text{Square variance:} \quad s^2 = 1.996 \cdot 10^{-14} \ [\text{m}]$$

The result is:

$$u_1 = \sqrt{s^2} = 1.41 \cdot 10^{-7} [\text{m}] \tag{2}$$

The component u_2 has been obtained by repeating the repeatability test with two different operators and calculating the difference between the largest and smallest mean. The value of the mean difference $\Delta\bar{x}$ had been obtained.

$$\Delta\bar{x} = 2 \quad [\mu\text{m}]$$

Considering a rectangular distribution, where all the values have the same probability, and considering the half-amplitude of the error, the uncertainty of operators, u_2, becomes:

$$u_2 = \frac{\Delta\bar{x}}{2\sqrt{3}} = 5.8 \cdot 10^{-7} \times L \quad [\text{m}] \tag{3}$$

The component u_3, due to temperature difference, can be estimated by using the following procedure. The instrument is considered at 20°C, and the maximum temperature difference (evaluated during shop-floor tests and adopting operators procedures), around 4°C, is supposed to be on the instrument, this can be done because the thermal dilation coefficient is the same. The error et is:

$$et = L_{20} \cdot a \cdot dt = 4.6 \cdot 10^{-5} \cdot L \quad [\text{m}] \tag{4}$$

Where L_{20} is the length, in meters, of the standard at 20°C, a is the thermal dilation coefficient ($11.5*10^{-6}$[m^2°C]), and dt is temperature difference, 4°C.

Supposing a rectangular distribution, and considering the half-amplitude of error, the uncertainty becomes as follows:

$$u_3 = \frac{et}{2\sqrt{3}} = 1.33 \cdot 10^{-5} \cdot L \quad [\text{m}] \tag{5}$$

The calibration uncertainty u_4 has been estimated to be $1.79*10^{6}+2.7*10^{-6}*L$ [m], considering a cover factor of 2. To estimate the calibration uncertainty, calibration machine uncertainty, standard uncertainty and temperature uncertainty had been considered. In order to add u_4 to uncertainty, it must be subdivided by cover factor as follows:

$$u_4 = 8.9 \cdot 10^{-7} + 1.35 \cdot 10^{-6} \cdot L \quad [\text{m}] \tag{6}$$

4.2 UNCERTAINTY COMPONENTS ADDITION

The constant and the proportional components are calculated separately, every component is squared and the components are added in order to obtain the combination of the uncertainties. To obtain the square root of the result is then calculated. It had been obtained:

$$u'(a) = 9.01 \cdot 10^{-7} \tag{7}$$

$$u'(b) = 1.33 \cdot 10^{-5} \cdot L \qquad [m] \tag{8}$$

$$U' = 9.0 \cdot 10^{-7} + 1.33 \cdot 10^{-5} \cdot L \tag{9}$$

This expression is linearized and it is superimposed that the values of the linear and quadratic coincide with extremes of the micrometer measuring range, where:

L1 = minimum length = 0.175 [m]
L2 = maximum length = 0.2 [m]

Then, the

$$u(b) = \frac{\sqrt{u'(a)^2 + u'(b)^2 L_2^2} - \sqrt{u'(a)^2 + u'(b)^2 L_1^2}}{L_2 - L_1} = 12.5 \cdot 10^{-6} \ [m] \tag{10}$$

$$u(a) = \sqrt{u'(a)^2 + u'(b)^2 L_1^2} - u(b) L_1 = 0.31 \cdot 10^{-6} \ [m] \tag{11}$$

We assume 1 as cover factor to obtain a confidence level of 68.26%, extended uncertainty becomes as follows:

$$U = 0.31 + 12.5L \ [\mu m] \text{ and L is expressed in [m]}$$

5 CONCLUSION

From the results, the micrometer uncertainties at the extremes on the measuring range are:

L = 175 mm U = 2.5 μm
L = 200 mm U = 2.8 μm

It had been established during the design phase of the uncertainty management system that uncertainties must be compared to design measures of the parts involved in measuring. Uncertainties must be at maximum 25% of the tolerance zone. Thus, for this micrometer, measurable limits of tolerance are:

L = 175 mm U_{175mm}*8 = 20.0 μm equivalent to ISO IT6
L = 200 mm U_{200mm}*8 = 22.4 μm equivalent to ISO IT6

In this study, it was possible to determine a method to define the adequacy of a micrometer for a determined measure. This study is useful both for operators and productive process designers.
Using the uncertainty estimation activity in the critical measure cases allows to use the instrumentation at best and determine the conformance of a product to the design tolerances. Uncertainty analysis on critical measures, related to normalized tolerance levels, have been indicated to designers, in order to make clear the measuring capabilities of Danieli & C. shop floors. This provided for the designers a new parameter for production cost estimation: tolerances above the measuring capabilities require investments, to improve the instrumentation or to investigate how to reduce the measure variability.
It is possible to summarize the advantages acquired from the activity of uncertainty definition and instrumentation selection in three points:

- The production process designer can use a precise criteria to choose the instrumentation, thus he can optimise the instrumentation management
- It is possible to furnish to the customer a document on conformance of the measures
- Uncertainty analysis is a fundamental parameter for production cost evaluation

REFERENCES

1. SINAL, 2000, Guida per la valutazione e la espressione dell'incertezza nelle misurazioni, www.sinal.it

2. BIMP, IEC, ISCC, ISO, IUPAC, IUPAP, OIML, (1993), Guide to the expression of uncertainty in measurement, Geneva, Switzerland, ISO Publication

3. D.C.Montgomery, (1996), Introduction to Statistical Quality Control, New York, John Wiley&Sons

4. S.Sartori, (1999), Un metodo approssimato per la stima dell'incertezza di una misura, XVIII Congresso SIT "Metrologia&Qualità", Torino

5. G.Barbato, M.Galetto, (2000), Uso della UNI CEI 9 per la pianificazione delle misure con le macchine di misura a coordinate, Sperimentare per la qualità, FrancoAngeli

6. UNI 9052, (1987), Criteri di base per le procedure di taratura dimensionale

7. UNI CEI ENV 13005, (2000), Guida all'espressione dell'incertezza di misura

8. ISO 14253-1, (1998), GPS – Inspection by Measurement of Workpieces and Measuring Instruments – Decision Rules for Proving Conformance or Non-Conformance with Specifications

9. ISO 10021-1, (1993), Metrological Confirmation System for Measuring Equipment

MICROMILLING – HIGH POTENTIAL TECHNOLOGY FOR MICROMECHANICAL PARTS

D. Spath[1], H. Tritschler[1], L. Bischoff[1] , W. Schulz[2]

[1] Institute of Machine Tools and Production Science, University of Karlsruhe (TH), Germany
[2] Institute of Ion-Beam Physics and Materials Research, Forschungszentrum Rossendorf e.V., Germany
[3] Laserpluss AG, Germany

KEYWORDS: Cutting, Micro-machining, Cutting tool

ABSTRACT. Many microfabrication processes lack the ability to structure wear resistant materials and to generate three dimensional geometries for small and medium lot sizes. Micromilling is a promising approach to overcome these limits of common microfabrication processes. In this paper it will be demonstrated that micromilling of steel with tungsten carbide tools shows good results if prerequisites regarding the tool and workpiece material structure and properties are fulfilled. For further miniaturization of structure details down to dimensions of 30-50 μm the diameters of the tools have to be reduced. This can no longer be done by grinding. In consequence new forceless processes like laser or ion beam machining are investigated. First results of structured end mills as well as milled structures will be presented.

1 MOTIVATION

Since machining steel using single point diamond tools brings about substantial difficulties in terms of tool wear, the utilization of ultra-fine grain hard metal tools represents a promising alternative for micro-cutting of steel [1, 2]. In the first step, we chose tungsten carbide end mills in order to manufacture steel microstructures. Micro injection molding as the main scope of application for micro-cutting imposes high demands on the surface quality – 0.5 μm Rz is desirable - and does not tolerate burr formation. Since commercially available micro cutting tools are mostly down-scaled from macroscopic cutting tools, we varied the tool geometry in order to identify suitable geometries for micro cutting.

2 EXPERIMENTAL EQUIPMENT

In order to obtain a tailored experimental equipment, a three axes micro-milling-machine has been designed, built, run-in and evaluated complementarily to the cutting experiments. The table of the machine is driven by AC servo motors connected to planetary thread screws with a machining envelope of 400 mm by 150 mm by 220 mm. As milling spindle, mainly an ultra-high speed spindle by The Precise Corporation, rotated with 160,000 rpm is used. This kind of extremely high spindle rotation is essential in end-milling with small tool diameters, because otherwise the cutting speed is insufficient. The high-speed spindle is equipped with hybrid ball bearings, i.e. steel bearings with ceramic balls, which provides for higher stiffness compared to air bearings and therefore makes the spindle very suitable for the machining of hard materials like steel. The evaluation of the machine proved that the positioning error as well as the minimum step width is better than 1 μm [4].

Published in: E.Kuljanic (Ed.) *Advanced Manufacturing Systems and Technology*,
CISM Courses and Lectures No. 437, Springer Wien New York, 2002.

3 WORKPIECE MATERIALS

Aside from hot-work tool steels like AISI H11 (X 38 CrMoV 5 1) and AISI H13 (X 40 CrMoV 5 1) [6, 7], which were examined in earlier experiments, recently also AISI O2 (90 MnCrV 8) was machined in different heat treatment states. It represents an oil-hardening cold-work tool steel with particularly high wear resistance against abrasive wear and a comparatively good machinability at the same time. The main application is mold and die making for injection molding and even forming. Like in the earlier experiments, special attention was given to the heat treatment, so that the grain size and – particularly important with a higher content of carbon – the distribution of carbide grains is below 10 µm and as homogenous as possible. Material hardness after heat treatment was varied mostly between 57 HRC and 64 HRC.

4 CAPABILITY OF COMMERCIAL TOOLS

With commercially available end milling tools, micro structures in hardened tool steel can be manufactured with an accuracy of better than 0.01 mm and a surface quality better than 0.3 µm Rz in less than an hours machining time (e.g. Figure 1).

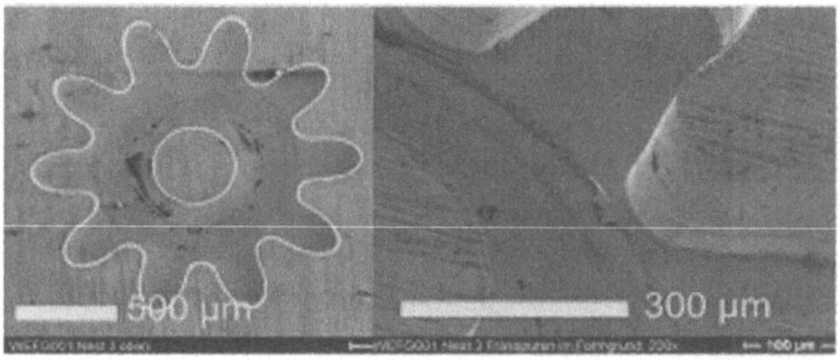

FIGURE 1. Steel mold cavity (53 HRC) for micro gear (l.) and detail of milled edge (r.)

Suitable process parameters, strategy and tool geometry provided, little or no burr formation occurs. One of the major drawbacks is that concave structures cannot be machined smaller than 0.1 mm due to the available tool diameters.

5 MANUFACTURING METHODS

The common manufacturing method for tungsten carbide cutting tools is grinding using diamond. With grinding wheels of micrometer-size diamond grain quality available, minute structuring of the tool geometry down to diameter 0.1 mm has become possible and is being marketed comparatively successfully. Due to the grinding forces and vibrations, though, the yield is dissatisfying. In order to obtain tools of even smaller diameter and also taylor-made geometry, tool manufacturing methods other than grinding were employed, namely laser machining and focused ion beam (FIB) milling.

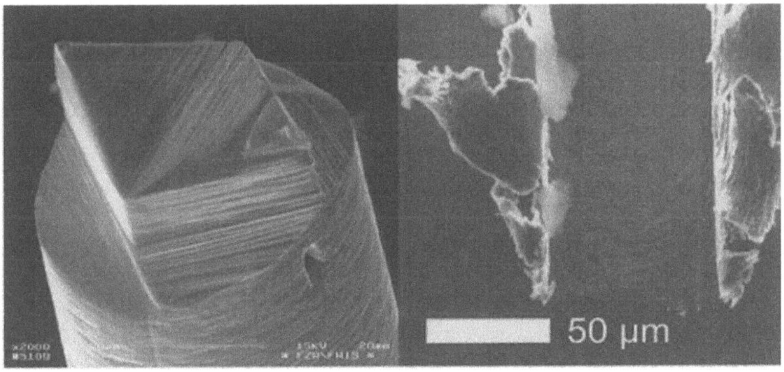

FIGURE 2. Ion milled two fluted end mill (l.) and detail of manufacturing sample (r.)

As has been previously proven [3, 4], ion milling is a suitable method to obtain minute cutting tools made of high speed steel and even tungsten carbide. As can be seen in Figure 2, ion milled end milling tools made of tungsten carbide hard metal are in principle eligible for machining steel. In this case, the material hardness was below 30 HRC, though. The manufacturing results are still dissatisfying regarding the extreme burr formation. With commercial micro end mills this is only met when applying unsuitable process parameters like insufficient feed rate, or with worn cutting edges. Aside from the influence of the comparatively soft material state, the latter probably applies in this case, since no tool coating was applied here.

6 VARIATION OF GEOMETRY

As was established earlier [7], single flute end milling tools appear more reasonable for very small tool diameters due to the disproportionateness of added up roundness errors of the tool, spindle, clamping etc. Therefore, both by FIB and laser machining, single flute end mills were manufactured. While for the laser-machined tool a simple engraving tool geometry was chosen, the FIB-machined tool was furnished with a rather intricate geometry providing for a better stability. The former was machined with a Laserpluss AG UV-laser source of 1 W at 256 nm, while the latter was machined with Au-ions at 35 keV, a current density of 1.2 A / cm^2 and a spot size of ~0.3 μm.

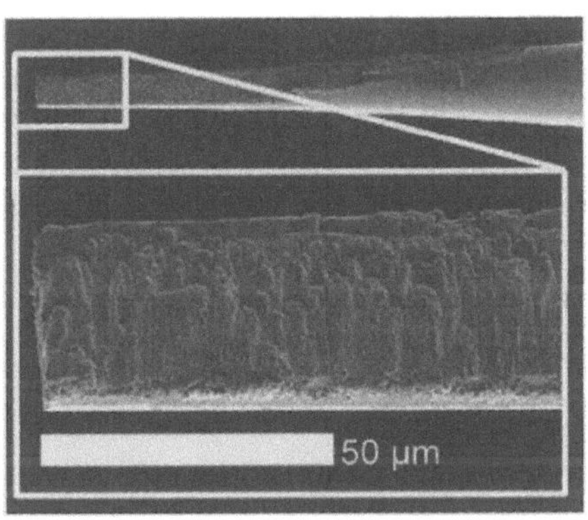

FIGURE 3. Laser machined single fluted end milling tool

As can be seen from the images in Figure 3, the surface quality of the laser machined tool still requires massive process optimization, whereas the FIB-machined tools are suitable for cutting experiments. The elaborate pre-machining preparations necessary with the currently used FIB-equipment, however, do not allow for statistically based experiments but rather for a mere investigation of feasibility.

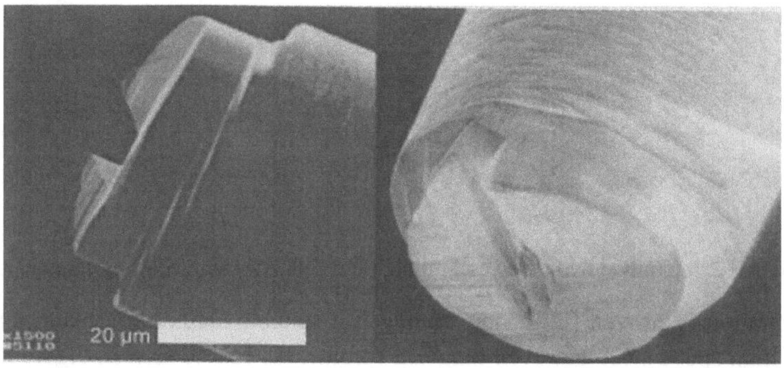

FIGURE 4. Ion milled single fluted tungsten carbide end mill

Since the FIB equipment currently used in this research is designed to control material removal in a slitting manner rather than in an excavating manner, the manufacturing of a geometry as presented in figure 4 is far more challenging than the one presented in Figure 2 and 3. By initial cutting experiments the fundamental ability of such type of cutting tool was established. Burr formation, though, was exceedingly strong with an average burr height of more than 10 µm. Future optimization is focused on burr reduction , which will be pursued through application of a higher rake angle and eventually through application of a sharper cutting edge using a different tool material.

7 VARIATION OF TOOL MATERIAL AND IMPROVEMENT OF CUTTING EDGE

Since both surface quality and – through influence on the cutting forces – the wear life of the tool strongly depend on the micro geometry of the cutting edge, care should be taken to its formation. An extremely sharp and at the same time durable edge is desirable. Depending on the grain size, tungsten carbide tools are in this respect limited to a minimum cutting edge roundness of 2 ~ 5 µm. Exceeding this limitation is only feasible by variation of the tool material. While single crystal cubic boron nitride is hardly available in grain sizes suitable for single point cutting tools, in this research artificial sapphire was used to obtain a sharper cutting edge.

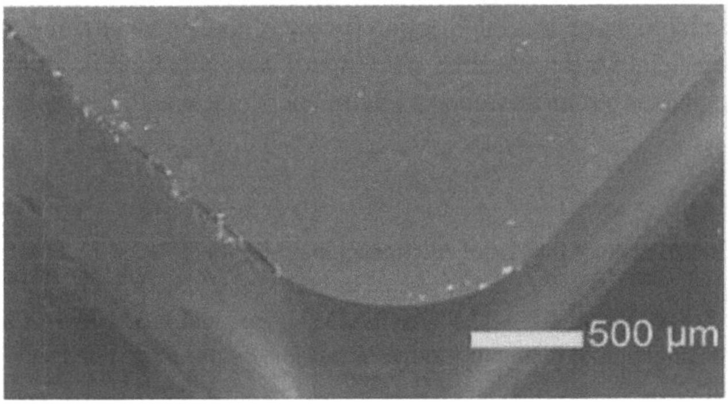

FIGURE 5. Artificial sapphire tool

Judging from SEM images (e.g. Figure 5), the edge appears comparable to single crystal diamond tools. Preliminary cutting experiments on face-turning of O2 tool steel (90 MnCrV 8) of 58 HRC have been successful. The persistance of the tool material in discontinuous cutting is yet to be assured.

8 CONCLUSIONS AND OUTLOOK

With the preliminary cutting experiments with end mills of about. 50 µm diameter and complex geometry having been successful, the next step will be the optimization of the tool geometry in order to obtain less burr formation. Further, PVD-coating for an economically reasonable tool wear life appears essential. Long ranged research has to focus on the evaluation of the feasibility of transferring the initial results in facing of tool steel using artificial sapphire towards micro end milling. The major task in ion milling of tungsten carbide tools lies in the optimization of the machining time, since the presently needed time of several hours is hampering any thought of commercialization. The currently upgraded ion milling facility in Rossendorf is planned to meet this demand. The laser machining process is currently falling back in terms of surface quality of the cutting tool and still needs substantial optimization efforts in this respect.

Since three years now the 3-axis machine tool has been a useful experimental equipment comparing to other conventional machine tools, although in terms of machining accuracy and particularly in the dynamic behavior of the x/y-table it lies by a factor 10 below the performance of commercially available ultra-precision machine tools. Future work, especially with tools of diameters below 0.1 mm or extremely brittle tool materials will be carried out on a Kugler ultra-precision milling center. This justifies a future hope for better tool wear life due to the enhanced smoothness in machine table motion.

ACKNOWLEDGEMENTS

The work presented here are part of a cooperation of various institutes at the University of Karlsruhe and the Research Center of Karlsruhe (FZK) and are financially funded by the Deutsche Forschungsgemeinschaft DFG ("german society for research"). The close cooparation with the Institute of Material Science and Engineering (iwk1) of the University of Karlsruhe (TH) has been particularly beneficial for this research. The experiments with modifications of the tool were carried out at Forschungszentrum Rossendorf and in cooperation with Laserpluss AG and HAM GmbH, Germany.

REFERENCES

1. Masuzawa, T., (2000), State of the Art of Micromachining Keynote Paper in: *Annals of the CIRP*, Vol 49/2
2. Paul, E., Evans, J.C., Mangamelli, A., McGlauflin, M.L., Polvani, R.S. (1996), Chemical aspects of tool wear in single point diamond turning, *Precision Engineering* Vol 18 (1996), Elsevier Science, 4-19
3. Adams, D.P., Vasile, M.J., Krishnan, A.S.M., (2000), Microgrooving and microthreading tools for fabricating curvilinear features, *Precision Engineering* Vol 24 (2000), Elsevier Science, 347-356
4. Weule, H., Schmidt, J., Hüntrup, V., (1999), Micromilling of Ferrous Materials, *Production Engineering* Vol. VI/2, 1999, p. 17-20
5. Spath, D., Hüntrup, V., (1999), Micro-Milling of Steel for Mould Manufacturing – Influences of Material, Tools and Process Parameters, *Proceedings of the 1st international conference of the euspen* Vol. 1, 203-206
6. Spath, D., Löhe, D., Tritschler, H., Schulze, V., Peichl, A. (2000), Microcutting of Steels – Interaction of Material Properties and Process Parameters, *Proceedings of the Micro.tec 2000* Vol. 2, 7-11
7. Schmidt, J., Tritschler, H., Haberer, H. (2001), Cutting Tools and Material Conditioning for Micro End Milling of Tool Steel, *Proceedings of the 2nd International conference of the euspen* Vol. 1
8. Weule, H., Hüntrup, V., Tritschler, H. (2001), Microcutting of Steel to Meet New Requirements in Miniaturization, *Annals of the CIRP* Vol. 50/1

FEM ANALYSIS IN A MACHINING CENTER

R. Pavan, M. Tellan, V. Ruffati

FAMUP Machining Center S. Quirino (PN) Italy

KEYWORDS: Machining centers, FEM analysis, design methodology

ABSTRACT: This paper analyses a methodology that may be followed for the practical design of the structural elements that compose a numeric-control machining centre, using a FEM (Finite Element Method) calculations. Detailed calculations are presented in the case of an important part of a machine tool such as the spindle-head holder.

1 THEORETICAL BASIS

The basic principle of FEM applied to continuous bodies consists in the fact that a structure with infinite unknown terms can be considered as a set of structural elements, connected at a discrete number of nodes and therefore with a finite number of unknown terms. If the relationship force-displacement of the single element is known, it is possible to determine its properties and analyse the behaviour of the assembled structure. The hypothesis is made that fictitious forces, that represent the distributed stresses that in reality are present in the structure, weigh upon the nodes. If this idealisation is allowed, the problem is reduced to the one of a conventional structure and then suitable for a numeric solution.

The Finite Element Method applied to continuous bodies is characterised by a series of essential steps:

1) The continuum (any part of the tooling machine that has to be analysed) is divided in imaginary (tetragonal) solid elements called finite elements;

2) The assumption is made that elements are connected among themselves at a discrete number of nodal points on their border. The displacements of these nodal points are the basic unknown quantities of the problem;

3) A set of functions is chosen in order to define univocally the displacements within each finite element, in terms of displacements at the nodes.

4) Such functions univocally define the strain status of each element. These strains, together with an eventual initial strain and the constitutive properties of the material, define the stress condition.

5) A system of forces is determined. These forces are concentrated at the nodes and equilibrate the forces at the contours and the distributed loads. The result is an equation of the following type:

$$\{F\}^e = [k]^e\{\delta\}^e + \{F\}^e_p + \{F\}^e_n \tag{1}$$

Published in: E.Kuljanic (Ed.) *Advanced Manufacturing Systems and Technology*,
CISM Courses and Lectures No. 437, Springer Wien New York, 2002.

where $[k]^e \{\delta\}^e$ represent the forces related to the node displacements, $\{F\}_p^e$ represents the nodal forces required to balance the distributed loads on the element; $\{F\}_n^e$ the nodal forces required to equilibrate the initial strain.
At this point the *displacement function* is defined in order to determine the deformation of a solid object subjected to external forces. For simplicity the analysis of a surface element can be considered, neglecting the three dimensional aspect (tetragonal solid elements) that represents only an extension of the above mentioned equations. A typical element is defined by the nodes i, j, m, etc. and by contour straight lines. The displacements for every internal point of the element are defined by the vector {f (x,y)}:

$$\{f\} = [N]\{\delta\}^e = [N_i, N_j, N_m, \ldots] \cdot \begin{Bmatrix} \delta_i \\ \delta_j \\ \delta_m \\ \cdot \\ \cdot \\ \cdot \end{Bmatrix} \tag{2}$$

where the components of [N] are in general function of the position and $\{\delta\}^e$ represents a list of nodal displacements for a particular element.
In *plane stress* Figure 1 represents the horizontal and vertical (two-dimensional) movements of a typical point internal to an element.

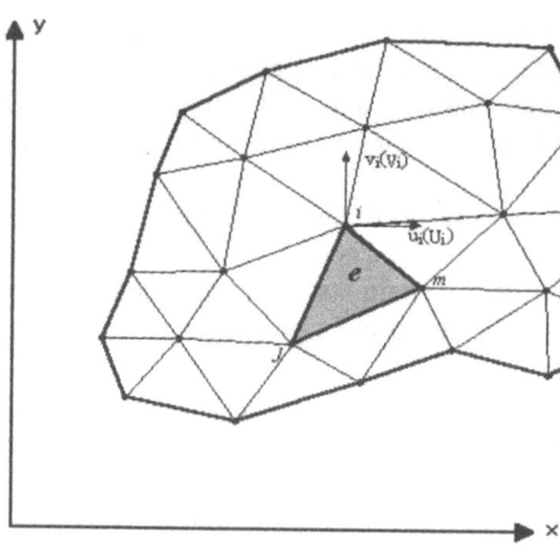

FIGURE 1. Plane stress region divided in finite elements.

The following vector represents the corresponding displacements of the nodes

$$\{\delta_i\} = \begin{Bmatrix} u_i \\ v_i \end{Bmatrix} \tag{3}$$

The strain can be determined for all points internal to the element starting from the known displacements, by means of an equation that in matricial form is:

$$\{\varepsilon\} = [B]\{\delta\}^e \tag{4}$$

In the case of plain stress the interesting strains are those in the plane and are defined in terms of the displacements by the well known equations:

$$\{\varepsilon\} = \begin{Bmatrix} \varepsilon_x \\ \varepsilon_y \\ \gamma_{xy} \end{Bmatrix} = \begin{Bmatrix} \frac{\partial u}{\partial x} \\ \frac{\partial v}{\partial y} \\ \frac{\partial u}{\partial y} + \frac{\partial v}{\partial x} \end{Bmatrix} \tag{5}$$

In general, the material comprehended in the elements can be subjected to internal strains like those due to temperature variations, contractions, crystal growth and so on. If such strains are indicated as zero-point, then the stresses will be calculated from the difference between effective strain and initial strain.
These stresses can be simply added to the general definition, so that, assuming a general elastic behaviour, the relationship between stress and strain will be linear and of the following type:

$$\{\sigma\} = [D](\{\varepsilon\} - \{\varepsilon_0\}) + \{\sigma_0\} \tag{6}$$

where [D] is an elasticity matrix containing the suitable properties of the material.
About *equivalent nodal forces,* accepted that

$$\{F\}^e = \begin{Bmatrix} F_i \\ F_j \\ F_m \\ \cdot \\ \cdot \\ \cdot \end{Bmatrix} \tag{7}$$

defines the nodal forces that are statically equivalent to the stress at the contours and to the loads distributed on the element, each one of the forces {**Fi**} must have the same components

number of the corresponding nodal displacements {i} and must be directed in the suitable corresponding direction.
The distributed loads {**p**} are defined as those acting on the volume unit of material internal to the element in the directions corresponding to those of the displacements {**f**} in that point.
In the particular case of plane stress the nodal forces are:

$$\{F_i\} = \begin{Bmatrix} U_i \\ V_i \end{Bmatrix} \tag{8}$$

with the components U and V corresponding to the directions of the displacements u and v, while the distributed load is:

$$\{p\} = \begin{Bmatrix} X \\ Y \end{Bmatrix} \tag{9}$$

where X and Y are the components of the body force, i.e. the gravity.
To accomplish the fact that nodal forces have to be statically equivalent to the real stresses at the border and to the distributed loads, the simplest procedure is to impose arbitrary nodal (virtual) displacements and to equalise the internal work to the external work done by the different forces and by stresses during the deformation.
$d\{\delta\}^e$ are the virtual displacements at the nodes that generate the following displacements and strains in the inner of the element:

$$d\{f\} = [N]d\{\delta\}^e \tag{10}$$

and

$$d\{\varepsilon\} = [B]d\{\delta\}^e \tag{11}$$

2 INDUSTRIAL APPLICATION

We can apply the above mentioned theory to tooling machines: in machining centres the main constitutive elements are designed in function of the manufacturing processes that the machining centre itself is capable to carry out. One of the most delicate parts of a machine tool is the spindle-head holder: this essential element is subjected both to its own weight and to external forces. These are due to all the elements that make part of the head and to machining forces (drilling, milling). The design of the head requires particular care because it must be done in such a way that the yielding it suffers during mechanical machining is confined in the material's elastic range and moreover respects the value limits imposed by firm's specifications [3].These requirements ensure the suitable precision in the operations that the machine will carry out. The first step is to establish machine's limits, intended as the maximum values of the cutting parameters that can be sustained, in order to choose the size of the engine that has to be applied to the spindle. Then a first approximated model can be developed using 3D CAD software.

For example we can consider one of the machines built by the firm Famup, in particular the head of the MMV200/80 model.

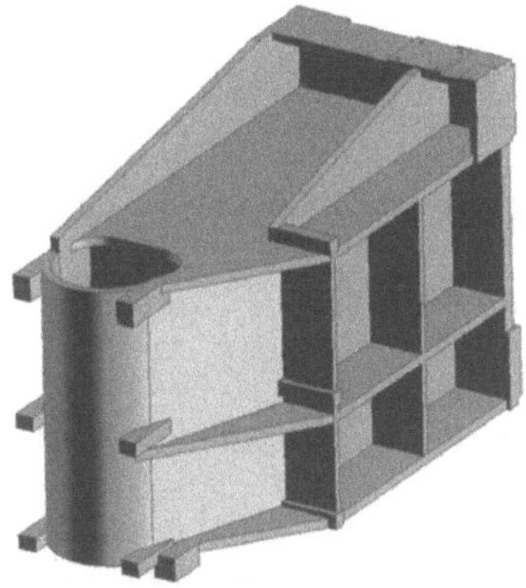

FIGURE 2. MMV200/80 head.

FIGURE 3. Mesh of MMV200/80 head.

Fig.2 represents the starting model for the first FEM analysis. The material used to build the head is cast-iron G25 ($\rho = 7250$ Kg/m^3).
The initial phase of the finite element analysis consists in the discretisation of the solid body as previously described: the software automatically generates a mesh [5]. The mesh is constituted by tetragonal elements and the total number of nodes is 125130, while the number of elements is 83362. In Fig.3 the mesh of the model is shown.

For a correct analysis attention must be paid to the constraints and to the loads applied to the structure, since this allows the comparison of results in case of analysis with different models. Fig. 4 shows how the constraints are applied.

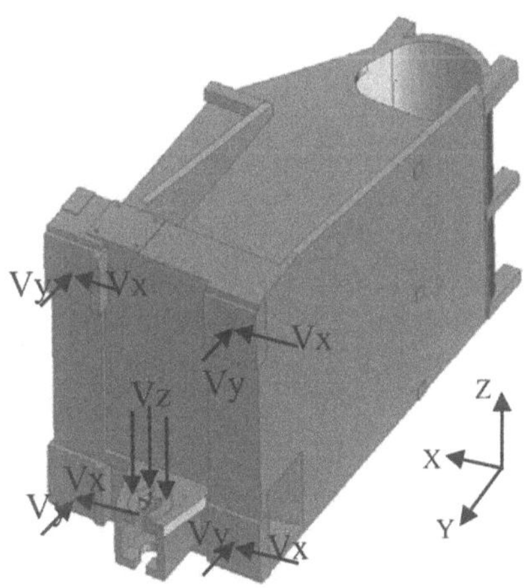

FIGURE 4. Constraints for MMV200/80 head.

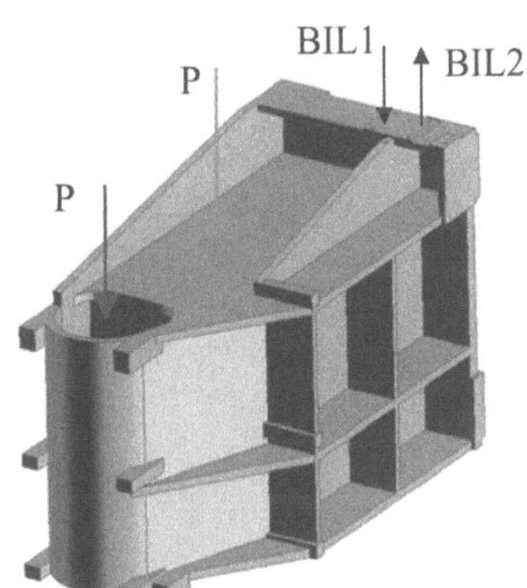

FIGURE 5. External forces for MMV200/80 head.

The structure has been constraint on the linear guideway and on the stock rolled ballscrews.
In the case of the head, the constraints to be applied are the following

1) Sliding blocks: the displacement along X axis and Y axis is blocked
2) Stock rolled ballscrew: the displacement along the Z axis is blocked

In Fig. 5 the load set is introduced: the forces that have to be applied are divided in two categories:

1) One is related with the external loads: the so called external forces , i.e. the weight **P**, the force due to the electrospindle and the protection carter **(Pt)**and the head balancing system **(BIL1, BIL2)**
2) The other one is related to the machining forces: drilling and milling.

The forces shown in Fig.6 refer to the tooling operation

1) Drilling: F(Z) represents the force of the tool in the feeding direction;
2) Milling: F(X) and F(Y) represent the cutting force, relative to both positive and negative sense of each axis.

After all data have been introduced in the finite element code, the analysis of the head can be carried out for different load conditions, which can be summarised as follows: Static condition; Drilling; Milling +X; Milling –X; Milling +Y; Milling –Y

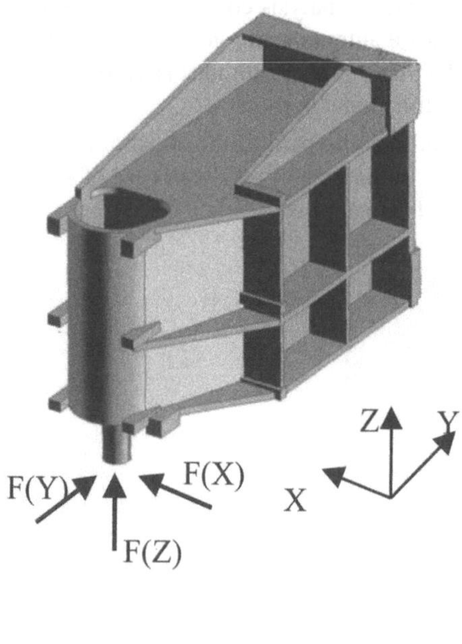

FIGURE 6. Machining forces for MMV200/80 head.

3 RESULTS

The following are the results obtained from FEM analysis.

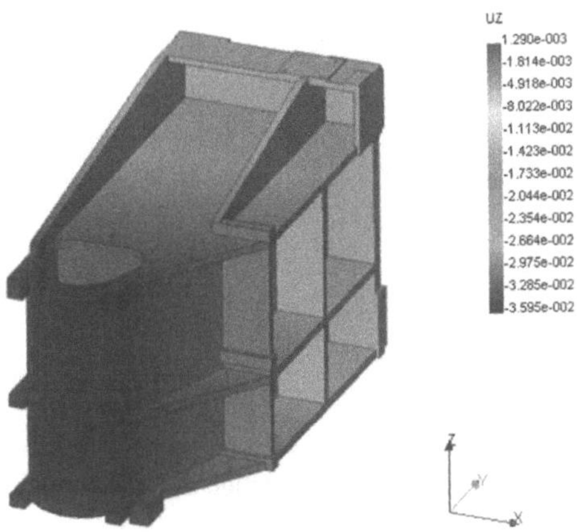

FIGURE 7. Yielding in Z direction in static conditions.

Fig. 7 shows the yielding of the structure in Z direction in the static condition. The same representation can be obtained for the other loading conditions listed above (omitted for conciseness). As from a first analysis the yielding of the structure is too large, thus an optimisation procedure is required.With "optimisation" we intend a redefinition of the geometry of the structure, in order to redistribute the material in strategic points with consequent decrease of the weight. In fact, if the structure is lighter, less material is required for the construction and therefore less power is required at the engine to move the masses. The investigation of the regions to lighten is not a random process: it is confirmed by the FEM analysis, so that after successive iterations the geometry of Fig. 8 is obtained.

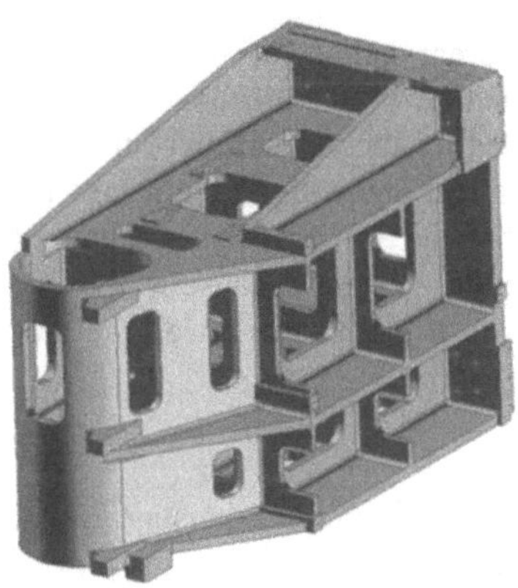

FIGURE 8. Final model of the head.

We can state that a study concerning the weight reduction of the structure has led to the following conclusion: a 25% decrease with respect to the initial weight has been achieved, with consequent reduction of the yielding, that is now within the limits of firm's specifications.
The results of the FEM analysis of the new model are shown in Fig. 9:

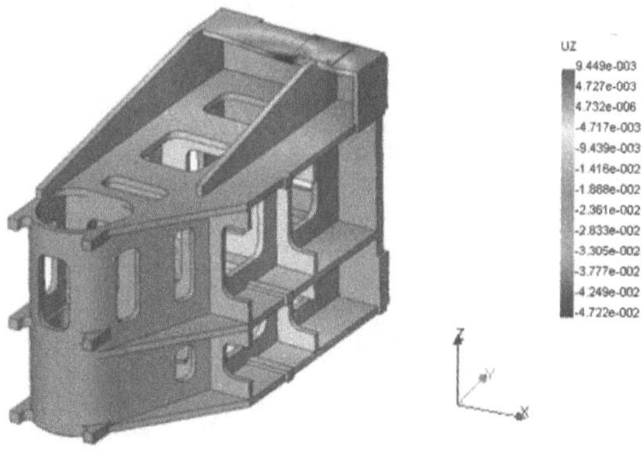

FIGURE 9. Yielding in Z direction in static condition.

It represents the displacements in the Z-direction in static conditions. The obtained values lead to the conclusion that the head has been optimised following correct criterion, respecting firm's specifications. Beside the structural study, a **vibration analysis** of the head should be carried out in order to determine the vibration modes of the structure.

4 CONCLUSIONS

This work reports about a methodology which has been followed by the firm FAMUP Machining Center. It must be considered essential for the correct design of any component of a tooling machine. The fundamental points of this study can be summarised in the following scheme: Identification of the data required for the project; Definition of the geometry of the model; FEM analysis of the structure; Optimisation procedure; FEM final analysis This operations sequence leads to the following conclusions: the starting point of a study is given by firm's specifications, that are necessary to give meaning to the FEM analysis. Indeed these specifications impose the maximal values of the structural yielding that can occur in a component of the tooling machine in the different working conditions and represent the reference that must be considered for the evaluation of the results. This allows advancing from a mathematical solid model to a real one without the necessity of a real experimental phase. In conclusion it can be stated that the FEM analysis represents a fundamental technological innovation in the field of structural design, because it allows to reduce time and costs, which are very important parameters in a global market as the present one, characterised by high competitiveness.

5 REFERENCES

1. Cosmos/M, Structural Research and Analysis Corporation (SRAC), 2001;
2. R. Pavan M. Tellan FEM Internal Documentation, 2002, UTEC, FAMUP Machining Center;
3. Cvdesign, Calcolo automatico delle strutture: gli elementi finiti, 2001;

AUTHORS INDEX

MIX
Papier aus verantwortungsvollen Quellen
Paper from responsible sources
FSC® C105338

If you have any concerns about our products,
you can contact us on
ProductSafety@springernature.com

In case Publisher is established outside the EU,
the EU authorized representative is:
Springer Nature Customer Service Center GmbH
Europaplatz 3, 69115 Heidelberg, Germany

Printed by Libri Plureos GmbH
in Hamburg, Germany